Michael Z. Hauschild · Ralph K. Rosenbaum
Stig Irving Olsen
Editors

Life Cycle Assessment

Theory and Practice

Springer

Editors
Michael Z. Hauschild
Department of Management Engineering
Technical University of Denmark
Kongens Lyngby
Denmark

Stig Irving Olsen
Department of Management Engineering
Technical University of Denmark
Kongens Lyngby
Denmark

Ralph K. Rosenbaum
IRSTEA, UMR ITAP, ELSA Research group and ELSA-PACT
Environmental and Social Sustainability Assessment
Montpellier
France

ISBN 978-3-319-56474-6 ISBN 978-3-319-56475-3 (eBook)
DOI 10.1007/978-3-319-56475-3

Library of Congress Control Number: 2017946049

Printed on acid-free paper

This Springer imprint is published by Springer Nature
The registered company is Springer International Publishing AG
The registered company address is: Gewerbestrasse 11, 6330 Cham, Switzerland

Preface

It is an old observation that 'What gets measured gets managed', and that what is not measured or measurable runs the risk of being neglected. It is therefore important that we have tools for assessing the sustainability of our choices when we develop the technologies and systems that shall help us determine and meet the needs of the present generations in a way that does not compromise the ability of our descendants to meet their needs in the future.

As you will learn from this book, we must take a life cycle perspective when we want to assess the sustainability of the solutions that lie in front of us. You will be presented with many examples of problem shifting where solutions that improve or solve a targeted problem unintentionally create other problems of environmental, economic or social nature somewhere else in the systems of processes and stakeholders affected by our choice. If we do not consider the totality of these systems in our analysis, we will fail to notice these unwanted consequences of our decision and we will not be able to take them into consideration. We also have to consider a broad range of potential impacts in our assessment, in fact all those is that the system can contribute to and that we consider relevant in the context of our decision-situation.

Life Cycle Assessment, LCA, offers this totality—it analyses the whole life cycle of the system or product that is the object of the study and it covers a broad range of impacts for which it attempts to perform a quantitative assessment. The focus of LCA has mainly been on the environmental impacts although both social and economic impacts can be included as well. It is an important assessment tool as demonstrated by the central role that it has been given in the environmental regulation in many parts of the world and certified by its ISO standardization and the strong increase in its use over the last decades by companies from all trades and all over the world.

Engineers and scientists who develop decision support, or make decisions where sustainability is a concern, should understand the need to view the solutions in a life cycle perspective and to consider possible trade-offs between environmental impacts and between the three sustainability dimensions. Designers and engineers who design and develop products and technical systems should be able to critically read and evaluate life cycle assessment information about the alternatives that they are considering, and the environmental sustainability specialists among them should also be able to perform the LCA studies.

Why this Book?

It is the purpose of this book to offer the reader the theory and practice of LCA in one volume comprising:

- A textbook, explaining the LCA methodology and the theory behind it in a pedagogical way with a meaningful balance between depth and accessibility
- A cookbook offering recipes with concrete actions needed to perform an LCA
- A repository of information about experience with the use and adaptation of LCA and LCA-based approaches within policy-making, decision support and life cycle engineering and management, and a collection of chapters presenting results and methodological challenges from the use of LCA in some of the central technological application areas of LCA

Focus is on environmental impacts but life cycle sustainability assessment is considered through introductory chapters on social LCA and on life cycle costing.

Who is the Target Audience?

The book was written to support the LCA learning of

- University students, from undergraduate to Ph.D. level
- Researchers and (university) teachers
- Professionals looking to get started on LCA and quantitative (environmental) sustainability assessment
- LCA practitioners looking to deepen their knowledge of specific aspects of LCA methodology (e.g. uncertainty management) and LCA practice in specific areas (e.g. electro-mobility, buildings, biomaterials, etc.) and looking for relevant literature for further reading.

The structure of the book with separate and comprehensive parts on LCA methodology (theory), LCA cookbook (own practice) and LCA applications (practice of others) allows it to cater to the needs of this rather broad group of potential users.

Who Wrote the Book?

A total of 68 authors contributed to the writing of this book (see short presentations of contributors at the end of each chapter). The core team consisted of researchers from the division for Quantitative Sustainability Assessment at the Department of Management Engineering at the Technical University of Denmark, where the three editors have or have had their employment (Ralph Rosenbaum now is an Industrial Chair for Environmental and Social Sustainability Assessment at the French National Research Institute of Science and Technology for Environment and Agriculture (Irstea) in Montpellier, France). Other contributions were solicited from leading experts within each field from the rest of the world, in particular for discussion of the different applications of LCA.

Who made it Possible?

A book like this requires much work apart from the writing of the text before your eyes, and it had never reached your hands without the indispensable contributions from staff of the division for Quantitative Sustainability Assessment at the Department of Management Engineering at the Technical University of Denmark.

We also wish to thank all contributing authors for their timely and fine contributions, their constructive collaboration and not least their patience with a production process that lasted far beyond what was planned when we started.

We hope that this book will find a broad audience worldwide and strengthen the assessment of sustainability in the future, because what gets measured gets managed…

Kongens Lyngby, Denmark — Michael Z. Hauschild
Montpellier, France — Ralph K. Rosenbaum
Kongens Lyngby, Denmark — Stig Irving Olsen

Contents

Editors and Contributors

About the Editors

Michael Z. Hauschild is Professor in Life Cycle Assessment and Head of the division for Quantitative Sustainability Assessment at the Department of Management Engineering, Technical University of Denmark. He has been overall responsible for the department's life cycle engineering research activities, teaching and professional training for more than a decade. A chemical engineer and ecotoxicologist of training, he entered the field of life cycle assessment method development and application with the EDIP project (Environmental Design of Industrial Products) 1992–1997. Together with colleagues he developed and documented one of the first full life cycle assessment methods and received the Great Environmental Prize of the Nordic Council of Ministers 1997 for this work. He has worked internationally in various scientific working groups and held the chair of the SETAC-Europe task force on ecotoxicity assessment in LCIA 1998–2002. 2002–2006 he chaired the UNEP/SETAC Life Cycle Initiative task force on Assessment of Toxic Impacts in LCIA facilitating the development of the UNEP/SETAC consensus model USEtox for evaluation of human and ecotoxicity in LCA, and since 2017 he has chaired the task force on ecotoxic impacts. He has been a member of the editorial board of The International Journal of Life Cycle Assessment since 1998, subject editor for LCIA of human and ecotoxic impacts since 2008, and he has been subject editor on LCA for the Journal of Industrial Ecology since 2010. As a consultant he has assisted in the development of the European Comission's International Life Cycle Data System (ILCD) guideline for LCA and the development of recommendations for life cycle impact assessment under the ILCD system. Furthermore, he is the founding Chair of the Nordic Life Cycle Association, NorLCA, aimed at broad dissemination of life cycle thinking in the Nordic countries and has been active in the International Academy for Production Engineering (CIRP) in agenda setting and support of life cycle engineering activities. He has been teaching LCA methodology and application to university students and professionals in industry and administration during his whole career; at the Technical University of Denmark close to 1000 master of engineering students since the late 1990s have graduated from the LCA course that he has been active in developing, renewing and running through around 20 years.

Ralph K. Rosenbaum is Head of the Industrial Chair for Environmental and Social Sustainability Assessment "ELSA-PACT" at the French National Research Institute of Science and Technology for Environment and Agriculture (Irstea) in Montpellier. Originally from Germany, he received his Environmental Engineering degree (Diplomingenieur) from the Technical University Berlin in 2003. He then pursued his Ph.D. thesis entitled "Multimedia and Food Chain Modelling of Toxics for Comparative Risk and Life Cycle Impact Assessment" at the Swiss Federal Institute of Technology Lausanne (EPFL) until 2006. In early 2007 he joined the team of CIRAIG at the École Polytechnique Montreal, Canada as researcher and lecturer. Before becoming affiliated with Irstea in 2014, he was appointed Associate Professor at the Technical University of Denmark (DTU) in Copenhagen in 2010. In 2015 he defended his Habilitation ("Habilitation à Diriger des Recherches"—HDR), entitled: "Increasing precision and applicability of life cycle impact assessment in the context of comparative environmental sustainability studies" at the University of Montpellier, France. Passionate about quantitative environmental sustainability assessment including Life Cycle Assessment (LCA) since 1997, Ralph Rosenbaum is an expert in environmental modelling, as well as application and development of LCA methodology and teaching related to sustainability and environmental assessment. He is co-author of the UNEP-SETAC consensus model for the evaluation of comparative toxicity USEtox and the LCIA methods IMPACT 2002+, Impact World+ and LC-Impact. Since 2007 he has been a subject editor of The International Journal of Life Cycle Assessment for impacts of chemicals on human health. He has been active in several international expert working groups of the UNEP-SETAC Life Cycle Initiative since its launch in 2002. He was member of the SETAC North America LCA Steering Committee from 2008 to 2010 and the SETAC Europe LCA Steering Committee from 2012 to 2018 and appointed to the LCIA Method Developers Advisory Group of the European LCA Platform project (ILCD) of the EU Commission in 2007. Since 2007 he has been developing and running courses on sustainability, LCA and related concepts and methods, teaching hundreds of professionals from industry, academia and government, as well as more than 600 students from Bachelor to Ph.D. level on three continents, and supervising numerous masters, Ph.D. and postdoc projects.

Stig Irving Olsen is Associate Professor at the division for Quantitative Sustainability Assessment at DTU Management Engineering at Technical University of Denmark. He graduated as environmental biologist from University of Copenhagen with a postgraduate education in Toxicology from the same university. He has a total of 9 years experience as a consultant in the field of toxicology and ecotoxicology. After some years as a consultant, he did an industrial Ph.D. working on "Life Cycle Assessment of basic chemicals" from Technical University of Denmark with five industrial partners, the main one being Novo Nordisk. He combined his knowledge and entered into methods development for life cycle impact assessment of toxic impacts, an area in which he was chairman for SETAC working groups during 1999–2001 and later member of an ensuing WG. He found interest in nanotechnology with the increasing societal focus, the potential environmental benefits, and the potential risks of nanoparticles and he became a member of a working group at the Danish board of technology and was invited to a number of governmental workshops in the USA and EU. He has also studied other emerging technologies such as third-generation biofuels. He has served as a reviewer of research proposals for EU, Sweden, Germany, Portugal and Switzerland. He is senior editor and member of founding board for the journal Integrated Environmental Assessment and Management and submission editor for The International Journal of Life Cycle Assessment. He has been teaching LCA and particularly application of simple LCA for 15 years in four different courses and has supervised numerous bachelors, masters, and Ph.D. students in the field of LCA.

Contributors

Naeem Adibi PLATEFORME [avniR]-cd2e, Loos-en-Gohelle, France; WELOOP, Lens, France

Karen Allacker Department of Architecture, Faculty of Engineering Science, KU Leuven, Leuven, Belgium

Per Dannemand Andersen Division for Technology and Innovation Management, Department of Management Engineering, Technical University of Denmark, Kongens Lyngby, Denmark

Assumpció Antón IRTA, Food and Agricultural Research Institute, Centre de Cabrils, Barcelona, Spain

Ioannis Bakas Division for Quantitative Sustainability Assessment, Department of Management Engineering, Technical University of Denmark, Kongens Lyngby, Denmark

Claudine Basset-Mens CIRAD, UPR Hortsys, ELSA—Research Group, Montpellier Cedex 5, France

Niki Bey Division for Quantitative Sustainability Assessment, Department of Management Engineering, Technical University of Denmark, Kongens Lyngby, Denmark

Morten Birkved Division for Quantitative Sustainability Assessment, Department of Management Engineering, Technical University of Denmark, Kongens Lyngby, Denmark

Anders Bjørn Division for Quantitative Sustainability Assessment, Department of Management Engineering, Technical University of Denmark, Kgs. Lyngby, Denmark; CIRAIG, Polytechnique Montréal, Montreal, QC, Canada

Anne-Marie Boulay LIRIDE, Sherbrooke University, Sherbrooke, Canada; CIRAIG, Polytechnique Montreal, Montreal, Canada

Alexandra Bonou Division for Quantitative Sustainability Assessment, Department of Management Engineering, Technical University of Denmark, Kongens Lyngby, Denmark

Mads Borup Division for Technology and Innovation Management, Department of Management Engineering, Technical University of Denmark, Kongens Lyngby, Denmark

Kossara Petrova Bozhilova-Kisheva Division for Quantitative Sustainability Assessment, Department of Management Engineering, Technical University of Denmark, Kgs.Lyngby, Denmark

Heidi B. Bugge Ecolabelling Denmark, Danish Standards Foundation, Nordhavn, Denmark

Sónia M. Carvalho Division for Quantitative Sustainability Assessment, Department of Management Engineering, Technical University of Denmark, Kongens Lyngby, Denmark

Felipe Cerdas Chair of Sustainable Manufacturing and Life Cycle Engineering, Institute for Machine Tools and Production Technology, Technische Universität Braunschweig, Brunswick, Germany

Julie Clavreul Residual Resources Engineering, Department of Environmental Engineering, Technical University of Denmark, Kongens Lyngby, Denmark

Andrea Corona Division for Quantitative Sustainability Assessment, Department of Management Engineering, Technical University of Denmark, Kongens Lyngby, Denmark

Teunis J. Dijkman Division for Quantitative Sustainability Assessment, Department of Management Engineering, Technical University of Denmark, Kongens Lyngby, Denmark

Patricia Egede Chair of Sustainable Manufacturing and Life Cycle Engineering, Institute for Machine Tools and Production Technology, Technische Universität Braunschweig, Brunswick, Germany

Alexi Ernstoff Division for Quantitative Sustainability Assessment, Department of Management Engineering, Technical University of Denmark, Kongens Lyngby, Denmark; EPFL Innovation Park, Quantis International, Lausanne, Switzerland

Nieves Espinosa Department of Energy Conversion and Storage, Technical University of Denmark, Roskilde, Denmark; Product Policy Bureau, European Commission Joint Research Centre, Sevilla, Spain

Peter Fantke Division for Quantitative Sustainability Assessment, Department of Management Engineering, Technical University of Denmark, Kongens Lyngby, Denmark

Matthias Finkbeiner Chair of Sustainable Engineering, Department of Environmental Technology, Technische Universität Berlin, Berlin, Germany

Jeppe Frydendal Danish Standards Foundation, Nordhavn, Denmark

Eskinder Gemechu ISM-CyVi, UMR 5255, University of Bordeaux, Talence, France

Emmanuel Gentil Copenhagen Resource Institute, Copenhagen K, Denmark

Stylianos Georgiadis Department of Applied Mathematics and Computer Science, Technical University of Denmark, Kongens Lyngby, Denmark; Global Decision Support Initiative, Technical University of Denmark, Kongens Lyngby, Denmark

Berit Godskesen HOFOR (Greater Copenhagen Utility), Copenhagen, Denmark

Benjamin Goldstein Division for Quantitative Sustainability Assessment, Department of Management Engineering, Technical University of Denmark, Kongens Lyngby, Denmark

Lisbeth Engel Hansen Danish Standards Foundation, Nordhavn, Denmark

Sune Balle Hansen Division of Quantitative Sustainability Assessment, Department for Management Engineering, Technical University of Denmark, Kongens Lyngby, Denmark

Michael Z. Hauschild Division for Quantitative Sustainability Assessment, Department of Management Engineering, Technical University of Denmark, Kongens Lyngby, Denmark

Christoph Herrmann Chair of Sustainable Manufacturing and Life Cycle Engineering, Institute for Machine Tools and Production Technology, Technische Universität Braunschweig, Brunswick, Germany

Susanne Vedel Hjuler Division of Quantitative Sustainability Assessment, Department for Management Engineering, Technical University of Denmark, Kongens Lyngby, Denmark; Novozymes A/S, Bagsværd, Denmark

Leise Jebahar Division for Quantitative Sustainability Assessment, Department of Management Engineering, Technical University of Denmark, Kongens Lyngby, Denmark

Louise Laumann Kjær Section of Engineering Design and Product Development, Department of Mechanical Engineering, Technical University of Denmark, Kongens Lyngby, Denmark

Henrik Fred Larsen Danish Road Directorate, Hedehusene, Denmark

Alexis Laurent Division for Quantitative Sustainability Assessment, Department of Management Engineering, Technical University of Denmark, Kongens Lyngby, Denmark

Julia Martínez-Blanco Chair of Sustainable Engineering, Department of Environmental Technology, Technische Universität Berlin, Berlin, Germany

Tuomas J. Mattila Finnish Environment Institute SYKE, Luonnonkoneisto Engineering Co-operative, Helsinki, Finland

Tim C. McAloone Section of Engineering Design and Product Development, Department of Mechanical Engineering, Technical University of Denmark, Kongens Lyngby, Denmark

Noa Meron The Porter School of Environmental Studies, Tel-Aviv University, Tel-Aviv, Israel

Mirko Miseljic Division for Quantitative Sustainability Assessment, Department of Management Engineering, Technical University of Denmark, Kgs. Lyngby, Denmark

Christine Molin Division for Quantitative Sustainability Assessment, Department of Management Engineering, Technical University of Denmark, Kgs. Lyngby, Denmark

Andreas Moltesen Division for Quantitative Sustainability Assessment, Department of Management Engineering, Technical University of Denmark, Kongens Lyngby, Denmark

Monia Niero Division for Quantitative Sustainability Assessment, Department of Management Engineering, Technical University of Denmark, Kongens Lyngby, Denmark

Montserrat Núñez IRSTEA, UMR ITAP, ELSA Research Group and ELSA-PACT Industrial Chair for Environmental and Social Sustainability Assessment, Montpellier, France

Stig Irving Olsen Division for Quantitative Sustainability Assessment, Department of Management Engineering, Technical University of Denmark, Kongens Lyngby, Denmark

Mikołaj Owsianiak Division for Quantitative Sustainability Assessment, Department of Management Engineering, Technical University of Denmark, Kongens Lyngby, Denmark

Aris Pagoropoulos Section of Engineering Design and Product Development, Department of Mechanical Engineering, Technical University of Denmark, Kongens Lyngby, Denmark

Rana Pant European Commission, Directorate D: Sustainable Resources, Bioeconomy Unit, Joint Research Centre, Ispra, Italy

Daniela C.A. Pigosso Section of Engineering Design and Product Development, Department of Mechanical Engineering, Technical University of Denmark, Kongens Lyngby, Denmark

Freja Nygaard Rasmussen Faculty of Engineering and Science, Danish Building Research Institute, Aalborg University, Copenhagen, Denmark

Jon Rasmussen Division for Quantitative Sustainability Assessment, Department of Management Engineering, Technical University of Denmark, Kongens Lyngby, Denmark

Jan-Markus Rödger Division for Quantitative Sustainability Assessment, Department of Management Engineering, Technical University of Denmark, Kongens Lyngby, Denmark

Ralph K. Rosenbaum IRSTEA, UMR ITAP, ELSA Research Group and ELSA-PACT—Industrial Chair for Environmental and Social Sustainability Assessment, Montpellier, France

Morten W. Ryberg Division for Quantitative Sustainability Assessment, Department of Management Engineering, Technical University of Denmark, Kgs. Lyngby, Denmark

Martin Rygaard Department of Environmental Engineering, Technical University of Denmark, Kongens Lyngby, Denmark

Serenella Sala European Commission, Directorate D: Sustainable Resources, Bioeconomy Unit, Joint Research Centre, Ispra, Italy

Anna Bernstad Saraiva SAGE/COPPE, Federal University of Rio de Janeiro, Rio de Janeiro, Brazil

Erwin M. Schau European Commission, Directorate D: Sustainable Resources, Bioeconomy Unit, Joint Research Centre, Ispra, Italy

Gitte Lemming Søndergaard Department of Environmental Engineering, Technical University of Denmark, Kongens Lyngby, Denmark

Guido Sonnemann ISM-CyVi, UMR 5255, University of Bordeaux, Talence, France

Sonia Valdivia World Resources Forum, St. Gallen, Switzerland

Marisa Vieira PRé Consultants bv, Amersfoort, The Netherlands

Arne Wangel Division for Quantitative Sustainability Assessment, Department of Management Engineering, Technical University of Denmark, Kongens Lyngby, Denmark

Caroline M. White Division for Quantitative Sustainability Assessment, Department of Management Engineering, Technical University of Denmark, Kongens Lyngby, Denmark

Part I
Introduction

Chapter 1
About This Book

Michael Z. Hauschild, Ralph K. Rosenbaum and Stig Irving Olsen

Abstract To reach the UN sustainable development goal, there is a need for comprehensive and robust tools to help decision-making identify the solutions that best support sustainable development. The decisions must have a system perspective, consider the life cycle, and all relevant impacts caused by the solution. Life Cycle Assessment (LCA) is a tool that has these characteristics and the ambition with this book is to offer a comprehensive and up-to-date introduction to the tool and its underlying methodological considerations and potential applications. The book consists of five parts. The first part introduces LCA. The second part is a text book aiming at university students from undergraduate to PhD level, and professionals from industry and within policy making. It follows ISO 14040/14044 structure, draws upon a variety of LCA methods published over the years, especially the ILCD, and offers prescriptions and recommendations for all the most important methodological choices that you meet when performing an LCA. The third part introduces applications of LCA and life cycle thinking by policy- and decision-makers in government and industry. The fourth part is a Cookbook guiding you through the concrete actions to undertake when performing an LCA. The fifth part contains some appendices. The book can be used as a text book, the chapter can be read as stand alone, and you can use the Cookbook as a manual on how to perform an LCA.

M.Z. Hauschild (✉) · S.I. Olsen
Division for Quantitative Sustainability Assessment, Department of Management Engineering, Technical University of Denmark, Kongens Lyngby, Denmark
e-mail: mzha@dtu.dk

R.K. Rosenbaum
IRSTEA, UMR ITAP, ELSA Research Group and ELSA-PACT—Industrial Chair for Environmental and Social Sustainability Assessment, Montpellier, France

M.Z. Hauschild et al. (eds.), *Life Cycle Assessment*,
DOI 10.1007/978-3-319-56475-3_1

1.1 Introduction

Our generation is facing daunting challenges of a changing climate and an overall increasing pressure on the environment, challenges that are under the influence of human-made activities. Reflecting on these environmental conditions and their relationship to social and economic challenges that we face, a sustainable development was coined in 1987 by UN's World Commission for Environment and Development as a development that "… meets the needs of the present generations without compromising the ability of future generations to meet their own needs" (UN WCED 1987). In 2015 the 193 member states of the United Nations adopted 17 goals to 'end poverty, protect the planet, and ensure prosperity for all as part of a new sustainable development agenda' by 2030, setting targets for the way in which the present generations can meet their needs (UN 2017). To meet the goals and targets, sustainability must gain strong prominence in decision support for professionals who are responsible for creating solutions for the future, but also for everybody else who, in today's global economy, is both a stakeholder and a decision-maker with a role to play concerning sustainability as a consumer, as member of a local community, or as a voter. Each individual needs answers and information based on comprehensive and robust tools to help them decide what best supports a sustainable development, from small- to large-scale decisions. To avoid the often seen problem shifting where solutions to a problem creates several new and often ignored problems, these decisions must take a systems perspective. They must consider what in this book is referred to as the life cycle of the solution, and they need to consider all the relevant impacts caused by the solution. Life Cycle Assessment (LCA) is a tool that has these characteristics, and there is a strong and growing need for professionals who understand or even master this tool and who know how to critically appraise and use the information that it provides. It is our ambition with this book to offer a comprehensive and up-to-date introduction to the tool and its underlying methodological considerations and potential applications.

1.2 Structure of This Book

The book consists of five parts.

The **first part** sets the scene. First, if you are a newcomer to LCA, you get a short introduction to important characteristics of LCA and some of its strengths and weaknesses, illustrated through a collection of questions that LCA can—or cannot be used for answering. This short introduction is followed by a presentation of the history of LCA from its early beginnings half a century ago to today, with a focus on methodological developments, growth in number and variety of applications and international harmonization and consensus building. Finally, LCA is positioned in the context of sustainability and its use as a tool for quantitative sustainability assessment is discussed.

The **second part** is a textbook aiming at university students from undergraduate to PhD level, and professionals from industry and within policy making who need a thorough and pedagogical introduction to LCA methodology. The textbook has been developed based on a cumulated experience from more than three decades with teaching LCA to engineering students at undergraduate and master level courses at Technical University of Denmark and Polytechnique Montréal, Canada, and it is intended to provide a complete curriculum for such courses.

The structure of the introduction to the LCA methodology follows the ISO framework (as presented and elaborated in the ISO 14040 and ISO 14044 standards (ISO 2006a, b)), and we have strived to keep the use of technical terms in accordance with the ISO terminology. When it comes to the methodological details, the ISO standards refrain from prescriptions or recommendations for many of the detailed decisions and choices that must be made by a practitioner who wants to perform an LCA. Here, we have sought inspiration in LCA methods published over the years, including the EDIP method (Wenzel et al. 1997; Hauschild and Wenzel 1998), the Ecoinvent methodologies (Weidema et al. 2013), the Consequential LCA (Ekvall and Weidema 2004), as well as more recent projects within the UNEP/SETAC Life Cycle Initiative and the development of the IMPACT World+ (http://www.impactworldplus.org), the LC-Impact (http://www.lc-impact.eu), or the ILCD life cycle impact assessment methods (Hauschild et al. 2013), and not least in the detailed guidance offered by the General guide for Life Cycle Assessment, the 'ILCD Handbook' that was elaborated by the European Commission to serve as the methodological backbone of its International Reference Life Cycle Data System (EC-JRC 2010). The ILCD Handbook was developed through a broad international consultation process with LCA experts, stakeholders and the public from all over the world with the ambition to minimize ambiguity in LCA studies and provide governments and businesses with a basis for assuring quality and consistency of life cycle data, methods and assessments (EC-JRC 2010; Pennington et al. 2010; Sala et al. 2012). Building on methodological elements from previously published LCA methods, it offers prescriptions and recommendations for all the most important methodological choices that you meet when performing an LCA. We use the ILCD method as a solidly founded, well documented, and detailed reference methodology that is in full accordance with the ISO standards and details methodology descriptions far beyond them.

This part of the textbook offers separate chapters on each phase of the LCA methodology and additional chapters on life cycle costing and social life cycle assessment as well as chapters on central methodological aspects like uncertainty management and sensitivity analysis, and use of input–output analysis in LCA.

The **third part** of the book offers a collection of chapters introducing applications of LCA and life cycle thinking by policy- and decision-makers in government and industry, written by authors who are experts in the field of their chapter. They start out with policy applications around the world and organizational LCA, then move on to industrial applications, life cycle management, ecodesign, environmental labels and declarations, and the Cradle to cradle concept. The focus then moves on to the application of LCA to different technological areas like energy systems, buildings, food and waste management. Eleven chapters present, within each their technological area, the main types of findings from published LCA studies, identifying methodological considerations that are particularly relevant and highlighting potential pitfalls when performing or using LCA studies within that area.

The **fourth part** consists of a Cookbook which takes you through all the phases of the LCA once more, but this time with concrete actions to undertake when performing an LCA. The ambition with the cookbook is to provide you with the recipes for performing an LCA. Where Part II answers the numerous 'why' questions, the Cookbook answers the 'what' and 'how' questions. It is intended to guide you through the many steps, activities and decisions that are needed to perform an LCA. The Cookbook follows the main structure of the ISO 14044 standard and gives detailed instructions on all the central activities, based on selection of those provisions and actions in the ILCD Handbook that are generally needed in order to perform an LCA.

The **fifth part** of the book is an appendix collection with supporting material for use in LCA teaching like a reporting template offering the student a recommended structure for an LCA report, an example of a complete LCA report on a case study based on student results from an LCA course, and an overview and comparison of existing life cycle impact assessment methods to compliment the methodology chapter on this phase of the LCA.

1.3 How to Use This Book?

As you will see, you may use this book *as a textbook*, focusing on the description of the theory in Part II. All the basic elements of the methodology are presented in chapters with clearly defined learning objectives. An exemplary LCA case study weaves through the methodology chapters and is used, where relevant, to give practical examples of the presented methodological elements. The case study is compiled at the end in a full LCA report in Part V of the book, illustrating the use of the reporting template and serving as an example for students of how a good student LCA report may look. You can select chapters from Part III of the book on the LCA applications that are relevant in your didactic context, and you can use the Cookbook in Part IV and the reporting template and example LCA report in Part V for support to perform a real LCA if this is part of your learning. Each chapter of the book was written in a way that allows it to also function as stand-alone material for studying the respective aspects that it presents. The chapters can thus also be read on their own in order to deepen your knowledge on their specific topics.

Once you have taken the learning from the book, you can use the Cookbook **as a manual on how to perform an LCA**. The cookbook is based on the ILCD guideline and it is thus not a universally endorsed LCA method—in fact, such a method does not exist beyond the ISO standards. We have, however, found that this guideline is useful as a reference because of its very detailed prescriptions. In cases where you disagree with certain provisions or where a different approach is more relevant for the study that you perform, it will still serve as a reference for transparently and efficiently reporting about the method that you have used by specifying the points where you have chosen a different approach.

Whether you aspire to be a practitioner of LCA or a user of LCA information, the textbook will also serve **as a repository of LCA experience** with the wealth of information on the many application areas presented in Part III of the book.

We wish you a fruitful learning with the book and success with your future LCA activities!

References

EC-JRC: European Commission—Joint Research Centre—Institute for Environment and Sustainability: International Reference Life Cycle Data System (ILCD) Handbook—General guide for Life Cycle Assessment—Detailed guidance. First edition March 2010. EUR 24708 EN. Luxembourg. Publications Office of the European Union (2010)

Ekvall, T., Weidema, B.P.: System boundaries and input data in a consequential life cycle inventory analysis. Int. J. LCA **9**(3), 161–171 (2004). doi:10.1007/BF02994190

Hauschild, M.Z., Wenzel, H.: Environmental Assessment of Products. Vol. 2—Scientific Background, 565 pp. Chapman & Hall, UK, Kluwer, Hingham. ISBN 0412 80810 2 (1998)

Hauschild, M.Z., Goedkoop, M., Guinée, J., Heijungs, R., Huijbregts, M., Jolliet, O., Margni, M., De Schryver, A., Humbert, S., Laurent, A., Sala, S., Pant, R.: Identifying best existing practice for characterization modeling in life cycle impact assessment. Int. J. Life Cycle Assess. **18**, 683–697 (2013)

ISO: Environmental Management—Life Cycle Assessment—Principles and Framework (ISO 14040). ISO, the International Organization for Standardization, Geneva (2006a)

ISO: Environmental Management—Life Cycle Assessment—Requirements and Guidelines (ISO 14044). ISO, the International Organization for Standardization, Geneva (2006b)

Pennington, D.W., Chomkhamsri, K., Pant, R., Wolf, M.-A., Bidoglio, G., Kögler, K., Misiga, P., Sponar, M., Lorz, B., Sonnemann, G., Masoni, P., Wang, H., Ling, L., Castanho, C., Sau Soon, C., Fieschi, M., Filareto, A., Hauschild, M.: ILCD handbook public consultation workshop—International Reference Life Cycle Data System (ILCD). Int. J. LCA **15**(3), 231–237 (2010)

Sala, S., Pant, R., Hauschild, M., Pennington, D.: Research needs and challenges from science to decision support. Lesson learnt from the development of the International Reference Life Cycle Data System (ILCD) recommendations for life cycle impact assessment. Sustainability **4**(7), 1412–1425 (2012)

UN: http://www.un.org/sustainabledevelopment/sustainable-development-goals/ (2017). Accessed 4 March 2017

UN WCED: Our Common Future. Oxford University Press, Oxford (1987)

Weidema, B.P., Bauer, C., Hischier, R., et al.: The Ecoinvent Database: Overview and Methodology, Data Quality Guideline for the Ecoinvent Database Version 3 (2013). www.ecoinvent.org

Wenzel, H., Hauschild, M.Z., Alting, L.: Environmental Assessment of Products. Vol. 1—Methodology, Tools and Case Studies in Product Development, 544 pp. Chapman & Hall, UK, Kluwer, Hingham (1997). ISBN 0 412 80800 5

Author Biographies

Michael Z. Hauschild involved in development of LCIA methodology since the early 1990s. Has led several SETAC and UNEP/SETAC working groups and participated in the development of the ISO standards and the ILCD methodological guidelines. Main LCA interests are chemical impacts, spatial differentiation and science-based boundaries in LCIA.

Ralph K. Rosenbaum LCA expert and environmental modeller focusing on LCIA development since early 2000s. Contributed to several UNEP/SETAC working groups towards global harmonisation of LCA methodology. Interested in LCIA modelling of emissions and water/soil resource use, operationalisation of uncertainty management and spatial differentiation.

Stig Irving Olsen LCA expert both as researcher and as consultant. Involved in the development of LCA methodologies since mid 1990s. Contributed to UNEP/SETAC working groups on LCIA methodology. Main LCA interest is human toxicity impacts, emerging technologies, and decision making.

Chapter 2
Main Characteristics of LCA

Anders Bjørn, Mikołaj Owsianiak, Christine Molin and Alexis Laurent

Abstract Life cycle assessment (LCA) has a number of defining characteristics that enables it to address questions that no other assessment tools can address. This chapter begins by demonstrating how the use of LCA in the late 2000s led to a drastic shift in the dominant perception that biofuels were "green", "sustainable" or "carbon neutral", which led to a change in biofuel policies. This is followed by a grouping of the LCA characteristics into four headlines and an explanation of these: (1) takes a life cycle perspective, (2) covers a broad range of environmental issues, (3) is quantitative, (4) is based on science. From the insights of the LCA characteristics we then consider the strengths and limitations of LCA and end the chapter by listing 10 questions that LCA can answer and 3 that it cannot.

Learning objectives After studying this chapter the reader should be able to:

- Explain the relevance of LCA as a tool for environmental management.
- Explain four main characteristics of LCA.
- Demonstrate an understanding of strengths and limitations of LCA by providing examples of environment-related questions that LCA can answer and questions that LCA cannot answer.

A. Bjørn · M. Owsianiak · C. Molin · A. Laurent
Division for Quantitative Sustainability Assessment,
Department of Management Engineering,
Technical University of Denmark, Kongens Lyngby, Denmark

A. Bjørn (✉) · C. Molin · A. Laurent
CIRAIG, Polytechnique Montréal, 3333 Chemin Queen-Mary,
Montréal, QC, Canada
e-mail: anders.bjoern@polymtl.ca

M.Z. Hauschild et al. (eds.), *Life Cycle Assessment*,
DOI 10.1007/978-3-319-56475-3_2

2.1 Why Is LCA Important? Biofuel Case

LCA has a number of defining characteristics. Before elaborating on these characteristics a real life case is presented to show how the use of LCA provided new insights and led to major changes in policy. This is the case of first generation biofuels used in the transport sector.

The use of biofuels is not a new trend. They were used in the form of wood and peat before the industrialisation and were pretty much the only source of fuels then. This changed with the emergence of cheap fossil fuels, first in the form of coal, later followed by oil and natural gas. By the end of the twentieth-century fossil fuels had become the dominating source for meeting the world's primary energy demand. At the same time the transportation sector of developed nations was responsible for an increasing share of the total national energy demands [e.g. EC (2012)]. While electricity and heat increasingly were supplied by other sources than fossil fuels, a similar transition could not be observed for transportation energy (IEA 2015).

The 2000s witnessed a renewed interest in using biofuels in the transportation sector, spurred by increasing oil prices, the question of energy security and concerns over climate change. Biofuels were seen as potentially cost competitive with gasoline and diesel and they were considered means to reduce dependencies on large exporters of oil, many of which were (and are) located in politically unstable regions of the world. In the early 2000s biofuels in the transportation sector were also generally considered much better for the climate than fossil fuels. The reasoning was that the CO_2 emitted from the combustion of biofuels has a "neutral" effect on climate change, because it belongs to the biogenic carbon cycle, meaning that it used to be in the atmosphere before being taken up, via photosynthesis, by the plants that were the sources of the biofuel and that it will be taken up by new plants again. By contrast, CO_2 emitted from the combustion of fossil fuels originates in carbon that belongs to the much slower geological carbon cycle and can be considered as effectively isolated from the atmosphere, because it would have stayed in the ground for millions of years, had it not been extracted to be used as fuel.

While the distinction between biogenic and fossil CO_2 is important, LCA studies (Zah and Laurance 2008; Fargione et al. 2008; Searchinger et al. 2008) have shown that it was a mistake to:

(1) consider the use of biofuels in the transport sector inherently "climate neutral"
(2) disregard potential increases in environmental problems other than climate change from a transition from fossil fuels to biofuels.

Regarding the first point, LCA takes a life cycle perspective when evaluating environmental impacts of a product or a system. In this case it means not only considering the use stage of the biofuel, i.e. where its chemical energy is transformed to kinetic energy in a vehicle's combustion engine, but also considering the industrial and agricultural processes prior to the delivery of the biofuel to the fuel tank of the vehicle (see Fig. 2.1).

Fig. 2.1 Graphic representation of the biofuels life cycle from feedstock to end user (Icons made by Flaticon from www.flaticon.com)

When taking a life cycle perspective it is clear that no biofuel is "climate neutral", because of the inputs of fossil fuels needed in industrial processes prior to the use stage. In addition, a consequence of the increased demand for biofuel crops may be the conversion of natural land (such as forest) to cultivated land and this releases the carbon bound in the natural biomass (e.g. wood) and the soil as CO_2. Sometimes the conversion of natural land happens as an indirect consequence, i.e. forest is being cleared to make room for the crops that used to be cultivated at the piece of land now used for biofuel crops. This means that a country that increases its production of biofuel crops, at the expense of a decrease in food crops may indirectly contribute to a loss of natural land (e.g. forest) somewhere else, possibly on a different continent, due to the mechanisms of international trade.

Regarding the second point, LCA considers multiple environmental issues (and sometimes social issues, see Chap. 16) when evaluating a product or a system. This is an important attribute in the case of biofuels because the release of nutrients from fertilizer use and synthetic chemicals from pesticide use, lead to eutrophication and toxic effects on freshwater ecosystems and elsewhere, and because the cultivation requires large amounts of land and water for irrigation, which can lead to biodiversity loss and water scarcity. Social impacts from an increased production of biofuels have also been reported in the form of increasing food prices.

The insights provided by LCA were a key reason for the rapid change in perspective on biofuels by policy-makers and media that began around 2008. For example, in 2010 the European Commission amended its legislation on biofuels by introducing a set of sustainability criteria, which relates to life cycle emissions of greenhouse gases and prohibits the conversion of land with previously "high carbon stock" and "high biodiversity" for the production of biofuels (EC 2010).

With the above text, we are not arguing that the transportation sector should abandon biofuels as a strategy to reduce its use of fossil fuels and climate impacts. We are merely trying to show that the world is not black and white and that a more holistic perspective is required when evaluating and guiding technological changes.

2.2 Main Characteristics

Having made a case for LCA with the topic of biofuels, we now turn to describing its main characteristics in slightly more technical terms and end the chapter by listing its strengths and limitations.

2.2.1 Takes a Life Cycle Perspective

The life cycle metaphor is borrowed from the field of biology. For example, the life cycle of a butterfly starts with an egg, which bursts and lets a caterpillar out that turns into a pupa from which a butterfly emerges that eventually dies after laying eggs for the cycle to be repeated. In much the same way a man-made object starts its lifecycle by the harvesting and extraction of resources, followed by production, use and eventually management of the object as waste, which marks the end of the life cycle. Recycling or reuse can be seen as "new eggs" for the life cycles of other man-made objects. The objects studied in LCA are often physical products and the term "product system" signals that a life cycle perspective is taken, i.e. that all the processes required to deliver the function of the product are considered. For example, the function of a car fuel is to propel a car. As illustrated in the case above, the delivery of this function requires a number of industrial and agricultural processes that can be conceptually organised in stages of the life cycle of a biofuel (see Fig. 2.1). The core reason for taking a life cycle perspective is that it allows identifying and preventing the burden shifting between life cycle stages or processes that happens if efforts for lowering environmental impacts in one process or life cycle stage unintentionally create (possibly larger) environmental impacts in other processes or life cycle stages. As shown above, the substitution of fossil fuels with biofuels reduces impacts on climate change from the use stage but increases climate change impacts from the harvest and extraction stage. Although LCA is mostly used to study product systems, it can also be used to study more complex man-made objects, such as companies (see Chap. 22), energy-, transport- or waste management systems (see Chaps. 26, 27 and 35) and infrastructure and cities (see Chap. 28). In all applications the assessment takes a life cycle perspective having the function of the studied entity as focal point.

2.2.2 Covers a Broad Range of Environmental Issues

In LCA, the comprehensive coverage of processes over the life cycle is complemented by a comprehensive coverage of environmental issues. Rather than focusing exclusively on, say, climate change, which generally receives most attention these days, LCA covers a broad range of environmental issues, typically around fifteen (see Chap. 10). These issues include climate change, freshwater use, land occupation and transformation, aquatic eutrophication, toxic impacts on human health, depletion of non-renewable resources and eco-toxic effects from metals and synthetic organic chemicals. The core reason for considering multiple environmental issues is to avoid burden shifting, which is also why a life cycle perspective is taken. Here burden shifting happens if efforts for lowering one type of environmental impact unintentionally increase other types of environmental impacts.

As shown above, decreasing impacts on climate change by substituting fossil fuels with biofuels has the potential to cause an increase in other environmental issues such as water scarcity, eutrophication, land occupation and transformation.

2.2.3 *Is Quantitative*

LCA results answer the question "how much does a product system potentially impact the environment?" Part of the answer may be "the impact on climate change is 87 kg of CO_2 equivalents". The quantitative nature of LCA means that it can be used to compare environmental impacts of different processes and product systems. This can, for example, be used to judge which products or systems are better for the environment or to point to the processes that contribute the most to the overall impact and therefore should receive attention. LCA results are calculated by (1) mapping all emissions and resource uses and, if possible, the geographical locations of these, and (2) use factors derived from mathematical cause/effect models to calculate potential impacts on the environment from these emissions and resource uses. The first step often involves thousands of emissions and resource uses, e.g. "0.187 kg CO_2, 0.897 kg nitrogen to freshwater, 0.000000859 kg dioxin to air, 1.54 kg bauxite, 0.331 m^3 freshwater…". In the second step the complexity is reduced by classifying these thousands of flows into a manageable number of environmental issues, typically around fifteen (see above). Quantifications generally aim for the "best estimate", meaning that average values of parameters involved in the modelling are consistently chosen (see Chap. 10).

2.2.4 *Is Based on Science*

The quantification of potential impacts in LCA is rooted in natural science. Flows are generally based on measurements, e.g. water gauges or particle counters at industrial sites or mass balances over the processes. The models of the relationships between emission (or resource consumption) and impact are based on proven causalities, e.g. the chemical reaction schemes involving nitrogen oxides and volatile organic compounds in the formation of atmospheric ground level ozone (smog) or on empirically observed relationships, e.g. between the concentration of phosphorous in a lake and the observed numbers of species and their populations. On top of its science core, LCA requires value judgement, which is most evident in the optional step of assigning weights to different types of environmental problems to evaluate the overall impact of a product system. LCA strives to handle value judgement consistently and transparently and in some cases allows practitioners to make modelling choices based on their own values, for example with respect to the number of years into the future that environmental impacts should be considered in the assessment.

2.3 Strengths and Limitations of LCA

A main strength of LCA is its comprehensiveness in terms of its life cycle perspective and coverage of environmental issues. This allows the comparison of environmental impacts of product systems that are made up of hundreds of processes, accounting for thousands of resource uses and emissions that are taking place in different places at different times. However, the comprehensiveness is also a limitation, as it requires simplifications and generalisations in the modelling of the product system and the environmental impacts that prevent LCA from calculating *actual* environmental impacts. Considering the uncertainties in mapping of resource uses and emissions and in modelling their impacts and the fact that calculated impacts are aggregated over time (e.g. tomorrow and in 20 years) and space (e.g. Germany and China) it is more accurate to say that LCA calculates impact *potentials*.

Another strength in the context of comparative assessments is that LCA follows the "best estimate" principle. This generally allows for unbiased comparisons because it means that the same level of precaution is applied throughout the impact assessment modelling. A limitation related to following the "best estimate" principle is, however, that LCA models are based on the average performance of the processes and do not support the consideration of risks of rare but very problematic events like marine oil spills or accidents at industrial sites. As a consequence, nuclear power, for example, appears quite environmentally friendly in LCA because the small risk of a devastating disaster, like the ones that happened in Chernobyl, the Ukraine or Fukushima, Japan, is not considered.

A final limitation worth keeping in mind is that, while LCA can tell you what (product system) is better for the environment, it cannot tell you if better is "good enough". It is therefore wrong to conclude that a product is environmentally sustainable, in absolute terms, with reference to an LCA showing that the product has a lower environmental impact than another product. Chapter 5 elaborates on the relationship between LCA and sustainability.

The above characteristics mean that LCA is suitable for answering some questions and unsuitable for answering others. Box 2.1 provides examples of questions that LCA can and cannot answer.

Box 2.1. What LCA can and cannot answer

Examples of questions LCA can answer:

1. Is paper, plastic or textile bags the most environmentally friendly option for carrying groceries back from the supermarket?
2. From an environmental point of view should we use glass fibre composite or steel for the car body?
3. How can the overall environmental impact of a refrigerator be minimised with the least effort?

4. What is the most environmentally friendly way to package and transport food?
5. From an environmental perspective, should plastics be incinerated or recycled and which parameters do the conclusion depend on?
6. Where is the environmental optimum in the trade-off between minimising heat loss and minimising the use of impact-intensive materials in a window (see illustrative case on window frames in Chap. 39)?
7. Should a plastic zipper be added to cheese packaging to reduce household food waste and thereby reduce the overall environmental impacts of cheese?
8. Is it more environmentally friendly to do the dishes manually or using a dishwasher?
9. Should a company target its own processes, its suppliers, its customers or the waste management sector in the effort of reducing the environmental impact of its products?
10. Are electric cars more environmentally friendly than conventional internal combustion engine cars and what are the important parameters deciding this (see Chap. 27)?

Examples of questions LCA cannot answer:

1. Should taxes on old diesel cars be increased to reduce emissions of particles and thereby reduce hospital spending on treating lung diseases?

Explanation: LCA cannot be used to compare the societal disadvantages of higher taxes with advantages of less pollution. Cost benefit analysis combined with Health Assessment Studies would be a better tool for answering this question.

2. Do current emissions from a specific factory lead to pollutant concentrations above regulatory thresholds in nearby aquatic ecosystems?

Explanation: LCA is not designed to evaluate impacts of a single emission source on local ecosystems and contains no information on regulatory thresholds. Chemical risk assessment is a more appropriate tool for answering this question.

3. Do total global emissions of endocrine disruptors cause polar bears to become hermaphrodites?

Explanation: LCA is not designed to assess a specific effect on a specific organism from a specific group of chemicals. It would be more meaningful to measure the concentration of endocrine disruptors in (deceased) polar bears and compare those measurements with observed occurrences of hermaphrodite individuals.

References

EC: Communication from the Commission on the practical implementation of the EU biofuels and bioliquids sustainability scheme and on counting rules for biofuels. Brussels (2010)

EC: EU transport in figures. Statistical Pocketbook 2012. Brussels (2012)

Fargione, J., Hill, J., Tilman, D., et al.: Land clearing and the biofuel carbon debt. Science **319**, 1235–1238 (2008). doi:10.1126/science.1152747

IEA: Recent energy trends in OECD. Excerpt from energy balances of OECD countries (2015)

Searchinger, T., Heimlich, R., Houghton, R.A., et al.: Use of U.S. croplands for biofuels increases greenhouse gases through emissions from land-use change. Science **319**, 1238–1240 (2008). doi:10.1126/science.1151861

Zah, R., Laurance, W.: Biofuel's impact on environment. Ind. Bioprocess **30**, 9 (2008)

Author Biographies

Anders Bjørn part of the LCA community since the early 2010s. Main focus is interpretations of sustainability and integration of sustainability targets in LCA to enable absolute sustainability assessments.

Mikołaj Owsianiak involved in development and application of life cycle impact assessment methods in sustainability assessment of technologies. Has worked on issues associated with: soils (remediation), metals (toxic impact assessment), biodiesel (fate in the environment), and carbonaceous materials (biochar and hydrochar).

Christine Molin active in the field of LCA since 1992. Special interest in the development and dissemination of LCA and in the use of LCA in small and medium sized enterprises.

Alexis Laurent working with LCA since 2010 with a strong LCIA focus, particularly on normalisation aspects. Main LCA interests include development of LCIA methods, LCA applications and footprinting of large-scale systems for policy-making (e.g. nations, sectors), and LCA applied to various technology domains, including energy systems.

Chapter 3
LCA History

Anders Bjørn, Mikołaj Owsianiak, Christine Molin and Michael Z. Hauschild

Abstract The idea of LCA was conceived in the 1960s when environmental degradation and in particular the limited access to resources started becoming a concern. This chapter gives a brief summary of the history of LCA since then with a focus on the fields of methodological development, application, international harmonisation and standardisation, and dissemination. LCA had its early roots in packaging studies and focused mainly on energy use and a few emissions, spurring a largely un-coordinated method development in the US and Northern Europe. Studies were primarily done for companies, who used them internally and made little communication to stakeholders. After a silent period in the 1970s, the 1980s and 1990s saw an increase in methodological development and international collaboration and coordination in the scientific community and method development increasingly took place in universities. With the consolidation of the methodological basis, application of LCA widened to encompass a rapidly increasing range of products and systems with studies commissioned or performed by both industry and governments, and results were increasingly communicated through academic papers and industry and government reports. To this day, methodological development has continued, and increasing attention has been given to international scientific consensus building on central parts of the LCA methodology, and standardisation of LCA and related approaches.

Learning Objectives
After studying this chapter, the reader should be able to:

- Explain how LCA emerged and what characterised the early years of development.

A. Bjørn (✉) · M. Owsianiak · C. Molin · M.Z. Hauschild
Division for Quantitative Sustainability Assessment, Department of Management Engineering, Technical University of Denmark, Kongens Lyngby, Denmark
e-mail: anders.bjoern@polymtl.ca

A. Bjørn · C. Molin · M.Z. Hauschild
CIRAIG, Polytechnique Montréal, 3333 Chemin Queen-Mary, Montréal, QC, Canada

M.Z. Hauschild et al. (eds.), *Life Cycle Assessment*,
DOI 10.1007/978-3-319-56475-3_3

- Outline the history of LCA from the 1970s to the present in terms of methodological development, application, international harmonisation and standardisation and dissemination.

3.1 Introduction

Concerns over environmental pollution and energy and material scarcity have motivated the development of life-cycle-oriented approaches for environmental profiling of products. Life Cycle Assessment (LCA) has experienced a strong development both in methodology and applications since the first life-cycle-oriented methods were proposed in the 1960s. Today LCA is defined as “a tool to assess the potential environmental impacts and resources used throughout a product’s life cycle, i.e. from raw material acquisition, via production and use stages, to waste management” (ISO 2006b). In this chapter, we present a brief account of the history of LCA in terms of methodological development, standardisation and regulation, application, and education and dissemination. Important elements of the history are summarised chronologically in Table 3.1.

3.2 Methodological Development

Life-cycle-oriented methods that were precursors of today’s LCA were developed in the 1960s in collaboration between universities and industry. They were known as Resource and Environmental Profile Analysis (REPA) (Hunt et al. 1992) or Ecobalances until the term LCA became the norm in the 1990s. The method development initiated in the US and mainly took place there and in Northern Europe. Early methods could be characterised as material and energy accounting and were inspired by material flow accounting, as they were focused on inventorying energy and resource use (crude oil, steel, etc.), emissions and generation of solid waste, from each industrial process in the life cycle of product systems.

As inventories got more complex, the initial focus on accounting the physical flows in a product life cycle was gradually extended with a translation of the inventory results into environmental impact potentials. In other words, from a list of resource uses and emissions a set of indicator scores for an assessed product was calculated, representing contributions to a number of impacts categories, such as climate change, eutrophication and resource scarcity.

In the early years of the LCA history, environmental concerns addressed by the methods tended to shift with public concerns, and there was no consistency or harmonisation of the applied methods. In some years, the focus was on the generation of solid waste, which was considered problematic, especially in the US, where landfilling was the dominant waste management practice. In other years,

Table 3.1 Selected events in LCA history

Event	Year	Note
The (perhaps) first LCA-oriented study was presented on energy requirements for the production of chemical intermediates and products	1963	World Energy Conference, Harold Smith
Coca Cola commissions its first study comparing beverage containers	1969	Not public
The methodological foundation for environmentally extended input/output analysis is made	1970	Leontief (1970)
Publication of the first public and peer-reviewed LCA study "Resource and Environmental Profile Analysis of Nine Beverage Container Alternatives", commissioned by the US EPA	1974	EPA (1974)
First impact assessment method based on critical volumes introduced	1984	BUS (1984)
The first widely used commercial LCA software, GaBi, was released in its first version	1989	Thinkstep (2016)
SimaPro, another widely used commercial LCA software, was released in its first version	1990	PRé (2016)
The term "life cycle assessment" was coined	1990	SETAC (1991)
Emergence of a number of LCI databases managed by different institutions	Early 1990s	
First environmental theme-oriented impact assessment methodology, CML92	1992	Heijungs et al. (1992)
SETAC Code of Practice published in effort to harmonise LCA framework, terminology and methodology	1993	SETAC (1993b)
The academic journal fully dedicated to LCA, The International Journal of Life Cycle Assessment, was born	1996	
ISO 14040 standard on LCA principles and framework released	1997	ISO 14040
ISO 14041 standard on goal and scope definition released	1998	ISO 14041
Damage-oriented methodology Eco-indicator 99 emerges	1999	Goedkoop and Spriensma (2000)
ISO 14042 standard on life cycle impact assessment released	2000	ISO 14042
ISO 14043 standard on life cycle interpretation released	2000	ISO 14043
UNEP/SETAC Life Cycle Initiative launched	2002	
The LCI database ecoinvent version 1.01 is released	2003	Ecoinvent (2016)
Establishing of a general methodological framework and guideless for LCA through ISO 14040 and ISO 14044	2006	
A framework for Life Cycle Sustainability Analysis was proposed	2008	Klöpffer (2008)
ILCD handbook published	2010	EC (2010)
PEF and OEF guidelines published	2012 and later	

when the price on oil was fluctuating or high, energy use was the focus of early studies. Public concerns also shifted with respect to emissions, which in some periods were deemed to be sufficiently controlled by regulation and voluntary measures by industry, but at other times considered very problematic. Early impact assessment methods tended to represent impacts from emissions in the form of dilution volumes of air or water needed to dilute the emissions to safe levels, or below regulatory thresholds [e.g. the Swiss Ecopoint method from the 1980s (Ahbe et al. 1990)].

During the 1990s many impact assessment methods evolved, and the ambition has since then been to quantify all relevant environmental impacts, independent of shifting public concerns, with the goal of avoiding burden shifting. The first impact assessment methodology to cover a comprehensive set of midpoint impact categories, as we know them today, was CML92 (Heijungs et al. 1992). It was released in 1992 by the Institute of Environmental Sciences at Leiden University in the Netherlands. The Swedish EPS method (Steen 1999a, b) looking at the damages caused took a different approach focusing on the damages to ecosystems and human health, rather than midpoint impacts, an approach that was followed by the Dutch Eco-indicator 99 methodology released in 1999 with a more science-based approach to the damage modelling (Goedkoop and Spriensma 2000). The early 1990s also saw the birth of a number of life cycle inventory databases managed by different institutes and organisations and covering different industrial sectors. Due to differences in data standards and quality, the resource uses and emissions of a single industrial process could, however, differ substantially in the different databases, but at this point in the development, the focus was on expanding the coverage and for many processes, there were no data at all. This situation was improved in 2003 with the release of the first ecoinvent database (v 1.01) covering all industrial sectors and aiming for consistent data standards and quality (ecoinvent 2016).

In parallel to this development in process-based LCA, a "top-down" approach was developed based on the work of the economist Wassily Leontief on input-output analysis of economies (Leontief 1970). This "top-down" approach to constructing an inventory is based on combining the national statistics of the trade between sectors with information on sector-specific environmental loads to arrive at an environmentally extended input/output analysis (EEIOA see more in Chap. 14).

Inherent in the discussion of LCI data was also a more fundamental difference in the perception of the product life cycle and LCA and its potential application. The attributional perspective aims to quantify the environmental impacts that can be attributed to the product system based on a mapping of the emission and resource flows that accompany the product as it moves through its life cycle, applying representative average data for all processes involved in the life cycle in a book keeping approach. The consequential perspective is concerned with the potential consequences of the decision based on the results of the LCA, and involves modelling of the broader economic system that the decision affects (see Sect. 8.5).

The modelling of increasingly complex product systems and the proliferation of LCI data and impact assessment methodologies created a need for dedicated LCA software and the first versions of both SimaPro and GaBi, two widely used software, were released around 1990 (Thinkstep 2016; PRé 2016).

In the twenty-first century, impact assessment methods have continuously been refined and several methodologies have emerged and are frequently being updated. The first impact assessment methods took into account the often large differences in the environmental hazards of the individual emissions. The realisation that there can be very large differences also in the sensitivity of the environment receiving the impacts lead to the release of the EDIP2003 method (Hauschild and Potting 2005) with spatially differentiated impact assessment methods covering non-global impacts like eutrophication and acidification. With the globalisation of production and an increased focus on biobased products in LCA, methods for impact assessment of extraction-related impacts like water use and land use have seen a lot of activity in the 2000s and 2010s. Hybrid LCA has emerged to reap the benefits of process-based and input/output based inventory analysis. Acknowledging that sustainability also has a social dimension, a growing activity has attempted to develop methods for Social LCA to quantify social impacts of product life cycles. A framework for life cycle sustainability assessment (LCSA) has emerged for performing assessments and aims to take into account an environmental, social and economic dimension of sustainability (see Chap. 5).

3.3 Application

Many of the early process-based LCA studies analysed packaging, which was a great consumer concern around the 1970s. For example, moulded pulp trays were compared to plastic trays and plastic bottles were compared to refillable glass bottles. Studies were typically commissioned by companies producing or using the packaging, such as Coca Cola in a pioneering study in 1969. Rather than disclosing studies directly to consumers, the results were mainly used for internal purposes, such as guiding reduction of life cycle impacts.

LCA also caught the interest of government early on. For example, the US EPA commissioned a large peer reviewed study, which was published in 1974, with the aim of informing regulation on packaging (US EPA 1974). However, at that time the EPA decided that using LCA as a direct regulatory tool was impractical, because it was thought to require LCAs to be carried out on thousands of products followed by extensive micro-managing of private businesses.

During the 1980s, life-cycle-related tools received little attention in North America, but in Europe, a revival started around the middle of the decade with an increased interest in the impacts of milk packaging that inspired a number of large LCA studies performed in different European countries. All studies compared alternative packaging systems for milk distribution to private consumers (Bundesamt für Umweltschutz 1984; Franke 1984; Lundholm and Sundström 1985;

Mekel and Huppes 1990; Pommer et al. 1991). A comparison of the studies shows that although they aimed to answer the same question (is returnable packaging or milk cartons preferable from an environmental and resource perspective?), and although they compared more or less the same packaging technologies, they reached very different conclusions. Rather than disqualify LCA as a serious decision support tool, these findings triggered an international collaboration among scientists and LCA practitioners from industry and consultancy on furthering LCA methodology development and harmonisation, as reflected in the strong international development work and standardisation in the 1990s. Concurrent with the fast methodological development of the 1990s the application of LCA expanded to include numerous other types of products during this decade as reflected in the proliferation of LCA-based ecolabels. The first LCA-supported Nordic Ecolabel was initiated in 1989 to guide consumers towards products with the lowest environmental impacts, and the number of product categories covered by criteria grew rapidly under this and other ecolabels like the European Flower label and the German Blaue Engel (see Chap. 24 on Eco-labelling and environmental product declarations). Several European countries launched national product-oriented environmental strategies with LCA as the methodological backbone, presaging the European Integrated Product Policy(IPP) to be adopted at EU level in 2003 with policy instruments like the aforementioned ecolabels, environmental product declarations, green public purchase and integration of environmental aspects into standards development.

After the turn of the century, product applications continued to grow in number and broaden in scope, also inspired by the increased political focus on LCA in EU and other parts of the world. LCA studies were increasingly used to analyse questions on the macro scale related to, for example, national energy systems and waste management systems. A 2006 survey of LCA practitioners found that LCA results were primarily used in business strategy, research and development and product or process design, but that education, policy development and labelling/product declarations were also frequent uses (Smith Cooper and Fava 2006). A similar survey from 2011 found that most practitioners made LCA studies in the agriculture (56%) and food sectors (62%), while practitioners working with other consumer goods (38%) and energy (37%) industries were somewhat less frequent (Teixeira and Pax 2011). The growth in the private sector's use of LCA in the period is reflected in Fig. 3.1 which shows the development in the total annual number of corporate responsibility reports mentioning LCA.

The year 2008 became an important year in the history of LCA for policy support, as the European Commission initiated its Sustainable Consumption and Production and Sustainable Industrial Policy (SCP/SIP) Action Plan, incorporating the previous IPP and waste and resource strategies and having LCA as the analytical backbone, but this time without the micromanagement regulation scope explored by the US EPA three decades earlier. The use of LCA in policy development is discussed in Chap. 18.

In 2009, The Sustainability Consortium was formed with the US retailer Walmart as a central partner with the mission to create a more sustainable consumer

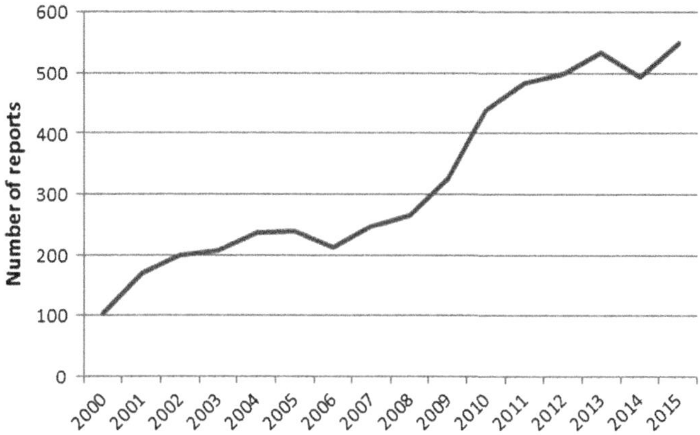

Fig. 3.1 Development in number of published corporate responsibility reports mentioning LCA ("Life cycle analysis" or "life cycle assessment") per year from 2000 to 2015. Based on a search in the PDF Search Tool of CorporateRegister (2016) carried out on April 25th, 2016

goods industry through the implementation of credible, transparent, and LCA-based reporting systems in the value chains of consumer products, targeting both environmental and social impacts. The activities of the sustainability consortium have the potential to strengthen the applicationof LCA further in the main regions supplying consumer products to the North American market, notably China and Southeast Asia.

3.4 International Harmonisation and Standardisation

With the awakening interest in LCA in the late 1980s, it soon became clear that there was a strong need for developing the methodology and harmonising the evolving methods to ensure consistency between studies.

3.4.1 Scientific Collaboration and Consensus Building

The global Society of Environmental Toxicology and Chemistry organised a workshop on "A Technical Framework for Life Cycle Assessment" in 1990 (SETAC 1991). This first event was followed by a series of workshops targeting central elements of the LCA methodology: in Leiden in the Netherlands (1991) (SETAC 1992), Sandestin Florida (SETAC 1993a) and Wintergreen (1992) (SETAC 1994) where central elements of LCA methodology were discussed with the aim of developing a common framework and agree on principles and research needs. The series culminated in a Code of Practice workshop held in Sesimbra,

Portugal, in 1993 leading to the development of the first official guidelines for LCA (SETAC 1993b)—a Code of practice for LCA. Through the rest of the 1990s SETAC working groups in Europe and North America further discussed the methodological elements with particular focus on inventory modelling and life cycle impact assessment, regularly publishing their recommendations in SETAC working group reports presenting the agreed state of the art and delivering recommendations for further research. The working groups helped coordinate the method development and strengthen the collaboration between the different research teams developing the LCA methods and they played an important role in the strong developments in LCA methodology through the 1990s. The work in these international fora was building on several important national and regional methodology development projects like the Nordic LCA Guideline project (Nordic Council of Ministers 1992; Lindfors et al. 1995), The Dutch LCA Handbook (Guinée et al. 2002) and the Danish EDIP project (Wenzel et al. 1997; Hauschild and Wenzel 1998)

In the late 1990s, leading researchers from the SETAC working group on life cycle impact assessment reached out to the United Nations Environmental Program (UNEP) to create a partnership to ensure further development of good LCA practice and global dissemination beyond Europe, North America and Japan, which had thus far been the main activity centres. The UNEP/SETAC Life Cycle Initiative was launched in 2002 and its changing working groups have taken over the method development activities of SETAC and increasingly focused on the dissemination of life cycle practices to the emerging economies through development of training materials and support with access to tools and data. The methodological recommendations have gained a more authoritative status with a formalised review procedure under the UNEP/SETAC Life Cycle Initiative.

3.4.2 International Standardisation

Taking off after the development of the SETAC code of practice for LCA in 1993, a formal standardisation process was initiated under the auspices of the International Organization of Standardization (ISO) to develop a global standard for LCA, building on the previous years' accomplishments in the scientific consensus building. The standard was to meet concerns from industry who increasingly wanted to use LCA for product development and marketing of greener products, but experienced that the lack of a standardised methodology meant that different studies of the same product could give opposite results depending on the concrete methodological choices. The standard development resulted in the adoption and release of four standards over the next seven years, addressing the principles and framework (ISO 14040), the goal and scope definition (ISO 14041), the life cycle impact assessment (ISO 14042) and the life cycle interpretation (ISO 14043). In a 2006 revision, the latter three were compiled in the ISO 14044 standard detailing the requirements and guidelines, without changing any requirements in the standards.

The ISO 14040 series standards concern the LCA methodology, but in the ISO 14000 series of Environmental Management standards, there are also standards and technical guidance reports on the applications of LCA for e.g. eco-design (ISO 14062, ISO 14006), communication of environmental performance (ISO 14020 series on ecolabels and ISO 14063), and greenhouse gas reporting and reduction (ISO 14064).

3.4.3 Standardisation of Methodology Beyond the ISO Standards: The European ILCD

LCA methodology was very young and rather immature while the ISO standardisation process took place in the 1990s, and the resulting standards are therefore not very detailed on specific methodological choices but rather focused on the framework and the fundamental principles of LCA. This is one of the reasons why the work of the UNEP/SETAC Life Cycle Initiative was needed to evaluate alternative practices and develop recommendations from a scientific point of view. It was also the background for a process initiated by the European Commissionin the mid-2000s to develop an International Life Cycle Data System (ILCD) with a database of life cycle inventory data and a series of methodological guidelines. With the development of the Integrated Product Policy and the action plan for Sustainable Consumption and Production, there was a need for a strong methodological basis of the LCA which was the method used for judging alternatives and communicating on the impacts of products and consumption. The ISO standards left too many possibilities for ambiguities in the applied methodology and in a consultation process, the EU Commission's Joint Research Centre's Institute for Environment and Sustainability developed a comprehensive guideline in LCA (EC-JRC 2010) that builds on the ISO 14040 and 14044 standards, and over 394 pages specifies the majority of the methodological choices that are left open by the ISO standards. Adherence to the ILCD guideline is intended to ensure more consistent and reproducible results of LCAs performed by different practitioners and hence increase comparability of LCA results from different studies. We have compiled the central provisions of the ILCD guideline as a Cookbook for LCA in Chap. 37 and the core methodological Chaps. (7–13) are inspired by and consistent with the ILCD guidelines. The ILCD work also involved a comparative analysis of all available LCIA methodologies (around 2008) comparing their approaches to assessment of the different midpoint and endpoint impact categories and identifying a recommendable practice for each impact category. The collection of best practices for each impact category was compiled as the ILCD impact assessment method (EC-JRC 2011). After the release of the ILCD guidelines in 2012, the EU Commission launched the Product Environmental Footprint (PEF) and Organisational Environmental Footprint (OEF) Guidelines as abbreviated and slightly revised versions of the ILCD guidelines targeting different categories of products or services to be applied by companies and organisations reporting on their environmental performance.

3.5 Dissemination

Early studies commissioned by companies were often not published due to confidential information on industrial processes and the difficulty of communicating results in non-technical language. The first peer-reviewed LCA-like study was the packaging study commissioned by the US EPA (see Sect. 3.2) published in 1974. After the development of the ISO 14040 series standards on LCA, starting in 1997, it became a common practice for companies to publish peer-reviewed LCA reports to document environmental claims, although full disclosure of underlying data is still rare due to confidentiality issues. Academic journals have become an important medium for the dissemination of LCA studies, whether made to support decisions in, e.g. companies, or for research purposes. In 1996, the first academic journal fully dedicated to LCA was born, The International Journal of Life Cycle Assessment. This journal and other journals have seen a sharp increase in the number of published papers related to life cycle assessment, from less than 100 in 1998 to more than 1300 in 2013 as illustrated in Fig. 3.2, which indicates an exponential development of the number of publications in this period. The publication of LCA reports outside academic journals is difficult to map, but is likely to have seen a similar development as indicated by the increase in company use of LCA illustrated in Fig. 3.1.

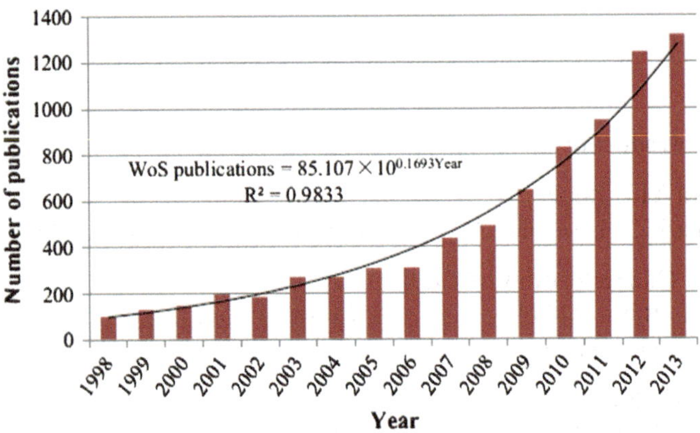

Fig. 3.2 Development in number of published LCA-related academic articles in English per year according to Web of Science (WoS) (Chen et al. 2014). The high R^2 value for the fitted exponential function indicates an exponential development. Reprinted with permission of Springer

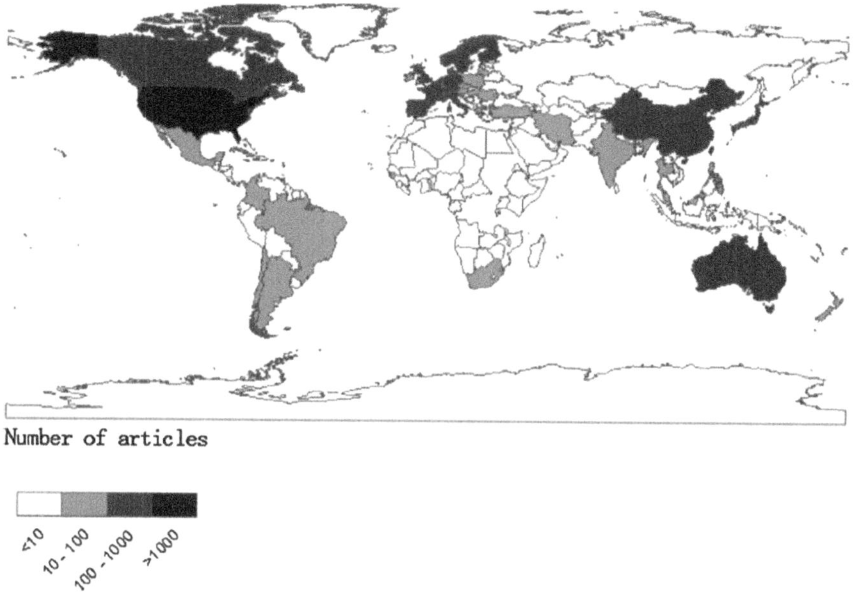

Fig. 3.3 Geographical distribution of articles published from 1998 to 2013 considering primary authors only (Hou et al. 2015). Reprinted with permission of Springer

Figure 3.3 shows that many of the English language LCA-related academic papers originate in the US and Europe, but that countries like Japan, China and South Korea have also had a noticeable publication activity. The limited activity on LCA in most emerging economies is clearly visible. Reasons for this are discussed in Chap. 19 on Globalisation and mainstreaming of LCA. Note, however, that LCA studies published in other languages than English are not included in Fig. 3.3, which therefore may lead to an underestimation of academic publications from emerging economies.

3.6 Concluding Remarks

LCA is a young discipline with 50 years of history and less than 30 years of intense development and application. Over the years, the methodology and applications have matured in the sense that scientific consensus and standards have emerged on how to perform LCA. The field has expanded in other ways when considering the number of publications, application domains and the geographical distribution of LCA competences. Table 3.1 summarises some of the important events in the history of LCA.

References

Ahbe, S., Braunschweig, A., Müller-Wenk, R.: Methodology for ecobalances based on ecological optimization. In: BUWAL (SAFEL) Environment Series No. 133, Bern (1990)

Bundesamt für Umweltschutz: Ökobilanzen von Packstoffen (Life-cycle assessment of packaging materials, in German). Bern, Switzerland. (Schriftenreihe Umweltschutz No. 24) (1984)

Chen, H., Yang, Y., Yang, Y., Jiang, W., Zhou, J.: A bibliometric investigation of life cycle assessment research in the web of science databases. Int. J. Life Cycle Assess. **19**(10), 1674–1685 (2014)

CorporateRegister. (2016). http://www.corporateregister.com. Accessed 25 Apr 2016

EC-JRC: European Commission-Joint Research Centre—Institute for Environment and Sustainability: International Reference Life Cycle Data System (ILCD) Handbook—General guide for Life Cycle Assessment—Detailed guidance. First edition March 2010. EUR 24708 EN. Publications Office of the European Union, Luxembourg (2010)

EC-JRC: European Commission—Joint Research Centre—Institute for Environment and Sustainability: International Reference Life Cycle Data System (ILCD) Handbook—Recommendations for Life Cycle Impact Assessment in the European context—Based on Existing Environmental Impact Assessment Models and Factors. First edition 2011, EUR 24571 EN. Publication Office of the European Union, Luxemburg (2011)

Ecoinvent: History. http://www.ecoinvent.org/about/history/history.html. (2016). Accessed 16 June 2016

Franke, M.: Umweltauswirkungen durch Getränkeverpackungen—Systematik zur Ermittlung der Umweltauswirkungen von komplexen Prozessen am Beispiel von Einweg-und Mehrweg-Getränkebehältern (Environmental impacts from beverage packaging—methodology for analysis of environmental impacts from complex processes with the example of oneway and returnable beverage containers, in German). EF-Verlag für Energie- u. Umwelttechnik, Berlin, Germany (1984)

Goedkoop, M.J., Spriensma, R.: Eco-indicator 99, a damage oriented method for lifecycle impact assessment, methodology report (update April 2000) (2000)

Guinée, J.B., Gorrée, M., Heijungs, R., Huppes, G., Kleijn, R., de Koning, A., van Oers, L., Wegener Sleeswijk, A., Suh, S., Udo de Haes, H.A., de Bruijn, J.A., van Duin, R., Huijbregts, M.A.J. (eds.) Handbook on Life Cycle Assessment: Operational Guide to the ISO Standards. Series: Eco-efficiency in industry and science. Kluwer Academic Publishers, Dordrecht. Hardbound. ISBN 1-4020-0228-9; Paperback, ISBN 1-4020-0557-1 (2002)

Hauschild, M.Z., Potting, J.: Spatial Differentiation in LCA Impact Assessment—The EDIP 2003 Methodology, Environmental News No. 80; Danish Environmental Protection Agency, Copenhagen, Denmark (2005)

Hauschild, M.Z., Wenzel, H.: Environmental assessment of products. Volume 2—Scientific Background, 565 pp. Chapman & Hall, UK/Kluwer Academic Publishers, Hingham. ISBN 0412 80810 2 (1998)

Heijungs, R., Guinée, J.B., Huppes, G., Lankreijer, R.M., Udo de Haes, H.A., Wegener Sleeswijk, A., Ansems, A.M.M., Eggels, P.G., van Duin, R., de Goede, H.P.: Environmental Life Cycle Assessment of products. Guide and Backgrounds. Centre of Environmental Science (CML), Leiden University, Leiden (1992)

Hou, Q., Mao, G., Zhao, L., Du, H., Zuo, J.: Mapping the scientific research on life cycle assessment: a bibliometric analysis. Int. J. Life Cycle Assess. **20**(4), 541–555 (2015)

Hunt, R.G., Sellers, J.D., Franklin, W.E.: Resource and environmental profile analysis: a life cycle environmental assessment for products and procedures. Environ. Impact Assess. Rev. **12**(12), 245–269 (1992)

ISO: Environmental Management—Life Cycle Assessment—Requirements and Guidelines (ISO 14044). ISO, The International Organization for Standardization, Geneva (2006b)

Klöpffer, W.: Life cycle sustainability assessment of products. Int. J. Life Cycle **13**(2), 89–94 (2008)

Leontief, W.: Environmental repercussions and economic structure. An input-output approach. Rev. Econ. Stat. **52**, 262–271 (1970)
Lindfors, L.-G., Christiansen, K., Hoffman, L., Virtanen, Y., Juntilla, V., Hanssen, O.-J., Rønning, A., Ekvall, T., Finnveden, G.: Nordic Guidelines on Life-Cycle Assessment. Nord1995:20, Nordic Council of Ministers, Copenhagen (1995)
Lundholm, M.P., Sundström, G.: Resource and Environmental Impact of Tetra Brik Carton and Refillable and Non-refillable Glass Bottles. Tetra Brik Aseptic Environmental Profile, AB Tetra Pak, Malmö (1985)
Mekel, O.C.L., Huppes, G.: Environmental Effects of Different Package Systems for Fresh Milk. Centrum voor Milieukunde, Rijksuniversiteit, Leiden, NL. (CML-Mededelingen no. 70) (1990)
Nordic Council of Ministers: Product Life Cycle Assessment—Principles and Methodology. Nord 1992:9, Nordic Council of Ministers, Copenhagen (1992)
Pommer, K., Hillersborg, A.S., Hirsbak, S., Hoffman, L., Schmidt, A.: Environmental Assessment of Milk Packaging. Return Bottles in Polycarbonate Compared to Disposable Cartons (in Danish). Danish Environmental Protection Agency, Copenhagen, Denmark. (Environmental project No. 168) (1991)
PRé:. Company History. https://www.pre-sustainability.com/company-history. Accessed 16 June 2016
SETAC: In: Fava, J., Denison, R., Jones, B., Curran, M.A., Vigon, B., Selke, S., Barnum, J. (eds.) SETAC Workshop Report: A Technical Framework for Life-Cycle Assessment. Smugglers Notch, Vermont, August 18–23 1990. SETAC Press (1991)
SETAC: Life-Cycle Assessment—Inventory, Classification, Valuation and Data Bases. 2–3 December 1991, Leiden, The Netherlands (1992)
SETAC: In: Fava, J., Consoli, F., Denison, R., Dickson, K., Mohin, T., Vigon, B. (eds.) Conceptual Framework for Life Cycle Impact Assessment. Workshop Report Sandestin FL, February 1–7, 1992. SETAC Press (1993a)
SETAC: Guidelines for Life-cycle Assessment: A "code of Practice": from the SETAC Workshop Held at Sesimbra, Portugal, 31 March-3 April 1993. SETAC Brussels and Pensacola (1993b)
SETAC: In: Fava, J., Jensen, A.A., Lindfors, L.-G., Pomper, S., De Smet, B., Warren, J., Vigon, B. (eds.) Life-Cycle Assessment Data Quality: A Conceptual Framework. Workshop Report Wintergreen, Virginia, October 1992. SETAC Press (1994)
Smith Cooper, J., Fava, J.A.: Life-cycle assessment practitioner survey—summary of results. J. Ind. Ecol. **10**(4), 12–14 (2006)
Steen, B.: A Systematic Approach to Environmental Priority Strategies in Product Development (EPS). Version 2000-general system characteristics; CPM report 1999:4, Chalmers University of Technology, Gothenburg, Sweden (1999a)
Steen, B.: A Systematic Approach to Environmental Priority Strategies in Product Development (EPS). Version 2000-Models and data of the default method; CPM report 1999:5, Chalmers University of Technology, Gothenburg, Sweden (1999b)
Teixeira, R., Pax, S.: A survey of life cycle assessment practitioners with a focus on the agri-food sector. J. Ind. Ecol. **15**, 817–820 (2011). doi:10.1111/j.1530-9290.2011.00421.x
Thinkstep: A brief history of Life Cycle Assessment (LCA). http://www.gabi-software.com/news/news-detail/article/a-brief-history-of-life-cycle-assessment-lca/. (2016). Accessed 16 June 2016
US EPA: Resource and Environmental Profile Analysis of Nine Beverage Container Alternatives —Final Report. EPA/530/SW-91c. U.S. Environmental Protection Agency (1974)
Wenzel, H., Hauschild, M.Z., Alting, L.: Environmental Assessment of Products. Volume 1—Methodology, Tools and Case Studies in Product Development, 544 pp. Chapman & Hall/UK, Kluwer Academic Publishers/Hingham. ISBN 0 412 80800 5

Author Biographies

Anders Bjørn Part of the LCA community since the early 2010s. Main focus is interpretations of sustainability and integration of sustainability targets in LCA to enable absolute sustainability assessments.

Mikołaj Owsianiak Involved in development and application of life cycle impact assessment methods in sustainability assessment of technologies. Has worked on issues associated with: soils (remediation), metals (toxic impact assessment), biodiesel (fate in the environment), and carbonaceous materials (biochar and hydrochar).

Christine Molin Active in the field of LCA since 1992. Special interest in the development and dissemination of LCA and in the use of LCA in small and medium sized enterprises.

Michael Z. Hauschild Involved in development of LCIA methodology since the early 1990s. Has led several SETAC and UNEP/SETAC working groups and participated in the development of the ISO standards and the ILCD methodological guidelines. Main LCA interests are chemical impacts, spatial differentiation and science-based boundaries in LCIA.

Chapter 4
LCA Applications

Mikołaj Owsianiak, Anders Bjørn, Alexis Laurent, Christine Molin and Morten W. Ryberg

Abstract The chapter gives examples of applications of LCA by the central societal actors in government, industry and citizens, and discusses major motivations and challenges for the use of LCA to support science-based decision-making from their respective perspectives. We highlight applications of LCA in policy formulation, implementation and evaluation, present different purposes of LCA application in industry at both product and corporate levels, and discuss challenges for LCA applications in small- and medium-sized enterprises. Our synthesis demonstrates the importance of LCA as a tool to quantify environmental impacts of products and systems and support decisions around production and consumption and highlights factors that prevent its even more widespread application.

Learning Objectives
After studying this chapter the reader should be able to:

- Explain the main motivations for use of LCA by governments, industry, and citizens and their main types of LCA applications.
- Demonstrate an understanding of the challenges and opportunities in the different types of LCA applications.

M. Owsianiak (✉) · A. Bjørn · A. Laurent · C. Molin · M.W. Ryberg
Division for Quantitative Sustainability Assessment,
Department of Management Engineering, Technical University
of Denmark, 2800 Kgs. Lyngby, Denmark
e-mail: miow@dtu.dk

A. Bjørn
CIRAIG, Polytechnique Montréal, 3333 Chemin Queen-Mary,
Montréal, QC, Canada

M.Z. Hauschild et al. (eds.), *Life Cycle Assessment*,
DOI 10.1007/978-3-319-56475-3_4

4.1 Background

Recent decades have witnessed numerous applications of LCA to support decisions in an environmental sustainability context (see Chap. 3). Much efforts have been made to facilitate the application of LCA and life cycle thinking in society ranging from the regulatory and governmental level, through industry and production to the level of citizens and consumers. The dissemination of LCA has been aided by a number of initiatives for supporting and harmonizing the application of the tool. In 1997 the first version of the ISO 14040 standard (later updated as ISO 2006a) was published in an attempt to harmonize the framework and principles of LCA and to increase transparency and comparability of LCA studies. In 2001, The United Nations Environment Programme (UNEP) and the Society for Environmental Toxicology and Chemistry (SETAC) joined forces in the launch of a global Partnership to strengthen the dissemination and use of LCA worldwide, known as the Life Cycle Initiative (LCI). The purpose of the initiative was to "enable users around the world to put life cycle thinking into effective practice". Another initiative supporting a more widespread application of LCA was The European Commission's project, The European Platform of Life Cycle Assessment, launched in 2005. Its objective was to "promote life cycle thinking in business and in policy making" in the European Union by focusing on underlying data and methodological needs. The homepages of these initiatives provide a wide palette of information, tools and support (http://www.lifecycleinitiative.org; http://eplca.jrc.ec.europa.eu/).

In parallel, many initiatives have been launched at the national level to facilitate and support the application of LCA, often under the auspices of governmental institutions such as environmental protection agencies (see Chap. 3), inspiring numerous private and public LCA consultancies to emerge in assistance to companies or institutions without the in-house LCA expertise. Recent widespread LCA-related services are an elaboration of Environmental Product Declarations (EPDs) or performance of Greenhouse Gasaccounting. Moreover, universities, research institutions and private companies often enter into close collaboration on LCA methodology development and application of LCA via, e.g. commercial projects or industrial PhDs.

Here, we present examples of applications and discuss major motivations and challenges for the use of LCA to support decision-making from the perspectives of decision-makers within governments, industry and citizens. More details are given in Part III of the book with chapters dedicated to different stakeholders and multiple examples of the use of LCA within different technology domains. Chapter 18 gives a more detailed introduction to the use of LCA and life cycle thinking in policy-making in different parts of the world, and Chap. 19 discusses the globalization of the use of LCA. Life cycle management (LCM) within business and industry is the topic of Chap. 22, while Chap. 24 introduces the use of LCA in the development and management of environmental labels and declarations.

4.2 Government Perspective

Application of LCA and life cycle based approaches can support policy formulation, policy implementation and regulation imposed by policies, and can be used to perform evaluation of policies. As part of the pan-European project CALCAS (Coordination Action for innovation in Life Cycle Analysis for Sustainability), reviews were conducted in mid-2000s to identify LCA applications to support different stages of the policy cycle, i.e. their formulation, implementation and evaluation (CALCAS 2008). Table 4.1 presents and overview of such applications. Since then, the pressing need to move towards more sustainable societies has made LCA increasingly recognized in high policy-level, and its role in the policy cycle has been formalized in some countries or regions. For example, in Europe, the European Commission has listed LCA as one of the reference models for the impact assessment of policies in the European Union (EU) within its "better regulation guidelines" document published in 2015 (European Commission 2015). This holds a potential to increase the use of LCA in retrospective assessments of existing policy frameworks (i.e. evaluations or fitness checks) and prospective assessments of future possible policy options (policy development).

Table 4.1 Examples of LCA applications at different stages of the policy cycle

Topic	Initiation year and/or geographical scope
LCA as a knowledge tool in policy formulation	
Environmental technologies action plan (ETAP)	2004; EU
Integrated product policy (IPP)	2003; EU
Directive on the eco-design of energy using products (EuP)	2005; EU
Strategy for the sustainable use of natural resources	2005
Sustainable production and consumption action plan (SCP)	2007; EU
Biofuels	Germany
Application of pesticides	Costa Rica
Supporting the implementation of information based instruments: LCA & policy implementation	
Eco-labelling	Various countries
Environmental product declarations (EPD)	Various countries
Strategic environmental assessment directive	2004
Public procurement	EU, Japan
Construction products directive	1989; EU
Ordinance on the avoidance and recovery of packaging wastes	Germany
Waste management	France, Mexico, japan
LCA as a tool for policy evaluation	
Thematic strategy on prevention and recycling of waste & Waste framework directive	2005; EU
Waste oil directive	2000; EU

Based on CALCAS (2008)

4.2.1 Policy Formulation

As an example of LCA used for policy formulation, the European Commission has promoted Integrated Product Policy (IPP) to minimize environmental impacts of products by considering all stages of their life cycle, from the cradle to grave (Mudgal 2008). The IPP comprises various instruments and tools, ranging from soft instruments that act through influencing the market (like environmental labelling or green taxation), through subsidies to industries (e.g. financial support to pioneers), to hard regulation such as the Eco-design Directive for Energy-related Products (ErP), which establishes a regulatory framework for eco-design of products that use energy and products that allow for generation, transfer and measurement of energy (Directive 2009/125/EC). This directive is an example of how life cycle thinking has guided policymaking within the EU, where the focus has shifted from manufacturing processes, to a focus on the use of products and their disposal (Wenzel et al. 1997; Azapagic and Perdan 2000). Many other examples of the use of LCA in policy formulation are given in Chap. 18.

A major challenge to the applicationof LCA in these contexts is the communication of environmental performance of products. It is often done using different approaches to life cycle inventory modelling and life cycle impact assessment, which may lead to inconsistent and sometimes misleading results. To facilitate the communication of reliable and reproducible information about the environmental performance of products and organizations, the European Commission has elaborated LCA-based methods for product environmental footprint (PEF), and organization environmental footprint (OEF) (Finkbeiner 2014; Galatola and Pant 2014) (see also Chap. 24).

4.2.2 Policy Implementation and Evaluation

Governments may use LCA as decision support to advice the introduction of novel technologies in the market (e.g. the use of biofuels, or introduction of electric cars) or the selection of waste management systems (e.g. EU Waste Framework Directive 2008/98/EC imposing "to handle waste in a way that does not have a negative impact on the environment or human health" and requiring the need for life cycle thinking in waste management) (European Parliament and Council 2008; Meylan et al. 2014). In Denmark, LCA was used in the 1990s to guide the development of the current Danish collection system for beverage containers (glass and plastic bottles and aluminium cans) and it has been used for assessment of recycling strategies for various waste fractions. The country has also operated with panels of key actors along the product life cycle who were consulted in the development of product-oriented policy initiatives. In Switzerland, findings from an LCA study were used to justify compensation rates to municipalities according to how waste glass packaging is collected and what disposal option is chosen by the municipality

(Meylan et al. 2014). In Sweden LCA was used to assess environmental impacts of introducing waste incineration tax, considered to "encourage waste reduction and increase materials recycling and biological treatment" (Björklund and Finnveden 2007). While the proposed design of such a tax would result in increased recycling, the LCA found that this would lead to only small environmental improvements. Thus, it was proposed that the design of the tax should include the fossil carbon content of the waste. Such examples can also be found outside Europe. In the United States, the California Oil Recycling Enhancement Act was initiated in 2009 to support management of used oil and support selection of least-polluting options (refining and reuse, distillation or combustion with energy recovery) by the state (Reed 2012). This act "requires that the Department of Resources Recycling and Recovery coordinate, with input from representatives of all used oil stakeholders, a comprehensive life cycle assessment of California's used lubricating and industrial oil management process" (CalRecycle 2012).

4.3 Industry Perspective

The application of LCA in enterprises can be classified into five main purposes: (i) decision support in product and process development; (ii) marketing purposes (e.g. Eco-labelling); (iii) development and selection of indicators used in monitoring of environmental performance of products or plants; (iv) selection of suppliers or subcontractors; and (v) strategic planning (Huang and Hunkeler 1995; Bültmann 1997; Hanssen 1999; Baumann 2000; Heiskanen 2000; Frankl and Rubik 2000; Ekvall 2012). We note that LCA applications within industry may well serve more than one purpose, and often the same LCA can be used for different purposes within acompany (e.g. product development is often combined with marketing efforts). Furthermore, as experience with using LCA grows in an enterprise, one application can trigger another (e.g. insights gained from an LCA into product environmental performance can lead to decisions about selection of suppliers or setting strategies). We also note that although LCA has traditionally been developed as a tool to be used at product level, and is still used as such, there is an increasing interest in using LCA at the corporate level to reflect the performance of the company or individual plants in a life cycle perspective. This is particularly relevant for (but not limited to) large enterprises and for applications related to monitoring of environmental performance and strategic planning.

4.3.1 Applications at Product Level

At product level, LCA is often used during product development and for identifying environmental hotspots of a product or process either within the organization or in its supply chain. For instance, a survey showed that the German industry in the

1990s mainly used LCA internally, to identify hotspots in products and systems, followed by product and process optimization (Bültmann 1997; Frankl and Rubik 2000). Another survey showed that large Danish companies, represented by 39 companies considered to cover 90–100% of Danish enterprises having practical experience with LCA in the 1990s, indicated that LCA had revealed new environmental aspects of their products that they had not anticipated. In 79% of the cases, this led to setting new priorities for environmental efforts, including changes in products and processes, like saving or substituting materials (Broberg and Christensen 1999).

In parallel to application in product and process development, LCA is often used for marketing purposes at different levels. As public concerns about the state of the environment have become increasingly pronounced and consumers more environmentally conscious, enterprises have also placed a larger focus on quantifying their environmental performance, using LCA and communicating this to the public as a way to brand their enterprise as green. Here, the major company expectations to the use of LCA are to get a competitive advantage and increase the company image or reputation (Broberg and Christensen 1999). Ecolabels or environmental product declarations (Chap. 24) can signal good environmental performance and be used to make a given product more appealing for environmentally conscious consumers.

4.3.2 Applications at Corporate Level

The use of LCA to document and monitor environmental performance at the corporate level is today often limited to a few selected impact categories, typically footprint indicators (see Sect. 10.4) like carbon footprint and blue water footprint. This situation may change in the future together with the development of guidelines for organization environmental footprint (OEF) (Dubois and Humbert 2015). At the corporate level, industry can also use LCA for setting strategic objectives. For example, Unilever set a target of halving their environmental impact by 2030, considering the life cycle of their products (Unilever 2015). Similarly, companies may want to carry out LCA to better understand their environmental performance in an effort to implement environmental management system (EMS) (Lewandowska et al. 2013, 2014). EMS is "a tool to implement a structured program of continual improvement in environmental performance" and "a tool to manage and communicate an enterprise's environmental performance to internal and outside parties" (Lombardo 2012). EMS standards nowadays often require a life cycle perspective in order to avoid greenwashing by companies outsourcing parts of their production to suppliers. There is thus often a relationship between the implementation of EMS and the implementation of LCA within companies. For example, among Spanish automotive supplier companies who have received the EMS ISO 14001 certification and have a certified eco-management and audit scheme (EMAS), the use of LCA is a common practice (Gonzalez et al. 2008). Organizations who have implemented a certified EMS impose higher demands on their suppliers to adopt environmentally friendly practices (Gonzalez et al. 2008). The contributions made by LCA to EMS range from the identification of overall environmental aspects and identification of the activities in the life cycle that have the largest environmental burdens, to a comparison of alternative manufacturing routes (Stewart et al. 1999). A major challenge in this context seems to be putting the results into practice, mainly due to lack of power or information of stakeholders along the product supply chain (Nakano and Hirao 2011).

4.3.3 Challenges of Small- and Medium-Sized Enterprises

Small- and medium-sized enterprises (SME) can use LCA for the same reasons as large companies. Yet, small- (10–49 employees) and medium-sized (50–249 employees) enterprises generally lag behind large companies in the implementation of LCA (Johnson and Schaltegger 2015). The major reasons are thought to be the cost of an LCA, the need for changes in workplace routines, perceived complexity of the LCA methodology and shortage of qualified personnel to carry out an LCA (Kurczewski 2013). A study of 10 SMEs revealed that a downside of LCA is that it becomes too comprehensive and too complex to be easily understood, leaving an impression in some companies of LCA as a 'black box' (Zackrisson et al. 2008). A closer collaboration with an experienced LCA practitioner and an expert was found to resolve this problem in some of the cases (Zackrisson et al. 2008). Similarly, based on a comprehensive literature review, Johnson and Schaltegger (2015) reported that major barriers for implementation of sustainability management tools (including LCA) by SMEs were (i) lack of awareness of sustainability issues; (ii) absence of perceived benefits; (iii) lack of knowledge and expertise on sustainability issues; (iv) lack of human and financial resources; (v) insufficient external drivers and incentives; (vi) unsuitability of formal management tools to fit the often informal and flexible SME structure; and (vii) complexity of tools.

While the use of LCAs by SMEs was considered marginal (as of 2012), it is however reported to become more and more common (Baumann et al. 2012; Schischke et al. 2012; Kurczewski 2013). This may be due to the increased legislative focus on environmental performance, and the potential market benefits from having an environmentally friendly profile, not least through a market pull from large companies that are often important costumers. This is reflected by a survey of 146 European SMEs which revealed that most SMEs have limited knowledge of LCA, and have little internal knowledge of environmental assessments and their communication (Pamminger 2011). The main drivers for SMEs to start using environmental assessment tools have been the customer demand or the pressure from legislation (Pamminger 2011; Schischke et al. 2012). However, industries focusing on emerging renewable resource technologies, such as bio-based plastic, had more knowledge and were, in fact, keen on using LCA for communicating the environmental performance and benefits of their technology compared to conventional technologies (Pamminger 2011). The authors' experience with LCA application by SMEs in Western Europe shows that SMEs are eager to contribute to an LCA (e.g. through provision of data) when a dedicated and sufficient budget is available, e.g. through the involvement in a larger research project. Experience also shows that SMEs typically find interest in identifying impact reduction opportunities, particularly those stemming from activities in the life cycle on which they themselves exert some influence. Similar findings were reported in European countries where the tradition of using LCA has historically not been that strong (Kurczewski 2013; Witczak et al. 2014).

4.4 Citizen Perspective

LCA results can also serve as decision support for individuals, be it in their capacity of citizens or consumers. In many cases, these decisions relate to the private consumption of goods and services. Consumers are knowingly or unknowingly exposed to LCA results, or conclusions drawn from LCA results, through ecolabels (see Chap. 24) or other consumer information from producers (e.g. printed on packaging) and media reporting academic findings, and they hold some power through their influence in the market of consumer products. Consumer decisions that may be supported by an LCA can range from choosing the product with the lowest environmental impact amongst a group of similar products (e.g. the more environmentally friendly vacuum cleaner), over choosing the most environmentally sound way of fulfilling a function (e.g. washing dishes by hand or in a dishwasher) to most effectively reducing the total personal environmental impact (e.g. reduce meat consumption, hot showers or car driving).

Besides decisions related to private consumption, citizens may also indirectly be affected by LCA results when following political discussions on large infrastructure-related decisions where LCA provides the underlying decision support. For example, municipalities often use LCA to support decisions on waste management infrastructure (European Commission 2008). If a political decision is made about increasing recycling and reducing landfilling or incineration, this will affect citizens, as they will have to sort their waste into recyclable fractions rather than throw all their waste into the same bin. Chapter 35 deals with the use of LCA in waste management.

4.5 Concluding Remarks

LCA is an important and useful tool to map environmental impacts and support policy development and concrete decisions, and for a company it can support the development of a positive image. There are, however, factors that hamper its more widespread application. This chapter has mainly addressed LCA applications in developed countries because this is where LCA has been applied the most and the needed data has been most available. However, large differences exist in the application of LCA between developed and developing countries in terms of both frequency and incentives. These differences and the challenges that they pose for a global dissemination of LCA and life cycle thinking are discussed in Chap. 19 on globalization and mainstreaming of LCA. The next chapter takes a closer look at the relationship between LCA and sustainable development.

References

Azapagic, A., Perdan, S.: Indicators of sustainable development for industry: a general framework. Trans. IChemE **78**, 243–261 (2000). doi:10.1205/095758200530763

Baumann, H.: Introduction and organisation of LCA activities in industry. Int. J. Life Cycle Assess. **5**, 363–368 (2000). doi:10.1007/BF02978673

Baumann, M., Held, M., Herrmann, C., et al.: Ecodesign tool for SMEs in the electronics sector. In: Electronics Goes Green 2012+, ECG 2012—Joint International Conference Exhibition Proceedings (2012)

Björklund, A.E., Finnveden, G.: Life cycle assessment of a national policy proposal—the case of a Swedish waste incineration tax. Waste Manag. **27**, 1046–1058 (2007). doi:10.1016/j.wasman.2007.02.027

Broberg, O., Christensen, P.: LCA experiences in Danish industry—results of a survey. Int. J. LCA **4**, 257–262 (1999)

Bültmann, A.: Produktökobilanzen und ihre Anwendung in Deutschen Unternehmen. Schriftenreihe des IÖW. **112**, 97 (1997)

CALCAS (Co-ordination Action for Innovation in Life-Cycle Analysis for Sustainability): Critical Review of the Current Research Needs and Limitations Related to ISO-LCA Practice: Deliverable D7 of CALCAS Work Package 5, by Zamagni, A., P. Buttol, P. L. Porta, R Buonamici, P. Masoni, J. Guinée, R. Heijungs, T. Ekvall, R. Bersani, A. Bieńkowska, and U. Pretato (2008). http://www.calcasproject.net/

CalRecycle: Used oil recycling program (2012)

Directive 2009/125/EC: Directive 2009/125/EC of the European Parliament and of the Council of 21 October 2009 establishing a framework for the setting of ecodesign requirements for energy-related products (recast). Off J Eur Union 10–35. doi:10.1016/j.cirp.2012.03.121 (2009)

Dubois, C., Humbert, S.: How to calculate the LCA of a retailer. LCA Discuss. Forum **31**(03), 2015 (2015)

Ekvall, T.: Life cycle assessment. In: Ahmed K (ed) Getting to green: a sourcebook of pollution management policy tools for growth and competitiveness. World Bank, Washington, DC. http://documents.worldbank.org/curated/en/560021468330349857/Getting-to-green-a-sourcebook-of-pollution-management-policy-tools-for-growth-and-competitiveness (2012)

European Commission: Better Regulation Guidelines. Commission staff working document. COM (2015) 215 final; SWD(2015) 111 final. May 2015. EU Commission. Strasbourg, FR. http://ec.europa.eu/smart-regulation/guidelines/docs/swd_br_guidelines_en.pdf (better regulation guidelines; accessed 26/10/2015) and http://ec.europa.eu/smart-regulation/guidelines/toc_tool_en.htm (better regulation "toolbox"; accessed 26/10/2015) (2015)

European Commission: Directive 2008/98/EC of the European Parliament and of the Council of 19 November 2008 on waste and repealing certain Directives (Text with EEA relevance) (2008)

European Parliament and Council: Directive 2008/98/EC of the European Parliament and of the Council on waste and repealing certain Directives (2008)

Finkbeiner, M.: Product environmental footprint—breakthrough or breakdown for policy implementation of life cycle assessment? Int. J. Life Cycle Assess. **19**, 266–271 (2014). doi:10.1007/s11367-013-0678-x

Frankl, P., Rubik, F.: Life cycle assessment in industry and business. Adoption patterns, applications and implications. In: IGARSS 2014 (2000). doi:10.1007/s13398-014-0173-7.2

Galatola, M., Pant, R.: Reply to the editorial "product environmental footprint—breakthrough or breakdown for policy implementation of life cycle assessment?" Written by Prof. Finkbeiner (Int J Life Cycle Assess 19(2):266–271). Int. J. Life Cycle Assess. **19**, 1356–1360 (2014). doi:10.1007/s11367-014-0740-3

Gonzalez, P., Sarkis, J., Adenso-Diaz, B.: Environmental management system certification and its influence on corporate practices. Evidence from the automotive industry. Int. J. Oper. Prod. Manag. **28**, 1021–1041 (2008)

Hanssen, O.J.: Status of life cycle assessment (LCA) activities in the nordic region. Int. J. Life Cycle Assess. **4**, 262–262 (1999). doi:10.1007/BF02979177

Heiskanen, E.: Managers' interpretations of LCA: enlightenment and responsibility or confusion and denial? Bus. Strateg. Environ. **9**, 239–254 (2000). doi:10.1002/1099-0836(200007/08)9:4<239:AID-BSE250>3.0.CO;2-6

Huang, E., Hunkeler, D.: Life cycle analysis: summary of a fortune 500 survey and a Japanese comparison. In: Vanderbilt Univ. US–Japan Cent (1995)

ISO: Environmental management—life cycle assessment—principles and framework (ISO 14040). ISO, the International Organization for Standardization, Geneva (2006a)

Johnson, M.P., Schaltegger, S.: Two decades of sustainability management tools for SMEs: how far have we come? J. Small Bus. Manag. (2015). doi:10.1111/jsbm.12154

Kurczewski, P.: Life cycle thinking in small and medium enterprises: the results of research on the implementation of life cycle tools in Polish SMEs-part 1: background and framework. Int. J. Life Cycle Assess. (2013). doi:10.1007/s11367-013-0675-0

Lewandowska, A., Kurczewski, P., Kulczycka, J., Joachimiak, K., Matuszak-Flejszman, A., Baumann, H., Ciroth, A.: LCA as an element in environmental management systems—comparison of conditions in selected organisations in Poland, Sweden and Germany: part 1: background and initial assumptions. Int. J. Life Cycle Assess. **18**, 472–480 (2013). doi:10.1007/s11367-012-0480-1

Lewandowska, A., Matuszak-Flejszman, A.: Eco-design as a normative element of Environmental Management Systems—the context of the revised ISO 14001:2015. Int. J. Life Cycle Assess. **19**, 1794–1798 (2014). doi:10.1007/s11367-014-0787-1

Lombardo, P.: Environmental management systems. In: Ahmed, K. Getting to Green: A Sourcebook of Pollution Management Policy Tools for Growth and Competitiveness. World Bank, Washington, DC (2012). http://documents.worldbank.org/curated/en/560021468330349857/Getting-to-green-a-sourcebook-of-pollution-management-policy-tools-for-growth-and-competitiveness

Meylan, G., Stauffacher, M., Krütli, P., et al.: Identifying stakeholders' views on the eco-efficiency assessment of a municipal solid waste management system. J. Ind. Ecol. (2014). doi:10.1111/jiec.12192

Mudgal, S.: Reporting on the Implementation of Integrated Product Policy (IPP). European Commission, DG Environment. Service Contract 0703307/2007/481297/G.4. http://ec.europa.eu/environment/ipp/pdf/bio_ipp.pdf. (2008). Accessed 11 August 2017

Nakano, K., Hirao, M.: Collaborative activity with business partners for improvement of product environmental performance using LCA. J. Clean. Prod. **19**, 1189–1197 (2011). doi:10.1016/j.jclepro.2011.03.007

Pamminger, R.: Technical report on needs and demands of SMEs. Deliverable D1.1 LCA to go project within the European Commission's Seventh Framework Programme. http://www.LCA2go.eu. (2011). Accessed 11 August 2017

Reed, D.L.: Life-Cycle Assessment in Government Policy in the United States. PhD Thesis. University of Tennessee. http://trace.tennessee.edu/utk_graddiss/1394. (2012). Accessed 11 August 2017

Schischke, K., Nissen, N.F., Sherry, J. et al.: Life cycle thinking in small and medium sized enterprises—status quo and strategic needs in the electronics sector. In: Electronics Go Green 2012 (2012)

Stewart, J.R., Collins, M.W., Anderson, R., Murphy, W.R.: Life Cycle Assessment as a tool for environmental management. Clean Technol. Environ. Policy **1**, 73–81 (1999). doi:10.1007/s100980050013

Unilever: Unilever sustainable living plan. Summary of progress 2015. Mobilising Collective Action. https://www.unilever.com/Images/uslp-mobilising-collective-action-summary-of-progress-2015_tcm244-424809_en.pdf. (2015). Accessed 11 August 2017

Wenzel, H., Hauschild, M.Z., Alting, L.: Environmental assessment of products. In: Vol. 1—Methodology, Tools and Case Studies in Product Development. Kluwer Academic Publishers, Hingham, p. 544. ISBN 0 412 80800 5 (1997)

Witczak, J., Kasprzak, J., Klos, Z., et al.: Life cycle thinking in small and medium enterprises: the results of research on the implementation of life cycle tools in Polish SMEs-part 2: LCA related aspects. Int. J. Life Cycle Assess. (2014). doi:10.1007/s11367-013-0687-9

Zackrisson, M., Rocha, C., Christiansen, K., Jarnehammar, A.: Stepwise environmental product declarations: ten SME case studies. J. Clean. Prod. **16**, 1872–1886 (2008). doi:10.1016/j.jclepro.2008.01.001

Author Biographies

Mikołaj Owsianiak involved in development and application of life cycle impact assessment methods in sustainability assessment of technologies. Has worked on issues associated with: soils (remediation), metals (toxic impact assessment), biodiesel (fate in the environment), and carbonaceous materials (biochar and hydrochar).

Anders Bjørn part of the LCA community since the early 2010s. Main focus is interpretations of sustainability and integration of sustainability targets in LCA to enable absolute sustainability assessments.

Alexis Laurent working with LCA since 2010 with a strong LCIA focus, particularly on normalisation aspects. Main LCA interests include development of LCIA methods, LCA applications and footprinting of large-scale systems for policy-making (e.g. nations, sectors), and LCA applied to various technology domains, including energy systems.

Christine Molin active in the field of LCA since 1992. Special interest in the development and dissemination of LCA and in the use of LCA in small and medium sized enterprises.

Morten W. Ryberg environmental engineer working with LCA since 2013. Primary focus on advancing absolute sustainability assessments by integrating and implementing environmental sustainability references in LCA. Other works include environmental assessment of coal cleaning technologies and of new carbonaceous materials.

Chapter 5
LCA and Sustainability

Andreas Moltesen and Anders Bjørn

Abstract LCA is often presented as a sustainability assessment tool. This chapter analyses the relationship between LCA and sustainability. This is done by first outlining the history of the sustainability concept, which gained momentum with the Brundtland Commission's report 'Our Common Future report' in 1987, and presenting the most common interpretations of the concept, which generally comprise four dimensions: (1) measures of welfare, (2) inter-generational equity, (3) intra-generational equity and (4) interspecies equity. The relevance of environmental protection for dimensions 2 and 4 is then demonstrated, and the strategy of LCA to achieving environmental protection, namely to guide the reduction of environmental impacts per delivery of a function, is explained. The attempt to broaden the scope of LCA, beyond environmental protection, by so-called life cycle sustainability assessment (LCSA) is outlined. Finally, the limitations of LCA in guiding a sustainable development are discussed.

Learning Objectives
After studying this chapter the reader should be able to:

- Explain the most common interpretations of the definition of sustainable development from Our Common Future.
- Account for the relevance of environmental protection to sustainability.
- Describe the type of sustainability strategy that LCA may support and discuss its limitations.

A. Moltesen (✉) · A. Bjørn
Division for Quantitative Sustainability Assessment, Department of Management Engineering, Technical University of Denmark, Kgs. Lyngby, Denmark
e-mail: andreasmoltesen@gmail.com

A. Bjørn
CIRAIG, Polytechnique Montréal, 3333 Chemin Queen-Mary, Montreal, QC, Canada

M.Z. Hauschild et al. (eds.), *Life Cycle Assessment*,
DOI 10.1007/978-3-319-56475-3_5

5.1 Introduction

In 1987, the United Nations' World Commission on Environment and Development published its report *Our Common Future*, which is sometimes referred to as the *Brundtland Report* after its chairperson, Gro Harlem Brundtland (WCED 1987). The report was a response; on the one hand to the growing disparity between North and South and on the other hand to the increased awareness that many of the natural systems on which we depend were under increasing stress. Development of the South was seen as urgently needed, but the development had to be achieved in an environmentally sound way which would allow for a continued thriving of the world's population—also in the future. The development in other words had to be sustainable. While the term "sustainable development" was already introduced in 1980 by the International Union for the Conservation of Nature, the publication of *Our Common Future* created a widespread awareness of sustainable development and provided its most well-known definition: "… development that meets the needs of the present without compromising the ability of future generations to meet their own needs". By coupling the concern for the present and future generations, the concept of sustainable development, as defined in *Our Common Future,* provided a framework for thinking these two increasingly pressing global challenges together in one immensely influential term.

The ability of present and future generations to meet their needs depends strongly on the life support functions of the earth and inherent in the definition of sustainable development is thus a concern for the health of the environment. The development of LCA can in many regards be seen as stemming from the same concern for environmental protection (see Chap. 3). A natural question may therefore be; How does LCA and sustainable development relate, and to what extent can LCA be used as a methodology for informing decisions towards sustainability?

To answer these questions we will start by giving an overview of how sustainable development is understood in literature, followed by an analysis of the possibilities and limitations for LCA to support it.

5.2 What Is Sustainability?

Since the publication of *Our Common Future*, many different definitions of "sustainable development" or the related term "sustainability" have been presented. In this chapter we will use these two terms interchangeably, but it should be mentioned that in literature, these concepts can be used with different connotations. It is, for example, sometimes asserted that sustainable development is primarily about development (sometimes seen as synonymous with economic growth), whereas sustainability gives priority to the environment. Others have argued that the difference is rather that sustainable development should be seen as the process or journey to achieving sustainability.

Proposals for definitions of sustainable development have been booming after the publication of *Our Common Future*, and have added several nuances and potential modifications to this definition. For example, some have argued against the one-sided focus on human needs. In the definition of sustainable development given above, there is little room for considering other living species than humans, unless these species directly serve as means to meet these human needs. In line with this, it has been argued that the definition is too narrow, and that other living species should be considered as well.

Others have debated the word “need”, and suggested several others and in many regards related words such as “wellbeing”, “utility”, “welfare” and “aspiration”.

Finally, it should be mentioned that the researchers, especially within the economic discipline, have omitted the focus on the needs of the present and claimed that sustainability is simply about ensuring that the total utility or welfare of a society can be maintained over an infinite time horizon (Pezzey 1992).

Despite these variations, there is a large degree of common ground in definitions of sustainability. Sustainability can be seen as comprising by the following four dimensions, with varying emphasis:

1. The first dimension relates to *measures of welfare* that is to be achieved in the population comprised by the definition (see Dimensions 2–4). This measure of welfare comprises several different concepts, such as “need”, “utility”, “happiness” and “aspiration”. Several others can be found in literature.
2. The second dimension relates to the concern for *inter-generational equity*, i.e. a concern for the equity in the welfare (as defined by the first dimension) between this and future generations. In most cases, these future generations comprise anyone born in the future, i.e. from tomorrow till infinite time has passed. This concern, together with some version of the first dimension, is found in all definitions of sustainability.
3. The third dimension relates to *intra-generational equity*. Within this dimension, we consider the extent to which the measures of welfare are equally distributed within a generation both on a macro-scale (i.e. among developed and developing nations) and on a micro-scale (i.e. the equality within a given nation, region or local community). As noted above, there is a large difference in the definitions with regards to whether this dimension is considered at all.
4. The fourth and final dimension relates to *interspecies equity*, relating to whether it is only the welfare (however defined) of humans which is a goal, or whether also the thriving of other living organisms (independent of their potential to contribute to human welfare) is considered. It should be noted that most definitions (including the original definition given in *Our Common Future*) are anthropocentric (i.e. human centred) and therefore do not include this dimension.

5.3 Sustainability and the Environmental Concern

Except from the fourth dimension of sustainability, which is typically not considered, there is no explicit consideration of environmental conservation in most definitions of sustainability. It may therefore seem odd that environmental protection is often seen as being more or less synonymous with sustainability. The reason should primarily be found in the concern for inter-generational equity. The rationale behind protecting the environment from a concern for inter-generational equity is that the natural resources and the services that nature provides are seen as the foundation for society. Without a functioning environment we will not be able to cultivate crops, secure clean air, be protected from ultraviolet radiation from the sun, etc. The idea is thus that protecting the environment is necessary to give future generations the same possibilities for achieving the levels of welfare that current generations are experiencing.

Thus, besides the concern for intra-generational equity, which is not ensured simply by protecting the environment, but which calls for initiatives related to combating poverty, sustainability includes a concern for environmental protection. The extent to which the environment should be protected as a condition for the inter-generational equity dimension of sustainability is, however, not clear-cut. Clearly, human needs cannot be met if humans cannot breathe due to air pollution or lack of oxygen. But the more detailed dependency of human needs on specific functions or qualities of the environment is disputed. For example, will the potential for meeting human needs be violated if the panda bear becomes extinct? And to what extent can technology replace the services and functions provided by ecosystems?

While keeping this discussion in mind, researchers have attempted to quantify carrying capacities of ecosystemsthat must not be exceeded to maintain functions and other ecosystem aspects of interest. For example, the carrying capacities of different terrestrial ecosystems in Europeand elsewhere towards deposition of acidifying compounds (sometimes termed critical loads) have been calculated (Hettelingh et al. 2007). At the global scale planetary boundaries have been proposed and tentatively quantified. Planetary boundaries can be interpreted as carrying capacities for the entire Earth System towards various anthropogenic pressures, such as greenhouse gases and interference with nutrient cycles. If exceeded there is a substantial risk that the Earth System will change from its well-known and relatively stable state that has characterized the Holocene geological epoch in the past 12,000 years to an unknown state (Rockström 2009; Steffen et al. 2015a). According to estimates, this exceedance has already happened for four of the nine proposed planetary boundaries, as shown in Fig. 5.1.

As this chapter is about the role of LCA in the environmental protection needed to achieve sustainability we will only address the part of the sustainability definition pertaining to the environment. Chapter 16 addresses the development of what has been termed Social LCA, addressing the social dimension of sustainability.

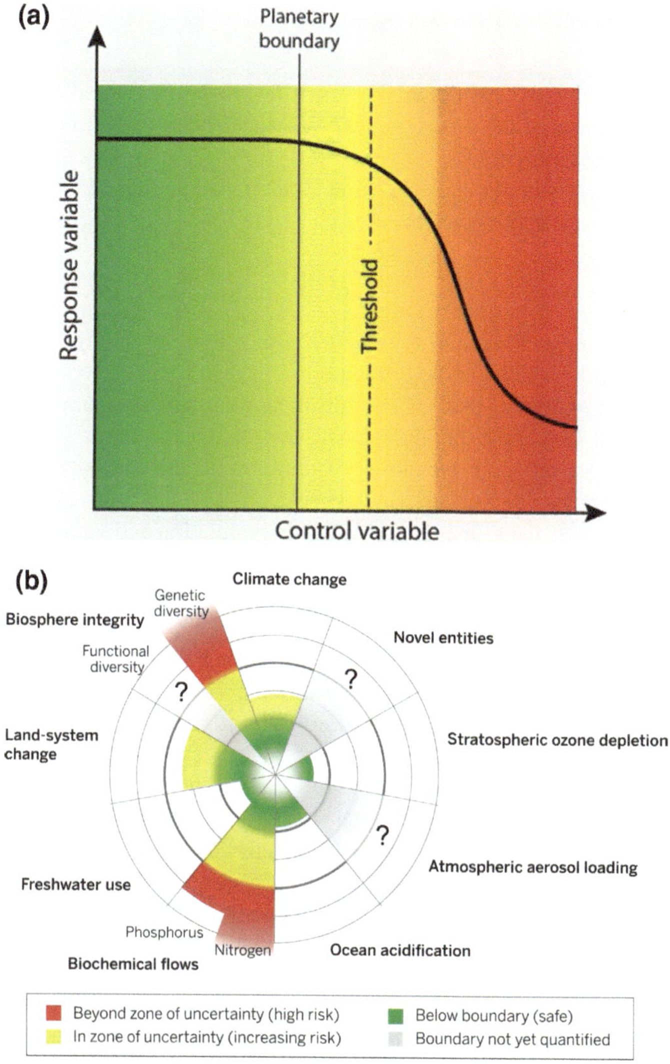

Fig. 5.1 Planetary boundaries. **a** Illustrates the concept of thresholds and boundaries in relation to an ecosystem's response to increasing human pressure. **b** Shows the proposed nine boundaries (two of them subdivided for specific pressures) and that mankind has currently exceeded four of them, two beyond the zone of uncertainty (Steffen et al. 2015a). Reprinted with permission from AAAS

5.4 Sustainability and LCA

If sustainability entails that the environment has to be conserved, the question is How can we conserve the environment? What are the overall drivers that lead to environmental deterioration?

These questions were first addressed in Holdren and Ehrlich (1974), whose work in a modified form lead to the formulation of the so-called IPAT equation, or

$$I = P \cdot A \cdot T \tag{5.1}$$

where (I) is the environmental impact, (P) is the population, (A) is the per capita affluence and (T) is the technology factor.

The formula expresses that the overall impact on the environment is controlled by the number of people on the planet, their affluence, expressed in material affluence per person, and technology's environmental intensity, expressed in environmental impact per material affluence.

Figure 5.2 shows the global development in population and various indicators of affluence, such as GDP, transportation and paper production, along with indicators of environmental pressures and impacts from 1750 to 2010. Figure 5.2a shows that while the world population has almost tripled from 1950 to 2010, all the indicators of affluence have increased at higher rates, meaning that the per capita affluence ("*A*" in the IPAT equation) has increased in the period (note that this increase has been unequal—income differences between and within countries have increased in the period). Figure 5.2b shows that the combined effect of an increasing population and increasing per capita affluence ("*P*" and "*A*" in the IPAT equation) has led to increases in environmental pressure and impacts ("*I*" in the IPAT equation). This means that technological improvements in environmental impact per material affluence ("*T*" in the IPAT equation) have been insufficient for maintaining environmental pressures and impacts at a status quo, let alone for decreasing them.

With the historical development in mind, the IPAT equation shows us that we, in theory, have three overall knots and handles to manipulate to ensure that loads on the environment do not exceed carrying capacities. Two of these three parameters, the number of people and their affluence, have been difficult to handle. In relation to the number of people, this can either be regulated by increasing mortality or reducing fertility, and in most parts of the world issues like these are not on the political agenda. In some parts of the world, for example in the EU, Russia and Japan, it is even seen as a political aim to increase fertility. However, despite this, projections show that the world population may stabilize around 10 billion in 2050.

With regards to the affluence, we have already established above that to increase the intra-generational equity, there is a need for increasing the affluence of the ones mostly in need. Reducing the overall affluence while increasing the affluence of the poorest inevitably calls for a decrease in the affluence of the richest part of the world population which is a difficult program for a political party striving for (re-)election

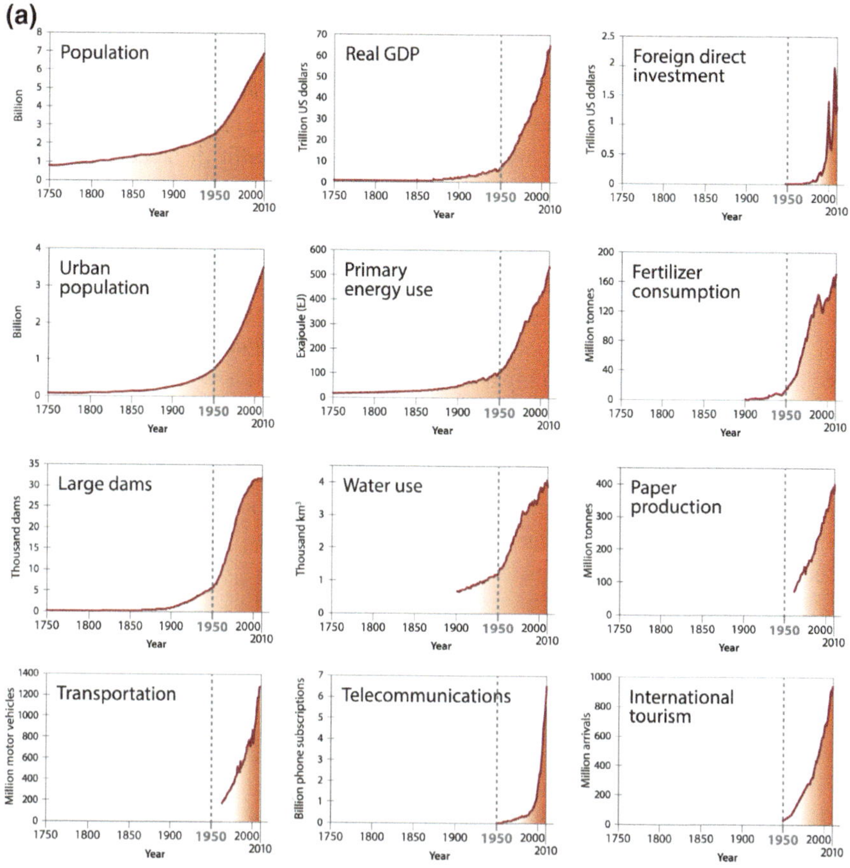

Fig. 5.2 Global development in a selection of **a** socio-economic indicators and **b** pressures and impacts on the environment from 1750 to 2010 (Steffen et al. 2015b). Reprinted by Permission of SAGE Publications, Ltd.

in a liberal democracy as found in most affluent societies today. The "*A*" in the IPAT equation above is therefore expected to increase over time.

What is left is the development of technology, which can allow us to regulate the environmental impact per consumed unit (the '*T*' factor in the IPAT equation). To increase the output or functionality while keeping a constant environmental impact corresponds to increasing what is often termed eco-efficiency. According to the World Business Council of Sustainable Development "eco-efficiency is achieved by the delivery of competitively priced goods and services that satisfy human needs and bring quality of life while progressively reducing environmental impacts of goods and resource intensity throughout the entire life cycle to a level at least in line with the Earth's estimated carrying capacity" (WBCSD 2000). By increasing the eco-efficiency of existing products and technologies, the idea is thus that we will be able to consume the same, or more, while at the same time lowering the overall

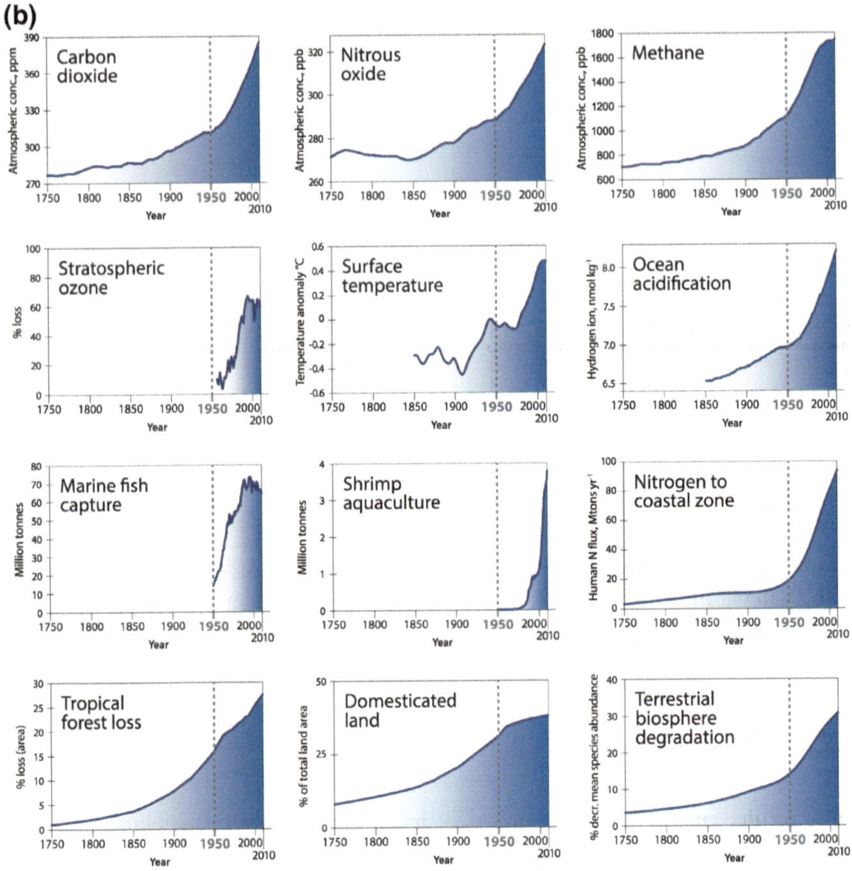

Fig. 5.2 (continued)

environmental burden of this consumption to a level that does not exceed carrying capacities.

As outlined in the chapters above, and as will be further detailed in the remaining parts of this book, LCA shows how a specific functionality can be achieved in the most environmentally friendly way among a predefined list of alternatives, or in which parts of the life cycle it is particularly important to improve a product to reduce its environmental impacts, in other words, increase its eco-efficiency. LCA can therefore be seen as a methodology that can guide decisions towards improving one of the three dimensions in the IPAT equation, namely the technology ("*T*") dimension.

5.5 A Note on Life Cycle Sustainability Assessment

It has been proposed to expand LCA into life cycle sustainability assessment (LCSA) to also encompass socialand economic aspects, in addition to environmental aspects of sustainability when analysing product life cycles (Kloepffer 2008; Zamagni 2012). The idea of LCSA builds on the so-called "three pillars" (or three dimensions) interpretation of sustainability, according to which sustainability is composed of an environmental, social and economic pillar. This interpretation gained momentum with the concept of the "Triple bottom line" by Elkington (1997), who proposed that businesses should manage environmental, social and economic aspects of sustainability in the same quantitative way that financial aspects are typically managed inaccounting. Accordingly, Kloepffer (2008) proposed the following scheme for LCSA:

$$\mathrm{LCSA} = \mathrm{LCA} + \mathrm{LCC} + \mathrm{SLCA} \tag{5.2}$$

LCC is an abbreviation for life cycle costingwhich aims to quantify all costs associated with the life cycle of a product that is directly covered by one or more of the actors in that life cycle. S-LCA is an abbreviation for social life cycle assessment, which has the goal of assessing the social impacts of a product over its life cycle. LCC and S-LCA are detailed in Chaps. 15 and 16 of this book. An important requirement of LCSA is that the three pillars of sustainability must be assessed using the same system boundaries, i.e. that the same elements of a product life cycle are considered in all three assessments (Kloepffer 2008) (see Chap. 8, for an elaboration on system boundaries).

While LCSA is much less mature than LCA and there is a little agreement of how to actually perform it, two fundamental aspects of LCSA deserve highlighting in this chapter:

1. LCSA seems to be based on the assumption that sustainability is something that can be balanced between an environmental, social and economic dimension. This is hinted by the scheme proposed by Kloepffer (2008), according to which a decrease in one sustainability dimension (e.g. environmental) can be compensated by an increase in another dimension (e.g. social). This conflicts with the concept of carrying capacity, according to which the meeting of human needs depends on a minimum level of environmental protection, as mentioned in Sect. 5.2. In our view it would therefore be misleading to assess a product that has a relatively good performance in an LCC and an S-LCA, but a relatively poor performance in an LCA, to be overall sustainable, because the bad performance in an LCA may be contributing to the exceedances of carrying capacities, which in the long term threatens the meeting of human needs and thus social (and economic) sustainability. This perspective is reflected by a popular quote, attributed to Dr. Guy McPherson: "If you really think that the environment is less important than the economy, try holding your breath while you count your money" (McPherson 2009).

2. LCSA includes an economic dimension of sustainability. This is consistent with the common "three pillar" interpretation of sustainability, but it can be questioned how relevant LCC is for sustainability assessments. This is because the costsquantified by LCC are only relevant to sustainability if these costs apply to the poor, which are of concern to the intra-generational equity dimension of sustainability (Jørgensen et al. 2013). Yet, quantifying the monetary gains or losses for the poor is already an aspect commonly included in S-LCA (see Chap. 16).

5.6 Limitations to the Strategy for Achieving Sustainability Through LCA

Even though LCA gives us the very valuable possibility of choosing the most eco-efficient way of achieving a specific functionality or service, this approach has some important limitations in regards to ensuring (environmental) sustainability.

Following the IPAT equation, and knowing the projections for the population growth and the goals for the increase inaverage affluence, it has been estimated that a factor 4, or higher, increase in the eco-efficiency of technologies or products is needed just to ensure a status quo with regards to our impacts on the environment (Reijnders 1998). But as shown in Fig. 5.1, status quo, with regards to some environmental impacts, is not good enough if we are to guarantee a sustainable development, because a number of planetary boundaries have already been exceeded. For some technologies and products an increase in "*T*" closer to a factor 10 may therefore be required.

It is evident that a factor of 10 increase in the eco-efficiency of technologies or products in many cases will be difficult to achieve. For example, even the most eco-efficient cars are far from a factor 10 more efficient than the average car, both regarding energy consumption during use and material consumption during production (Girod et al. 2014). In other cases, however, a factor 10 increase in the eco-efficiency of products has been achieved in isolated areas. Freon and other ozone depleting gases used in for example refrigerators have more or less been phased out as a result of the Montreal Protocol, leading to an eco-efficiency increase on this isolated area, far better than a factor of 10 (WMO 2014).

However, one thing is to increase the eco-efficiency of the product, another is how we administer the gains achieved through the increased efficiency. History has demonstrated that the level of services that we want from products and technologies is not static. As soon as new possibilities evolve we tend simply to expand our wants and expectations (which might not be the same as needs, depending on the interpretation of sustainability). Evidence suggests that increases in eco-efficiencies in some cases due to changes in wants and expectations lead to so-called "rebound effects". An example of a rebound effect could be if an increase in eco-efficiencyof the car engine leads the producer to increase the power of the motor, add extra comfort to the car, or if costumers travel longer distances due to an improved fuel economy, reducing or eliminating the effect of the increase in eco-efficiency. Another example is seen in the lighting technologies: Since the light bulb was invented there has been an enormous increase in the energy efficiency, which has equally lead to a dramatic decrease in the price of light. But as our appetite for more light seems insatiable this increase in eco-efficiency has been met by a corresponding increase in demand—with no signs of saturation. In fact, it has been found that the fraction of GDP spent on light has remained almost constant, close to 1% over the last three centuries in the UK and that this fraction is similar in other countries spanning diverse temporal, geographic, technological and economic circumstances (Tsao and Waide 2010).

In sum, this implies that while LCA may help identify the most eco-efficient solution among a range of alternatives, the actual eco-efficiency that we may achieve through redesign and technological inventions is in many cases insufficient. Furthermore, the increases that are gained in eco-efficiency on the product or technology level may be counterbalanced by increases in demand. Impacts on the environment quantified using LCA can be put into a sustainability perspective by relating them to environmental carrying capacities (Bjørn et al. 2015). This can facilitate an absolute evaluation of whether a studied product can be considered environmentally sustainable, and if not, how much further environmental impacts must be reduced for this to come true. Such an absolute perspective can complement the common relative perspective of LCA which is about identifying the product system that is better for the environment, but that might not be good enough from a sustainability perspective.

Yet, even when an absolute perspective is taken LCA cannot, by itself, cover all relevant aspects of sustainability. Many sustainability researchers have argued that the narrow focus on eco-efficiency simply will not suffice. They propose that we have to look at the necessity of the services, and not only at providing the services in the most eco-efficient way. In other words, these researchers talk about the necessity to adjust the "*A*", the affluence, in the IPAT equation. In this relation, the LCA falls short—it is a tool to find the most eco-efficient way to deliver this service among a list of predefined alternatives—not a tool for identifying the importance of various services.

Increases in eco-efficiency are high on the agenda in many companies, not least because of the often accompanying cost reductions, and on this journey there is no doubt that the LCA will be an invaluable tool to show the way. However, at the same time, we have to be open to the possibility that we may need to discuss not only how different services should be provided, but also the more sensitive and political question—whether a service should be provided at all, if we are to ensure that the future generations are given the same possibilities for meeting their needs as we were given.

References

Bjørn, A., Diamond, M., Owsianiak, M., et al.: Strengthening the link between life cycle assessment and indicators for absolute sustainability to support development within planetary boundaries. Environ Sci Technol **49**, 6370–6371 (2015). doi:10.1021/acs.est.5b02106

Elkington, J.: Cannibals with Forks: The Triple Bottom Line of 21st Century Business. Capstone Publ, Oxford (1997)

Girod, B., van Vuuren, D.P., Hertwich, E.G.: Climate policy through changing consumption choices: Options and obstacles for reducing greenhouse gas emissions. Glob. Environ. Chang. **25**, 5–15 (2014). doi:10.1016/j.gloenvcha.2014.01.004

Hettelingh, J.-P., Posch, M., Slootweg, J., et al.: Critical loads and dynamic modelling to assess European areas at risk of acidification and eutrophication. Water Air Soil Pollut. Focus **7**, 379–384 (2007). doi:10.1007/s11267-006-9099-1

Holdren, J.P., Ehrlich, P.R.: Human population and the global environment. Am. Sci. **62**, 282–292 (1974)

Jørgensen, A., Herrmann, I.T., Bjørn, A.: Analysis of the link between a definition of sustainability and the life cycle methodologies. Int. J. Life Cycle Assess. **18**, 1440–1449 (2013). doi:10.1007/s11367-013-0617-x

Kloepffer, W.: Life cycle sustainability assessment of products. Int. J. Life Cycle Assess. **13**, 89–95 (2008). doi:10.1065/lca2008.02.376

McPherson, G.: Time for a revolution. http://guymcpherson.com/2009/05/time-for-a-revolution/ (2009). Accessed 1 June 2016

Pezzey, J.: Sustainable Development Concepts—An Economic Analysis. World Bank Environment Paper No. 2. The World Bank, Washington DC (1992)

Reijnders, L.: The factor X debate: setting targets for eco-efficiency. J. Ind. Ecol. **2**, 13–22 (1998)

Rockström, J.: Planetary Boundaries: Exploring the Safe Operating Space for Humanity. Ecology and Society. **14**(2), art. 32 (2009)

Steffen, W., Richardson, K., Rockström, J., et al.: Planetary boundaries: Guiding human development on a changing planet. Science (2015a). doi:10.1126/science.1259855

Steffen, W., Broadgate, W., Deutsch, L., Gaffney, O., Ludwig, C.: The trajectory of the Anthropocene: the great Acceleration. Anthrop. Rev. **2**(1), 81–98 (2015b)

Tsao, J.Y., Waide, P.: The world's appetite for light: empirical data and trends spanning three centuries and six continents. Leukos **6**, 259–281 (2010)

WBCSD: Eco-efficiency—Creating More Value With Less Impact. World Business Council for Sustainable Development, Washington DC (2000)

WCED: Report of the World Commission On Environment And Development: Our Common Future. Oxford University Press, Oxford (1987)

WMO: Scientific Assessment Of Ozone Depletion: World Meteorological Organization, Global Ozone Research and Monitoring Project, Geneva (2014) doi:10.1007/s11367-012-0389-8

Zamagni, A.: Life cycle sustainability assessment. Int. J. Life Cycle Assess. **17**, 373–376 (2012). doi:10.1007/s11367-012-0389-8

Author Biographies

Andreas Moltesen has been working with LCA since 2006 with a particular focus on social life cycle assessment. He has later worked on life cycle assessments of biofuels and is currently particularly involved with life cycle assessments of transport systems.

Anders Bjørn part of the LCA community since the early 2010s. Main focus is interpretations of sustainability and integration of sustainability targets in LCA to enable absolute sustainability assessments.

Part II
Methodology

Chapter 6
Introduction to LCA Methodology

Michael Z. Hauschild

Abstract In order to offer the reader an overview of the LCA methodology in the preparation of the more detailed description of its different phases, a brief introduction is given to the methodological framework according to the ISO 14040 standard and the main elements of each of its phases. Emphasis is on the iterative nature of the LCA process with its many feedback loops between the different phases. It is explained how the integrated use of sensitivity analysis helps identify key assumptions and key data and thus ensure effectiveness by directing the focus of the LCA practitioner to those parts of the study where additional work contributes most to strengthen the results and conclusions of the study.

Learning Objectives
After studying this chapter, the reader should be able to

- Draw and explain the methodological framework for LCA.
- Present an overview of the phases of LCA, their purpose and main elements.
- Explain the iterative nature of LCA and its rationale in terms of helping the LCA practitioner focus on what matters most for the results and conclusions of the study.

6.1 Introduction

As described in Chap. 3, the need for agreement on common principles for how to perform an LCA was realised back in the 1980s. An international discussion of methodological issues took off around 1990 under the auspices of SETAC leading

M.Z. Hauschild (✉)
Division for Quantitative Sustainability Assessment, Department of Management Engineering, Technical University of Denmark, Kongens Lyngby, Denmark
e-mail: mzha@dtu.dk

M.Z. Hauschild et al. (eds.), *Life Cycle Assessment*,
DOI 10.1007/978-3-319-56475-3_6

to publication of state-of-the-art reports and codes of conduct for different parts of the LCA methodology throughout the 1990s and feeding into the standardisation process that went on in parallel. Although many methodological aspects are still under discussion and development continues today, the fundamental structure has been stable since the appearance of the first ISO 14040 standard in 1997, and it is also applied in major LCA methodologies like the CML (Guinée 2002), EDIP97 (Wenzel et al. 1997), and by the ILCD guidelines from the EU Commission (EC-JRC 2010).

The methodology chapters in Part II of this book give a detailed presentation of the LCA methodology structured according to the ISO framework and referring to the recommendations and requirements given by the ILCD guidelines. References are not given consistently to these sources throughout the chapters but unless otherwise mentioned, they are the basis of the presented methodology.

The European ILCD guidelines for LCA (EC-JRC 2010) are strongly founded in the framework and methodological requirements of the ISO LCA standards (ISO 2006a, b) but they go further and offer methodological guidance at a much more detailed level than the standards do. They are the outcome of a comprehensive consultation process involving hearings of experts and stakeholders, and on this basis, we have chosen them as a useful reference for discussing LCA methodology and specifying methodological choices. In Chap. 37 the most important methodological actions and requirements of the ILCD guideline are presented in the form of a cookbook or checklist that you can refer to as a reference methodology to follow, or to deviate from at specific and transparently documented points of the methodology.

6.2 The Phases of LCA

We begin in this introductory chapter with a brief description of the main methodological phases and the way in which their results are assessed and refined in a focused iterative process. This will give you an overview of the methodology before you dig into the details and peculiarities of its different phases and elements, and it will introduce you to the iterative approach, which is fundamental for performing a successful LCA.

As illustrated in Fig. 6.1, the ISO standard distinguishes the methodological framework of LCA from its different applications, which are multiple such as product development, Ecolabelling, carbon footprint and other footprints (see Part III of the textbook for examples). Applications of LCA are treated in separate publications from the standard organisation. The LCA framework operates with four separate phases, Goal and scope definition, Inventory analysis, Impact assessment and Interpretation.

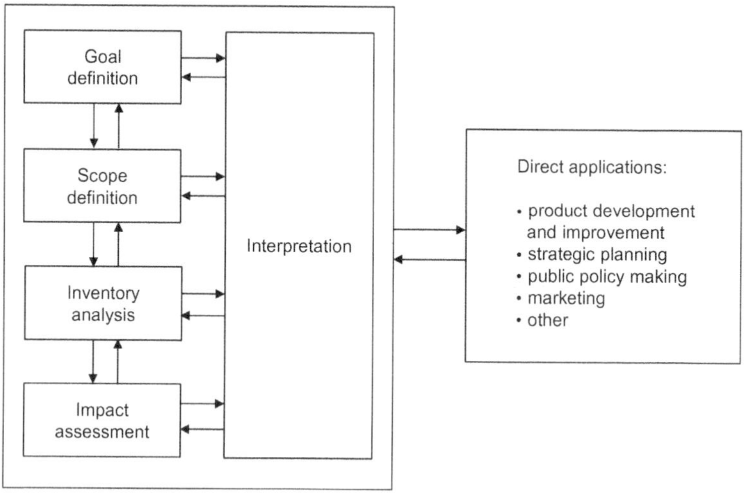

Fig. 6.1 Framework of LCA modified from the ISO 14040 standard

6.2.1 Goal and Scope Definition

An LCA starts with a well-considered and deliberate definition of the goal of the study (see Chap. 7). Why is this study performed? Which question(s) is it intended to answer and for whom is it performed? The goal definition sets the context of the LCA study and is the basis of the scope definition (see Chap. 8) where the assessment is framed and outlined in accordance with the goal definition, primarily in terms of

- Defining the functional unit: a quantitative description of the function or service for which the assessment is performed, and the basis of determining the reference flow of product that scales the data collection in the next LCA phase, the inventory analysis.
- Scoping the product system, deciding which activities and processes belong to the life cycle of the product that is studied.
- Selecting the assessment parameters, i.e. the impacts that shall be assessed in the study.
- Selecting the geographical and temporal boundaries and settings of the study and the level of technology that is relevant for the processes in the product system.
- Deciding the relevant perspective to apply in the study: should it be a consequential study assessing the impacts that can be expected as a consequence of choosing one alternative over another, or should it be an attributional study assessing the impacts that are associated with the studied activity?
- Identifying the need to perform critical review, in particular if the study is a comparative assertion intended to be disclosed to the public.

The goal definition and the ensuing scope definition are very important to consider when the results of the study are interpreted since these definitions involve choices that determine the collection of data and the way in which the system is modelled and assessed. They therefore have a strong influence on the validity of the conclusions and recommendations that are based on the results of the LCA.

6.2.2 *Inventory Analysis*

Following the definition of goal and scope, the inventory analysis collects information about the physical flows in terms of input of resources, materials, semi-products and products and the output of emissions, waste and valuable products for the product system (see Chap. 9). The analysis studies all the processes that were identified as belonging to the product system, and the flows are scaled in accordance with the reference flow of product that is determined from the functional unit. Due to the comprehensiveness of most product systems, the inventory analysis often relies on generic data for many processes originating from databases with unit processes or cradle-to-gate data, presenting the in- and output flows for one unit process, e.g. for production of a material, generation of heat or electricity, transportation or waste management. Environmentally extended input–output analysis can be used to support and qualify the collection of inventory data as discussed in Chap. 14.

The outcome of the inventory analysis is the life cycle inventory, a list of quantified physical elementary flows for the product system that is associated with the provision of the service or function described by the functional unit.

6.2.3 *Impact Assessment*

Taking the life cycle inventory as a starting point, the impact assessment translates the physical flows and interventions of the product system into impacts on the environment using knowledge and models from environmental science (see Chap. 10). The impact assessment consists of five elements of which the first three are mandatory according to the ISO 14040 standard:

1. *Selection* of impact categories representative of the assessment parameters that were chosen as part of the scope definition. For each impact category, a representative indicator is chosen together with an environmental model that can be used to quantify the impact of elementary flows on the indicator.
2. *Classification* of elementary flows from the inventory by assigning them to impact categories according to their ability to contribute by impacting the chosen indicator.

3. *Characterisation* using environmental models for the impact category to quantify the ability of each of the assigned elementary flows to impact the indicator of the category. The resulting characterised impact scores are expressed in a common metric for the impact category. This allows aggregation of all contributions into one score, representing the total impact that the product system has for that category. The collection of aggregated indicator scores for the different impact categories (each expressed in its own metric) constitutes the characterised impact profile of the product system.
4. *Normalisation* is used to inform about the relative magnitude of each of the characterised scores for the different impact categories by expressing them relative to a common set of reference impacts—one reference impact per impact category. Often the background impact from society is used as a reference. The result of the normalisation is the normalised impact profile of the product system in which all category indicator scores are expressed in the same metric.
5. *Grouping or weighting* supports comparison across the impact categories by *grouping* and possibly ranking them according to their perceived severity, or by *weighting* them using weighting factors that for each impact category gives a quantitative expression of how severe it is relative to the other impact categories. Quantitative weighting allows aggregation of all the weighted impact scores into one overall environmental impact score for the product system, which may be useful when the results of the LCA are used in decision support together with other condensed information like the economic costs of the alternatives.

The main focus of this book is the traditional environmental LCA focusing on the environmental impacts of the product system, but for sustainability assessment, also social and economic impacts need to be considered. For these other dimensions of sustainability, a life cycle perspective is as relevant as it is for the environmental dimension and in a life cycle sustainability assessment (LCSA—See Chap. 5) they may be addressed through a social LCA (S-LCA) and a life cycle costing analysis (LCC). Both of these assessment techniques have their own distinct methodological foundation which shares the fundamental framework of environmental LCA but has many distinct elements in all phases of the methodology as introduced in Chaps. 15 (LCC) and 16 (S-LCA).

Interpretation The results of the study are interpreted in order to answer the question(s) posed as part of the goal definition (see Chap. 12). The interpretation considers both results of the inventory analysis and the impact assessment elements characterisation and, possibly, normalisation and weighting. The interpretation must be done with the goal and scope definition in mind and respect the restrictions that the scoping choices impose on a meaningful interpretation of the results, e.g. due to geographical, temporal or technological assumptions.

Sensitivity analysis and uncertainty analysis are applied as part of the interpretation to guide the development of conclusions from the results, to appraise the robustness of the conclusions, and to identify the focus points for further work in order to further strengthen the conslusions.

6.3 The Iterative Nature of LCA

In Fig. 6.1 a number of arrows indicate that rather than a linearly proceeding process, LCA involves many feedback loops between the different phases of the LCA. Insights from the impact assessment are used in refining the inventory analysis and insights from both of these phases may feed back to the scope definition, e.g. in the setting of the boundaries of the product system, what to include and what to exclude. Sensitivity and uncertainty analysis are thus not just performed in the interpretation at the end but throughout the study as part of both inventory analysis and impact assessment in order to identify the key figures or key assumptions of the study and the data that are associated with the largest uncertainties (see Chap. 11). Each phase of the methodology provides feedback to the previous phases of the study and helps target the next iteration of the LCA. The best precision is obtained with minimum work effort if the focus is on improving the key figures wherever possible and needed, and on reducing the largest uncertainties.

In practice, the first iteration will often be a screening that covers the full life cycle, but in terms of inventory data largely is based on easily accessible data from available databases. Following the impact assessment, the parts of the product system that contribute most strongly to the total results can be identified, and the chosen boundaries of the product system can be tested. As a consequence, the scoping may have to be refined. The impact assessment results also allow identifying those inventory data or assumptions made in the inventory analysis that have the largest influence on the overall results or for which the uncertainties are so large that they potentially could be key figures. These data should be the target of the next iteration, where effort should be focused on testing and refining these assumptions or data and get more representative or recent data. Based on the revised inventory a new impact assessment is performed, and the sensitivity analysis is performed once more to see which are now the key figures and key assumptions. Large uncertainties may also accompany the factors applied in the characterisation of some of the inventory flows in the impact assessment, and if the sensitivity analysis indicates that such uncertainties may have a decisive influence on the results, these factors will also be the target of a consecutive iteration. Figure 6.2 illustrates the iterative approach to performing an LCA.

As illustrated by the narrowing spiral in Fig. 6.2, the uncertainty of the LCA results is reduced through the repeated iterations, and these are carried on until the remaining uncertainty of the results is sufficiently small to meet the goal of the study. If the goal is to identify which among several alternatives has the lowest environmental impacts, the number of needed iterations may be low if the alternatives show large differences in their impacts, while a higher number of iterations will be needed if the alternatives are more similar. An LCA performed to support an environmental product declaration with a general requirement to the uncertainty of the impact scores can require a high number of iterations before all impact scores are determined within the stipulated level of uncertainty.

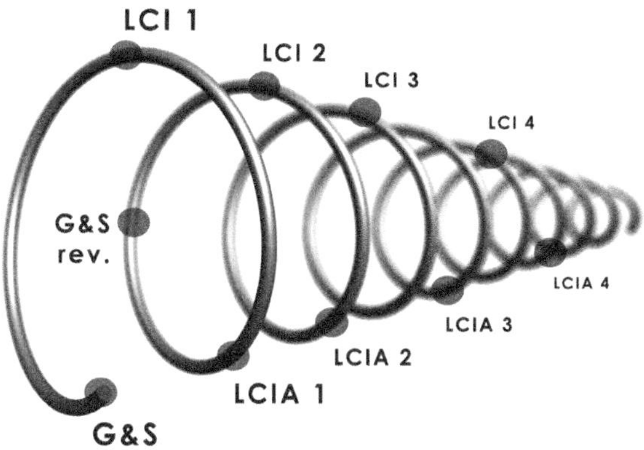

Fig. 6.2 Using sensitivity analysis and uncertainty analysis as integrated tools, the phases of the LCA methodology are repeated with focus on improving and strengthening the identified key figures and assumptions in consecutive iterations until the strength of the conclusions meets the requirements posed by the goal and scope definition

With this overview of the LCA framework, its interconnected phases and how iteration is used to ensure effectiveness when performing an LCA, you are now prepared for diving into the intricate details of the many elements of the LCA methodology. Enjoy!

References

This chapter is to a large extent based on the ILCD handbook and the ISO standards 14040 and 14044. Due to the scope of this chapter, some details have been omitted, and some procedures have been rephrased to make the text more relevant to students. For more details, the reader may refer to these texts:

EC-JRC (2010) European Commission—Joint Research Centre—Institute for Environment and Sustainability: International Reference Life Cycle Data System (ILCD) Handbook—General guide for Life Cycle Assessment—Detailed guidance. First edition March 2010. EUR 24708 EN. Luxembourg. Publications Office of the European Union.

ISO: Environmental management—life cycle assessment—principles and framework (ISO 14040). ISO, The International Organization for Standardization, Geneva (2006a)

ISO: Environmental management—life cycle assessment—requirements and guidelines (ISO 14044). ISO, The International Organization for Standardization, Geneva (2006b)

Additional References Quoted in the Text

Guinée, J.B., Gorrée, M., Heijungs, R., Huppes, G., Kleijn, R., van Oers, L., Wegener Sleeswijk, A., Suh, S., Udo de Haes, H.A., de Bruijn, H., van Duin, R., Huijbregts, M.A.J.: Handbook on Life Cycle Assessment: Operational Guide to the ISO Standards. Kluwer Academic Publishers, Dordrecht (2002). ISBN 1-4020-0228-9

Wenzel, H., Hauschild, M.Z., Alting, L.: Environmental Assessment of Products, vol. 1—Methodology, Tools and Case Studies in Product Development, Kluwer Academic Publishers, Hingham, MA. USA (1997). ISBN 0 412 80800 5

Author Biography

Michael Z. Hauschild Involved in development of LCIA methodology since the early 1990s. Has led several SETAC and UNEP/SETAC working groups and participated in the development of the ISO standards and the ILCD methodological guidelines. Main LCA interests are chemical impacts, spatial differentiation and science-based boundaries in LCIA.

Chapter 7
Goal Definition

Anders Bjørn, Alexis Laurent, Mikołaj Owsianiak and Stig Irving Olsen

Abstract The goal definitionis the first phase of an LCA and determines the purpose of a study in detail. This chapter teaches how to perform the six aspects of a goal definition: (1) Intended applications of the results, (2) Limitations due to methodological choices, (3) Decision context and reasons for carrying out the study, (4) Target audience, (5) Comparative studies to be disclosed to the public and (6) Commissioner of the study and other influential actors. The instructions address both the conduct and reporting of a goal definition and are largely based on the ILCD guidance document (EC-JRC in European Commission—Joint Research Centre—Institute for Environment and Sustainability: International Reference Life Cycle Data System (ILCD) Handbook—General Guide for Life Cycle Assessment —Detailed Guidance. Publications Office of the European Union, Luxembourg 2010).

Learning Objectives

After studying this chapter, the reader should be able to:

- Define the goal of any LCA study.
- Explain the six goal aspects and their relevance for the subsequent LCA phases.

A. Bjørn (✉) · A. Laurent · M. Owsianiak · S.I. Olsen
Division for Quantitative Sustainability Assessment,
Department of Management Engineering, Technical University of Denmark,
Kongens Lyngby, Denmark
e-mail: anders.bjoern@polymtl.ca

A. Bjørn
CIRAIG, Polytechnique Montréal, 3333 Chemin Queen-Mary,
Montreal, QC, Canada

M.Z. Hauschild et al. (eds.), *Life Cycle Assessment*,
DOI 10.1007/978-3-319-56475-3_7

7.1 Introduction

The goal definitionis the first phase of any LCA. Here, the purpose of the study is elaborately defined and described. This greatly influences the LCA because decisions made in later LCA phases (Chaps. 8–12) must be consistent with the goal definition. The influence may also go the other way, for example, if unforeseen data limitations in the inventory analysis (Chap. 9) necessitate a revision of the goal definition. Such a revision is an example of the iterative nature of LCA (see Chap. 6).

The goal definition based on the ISO standard requirements generally contains six aspects:

1. Intended applications of the results
2. Limitations due to methodological choices
3. Decision context and reasons for carrying out the study
4. Target audience
5. Comparative studies to be disclosed to the public
6. Commissioner of the study and other influential actors.

Each aspect must be considered when performing an LCA. Aspects 1 and 3 are central for *doing* an LCA because they have pervasive influence on decisions made in later LCA phases. On the other hand, aspects 2, 4, 5 and 6 mainly relate to *communicating the results* of an LCA. For these aspects, we further refer to Chaps. 13, 37–39, which provide specific guidance on and examples of the reporting and reviewing of LCA results.

7.2 Intended Applications of the Results

All LCAs involve studying one or more product systems and this can be used in several applications, such as

- Comparing environmental impacts of specific goods or services.
- Identifying the parts of a product system that contribute most to its environmental impact (i.e. "hot spot identification", focusing in product development).
- Evaluating improvement potentials from changes in product designs(analysis and 'what-if' scenarios in eco-design).
- Documenting the environmental performance of products (e.g. in marketing using environmental product declarations or other types of product environmental footprints).
- Developing criteria for an eco-label.
- Developing policies that consider environmental aspects.

It is important to determine the intended application(s) of the LCA results at the onset, because it influences later phases of an LCA, such as the drawing of system boundaries (Chap. 8), sourcing of inventory data (Chap. 9) and interpretation of results (Chap. 12). Often, several separate applications are intended in a study. For example, the intended applications of the results of the illustrative case on window frames in Chap. 39 were both to benchmark a new window design against three windows already on the market and to identify hot spots in the life cycle of the compared windows with the aim of guiding future impact reduction efforts.

7.3 Limitations Due to Methodological Choices

This aspect can be seen as a critical reflection of what the LCA results can and cannot be used for. If a study only covers climate change (often referred to as a "carbon footprint" study) it is, for example, important to stress that results cannot be used to claim a general environmental superiority of a studied product or conclude anything about its overall "environmental friendliness". Also, if a comparative study disregards one or more life cycle stages, it is important to stress how that limits the interpretation of results. For example, a study comparing the production of 1 tonne aluminum to the production of 1 tonne steel from mining to ingot cannot be used to identify the environmentally soundest material for use in a car, because the density difference of the two metals leads to differences in the amount of metal used for the car body and differences in the car mileage (fuel consumption per kilometre), causing different environmental impacts in the use stage and finally also in the disposal stage. In the illustrative window frame case study (Chap. 39) a stated limitation of the study was that a site-generic LCIA approach was taken in spite of impacts being concentrated around Scandinavia, where the natural environments, for some impact categories, do not correspond to the global average (e.g. Scandinavian soils show a higher sensitivity to depositions of acidifying compounds). Note that the limitations stated here should only relate to the choices made in the goal and scope phases of an LCA (this chapter and Chap. 8). These choices all relate to the *planning* and *use* of an LCA. On the contrary, choices made during the inventory and impact assessment phases of an LCA (Chaps. 9 and 10) relate to *unforeseen constraints and assumptions* (for example with respect to data availability) and must be documented at a later point in an LCA report, for example, in the inventory analysis part (Chap. 9) or in the interpretation part of a report (see Chap. 12).

7.4 Decision Context and Reasons for Carrying Out the Study

This is an important aspect of the goal definition because it strongly influences the appropriate elaboration of a life cycle inventory (Chap. 9). First, the reasons for carrying out a study must be understood. The reasons should be clearly connected

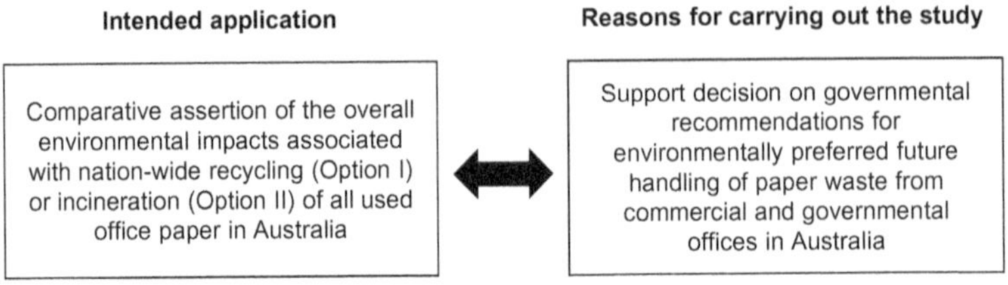

Fig. 7.1 Example of reasons for carrying out a study in continuation of the intended application

to the intended application of results (Sect. 7.2) and specifically address drivers and motivations with respect to decision-making. Figure 7.1 provides an example of reasons for carrying out a study in continuation of the intended applications.

Note that there is some ambiguity about the differences between "Intended application" and "Reasons for carrying out the study" in the ILCD guideline. As a rule of thumb the former should describe *what* a study does, while the latter should address *why* a study is made. The reasons for carrying out at study help understanding its decision context. In the example shown in Fig. 7.1 the study is motivated by a need for decision supporton governmental recommendations of paper waste handling. This means that the results and recommendations of the study can be expected to lead to changes in the analysed system. These changes may, in turn, lead to so-called "structural changes" in other systems that the studied product system interacts with. A structural change occurs when a change in one product system has such a large influence on the demand for a good or a service that it leads to new equipment being installed (increase in production capacity) or existing equipment being prematurely taken out of use (decrease in production capacity). As a rule of thumb, structural changes can be assumed to take place if the analysed decision leads to an additional demand or supply of a product that exceed the average percentage of annual replacement of total capacity (100% divided by the average equipment lifetime in years, e.g. 20). Structural changes result in qualitative and quantitative differences of industries and this must be considered in the inventory modelling (Chap. 9). In combination the above considerations help identify three different decision context situations and any LCA should be classified into one of these as part of the goal definition. Box 7.1 presents these three decision contexts and Fig. 7.2 presents a decision tree for how to determine the correct decision context of an LCA study.

Box 7.1 The Three Types of Decision Contexts

Situation A (Micro-level decision support): The study results are intended used to support a decision, but the small scale of the studied product system means that regardless the decision made, it will not cause structural changes in the systems that the studied product system interacts with. Many studies that intend to compare individual product systems, identify hotspots within these (see Sect. 7.2) or document the environmental performance of a product in the form of an environmental product declaration fall into this decision context. The decision support of the LCA study may lead to limited changes in other systems, e.g. a reduced demand for electricity, but the changes are not of a structural nature, e.g. no electricity production equipment will be prematurely taken out of use.

Situation B (Meso/macro-level decision support): The study results are intended used to support a decision, and the scale of the studied product system is such that the decisions that are made are expected to cause structural changes in one or more processes of the systems that the studied product system interacts with. An example of a study that would be classified as belonging to this type of decision context is a study intended as decision support for policy development on potential nationwide substitution of diesel derived from oil with biodiesel for private cars. Such a decision will lead to structural changes in the biodiesel industry in the form of new equipment being installed to respond to the substantially increased demand for biofuels.

Situation C (Accounting): The study is not to be used to support decisions and is of a purely descriptive nature. It is documenting what has already happened, or what will happen due to a decision that has already been taken. Therefore, the presence of the LCA study will not lead to changes (small or structural) on other systems. Interactions with other systems (whether taking place in the past or in the future), e.g. through energy generated from waste incineration, can either be included in the product system model (**Situation C1**) or considered partially in the LCA through allocation (see Chap. 8) (**Situation C2**). C1 is used unless C2 is specifically prescribed by the commissioner's goal of the study.

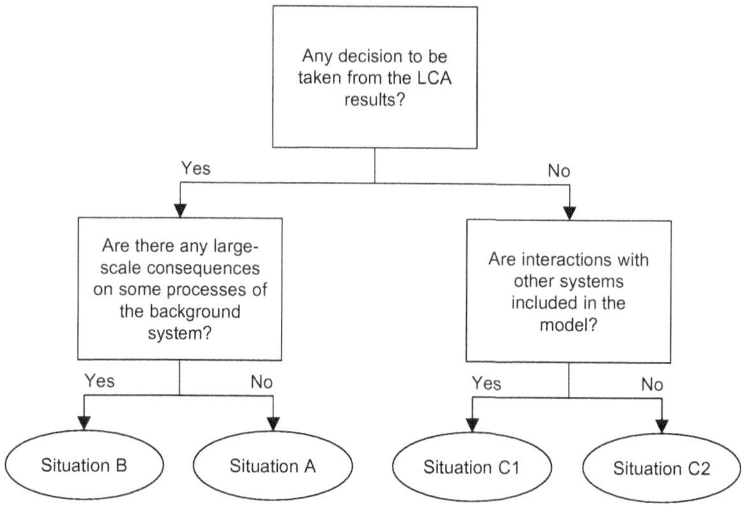

Fig. 7.2 Decision tree for how to identify the correct decision context

Figure 7.2 shows that the identification of the decision context depends on:

- Whether the study is intended as decision support
- Whether structural changes in interacting systems are expected from a decision supported by the study.
- Whether it is chosen to model interactions with other systems as part of the product system model or to handle them partially through allocation (see Chap. 8).

In the illustrative case of the window frames, the reason to carry out the study was to attract environmentally conscious consumers, through the use of an eco-label that the LCA results would help obtain. The study is thus to be used for decision support, but since it is concerned with a single product, this decision support is not expected to lead to structural changes in other systems. The decision context of the study is therefore Situation A (Micro-level decision support).

7.5 Target Audience

The goal definition must state the target audience of the study, i.e. to whom the results of the study are intended to be communicated. The target audience may be consumers, consumer organisations, companies (managers, product developers, etc.), government, NGOs and others. The target audience greatly influences the extent to which details of the study should be documented, the technical level of reporting (Chap. 8) and the interpretation of results (Chap. 12). In the illustrative window frame case study, the employees of the window producer NorWin's environmental and design departments are the target audience. Since this audience is unfamiliar with LCA, the content of the report was presented pedagogically by explaining technical terms that the readers could not be expected to be familiar with. When the readers are unfamiliar with LCA it may also be appropriate to provide brief background information about LCA of the type given in Chap. 2 of this book.

7.6 Comparative Studies to Be Disclosed to the Public

The goal definition should explicitly state whether the LCA study is of a comparative nature (see Sect. 7.2) and if it is intended to be disclosed to the public. If this is the case, the ISO standard specifies a number of requirements on the conduct and documentation of the study and an external review process, due to the potential consequences that the communication of the results of the study may have for external companies, institutions, consumers and other stakeholders. The ISO requirements are detailed in Chap. 8 and are basically meant to ensure transparency and good quality of a study.

7.7 Commissioner of the Study and Other Influential Actors

The goal definition should also explicitly state who commissioned the study, who financed it (usually the commissioning organisation) and other organisations that have influence on the study, including those of the LCA experts conducting the study. This step of the goal definition is meant to highlight potential conflicts of interest to readers of the study. Such conflict of interest may occur if a key provider of data has an economic interest in particular LCA results and interpretations. In comparative studies, it may also lead to an unintentional bias of the data collection. The commissioner of the study will normally provide data that is up to date and reflects the current performance of the technology for the commissioner's own product. In contrast, the data collection for the other product(s) in the comparison will typically have to be based on literature and databases and hence, due to the delay involved in publishing the data, represent the state of the art several years ago.

References

This chapter is to a large extent based on the ILCD handbook and the ISO standards 14040 and 14044. Due to the scope of this chapter, some details have been omitted, and some procedures have been rephrased to make the text more relevant to students. For more details, the reader may refer to these texts:

EC-JRC.: European Commission—Joint Research Centre—Institute for Environment and Sustainability: International Reference Life Cycle Data System (ILCD) Handbook—General Guide for Life Cycle Assessment—Detailed Guidance. First edition March 2010. EUR 24708 EN. Publications Office of the European Union, Luxembourg (2010)

ISO.: Environmental Management—Life Cycle Assessment—Principles and Framework (ISO 14040). ISO, the International Organization for Standardization, Geneva (2006a)

ISO.: Environmental Management—Life Cycle Assessment—Requirements and Guidelines (ISO 14044). ISO, the International Organization for Standardization, Geneva (2006b)

Author Biographies

Anders Bjørn part of the LCA community since the early 2010s. Main focus is interpretations of sustainability and integration of sustainability targets in LCA to enable absolute sustainability assessments.

Alexis Laurent working with LCA since 2010 with a strong LCIA focus, particularly on normalisation aspects. Main LCA interests include development of LCIA methods, LCA applications and footprinting of large-scale systems for policy-making (e.g. nations, sectors), and LCA applied to various technology domains, including energy systems.

Mikołaj Owsianiak involved in development and application of life cycle impact assessment methods in sustainability assessment of technologies. Has worked on issues associated with: soils (remediation), metals (toxic impact assessment), biodiesel (fate in the environment), and carbonaceous materials (biochar and hydrochar).

Stig Irving Olsen LCA expert both as researcher and as consultant. Involved in the development of LCA methodologies since mid 1990's. Contributed to UNEP/SETAC working groups on LCIA methodology. Main LCA interest is human toxicity impacts, emerging technologies, and decision making.

Chapter 8
Scope Definition

Anders Bjørn, Mikołaj Owsianiak, Alexis Laurent, Stig Irving Olsen, Andrea Corona and Michael Z. Hauschild

Abstract The scope definition is the second phase of an LCA. It determines what product systems are to be assessed and how this assessment should take place. This chapter teaches how to perform a scope definition. First, important terminology and key concepts of LCA are introduced. Then, the nine items making up a scope definition are elaborately explained: (1) Deliverables. (2) Object of assessment, (3) LCI modelling framework and handling of multifunctional processes, (4) System boundaries and completeness requirements, (5) Representativeness of LCI data, (6) Preparing the basis for the impact assessment, (7) Special requirements for system comparisons, (8) Critical review needs and (9) Planning reporting of results. The instructions relate both to the performance and reporting of a scope definition and are largely based on ILCD.

Learning Objectives

After studying this chapter, the reader should be able to:

- Define the scope of any LCA study.
- Explain each of the nine scope items and their relevance for the subsequent LCA phases.
- Define a functional unit for any kind of LCA study.
- Explain the fundamental characteristics of an attributional and a consequential modellingapproach and how the decision context determines the choice between them.
- Explain how the iterative approach to LCA helps getting the system boundaries and completeness right.

A. Bjørn (✉) · M. Owsianiak · A. Laurent · S.I. Olsen · A. Corona · M.Z. Hauschild
Division for Quantitative Sustainability Assessment, Department of Management Engineering, Technical University of Denmark, Kgs Lyngby, Denmark
e-mail: anders.bjoern@polymtl.ca

A. Bjørn
CIRAIG, Polytechnique Montréal, 3333 Chemin Queen-Mary, Montreal, QC, Canada

M.Z. Hauschild et al. (eds.), *Life Cycle Assessment*,
DOI 10.1007/978-3-319-56475-3_8

8.1 Introduction

The scope definition determines what product systems are to be assessed and how this assessment should take place. Together with the goal definition (Chap. 7) the scope definition serves as a firm guide for how the ensuing LCA phases should be performed (Inventory analysis, Impact assessment and Interpretation, including uncertainty and sensitivity analysis) and for how the LCA should be reported. An overarching aim of the scope definition is to ensure and document the consistency of methods, assumptions and data and strengthen the reproducibility of the study.

A scope definition consists of the following nine scope items:

1. Deliverables
2. Object of the assessment
3. LCI modelling framework and handling of multifunctional processes
4. System boundaries and completeness requirements
5. Representativeness of LCI data
6. Preparation of the basis for the impact assessment
7. Special requirements for system comparisons
8. Needs for critical review
9. Planning reporting of results.

Each item must be considered when performing an LCA. Items 2–6 are central for *doing* an LCA because these have a pervasive influence on decisions made in later LCA phases. Aspects 1, 7, 8 and 9 mainly relate to *reporting and communicating* an LCA study. For these items, we further refer to Chaps. 13, 37–39, which provide specific guidance on the reviewing and reporting of LCAs. Note that the aspect of data quality requirements, which ILCD proposes as a separate scope item, is here considered under scope items 4 and 5.

8.2 Terminology and Key Concepts

Before explaining the nine scope items, we present the terminology and key concepts that are used in this chapter.

8.2.1 Unit Process and Flows

A unit process is the smallest element considered in a life cycle inventory model (see below) for which input and output data are quantified. Unit processes can therefore be considered the building blocks of a life cycle inventory model that are "glued together" by input and output data, which can be organised into six categories of physical flows:

Input flows:

1. Materials
2. Energy
3. Resources

Output flows:

4. Products
5. Waste to treatment
6. Emissions.

Figure 8.1 shows a unit process of steel sheet rolling with an example of flows for each of the six categories.

In practice, a unit process can represent a single process, e.g. the rolling of steel, but it can also represent an entire facility that contains many different processes, e.g. a slaughterhouse, if this offers the sufficient level of detail for the inventory modelling. The latter type of unit process may be physically subdivided into two or more new unit processes in a life cycle inventory model, see Sect. 8.5.4. Generally, unit processes do not gain or lose mass over time and the sum of all input flows-should therefore be equal to the sum of all output flows at the level of elements (e.g. copper) and in aggregation.

Output flows belonging to the *product* or *waste to treatment* categories from one unit process can act as input flows belonging to the categories *materials* and *energy* for other unit processes and this is how unit processes are linked in a life cycle inventory model. By comparison, resources and emission flows are not exchanged between unit processes. They are referred to as *elementary flows*, and defined by ILCD (using a slight modification of the ISO definition) as "single substance or energy entering the system being studied that has been drawn from the ecosphere without previous human transformation, or single substance or energy leaving the system being studied that is released into the ecosphere without subsequent human transformation". The ecosphere can be understood as "the environment" and is elaborated below. Note that a single substance should be seen as an ideal and that some elementary flows in existing LCA practice are heterogeneous materials (such

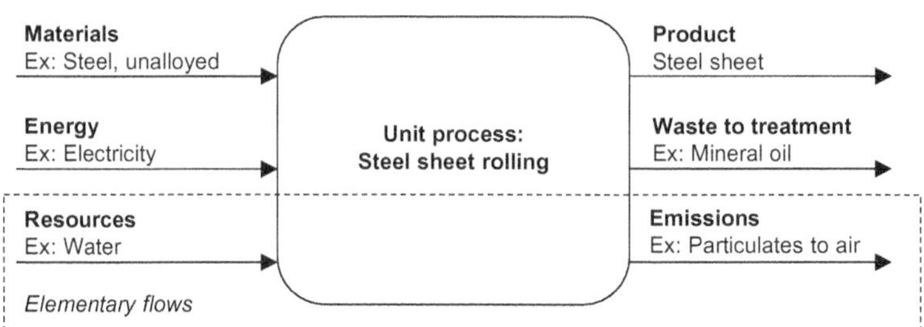

Fig. 8.1 The unit process of steel sheet rolling and examples of flows. The actual unit process contains 86 flows [inspired by: ecoinvent v3 (Weidema et al. 2013)]

as the elementary flow bauxite which contains different minerals, some of which, e.g. $Al(OH)_3$, are sources of aluminium) or cover a group of individual substances (such as the elementary flow VOCs, volatile organic compounds).

What makes resource flows differently from material and energy flows is that they have been "drawn from the ecosphere without previous human transformation". This means that resource flows are not outputs from other unit processes. In the steel sheet example of Fig. 8.1, the resource flow "water" may be sourced directly from a river close to the location of the steel sheet rolling process (i.e. no previous human transformation), whereas unalloyed steel (a material flow) is the product flow of another unit process and acts as a material flow to the steel sheet rolling unit process. Also, in the example of a unit process composed of an entire slaughterhouse, solar influx may be harvested directly in photovoltaic panels on the roof of the slaughterhouse to produce electricity and the solar influx is then a resource flow to the unit process because it has not undergone a previous human transformation. If the slaughterhouse instead was purchasing electricity from the grid, this electricity would be an energy flow to the slaughterhouse unit process because it has undergone previous human transformation, meaning that it is a product flow of another unit process (e.g. a coal-fired power plant). Similarly, what makes emission flows differently from waste flows is that they are "released into the ecosphere without subsequent human transformation". This means that emissions are not inputs to other unit processes. In the steel sheet example shown in Fig. 8.1, particulates (emission flow) are emitted directly into the air, whereas mineral oil will go through treatment, i.e. be a material input for another unit process. Chapter 9 will further explain how these concepts are used to model an LCI.

8.2.2 The Technosphere and the Ecosphere

LCA divides the world into a technosphere and an ecosphere, see Fig. 8.2.

The *technosphere* can be understood as everything that is intentionally "man-made" and also includes processes that are natural in origin, but manipulated by humans, such as photosynthesis when part of an agricultural system. All unit processes of an LCI model belong to the technosphere.

The *ecosphere* is sometimes referred to as "the environment" or "nature" in layman's terms and can be understood as everything which is not intentionally "man-made". In the ecosphere reside those qualities that LCA has been designed to protect, i.e. ecosystems, human health and resource availability. These qualities are called Areas of Protection or damage categories in the field of LCA (see Chap. 10). Changes to the ecosphere can be considered unintentional "man-made" consequences of activities in the technosphere. Note that the ecosphere also undergoes natural changes, for example, via ice age cycles or natural ecological successions, which means that it can be difficult to choose an appropriate natural reference state against which human impacts should be measured, see Chap. 10.

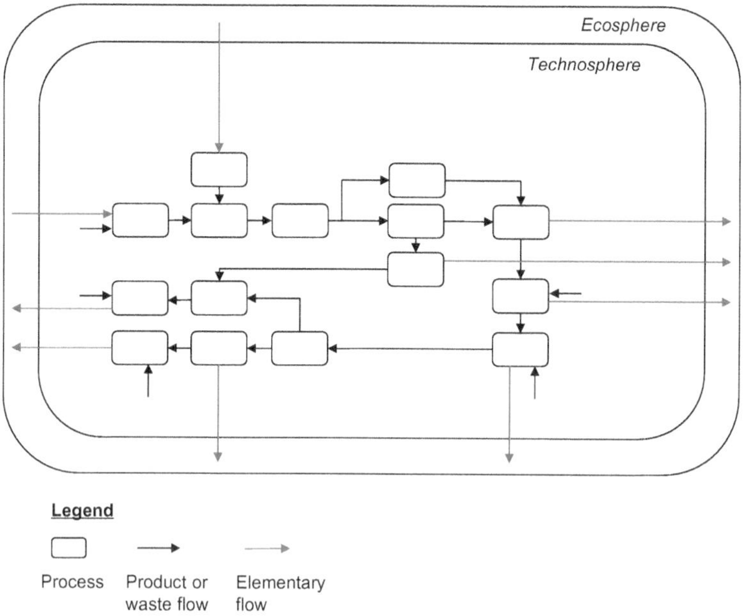

Fig. 8.2 Division between ecosphere and technosphere for a generic product system. Elementary flows are represented by *blue arrows*, while flows within the technosphere are in *black*

Elementary flows are per definition the only flows that go across the boundary between the technosphere and the ecosphere (see Sect. 8.2.1) and it is because of these flows that the Areas of Protections are potentially impacted by the product systems assessed in LCA. Note that there is no clear-cut large-scale spatial separation between the technosphere and the ecosphere. The two spheres are in fact largely intermingled and therefore quite abstract. Surely, natural reserves and undeveloped land largely belong to the ecosphere, but the transportation and tourism infrastructure (roads, trash bins, etc.) going through them belong to the technosphere. In addition, though cities may appear like they belong 100% to the technosphere, the outdoor or indoor air that the population inhales belongs to the ecosphere, because human health can be impacted through air pollution. Note also that the exact location of the boundary between the technosphere and the ecosphere is often debated in the LCA community, for example, with regards to agricultural systems (see Chaps. 29 and 30).

8.2.3 Foreground and Background System

Often hundreds of unit processes are required to deliver the product studied in an LCA. It is useful to distinguish between unit processes belonging to the *foreground*

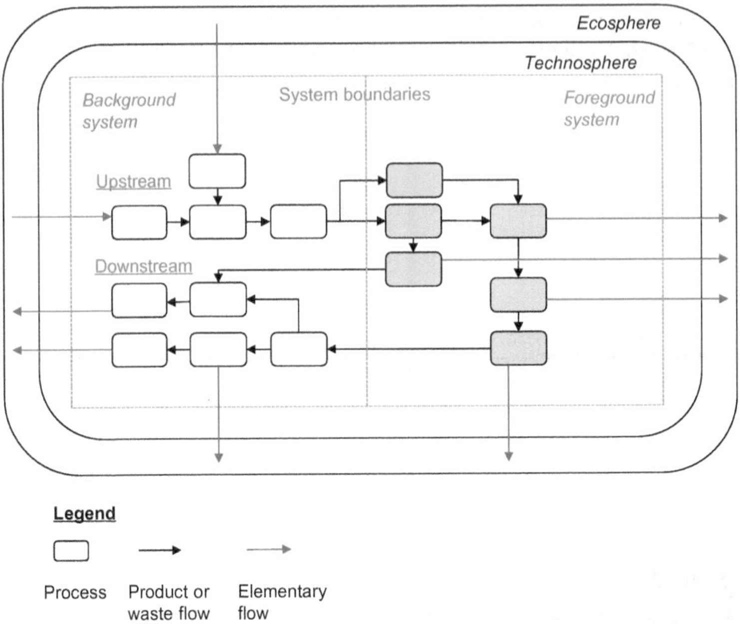

Fig. 8.3 LCI model for the generic product system from Fig. 8.2. The *green box* represents the boundaries of the product system with the division between foreground and background systems indicated. Unit processes with *grey shading* belong to the foreground processes, while unit processes without shading belong to the background system. Part of the background system lies upstream in the value chain and feeds into the foreground system. Another part lies downstream and receives input from the foreground system. *Black arrows* between unit processes indicate material, energy, product or waste flows. *Blue arrows* to and from each unit process represent elementary flows (resources and emissions)

and *background system*. The foreground and background systems are indicated in Fig. 8.3 for a generic product system.

The *foreground system* is commonly defined as comprising those processes of a product system that are specific to it. These processes are in the study of a product typically some of the tier-one suppliers, but may also be suppliers further up the supply chain (e.g. tier-two or tier-three) if these are known by the producer, e.g. through a system of material certification. The foreground system is largely modelled using primary data, i.e. data collected first-hand by the LCA practitioner, e.g. obtained through the commissioner of the study. From a management perspective, processes in the foreground system can often be changed by the decision-maker commissioning a study (e.g. a company), either because they are directly operated by the decision-marker (e.g. at the production site) or because the decision-maker has the power to change or influence the processes, e.g. via purchase decisions or consumer information. In this context, a change can be choosing another supplier (introducing a different unit process in the model) or influencing the way a unit process is operated, thereby changing all or some of its six types of flows qualitatively and quantitatively.

The *background system*, in contrast, is commonly defined as those processes of a system that are not specific to it. Such processes take part in numerous product systems besides the one studied. Examples are society's electricity supply, the production of metallic copper, or the waste management systems. Neither of these is specific to the product under study, but typically purchased in a market without possibility to choose between specified individual suppliers. The background system is typically modelled using LCI databases, which contain average industry data representing the process in specific nations or regions. From a management perspective, processes in the background system can typically not be structurally changed by the decision-maker commissioning a study (e.g. a company), because the decision-maker is only a minor customer and therefore can only exert limited power or because the suppliers are anonymous to the customer like the case of copper which is bought on the global metal market (an exception is Situation B studies where the decision-maker has influencing power on the background system, see Chap. 7). The distinction between foreground and background system is especially useful for planning data collection for the inventory analysis (see Chap. 9) and for making recommendations as part of the interpretation of LCA results (see Chap. 12).

8.2.4 Life Cycle Inventory Model and Results

A life cycle inventory (LCI) model aims to link all unit processes that are required to deliver the product(s) studied in an LCA (glueing together the product system). Figure 8.3 shows an example of an LCI model for a generic product.

An LCI result is an inventory of the aggregated quantities of elementary flows, separated into resources and emissions, from all the unit processes within the system boundary. These elementary flow quantities must be correctly scaled to the assessed product by considering the extent to which the function of each unit process is required to deliver the studied product (see Chap. 9).

8.2.5 Life Cycle Impact Assessment

LCIA is composed of selection of impact categories, classification and characterisation, normalisation and weighting (the latter two are optional steps according to ISO). Chapter 10 details these steps and only their main characteristics and purposes are presented here.

Selection of Impact Categories, Classification and Characterisation

The first step of LCIA involves *selecting the impact categories* that are relevant to consider in the LCA (considering the goal and scope of the study) and *classifying* the elementary flows of the LCI results into these impact categories.

The classification is based on the identification of the environmental issues that each elementary flow can contribute to, such as water depletion, non-renewable resource depletion, climate change or freshwater eutrophication. The purpose of the next step, *characterisation*, is to translate the LCI results (quantities of elementary flows aggregated across all unit processes of an LCI model) into indicator scores for the different impact categories. This essentially reduces a list of hundreds of quantified flows (the LCI results) to a manageable number of indicator scores (typically around 10 or fewer) with a clear environmental meaning, which is practical when comparing the environmental performance of two or more products.

Normalisation
Normalisation is an optional step under ISO 14044:2006 to support the interpretation of the impact profile from the characterisation. Normalisation means that indicator scores for all impact categories are expressed in a common metric, typically the annual contributions to total environmental impacts of an average person. This serves mainly three purposes: (1) for decision-makers to better understand the magnitude of characterised results by relating them to a common familiar and external reference, (2) to check for errors in the assessment resulting in unreasonably low or high normalised results and (3) to pave the road for weighting.

Weighting
Like normalisation, weighting is an optional step under ISO 14044:2006 to support the interpretation of the impact profile. In weighting, the (typically normalised) indicator scores for the different impact categories are made comparable by assigning weights to each impact category that is intended to reflect their relative importance. This relative importance is inherently subjective and can be based on the opinion of experts, policymakers or the general public (or a combination of these). Weighting allows calculating a single indicator score by summing all the weighted impact scores. This is often considered useful by decision-makers wanting to understand which product system performs best "overall" in a comparison.

The detailed choices on impact assessment methods and factors are made in the impact assessment phase of the LCA but it is necessary to select the impact categories in the scoping phase to ensure that the inventory analysis collects data on all elementary flows of potential relevance for the selected impact categories.

8.3 Deliverables

The types of deliverables should directly reflect the intended applications of results, as defined in the goal definition. To be compatible with the ISO 14044 standard an LCA study must include an impact assessment, and most LCA studies have two deliverables, the LCI results and the LCIA results. Some LCA studies (e.g. collection of data for unit process databases) only involve the construction of a life cycle inventory (LCI), in which case the only deliverable is the LCI results. In any case, LCI results should be documented with full transparency (see Sect. 9.7) to ensure reproducibility of the LCA study and potentially allow elements of the underlying LCI model to be used as data sources for other LCA studies, if results are publicly released. LCIA results must be documented by the numerical values of the characterised results for each impact category covered. If normalisation and weighting of characterised results is carried out (see Sect. 8.2.5) the results of these steps must also be documented numerically.

8.4 Object of Assessment

8.4.1 Functions

All LCAs study one or more product systems composed of many unit processes that are active throughout the life cycles of the product system(s). To study these systems the functions they provide must be understood. Indeed, LCA is the environmental assessment of needs fulfilment focusing on functions first and then on the products needed to provide these functions. An LCA study should thus first define the functions from the perspective of the user (later the perspective will change when secondary functions are to be defined, see Sect. 8.5). For example, two different energy technologies may be compared on the basis of the function they provide of enabling the delivery of electricity to households (through a common distribution system). Functions are especially important to understand when comparing two or more product systems because a comparison is only fair and meaningful if the compared systems provide (roughly) the same function(s) to the user. For example, a tablet and a newspaper both provide the function of a news media, but because the tablet provides more functions (access to other websites, word processing and other software) a direct comparison of environmental impacts of a newspaper and a tablet would not be meaningful. An LCA must therefore always be anchored in a precise, quantitative description of the function(s) provided by the analysed product system. In the illustrative case on window frames in Chap. 39, the windows are compared based on their function of allowing daylight into a building.

8.4.2 Functional Unit

To support a fair and relevant quantitative comparison of alternative ways of providing a function, knowledge of the functions provided by the alternative product systems must be used to define a *functional unit*. A functional unit defines the qualitative aspects and quantifies the quantitative aspects of the function, which generally involves answering the questions "what?", "how much?", "for how long/how many times?", "where" and "how well?". For example, a comparison of outdoor paints may be based on the functional unit: "Complete coverage of 1 m^2 primed outdoor wall for 10 years in Germany in a uniform colour at 99.9% opacity". This is not to say that all LCAs on paint should have this functional unit. In other cases, for example, a particular colour or sheen may be considered an important function and should be included in the functional unit. It is important to understand that the functional unit should always include a function and not simply be a physical quantity, such as 1 kg, 1L or 1 MJ. For example, it would be wrong to compare paints on the basis of a functional unit of "1L paint", since an identical quantity of different paints may deliver different functions, e.g. in terms of area of wall that can be covered, or the quality and duration of the coverage. Figure 8.4 illustrates how this functional unit is composed of answers to the five questions presented above.

It is important to define the functional unit right because it significantly influences the way LCA is performed, its results and interpretation, especially in comparative studies (see Sect. 8.9). This is because the functional unit serves as a reference point for deciding which unit processes to include and to what extent they are drawn upon. It is therefore essential to ensure that the functional unit fully captures the relevant functional aspects of the studied systems. In the following paragraphs, we provide some guidance for defining a correct functional unit.

To get started, two concepts from the product development field are generally useful. These are *obligatory* properties and *positioning* properties. The obligatory properties are features that the product must possess for any user to perceive it as a product (e.g. ability to cover and protect the wall against the weather for an outdoor wall paint) and may also include legally required features (e.g. a car must have seat belts). These can usually be expressed in technical terms. The positioning properties, on the other hand, are optional features of a product, which can be used to position it as more attractive to the consumer in the competition with other similar products. Examples include price, colour, comfort, convenience, image, fashion and aesthetic aspects of the product. Positioning properties often vary from consumer to consumer as opposed to obligatory properties. Tables 8.1 and 8.2 show an example of obligatory and positioning properties for an outdoor wall paint and the window frame case study (Chap. 39), respectively.

After having listed the obligatory and positioning properties they need to be transformed into the functional unit, i.e. they should be used to address the

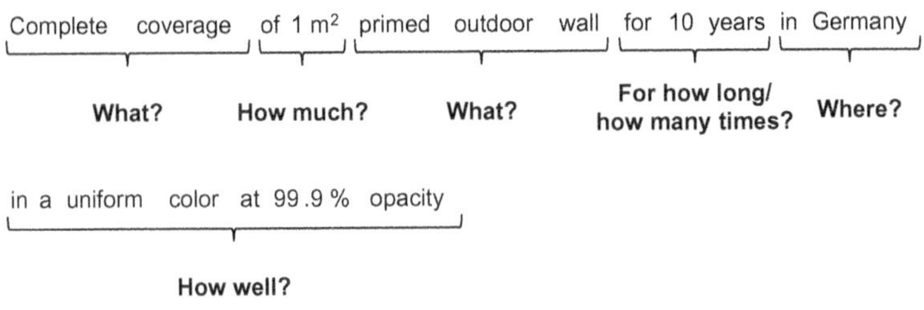

Fig. 8.4 Example of a functional unit composed of five questions

Table 8.1 Derivation of functional unit on the basis of obligatory and positioning properties of an outdoor wall paint

Obligatory properties	Positioning properties
Cover wall with uniform colour Protect wall against rain, sun and microalgae Provide surface that is easy to clean Meet health requirements for application	Drip-free application Many different colour tones to select from Water-based Well covering (needs only one application)
Functional unit Complete coverage of 1 m^2 primed outdoor wall for 10 years in Germany at a uniform colour at 99.9% opacity *Reference flow* 0.67 L of water-based paint A (needs two applications and a re-paint every 2½ years) 0.15 L of water-based paint B (low content of water, only needs one application and lasts 5 years until re-paint is required)	

Table 8.2 Derivation of functional unit on the basis of obligatory and positioning properties of windows

Obligatory properties	Positioning properties
Allow daylight into a building through a physical barrier	Protection from outdoor climate (thermal and noise insulation) Allow ventilation between indoor and outdoor Provide aesthetic functionality to the building Protection against breaking into the building
Functional unit Allow daylight into a building through a physical barrier, equivalent to light being transmitted through an area of 1.23 × 1.48 m^2 with visible light transmittance of at least 0.7, for 20 years *Reference flow* 0.5–0.67 window frames, depending on material 1 window pane Paint for maintaining surface of window pane (dependent on frame material)	

questions "what?", "how much?", "for how long/how many times?", "where" and "how well?", as in the example of Fig. 8.4. When defining the functional unit it is useful to distinguish between its quantitative and qualitative aspects.

The *quantitative* aspects always make up the answers to the "how much?" and "for how long/how many times?" questions and often take part of the answer to the "how well?" question. In the example of an LCA on shopping bags quantitative functional aspects may be the volume ("how much?"), the number of shopping trips that the bag should be used for ("how long/many times?") and strength, i.e. the weight that can be carried ("how well?"). For products that are continually in use (e.g. a fridge or a paint) the "how long/many times?" question should be addressed in the form of the time during which the product is in function (as in the paint example of Fig. 8.4). For products that are not in use all the time (e.g. clothes, mobile phones) the "how long/many times?" question should instead be addressed by specifying the intensity of the use, either as the total duration of use (e.g. 1000 h) or the number of times that the function is provided (e.g. 50 shopping trips for the shopping back example above). In the window frame case study the "how well?" question was partly addressed quantitatively by defining a visible light transmittance (the fraction of light that a window allows into the building) of at least 0.7 in the functional unit. The magnitude of the quantitative aspects in the functional unit can be chosen more or less arbitrarily. However, for the users of an LCA, it often makes the most sense to relate it to the magnitudes of typical use by a person, a family or a community. In the example of Fig. 8.4 it would be less intuitive to relate to a functional unit involving the complete coverage of 1 km^2 primed outdoor wall, while a good magnitude in the functional unit for a study of waste incinerators could be the household waste generated by the municipality in one year.

The *qualitative* aspects cover the way in which the function is provided and are often not easily quantifiable and sometimes not even clear-cut. The "what?" and "where" questions require qualitative answers. In the example of Fig. 8.4 the "what?" question is answered by "complete coverage of primed outdoor wall" and the "where?" question by "Germany". Other qualitative aspects are often used to answer the "how well?" question. These could be legal requirements, e.g. fire safety measures in a car or an office building, or technical standards, e.g. RAL code 3020 for the colour of paint. References to relevant legal requirements and technical standards in the functional unit are helpful, because they ensure comparability through adherence to the standard. To fully address the "how well?" question subjective or ambiguous elements related to user perception (e.g. fashion) are often important to include, to ensure comparability of different products. For example, products may be discarded by users although they still fulfil their technical functions because they are no longer perceived as fashionable. For this reason, it is important to understand which aspects of a studied product's function, including non-technical aspects such as fashion, that are perceived as important by users. LCA practitioners carrying out a study are therefore advised to consult the users of the product or service that is studied to ensure that the definition of the functional unit captures their perception of the product's functionality. Those non-technical aspects that differ between compared products should either be included in the functional unit or considered separately in the interpretation phase of the LCA (see Chap. 12).

The authors of this chapter have over the years encountered many types of mistakes in the definition of functional units. Box 8.1 provides selected examples of such mistakes and explains what is wrong with them and what needs to be considered to prevent making them.

Box 8.1: Common Types of Mistakes when Defining the Functional Unit

1. **Assuming that same physical quantity of product equals the same function**:

   Example: "1 kg of packaging material"

   Explanation: A physical quantity, such as mass, is not a function. The mass required to provide a packaging function often depends on the material.

As an example, glass and PET in beverage packaging have different densities and physical properties, and different masses will therefore be required for providing the same function. To prevent mistakes like this, the functionality of the product should be considered (for example, what is the functionality of packaging?).

2. **Being overly restrictive**:

Example: "Enable watching of television with a 30 W power consumption for 10,000 h"

Explanation: A fixed power consumption is (except in special cases) not relevant to the user of a television and means that only televisions with that exact power consumption can be included in a study. To prevent mistakes like this, it must be ensured that the functional unit only covers what relates to the function of the product (to watch television).

3. **Incorrect use of technical standards or legal requirements**:

Example: "Driving 1000 average person-kilometres in a diesel passenger car that fulfils the Euro 6 standard and therefore emitting less than 0.08 g NO_X per kilometre (Euro 6 standard) during use"

Explanation: Often products can demonstrate compliance with the law or a voluntary standard when completing a test that does not represent the actual conditions of the product's use. A passenger car complying with the Euro 6 standard may emit more NO_X than 0.08 g/km, depending on the driving pattern, climate, etc. A misinterpretation of a technical standard in the functional unit can therefore lead to mistakes in the LCI (in this case, underestimated NO_X emissions). To prevent mistakes like this, the condition of the use must be considered. Generally, a reference to a technical standard in the functional unit does not need to be accompanied by the exact meaning of the technical standard, as this will be dealt with in the LCI modelling step.

It must be stressed that a solid insight in the relevant technological domain is required to define a meaningful functional unit. For example, good knowledge about biofuels, nanomaterials or remediation of contaminated sites is required to define meaningful functional units for these technologies. Chapters 26–36 discuss the application of LCA, including the definition of functional units, for a wide range of technological domains.

8.4.3 Reference Flows

When the functional unit has been defined, the *reference flows* can be determined. A reference flow is the product flow to which all input and output flows for the processes in the product system must be quantitatively related. In other words, the reference flow is the amount of product that is needed to realise the functional unit. For example, as shown in Table 8.1, 0.67 L of paint A is required to realise the functional unit in Fig. 8.4, while the same functional unit is realised with 0.15 L of paint B. The reference flow is typically different qualitatively and quantitatively for different products compared on the basis of a functional unit, due to differences in product properties and characteristics (e.g. viscosity and tear resistance of a paint). The reference flow is the starting point for the ensuing LCI analysis phase of an LCA (see Chap. 9), because it determines all the product flows required throughout the life cycle of the product system studied and their associated elementary flows (resource uses and emissions). It is very important not to confuse a reference flow with a functional unit (see Example 1 in Box 8.1). The former can only be known when the latter is correctly defined. One should, for example, never base an LCA on the comparison of 1 L of two different paints, unless a correctly defined functional unit has shown that the reference flows of the compared paints are quantitatively identical. It is important to understand the use situation in order to correctly define reference flows. For example, to define reference flows in a comparison of a disposable cardboard cup and a ceramic cup, the LCA practitioner must understand the number of times the two cups are used before they are discarded and how the ceramic cup is cleaned (by hand or dishwasher, and the associated consumption of detergent and water and its temperature). Tables 8.1 and 8.2 include functional unit and corresponding reference flows for the example of outdoor wall paint and the window frame case study (Chap. 39), respectively.

8.5 LCI Modelling Framework and Handling of Multifunctional Processes

This part of the scope definition deals with the choice of an appropriate LCI modelling framework and ways to handle multifunctional processes. These choices must be made in accordance with the goal definition, particularly the identified decision context (Situation A, B or C, see Sect. 7.3), and they have a strong influence on the inventory analysis, the LCA results and their interpretation.

8.5.1 Secondary Functions and Multifunctional Processes

To understand why different LCI modelling frameworks exist we first need to consider that a product system often delivers other types of function than the type dealt with in Tables 8.1 and 8.2. The functions of Tables 8.1 and 8.2 all relate to obligatory or positioning properties and are intended functions made available to product users by, e.g. companies selling the products. They are called *primary functions*. In addition to those, *secondary functions* can also emerge in the life cycle of a product system. Secondary functions are unintended functions that usually have low or no relevance to the users of a product, meaning that they are not contributing to the obligatory or positioning properties. Instead, secondary functions are relevant to other systems of the technosphere that the studied product system interacts with. The existence of secondary functions reflects the fact that some processes are *multifunctional*. A process is multifunctional when it provides more than one function, meaning that it either delivers more than one product output and/or provides more than one service. An example of a multifunctional process that delivers more than one product output is animal husbandry where the cow may deliver both milk, meat, hide, bone meal and other products with an economic value. The production of the hide is an example of a secondary function of the husbandry from the perspective of the user of a bottle of milk, since hide is neither an obligatory nor a positioning property of the milk. An example of a multifunctional process that both deliver more than one product output and provide more than one service is waste incineration. It provides the multiple services of getting rid of many different types of wastes (the obligatory property) and can deliver both electricity and heat while doing so. Thus, secondary functions of a product that is disposed of by incineration are the production of heat and electricity. These secondary functions are relevant from the perspective of the energy system that the product system interact with because a change in the volume of discarded products that is incinerated leads to a change in the amount of energy generated from incineration.

Multifunctional processes constitute a methodological challenge in LCA, which is based on the idea of analysing individual product systems based on the primary functions they provide in order to determine the environmental impact from the product. In the real world, there is hardly any product system that exists in isolation. As soon as a by-product arises from a multifunctional process (e.g. animal husbandry), it is economic common sense to try to utilise it, often in a different context from the product system being analysed in the LCA. This means that the process becomes part of another product system as well, and that the environmental impacts from the process can no longer be fully ascribed to the product system studied.

8.5.2 The ISO 14044 Hierarchy to Solving Multifunctionality

In order to solve multifunctionality issues, the ISO 14044 standard presents a hierarchy of solutions. These solutions can both be used to make different product systems functionally comparable and to represent a single product system in a hotspot analysis. The levels of the hierarchy are presented below and the hierarchy is summarised as a decision tree in Fig. 8.5. Chapter 9 shows how to use to ISO hierarchy in practice when constructing an LCI.

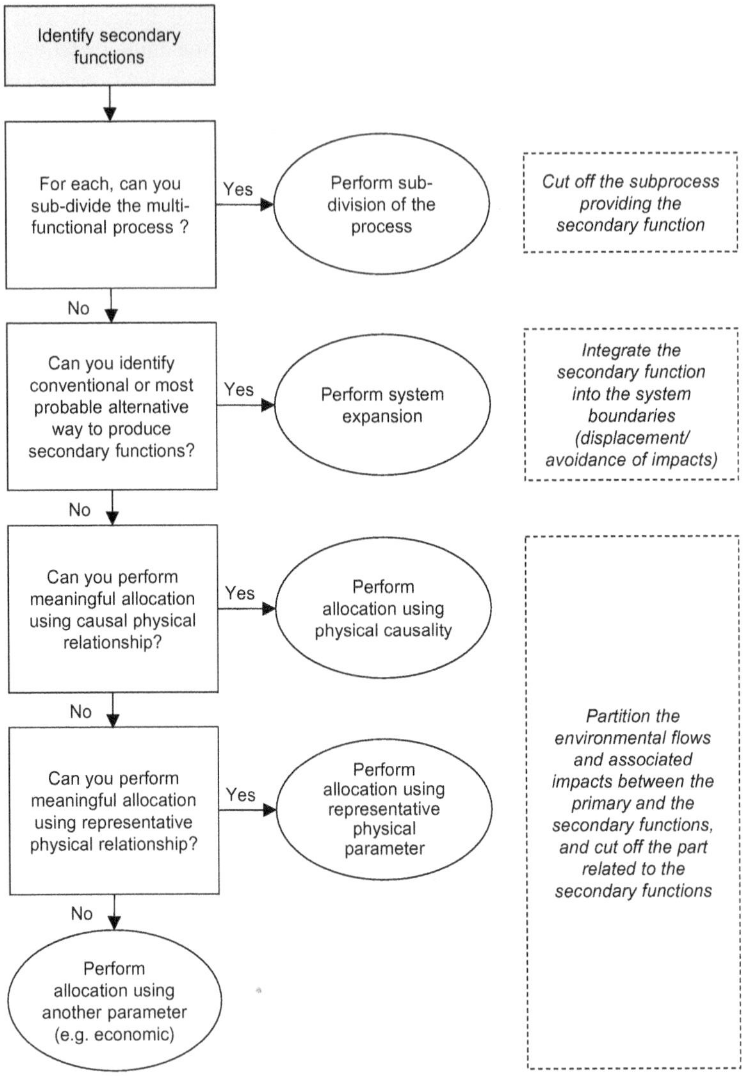

Fig. 8.5 ISO hierarchy for solving multifunctionality presented in a decision tree

Subdivision of Unit Process

First choice is to try to solve this problem through increasing the resolution of the modelling by dividing the multifunctional unit process into minor units to see whether it is possible in this way to separate the production of the product from the production of the co-product, and if so exclude the subprocesses that provide the additional functions from the product system, see Fig. 8.6.

An example of subdivision is when a factory produces two products. Here, the subdivision approach may lead to the realisation that the factory actually contains a

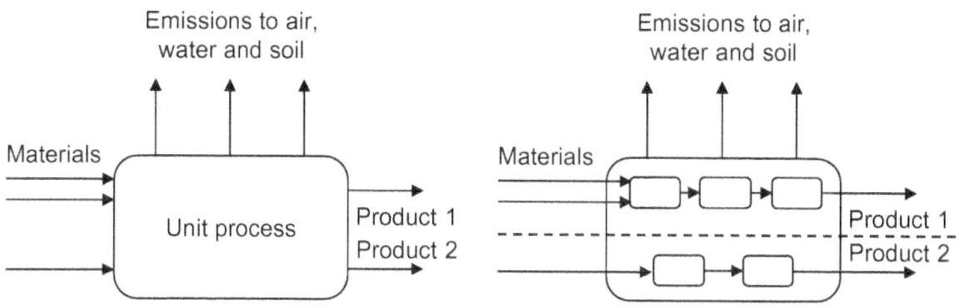

Fig. 8.6 Solving the multifunctionality problem by increasing the modelling resolution and sub-dividing the process into minor units which can unambiguously be assigned to either of the functional outputs

number of processes and that the processes needed for the production of the first product are physically separated from the processes needed for the production of the second product. This approach to solving multifunctionality does not always work. Even if you zoom to the molecular level of a cow, it is not possible to physically separate the metabolic processes in the cow that lead to the production of milk from the ones that lead to the production of meat or hide.

System Expansion

If subdivision fails to solve the multifunctionality problem, the ISO standard recommends trying to solve the problem by system expansion. In a comparison of two processes, this means expanding the second process with the most likely alternative way of providing the secondary function of the first process. In the comparison of power plant 1, which has district heating with co-generated heat as a secondary function, with a power plant 2, which only produces electricity, this means expanding the system of plant 2 with the most likely alternative way or combination of ways of providing district heat in that region (see Fig. 8.7).

Expansion of system 2 with the alternative way to produce the secondary function of system 1 is equivalent to subtracting the alternative way from system 1 (which provides the function). This is also called to credit system 1 with the inputs and outputs which are avoided when its secondary service replaces this alternative production. In the case of district heating being the secondary function, system expansion would thus be the same as crediting the power plant, which produces the district heat, through subtracting the impacts from the most likely alternative way of producing this heat as illustrated in Fig. 8.6.

In Fig. 8.6 equation B follows from equation A by subtraction of the alternative way of district heating from both sides of the equal sign. The approach of system expansion is thus mathematically equivalent to crediting for avoided production. Crediting for avoided production is typically used to account for secondary functions in a hotspot analysis where there is not a comparison of two alternative systems. For example, a product system that includes incineration can be credited for the avoided impacts from the production of heat and electricity by subtracting the avoided elementary flows in the inventory of the process (see Chap. 9 for technical details). In the milk example, system expansion can be performed by

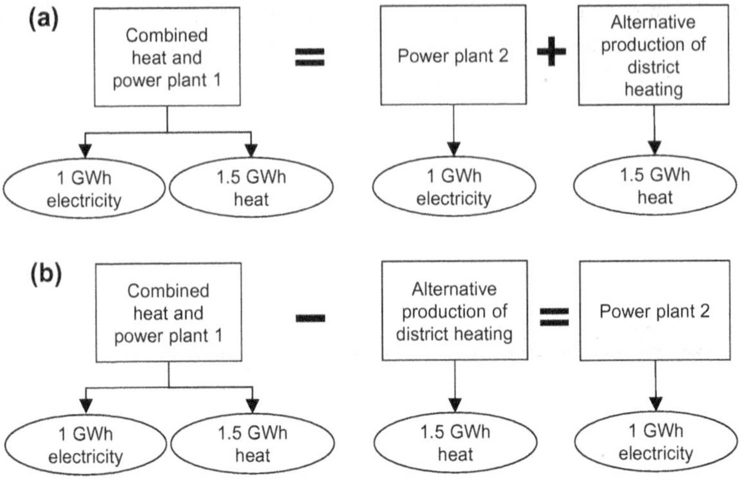

Fig. 8.7 Equivalent modelling approach when dealing with multifunctionality. **a** System expansion: to ensure equal functionality system 2 is expanded to include the secondary function of system 1. **b** Crediting: system 1 is credited for the production of the secondary function, in order to have equal functionality of system 2

crediting the milk for the avoided impacts from alternative production of beef and other co-products. This alternative production might be the raising of cattle in a pure beef production system (which includes hides and other low-value co-products). Note that quality differences between dairy cow meat and cattle meat means that they may not be functionally equivalent. This may require the application of a value correction factor to the crediting.

An important task in system expansion is to identify the process (or combination of processes) which is superseded by the co-product. This relates to the decision context (Situation A, B or C1/C2) identified in the goal definition (Sect. 7.4) and will be dealt with in Sect. 8.5.3.

Allocation

Sometimes it is not feasible to obtain complete functional equivalency between the compared systems or to isolate the primary function of a process from the secondary functions through system expansion. This may be the case when there is no alternative way to produce the secondary functions. A classic example of such a multi-output process is a petrochemical refinery with a variety of different organic substances as output without any mainstream alternative routes of production for these. It may also be the case when the most likely alternative route also has secondary functions, creating the need for further system expansion introducing alternative routes for the new level of secondary functions, which again may have secondary functions, creating the need for further system expansion and so on. In the milk example, the alternative production of meat from raising of cattle for example leads to the co-production of horn (for example used in jewellery production), which cannot be produced in isolation and for which there may not exist a functionally equivalent material.

When system expansion is not feasible, or when it is in conflict with the goal definition (for Situation C2, see below), the ISO 14044 standard recommends dividing the inputs and outputs of the multifunctional process or system between the different products or functions. This is called allocation.

If possible, the allocation should be performed in accordance with the underlying causal physical relationships between the different products or functions, reflecting the way in which the input and output quantities are affected by changes in the quantities of products or functions delivered by the process or system. For example, in the hypothetical example of a waste incineration plant that incinerates two waste inputs, batteries and plastic, emissions of the toxic metal cadmium from the process will originate entirely from the batteries, given that the plastic stream contains no cadmium and that cadmium cannot be formed in the waste incineration process. This conclusion on the origin of cadmium, based on deductive reasoning, could also have been reached empirically by measuring changes in cadmium emissions in response to changes in waste inputs (e.g. a doubling of cadmium emissions would be expected from a doubling of battery inputs). A causal physical relationship can thus be established and cadmium emissions can be allocated 100% to the batteries. In the case of the milk example, the International Dairy Association recommends that physical allocation be based on the different physiological feed requirements for an animal to produce milk and meat (IDF 2010). In the absence of a causal physical relationship between the products, the ISO standard recommends performing the allocation according to representative parameters. This is possible when co-products provide identical or similar functions. In the case of a waste incineration plant that delivers both heat and electricity as output, the exergy content of the two flows may, depending on the study context, be used as a representative physical parameter or allocation key, because it reflects the potential of each energyform to perform mechanical work. Here, it is important that the representative physical parameter actually represents a common function of the co-products. In the example of an agricultural process that produces both wheat and straw, the energy content of the two flows can only be used as a representative parameter if they are both intended as animal fodder (a common function). If instead, the wheat is intended as food for humans this choice of representative parameter would be wrong (food for humans deliver many more functions than energy, e.g. vitamins and taste).

When no common representative physical parameter can be identified for the different outputs, another relationship must be found between them. As an example, the ISO standard mentions an economic relationship, and indeed, this is a frequently applied allocation parameter. In economic allocation the inputs and outputs of the process or system are divided between its products according to their respective economic values, e.g. determined as their long-term average market prices, or some shadow price in cases where there is no market, e.g. for intermediary products. A justification for the use of economic allocation is that products are produced due to an incentive of financial income, and that a co-product with a market value close to 0 should be allocated a correspondingly low share of the non-product flows of a process, compared to a primary product with a high market value. In the extreme situation where the value of the co-product is zero, its allocated share of the inputs and outputs also becomes zero in accordance with the fact that a zero-value output is not a co-product but waste and should be modelled as such.

8.5.3 LCI Modelling Framework: Attributional and Consequential LCA

Traditionally, there have been two main LCI modelling frameworks: *attributional* and *consequential* modelling. In the ILCD guidelines, these were adapted to match the four decision context situations (i.e. A, B, C1 and C2). Understanding the difference between attributional and consequential modelling and when to use what has been one of the most difficult aspects of LCA, and there is still no consensus on this issue within the LCA community. In addition, some aspects of the terminology defined in the ILCD guidelines, in particular with regard to the definition and settings of attributional modelling, are inconsistent with the traditional views within the LCA community, thus adding more confusion to the matter (Ekvall et al. 2016). Below we first offer an explanation of the two modelling frameworks, including their handling of multifunctional processes and the use of average or so-called marginal LCI data (to be explained below). Where relevant we specify discrepancies between the ILCD guidelines and the traditional views. Table 8.3 summarises the explanation and discrepancies. We then provide guidance in compliance with the ILCD guidelines for selecting the LCI modelling framework with consideration to the goal definition.

Attributional LCI modelling was initially the common practice when LCA development caught pace in the early-mid nineties. The overall aim of attributional modelling is to represent a product system in isolation from the rest of the technosphere or economy. The question addressed by attributional LCA can be said to

Table 8.3 The meaning of the attributional and consequential modelling frameworks and their handling of multifunctionality

LCI modelling framework	Question to be answered	Handling of multifunctional processes when subdivision is not possible		Modelling of background system
		Before ILCD	ILCD	
Attributional	What environmental impact can be attributed to product X?	Allocation	System expansion or allocation	Average processes
Consequential	What are the environmental consequences of consuming X?	System expansion	System expansion	Marginal processes

be "what environmental impact can be *attributed* to product X?" or "what environmental impact is product X *responsible* for?" As hinted by these questions, there is an element of subjectivity involved in attributing impacts to a product system or deciding the impact responsibility of a product system. This subjectivity arises in the act of artificially separating the studied product system from the rest of the economy. This separation is artificial because many, if not most, product systems interact with other products systems through multifunctional processes, meaning that they, as explained in the previous section, cannot be described as physical entities in isolation. For example, from a strict physical perspective, the product system of a bottle of milk cannot be described in isolation and the assignment of processes that the product system is seen as "responsible for" therefore involves choices. Before the ILCD guidelines came into place attributional modelling was generally associated with allocation as the approach to solving the issue of multifunctional processes, provided that subdivision (the preferred solution of the ISO hierarchy) was not possible. By contrast, ILCD in some cases recommends solving multifunctionality by system expansion within an attributional modelling framework (see below).

Besides the issue of multifunctionality, attributional LCA is also associated with the use of average processes in the background system, which reflects the modelling of an average supply chain. In practice, this means that a market mix is used. This could be for the global aluminium market or the electricitymarket of a nation. The former is composed of a range of bauxite mines with different ore grades and processing facilities that employ different production technologies, while the latter is composed of different energy conversion technologies, such as the combustion of coal, natural gas, oil and biomass, the harvesting of wind and solar power and the use of nuclear power. As an example, Fig. 8.8 shows the Danish electricity consumption mix in 2014.

Consequential LCI modelling was developed around the year 2000 to eliminate the weakness inherent in the attributional LCA modelling framework due to the attempt to artificially separate a product from the rest of the economy. Its overall aim is to describe the changes to the economy caused by the introduction of the studied product system, i.e. the product system's consequence. Consequential LCI modelling thus aims to answer the question "*What are the environmental consequences of consuming X?*" For example, a consequential LCA of a bottle of milk would attempt to model how the market responds to the change in demand for milk represented by the functional unit of the study (e.g. involving a milk volume of 1 L or a specified nutritional value). This is a very different approach than attributional modelling because the change in the economy can look very different than the representation of the isolated bottle of milk system. For example, the increased demand for milk may lead to an increase in the capacity for milk production (i.e. the numbers of cows giving milk), which in turn may lead to a reduction in the production of some meat (e.g. beef from raising cattle) due to the increasing supply of meat from dairy cows. This corresponds to handling the multifunctional process of milk production by system expansion. A consequence of increased consumption of milk may therefore be a reduction in environmental impacts from the avoided

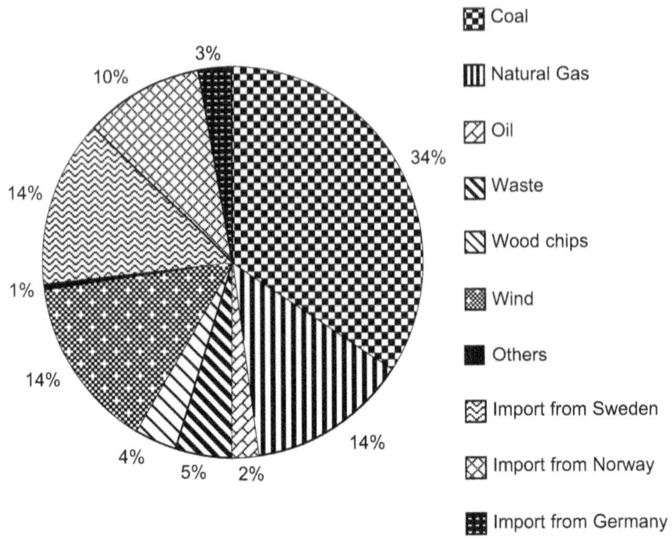

Fig. 8.8 Danish electricity consumption mix in 2014 (low voltage, e.g. for domestic consumption). Imports from neighbouring countries can be further broken down into energy sources (Treyer and Bauer 2013)

production of beef from cattle, which is somewhat counterintuitive. The market may also be influenced by an increased demand for a product in other cases than multifunctionality. For example, if an additional kg is demanded of a fish species that is already fished at its maximum level permitted by regulation (a production constraint) a consequence may be an increase in the production of another protein source that is not constrained, such as chicken, and the environmental impacts following this increase. The examples show that consequential modelling to a large extent relies on a good understanding of and ability to model the dynamics of the economic system, which requires a markedly different way of thinking than the engineering perspective on product supply chains that historically has been in the core of LCA (see Chap. 3).

Contrary to attributional LCA, consequential LCA is not associated with the use of average processes for modelling the background system, but instead with the use of marginal processes. These are the processes that are employed or taken out of use as a response to an increase or decrease in the demand for a product, respectively. In the example of the Danish electricity system, the short-term marginal process will never be solar or wind, because solar irradiance or the wind are natural processes that cannot be "turned up or down" in response to a short-term change in electricity demand. Instead, the short-term marginal process in this example is a combustion process because it is possible to quickly adjust the rate at which something (e.g. coal or natural gas) is combusted in response to a change in electricity demand. The short-term marginal is often the combustion of natural gas, because this is a more expensive way of generating electricity than coal and thus sensitive to changes in electricity prices caused by changes in electricity demand (often, natural gas is only used during peak demand when a relatively high electricity price makes this technology economically viable). However, the relevant marginal processes to include in an LCI model are not always the ones that are affected as an immediate consequence of a decision, i.e. short-term marginal processes. Long-term marginal processes may be more relevant if a decision leads to large changes in supply or demand. Long-term marginal processes represent changes in the installed production capacity in response to the projected development of electricity demand. Often it is difficult to identify a single long-term marginal process, which is why a mix of potential long-term marginal processes is often used. Figure 8.9 shows such a mix for the long-term marginal electricity technology in the Danish market. See Chap. 9 on the identification of short- and long-term marginal processes.

It can be seen that fewer electricity production processes are part of the mix in Fig. 8.9 for consequential modelling than the mix in Fig. 8.8 for attributional modelling. For example, waste as an electricity source is not part of the consequential mix and this is because the long-term planning of waste incineration is thought to consider projections in future waste volumes (the primary function of waste incineration is to "get rid of" solid waste) rather than projections in future electricity demand. On the other hand, the construction of new wind turbines and coal-fired power plants (and to a very small extent, hydropower plants and rooftop photovoltaic panels) are thought to consider projections in future Danish electricity demand. When to consider short- versus long-term marginal processes in consequential LCA and how to identify these are still being debated in the LCA community.

Note that while the background system is modelled differently in attributional and consequential LCA, the foreground system is overall modelled in the same way, the only exception being the handling of multifunctional processes.

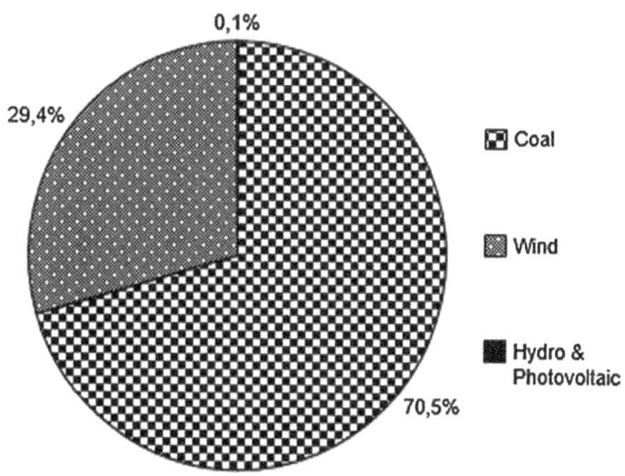

Fig. 8.9 Danish market mix of long-term marginal electricity processes (low voltage, e.g. for domestic consumption) (Treyer and Bauer 2013)

8.5.4 Recommended Modelling Choices for the Identified Decision Context

ILCD provides recommendations for model choices for each of decision contexts (Situation A, B, C1 and C2) identified as part of the goal definition (see Chap. 7). These recommendations are the outcome of a comprehensive consultation process within the LCA community. Since different actors with different views have had a saying in this consensus process leading up to the ILCD recommendations, they are somewhat internally inconsistent, as pointed out by, for example, Ekvall et al. (2016). Below we present the recommendations for each decision context and make notes about the parts that are disputed. Table 8.4 summarises the recommendations.

Situation A

Situation A concerns micro-level decision support (see Chap. 7) and the consequence of a decision (e.g. the introduction of a new product on the market) is therefore of interest. Ideally, the marginal process should therefore be identified and used for all background processes (such as electricity supply) and cases of multifunctionality (e.g. of an incineration process) should be handled by system expansion with marginal processes, provided that subdivision is not possible (see Sect. 8.5.2). This ideal for Situation A is logically consistent with a consequential modelling framework. Yet, ILCD recommends using an average market consumption mix for background processes and in cases of system expansion in the background system. ILCD terms this attributional modelling, although system expansion was previously associated with consequential modelling, as mentioned above. The main reason for diverging from the ideal is that for the small changes studied under Situation A it can be very difficult to identify marginal processes, i.e. to understand the long- and short-term consequences on the market of introducing a small change in its composition of product systems. The actual market behaviour in response to small changes may also not be well-represented by simple mathematical

Table 8.4 Summary of ILCD recommendations on LCI modelling choices

Decision context	LCI modelling framework (ILCD terminology)	Handling of multifunctional processes when subdivision is not possible	Modelling of background system
Situation A	Attributional	System expansion	Average processes
Situation B	Mix of attributional and consequential	System expansion	Mix of long-term marginal processes for processes structurally changed. Average processes in all other cases
Situation C1	Attributional	System expansion	Average processes
Situation C2	Attributional	Allocation	Average processes

equations, which makes it difficult to model what will actually happen, short-term and long-term, when, for example, a light-bulb is turned on, compared to a situation where it is not turned on. There is therefore a risk of using wrong marginal processes and this is problematic because LCA results are often quite sensitive to the choices of marginal process (e.g. natural gas vs. wind for electricity supply). These considerations have lead ILCD to pragmatically recommend using average processes in the background system. It must be mentioned that some LCA experts prefer to pursue the ideal by using marginal processes in Situation A studies, which conflicts with the presented ILCD recommendations.

Situation B
Situation B concerns meso/macro-level decision support (see Chap. 7). ILCD recommends the same modelling choices as for Situation A, with the exception that background processes in the studied product system that have been identified as being affected by structural changes as consequence of the analysed decision are recommended to be modelled as mix of the long-term marginal processes. The logic behind this exception is that marginal processes for suppliers that experience structural changes are easier to identify than marginal processes for suppliers that just experience changes in terms of the volume of products they deliver. The reason for the focus on the long-term marginal is that the consequences studied under Situation B are generally long term. Still, identifying the correct long-term marginal processes in Situation B can be challenging and this is why it is pragmatically recommended to use a mix of possible long-term marginal processes, rather than actual long-term marginal processes, such as the mix for electricity shown in Fig. 8.9. Chapter 9 addresses the calculation of such a mix. In light of the uncertainty involved, we advise to model the LCI using a range of different mixes to analyse how sensitive results are to the estimated mix (see Chap. 12). As for Situation A studies, some LCA experts prefer to pursue the ideal of using a fully consequential approach by only using marginal processes (either single process or a mix) in Situation B studies, which conflicts with the presented ILCD recommendations.

Situation C
Situation C relates to accounting, meaning that studies are not to be used to directly support decisions and are of purely descriptive nature, often describing what has already happened. Situation C1 considers interactions with other systems and ILCD recommends handling this interaction via system expansion (for solving multifunctional processes where subdivision is not possible) and use of average processes in the background system. This means that the recommendations of ILCD in practice are similar for Situation A and C1, even though the modelling ideals of Situation A and C1 are different. Situation C2 disregards interactions with other systems and ILCD therefore recommends that allocation be systematically used to solve multifunctional processes, provided that subdivision is not possible. Note that this conflicts with the ISO hierarchy, according to which system expansion should be performed when possible instead of allocation.

8.6 System Boundaries and Completeness Requirements

System boundaries demarcate the boundaries between the studied product system and (1) the surrounding economy (technosphere) and (2) the environment (ecosphere). "Completeness requirements" is a related concept that can be used to determine what processes should be included within the system boundaries to reach the degree of completeness in the product system modelling that is needed to be in agreement with the goal of a study (see details below). The setting of the system boundaries can have a large influence on LCA results because they determine the unit processes from which environmental impacts should be quantified. At this point in the scope definition, the system boundaries should be represented in a diagram that provides an overview of which parts of the studied product system(s) that are included and which are excluded. An appropriate level of detail in this diagram is the life cycle stages (such as production, manufacturing, transportation, retail, use and disposal) or the main processing steps. It is often useful to start with the process or life cycle stage that delivers the reference flow and then expand upstream and downstream. See Fig. 8.10 for an example diagram for the study of a steel sheet used to prevent accidents during roadworks. Note that the diagram does not need to contain individual unit processes, as this full level of detail will only be achieved in the actual construction of the inventory model (Chap. 9).

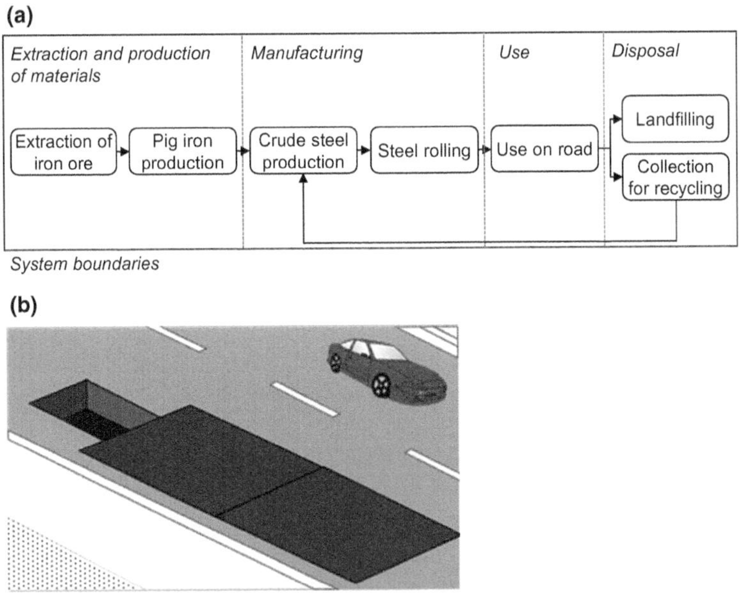

Fig. 8.10 **a** Example of system boundaries diagram for the life cycle of a steel sheet used to prevent accidents during roadworks. Only the main process steps in the life cycle are shown. **b** Illustration of steel sheets in use

8.6.1 Ideal System Boundaries

Ideally, within the system boundaries should be all the unit processes required to deliver the reference flow(s) defined by the functional unit. In cases where multi-functionality is handled by system expansion, this also includes processes from other systems that interact with the studied system. System boundaries should ideally be set so that all flows crossing them are elementary flows (resources and emissions). In other words, no material, energy, product or waste to treatment flows should cross the system boundaries. Ideal system boundaries thereby contain all the unit processes used to deliver the reference flow(s) by (1) generating energy and products (materials for other unit processes) from extracted resources and (2) treating waste flows to the point where the only outputs are emissions. Figure 8.11 illustrates an ideal system boundary for a simple hypothetical product system containing just fifteen unit processes. In this case, the inventory model is fully complete, because all unit processes needed to deliver the reference flows are inside the system boundaries.

Outside the system boundaries lies the rest of the *technosphere* (not shown in Fig. 8.9), i.e. the total body of other product systems in the global economy, and the *ecosphere*, i.e. which is affected by resource uses and emissions from the technosphere.

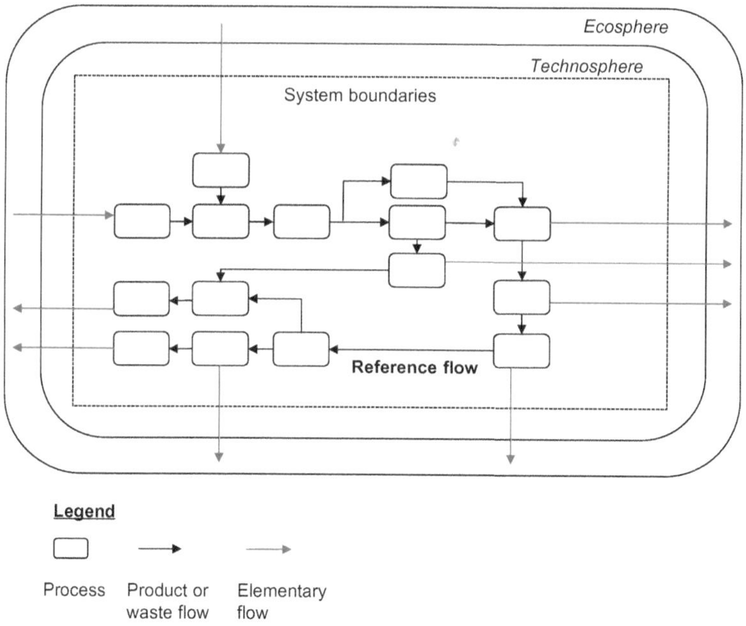

Fig. 8.11 Setting of system boundaries for a simple hypothetical product system. The boundary contains all the unit processes required to deliver the reference flow (*bold*), and the only flows crossing the system boundaries are elementary flows (*blue*). Note that the rest of the technosphere is not shown

8.6.2 Reasons to Divert from Ideal System Boundaries

There are three reasons to divert from working with ideal system boundaries:

First, if a study does not take a full life cycle perspective the rule of only allowing elementary flows to cross the system boundary does not apply. A study taking a full life cycle perspective aims to cover all the processes that are needed to deliver the function(s) of interest upstream (extraction and production of raw materials and manufacturing) and downstream (disposal) to the use stage. By contrast, a so-called "cradle-to-gate" study is an example of a study not taking a full life cycle perspective because the system boundary ends at the gate of the factory where the studied product is produced. In this case, the product flow thus crosses the system boundary, as shown in Fig. 8.12 (based on the simple hypothetical product system shown in Fig. 8.11). The goal definition's intended applications of results decides whether a full life cycle perspective should be taken (see Chap. 7). This decision is usually also reflected by the functional unit (see Sect. 8.4.2).

Second, in comparative studies it is justified to exclude identical processes if they deliver identical quantities of services (energy, materials or treatment of waste) in the systems studied. For example, in the illustrative case study on window frames (Chap. 39) comparing four windows, the processes involved in cleaning the

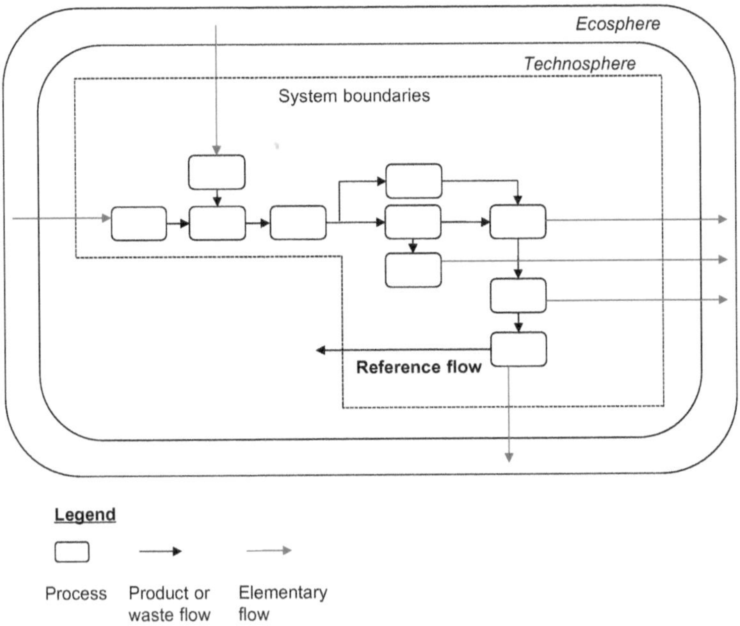

Fig. 8.12 Setting of system boundaries for a simple hypothetical product system in a cradle-to-gate assessment. In this case, the reference flow (*bold*) is crossing the system boundaries, to the rest of the technosphere; in addition to the elementary flows (*blue*) entering or leaving the ecosphere

windows throughout their use stages were excluded from the system of each window because users are expected to clean the windows, that all have the same surface area, in the same way, using the same amount of water and detergent. While this kind of exclusion is allowed for comparisons between systems, it prevents a proper hotspot analysis because it is unknown how much the omitted processes contribute to overall environmental impacts.

Third, constructing an LCI model with ideal boundaries is practically impossible. This is because the number of unit processes actually required to deliver a reference flow is often, even for simple products, enormous: Typically, unit processes require around 5–10 material or energy inputs that each needs to be produced by a unit process that in itself requires around 5–10 material or energy inputs, etc. Furthermore, many product systems include examples of infinite loops where one process A requires input from another process B to deliver an output that is needed by process B to produce the input to process A. Every step back in a value chain represents a step back in time and ideal system boundaries would therefore need to encompass a large part of industrial history, which is not practically possible to model. Yet, amongst the enormous number of unit processes that should ideally be included in the system boundaries, only a minority actually have a quantitatively relevant contribution to the environmental impacts of the studied product system. For example, the ballpoint pens used by employees at a coal-fired power plant obviously have an insignificant contribution to the environmental impacts of a unit of power generation.

Therefore, all LCA studies in practice cut-off some unit processes that are actually needed (although to a very limited extent) to deliver the reference flow. This presents a dilemma of the system scoping. You should include within your system boundaries the processes that matter, i.e. contribute significantly to the overall impacts from the product system, but how can you determine whether a process matters before you know what the total impacts are and can relate the impacts from the process to this number? The solution to this dilemma lies in the iterative approach to LCA that was introduced in Sect. 6.3 and presented for inventory modelling in Sect. 9.3. Figure 8.13 shows examples of excluded product flows.

8.6.3 Completeness Requirements: Quantitative or Qualitative?

Completeness requirements are understood quantitatively by the ILCD guideline as the share (%) of a product's actual environmental impact that a study aims to capture. From this understanding, completeness requirements would, for example, be lower for a study that intends to provide an initial screening of hot spots for a company to familiarise itself with the concept of life cycle thinking (e.g. 70%), than for a study that intends to provide an environmental product declaration (EPD) for

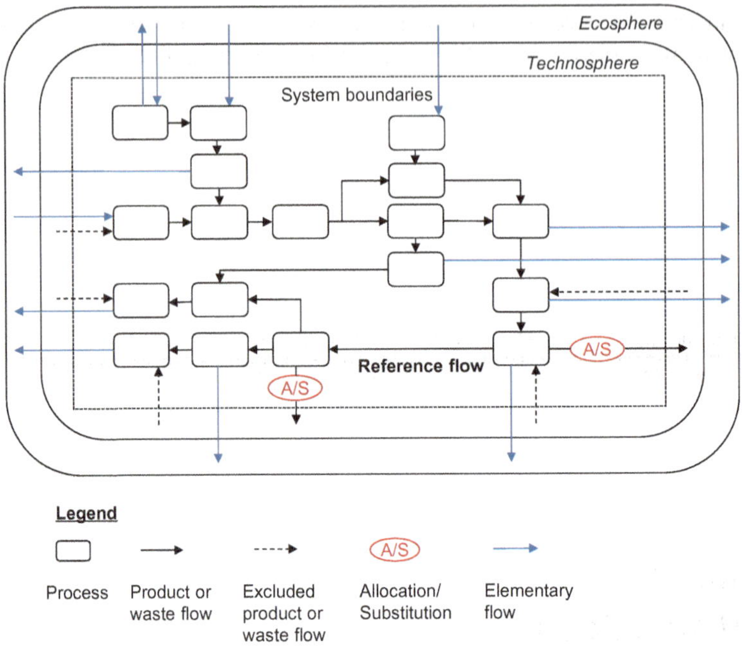

Fig. 8.13 Setting of system boundaries for a realistic product system. In this case, some processes are not included within the system boundaries (cut-off), as illustrated by the excluded product and waste flows. The exact system boundaries depend on whether allocation or substitution is performed in the handling of multifunctional processes

consumers to consider environmental aspects of their purchasing (e.g. 90%). In an LCA guiding the choice between two product designs, the completeness requirement depends on the difference in impact between the product designs. If there is a large (expected) difference, the requirement to completeness would be lower than if the product designs have very similar impacts. In practice, it is often difficult to derive a quantitative completeness requirement from the, generally qualitative, goal definition. In addition, a quantitative completeness requirement is often not helpful for deciding whether a specific process should be included in the system or can be cut-off. To know whether a process can be cut-off one must know how much that process contributes to the total LCIA results for the product system. In other words, one must include the process to figure out if it can be excluded. To circumvent this paradox, some LCA practitioners take a more practical approach by deriving a mass-based cut-off criterion,such as 0.1% from quantitative completeness requirements. This would mean that processes delivering flows with a mass of less than 0.1% of the reference flow can be cut-off. We do not recommend following this approach blindly, because flows that are quantitatively small may still lead to large impacts and therefore have to be included in the modelling. For example, a low mass share of gold in a laptop can account for relatively large impacts due to mining activities, and a small quantity of radioactive waste, e.g. from hospital equipment, can require extensive waste treatment, and therefore be associated with environmental impacts that should not be neglected.

Due to the limitations of working with quantitative completeness requirements, we here advocate a qualitative approach. This means specifying the parts of a life cycle that must be included in the system boundaries and arguing why cutting off other parts is acceptable. For example, an LCA practitioner may know from similar LCA studies or previous experience that transportation between the use stage and the waste management stage, or business trips of the employees of a tier-one supplier are negligible. For new LCA practitioners, it can be difficult to create reasonable completeness requirements and it is therefore always important to explicitly report and justify them. Applying the iterative approach, the omission of any processes should be justified in a sensitivity analysis after the inventory analysis and impact assessment. If the sensitivity analysis indicates that the process may be important with the chosen completeness requirements, it should be included (and perhaps refined) in the next iteration. We stress that an LCA practitioner should not blindly apply "default" qualitative completeness requirements, such as disregarding the production and maintenance of infrastructure, to any study, but always base the requirements on a case-specific assessment. This is to avoid cutting off parts of a life cycle that are important in the specific study, although they may typically not be important.

As with most items of the scope definition, completeness requirements are meant to guide the initial LCI analysis, but during this analysis unforeseen limitations may mean that the requirements are in practice not possible to follow. The LCA practitioner can either handle this situation by modifying the completeness requirements in a new iteration of the scope definition or by explicitly documenting in the LCI analysis the parts of the LCI model that do not fulfil the completeness requirements.

8.7 Representativeness of LCI Data

It is the aim of LCA to reflect physical reality. This means that the model should represent what actually happens or has happened to the extent possible, and the unit processes applied to model the product system must be representative of the processes which are actually used in the analysed product system (in case of attributional LCA) or affected due to the introduction of the assessed product on the market (in the case of consequential LCA).

Typically, parts of a foreground system will be based on data (elementary flows, etc.) collected first-hand by the LCA practitioner, e.g. from the company commissioning the study. This primary data is, provided that it contains no errors, per definition representative of the specific process occurring at the time that the data was collected. Other parts of the foreground system and the entire background system, on the other hand, are constructed from other than first-hand data sources and when doing so it is important to consider how representative the chosen or constructed unit processes are of the actual unit processes that they are models for. Representativeness of LCI data can be understood in three interrelated dimensions: geographical, time-related and technological. Based on the goal definition and knowledge about the studied product system, the scope definition must provide guidance and requirements for the inventory analysis with respect to representativeness of LCI data, as explained below for each dimension of representativeness. Besides serving as a guide for carrying out the inventory analysis, the representativeness of data should also be used in the interpretation of the results to reflect upon the extent to which the product system model corresponds to reality (Chap. 12).

8.7.1 Geographical Representativeness

The geographical representativeness reflects how well the inventory data represents the actual processes regarding location-specific parameters. Geographical representativeness is important to consider because two processes delivering the same product output, but taking place in two different locations (e.g. nations), can be quite different in terms of the other flows (elementary flows, energy flows, material flows and waste to treatment). Differences between unit processes can be caused by geographical differences, such as local climate and proximity to natural resources, and regulatory differences, such as energy taxes and emission thresholds. In addition, when a mix of processes (market mix for attributional LCA and mix of marginal processes for consequential LCA) is used to model the background system or perform system expansion, the location of the mix used in the model versus the actual location of the mix must be considered. For example, the electricity mixes of Denmark (mainly coal and wind power) and Sweden (mainly nuclear and hydropower) vary quite a lot; despite the close proximity of the two countries, see Fig. 8.14.

This can in part be explained by geographical differences (Sweden has mountains and therefore a potential for generating hydropower—Denmark is flat) and in part from social and political differences (Sweden has nuclear power plants—Denmark does not, largely due to public resistance).

Due to the importance of geographical representativeness the LCA practitioner must in the scope definition define the geographical scope of the processes, or combinations of processes, taking place in the product system. The starting point should be the foreground system, where the locations of processes are typically known with high certainty. The LCA practitioner can then proceed to defining the geographical scope of upstream and downstream processes that typically are more uncertain the more "process steps" from the key processes they are in the model. The appropriate resolution of the geographical scopes (e.g. city, region, nation or continent) depends on factors such as the spatial coverage of regulation (typically following national borders), geographical variations (e.g. weather, climate) and the spatial extent of markets (some markets are very local, while others are global).

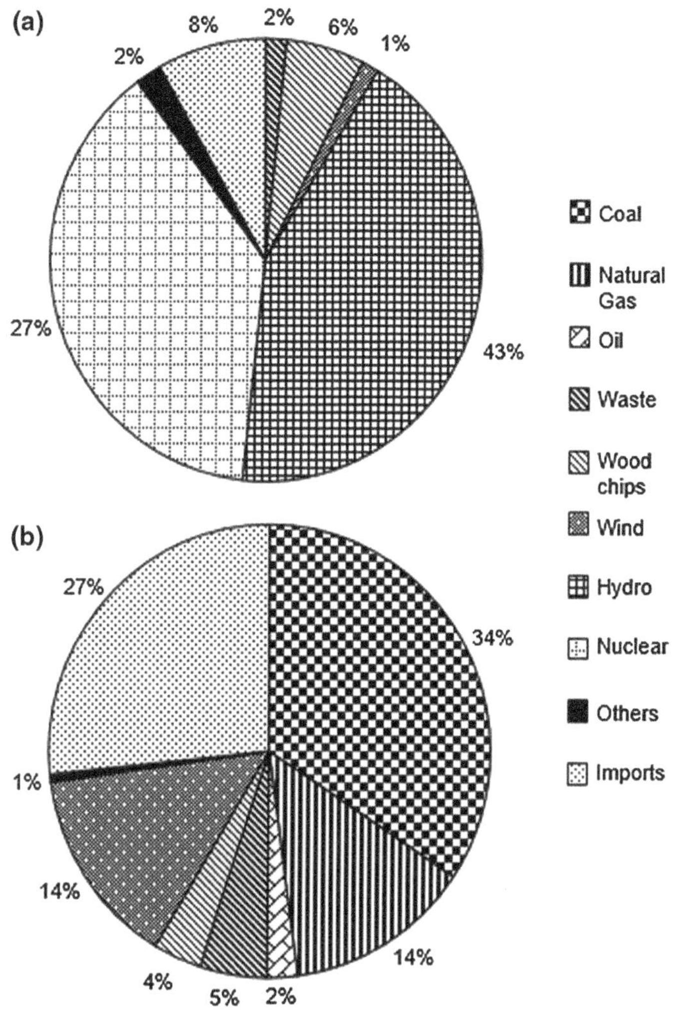

Fig. 8.14 Swedish (**a**) and Danish (**b**) electricity consumption mix in 2014 (low voltage, e.g. for domestic consumption). Imports from neighbouring countries can be further broken down into energy sources. 'Others' is an aggregation of all energy sources contributing less than 1% to the mix (Treyer and Bauer 2013)

Table 8.5 shows the geographical scope for life cycle stages and processes in the illustrative case of window frames (Chap. 39).

During an inventory analysis, it is common that some unit processes cannot be obtained for the described location, such as a specific country. In these cases, the LCA practitioner must choose the most representative unit process to approximate the actual unit process based on his or her knowledge of geographical variations in central factors such as climate, regulation and markets. For example, in a study

Table 8.5 Geographical scope for life cycle stages and central unit processes in the window frames case study

Stage	Window type			
	Wood	Wood/aluminium	PVC	Wood/composite
Materials	Metal ores: not known			
	Crude oil: Norway, Russia, Middle East			
	Forestry: Finland		–	–
Manufacturing	Glass pane: Sweden			
	Wood frame: Scandinavia	Wood frame: Scandinavia	PVC frame: Germany	Composite frame: Germany
	Other elements: Europe	Other elements: mainly Europe	Other elements: mainly Europe	Other elements: mainly Europe
	Assembly: Denmark			
Use (heat supply)	Mainly Scandinavia, Germany	Mainly Scandinavia, Germany	Europe	Mainly Scandinavia
Disposal	The same as the use stage			

involving clothes washing in Vietnam the unit process for a certain waste water treatment process in Thailand may be a good approximation for a Vietnamese unit process if the treatment efficiency is the same, because of the climatic similarities between the two countries. If needed, the proxy process may be adjusted to better represent the actual process of the product system. Chapter 9 elaborates on the choices related to geographical representativeness when constructing an inventory model. The influence of a low geographical representativeness on the conclusionsof the study must be evaluated in the interpretation of the LCA results (see Chap. 12).

8.7.2 Time-Related Representativeness

Just as two processes delivering the same product output can be different if they occur in different locations, they can also be different if they occur at different times. This is due to technological innovation and development, which often tends to lead to more efficient processes over time, meaningless input (energy, material and resource flows) and sometimes also less unwanted by-products (waste to be treated and emissions) per unit of output. The time-related representativeness reflects how well the inventory data represents the actual processes regarding the time (e.g. year) they occur. Technological innovation is "faster" in some sectors than others. Therefore, a unit process that reflects the situation 10 years prior to the occurrence of the process in the product system can have a high time-related representativeness if it belongs to a mature sector with little technological

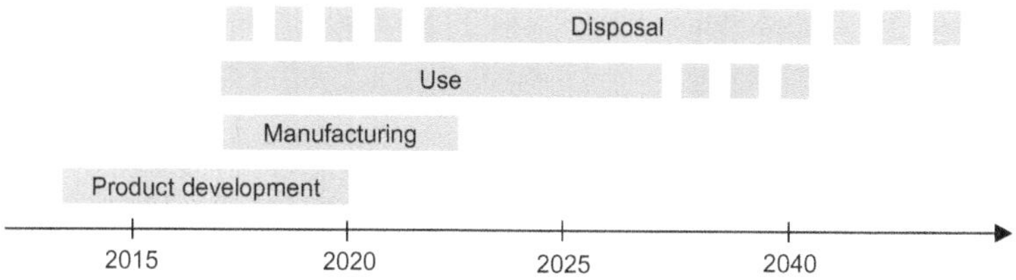

Fig. 8.15 Example of time frames expressed for different life cycle stages

innovation (such as the pulp and paper industry); but it can have a low representativeness if it is part of a sector with rapid technological development, such as IT, energy (with the growing focus on decarbonisation) and waste treatment (with the focus on waste avoidance and recycling of materials).

In line with the requirements to define the geographical scope of processes, LCA practitioners must in the scope definition define the time frame of the processes in the different stages of the life cycle. Figure 8.15 gives an example of how time frames can be represented.

These times are largely influenced by the expected lifetime of the studied product(s). For example, in a study involving furniture the expected lifetime, from consumer purchase to disposal, is decisive for the time at which waste treatment can be expected to occur. In other cases, the lifetime of installed capacity in the foreground system has a great influence on the time frames. For example, in a study involving a decision to construct a new incineration plant, the number of years that it is planned to operate (typically 20–30 years) is decisive for the timing of the involved unit processes. In all cases, the intended application of results and reasons for carrying out a study, as stated in the goal definition, can guide the time-related requirements. In the illustrative case study on window frames, the time frame of the manufacturing and use stage is estimated to be 5 and 20 years, respectively.

Following the formulation of the time-related requirements, the LCA practitioner must attempt to obtain the highest overall possible time-related representativeness when constructing the inventory model, within the time or budget constraints of the LCA study. When comparing the time aspects of the obtained inventory data with the time-related requirements it must be noted that the time at which a dataset was published is usually not equivalent to the time for which its data is valid (several years may pass between the first-hand collection of data and the publication of the data). In the foreground system the focus should be on those processes taking place in the future that the results of an initial iteration show to be important and that is also expected to change relatively rapidly (see above). The available current or past data for these processes can be used to project how they will evolve in the future. For example, the electricity mix of the future might be projected from past trends along with plans issued by public authorities that govern the electricity system. See also Chap. 21 on prospective LCAs and technological foresight. Regarding the background system, LCA practitioners usually have to make do with the most recent process contained in the LCI database used, while considering any trade-offs with geographical representativeness. The influence of a low time-related representativeness on the conclusions of the study must be evaluated in the interpretation of the LCA results (see Chap. 12).

In comparative studies it is important to investigate whether there is a risk that differences in time-related representativeness for the compared alternatives can lead to a bias that favours one product system over the others. This could, for example, be the case in a comparison of two technologies if the data of one technology is older (in terms of the year they are valid for) than the data of the other technology.

Just as some LCIA methods are spatially differentiated, there are also LCIA methods that are temporarily differentiated, meaning that their results are affected by the timing of elementary flows (see Chap. 10). So far, this LCIA practice has been limited to mainly distinguishing between "short-term" and "long-term" emissions, which is, for example, relevant when including landfilling processes, from which some emissions are projected to occur hundreds or even thousands of years after the landfilling of a given material. In addition, some climate change indicators consider when an emission occurs, which, for example, enables quantification of the benefits of temporary carbon storage. In specific cases, the difference of inventory data in the course of the year (especially hot and cold season) and the day (daytime/night) are relevant for a study. It is to be checked along the goal of the study whether such intra-annual or intra-day specific data might be needed (e.g. on night-time electricity base-load data for charging electric car batteries overnight). In all cases, the time-related information for elementary flows required by the LCIA methods chosen in the previous step of the scope definition should guide the data collection and output format of the inventory analysis.

8.7.3 Technological Representativeness

Two identical products can be produced using two different technologies and thereby be associated with different (sets of) unit processes and related flows. For example, crude steel can be produced using an electric arc furnace (EAF) or a basic oxygen furnace (BOF), which are two very different technologies involving different inventory flows. Technological representativeness reflects how well the inventory data represents the actual technologies involved in the studied product system. Technological representativeness is interlinked with geographical and temporal representativeness. For example, the technology mix involved in the production of electricity (coal power, natural gas, nuclear power, windmills, etc.) varies in space (e.g. from country to country) and over time. The LCA practitioner must use his or her knowledge about the product system to ensure (to the highest degree possible) that it is modelled using unit processes that reflect the actual technologies involved. It is important to ensure that the unit processes modelled in the system are in fact internally technologically compatible, meaning that the product output of one process should meet the quality requirements for input materials of the next process in the system. For example, if a unit process requires steel that is stainless and heat resistant as material input, then it is incompatible with the product of a unit process producing basic grade steel without these properties.

The scope definition should therefore contain a list of technologies that are known to be involved in the foreground system and in those parts of the background system for which such knowledge exists (typically energysupply, waste management and transportation), specifying representativeness requirements. This list should be partly based on the outcome of the geographical scope and time frames in terms of where and when processes are taking place.

8.8 Preparing the Basis for the Impact Assessment

The planning of the impact assessment in the scope definition has two main purposes. The first is to ensure that it is done in accordance with the goal definition and the second is to prepare for the inventory analysis where the elementary flows (resources and emissions) that should be included depend on the impact categories to be covered in the LCIA. These elementary flows may also depend on the particular LCIA methods that are used to model these impact categories because different LCIA methods can cover different elementary flows. Planning how to perform the LCIA prior to the life cycle inventory analysis therefore helps ensuring that the right data is being collected in the cycle inventory analysis. A brief guidance on the planning of the LCIA is given in the following sections. Chapter 10 gives a comprehensive introduction to the science behind LCIA and how to report results.

8.8.1 Selection of Impact Coverage

According to the ISO 14044 standard for LCA, the selection of impact categories to be covered by an LCA "shall reflect a comprehensive set of environmental issues related to the product system being studied, taking the goal and scope into consideration". This means that all environmental impacts where the product system has relevant contributions must be included in the impact assessment, unless the goal definition explicitly states otherwise. The latter is the case, e.g. in carbon or water footprinting studies, and in such studies the limitations imposed by the narrow impact coverage should be stressed in the goal definition and addressed the interpretation of results. Other valid reasons to exclude one or more impact categories from the assessment is when an initial iteration of the LCA shows that they do not contribute to the differentiation between the alternatives in a comparative LCA, or when they have a negligible contribution to the overall impacts, estimated by aggregating indicator scores for different impact categories to a single score following normalisation and weighting (see Sect. 8.2.5). In this case, the excluded impact categories must be listed as deliberately omitted in the scope definition of the LCA report with reference to the outcome of the initial iteration. Transparency on the selection of impact categories is essential to avoid an "interest-driven" selection of impact categories where impact categories are excluded, e.g. because they disfavour the product produced by the commissioner of a study in a comparative analysis.

8.8.2 Selection of LCIA Methods

To support the choice between alternative LCIA methods that can be used to calculate an indicator score for the same impact category, ILCD has developed six criteria for evaluating the methods:

1. Completeness of scope: how well does the indicator and the characterisation model cover the environmental mechanisms associated with the impact category under assessment?
2. Environmental relevance: to what extent are the critical parts of the impact pathway included and modelled in accordance with the current state of the art?
3. Scientific robustness and Certainty: how well has the model been peer reviewed, does it represent state of the art, can it be validated against monitoring data and are uncertainties reported?
4. Documentation, Transparency and Reproducibility: how accessible are the model, the model documentation, the characterisation factors and the applied input data?
5. Applicability: are characterisation factors provided for the important elementary flows for this impact category in a form that is straightforward to apply?
6. Stakeholders' acceptance: has the model been endorsed by competent authorities, are the model principles and applied metric understandable for users of the LCA results in a business and policy contexts?

These criteria can be difficult to apply for LCA practitioners that are not experts in LCIA modelling, but further insight can be gathered in Chaps. 10 and 40 gives an overview of available LCIA methods, discusses their main differences and how they perform on the six criteria.

In practice, an LCA practitioner will often rely on the use of software to model the product system and perform the impact assessment and then simply calculate LCIA scores for all the impacts categories that are made available in the software as part of an LCIA method. An LCIA method is a collection of impact categories that aims to have a broad coverage of environmental issues, and it is typically developed by one research group (Hauschild et al., 2013). If several LCIA methods are available, it may be useful to apply more than one to test the sensitivity of the results to the choice of LCIA method (see Chap. 11). This is an easy way to explore the sensitivity of LCIA results because calculating results for multiple impact categories in LCA software essentially takes the same time as calculating results for a single impact category.

For some LCA studies, no LCIA method may cover an environmental impact that is considered relevant. In such cases, the LCA practitioner can choose to develop an LCIA method on their own and this development should be guided by the six criteria above. Often, however, the development of a new impact category is not feasible for an LCA practitioner, due to budget constraints and limited knowledge of the impact pathway. The potentially relevant environmental impacts that are not covered by the impact assessment should be highlighted in the scope definition and considered qualitatively in the interpretation of results (Chap. 12).

An important aspect related to compatibility between the collected elementary flow of the life cycle inventory analysis and the ensuing LCIA is the degree of spatial differentiation of the LCA study. Spatial differentiation essentially means taking into account where an elementary flow occurs. This information is relevant for many impact categories, because the sensitivity of the environment towards 1 unit of elementary flow differs from place to place (see more details in Chap. 10). Many popular LCIA methods are (still) spatially generic. Yet, spatially differentiated methods have over the years increased in numbers and quality and their use may therefore increase in the future. If it is chosen to use spatially differentiated methods it is important to collect spatial information for the elementary flows in the life cycle inventory analysis (e.g. name of nation, watershed ID or grid cell defined by GIS coordinates) that is compatible with these methods.

Normalisation and weighting are optional LCIA steps under ISO 14044:2006, and as part of the scope definition the LCA practitioner should decide whether normalisation and weighting is needed. Are the steps relevant for the intended application(s) and target audience of the LCA study (see Goal definition in Chap. 7)? Normalisation is usually beneficial to aid the understanding of results if the target audience are not experts, and weighting is required if an aggregation of impact scores across the environmental impact categories is intended. On top of normalisation, an LCA practitioner may thus choose to include weighting, if the commissioner of a study has specifically asked for single score results. The decision to perform normalisation and weighting can also influence the choice of LCIA method since not all methods support these steps. A detailed description of normalisation and weighting is given in Chap. 10.

8.9 Special Requirements for System Comparisons

Many LCA studies compare systems, e.g. when two or more products fulfil the same function as captured in the functional unit. The ISO 14044 standard poses a number of special requirements for the scope definition of comparative studies to ensure that the systems can actually be compared: "Systems shall be compared using the same functional unit and equivalent methodological considerations, such as performance, system boundary, data quality, allocation procedures, decision rules on evaluating inputs, and outputs and impact assessment. Any differences between systems regarding these parameters shall be identified and reported". When a comparative study is intended to conclude on the superiority or equivalence of the compared alternatives in terms of their environmental performance, and to make these conclusions publically available, the standard identifies it as a "comparative assertion intended to be disclosed to the public". For such applications of LCA, the standard requires that these points shall be evaluated in a critical review performed by a panel of interested parties (see Sect. 8.10 and Chap. 13).

These special requirements reflect the consequences that the comparative use of LCA results may have for other companies, institutions and stakeholders that are not directly involved in the study and they are intended to prevent the misuse of LCA in market competition.

To prevent misleading LCA results and the misuse of LCA in comparative assertions, the ILCD guideline furthermore requires that:

- The uncertainties involved must be evaluated and communicated when one product system appears to have a lower environmental impact for one or more impact categories than another, see Chaps. 11 and 12 for details.
- In the case where the goal definition prescribes a comparison based on a single indicator (e.g. carbon footprint) the LCA study must highlight that the comparison is not suitable to identify environmental preferable alternatives, as it only covers the considered impact(s) (e.g. climate change). This applies unless it can be sufficiently demonstrated that the compared alternatives do not differ in other relevant environmental impacts to a degree that would change the conclusions of the comparison if those other impacts would be included in the analysis. Such demonstrations may be in the form of other LCA studies available for sufficiently similar systems.

8.10 Need for Critical Review

A critical review is performed by experts not involved in making a study. A critical review is sometimes required (e.g. for studies with the intended publication of results), but even when there is no formal requirement a critical review is useful for improving the quality and credibility of a study.

Chapter 13 deals specifically with the critical review stage of an LCA, presents the different types of critical review and explains for what kind of LCA studies (with reference to the goal definition) these are needed. It is, however, useful already during the scope definition to decide whether a critical review is needed or intended. If a review is required or intended, the scope definition should furthermore specify the form of the review in order to allow the documentation and reporting of the study to be tailored to meet the later requirements from the peer reviewers. It should also, in the scope definition, be decided whether the review should be performed on the final draft of the LCA report or whether it should be done in an interactive process throughout the performance of a study. In this case, the reviewers are given the opportunity to comment on the goal and scope definition prior to the onset of the inventory analysis, and possibly on interim results of the impact assessment and interpretation before the final reporting so that their comments can guide the process of the LCA.

8.11 Planning the Reporting of Results

Product systems can be very complex, and choices are often made during the LCA that can influence the conclusions. To reduce the risk of erroneous and misleading use of the LCA, it is essential that the reporting is clear and transparent with a clear indication of what has and what has not been included in the study and which conclusions and recommendations the outcome supports.

The reporting of an LCA study should target the audience as it is specified in the goal definition. Depending on whether the study is comparative and public, the ILCD guideline identifies three reporting levels:

1. Internal use by the commissioner of study;
2. External use by the third party, i.e. a limited, well-defined list of recipients with at least one organisation that has not participated in the study.
3. Comparative studies to be disclosed to the public.

Due to the sensitive nature of comparative assertions based on LCA, there are a number of additional reporting requirements to level 3 studies. No formal requirements apply to level 1, but it is recommended to follow the requirements for level 2. Chapter 38 shows all the elements that an LCA report should cover, according to level 2 and 3, and proposes a sequence of these elements, and the reporting of the case study on window frames in Chap. 39 demonstrates the application of the template in a comparative study.

References

This chapter is to a large extent based on the ILCD handbook and the ISO standards 14040 and 14044. Due to the scope of this chapter, some details have been omitted, and some procedures have been rephrased to make the text more relevant to students. For more details, the reader may refer to these texts:

EC-JRC.: European Commission—Joint Research Centre—Institute for Environment and Sustainability: International Reference Life Cycle Data System (ILCD) Handbook—General Guide for Life Cycle Assessment—Detailed Guidance. First edition March 2010. EUR 24708 EN. Publications Office of the European Union, Luxembourg (2010)

ISO.: Environmental Management—Life Cycle Assessment—Principles and Framework (ISO 14040). ISO, the International Organization for Standardization, Geneva (2006a)

ISO.: Environmental Management—Life Cycle Assessment—Requirements and Guidelines (ISO 14044). ISO, the International Organization for Standardization, Geneva (2006b)

Additional References Used in the Text

Ekvall, T., Azapagic, A., Finnveden, G., et al.: Attributional and consequential LCA in the ILCD handbook. Int. J. Life Cycle Assess. **21**, 293–296 (2016). doi:10.1007/s11367-015-1026-0

Hauschild, M.Z., Goedkoop, M., Guinée, J., et al.: Identifying best existing practice for characterization modeling in life cycle impact assessment. Int. J. Life Cycle Assess. **18**, 683–697 (2013). doi:10.1007/s11367-012-0489-5

IDF.: A common carbon footprint approach for dairy. The IDF guide to standard life cycle assessment methodology for the dairy sector. http://www.idf-lca-guide.org/Public/en/LCA+Guide/LCA+Guidelines+overview (2010). Accessed 2 Aug 2016

Treyer, K., Bauer, C.: Life cycle inventories of electricity generation and power supply in version 3 of the ecoinvent database-part I: electricity generation. Int. J. Life Cycle Assess. **3**, 1–19 (2013). doi:10.1007/s11367-013-0665-2

Weidema, B.P., Bauer, C., Hischier, R., et al.: The ecoinvent database: overview and methodology. Data quality guideline for the ecoinvent database version 3. www.ecoinvent.org (2013)

Author Biographies

Anders Bjørn part of the LCA community since the early 2010s. Main focus is interpretations of sustainability and integration of sustainability targets in LCA to enable absolute sustainability assessments.

Mikołaj Owsianiak involved in development and application of life cycle impact assessment methods in sustainability assessment of technologies. Has worked on issues associated with: soils (remediation), metals (toxic impact assessment), biodiesel (fate in the environment), and carbonaceous materials (biochar and hydrochar).

Alexis Laurent working with LCA since 2010 with a strong LCIA focus, particularly on normalisation aspects. Main LCA interests include development of LCIA methods, LCA applications and footprinting of large-scale systems for policy-making (e.g. nations, sectors), and LCA applied to various technology domains, including energy systems.

Stig Irving Olsen LCA expert both as researcher and as consultant. Involved in the development of LCA methodologies since mid 1990's. Contributed to UNEP/SETAC working groups on LCIA methodology. Main LCA interest is human toxicity impacts, emerging technologies, and decision making.

Andrea Corona materials engineer with a focus on sustainability assessment of biobased products and composite materials. Working with LCA from 2012. Main interests include life cycle engineering, product development and eco-design.

Michael Z. Hauschild involved in development of LCIA methodology since the early 1990s. Has led several SETAC and UNEP/SETAC working groups and participated in the development of the ISO standards and the ILCD methodological guidelines. Main LCA interests are chemical impacts, spatial differentiation and science-based boundaries in LCIA.

Chapter 9
Life Cycle Inventory Analysis

Anders Bjørn, Andreas Moltesen, Alexis Laurent, Mikołaj Owsianiak, Andrea Corona, Morten Birkved and Michael Z. Hauschild

Abstract The inventory analysis is the third and often most time-consuming part of an LCA. The analysis is guided by the goal and scope definition, and its core activity is the collection and compilation of data on elementary flows from all processes in the studied product system(s) drawing on a combination of different sources. The output is a compiled inventory of elementary flows that is used as basis of the subsequent life cycle impact assessment phase. This chapter teaches how to carry out this task through six steps: (1) identifying processes for the LCI model of the product system; (2) planning and collecting data; (3) constructing and quality checking unit processes; (4) constructing LCI model and calculating LCI results; (5) preparing the basis for uncertainty management and sensitivity analysis; and (6) reporting.

Learning Objectives
After studying this chapter the reader should be able to:

- Collect and critically evaluate the data quality of an LCI.
- Construct a unit process from first-hand gathered data.
- Build an LCI model using either attributional or consequential approach and explain the differences between the two approaches.
- Explain what data is required for uncertainty and sensitivity analyses and how to collect these data.
- Document an LCI model, including unit processes and LCI results.

A. Bjørn (✉) · A. Moltesen · A. Laurent ·
M. Owsianiak · A. Corona · M. Birkved · M.Z. Hauschild
Division for Quantitative Sustainability Assessment, Department of Management Engineering, Technical University of Denmark, Kongens Lyngby, Denmark
e-mail: anders.bjoern@polymtl.ca

A. Bjørn
CIRAIG, Polytechnique Montréal, 3333 Chemin Queen-Mary, Montréal, QC, Canada

M.Z. Hauschild et al. (eds.), *Life Cycle Assessment*,
DOI 10.1007/978-3-319-56475-3_9

9.1 Introduction

During the life cycle inventory (LCI) analysis phase of an LCA the collection of data and the modelling of the flows to, from and within the product system(s) is done. This must be in line with the goal definition (see Chap. 7) and (to the extent possible) meet the requirements derived in the scope definition (see Chap. 8). The LCI result is a list of quantified elementary flows crossing the system boundary of the studied life cycle and it is used as input to the subsequent LCIA phase (see Chap. 10). Insights that the LCA practitioner gains when conducting the LCI analysis are also commonly used to adjust the requirements of the scope definition, e.g. when unforeseen data limitations lead to the need for a modification of the completeness requirements (see Sect. 8.6.3). Typically, the LCI analysis is the phase that requires the most efforts and resources from the LCA practitioner, and it is rarely practically possible to collect the highest quality of data for all processes of the LCI due to the unreasonable high cost that would be involved. Fortunately, it is also rarely needed in order to meet the goal and support the intended applications of the LCA. Therefore, the inventory analysis requires a structured approach to ensure that time is being spent on collection of data for those parts of the product's life cycle that are most important for the overall impacts from the product system. Several iterations between the LCI and LCIA phase are normally needed to meet the goal of the study, with each iteration providing insight into which inventory data are the most important for the LCA results (see Chap. 6).

In this chapter, we provide practical guidance on how to perform an LCI analysis using an iterative approach to LCA. We will focus on providing detailed guidance for the four decision contexts (A, B, C1 and C2) in line with the ILCD guideline. The chapter is structured around six steps of an LCI analysis:

1. Identifying processes for the LCI model
2. Planning and collecting data
3. Constructing and quality checking unit processes
4. Constructing LCI model and calculating LCI results
5. Preparing the basis for uncertainty management and sensitivity analysis
6. Reporting.

Before digging into the details, we note that this chapter teaches how to construct an LCI using knowledge about the industrial processes taking part in a life cycle and the physical flows connecting them. This is called a process-based (or bottom-up) approach to inventory modelling. A complementary approach to constructing an LCI is to model the life cycle inventory for the product from a macroscale perspective by drawing on a combination of (1) information on elementary flows associated with one unit of economic activity in different sectors and (2) national statistics on the trade of products and services between sectors. This is called environmentally extended input–output analysis (EEIO) and in contrast to the process-based approach it can be seen as a top-down approach to inventory modelling. The strength of EEIO is that a completeness of 100%, in theory, can be achieved in the sense that no processes need to be cut-off due to missing data or budget constraints. The two main weaknesses of the EEIO approach are (1) that the coverage of elementary flows is rather limited, compared to the process-based approach and that (2) the resolution of many products and services is quite low due to the heterogeneous nature of many sectors, as defined by national trade statistics. Chapter 14 deals with IO-LCA and in particular how to use EEIO to complement and guide process-based LCA. This chapter will make references to EEIO, when the approach can complement the process-based approach.

9.2 Identifying Processes for the LCI Model

This first step of *the LCI details the coarse* initial system diagram made under the scope item System boundaries (see Sect. 8.6) and draws upon the related completeness requirements. The outcome of the step is a detailed depiction of the foreground system, i.e. all the processes it is composed of and their links, and the processes of the background system 'neighbouring' the foreground system, i.e. where links to LCI database processes will be established.

9.2.1 Detailing the Physical Value Chain

For all decision contexts (A, B, C1 and C2—see Sect. 7.4) the approach to identifying processes is to start with the reference flow and construct the entire foreground system process by process:

0. The unit process having the reference flow, as product output, should first be identified (or unit processes, in the case of more than one reference flow). This is termed a 'level 0' process. In a study where a window is the reference flow, the level 0 process is the assembly of the window.
1. The processes required to deliver flows that will be *physically embodied* in the reference flow should then be identified. These are termed "level 1" processes. In the window example, examples of level 1 processes are the production of glass and the window frame.
2. The processes required to deliver flows that perform a *supporting function* to the level 0 process (i.e. not becoming physically embodied in its output) should then be identified. These are termed 'level 2' processes. In the window example, examples of level 2 processes are the supply of electricity used in the assembly of the window or the transportation needed to deliver the flows of the level 1 processes to the level 0 process.
3. The processes required to *deliver services* to the level 0 processes should then be identified. These are termed 'level 3' processes. In the window example, examples of level 3 processes are administration and marketing.

4. The processes required to produce and maintain the *infrastructure* that enables the level 0 process should then be identified. These are termed 'level 4' processes. In the window example, examples of level 4 processes are production and maintenance (oiling, replacing and repairing parts) of the assembly machines.

After having identified level 1, 2, 3 and 4 processes belonging to the level 0 process (the reference flow), Step 1–4 is then repeated for each these processes. This procedure is illustrated in Fig. 9.1 for the window example.

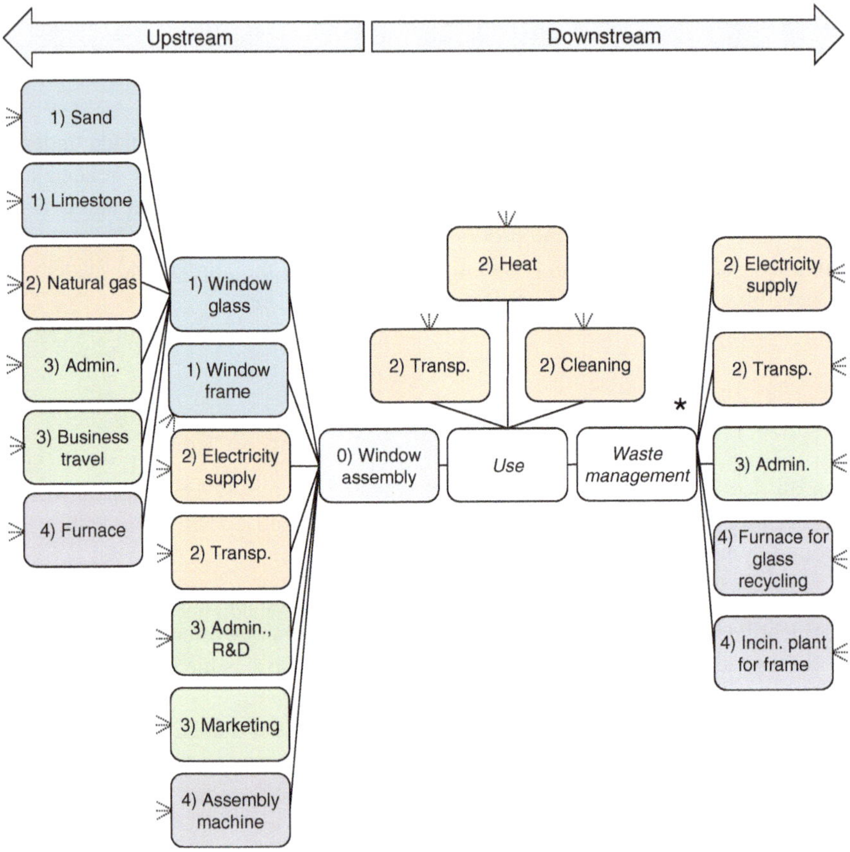

Fig. 9.1 Procedure for identifying processes of the foreground system, exemplified in the study of the life cycle of a window. The starting point is the process that delivers the reference flow, '0) Window assembly'. The foreground system is then populated process by process by proceeding upstream and downstream from the reference flow. *Unlinked arrows* present on some processes indicate the existence of other processes that were not included in the figure. Use and Waste management are in *italic*, because they represent life cycle stages, rather than actual processes. The *star* at 'Waste management' indicates the existence of multifunctional processes, i.e. glass recycling and incineration of window frame. *Abbreviations* in the figure: *Incin* incineration, *Transp* transportation, *Admin* administration, *R&D* research and development. *Numbering* and *colour code*, identify the process level for the different foreground processes

Processes downstream, i.e. in the use and waste management stages, should be identified in a similar fashion. The procedure is, in principle, repeated until the foreground system is completed and can be linked to LCI database processes of the background system, as described later in this chapter. When carrying out this procedure, the LCA practitioner should identify all multifunctional processes, because they have to be handled next.

Note that the step of identifying processes for the LCI model and the step of planning and collection of data are somewhat interrelated. For example, data collected for a given process may lead to the realisation that one or more upstream processes are different than the ones previously (assumed) identified. During data collection the LCA practitioner may, for example, realise that a plastic component is actually produced from biomaterials rather than petrochemicals, as was initially assumed. The identified processes in this first inventory step should therefore be considered preliminary.

In practice, many processes belonging to level 3 and 4 will end up being entirely omitted from an LCI model, because their individual contribution to the indicator score is expected to be insignificant and because data can be hard to find, at least when using the 'bottom-up' (=process-based) approach to constructing inventories. In such cases, the environmental impacts of product systems are systematically underestimated by various degrees. It is an important task of the inventory analysis and consecutive impact assessment to ensure that this underestimation does not violate the completeness requirements for the study. Chapter 14 shows how IO-LCA can complement process-based LCA to better cover the impacts from level 3 and 4 processes.

9.2.2 Handling of Multifunctional Processes

Section 8.5.2 presented the ISO hierarchy for solving multifunctionality, i.e. processes in the product system that deliver several outputs or services of which not all are used by the reference flow of the study. According to this hierarchy, the preferred solution is subdivision of the concerned process, and if this is not possible, system expansion and, as a last resort, allocation. Below, examples are given for how to carry out each solution in practice. This guidance is primarily relevant for the foreground system because multifunctionality has typically already been handled for the processes in the LCI databases that are used to construct the background system. Some LCI databases exist in different versions, according to how multifunctionality has been solved (see Sect. 9.3 below). For the background system this reduces the job of the LCA practitioner to just source processes from the appropriate version of the LCI databases. Yet, even in the background system, the LCA practitioner may sometimes have to solve multifunctionality manually when no appropriate solutions exist in the used LCI databases. We note that many waste treatment processes are multifunctional because they both offer the function of managing (often heterogeneous) waste streams and the function of providing product flows, such as recycled materials or electricity. We refer to Chap. 35 on application of LCA to solid waste management systems for more details on how these special cases of multifunctionality are solved in LCA practice.

Subdivision

When possible, subdivision should always be the solution to multifunctionality. Unit processes can be defined at many levels of detail and for the use in LCA there is no point in detailing them beyond what is needed for the modelling purpose in the LCA. This may mean that by increasing the detail applied in the modelling, the multifunctionality may be revealed as artificial. For example, a process that encompasses an entire factory producing two different products may have been identified from the procedure detailed in Sect. 9.2.1. If this factory is in fact using different and independent machines and work stations for manufacturing the two products, the initial process can, by introducing additional detail in the modelling of the process, be subdivided into two or more processes that each contribute to the production of only one of the products, see Fig. 9.2. Note that it is often not possible to fully physically divide a process according to the co-products. In the factory example room lighting, room heating and administration (all level 3 processes, according to Sect. 9.2.1) may not be possible to divide between the co-products. In such cases, subdivision needs to be supplemented with or replaced by another solution to multifunctionality. Note also, that in practice data availability often determines whether subdivision is possible. In the factory example, it may be that data only exist for the electricity consumption of the entire factory, i.e. the consumption of each machine is unknown and in this case, subdivision would be practically impossible. In addition, there are many situations where the creation of the co-products is integrated into the process in a way that impedes the multifunctionality to be addressed by subdivision. This is the case for many biological and chemical processes.

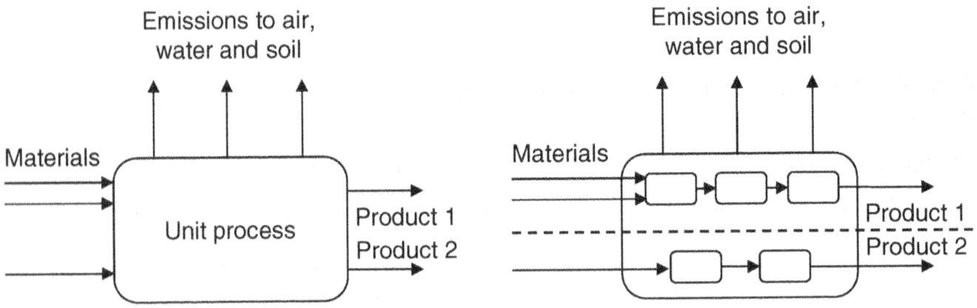

Fig. 9.2 Solving the multifunctionality problem by increasing the modelling resolution and sub-dividing the process into minor units which can unambiguously be assigned to either of the functional outputs

System Expansion

System expansion is second in the ISO hierarchy. As explained in Sect. 8.5.2, system expansion is mathematically identical to crediting the studied product system with the avoided production of the secondary function(s) that would alternatively have been produced and delivered somewhere else in the technosphere. When modelling a life cycle inventory, the technique used to perform crediting varies between LCA software (see Sect. 9.5). The identification of avoided processes depends on the decision context. For Situation A and C1 a market mix is used, which corresponds to the average process used to supply the entire market (see Sect. 8.5.4). To calculate a market mix, one needs to know the amount of product or service that is produced and delivered to the relevant market by each process at the time when the secondary function is delivered by the studied product system (see Fig. 8.13). So, for example, if recycled steel is a co-product of a studied life cycle and the two processes for producing steel, electric arc furnace (EAF) and a basic oxygen furnace (BOF), delivered 60 and 140 million tonnes, respectively, in the relevant market and reference year, then the market mix would be 30% EAF and 70% BOF (World Steel 2015). The LCI model should thus be credited with a constructed process composed of 30% of the flow quantities associated with the production of 1 unit of EAF steel and 70% of the flow quantities associated with the production of 1 unit of BOF steel. It is important to identify the correct market for each system expansion. The correct market must reflect the geographical and temporal scope (see Sect. 8.7). Note that some goods and services are sold in global markets due to the low cost of transportation relative to their value (e.g. gold), while other goods and services are sold on local or regional markets due to high transportation cost (e.g. some biomaterials and water) or regulation. Information on volumes produced and delivered to markets can often be obtained from reports or databases of industry organisations (e.g. the World Steel Association in the example of recycled steel). In consequential modelling (parts of Situation B, see Sect. 8.5.4), the avoided process is not a market mix, but the marginal process (or a mix of marginal processes) and its identification is explained in Sect. 9.2.3.

Allocation

Allocation is the third and last option in the ISO hierarchy. As mentioned in Sect. 8.5.2 allocation should, when possible, be based on (1) causal physical relationship, followed by (2) a common representative physical parameter and, as a last resort, (3) economic value.

The causal physical relationship approach is possible when the ratio between quantities of co-products can be changed. Consider again the above example of a factory producing two products (x and y), where only the total electricity consumption is known. Here it would be possible to derive the electricity consumption of x and y by collecting data on production volumes and total electricity consumption at two points in time, where the relationship between the produced quantities are different. This could lead to the following simple system of equations:

$$\text{Time 1}: \ 10\,\text{tonnes} * X + 20\,\text{tonnes} * Y = 10.000\,\text{kWh} \tag{9.1}$$

$$\text{Time 2}: \ 10\,\text{tonnes} * X + 40\,\text{tonnes} * Y = 12.000\,\text{kWh} \tag{9.2}$$

Here, X and Y represent the electricity consumption of product x and y (kWh/tonne) and by solving the equation system, one finds that X is 800 kWh/tonne and Y is 100 kWh/tonne. If time 1 is representative for the unit process to be applied in the LCI model, then 80% (10 tonnes * 800 kWh/tonne divided by 10.000 kWh) of the factory's electricity consumption should be allocated to product x. Note that this 80% allocation factor should not blindly be applied to allocate the remaining flows (e.g. consumption of heat and emissions of NO_x) between product x and y, for which the causal physical relationships may be different. Note also that allocation according to a causal physical relationship is in many cases not possible, because the ratio between co-products or co-services for many processes cannot be changed. For example, it is not for practical purposes possible to reduce or increase the production of straw, while keeping the production of wheat constant.

The representative physical parameter approach is possible when co-products provide a similar function. For example, in the case of a fractional distillation process of crude oil, a similar function of many of the co-products (e.g. diesel, petrol, kerosene, propane and bunker oil) is to serve as a fuel to drive a process performing mechanical work, and therefore exergy, which can be interpreted as the maximum useful work, is an appropriate representative physical parameter. The parameter values of each co-product can typically be obtained from physical or chemical compendiums (e.g. in the case of exergy values). Once the values have been obtained, calculating the allocation factor is straightforward. For example, if co-products x, y and z are produced in quantities 1, 3 and 6 kg and if their representative physical parameter values are 10, 1, and 0.5 per kg, then the total parameter value would be 16 i.e. (1 * 10 + 3 * 1 + 6 * 0.5) and the allocation factor for product x would be 62.5% (1 * 10 divided by 16) and so on. Note that in the distillation process case, the functions of the co-products are not entirely identical. Airplanes cannot fly on bunker oil, and bitumen, one of the co-products, cannot be used as a fuel. Allocating according to a representative physical parameter is therefore not ideal, but may be the best solution, compared to other allocation approaches. This example illustrates that there is often not a single correct allocation approach and the choice of approach therefore depends on the judgement of the LCA practitioner. The sensitivity of the LCA results to this judgement may be investigated in a sensitivity analysis applying different possible allocation factors, as explained in Sect. 9.6. Note that it is very important to choose a representative parameter that is actually representative for the function of all co-products. For example, mass is not a representative parameter for the co-production of milk and meat from dairy cows because the functions of milk and meat are not their mass. In this case, some measure of nutritional value would be a more representative parameter.

The economic value approach is recommended as a last resort and is generally easy to carry out due to the abundance of price data on goods and services. Prices may be obtained by contacting the company running the multifunctional process in question or from the stock exchange in case of global markets, e.g. for some metals. For some co-products there may not be a market because they need to go through additional processing before they are sold. In that case, the LCA practitioner should calculate a shadow price. For example, straw, a co-product of wheat production, needs to be baled before it is sold, and the economic value of baled straw must therefore be subtracted the cost to the farmer of baling the straw to calculate the shadow price of the unbaled straw leaving the multifunctional process of wheat production. Note that the prices of most goods and services are volatile to varying degrees. It is therefore recommended to calculate average values for the time period that is relevant to the temporal scope of the study (see Sect. 8.7.2). Once the economic values have been determined, allocation factors are calculated in the same way as the above generic example for the representative physical parameter approach.

It should be noted that although allocation by economic value is the last resort according to ISO, it is widely used in practice. This is because the other solutions to the handling of multifunctional processes are often not possible due to the nature of the multifunctional process or due to lack of the required information and data to identify the relevant process for a system expansion or to determine a causal physical relationship, or a common representative physical parameter. By contrast, the price data needed to carry out economic allocation is generally available. For this reason, economic allocation is done by some LCA researchers recommended as a default solution to multifunctionality, e.g. by the Dutch CML Guideline (Guinée et al. 2002), and the LCI database ecoinvent comes in a version where allocation by economic value is systematically applied to all multifunctional processes (see Sect. 9.3.2 below).

9.2.3 Consequential Modelling

In most cases, a consequential LCI will include other processes than an attributional LCI for the same product system. The attributional LCI includes the processes which the assessed product 'sees' from its journey from the cradle to the grave. If, for example, the assessed product is a plastic cup, the start of the journey will be some extracted crude oil, which through a sequence of production processes will be processed into plastic. This will then be transported to the shop, be bought by a user, who will use it once and then discard it, after which it will be transported to, say, an incinerator and burned. In the attributional LCI, each of these processes: the production of crude oil, the conversion into plastic, the transport and incineration will be included.

The consequential LCI is different; the goal of the assessment is to identify the environmental impacts caused by a decision, for example the decision to buy a plastic cup. The processes that change due to a decision may not be the same that a product 'sees' throughout its product life (see Fig. 9.1). The following example may make this easier to understand.

Assume now for the sake of the example that we have reached the peak in oil production: we simply cannot economically extract more oil than we are already doing. This implies that the decision to, say, use this plastic cup will not result in an increase in the production of oil, as this is already at its maximum. What happens instead may be that the price of oil will go up due to the increase in demand (which in this example is going to be extremely small due to the small amount of oil needed to produce the cup. However, here it is the principle that is of interest). The increase in price may cause other users of oil to reduce their use, or find a substitute for their use of oil. In this example we will assume that some oil users will find natural gas a suitable substitute and these users will therefore increase their demand for natural gas to compensate for the decreased availability of oil. This implies that, given these assumptions, an increase in the demand for oil created from the increase in demand for plastic cups will not result in an increase in the production of oil, but rather in the production of natural gas. The consequential LCI will therefore not include an extraction of oil, but rather an increased extraction of natural gas. This line of thinking obviously does not only relate to the oil used in the production of the plastic, but to all the inputs used when the plastic cup is produced.

Another very important difference between the attributional and consequential LCI is that in an attributional LCI the normal procedure is to assume that the electricity consumed in the production of the plastic cup is produced by all the suppliers on the market, depending on their market share. In a consequential LCI, this is different: If we increase the demand for electricity in the market, it is most likely that not all the suppliers are going to increase their production to meet the increase in demand. The reason is that the most cost-efficient producers will already produce at full capacity. This is for example going to be the case for nuclear power plants. This means that if we increase the demand for electricity, we will not influence the extent of the production from the nuclear power plants. Rather, we will influence other types of power plants, for example natural gas power plants, which are more expensive to operate (per kWh), and which will therefore only produce during peak load situations (when electricity prices are higher). The same thinking is applied when studying the effect of increasing or decreasing demands for other products than electricity. Rather than including an average of the producers in the market in the LCI, as is done in the attributional LCI, it is the 'marginal' producers, which are included in the consequential LCI. A marginal producer is a producer who will change its supply due to small changes in demand.

A final important difference between the attributional and consequential LCI lies in the handling of multifunctional processes (see Sect. 8.5.2). In a consequential LCI, the multi-output processes are always handled by system expansion (if subdivision is not possible).

Based on the outline above, there are generally three different tasks in a consequential LCI:

1. To identify whether an increase or decrease in demand for a product will actually lead to corresponding increases or decreases in supply for that product. As illustrated with the oil and gas example above, this is not necessarily the case.
2. To identify which production technology will be affected by the change in supply of products. This is most likely not going to be an average of the production technologies on the market, but rather one or a few operating on the margin.
3. To identify which product substitutes which. This is relevant when changing demands for a product whose production is constrained, such as oil in the example above. It is also relevant for the handling of multi-output processes, where it involves identifying the product that will be affected (substitute or be substituted) by a co-product from a multi-output process.

From the discussion above, it can be seen that if we want to perform an attributional LCI, we can do so simply on the basis of knowledge about the product and the parts that it includes: we need to know about how plastic cups are made, used and discarded. However, if we want to perform a consequential LCI, besides the technical knowledge about how the plastic cup is produced, used and discarded, we also need knowledge about how the market reacts to an increase (or decrease) in demand and supply.

As can be imagined, answering how the market reacts is easier said than done. What will actually happen if I increase the demand for this or that? Modelling the reactions of the market is a very complex task—just ask any stockbroker! Outlining what will happen is therefore necessarily somewhat uncertain, especially if the assessment addresses decisions in the more distant future. However, to ease the answering of these questions we will in this chapter outline a range of 'rules of thumb' developed for identifying the processes that are likely to change due to a decision.

As outlined above, the goal of the consequential LCA is to answer questions of the type: *"What are the environmental consequences if …?"*. As the environmental consequences that are considered arise from changes in production of products, this overall question answered in the consequential LCA can basically be translated to "What changes in the production of goods if we demand/supply more/less of *X*(, *Y*, *Z*, …)?". We continue asking this question until we have covered all induced changes. For example, in the case where we want to assess what happens if we use a plastic cup, we basically want to increase the demand for plastic cups. We therefore start by asking: *"What will happen if I increase the demand for plastic cups?"* If what happens most likely turns out to be that additional cups will be produced, then the follow-up question will be: *"What will happen if we produce additional plastic cups?"* The overall approach of identifying processes to include in a consequential modelling of the product system is to repeatedly ask this question for each step upstream and downstream from the reference flow (see Sect. 9.2.1) until all changes have been covered.

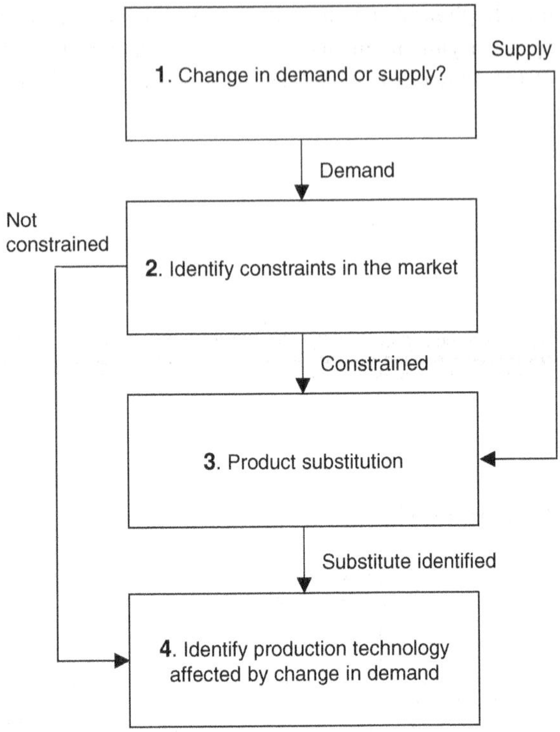

Fig. 9.3 4-Step approach for identifying affected process in a full consequential LCA

We recommend solving this task by following a 4-step procedure shown in Fig. 9.3. Depending on the concrete case, one or more steps can be skipped (as will be explained below).

Step 1: Change in demand or supply?

When performing a consequential LCI, full elasticity of supply is generally assumed. This implies that a change in demand for some function will lead to a change in supply of products that can fulfil this demand, but that change in supply, will not lead to a change in demand. There will therefore be a difference between the market effects of changing demand and changing supply, as will be visible in the steps below.

First step in the procedure is therefore to consider whether the question at hand addresses a change in demand or supply; e.g. are we assessing the question: "What happens if I *demand* more/less of *X*?" or the question "What happens if I *supply* more/less of *Y*?" Note that handling a co-product from a multi-output process in the studied life cycle relates to changes in supply of this co-product, and therefore is related to the latter type of question.

If the assessed decision relates to changes in demand, go to Step 2. If it relates to changes in supply, go to Step 3.

Step 2: Identify constraints in the market

If we increase (or decrease) our demand for *X*, the market will, according to standard economic theory, respond by increasing (or decreasing) the supply of *X*. In many cases, at least on a short term, there will not be a one-to-one relationship between increases in demand and supply. The reason is that an increase in demand will often result in an increase in price, implying that some users may stop using the product and potentially find a cheaper substitute product. Hereby, the supply and demand will reach a new steady state, which will often not entirely correspond to the initial demand plus the increase. Despite that these thoughts about price elasticity have been introduced in LCA literature, for simplicity, the default assumption here will be that the increase (or decrease) in demand will spur an equally large increase (or decrease) in supply, which is also the most common assumption in consequential LCI.

However, in many cases markets face various constraints and other market imperfections. An increase (or decrease) in demand will therefore not always lead to an increase (or decrease) in supply. Market limitations may be of a legal, economical, technical or physical nature. For example, straw used for co-firing in power plants is, due to the transport cost to value ratio, not transported far from the production site. Moreover, as there is limited production capacity of straw in a given area, an increase in demand within this area will in many cases not result in an increase in supply. Another example may be the demand for recycled metals, which are often constrained by the amount of waste input to recycling processes, in which case an increase in demand will not result in an increase in supply of recycled metals. Other constraints may be due to legally set boundaries for how much of a certain good may be produced. If the production already fills the boundaries, a small increase or decrease in demand will also not have any effect on supply.

There are thus a number of situations where the default assumption—that an increase or decrease in demand results in an increase or decrease in supply—may not hold true. In these situations, *the market is constrained*, and a central task will be to identify how existing or potential users will handle the increase or decrease in demand. In the example above with an increased demand for recycled metal, a reasonable assumption may be that existing users of the recycled metal will use virgin metal instead. In other words, an increase in the demand for a product already produced at maximum will not lead to an increase in supply, but more likely make existing users find a substitute. A guideline for identification of which products can substitute which is provided under Step 3.

The assessed decision may also lead to a decrease in the demand for a product, which is only produced to a certain amount. If this product is already fully utilised, a reasonable assumption may be that a decrease in demand for the product in question will not lead to a decrease in the supply of this product, as other users will use up the freed supply. In the metal example above, this could imply that the freed supply of recycled metal will be used up by a user of virgin metal, in total lowering the demand of virgin metal, while keeping the utilisation of recycled metal at the same level.

It may also happen that the freed supply resulting from a decreased demand does not lead other users to utilise the product. If this is the case, it can be assumed that less will be produced of the product, or if the product is a co-product of another and more valuable product, and its production therefore bound, it may end up as waste, implying that a decrease in the demand for the product will simply imply more waste.

It may seem an enormous task to try to identify whether all the commodities included in the life cycle are constrained in their production. However, in practice the assumption will often be that a product is constrained if:

- It is a co-product from a process that has another more valuable product, as it will never be the less valuable product that will control the overall output of the production (e.g. waste from a slaughterhouse that may be utilised for biodiesel production is constrained by the amount of meat produced).
- Its production is limited by regulation (e.g. regulation may set a limit for the overall annual catch of commercial fish species).
- Its production is limited physically (for example the production of wood on an island is restricted by e.g. forest area and a high cost of transportation may mean that import is not an economic option).

Identifying whether a commodity is produced as a less valuable by-product will often be quite easy, but the identification of both regulatory and physical constraints may be more difficult. It will in many cases require knowledge about the specific market in which the change in demand is made, which will often require advice from experts. Furthermore, one must know whether the production capacity for the product, for which demand is changed, is already filly utilised. Figure 9.4 presents a decision tree for dealing with potential market constraints.

In case of unconstrained markets, go directly to Step 4. Constrained markets must in some cases (see Fig. 9.4) be studied in Step 3 first to identify what other users prefer as a substitute (in the case of increased demand) or which substitute other users stop using (in the case of decreased demand).

Step 3: Product substitution

As noted in Step 1 it is commonly assumed in consequential LCA that supply follows demand. This implies that if we change supply, we will not change the demand but rather affect the competition between suppliers to cover the demand. For example, if we reduce the supply of crude oil on the market, it is assumed that the crude oil users will attempt to find a substitute for the crude oil, creating a demand for other products satisfying the same service as offered by the crude oil. The demand for the service that the crude oil is providing is thereby assumed to be constant, but there is a change in the way the demand is met.

Following this assumption about demand driven consumption, changes in how the demand is met may arise if the supply of a product is changed, or if we change demand for a product whose production is constrained (as explained in Step 2).

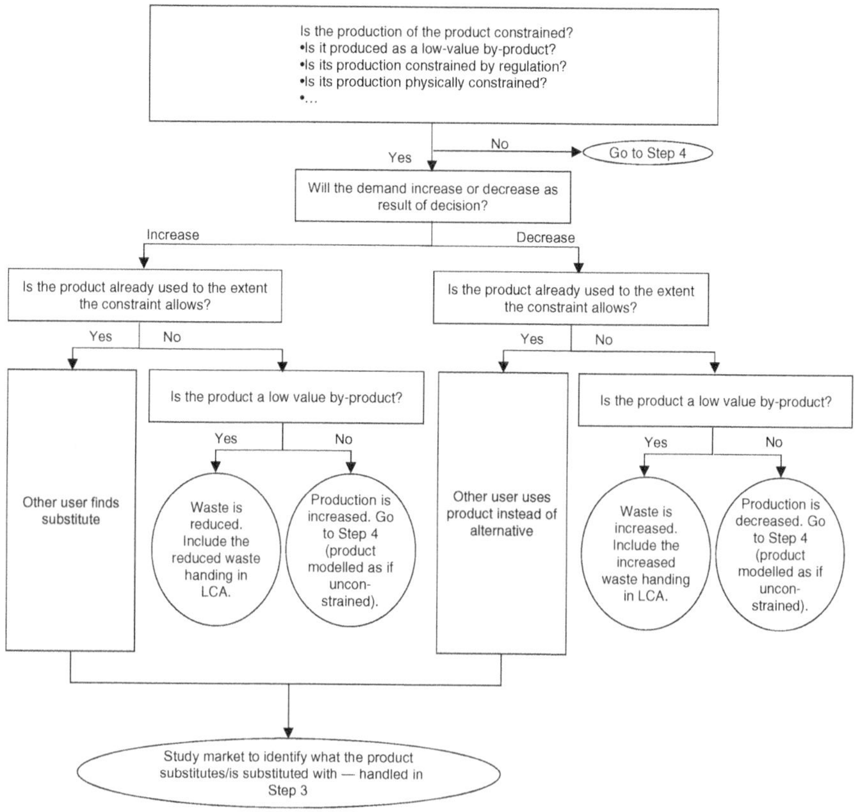

Fig. 9.4 How to identify constrained production and how to handle it

In each of these cases, we need to identify the substitutions that occur in the market, like in the above example where gas substitutes oil. The question that we will address in this step of the consequential LCI is: *"How do we identify which product substitutes which?"*

In order to identify which products can substitute which, there are two aspects that we have to consider:

- The products must deliver the same service(s) for the product user.
- The product working as a substitute has to be available.

Below we will address each of these two issues.

Identifying a satisfying substitute for the product user

A product may provide different services for different users, implying that one product may be a fully satisfying substitute for one user, but completely useless for another. Thus, to identify which product can substitute which, we first need to identify the product user who is likely to find a substitute due to an increase in demand or decrease in supply or who decides to use the product instead of a substitute due to a decrease in demand or increase in supply, i.e. *the marginal user*. However, in reality, identifying the marginal user may be very difficult. Therefore, if a market analysis shows that the product is used in significant amounts for several different purposes, it is advised to make different scenarios for each of these potential substitutions. This can feed into sensitivity and uncertainty analysis of an LCA (see Sect. 9.6 and Chap. 11). In this case, this step should be followed for each of the scenarios.

Having identified the marginal product users and what they use the product for, the next step is to identify what can be used as a substitute for the product by the different marginal users.

Identifying what will be a satisfying substitute for a specific user will in most cases require a large amount of background information about the market where the substitution will take place, and hence involve some elements of uncertainty. However, for a product to work as a substitute, it needs to fulfil the same functions for the user. As outlined in Weidema (2003), these may relate to:

- *Functionality*, related to the main function of the product
- *Technical quality*, such as stability, durability, ease of maintenance
- *Costs* related to purchase, use and disposal
- *Additional services* rendered during use and disposal
- *Aesthetics*, such as appearance and design
- *Image* (of the product or the producer)
- *Specific health and environmental properties*, for example non-toxicity.

Apart from the basic functionality of the product, which can be seen as an obligatory property of the product (see Chap. 8), the importance of these properties will to a large extent depend on the product user. If the product user is a company using the product in its production, the functionality and technical quality will normally be the most important, for some companies accompanied by health and environmental issues. For consumers, on the other hand, issues like aesthetics and image may have a high priority.

It should be noted that there may be not one but several products that work as a substitute for a product. If it is possible to identify the distribution between the alternative product substitutes, the consequential LCI should be based on this. If this is not possible, it may be necessary to develop several scenarios for each of the likely substitutes.

Product availability

Ensuring that the substitute has the necessary functionality, however, is not enough. The substitute also has to be available. A substitute is unavailable if constrained and already used to the extent that the constraint allows. To identify whether the substitute is available, we need to perform parts of Step 2 (included in the decision tree below), which also had as a goal to identify the availability of a product. As the discussion of how to perform this identification is going to be the same as under Step 2, the reader is referred to this section for further explanation.

Figure 9.5 presents a decision tree for identifying product substitutes.

As an additional consideration, it should be noted that in some cases one product will not substitute another directly. For example, the production of biodiesel leads to the co-production of glycerol which contains salts and other impurities. Before it

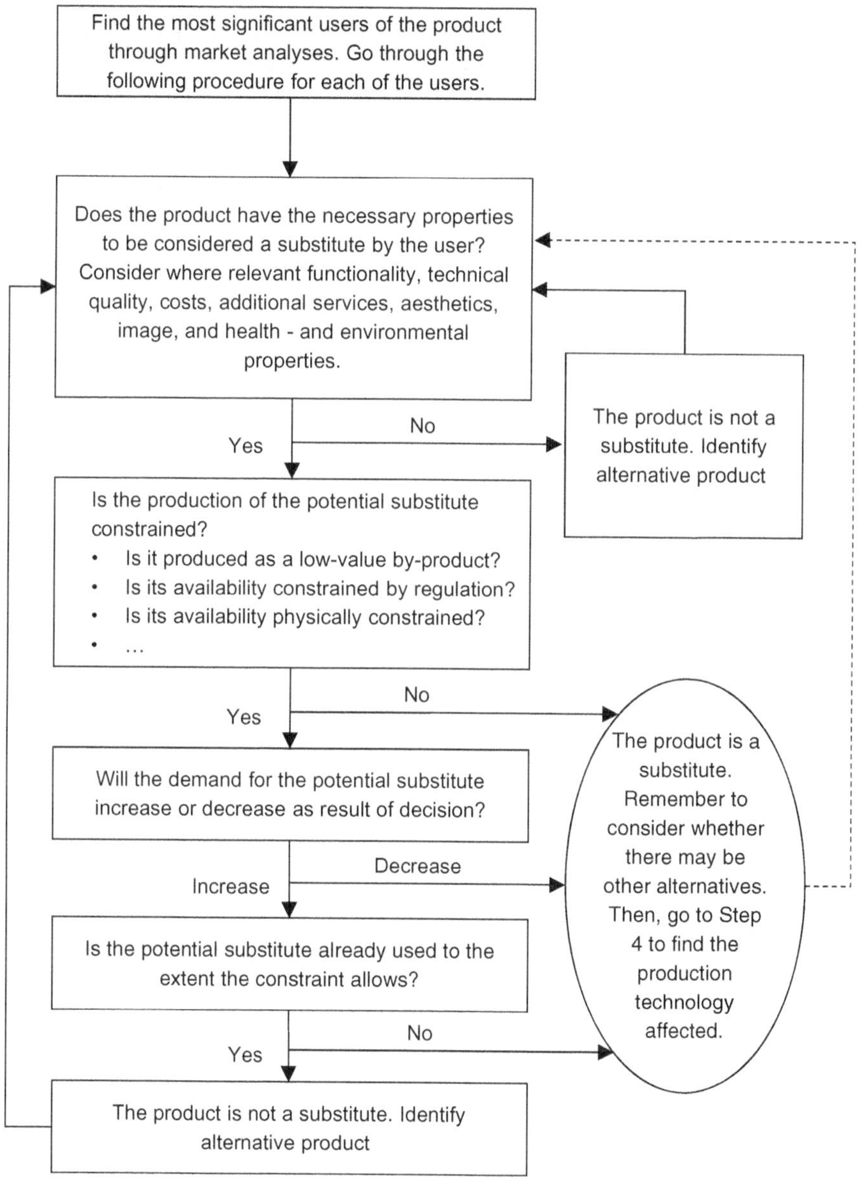

Fig. 9.5 Procedure for identifying possible substitutions of products as consequence of changes in supply or demand

can be sold on the glycerol market, it therefore needs to be distilled. In this case, and in others where additional treatment is needed for the product to be considered as a substitute, these additional treatments need to be included in the LCI. Also, it should be noted that in some cases a product substitution may create a cascading substitution effect (not captured by Fig. 9.5 for simplicity). E.g. if a decrease in demand for product A leads to other users using product A instead of B (substitutes), which is a waste product that is fully used (nothing goes directly to waste management), this can lead to other users using product B instead of C (substitutes), and so on and so forth.

Step 4: Identify production technology affected by change in demand

If the product for which the demand is changed is not limited in supply, it will normally be assumed in a consequential LCI that supply follows demand in a one-to-one relationship. The question is, however, which production technology will be affected by the change in demand. Identifying this technology is the purpose of this step.

In many cases, similar products can be produced with very different environmental impacts. Just think of electricity that may be produced from wind turbines or coal fired power plants. It is therefore in many cases important to identify not only that the production of a certain product will change as a result of the assessed decision, but also to identify as accurately as possible which supplier, and hereby which production technology will be affected by the change in demand.

For doing so, three issues need consideration: The size of the change in demand created by the decision, the trend in the market and whether the assessed decision leads to an increase or a decrease in demand. These issues will be discussed below.

<u>Size of change</u>

When identifying which technology will be affected by the change in demand, it is important to distinguish two different perspectives: The immediate production perspective and the perspective relating to changes in production technologies in the market. Consider the following example of electricity generation: Some technologies cost more to run than others. The production of electricity from gas turbines is, for example, often more expensive than electricity produced from coal. This implies that only coal power will be used, when the capacity of the installed coal power plants is sufficient to cover the demand. However, when the demand increases above what can be supplied by the coal power plants, gas power plants will start to produce. From an immediate production perspective, the concrete technology that will supply the demand will depend on the cost efficiency of the production technologies with available production capacity—the least cost efficient will be only be used to supply peak load.

However, this is only the immediate consequence of the decision. If the electricity consumption in the given market in general is increasing or stable, a decision leading to an increase in demand will push for an increase in the installed power production capacity. In other words, the decision will have an effect on installed capacity. Assume now that the planned implementations of power plants in the market are wind turbines. The long-term effects of increasing the demand will then be a corresponding increased implementation of wind turbines.

The difference between the immediately affected production technology, known as the 'short-term marginal' and the effect on the installed production technology, known as the 'long-term marginal' may be very large—in the example above, the difference was between coal and gas power and wind. It can therefore be a very important decision for the results of the LCA whether the short or long-term marginal is used in the LCI. The general rule has been to use the long-term marginal when the assessed decision is creating large changes in demand, and use short-term marginal when the assessed decision creates small changes in demand. A change in demand is in this context considered small, if it is smaller than the average percentage of annual replacement of capacity (often around 5%, see below). The argument is that these small changes will be part of the general trend in the market and therefore be handled by the trend in the market. The signal they send is therefore considered too small to overcome the threshold for a structural change in production capacity. The difference in the size of changes assumed in the LCA is in fact what makes Situation A and B studies different in the ILCD classification (see Sect. 7.4).

Trend in the market

The electricity example above relates to the situation where the market trend points towards a stable or increasing demand. However, if the market trend is rapidly decreasing, the long-term marginal response to a decision that leads to an increase in demand will not be an increase in the implementation of more wind turbines but rather the continued use of coal or gas power plants that would otherwise have been taken out of operation. In this market, the demand caused by the assessed decision will thereby make the existing least competitive technology stay longer on the market.

The distinction between whether the trend in a decreasing market is slowly decreasing or rapidly decreasing depends on whether the decrease happens below or above the average replacement rate for the production technology. For example, a market trend would be characterised as rapidly decreasing if it decreases by 10% per year, while the average replacement rate for the production technology is 5%. Note that a replacement rate of 5% means that production plants are designed to operate for 20 years, which is quite common, depending on the technologies involved. The reason for making this distinction in market trends is that for increasing, stable or slowly decreasing market trends there is a need for implementation of new production technology, and changes in demand will therefore affect this implementation rate. For rapidly decreasing market trends, however, the decrease is faster than the decommissioning rate for the technology, implying that production plants would be taken out of use before their design life time. In such cases (small) changes in demand will not lead to changes in implementation of new technology (e.g. wind turbines), but merely to the changes in the speed of decommissioning (e.g. coal or gas power plants).

Increase or decrease in demand

The electricity generation example above relates to the situation where the assessed decision leads to an *increase in demand*, and the general trend in the market is either on the increase or decrease. However, the assessed decision may also lead to a *decrease in demand*. If the assessed decision leads to a large decrease in demand in a market with an increasing market trend, the implementation of new technologies will be postponed, implying that existing least cost effective technologies will continue to be used for a longer time.

As showed in the discussions above, there are three aspects that need to be considered, and since each of them has two possible outcomes, there is a total of eight possible combinations. Not all combinations were discussed above, but they follow the same logic. Table 9.1 summarises the discussions above and gives an outline of how to perform the identification of which technology is affected by a change in demand for all eight combinations.

Table 9.1 Identification of the technology which will be affected by a change in demand (i.e. the marginal technology)

	Long-term marginal	Short-term marginal
Decision leads to *increase* in demand	*Increasing* market trend: Implementation of new production technology is promoted—increase in demand is supplied by the production technology *to be implemented* in the context	*Increasing* market trend: Less cost-efficient technology will be used to supply increase in demand—increase in demand is supplied by *least* cost effective technology available on the market
Decision leads to *increase* in demand	*Decreasing* market trend: Decommissioning of least competitive technology is delayed. Increase in demand is supplied by *least* cost effective technology available on the market	*Decreasing* market trend: Less cost-efficient technology will be used to supply increase in demand. Increase in demand is supplied by *least* cost effective technology available on the market
Decision leads to *decrease* in demand	*Increasing* market trend: Implementation of new production technology is delayed. Decrease in demand saves the supply from production technology *to be implemented* in the context	*Increasing* market trend: The least cost-efficient technology is no longer needed because of reduced demand. Decrease in demand saves the supply from *least* cost effective technology available on the market
Decision leads to *decrease* in demand	*Decreasing* market trend: Decommissioning of least competitive technology is promoted—decrease in demand saves the supply from *least* cost effective technology available on the market	*Decreasing* market trend: The least cost-efficient technology is no longer needed because of reduced demand—decrease in demand saves the supply from *least* cost effective technology available on the market
After identifying the production technology affected by the change in demand, go to Step 1 again to address other changes created by the assessed decision, if all changes are not already handled.		

The table shows that, depending on the combination of the three aspects, the marginal technology can either be the least cost effective technology available on the existing market (6 of the combinations) or the future production technology to be implemented (2 of the combinations). In practice, the marginal technology, especially long term, can be difficult to identify, and this is a potential source of considerable uncertainty in the inventory analysis. The importance of this uncertainty may be investigated by sensitivity scenarios for the different potential marginal technologies. Furthermore, it is possible to create a mix of potential marginal processes, which means that the inventory data becomes a mix of data from the different potential marginal processes, as demonstrated in Sect. 9.5. This approach is used in the ecoinvent database in its version 3 (and higher).

Note that in the discussions above, we have mentioned 'the market' as one entity. However, in reality, there may be many markets for one product, e.g. when the product has high transportation costs compared to the value of the product. In cases where there are many small markets for the same product, the market trend has to be identified in the affected local market. For other products where the transportation costs are lower, there may be only one global market. The spatial nature of a market has to be established as a first task when identifying changes in supply.

Secondary consequences and concluding remarks

In the presented 4-step guidance we have only addressed the rather 'direct consequences' of increased or decreased demands and supplies. However, several derived effects or secondary consequences of these direct consequences may be found. Depending on the size of these consequences and the scope of the assessment, these may be relevant to consider. Common for each of them is that they are difficult to foresee and even more difficult to quantify. We therefore cannot establish a general procedure for identifying and quantifying these, more than stating that in-depth knowledge on the topic of concern in most cases will be necessary. A few examples of the types of secondary consequences are given below.

Additional or reduced production of a product may affect market prices for the product hereby affecting the broader demand for the product. For example, if the assessed decision will lead to the increase in the cost of, say, wheat, the behaviour of other consumers may be to consume less wheat due to this increase. It may also be that due to the increase in price, some consumers will begin to use, e.g. corn instead of wheat, hereby increasing the demand for corn.

Changes in market prices may not only affect the consumers but also the producers. In the example above, increases in wheat prices may cause producers to increase the intensity of their production, typically done through increasing the fertiliser use, or through increasing the agricultural area (for more discussions about secondary consequences specifically related to biomaterial production, see Chap. 30). However, it may also be imagined that the increase in price of wheat may cause producers to intensify research related to yield increase, potentially leading to, say, a decrease in area/fertiliser/pesticide use per produced unit of wheat.

Other types of 'secondary consequences' related not to prices of products but to the time consumption of products can also be imagined for some products. For example, a washing machine may lead to significant time savings for the user. The question is then what this time will be used for. In some cases, what is gained in terms of time savings by various household appliances is to some extent used on other 'time-consuming' household appliances, such as TV or videogames. When assessing a washing machine, it may therefore in some cases make sense to include an increase in power consumption from the TV set, or something similar.

As may be obvious from the example above, identifying the secondary consequences will in many cases be very difficult and associated with considerable uncertainties. Furthermore, these effects are typically far from linear and when certain thresholds are passed a complete shift of parts of the market can be the consequence (e.g. the point where the production cost of wind power makes it fully competitive in certain market segments).

Whenever these effects are considered in an LCA, it will often be advisable to make several different scenarios where various realistic possibilities are addressed in order to assess the potential variability of the results (see Sect. 9.6).

However, despite the problems of identifying these secondary consequences, it is evident that if the goal of the assessment is to get as complete an overview of the consequences of a decision, none of these should be omitted a priori, but should be included if at all considered to be practically possible and important for the outcome of the study.

This concludes the introduction and guide to consequential LCA. Readers are invited to consult the Appendix for an example of how to use the 4-step guideline in a case study of the consequences of increasing the supply of biodiesel from poultry fat. As we hope to have demonstrated, consequential LCA is conceptually appealing because it aims to address the consequences of a potential decision. After all, why bother making an LCA study (or paying for one) if its outcomes are not expected to have a consequence on the physical world? We also hope to have demonstrated that the answers to the many questions that need to be addressed throughout the 4-step guide are often associated with large uncertainties. Even advanced economic models generally do a poor job at predicting concrete consequences in markets following some sort of perturbation (consider how global financial crises tend to take also financial analysts by surprise) and simplifying assumptions have to be applied. These uncertainties are one reason why many LCA practitioners prefer an attributional approach. Its use of average process data and frequent use of allocation is theoretically difficult to defend when the goal of an LCA study is to support decisions (i.e. study the consequence of decisions), which is the case for Situation A and B studies in the terminology of ILCD (see Sect. 7.4). Yet, attributional LCA does not suffer from uncertainties related to economic modelling and is preferred by some LCA practitioners for this reason and considered to be 'on average more correct than consequential LCA'. This is also part of the reason why ILCD recommends an attributional approach even for goal situation

A where the LCA supports a decision but the scale is small and market elasticities make identification of the marginal product or technology uncertain in many cases.

9.3 Planning and Collection of Data

Based on the scope definition and the processes identified to belong within the system boundaries, the collection of data for these processes has to be planned and carried out. The planning has the purpose of balancing the effort of data collection by the relevance of the respective data and information. This is essential in order to avoid wasting time on collecting high-quality data that have a low relevance for the LCA results and/or spend too little time on collecting high-quality data where it is highly relevant for the results. Planning and collection of data are iterative processes, which is why they are addressed together in this section. These processes are an integrated part of the iterative approach to LCA that also involves the calculation of LCIA results. For example, the first iteration of LCIA results may guide the practitioner about which data are particularly relevant to focus on in a second iteration.

As starting point for data collection, we encourage practitioners to create a table that outlines a plan for the data collection for each process or single data point, see template in Table 9.2 (elements of the table are explained below). Note that the data eventually collected by the practitioner will often diverge from the initial plan due to unforeseen limitations and results of early iterations of LCIA phase that may lead to changes in the data specificity that the practitioner aims at for each individual process or single data point. The table can therefore be adapted accordingly at each iteration and be used in its final version (i.e. final iteration of the study) to document the metadata behind the LCI data (see Sect. 9.7).

The initial planning should be based on the requirements to data representativeness from the scope definition, as well as on the efforts that are expected in order

Table 9.2 Template for planning and collection of data

Process or single data point	Specificity					Type	Source	Access
	Very high	High	Medium	Low	Very low			
X	X					Concentration	Process engineer	Questionnaire
Y		X				Kg/year	Academic paper	Online search
Z				X		Unit process	ecoinvent	Database search

The structure of the table can follow life cycle stages of the product. Based on Wenzel et al. (1997)

to obtain data of a given quality. Data quality is here classified into one of five categories of data specificity shown in Table 9.3.

The efforts required to obtain data of a given quality can be estimated for each data point (e.g. a flow quantity) by considering three additional dimensions of the data in Table 9.2: data type, data source and data access. Examples are given for each of these in Table 9.4. The following sub-sections are structured according to the collection of data for each of the five data specificity levels and address challenges that the LCA practitioner commonly faces for each of the three dimensions of Table 9.4.

Table 9.3 Classification of data specificity (inspired by Wenzel et al. 1997)

Data specificity	Explanation
Very high	Measured directly at specific process site or scaled from measurement
High	Derived from measurements at specific process site via modelling
Medium	LCI database process or data from literature specific to actual process, e.g. according to best available technology standard or country average. Specificity may be improved by modifying a process with site-specific data
Low	Generic LCI database process or data from literature, e.g. covering a mix of technologies in a country or region
Very low	Judgement by expert or LCA practitioner

Table 9.4 Three dimensions influencing the effort required to obtain data

		Examples and notes
Data type	Complete unit process	Includes all flows scaled to 1 unit of reference flow for process
	Individual flow to/from process per unit of time	*X* kg/year, covers elementary flows and other flow types
	Technical or geographic parameters	Process pressure, temperature, soil pH, precipitation
	Concentrations	*X* g/m^3 flue gas or waste water to treatment
	Quantities of products bought per year	*X* kg steel of specified grade (i.e. material flow to process)
	Use characteristics	Temperature of clothes washing, driving pattern of car
	Sector statistics	Sector-average data
	Economy-wide statistics	Infrastructure data, trade data
Data source	*Experts internal to commissioner*	
	Process engineers	Flow data on internal processes
	Purchasing department	Supplier data
	Research and development or design	Data on product concepts, not yet marketed
	Experts external to commissioner	
	Researchers	Expert in relevant technological domain

(continued)

Table 9.4 (continued)

		Examples and notes
	Consultants	Person having long experience with conducting similar studies
	Industry representatives	Person with broad overview of relevant industry
	Public	
	Other LCA studies	Academic literature, reports commissioned by companies
	LCI databases	ecoinvent, LCAfood
	LCI models	PestLCI
	Company CSR reports	Mentioning of key environmental figures
	Industry association reports and databases	Volumes produced, average elementary flows
	Legal documents	Details on best available technologies, regulatory thresholds
	National or supranational statistical agencies	Mixes of waste treatment, transport, energy, etc.
	Consumer organisations	Average life time of products
Data access	Online search	Google, databases, websites
	Questionnaire	Employees at commissioning company or suppliers
	Direct dialogue	Physical visits to site, email or telephone contact
	First-hand gathering by LCA practitioner	Measurements at site with own equipment

The points listed under each dimension are illustrative and not exhaustive

9.3.1 Very High and High Data Specificity

The data type to be prioritised is always complete unit processes, because these form the basis of the LCI results. However, for very high and high data specificity, complete site-specific unit processes often do not exist and therefore must be constructed by the practitioner from single data points.

For very high specificity, these data points are directly measured input and output flows, i.e. elementary flows from/to the ecosphere and other flows from/to other processes in the technosphere. Ideally, elementary flow data should be gathered in the physical unit matching the characterisation factors to be applied in LCIA (usually 'kg') per specific reference flow of the unit process (usually the primary product output). For a CO_2 emission (i.e. elementary flow) from an electricity generation process, this would mean an amount of kg CO_2 per kWh electricity produced. Often, a directly measured flow will not be available in this form, but rather as a quantity per unit of time (e.g. kg per year). In this case, the flow needs to be scaled to one unit of reference flow. Figure 9.6 shows an example of how to do this in practice.

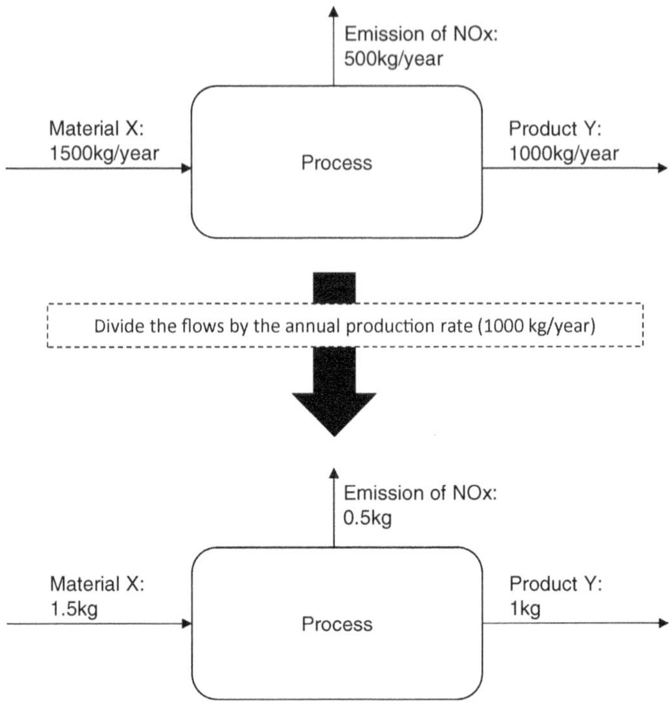

Fig. 9.6 Example of the scaling of three annual flows to one unit (kg) of reference flow (product *Y*)

Often, a company will not possess all the relevant data required for a unit process, due to the cost of systematically measuring all inputs and outputs. When direct site measurements of flows are not available, the flows can be modelled from other site-specific data, in which case the data quality is high, as opposed to very high. Such other site-specific data can be the concentration of pollutants in effluents (typically wastewater or flue gas). Figure 9.7 shows an example of how to calculate copper emissions to untreated wastewater from the concentration of copper in the wastewater.

Another approach is to calculate output flows from site-specific measurements of input flows using a mass balance. Since unit processes, in general, do not gain or loose mass (or energy) over time, the mass of inputs should equal the mass of outputs. So, if a company consumes 10 m^3 of natural gas per year, the CO_2 emissions can be estimated from the mass of natural gas (calculated using its density) and the stoichiometry of the combustion reaction (natural gas is mainly composed of methane, CH_4). A mass balance approach can also be applied to modelling at the level of elements. If for example, a company consumes 950 g copper per unit of reference flow, but one unit of reference flow only contains 928 g copper, then the remaining 22 g per unit of reference flow must leave the process as

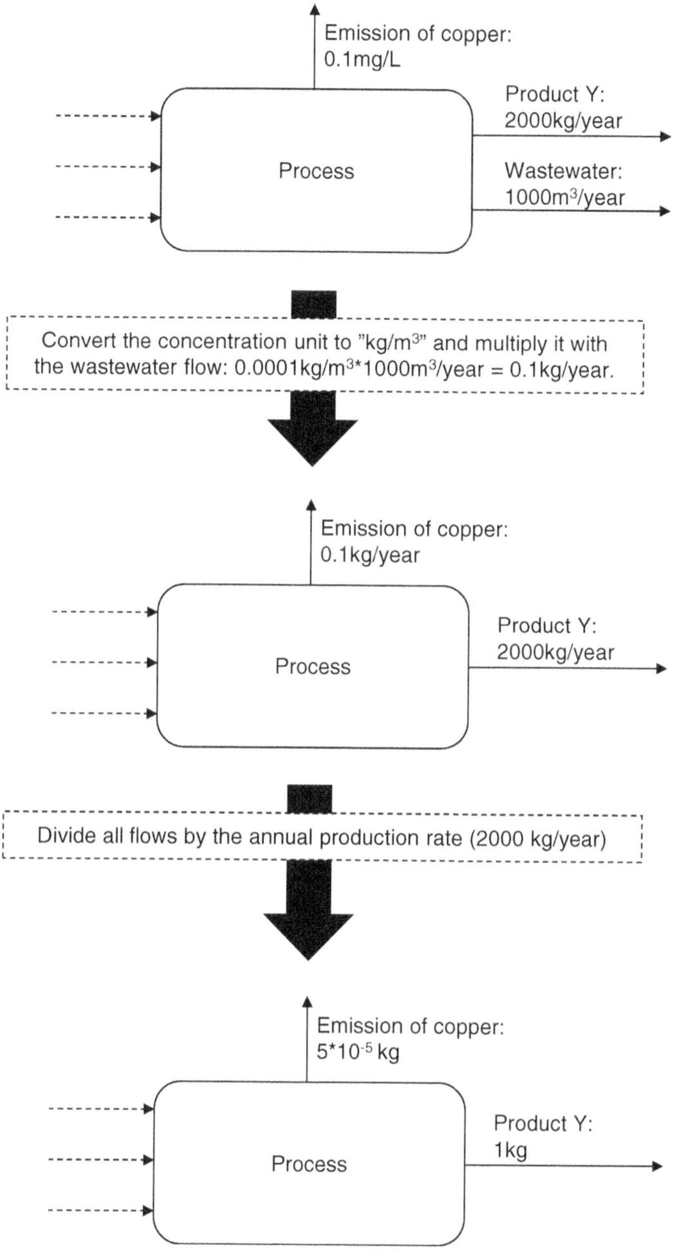

Fig. 9.7 Example of the calculation of an emission per reference flow (1 kg of product *Y*) from a wastewater concentration. *Dotted arrows* indicate input flows not considered in the example

a waste flow to treatment or as an emission. In many cases, a simple back-of-the-envelope mass balance calculation will not suffice, because the relationship between input and output flows is complicated and dependent on many parameters, and it is more appropriate to apply dedicated LCI models. For example, the LCI model PestLCI (Birkved and Hauschild 2006) calculates emissions of pesticides from field applications via different routes (e.g. evaporation, air drift, emissions through drainage pipes and groundwater leaching) based on the application of a specific pesticide to the field, its physical and chemical properties, and a large number of context-specific parameters, such as crop type, time of application, soil pH and slope. Note that the data specificity obtained from LCI models can only be characterised as high if all inputs and parameters are in fact site-specific (and when relevant, time-specific). If this is not the case, the data specificity is lower.

High and very high specificity data (e.g. on elementary flows such as CO_2 emissions and freshwater use) are sometimes available in reports, e.g. 'green accounts' or CSR reports, published by the company operating the process of interest, but often the source of such data is employees working with or operating the process. These may be process engineers monitoring flow data as part of their daily routine, or they may work in the purchasing department and thereby have knowledge about the amounts of input flows (materials and energy) purchased and the identify of suppliers. The latter may be used to contact suppliers for data specific to processes at their sites and the procedure can, in principle, be repeated several times to obtain company internal data further upstream in the foreground system.

Company internal data may be accessed by asking the employees to fill out questionnaires combined with a physical visit to the site, email or telephone contact. This way of obtaining data can be straightforward or require lots of effort depending on the willingness of the employees possessing the data to share them in a relevant format. From our experience, this willingness is generally higher when the commissioner of a study is part of the same company and department as the employee holding the data or if the LCA study has been given attention by the management level in a company. It should be noted that company internal data are sometimes confidential. In some cases, they are not possible to obtain, but in other cases the confidentiality issues may be handled by the LCA practitioner signing a non-disclosure agreement and reporting any confidential data of importance to the study in a special appendix to the report that is only accessible to a selected group of people (typically including members of a peer review panel if the study is peer reviewed).

9.3.2 Medium and Low Specificity Data

For reasons given above it is in practice rarely feasible (nor necessary) to obtain all foreground system data from site measurements, i.e. with high or very high specificity. A large part of the data collection therefore usually takes place online by searching, identifying and accessing publicly available sources, such as other LCA studies, industry association reports and national statistics. It is also possible to identify, via online searching, data for a process that is very similar to the actual process to be modelled, either because the reference flow of the processes is the same (e.g. the incineration of polypropylene) or similar (e.g. the incineration of polypropylene versus polyethylene). The strategy of extrapolation from data for similar processes is especially useful to 'fill out gaps' in a preliminary unit process, but the LCA practitioner must carefully check the representativeness of the process used for extrapolation. For example, if the initial data collection effort has led to a handful of high or very high specificity emission data, but no resource inputs for a process, the remaining flows may be quantified by extrapolation from a similar process. Such similar process can be sourced from scientific papers or other sources, which can document sufficient representativeness (technology, geography, time) and disclose sufficient data to check the agreement with the existing handful of high specificity emission data for the original unit process. A special case of extrapolation is for novel technologies that may not yet operate at industrial scale anywhere at the point in time where the study is to be conducted. Here, an obvious source of extrapolation is laboratory scale processes. It is, however, important to consider how the relationships between the flows of a process changes from laboratory to industrial scale. Often the technology of the process will change, not just in size, at the upscaling from lab scale to commercial scale, and this typically leads to increased efficiency (e.g. less input per reference flow output) and changes in the quality of flows.

The effort required to access data via online searching depends on the expertise of the practitioner (e.g. familiarity with the terminology of the concerned technical domain) and on how well-studied the phenomena behind the data is. For example, there is generally more publically available data on greenhouse gas emissions than on emissions of synthetic chemicals used for very specific industrial purposes and produced in low volumes. The effort also depends on the number of data points that can be accessed from each source. A unit process is often composed of more than 100 flows (the majority often being elementary flows). Some sources, such as LCI databases, contain data for all flows making up a unit process, while other sources, e.g. statistical agencies, may only cover a few elementary flows.

LCI databases are used to source data for the background system and for the parts of the foreground system where more specific data can or will not be obtained. Table 9.5 presents a non-exhaustive list of LCI databases.

Table 9.5 List of process-based LCI databases (not exhaustive)

Name	Description	References
ecoinvent	Swiss database that contains approximately 12,500 processes (version 3) organised under different themes like transport, energy, material production, agriculture, etc. All processes are available as unit- and system-processes and all processes are documented in detail. Updated regularly	ecoinvent; www.ecoinvent.org
ELCD	Database of the JRC of the European Commission, contains more than 300 datasets on energy, material production, disposal and transport	Joint Research Centre of the European Commission; eplca.jrc.ec.europa.eu/ELCD3/index.xhtml
Agri-footprint	A comprehensive LCI database of feed, food and biomass, containing around 3500 products and processes	Blonk Consultants; www.agri-footprint.com
LCA Food	Danish database containing more than 600 data sets on basic food products and related processes from agriculture, aquaculture, fishery, industry, wholesale and supermarket, including waste treatment processes	2.-0 LCA Consultants and Aarhus University; www.lcafood.dk
Swedish National LCA database	Contains more than 500 well-documented LCI data sets in SPINE format for a wide range of industrial processes and household goods and services	Competence Centre for Environmental Assessment of Product and Material Systems of Chalmers University of Technology; cpmdatabase.cpm.chalmers.se
GaBi databases	Separate databases mainly based on primary data collection. Cover sectors from agriculture to electronics and automotive industries, textiles and retail, through to services. Contains more than 10,000 Life Cycle Inventory profiles	GaBi; www.gabi-software.com/international/databases/gabi-databases/
LC-inventories	Over 1000 process data sets, which are corrections, updates or extensions of ecoinvent v2.2 database, created by ESU-Services and other authors	The Swiss Federal Office for the Environment and ESU-services; www.lc-inventories.ch
NEEDS	Database designed for long-term environmental assessment. Contains around 800 processes of future energy supply systems, future material supply, and future transport services	Members of a European research project; www.needs-project.org/needswebdb/index.php
NREL	US-American database with around 300 datasets related to the production of materials, components, or assembly in the U.S.	National Renewable Energy Laboratory; www.nrel.gov/lci
ProBas	Comprises more than 8000 datasets on energy, material production, transport and disposal, different data sources and	German Federal Environmental Agency; www.probas.umweltbundesamt.de/php/index.php

(continued)

Table 9.5 (continued)

Name	Description	References
	data quality. Focuses on processes within Germany	
LCA Commons	More than 18,000 datasets for U.S. agriculture production and agriculturally derived products	USDA; www.lcacommons.gov
Ökobaudat	German database with around 950 environmental product declaration datasets for building materials, building processes and transport processes	Federal Ministry for the Environment, Nature Conservation, Building and Nuclear Safety; http://www.oekobaudat.de/en.html

While a number of LCI databases are available and some of them contain high-quality data for specific technologies or industries as shown in Table 9.5, the most comprehensive, and probably most widely used, database is ecoinvent and in the following section we there focus on this database and encourage the reader to look for similar information about other databases using the references given in Table 9.5 as relevant. ecoinvent version 3 contains approximately 12,500 unit processes and each process exists in an 'allocation, default' (or APOS: allocation at the point of substitution), an 'allocation, recycled content' (or 'cut-off') and a 'consequential' version. The 'allocation, default' version uses price as allocation key as a rule, except for a few processes, where representative physical parameters are used (such as for processes involving co-production of electricity and heat) where markets are judged distorted by, e.g., regulation, and also corrects for fluctuating prices by applying three-year, historical average prices for some processes (Weidema et al. 2013). The cut-off version is identical to the default allocation version, except for the handling of recyclable materials that are cut-off before being sent to recycling. This means that recyclable materials do not bring any benefits to the primary user of the materials and are considered available 'burden-free' to recycling processes, and that the impacts attributed to secondary recycled materials are only those of the recycling processes and the associated transportation. By contrast, in the default allocation version secondary recycled materials are also allocated a share of the material's previous life cycle impacts (based on economic allocation). The existence of the two allocation approaches for recyclable materials in ecoinvent ('default' and 'cut-off') reflects the fact that there is little consensus on how to perform such allocation in the most reasonable way. The cut-off allocation is the recommended approach in the European Product Environmental Footprint guideline (EC-JRC 2012).

The consequential version of ecoinvent uses the long-term marginal technology, which is identified by considering whether a market is increasing (or stable, or slowly decreasing) or rapidly decreasing, in line with Table 9.1. The ecoinvent centre advocates the use of the consequential version of the database not only for large-scale decisions (studied in Situation B studies, according to ILCD), but also for small-scale decisions, which are by definition too small to cause structural changes outside the foreground system, i.e. too small to lead to new equipment being installed (increase in production capacity) or existing equipment being prematurely taken out of use (decrease in production capacity). Yet the ecoinvent centre argues that the consequential version of the database (which is based on the long-term marginal technology) is "applicable to study the effect of small, short-term decisions, since each individual short-term decision contributes to the accumulated trend in the market volume, which is the basis for decisions on capital investment" (Weidema et al. 2013). In relation to the ILCD-defined decision context situations, the ecoinvent 3 database can, strictly speaking, only be used to model consistently the parts of Situation B studies involving structural changes (using the consequential database) and Situation C2 studies (using the allocation default or cut-off database). However, as noted in Sect. 9.2.2, economic allocation is often the only practical solution to multifunctionality, irrespective of decision context. We therefore advise that one of the two allocation versions of ecoinvent is used for Situation A, B (only non-structural changes), C1 and C2. However, the LCA practitioner should check for any multifunctional processes that have high contributions to early iteration LCA results and, where appropriate and technically feasible, manually change the multifunctionality solution in accordance with the scope definition of the study to test its influence on LCA results.

Whenever data is sourced by online searches or LCI databases it is important to pay attention to the available metadata describing the characteristics and conditions of the process to evaluate how representative the data is for the actual data needed. Metadata usually specifies the exact technology (or mix of technologies, in the case of average or generic data) involved in a process, the location (e.g. country) of the unit process, the time during which the data applies and relevant operating conditions (e.g. climate). The metadata allows distinguishing between medium and low data specificity (see Table 9.4). Relevant metadata for foreground processes should be reported by the LCA practitioner (see Sect. 9.7) and furthermore considered in the later sensitivity analysis and uncertainty management (see Sect. 9.6).

When using a unit process from an LCI database in the foreground system it is preferable to adapt it to make it more representative of the actual process to be modelled to the extent that this is possible (see Sect. 8.7). One improvement of the representativeness that is usually possible is to manually change the electricity grid mix that fuels the process to a mix that matches the geographical and temporal scope of the study. Note that such adaptation is not possible if a unit process is 'aggregated', meaning that the elementary flows of all processes upstream and downstream have been aggregated, so the reference flow is the only output of the aggregated process (or input, in the case of waste treatment processes) apart from the elementary flows.

Aggregated unit processes are often preferred for constructing the background system because the LCA practitioner only needs to include the aggregated processes that link to the foreground processes of an LCI model.

9.3.3 *Very Low Specificity Data*

If efforts to obtain data have been fruitless, one may rely on expert judgement. People may qualify as experts if they are knowledgeable in the technical domain relevant for the data (e.g. plastic moulding) or if they have conducted similar LCA studies themselves in the past. If no expert is available, a last resort is to use a 'reasonable worst case' for the calculation of the first iteration of LCA results. A reasonable worst case value may be derived from knowledge of similar or related processes or from correlation or calculation from other flows of the process or other processes. The results will then show if the data is potentially important or negligible (judging against the cut-off criteria identified in the scope definition). In the first case, the practitioner may try again to obtain data of better quality or address the issue in the interpretation of results. In the latter case, the reasonable worst case data may either be kept in the model or removed. Whatever option is chosen it should be reported (see Sect. 9.7) for the sake of other LCA practitioners wanting to use (parts of) the inventory model in future LCA studies.

9.4 Constructing and Quality Checking Unit Processes

The data that is collected should represent full operation cycle of the process, including preparatory activities like heating, calibration (with potential loss of materials and products as scrap), operation, idling, cleaning and maintenance. It should also take into account typical scrap rates during operation. This means that the data collection should be based on a longer period of operation, ideally covering several production cycles, perhaps one year's production. Sometimes also the impacts from the manufacturing and end-of-life stage of the production equipment are important and then they should also be included in the data collection. When the data has been collected, it is time to construct unit processes. As mentioned, the type of data collected can vary (see Table 9.3) and it is important to ensure that all the data has the right format for a unit process. To reiterate, all data must be in the form of flows. Elementary flows must be in a unit that matches that of the characterisation factors to be applied ('kg' in many cases), and all flows should be scaled to 1 unit of the reference flow of a unit process (see Figs. 9.6 and 9.7). Note that unit processes obtained from LCI databases already have the right format and are therefore ready to incorporate in an LCI model (see Sect. 9.5).

9.4.1 *Quality Check of a Unit Process*

When constructing unit processes there is a risk that they are incomplete and that there are errors in the flow quantities. Incompleteness may be caused by the fact that some flows are not monitored or reported. Errors in flow quantities may be caused by errors in reported measurements (e.g. a technician writing 'g' instead of 'mg') or errors in the calculation of flows and conversions of units (e.g. if one had forgotten to convert the concentration unit in the example of Fig. 9.7). To avoid (critical) incompleteness and quantitative errors, constructed unit processes should be checked before they are used in an LCI model. Such a quality check can be supported by calculation and interpretation of first iteration LCIA results, e.g. through the identification of the most contributing process and substances.

Completeness of flows

There are three complementary approaches for validating the completeness of flows.

1. Knowledge of similar processes can help identifying potentially missing flows. For example, the LCA practitioner may suspect one or more missing flows, if a unit process for a specific paper production process contains no chlorine containing compounds in the wastewater to treatment and the practitioner knowns from previous experience that chlorine compounds are typically present in the effluent of paper production processes.
2. Knowledge of the nature of a physical transformation in a process can hint what emissions or waste flows to treatment may be missing. For example, NO_x gases are known to be formed whenever a combustion process occurs in the presence of nitrogen, the major constituent of atmospheric air. Filters can capture large fractions of generated NO_x before it becomes an emission, but usually not every single molecule.
3. A qualitative comparison of input and output flows can show if there is disagreement between the elements entering a process and the elements leaving a process. For example, a process cannot emit large quantities of CO_2, without inputs of carbon sources in the form of fossil fuels (e.g. coal, natural gas or oil). While using this validation technique it should be kept in mind that some flows entering and leaving a process are elementarily heterogeneous. For example, mercury is a common emission from the combustion of coal due to the mercury content (typically in the order of 0.00001%) of the coal entering the process as a heterogeneous material flow. In this case, the mercury input is 'hidden' in the coal input and it would therefore be wrong to assume that a homogenous input of mercury is missing on account of the emission of mercury.

Flow quantities

A unit process should obviously not only contain the right flows, but also the right quantities of these flows. A number of validation approaches exist for checking flow quantities.

A mass balance is a universal approach because the sum of flows entering a process should amount to the same number as the sum of flows leaving a process since no accumulation occurs inside the process. A mass balance is therefore an efficient way of spotting errors, for example if the mass of outputs is on the order of

1000 times the mass of inputs. Note, however, that flow quantities may be correct, even if the law of conservation of mass seems to be violated. This is because most of the constituents of atmospheric air, e.g. oxygen and nitrogen, are generally not counted as resource inputs in unit processes, in which case the mass of outputs appear larger than the mass of inputs (e.g. due to combustion products such as CO_2, H_2O and NO_x). A mass balance can also be applied at the level of individual elements, but one should be aware of 'hidden' elements in heterogeneous flows, as described above. Energy balances can in principle also be used as a validation approach, but this would require calculations of the chemical energy stored in inputs and outputs and quantification of heat lost to the environment, which is often not reported as an emission in a unit process.

Following validation based on mass balance a complementary validation based on stoichiometry can be carried out if the process to be validated involves one or more chemical reactions. This serves to check if the ratio between inputs and outputs involved in a chemical reaction is correct. For example, stoichiometry gives us the correct ratio between inputs and outputs in the electrolysis of water in the presence of sodium chloride: $2NaCl + 2H_2O \rightarrow 2NaOH + H_2 + Cl_2$. The mass (g) of each molecule can then be calculated by multiplying its stoichiometric coefficient (mole) and its molar mass (g/mole).

Other validation approaches rely on comparisons to external information. This could be information for similar processes that are expected to contain flows of similar magnitudes as the process to be validated. The external information could also be legal limits. For example, if an emission of nitrogen dioxide corresponds to 100 times a regulatory emission limit, it is a strong indication that there is an error in the emission quantity (note however that many regulatory limits are given as concentrations rather than mass flows, in which case a conversion is needed).

Yet another validation approach relies on the first iteration of LCIA results. These are useful for identifying erroneously high flow quantities. For example, if the contribution from a single elementary flow of a single unit process contributes with 99.9% of the impact for an impact category, this is a strong indication that the flow quantity is too high (e.g. due to a factor 1000 unit conversion mistake in a calculation or data entry in the LCA software). This validation approach can also be used to check for mistakes in the ID of an elementary flow, such as mistakenly using the name 'dioxin' for an emission of 'carbon dioxide' (dioxin is a group of extremely toxic chemicals).

9.4.2 Using Flow Names Compatible with LCA Software

To prepare a unit processes for use in an LCI model it is important that the LCA software used 'understands' the identity of the flows of the unit process. If this is not the case, a flow cannot be linked correctly to other processes or characterisation factors (in the case of elementary flows). There have been attempts at harmonising flow names across LCI databases and LCA software, but the LCA practitioner should always check the flow nomenclature of the software used (e.g. SimaPro, GaBi or OpenLCA) and follow this when naming the flows of constructed unit processes. Unit processes of LCI databases (see Table 9.5) are commonly integrated into LCA software, which ensures that their flow names are correct.

LCA practitioners may face a situation where an LCA software has no name for a given elementary flow or the CAS-number (Chemical Abstract System number—a unique identifier for a chemical) of an emitted chemical does not exist in the list of flow names in a software. In this case, the LCA practitioner should check if there is a characterisation factor (CF) for the chemical in the LCIA method to be applied in the ensuing LCIA step. If this is the case, the LCA practitioner should create a new flow in the LCA software with a name identical to the name of the CF, so the software can create the link. If there is no CF, the LCA practitioner can either calculate the CF on his/her own when guidelines to do so exist (e.g. for the USEtox model; see Chap. 40) or discuss the potential of that substance to contribute to the total environmental impact and to the resulting interpretation of the results. In the case of missing flows that are not elementary flows, these should also be created in the LCA software and used to link processes together. For example, in the case of a waste to treatment flow that is specific to the studied system (part of the foreground system), and therefore not existing in the LCA software, this flow should be created in the software and used to link the process having it as an output to the most appropriate waste treatment process that is available.

9.5 Constructing the LCI Model and Calculating LCI Results

When all unit processes have been constructed or collected from LCI databases the LCA practitioner can construct the LCI model. Each unit process can be seen as a 'building block' in the LCI model, the 'size' of which is ultimately decided by the study's reference flow derived from the functional unit in the scope definition (see Chap. 8). This is because the reference flow decides the quantity required of each unit process-specific reference flow. In other words, each unit process must be scaled to fit the LCI model. Figure 9.8 shows an example of how this is done manually for a simplified system composed of just three unit processes each having just 4 flows.

In Fig. 9.8, Process 1 is first scaled to match the reference flow of the study (100 kg of Product *X*). After the scaling of Process 1, 200 kg of Product *Y* is required, which Process 2 is scaled according to. This means that 240 kg of Product *Z* is required, which Process 3 is scaled according to, etc. In practice, LCA software can carry out the scaling automatically for the practitioner, when told what the reference flow of a study is.

In practice, inventory modelling is normally performed using a dedicated software which supports both the building of the product system model, connecting the

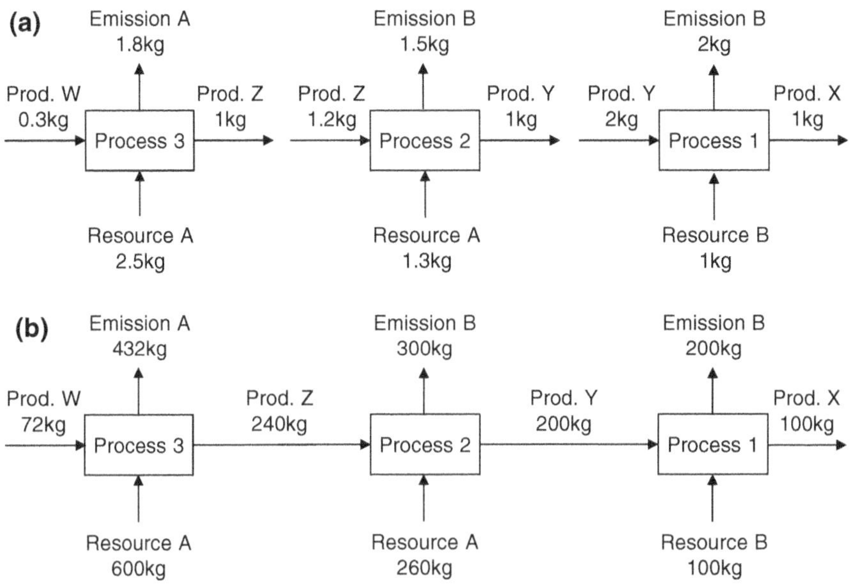

Fig. 9.8 Three simplified unit processes unconnected (**a**) and connected (**b**) based on a study reference flow of 100 kg of product *X* (the reference flow of process 1)

Table 9.6 Software for performing LCA (non-exhaustive list)

Name	Information
SimaPro	Pré Consultants; www.pre-sustainability.com/simapro
GaBi	Thinkstep; www.gabi-software.com/international/index/
OpenLCA	GreenDelta (open access); www.openlca.org/
Umberto	Ifu Hamburg; www.ifu.com/en/umberto/

relevant unit processes; the linking to available unit process databases and storing of own processes, and the linking of elementary flows in the inventory results to the relevant characterisation factors for the life cycle impact assessment. Table 9.6 shows some of the widely used software for LCA

9.5.1 *Database and Software Specific Aspects*

As mentioned in Sect. 9.3.2 processes from LCI databases exist in disaggregated and aggregated versions, the difference being that the latter scales all processes upstream and downstream according to the reference flow of the process and aggregates their elementary flows, so that the only output of the aggregated process (or input, in the case of waste treatment processes) that is not an elementary flow is its reference flow. In practice, some LCI databases only provide aggregated processes, which means that it is not possible to modify them to increase their representativeness for the study. When aggregating processes, the LCI database providers have made choices on how to handle multifunctional processes and how to cut-off the life cycle of the process' reference flow because including all processes is not practically achievable in process-based LCI modelling (see Sect. 9.1). These choices also relate to solving the issue of closed loops between processes, which occurs if two processes need each other's outputs as inputs. This issue is commonly solved by matrix inversion (Heijungs and Suh 2002).

The way system expansion is performed in the construction of the inventory model depends on the LCA software used, but it is usually simple to implement for the LCA practitioner. For example, in GaBi it is performed by connecting the avoided process as input but with a scaling factor of −1 so that it is computed negatively as a crediting. In SimaPro, system expansion is performed by making a direct link to the avoided process in the flow category 'Avoided products', and the software automatically accounts for it negatively when processing the assessment. In OpenLCA, which is a free LCA software, system expansion is modelled as an avoided output of a unit process, in practice marking an output flow as 'avoided product' by checking a mark in the process.

9.5.2 *Calculation of LCI Results*

The LCI results are the compilation of elementary flows over all the processes that are part of the LCI model (scaled to the reference flow of the functional unit). For the simplified product system in Fig. 9.8 the results would simply be the sum of each of the resources and emissions across all the processes, see Fig. 9.9 describing final LCI results.

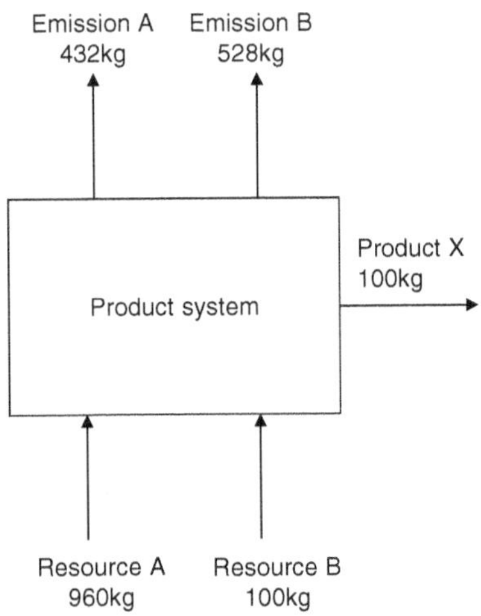

Fig. 9.9 LCI results for the product system in Fig. 9.8. The aggregated elementary flows of product *W* (72 kg), that are not shown in Fig. 9.8, are 100 kg of resource A and 28 kg of emission B

In practice, the number of flows and processes is normally huge, but no manual work is typically required from the LCA practitioner as the LCA software can calculate LCI results for a product system with one click of a mouse button. Such LCI results are the basis for the subsequent life cycle impact assessment phase (unless the goal of a study is to simply calculate the LCI results).

9.6 Data Needs for Uncertainty and Sensitivity Analysis

Uncertainty and sensitivity analysis is important for the interpretation of LCIA results because they can inform the LCA practitioner on how robust the conclusions of the study are and where future studies should focus to make results even more robust. Chapter 11 is dedicated to these matters and details the theoretical background and the practical use of uncertainty and sensitivity analyses. The following describes the data that needs to be collected during the inventory analysis as inputs for uncertainty and sensitivity analyses.

Uncertainty analysis allows for the quantification of uncertainties of the final result, as a consequence of the uncertainty of each parameter in the LCI model. To enable an uncertainty analysis, the practitioner must, for quantitative parameters in the foreground system, collect information on their statistical distribution (e.g. normal, log-normal or uniform) and corresponding statistical parameter values (e.g. mean and standard deviation for normally distributed parameters).

Sensitivity analysis allows for systematic identification of the parameters that have the highest influence on the LCIA results. The influence of parameters on results is calculated by changing them, one by one, and observing the changes in results. These changes in parameters should reflect uncertainties about the actual product system modelled. For quantitative parameters in the foreground system, the practitioner should aim to collect minimum and maximum values, or a low and a high percentile (e.g. 2.5th and 97.5th) when a parameter's statistical distribution is known (see above), in addition to the default value that is used in the LCI model. For example, a specific farmer may on average apply 2 kg of a specific pesticide to produce 1 tonne of potatoes, but this number may vary from 0.5 to 3 kg, depending on weather conditions. For discrete parameters or assumptions in the foreground system the practitioner should develop a number of sensitivity scenarios. For example, a part of the product system may be located in a different country than assumed in the LCI model and a sensitivity scenario would thus involve differences in energy mix, waste treatment technologies, etc. Note that the data requirements for sensitivity and uncertainty analysis overlap and data collection can therefore be performed in parallel by the practitioner.

It often takes more time to collect sensitivity and uncertainty data for some parameters in the foreground system than for others and it may not be necessary to collect data for all processes, depending on the outcome of the first iteration of the analysis. For example, if a process is found to contribute to less than 0.1% of total impacts, then its sensitivity and uncertainty data should generally not be a high priority as illustrated by Fig. 12.3.

For the background system, many LCI databases include uncertainty information on processes, which can feed into uncertainty and sensitivity analysis in LCA software. The practitioner therefore needs not to bother about such data in the inventory analysis.

9.7 Reporting

The reporting of the inventory analysis should contain six elements:

1. Documentation of LCI model at system level.
2. Documentation of each unit process.
3. Documentation of metadata.
4. Documentation of LCI results.
5. Assumptions for each life cycle stage.
6. Documentation of data collected for uncertainty and sensitivity analysis.

Elements 1, 2 and 3 should allow the reader to recreate the LCI results, which are documented in Element 4 (i.e. exigence of reproducibility of the study). Element 5 should allow the reader to judge the reasonability of all assumptions performed (i.e. exigence of transparency) and Element 6 should allow the reader to recreate the uncertainty and sensitivity analysis (exigence of reproducibility and consistency). Below we elaborate on each element and we further refer to the illustrative case on window frames in Chap. 39 for an example of how the inventory analysis may be reported.

9.7.1 Documentation of LCI Model at System Level

We propose to use a flowchart that contains all the linked processes in the foreground system for each studied product system and shows their links to processes in the background system. Each process should be named and, depending on the size of the foreground system, flow names and quantities may also be given (this information is, however, not essential, as it will also be given in second reporting element). Figure 9.10 illustrates how to document a flow chart for a simple, hypothetical LCI model (flow names and quantities not shown). Flow chart should be reported in the main part of an LCA report.

Note that only the unit processes of the background system that are linked to ('neighbouring') the foreground system needs to be included in the flow chart. These are processes UP1 to UP8 in Fig. 9.10. From this information, the reader

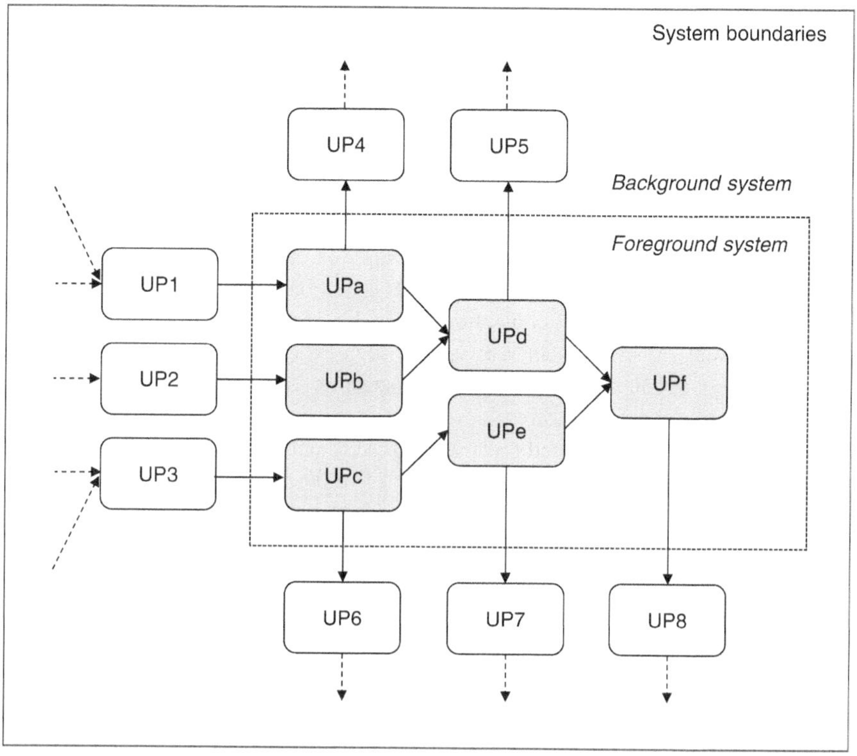

Fig. 9.10 Documentation of LCI model in flow chart. *Arrows* between unit processes (UPs) indicate material, energy, product or waste flows. Unit processes belonging to the foreground and background system are identified with a *letter* or with a *number* respectively. Only the background unit processes neighbouring the foreground system should be included and *dotted arrows* to and from these processes indicate the existence of additional background processes.

may reconstruct the remaining background system on his/her own by using aggregated versions of the reported 'neighbouring' background processes from the relevant LCI database(s).

9.7.2 Documentation of Each Unit Process

We recommend the creation of a table for each unit process in the foreground system that contains its name (identical to the one used in the flow chart of the first reporting element) and the names and quantities of all flows (materials, energy, resources, products, waste to treatment and emissions—same units as used in LCA software), see scheme in Table 9.7. We also advocate providing the source of a process or flow (e.g. name of the database where process is from), a reference to the

Table 9.7 Scheme for documenting foreground processes

Outputs	Quantity	Unit	Source/note
Reference flow (main product or function)			
Reference flow	1	kg	E.g.: input and output flows are not scaled to the functional unit of the product system
Other outputs (avoided product or function; waste to treatment)			
Avoided product 1	–	kg	E.g. please see Table A1 for the corresponding unit process
Waste 1	–	kg	E.g. ecoinvent ver. 3.0
Waste 2	–	m^3	E.g. ecoinvent ver. 3.0
Waste *n*	–	m^3	E.g. ecoinvent ver. 3.0
Emissions (to air; water; soil)			
Emission 1	–	kg	E.g. please see "Appendix" for details on calculation of emissions
Emission 2	–	kg	E.g. please see "Appendix" for details on calculation of emissions
Emission *n*	–	m^3	E.g. please see "Appendix" for details on calculation of emissions

Inputs	Quantity	Unit	Source/note
Materials			
Material 1	–	kg	E.g. please see Table A2 for the corresponding unit process
Material 2	–	kg	E.g. please see Table A3 for the corresponding unit process
Material *n*	–	m^3	E.g. ecoinvent ver. 3.0
Energy			
Energy 1	–	MJ	E.g. ecoinvent ver. 3.0; see Table 1 in the main report for the source of values
Energy 2	–	MJ	E.g. ecoinvent ver. 3.0; see Table 1 in the main report for the source of values
Energy n	–	MJ	E.g. please see Table A4 for the corresponding unit process
Resources			
Resource 1	–	kg	E.g. ecoinvent ver. 3.0
Resource 2	–	kg	E.g. ecoinvent ver. 3.0
Resource *n*	–	m^3	E.g. ecoinvent ver. 3.0
			E.g. ecoinvent ver. 3.0

Units are illustrative. *Note* that for waste treatment processes the reference flow is usually a material input. *Note* that the column 'source/note' is based on fictive examples and references included therein are not a part of this textbook chapter

section of a report where details of a calculations (e.g. emissions) are provided; and finally, reference to other unit process tables that are input or output to the process of interest. Because the number of processes to be documented is often large, tables like Table 9.7 are usually best reported in an appendix to the LCA report.

The flow quantities of process tables should either be scaled to 1 unit of the reference flow of the process (as shown in Table 9.7) or scaled to the quantity of process reference flow required to meet the reference flow of the study (derived from the functional unit). For neighbouring background processes (UP1 to UP8 in Fig. 9.10), the name of the process and the name and version of the database it was sourced from is sufficient, because the reader may use this information to recreate the remainder of the background system.

Note that inventory data in the foreground system are sometimes confidential, for example when a manufacturer wants to prevent the details of the production processes to be disclosed to the public or competitors. In terms of documenting LCI results, confidentiality issues can be handled by placing the process tables containing confidential data in an appendix that is only made available to groups of people that are cleared by the supplier of the data (e.g. employees of the organisation commissioning a study and an external critical reviewer).

9.7.3 Documentation of Metadata

We recommend reporting metadata according to specificity, type, source and access using the structure of Table 9.2 (introduced for data planning and collection). For easy overview, the rows of the table should be grouped into life cycle stages. The data specificity classification (from very low to very high) for each data point should be transparent, i.e. by writing in the relevant cell why a data point was classified to a given specificity, rather than simply making a cross. The documentation of these metadata should be consistent with the documentation of unit processes described in Table 9.7, and cross-references between the two should be made (e.g. notes and data sources reported in tables documenting unit processes may readily refer to the table with metadata) We advocate reporting metadata in the main part of the LCA report.

9.7.4 Documentation of LCI Results

The LCI results should simply be documented as a list of quantified elementary flows, divided into resources and emissions, i.e. as in Table 9.7. This typically consists of an extensive table, which can be documented as an appendix for readability of the LCA report.

9.7.5 Assumptions for Each Life Cycle Stage

Due to lack of information and budget constraints, it is common to make several assumptions when constructing an LCI model. For example, data originally planned to be collected in medium or high specificity may end up being collected in low specificity. Thereby assumptions need to be made on what low-quality data can best represent the actual data. For example, should a wastewater treatment process in Vietnam, for which data could not be obtained, be approximated by a process in Thailand, possibly correcting for the Vietnamese electricity mix, or should it rather be approximated by an average process for the entire South East Asian region? All assumptions made during the construction of the LCI model should be transparently documented. We recommend that major assumptions are indicated, when describing the data collection and modelling of each individual life cycle stage, to facilitate cross-comparison with the documentation of metadata. Major assumptions may also be included directly in the table containing metadata. References to the sensitivity analysis should be given for assumptions whose influence on LCIA results are tested by the creation and analysis of sensitivity scenarios (see next subsection). We also recommend that a list of all assumptions, minor and major, be placed in an 'Appendix'.

9.7.6 Documentation of Data Collected for Uncertainty and Sensitivity Analysis

For sensitivity analyses, the LCA report must state which parameters are analysed and whether this is done by calculating normalised sensitivity coefficients (for parameters of a continuous nature) or by the construction of sensitivity scenarios (for parameters of a discrete nature). In the former case, the perturbed values for each parameter must be documented and the basis of these explained (e.g. reported min/max-values, 2.5/97.5 percentiles, or an arbitrary value, such as $\pm 10\%$). In the latter case, the sensitivity scenarios should be documented and references to the assumptions they are based on made (see previous subsection).

For uncertainty analyses, the best practice is to use statistical distributions of parameter values as input to Monte Carlo analysis (see Sect. 9.6), in which case the distributions (e.g. uniform, normal or log-normal) and statistical parameters (e.g. standard deviation) must be documented for each parameter value covered in the uncertainty analysis. If, due to lack of such data, the Pedigree approach is taken, the underlying uncertainty factors and calculated geometric standard deviation for process must be documented. An example was given earlier in Table 9.6.

Appendix: Example of Consequential LCA on Biodiesel Made from Poultry Fat

To help you get an overview of the 4-step procedure for performing a consequential LCI (presented in Sect. 9.2.3), an example is here presented, which shows some parts of a consequential LCI looking at the decision to supply additionally 200 tonnes of biodiesel based on poultry fat. It should be noted that this is a constructed example and that the factual claims made may not be completely accurate.

To start the procedure, we go to Step 1. Here we are asked to consider whether the assessed decision leads to changes in demand or supply. Clearly, this decision leads to changes in supply. This implies that we move directly to Step 3.

Step 3 is based on the assumption that demand is constant, and given that we increase supply of poultry fat biodiesel we therefore have to consider what other products it substitutes. According to the procedure given in Step 3, we need to identify a user and a satisfying substitute for the user which fulfils the same functions terms of functionality, technical quality, costs, etc.

Biodiesel is only used by drivers of diesel vehicles and can be blended with petrochemical diesel or used as a full substitute for petrochemical diesel in ordinary diesel engines. As it is often sold under favourable tax conditions, it seems reasonable to assume that it will substitute ordinary diesel. However, another scenario which may also in some cases be realistic to consider is that it will substitute other types of biodiesel (e.g. based on other substrates). Ordinary diesel and other types of biodiesel can both be produced without constraints (the answer to the second question of the decision tree in Fig. 9.5 is 'no') and can therefore both be considered reasonable alternatives. In this example, however, we will only consider the former.

Having found petrochemical diesel as a substitute, we go to Step 4 to identify which technology will produce the diesel, which is substituted. Here, we need to consider the trend in the market, the scope of the decision, and whether the decision leads to an increase or decrease in demand. Having addressed these issues, we find that the substituted diesel is produced by the least cost-efficient technology supplying the market at the time of our decision, which we find to be crude oil produced from tar sand.

Biodiesel does not contain the same amount of energy per weight unit as ordinary diesel, implying that we will need more biodiesel than diesel to drive a certain distance. The ratio is around 37:42, implying that for each kg of poultry fat biodiesel we produce and use extra, we will reduce the production and use of diesel made from tar sand by 37/42 kg.

The production of biodiesel inevitably leads to the co-production of glycerol. When we decide to increase the production of biodiesel by 200 tonnes, we will also increase the production of glycerol by approximately 20 tonne. As this is a result of our decision to produce more biodiesel, it needs to be included in the assessment. We therefore start again in Step 1 by asking the question: "What happens if we increase the supply of glycerol by 20 tonnes?" Being a supply oriented question, we go directly to Step 3, where we are asked to identify products for which glycerol can serve as a substitute, based on relevant functionality, technical quality, costs, etc. Through analysing the biodiesel market, for example through biodiesel journals and experts in the field, we find that glycerol from biodiesel can be used by producers of chemicals, especially for the production of propylene glycol. Hereby glycerol can, after distillation and processing, substitute other feedstock in the production of propylene glycol. Having identified a substitute, we go to Step 4, to identify the propylene glycol production technology affected by the change in feedstock to glycerol. This procedure (not detailed here) allows us to include the avoided production of propylene glycol in our LCI. When doing so, it is important to identify the processes needed to convert the crude glycerol to propylene glycol and remember to take into consideration the conversion rate.

Having considered both the substitution of diesel with biodiesel and conventional propylene glycol with propylene glycol made from glycerol, we have now considered all the downstream parts of the life cycle. However, our decision to supply more poultry fat biodiesel will also create changes in the upstream part of the life cycle: If we want to supply more poultry fat biodiesel, we need more of the constituents included for producing the biodiesel. The demand for these constituents thereby increases. In the concrete case, biodiesel is made from poultry fat and methanol, which are brought to react using a strong base, often sodium hydroxide. For the sake of simplicity, we will here only consider the increased demand for poultry fat and methanol.

Thus, we return to Step 1 and ask: "What happens if I increase the demand for poultry fat?" As this is clearly a question that relates to demand, we go to Step 2.

The first part of the decision tree in Step 2 (Fig. 9.4) asks us to consider whether the production of the product is constrained. In this case, this is actually the case, since poultry fat is a low value by-product from the production of other poultry products, mainly meat. The production of poultry fat therefore follows the demand for poultry meat, and additional demand for poultry fat will not result in an additional supply of poultry fat. As the assessed decision will lead to an increase in the demand for poultry fat, and as market analysis shows us that poultry fat is already used to the extent the constraint of being a co-product allows (in other words, no poultry fat is wasted), we go to Step 3, to find out which product can substitute our use of poultry fat. Poultry fat is mainly used in the feed industry and through contacts to feed producers we find that they are able to use palm and soybean oil in a certain relationship instead of poultry fat. This implies that if we decide to produce more biodiesel from poultry fat and thereby demand more poultry fat, we will not increase the supply of poultry fat but rather increase the demand for palm and soybean oil. To identify the consequences of the increased demand for these oils, we go through the relevant Steps 2–3 for each of these, but to keep this example relatively simple, we will not go further into documenting these steps.

Assuming that we have now fully outlined the processes that change as a result of our increase in demand for palm and soybean oil, we turn to the other main constituent of biodiesel, namely methanol. As noted above, we also increase the demand for methanol. We therefore again start in Step 1 by asking the question: "What happens when I increase the demand for methanol?" As this is a demand oriented question, we go to Step 2. Here we are first asked whether methanol can be produced without constraints. As this is the case, we go to Step 4. Here we are asked to consider the overall trend in the market, the scope of the decision in comparison to the overall market for methanol, and whether the decision leads to an increase or decrease in demand. Through market studies we find that the trend in the market, which can be considered global, is an increasing production. Secondly, the size of the decision, which in this case is to produce a few hundred extra tonnes of poultry fat biodiesel will amount to very little compared to the overall market volume for methanol. We should therefore identify the *short-term* marginal producer.

Given that our decision leads to an increase in demand, we are told by Table 9.1, that the methanol will be produced by the least competitive producer on the market. As there are more or less only producers making methanol from synthetic gas, we assume that the methanol will be produced using this technology.

Other inputs and outputs to and from the biodiesel process are handled in a similar way, but to keep the example relatively short, these will not be discussed here.

As the example shows, creating a consequential LCI is in many cases a rather laborious task as detailed knowledge is needed about the markets affected by the decision, as for example establishing knowledge about potential substitutes for poultry fat in the feed industry in the example above. Much of the time spent making the LCA will therefore often be used in preparing the consequential LCI.

References

This chapter is to a large extent based on the ILCD handbook and the ISO standard 14040 and 14044. Due to the scope of this chapter, some details have been omitted, and some procedures have been rephrased to to make the text more relevant to students. For more details, the reader may refer to these texts:

EC-JRC: European Commission—Joint Research Centre—Institute for Environment and Sustainability: International Reference Life Cycle Data System (ILCD) Handbook—General guide for Life Cycle Assessment—Detailed guidance, 1st edn. March 2010. EUR 24708 EN. Luxembourg, Publications Office of the European Union (2010)

ISO: Environmental Management—Life Cycle Assessment—Principles and Framework (ISO 14040). ISO, the International Organization for Standardization, Geneva (2006a)

ISO: Environmental Management—Life Cycle Assessment—Requirements and Guidelines (ISO 14044). ISO, the International Organization for Standardization, Geneva (2006b)

Additional References Used in the Text

Birkved, M., Hauschild, M.Z.: PestLCI—a model for estimating field emissions of pesticides in agricultural LCA. Ecol. Model. **198**, 433–451 (2006). doi:10.1016/j.ecolmodel.2006.05.035

Ciroth, A.: Refining the pedigree matrix approach in ecoinvent: towards empirical uncertainty factors. Presentation at the LCA Discussion Forum Zürich, 13, September 2013 (2013)

EC-JRC: Product Environmental Footprint (PEF) Guide. Deliverable 2 and 4A of the Administrative Arrangement between DG Environment and the Joint Research Centre No N 070307/2009/552517, including Amendment No 1 from December 2010. Ispra, Italy (2012)

Goedkoop M, Oele M, Leijting J, et al.: Introduction to LCA with SimaPro. Report version 5.2, January 2016 (c) 2002–2016 Pré. (2016). www.pre-sustainability.com

Guinée, J.B., Gorrée, M., Heijungs, R., Huppes, G., Kleijn, R., van Oers, L., Wegener Sleeswijk, A., Suh, S., Udo de Haes, H.A., de Bruijn, H., van Duin, R., Huijbregts, M.A.J.: Handbook on Life Cycle Assessment: Operational Guide to the ISO Standards. Kluwer Academic Publishers, Dordrecht (2002). ISBN 1-4020-0228-9

Heijungs, R., Suh, S.: The computational structure of life cycle assessment. Kluwer Academic Publishers, Dortrecht (2002)

Weidema, B.P.: Market Information in Life Cycle Assessment. Environment Project No. 863, vol. 863, p. 147 (2003)

Weidema, B.P., Bauer, C., Hischier, R., et al.: The ecoinvent database: overview and methodology. Data Quality Guideline for the Ecoinvent Database Version 3 (2013). www.ecoinvent.org

Wenzel, H., Hauschild, M.Z., Alting, L.: Environmental assessment of products. In: Methodology, Tools and Case Studies in Product Development, vol. 1, 544 pp. Chapman & Hall, Kluwer Academic Publishers, Hingham, MA (ISBN:0 412 80800 5) (1997)

World Coal Association: Coal and steel statistics 2012 (2012)

World Steel: World steel in figures 2015 (2015)

Author Biographies

Anders Bjørn Part of the LCA community since the early 2010s. Main focus is interpretations of sustainability and integration of sustainability targets in LCA to enable absolute sustainability assessments.

Andreas Moltesen Has been working with LCA since 2006 with a particular focus on social life cycle assessment. He has later worked on life cycle assessments of biofuels and is currently particularly involved with life cycle assessments of transport systems.

Alexis Laurent Working with LCA since 2010 with a strong LCIA focus, particularly on normalisation aspects. Main LCA interests include development of LCIA methods, LCA applications and footprinting of large-scale systems for policy-making (e.g. nations, sectors), and LCA applied to various technology domains, including energy systems.

Mikołaj Owsianiak Involved in development and application of life cycle impact assessment methods in sustainability assessment of technologies. Has worked on issues associated with: soils (remediation), metals (toxic impact assessment), biodiesel (fate in the environment), and carbonaceous materials (biochar and hydrochar).

Andrea Corona Materials engineer with a focus on sustainability assessment of biobased products and composite materials. Working with LCA from 2012. Main interests include life cycle engineering, product development and eco-design

Morten Birkved With a background in environmental chemistry and LCA, Morten's primary research activities are focused on the development of quantification methods for ecosphere-technosphere exchanges, general LCI, scope modelling, assessment of buildings and built environments, system modelling and fused assessment forms.

Michael Z. Hauschild Involved in development of LCIA methodology since the early 1990s. Has led several SETAC and UNEP/SETAC working groups and participated in the development of the ISO standards and the ILCD methodological guidelines. Main LCA interests are chemical impacts, spatial differentiation and science-based boundaries in LCIA.

Chapter 10
Life Cycle Impact Assessment

Ralph K. Rosenbaum, Michael Z. Hauschild, Anne-Marie Boulay, Peter Fantke, Alexis Laurent, Montserrat Núñez and Marisa Vieira

Abstract This chapter is dedicated to the third phase of an LCA study, the Life Cycle Impact Assessment (LCIA) where the life cycle inventory's information on elementary flows is translated into environmental impact scores. In contrast to the three other LCA phases, LCIA is in practice largely automated by LCA software, but the underlying principles, models and factors should still be well understood by practitioners to ensure the insight that is needed for a qualified interpretation of the results. This chapter teaches the fundamentals of LCIA and opens the black box of LCIA with its characterisation models and factors to inform the reader about: (1) the main purpose and characteristics of LCIA, (2) the mandatory and optional steps of LCIA according to the ISO standard, and (3) the science and methods underlying the assessment for each environmental impact category. For each impact category, the reader is taken through (a) the underlying environmental problem, (b) the underlying environmental mechanism and its fundamental modelling principles, (c) the main anthropogenic sources causing the problem and (d) the main methods available in LCIA. An annex to this book offers a comprehensive qualitative comparison of the main elements and properties of the most widely used and also the latest LCIA methods for each impact category, to further assist the advanced practitioner to make an informed choice between LCIA methods.

R.K. Rosenbaum (✉) · M. Núñez
IRSTEA, UMR ITAP, ELSA Research Group and ELSA-PACT Industrial Chair for Environmental and Social Sustainability Assessment,
361 rue Jean-François Breton, BP 5095, 34196 Montpellier, France
e-mail: ralph.rosenbaum@irstea.fr

M.Z. Hauschild · P. Fantke · A. Laurent
Division for Quantitative Sustainability Assessment, Department of Management Engineering, Technical University of Denmark, Kongens Lyngby, Denmark

A.-M. Boulay
LIRIDE, Sherbrooke University, Sherbrooke, QC, Canada

M. Vieira
PRé Consultants bv, Amersfoort, The Netherlands

A.-M. Boulay
CIRAIG, Polytechnique Montreal, Montreal, QC, Canada

M.Z. Hauschild et al. (eds.), *Life Cycle Assessment*,
DOI 10.1007/978-3-319-56475-3_10

Learning Objectives

After studying this chapter, the reader should be able to:

- Explain and discuss the process and main purposes of the LCIA phase of an LCA study.
- Distinguish and explain the mandatory and optional steps according to international standards for LCA.
- Differentiate and describe each of the impact categories applied in LCIA regarding:
  - the underlying environmental problem,
  - the environmental mechanism and its fundamental modelling principles,
  - the main anthropogenic sources causing the problem,
  - the main methods used in LCIA.

10.1 Introduction

In practice, the Life Cycle Impact Assessment (LCIA) phase is largely automated and essentially requires the practitioner to choose an LCIA method and a few other settings for it via menus and buttons in LCA software. However, as straightforward as that may seem, without understanding a few basic, underlying principles and the meaning of the indicators, neither an informed choice of LCIA method nor a meaningful and robust interpretation of LCA results are possible. However, the important extent of science and its inherent multidisciplinarity frequently result in a perceived opacity of this phase. This chapter intends to open the black box of LCIA with its characterisation models and factors, and to accessibly explain (1) its main purpose and characteristics, (2) the mandatory and optional steps according to ISO and (3) the meaning and handling of each impact category. While this chapter is a pedagogical and focused introduction into the complex and broad aspects of LCIA, a more profound and in-depth description, targeting experienced LCA practitioners and scientists, can be found in Hauschild and Huijbregts (2015).

Once the Life Cycle Inventory (LCI) is established containing all elementary flows relevant for the product system under assessment, the next question to answer will be something like: How to compare 1 g of lead emitted into water to 1 g of CO_2 emitted into the air? In other words, how to compare apples with pears? Life Cycle Impact Assessment is a phase of LCA aiming to assess the magnitude of contribution of each elementary flow (i.e. emissions or resource use of a product system) to an impact on the environment. Its objective is to examine the product system from an environmental perspective using impact categories and category indicators in conjunction with the results of the inventory analysis. This will provide information useful in the interpretation phase.

As the focal point of this phase of an LCA (and also of this chapter), it is a relevant question to ask what is an environmental impact? It could be defined as a set of environmental changes, positive or negative, due to an anthropogenic intervention. Such impacts are studied and assessed using a wide range of quantitative and qualitative tools, all with specific aims and goals to inform or enable more sustainable decisions. In LCA this is an important phase, as it transforms an elementary flow from the inventory into its potential impacts on the environment. This is necessary since elementary flows are just quantities emitted or used but not directly comparable to each other in terms of the importance of their impact. For example, 1 kg of methane emitted into air does not have the same impact on climate change as 1 kg of CO_2, even though their emitted quantities are the same (1 kg) since methane is a much stronger greenhouse gas (GHG). LCIA characterisation methods essentially model the environmental mechanism that underlies each of the impact categories as a cause–effect chain starting from the environmental intervention (emission or physical interaction) all the way to its impact. However, the results of the LCIA should neither be interpreted as predicted actual environmental effects nor as predicted exceedance of thresholds or safety margins nor as risks to the environment or human health. The results of this LCA phase are scores that represent potential impacts, a concept that is explained further on.

The ISO 14040/14044 standards (ISO 2006a, b) distinguish mandatory and optional steps for the LCIA phase, which will all be explained further in this chapter:

Mandatory steps:

- Selection of impact categories, category indicators and characterisation models (in practice typically done by choosing an already existing LCIA method) →*Which impacts do I need to assess?*
- Classification (assigning LCI results to impact categories according to their known potential effects, i.e. in practice typically done automatically by LCI databases and LCA software) →*Which impact(s) does each LCI result contribute to?*
- Characterisation (calculating category indicator results quantifying contributions from the inventory flows to the different impact categories, i.e. typically done automatically by LCA software) →*How much does each LCI result contribute?*

Optional steps:

- Normalisation (expressing LCIA results relative to those of a reference system) →*Is that much?*
- Weighting (prioritising or assigning weights to the each impact category) →*Is it important?*
- Grouping (aggregating several impact indicator results into a group)

As already mentioned, it is important to keep in mind that the impacts that are assessed in the LCIA phase should be interpreted as impact potentials, not as actual impacts, nor as exceeding of thresholds or safety margins, or risk, because they are:

- Relative expressions of potential impacts associated with the life cycle of a reference flow needed to support a unit of function (=functional unit)
- Based on inventory data that are integrated over space and time, and thus often occurring at different locations and over different time horizons
- Based on impact assessment data which lack information about the specific conditions of the exposed environment (e.g. the concomitant exposure to substances from other product systems)

Terminology and definitions are given in Table 10.1.

Table 10.1 Essential terminology and definitions

Term	Definition	Source
Area of protection	A cluster of category endpoints of recognisable value to society. Examples are human health, natural resources and natural environment.	Hauschild and Huijbregts (2015)
Category indicator	Quantifiable representation of an impact category	ISO (2006b)
Category endpoint	Attribute or aspect of natural environment, human health or resources, identifying an environmental issue giving cause for concern	ISO (2006b)
Characterisation model	Reflect the environmental mechanism by describing the relationship between the LCI results, category indicators and, in some cases, category endpoint(s). The characterisation model is used to derive the characterisation factors.	ISO (2006b)
Characterisation factor	Factor derived from a characterisation model which is applied to convert an assigned life cycle inventory analysis result to the common unit of the category indicator	ISO (2006b)
Ecosphere	The biosphere of the earth, especially when the interaction between the living and non-living components is emphasised	Oxford Dictionary of English
Elementary flow	Material or energy entering the system being studied that has been drawn from the environment without previous human transformation, or material or energy leaving the system being studied that is released into the environment without subsequent human transformation	ISO (2006b)
Environmental impact	Potential impact on the natural environment, human health or the depletion of natural resources, caused by the interventions between the technosphere and the ecosphere as covered by LCA (e.g. emissions, resource extraction, land use)	EC-JRC (2010a)
Environmental mechanism	System of physical, chemical and biological processes for a given impact category, linking the life cycle inventory analysis results to category indicators and to category endpoints	ISO (2006b)

(continued)

Table 10.1 (continued)

Term	Definition	Source
Environmental relevance	Degree of linkage between category indicator result and category endpoints	ISO (2006b)
Impact category	Class representing environmental issues of concern to which life cycle inventory analysis results may be assigned	ISO (2006b)
Impact pathway	Cause–effect chain of an environmental mechanism	
LCIA method	Collection of individual characterisation models (each addressing their separate impact category)	Hauschild et al. (2013)
Midpoint indicator	Impact category indicator located somewhere along the impact pathway between emission and category endpoint	Hauschild and Huijbregts (2015)
Potential impact	Relative performance indicators which can be the basis of comparisons and optimisation of the system or product	Hauschild and Huijbregts (2015)
Technosphere	The sphere or realm of human technological activity; the technologically modified environment	Oxford Dictionary of English

10.2 Mandatory Steps According to ISO 14040/14044

10.2.1 Selection of Impact Categories, Category Indicators and Characterisation Models

The contents of this section have been modified from Rosenbaum, R.K.: Selection of impact categories, category indicators and characterisation models in goal and scope definition, appearing as Chapter 2 of Curran M. A. (ed.) LCA Compendium—The Complete World of Life Cycle Assessment—Goal and scope definition in Life Cycle Assessment pp 63–122, Springer, Dordrecht (2017).

The objective of selecting impact categories, category indicators and characterisation models is to find the most useful and needed ones for a given goal. To help guide the collection of information on the relevant elementary flows in the inventory analysis, the selection of impact categories must be in accordance with the goal of the study and is done in the scope definition phase prior to the collection of inventory data to ensure that the latter is targeted towards what is to be assessed in the end (see Chaps. 7 and 8 on Goal and Scope definition). A frequent difficulty is the determination of the criteria that define what is useful and needed in the context of the study. Some criteria are given by ISO 14044 (2006b), either as requirements or as recommendations. The requirements are obligatory for compliance with the ISO standard, and will therefore be among the focus points of a Critical Review (see Chap. 13 on Critical Review). Some of these requirements and recommendations concern LCA practitioners and LCIA method developers alike, while others are most relevant for developers of LCIA methods and of LCA software. The focus is here on the former, i.e. requirements concerning LCA practitioners.

ISO 14044 (2006b) states that the choice of impact categories needs to assure that they

- Are not redundant and do not lead to double counting
- Do not disguise significant impacts
- Are complete
- Allow traceability

Furthermore, this list is complemented with a number of obligatory criteria, requiring that the selection of impact categories, category indicators and characterisation models shall be:

- Consistent with the goal and scope of the study (when, for example, environmental sustainability assessment is the goal of a study, the practitioner cannot choose a limited set of indicators, or a single indicator footprint approach, as this would be inconsistent with the sustainability objective of avoiding burden-shifting among impact categories)
- Justified in the study report
- Comprehensive regarding environmental issues related to the product system under study (essentially meaning that all environmental issues—represented by the various impact categories—which a product system may affect need to be included, again in order to reveal any problem-shifting from one impact category to another)
- Well documented with all information and sources being referenced (in practice it is normally sufficient to provide name and version number of the LCIA method used together with at least one main reference, which should provide all primary references used to build the method)

ISO 14044 (2006b) *recommendations* for the selection of impact categories, category indicators and characterisation models by a practitioner include:

- International acceptance of impact categories, category indicators and characterisation models, i.e. based on an international agreement or approved by a competent international body
- Minimisation of value-choices and assumptions made during the selection of impact categories, category indicators and characterisation models
- Scientific and technical validity of the characterisation model for each category indicator (e.g. not based on unpublished or outdated material)
- Being based upon a distinct, identifiable environmental mechanism and reproducible empirical observation
- Environmental relevance of category indicators

Numerous further criteria but also practical constraints beyond ISO 14044 exist and are applied, consciously or unconsciously, often based on experience or recommendations from colleagues. In practice the selection of impact categories, category indicators and characterisation models usually boils down to selecting an

LCIA method (or several) available in the version of the LCA software that the practitioner has access to.

External factors for this choice will be among other:

- Requirements following from the defined goal (see Chap. 7) and specified in the scope definition of the LCA (see Chap. 8)
- Requirements by the commissioner of an LCA
- Fixed requirements, e.g. for Environmental Product Declarations (EPDs) or Product Environmental Footprints (PEFs) from underlying sector-based Product Category Rules (PCRs) or from labelling schemes (see Chap. 24)

Practical constraints may, for example, consist of:

- Availability, completeness and quality of LCI results required for a specific impact category
- Availability, completeness and quality of characterisation models and factors for a specific impact category, including the need to consider specific rare or new impact categories, such as noise, which may only be supported by one or two LCIA methods if at all
- If normalisation is required, availability, completeness and quality of normalisation factors for a specific impact category or LCIA method

If practical constraints prevent the practitioner from including what has been identified as relevant impact categories, this needs to be made clear in the discussion and interpretation of the LCA results and comments need to be made on whether it may change the conclusions. In the illustrative case on window frames in Chap. 39, the method recommended for characterisation by the International Life Cycle Data system (ILCD) is chosen as life cycle impact assessment method (EC-JRC 2011), and all impact categories covered by the method are included in the study.

In common LCA practice, a number of category indicators, based on specific characterisation models is combined into predefined sets or methods, often referred to as life cycle impact assessment methods or simply LCIA methods (EC-JRC 2011; Hauschild et al. 2013), available in LCA software under names such as ReCiPe, CML, TRACI, EDIP, LIME, IMPACT 2002+, etc. However, with an increasing number of LCIA methods and indicators available, the task of choosing one requires a tangible effort from the practitioner to understand the main characteristics of these methods and to keep up-to-date with the developments in the field of LCIA. A qualitative and comparative overview of the main characteristics of current LCIA methods can be found in Chap. 40 of the Annex of this book.

10.2.1.1 How to Choose an LCIA Method?

A number of LCIA methods have been published since the first one appeared in 1984. Figure 10.1 shows the most common methodologies published since 2000

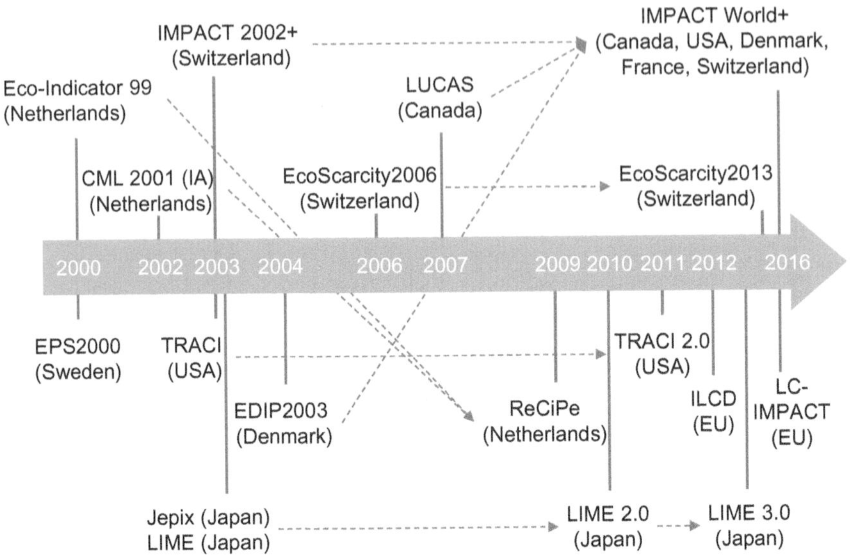

Fig. 10.1 LCIA methods published since 2000 with country/region of origin in *brackets*. *Dotted arrows* represent methodology updates (Rosenbaum 2017)

that all meet the requirements of ISO 14044. A more detailed overview of these methods can be found in Chap. 40.

When selecting an LCIA method, the requirements, recommendations, external and internal factors and constraints discussed above all need to be considered. This leads to a number of questions and criteria that should be answered in order to systematically identify the most suitable one. Here is a non-exhaustive list of relevant questions to address:

- Which impact categories (or environmental problems) do I need to cover and can I justify those that I am excluding?
- In which region does my life cycle (or its most contributing processes) take place?
- Do I need midpoint or endpoint assessment, or both?
- Which elementary flows do I need to characterise?
- Are there any recommendations from relevant organisations that can help me choose?
- How easily can the units of the impact categories be interpreted (e.g. absolute units, equivalents, monetary terms, etc.)?
- How well is the method documented?
- How easily can the results (units, aggregation into specific indicator groups, etc.) be communicated?
- Do I need to apply normalisation and if yes for which reference system (in most cases it is not recommendable to mix characterisation and normalisation factors

from different LCIA methods due to the difference in characterisation modelling, units, numerical values, etc..)?
- When was the method published and have there been important scientific advances in the meantime?
- Do I have the resources/data availability to apply a regionalised methodology (providing more precise results)?
- Do I need to quantify the uncertainty of both LCI and LCIA and does the LCIA method support that?

ISO 14040/14044 by principle do not provide any recommendations about which LCIA method should be used, but some organisations do recommend the use of a specific LCIA method or parts of it. The European Commission has established specific recommendations for midpoint and endpoint impact categories by systematically comparing and evaluating all relevant existing approaches per category, leading to the recommendation of the best available approach (EC-JRC 2011). This effort resulted in a set of characterisation factors, which is directly available in all major LCA software as the ILCD method. Some methods with a stronger national focus are recommended by national governmental bodies for use in their respective country, such as LIME in Japan, or TRACI in the US.

Given the amount of LCIA methods available and the amount of time required to stay informed about them, it may be tempting to essentially stick to the method(s) that the LCA practitioner knows best or has used for a long time, or that was recommended by a colleague, or simply choosing a method requested by the client to allow comparison with results from previous studies. It is however beneficial to apply a more systematic approach to LCIA method selection that in combination with the LCIA method comparison in Chap. 40 allows to determine the relevant selection questions and criteria, thus optimising the interpretability and robustness of the results of the study. The following properties are compared in Chap. 40 per impact category and for both midpoint and endpoint LCIA methods:

- Aspects/diseases/ecosystems (which kinds of impacts) that are considered
- Characterisation model used
- Selected central details about fate, exposure, effect and damage modelling
- Reliance on marginal or average indicator
- Emission compartments considered
- Time horizon considered
- Geographical region modelled
- Level of spatial differentiation considered
- Number of elementary flows covered
- Unit of the indicator

Not all of these properties may be of equal relevance for choosing an LCIA method for each practitioner or study, but they are identified here as relevant and fact-based properties.

Further details on the selection of impact categories, category indicators and characterisation models can be found in Rosenbaum (2017) and Hauschild and Huijbregts (2015).

10.2.2 Classification

In this step, the elementary flows of the LCI are assigned to the impact categories to which they contribute; for example an emission of CO_2 into air is assigned to climate change or the consumption of water to the water use impact category, respectively. This is not without difficulty because some of the emitted substances can have multiple impacts in two modes:

- In parallel: a substance has several simultaneous impacts, such as SO_2 which causes acidification and is toxic to humans when inhaled.
- In series: a substance has an adverse effect which itself becomes the cause of something else, such as SO_2 which causes acidification, which then may mobilise heavy metals in soil which are toxic to humans and ecosystems.

This step requires considerable understanding and expert knowledge of environmental impacts and is therefore typically being handled automatically by LCA software (using expert-based, pre-programmed classification tables) and not a task that the LCA practitioner needs to undertake.

10.2.3 Characterisation

In this step, all elementary flows in the LCI are assessed according to the degree to which they contribute to an impact. To this end, all elementary flows *E*, classified within a specific impact category *c* (representing an environmental issue of concern), are multiplied by their respective characterisation factor CF and summed over all relevant interventions *i* (emissions or resource extractions) resulting in an impact score IS for the environmental impact category (expressed in a specific unit equal for all elementary flows within the same impact category):

$$\mathrm{IS}_c = \sum_i (\mathrm{CF}_i \cdot E_i) \tag{10.1}$$

For each impact category, the indicator results are summed to determine the overall results for the category. In the following sections, the general principles of how CFs are calculated and interpreted will be discussed. In order to provide a better understanding of what CFs in each impact category represent and how they are derived, Sects. 10.6–10.16 will, for each impact category, explain the corresponding (1) problem observed, (2) principal environmental mechanism, (3) main causes and (4) most widely used characterisation models.

10.2.3.1 What Is a Characterisation Factor?

A characterisation factor (CF) represents the contribution per quantity of an elementary flow to a specific environmental impact (category). It is calculated using (scientifically valid and quantitative) models of the environmental mechanism representing as realistically as possible the cause–effect chain of events leading to effects (impacts) on the environment for all elementary flows which contribute to this impact. The unit of a CF is the same for all elementary flows within an impact category. It is defined by the characterisation model developers and may express the impacts directly in absolute terms (e.g. number of disease cases/unit toxic emission) or indirectly through relating them to the impact of a reference elementary flow (e.g. CO_2-equivalents/unit emission of greenhouse gases).

10.2.3.2 How Is It Calculated?

The modelling of a characterisation factor involves the use of different models and parameters and is typically conducted by experts for a particular impact category and its underlying impact pathway or environmental mechanism. Various assumptions and methodological choices are involved and this may affect the output as reflected in the differences in results that may be observed for the same impact category when applying different LCIA methods. This must be considered when interpreting the result of the LCIA phase. The first step when establishing an impact category is the observation of an adverse effect of concern in the environment, leading to the conclusion that we need to consider such effects in the context of decisions towards more sustainable developments. Once accepted as an effect of concern, the focus will be on how to characterise (quantify) the observed effect in the framework of LCA.

The basis and starting point of any characterisation model is always the establishment of a model for the environmental mechanism represented by a cause–effect chain. Its starting point is always the environmental intervention (represented by elementary flows), essentially distinguishing two types based on the direction of the relevant elementary flows between technosphere and ecosphere:

- An emission into the environment (=elementary flow from the technosphere to the ecosphere),
  or
- A resource extraction from the environment (=elementary flow from the ecosphere to the technosphere).

10.2.3.3 Emission-Related Impacts

For the first type, an emission into the environment, the principal cause–effect chain may be divided into the following main steps:

- Emission: into air, water or soil (for some product systems also other compartments may be relevant such as groundwater, indoor air, etc.)
- Fate: environmental processes causing transport, distribution and transformation of the emitted substance in the environment. Depending on the physical and chemical properties of the substance and the local conditions at the site of emission, a substance may be transferred between different environmental compartments, be transported over long distances by wind or flowing water, and be undergoing degradation and transformation into other molecules and chemical species.
- Exposure: contact of the substance from the environment to a sensitive target like animals and plants, entire ecosystems (freshwater, marine, terrestrial or aerial) or humans. Exposure may involve processes like inhalation of air, ingestion of food and water or dermal contact via skin and other surfaces.
- Effects: observed adverse effects in the sensitive target after exposure to the substance, e.g. increase in the number of disease cases (ranging from reversible temporary problems to irreversible permanent problems and death) per unit intake in a human population or number of species affected (e.g. by disease, behaviour, immobility, reproduction, death, etc.) after exposure of an ecosystem
- Damage: distinguishing the severity of observed effects by quantifying the fraction of species potentially disappearing from an ecosystem, or for human health by giving more weight to death and irreversible permanent problems (e.g. reduced mobility or dysfunctional organs) than to reversible temporary problems (e.g. a skin rash or headache)

These steps together constitute the environmental mechanism of the impact category and their specific features will vary depending on the impact category we are looking at.

10.2.3.4 Extraction-Related Impacts

For the second type of elementary flow, a resource extraction from the environment, the principal cause–effect chain may comprise some or all of the following main steps (with significant simplifications possible for some resources where not all steps may be relevant, e.g. minerals):

- Extraction or use: of minerals, crude oil, water or soil, etc.
- Fate: (physical) changes to local conditions in the environment, e.g. soil organic carbon content, soil permeability, groundwater level, soil albedo, release of stored carbon, etc.

- Exposure: change in available quantity, quality or functionality of a resource and potential competition among several users (human or ecosystems, with different degrees of ability to adapt and/or compensate), e.g. habitat loss, dehydration stress, soil biotic productivity, etc.
- Effects: adverse effects on directly affected users that are unable to adapt or compensate (e.g. diseases due to lower water quality, migration or death of species due to lack of water or habitat, malnutrition, etc.) and contributions to other impact pathways (e.g. global warming due to change in soil albedo or released soil carbon)
- Damage: distinguishing the severity of observed effects by quantifying the reduction of biodiversity, or human health of a population affected (although not yet common practice, this may even go as far as including social effects such as war on water access)

This mechanism will have specific features and may vary significantly between impact categories, but the principle remains valid for all extraction-related impact categories, currently being:

- Land Use (affecting biotic productivity, aquifer recharge, carbon sequestration, albedo, erosion, mechanical and chemical filtration capacity, biodiversity, etc.)
- Water use (affecting human health, aquatic ecosystems, terrestrial ecosystems)
- Abiotic resource use (fossil and mineral) affecting the future availability of the non-renewable abiotic resources
- Biotic resource use (e.g. fishing or wood logging) affecting the future availability of the renewable biotic resources and the ecosystems from which they are harvested.

10.2.3.5 The Impact Indicator

The starting point of the environmental mechanism is set by an environmental intervention in the form of an elementary flow in the LCI, and the contribution from the LCI flow is measured by the ability to affect an indicator for the impact category which is selected along the cause–effect chain of the impact category. Apart from the feasibility of modelling the indicator, this selection should be guided by the environmental relevance of the indicator. For example, there is limited relevance in choosing human exposure to the substance as an indicator for its human health impacts, because even if a substance is taken in by a population (i.e. exposure can be observed and quantified), it might not cause any health effect due to a low toxicity of the substance, and this would be ignored if a purely exposure-based indicator was chosen. In general, the further down the cause–effect chain an indicator is chosen, the more environmental relevance (and meaning) it will have.

However, at the same time the level of model and parameter uncertainty may increase further down the cause–effect chain, while measurability decreases (and hence the possibility to evaluate and check the result against observations that can be directly linked to the original cause). Contrary to a frequent misconception, that does not mean that the total uncertainty (i.e. including all its sources, not just parameter and model uncertainty) of an indicator increases when going further down the cause–effect chain, because the increase in parameter and model uncertainty is compensated by an increase in environmental relevance. If the latter is low (as is the case for indicators placed early in the cause–effect chain) the relationship of an indicator to an environmental issue is assumed but not modelled and thus hypothetical and therefore uncertain. A detailed discussion on these issues can be found in Chap. 11.

To select the impact indicator, developers must therefore strike a compromise between choosing an indicator of impact:

1. Early in the environmental mechanism, giving a more measurable (e.g. in the lab) result but with less environmental relevance and more remote from the concerns directly observable in the environment

   Versus

2. 2. Downstream in the environmental mechanism, giving more relevant but hardly verifiable information (e.g. degraded ecosystems, affected human lifetime)

This has led to the establishment of two different types of impact categories, applying indicators on two different levels of the environmental mechanism: midpoint impact indicators (representing option 1 from above) and endpoint impact indicators (representing option 2).

10.2.3.6 Midpoint Impact Indicators

When the impact assessment is based on midpoint impact indicators, the classification gathers the inventory results into groups of substance flows that have the ability to contribute to the same environmental effect in preparation for a more detailed assessment of potential impacts of the environmental interventions, applying the characterisation factors that have been developed for the concerned impact category. For example, all elementary flows of substances that may have a carcinogenic effect on humans will be classified in the same midpoint category called “toxic carcinogen” and the characterisation will calculate their contribution to this impact. Typical (*and emerging*) midpoint categories (including respective sub-categories/impact pathways) are:

- Climate change
- Stratospheric ozone depletion
- Acidification (terrestrial, freshwater)
- Eutrophication (terrestrial, freshwater, marine)
- Photochemical ozone formation
- Ecotoxicity (terrestrial, freshwater, marine)
- Human toxicity (cancer, non-cancer)

- Particulate matter formation
- Ionising radiation (human health, aquatic and terrestrial ecosystems)
- Land Use (biotic productivity, aquifer recharge, carbon sequestration, albedo, erosion, mechanical and chemical filtration capacity, biodiversity)
- Water use (human health, aquatic ecosystems, terrestrial ecosystems, ecosystem services)
- Abiotic resource use (fossil and mineral)
- *Biotic resource use* (e.g. fishing or wood logging)
- *Noise*
- *Pathogens*

The characterisation at midpoint level of the elementary flows in the life cycle inventory results in a collection of midpoint impact indicator scores, jointly referred to as the characterised impact profile of the product system at midpoint level. This profile may be reported as the result of the life cycle impact assessment, and it may also serve as preparation for the characterisation of impacts at endpoint level.

10.2.3.7 Endpoint Impact Indicators

Additional modelling elements are used to expand or link midpoint indicators to one or more endpoint indicator (sometimes also referred to as damage or severity). These endpoint indicators are representative of different topics or "Areas of Protection" (AoP) that "defend" our interests as a society with regards to human health, ecosystems or planetary life support functions including ecosystem services and resources, for example. As discussed, endpoint indicators are chosen further down the cause–effect chain of the environmental mechanism closer to or at the very endpoint of the chains—the Areas of Protection. The numerous different midpoint indicators therefore all contribute to a relatively small set of endpoint indicators as can be observed in Fig. 10.2. Although, different distinctions are possible and exist, typical endpoint indicators are:

- Human health
- Ecosystem quality or natural environment
- Natural resources and ecosystem services

Therefore, the same list of impact categories as for midpoint indicators (see above) applies to endpoint indicators but with a further distinction regarding which of the three AoPs are affected (e.g. climate change usually has one midpoint indicator, but two endpoint indicators, one for human health and one for ecosystem quality—see Fig. 10.2). All endpoint indicators for the same AoP have a common unit and can be summed up to an aggregated impact score per AoP (assuming equal or different weighting of each endpoint indicator). Before aggregation, however, an environmental profile on endpoint level is as detailed as on midpoint level and allows for a contribution analysis of impact categories per AoP (e.g. which impact category contributes the most to human health impacts). On midpoint level,

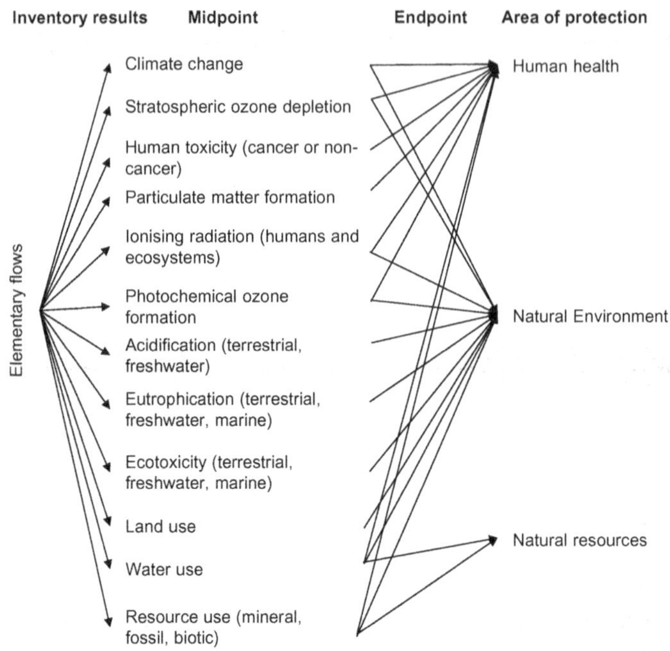

Fig. 10.2 Framework of the ILCD characterisation linking elementary flows from the inventory results to indicator results at midpoint level and endpoint level for 15 midpoint impact categories and 3 areas of protection [adapted from EC-JRC (2010b)]

aggregation and contribution analysis of multiple impact categories are only possible after applying normalisation and weighting.

There are three frequent misconceptions related to that:

1. *Misconception*: Applying normalisation, weighting and aggregation to midpoint indicator results is the same as calculating endpoint indicator results. Or in other words, midpoint indicator results that are normalised, weighted and aggregated into one impact score per AoP have the same unit as endpoint indicator results aggregated into one impact score per AoP. Therefore, both results are identical.
   Fact: Even though the unit of both aggregated indicators is the same, their numerical value and their physical meaning are completely different. They are not identical and cannot be interpreted in the same way.
2. *Misconception*: Changing from midpoint to endpoint characterisation implies a loss of information due to aggregation from about 15 midpoints into only three endpoint indicators.
   Fact: Before aggregation is applied, endpoint indicators are constituted for the same amount of impact categories as on midpoint level, but not every impact category contributes to each AoP (e.g. mineral resource depletion does not contribute to human health impacts). Therefore, the same analysis of contribution

per impact category is possible as for normalised and weighted midpoint indicators while avoiding the need for normalisation and weighting and the associated increased uncertainty and change in meaning.

3. *Misconception*: Endpoint characterisation is more uncertain than midpoint characterisation.
   Fact: This may be the case when looking at a limited set of sources of uncertainty and how they contribute to the uncertainty of the value of the indicator. However, when considering all relevant sources of uncertainty and the relevance of the indicator for the decision at hand, the choice of indicator has no influence on the uncertainty of the consequences of the decision. This is discussed in detail in Chap. 11.

To go from midpoint to endpoint indicator scores, additional midpoint-to-endpoint characterisation factors (sometimes also referred to as severity or damage characterisation factors) are needed, expressing the ability of a change in the midpoint indicator to affect the endpoint indicator. In contrast to the midpoint characterisation factors which reflect the properties of the elementary flow and hence are elementary flow-specific, the midpoint-to-endpoint characterisation factors reflect the properties of the midpoint indicator and there is hence only one per midpoint impact category. Some LCIA methods only support endpoint characterisation and here the midpoint and midpoint-to-endpoint characterisation is combined in one characterisation factor.

10.2.3.8 Midpoint or Endpoint Assessment?

Next to the relationship between environmental relevance and various sources of uncertainty discussed above (and in more detail in Chap. 11), the possibility to aggregate information from midpoint to endpoint level while avoiding normalisation has the advantage of providing more condensed information (fewer indicator results) to consider for a decision, while still being transparent as to which impact pathway(s) are the main causes of these damages. Instead of perceiving midpoint and endpoint characterisation as two alternatives to choose from, it is recommended to conduct an LCIA on both midpoint and endpoint level (using an LCIA method that provides both) to support the interpretation of the results obtained and which complement each other respectively.

10.2.3.9 Time Horizons and Temporal Variability?

Environmental impacts caused by an intervention will require different amounts of time to occur, depending on the environmental mechanism and the speed at which its processes take place. This means that next to the fact that the numerous elementary flows of an LCI may occur at different moments in time during the life cycle of the product or service assessed (which may be long for certain products like buildings for example), there is also a difference in the lag until their impacts occur. However, the way LCA is currently conducted, potential impacts are assessed as if interventions and potential impacts were happening instantly, aggregating them over time and over the entire life cycle. This means that these potential impacts need to be interpreted as a "backpack" of potential impacts attributable to the product or service assessed.

Next to such temporal variability, another potential source of time-related inconsistency in LCA is the problem of applying different time horizons for different impact categories. These time horizons are sometimes explicit (e.g. the 20 and 100 years' time horizons for global warming potentials), but in most cases implicit in the way the environmental mechanism has been modelled (e.g. over what time horizon the impact has been integrated). This may result in a mixing of different time horizons for different impacts in the same LCIA, which may have implications for the interpretation of LCA results. For example, methane has a lifetime much shorter than CO_2. Therefore, depending on the time horizon chosen, the characterisation of methane will change. This is directly connected to the question of how to consider potential impacts affecting current and immediate future generations versus those affecting generations in a more distant future.

Another issue concerns the temporal course of the emission and its resulting impacts. While some impacts may be immediately (i.e. within a few years) tangible and directly affecting a larger number of individuals (human or not), some impacts may be very small at any given moment in time, but permanently occurring for tens to hundreds of thousands of years (e.g. impacts from heavy metal emissions from landfills or mine tailings). Between these two illustrative extremes, lies any possible combination of duration versus severity.

10.2.3.10 Spatial Variability and Regionalisation?

Some impacts are described as global because their environmental mechanism is the same regardless where in the world the emission occurs. Global warming and stratospheric ozone depletion are two examples. Other impacts, such as acidification, eutrophication or toxicity may be classified as regional, affecting a (sub-)continent or a smaller region surrounding the point of emission only. Impacts affecting a small area are designated as local impacts, water or direct land-use impacts on biodiversity for example. Whereas for global impact categories the site where the intervention takes place has no considerable influence on the type and magnitude of its related potential impact(s), for regional or local impacts this may influence the magnitude of the potential impact(s) up to several orders of magnitude (e.g. a toxic emission taking place in a very large and densely populated city or habitat versus somewhere remote in a large desert). This spatial variability can be dealt with in two ways:

- Identification and modelling of archetypal emission/extraction situations and their potential impacts (e.g. toxic emission into urban air, rural air or remote air) or spatialized archetypes (e.g. city-specific emissions, formation and background concentrations of particulate matter and related mortality rates)

Or

- Modelling impacts with a certain degree of spatial resolution (e.g. sub-continental, country-level, sub-water-shed level or GPS grid-based), allowing for a characterisation which can be specific to any given place of emission or extraction

Both solutions require that the place of emission/extraction is known for each flow in the inventory—either explicitly (e.g. by country or geographical coordinates such as latitude and longitude) or regarding the most representative archetype. In order to support a spatially differentiated impact assessment, the life cycle inventory must thus not be aggregated to present one total intervention per elementary flow since this will lose the information about location of the interventions which is needed to select the right CF. Otherwise, generic global average CFs need to be used, leading to a higher uncertainty due to the spatial variability not considered in the characterisation. In contrast to the site-generic LCIA method, which provides one CF per combination of elementary flow and intervention/emission compartment, the spatially differentiated characterisation method provides one CF per combination of elementary flow, intervention/emission compartment and spatial unit. For grid-based methods, this may amount to thousands of CFs for each contributing elementary flow.

It depends on the impact category and emission situation to evaluate whether a spatial or archetypal setup will give the more accurate solution (e.g. urban/rural differences in particulate matter-related health effects might not be captured by spatial models with typical resolutions lower than $10 \times 10\ \text{km}^2$ at the global scale, whereas an archetypal model distinguishing between urban and rural emission situations would capture such differences). It should be noted that country-based characterisation is not meaningful from a scientific point of view, as most impacts are not influenced by political borders, although from a practical data-availability point of view this currently not unusual practice is understandable and normally an improvement to not considering the spatial variation at all. It should furthermore be noted that most currently available LCA software fails to support spatially differentiated characterisation, and therefore most LCAs are performed using the site-generic CFs.

10.2.3.11 The Units?

The unit of CFs for midpoint impact categories is specific for each category and LCIA method chosen, and therefore discussed in detail in the corresponding section dedicated in detail to each LCIA method in Chap. 40. However, two different approaches can be identified—expression in absolute form as the modelled indicator result (e.g. area of ecosystem exposed above its carrying capacity per kg of substance emitted for acidification) or expression in a relative form as that emission of a reference substance for the impact category which would lead to the same level of impact (e.g. kg CO_2-equivalents/kg of substance emitted for climate change).

In contrast, endpoint CFs are typically expressed in absolute units and the units are relatively common between those LCIA methods that cover endpoint modelling:

Human health: [years] expressed as DALY (Disability-Adjusted Life Years). This unit is based on a concept proposed by Murray and Lopez (1996) and used by the World Health Organisation. It considers different severity contributions defined as "Years of Life Lost per affected Person" YLL_p [year/disease case] and "Years of Life lived with a Disability per affected Person" YLD_p [years/disease case]. These statistical values are calculated on the basis of number and age of deaths (YLL) and disabilities (YLD) for a given disease. This information can be combined into a single indicator using disability weights for each type of disability to yield the "Disability Adjusted Life Years per affected Person" $DALY_p$ [year/person].

Ecosystem quality or Natural environment: [m^2 year] or [m^3 year] expressed as Potentially Disappeared Fraction (PDF). It can be interpreted as the time and area (or volume) integrated increase in the disappeared fraction of species in an ecosystem [dimensionless] per unit of midpoint impact indicator increase. It essentially quantifies the fraction of all species present in an ecosystem that potentially disappears (regardless whether due to death, reduced reproduction or immigration) over a certain area or volume and during a certain length of time. Different ecosystems have different numbers of species that can be affected by the impact and it is necessary to correct for such differences when aggregating the potentially disappeared fractions of species across the different impact categories at endpoint (i.e. a PAF of 0.5 for 10 species represents 5 species potentially lost, whereas the same PAF for 1000 species represents 500 species potentially lost).

Resource depletion and ecosystem services: Different approaches exist and since there is still no common perception of what the area of protection for resources is (Hauschild et al. 2013), there is also no consensus forming on how to model damage in the form of resource depletion. Some proposals focus on the future costs for extraction of the resource as a consequence of current depletion, and these divide into costs in the form of energy or exergy use for future extraction (measured in MJ) or monetary costs (measured in current currency like USD, Yen or Euro).

10.2.3.12 Uncertainties?

Uncertainties can be important in LCIA and contribute substantially to overall uncertainty of an LCA result. For some impact categories, this contribution may be much larger than that of the LCI. At the same time, it is also crucial to be aware that large uncertainty is by no means a valid reason to exclude an impact category from the assessment. One of the more uncertain impact categories is human toxicity and it has to be capable of dealing with hundreds to thousands of different elementary flows, which may differ by more than 20 orders of magnitude in their impact potential, due to the sheer number of substances (i.e. elementary flows) that may be assigned to this category and the variation in their environmental persistence and potential toxicity. It is much more certain to consistently characterise an impact category to which only a handful of elementary flows are assigned showing impact potentials that range only three or four orders of magnitude from the least to the most impacting elementary flow (e.g. eutrophication, acidification or global warming).

With the exception of photochemical ozone formation, there is no other impact category that covers even 100 different elementary flows. In this respect, there is hence a factor of >1000 between other impact categories and the toxicity categories (human health and ecotoxicity). This means that due to the large variety of substances with a toxicity potential, there will always be a very large uncertainty inherent in these categories, although developers will eventually be able to lower some of the model and parameter uncertainties currently observed. Excluding them from the assessment because of their uncertainty would therefore mean that toxicity would never be considered in LCA, which clearly risks violating the goal of LCA to avoid problem-shifting from one impact category to another. Besides, the uncertainty of assigning a zero-impact to a potentially toxic elementary flow by neglecting the toxicity impact categories is certainly higher than the inherent uncertainty of the related characterisation factors.

The solution rather lies in the way we interpret such inherently uncertain impact potentials, whereas a more certain impact indicator may allow for identifying the exact contribution of each elementary flow to the total impact in this category, toxicity indicators allow for identifying the (usually 5–20) largest contributing elementary flows, which will constitute >95% of the total impact. A further distinction between these will not be possible due to their uncertainty. Assuming that an average and complete LCI may contain several hundreds of potentially toxic elementary flows, one can then disregard all the remaining (several hundred) flows due to their low contribution to total toxicity. A further discussion and recommendations can be found in Rosenbaum et al. (2008).

Overall uncertainty in LCA is comprised of many different types of uncertainty as further discussed in Chap. 11. Variability (e.g. spatial or temporal/seasonal) may also be an important contributor, which should by principle be considered separately, as its contribution can be reduced to a large extent by accounting for it in the characterisation as discussed above for spatial variability and regionalised LCI and LCIA. Uncertainty in LCIA can only be reduced by improved data or model quality, essentially coming from updated LCIA methods, which is a good reason for a practitioner to keep up with the latest developments in LCIA, which may well lead to less uncertain results than the method one has been using for ten years. Most existing LCIA methods do not present information about the uncertainty of the characterisation factors.

10.2.3.13 What Are the Main Assumptions?

In current LCIA methods, some assumptions are considered as a basic requirement in the context of LCA:

- Steady-state: Although exceptions exist, LCIA models are usually not dynamic (i.e. representing the variation of an environmental system's state over time and for specific time steps), but represent the environment as a system in steady state, i.e. all parameters which define its behaviour are not changing over time.
- Linearity: As life cycle inventory (LCI) data are typically not spatially and/or temporally differentiated, integration of the impact over time and space is required. In LCIA, this leads to the use of characterisation models assuming steady-state conditions, which implies a linear relationship between the increase in an elementary flow and the consequent increase in its potential environmental impact. In other words, e.g. doubling the amount of an elementary flow doubles its potential impact.
- Marginal versus average modelling: These terms are used in different ways and meanings in the LCA context; here they describe two different impact modelling principles or choices: a marginal impact modelling approach represents the additional impact per additional unit emission/resource extraction caused by the product system on top of the existing background impact (which is not caused by the modelled product system). This allows, e.g. considering nonlinearity of impacts depending on local conditions like high or low background concentrations to which the product systems adds an additional emission). An average impact modelling approach is strictly linear and represents an average impact independent from existing background impacts, which is similar to dividing the overall impact by the overall emissions. This is further discussed by Huijbregts et al. (2011). Note that marginal and average modelling are both suitable for small-scale interventions such as those related to a product or service. However, when medium-scale or large-scale interventions (or consequences) are to be assessed, the characterisation factors should represent non-marginal potential impacts and may also have to consider nonlinearity.
- Potential impacts: LCIA results are not actual or predicted impacts, nor exceedance of thresholds or safety margins, or risk. They are relative expressions of impacts associated with the life cycle of a reference unit of function (=functional unit), based on inventory data which are integrated over space and time, representing different locations and time horizons and based on impact assessment data which lack information about the specific conditions of the exposed environment.
- Conservation of mass/energy and mass/energy balance: Mass/energy cannot be created or disappear, it can only be transferred. Following this principle, processes of transport or transformation of mass or energy are (or at least should be) modelled assuming that the mass/energy balance is conserved at all times.
- Parsimony: This refers to the basic modelling principle of "as simple as possible and as complex as necessary", an ideal balance that applies to LCIA characterisation models as well as to the entire LCA approach.
- Relativity: LCA results are relative expressions of impacts that relate to a functional unit and can be compared between different alternatives providing the same function (e.g. option A is more environmentally friendly than option B).

An absolute interpretation of LCA results (e.g. option A is sustainable, option B is not) is not advisable as it requires a lot of additional assumptions.

- Best estimates: A fundamental value choice in LCA is not to be conservative, precautionary or protective, but to focus on avoiding any bias between compared scenarios by assuming average conditions, also referred to as best estimates. Products or services assessed in LCA are typically not representing one specific example (e.g. with a serial number or from a specific date), but an average, often disregarding whether a specific life cycle process took place in summer or winter, during the day or night, etc. As discussed by Pennington et al. (2004), LCA is a comparative assessment methodology. Direct adoption of conservative regulatory methodology and data is often not appropriate, and should be avoided in LCIA in order not to bias comparison between impact categories where different levels of precaution may be applied.

10.3 Optional Steps According to ISO 14040/14044

10.3.1 Normalisation

The indicator scores for the different midpoint indicators are expressed in units that vary between impact categories and this makes it unfeasible to relate them to each other and to decide which of them are large and which small. To support such comparisons, it is necessary to put them into perspective, and this is the purpose of the normalisation step, where the product system's potential impacts are compared to those of a reference system like a country, the world or an industrial sector. By relating the different impact potentials to a common scale they can be expressed in common units, which provide an impression of which of the environmental impact potentials are large and which are small, relative to the reference system. Normalisation can be useful for:

- Providing an impression of the relative magnitudes of the environmental impact potentials
- Presenting the results in a form suitable for a subsequent weighting
- Controlling consistency and reliability
- Communicating results

Typical references are total impacts per impact category per:

- Geographical zone which can be global, continental, national, regional or local
- Inhabitant of a geographical zone (e.g. expressing the "environmental space" occupied per average person)
- Industrial sector of a geographical zone (e.g. expressing the "environmental space" occupied by this product system relative to similar industrial activities)

- Baseline reference scenario, such as another product system (e.g. expressing the "environmental space" occupied by this product system relative to a similar reference system using best available technology)

Using one of the first three reference systems listed above is also referred to as external normalisation. Using the last reference system in the list is also called internal normalisation when the reference scenario is one of the compared alternatives, such as the best or worse of all compared options or the baseline scenario representing, e.g. a current situation that is intended to be improved or a virtual or ideal scenario representing a goal to be reached. Normalised impact scores, when using internal normalisation, are often communicated as percentages relative to the reference system. In the illustrative case on window frames in Chap. 39 an internal normalisation is applied using the wooden frame window as reference (indexing it to 100%) to reveal how the studied alternatives compare to this baseline choice. The study also applies external normalisation in order to compare the size of the different midpoint impact scores with the European person equivalent impact scores that is provided as default normalisation references for the LCIA method applied in the study (the ILCD method).

In practice, an LCIA method generally provides normalisation factors for use with its characterisation factors. The normalisation factors should be calculated using the same characterisation factors for the reference inventory as used for the inventory of the product system. Normalisation factors from different LCIA methods thus cannot be mixed or combined with characterisation factors from another LCIA method. This means that as an LCA practitioner you are usually limited to the reference system chosen by the LCIA method developers. Normalisation is applied using normalisation factors (NF). These are essentially calculated per impact category by conducting an LCI and LCIA on the reference system, i.e. quantifying all environmental interventions E for all elementary flows i for the reference system and applying the characterisation factors CF per elementary flow i, respectively, for each impact category c. Although not obligatory, the normalisation reference is typically divided by the population P of the reference region r, in order to express the NF per average inhabitant of the reference region (per capita impacts or "person equivalents"). This way, a total impact of the reference system per impact category is calculated, resulting in one NF per impact category c:

$$\mathrm{NF}_c = \left(\frac{\sum_i (\mathrm{CF}_i \cdot E_i)}{P_r}\right)^{-1} \tag{10.2}$$

Ensuring consistency, the LCI data used to calculate a NF need to represent a common reference year and duration of activity (typically one year, being the reference year) for all impact categories. This results in NF having a unit expressing an impact per person and year, also referred to as person equivalent. A normalised impact score NS for a product system is calculated by multiplying the calculated impact score IS for the product system by the relevant NF per impact category c:

$$NS_c = IS_c \cdot NF_c \tag{10.3}$$

Two different approaches exist for collection of inventory data for the calculation of NFs (with the exception for global NFs, where both approaches give equal results):

- Production-based (or top-down), representing the interventions taking place in the reference region as result of the total activities in the region
- Consumption-based (or bottom-up), representing the interventions that are caused somewhere in the world as consequence of the consumption taking place in the reference region (and thus representing the demand for industrial and other activities within and outside the reference region)

Other ways to derive NF (although somewhat bordering to weighting already) are to base them on a conceptual "available environmental space". This can be determined using, e.g. political targets for limits of environmental interventions or impacts for a given duration and reference year (i.e. "politically determined environmental space" being the average environmental impact per inhabitant if the political reduction targets are to be met), or a region's or the planet's carrying capacity (i.e. "environmental space" being the amount of environmental interventions or impacts that the region or planet can buffer without suffering changes to its environmental equilibrium within each impact category). The latter would require knowing the amount of impact that a region or the planet can take before suffering permanent damage, which is a concept associated with much ambiguity and hence very uncertain to quantify. There is increasing focus on science-based targets in the environmental regulation with the 2 °C ceiling for climate change as the most prominent example, and this may lead to future consensus building on science-based targets also for some of the other impacts that are modelled in LCIA. Political targets are often determined at different times and apply to different periods of time. In order to ensure a consistent treatment of each impact category, it is necessary to harmonise the target values available so that all targets for any given intervention are converted to apply to the same period and reference year. The targets can be harmonised by interpolating or extrapolating to a reduction target for a common target year, computed relative to interventions in the reference year. More details can be found in Hauschild and Wenzel (1998).

Caution is required when interpreting normalised LCA results! Applying normalisation harmonises the metrics for the different impact potentials and brings them on a common scale, but it also changes the results of the LCA and consequently may change the conclusions drawn from these. Since there is no one objectively correct choice of reference systems for normalisation, the interpretation of normalised LCA results must therefore always be done with due consideration of this choice of normalisation reference. A few main issues that need to be considered when interpreting normalised LCA results are:

- Depending on the size of and activities reflected in the reference system, different biases may be introduced in the comparison of the impact scores of a product system. As a general principle, the larger the reference system, the lesser the risk of such bias when normalising against the background activities of society.

- While supporting comparison of results across impact categories, normalised LCA results cannot be interpreted as reflecting a weight or importance of one impact category relative to others. Normalisation helps to identify the impacts from the product system that are large compared to the chosen reference system, but large is not necessary the same as important. It is therefore not suitable as the only basis for identification of key issues/impacts in a product system, unless explicitly required by the goal and scope definition (e.g. evaluating the environmental impact contribution of a product system to a reference system which it is part of).
- Unless (a) the reference system is global or (b) all environmental interventions of the product system assessed take place in the same region as those of the reference system, the direct interpretation of normalised impacts as contributions to or fractions of the reference system is misleading because parts of the life cycle of the product or service take place in different regions of the world, including outside the reference system.

By expressing the different impact scores on a common scale, normalisation can also help checking for potential errors in the modelling of the product system. If the results are expressed in person equivalents, it is possible to spot modelling errors leading to extremely high or low impacts in some of the impact categories—like frequent unit errors when emissions are expressed in kg instead of g. Looking across the impact category results in a normalised impact profile, it is also possible for the more experienced LCA practitioner to check whether they follow the pattern that would be expected for this type of product or service.

Although characterisation and aggregation at endpoint level leads to fewer impact scores (typically three), normalisation may still be useful with the same purposes as normalisation at midpoint level. The calculation and application of the endpoint normalisation references follows the same procedure as for midpoint normalisation, just applying combined midpoint and endpoint characterisation factors in Eq. 10.2.

10.3.2 Weighting (and Aggregation)

Weighting can be used to determine which impacts are most important and how important they are. This step can only be applied after the normalisation step and allows the prioritisation of impact categories by applying different or equal weights to each category indicator. It is important to note that there is no scientific or objective basis for this step. This means that, no matter which weighting method or scheme is applied, it will always be based on the subjective choices of one person or a group of individuals. Weighting can be useful for:

- Aggregating impact scores into several or one single indicator (note that according to ISO 14040/14044 there is no scientific basis on which to reduce the results of an LCA to a single result or score because of the underlying ethical value-choices)
- Comparing across impact categories
- Communicating results applying an underlying prioritisation of ethical values

Note that in all of these cases weighting is applied, either implicitly or explicitly! Even when applying no explicit weighting factors in the aggregation by simply summing up impact scores, there is always an implicit equal weighting (all weighting factors = 1) inherently applied when doing any of the above. According to ISO 14044, weighting is not permitted in a comparative assertion disclosed to the public and weighted results should always be reported together with the non-weighted ones in order to maintain transparency. The weighting scheme used in an LCA needs to be in accordance with the goal and scope definition. This implies that the target group including their preferences and the decisions intended to be supported by the study need to be considered, making shared values crucial for the acceptance of the results of the LCA. This can pose important problems due to the variety of possible values among stakeholders, including:

- Shareholders
- Customers
- Employees
- Retailers
- Authorities
- Neighbours
- Insurance companies
- NGOs (opinion leaders)
- …

It may not be possible to arrive at weighting factors that will reflect the values of all stakeholders so focus will typically have to be on the most important stakeholders, but is it possible to develop one set of weighting factors that they will all agree on? If this is not the case, several sets of weighting factors may have to be applied, representing the preferences of the most important stakeholder groups. Sometimes the use of the different sets will lead to the same final recommendations which may then satisfy all the main stakeholders. When this is not the case, a further prioritisation of the stakeholders is needed, or the analysed product system (s) must be altered in a way that allows an unambiguous recommendation across the applied weighting sets.

The weighting of midpoint indicators should not be purely value-based. More, to some extent, science-based criteria for importance of environmental impacts may be:

- Probability of the modelled consequences, how certain are we on the modelled cause–effect relations?
- What is the resilience of the affected systems?

- Existence of impact thresholds—in the characterisation modelling we typically assume linear cause–effect relationships for the small interventions in the product system, but in the full environmental scale, there may be impact levels that represent tipping points beyond which much more problematic effects occur.
- If so, then how far are we from such critical impact levels—is this an important concern in the near future?
- Severity of effect and gravity of consequences—disability, death, local extinction, global extinction
- Geographical scale
- Population density is essential for the impacts on human health.
- Possibility to compensate/adapt to impact
- Temporal aspects of consequences—when will we feel the consequences, and for how long?
- Is the mechanism reversible, can we return to current conditions if we stop the impacts?

Indeed, many of these science-based criteria are attempted to be included in the environmental modelling linking midpoint indicators to endpoint indicators, and midpoint-to-endpoint characterisation factors may thus be seen as science-based weighting factors for the midpoint impact categories.

Different principles applied to derive weighting factors are:

- Social assessment of the damages (expressed in financial terms like willingness to pay), e.g. impact on human health based on the cost that society is prepared to pay for healthcare (e.g. used in EPS and LIME LCIA methods)
- Prevention costs (to prevent or remedy the impact through technical means), e.g. the higher the costs, the higher the weighting of the impact
- Energy consumption (to prevent or remedy the impact through technical means), e.g. the higher the energy consumption, the higher the weighting of the impact
- Expert panel or stakeholder assessment, e.g. weight attributed based on the relative significance, from a scientific perspective (subjective to each expert), of the different impact categories
- Distance-to-target (politically or scientifically defined): degree at which the targeted impact level is reached (distance from the target value), the greater the distance, the more weight is assigned to the impact (e.g. used in EDIP, Ecopoints and Swiss Ecoscarcity LCIA methods).
- Social science-based perspectives, not representing the choices of a specific individual, but regrouping typical combinations of ethical values and preferences present in society into a few, internally consistent profiles (e.g. used in ReCiPe and Ecoindicator99 LCIA methods).

The latter approach is relatively widely used and applies three cultural perspectives, the Hierarchist, the Individualist and the Egalitarian (a forth perspective, the Fatalist is not developed for use in LCA since the fatalist is expected not to be represented among decision-makers, targeted by an LCA). For each cultural perspective coherent choices are described in Table 10.2 for some of the central

Table 10.2 Cultural perspectives represented by preference with coherent choices (Hofstetter 1998)

	Time perspective	Manageability	Required level of evidence
H (Hierarchist)	Balance between short and long term	Proper policy can avoid many problems	Inclusion based on consensus
I (Individualist)	Short term	Technology can avoid many problems	Only proven effects
E (Egalitarian)	Very long term	Problems can lead to catastrophe	All possible effects

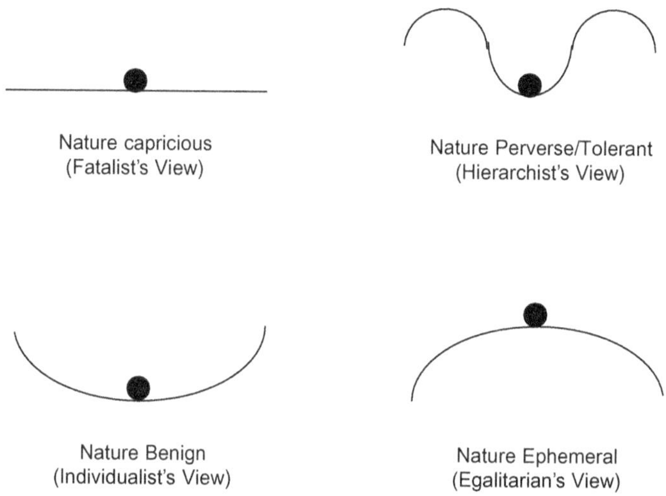

Fig. 10.3 Different archetypal perceptions of nature [adapted from Thompson (1990)]

assumptions made in the characterisation modelling and in the development of a set of consistent weighting factors for each archetype.

The different archetypal views on nature and the related risk perceptions are illustrated in Fig. 10.3. The dot represents the state of nature as a rolling ball, shifted by human activities along the curve representing nature's reaction to a shift. Its position in the figures indicates the state of harmony between humans and nature according to the four archetypal views.

10.3.3 Grouping

This step consists in placing the impact categories in one or several groups or clusters (as defined in goal and scope) and can involve sorting or ranking, applying one of two possible methods:

- Sorting and clustering midpoint impact categories on a nominal basis (e.g.: by characteristics such as emission-related and resource-related, or global, regional or local spatial scales)
- Ranking the impact categories according to a set (subjective—based on ethical value-choices) hierarchy (e.g.: high, medium or low priority)

10.4 Footprints Versus LCA

"I was exceedingly surprised with the print of a man's naked foot on the shore, which was very plain to be seen in the sand." (Daniel Defoe, Robinson Crusoe, 1719). The meaning of the term "footprint" has largely evolved since Daniel Defoe's famous novel and is currently used in several contexts (Safire 2008). Its appearance in the environmental field can be tracked back to 1992 when William Rees published the first academic article on the thus-termed "ecological footprint" (Rees 1992), which was further developed by him and Mathis Wackernagel in the following years. Its aim is to quantify the mark left by human activities on natural environment.

Since then, the mental images created by the word have contributed to its use as an effective way of communicating on different environmental issues and raising environmental awareness within the scientific community as well as among policy communities and the general public. Since the early 2000s, several footprints have thus emerged within the environmental field with different definitions and meanings, ranging from improved ecological footprint methodologies to the representation of specific impacts of human activities on ecosystems or human health to a measure of a specific resource use. Prominent examples are:

- Ecological footprint focusing on land use (http://www.footprintnetwork.org)
- Cumulative Energy Demand (CED) focusing on non-renewable energy
- Material Input Per unit of Service (MIPS) focusing on material use
- Water footprint focusing on water use volumetric accounting (http://waterfootprint.org)
- Water footprint focusing on water use impacts including pollution (ISO 14046)
- Carbon footprint focusing on climate change (ISO 14064, ISO/TS 14067, WRI/WBCSD GHG protocol, PAS 2050)

Later developments focused on the introduction of new environmental concerns or enlarging the scope of footprints. Examples for such emerging footprints are:

- Chemical footprint focusing on toxicity impacts
- Phosphorus depletion footprint

As illustrated in Fig. 10.4, all footprints are fundamentally based on the life cycle perspective and most of them focus on one environmental issue or area of concern.

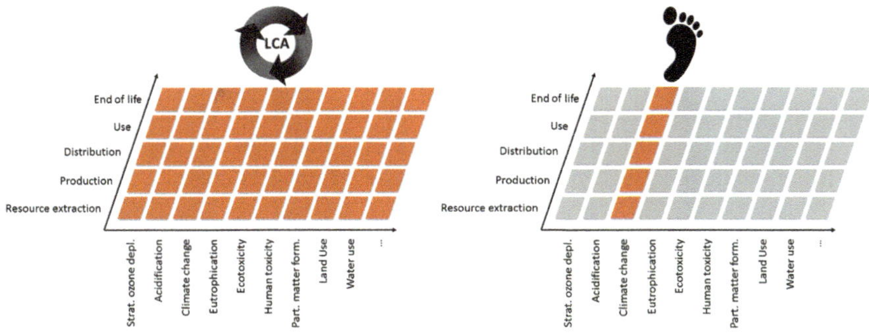

Fig. 10.4 The fundamental difference in scope and completeness between LCA and footprints while both apply the life cycle perspective

They can be applied to a large variety of assessment targets like products, services, organisations, persons and populations, sites and regions, even countries or the entire world. Their success in the last decades lies in their particular strengths:

- Easily accessible and intuitive concept
- Easy to communicate about specific environmental issues or achievements with non-environmental experts (policy and decision-making communities, general public)
- Availability of data
- Easy to perform
- Wide range of assessment targets can easily be assessed

These strengths, however, also come with a number of important limitations:

- Their focus on one environmental issue does not inform about a potential burden-shifting from one environmental issue (e.g. climate change) to another (e.g. water availability). Therefore, while they allow for identification of the best option for one environmental problem, they are not suitable to support decisions regarding environmental sustainability, which need to consider all potential environmental problems.
- Some footprints only assess the quantity of a resource used (e.g. ecological footprint, CED, MIPS and volumetric water footprint), which is comparable to the accounting of quantities used or emitted in the life cycle inventory (see Chap. 9). Such footprints therefore do not inform about the associated environmental consequences of the resources used or emissions accounted, and they do not quantify potential impacts on a given area of protection. Among other, this limitation compromises the comparability of footprints for different options to choose from.
- Impact-based footprints (e.g. carbon footprint), at least historically, assess impacts on midpoint level and hence do not reflect damages, which has implications on their environmental relevance. However, with an increasing

range of endpoint impact indicators available, this may be solved with science advancing further.
- Different footprints can usually not be combined to enlarge their environmental scope because their system boundaries (see Chaps. 8 and 9) are not aligned and double counting of impacts becomes likely, which increases the risk of bias to the comparison, the same way the omission of impacts does.

As mentioned above, the focus on single environmental problems has important implications regarding the risks of using footprints in decision-making processes. A study by Huijbregts et al. (2008) calculated 2630 product-specific ecological footprints of products and services (e.g. energy, materials, transport, waste treatment, etc.). They concluded that "Ecological footprints may [...] serve as a screening indicator for environmental performance... [and provide] a more complete picture of environmental pressure compared to non-renewable CED [Cumulative Energy Demand]", while also observing that "There are cases that may [...] not be assessed in an adequate way in terms of environmental impact. For example, a farmer switching from organic to intensive farming would benefit by a smaller footprint for using less land, while the environmental burdens from applying more chemicals [i.e. pesticides and fertilisers] would be neglected". Thus, the usefulness of the ecological footprint as a stand-alone indicator may often be limited (Huijbregts et al. 2008).

The limitations of carbon footprints (i.e. the climate change impact indicator in LCA) as environmental sustainability indicators was investigated by a study from Laurent et al. (2012), who assessed the carbon footprint and 13 other impact scores from 4000 different products, technologies and services (e.g. energy generation, transportation, material production, infrastructure, waste management). They found "that some environmental impacts, notably those related to emissions of toxic substances, often do not covary with climate change impacts. In such situations, carbon footprint is a poor representative of the environmental burden of products, and environmental management focused exclusively on [carbon footprint] runs the risk of inadvertently shifting the problem to other environmental impacts when products are optimised to become more "green". These findings call for the use of more broadly encompassing tools to assess and manage environmental sustainability" (Laurent et al. 2012).

This problem is demonstrated in Fig. 10.5, which shows the carbon footprint, ecological footprint, volumetric water footprint and the LCA results for an illustrative comparison of two products A and B. If one had to choose between option A and B, the decision would be different and thus depending on, which footprint was considered, whereas LCA results provide the full range of potential impacts to consider in the decision.

The large variety in footprints and their definitions and methodological basis in combination with their wide use in environmental communication and marketing claims, has resulted in confusing and often contradictory messages to buyers. This ultimately limited the development and functioning of a market for green products (Ridoutt et al. 2015, 2016). In response, a group of experts established under the

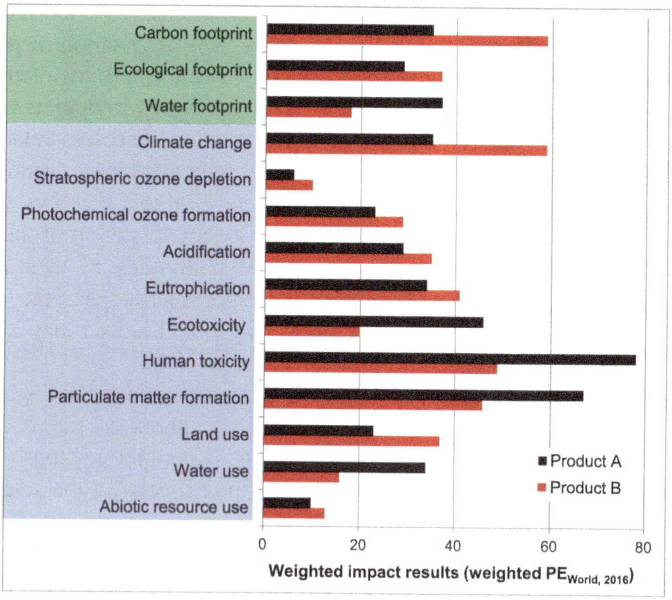

Fig. 10.5 Comparing two products, which alternative would you choose? Examples of footprints are indicated in *green shading*; impact categories commonly assessed in LCA are indicated in *blue shading*

auspices of the UNEP/SETAC Life Cycle Initiative defined footprint as "Metric used to report life cycle assessment results addressing an area of concern [the latter specified as an] Environmental topic defined by the interest of society" (Ridoutt et al. 2016). This definition underpins a footprint's focus on environmental issues particularly perceived by society (e.g. climate change or water scarcity) and allows for a clear distinction to LCA, which is primarily oriented "toward stakeholders interested in comprehensive evaluation of overall environmental performance and trade-offs among impact categories" (Ridoutt et al. 2016) and related areas of protection. This definition also recognises the inherent complexity of an environmental performance profile resulting from an LCA study, which requires a certain expertise to be correctly interpreted.

In conclusion, footprints are life cycle-based, narrow-scoped, environmental metrics focusing on an area of concern. They are widely and easily applicable, as well as easily understood by non-environmental experts and therefore straightforward to communicate. They are particularly useful for communication of environmental problems or achieved improvements, as long as their use is restrained to their coverage of environmental concerns and care is taken when interpreting them (burden-shifting), particularly when results are disclosed to non-expert audiences (e.g. public opinion). A footprint's life cycle perspective can be an inspiring first contact with the concept of life cycle thinking for the general public, and for policy and decision-makers it often serves as an entry-door into the concept and methodology of LCA. Footprints have the ability to raise environmental awareness and therefore are springboards towards the use of more-encompassing assessment tools such as LCA. They can constitute a first step for organisations or companies, who can already implement procedures as a preparation for full environmental assessments. However, due to a footprint's narrow scope and limited representativeness for a comprehensive set of environmental indicators, they are not suitable for decision-support of any kind including product labels, ecodesign, policy-support and the like.

10.5 Detailed Description of Impact Categories Currently Assessed in LCA

The following sections document how the most commonly considered environmental problems (i.e. impact categories) are handled in life cycle impact assessment. Ionising radiation is also a commonly addressed impact category in LCA, but was not included in the detailed overview here due to its specificity to a limited number of processes in the LCI. The impact categories are dealt with in sequence going from global over regional towards local and addressing first the emission-related and then the extraction-related categories. The common structure of the sections is:

- What is the problem?
- What is the underlying environmental mechanism and how is it modelled in LCIA?
- What are the human activities and elementary flows contributing most to the problem? (emission-based categories only)
- What are the most widely used, existing LCIA characterisation models?

Beyond the classic list of impact categories discussed hereafter, there is a number of emerging categories currently in the stage of research and development. Though potentially relevant they have not yet reached sufficient methodological maturity to be operational for the majority of practitioners and no or only few LCIA methods have included them in their indicator set. Some examples are:

- Biotic resources such as fish or wood
- Noise
- Pathogens
- Salinization
- Accidents
- Impacts of Genetically Modified Organisms (GMO)

A profound comparison of existing LCIA methods was performed by Hauschild et al. (2013) for the establishment of recommended LCIA models for the European context. Taking Hauschild et al.'s work as a starting point, the tables in Chap. 40

provide a complete and updated qualitative comparison of widely used LCIA methods available in current LCA software.

10.6 Climate Change

10.6.1 Problem

The greenhouse effect of our atmosphere, discovered and explored from the early 19th century, is vital to life on our planet and has always existed since the dawn of life on Earth. Without it the global average temperature of our atmosphere near the ground would be −18 °C instead of currently 15 °C. Hence, there are natural drivers and sources keeping it in balance (with periodical imbalances leading to natural events such as ice ages). In addition to those, anthropogenic activities also contribute to this effect increasing its intensity and creating *global warming*, which refers to the phenomenon of rising surface temperature across the planet averaged over longer periods of time. The Intergovernmental Panel on Climate Change (2014a) (IPCC) defines *climate change* as "a change in the state of the climate that can be identified (e.g. using statistical tests) by changes in the mean and/or the variability of its properties, and that persists for an extended period, typically decades or longer". IPCC observed an acceleration of the rise in planetary surface temperature in the last five to six decades, with the highest rates at the very northern latitudes of the Arctic. Ocean temperatures are also on the rise down to a depth of at least 3000 m and have so far absorbed most of the heat trapped in the atmosphere. Tropospheric temperatures are following similar trends as the surface. Although, still debated by few sceptics, most scientists agree on the presence of this effect with anthropogenic activities as the main cause. These are also the focal point of LCIA methodology and hence of this chapter.

Effects observed by IPCC with varying degrees of confidence based on statistical measures (IPCC 2014a):

- Rise of atmospheric temperature with the last three decades from 1983 to 2012 being very likely the warmest 30-year period of the last 800 years in the Northern Hemisphere and likely the warmest 30-year period of the last 1400 years
- Rise of ocean temperature in the upper 75 m by a global average of 0.11 °C per decade from 1971 to 2010
- Melting of glaciers, snow and ice caps, polar sea ice and ice packs and sheets (≠polar sea ice) and permafrost soils
- Rise in global mean sea levels by 0.19 m over the period 1901–2010 (due to thermal expansion and additional water from melting ice)
- Increase in frequency and intensity of weather-based natural disasters, essentially due to increased atmospheric humidity and consequent changes in

atmospheric thermodynamics (i.e. energy absorption via evaporation and condensation) and cloud formation
- Intense tropical cyclone activity increased in the North Atlantic since 1970
- Heavy precipitation and consequent flooding (North America and Europe)
- Droughts
- Wildfires
- Heat waves (Europe, Asia and Australia)
- Alteration of hydrological systems affecting quantity and quality of water resources
- Negative impacts of climate change on agricultural crop yields more common than positive impacts
- Shifting of geographic ranges, seasonal activities, migration patterns, abundances and species interactions (including in biodiversity) by many terrestrial, freshwater and marine species
- Changes in infectious disease vectors

The continuation and intensification of already observed effects as well as those not yet observed (but predicted by models as potential consequences of further global warming) depend on the future increase in surface temperature which is predicted using atmospheric climate models and a variety of forecasted emission scenarios ranging from conservative to optimistic. Given the inertia of atmospheric and oceanic processes and the global climate, it is expected that global warming will continue over the next century. Even if emissions of GHGs would stop immediately, global warming would continue and only slow down over many decades. The following effects are not yet observed and highly debated in the scientific community; hence consensus or general agreement regarding their likelihood is not established. Nevertheless, they are possible impacts and should be seen as part of the possible effects of global warming, especially when considering longer time horizons.

- Slowing down of the thermohaline circulation of cold and salt water to the ocean floor at high latitudes of the northern hemisphere (e.g. Gulf stream), among other things responsible for global heat distribution, oceanic nutrient transport, the renewal of deep ocean water, and the relative mildness of the European climate. This circulation as shown in Fig. 10.6 is driven by differences in the density of water due to varying salinity and differences in water temperature, and might be affected by freshwater inflow from melting ice, decreasing sea water salinity and consequently reducing its density and the density gradient between different oceanic zones.
- Increasing frequency and intensity of "El Niño" events while decreasing that of its counterpart "La Niña" might be possible, although it is unclear to what extent this is influenced by global warming. One possibility is that this effect only occurs in the initial phase of global warming, while weakening again later when the deeper layers of the ocean get warmer as well. Dramatic changes cannot be fully excluded based on current evidence; therefore, this effect is considered a potential tipping element in our climate.

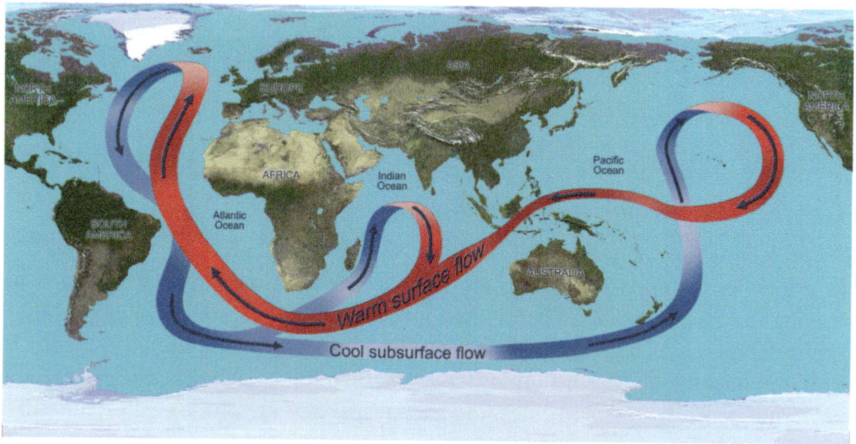

Fig. 10.6 "The big loop" takes 1500 years to circumnavigate the globe (NASA/JPL 2010, public domain, http://www.jpl.nasa.gov/news/news.php?release=2010-101)

- Mobilisation and release of oceanic methane hydrate (water ice containing large amounts of methane in its crystal structure) present in deep ocean sediments and permafrost, could lead to further global warming and significantly affect the atmospheric oxygen content. There is large uncertainty regarding the amounts and size of reserves found under sediments on the ocean floors, but a relatively sudden release of large amounts of methane hydrate deposits is believed to be a main factor in the global warming of 6 °C during the end-Permian extinction event (Benton and Twitchet 2003) when 96% of all marine species became extinct 251 million years ago.
- Effects on Earth's primary "lung": phytoplankton which produces 80% of terrestrial oxygen and absorbs a significant share of CO_2.
- In addition to the environmental effects discussed above, the human population is likely to be affected by further severe consequences should other adaptation strategies prove inefficient: disease, malnutrition and starvation, dehydration, environmental refugees, wars and ultimately death.
- Nonlinearity of cause–effect chains, feedback and irreversible tipping points: Although, in LCIA models, linearity of cause–effect chains is assumed, the above discussed effects present several examples of mechanisms that are unlikely to depend linearly on the temperature increase, i.e. they will not change proportionally in frequency and/or intensity per degree of change in global temperature. Furthermore, they are likely to directly or indirectly influence each other, causing feedback reactions adding further nonlinearity. Additionally, some of these effects will be irreversible, changing the climate from one stable state to another. This phenomenon is referred to as tipping points, and the above-mentioned release of methane from methane hydrates and the alteration of the Gulf stream are examples. Lenton et al. (2008) discuss a number of additional potential tipping points.

- Forest dieback (Boreal forest, Amazon rainforest).
- Area encompassed by monsoon systems will increase with intensified precipitation.

10.6.2 Environmental Mechanism

In principle, the energy reaching the Earth's atmosphere from solar radiation and leaving it again (e.g. via reflection and infrared radiation) is in balance, creating a stable temperature regime in our atmosphere. As shown in Fig. 10.7, from the sunlight reaching the Earth's atmosphere, one fraction (~28%) is directly reflected back into space by air molecules, clouds and the surface of the earth (particularly oceans and icy regions such as the Arctic and Antarctic): this effect is called albedo. The remainder is absorbed in the atmosphere by greenhouse gases (GHG) (21%) and the Earth's surface (50%). The latter heats up the planetary surface and is released back into the atmosphere as infrared radiation (black body radiation) with a longer wave length than the absorbed radiation. This infrared radiation is partially absorbed by GHGs and therefore kept in the atmosphere instead of being released into space, explaining why the temperature of the atmosphere increases with its content of GHGs.

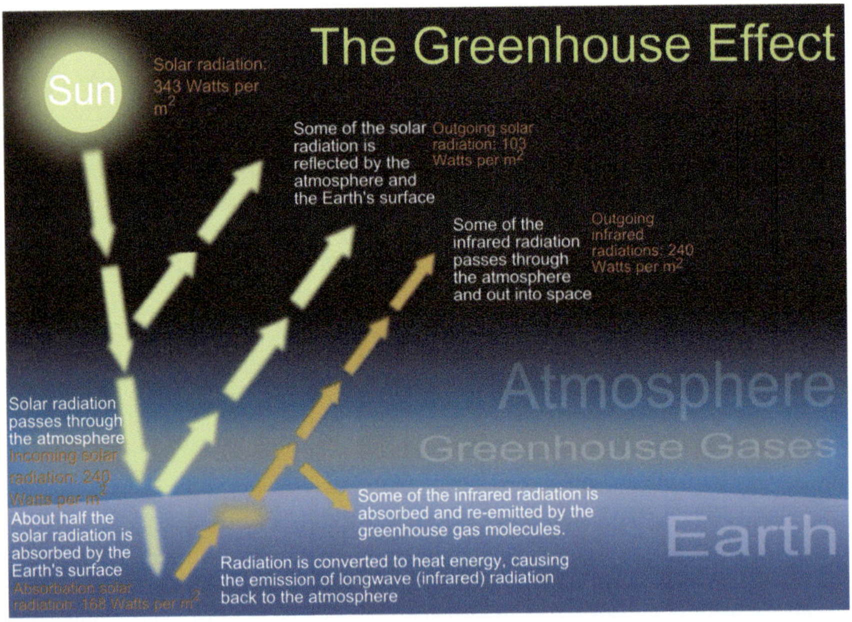

Fig. 10.7 The greenhouse effect (©User: ZooFari/Wikimedia Commons/CC-BY-SA-3.0)

A cause–effect chain for climate change is shown in Fig. 10.8 and can be summarised as follows:

1. GHG emissions
2. Transport, transformation and distribution of GHG in the atmosphere
3. Disturbance of the radiation balance—radiative forcing (primary effect, midpoint)
4. Increase in global temperatures of atmosphere and surface
5. Increase in sea level due to heat expansion and the melting of land-based ice
6. Increased water vapour content of the atmosphere causing more extreme weather
7. Negative effects on the ecosystems and human health (endpoint)

Until now the unanimously used climate change indicator on midpoint level in LCA has been the Global Warming Potential, an emission metric first introduced in the IPCC First Assessment Report (IPCC 1990) and continuously updated by IPCC

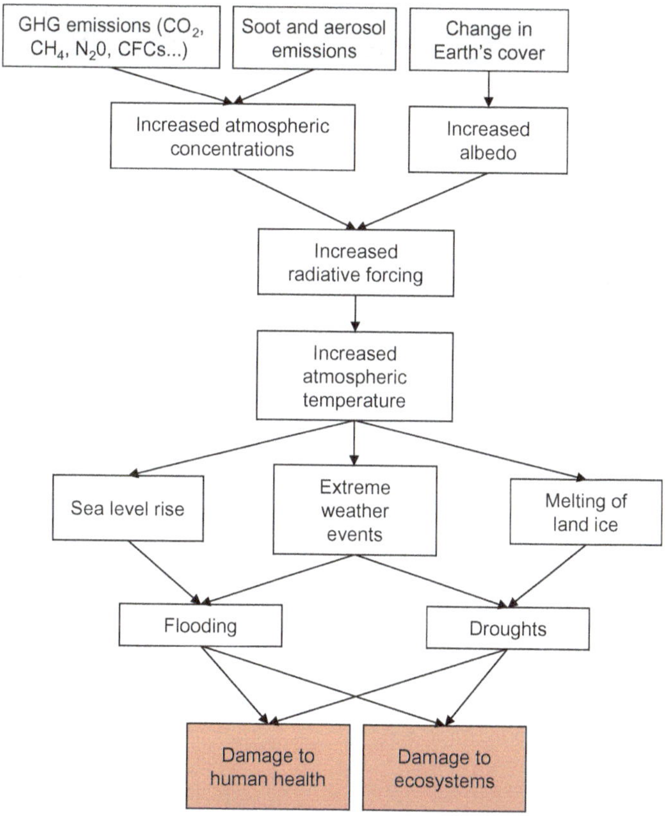

Fig. 10.8 Impact pathway for climate change

since then with the latest version in the Fifth Assessment Report (IPCC 2013). Global warming potentials are calculated for each GHG according to:

$$\mathrm{GWP}_i = \frac{\int_0^T a_i \cdot C_i(t)\mathrm{d}t}{\int_0^T a_{\mathrm{CO}_2} \cdot C_{\mathrm{CO}_2}(t)\mathrm{d}t} \tag{10.4}$$

where

- a_i: thermal radiation absorption (instant radiative forcing) following an increase of one unit in the concentration of gas i
- $C_i(t)$: Concentration of gas i remaining at time t after emission
- T: number of years for which the integration is carried out (e.g. 20 or 100 years)

GWP100-year is directly used in LCIA as the characterisation factor. As shown above, it is the ratio of the cumulated radiative forcing over 100 years of a given GHG and that of CO_2, with the unit of kg CO_2-eq/kg GHG. Therefore, GWP for CO_2 is always 1 and a GWP100 for methane of 28 kg CO_2-eq/kg methane (see Table 10.3) means that methane has 28 times the cumulated radiative forcing of CO_2 when integrating over 100 years. The difference in GWP20 and GWP100 for methane shown in Table 10.3 is due to the fact that methane has a relatively short atmospheric lifetime of 12 years compared to CO_2's lifetime which is at least one order of magnitude higher, which means that methane's GWP gets lower the longer the time horizon over which it is integrated (i.e. sort of a 'dilution' of its effect over a longer time). On the other hand a more persistent GHG such as nitrous oxide with 120 years lifetime has a similar value when integrating over 20 and 100 years and the 'time-dilution' effect would only become visible when integrating over time periods significantly longer than 120 years.

10.6.3 Emissions and Main Sources

Many greenhouse gases are naturally present in the atmosphere and contribute to the natural greenhouse effect. Estimated main contributors to the natural greenhouse effect are:

Table 10.3 Excerpt from the list of GWP (IPCC 2014a)

Substance	Molecule	Atmospheric lifetime (years)	Radiative efficiency (W/(m^2 ppb))	GWP (kg CO_2-eq/kg GHG)	
				20 years	100 years
Carbon dioxide	CO_2		1.37E−05	1	1
Methane	CH_4	12	3.63E−04	84	28
Nitrous oxide	N_2O	121	3.00E−03	264	265

- Water vapour: ~55%
- Carbon dioxide (CO_2): 39%
- Ozone (O_3): 2%
- Methane (CH_4): 2%
- Nitrous oxide (N_2O): 2%

Anthropogenic water vapour emissions do not contribute to climate change as the presence of water vapour is a function of atmospheric temperature and evaporation surfaces. For the other constituents however, anthropogenic sources for CO_2, CH_4 and N_2O do contribute to increasing the greenhouse effect beyond its natural state. Further relevant GHG emissions also include industrial volatile and persistent halocarbons (chlorinated fluorocarbons including CFCs ("freons"), HCFCs and perfluoromethane) and sulphur hexafluoride (SF_6). GHG emissions are attributable to almost any human activity. The most important contributing activities are: burning of fossil fuels and deforestation (including releasing carbon from soil and change in albedo). Figure 10.9 shows the global contributions to GWP from five major economic sectors for the year 2010. Industry, agriculture, housing and transport are the dominating contributors to GHG emissions.

In addition to the greenhouse gases which all exert their radiative forcing in the atmosphere over timespans of years to centuries, there are also more short-lived radiative forcing agents that are important for the atmospheric temperature in a more short-term perspective. These include:

- Sulphate aerosols (particulate air pollution caused by the emission of sulphur oxides from combustion processes) that reduce the incoming radiation from the sun and thus have a negative contribution to climate change
- Nitrogen oxides NO and NO_2 (jointly called NO_x) and VOC from combustion processes, that contribute to photochemical formation of ozone (see Sect. 10.10) which is a strong but short-lived radiative forcing gas

The radiative forcing impact of short-lived agents like these is very uncertain to model on a global scale, and their contribution to climate change is therefore not currently included in LCIA.

10.6.4 Existing Characterisation Models

All existing LCIA methods use the GWP (Eq. 10.4) for midpoint characterisation. In terms of time horizon most use 100 years, which has been recommended by IPCC as the best basis for comparison of GHGs, while some methods use a 500 year time horizon to better incorporate the full contribution from the GHGs. As mentioned, the longer time perspective puts a higher weight on long-lived GHGs like nitrous oxide, CFCs and SF_6 and a lower weight on short-lived GHGs like methane.

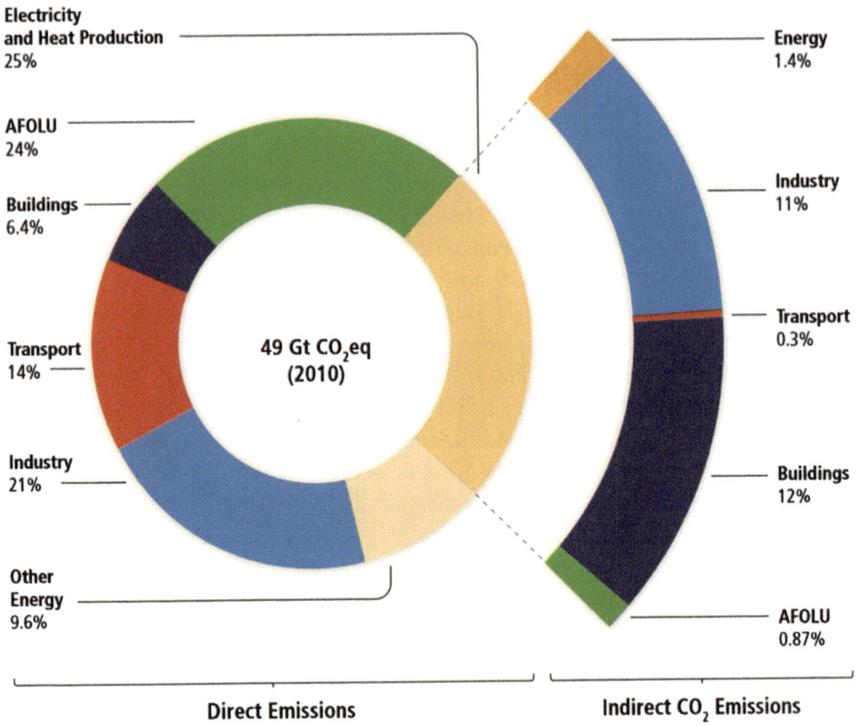

Fig. 10.9 Direct GHG emission shares (% of total anthropogenic GHG emissions) of five major economic sectors in the world in 2010. The pull-out shows how indirect CO_2 emission shares (in % of total anthropogenic GHG emissions) from electricity and heat production are attributed to sectors of final energy use. 'Other Energy' refers to all GHG emission sources in the energy sector other than electricity and heat production. 'AFOLU' stands for Agriculture, Forestry, and Other Land Use [taken from IPCC (2014b)]

So far radiative forcing agents with shorter atmospheric lifetime than methane are not considered in LCIA even though they also contribute to changing temperatures. However, a UNEP-SETAC expert workshop in 2016 recommended that climate change assessment at midpoint should be split into two sub-categories, respectively, focusing on the long-term climate change contributions and on the rate by which temperature changes occur. The two would be expressed in different metrics and not aggregated at midpoint level. It is expected that the distinction into two midpoint categories will cater better for the damage modelling since both rate of change and magnitude of the long-term temperature increase are important.

Endpoint characterisation of climate change is a challenge due to the complexity of the underlying environmental mechanisms with multiple feedback loops of which many are probably unknown, the global scale and the very long time perspective. In particular damages to human health are also strongly affected by local and regional differences in vulnerability and ability of societies to adapt to changing climate conditions. Some endpoint methods have proposed endpoint characterisation factors (e.g. Ecoindicator99, ReCiPe, LIME, IMPACT World+ and LC-IMPACT), but due to the state of current climate damage models, they inevitably miss many damage pathways and are accompanied by very large uncertainties, where even the size of these uncertainties is difficult to assess. This is why other endpoint methods (e.g. IMPACT 2002+) refrain from endpoint modelling for this impact category and present the midpoint results for climate change together with the endpoint results for the rest of the impact categories. In any case, endpoint results for climate change must be taken with the greatest caution in the interpretation of results. For further details see Chap. 40 and Hauschild and Huijbregts (2015).

10.7 Stratospheric Ozone Depletion

10.7.1 Problem

Ozone (O_3) is a highly reactive and unstable molecule consisting of three oxygen atoms and forms a bluish gas at normal ambient temperature with a distinct somewhat sharp odour. This molecule is present in lower atmospheric layers (tropospheric ozone as a consequence of photochemical ozone formation) and in larger concentrations (about 8 ppmv) also in higher altitudes between 15 and 40 km above ground (stratospheric ozone). Tropospheric, ground-level ozone is considered a pollutant due to its many harmful effects there on humans, animals, plants and materials (see Sect. 10.10). However, as a component of stratospheric atmospheric layers, it is vital to life on planet Earth, due to its capacity to absorb energy-rich UV radiation, thus preventing destructive amounts of it from reaching life on the planet's surface.

Stratospheric ozone depletion refers to the declining concentrations of stratospheric ozone observed since the late 1970s, which are observed in various ways: (1) As the 'ozone depletion area' or 'ozone hole' (an ambiguous term often used in public media referring to an area of critically low stratospheric ozone concentration), a recurring annual cycle of relatively extreme drops in O_3 concentrations over the poles which start to manifest annually in the late winter/early spring of each hemisphere (i.e. from around September/October over the South pole and March/April over the North pole) before concentrations recover again with increasing stratospheric temperatures towards the summer. 'Ozone holes' have been observed over Antarctic since the early 1980s as shown in Fig. 10.10. (2) A general decline of several percent per decade in O_3 concentrations in the entire stratosphere. Ozone concentration is considered as critically low when the value of the integrated ozone column falls below 220 Dobson units (a normal value being about 300 Dobson units). Dobson Units express the whole of ozone in a column from the ground passing through the atmosphere.

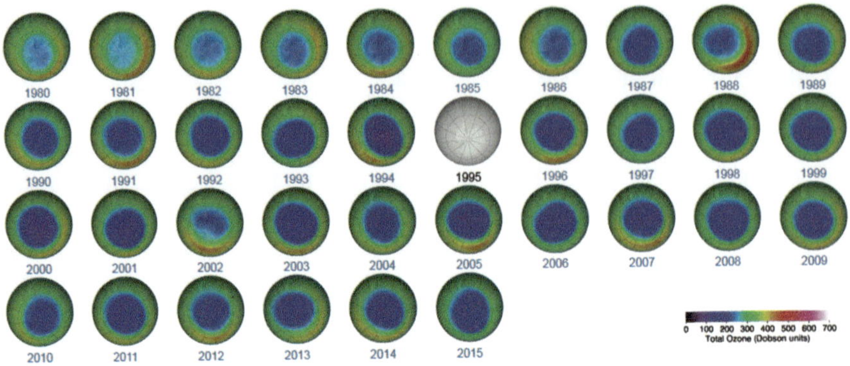

Fig. 10.10 Evolution of the hole in the ozone layer over Antarctica in September from 1980 to 2015 (*Source* NASA Ozone Watch 2016, public domain, http://ozonewatch.gsfc.nasa.gov/monthly/climatology_09_SH.html)

Data for Europe for example show a decline of 5.4% of stratospheric O_3 concentration per decade since the 1980s when measured in winter and spring, with an improving trend over the period 1995–2000. However, in later years low concentration records were broken on an almost annual basis. To date, the largest 'ozone hole' in human history was observed in 2006 with 29.5 million km^2 over Antarctica, but even in 2015 its largest spread still reached 28.2 million km^2. The largest Arctic 'ozone hole' ever was observed in 2011.

Impacts of stratospheric ozone depletion are essentially linked to reduced absorption of solar radiation in the stratosphere leading to increased UV radiation intensities at the planet surface, of which three broad (wavelength) classes are distinguished: UV-C, UV-B and UV-A. The impact of UV radiation on living organisms depends on its wavelength, the shorter the more dangerous. UV-C is the most dangerous wavelength range, but almost completely filtered by the ozone layer. UV-B (wavelengths 280–315 nm) is of the most concern due to ozone layer depletion, while UV-A is not absorbed by ozone.

Depending on duration and intensity of exposure to UV-B, impacts on human health are suspected to include skin cancer, cataracts, sun burn, increased skin cell ageing, immune system diseases, headaches, burning eyes and irritation to the respiratory passages. Ecosystem effects are linked to epidermal damage to animals (observed e.g. in whales), and radiation damage to the photosynthetic organs of plants causing reduced photosynthesis, leading to lower yields and crop quality in agricultural produce and loss of phytoplankton, the primary producers of aquatic food chains, particularly in the polar oceans. Additionally, UV-B accelerates the generation of photochemical smog, thereby stimulating the production of tropospheric ozone, which is a harmful pollutant (see Sect. 10.10).

10.7.2 Environmental Mechanism

Stratospheric ozone concentrations result from a balance between O_3 formation and destruction under the influence of solar (UV) radiation, temperature and the presence of other chemicals. The annual cycle of ozone destruction over the poles develops under the presence of several influencing factors with its intensity directly depending on their combined intensity: (1) meteorological factors (i.e. strong stratospheric winds and low temperature) and (2) the presence of ozone depleting chemicals.

Meteorological factors involve the formation of the "polar vortex", a circumpolar stratospheric wind phenomenon, in the polar night during the polar winter, when almost no sunlight reaches the pole. This vortex isolates the air in polar latitudes from the rest of Earth's atmosphere, preventing ozone and other molecules from entering. As the darkness continues, the air inside the polar vortex gets very cold, with temperatures dropping below −80 °C. At such temperatures a special type of clouds, called Polar Stratospheric Clouds (PSC), begins to form. Unlike tropospheric clouds, these are not primarily constituted of water droplets, but of tri-hydrated nitric acid particles, which can form larger ice particles containing dissolved nitric acid in their core as temperature continues to drop. The presence of PSC is crucial for the accelerated ozone depletion over the polar regions because they provide a solid phase in the otherwise extremely clean stratospheric air on which the ozone-degrading processes occur much more efficiently.

Chemical factors involve the presence of chlorine and bromine compounds in the atmosphere as important contributors to the destruction of ozone. The majority of the chlorine compounds and half of the bromine compounds that reach the stratosphere stem from human activities.

Due to their extreme stability, chlorofluorocarbons (CFCs) are not degraded in the troposphere but slowly (over years) transported into the stratosphere. Here, they are broken down into reactive chlorine radicals under the influence of the very energy-rich UV radiation at the upper layers of the ozone layer. One chlorine atom can destroy very high numbers of ozone molecules, before it is eventually inactivated through reaction with nitrogen oxides or methane present in the stratosphere. The degradation and inactivation scheme is illustrated in a simplified form for a CFC molecule in Fig. 10.11.

When they are isolated in the polar vortex and in the presence of PSC, these stable chlorine and bromine forms come into contact with heterogeneous phases (gas/liquid or gas/solid) on the surface of the particles forming the PSC, which breaks them down and release the activated free chlorine and bromine, known as "active" ozone depleting substances (ODS). These reactions are very fast and, as explained, strongly enhanced by the presence of PSC, a phenomenon which was neglected before the discovery of the 'ozone hole'.

While this describes the fate mechanism leading to stratospheric ozone reduction, Fig. 10.12 shows the impact pathway leading to ozone depletion in the

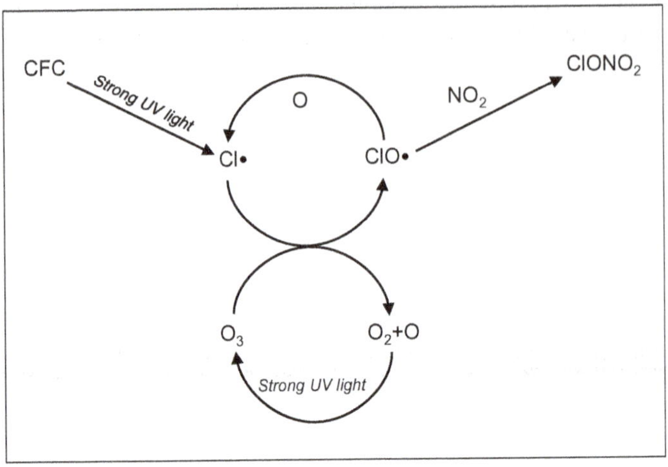

Fig. 10.11 Degradation of ozone catalysed by chlorine in the stratosphere (simplified)

Fig. 10.12 Impact pathway for stratospheric ozone depletion

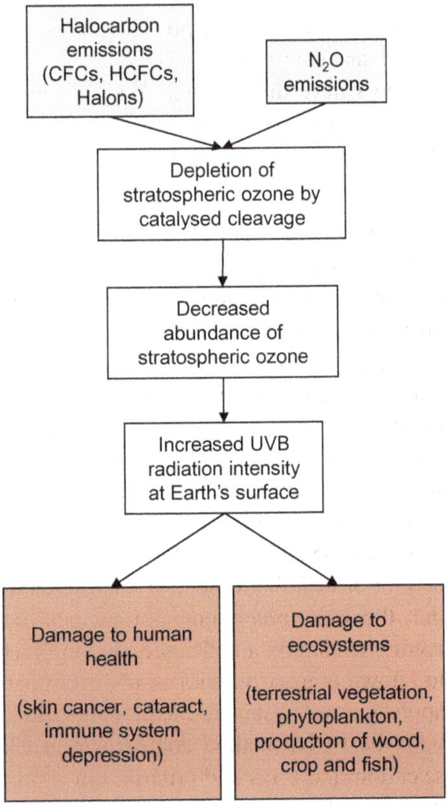

stratosphere from man-made emissions of long-lived halocarbons and nitrous oxide as used by most LCIA methods.

The midpoint indicator used without exception in all LCIA methods to calculate characterisation factors is the Ozone Depletion Potential (ODP). In a similar manner as the Global Warming Potential (GWP), it evaluates the potential of a chemical to destroy the ozone layer based on a model from the World Meteorological Organization (WMO 2014). The ODP essentially expresses the global reduction in stratospheric O_3 concentration C_{O_3} due to an ozone depleting substance *i* relative to the global reduction of stratospheric O_3 concentration C_{O_3} due to 1 kg of CFC-11 ($CFCl_3$), and is hence expressed in CFC-11 equivalents:

$$\mathrm{ODP}_i = \frac{\Delta C_{\mathrm{O_3}}(i)}{\Delta C_{\mathrm{O_3}}(\mathrm{CFC}-11)} \tag{10.5}$$

10.7.3 Emissions and Main Sources

The halogen compounds in the stratosphere are mostly originating from very stable industrial halocarbon gases used as solvents or refrigerants (the chlorinated CFCs or freons), or fire extinguishers (the brominated halons). Groups of anthropogenic ODS are: bromochloromethanes (BCM), CFCs, carbon tetrachloride, hydrobromofluorocarbons (HBFCs), hydrochlorofluorocarbons (HCFCs), tetrachloromethane, 1,1,1-trichloromethane, methyl bromide, methyl chloride and halons. The main uses of ODS during the last century were: fire extinguishing systems (halon), plastic foams, propellant gas in spray cans, fumigate and pesticides (methyl bromide), metered-dose inhalers (MDIs), refrigeration and air-conditioning and solvent degreasing.

Natural ozone depleting substances are CH_4, N_2O, H_2O and halogenated substances with sufficient stability and/or release rates to allow them to reach the stratosphere. All ozone depleting substances have two common characteristics, being:

- Chemically very stable in the lower atmosphere
- Capable of releasing chloride or bromide under UV radiation (photodissociation)

The phasing-out of production and use of the concerned substances has been successfully enforced under the Montreal protocol, which was signed in 1987 and led to phasing-out of consumption and production of ODS by 1996 in developed countries and by 2010 in developing countries. If continuously respected, this effort should lead to the cessation of the annual appearance of the 'ozone hole' around 2070, the delay being due to the facts that (1) we are still emitting decreasing amounts of relevant substances (mostly during the end-of-life treatment of old refrigeration and air-conditioning systems) and (2) they are very persistent and may take decades to reach the poles and hence continue their adverse effects for a prolonged time. When significant emissions or dominating impacts of ODS are observed in LCIs or LCA results nowadays, it is likely because the data originate from references before the phase-out and hence it is most likely an artefact due to obsolete data, unless the end-of-life treatment of old refrigeration and air-conditioning systems are an important component of the LCA.

10.7.4 Existing Characterisation Models

Without any exception, all existing LCIA methods use the ODP as midpoint indicator (although not all of them have the most recent version). For endpoint characterisation, different midpoint-to-endpoint models are applied that relate ozone depletion to increased UV radiation and ultimately to skin cancer and cataract in humans. All endpoint LCIA methods characterise impacts on human health, but only the Japanese method LIME additionally considers impacts on Net Primary Productivity (NPP) for coniferous forests, agriculture (soybean, rice, green pea, mustard) and phytoplankton at high latitudes. For further details see Chap. 40 and Hauschild and Huijbregts (2015).

10.8 Acidification

10.8.1 Problem

During the 1980s and 90s, the effects of acidification of the environment became clearly visible in the form of a pronounced lack of health especially among conifers in many forests in Europe and the USA, resulting locally in forest decline, leading to accelerated clearing of whole forests. Clear acidic lakes without fish go right back to the beginning of the twentieth century, occurring locally for example in Norway and Sweden as a result of human activities, but the extent of the problem increased dramatically in more recent times, and during the 1990s there was serious acidification in more than 10,000 Scandinavian lakes. Metals, surface coatings and mineral building materials exposed to wind and weather are crumbling and disintegrating at a rate which is unparalleled in history, with consequent major socio-economic costs and loss of irreplaceable historic monuments in many parts of the industrialised world.

The acidification problems were one of the main environmental concerns in Europe and North America in the 1980s and 90s but through targeted regulation of the main sources in the energy, industry and transportation sectors followed by liming to restore the pH of the natural soils and waters, it is no longer a major concern in these regions. In China, however, acidification impacts are dramatic in some areas due to the extensive use of coal-fired power generation using sulphur-rich coal.

10.8.2 Environmental Mechanism

Acidification of soil or aquatic ecosystems can be defined as an impact which leads to a fall in the system's acid neutralising capacity (ANC), i.e. a reduction in the quantity of substances in the system which are able to neutralise hydrogen ions added to the system.

ANC can be reduced by:

1. Addition of hydrogen ions, which displace other cations which can then be leached out of the system
2. Uptake of cations in plants or other biomass which is collected and removed from the system

Particularly the former is relevant for acidification impacts in LCA. Acidification occurs naturally over time, but it is greatly increased by man-made input of hydrogen ions to soil and vegetation. The main source is air-borne emissions of gases that release hydrogen when they are degraded in the atmosphere or after deposition to soil, vegetation or water. Deposition is increased during precipitation events where the gases are dissolved in water and come down with rain, which can be rather acidic with pH values down to 3–4 in cases of strong air pollution ("acid rain").

The most important acidifying man-made compounds are:

Sulphur oxides, SO_2 and SO_3 (or jointly SO_x), the acidic anhydrides of sulphurous acid H_2SO_3 and sulphuric acid H_2SO_4, respectively, meaning that upon absorption of water from the atmosphere they form these very strong acids which both release two hydrogen ions when deposited:

$$SO_2 + H_2O \rightarrow H_2SO_3 \rightarrow 2H^+ + SO_3^{2-}$$

$$SO_3 + H_2O \rightarrow H_2SO_4 \rightarrow 2H^+ + SO_4^{2-}$$

Nitrogen oxides, NO and NO_2 (or jointly NO_x) that are also acidic anhydrides as they can be converted to nitric and nitrous acids by oxidation in the troposphere. NO is oxidised to NO_2 primarily by reaction with ozone (see Sect. 10.10):

$$NO + O_3 \rightarrow NO_2 + O_2$$

NO_2 can be oxidised to nitric acid, HNO_3 or $HONO_2$:

$$NO_2 + OH + M \rightarrow HONO_2 + M$$

where OH is hydroxyl radical present in the atmosphere and M is an inactive body which can remove surplus energy.

Ammonia, which is in itself a base (absorbing hydrogen ions via the reaction $NH_3 + H^+ \rightarrow NH_4{}^+$), but upon complete mineralisation through nitrite, $NO_2{}^+$, to nitrate, $NO_3{}^-$ releases one net proton:

$$NH_3 + 2O_2 \rightarrow H^+ + NO_3{}^- + H_2O$$

Strong acids like hydrochloric acid, HCl or sulphuric acid, H_2SO_4, which release their content of hydrogen ions as soon as they are dissolved in water and thus also are strongly acidifying.

Because of their high water solubility, the atmospheric residence time of these acidifying substances is limited to a few days, and therefore acidification is a regional effect with its extent limited to the region around the point of emission.

When acidifying compounds deposit on plant leaves or needles, they can damage these vital plant organs and through this damage the plants. When the acidifying compounds reach the soil, protons are released in the soil where they may lower the pH of the soil water and cause release of metal ions bound in the soil. Some of these metals are toxic to the plants in the soil, others are essential for plant growth, but after their release, they wash out, and the availability of these metals to plants may then become limiting for plant growth. The result is stress on the plants through root and leaf damage and after prolonged exposure the plants may die as a direct consequence of this or through diseases or parasites that benefit from the weakened constitution of the plant. Lakes are also exposed to the acidification, in particular through the acidified soil water leaching to the lake. When the pH of a lake drops, the availability of carbon in the water in its dominating form around neutral pH, which is $HCO_3{}^-$, is converted to dissolved CO_2. The solubility of toxic metals is increased, in particular aluminium which may precipitate on the gills of fish at pH 5. The phytoplankton and macrophyte flora gradually change and also the fauna is affected. Humic acids that give the lakewater a brown colour are precipitated, and the acidified lakes appear clear and blue.

The sensitivity to acidification is strongly influenced by the geology and nature of the soil. Calcareous soils with a high content of calcium carbonate are well buffered meaning that they will resist the change in pH by neutralising the input of hydrogen ions with the basic carbonate ions:

$$H^+ + CaCO_3 \rightarrow Ca^{2+} + HCO_3{}^-$$

$$H^+ + HCO_3{}^- \rightarrow H_2O + CO_2$$

As long as there is calcium carbonate in the soil, it will thus not be acidified.

Soils that are rich in clay are also resistant to acidification through their ability to adsorb the protons on clay mineral surfaces under release of metal ions, while sandy soils are more sensitive to acidification. The sensitivity of an ecosystem towards acidification can be described by its critical load—"A quantitative estimate of an exposure to one or more pollutants below which significant harmful effects on specified sensitive elements of the environment do not occur according to present knowledge" (Nilsson and Grennfelt 1988). Critical loads are high in calcareous regions like the Mediterranean and low in e.g. granite rock regions like most of Scandinavia.

Incorporating the environmental mechanism described above, the impact pathway of acidification is illustrated in Fig. 10.13.

Oceanic acidification is the process of dissolution of CO_2 into seawater leading to a slight lowering of the pH in the open oceans as a consequence of increasing concentrations of CO_2 in the atmosphere. Dissolution of CO_2 in water generates carbonic acid, a rather weak acid (think soda water), which releases protons according to

$$CO_2 + H_2O \rightarrow H_2CO_3 \rightarrow HCO_3^- + H^+$$

The slightly lowered pH is deleterious to coral reefs, which should be included in endpoint characterisation. CO_2 is the only important contributor to oceanic acidification and inclusion of this impact category on midpoint level therefore offers little additional information to the LCIA that already considers climate change, we will hence not discuss it further here.

10.8.3 Emissions and Main Sources

Sulphur dioxides and nitrogen oxides are the man-made emissions that contribute the most to acidification. Historically metal smelters of the mining industry have been strong sources of local acidification with large localised emissions of sulphur oxides. Today, the main sources of both SO_x and NO_x are combustion processes in thermal power plants, combustion engines, waste incinerators and decentralised furnaces. For sulphur oxides, the level of emissions depends on the sulphur content of the fuels. Since nitrogen is abundant in the atmosphere and hence in all combustion processes using air, emissions of nitrogen oxides are mainly determined by conditions of the combustion process and possible treatment of the flue gases through catalysers and filters. As response to the serious problems with acidification in Europe and North America in previous times, regulation now ensures that sulphur content is removed from the fuels, that important combustion activities like thermal power plants and waste incinerators have an efficient neutralisation of the flue gases before they are released, and that combustion engines have catalysers lowering the NO_x content of the exhaust gases.

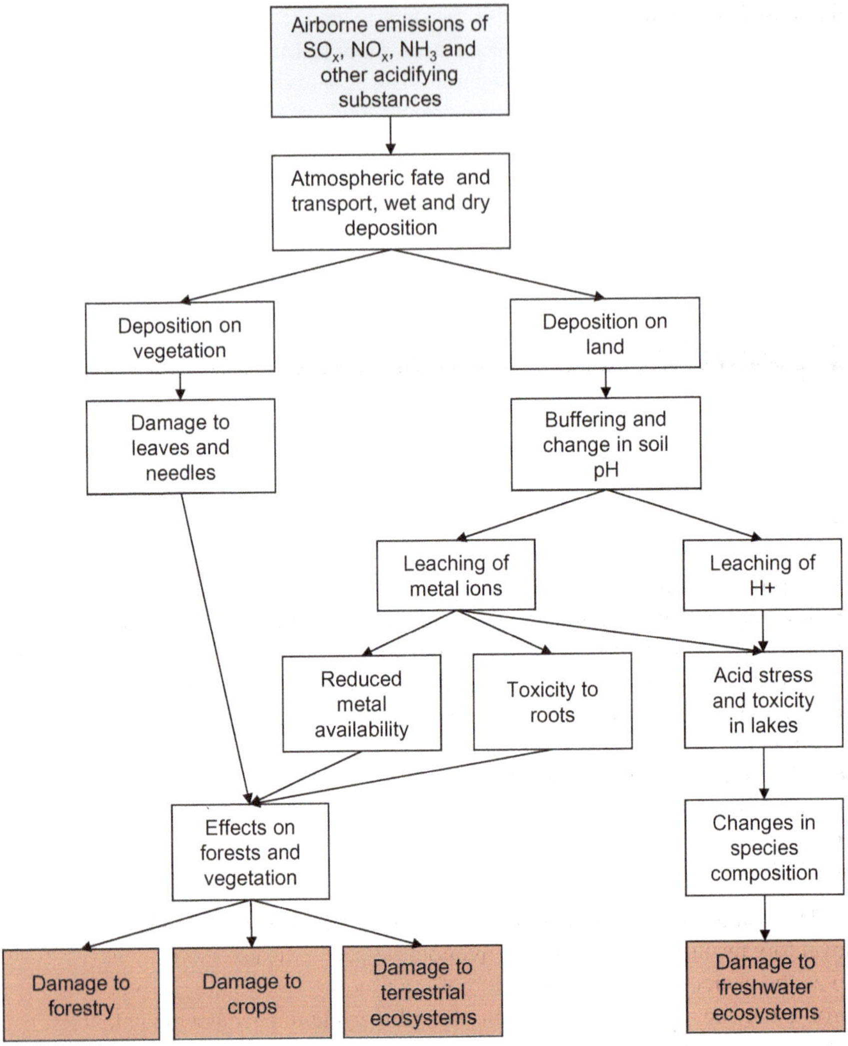

Fig. 10.13 Impact pathway for acidification

Ammonia is also an important contributor to acidification in some regions and the main sources are all related to agriculture using NH_3 as a fertiliser, and to animal husbandry, in particular pig and chicken farms, with ammonia emissions from stables and dispersion of manure.

Mineral acids like HCl and H_2SO_4 rarely appear as elementary flows in life cycle inventories but they may be emitted from some industrial processes and also from waste incinerators with inefficient flue gas treatment.

10.8.4 *Existing Characterisation Models*

The acidification potential depends both on the potency of the emitted gas and on the sensitivity of the receiving environment in terms of buffering capacity of the soils and sensitivity of the ecosystems to acidification as expressed by their critical load. While the difference between the contributing gases is modest—within a factor 5–10 across substances, the difference between sensitivities in different locations can be several orders of magnitudes depending on the geology and soil characteristics. Early characterisation models were site-generic and only incorporated the difference in ability to release protons, but newer models incorporate more and more of the cause–effect chain in Fig. 10.13 and model e.g. the area of ecosystem in the deposition area that becomes exposed above its critical load. This requires a site-dependent LCIA approach where the characterisation factor is determined not just per emitted substance but also per emission location. Characterisation factors may be expressed as absolute values or as an equivalent emission of a reference substance which in that case is usually SO_2. For further details see Chap. 40 and Hauschild and Huijbregts (2015).

10.9 Eutrophication

10.9.1 *Problem*

Nutrients occur naturally in the environment, where they are a fundamental precondition for the existence of life. The species composition and productivity of different ecosystems reflect the availability of nutrients, and natural differences in the availability of nitrogen and phosphorus are thus one of the reasons for the existing multiplicity of species and of different types of ecosystems. Ecosystems are dynamic, and if they are affected by a changed availability of nutrients, they simply adapt to a new balance with their surroundings. Originally, eutrophication of aquatic environments, such as rivers or lakes, describes its eutrophic character (from the Greek word "eu"—good or true, and "trophein"—feed), meaning nutrient-rich. From the 1970s the term was used to describe the slow suffocation of large lakes. It now has a meaning close to dystrophic, i.e. poor conditions and low in oxygen, supporting little life. An aquatic ecosystem in strong imbalance is named hypertrophic, when close to a natural equilibrium it is called mesotrophic, and when healthy it is called oligotrophic.

The perhaps most prominent effect of eutrophication in lakes, rivers and the coastal sea are lower water quality including low visibility or for stronger situations massive amounts of algae in the surface layers of those waters. Eutrophication essentially describes the enrichment of the aquatic environment with nutrient salts leading to an increased biomass production of planktonic algae, gelatinous zooplankton and higher aquatic plants, which results in the degradation of (organoleptic) water quality (e.g. appearance, colour, smell, taste) and an altered species composition of the ecosystem. It may also lead to the development of toxic phytoplankton, dynophysis, cyanobacteria or blue-green algae. When the algae die, they sink to the bottom where they are degraded under oxygen consumption. As a consequence, the concentration of dissolved oxygen decreases (hypoxia), which results in biodiversity loss (flora and fauna). Ultimately, if the process is not stopped, this will turn a lake into a swamp, that will gradually become grassland and forest. This process occurs naturally but over a much longer time horizon.

For terrestrial systems, the most significant environmental problem in relation to nitrogen compound loading is changes in the function and species composition of nitrogen-poor (and nitrogen limited) ecosystems in heathlands, dune vegetation, commons and raised bogs as a result of the atmospheric deposition of nitrogen compounds. Forestry and agriculture may also be affected by reduced yields via damage to forests and crops. This section however focuses on aquatic eutrophication.

10.9.2 Environmental Mechanism

The food chain in aquatic ecosystems can be distinguished into three trophic levels: primary producers (algae and plants producing biomass via photosynthesis), primary consumers (species consuming algae and plants, the vegetarians) and secondary consumers (species consuming primary consumers, the carnivores). In addition to sunlight, growth of primary producers (algae and higher plants) requires all of the elements which enter into their anabolism (i.e. their synthesis of the molecules which constitute the organisms' cells). A molecular formula for the average composition of an aquatic organism is $C_{106}H_{263}O_{110}N_{16}P$ (Stumm and Morgan 1981). Apart from the elements represented in this formula, minor quantities of a large number of other elements are required, e.g. potassium, magnesium, calcium, iron, manganese, copper, silicon and boron (Salisbury and Ross 1978). In principle, the availability of any of these elements can determine the potential extent of the growth of the primary producers in a given system. The elements entering in greatest quantities into the primary producers (as in all other living organisms) are carbon, C, hydrogen, H and oxygen, O. The availability of water can limit growth in terrestrial plants, but the availability of one of the three basic elements is rarely a limiting factor in the growth of primary producers.

The other elements which enter into the construction of the primary producers are nutrients, as the availability of these elements in sufficient quantities is necessary to ensure growth. The nutrients are classified as macronutrients (>1000 μg/g dry matter in plants) and micronutrients (<100 μg/g dry matter in plants) (Salisbury and Ross 1978). In rare cases, growth is limited by the availability of one of the micronutrients, but very small quantities of these elements are required by the primary producers, and these elements are therefore limiting only on very poor soils. Of the macronutrients, sulphur is added to all ecosystems in fair quantities in most of the industrialised world by the atmospheric deposition of sulphur compounds from flue gases resulting from energy conversion based on fossil resources. Calcium, potassium and magnesium occur in lime and clay, respectively, which exist in large quantities in soils.

In practice, one of the two last macronutrients, nitrogen and phosphorus, is therefore almost always the limiting element for the growth of primary producers, and it is therefore reasonable to regard only the elements nitrogen and phosphorus as contributors to nutrient enrichment. In many lakes, phosphorus deficiency, or a combination of nitrogen and phosphorus deficiencies, is typically limiting growth, and their addition promotes algal growth. In coastal waters and seas, nitrogen is often the limiting nutrient. *Substances which contain nitrogen or phosphorus in a biologically available form are therefore classified as potential contributors to nutrient enrichment.* As is evident from the formula for the average composition of aquatic organisms, the ratio of nitrogen to phosphorus is of the order of 16. If the concentration of bioavailable nitrogen is significantly more than 16 times the concentration of bioavailable phosphorus in an ecosystem, it is thus reasonable to assume that phosphorus is the limiting nutrient, and vice versa. Since most of the atmosphere consists of free molecular nitrogen, N_2, further addition of N_2 will not have any effect, and it is also not directly bioavailable. N_2 is therefore not classified as contributing to nutrient enrichment.

For aquatic eutrophication, the starting point of the cause–effect chain is the emission of a compound containing either nitrogen (N) or phosphorus (P). Increased availability of nutrients will primarily increase the growth of algae and plants, especially in summer with abundant sunlight. This algae growth is visible as rivers, lakes or coastal waters turn turbid in summer. Eventually, the algae will sink to the bottom where they are decomposed by degraders like bacteria under consumption of oxygen in the bottom layer. With the sunlight being increasingly blocked from reaching deeper water layers, the build-up of a temperature gradient causes stratification in deep lakes and some coastal waters in the summer months. In the marine environment, stratification is determined by density differences between salt water flowing in from the sea and brackish water flowing out from river deltas and fjords. Such stratification prevents effective mixing of the water column. If fresh oxygen-rich water from the surface does not find its way to the bottom layers, the oxygen concentration near the bottom will gradually be reduced until the bottom-dwelling organisms move away or die. As the oxygen concentration approaches zero, poisonous substances such as hydrogen sulphide, H_2S, are formed in the sediments, where they accumulate in gas pockets which, when released again, kill those organisms exposed to them.

The main cause–effect chain as shown in Fig. 10.14 can be summarised as:

- Emission of N or P containing substances
- Growth and blooming of algae and higher plants increases
- Sunlight no longer reaches lower water layers, which creates a temperature gradient with increasing depth
- This supports a stable stratification of water layers reducing the transport of fresh oxygen-rich surface water to deeper layers

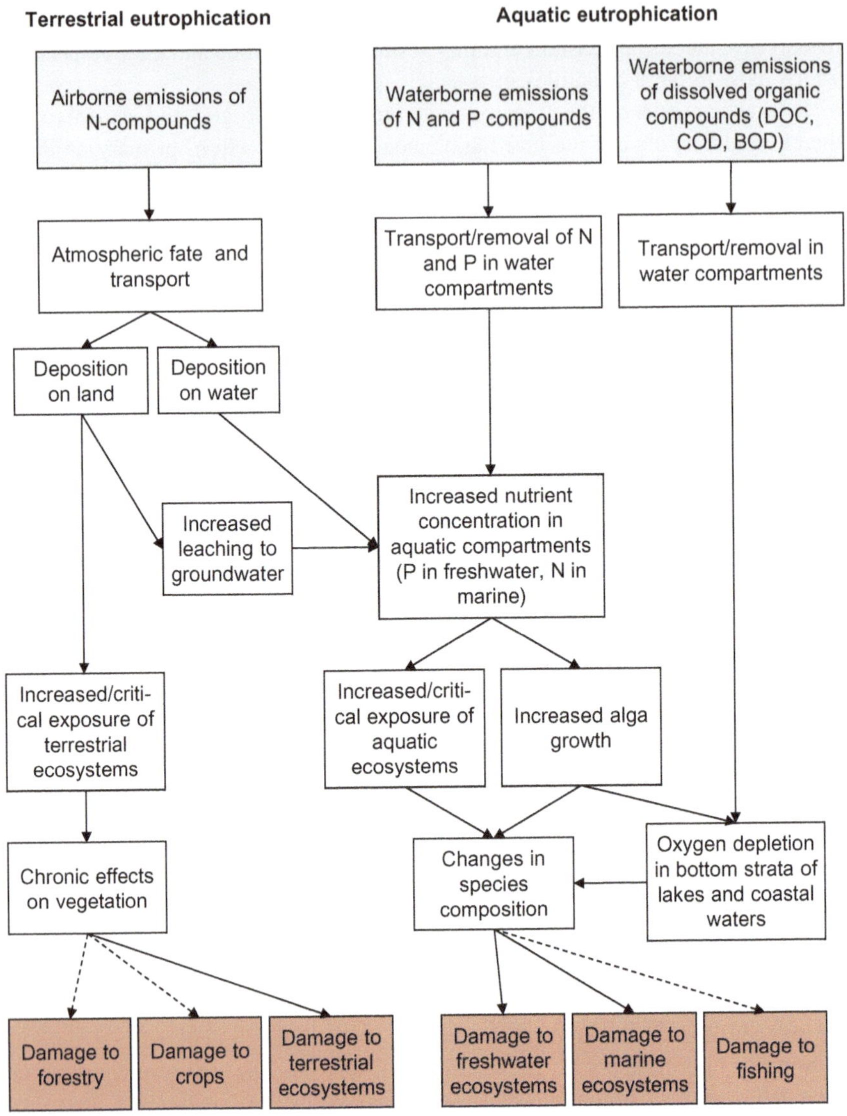

Fig. 10.14 Impact pathways for terrestrial and aquatic (freshwater and marine) eutrophication [adapted from EC-JRC (2011)]

- Oxygen is steadily depleted in bottom layers, which leads to suffocation of bottom-dwelling species and fish
- This is additionally accelerated by the oxygen consuming decomposition of the dead species and sedimented dead algae
- The aquatic medium becomes hypoxic and finally anoxic, favouring the formation of reducing compounds and noxious gases (mercaptans, methane)

In a tripartite division of environmental impact categories into global, regional and local, eutrophication is considered a local to regional impact. As a consequence of the above explanations, impact potentials are highly dependent on local conditions, e.g. whether the recipient of the emission will support the requisite conversion of the emission (e.g. mineralisation of organic nitrogenous compounds), or whether the recipient is limited in nitrogen or phosphorus, while both elements are always considered potential contributors to eutrophication.

The calculation of characterisation factors for a nutrient enriching substance consists of an assessment of the number of moles of nitrogen or phosphorus which can be released into the environment from one mole of the substance emitted. This can be expressed in the form of two nutrient enrichment equivalents, as kg N-equivalents and kg P-equivalents. The possible consequences of eutrophication are often irrespective of whether nitrogen or phosphorus is the causing agent. In some situations it can therefore be desirable to reduce the complexity of the results of the environmental assessment by expressing eutrophication as one equivalent, so that the contributions for nitrogen and phosphorus are aggregated. In this case the impact potential may also be expressed as an equivalent emission of a reference substance (e.g. NO_3^- as one of the most important nutrient enriching substances). Aggregation of N and P potentials requires an assumption concerning the magnitude of the ratio N/P between these two elements in living organisms. As explained above a molar ratio of 16 can be used for nitrogen:phosphorus in living material. One mole of phosphorus (in an area where the availability of phosphorus limits growth) therefore contributes as much to eutrophication as 16 mol of nitrogen (in an area where the availability of nitrogen limits growth). The aggregate nutrient enrichment potential for nitrogenous substances is then calculated as the emission's N potential multiplied by the gram/mol molecular weight of the reference substance (e.g. NO_3^- of 62 g/mol). The P potential for phosphorous-containing substances is multiplied by 16 times the gram/mol molecular weight of the reference substance.

The primary receiving compartment for agricultural emissions is mainly freshwater where some of the nitrogen may be removed on the way to the marine systems by denitrification in rivers and lakes converting the nitrogen into molecular N_2 which is released to the atmosphere. Loading of freshwater with nitrogen is thus greater than the quantity conveyed to the marine areas via rivers and streams. Phosphorous compounds do not undergo this kind of conversion but phosphate forms insoluble salts with many metals and this may lead to some removal through accumulation of phosphorus in lake sediments. Phosphorus accumulated in the sediments of rivers and streams during drier periods may later be washed out into the marine environment when the water flow increases, e.g. after a thunderstorm.

10.9.3 Emissions and Main Sources

Due to the use of inorganic fertilisers and manure, agriculture is a significant source of phosphorus and nitrogen emissions in the form of phosphates and nitrates, respectively, affecting groundwater via percolation and surface water via runoff and leaching processes, and of ammonia emitted to air and deposited on land nearby. Oxides of nitrogen may be emitted from incineration processes. Point sources in the form of wastewater treatment plants for households (e.g. from polyphosphates in detergents) and industry as well as fish farming are important sources of phosphorus and nitrates. Apart from man-made emissions, natural sources include leaching and runoff of nitrogen and phosphates. The natural addition of nutrients to *terrestrial areas* is believed to consist mainly of atmospheric deposition of oxides of nitrogen and ammonia while some natural plant species also possess the ability to fixate atmospheric nitrogen.

Emissions of organic materials can lead to oxygen consumption by bacteria degrading this organic matter and thus contributing to oxygen depletion similarly to what is observed as a result of the nutrient enrichment of lakes and coastal waters. However, this is a primary effect and is strictly speaking not part of the nutrient enrichment mechanism. Therefore, emissions of BOD (biological oxygen demand—substances which consume oxygen on degradation) or COD (chemical oxygen demand) may additionally be characterised by some LCIA methods considering oxygen depletion (hypoxia) in water as a common midpoint for both mechanisms. Most LCIA methods are currently based on the N/P ratio and typically do not classify BOD or COD as contributing to nutrient enrichment and thus eutrophication. In large parts of the industrialised world organic matter emissions are only of local significance in watercourses and for occasional emissions of untreated effluent.

10.9.4 Existing Characterisation Models

The essential evolutions during the last decade were related to improved fate modelling, distinguishing P-limited (freshwater) and N-limited (marine) ecosystems, introduction of a midpoint effect factor in the more recent methods, and characterisation models becoming global and spatially more detailed. Midpoint LCIA methods usually propose units in P- and N-equivalents such as kg P-eq or kg PO_4^{3-}-eq and kg N-eq or kg NO_3^{-}-eq. For endpoint characterisation most models use Potentially Disappeared Fraction of species (PDF) in [m^2 years], except LIME which uses Net Primary Productivity (NPP) loss. For further details see Chap. 40 and Hauschild and Huijbregts (2015).

10.10 Photochemical Ozone Formation

This impact category appears under a number of different names in the various LCIA methods: (tropospheric) ozone formation, photochemical ozone formation or creation, photo oxidant formation, photosmog or summer smog. There are minor differences, but in essence they all address the impacts from ozone and other reactive oxygen compounds formed as secondary contaminants in the troposphere by the oxidation of the primary contaminants volatile organic compounds (VOC), or carbon monoxide in the presence of nitrogen oxides (NO_x) under the influence of light. VOCs are here defined as organic compounds with a boiling point below 250 °C (WHO 1989). NO_x is a joint name for the nitrogen monoxide NO and nitrogen dioxide NO_2.

10.10.1 Problem

The negative impacts from the photochemically generated pollutants are due to their reactive nature which enables them to oxidise organic molecules in exposed surfaces. Impacts on humans arise when the ozone and other reactive oxygen compounds, which are formed in the process, are inhaled and come into contact with the surface of the respiratory tract, where they damage tissue and cause respiratory diseases. Impacts on vegetation arise when the reactive compounds attack the surfaces of plants or enter plant leaves and cause oxidative damage on their photosynthetic organs. Impacts on man-made materials are caused by oxidation and damage many types of organic materials which are exposed to ambient air. It is thus not the VOCs per se which cause the environmental problems associated with photochemical ozone formation, but the products of their transformation in the troposphere which is the lower stratum of the atmosphere, from the surface of the earth to the tropopause 8–17 km above us. Direct toxic effects on humans from VOCs are treated separately in the impact category human toxicity (see Sect. 10.12). Apart from a general increase in the tropospheric ozone concentration, photochemical ozone formation may cause smog episodes on a more local scale in and around cities with a combination of large emissions and the right meteorological conditions. During smog episodes, the concentrations of ozone and other photooxidants reach extreme levels causing immediate damage to human health.

10.10.2 Environmental Mechanism

The photochemical formation of ozone and other reactive oxygen compounds in the troposphere from emissions of VOCs and NO_x follows rather complex reaction schemes that depend on the nature of the specific organic compound emitted. A simplified presentation of the fundamental elements of the schemes is given in Fig. 10.15 and can be summarised as:

1. VOCs (written as RH) or CO react with hydroxyl radical $OH^{\bullet}$ in the troposphere and form peroxy radicals, $ROO^{\bullet}$
2. The peroxy radicals oxidise NO to NO_2

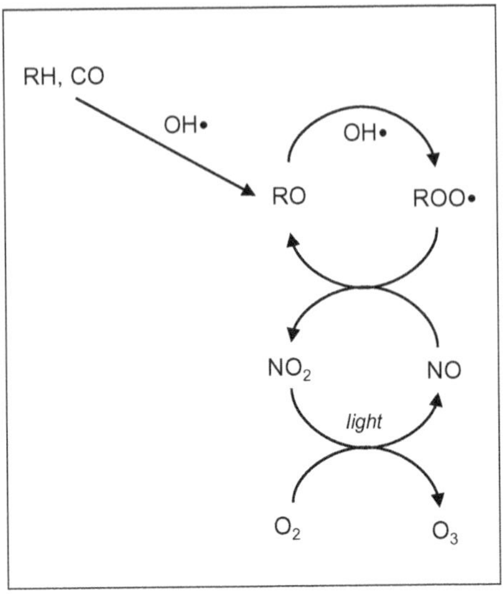

Fig. 10.15 Simplified presentation of the photochemical formation of ozone

3. NO_2 is split by sunlight with formation of NO and release of free oxygen atoms
4. Free oxygen atoms react with molecular oxygen O_2 to form ozone

Both VOCs and nitrogen oxides are thus needed for the photochemical ozone formation and both contribute to the formation of ozone and other oxidants. VOC and NO_x sources are very heterogeneously distributed across Europe. VOC emissions involve hundreds of different organic compounds, depending on the nature of the source and activity causing the emission. This means that at the regional level, photochemical formation of ozone is highly non-linear and dynamic with the influence of meteorological conditions and on top of this the interaction between the different VOCs from both anthropogenic and natural sources like forests, and a large number of different reaction products. A further complication arises because NO may react with the formed ozone, abstracting an oxygen atom to give oxygen and NO_2. This means that depending on the conditions, NO may locally have a negative ozone formation potential and hence a negative characterisation factor for this impact category. Rather than a permanent removal of ozone this reaction of NO leads to a geographic displacement of the ozone formation since the NO_2 thus formed can later cause ozone formation again following the scheme in Fig. 10.15, just in a different location.

The ozone formation requires the reaction between a hydroxyl radical and a bond between carbon and hydrogen or another carbon atom in a VOC molecule. The relative strength of a volatile organic compound in terms of ozone formation potential per unit weight thus depends on how many such bonds it contains. The strength grows with the number of double or triple bonds and declines with the content of elements other than carbon and hydrogen. The following general ranking can be given from high to low ozone formation potential:

1. Alkenes (decreasing with chain length) and aromatics (increasing with the degree of alkyl substitution, decreasing with the length of the chain in the substituted alkyl group)
2. Aldehydes (the strongest is formaldehyde; benzaldehyde has no or even a negative ozone formation potential)
3. Ketones
4. Alkanes (almost constant from a chain length of three carbon atoms and upwards), alcohols and esters (the more oxygen in the molecule, the weaker)
5. Halocarbons (decreasing with the degree of halogen substitution and the weight of the halogen element)

Animals and humans are mainly exposed to the photochemical oxidants through inhalation of the surrounding air, and the effects therefore appear in their respiratory organs. Ozone is detectable by its odour at a concentration of ca. 20 ppb in pure air, but only at somewhat higher concentrations we start to see acute symptoms like increased resistance of the respiratory passages and irritation of the eyes, followed at even higher concentrations by more serious effects like oedema of the lungs, which can lead to long-term incapacity. Smog episodes with extreme concentrations of photochemical oxidants in urban areas are known to cause increased mortality. Chronic respiratory illness may result from long-term exposure to the photochemical oxidants.

Plants rely on continuous exchange of air between their photosynthetic organs (leaves or needles) and the atmosphere to absorb the carbon dioxide which is needed for photosynthesis. Ozone and other photooxidants enter together with the air and through their oxidative properties damage the photosynthetic organelles, leading to discolouration of the leaves followed by withering of the plant. The sensitivity of the plant varies with the season and also between plant species, but considerable growth reductions are observed in areas with high ozone concentrations during the growth season. Agriculture yield losses of 10–15% have been estimated for common crop plants.

Figure 10.16 summarises the impact pathway for photochemical ozone formation linking emissions of VOCs, CO and NO_x to the resulting damage to the areas of protection.

10.10.3 Emissions and Main Sources

In some cases the emissions of individual substances are known, but in the case of oil products the emissions will often be composed of many different substances and will be specified under collective designations like VOCs or nmVOCs (non-methane VOCs, i.e. VOCs apart from methane which is typically reported separately due to its nature as a strong greenhouse gas) and sometimes also HCs (hydrocarbons), or nmHCs (non-methane hydrocarbons, i.e. hydrocarbons excluding methane).

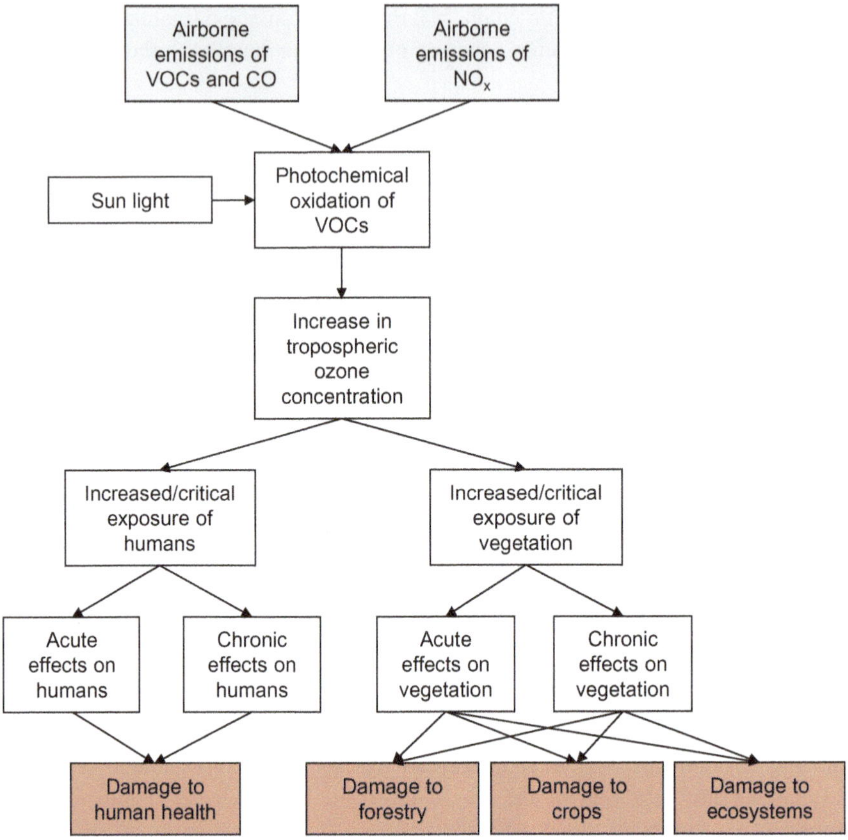

Fig. 10.16 Impact pathway for photochemical ozone formation [adapted from EC-JRC (2011)]

The most important man-made emissions of VOCs derive from road traffic and the use of organic solvents, which during 2000–2010 in Europe amounted to around 40% of the total man-made nmVOC emissions. A further 7% derives from industrial processes and 10% are fugitive emissions (Laurent and Hauschild 2014). VOCs are also emitted in large quantities from vegetation, in particular forests, but unless a man-made manipulation of the natural system affects its emissions of VOCs, these will not be reported in an LCI and hence not dealt with in the impact assessment. Carbon monoxide is emitted from combustion processes with insufficient oxygen supply. These include road traffic and various forms of incomplete combustion of fossil fuels or biomass in stationary systems. Nitrogen oxides are also emitted from combustion processes in transport, energy- and waste incineration systems.

10.10.4 Existing Characterisation Models

The complexity of the underlying reaction schemes and the high number of individual contributing substances for which photochemical ozone formation characterisation factors must be calculated calls for simplification in the characterisation modelling. Existing characterisation models apply one of two approaches:

The first alternative is to simplify the non-linear and dynamic behaviour of the photochemical oxidation schemes by modelling one or a few typical situations in terms of meteorology, atmospheric chemistry and concomitant emissions of other air pollutants. For each individual VOC, characterisation factors may then be presented for each situation or in the form of a weighted average across the situations.

The second alternative is to ignore the variation between individual VOCs and concentrate on getting the spatial and temporal specificities well represented in the characterisation model. This approach leads to spatially (and possibly temporally) differentiated characterisation factors for VOCs (as a group, ignoring variation in strength between individual substances), CO and NO_x. Often methane is treated separately from the rest of the VOCs (which are then termed non-methane VOCs or nmVOCs) due to its very low characterisation factor which really distinguishes it from the majority of the other VOCs.

The first approach is adopted in characterisation models based on the POCP (Photochemical Ozone Creation Potential) or MIR (Maximum Incremental Reactivity) concepts. The second approach is adopted in regionally differentiated models which attempt to capture the non-linear nature of the ozone formation with its spatially and temporally determined differences. For further details see Chap. 40 and Hauschild and Huijbregts (2015).

10.11 Ecotoxicity

The contents of this section have been modified from Rosenbaum, R.K.: Ecotoxicity, appearing as Chapter 8 of Hauschild M. Z. and Huijbregts M. A. J. (eds.) LCA Compendium—The Complete World of Life Cycle Assessment—Life Cycle Impact Assessment, pp 139–162. Springer, Dordrecht (2015).

10.11.1 Problem

About 500 years ago Paracelsus stated that ‘All substances are poisons; there is none which is not a poison. The right dose differentiates a poison and a remedy’. Today’s toxicology science still agrees and adheres to this principle and in consequence any substance emitted may lead to toxic impacts depending on a number of driving factors: (1) emitted quantity (determined in the LCI), (2) mobility,

(3) persistence, (4) exposure patterns and bioavailability and (5) toxicity, with the latter four considered by the characterisation factor.

This shows that toxicity is not the only parameter that determines the potential ecotoxic impact of a chemical in the environment as it first has to reach and enter a potential target organism. For example, a substance may be very toxic, but never reach any organism due to its short lifetime in the environment (e.g. rapid degradation) or because it is not sufficiently mobile to be transported to a target organism and ends up bound to soil or buried in sediment, in which case it contributes little to ecotoxic impacts. On the other hand, another substance may not be very toxic, but if it is emitted in large quantities and over prolonged periods of time or has a strong environmental persistence, it may still cause an ecotoxic impact.

Chemical emissions into the environment will affect terrestrial, freshwater, marine and aerial (i.e. flying and gliding animals) ecosystems depending on the environmental conditions of the place and time of emission and the characteristics of the substance emitted. They can affect natural organisms in many different ways, causing increased mortality, reduced mobility, reduced growth or reproduction rate, mutations, behavioural changes, changes in biomass or photosynthesis, activity etc.

10.11.2 Environmental Mechanism

As shown in Fig. 10.17, the environmental mechanism of ecotoxic impacts of chemicals in LCA can be divided into four consecutive steps.

1. Fate modelling estimates the increase in concentration in a given environmental medium due to an emission quantified in the life cycle inventory
2. The exposure model quantifies the chemical's bioavailability in the different media by determining the bioavailable fraction out of the total concentration
3. The effect model relates the amount available to an effect on the ecosystem. This is typically considered a midpoint indicator in LCA, as no distinction between the severity of observed effects is made (e.g. a temporary/reversible decrease in mobility and death are given the same importance)
4. Finally, the severity (or damage) model translates the effects on the ecosystem into an ecosystem population (i.e. biodiversity) change integrated over time and space

All four parts of this environmental mechanism are accounted for in the definition of the substance-specific and emission compartment-specific ecotoxicity characterisation factor CF_{eco}:

$$CF_{eco} = FF \times XF_{eco} \times EF_{eco} \times SF_{eco} \tag{10.6}$$

where FF is the fate factor, XF_{eco} the ecosystem exposure factor, EF_{eco} the ecotoxicity effect factor (midpoint effects), and SF_{eco} the ecosystem severity factor (endpoint effects). Each of these four elements of the environmental mechanism of

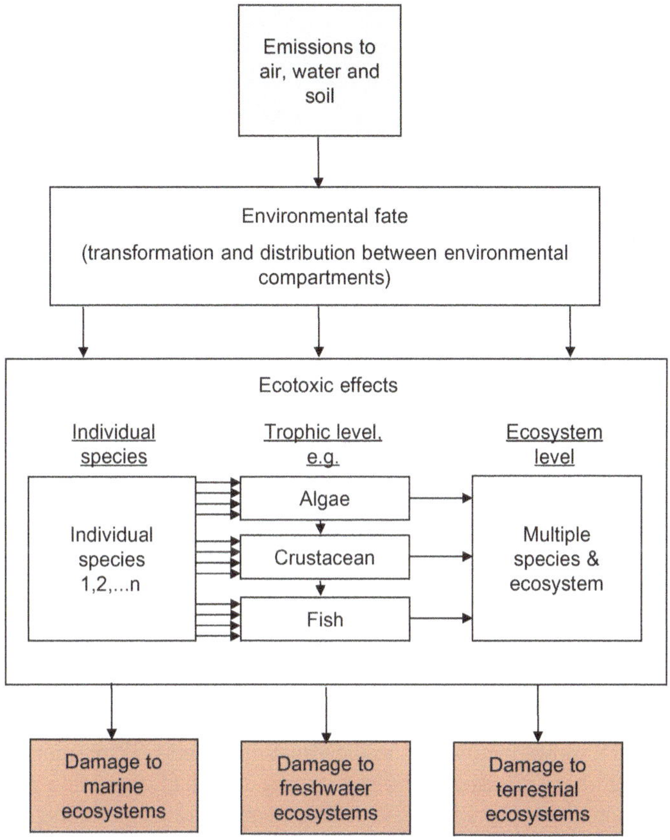

Fig. 10.17 General scheme of the Impact pathway for ecotoxicity [adapted from EC-JRC (2011)]

ecotoxicity, and thus its characterisation factor, is described in the following sections. Some LCIA methods also directly combine EF_{eco} and SF_{eco} into a single damage factor, directly calculating an endpoint characterisation factor. For midpoint characterisation, SF_{eco} is simply omitted and CF_{eco} is then the midpoint *ecotoxicity* characterisation factor.

A method for toxic impact assessment of chemicals in the framework of LCA must be able to cover the very large number of potentially toxic substances in the inventory in terms of available characterisation factors. It must also be based on integration of the impact over time and space as LCI data are typically not spatially and/or temporally differentiated, and the characterisation factor must relate to a mass flow and not require any information about concentrations of the substance as this information is not available in the LCI. To be compatible with the effect model, the fate model must translate chemical emissions calculated in the life cycle inventory into an increase in concentration in the relevant medium. In the characterisation modelling this leads to the use of fate models assuming steady-state conditions.

The *fate* model predicts the chemical behaviour/distribution in the environment accounting for multimedia (i.e. between environmental media and compartments) and spatial (i.e. between different zones but within the same compartment or medium) transport between environmental compartments (e.g. air, water, soil). This is accomplished via modelling of (thermodynamic) exchange processes such as partitioning, diffusion, sorption, advection, convection—represented as arrows in Fig. 10.18—as well as biotic and abiotic degradation (e.g. biodegradation, hydrolysis or photolysis), or burial in sediments. Degradation is an important loss process for most organic substances, but may also lead to toxic breakdown compounds. The rate by which the degradation occurs can be derived from the half-life of the substance in the medium and it depends both on the properties of the substance and on environmental conditions such as temperature, insolation or presence of reaction partners (e.g. OH radicals for atmospheric degradation). The basic principle underlying a fate model is a mass balance for each compartment leading to a system of differential equations which are solved simultaneously, which can be done for steady-state or dynamic conditions. A life cycle inventory typically reports emissions as masses emitted into an environmental compartment for a given functional unit. The mathematical relationship between the steady-state solution for a continuous emission and the time-integrated solution for a mass of chemical released into the environment has been demonstrated (Heijungs 1995; Mackay and Seth 1999).

Figure 10.18 shows the overall nested structure of the USEtox model which is a widely used global scientific consensus model for characterisation modelling of human and ecotoxic impacts in LCA. Further details on fate modelling principles in the USEtox model can be found in Henderson et al. (2011) and Rosenbaum et al. (2008).

Exposure is the contact between a target organism and a pollutant over an exposure boundary for a specific duration and frequency. The exposure model accounts for the fact that not necessarily the total ('bulk') chemical concentration present in the environment is available for exposure of organisms. Several factors and processes such as sorption, dissolution, dissociation and speciation may influence (i.e. reduce) the amount of chemical available for ecosystem exposure. Such phenomena can be defined as bioavailability ("freely available to cross an organism's cellular membrane from the medium the organism inhabits at a given time"), and bioaccessibility ("what is actually bioavailable now plus what is potentially bioavailable").

The *effect* model characterises the fraction of species within an ecosystem that will be affected by a certain chemical exposure. Effects are described quantitatively by lab-test derived concentration-response curves relating the concentration of a chemical to the fraction of a test group that is affected (e.g. when using the EC50—the Effect Concentration affecting 50% of a group of individuals of the same test species compared to a control situation). Affected can mean various things, such as increased mortality, reduced mobility, reduced growth or reproduction rate,

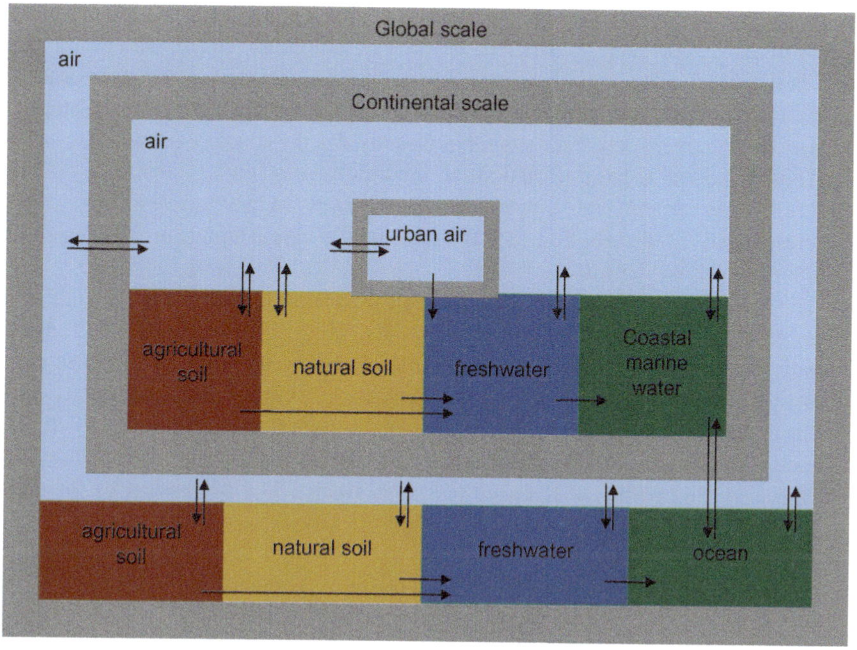

Fig. 10.18 The USEtox fate model [taken from Rosenbaum et al. (2008)]

mutations, behavioural changes, or changes in biomass or photosynthesis. These are the effects that may be observed during standardised laboratory-based ecotoxicity tests, and the results are specific for each combination of substance and species. Toxic effects are further distinguished into acute, sub-chronic and chronic toxicity (including further sub-groups like sub-acute, etc.). Acute toxicity describes an adverse effect after a short period of exposure, relative to the lifetime of the animal (e.g. <7 days for vertebrates, invertebrates or plants and <4 days for algae). Chronic toxicity is based on exposure over a prolonged period of time covering at least one life cycle or one sensitive period (e.g. ≥ 32 days for vertebrates, ≥ 21 days for invertebrates, ≥ 7 days for plants and ≥ 4 days for algae).

When relating to freshwater ecosystems, the question arises what exactly we mean by that. In LCIA, a freshwater ecosystem is typically seen as consisting of at least three trophic levels:

1. Primary producers, converting sunlight into biomass via photosynthesis (i.e. phytoplankton, algae)
2. Primary consumers, living off primary producers (i.e. zooplankton, invertebrates, planktivorous fish)
3. Secondary consumers at the upper end of the aquatic food chain (i.e. piscivorous fish)

It should be noted that only impacts on cold-blooded species in freshwater ecosystems are currently considered. There is no minimum requirement established, which trophic levels should be covered by a characterisation factor for terrestrial or marine ecosystems and available methods usually extrapolate from freshwater data or use the relatively few data available directly for these ecosystems.

There is often a large variation of sensitivity to a given substance between different species in the freshwater ecosystem. This is described by a species-sensitivity distribution (SSD) curve, which hence represents the sensitivity of the entire ecosystem to a substance—see Fig. 10.19.

The SSD is constructed using the respective geometric mean of all available and representative toxicity values for each species. This curve represents the range of sensitivities to exposure to a given substance among the different species in an ecosystem from the most sensitive to the most robust species. The ecotoxicity effect factor is then calculated using the HC50—Hazardous Concentration at which 50% of the species (in an aquatic ecosystem) are exposed to a concentration above their EC50, according to the SSD curve (see Fig. 10.19). The dimension of the effect factor is PAF—Potentially Affected Fraction of species, while the unit is typically m^3/kg.

The ecotoxicological effect factor of a chemical is calculated as:

$$EF_{eco} = \frac{0.5}{HC50} \tag{10.7}$$

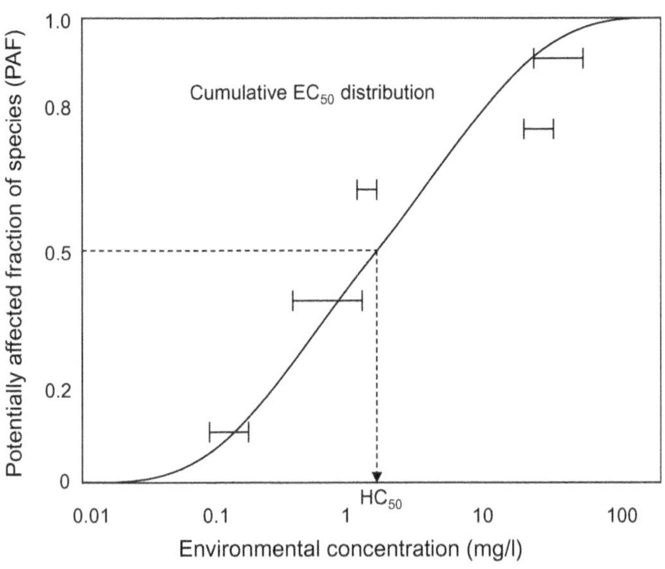

Fig. 10.19 Species-sensitivity distribution (SSD) curve representing the sensitivity of the ecosystem to a chemical substance

The HC50 value can be determined from the SSD curve but is often, more conveniently, calculated as the geometric mean of the EC50 values per species s, respectively:

$$\log \mathrm{HC50} = \frac{1}{n_{\mathrm{s}}} \cdot \sum_{\mathrm{s}} \log \mathrm{EC50}_{\mathrm{s}} \tag{10.8}$$

where n_s is the number of species.

A *damage* model, incorporating the severity of the effect, goes even further along the cause–effect chain and quantifies how many species are disappearing (instead of 'just' affected) from a given ecosystem. Disappearance may be caused by mortality, reduced proliferation or migration, for example.

10.11.3 Emissions and Main Sources

Chemicals are a main pillar of our industrialised economy, they are used in virtually any product around the globe and therefore numerous, used in large quantities and emitted from nearly all processes that an LCI may contain. Ecotoxity is very different from any other (non-toxicity) impact category when it comes to the number of potentially relevant elementary flows. Whereas no other (non-toxicity) impact category—with the exception of photochemical ozone formation—exceeds 100 contributing elementary flows (and related characterisation factors), the toxicity categories are facing the challenge of having to characterise several tens of thousands of chemicals with huge differences in their abilities to cause toxic impacts. The CAS registry currently (end of 2016) contains more than 124 million unique organic and inorganic structures (www.cas.org/about-cas/cas-fact-sheets) of which roughly 200,000 may play an industrial role as reflected by the ever increasing number of more than 123,000 substances registered in the European Classification and Labelling Inventory Database which contains REACH (Registration, Evaluation, Authorisation and Restriction of Chemical substances) registrations and CLP (Classification, Labelling and Packaging of substances and mixtures) notifications so far received by the European Chemicals Agency (ECHA: http://echa.europa.eu/information-on-chemicals/cl-inventory-database). Current LCIA models cover around 3000 substances for aquatic ecotoxicity.

10.11.4 Existing Characterisation Models

Characterisation methods like EDIP account for fate and exposure relying on key properties of the chemical applied to empirical models. Mechanistic models and methodologies have been published accounting for fate, exposure and effects providing cardinal impact measures. Among these methods are IMPACT 2002 (used in IMPACT 2002+) and USES-LCA (used in CML and ReCiPe). All these methods adopt environmental multimedia, multipathway models employing mechanistic cause–effect chains to account for the environmental fate, exposure and effects processes. However, they do not necessarily agree on how these processes are to be modelled, leading to variations in results of LCA studies related to the choice of LCIA method. Based on an extensive comparison of these models followed by a consensus-building process, the scientific consensus model USEtox (UNEP/SETAC toxicity consensus model) was developed with the intention to solve this situation by representing a scientifically agreed consensus approach to the characterisation of human toxicity and freshwater ecotoxicity (Hauschild et al. 2008; Rosenbaum et al. 2008; Henderson et al. 2011). It has been recommended and used by central international organisations like the United Nations Environment Program UNEP, Society of Environmental Toxicology and Chemistry SETAC, the European Commission and US-EPA to characterise human and ecotoxicity in LCIA.

Among the existing characterisation models on midpoint level, three main groups can be distinguished: (1) mechanistic, multimedia fate, exposure and effect models, (2) key property-based partial fate models and (3) non-fate models (EC-JRC 2011). According to ISO 14044 (2006b) "Characterisation models reflect the environmental mechanism by describing the relationship between the LCI results, category indicators and, in some cases, category endpoints. […] The environmental mechanism is the total of environmental processes related to the characterisation of the impacts." Therefore, ecotoxicity characterisation models falling into categories (2) and (3) do not completely fulfil this criterion. Caution is advised regarding their use and most importantly the interpretation of their results, which should not be employed without prior in-depth study of their respective documentation. Having said that, depending on the goal and scope of the LCA, they may still be an adequate choice in some applications, and indeed may agree quite well with the more sophisticated multimedia-based models.

Ecotoxicity endpoint modelling is still in an early state and much research needs to be performed before maturity is reached. The authors of the ILCD LCIA handbook concluded that "For all the three evaluated endpoint methods (EPS2000, ReCiPe, IMPACT 2002+), there is little or no compliance with the scientific and stakeholder acceptance criteria, as the overall concept of the endpoint effect factors is hardly validated and the endpoint part of the methods is not endorsed by an authoritative body. […] No method is recommended for the endpoint assessment of ecotoxicity, as no method is mature enough." (EC-JRC 2011).

When interpreting the results of existing methods, it is important to keep in mind that many aspects are not or only very insufficiently covered. This includes elements like terrestrial and marine ecotoxicity as well as toxicity of pesticides in pollinators.

For further details see Chap. 40 and Hauschild and Huijbregts (2015).

10.12 Human Toxicity

As explained in Sect. 10.11, both toxicity impact categories have a number of things in common, like main emissions and sources, modelling principles, model structure and even some of the models used in the characterisation are identical between the human toxicity and ecotoxicity impact categories. Notably the fate model used is the same in LCIA methods using mechanistic characterisation modelling, which is the majority of existing methods. Therefore, only those parts that are specific for human toxicity and different from ecotoxicity will be discussed here. It is recommended to first read Sect. 10.11 in order to understand the main underlying principles not repeated hereafter.

10.12.1 Problem

Human toxicity in LCA is based on essentially the same driving factors as ecotoxicity: (1) emitted quantity (determined in the LCI), (2) mobility, (3) persistence, (4) exposure patterns and (5) human toxicity, with the latter four considered by the characterisation factor. The respective mechanisms and parameters are certainly different and specific for human toxicity, notably for the exposure modelling, where many factors capturing human behaviour, such as dietary habits, influence human exposure pattern.

Chemical exposure of humans can result from emissions into the environment which will affect the whole population, but also from the many chemical ingredients in products released during their production, use, or end-of-life treatment and thus affecting workers or consumers. Chemical emissions are responsible for, or contribute to, many health impacts such as a wide range of non-cancer diseases as well as increased cancer risks for those chemicals that are carcinogenic.

10.12.2 Environmental Mechanism

Modelling the toxicological effects on human health of a chemical emitted into the environment, whether released on purpose (e.g. pesticides applied in agriculture), as a by-product from industrial processes, or by accident, implies a cause–effect chain, linking emissions and impacts through four consecutive steps as depicted in Fig. 10.20.

The cause–effect chain links the emission to the resulting mass in the environmental compartments (fate model) and on to the intake of the substance by the overall population via food and inhalation exposure pathways (human exposure model), and to the resulting number of cases of various human health risks by comparison of exposure with the known dose-response relationship for the

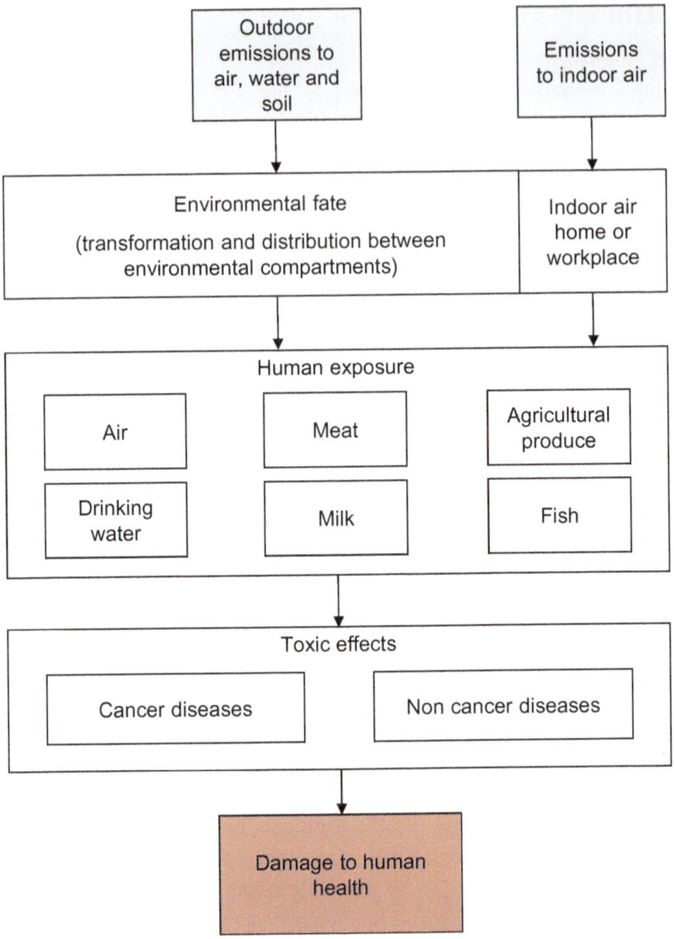

Fig. 10.20 General scheme of the impact pathway for human toxicity [adapted from EC-JRC (2011)]

chemical (toxic effect model) and finally their damage to the health of the overall population. In the characterisation modelling, the links of this cause–effect chain are expressed, similarly to Eq. 10.6, as factors corresponding to the successive steps of fate, exposure, effects and severity:

$$CF_{hh} = FF \times XF_{hh} \times EF_{hh} \times SF_{hh} \tag{10.9}$$

where CF_{hh} is the human health characterisation factor, FF the fate factor, XF_{hh} the human exposure factor, EF_{hh} the human toxicity effect factor (midpoint effects) and SF_{hh} the human health severity factor (endpoint effects). Some LCIA methods also directly combine EF_{hh} and SF_{hh} into a single damage factor, directly calculating an endpoint characterisation factor. For midpoint characterisation, SF_{hh} is simply omitted and CF_{hh} is then the midpoint human *toxicity* (i.e. not human *health*) characterisation factor.

The *midpoint human toxicity characterisation factor* [number of cases/$kg_{emitted}$] expresses the toxic impact on the global human population per mass unit emitted into the environment and can be interpreted as the increase in population risk of disease cases due to an emission into a specific environmental compartment. The *endpoint human health characterisation factor* [DALY/$kg_{emitted}$] quantifies the impact on human health in the global population in Disability-Adjusted Life Years (DALY) per mass unit emitted into the environment. DALY is a statistical measure of population life years lost or affected by disease (or other influences) and is used among other by the World Health Organisation.

The *fate* model is, without exception, the same as for ecotoxicity. Logically, the environment in which a chemical is transported, distributed and transformed is the same, no matter who will be affected. Therefore, for the sake of consistency, all LCIA methods that cover human toxicity are using the same fate model as for ecotoxicity, but of course different exposure and effect models, as this will be specific for the targeted organism (human or ecosystem species). The fate model is therefore the same as described in Sect. 10.11.

The *exposure* model relates the amount of chemical in a given environmental compartment to the chemical intake by humans (exposure rates). It can be differentiated into direct intake (e.g. by breathing air and drinking water), indirect intake through bioconcentration processes in animal tissues (e.g. meat, milk and fish) and intake by dermal contact. An exposure pathway is defined as the course a chemical takes from the environment to the exposed population, for example through air, meat, milk, fish, water or vegetables. Exposure pathways can be further aggregated into exposure routes, such as inhalation of air, ingestion of food including drinking water and other matter such as soil particles and dermal exposure. The human exposure model is designed for assessing human exposure to toxic chemical emissions applying realistic exposure assumptions and being adapted to take spatial variability into account. In LCIA human exposure is always assessed at the population level.

The *intake fraction* iF is calculated as the product of fate and exposure factor ($iF = FF * XF_{hh}$ [$kg_{intake}/kg_{emitted}$]) and it can be interpreted as the fraction of an emission that is taken in by the overall population through all exposure routes, i.e. as a result of food contamination, inhalation and dermal exposure. A high value, such as iF = 0.001 for dioxins, reflects that humans will take in 1 part out of 1000 of the mass of a chemical released. Dioxins are very efficient in exposing humans as reflected by the high intake fraction. For other chemical emissions, intake fraction values typically lie in the range of 10^{-10} to 10^{-5}.

The *effect* model relates the quantity of a chemical taken in by the population via a given exposure route (inhalation and ingestion, respectively, dermal uptake is not currently modelled in LCIA) to the toxic effects of the chemical once it has entered the human organism and can be interpreted as the increase in the number of cases of a given human health effect (e.g. cancer or non-cancer diseases) in the exposed population per unit mass taken in. The two general effect classes, cancer and non-cancer, each cover a multitude of different diseases, so this is a simplification reflecting the fact that it is very difficult to predict the many underlying human toxicity endpoints from the animal dose-response curves from laboratory experiments with test animals which are normally the basis of the effect factor.

The *severity* factor represents adversely affected life years per disease case (DALY/case), distinguishing between differences in the severity of disabilities caused by diseases in terms of affected life years, e.g. discriminating between a lethal cancer and a reversible skin irritation. It is quantified by the statistically determined, population-based years of life lost (YLL) and years of life disabled (YLD) due to a disease.

10.12.3 Emissions and Main Sources

The relevant emissions and main sources are identical to those of the ecotoxicity impact category and discussed in Sect. 10.11.

10.12.4 Existing Characterisation Models

Again here, Sect. 10.11 contains a discussion on existing characterisation models, which largely applies also to the human toxicity impact category.

In USEtox, the units of the two human toxicity midpoint indicators for non-cancer and cancer are Comparative Toxic Unit for humans CTU_h [disease cases]. They can be added up to a single human health indicator, but then the interpretation needs to consider that this intrinsically assumes equal weighting between cancer and non-cancer effects (which includes equal weighting between e.g. a reversible skin rash and non-reversible death). Human health endpoint indicators in USEtox are given in the Comparative Damage Unit for human health CDU_h [DALY]. In accordance with the purpose of endpoint modelling, this indicator better represents the distinction of the severity of different effects.

When interpreting human toxicity indicators from existing methods, it is important to be aware that these only provide indicators for global population exposure to outdoor and indoor emissions, while human toxicity for occupational exposure of workers or direct exposure related to product use for consumers are not yet covered by USEtox and the other characterisation models, despite their very high relevance. Products of special interest in this context are cosmetics, plant protection products, textiles, pharmaceuticals and many others, that may in particular contain substances having toxic properties and have the potential to cause mutagenic, neurotoxic or endocrine disrupting effects. This is the subject of ongoing research and will be included in LCIA methods once the models are mature and operational.

For further details see Chap. 40 and Hauschild and Huijbregts (2015).

10.13 Particulate Matter Formation

In existing LCIA methods, health impacts from exposure to particulate matter (PM) as impact category is referred to by different terms (e.g. 'particulate matter/respiratory inorganics' in ILCD, 'respiratory effects' in IMPACT 2002+, 'human health criteria pollutants' in TRACI, or 'particulate matter formation' in ReCiPe). Although causing mainly toxicity-related health effects, exposure to PM is considered a separate impact category in most LCIA methods. This is mainly due to a number of important differences between the characterisation of PM formation and that of human toxicity. These differences include the complex atmospheric chemistry involved in the formation of secondary PM from different precursor substances which requires a different fate model. Furthermore, different emission heights are important to consider, global monitoring data for PM air concentrations are used, and the effect assessment is based on exposure-response functions mostly derived from epidemiological evidence, which is not possible for most toxic chemicals due to missing emission locations and exposure- or dose-response information.

10.13.1 Problem

A large number of studies including the global burden of disease (GBD) study series consider particulate matter (PM) to be a leading environmental stressor contributing to global human disease burden (i.e. all diseases around the world) via occupational and household indoor exposure as well as urban and rural outdoor (ambient) exposures. In 2013, outdoor PM pollution accounted for 2.9 million deaths and 70 million DALY, and household PM pollution from solid fuels accounted for 2.9 million deaths and 81 million DALY (Forouzanfar et al. 2015). With that, outdoor and household PM pollution combined contributed in 2013 with 71% to premature deaths attributable to all environmental risk factors and with 19% to premature death attributable to all risk factors (i.e. including behavioural etc.). This means that exposure to PM accounts on average for 1 out of 5 premature deaths worldwide. Thereby, exposure to PM is associated in epidemiological and toxicological studies with various adverse health effects and reduction in life expectancy including chronic and acute respiratory and cardiovascular diseases, chronic and acute mortality, lung cancer, diabetes and adverse birth outcomes (Fantke et al. 2015).

PM can be distinguished according to formation type (primary and secondary) and according to aerodynamic diameter (respirable, coarse, fine and ultrafine). Primary PM refers to particles that are directly emitted, e.g. from road transport, power plants or farming activities. Secondary PM refers to organic and inorganic particles formed through reactions of precursor substances including nitrogen oxides (NO_x), sulphur oxides (SO_x), ammonia (NH_3), semivolatile and volatile organic compounds (VOC). Secondary particles include sulphate, nitrate and organic carbonaceous materials and can make up to 50% of ambient PM concentrations. Respirable particles (PM_{10}) have an aerodynamic diameter less than 10 μm, coarse particles ($PM_{10–2.5}$) between 2.5 and 10 μm, fine particles ($PM_{2.5}$) less than 2.5 μm, and ultrafine particles (UFP) less than 100 nm (WHO 2006). $PM_{2.5}$ is often referred to as the indicator that best describes the component of PM responsible for adverse human health effects (Lim et al. 2012; Brauer et al. 2016).

10.13.2 Environmental Mechanism

Characterising health impacts from exposure to PM associated with emissions of primary PM or secondary PM precursor substances builds on the general LCIA framework for characterising emissions of air pollutants (see Fig. 10.2). The impact pathway for health impacts from PM emissions is illustrated in Fig. 10.21 and starts from primary PM emissions or secondary PM precursor substances emitted into air.

As for the toxicity impact categories, combining all factors from emission to health impacts or damages yields the characterisation factor for particulate matter formation (CF) with units [disease cases/$kg_{emitted}$] at midpoint level (i.e. excluding SF) and [DALY/$kg_{emitted}$] at endpoint level:

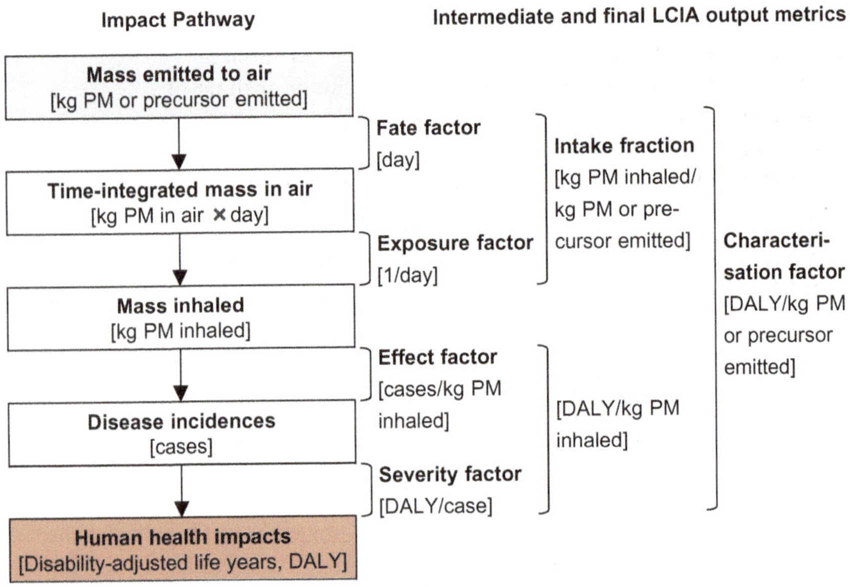

Fig. 10.21 Schematic impact pathway and related output metrics for characterising health impacts from particulate matter (PM) exposure in life cycle impact assessment [adapted from Fantke et al. (2015)]

$$CF = FF \times XF \times EF \times SF \tag{10.10}$$

Emissions are expressed as mass of PM or precursor substance released into air. From there, the impact pathway follows different distribution processes within and between air compartments and/or regions (indoor, outdoor, urban, rural, etc.) yielding a time-integrated mass of PM in the different air compartments and/or regions. Relating the time-integrated PM mass in air to the mass of PM or precursor substance emitted yields the fate factor (FF) with unit kg in air integrated over one day per kg emitted. A certain fraction of PM mass in air is subsequently inhaled by an exposed human population. This fraction is expressed by the exposure factor (XF) describing the rate at which PM is inhaled with unit kg PM inhaled per kg PM in air integrated over one day. Multiplying FF and XF yields the cumulative PM mass inhaled by an exposed population per kg PM or precursor emitted expressed as human intake fraction (iF). Inhaling PM mass may then lead to a cumulative population risk referred to as expected disease incidences in the exposed human population and typically assessed based on PM air concentration. Relating PM concentration in air to cumulative population risk yields the exposure-response or effect factor (EF) with unit disease cases (e.g. death for mortality effects) per kg PM inhaled. Finally, disease incidences are translated into human health damages by accounting for the disease severity expressed as disability-adjusted life years (DALY) that include mortality and morbidity effects. Linking health damages to disease incidences yields the severity (or damage) factor (SF) with unit DALY per disease case.

For characterising health impacts from emissions of PM or precursor substances, several aspects influence emission, fate, intake and health effects. Regardless of the modelling setup (spatial vs. archetypal; including or disregarding indoor sources and/or secondary PM formation, etc.), main influential aspects are spatiotemporally variable population density and activity patterns, background PM concentration in air, background disease rate and background severity, emission location (e.g. indoor vs. outdoor or urban vs. rural) and emission height, as well as potential nonlinearity in the disease-specific exposure-response relationship. The effect of using a non-linear exposure-response curve in the calculation of CFs following the marginal and average approach is illustrated in Fig. 10.22 for two distinct background concentration scenarios, where the difference between marginal and average approach is increasing with increasing background concentration for an exposure-response curve of supralinear shape.

10.13.3 Emissions and Main Sources

Substances considered in the different LCIA methods to contribute to health impacts from PM are typically one or more PM fractions (PM_{10}, $PM_{10-2.5}$, $PM_{2.5}$)

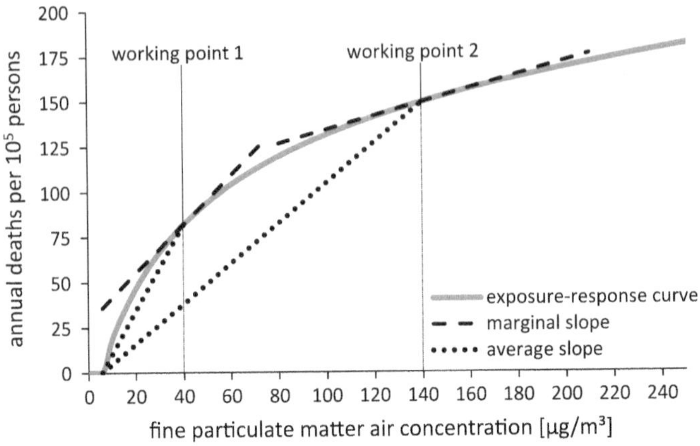

Fig. 10.22 Illustration of using a non-linear exposure-response curve for health effects from fine particulate matter exposure with *dashed* and *dotted lines* as approaches for calculating marginal and average (between working point and theoretical minimum-risk concentration) characterisation factors, respectively, at different background concentrations in air as working points. Exposure-response curve based on data from Apte et al. (2015)

and PM precursor substances (mostly NO_x, SO_2 and NH_3) and in some cases also carbon monoxide (e.g. IMPACT 2002+) or non-methane volatile organic compounds (e.g. ReCiPe). Relevant emission sources of PM (and/or precursors) are for example road traffic, stationary emissions from coal/gas-fired power plants or indoor emissions from solid fuels combustion. Several emission sources are ground-level sources (e.g. road traffic and household combustion), while others are considered to occur at higher stack levels (typically stationary emission sources, e.g. power plants).

10.13.4 Existing Characterisation Models

In LCIA, archetypal impact assessment scenarios (e.g. urban, rural) are often used instead of spatialized or site-specific scenarios, especially when emission locations are unknown or fate, exposure and/or effect data do not allow for spatial differentiation. Such archetypal approach and related intake fractions were proposed by Humbert et al. (2011) with population density (urban, rural and remote) and emission height (ground-level, low-stack and high-stack emissions) as main determinants of PM and precursor impacts. The UNEP/SETAC Life Cycle Initiative established a task force to build a framework for consistently quantifying health effects from PM exposure and for recommending PM characterisation factors for application in LCIA with fine particulate matter ($PM_{2.5}$) as representative indicator. First recommendations from this task force focus on the integration of indoor and outdoor environments, the archetypal approach capturing best the dominating differences between urban and rural areas and a number of other improvements (Fantke et al. 2015).

Most LCIA characterisation methods addressing particulate matter formation follow the framework described in this section. There are some methods, however, that characterise impacts from particles as part of the 'human toxicity' impact category (e.g. CML 2002 and EDIP 2003), while most methods (including all methods developed after 2010) characterise human toxicity impacts from chemicals and impacts from particles as separate impact categories, mainly due to the differences in available data that allow using more refined models and less generic assumptions for the impact assessment of particle emissions.

The most recent characterisation models—all damage-oriented—include work by van Zelm et al. (2008) providing characterisation factors for primary and secondary PM_{10} for Europe based on a source receptor model, work by Gronlund et al. (2015) giving archetypal characterisation factors for primary $PM_{2.5}$ and secondary $PM_{2.5}$ precursors based on US data and work by van Zelm et al. (2016) proposing averaged primary and secondary $PM_{2.5}$ characterisation factors for 56 world regions based on a global atmospheric transport model. However, none of the currently available approaches includes indoor sources, is able to distinguish emission situations at the city level or considers the non-linear nature of available exposure-response curves, which is why further research is needed for this impact category. For further details see Chap. 40 and Hauschild and Huijbregts (2015).

10.14 Land Use

10.14.1 Problem

Land use refers to anthropogenic activities in a given soil area. Examples of land use are agricultural and forestry production, urban settlement and mineral extraction. The land use type in a specific area can be identified by the physical coverage of its surface, for example tomato crop grows in open-field orchards or under greenhouses, artificial surfaces with infrastructure are the expression of human settlements and open-pits are a sign of ore extraction. There is thus a direct link between land use and land cover, which is used to analyse land use dynamics and landscape change patterns.

Soil is a finite resource, which contributes to the environmental consequences of its use. Soil loss actually occurs quantitatively with the average soil formation rate being extremely low compared to the soil depletion rate. It also affects qualitative soil attributes, because degrading takes place via unsustainable management practices for the highest quality soils, which are those able to fulfil a greater diversity of purposes. As soil or land surface available at a given time is limited, land-use competition between resource users for occupying the same space often arises. This drives continuous changes in land uses. Croplands, pastures, urban areas and other land-use-intensive, human activities have expanded worldwide in the last decades at the expense of natural areas to satisfy our growing society's needs for food, fibre, living space and transport infrastructure. Such changes transform the planet's land surface and lead to large and often irreversible impacts on ecosystems and human quality of life (EEA 2010). For example, forest clearing contributes to climate change with the release of carbon from the soil to the atmosphere. The loss, fragmentation and modification of habitats lead to biodiversity decline. Land use change alters the hydrological cycle by river diversion and by modifying the portion of precipitation into runoff, infiltration and evapotranspiration flows (Foley et al. 2005). After soil surface conversion, inappropriate management practices on human-dominated lands can also trigger a manifold of environmental effects on soil physical properties. In agricultural lands, mechanised farming can induce soil compaction, which affects aquifer recharge and the natural capacity of the soil to remove pollutants. Erosion is also a spread environmental concern of intensive agricultural practices. In urban and industrial areas, soil has been replaced by concrete surfaces and all its functions annulled.

The Millennium Ecosystem Assessment (2005) provides a comprehensive description of how human land-use activities affect biodiversity and the delivery of ecological functions. Some ecological effects of land use are:

- Biodiversity decrease at the ecosystem, species and genetic levels
- Impacts on local and regional climate regulation due to changes in land cover and albedo, e.g. tropical deforestation and desertification may locally reduce precipitation
- Regional decline in food production per capita due to soil erosion and desertification, especially in dry lands
- Rise in flood and drought risks through loss of wetlands, forests and mangroves
- Change in the water cycle by river diversion and by greater appropriation of freshwater from rivers, lakes and aquifers to be used for irrigation of areas converted to agriculture

To sum up, land-use activities (including land conversion and land use itself) cause noticeable damages on biodiversity and on the performance of soil to provide ecological functions as illustrated in Fig. 10.23. These ecological functions upon which human well-being depends are also referred to as ecosystem services (Millenium Ecosystem Assessment 2005), and together with biodiversity loss are the focus of the LCIA land-use impact category.

10.14.2 Environmental Mechanism

The LCIA land-use impact category covers a range of consequences of human land use, being a receptacle (or 'bulk') category for many impact indicators. It does not

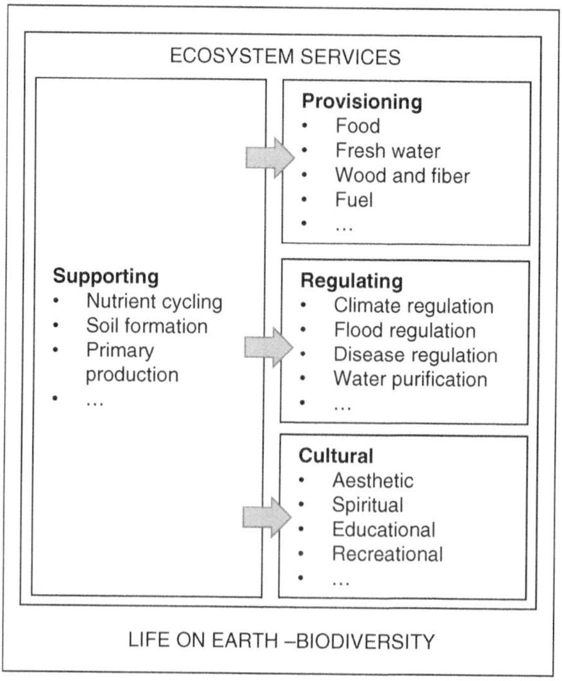

Fig. 10.23 The land use impact category focuses on damage to biodiversity—which represents the foundation of ecosystems—as well as on the provision of ecosystem services, due to land conversion and land use [adapted from Millenium Ecosystem Assessment (2005)]

assess nutrients, pesticides and any other types of emission to the ecosphere which are characterised by the corresponding emission-based impact category (e.g. eutrophication for emission of nutrients, ecotoxicity for emission of pesticides). Their inclusion in the land-use category would lead to double counting of the same impact.

The general land-use environmental mechanism follows the model of Fig. 10.24. It shows the cause–effect chain from the elementary flow (i.e. land transformation or land occupation) to the endpoint damages on human health and ecosystems as well as available soil resources. Land transformation refers to the conversion from one state to another (also known as land use change, LUC) and land occupation to the use of a certain area for a particular purpose (also known as land use, LU). The figure should be read as follows, giving an example of the depicted impact pathways: land occupation leads to physical changes to soil, which leads to an altered soil function and affects habitats and net primary production which eventually leads to damage on ecosystem quality. The picture provides a good display of the complexity involved in land-use modelling. For some of the presented impacts, such as warming effect due to albedo change or landscape impairment, characterisation models have yet to be developed.

The same type of human activity may cause different land-use related impacts depending on the region of the world where the activity takes place. This variation is due to the strong influence of climate, soil quality, topography and ecological quality on the magnitude of the impact. For example, deforestation of a forest area

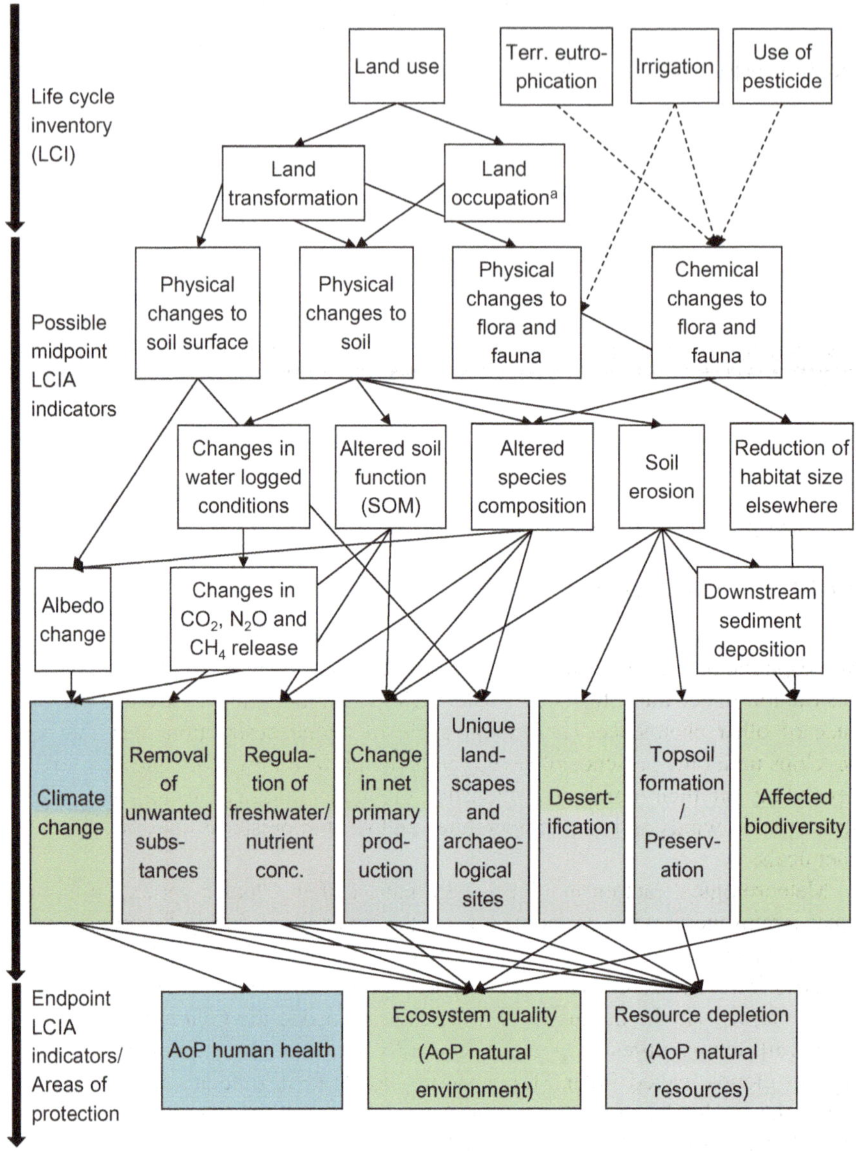

Fig. 10.24 Impact pathway for land use impacts; *dashed arrows* indicate impacts covered by emission-related impact categories and by water use in the case of irrigation [adapted from EC-JRC (2011)]. [a]Land occupation will not cause changes but will contribute to prolong the changed conditions

for use in agriculture in the Brazilian Amazon has a greater impact in terms of number of species affected than forest clearing in an ecologically poorer European region. Because land use impacts depend on-site-specific conditions, land use is considered a local impact category in LCA, in opposition to other impact categories of global geographic scope such as climate change, whose environmental effects (in terms of radiative forcing) are independent of the location of the emission.

As a consequence of the above explanation, methods that focus on land-use impacts should include geospatial data both in the LCI and the LCIA phases. The inventory must contain information on the geographic location of the human intervention, with a level of detail that may vary from the exact coordinates to coarser scales (e.g. biome, country, continent), depending on the goal and scope of the study and if the inventory refers to the foreground or to the background system (see Chap. 9). In the LCIA, characterisation factors for a given impact indicator must capture the sensitivity of the habitat to the impact modelled. For example, characterisation factors for soil erosion may include information on the soil depth in the specific location of the activity under evaluation, as the impact of soil loss will depend on the soil stock size, i.e. thinner soils are more vulnerable than thicker soils (Núñez et al. 2013). Every geographic unit of regionalised impact assessment methods has its own characterisation factor. Within the boundary of such a unit, it is assumed that an activity triggers the same impacts on land.

10.14.3 Existing Characterisation Models

Characterisation of land use in LCA has been extensively discussed over the last decades but is far from being settled, because the first operational methods have only been available since 2010. Until then, land use was only an inventory flow-counted in units of surface occupied and time of occupation (m^2 and years) and surface transformed (m^2), without any associated impact. The main reason for this "late development" is that land-use related impacts rely on spatial and temporal conditions where the evaluated activity takes place, whereas traditional LCA is site-generic. During the last few years, the release of geographical information system (GIS) software and data sets have brought new opportunities in LCA to model land-use impacts and in general, any other spatially dependent impact category.

Today, there are LCIA methods to evaluate impacts on biodiversity and impacts on several ecosystem services. From the long list of services provided by terrestrial ecosystems (24 acknowledged in the Millennium Ecosystem Assessment international work programme (2005), LCA focuses on those which are recognised as being more environmentally relevant (i.e. educational and spiritual values are excluded). A non-exhaustive list of methods is provided below. For completeness, see Milà i Canals and de Baan (2015):

- Impacts on biodiversity: Biodiversity should be preserved because of its intrinsic value. The most commonly applied indicator is based on species richness, given the availability of data (Scholz 2007; Koellner and Scholz 2008; de Baan et al. 2013a, b). Damage on biodiversity is commonly expressed in quantity of species biodiversity loss, either in relative terms (potentially disappeared fraction of species times surface, PDF.m^2) or in absolute species loss. Existing indicators for biodiversity are at the endpoint level (in Fig. 10.24, Ecosystem quality-AoP natural environment box in the lower row). The UNEP-SETAC Life Cycle Initiative project on global guidance for LCIA indicators and methods provisionally recommended characterisation factors from Chaudhary et al. (2015) representing global potential species loss from land use to assess impacts on biodiversity due to land use and land-use change as hotspot analysis in LCA only (not for comparative assertions nor eco-labelling). Further testing of the CFs as well as the development of CFs for further land-use types are required to provide full recommendation.

- Impacts on ecosystem services: Includes a range of indicators for life support functions that ecosystems provide. Ecosystem services are hardly covered in LCIA and proposals are still incipient. All available methods are on the midpoint level (in Fig. 10.24, boxes between the LCI and the endpoint), which means that comparison or aggregation with damages on biodiversity is not possible so far. The recent draft review of land-use characterisation models for use in Product and Organisation Environmental Footprint (PEF/OEF) provisionally (i.e. "apply with caution") recommended characterisation factors from LANCA (Bos et al. 2016) to assess impacts on ecosystem services (EC-JRC 2016). Currently, there are LCA methods for the following ecosystem services:
  - Biotic production potential: capacity of ecosystems to produce and sustain biomass on the long term. Available indicators are based on the soil organic matter (or carbon) content (Brandão and Milà i Canals 2013), the biotic production (Bos et al. 2016) and the human appropriation of the biotic production (Alvarenga et al. 2015)
  - Carbon sequestration potential: capacity of ecosystems to regulate climate by carbon uptake from the air. The size of the climatic impact is determined by the amount of CO_2 transfers between vegetation/soil and the atmosphere in the course of terrestrial release and re-storage of carbon (Müller-Wenk and Brandão 2010)
  - Freshwater regulation potential: capacity of ecosystems to regulate peak flow and base flow of surface water. Available indicators refer to the way a land-use system affects average water availability, flood and drought risks, based on the partition of precipitation between evapotranspiration, groundwater infiltration and surface runoff (Saad et al. 2013; Bos et al. 2016)
  - Water purification potential: mechanical, physical and chemical capacity of ecosystems to absorb, bind or remove pollutants from water. Site-specific soil properties such as texture, porosity and cation exchange capacity are used as the basis for the assessment (Saad et al. 2013)
  - Erosion regulation potential: capacity of ecosystems to stabilise soils and to prevent sediment accumulation downstream. The soil performance is determined by the amount of soil loss (Saad et al. 2013; Bos et al. 2016) and how this soil loss reduces the on-site soil reserves and the biotic production (Núñez et al. 2013)

- Desertification regulation potential: capacity of dry lands to resist irreversible degradation on the human time-frame. A multi-indicator system of four variables, namely climate aridity, soil erosion, aquifer exploitation and fire risk, determines the desertification ecosystem vulnerability (Núñez et al. 2010)

The land-use impact category is likely the LCA category most affected by potential problems of double counting. This is because methods for emissions and methods for land use have been developed under two different, incompatible approaches. Emission models are bottom-up: the starting point is the elementary flowin the LCI and the impact model describes stepwise all the mechanisms that link the cause (the LCI) to the consequence (midpoint or endpoint impact). Land-use models, in contrast, are top-down. This means that they are based on empirical observations of the state of the environment, but there is no evidence of the connection between the consequence and the (supposed) cause. For example, methods to evaluate biodiversity damage are based on databases of the species present under different land-use types. The reduction in species richness from e.g. a forest to an arable intensive agricultural land is driven by many reasons that partially add to each other: cut down of trees and replacement for crops, use of tractor and other agricultural machinery, emission of pesticides and fertilisers, etc. However, how and how much each of the reasons above contributes to the actual biodiversity loss observed in the agricultural land is not known. The development of mechanistic models such as the ones used to characterise emissions, have the potential to resolve the issue of double counting. For further details see Chap. 40 and Hauschild and Huijbregts (2015).

10.15 Water Use

10.15.1 Problem

Water is a renewable resource which, thanks to the water cycle, does not disappear. It is a resource different from any other for two main reasons: (1) it is essential for human andecosystem life and (2) its functions are directly linked to its geographic and seasonal availability, since transporting it (and to a lesser extent, storing it) is often impractical and costly. There is sufficient water on our planet to meet current needs of ecosystems and humans. About 119,000 km^3 are received every year on land in different forms of precipitation, out of which 62% are sent back directly to the atmosphere via evaporation and plant transpiration. Out of the 38% remaining, humans use only about 3%, out of which 2.1% for agriculture, 0.6% for industrial uses and 0.3% for domestic uses. However, despite these small fractions, there are still important issues associated with water availability. Many important rivers are running dry from overuse (including the Colorado, Yellow and Indus), greatly affecting local aquatic and terrestrial ecosystems. Humans compete for the use of water in some regions, sometimes leading to the exchange of water rights on the market or to the exacerbation of tensions between nations. The World Water Council described the problem well by stating: "There is a water crisis today. But the crisis is not about having too little water to satisfy our needs. It is a crisis of managing water so badly that billions of people—and the environment—suffer badly". In addition to the current mismanagement of the water, which is strongly linked to a competing demand for human uses and ecosystems for a limited renewable resource, the human demand is only increasing, namely due to a growing population and changing diets (with increasing meat consumption). Water availability is also changing due to climate change, aggravating droughts and flooding and hence further increasing the gap between the demand and availability in many highly populated regions around the world. Since the problems associated with water are dependent on where and when water is available, as well as in which quality, it is these aspects that also need to be considered when we assess potential impacts of human freshwater use on the environment (including human health) in LCA.

10.15.2 Environmental Mechanism

Before diving into the assessment of potential impacts associated with water, some concepts are important to establish first.

- Types of water use: Water can be used in many different manners and the term water use represents a generic term encompassing any type of use. Consumptive and degradative use are the two main types of use and all other types of use (borrowing, turbinated, cooling, etc.) can generally be defined by one or a combination of the following three terms:
  - Water withdrawal: "anthropogenic removal of water from any water body or from any drainage basin either permanently or temporarily" (ISO 2014)
  - Consumptive use/water consumption: water use where water is evaporated, integrated in a product or released in a different location then the source
  - Degradative use/water degradation: Water that is withdrawn and released in the same location, but with a degraded quality. This includes all forms of pollution: organic, inorganic, thermal, etc. (ISO 2014)
- Sources of water: Different sources of water should be distinguished as impacts from using them will often differ. In general, the following main sources are differentiated: surface water, groundwater, rainwater, wastewater and sea water. Some more specific descriptions can include brackish water (saline water with lower salinity than sea water, generally between 1000 and 10,000 mg/l) or fossil water (non-renewable groundwater)
- Water availability: when used as an indicator, this describes the "extent to which humans and ecosystems have sufficient water resources for their needs", with a note that "Water quality can also influence availability, e.g. if quality is not

sufficient to meet users' needs. If water availability only considers water quantity, it is called water scarcity". (ISO 2014). However, this term (water availability) is also used to refer to the renewable water volume that is available in a specific area during a specific time, most typically annually or monthly over a watershed (m^3/year or m^3/month)

- Water Scarcity: Different definitions exist for water scarcity, but in LCA the following standardised one is retained: "extent to which demand for water compares to the replenishment of water in an area, e.g. a drainage basin, without taking into account the water quality" (ISO 2014)
- Watershed (also called drainage basin): "Area from which direct surface runoff from precipitation drains by gravity into a stream or other water body" (ISO 2014). In general the main watershed is taken as the reference geographical area to define the same location, as countries are often too large to represent local water issues and smaller areas would lack data and relevance

As mentioned above, freshwater is received from precipitation and a fraction of it (about 38%) is made available as "blue water", or flowing water which can be used by humans and ecosystems via lakes, rivers or groundwater. Some freshwater is also present in deep fossil aquifers, which are not renewable (not recharged by precipitation), and can be used by humans if pumped out. Groundwater aquifers can recharge lakes and rivers, and vice versa, depending on the topology, soil porosity, etc. Surface water is used by humans, aquatic ecosystems and terrestrial ecosystems, whereas groundwater can be used by some terrestrial ecosystems and humans.

Water use impact assessment at midpoint level typically focuses on water deprivation. Although water is renewed, there is a limited amount available in an area at any point in time, and different users must share, or compete for, the resource. Consuming a certain volume of water will lower its availability for users downstream and may also affect groundwater recharge for example. Users depending on this water may be deprived and suffer consequences. The extent to which they will be deprived will depend on the water scarcity in a region (Fig. 10.25). The higher the demand in comparison to the availability, the more likely a user will be deprived. This user can be (1) humans (present and future

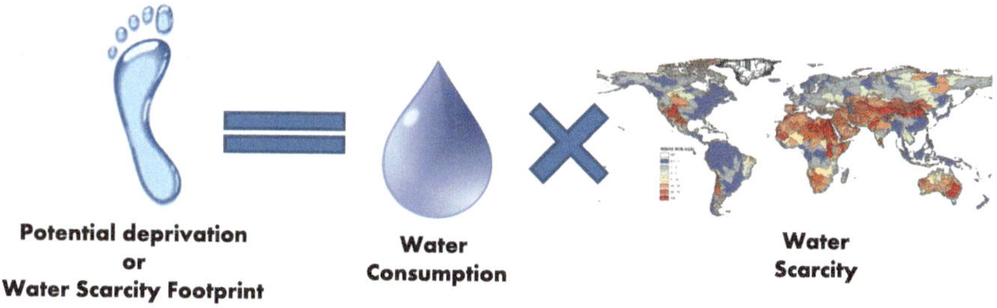

Fig. 10.25 The potential deprivation caused by an additional water consumption in a region is assessed by multiplying this water consumption with a local water scarcity factor. The result is also called a water scarcity footprint

generations) and (2) ecosystems (Bayart et al. 2010). Quantifying "the potential of a user (humans or ecosystems) to be deprived when water is consumed in a region" (Boulay et al. 2017) is the question normally answered at the midpoint level using for example a scarcity indicator (or user-specific deprivation potential if they exist), whereas assessing the potential damages from this deprivation on human health and ecosystem quality is an endpoint assessment.

At the endpoint level, water use impact assessment is focused on the consequences of the water deprivation for humans and ecosystems. The higher the scarcity (and competition between human users), the larger the fraction of an additional water consumption that will deprive another user. Which human user is affected will depend on the share of each water user in a region, as well as their ability to adapt to water deprivation. If the deprived users have access to sufficient socio-economic resources, they may adapt and turn towards a backup technology like desalinisation of seawater or freshwater import to meet their needs. Impacts from human deprivation are then shifted from being solely on human health to all impact categories that are affected by the use of this backup technology. However, if socio-economic means are not sufficient to adapt to lower water and/or food availability, deprivation may occur. Since the potential impacts associated with water deprivation for humans assessed in LCA are on human health, deprivation of water for domestic use, agriculture and aquaculture/fisheries are relevant. Domestic users which already compete for water and have no means to compensate lower water availability via purchasing or technological means will suffer from freshwater deprivation, which is associated to water-related diseases caused by the use of improper water sources and change of behaviour. Agricultural users that are deprived of water for irrigation may produce less, which in turn will lead to lower food availability, either locally or internationally through trade, which may increase health damages associated with malnutrition. Similarly, lower freshwater availability for aquaculture or fisheries could lower fish supply and also contribute to malnutrition impacts, although this was shown to be negligible in comparison to other users' deprivation. This impact pathway, leading to damages on human health, is shown in Fig. 10.26.

Consuming water can also affect water availability for aquatic and terrestrial ecosystems. If the flow of the river is altered, or the volume of the lake is reduced, aquatic ecosystems have less habitat space and may either have to adapt or suffer a change in species density. Since water compartments are strongly interconnected, consuming water in a lake can affect the groundwater availability and vice versa, and each change in availability can lead to a loss of species. Consuming water can also alter the quality by reducing the depth of the water body for example, increasing temperature or concentrating contaminants. Aquatic ecosystems are dependent not only on a minimum volume for their habitat, but also on the flow variations which are naturally influenced by seasons. Human interference with this flow variation can also cause potential species loss. The groundwater table in some regions directly feeds the roots of the vegetation and lowering the aquifer's level can mean that shorter roots species no longer reach their source of water. The relevant mechanisms are summarised in Fig. 10.27. These impact pathways appear

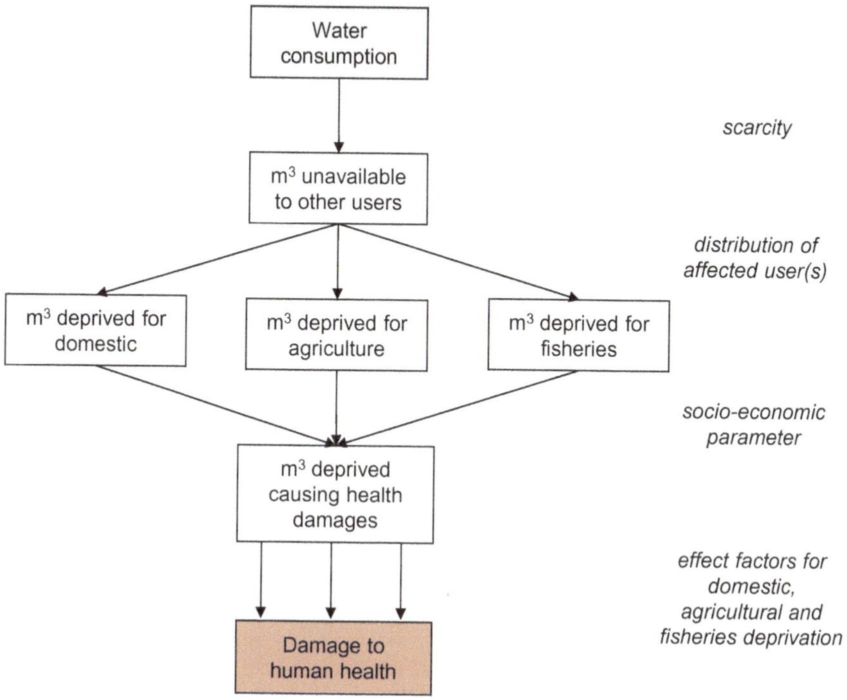

Fig. 10.26 Impact pathway from water consumption to water deprivation for human users leading to potential impacts on human health in Disability Adjusted Life Years (DALY) [adapted from Boulay et al. (2015)]

to be complementary, however more research is needed to determine how they should be used together and to provide one harmonised methodology.

10.15.3 Existing Characterisation Models

A stress/scarcity index (here used interchangeably) is the most commonly used midpoint, even if it does not necessarily represent an actual point on the impact pathway of all endpoint categories. A scarcity index is based on the comparison between water used and renewable water available, and represents the level of competition present between the different users (ideally human users and ecosystems). Early indicators (Frischknecht et al. 2008; Pfister et al. 2009) are based on withdrawal-to-availability (WTA) ratios as these were the data available at the time. Since water that is withdrawn but released into the same watershed (within a reasonable time-frame) does not contribute to scarcity, indicators emerged which were based on consumption-to-availability (CTA) ratios instead of withdrawals, when the needed data became available (Boulay et al. 2011; Hoekstra et al. 2012;

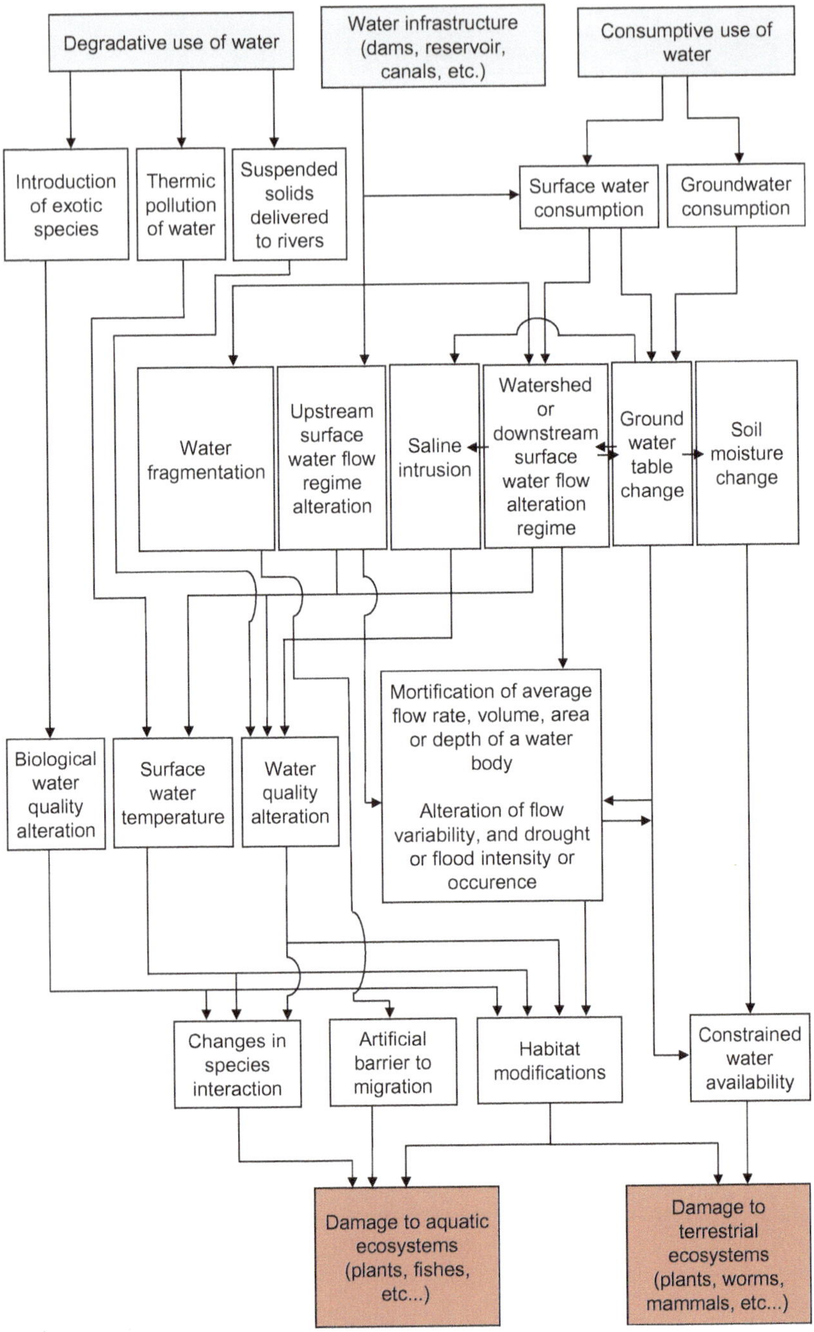

Fig. 10.27 Impact pathways affecting ecosystem quality [adapted from Núñez et al. (2016)]

Berger et al. 2014). Further development led to the inclusion of environmental water requirements as part of the water demand in order to better represent the total water demand from all users, including ecosystems, and resulted in a ratio based on demand-to-availability (DTA) being proposed (Boulay et al. 2014).

However, one important information was lost in all these indicators: the absolute availability. A ratio of 0.5 may indicate that half of the available water is currently withdrawn, consumed or demanded, but it does not inform on the magnitude of this water volume (i.e. is it 1 or 1000 m^3?). Regions differ largely in terms of absolute water availability (or aridity) and this information should not be discarded by only looking at the fraction of available water that is being used. In 2016, the WULCA group (see below) proposed the area-specific Available Water Remaining indicator (based on availability minus demand), AWARE, inverted and normalised with the world average (Boulay et al. 2017). Ranging between 0.1 and 100, this index assesses the potential to deprive another user (human or ecosystem) of water, based on the relative amount, comparing to the world average, of water remaining per area once the demand has been met. The more water remaining compared to the average, the lower the potential to deprive another user, and vice versa.

It should be noted that some midpoints also propose to include quality aspects, allowing the quantification of lower availability being caused by both consumptive and degradative use. This is either done through the use of water quality categories and the assessment of their individual scarcity (Boulay et al. 2011), or through a distance-to-target approach, or dilution volume equivalent, in relation to a reference standard (Ridoutt and Pfister 2010; Bayart et al. 2014).

As mentioned above, human water deprivation can cause health damage by depriving three users: domestic, agriculture or aquaculture/fisheries. Domestic deprivation has been assessed in two methods (Motoshita et al. 2011; Boulay et al. 2011) which quantify the impact pathways described above, either mechanistically or statistically. Both provide characterisation factors in DALY/m^3 consumed and the details of the differences between the methods are described in Boulay et al. (2015).

Agricultural deprivation has been assessed in three methods (Pfister et al. 2009; Boulay et al. 2011; Motoshita et al. 2014). Differences are based on the user competition factor (scarcity) used, the underlying sources of data, the parameter upon which to base the capacity of users to adapt to water deprivation or not, the calculation of the effect factor and, most importantly, the inclusion or not of the trade effect, i.e. the ripple effect of lower food production to lower income and importing countries. Analysis of these methods and modelling choices is provided in Boulay et al. (2015) and at time of writing a consensus was built based on these three models and is described in the Pellston Workshop report from Valencia, 2016.

For the damage that water use may cause on ecosystems, several methods exist that attempt to quantify a part of the complex impact pathways between water consumption and loss of species, i.e. ecosystem quality impacts. An overview of these methods was prepared by Núñez et al. (2016) who analysed in details the existing models, assumptions and consistency. The large majority of them have not yet found their way into LCA practice. None of these endpoint models use water scarcity as a modelling parameter, and hence scarcity does not represent a "true midpoint" for ecosystem quality.

The assessment of impacts on the impact category *resources*, or *ecosystem services* and resources, is still subject to debate and development. The main question pending being "what exactly are we trying to quantify?". For the case of water, this can be answered in different ways: future generation deprivation, resource-equivalent approach or monetarisation, but these still require further development. The use of non-renewable sources of water fromfossil aquifers would fall in this category.

For further details see Chap. 40 and Hauschild and Huijbregts (2015). Water is a precious resource for humans and ecosystems and our attempts to protect it come in different forms and from different angles. Numerous initiatives exist and indicators of all kinds are emerging regularly and, for the time being, continuously evolving. This should not be perceived as a problem or a sign of lesser value for these indicators; it simply reflects the fact that potential issues associated with water are diverse and so are the approaches to quantify and minimise them. The LCA approach aims to quantify potential impacts associated with human activities (a product, a service or an organisation) on specific areas of protection. Water-related indicators developed within the LCA framework are aligned with this goal, and efforts have been made to build consensus on these methodologies. The WULCA (water use in LCA) expert working group of the UNEP-SETAC Life Cycle Initiative has fostered the development and global harmonisation through international consensus of the water-related impact assessment methods in LCA. For further information on the existing methods, the reader is encouraged to explore the website: www.wulca-waterlca.org.

10.16 Abiotic Resource Use

10.16.1 Problem

Natural resources constitute the material foundation of our societies and economies and, paraphrasing the definition of sustainability by the United Nation's Commission on Environment and Development (the Brundtland Commission), they are as such fundamental for our abilities to fulfil our needs as well as for future generations' possibilities to fulfil their own needs. Since we don't know with any certitude what the needs of future generations for specific resources will be, and in order to respect the principle of sustainability, we have to ensure that the future resource availability is as good as possible compared to the current generation's situation, i.e. we have to consider the future availability for all resources that we know and dispose of today.

The definition of natural resources has an anthropocentric starting point. What humans need from nature in order to sustain their livelihood and activities is a resource. For the context of LCA, Udo de Haes et al. (1999) thus define natural resources as: "… those elements that are extracted for human use. They comprise both abiotic resources, such as fossil fuels and mineral ores, and biotic resources, such as wood and fish. They have predominantly a functional value for society."

Although water and land are also resources, their use causes direct impacts on the environment. In this respect they differ from the other resources and they are therefore treated as individual impact categories and described in separate sections. Currently, the resource use impact category covers mostly fossil fuels, minerals and metals so this will also be the focus here.

In terms of future availability of a resource the issue is not the current extraction and use of the resource per se but the depletion or dissipation of the resource. Similar to the use of land, the use of resources can be viewed from an occupation perspective and a transformation perspective. While a resource is used for one purpose it is not available for other purposes, and there is thus a competition situation. When resources are used in a way that caters to their easy reuse at the end of the product life, they are still occupied and not immediately available to other use, but they are in principle available to future use for other purposes. This is the case for many uses of metals today. The occupation perspective is normally not addressed in LCIA of resources today [with the exception of Schneider et al. (2011)]. Rather than resource *use* the focus of the impact assessment is usually on the resource *loss* that occurs throughout the life cycle.

Resource loss occurs through transformation of the resource when the use is either consumptive or dispersive. *Consumptive resource* use converts the resource in a way so that it no longer serves as the resource it was. An example is the use of fossil resources as fuels, converting them in the combustion process into CO_2 and water. The transformation occurring in *dispersive resource use* does not lose the resource but uses it in a way that leads to its dispersal in the technosphere or ecosphere in forms that are less accessible to human use than the original resource was. Dispersive use occurs for most of the metals.

There is still much debate about what the issue of concern of natural resources is and about how this should be addressed in LCIA (Hauschild et al. 2013). This may be explained by the difference in functional values of natural resources on the one hand, and intrinsic or existence values of other impact categories, assessing impacts on human health and ecosystem quality, on the other hand. Steen (2006) summarised different perceptions of the problem with abiotic resources in LCIA as: "… (1) assuming that mining cost will be a limiting factor, (2) assuming that collecting metals or other substances from low-grade sources is mainly an issue of energy, (3) assuming that scarcity is a major threat and (4) assuming that environmental impacts from mining and processing of mineral resources are the main problem."

The extraction of resources and their conversion into materials that are used in product systems are accompanied by energy use and direct emissions that make the raw material extraction sector an important contributor to environmental impacts and damages in many parts of the world. These impacts are addressed by the other impact categories which are considered in LCA, and hence not treated under the resource depletion impact category.

10.16.2 Environmental Mechanism

With a focus on resource availability for current and future generations, the environmental mechanism may look as shown in Fig. 10.28. It is assumed that resources with easy and/or cheap access and with high concentration or quality are extracted first. Consequently, today's resource extraction will lead future generations to extract lower concentration or lower value resources. This results in additional efforts for the extraction of the same amount of resource which can be translated into higher energy or costs. The endpoint of the impact pathway for resource use is often assessed as the future consequences of resource extraction. Schneider et al. (2014) went further in the pathway with the development of a new model for the assessment of resource provision including economic aspects that influence the security of supply and affect the availability of resources for human use.

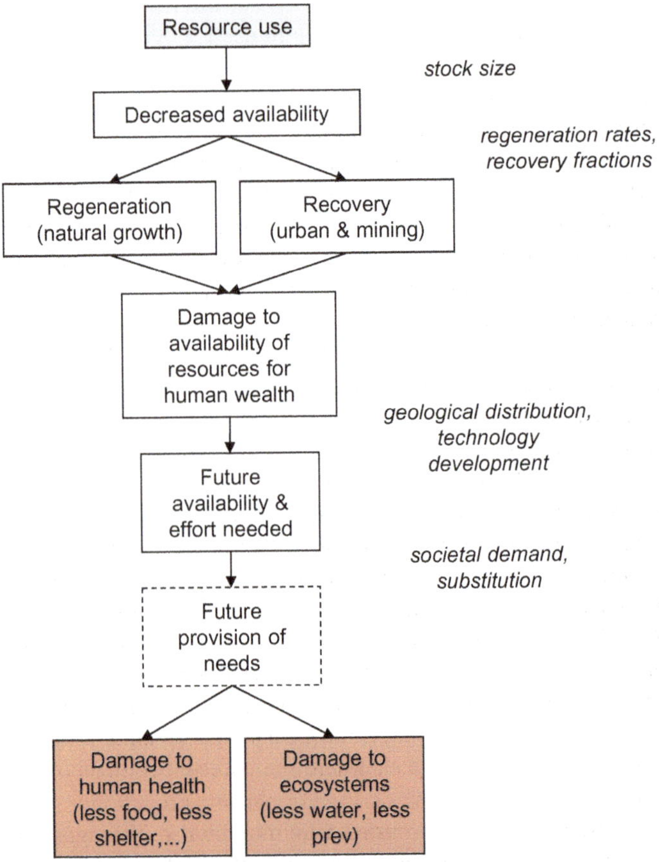

Fig. 10.28 Impact pathway for resource depletion [adapted from EC-JRC (2010b)]

Several classification schemes exist for resources (Lindeijer et al. 2002), classifying them according to their origin into *Abiotic resources* (inorganic materials—e.g. water and metals, or organic materials that are non-living at the moment of their extraction—fossil resources) and *Biotic resources* (living at least until the time of their extraction or harvest from the environment, and hence originating in the biomass). A further classification may be done according to the ability of the resource to be regenerated and the rate by which it may occur. Here resources are classified into:

- Stock resources exist as a finite and fixed amount (reserve) in the ecosphere and are not regenerated (metals in ores) or regenerated so slowly that for practical purposes the regeneration can be ignored (fossil resources)
- Fund resources regenerate but can still be depleted (like the stock resources) if the rate of extraction exceeds the rate of regeneration. Depletion can be temporary if the resource is allowed to recover but it can also be permanent for biotic fund resources where the species underlying the resource becomes extinct. Biotic resources are fund resources but there are also examples of abiotic resources like sand and gravel where the regeneration rate is so high that it is meaningful to classify them as fund resources
- Flow resources are provided as a flow (e.g. solar radiation, wind and to some extent freshwater) and can be harvested as they flow by. Flow resources cannot be globally depleted but there may be local or temporal low availability (notably for freshwater—see Sect. 10.15)

Stock resources are also referred to as *non-renewable resources* while fund and flow resources jointly are referred to as *renewable resources*. Resources may also be classified as *exhaustible*, i.e. they can be completely used up, and *inexhaustible*, which are unlimited.

10.16.3 Existing Characterisation Models

Impacts resulting from resource use are often divided into three categories following the impact pathway (see Fig. 10.28):

1. Methods aggregating natural resource consumption based on an inherent property
2. Methods relating natural resource consumption to resource stocks or availability
3. Methods relating current natural resource consumption to consequences of future extraction of natural resources (e.g. potential increased energy use or costs).

Category 1 methods focus for example on exergy [expressing the maximum amount of useful work the resource can provide in its current form, (Dewulf et al. 2007)], energy (Frischknecht et al. 2015) and solar energy (Rugani et al. 2011).

While being very reproducible and also easy to determine, the relevance of exergy loss to the scarcity and future availability of the resource is not obvious and therefore these methods are not recommended by the European Commission (EC-JRC 2011). However, the cumulative energy demand (CED) method (Frischknecht et al. 2015) is still used frequently as a resource accounting method in LCA studies and is also part of various comprehensive LCIA methods like CML-IA for fossil fuels (Guinée et al. 2002), ReCiPe (Goedkoop et al. 2012) and the Ecological Scarcity method (Frischknecht and Büsser Knöpfel 2013).

Viewing resource use from a sustainability perspective, the characterisation at midpoint level in the environmental mechanism (Fig. 10.28) should address its impact on the future availability of the resource for human activities. Several category 2 methods do this through incorporating a measure of the scarcity of the resource, expressed by the relationship between what is there and what is extracted, i.e. between the size of the stock or fund and the size of the extraction. However, there are different measures to determine the size of the stock or fund yet to be extracted.

Figure 10.29 shows a terminology for classifying a stock resource into classes according to their economic extractability and whether they are known or unknown. Here we will describe those most used in LCIA. The *reserves* are the part of the resource which are economically feasible to exploit with current technology. The *reserve base* is the part of the demonstrated resource that has a reasonable potential to become economically and technically available if the price of the resource increases or if more efficient extraction technology becomes available. *Ultimate reserves* are the resources that are ultimately available in the earth's crust, which include nonconventional and low-grade materials and common rocks. This reserve

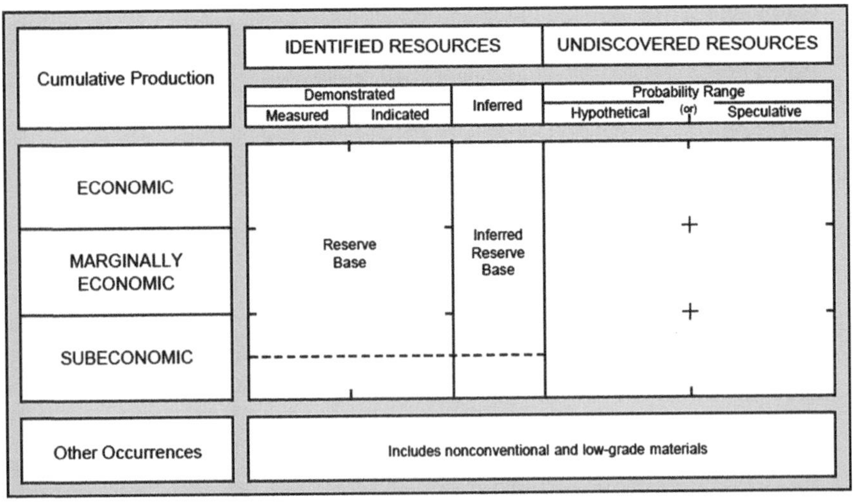

Fig. 10.29 Resource/reserve classification for minerals [taken from U.S. Geological Survey (2015)]

estimate refers to the quantity of resources that is ultimately available, estimated by multiplying the average natural concentration of the resources in the earth's crust by the mass of the crust. Lately, the *extractable geologic resource*, also called *ultimate recoverable resource* and *ultimately extractable reserves*, has also been adopted by a few LCIA methods. This reserve type is the amount of a given metal in ore in the upper earth's crust that is judged to be extractable over the long term, e.g. 0.01% (UNEP International Panel on Sustainable Resource Management 2011).

Each reserve estimate has pros and cons. *Reserves* are known and economically viable to extract, but this amount can fluctuate considerably with changes in prices and discoveries of new deposits. *Reserve base* has not been reported by the US Geological Survey since 2009 because its size also increases and decreases based on technological advances, economic fluctuations and new discoveries, etc. Consequently, basing the characterisation factoron *reserves* or *reserve base* has the problem that it changes with time. *Ultimate reserves* are calculated on basis of the average concentration of metals in the earth's crust so they are more stable but this is not a good indicator of the quantity of the resource that can realistically be exploited. Finally, the *extractable geologic resource* seems to be a quite certain reserve estimate but authors are still debating how to quantify it (Schneider et al. 2015).

From the *category 2 methods*, CML-IA and EDIP are the most widely used. The CML-IA method for characterisation of abiotic stock resources defines an Abiotic Depletion Potential, ADP with a characterisation factor based on the annual extraction rate and the reserve estimates. In Guinée et al. (2002) only the *ultimate reserves* are included, but Oers et al. (2002) defined additional characterisation factors on the basis of *reserves* and *reserve base* estimates. CML-IA using *reserve base* estimates is the method recommended in the ILCD Handbook for LCIA in the European context (EC-JRC 2011).

An alternative approach inspired by the EDIP method (Hauschild and Wenzel 1998) bases the assessment for the abiotic stock resources on the *reserve base* and defines the characterisation as the inverse person reserve, i.e. the amount of *reserve base* per person in the world. For renewable resources, the EDIP inspired characterisation is based on the difference between the extraction rate and the regeneration rate. If the regeneration rate exceeds the extraction rate, it is considered that there is no resource availability issue, and the characterisation factor is given the value 0.

Further, down the impact pathway, *category 3 methods* have been developed expressing the future consequences of current resource consumption. Some methods quantify these consequences as additional energy requirements: Eco-Indicator 99, IMPACT 2002+; some methods quantify this effort as additional costs: ReCiPe and Surplus Cost Potential on basis of relationships between extraction and cost increase (Ponsioen et al. 2014; Vieira et al. 2016b), EPS 2000 and the Stepwise method based on willingness to pay; and some methods quantify this effort as additional ore material that has to be dealt with: Ore Requirement Indicator ORI (Swart and Dewulf 2013) and Surplus Ore Potential SOP (Vieira et al. 2016a) used in the LC-IMPACT LCIA method. These methods suffer from a strong dependency on rather uncertain assumptions about the future efficiencies and energy needs of mining and extraction technologies, but they seem to better capture the issue of concern which is assuring a supply of resources to future generations.

Schneider et al. (2014) defined a semi-quantitative method expressed as the economic resource scarcity potential (ESP) for evaluating resource use based on life cycle assessment. This method includes elements typically used in the discipline of raw materials criticality, like governance and socio-economic stability, trade barriers, etc., for which each element are scaled to the range 0–1.

For metal resources, characterisation factors are mostly applied to the metal content in the ore, not the mineral that is extracted. The relevant inventory information is thus the amount of metal used as input, not the amount of mineral. This is also how life cycle inventory (LCI) databases model elementary flows of mineral and metal resources. Schneider et al. (2015) considers not only the geological stock not yet extracted, but also the anthropogenic stock in circulation in products and goods.

The geographic scale at which it is relevant to judge the availability and depletion of a resource depends on the relationship between the price and the density/transportability of the resource. The scale is global for the valuable and dense stock and fund resources that are easy to transport and hence traded on a world market (metals, oil, coal, tropical hardwood), while it is regional for the less valuable and/or less dense stock and fund resources that are used and extracted regionally (natural gas, sand and gravel, limestone) or even locally.

For further details see Chap. 40 and Hauschild and Huijbregts (2015).

References

Alvarenga, R., Erb, K.-H., Haberl, H., Soares, S., van Zelm, R., Dewulf, J.: Global land use impacts on biomass production—a spatial-differentiated resource-related life cycle impact assessment model. Int. J. Life Cycle Assess. **20**, 440–450 (2015). doi:10.1007/s11367-014-0843-x

Apte, J.S., Marshall, J.D., Cohen, A.J., Brauer, M.: Addressing global mortality from ambient PM2.5. Environ. Sci. Technol. **49**, 8057–8066 (2015). doi:10.1021/acs.est.5b01236

Bayart, J.-B., Margni, M., Bulle, C., Deschênes, L., Pfister, S., Koehler, A., Vince, F.: Framework for assessment of off-stream freshwater use within LCA. Int. J. Life Cycle Assess. **15**, 439–453 (2010). doi:10.1007/s11367-010-0172-7

Bayart, J.-B., Worbe, S., Grimaud, J., Aoustin, E.: The Water Impact Index: a simplified single-indicator approach for water footprinting. Int. J. Life Cycle Assess. **19**, 1336–1344 (2014). doi:10.1007/s11367-014-0732-3

Benton, M.J., Twitchet, R.J.: How to kill (almost) all life: the end-Permian extinction event. Trends Ecol. Evol. **18**, 358–365 (2003). doi:10.1016/S0169-5347(03)00093-4

Berger, M., van der Ent, R., Eisner, S., Bach, V., Finkbeiner, M.: Water Accounting and Vulnerability Evaluation (WAVE): considering atmospheric evaporation recycling and the risk of freshwater depletion in water footprinting. Environ. Sci. Technol. **48**, 4521–4528 (2014). doi:10.1021/es404994t

Bos, U., Horn, R., Beck, T., Lindner, J., Fischer, M.: LANCA®—Characterization Factors for Life Cycle Impact Assessment, Version 2.0. Fraunhofer Verlag, Stuttgart (2016)

Boulay, A.-M., Bulle, C., Bayart, J.-B., Deschenes, L., Margni, M.: Regional characterization of freshwater use in LCA: modeling direct impacts on human health. Environ. Sci. Technol. **45**, 8948–8957 (2011). doi:10.1021/es1030883

Boulay, A.-M., Bare, J., Benini, L., Berger, M., Klemmayer, I., Lathuilliere, M., Loubet, P., Manzardo, A., Margni, M., Ridoutt, B.: Building consensus on a generic water scarcity indicator for LCA-based water footprint: preliminary results from WULCA. LCA Food 2050 (2014)

Boulay, A.-M., Motoshita, M., Pfister, S., Bulle, C., Muñoz, I., Franceschini, H., Margni, M.: Analysis of water use impact assessment methods (part A): evaluation of modeling choices based on a quantitative comparison of scarcity and human health indicators. Int. J. Life Cycle Assess. **20**, 139–160 (2015). doi:10.1007/s11367-014-0814-2

Boulay, A.-M., Bare, J., Benini, L., Berger, M., Lathuilliere, M., Manzardo, A., Margni, M., Motoshita, M., Núñez, M., Pastor, A.V., Ridoutt, B.G., Oki, T., Worbe, S., Pfister, S.: The WULCA consensus characterization model for water scarcity footprints: assessing impacts of water consumption based on available water remaining (AWARE). Int. J. Life Cycle Assess. In Press. (2017). doi:10.1007/s11367-017-1333-8

Brandão, M., Milà i Canals, L.: Global characterisation factors to assess land use impacts on biotic production. Int. J. Life Cycle Assess. **18**, 1243–1252 (2013). doi:10.1007/s11367-012-0381-3

Brauer, M., Freedman, G., Frostad, J., van Donkelaar, A., Martin, R.V., Dentener, F., van Dingenen, R., Estep, K., Amini, H., Apte, J.S., Balakrishnan, K., Barregard, L., Broday, D., Feigin, V., Ghosh, S., Hopke, P.K., Knibbs, L.D., Kokubo, Y., Liu, Y., Ma, S., Morawska, L., Sangrador, J.L.T., Shaddick, G., Anderson, H.R., Vos, T., Forouzanfar, M.H., Burnett, R.T., Cohen, A.: Ambient air pollution exposure estimation for the global burden of disease 2013. Environ. Sci. Technol. **50**, 79–88 (2016). doi:10.1021/acs.est.5b03709

Chaudhary, A., Verones, F., de Baan, L., Hellweg, S.: Quantifying land use impacts on biodiversity: combining species-area models and vulnerability indicators. Environ. Sci. Technol. **49**, 9987–9995 (2015). doi:10.1021/acs.est.5b02507

de Baan, L., Alkemade, R., Koellner, T.: Land use impacts on biodiversity in LCA: a global approach. Int. J. Life Cycle Assess. **18**, 1216–1230 (2013a). doi:10.1007/s11367-012-0412-0

de Baan, L., Mutel, C.L., Curran, M., Hellweg, S., Koellner, T.: Land use in life cycle assessment: global characterization factors based on regional and global potential species extinction. Environ. Sci. Technol. **47**, 9281–9290 (2013b). doi:10.1021/es400592q

Dewulf, J., De Meester, B., Van der Vorst, G., Van Langenhove, H., Bösch, M., Hellweg, S., Huijbregts, M.A.J.: Cumulative Exergy Extraction from the Natural Environment (CEENE): a comprehensive Life Cycle Impact Assessment method for resource accounting. Environ. Sci. Technol. **41**, 8477–8483 (2007). doi:10.1021/es0711415

EC-JRC—European Commission-Joint Research Centre—Institute for Environment and Sustainability: International Reference Life Cycle Data System (ILCD): Handbook—General Guide for Life Cycle Assessment—Detailed Guidance, 1st edn March 2010. EUR 24708 EN. Publications Office of the European Union, Luxembourg (2010a)

EC-JRC—European Commission-Joint Research Centre—Institute for Environment and Sustainability: International Life Cycle Data System (ILCD) Handbook—Framework and Requirements for Life Cycle Impact Assessment Models and Indicators, 1st edn. European Commission, Joint Research Centre, Institute for Environment and Sustainability, Ispra, Italy (2010b)

EC-JRC—European Commission-Joint Research Centre—Institute for Environment and Sustainability: International Reference Life Cycle Data System (ILCD): Handbook—Recommendations for Life Cycle Impact Assessment in the European Context—Based on Existing Environmental Impact Assessment Models and Factors, 1st edn 2011, EUR 24571 EN. Publication Office of the European Union, Luxemburg (2011)

EC-JRC: Environmental Footprint—Update of Life Cycle Impact Assessment Methods; DRAFT for TAB (Status: May 2, 2016) Resources, Water, Land. Ispra, Italy (2016)

EEA: The European Environment. State and Outlook 2010. Land Use (2010)

Fantke, P., Jolliet, O., Evans, J.S., Apte, J.S., Cohen, A.J., Hänninen, O.O., Hurley, F., Jantunen, M.J., Jerrett, M., Levy, J.I., Loh, M.M., Marshall, J.D., Miller, B.G., Preiss, P., Spadaro, J.V., Tainio, M., Tuomisto, J.T., Weschler, C.J., McKone, T.E.: Health effects of fine particulate matter in life cycle impact assessment: findings from the Basel Guidance Workshop. Int. J. Life Cycle Assess. **20**, 276–288 (2015). doi:10.1007/s11367-014-0822-2

Foley, J.A., Defries, R., Asner, G.P., Barford, C., Bonan, G., Carpenter, S.R., Chapin, F.S.: Global consequences of land use. Science **309**, 570–574 (2005). doi:10.1126/science.1111772

Forouzanfar, M.H., Alexander, L., Anderson, H.R., Bachman, V.F., Biryukov, S., Brauer, M., Murray, C.J., et al.: Global, regional, and national comparative risk assessment of 79 behavioural, environmental and occupational, and metabolic risks or clusters of risks in 188 countries, 1990–2013: a systematic analysis for the Global Burden of Disease Study 2013. Lancet **386**, 2287–2323 (2015). doi:10.1016/S0140-6736(15)00128-2

Frischknecht, R., Büsser Knöpfel, S.: Swiss eco-factors 2013 according to the ecological scarcity method. In: Methodological Fundamentals and their Application in Switzerland. Environmental Studies No. 1330, Bern, Switzerland (2013)

Frischknecht, R., Steiner, R., Braunschweig, A., Egli, N., Hildesheimer, G.: Swiss Ecological Scarcity Method: The New Version 2006 (2008)

Frischknecht, R., Wyss, F., Büsser Knöpfel, S., Lützkendorf, T., Balouktsi, M.: Cumulative energy demand in LCA: the energy harvested approach. Int. J. Life Cycle Assess. **20**, 957–969 (2015). doi:10.1007/s11367-015-0897-4

Goedkoop, M., Heijungs, R., Huijbregts, M.A.J., De Schryver, A., Struijs, J., van Zelm, R., Ministry of Housing SP and E (VROM): ReCiPe 2008—A Life Cycle Impact Assessment Method Which Comprises Harmonised Category Indicators at the Midpoint and the Endpoint Level, 1st edn revised. Ministry of Housing, Spatial Planning and Environment (VROM) (2012)

Gronlund, C., Humbert, S., Shaked, S., O'Neill, M., Jolliet, O.: Characterizing the burden of disease of particulate matter for life cycle impact assessment. Air Qual. Atmos. Health **8**, 29–46 (2015). doi:10.1007/s11869-014-0283-6

Guinée, J.B., Gorrée, M., Heijungs, R., Huppes, G., Kleijn, R., van Oers, L., Wegener Sleeswijk, A., Suh, S., Udo de Haes, H.A., de Bruijn, H., van Duin, R., Huijbregts, M.A.J.: Handbook on Life Cycle Assessment: Operational Guide to the ISO Standards. Kluwer Academic Publishers, Dordrecht (2002). ISBN 1-4020-0228-9

Hauschild, M.Z., Huijbregts, M.A.J. (eds.): Life Cycle Impact Assessment. In: Klöpffer, W., Curran, M. (eds.) LCA Compendium—The Complete World of Life Cycle Assessment, p. 339. Springer, Dordrecht (2015). doi:10.1007/978-94-017-9744-3

Hauschild, M., Wenzel, H.: Environmental Assessment of Products, Volume 2: Scientific Background. Kluwer Academic Publishers, Hingham (1998)

Hauschild, M.Z., Huijbregts, M.A.J., Jolliet, O., MacLeod, M., Margni, M., Van de Meent, D., Rosenbaum, R.K., McKone, T.E.: Building a model based on scientific consensus for life cycle impact assessment of chemicals: the search for harmony and parsimony. Environ. Sci. Technol. **42**, 7032–7037 (2008). doi:10.1021/es703145t

Hauschild, M., Goedkoop, M., Guinée, J.B., Heijungs, R., Huijbregts, M., Jolliet, O., Margni, M., Schryver, A., Humbert, S., Laurent, A., Sala, S., Pant, R.: Identifying best existing practice for characterization modeling in life cycle impact assessment. Int. J. Life Cycle Assess. **18**, 683–697 (2013). doi:10.1007/s11367-012-0489-5

Heijungs, R.: Harmonization of methods for impact assessment. Environ. Sci. Pollut. Res. **2**, 217–224 (1995). doi:10.1007/BF02986769

Henderson, A., Hauschild, M.Z., Van de Meent, D., Huijbregts, M.A.J., Larsen, H.F., Margni, M., McKone, T.E., Payet, J., Rosenbaum, R.K., Jolliet, O.: USEtox fate and ecotoxicity factors for comparative assessment of toxic emissions in life cycle analysis: sensitivity to key chemical properties. Int. J. Life Cycle Assess. **16**, 701–709 (2011). doi:10.1007/s11367-011-0294-6

Hoekstra, A.Y., Mekonnen, M.M., Chapagain, A.K., Mathews, R.E., Richter, B.D.: Global monthly water scarcity: blue water footprints versus blue water availability. PLoS ONE (2012). doi:10.1371/journal.pone.0032688

Hofstetter, P.: Perspectives in Life Cycle Impact Assessment: A Structure Approach to Combine Models of the Technosphere, Ecosphere and Valuesphere. Kluwer Academic Publishers, Dordrecht (1998)

Huijbregts, M.A.J., Hellweg, S., Frischknecht, R., Hungerbühler, K., Hendriks, A.J.: Ecological footprint accounting in the life cycle assessment of products. Ecol. Econ. **64**, 798–807 (2008). doi:10.1016/j.ecolecon.2007.04.017

Huijbregts, M.A.J., Hellweg, S., Hertwich, E.: Do we need a paradigm shift in life cycle impact assessment? Environ. Sci. Technol. **45**, 3833–3834 (2011). doi:10.1021/es200918b

Humbert, S., Marshall, J.D., Shaked, S., Spadaro, J.V., Nishioka, Y., Preiss, P., McKone, T.E., Horvath, A., Jolliet, O.: Intake fractions for particulate matter: recommendations for life cycle assessment. Environ. Sci. Technol. **45**, 4808–4816 (2011). doi:10.1021/es103563z

IPCC: Climate Change: The IPCC Scientific Assessment. Cambridge University Press, Cambridge (1990)

IPCC: Climate change 2013: the physical science basis. In: Stocker, T.F., Qin, D., Plattner, G.-K., Tignor, M., Allen, S.K., Boschung, J., Nauels, A., Xia, Y. (eds.) Contribution of Working Group I to the Fifth Assessment Report of the Intergovernmental Panel on Climate Change, Cambridge, UK (2013)

IPCC: Climate change 2014: synthesis report. In: Core Writing Team, Pachauri, R.K., Meyer, L.A. (eds.) Contribution of Working Groups I, II and III to the Fifth Assessment Report of the Intergovernmental Panel on Climate Change. IPCC, Geneva (2014a)

IPCC: Climate change 2014: mitigation of climate change. In: Edenhofer, O., Pichs-Madruga, R., Sokona, Y., Farahani, E., Kadner, S., Seyboth, K., Adler, A. (eds.) Contribution of Working Group III to the Fifth Assessment Report of the Intergovernmental Panel on Climate Change, Cambridge, UK (2014b)

ISO: Environmental Management—Life Cycle Assessment—Principles and Framework (ISO 14040). ISO, the International Organization for Standardization, Geneva (2006a)

ISO: Environmental Management—Life Cycle Assessment—Requirements and Guidelines (ISO 14044). ISO, the International Organization for Standardization, Geneva (2006b)

ISO: Environmental Management—Water Footprint—Principles, Requirements and Guidelines (ISO 14046). ISO, the International Organization for Standardization, Geneva (2014)

Koellner, T., Scholz, R.W.: Assessment of land use impacts on the natural environment. Part 2: Generic characterization factors for local species diversity in Central Europe. Int. J. Life Cycle Assess. **13**, 32–48 (2008). doi:10.1065/lca2006.12.292.2

Laurent, A., Hauschild, M.Z.: Impacts of NMVOC emissions on human health in European countries for 2000–2010: use of sector-specific substance profiles. Atmos. Environ. **85**, 247–255 (2014). doi:10.1016/j.atmosenv.2013.11.060

Laurent, A., Olsen, S.I., Hauschild, M.Z.: Limitations of carbon footprint as indicator of environmental sustainability. Environ. Sci. Technol. **46**, 4100–4108 (2012). doi:10.1021/es204163f

Lenton, T.M., Held, H., Kriegler, E., Hall, J.W., Lucht, W., Rahmstorf, S., Schellnhuber, H.J.: Tipping elements in the Earth's climate system. Proc. Natl. Acad. Sci. **105**, 1786–1793 (2008). doi:10.1073/pnas.0705414105

Lim, S.S., Vos, T., Flaxman, A.D., Danaei, G., Shibuya, K., Adair-Rohani, H., Memish, Z.A., et al.: A comparative risk assessment of burden of disease and injury attributable to 67 risk factors and risk factor clusters in 21 regions, 1990–2010: a systematic analysis for the Global Burden of Disease Study 2010. Lancet **380**, 2224–2260 (2012). doi:10.1016/S0140-6736(12)61766-8

Lindeijer, E., Müller-Wenk, R., Steen, B.: Impact assessment of resources and land use. In: Udo de Haes, H.A., Finnveden, G., Goedkoop, M., et al. (eds.) Life Cycle Impact Assessment—Striving Towards Best Practice, pp. 11–64. SETAC, Pensacola, USA (2002)

Mackay, D., Seth, R.: The role of mass balance modelling in impact assessment and pollution prevention. In: Sikdar, S.K., Diwekar, U. (eds.) Tools and Methods for Pollution Prevention, pp. 157–179. Kluwer Academic Publishers, Dordrecht (1999)

Mila i Canals, L., de Baan, L.: Land use. In: Hauschild, M.Z., Huijbregts, M.A.J. (eds.) LCA Compendium—The Complete World of Life Cycle Assessment—Life Cycle Impact Assessment, pp 197–222. Springer, Dordrecht (2015). doi:10.1007/978-94-017-9744-3

Millenium Ecosystem Assessment: Ecosystems and human well-being: biodiversity Synthesis. World Resources Institute, Washington, DC (2005)

Motoshita, M., Itsubo, N., Inaba, A.: Development of impact factors on damage to health by infectious diseases caused by domestic water scarcity. Int. J. Life Cycle Assess. **16**, 65–73 (2011). doi:10.1007/s11367-010-0236-8

Motoshita, M., Ono, Y., Pfister, S., Boulay, A.-M., Berger, M., Nansai, K., Tahara, K., Itsubo, N., Inaba, A.: Consistent characterisation factors at midpoint and endpoint relevant to agricultural water scarcity arising from freshwater consumption. Int. J. Life Cycle Assess. (2014). doi:10.1007/s11367-014-0811-5

Müller-Wenk, R., Brandão, M.: Climatic impact of land use in LCA—carbon transfers between vegetation/soil and air. Int. J. Life Cycle Assess. **15**, 172–182 (2010). doi:10.1007/s11367-009-0144-y

Murray, C.J., Lopez, A.D.: The Global Burden of Disease. Harvard School of Public Health, World Bank, and World Health Organization, Geneva, Switzerland (1996)

Nilsson, J., Grennfelt, P.: Critical loads for sulphur and nitrogen. Report from a Workshop held at Skokloster, Sweden, 19–24 March 1988. Environmental Report 1988:15, Copenhagen, Denmark (1988)

Núñez, M., Civit, B., Muñoz, P., Arena, A.P., Rieradevall, J., Antón, A.: Assessing potential desertification environmental impact in life cycle assessment. Part 1: methodological aspects. Int. J. Life Cycle Assess. **15**, 67–78 (2010). doi:10.1007/s11367-009-0126-0

Núñez, M., Antón, A., Muñoz, P., Rieradevall, J.: Inclusion of soil erosion impacts in life cycle assessment on a global scale: application to energy crops in Spain. Int. J. Life Cycle Assess. **18**, 755–767 (2013). doi:10.1007/s11367-012-0525-5

Núñez, M., Bouchard, C., Bulle, C., Boulay, A.-M., Margni, M.: Critical analysis of life cycle impact assessment methods addressing consequences of freshwater use on ecosystems and recommendations for future method development. Int. J. Life Cycle Assess. **21**, 1799–1815 (2016). doi:10.1007/s11367-016-1127-4

Pennington, D.W., Rydberg, T., Potting, J., Finnveden, G., Lindeijer, E., Jolliet, O., Rebitzer, G.: Life cycle assessment Part 2: Current impact assessment practice. Environ. Int. **30**, 721–739 (2004). doi:10.1016/j.envint.2003.12.009

Pfister, S., Koehler, A., Hellweg, S.: Assessing the environmental impacts of freshwater consumption in LCA. Environ. Sci. Technol. **43**, 4098–4104 (2009). doi:10.1021/es802423e

Ponsioen, T.C., Vieira, M.D.M., Goedkoop, M.J.: Surplus cost as a life cycle impact indicator for fossil resource scarcity. Int. J. Life Cycle Assess. **19**, 872–881 (2014). doi:10.1007/s11367-013-0676-z

Rees, W.E.: Ecological footprints and appropriated carrying capacity: what urban economics leaves out. Environ. Urban **4**, 121–130 (1992). doi:10.1177/095624789200400212

Ridoutt, B.G., Pfister, S.: A revised approach to water footprinting to make transparent the impacts of consumption and production on global freshwater scarcity. Glob. Environ. Change **20**, 113–120 (2010). doi:10.1016/j.gloenvcha.2009.08.003

Ridoutt, B.G., Fantke, P., Pfister, S., Bare, J., Boulay, A.-M., Cherubini, F., Frischknecht, R., Hauschild, M., Hellweg, S., Henderson, A., Jolliet, O., Levasseur, A., Margni, M., McKone, T., Michelsen, O., Milà i Canals, L., Page, G., Pant, R., Raugei, M., Sala, S., Saouter, E., Verones, F., Wiedmann, T.: Making sense of the minefield of footprint indicators. Environ. Sci. Technol. **49**, 2601–2603 (2015). doi:10.1021/acs.est.5b00163

Ridoutt, B.G., Pfister, S., Manzardo, A., Bare, J., Boulay, A.-M., Cherubini, F., Fantke, P., Frischknecht, R., Hauschild, M., Henderson, A., Jolliet, O., Levasseur, A., Margni, M., McKone, T., Michelsen, O., i Canals, L., Page, G., Pant, R., Raugei, M., Sala, S., Verones, F.: Area of concern: a new paradigm in life cycle assessment for the development of footprint metrics. Int. J. Life Cycle Assess. **21**, 276–280 (2016). doi:10.1007/s11367-015-1011-7

Rosenbaum, R.K.: Selection of impact categories, category indicators and characterisation models in goal and scope definition. In: Curran, M.A. (ed.) LCA Compendium—The Complete World of Life Cycle Assessment—Goal and scope definition in Life Cycle Assessment, pp 63–122. Springer, Dordrecht (2017). doi:10.1007/978-94-024-0855-3

Rosenbaum, R.K., Bachmann, T.M.K., Gold, L.S., Huijbregts, M.A.J., Jolliet, O., Juraske, R., Koehler, A., Larsen, H.F., MacLeod, M., Margni, M., McKone, T.E., Payet, J., Schuhmacher, M., Van de Meent, D., Hauschild, M.Z.: USEtox - The UNEP/SETAC-consensus model: recommended characterisation factors for human toxicity and freshwater ecotoxicity in Life

Cycle Impact Assessment. Int. J. Life Cycle Assess. **13**, 532–546 (2008). doi:10.1007/s11367-008-0038-4

Rugani, B., Huijbregts, M.A.J., Mutel, C., Bastianoni, S., Hellweg, S.: Solar energy demand (SED) of commodity life cycles. Environ. Sci. Technol. **45**, 5426–5433 (2011). doi:10.1021/es103537f

Saad, R., Koellner, T., Margni, M.: Land use impacts on freshwater regulation, erosion regulation, and water purification: a spatial approach for a global scale level. Int. J. Life Cycle Assess. **18**, 1253–1264 (2013). doi:10.1007/s11367-013-0577-1

Safire, W.: Footprint. New York Times Sunday Mag MM20 (2008)

Salisbury, F.B., Ross, C.W.: Plant Physiology, 3rd edn. Wadsworth Publishing Company, Belmont (1978)

Schneider, L., Berger, M., Finkbeiner, M.: The anthropogenic stock extended abiotic depletion potential (AADP) as a new parameterisation to model the depletion of abiotic resources. Int. J. Life Cycle Assess. **16**, 929–936 (2011). doi:10.1007/s11367-011-0313-7

Schneider, L., Berger, M., Schüler-Hainsch, E., Knöfel, S., Ruhland, K., Mosig, J., Bach, V., Finkbeiner, M.: The economic resource scarcity potential (ESP) for evaluating resource use based on life cycle assessment. Int. J. Life Cycle Assess. **19**, 601–610 (2014). doi:10.1007/s11367-013-0666-1

Schneider, L., Berger, M., Finkbeiner, M.: Abiotic resource depletion in LCA—background and update of the anthropogenic stock extended abiotic depletion potential (AADP) model. Int. J. Life Cycle Assess. **20**, 709–721 (2015). doi:10.1007/s11367-015-0864-0

Scholz, R.W.: Assessment of land use impacts on the natural environment. Part 1: An analytical framework for pure land occupation and land use change. Int. J. Life Cycle Assess. **12**, 16–23 (2007). doi:10.1065/lca2006.12.292.1

Steen, B.A.: Abiotic resource depletion different perceptions of the problem with mineral deposits. Int. J. Life Cycle Assess. **11**, 49–54 (2006). doi:10.1065/lca2006.04.011

Stumm, W., Morgan, J.J.: Aquatic chemistry—introduction emphasizing chemical equilibria in natural waters, 2nd edn. Wiley, New York (1981)

Swart, P., Dewulf, J.: Quantifying the impacts of primary metal resource use in life cycle assessment based on recent mining data. Resour. Conserv. Recycl. **73**, 180–187 (2013). doi:10.1016/j.resconrec.2013.02.007

Thompson, M., Ellis, R.J., Wildavsky, A.: Cultural Theory. Westview Press, Boulder (1990)

Udo de Haes, H.A., Jolliet, O., Finnveden, G., Hauschild, M., Krewitt, W., Mueller-Wenk, R.: Best available practice regarding impact categories and category indicators in life cycle impact assessment. Part 1. Int. J. Life Cycle Assess. **4**, 167–174 (1999). doi:10.1007/BF02979453

UNEP International Panel on Sustainable Resource Management: Estimating long-run geological stocks of metals (2011)

U.S. Geological Survey: Mineral Commodity Summaries 2015: U.S. Geological Survey (2015)

Van Oers, L., De Koning, A., Guinée, J., Huppes, G.: Abiotic Resource Depletion in LCA—Improving Characterization Factors for Abiotic Resource Depletion as Recommended in the New Dutch LCA Handbook. Road and Hydraulic Engineering Institute of the Dutch Ministry of Transport, Public Works and Water Management (V&W), Delft, The Netherlands (2002)

van Zelm, R., Huijbregts, M.A.J., den Hollander, H.A., van Jaarsveld, H.A., Sauter, F.J., Struijs, J., van Wijnen, H.J., van de Meent, D.: European characterization factors for human health damage of PM10 and ozone in life cycle impact assessment. Atmos. Environ. **42**, 441–453 (2008). doi:10.1016/j.atmosenv.2007.09.072

van Zelm, R., Preiss, P., van Goethem, T., Van Dingenen, R., Huijbregts, M.: Regionalized life cycle impact assessment of air pollution on the global scale: damage to human health and vegetation. Atmos. Environ. **134**, 129–137 (2016). doi:10.1016/j.atmosenv.2016.03.044

Vieira, M.D.M., Ponsioen, T.C., Goedkoop, M.J., Huijbregts, M.A.J.: Surplus ore potential as a scarcity indicator for resource extraction. J. Ind. Ecol. (2016a). doi:10.1111/jiec.12444

Vieira, M.D.M., Ponsioen, T.C., Goedkoop, M.J., Huijbregts, M.A.J.: Surplus cost potential as a life cycle impact indicator for metal extraction. Resources **5**, 2 (2016b). doi:10.3390/resources5010002

WHO: Indoor air quality: organic pollutants. Report on a WHO Meeting, Berlin, 23–27 August 1987. EURO Reports and Studies 111. Geneva, Switzerland (1989)

WHO: Health risks of particulate matter from long-range transboundary air pollution, Bonn, Germany (2006)

WMO: Scientific Assessment of Ozone Depletion: 2014, World Meteorological Organization, Global Ozone Research and Monitoring Project Report No. 55. Geneva, Switzerland (2014)

Author Biographies

Ralph K. Rosenbaum LCA expert and environmental modeller focusing on LCIA development since early 2000s. Contributed to several UNEP/SETAC working groups towards global harmonisation of LCA methodology. Interested in LCIA modelling of emissions and water/soil resource use, operationalisation of uncertainty management and spatial differentiation.

Michael Hauschild Involved in development of LCIA methodology since the early 1990s. Has led several SETAC and UNEP/SETAC working groups and participated in the development of the ISO standards and the ILCD methodological guidelines. Main LCA interests are chemical impacts, spatial differentiation and science-based boundaries in LCIA.

Anne-Marie Boulay LCA and water footprint expert focusing on water use impacts assessment in LCA since 2008. Chaired UNEP/SETAC and FAO working groups towards harmonisation of water use impact assessment methodologies, and has been involved in ISO standards development on water footprint, providing training on the topic for UNEP and ISO worldwide.

Peter Fantke Develops methods for LCIA, health impact assessment and chemical alternatives assessment since 2006. Has contributed to UNEP/SETAC LCIA working groups and is USEtox Manager. Interested in quantifying and characterising chemical emissions, uncertainty analysis, consumer exposure, chemical substitution and model parameterisation.

Alexis Laurent Working with LCA since 2010 with a strong LCIA focus, particularly on normalisation aspects. Main LCA interests include development of LCIA methods, LCA applications and footprinting of large-scale systems for policy-making (e.g. nations, sectors), and LCA applied to various technology domains, including energy systems.

Montserrat Núñez Environmental scientist with interest in LCA, agriculture, and modelling of environmental impacts from resource use in agricultural activities. Involved in development of LCIA methods to assess impacts of land use and water use since 2007.

Marisa Vieira Both as researcher and consultant developing and applying LCA, LCM and footprints since 2007. Main interests are the application of life cycle thinking in companies, branch associations and policy and capability development through training on LCA, SimaPro and Environmental Footprint.

Chapter 11
Uncertainty Management and Sensitivity Analysis

Ralph K. Rosenbaum, Stylianos Georgiadis and Peter Fantke

Abstract Uncertainty is always there and LCA is no exception to that. The presence of uncertainties of different types and from numerous sources in LCA results is a fact, but managing them allows to quantify and improve the precision of a study and the robustness of its conclusions. LCA practice sometimes suffers from an imbalanced perception of uncertainties, justifying modelling choices and omissions. Identifying prevalent misconceptions around uncertainties in LCA is a central goal of this chapter, aiming to establish a positive approach focusing on the advantages of uncertainty management. The main objectives of this chapter are to learn how to deal with uncertainty in the context of LCA, how to quantify it, interpret and use it, and how to communicate it. The subject is approached more holistically than just focusing on relevant statistical methods or purely mathematical aspects. This chapter is neither a precise statistical method description, nor a philosophical essay about the concepts of uncertainty, knowledge and truth, although you will find a little bit of both. This chapter contains (1) an introduction of the essential terminology and concepts of relevance for LCA; (2) a discussion of main sources of uncertainty and how to quantify them; (3) a presentation of approaches to calculate uncertainty for the final results (propagation); (4) a discussion of how to use uncertainty information and how to take it into account in the

R.K. Rosenbaum (✉)
IRSTEA, UMR ITAP, ELSA Research Group and ELSA-PACT—Industrial Chair for Environmental and Social Sustainability, 361 rue Jean-François Breton, BP 5095, 34196 Montpellier, France
e-mail: ralph.rosenbaum@irstea.fr

S. Georgiadis
Department of Applied Mathematics and Computer Science, Technical University of Denmark, Kongens Lyngby, Denmark

S. Georgiadis
Global Decision Support Initiative, Technical University of Denmark, Kongens Lyngby, Denmark

P. Fantke
Division for Quantitative Sustainability Assessment, Department of Management Engineering, Technical University of Denmark, Kongens Lyngby, Denmark

M.Z. Hauschild et al. (eds.), *Life Cycle Assessment*,
DOI 10.1007/978-3-319-56475-3_11

interpretation of the results; and finally (5) a discussion of how to manage, communicate and present uncertainty information together with the LCA results.

Learning Objectives

After studying this chapter, the reader should be able to:

- Explain the importance and usefulness of addressing uncertainty in LCA
- Distinguish types and sources of uncertainty and variability and explain important misconceptions of uncertainty in the context of LCA
- List the dominating sources of uncertainty in a typical LCA
- Explain the relevant concepts and vocabulary of uncertainty
- Analyse sensitivity, uncertainty and variability and use these insights to reduce overall uncertainty when performing an LCA
- Express and communicate uncertainty in an appropriate way, catering to the purpose of the analysis
- Apply uncertainty information in results interpretation and decision support

11.1 Introduction

The British mathematician, science historian, author and inventor Jacob Bronowski wrote that "Knowledge is an unending adventure at the edge of uncertainty". This is a perfect motto and inspiration for this chapter. Before learning how to deal with uncertainty in the context of LCA, how to quantify it, interpret and use it, or communicate it, which are the main objectives of this chapter, it is useful to truly understand the concept of uncertainty in a broader sense. It is for that reason that we have chosen to approach the subject much more holistically than just focusing on relevant statistical methods, mathematical aspects and the like. This chapter is neither a precise statistical method description, nor a philosophical essay about the concepts of uncertainty, knowledge and truth, although you will find a little bit of both.

First of all, uncertainty is always there, it is the elephant in the room no matter what we are doing or talking about. From individuals to the entire humanity, from a child to a stock market broker to the most accomplished Nobel laureate, many of our daily efforts are related to knowing more, doing better, being more precise and more accurate. Acquiring knowledge and information and reducing the uncertainty around them is a driving force behind all human advancement, mobilising incredible amounts of resources worldwide. It is in fact one (if not the) driving force behind most things we do.

Uncertainty is also often the elephant in the room when people talk about or apply LCA. It is always there but some may fear it and ignore it deliberately, some may use it to criticise or even discredit LCA. An oversimplified understanding of uncertainty is a good part of the problem's root in both cases. Uncertainty is indeed frequently perceived as potentially discrediting LCA and its results as being too uncertain, unreliable, and insufficiently capable of distinguishing the compared options. The often considerable resources required for quantifying and managing uncertainty in an LCA study is an important barrier for their adequate consideration. Nevertheless, the presence of uncertainties of different types and from numerous sources in LCA results is a fact and ignoring them may be more detrimental than managing them in an integrated manner which allows their meaningful use to quantify and improve the precision of a study and the robustness of its conclusions.

LCA practice sometimes suffers from an imbalanced perception of uncertainties and their use in justifying modelling choices and omissions (e.g. excluding impact categories due to their perceived uncertainty). Identifying prevalent misconceptions, in some cases "myths", around uncertainties is another central goal of this chapter. The ambition is to help balancing the discussions around uncertainty in LCA and establish a positive discourse that focuses on the advantages of uncertainty management. Proper uncertainty management allows for more robust results and conclusions in support of science-based decision-making, grounded on the (accurate) recognition and discussion of inevitable and ubiquitous uncertainties.

Consider the following conceptual and simplified example to illustrate how fundamentally useful uncertainty assessment and management are in LCA. Figure 11.1 shows the results of an LCA study, performing a comparison of two alternative options A and B, for a given impact category like water use for example. The point estimate (i.e. reproducible, single value output from the LCA model without considering variations in inputs) impact score is 4 for option A and 6 for option B, which may suggest that option *A* is preferable, i.e. less environmentally impacting, over option *B* by a factor of 1.5. However, considering the uncertainties (including correlations between both options), the impact scores can be shown as superposed distributions as demonstrated in Fig. 11.1 (even though this may not be the best way to compare scenarios as discussed later in this chapter). Where the distributions are overlapping, option *B* has certain chances to be preferable over option *A*, the opposite of the conclusion drawn above from only looking at the point estimates. The more the distributions overlap, the higher the chances that option *A* may not be preferable to option *B*. In the left plot, there is a relatively small overlap of both distributions, and hence a relatively low chance to take the wrong decision when preferring option *A* over option *B*. In the centre plot, it is essentially impossible to discern the impact scores of both options and the chances to make the wrong conclusion would be high, no matter which option is chosen. In the right plot, the dispersions of both options are different (which will usually be the case in practice) and need to be evaluated in order to derive more reliable results. How to deal with such cases is discussed further-on in the chapter. That means that if the uncertainty cannot be further reduced (e.g. by using more certain data or models), both options are basically equal in terms of their potential environmental impact on water use.

The consideration and communication of uncertainties related to results obtained via modelling and/or measurements is vital for their correct interpretation. This is

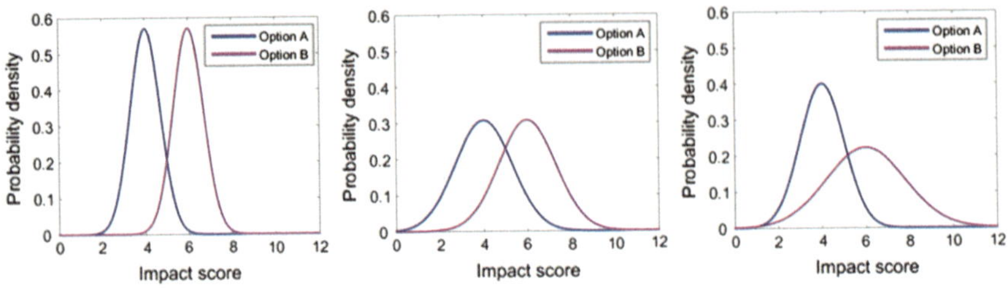

Fig. 11.1 Illustrative comparison of impact scores from two options *A* and *B* with the point estimates of 4 and 6, respectively, in all graphs, but with different uncertainties, low in the *left*, high in the *centre* and mixed in the *right* graph

often hampered by the difficulty to assign and propagate uncertainty information of the usually numerous parameters of a model as required by uncertainty assessment methods. This problem becomes even more apparent when modelling large systems as usually done in LCA, not to mention that there is more to overall uncertainty of a model result than just what parameters contribute. In current daily LCA practice, this often leads to complete omission of this important and integral aspect of any model result, while it may potentially influence or even change the conclusions of a study.

Uncertainty thus refers to everything we do not know and we cannot be certain about, regardless whether we are aware of it or not. In order to create a common basis of understanding when using technical terms and vocabulary around uncertainty, a thorough definition of important terms and concepts will be provided as starting point in the following section.

11.2 Essential Concepts and Definitions

In order to provide an accessible and operational angle on the subject, we have deliberately chosen to use simplified terminology and explanations that do not always capture everything there is to say. The focus of this chapter is on what is relevant for LCA students and practitioners, not on covering all aspects around statistical concepts, terms and definitions. For many concepts there may be multiple terms that are used synonymously in literature and in some cases there may not even be consensus on specific terms and their definition, such as what a sensitivity analysis exactly is for example. The implicit imprecision may be shocking to experts in statistics, but avoiding to capture the full complexity substantially helps getting a first grasp and understanding of the main concepts and how they are used in LCA practice, which is the main purpose of this chapter.

11.2.1 Defining Uncertainty, Variability and Sensitivity

The term **uncertainty** is used with a fairly large variation in its definition, including or excluding (somewhat) adjacent concepts like variability and sensitivity. It is therefore difficult if not impossible to give a universally valid and accepted definition of uncertainty. For the sake of defining a common understanding within the scope of this book, we use the definition of uncertainty as comprising everything we do not know, expressed as *the probability or confidence for a certain event to occur*. More precisely, the "unknown" includes both random and systematic errors (of estimating, measuring or collecting data), mistakes, and epistemological (or epistemic) uncertainty (i.e. lack of scientific knowledge and consequent misinterpretations). To put it a bit bluntly, uncertainty in principle describes the degree to which we may be off from the truth. In reality it is of course impossible for us to know that, otherwise we would not have to face uncertainty since we would know the truth (and we will avoid attempting to define what "truth" itself means). Therefore, in practice we define *reference points* that we assume to represent truth or at least to be close to it. A typical example for such a reference point would be a measurement. If we trust the measuring method and protocol we trust that a measurement represents a sort of truth at a specific point in space and time and the difference between a modelled estimate and a corresponding measured value is then used as an indicator for uncertainty. Ciroth et al. (2004) discuss and nicely illustrate this discrepancy between measured and true value and what uncertainty represents in that respect.

It is then important to keep in mind that the measured value inevitably comes with its own uncertainty due to possible measurement errors (and mistakes) and due to the uncertainty of how suitable the measurement method and how representative the sampling was regarding the actual "truth". Uncertainty can thus be quantified and reduced by knowing more, which usually requires us to invest more resources in order to gain more knowledge (e.g. by performing additional measurements or collecting more data and refining the model). However, no matter how many resources we have available, we can never be certain that we have eliminated (or at least minimised) uncertainty.

In order to define **variability**, let's take the example of body weight distributions in a human population. Many observations we can make will always have more than one value, as soon as we measure more than one sample (i.e. a sub-set of data points from a population of measured data), human body weight being an intuitive example. We are thus faced with a natural variability that simply represents the variety or spread in the data that we will always observe. With enough resources at hand that allow us to take every possible sample, we can perfectly well measure and quantify this variability, but we can never reduce it. In the context of LCA, we are typically faced with three different types of variability: (1) temporal variability (e.g. seasonal changes in temperature), (2) spatial or geographical variability (e.g. population density in different regions), and (3) inter-individual variability of humans, animals, other species (e.g. differences in diets) or technologies.

In LCA practice, the terms variability and uncertainty are often not distinguished or overarching one another (i.e. variability is often included as one aspect of uncertainty). However, for their important differences described before, it is recommendable and good practice to quantify and maintain both well separated as this will allow us to put this information to good use when interpreting and improving LCA results. We will come back to that later.

The **sensitivity** of a model describes the extent to which the variation of an input parameter or a choice (e.g. time horizon in the functional unit) leads to variation of the model result. A model is sensitive toward a parameter if a small change in this parameter will result in a large change in the model result, whereas a model is insensitive toward a parameter if any change in this parameter will have no (or negligible) effect on the model result (which in certain cases might indicate that this parameter may not be needed in the model, or at least that it is not an important input parameter for this particular value of the model result). Sensitivity may be analysed for both continuous and discrete input parameters, and it can also be analysed for choices leading to discrete sets of input values. For example, the choice of LCIA method is always a discrete choice between a certain number of fixed options (i.e. available methods). It is worth noting that the term sensitivity is used in various and inconsistent ways throughout literature and no agreement on its exact definition exists. Two main uses could be distinguished: (1) For some authors sensitivity includes the effect of uncertainty and thus considers the range of variation of input parameters as a function of their uncertainty (which hence needs to be known), varying them all at the same time. This is also called *global sensitivity analysis* and is essentially what this chapter refers to as uncertainty analysis. (2) Others define sensitivity solely as the effect of a certain change in input on the output applying a predefined variation without considering the uncertainty. This is analysed by varying one parameter at a time and also called *local sensitivity analysis*. In the context of this book and many publications in the LCA community, sensitivity only describes the variation of a result due to variation of an input or choice, without considering its uncertainty, i.e. local sensitivity.

11.2.2 Defining Accuracy and Precision in the LCA Context

When talking about uncertainty, a number of terms are often used in conjunction or interchangeably which seem to be synonyms but in fact are not. Two such terms are accuracy and precision. The definition of these terms in general English dictionaries varies to some extent, the Oxford English Dictionary for example defines accuracy as technical noun being "The degree to which the result of a measurement, calculation, or specification conforms to the correct value or a standard" and precision as technical noun being "Refinement in a measurement, calculation, or specification…". Therefore, both terms are independent and while accuracy refers to the correctness of a value, precision relates to the relationship among multiple measurements or calculation results. It is therefore useful to have a closer look at the actual meaning of these terms in a technical or scientific context and what they imply for LCA. Accuracy describes the closeness of a measured or modelled value to its "true" value. Precision represents the quality of being reproducible in amount or performance (i.e. any repetition of a calculation, experiment, model run, etc. gives a similar result when precise or a wide spread when imprecise), but a reproducible result does not necessarily have to be very accurate or even "true". In consequence, the accuracy of a model result may be high while its precision can be low as illustrated in Fig. 11.2. This means that the average of such model results will still represent meaningful information even though the results' spread (i.e. the standard deviation) may be large. In contrast, a very precise measurement or model

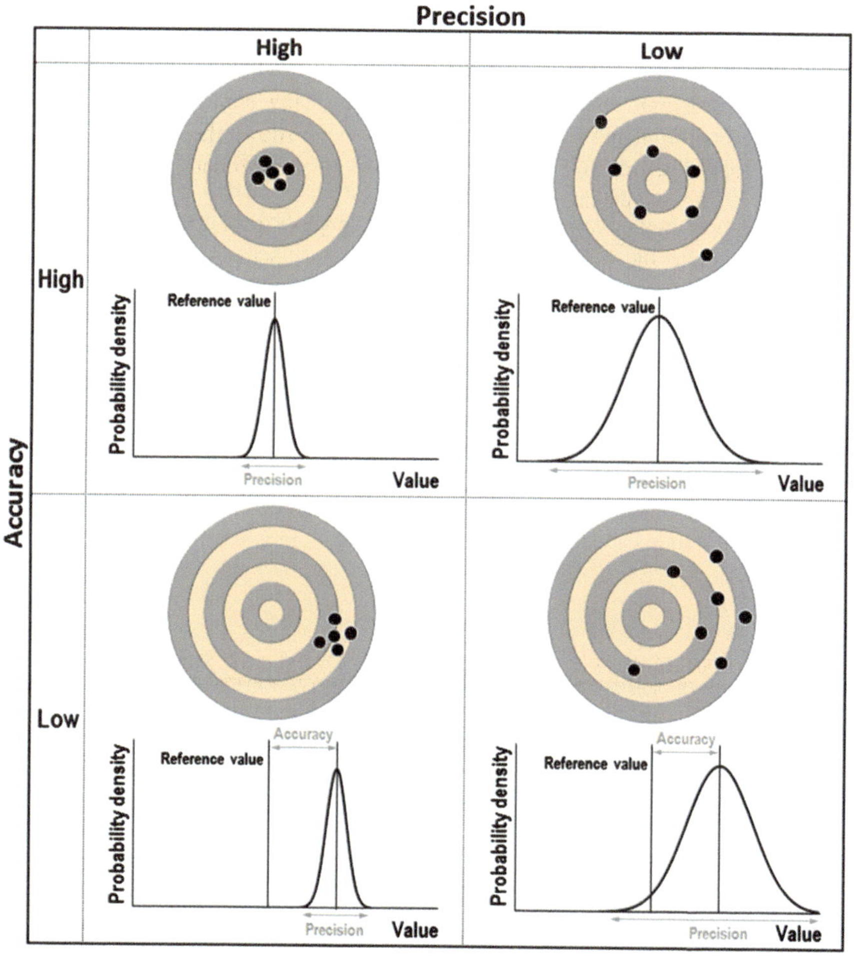

Fig. 11.2 Illustration of precision and accuracy

result (i.e. with a small standard deviation) is not necessarily meaningful if it comes with low accuracy regarding the information one is actually looking for.

In the LCA context, this can be illustrated using the different time horizons of the global warming potential (GWP). When intending to capture potential impacts from global warming of greenhouse gas (GHG) emissions, the GWP is integrated over 20, 100 and until the 4th IPCC Assessment Report (IPCC 2007) even over 500 years. It is intuitive that precision decreases with an increasing time horizon due to the assumptions necessary to model and predict far into the future, but does accuracy also automatically decrease with longer time horizons?

In order to answer that question, we need to consider that most GHGs stay much longer in the atmosphere than 20 years. GWP20 is a very precise and probably accurate indicator for the cumulative radiative forcing (i.e. the capacity to absorb energy, which can be measured in the lab) of a molecule during 20 years, but it neglects that this molecule may still be active long after. It is thus a very inaccurate indicator for the total potential contribution of the molecule to global warming, which is what we are usually interested in for an LCA study (unless the goal and scope definition requires a focus on short-term impacts). Therefore, implicitly assuming that GWP20 quantifies the (total) potential contribution of an emission to global warming bears a risk of interpreting LCA results wrongly in spite of using an indicator that is very precise, as it is inaccurate for the objective at hand (Fig. 11.3).

This example may seem somewhat obvious, but there are many other instances of exactly this type of confusion that can be found in current LCA practice. Another example is the comparison of the uncertainty of indicator results from different impact categories. The GWP is generally perceived as a fairly certain midpoint indicator whereas human toxicity is seen as a very uncertain midpoint indicator, an argument that is sometimes used to justify the omission of toxicity characterisation from an LCA study. It is worth reflecting whether this direct comparison of uncertainties makes sense by looking at the environmental relevance of what both indicators are actually quantifying.

We discussed in Chap. 10 that GWP is the time-integrated radiative forcing of a substance per unit mass emitted. The input data required to calculate it are relatively

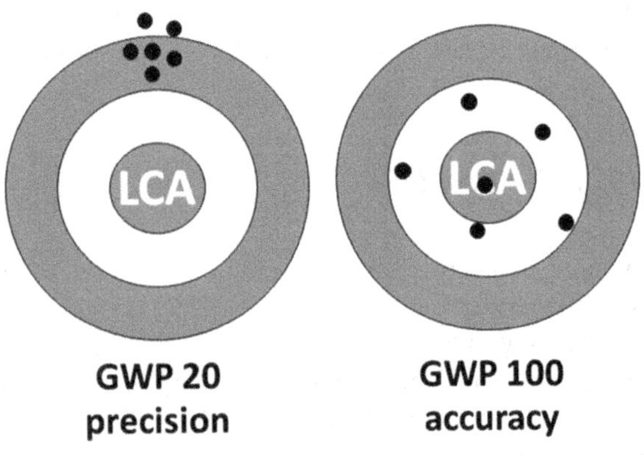

Fig. 11.3 GWP20 more precise but less accurate from an LCA perspective than GWP100

straightforward to measure and well reproducible in a laboratory, or in other words it is a *precise* indicator. It indicates the potential absorption of energy in molecules in the atmosphere, but it does not inform us on its impact on the environment or human health, or in other words it is *not accurate* regarding the goal of quantifying potential environmental impacts. Most toxicity midpoint indicators, however, quantify statistically how many disease cases (or affected species) may potentially occur in a human (or ecosystem) population per mass emitted. Therefore, toxicity indicators are much more representative regarding the consequences of a potential impact than GWP, or in other words a toxicity midpoint indicator has a higher environmental relevance than GWP and may thus actually be more accurate than GWP, while being less precise. Only the inherent, and most likely often unconscious, assumption of causal links between radiative forcing—increased temperature—melting polar caps—rising sea levels—more extreme weather events—loss of agricultural yield—increased competition for food—starvation and possibly even war—and thus effects on human health makes this indicator useful for LCA, but does it make it actually less uncertain for indicating a potential environmental or human health impact?

The argument of too high uncertainty of toxicity indicators thus refers to their precision (reproducibility), but not necessarily to their accuracy (in representing environmental impacts) and may hence be misleading. In addition, the spread between the highest and the lowest values for an indicator may differ widely between impact categories. Given that the toxicity-related impact categories cover several thousand elementary flows (i.e. chemical emissions) with different environmental mechanisms, related variability is higher by several orders of magnitude than for impact categories only covering a handful of elementary flows (e.g. climate change including ~50 chemicals). An example of the relationship between uncertainty around results for a single chemical and spread of results across chemicals is given in Chap. 31, Fig. 31.7.

In LCA, uncertainty should always be referring to what a study aims to quantify. The environmental relevance of indicators varies greatly among impact categories and is also a source of uncertainty towards the conclusions of a study. Just because this uncertainty is not quantified or even somewhat unconscious, that does not mean that it is not present. Hence, a direct comparison of purely precision-related uncertainty among midpoint indicators is not meaningful unless the compared indicators have a similar level of accuracy (i.e. environmental relevance).

This brings us to another common misconception about the uncertainty of LCA indicators, namely the choice of using midpoint or endpoint indicators. The typical trade-off between both options is that a midpoint indicator result will be more precise but less environmentally relevant, while it will be the opposite for an endpoint indicator (i.e. less precise but more environmentally relevant). Therefore, endpoint indicators are typically perceived as more uncertain based on their usually lower precision (due to a larger number of choices and hypotheses involved in their modelling compared to midpoint indicators). When considering environmental relevance as a measure of accuracy and a type of uncertainty (as discussed further below), it is important to keep in mind that midpoint indicators have a large portion of (unquantified or unperceived/unconscious) uncertainty due to their lower environmental relevance compared to endpoint indicators. As depicted in Fig. 11.4, overall uncertainty may increase or decrease from a midpoint to an endpoint indicator of a given impact category, depending on the uncertainty of models and parameters used for endpoint modelling.

However, Weidema (2009) pointed out that this "figure implies that it is possible to make a trade-off between relevance and uncertainty, in which the overall error is minimised … [and] … that the consequences of the decision will be less uncertain if the decision is taken at the point where the overall error is minimised—that is, at a midpoint […] (e.g., at the level of CO_2-equivalents)", which is a common perception among LCA practitioners and clients. Weidema then rightfully argues that "When the decision is implemented, however, the consequences occur not only at the level of the midpoint but also at the level of the endpoint (the decision will result in lost species and lost lives). This implies that the apparently low uncertainty of the decision at midpoint does not reduce the uncertainty of the consequences of the decision at endpoint level, which are still as uncertain as indicated at the bottom of [the] figure […]. If the consequences at endpoint level (e.g., lost species and lost lives) are what we really are interested in (as implied by the maximum relevance), then taking the decision at the midpoint level (e.g., CO_2-equivalents) is simply the same as ignoring the true uncertainty of the consequences of the decision." In other words, if minimal or avoided environmental consequences are the objective of a

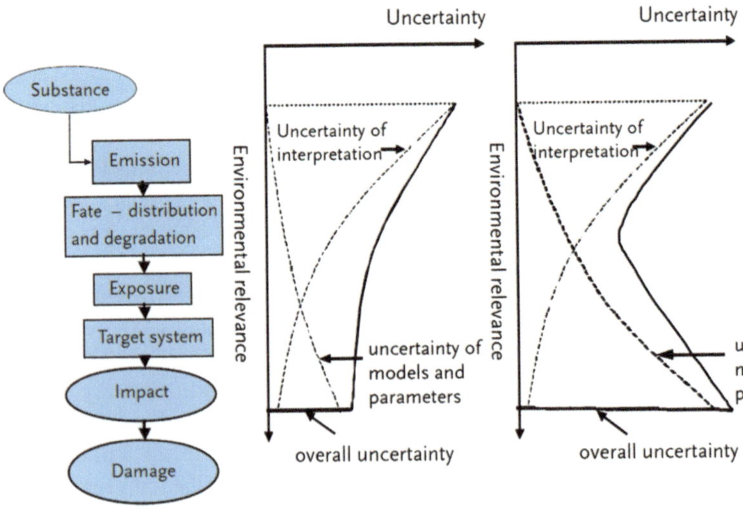

Fig. 11.4 Conceptual representation of how overall uncertainty may decrease (*middle*) or increase (*right*) from midpoint to endpoint (*damage*) in an impact pathway (*left*); uncertainty of interpretation and uncertainty of models and parameters contribute to different extents to overall uncertainty on midpoint (early in the impact pathway) and on endpoint/damage level (end of the impact pathway) while environmental relevance increases [taken from Hauschild and Potting (2005)]

decision, choosing midpoint indicators because they can be quantified with higher precision will still not avoid the uncertainty of that decision's environmental consequences since a midpoint indicator is less relevant (representative) for the environmental consequences to be avoided. Weidema (2009) entertainingly compares this flawed logic as being "representative of the situation of the drunk who, when asked why he was searching for his keys under the streetlight although he had lost them in the dark alley, responded that it was easier to see under the light". In consequence, the overall uncertainty of endpoint indicators may not (always) be much different to that of midpoint indicators from a decision-support perspective as indicated in Fig. 11.4 where the development in the "overall uncertainty" accompanying the decision may sometimes be lowest at the damage level, when the reduction in interpretation uncertainty, going from midpoint to damage, more than compensates the increase in model and parameter uncertainty of the applied characterisation model.

Hopefully, these examples illustrate that when discussing uncertainties between LCA indicators (of different impact categories or between midpoint and endpoint level), all types of uncertainty combined with the related concepts of precision and accuracy need to be considered or else the risk of oversimplifying and comparing apples and oranges is imminent, which may lead to unjustified and wrong conclusions.

The very purpose of any model is to represent a simplification of reality, but what is the right level of simplification? In order to establish a useful model, a meaningful level of complexity is required. As illustrated in Fig. 11.5 adapted from Ciroth (2004), the overall error (of representing reality) of a model is, among other, a function of the error due to an inaccurate representation of reality (too complex model with, e.g. too many input parameters and algorithms that introduce each their own uncertainty) and the error due to ignoring too much of the complexity of reality (too simplistic model). Accordingly, balancing both will yield the lowest overall model-related error. This is known as the parsimony principle, i.e. as simple as possible and as complex as necessary, and intuitively is a suitable leitmotif for LCA.

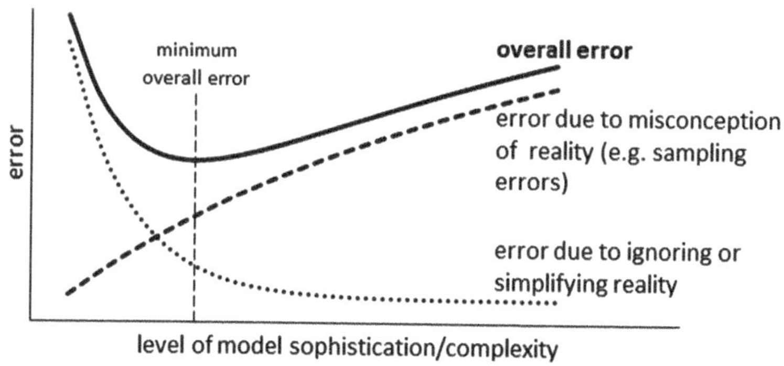

Fig. 11.5 Too complex modelling can have a similar error of representing reality as too simplistic modelling [modified from Ciroth (2004)]

Simplicity is often perceived as a desirable quality of a model making it easy to understand and less data demanding. Complexity on the other hand is frequently perceived as cumbersome, non-transparent and data intensive. However, rejecting complexity as such, without regarding its relevance and influence on the decision at hand, will of course be simpler and also lead to a decision, but it may not be a decision fulfilling the LCA objective of choosing an environmentally preferable option. In other words, it may be a more precise but less accurate and thus a potentially misleading decision. Given the inherent (i.e. unavoidable) complexity of environmental processes and our still limited knowledge of them, the principle of "It is better to be vaguely right than exactly wrong" (Read 1920) is a much cited and useful angle when discussing uncertainties in LCA, thereby also acknowledging that we should never design our models more complex than necessary to avoid "paralysis by analysis" potentially leading to no operational model at all and, hence, to no decision (support).

11.2.3 Representing Uncertainty

The probabilistic nature of uncertainty of the studied process or object is conceptualised by a probability distribution. The probability distribution of a continuous variable is described by a distribution function, usually the probability density function (PDF—not to be confused with the abbreviation PDF for Potentially Disappeared Fraction of species as used in Chap. 10). In practice, the PDF of an input parameter x is estimated by the values x_i measured over a sample, ranging from a minimum to a maximum value. Hence, the probability is approximated by the relative frequency when enough values are sampled. For example, when measuring the body weight of individuals in a human population of several thousand people, we will always find a range of values with a minimum value given by the lightest and a maximum value given by the heaviest individual(s) among those measured. Drawing the full range of measured values on the x axis and how often each of these values occurs (=their relative frequency) on the y axis results in a distribution function (a PDF) as illustrated in Fig. 11.6.

The shape of this function varies substantially depending on the frequency of the values of a variable. Many shape patterns have been clearly defined and termed, distinguishing continuous distributions such as normal, log-normal, or beta, and discrete ones such as binomial, Poisson, or hypergeometric, the latter being characterised by a probability mass function (PMF). When representing uncertainties, these names are used to describe the **type of distribution** and are an essential element when addressing the uncertainty of a (measured or estimated) parameter or the model output. Various methods exist to fit a continuous or a discrete distribution over a set of values.

Generally, important measures to describe uncertainties of an input parameter x or the model output are the *standard deviation* for the spread of a distribution, and for the central tendency of a distribution the arithmetic mean (or average), the

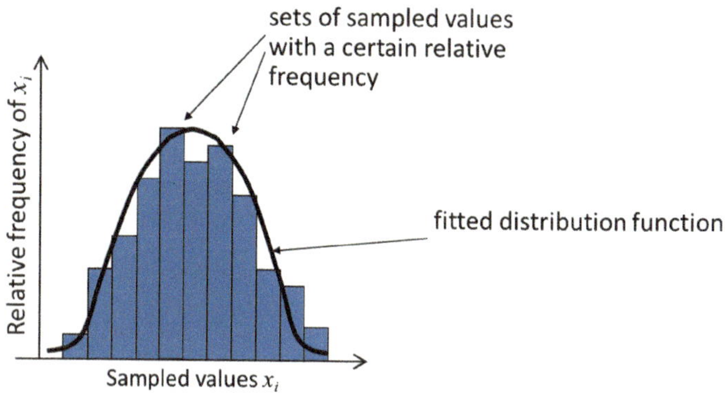

Fig. 11.6 Fitting of a distribution to a set of values for a variable

median, and (more rarely) the mode. The **arithmetic mean** or **average** of a sample is calculated as the sum of all values divided by the total count of all values. The **mode** is the most frequent (i.e. most probable) value within the dataset, and the **median** is the value separating the upper 50% and the lower 50% of all values when ranked in order of their magnitude. In a perfectly normally distributed dataset, the average, median, and mode are identical, whereas in any skewed distribution (e.g. a log-normal distribution) these central tendency measures have different values. However, the mean has the disadvantage to be very susceptible to outliers (unusually small or large values within a dataset) and skewed data. Therefore, the mean does not represent the best central value in skewed distributions (e.g. log-normal), whereas the median is less affected by the skewness of a dataset. The variation of the sample values is most commonly described by the (sample) **standard deviation.** The PDF or PMF are sufficient to fully characterise the distribution of an input parameter, but it is not always evident to derive these functions. Then the combined knowledge of the average (or median) and the (sample) standard deviation can provide a useful description of the behaviour of a parameter.

In-between the minimum and the maximum values of the range, we will find all sampled values and measures of central tendency for a probability distribution, like the *average* and the *median* body weight in the previous example. For the quantification of uncertainty, we usually do not use the entire range between these two extrema, but rather a sub-set of (more representative) values. Figure 11.7a represents a normal distribution for an input parameter x with known parameters μ (=mean) and σ (=standard deviation). Integrating under the curve of the normal distribution from negative to positive infinity, the area is 1 (i.e. 100%). Consequently, the probability for a value drawn from this distribution to fall in the range $\pm\infty$ is 100%. Obviously, this is not useful in terms of describing the uncertainty of a parameter.

In the context of environmental modelling (including LCA) the typically used uncertainty range is the 95% interval as given in Fig. 11.7a as shaded area for a

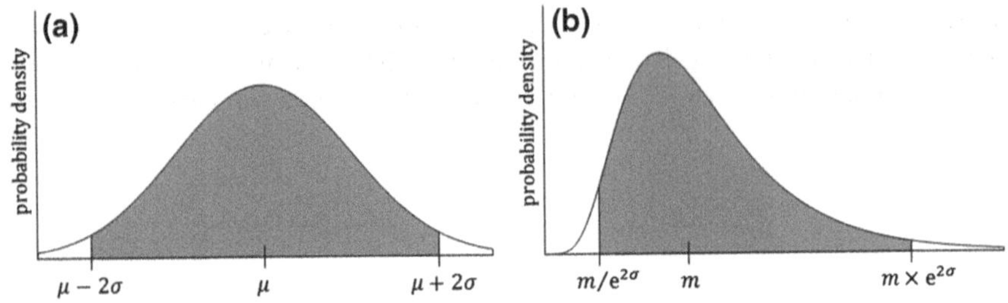

Fig. 11.7 **a** Normal distribution and **b** log-normal distribution with 95% uncertainty interval ranges *shaded in grey*

normally distributed input parameter. This 95% uncertainty interval can be interpreted as the range of values within which (approximately) 95% of all randomly measured values can be found. When the distribution function is known, we can also say that any sampled (or measured) value one may take in the future will fall within this range with 95% chances. Assuming a normal distribution for our example on body weight, this means that 95% of all measured weights from our population will fall within this range and that if picking randomly a person from that population, one will have 95% chances that this person has a body weight within this range of values and only 5% chances to pick a person lighter or heavier than that. The limits of the uncertainty interval are referred to via various names such as upper and lower bounds or **2.5th (lower bound) and 97.5th (upper bound) percentiles**. Other used uncertainty intervals for normally distributed variables are the 68 and the 99.7% intervals.

The link between measures of central tendency (especially the mean and median) and dispersion (standard deviation) of an input parameter x with the upper and lower uncertainty bounds is detailed in the following. Going back to the normal distribution in Fig. 11.7a with mean value μ and standard deviation σ, the 95% uncertainty interval (approximately) corresponds to the interval range between $\mu - 2\sigma$ and $\mu + 2\sigma$. The limits of this interval are the 2.5th percentile (2.5th %ile) as the lower bound at $\mu - 2\sigma$ and the 97.5th %ile as the upper bound at $\mu + 2\sigma$. Integrating over a range within $\pm\sigma$ from the mean value μ, the resulting value is 0.6826; hence, the probability for a value to fall within the range $\pm\sigma$ around the mean is approximately 68%. This range is called the 68% (sometimes 65%) uncertainty interval. You may have guessed it by now, the 99.7% uncertainty interval is then bounded by $\mu - 3\sigma$ on the lower and $\mu + 3\sigma$ on the upper end of the distribution.

If an input parameter x is log-normally distributed with population parameters μ and σ, it means that the natural logarithm of the parameter follows a normal distribution. This distribution is often observed for measurements of environmental input parameters and hence frequently used in environmental modelling. The median value m of the log-normal distribution is identical to the **geometric mean** e^{μ}, while the mean of the distribution is $e^{\mu + \sigma^2/2}$. The mean is larger than the median as this distribution is right skewed. For a log-normally distributed input parameter, the corresponding distribution and the 95% uncertainty interval are depicted in Fig. 11.7b. The 95% uncertainty interval (approximately) corresponds to the integration over the range $m/\mathrm{e}^{2\sigma}$ to $m \times \mathrm{e}^{2\sigma}$. The exponential term is thereby defined as the **squared geometric standard deviation**:

$$\mathrm{GSD}^2 \triangleq \mathrm{e}^{2\sigma}. \tag{11.1}$$

With that, the GSD^2 is used to define the 2.5th and 97.5th %iles, i.e. the 95% uncertainty interval bounds, of a log-normal probability distribution around the median m of x as

$$\mathrm{Probability}\left\{\frac{m}{\mathrm{GSD}^2} < x < m \times \mathrm{GSD}^2\right\} \approx 0.95. \tag{11.2}$$

The uncertainty intervals as discussed above should be distinguished from the confidence intervals. In practice, a population parameter (mean, median or standard deviation) is often unknown. In statistical data analysis, **confidence intervals** are usually calculated, that is the estimated range of values that frequently contains the "true" value of the unknown population parameter, if the sampling procedure is repeated. We need here to clarify some common misconceptions around the interpretation of confidence intervals. For our example on body weight, suppose a 95% confidence interval for the unknown true mean weight that ranges from a to b ($a < b$). The statements "95% of the population weighs between a and b kilograms" or "There is a 95% chance that the mean weight of the population lies between a and b kilograms" are false. The correct interpretation is "If we were to repeat the weight measurement over and over, then 95% of the time, on average, the confidence intervals contain the true mean weight." The latter does not refer directly to a property of the population parameter, but a property of the procedure itself. Two useful further readings on common misconceptions and misinterpretations of confidence intervals and other statistical methods and parameters are the papers from Greenland et al. (2016) and Hoekstra et al. (2014). For a further study of confidence intervals, and all the concepts presented in this section as well, the reader can refer to bibliography in probability and statistics, e.g. Walpole et al. (2012).

The type of distribution is an important element to precisely describe the uncertainty of a parameter. The simplifying assumption of a certain type of distribution (in LCA typically log-normal), instead of attempting to identify the exact distribution, is very useful when little or no information is available about a parameter or when using simplified, approximate analytical uncertainty propagation methods. However, this is sometimes met with criticism by practitioners who would like to integrate uncertainty information into their LCA studies using the exact distribution type. While from a purely statistical point of view this is the ideal, the very large number of variables and their distributions for individual inventory data and characterisation factors used to quantify the uncertainty of an impact score, will often result in a normal distribution of the impact score. This phenomenon is called the "central limit theorem" which states that the arithmetic mean of a sufficiently large number of independent values will be approximately normally distributed, regardless of the underlying input distributions (Pólya 1920). Although, this theorem requires certain conditions to be fulfilled (e.g. independence of the included parameters, existence of a finite expected value and standard deviation for each parameter), it is reasonable to assume these conditions to be fulfilled by most unit processes in LCI. This practical assumption offers several ways to significantly and parsimoniously simplify uncertainty quantification in the LCA context with a likely acceptable loss of precision when assuming one or only a few distribution types for LCA input parameters.

11.3 Addressing Uncertainty in LCA

11.3.1 Types and Sources of Uncertainty and Variability in LCA

There is no shortage of classifications of uncertainty types in literature, ranging from only two or three classes up to ten or more different types. A very useful classification for LCA was published by Huijbregts (1998) and comprises the following classes:

1. Temporal variability (e.g. seasons),
2. Spatial variability (e.g. population density, climate conditions),
3. Variability between objects (e.g. between different individuals),
4. Parameter uncertainty (e.g. inaccuracy, lack or non-representativeness of input data and model parameters),
5. Model (structure) uncertainty (e.g. algorithms in process and characterisation models),
6. Uncertainty due to choices (e.g. definition of functional unit and system boundaries, selection of LCIA method),

to which Björklund (2002) added:

7. Epistemological uncertainty (e.g. lack of relevant knowledge),
8. Mistakes (e.g. choosing the wrong substance or process due to similar names as references, unit conversions or unclear units like tons vs. metric tons/tonnes),

and to which we add:

9. Relevance uncertainty (e.g. environmental relevance, accuracy or representativeness of an indicator towards an area of protection).

Huijbregts (1998) also provided an illustrative list of examples of sources of uncertainty for each type and per LCA phase, which was slightly modified by

Björklund (2002) and by the authors of the present chapter and which is shown in Table 11.1. A classification of uncertainty types widely used in many fields of application distinguishes only three different types: parameter, model, and scenario uncertainty. Most of the nine uncertainty types listed above are essentially sub-classes of these three types as indicated in Table 11.1. Parameter uncertainty comprises variability and uncertainty in model input parameters. Model uncertainty indicates the uncertainty of the model itself via setup, initial and boundary conditions defined, variables/indicators taken into account, and equations used. Scenario uncertainty can be interpreted as uncertainty in the application and use of the model and its results under predefined conditions and assumptions. Whereas parameter and model uncertainty only contribute to the uncertainty of the numerical model results, scenario uncertainty may also contribute to uncertainty in the interpretation of the model results and, hence, that of a consequent decision as illustrated in Fig. 11.9.

For a number of reasons, parameter uncertainty and variability is the uncertainty type that is best considered in current LCA practice and it is what most people refer to when discussing uncertainty in LCA. With occasional, rare exceptions, the few published LCA studies that include uncertainty, essentially consider parameter uncertainty and variability. This kind of uncertainty is estimated in LCI databases such as ecoinvent and in some LCIA methods such as Impact World+ or LC-Impact, and LCA software allows to include the respective calculations in an LCA study. It is also a source of uncertainty that practitioners can address by improving data quality and representativeness, e.g. using primary data for foreground processes, or via spatialised LCA. This can be illustrated using three axes of data representativeness as discussed by Weidema et al. (2003), which constitute a three-dimensional space as shown in Fig. 11.8. LCI data may thus be too detailed, too un-specific, or too non-representative along one, two, or all three axes. Their distance on each axis to the range of data needed thereby represents their uncertainty.

It is important to keep in mind that most types of uncertainty and variability listed in Table 11.1 will contribute, to varying degrees, to the overall uncertainty of a quantitative LCA result (i.e. impact score). Just because parameter uncertainty is essentially the most accessible one and therefore the most frequently assessed or discussed type of uncertainty, it does not mean that it is always the most important (i.e. most contributing) one. The ninth type in the list above (uncertainty related to environmental relevance, accuracy or representativeness) refers to how completely all relevant processes are included in a model, notably to how completely an environmental mechanism is represented in a given characterisation model for a given category midpoint or endpoint (as illustrated in Fig. 11.4). Note that completeness and representativeness relate directly to the goal and scope of an LCA, e.g. the GWP model may be perfectly representative and complete if the goal of a study is to calculate a carbon footprint, while it may be incomplete and of low (environmental) relevance if the goal is to quantify the contribution of an activity to climate change-related human health impacts. For this reason, uncertainty related to environmental relevance or representativeness (i.e. termed here as relevance uncertainty in line with Paparella et al. 2013) cannot be part of the model

Table 11.1 Examples of sources of uncertainty and variability for each type of uncertainty per LCA phase

Uncertainty type		LCA phase				
		Goal and scope definition	Inventory analysis	Choice of impact categories and classification	Characterisation	Weighting and normalisation
Variability	Temporal variability	Inherent variations in the studied system	Differences in yearly emission factors	Inconsistent time horizons between impact categories	Time horizon or change in environmental characteristics over time	Change of social preference over time
	Spatial variability		Regional differences in emission factors	Regional differences in relevance of an impact category	Regional differences in environmental, ecological sensitivity or characteristics	Regional differences in distance to (political) targets
	Variability between objects		Differences in technology between factories which produce the same product	Differences among technologies in relevance of an impact category	Differences in environmental, ecological and human characteristics	Differences in individual preferences when using a panel method
Parameter uncertainty			Inaccurate, non-representative or no inventory data		Uncertainty in lifetimes of substances	Inaccurate normalisation data

(continued)

Table 11.1 (continued)

Uncertainty type		LCA phase				
		Goal and scope definition	Inventory analysis	Choice of impact categories and classification	Characterisation	Weighting and normalisation
Model uncertainty	Model structure uncertainty		Linear instead of nonlinear modelling, assuming continuous emissions	Impact categories are not known; Contribution of impact category is not known	Linear instead of nonlinear modelling, assuming steady-state conditions	Weighting criteria are not operational
	Uncertainty due to choices	Choice of system boundaries	Choice of allocation methods			
Scenario uncertainty		Choice of functional unit	Choice of technology level	Leaving out known impact categories	Choice of the characterisation method(s)	Choice of normalisation reference system or weighting method
Relevance uncertainty		Environmental relevance and representativeness required for decision		Completeness of (relevant) impact categories covered	Representativeness of an indicator regarding a given area of protection	Neglecting the influence of normalisation or weighting factors on results when interpreting them
Epistemological uncertainty		Ignorance about relevant aspects of studied system	Ignorance about modelled processes	Ignorance about relevant impact mechanisms	Ignorance about relevant environmental processes	Ignorance about relevant priorities
Mistakes		Any	Any	Any	Any	Any

Extended from Huijbregts (1998) and Björklund (2002)

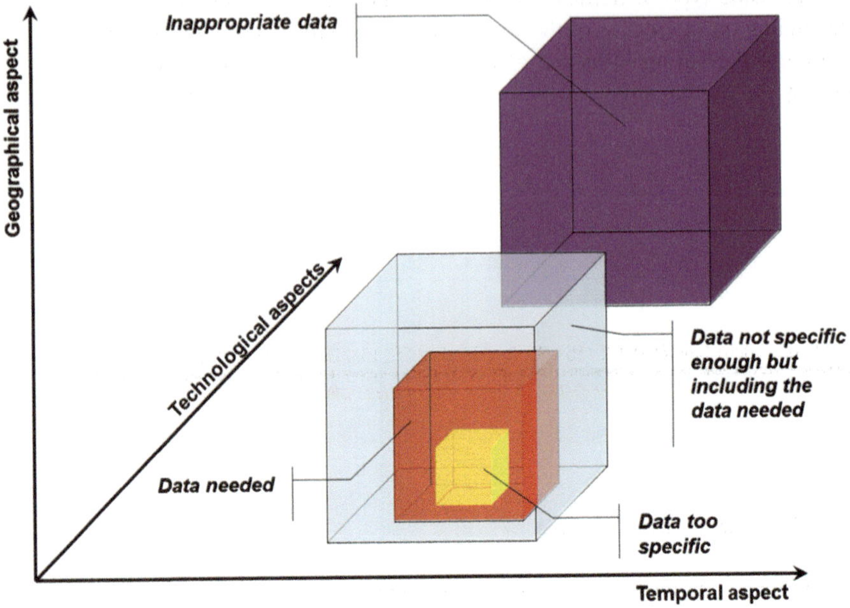

Fig. 11.8 Three aspects of data representativeness in LCA based on Weidema et al. (2003)

uncertainty, which is an intrinsic property of a model result and does not depend on how it is used or interpreted.

Uncertainties and variabilities are typically discussed regarding their importance for the uncertainty of a numerical model output or result, i.e. a number with its standard deviation and eventually a distribution function, which describes the uncertainty of the underlying tool and its result. This does however not consider what this result is being used for, which decision it supports and how it is being interpreted in the context of this decision. In order to also be able to represent and discuss additional sources of uncertainty related to results interpretation and the decision context, the concept of *relevance uncertainty* may be helpful. The more representative an indicator is for a given environmental (or social or economic) problem or damage, the lower the uncertainty on its interpretation, as discussed before. As shown in Fig. 11.9, this may be called the relevance uncertainty, which essentially contributes to the uncertainty of a conclusion or decision, but not to that of the numerical model result. Weidema (2009) pointed this out by stating that "Perhaps the cause of the logical error in the interpretation of (Fig. 11.4) … is that it requires that relevance (or uncertainty of interpretation) can be measured in the same unit as uncertainty of measurement, which is, in fact, not possible. Relevance is what we look for; uncertainty addresses the reliability of our measurement. When we are deciding how to measure what we look for, it is irrelevant how precisely we can measure what we do not look for".

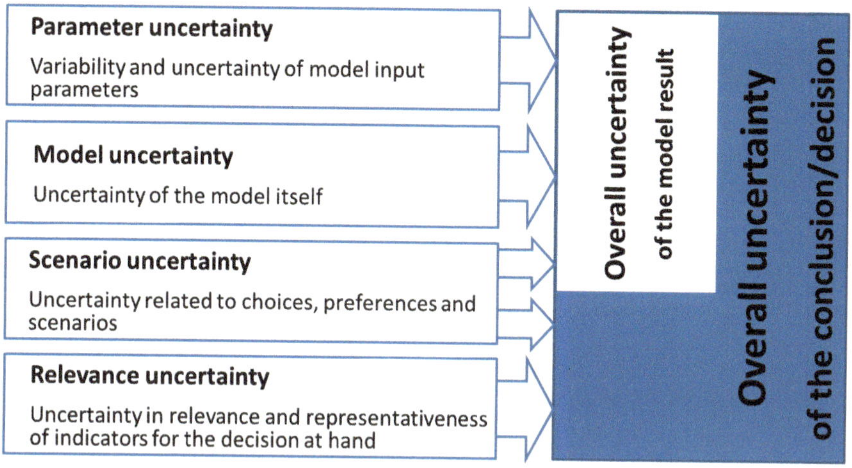

Fig. 11.9 Types of uncertainty and their contribution to result and decision uncertainty

This can be illustrated via a simple example on the use of indicators. Before leaving the house in the morning, many people check the outdoor temperature. What will be the uncertainty of this information? We can probably assume it to be low, so it is a very certain indicator value. However, the real question behind may not be what the value of the current temperature is, but what would be the adequate way to dress for the day. This decision requires a number of indicators, among other temperature, but also wind speed (or chill factor), rainfall and the predictions of those parameters for the rest of the day. Now the uncertainty of these indicator values is probably already a bit higher than that of the current temperature, but that's not all, since the decision at hand is the choice of clothes. This however comes with its own uncertainty on the interpretation of the link between preferable clothes and the available indicator values for temperature, wind speed, precipitation, all with their respective predictions and related uncertainties. The overall uncertainty of the decision is therefore dependent not only on the contributions from the indicator values but also on their interpretation and on how to conclude from them to choosing among a range of options for pants, jumpers, shoes, and jackets.

To translate this example into the world of LCA, one could ask "What is the GWP100 of 1 litre biodiesel?" and most practitioners will be able to answer this question (using a number of assumptions and choices) with reasonable certainty. However, a typical LCA goal is not the quantification of a given indicator, but the support of a decision like "Is biodiesel environmentally preferable to fossil diesel?". To answer that question, multiple indicators besides GWP100 such as land use, (pesticide-related) toxicity, eutrophication and others will have to be calculated. The resulting midpoint or endpoint indicator values, respectively, have their uncertainties and comparing them among both diesel types also adds uncertainty. However, the overall uncertainty of the answer to the question of preference is also affected by how the link between the differences of the indicator values and their representation of environmental consequences is interpreted.

Completeness refers to a parsimonious balance between simplicity and complexity (as discussed above in relation to Fig. 11.5), not to the need to include everything. The same parsimony principle also applies to the related balance between parameter and model uncertainty. In essence, a too simple model will be missing important processes and thus have high scenario and relevance uncertainty due to low environmental relevance, but will have low parameter and model uncertainty. A too complex model, in contrast, may need many (uncertain or unknown) parameters and may imprecisely represent some processes (high model uncertainty), but will also be more (environmentally) relevant, i.e. low uncertainty on representativeness. Similar to Fig. 11.5, overall uncertainty will thus, again, be lowest when both extremes are well balanced, the model being as simple as possible and as complex as necessary (i.e. following the parsimony principle), representing well all significantly influential processes (van Zelm and Huijbregts 2013). This is another example why the assumption of low uncertainty for a simple model, just because it needs few parameters, is incorrect and misleading.

When discussing uncertainty or error in LCA, it is also important to be aware of the implications of random versus systematic errors. In most fields where uncertainty assessment is addressed, the goal is to be precise on an absolute indicator, like temperature or weight for example, which aims to respectively indicate how hot or cold or how light or heavy something or someone is. With some exceptions, the goal in LCA is usually to compare (even a hotspot analysis is essentially a comparison between all processes within a product system) and provide a relative indicator of how much better or worse an option is compared to another, as opposed to indicating how good or how bad something is in absolute terms. In such a comparative context, a systematic error—affecting all compared objects in the same way—may have little importance for the interpretation of results and drawing conclusions. It will just shift all results up or down systematically. It thus affects the result in absolute terms (i.e. the numbers are all higher or lower), but not in relative terms (i.e. the quantitative difference between compared objects remains largely the same). This is frequently ignored when LCA is being criticised as too uncertain, essentially because people tend to interpret its uncertainty in absolute terms and compare it with the absolute uncertainty of other methods like quantitative risk assessment for example, whereas much of the absolute uncertainty does not contribute to the uncertainty of the difference between compared alternatives, which will be further discussed in Sect. 11.4.2. This is also related to why LCA results represent *potential impacts* and (usually) not predictions of observable impacts (see discussion and definition in Chap. 10).

11.3.2 Uncertainty Quantification and Propagation Methods

A quantitative uncertainty management is still a rare sight in LCA practice. If integrated, the most commonly considered types of uncertainty are parameter uncertainty and variability. Parameter uncertainty for example is captured in uncertainty estimates for inventory data such as given in the ecoinvent database.

The quantification of uncertainty refers to the task of establishing a quantitative measure of uncertainty for (1) a specific source of uncertainty in an LCA (e.g. a mean value, standard deviation, and distribution type for a variable or an other uncertain aspect), and (2) the overall uncertainty of an LCA as a result of the combination of specific sources of uncertainty. The latter is achieved using uncertainty propagation methods.

Having discussed the types and sources of uncertainty and variability that are relevant for LCA, the question arises how to quantify them in order to consider and manage them during the assessment process. Some uncertainty types may be more straightforward to quantify statistically than others (e.g. variability of measurable parameters, uncertainty due to some choices), some can be estimated but may be very difficult to quantify (e.g. model uncertainty) and those that relate to the unknown cannot be quantified at all (e.g. mistakes, epistemological uncertainty, and environmental relevance). The latter can (and should) be considered qualitatively during the interpretation of LCA results (see Chap. 12). In consequence, the quantitative overall estimated uncertainty of a model result is both incomplete and uncertain in itself. This however does not make this information useless, but it is essential to consider when interpreting results including their uncertainty.

Several methods to quantify the (quantifiable) uncertainty elements of an LCA have been proposed and implemented to some extent into LCA. Among these methods are reporting uncertainty intervals, analysing parameter variability and/or different scenarios, translating qualitative data quality 'pedigree criteria' into a numerical pedigree matrix, using fuzzy data sets, applying analytical uncertainty propagation, conducting numerical, probabilistic simulations based on e.g. Monte Carlo analysis, using Bayesian statistics, or a combination of some of these methods. The following sections describe three methods that are already used in LCA: (1) the semi-quantitative pedigree matrix approach used for example by ecoinvent for the quantification of variability and uncertainty of LCI data; (2) Monte Carlo simulation used in LCA software like SimaPro, GaBi, openLCA, and the more explorative/educational LCA tools CMLCA and Brightway 2; and (3) Taylor series expansion used in CMLCA. A broader overview of selected quantitative uncertainty propagation methods in the context of LCA or the comparison of specific methods can be found in Lloyd and Ries (2007), Heijungs and Huijbregts (2004), or Groen et al. (2014).

Pedigree Matrix Approach

Information about the uncertainty associated with elementary flows is often not available or difficult to quantify for the hundreds to thousands of flows in a typical LCI. To nevertheless address uncertainty related to LCI results, a simplified semi-quantitative procedure can be used and is implemented into the ecoinvent database and also used by ILCD (EC-JRC 2010). It quantifies (exclusively) parameter uncertainty via combining two different kinds of uncertainty:

(1) Basic uncertainty due to variation and stochastic error of the values for elementary flows, from measurement uncertainties, activity specific variations, temporal variations, etc. This is quantified either using statistical methods when sufficient data are available, or via a simplified approach assuming a log-normal distribution, establishing an approximation that reflects the lack of sufficient information to calculate a more precise estimate.

(2) Additional uncertainty based on data quality indicators using a qualitative assessment of "reliability", "completeness" and representativeness in terms of "temporal correlation", "geographical correlation", and "further technological correlation". These quality indicators are assigned different scores expressing for each value different degrees of data quality and uncertainty and are represented by a numerical value (1, 2, 3, etc.) for each data quality and uncertainty degree. The lower a score for any quality indicator, the higher is the data quality and/or the lower the data-related uncertainty. As illustrated in Fig. 11.10, combining data indicators in rows with the scores for each indicator in columns gives the so-called "pedigree matrix" considering additional uncertainty (uncertainty due to using imperfect data).

Originally, the semi-quantitative pedigree matrix approach was proposed by Funtowicz and Ravetz (1990) in a framework for managing "all sorts of uncertainty" and later adapted to LCI modelling by Weidema and Wesnæs (1996) as being integrated into ecoinvent. The concept of the pedigree matrix is shown for the data quality indicator "reliability" of the data sources in Fig. 11.10. Combining data quality indicators with their respective scores gives a set of uncertainty factors aggregated into (geometric) standard deviations based on assuming log-normally distributed data in ecoinvent 2. These uncertainty factors are based on expert judgment, without (documented) empirical foundation and have been updated with a more empirical approach by Ciroth et al. (2016) based on analysing LCA studies and data with focus on industrial processes separately for each data quality indicator. Furthermore, in ecoinvent 3 the mathematical framework has been developed to also calculate uncertainty factors for distributions other than log-normal from the coefficient of variation chosen as a universal measure of variability and defined as the ratio between the arithmetic standard deviation and mean for all distributions (Muller et al. 2016).

The pedigree matrix based approach was also applied in LCIA for estimating input data uncertainty for toxicity characterisation by Fantke et al. (2012). In this context, the matrix columns represent data-related base uncertainty and the matrix rows represent spatiotemporal data variability. This application and the framework

Qualitative scoring

indicator score / indicator	1	2	3	4	5 (default)
reliability	verified data based on measurements	Verified data partly based on assumptions or non-verified data based on measurements	Non-verified data partly based on qualified estimates	Qualified estimate (e.g. by industrial expert)	Non-qualified estimate

Uncertainty factors (ecoinvent 2) based on expert judgment

indicator score / indicator	1	2	3	4	5 (default)
reliability	1.00	1.05	1.10	1.20	1.50

Tentative uncertainty factors (ecoinvent 3) based on empirical foundation

indicator score / indicator	1	2	3	4	5 (default)
reliability	1.00	1.54 (interim)	1.61	1.69	(n.a.)

Fig. 11.10 Excerpt from the ecoinvent 3 pedigree matrix showing scores for the data quality indicator "reliability" of the data source [adapted from Ciroth et al. (2016)] and how the scores are translated into numerical uncertainty factors based on expert judgement (for ecoinvent 2) or based on empirical data (for ecoinvent 3). The full matrices contain scores for five different indicators

laid out by Muller et al. (2016) demonstrate that the semi-quantitative pedigree approach can be flexibly applied to different aspects of an LCA study based on a diversity of different data lacking fully quantifiable uncertainty information.

Numerical Uncertainty Propagation

The most widely used uncertainty propagation method is a numerical approach called Monte Carlo simulation (sometimes also referred to as Monte Carlo analysis). It is available in all major LCA software (although not in all respective versions). Its basic principle is the repetition of model calculations (i.e. iterations) using values for each input parameter sampled from its defined probability distribution. A Monte Carlo simulation is outlined as follows:

Step 1: generate samples of random values for all input variables;
Step 2: apply the model on the generated values to calculate the model output in terms of LCA results;
Step 3: analyse statistically the model output.

The model output can therefore be represented by a probability distribution instead of a single value. An insufficient number of iterations will result in an unreliable empirical distribution of the output variable that may neither consider the full (or at least sufficiently representative) range of output values possible, nor represent the true shape of the distribution. In consequence, the distribution type will not be stable and the uncertainty will therefore be imprecisely estimated.

The accuracy of a Monte Carlo analysis increases as the number of iterations becomes larger. However, there is no generic approach to determine when the number of iterations is 'large enough'. Consequently, the number of iterations may vary among practitioners and also among studies performed by the same practitioner. The number of simulations does not dependent on the number of input parameters, but, in practice, the more complex a (LCI) model is, the more time-consuming a Monte Carlo simulation becomes, which may affect the total number of iterations. Instead of pre-defining the necessary number of iterations, it may be more efficient to run a few tests using an increasing number of simulations until the uncertainty measures (mean, standard deviation or eventually the distribution shape) does not change above an 'acceptable' difference, when further increasing the number of iterations. With enough experience, the number of required iterations can be identified based on the type or complexity of a study.

While the basic, iterative principle is the same for any implementation of Monte Carlo simulation, the sampling method (i.e. how the values from the distribution of an input parameter are sampled) can vary. The simplest sampling method is called "simple random sampling" (SRS), sometimes also "Monte Carlo sampling", and it randomly samples a value from the entire distribution of a parameter, as many times as the number of iterations set. Another, more optimised sampling approach is "Latin hypercube sampling" (LHS), which is a stratified sampling method that first divides a distribution into segments of equal probability and randomly samples one value from each segment. Subsequently, for each iteration, one of these pre-sampled values is randomly selected and used as input parameter value. If correctly set up, this allows a better representation of extreme values (close to upper and lower bounds of the distribution) and can significantly reduce the amount of iterations required as it needs less iterations in order to create a sufficiently representative amount of combinations of the different input parameter values. There are several specialised, further optimised variants of this sampling technique, including for example Median Latin Hypercube sampling, which samples the median of each segment instead of a random value. For most LCA applications with its many distributions and multiple sources of variance contributing to the result's overall uncertainty, there will often be no difference or particular advantage in using LHS compared to SRS. Only when a small amount (typically less than five) of input parameters contributes most to the overall output uncertainty, the advantage of LHS may be tangible. For an overview on simulation and sampling approaches, see e.g. Ross (2012).

Since all inputs are assumed to vary independently and thus in principle any combination of input values is possible, Monte Carlo simulation as described above implies mutual independence of all input parameters. In LCA however, many input parameters are correlated, i.e. if one parameter has an increased value any correlated parameter will consequently have a value that is higher or lower by a specific factor. This dependency of two or more parameters can be expressed using covariance or a correlation coefficient, which can be incorporated into a Monte Carlo simulation so that no impossible combinations of input values are sampled. This will typically lead to a reduction (sometimes an increase) in output uncertainty that can be very large compared to assuming input parameter independence and it is therefore essential to consider. Note that, when correlations exist, appropriate conditional distributions are required. The difficulty in LCA practice is to identify and, even more so, to quantify input parameter correlations, which may be numerous and not typically provided in LCI databases. For a single scenario, Groen and Heijungs (2017) analysed the importance of correlation in uncertainty and sensitivity analysis in LCA. They compared two approaches to include correlation of input parameters and demonstrated that the risk of ignoring correlation can be quantified. They found that in some cases it may not be necessary to quantify and consider correlation and that the risk of ignoring it can be included in the uncertainty analysis and thus be considered for the quantification of the robustness of the results and the consequent decision. One possible way of identifying and managing input parameter correlation is described in the Supporting Information of Fantke et al. (2012).

A note to avoid confusion: LCA (and other) literature sometimes refers to Monte Carlo and Latin Hypercube (with or without further specification whether simulation, analysis or sampling is meant) as if they were two distinct alternative sampling methods. As described above however, both belong to the family of Monte Carlo simulations and the difference is the sampling method.

Analytical Uncertainty Propagation

The most classic, simple and well-established analytical approach to uncertainty analysis, which is widely used in physical sciences and engineering, is the first-order approximation or Gaussian approximation, named after its famous developer Carl Friedrich Gauss. Morgan and Henrion (1990) described how a first-order approximation can be derived from the Taylor series (i.e. the representation of a function as an infinite sum of terms calculated from its derivatives at a given point), a technique based on a Taylor series expansion of the function relating model input parameters to model results (output). They extended this to a number of special cases, essentially allowing a wider application. This method uses linear first-order equations within a fully multiplicative set of parameters assuming independence of all relevant inputs.

In this method, relative (normalised local) sensitivity coefficients $S_{\hat{x}}$ are defined for each input variable x, calculated from the change of model output y (∂ output) per relative change of input variable x (∂ input) and evaluated at the point $x = \hat{x}$:

$$S_{\hat{x}} \triangleq \frac{\partial\,\text{output}/\text{output}}{\partial\,\text{input}/\text{input}} = \left.\frac{\partial y/y}{\partial x/x}\right|_{x=\hat{x}} \tag{11.3}$$

Model output uncertainty, represented by the corresponding squared geometric standard deviation of model output, GSD_y^2, can be described by its variance, $\text{var}[\ln(y)]$, i.e. the variation around its mean value. Output variance depends on the variance of all model input variables, $\mathrm{var}[\ln(x_i)]$ (Morgan and Henrion 1990). If we use the fact that the variance of any input variable is related to its $\mathrm{GSD}_{x_i}^2$ by $\mathrm{var}[\ln(x_i)] = [\ln(\mathrm{GSD}_{x_i})]^2$, we can express model output uncertainty via its GSD_y^2 as a function of $\mathrm{GSD}_{x_i}^2$ of model input:

$$\mathrm{GSD}_y^2 = \exp\left(2 \times \sqrt{\sum_{i=1} \mathrm{var}[\ln(x_i)]}\right) = \exp\left(\sqrt{\sum_{i=1} \left[\ln\left(\mathrm{GSD}_{x_i}^2\right)\right]^2}\right) \quad (11.4)$$

The $\mathrm{GSD}_{x_i}^2$ for the different input variables need to be known or can be approximated e.g. for log-normally distributed data from the 95% uncertainty interval by $\mathrm{GSD}^2 = \sqrt{97.5\text{th}\%\text{ile}/2.5\text{th}\%\text{ile}}$ (see also Fig. 11.7b and Eqs. 11.1 and 11.2) to ultimately arrive at an overall model output uncertainty using this analytical uncertainty quantification approach.

In the LCA context, this method was first proposed for use in LCI (Heijungs 1996, 2002, 2010; Heijungs et al. 2005). Based on this, application to LCA was demonstrated by Ciroth et al. (2004) for a virtual case and by Hong et al. (2010) for the real case of the carbon footprint of a car part comparing several scenarios and considering the dependency of many LCI and LCIA parameters shared by the considered scenarios, which is essential when comparing them. Imbeault-Tétreault et al. (2013) applied it to a complete LCA comprising 881 unit processes with 689 elementary flows comparing two scenarios and considering their dependencies. Different implementations of this method are possible, dealing in different ways with the limitations of this approach based on different underlying assumptions as compared and critically discussed by Heijungs and Lenzen (2014).

Comparisons with the results from Monte Carlo simulation which is considered to be the reference method for uncertainty propagation in LCA, consistently found good accordance between both methods applied to LCA (Ciroth et al. 2004; Hong et al. 2010; Imbeault-Tetreault et al. 2013; Heijungs and Lenzen 2014).

The main advantages of the analytical approach are its relative simplicity and calculation speed. The uncertainty is instantly calculated, whereas Monte Carlo simulation may take several minutes for small systems and few iterations to hours or even days of calculation time for complex systems and many iterations. For a typical LCA and a reasonable number of iterations, half an hour up to several hours on a modern computer can be expected. This is a major drawback towards routine uncertainty assessment in LCA and a central motivation for the authors mentioned above to explore analytical approaches for use in LCA. On the other hand, analytical methods are limited to predominantly simple (i.e. linear and continuous) models. An overview of strengths and weaknesses of analytical versus numerical methods in LCA was derived by Heijungs and Lenzen (2014) and is summarised in Table 11.2. It is worth mentioning that the analytical approach does not provide information about the distribution type of its result, only the standard deviation, but in LCA, a log-normal distribution is often assumed for the output similarly to the

Table 11.2 Comparison of main strengths and weaknesses of analytical and numerical uncertainty propagation methods

	Analytical: Taylor series expansion	Numerical: Monte Carlo simulation
Uncertainty information required per parameter	Standard deviation	Standard deviation, distribution type, parameter(s) describing the distribution
Uncertainty information obtained for model result	Standard deviation	Standard deviation, distribution type, further statistical analysis (e.g. median, interquartile range, etc.)
Applicability	Linear (almost), continuous functions; small uncertainties; no covariance (unless considered in additional term)	Linear and nonlinear, continuous and discrete functions; small and large uncertainties; no covariance (unless considered in additional term)
Calculation time	Instantly	Several minutes to hours
Capturing correlation of input parameters	Possible	Possible
Advantages	• Fast calculation time (i.e. seconds) • Distribution type and parameters of inputs not required • Useful screening approach	• Distribution type and parameters or outputs determined • Flexible and widely applicable including to complex models
Disadvantages	• Distribution type and parameters or outputs not determined • Fairly rigid and limited to simple linear models • Less widely applicable than Monte Carlo	• Long (sometimes very long) calculation time (i.e. hours to days) • More input information required

input parameters. Heijungs and Lenzen (2014) concluded that both methods should be implemented in LCA software and used complementarily in LCA, in order to profit from their respective advantages.

Quantification of Sensitivity

Sensitivity can be quantified using perturbation analysis (although often also referred to as sensitivity analysis, a term which is not clearly defined and used in different ways in literature, including or excluding uncertainty). Perturbation analysis can be performed numerically by varying an input parameter (e.g. by a fixed amount, a percentage, a standard deviation, or between a minimum and a maximum) and observing the resulting change in model output relative to the result using the unchanged input parameter (Heijungs 1994). The sensitivity S is then the ratio of the relative change in output divided by the relative change in input as given in Eq. 11.3. There are also analytical approaches available to provide this analysis (Heijungs 1994, 2002, 2010). The illustrative case study of an LCA on window frames in Chap. 39) identifies sensitive parameters calculating sensitivity ratios using Eq. 11.3 and also runs two sensitivity scenarios to test the influence of central assumptions in the study concerning the choice of geographical location (Danish vs. average European residence) and the decision whether to go for a design with a two-layered or a three-layered window pane.

If the input is not a parameter but a discrete choice (e.g. system boundaries, allocation rules, functional unit, LCIA method), a so-called scenario analysis evaluates the change in the result for each alternative considered (or meaningful) for a given choice. In this case Eq. 11.3 cannot be applied and a change in a choice may entail a change in several (correlated or mutually independent) input parameters, such as the case for the choice of LCIA method, which will usually change all characterisation factors. The analysis of the influence of a choice on the result is therefore referred to as scenario analysis, with each choice representing a possible scenario. Although they are formally two different types of analysis, a scenario analysis can be seen as a sort of sensitivity analysis, but for discrete changes in (often multiple) inputs instead of variation of one continuous parameter value at a time. A scenario analysis is also often used to represent different possibilities, e.g. future developments or best-case/worst-case scenarios, of how a number of parameters may change.

11.4 Interpretation and Use of Uncertainty Information

Once the uncertainties of input parameters, models, choices, etc., have been quantified and propagated, so that the results are not calculated deterministically but probabilistically (i.e. accompanied with a standard deviation and eventually a distribution of output values), the obtained information on uncertainty in the result can be used to improve (i.e. reduce) the uncertainty of important inputs and to enhance the interpretation and the robustness of conclusions drawn. This can be done in several, mostly complementary ways discussed in the following sections. We first discuss how to interpret uncertainty, variability, and sensitivity information, respectively, as the results of an uncertainty assessment in the LCA context. Then, we discuss the combined use of them and how to use the information obtained to reduce the uncertainty of an LCA study and the robustness of its conclusions.

11.4.1 Interpreting Uncertainty, Variability, and Sensitivity

As discussed, the sensitivity analysis points out those input parameters that have an important influence on the result, while the uncertainty analysis (including variability) provides information on the spread of the result due to the spread in input data and other sources of uncertainty. An input parameter may be very uncertain, but if the model output is insensitive to this parameter, the uncertainty of the input parameter will not contribute to that of the result (since no change in its value changes the model result) and improving the certainty of this parameter with better input data will bring no improvement to the robustness of the result and would thus be wasted effort. On the other hand, the model result may be very sensitive to an input parameter that is very certain, in which case it would depend on their degree of certainty and sensitivity whether or not better data would still improve the result's robustness. From this illustration it is clear that neither the sensitivity nor the uncertainty of a parameter should be interpreted on their own, whereas the combination of both allows a meaningful judgement of the importance of a parameter regarding the model output. This can be illustrated plotting both aspects as illustrated in Fig. 11.11, which shows both cases described above plus the two cases of (1) complete insensitivity combined with complete certainty of a parameter, which makes it negligible regarding its importance for the output uncertainty, and (2) high sensitivity combined with high uncertainty of a parameter, which will identify the primary parameters to focus data collection and improvement (i.e. reducing parameter uncertainty) on in order to obtain the largest gains in result certainty. This basic concept is useful to keep in mind when identifying dominant sources of uncertainty for any model result.

The concept of identifying and ranking sources of uncertainty, like input parameters, unit processes, or characterisation factors, in terms of their contribution to uncertainty in LCA results is called 'identification of significant issues in the Interpretation phase (see Chap. 12) but also referred to as key issue analysis, importance analysis or uncertainty contribution analysis, which is not to be confused with the impact contribution analysis or dominance analysis frequently used in LCA that identifies the unit processes most contributing to an impact score. It is useful for identification of important sources of uncertainty, where better information or data would directly improve the certainty of the result and hence the robustness of the conclusion. It can be applied to focus data acquisition and model refinement, ensuring that additional effort in getting better data or improving models actually contributes to more robust results. This also relates back to the discussion above on precision and accuracy, confirming that improving precision

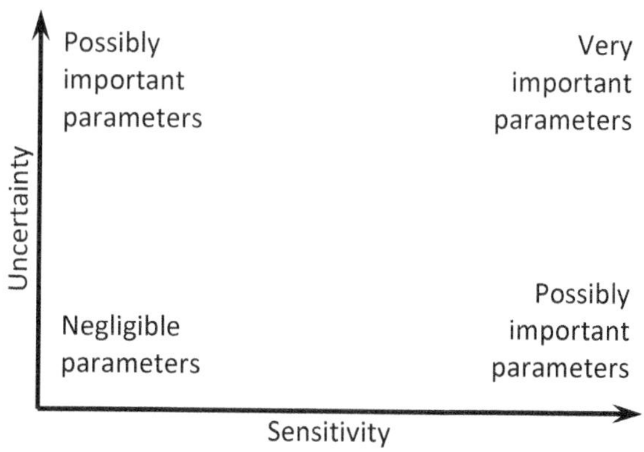

Fig. 11.11 Combining uncertainty of and sensitivity toward an input parameter to identify its importance in terms of contribution to overall uncertainty of the model result [based on Heijungs (1996)]; instead of parameters, any source of uncertainty in LCA could be identified, including for example characterisation factors

by finding more precise data does not automatically result in lower uncertainty if those data are not central to the impact score they are used to model.

Combining the information gathered via key issue analysis with that from impact contribution analysis, helps identifying unit processes that contribute significantly to (1) the impact score of a highly local or regional impact category (e.g. eutrophication, toxicity, land use, or water use, see Chap. 10) and (2) the uncertainty of that impact score. This can be used to apply a smart, partial regionalisation of the LCI model and its LCIA characterisation. Instead of using spatially resolved LCI and LCIA data for the entire LCA (which is resource intensive and thus usually prohibitive for both the practitioner and the LCA software used), only the identified unit processes and elementary flows are regionalised using primary input data and regionalised LCIA characterisation factors (or derived, representative archetypes of them). If the uncertainty and variability information of the underlying elementary flows and characterisation factors has been kept separate (i.e. not been combined into a single uncertainty distribution), this will result in a (substantially) lower overall uncertainty of the impact score, since the contribution from spatial variability will be eliminated (or at least reduced) by using spatially resolved data and characterisation factors for these processes. This method allows a parsimonious consideration of complexity due to spatial variability and rewards the practitioner's additional effort directly by a lower overall uncertainty and hence a more robust result and conclusion. The same approach can also be applied to temporal variability, i.e. using temporally explicit data instead of annual averages when it sufficiently influences the result's uncertainty, e.g. for water consumption.

11.4.2 Relevance of Uncertainty When Comparing Scenarios

So far we have discussed various aspects related to assessing the uncertainty of a single scenario, i.e. the environmental profile of one option, without comparing two or more alternative options, the latter being one of the most frequent applications of LCA. In the case of a comparative LCA however, there is an additional aspect to consider: the correlation of numerous input parameters between the compared scenarios, where many processes (i.e. electricity, fuel, transport, etc.) and almost all characterisation factors will be the same in several or all compared scenarios. When comparing two scenarios, the focal point is thus not on how large the value of an impact score is but what the difference (or the ratio) between two impact scores (i.e. between two scenarios or compared systems) is. Consequently, instead of the absolute uncertainty of a single impact score, the uncertainty of the difference (or ratio) between two impact scores needs to be assessed, because the uncertainty of correlated parameters will be the same in both scenarios and thus not contribute to the uncertainty of the difference between the scenarios. In other words, comparing two scenarios and their respective uncertainties (e.g. by simply overlaying both distributions) without considering correlation, the uncertainty will be (strongly) overestimated, which may be misleading and result in the wrong conclusion. Note that in the illustrative case study presented in Chap. 39, it was not possible to consider the correlations between the compared scenarios, due to software limitations. The technical possibilities for uncertainty analysis vary between available LCA software and may also evolve (i.e. improve) from older to newer versions. Choosing LCA software that supports the requirements of a proposer uncertainty analysis is therefore essential. The uncertainty analysis presented in the illustrative case study is a screening level analysis and considers in its interpretation that uncertainty in the comparison of scenarios is overestimated due to lacking consideration of correlations.

There are two frequently used ways to compare the impact scores of two scenarios A and B, calculating the difference $A - B$, or the ratio A/B. When using the difference, the result can be $A - B < 0$ when A has a lower impact score than B ($A < B$), it can be $A - B = 0$ when $A = B$, or it can be $A - B > 0$ when $A > B$. The second way works similarly, with $A/B < 1$ when $A < B$, $A/B = 1$ when $A = B$, or $A/B > 1$ when $A > B$. In both cases the environmentally preferable option for a given impact category (i.e. compared impact scores) is easily identified. The uncertainty of the difference or ratio can be quantified using covariance or correlation coefficients, which can be assessed with both numerical and analytical uncertainty propagation methods. When using Monte Carlo simulation, it is also straightforward to calculate the above difference or ratio pairing the results from the iterations from each scenario. This will result in a number of iterations where $A > B$ and some where $A < B$ unless one scenario is always better than the alternative over its entire range of uncertainty. This can then be interpreted as the frequencies of each case, i.e. x% of iterations where $A > B$ and y% of iterations where $A < B$, with x and y representing the respective probability given that enough iterations where calculated. This means that it is possible to calculate the probability of A being environmentally preferable over B and vice versa as illustrated in Fig. 11.12. For example, if A is better than B in 25% of the simulated cases, there will be 75% where B is better than A. The conclusion may thus be that B is better than A with 75% likelihood, or in other words with a 25% probability to be wrong.

In a decision support context, the probability of one alternative being preferable over another is an essential measure of robustness of a recommendation and eventually an information that only uncertainty assessment can provide. If the decision is to choose one option over all other alternatives, there is a substantial added value for the decision maker if the probability for this to be wrong can be quantified. It helps, among other, to provide perspective on the robustness of an environmental gain of a certain option relative to other measures such as costs for example. If a higher investment is required but the probability that this really is an environmentally preferable option is very high, the investment may be easier to justify. Several authors demonstrated how to apply this in LCA (Hong et al. 2010; Wei et al. 2016) and the following section provides an example where this approach was also used to enhance the interpretation of results and express the robustness of the conclusions.

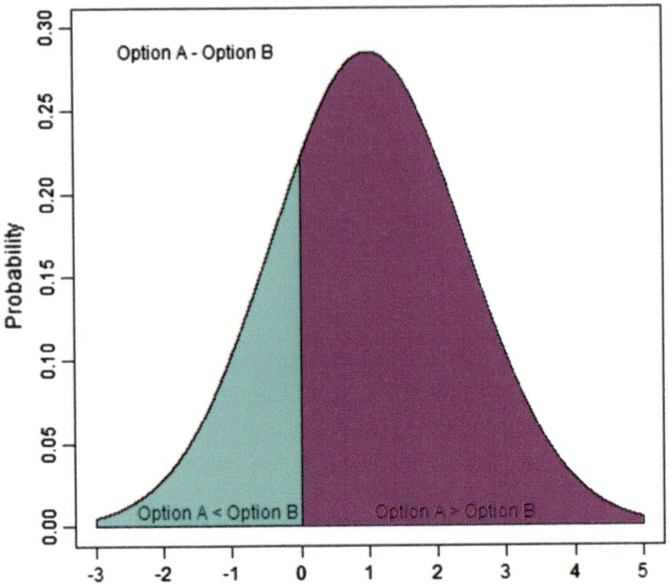

Fig. 11.12 Uncertainty of the difference between two scenarios *A* and *B*

The example is a real case comparing different, functionally equivalent solutions for hand dryers in public restrooms (Quantis 2009). The study compared the (1) XLERATOR Hand Dryer (high-speed air flow) to (2) conventional hand dryers (low-speed air flow), (3) paper towels with virgin paper, and (4) paper towels with 100% recycled paper. The functional unit was to dry 260,000 pairs of hands. The study was performed by Quantis' Boston office, commissioned by Excel Dryer Inc., underwent critical review according to ISO 14040/14044 and has been published (available via exceldryer.com). It is in many ways a classical LCA study, but what makes it stand out as an interesting example is that for climate change impacts, an uncertainty assessment was performed in order to determine the confidence in the conclusions regarding the preferable solution.

Using the analytical propagation method, output uncertainty was calculated for the climate change results of all four scenarios as shown in Fig. 11.13. Even though it may be tempting to compare the distributions directly, the latter do not consider dependency and thus overestimate the uncertainty of the difference between scenarios when comparing them. They do, however, indicate the uncertainty of each scenario individually with the XLERATOR showing the lowest spread, which is due to the fact that many primary data are used that the commissioner has direct access to, whereas the input data for alternative scenarios are estimated or taken from other sources and secondary data, thus increasing their uncertainty. The XLERATOR shows the lowest impact score and very little overlap with the uncertainty range of the alternative scenarios. This allows a first conclusion that it is very certain that this is the preferable alternative among the compared options,

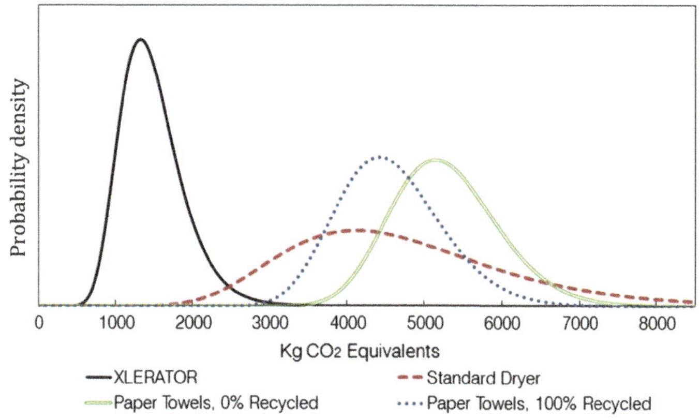

Fig. 11.13 Probability distributions (probability density functions) for the climate change impact scores of four compared alternatives to dry hands (Quantis 2009)

because the uncertainty range can only be smaller when considering the dependency of parameters between the alternatives.

In order to gain deeper insights into the uncertainty when comparing these alternatives, a paired comparison between two scenarios at a time was performed. A selection of the results is shown in Fig. 11.14. As discussed above, the ratio of two study results can be used to compare them and determine whether or not one of the two alternatives is environmentally preferable. It is clearly demonstrated that the XLERATOR consistently has the lowest impact score and that the probability that this is the wrong conclusion is virtually zero. In other words, according to the uncertainty analysis, it is 100% certain that the XLERATOR is the most preferable among all considered options regarding climate change impacts.

It is an important question to ask which aspects of uncertainty have been considered and how completely the uncertainty has been captured. If important sources of uncertainty that are independent between scenarios have been omitted, the uncertainty of the ratio between two scenarios may well be larger and the conclusion would be less robust. Assuring a complete consideration of important sources of uncertainty contributing to the difference between two scenarios is essential in order to fully trust the resulting measure of confidence in concluding the preference of one scenario over another.

The comparison of other scenarios provides examples of less certain outcomes. The comparison of standard dryer and virgin paper towels shows that a part of the resulting distribution of the ratio between both scenarios is larger than 1. According to the numerical results provided in the report, there is a 24% chance that virgin paper towels have a lower climate change impact than standard dryers. Consequently, there is a 76% chance that standard dryers are less impacting than virgin paper towels. When comparing standard dryer and recycled paper towels, the uncertainty distribution of the ratio between both is almost equally spread around 1, which means that there are about 50% chance for both possible conclusions. In that

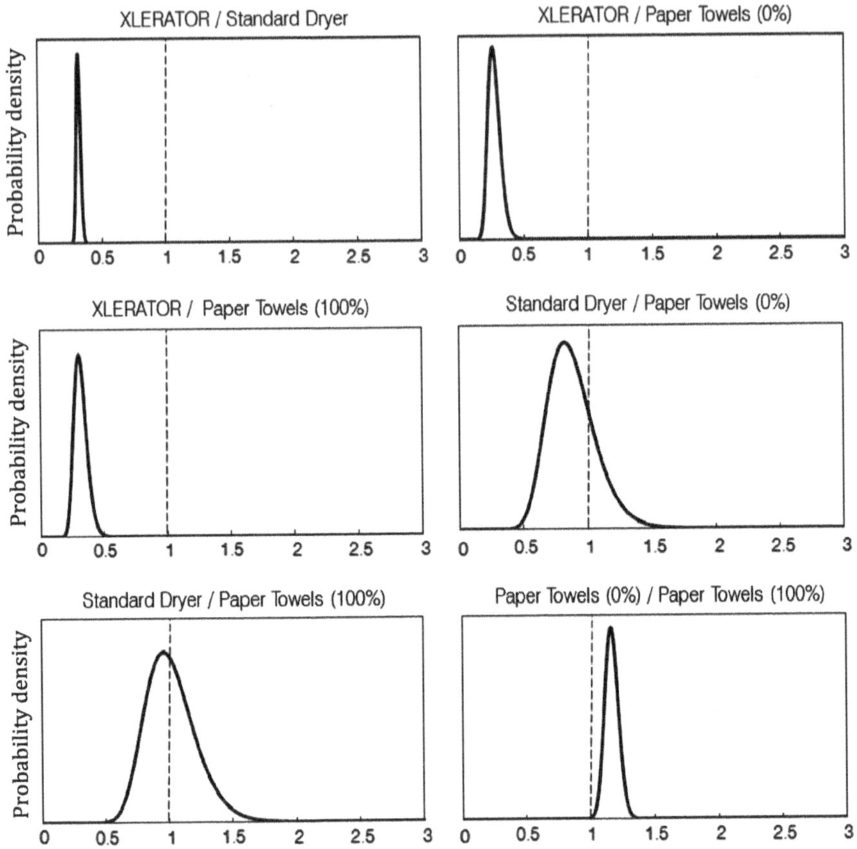

Fig. 11.14 Paired comparison of the climate change impact score ratio of alternative scenarios including uncertainty (Quantis 2009)

case, both scenarios have to be considered essentially equal and no conclusion regarding their (difference in) climate change impact can be drawn. Comparing the two paper towel options, it appears that recycled paper towels are the less impacting alternative, but the distribution of the ratio between both is close to 1. Additionally, the report states that a number of potentially important and independent uncertainties have not been quantified, such as "the methodological issues relating to allocating for recycled content and that the data used do not include impacts for the processing of the recycled paper". Hence, the range of the uncertainty distribution of the ratio between both scenarios may be larger and the conclusion of preference for the 100% recycled paper towels may be less robust.

In order to derive concrete conclusions if one option is better than (preferable to) another one, the observed differences need to be examined in statistical terms. The two most used statistical tools to examine if their difference is statistically significant are confidence intervals and hypothesis testing.

There is still room for improvement, but this study is an excellent example of how uncertainty analysis strengthens the robustness and the trust in the conclusions of an LCA. Besides showing the added value of uncertainty assessment and interpretation, it also illustrates the feasibility of quantifying and managing uncertainty in LCA.

11.5 Communication of Uncertainty

Besides quantifying and improving the robustness of an LCA and its conclusions, the question of how to communicate this beyond the practitioners directly involved in the study is fundamentally important and may often be more complex than anticipated. Like any communication it needs to be adapted to the target audience and will have to look very differently if targeting the general public, high-level decision/policy makers, or fellow practitioners and it will depend on the goal and scope of the LCA itself and, thus, differ if the goal was, e.g. to support the eco-design of a product or the overall environmental performance of a company. The following set of questions is useful to address in order to identify a meaningful uncertainty communication strategy:

1. Who is the target audience and how familiar is this audience with LCA and its aspects of uncertainty?
2. What exactly should be communicated in relation to uncertainty?
3. How should uncertainty results be represented?

11.5.1 Who? Identifying the Target Audience

Before choosing which uncertainty information should be conveyed, with how much detail and how exactly, it is essential to identify the target audience(s) of this information and adapt the communication strategy accordingly. Each potential target audience will understand and interpret uncertainty information differently in function of how familiar they are with underlying methodology, sources, types and meaning of uncertainty. There are many ways of classifying target audiences, but several main target groups (not necessarily always applying to all LCA reporting situations) may be:

- LCA experts, e.g. other practitioners, scientists, etc., who are very familiar with the subject. This may well be the easiest case, since little or no selection of uncertainty information, or a particularly adapted presentation will be required in most cases.
- Informed stakeholders with expertise regarding the LCA, the studied subject or the indicators considered (e.g. environmental, social, or economic), such as

NGOs, competitors, governmental agencies, etc. This target group will be able to access the core issues of an LCA and its uncertainty as long as some guidance and transparency regarding uncertainty are provided and the information is presented in a way that does not require in-depth expertise and routine.
- The general public, like NGOs, consumers, workers, or neighbours of a production site, will usually need as much pre-selection, pre-digestion and simplification of uncertainty information as possible.
- High-level decision makers in a company or national/international policy-context will not be familiar with technical details around the LCA study and uncertainty analysis. They have little time to spend on understanding any details and need to know quickly what the implications of the underlying uncertainties are for their decision(s). They may want to know which uncertainties are considered and how certain they can be regarding the robustness of the LCA results.
- Medium-level decision-makers such as regional or local policy-makers, or industrial production managers may require to be presented with uncertainty information somewhere in-between high-level decision-makers and the general public, depending on the context.
- The commissioner(s) of an LCA may fall into any of these groups and will have a particular interest in the uncertainty of its results.

It may well be that an LCA study needs to address several of these target groups and that a meaningful compromise needs to be found. A good way to deal with multiple target groups' needs is to prepare an adapted presentation for each target group, e.g. via an executive summary (for high-level) and a technical summary (for medium-level and informed stakeholders), or via dedicated reports or at least interpretation and discussion chapters for a given target group. The LCA report on window frames provided as an illustrative case study in Chap. 39 provides both an executive summary and a technical summary addressing different target groups for the report.

11.5.2 What? Selecting Which Information Is Relevant to Communicate

There are many aspects related to uncertainty that could be communicated but need to be selected depending on the target group of the information and what they can and need to do with it, but also considering the importance of transparency:

1. Assumptions and hypotheses underlying a study, including simplifications and generalisations;
2. Representativeness of information, models, and data used;
3. General level of scientific knowledge and understanding about important aspects of a study, particularly for new issues or approaches used;

4. Subjective, ethical or moral values and choices implicitly or explicitly included in the study;
5. Aspects that have not been considered (for whatever reason) but that may be important;
6. Types and sources of uncertainty that have been quantified;
7. Types and sources of uncertainty that have not been quantified but that are expected to be important contributors to overall uncertainty of results and/or conclusions;
8. How exactly uncertainties have been quantified and propagated;
9. Types of analyses that have been performed to consider uncertainty (e.g. sensitivity, uncertainty, uncertainty contribution, scenario analysis, etc.);
10. Uncertainty management and reduction strategies applied;
11. Robustness of the numerical results eventually including the quantitative uncertainty of some or all of them and a list of the most sensitive underlying assumptions and data;
12. Robustness of the conclusions and recommendations, eventually including quantitative measures and a list of the most sensitive underlying assumptions, data and choices;
13. Implications and consequences of the uncertainty underlying the results and/or conclusions.

It is important to keep in mind that communication of uncertainty does not necessarily imply its quantification using sophisticated methodology and substantial resources. The absolute minimum of a qualitative discussion of some or all aspects listed above can and should always be provided by a practitioner.

11.5.3 How? Representing Uncertainty Effectively

This section is largely inspired by a report from Wardekker et al. (2013), which nicely summarises the essential aspects around representing uncertainty. Although not specifically adapted to LCA, further details and insights beyond the selection in this chapter may be found there. When communicating LCA results, in which ever way, it is important to be aware that it is the responsibility of the author(s) (i.e. the practitioner, sometimes also the commissioner) to consider and adapt to the target audience. It is clearly insufficient to focus on a scientifically correct and complete presentation of results and related uncertainty, leaving the responsibility of their correct interpretation solely to the (target) audience. When choosing a way to express and represent uncertainty, it is thus important to keep in mind that the target audience may interpret it very differently than intended. Only using point estimates or deterministic results and conclusions, without mentioning any uncertainty, already bears the risk of unintentional interpretations. This will be even more the case when including uncertainty information, where it is well possible that referring to a low probability of an environmental consequence to occur may result in unintentional focus and unrest about this unexpected risk. It is also possible that evoking a high probability of adverse effects may not be noticed as an issue of concern, just because it was presented as something of a certain probability and not as the (almost) certain environmental consequence of an act or decision. In other words, the same uncertainty information may result in opposite interpretations by different readers. While this is difficult to fully foresee and avoid, paying attention to such details when preparing a presentation or report can help avoiding unintentional or wrong interpretation when considering the target audience's context and interpretation capacity.

Wording and phrasing are essential elements in this context. For example, a non-technical audience may not be familiar with the meaning and implication of terms like risk, probability or likelihood. The expression of uncertainty in a positive way versus a negative way can make an important difference. To illustrate this, the following two phrases express the same uncertainty information in an LCA comparing two alternatives *A* and *B*, but in a very different way: (1) "there is a 10% risk that choosing option *B* may be the wrong decision" versus "there is a chance of 90% that option *B* is the best option".

Besides paying attention to how an information is phrased (sent), it also plays an important role how the information is received, which Wardekker et al. (2013) describe via three effects of distortion:

- "Availability: matters that easily come to mind are generally regarded as occurring more frequently or more likely to occur than matters that are more obscure. A strong focus on a specific issue (in the media) may result in people regarding it as more likely to occur.
- Confirmation: once a view has been adopted, new information will be interpreted on the basis of this view. It is difficult to change people's views.
- Overconfidence: people are often too certain of their own judgement. This applies to the general public as well as to scientists."

The exact place in a report or presentation where uncertainty information is included is worth some consideration. Numerous options exist, but each solution may bear its particular risk of failure to communicate uncertainty, like a dedicated chapter stating all there is to state, may be easily ignored because it is little inviting to read, or an annex containing all relevant information may never be read, as it is not part of the main body of the report and therefore may not be considered relevant by some readers. It may be a good idea to spread uncertainty information meaningfully in different parts of the report, a concept referred to as progressive disclosure of information (PDI) which employs the concept of layers of information, distinguishing "outer layers (e.g. press release, summary, [oral presentations]) [that] refer to non-technical information, uncertainties integrated into the message, emphasis on context, implications and consequences" and "inner layers (e.g., appendices, background report, [or specific section like introduction, conclusion, recommendations]) [containing] detailed technical information, uncertainties discussed separately, emphasis on types, sources and the extent of uncertainty)"

(Wardekker et al. 2013). Different layers can be used that are adapted to specific target groups and uncertainty information to communicate. In any case, conclusions and recommendations should always directly include relevant and central information regarding uncertainty.

There are different, often complementary ways to present uncertainty information:

- Qualitatively (e.g. reporting sources of uncertainty and their potential influence on results);
- Descriptively (e.g. reporting central tendencies like mean and variability around the mean);
- Graphically (e.g. visualising uncertainty information in graphs);
- Numerically (e.g. reporting ranges, probability distributions of results values or statistical results).

Presenting uncertainty information in a verbal or descriptive way is useful, as it allows direct integration with the results and conclusions, especially for non-quantitative information and may be retained more easily than numerical information by most readers. It is particularly well suited for inclusion with outer layers (e.g. report summary). Such a description of uncertainties may be based on a quantified evaluation or even just on a qualitative appreciation of uncertainty. In any case, it is important to keep in mind that many terms typically used to describe uncertainty are quite imprecise and prone to vary in perception and interpretation among individuals, e.g. large, small, important, significant, etc. It is essential to use these terms consistently with the same meaning throughout a report and that they match numerical results, if available. They may even be explicitly defined, e.g. very likely = 90–99% probability, likely = 80–89% and so on.

If quantified, uncertainty information can also be communicated numerically, e.g. in tables, as standard deviations, minimum and maximum bounds, ranges, uncertainty and confidence intervals, probabilities, comparison with other studies or measurements, etc. This is useful especially for application in inner layers of information, such as a report appendix. A frequent mistake in this case is the communication of results and quantified uncertainties with a "false precision" showing too many digits. This practice suggests a very precise quantification of uncertainty that is most likely not defendable in an LCA context. For example, considering a typical standard deviation of a global warming impact score, a value of 2.49678 is essentially the same as 2.5 and in fact even the same as 3. The opposite may also exist, when a "false imprecision" is used to express numerical results so vaguely that they could mean anything, or are immune to criticism, but not very helpful for decision support.

Graphical representation of uncertainty can be provided in many different ways, e.g. using error (or uncertainty) bars or bands, box plots, probability distributions, coefficients of variation, confidence intervals, etc. This has the advantage that a lot of information can be aggregated and shown in a concise and structured way, allowing to capture a lot of uncertainty information in a short time and single graph.

This is illustrated in Fig. 11.15 which shows an example of a box plot (or whisker plot) of the spread of freshwater ecotoxicity characterisation factors for 2499 organic chemicals and 4 emission compartments from USEtox 2.02 (see Chap. 10 for further information regarding freshwater ecotoxicity characterisation factors). The boxes efficiently illustrate that 90% of the characterisation factors fall within the range of five to six orders of magnitude, whereas the difference between the lowest and highest characterisation factors (grey dots) is in the range of 16–19 orders of magnitude. Although the actual shape of the uncertainty distribution cannot the seen, it is visible that the distribution is skewed towards higher values with the median (the value at 50%) being in the upper range of values and not in the centre.

However, graphical representation of uncertainty also bears the risk of being suggestive, easily misinterpreted, or too complex. One of the most common ways to represent uncertainty is to plot the probability distributions of the output variables, as presented in Fig. 11.13 and discussed in Sect. 11.4.2. Alternatively to PDFs in Fig. 11.13, the Cumulative Distribution Functions (CDFs) of the outputs could be derived in order to characterise uncertainty. Another tool to represent uncertainty are the so-called probability boxes, based on a probability bounds approach (Karanki et al. 2009). The book "Environmental Decisions in the Face of Uncertainty" from the Institute of Medicine (IOM 2013) contains a useful overview and more in-depth discussion on graphical and other representations of uncertainty. For example, a frequent mistake when representing uncertainty in LCA is the use of error bars. Figure 11.16 illustrates this with error bars that we added to the original graph from the Quantis study discussed above, so that the resulting graph below represents uncertainty in an alternative way to Fig. 11.13. This representation of uncertainty can be seen in numerous LCA publications and presentations. The error bars here represent the absolute uncertainty of each compared option, but they do not consider interdependence of uncertainties between scenarios. However, by presenting them next to each other, Fig. 11.16 suggests that the error bars can be

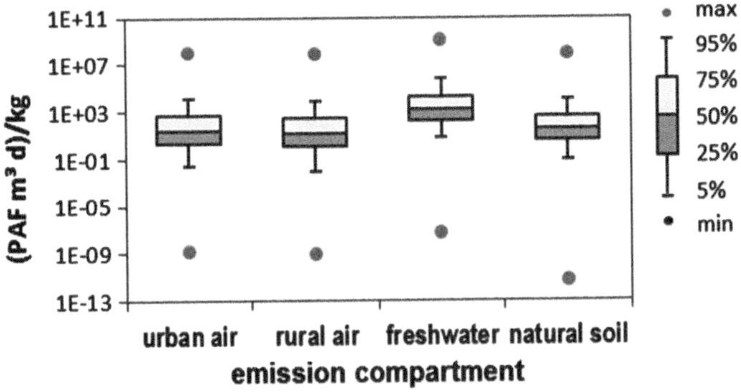

Fig. 11.15 *Box* or *whisker plot* of freshwater ecotoxicity characterisation factors for 2499 organic chemicals and 4 emission compartments from USEtox 2.02

directly compared among each other in order to determine if the uncertainty allows to visually distinguish these options. As discussed above, only the uncertainty of the difference (or ratio) for each paired comparison among these options (which will be smaller while only considering the uncertainty of the difference (or ratio) between two options) will truly allow to determine whether both options are distinguishable or essentially equal. The useful way of using error bars in this example would therefore be to present one for each pairing of these compared options, parallel to Fig. 11.14.

When using graphs, the scale of an axis should always reflect the underlying uncertainty. This is particularly important in LCA, where many impact scores may have an uncertainty spanning from one to several orders of magnitude, in which case it would be misleading to present them on a linear scale. In such cases, the results should preferably be shown using a log-scale, which will only emphasise larger differences between impact scores. Contrary to a frequent perception, this has nothing to do with data manipulation, since scores can still be identified by their exact value. It simply avoids over-exaggeration of very small differences that may look very large on a linear scale while (almost) disappear on a log-scale. A similar effect of over-exaggeration is achieved when zooming into a certain range of an axis, e.g. only showing the highest values from 80 to 100%, which will show differences between two points as much larger compared to the full range of the axis.

As indicated above, these approaches are complementary and should be used as such. Sometimes a repetition of the same (important) information via two different ways and at two different places in a report may be preferable over a concise, non-repetitive communication.

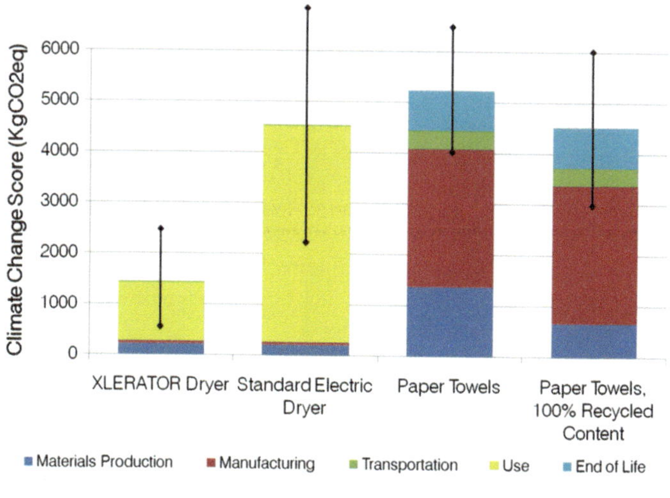

Fig. 11.16 Uncertainty bars for the climate change impact scores of four compared alternatives to dry hands (figure adapted from Quantis (2009) by adding error bars for illustrative purposes)

11.6 Management of Uncertainty

The strategy of how to consider and manage uncertainties in an LCA study depends on a number of factors that will determine what is feasible. The most important limitation is likely the availability of resources (time and/or budget) to collect additional information in order to quantify, represent and reduce uncertainty. Accessibility and level of operationalisation of the technical aspects of uncertainty assessment (e.g. databases providing default uncertainties for background LCI data and LCIA characterisation factors, LCA software providing ways to efficiently propagate uncertainties) is also frequently named as a potential barrier. In any case, there is always a minimum of uncertainty management that will be feasible without requiring important resources. In many scientific fields, uncertainty is managed using a tiered approach with each tier (or level of detail) progressively increasing the requirements and sophistication of uncertainty assessment and management. A particular advantage of such an approach is that it allows an iterative improvement and refinement of uncertainty management from a first qualitative listing of uncertainty sources, to a first quantitative estimation and screening, up to a sophisticated full uncertainty assessment as a study advances. This type of approach caters nicely to the iterative nature of LCA (see Sect. 6.3) and allows the LCA practitioner to adapt the extent of uncertainty management in a study to the available resources, instead of suggesting that uncertainty management always has to be done using the most complex approaches or not at all if resources are too limited to allow for a quantitative approach.

An example for such a tiered approach is the Guidance on Characterizing and Communicating Uncertainty in Exposure Assessment from the World Health Organisation (WHO 2008). It proposes four progressive tiers with increasing complexity from tier 0 (the absolute minimum) to tier 3 (the most sophisticated level):

Tier 0: Screening uncertainty analysis
Tier 1: Qualitative uncertainty analysis
Tier 2: Deterministic uncertainty analysis
Tier 3: Probabilistic uncertainty analysis

While the details of this framework are adapted to chemical exposure assessment, its underlying principle of iteratively increasing sophistication and complexity is a useful inspiration for LCA. Figure 11.17 shows the different levels of detail for each tier, from no uncertainty analysis (point estimate) at the bottom to probabilistic uncertainty analysis at the top.

An expert working group of the UNEP-SETAC Life Cycle Initiative on uncertainty management in LCA drafted a similar framework for LCA during a series of workshops between 2009 and 2012, which is a useful starting point

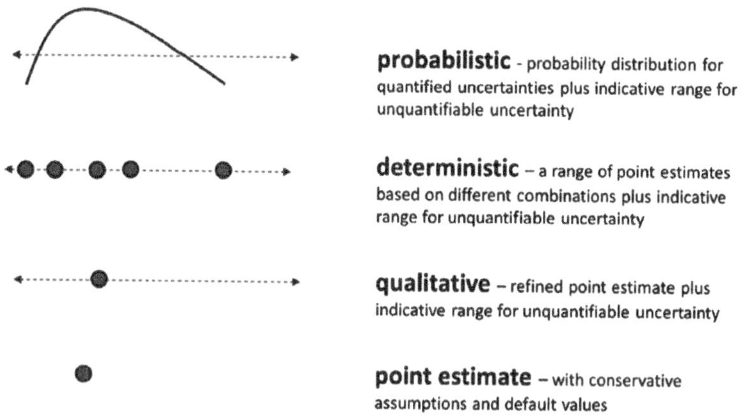

Fig. 11.17 Levels of detail for tiered uncertainty management strategies [taken from Paparella et al. (2013)]

towards integration of uncertainty management into LCA practice. They proposed five tiers:

Tier 0: Minimum transparency with a clear definition of what is considered a notable difference between scenarios for each impact category;

Tier 1: Screening level focusing on identification of important sources of parameter uncertainty providing information on importance and sensitivity of parameters, choices, assumptions, etc.;

Tier 2: Qualitative and semi-quantitative uncertainty assessment of important sources of uncertainty with systematic identification and description of uncertainties for all parameters, choices and assumptions including parameter and scenario uncertainty;

Tier 3: Quantitative uncertainty assessment of all sources of uncertainty with systematic quantification of uncertainties and variability for all parameters, choices and assumptions accounting for all quantifiable uncertainties;

Tier 4: Fully probabilistic LCA representing all relevant sources of influence by fully characterised uncertainty and variability separately.

In essence, different levels of sophistication are possible when establishing a strategy to integrate uncertainty management into a study and it is not always the most sophisticated level that is required. Compared to completely ignoring uncertainty, even a basic (e.g. qualitative) consideration is already better than nothing and a good and essential first step to pinpoint sources of uncertainty in the results of any LCA study. This helps to be conscious about potential pitfalls and misinterpretation when making a decision based on the conclusions of a study. It should also be noted that a tiered uncertainty assessment framework essentially serves as an orientation providing coherence for different levels of sophistication of uncertainty assessment.

It groups those elements of an uncertainty assessment that can be combined meaningfully on each level.

11.7 Perspectives

Uncertainty and variability are inherent properties of LCA, all its models, data, assumptions, and choices that are required when performing an LCA. Uncertainty is not the enemy, but it is unavoidable and its assessment can be helpful when put to good use for improving and interpreting LCA results. Uncertainty and its reduction is the very reason for the iterative nature of LCA and should hence be used as a guiding principle for the changes applied during each iteration of an LCA. Uncertainty and variability have many sources, some of which are quantifiable, while others are not, but all need to be considered when interpreting and discussing results and the robustness of a conclusion.

In order to be successfully applied in LCA, uncertainty assessment requires some knowledge of the underlying principles and methods as well as a set of tools supporting:

1. Quantification and storage of uncertainty, variability, and correlation or interdependence of inputs, models, assumptions, etc.
2. Propagation of input uncertainties to model output uncertainty
3. Tools for sensitivity, uncertainty, uncertainty contribution analysis and scenario comparison
4. Skilled interpretation and communication of relevant uncertainty information

Even though uncertainty assessment is an additional procedure to handle and provide resources for when conducting an LCA, it has multiple uses that will help ensuring that resources spent on the iterative improvement of the study actually contribute to a tangible improvement in uncertainty of the results and their enhanced interpretation in order to provide robust conclusions. Uncertainty assessment can notably be used to:

- Identify sources of uncertainty that dominantly contribute to the uncertainty of results
- Effectively target the iterative improvement of data, models and assumptions towards those elements that dominate the result(s) and their uncertainty
- Identify processes and elementary flows where archetypical or spatially explicit LCI and LCIA data will significantly reduce the uncertainty of the results due to the integration of spatial (or temporal) variability into the LCA
- Enhance the interpretation of results, e.g. which alternatives are truly different and which are not
- Quantify the confidence in the robustness of a conclusion or the probability of being wrong

Not only does the assessment and management of uncertainty in LCA provide a lot of opportunities and advantages, but ignoring it actually bears potentially important risks. For example, resources spent to improve the study may be inefficiently used when improving data and models with limited contribution to result uncertainty (e.g. when results are not sensitive to changes in inputs). Conclusions drawn from deterministic results may not only lack robustness but actually be misleading (e.g. when differences between results are not significant, i.e. falling within the uncertainty ranges of results).

From today's perspective, a lot can be done already to consider uncertainty, with many LCI databases and the first LCIA methods providing uncertainty estimates for their data, and most LCA software providing functionality to propagate those into the results. When exploring those options, it is important to be aware of the limitations that most if not all LCA software (while writing this book in 2016) does not provide the possibility to consider LCIA uncertainties, which may not always be obvious to the user. Running the uncertainty analysis will thus essentially propagate the uncertainties from the LCI database and result in a very incomplete quantification of uncertainty that may be missing many important sources on the LCIA side. Using this kind of uncertainty information to establish whether or not two alternatives have significantly different impact scores may still provide misleading conclusions and a false impression on their robustness due to its bias towards LCI uncertainty. To overcome this limitation, updates of LCIA methods will (increasingly) provide uncertainty estimates for characterisation factors (Bulle et al., in review).

With high uncertainties being a frequent, critical argument towards LCA, it is worth asking if LCA results are actually more uncertain than those from other assessment tools. No doubt that the precision of LCA results will be inferior to that of many other environmental assessment tools, especially the local and site-specific ones. This has a lot to do with scale, since LCA typically models entire supply chains that will usually be global, involve many processes about which little information is available, covering a broad range of environmental indicators and impact categories, and often spanning considerable time periods (defined in the duration of the functional unit) to be represented. The combination of large spatiotemporal scales and the complexity due to broad inventory flow and impact coverage, which is unique to LCA among environmental assessment tools, is the source of a lot of variability and uncertainty due to e.g. aggregating over larger spatial or temporal space and is thus simply a function of the space considered and data available. However, as discussed in this chapter, contrary to most environmental assessment tools, LCA does not attempt to predict absolute impacts, but rather focuses on the relative difference in potential impacts between alternatives, although exceptions exist, such as Environmental Product Declarations (EPD) which are "stand-alone" environmental profiles. Any systematic error and source of variability or uncertainty will usually have little influence on the uncertainty of the difference between alternatives. Therefore, the focus in LCA is accuracy and not necessarily precision.

While this chapter provides an overview of a range of aspects around uncertainty management in LCA, we recommend the cited literature for those readers looking for more in-depth insights into specific aspects. For further reading beyond literature cited above we recommend the following: Deeper insights on uncertainty representation in the context of LCA were published by Heijungs and Frischknecht (2005). For log-normally distributed parameters, Strom and Stansbury (2000) discuss the determination of distribution information from minimal literature information and provide a comprehensive overview on log-normal distributions. Heijungs and Kleijn (2001) further discuss contribution analysis, perturbation analysis, uncertainty analysis, comparative analysis, and discernibility analysis. De Schryver et al. (2011) explore how value choices in LCIA influence the uncertainty of (human health) characterisation factors. Clavreul et al. (2013) combine probability and possibility theories to represent stochastic and epistemic uncertainties in a consistent manner in LCA. Even though it does not discuss life cycle assessments and has a more risk-assessment based focus, a useful read regarding environmental decision making under uncertainty including aspects of communication and management of uncertainty is the book "Environmental Decisions in the Face of Uncertainty" from the Institute of Medicine (IOM 2013) which is freely available via The National Academies Press (NAP) website.

When discussing LCA indicators and results, we should be at least as critical, if not even more critical when presented with no or small uncertainties as we are when presented with large, but properly quantified uncertainties. Or to say it more eloquently with the words of physicist and Nobel laureate Richard P. Feynman: "What is not surrounded by uncertainty cannot be the truth".

Acknowledgements The authors gratefully acknowledge the support from Jon Dettling (Quantis) for allowing us to use their study as an example.

References

Björklund, A.: Survey of approaches to improve reliability in LCA. Int. J. Life Cycle Assess. **7**, 64–72 (2002). doi:10.1007/BF02978849

Bulle, C., Margni, M., Kashef-Haghighi, S., Boulay, A.-M., Bourgault, G., De Bruille, V., Cao, V., Fantke, P., Hauschild, M.Z., Henderson, A., Humbert, S., Kounina, A., Laurent, A., Levasseur, A., Liard, G., Patouillard, L., Rosenbaum, R.K., Roy, P.-O., Shaked, S., Jolliet, O.: IMPACT World+: A Globally Regionalized Life Cycle Impact Assessment Method (in review)

Ciroth, A.: Uncertainties in life cycle assessment. Int. J. Life Cycle Assess. **9**, 141–142 (2004). doi:10.1007/BF02994186

Ciroth, A., Fleischer, G., Steinbach, J.: Uncertainty calculation in life cycle assessments—A combined model of simulation and approximation. Int. J. Life Cycle Assess. **9**, 216–226 (2004). doi:10.1007/BF02978597

Ciroth, A., Muller, S., Weidema, B., Lesage, P.: Empirically based uncertainty factors for the pedigree matrix in ecoinvent. Int. J. Life Cycle Assess. **21**, 1338–1348 (2016). doi:10.1007/s11367-013-0670-5

Clavreul, J., Guyonnet, D., Tonini, D., Christensen, T.H.: Stochastic and epistemic uncertainty propagation in LCA. Int. J. Life Cycle Assess. **18**, 1393–1403 (2013). doi:10.1007/s11367-013-0572-6

De Schryver, A.M., van Zelm, R., Humbert, S., Pfister, S., McKone, T.E., Huijbregts, M.A.J.: Value choices in life cycle impact assessment of stressors causing human health damage. J. Ind. Ecol. **15**, 796–815 (2011). doi:10.1111/j.1530-9290.2011.00371.x

EC-JRC: European Commission-Joint Research Centre—Institute for Environment and Sustainability: International Reference Life Cycle Data System (ILCD): Handbook—General guide for Life Cycle Assessment—Detailed guidance. First edition March 2010. EUR 24708 EN. Publications Office of the European Union, Luxembourg (2010)

Fantke, P., Wieland, P., Juraske, R., Shaddick, G., Itoiz, E.S., Friedrich, R., Jolliet, O.: Parameterization models for pesticide exposure via crop consumption. Environ. Sci. Technol. **46**, 12864–12872 (2012). doi:10.1021/es301509u

Funtowicz, S.O., Ravetz, J.R.: Uncertainty and Quality in Science for Policy. Kluwer Academic Publishers, Dordrecht (1990)

Greenland, S., Senn, S.J., Rothman, K.J., Carlin, J.B., Poole, C., Goodman, S.N., Altman, D.G.: Statistical tests, P values, confidence intervals, and power: a guide to misinterpretations. Eur. J. Epidemiol. **31**, 337–350 (2016). doi:10.1007/s10654-016-0149-3

Groen, E.A., Heijungs, R.: Ignoring correlation in uncertainty and sensitivity analysis in life cycle assessment: what is the risk? Environ. Impact Assess. Rev. **62**, 98–109 (2017). doi:10.1016/j.eiar.2016.10.006

Groen, E.A., Heijungs, R., Bokkers, E.A.M., de Boer, I.J.M.: Methods for uncertainty propagation in life cycle assessment. Environ. Model Softw. **62**, 316–325 (2014). doi:10.1016/j.envsoft.2014.10.006

Hauschild, M.Z., Potting, J.: Spatial Differentiation in Life Cycle Impact Assessment: The EDIP2003 Methodology. The Danish Ministry of the Environment, Environmental Protection Agency, Copenhagen (2005)

Heijungs, R.: A generic method for the identification of options for cleaner products. Ecol. Econ. **10**, 69–81 (1994). doi:10.1016/0921-8009(94)90038-8

Heijungs, R.: Identification of key issues for further investigation in improving the reliability of life-cycle assessments. J. Clean. Prod. **4**, 159–166 (1996). doi:10.1016/S0959-6526(96)00042-X

Heijungs, R.: The use of matrix perturbation theory for addressing sensitivity and uncertainty issues in LCA. In: Proc. Fifth Int. Conf. EcoBalance—Pract. tools thoughtful Princ. Sustain., 6–8 Nov 2002, Tsukuba, Japan (2002)

Heijungs, R.: Sensitivity coefficients for matrix-based LCA. Int. J. Life Cycle Assess. **15**, 511–520 (2010). doi:10.1007/s11367-010-0158-5

Heijungs, R., Frischknecht, R.: Representing statistical distributions for uncertain parameters in LCA: relationships between mathematical forms, their representation in EcoSpold, and their representation in CMLCA. Int. J. Life Cycle Assess. **10**, 248–254 (2005). doi:10.1065/lca2004.09.177

Heijungs, R., Huijbregts, M.A.J.: A review of approaches to treat uncertainty in LCA. Trans. 2nd Bienn. In: Meet. Int. Environ. Model. Softw. Soc. iEMSs, Osnabrück, Ger, pp 332–339 (2004)

Heijungs, R., Kleijn, R.: Numerical approaches towards life cycle interpretation five examples. Int. J. Life Cycle Assess. **6**, 141–148 (2001). doi:10.1007/BF02978732

Heijungs, R., Lenzen, M.: Error propagation methods for LCA—a comparison. Int. J. Life Cycle Assess. **19**, 1445–1461 (2014). doi:10.1007/s11367-014-0751-0

Heijungs, R., Suh, S., Kleijn, R.: Numerical approaches to life cycle interpretation—The case of the Ecoinvent'96 database. Int. J. Life Cycle Assess. **10**, 103–112 (2005). doi:10.1065/lca2004.06.161

Hoekstra, R., Morey, R.D., Rouder, J.N., Wagenmakers, E.-J.: Robust misinterpretation of confidence intervals. Psychon. Bull. Rev. **21**, 1157–1164 (2014). doi:10.3758/s13423-013-0572-3

Hong, J., Shaked, S., Rosenbaum, R.K., Jolliet, O.: Analytical uncertainty propagation in life cycle inventory and impact assessment: application to an automobile front panel. Int. J. Life Cycle Assess. **15**, 499–510 (2010). doi:10.1007/s11367-010-0175-4

Huijbregts, M.A.J.: Application of uncertainty and variability in LCA Part I: a general framework for the analysis of uncertainty and variability in life cycle assessment. Int. J. Life Cycle Assess. **3**, 273–280 (1998). doi:10.1007/BF02979835

Imbeault-Tetreault, H., Jolliet, O., Deschênes, L., Rosenbaum, R.K.: Analytical propagation of uncertainty in LCA using matrix formulation. J. Ind. Ecol. **17**, 485–492 (2013). doi:10.1111/jiec.12001

IOM: Environmental Decisions in the Face of Uncertainty. The National Academies Press, Washington, DC (2013)

IPCC: Climate change 2007—the physical science basis. Intergovernmental Panel on Climate Change (2007)

Karanki, D.R., Kushwaha, H.S., Verma, A.K., Ajit, S.: Uncertainty analysis based on probability bounds (P-box) approach in probabilistic safety assessment. Risk Anal. **29**, 662–675 (2009). doi:10.1111/j.1539-6924.2009.01221.x

Lloyd, S.M., Ries, R.: Characterizing, propagating, and analyzing uncertainty in life-cycle assessment: A survey of quantitative approaches. J. Ind. Ecol. **11**, 161–179 (2007). doi:10.1162/jiec.2007.1136

Morgan, M.G., Henrion, M.: Uncertainty: A Guide Dealing with Uncertainty in Quantitative Risk and Policy Analysis. Cambridge University Press, Cambridge (1990)

Muller, S., Lesage, P., Ciroth, A., Mutel, C., Weidema, B.P., Samson, R.: The application of the pedigree approach to the distributions foreseen in ecoinvent v3. Int. J. Life Cycle Assess. **21**, 1327–1337 (2016). doi:10.1007/s11367-014-0759-5

Paparella, M., Daneshian, M., Hornek-Gausterer, R., Kinzl, M., Mauritz, I., Muhlegger, S.: Uncertainty of testing methods—what do we (want to) know? ALTEX **30**, 131–144 (2013)

Pólya, G.: Über den zentralen Grenzwertsatz der Wahrscheinlichkeitsrechnung und das Momentenproblem. Math. Z. **8**, 171–181 (1920). doi:10.1007/BF01206525

Quantis: Comparative Environmental Life Cycle Assessment of Hand Drying Systems: The XLERATOR Hand Dryer, Conventional Hand Dryers and Paper Towel Systems. Salem, MA (2009)

Read, C.: Logic: Deductive and Inductive, 4th edn. Simkin and Marshall, London (1920)

Ross, S.: Simulation, 5th edn. Academic Press, Cambridge (2012)

Strom, D.J., Stansbury, P.S.: Determining parameters of logs from minimal information. Am. Ind. Hyg. Assoc. J. **61**, 877–880 (2000). doi:10.1080/15298660008984601

van Zelm, R., Huijbregts, M.A.J.: Quantifying the trade-off between parameter and model structure uncertainty in life cycle impact assessment. Environ. Sci. Technol. **47**, 9274–9280 (2013). doi:10.1021/es305107s

Walpole, R.E., Myers, R.H., Myers, S.L., Ye, K.: Probability and Statistics for Engineers and Scientists, 9th edn. Prentice Hall, Englewood Cliffs (2012)

Wardekker, J.A., Kloprogge, P., Petersen, A.C., Janssen, P.H.M., van der Sluijs, J.P.: Guide for Uncertainty Communication, PBL Netherlands Environmental Assessment Agency, The Hague, The Netherlands (2013)

Wei, W., Larrey-Lassalle, P., Faure, T., Dumoulin, N., Roux, P., Mathias, J.-D.: Using the reliability theory for assessing the decision confidence probability for comparative life cycle assessments. Environ. Sci. Technol. **50**, 2272–2280 (2016). doi:10.1021/acs.est.5b03683

Weidema, B.P.: Avoiding or ignoring uncertainty. J. Ind. Ecol. **13**, 354–356 (2009). doi:10.1111/j.1530-9290.2009.00132.x

Weidema, B.P., Wesnæs, M.S.: Data quality management for life cycle inventories—an example of using data quality indicators. J. Clean. Prod. **4**, 167–174 (1996). doi:10.1016/S0959-6526(96)00043-1

Weidema, B.P., Frees, N., Petersen, E., Øllgaard, H.: Reducing uncertainty in LCI, Danish Environmental Protection Agency, Copenhagen, Denmark (2003)
WHO (2008) Guidance Document on Characterizing and Communicating Uncertainty in Exposure Assessment. International Programme on Chemical Safety (IPCS), World Health Organization (WHO), Geneva

Author Biographies

Ralph K. Rosenbaum LCA expert and environmental modeller focusing on LCIA development since early 2000s. Contributed to several UNEP/SETAC working groups towards global harmonisation of LCA methodology. Interested in LCIA modelling of emissions and water/soil resource use, operationalisation of uncertainty management and spatial differentiation.

Stylianos Georgiadis Background in Applied Mathematics, focusing on stochastic and statistical modelling with application in urban water systems, food risk assessment, reliability and queueing models. Interested in uncertainty quantification in risk management, life cycle assessment and decision analysis.

Peter Fantke Develops methods for LCIA, health impact assessment and chemical alternatives assessment since 2006. Has contributed to UNEP/SETAC LCIA working groups and is USEtox Manager. Interested in quantifying and characterising chemical emissions, uncertainty analysis, consumer exposure, chemical substitution and model parameterisation.

Chapter 12
Life Cycle Interpretation

Michael Z. Hauschild, Alexandra Bonou and Stig Irving Olsen

Abstract The interpretation is the final phase of an LCA where the results of the other phases are considered together and analysed in the light of the uncertainties of the applied data and the assumptions that have been made and documented throughout the study. This chapter teaches how to perform an interpretation. The process of interpretation starts with identification of potentially significant issues in the previous stages of goal and scope definition, inventory analysis and impact assessment, and examples of potential significant issues are given for each phase. The significance is then determined by checking completeness, sensitivity and consistency for each of these identified issues. The outcome is used to inform previous phases on the needs for strengthening the data basis of the study, and where this is not possible to reconsider the goal and scope definition of the study. Finally, guidance is given on how to draw conclusions based on the previous steps of the interpretation, qualify the conclusions in terms of their robustness, and develop recommendations based on the results of the study.

Learning objectives After studying this chapter, the reader should be able to:

- Explain the purpose of interpretation and its relationships to the other phases of the LCA.
- Explain what is meant by "significant issues" and give examples of potential significant issues from each of the methodological phases.
- Describe procedures to identify significant issues.
- Explain how sensitivity analysis and uncertainty information is used in combination to focus the data collection in previous phases of the LCA and to qualify the conclusions that are drawn from the results of the study.

M.Z. Hauschild (✉) · A. Bonou · S.I. Olsen
Division for Quantitative Sustainability Assessment, Department of Management Engineering, Technical University of Denmark, Kongens Lyngby, Denmark
e-mail: mzha@dtu.dk

M.Z. Hauschild et al. (eds.), *Life Cycle Assessment*,
DOI 10.1007/978-3-319-56475-3_12

12.1 Introduction

Interpretation is the phase of the LCA where the results of the other phases are considered together and analysed in the light of the uncertainties of the applied data and the assumptions that have been made and documented throughout the study. The outcome of the interpretation should be conclusions or recommendations that (1) respect the intentions of the goal definition and the restrictions that this imposes on the study through the scope definition and (2) take into account the appropriateness of the functional unit and system boundaries. The interpretation should present the conclusions of the LCA in an understandable way and help the users of the study appraise their robustness and potential weaknesses in light of any identified studylimitations.

Central elements of the interpretation phase such as sensitivity analysis and uncertainty analysis are also applied throughout the LCA process together with impact assessment tools as part of the iterative loops which are used in the drawing of boundaries and the collection of inventory and impact assessment data (see Chaps. 8–10). A more detailed presentation of these elements is given in Chap. 11.

The interpretation proceeds through three steps as illustrated in Fig. 12.1.

1. The significant issues (key processes and assumptions, most important elementary flows) from the other phases of the LCA are identified (see Sect. 12.2).

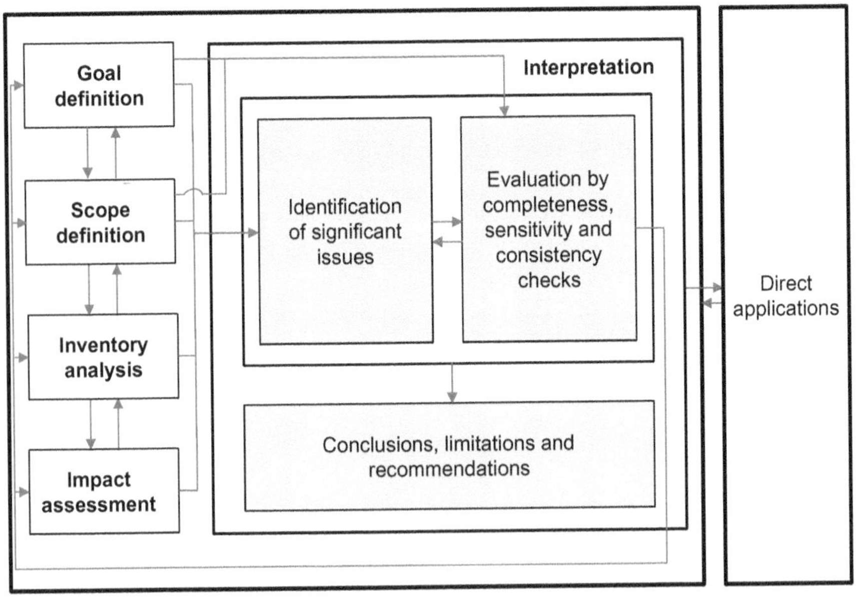

Fig. 12.1 The elements of the interpretation phase and their relations to each other and to the other phases of the LCA (revised from ISO 2006a, b)

2. These issues are evaluated with regard to their influence on overall results of the LCA and the completeness and consistency with which they have been handled in the study (see Sect. 12.3).
3. The results of the evaluation are used in the formulation of conclusions and recommendations from the study (Sect. 12.4).

In cases where the study involves comparison of two or more systems, there are additional considerations to be included in the interpretation (Sect. 12.5).

12.2 Identification of Significant Issues

The purpose of the first element of the life cycle interpretation is to analyse the results of earlier phases of the LCA in order to determine the most environmentally important issues, i.e. those issues that have the potential to change the final results of the LCA. The significant issues can be methodological choices and assumptions, inventory data for important life cycle processes, and/or characterisation, normalisation or weighting factors used in the impact assessment. The practitioner is encouraged to prepare a list of such choices during the practical execution of the LCA, the definition of goal and scope, the modelling of the product system and the impact assessment, to help with their identification (see for example reporting recommendations for life cycle inventory phase in Sect. 9.7). Table 12.1 provides examples of such influential issues.

As discussed in Chap. 11, sensitivity analysis can be performed as a contribution analysis where the contribution from each process or stage to the total results for an impact category is quantified and expressed. It can also be done as a dominance analysis, where the processes or stages are ranked according to their relative share in the total impact.

The identification of significant issues draws on the sensitivity analysis activities in the evaluation element of the interpretation phase in combination with information about potential key assumptions and uncertainty ranges for potential key numbers in inventory analysis and impact assessment. At the same time, the evaluation element takes the identified significant issues as an important input. The two elements are thus performed in iteration.

In the illustrative case on window frames in Chap. 39, life cycle impacts are dominated by the use stage in all impact categories for all four window frame designs. Parameters related to the use stage, such as the modelled heat loss, the assumed mix of heating sources, the LCI processes used to represent each heat conversion technology in the heat mix, and the relevant characterisation factors and normalisation references involved in the impact assessment were thus identified as significant issues.

Table 12.1 Examples of significant issues

What to look for	How to identify significant issues
Goal and Scope definition—methodological choices and assumptions	
Functional unit	Choice of functional unit, system expansion (assumption of alternative/replaced technologies), allocation model and setting of system boundaries are discrete choices that can be checked by running the different possibilities as scenarios and comparing the results to determine their influence on the final outcome and conclusions
Handling of multifunctional processes – System expansion – Allocation criteria	
Cut-off decisions and boundary settings	
Inventory analysis—data for product system processes	
Data for activities occurring in many parts of the product system, e.g. transportation or energy transformation processes	Sensitivity analysis is performed by varying the single issue, or in case of interdependency by joint variation of the concerned issues, and analysing their influence on the outcome of the study The range of variation applied for a given issue should reflect the uncertainty by which it is accompanied
Data for key processes: processes that contribute substantially to the environmental impact of the product system in one or more impact categories	
Data for key elementary flows: processes that contribute substantially to the overall results for an impact category	
Impact categories that dominate the total impacts from the product system	
Impact assessment factors	
Characterisation or normalisation factors used in the impact assessment	Sensitivity analysis is performed by varying the single issue, or in case of interdependency by joint variation of the concerned issues, and analysing their influence on the outcome of the study The range of variation applied for a given issue should reflect the uncertainty by which it is accompanied
Choice of impact assessment method and selection of impact categories	Other impact assessment methods and potentially omitted impact categories may be tested to see if they give different outcomes of the study

12.3 Evaluation

The evaluation element establishes the basis for the conclusions and recommendations that can be formulated in the final element of the interpretation (see Sect. 12.4). It is performed in an iterative interaction with the identification of key issues in order to determine the reliability and stability of the results from the identification element.

Like the identification of key issues, the evaluation covers the results from the earlier phases of the LCA, the inventory analysis and the impact assessment, in accordance with the goal and scope of the study, with focus on the significant issues identified among methodological choices and data.

The outcome of the evaluation is crucial to determine the strength of the conclusions and recommendations from the study, and it must therefore be presented in a way that gives the commissioner and user of the study a clear understanding of the outcome.

The evaluation involves:

- Completeness check.
- Sensitivity analysis in combination with uncertainty analysis.
- Consistency check.

12.3.1 Completeness Check

Completeness checks are performed for the inventory and the impact assessment in order to determine the degree to which the available data is complete for the processes and impacts, which were identified as significant issues. If relevant information is found to be missing or incomplete for some of the key processes or the most important elementary flows or impact categories, the necessity of such information for satisfying the goal and scope of the LCA must be investigated. If deemed necessary, the inventory and impact assessment phases must be revisited in order to fill the identified gaps. Alternatively, the goal and scope definition may have to be adjusted to accommodate the lack of completeness. If an important data deficiency cannot be remediated, this should be considered when formulating the limitations in the conclusions from the study (see Sect. 12.4). If the missing information is found to be of little importance, this should be documented in the reporting of the completeness check.

Taking the completeness check of the illustrative case on window frames (see Chap. 39) as an example, several gaps were identified. In relation to LCI, the applied heat mix was thus only representative for district heating (and hence not appropriate for situations with local heating sources), and the LCI unit processes used to model the energy technologies applied in the heat mix were geographically representative for Norway and Switzerland and hence not fully representative for Denmark. With regard to LCIA, the use of site-generic characterisation factors for some impact categories may not be fully representative for the specific impact pathways of environmental flows released in or close to Denmark. Once identified, those gaps therefore underwent the procedure described in Fig. 12.2 to be addressed in the study.

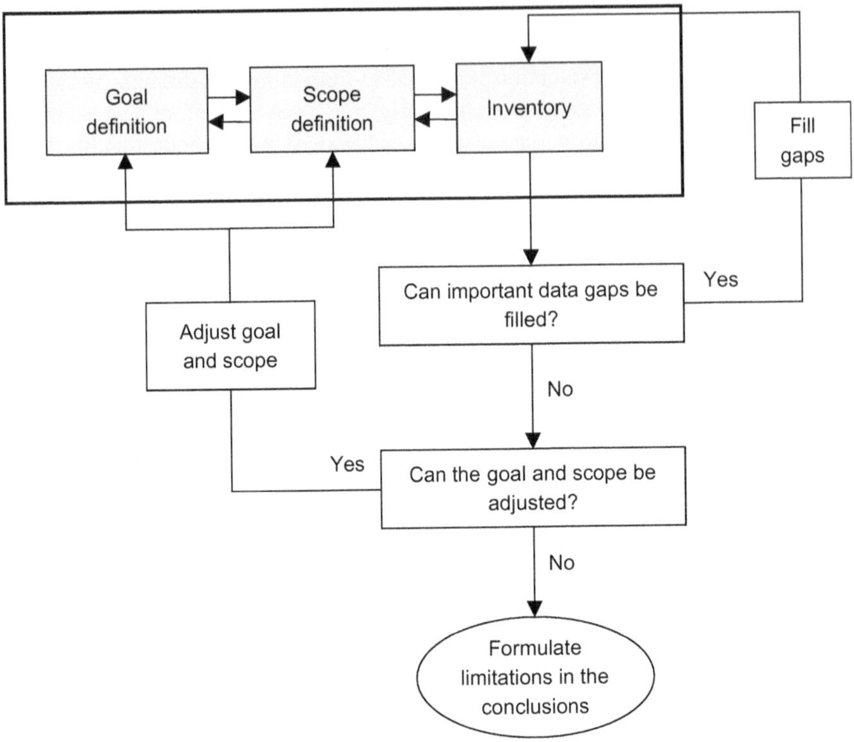

Fig. 12.2 Iterative interaction between completeness check and the earlier phases of the LCA

12.3.2 *Sensitivity Check*

Sensitivity check has the purpose of identifying the key processes and most important elementary flows as those elements that contribute most to the overall impacts from the product system. Sensitivity analysis can be performed and presented as a contribution analysis (which activities contribute to which environmental impact scores, by how much and through which elementary flows?) or a dominance analysis (which activities contribute most to which impacts or flows?). See Chap. 11 for a more detailed discussion of sensitivity analysis and how it is performed.

In the illustrative case on window frames, not all significant issues were covered in the sensitivity analysis due to lack of sufficient data and knowledge to construct sensitivity scenarios in some of the cases. A sensitivity scenario reflecting the EU27 heat mix was established and results showed that impacts for a few impact categories (mainly related to toxicity) were lower than in the baseline scenario, while most impacts were higher due to a larger share of oil and natural gas in the EU27 heat mix.

In support of the iterative approach applied in LCA, sensitivity analysis is also used as a steering activity in the iteration loops that are performed throughout the LCA in support of boundary setting for the product system, inventory data collection and impact assessment. The findings from these earlier sensitivity analyses are brought into the sensitivity check of the interpretation phase.

In the interpretation phase, sensitivity analysis is used together with information about the uncertainties of significant issues among inventory data, impact assessment data and methodological assumptions and choices to assess the reliability of the final results and the conclusions and recommendations which are based on them (Sect. 12.3) (Table 12.2).

Table 12.2 *Tools for sensitivity analysis*

Factors checked for sensitivity	Tools for sensitivity analysis
Data uncertainty	The influence of data uncertainty for key issues can be checked by allowing the data to vary within the limits given by the uncertainty estimates while modelling the product system and checking the results. If the information about the (stochastic) uncertainties of the individual elementary flows and characterisation factors allows it, it is also possible to calculate the uncertainty of the final results in terms of inventory and environmental impacts (e.g. simulating it using Monte Carlo techniques). See Chap. 11 for a more detailed discussion of uncertainty analysis and how it is performed
Methodological uncertainty	The influence of methodological (systematic) uncertainties can be checked by analysing different possible choices (e.g. of applied allocation principle) as scenarios and reporting the influence on the final results. Methodological choices which may be relevant to include in a sensitivity analysis include: handling of multifunctional processes (system expansion assumptions or allocation rules), cut-off criteria, boundary setting and system definition, and judgements and assumptions concerning data in the inventory; and for the impact assessment: selection of impact categories, assignment of inventory results (classification), calculation of category indicator results (characterisation), and normalisation and weighting of impact scores

The combination of sensitivity analysis and uncertainty analysis helps identify focus points for improved inventory data collection or impact assessment.

As illustrated in Fig. 12.3, data with a high uncertainty need not be a focus point for improvement if the sensitivity to this data is very low. In the same way, data which has a strong influence on the final results of the study may also not require further data collection effort if the representativeness of the data is high and its uncertainty negligible. The focus point for improvement of data quality should be data with a strong influence on the overall results and a high uncertainty or questionable. If such data cannot be improved, the result is a low precision which must be reported. If the precision is insufficient to meet the requirements from the intended application of the results, it may be necessary to revise the goal of the study. Figure 12.4 provides a decision tree for handling the sensitivity check.

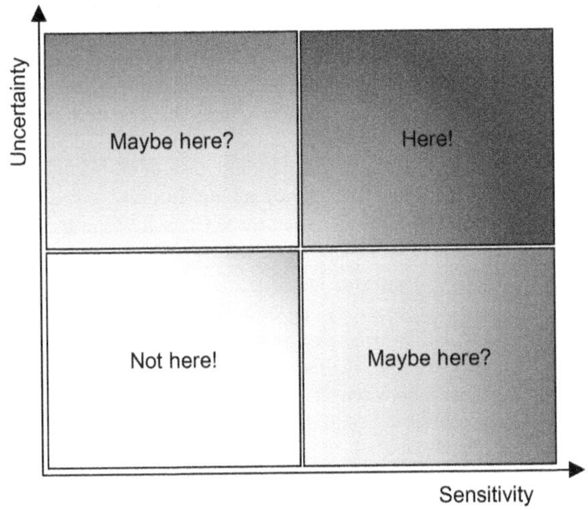

Fig. 12.3 Focusing collection of improved data by combining sensitivity and uncertainty information

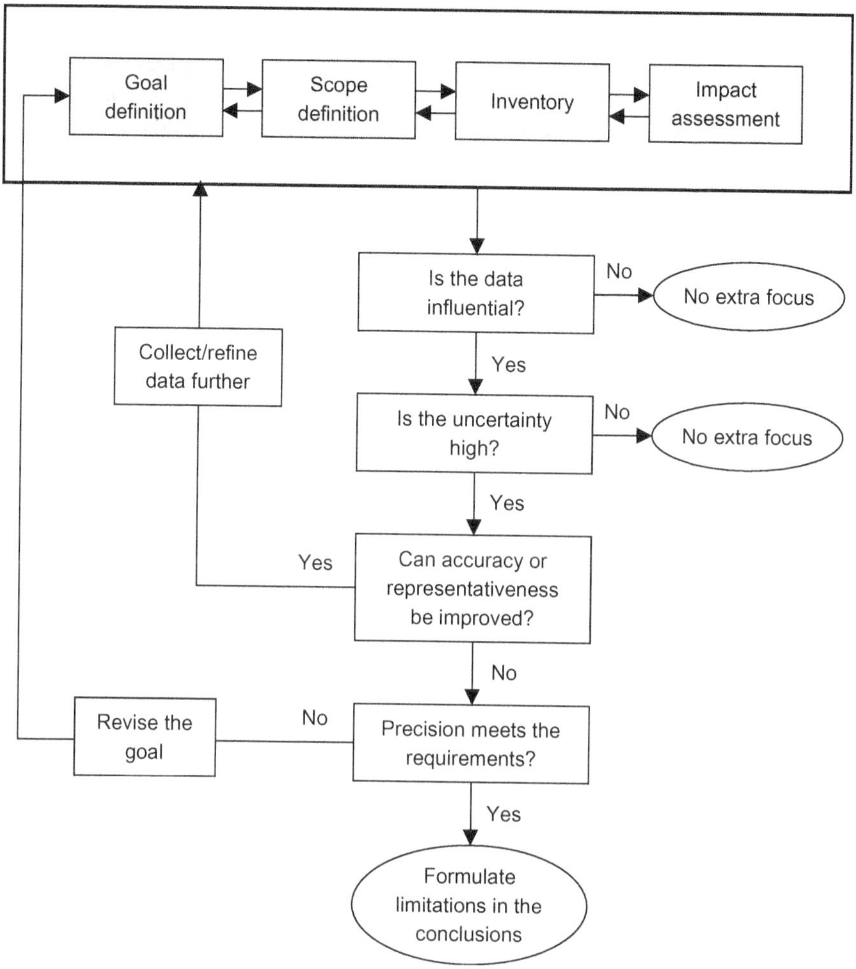

Fig. 12.4 Combination of sensitivity analysis and uncertainty information to focus improvement of the LCA data

12.3.3 Consistency Check

The consistency check is performed to investigate whether the assumptions, methods, and data, which have been applied in the study, are consistent with the goal and scope.

Are differences in the quality of inventory data along a product life cycle and between different product systems consistent with the significance of the processes, which the data represent and with the goal and scope of the study? Inventory data quality concerns both the time-related, the geographical, and the technological representativeness of the data, the appropriateness of the chosen unit process to represent the process of the product system, and the uncertainty of the data.

In case of comparison between different product systems, the consistency check also investigates whether allocation rules and system boundary setting as well as impact assessment have been consistently applied to all compared product systems.

When inconsistencies are identified, their influence on the results of the study is evaluated and considered to draw conclusions from the results.

Taking the window frame case as example again, the main identified inconsistency is between the goal and scope and the interpretation of the results which does not give due consideration to changes that may occur in particular in the background system within the time frame of the study (at least 20 years). Important changes are the Danish heat mix (for which the share of fossil fuels is expected to decrease) and the technological development in the heat supply technologies (see Chap. 39 for further details).

12.4 Conclusions, Limitations and Recommendations

Building on the outcome of the other elements of the interpretation, and drawing on the main findings from the earlier phases of the LCA, the final element of the interpretation has to draw conclusions and identify limitations of the study, and develop recommendations to the intended audience in accordance with the goal definition and the intended applications of the results.

The conclusions should be drawn in an iterative way: based on the identification of significant issues (Sect. 12.2) and the evaluation of these for completeness, sensitivity and consistency (Sect. 12.3), preliminary conclusions can be drawn. It is then checked whether these preliminary conclusions are in accordance with the requirements of the scope definition of the study (in particular data quality requirements, predefined assumptions and values, and limitations in methodology and study). If the conclusions are aligned with the requirements, they can be reported as final conclusions, otherwise they must be re-formulated and checked again.

Recommendations based on the final conclusions of the study should be logical and reasonable consequences of the conclusions. They should only be based on significant findings and relate to the intended application of the study as defined in the goal definition.

In the illustrative case on window frames (Chap. 39) it was concluded for example that the wood composite (W/C) window has the lowest impact among the four compared windows in all impact categories, and that impacts occurring in the use stage are generally dominating the total impacts and are caused by the demand for heat to compensate the heat losses that occur through the window. Albeit not visible in the results, due to the disregard of technological changes related to heat supply over the time frame of the study (likely going towards lower impacts), the dominance of the use stage impacts is likely to decrease with time, depending on what technological improvements are introduced in the other stages of the window life cycle, but it is still expected to remain significant in a foreseeable future. A follow-up study is recommended to further address these dynamics.

12.5 Interpretation for Comparative Studies

In studies that involve a comparison of product systems, the interpretation has to consider a number of additional points to ensure fair and relevant conclusions from the study.

- Significant issues must be determined for each of the systems, and special attention should be given to issues that differ between the systems and which have the potential to change the balance of the comparison.
- The completeness check must have specific focus on differences in the completeness of the treatment of some of the significant issues between the product systems. If there are differences that could influence the comparison results, these should be eliminated if possible and otherwise kept in mind in the formulation of conclusions.
- If an uncertainty analysis is performed to investigate whether the difference between two systems is statistically significant, the analysis should be performed on the difference between the systems (one system minus the other), which should be checked for a statistically significant difference from zero taking into account potential co-variation between processes of the two systems (e.g. processes which are the same). See the discussion of this point in Chap. 11.
- When an LCA is intended to be used in comparative assertions intended to be disclosed to the public, the ISO 14044 standard requires that the evaluation element include interpretative statements based on detailed sensitivity analyses. It is emphasised in the standard, that the inability of a statistical analysis to find significant differences between different studied alternatives does not automatically lead to the conclusion that such differences do not exist, rather that the study is not able to show them in a significant way.
- A consistency check must be performed of the treatment of the key assumptions and methodological choices in the different systems to avoid a bias and ensure a fair comparison.

  - Are differences in the quality of inventory data between the compared product systems acceptable, considering the relative importance of the processes in the product systems, and are the differences consistent with the goal and scope of the study? For example, if one study is based on specific and recent data with a high degree of representativeness for all the key processes while the other uses extrapolation from literature data, there is a bias in the inventory data that can make a comparison invalid.
  - Have allocation rules and system boundary setting been consistently applied to all product systems?

– Has the impact assessment been performed consistently for the systems, have the relevant impact categories been included for all systems, and have the impacts been calculated in the same way and with the same coverage of elementary flows for all the systems?

The influence of any identified inconsistencies on the outcome of the comparison should be evaluated, and taken into consideration when conclusions are drawn from the results.

References

This chapter is to a large extent based on the ILCD handbook and the ISO standards 14040 and 14044. Due to the scope of this chapter, some details have been omitted, and some procedures have been rephrased to make the text more relevant to students. For more details, the reader may refer to these texts:

EC-JRC: European Commission—Joint Research Centre—Institute for Environment and Sustainability: International Reference Life Cycle Data System (ILCD) Handbook—General guide for Life Cycle Assessment—Detailed guidance, 1st edn. March 2010. EUR 24708 EN, Luxembourg, Publications Office of the European Union (2010)

ISO: Environmental Management—Life Cycle Assessment—Principles and Framework (ISO 14040). ISO, the International Organization for Standardization, Geneva (2006a)

ISO: Environmental Management—Life Cycle Assessment—Requirements and Guidelines (ISO 14044). ISO, the International Organization for Standardization, Geneva (2006b)

Author Biographies

Michael Z. Hauschild Involved in development of LCIA methodology since the early 1990s. Has led several SETAC and UNEP/SETAC working groups and participated in the development of the ISO standards and the ILCD methodological guidelines. Main LCA interests are chemical impacts, spatial differentiation and science-based boundaries in LCIA.

Alexandra Bonou LCA expert and modeller focussing on ecodesign since 2009. Has worked on integrating life cycle thinking and environmental target setting in organisations of the private and public sector. Interested in life cycle management, product development, social life cycle assessment.

Stig Irving Olsen LCA expert both as researcher and as consultant. Involved in the development of LCA methodologies since mid 1990s. Contributed to UNEP/SETAC working groups on LCIA methodology. Main LCA interest is human toxicity impacts, emerging technologies, and decision making.

Chapter 13
Critical Review

Ralph K. Rosenbaum and Stig Irving Olsen

Abstract Manipulation and mistakes in LCA studies are as old as the tool itself, and so is its critical review. Besides preventing misuse and unsupported claims, critical review may also help identifying mistakes and more justifiable assumptions as well as generally improve the quality of a study. It thus supports the robustness of an LCA and increases trust in its results and conclusions. The focus of this chapter is on understanding what a critical review is, how the international standards define it, what its main elements are, and what reviewer qualifications are required. It is not the objective of this chapter to learn how to conduct a critical review, neither from a reviewer nor from a practitioner perspective. The foundation of this chapter and the basis for any critical review of LCA studies are the International Standards ISO 14040:2006, ISO 14044:2006 and ISO TS 14071:2014.

Learning Objectives
After studying this chapter, the reader should be able to:

- Explain when a critical review is needed and what is its purpose.
- Provide perspectives on the difference between critical review, scientific review and validation.
- Explain the principles, procedure, requirements, content, deliverables and options when conducting critical review and which international standards describe it.

R.K. Rosenbaum (✉)
IRSTEA, UMR ITAP, ELSA Research Group and ELSA-PACT—Industrial Chair for Environmental and Social Sustainability Assessment, 361 rue Jean-François Breton, BP 5095, 34196 Montpellier, France
e-mail: ralph.rosenbaum@irstea.fr

S.I. Olsen
Division for Quantitative Sustainability Assessment, Department for Management Engineering, Technical University of Denmark, Bygningstorvet, Building 115, 2800 Kgs. Lyngby, Denmark

M. Hauschild et al. (eds.), *Life Cycle Assessment*,
DOI 10.1007/978-3-319-56475-3_13

- Discuss the necessary qualifications of a reviewer and how they are selected and by whom.
- Describe the possible roles, obligations, tasks, and deliverables of a reviewer.

The focus of this chapter is on understanding what a critical review is and what its main elements are. It is important to note that it is NOT the objective of this chapter to learn how to conduct a critical review, neither from a reviewer nor from a practitioner perspective.

13.1 Introduction

Numerous LCA studies have been published and many of them based on the highest standards of quality and robustness, but there are also an alarming number of studies that contain either important mistakes or plain manipulations in order to obtain an intended result that would support a specific, pre-defined claim. These mistakes and manipulations may be subtle and difficult to detect but can also be immediately identifiable to the trained eye and a number of studies based on surprisingly blunt and evident manipulations have been published over the years. Especially some earlier studies have become classic and illustrative examples in LCA teaching of how not to do LCA (or comparative environmental claims in general) and they also nicely illustrate the purpose and need for critical review of published LCA studies. Two entertaining examples are:

(1) SUV versus hybrid car: A famous example is a study from the automotive marketing company CNW Marketing Research, Inc. from 2007 called "Dust to Dust: The Energy Cost of New Vehicles From Concept to Disposal". This study compared the life cycle energy costs of a number of automobiles from 2005 and had no hesitation to conclude (and widely communicate) that many large sport utility vehicles (SUVs) including GM's massive Hummer models H2 and H3 use less energy per mile driven than many smaller vehicles including the Toyota Prius hybrid car. Gleick (2007) analysed the information and commented "that the report's conclusions rely on faulty methods of analysis, untenable assumptions, selective use and presentation of data, and a complete lack of peer review. Even the most cursory look reveals serious biases and flaws: the average Hummer H1 is assumed to travel 379,000 miles and last for 35 years, while the average Prius is assumed to last only 109,000 miles over less than 12 years".

(2) Fast-food versus classic restaurant: A study from the 1990s comparing a fast-food restaurant with a normal restaurant that surprisingly concludes the environmental superiority of the fast-food option. When the study was redone by independent practitioners, they demonstrated that the system boundaries were chosen in a way that comparability of both options was not supported since important processes from the fast-food restaurant were excluded. Correcting these manipulations then yielded a different picture (Lang et al. 1994). The whole story can be found in Jolliet et al. (2015).

Product systems can be very complex, involving a high number of processes and locations. Their modelling in LCA builds on multiple data sources from measurements to unit process databases and involves influential assumptions, drawing on a diversity of expertise from process engineering to environmental and sometimes also social science. Results are often communicated to stakeholders and decision makers that cannot control the quality of the studies, and manipulation and mistakes in LCA studies are as old as the tool itself. The understanding of a need for an independent critical review of LCA studies thus came very early in the history of the methodology. The SETAC LCA "Code of practice" proposed it first in 1993 (Consoli et al. 1993) with more detailed procedural guidelines published later by Klöpffer (1997) and Weidema (1997), which still stand until today as essential references on how to conduct a critical review. As its superseded predecessor from the late 1990s, the revised international standard ISO 14044 (2006a) defines review procedures (although in much less detail than Klöpffer and Weidema, respectively) to ensure that an LCA study is conform to ISO requirements. As a further development from there, ISO published the technical standard ISO TS 14071 (2014) that aims to specify detailed ISO requirements for critical reviews. In consequence, this should ensure that all claims of a critically reviewed LCA study are well justified and supported by assumptions, methods and data used. Besides preventing misuse and unsupported claims, critical review may also help identifying mistakes and more justifiable assumptions as well as generally improve the quality of a study. It thus supports the robustness of an LCA and increases trust in its results and conclusions.

In general, there are different kinds of review processes associated with scientific and technical developments and they all fulfil different objectives and vary in their approach and process. Two different types of review processes are mainly relevant in the context of an LCA study: (1) scientific peer-review and (2) critical review according to ISO 14044 (2006a). While this chapter is focusing on the latter, there is much confusion between both and it is essential to clearly distinguish them and understand their differences. Table 13.1 provides a simplified overview of general tendencies for similarities and differences between both types of review.

Besides several similarities, the essential differences between these two review types are thus linked to their duration, depth, cost, transparency, confidentiality, content and objectives. As discussed by Curran and Young (2014), there is also an important and frequently ignored difference between the terms "critical review" and "verification" with their essential difference being that critical review relies on expert judgement whereas verification is based on comparison against objective evidence.

The focus of this chapter and the basis for any critical review of LCA studies are the International Standards ISO 14040 (2006b), ISO 14044 (2006a) and ISO TS 14071 (2014). However, it is worth noting that other review schemes exist that may be specified in more detail than in ISO 14044 and ISO TS 14071, while still being fundamentally based on them. These review schemes often have a very specific context of application and in most cases also a geographically limited relevance. One example is the International Reference Life Cycle Data System (ILCD) of the

Table 13.1 Similarities and differences between scientific peer-review and critical review according to ISO 14044 (note that this table represents a general tendency for each criteria, not an absolute truth as there will likely be cases of review processes that may differ on either side of the table)

<table>
<tr><th colspan="2"></th><th>Scientific peer-review</th><th>Critical review</th></tr>
<tr><td colspan="2">Timing relative to study</td><td>after</td><td>during or after</td></tr>
<tr><td colspan="2">Time spent per reviewer</td><td>1-2 days</td><td>2-7 days</td></tr>
<tr><td rowspan="9">Reviewer(s)</td><td>Accreditation required</td><td colspan="2">-</td></tr>
<tr><td>Number</td><td>1-3</td><td>1-5</td></tr>
<tr><td>Anonymity</td><td>✓</td><td>-</td></tr>
<tr><td>Selection</td><td>external (editor)</td><td>internal (commissioner/practitioner) or external (review panel chair)</td></tr>
<tr><td>Affiliation</td><td>external</td><td>external (internal)</td></tr>
<tr><td>Independence</td><td colspan="2">✓</td></tr>
<tr><td>Access to confidential information</td><td>-</td><td>✓</td></tr>
<tr><td>Remuneration</td><td>-</td><td>✓</td></tr>
<tr><td rowspan="11">Review objects</td><td>Objectives of an LCA</td><td>✓</td><td>-</td></tr>
<tr><td>Goal and scope definition</td><td colspan="2">✓</td></tr>
<tr><td>Interpretation and conclusions</td><td colspan="2">✓</td></tr>
<tr><td>Standardised review criteria</td><td>-</td><td>✓</td></tr>
<tr><td>Documents reviewed</td><td>Publication manuscript and supporting information</td><td>Full LCA report, primary input data, (computerised product system model and database(s))</td></tr>
<tr><td>Subjectivity of judgments</td><td colspan="2">✓</td></tr>
<tr><td>Methodology</td><td colspan="2">✓</td></tr>
<tr><td>Novelty/originality</td><td>✓</td><td>-</td></tr>
<tr><td>Scientific/technical validity</td><td colspan="2">✓</td></tr>
<tr><td>Input data quality and representativeness</td><td>(-)</td><td>✓</td></tr>
<tr><td>Conformity with ISO and other standards</td><td>-</td><td>✓</td></tr>
<tr><td rowspan="3"></td><td>Product system model</td><td>(-)</td><td>✓</td></tr>
<tr><td>Assumptions and choices</td><td colspan="2">✓</td></tr>
<tr><td>Mistakes</td><td colspan="2">✓</td></tr>
<tr><td colspan="2">Increased credibility of LCA study</td><td colspan="2">✓</td></tr>
<tr><td colspan="2">Increased quality of publication</td><td colspan="2">✓</td></tr>
<tr><td colspan="2">Increased quality of LCA study</td><td>(-)</td><td>✓</td></tr>
<tr><td colspan="2">Review comments/responses public (in case of publication)</td><td>-</td><td>✓</td></tr>
<tr><td colspan="2">Mandatory for publication</td><td>-
(✓ if published in scientific journal)</td><td>-
(✓ for comparative assertions)</td></tr>
<tr><td colspan="2">Consensus among reviewers</td><td>-</td><td>✓ (or minority statement)</td></tr>
</table>

European Commission with its series of ILCD handbooks, including one specifically dedicated to review schemes (EC-JRC 2010a) and another to reviewer qualifications for LCI datasets (EC-JRC 2010b) linked to the European reference Life Cycle Database ELCD. While these are valuable sources of information for the interested reader and we recommend them for further study, they will not be discussed in detail in this chapter.

13.2 Critical Review Process

As presented, critical review is a procedure intrinsically linked to ISO 14044 (2006a) which defines it as a "process intended to ensure consistency between a life cycle assessment and the principles and requirements of the International Standards on life cycle assessment" (Clause 3.45). However, critical review may also be performed just in order to improve the quality of the study and thus the trust in it. The following will detail why, how, and when critical reviews are performed.

13.2.1 Purpose

Critical review of an LCA study is useful in all cases where quality, robustness, and trust in results are wanted. Whether or not a review is required depends on the goal definition, i.e. the intended application and decision-context, the reasons for carrying out the study, and the intended audience. ISO 14044 recommends the use of critical reviews in general and makes it mandatory for "LCA studies where the results are intended to be used to support a comparative assertion intended to be disclosed to the public" (ISO 2006a). These mandatory critical reviews have to ("shall" in ISO terminology, which indicates an obligation) be performed by a panel of interested parties including at least three experts. A comparative assertion is defined by ISO as (ISO 2006a): an "environmental claim regarding the superiority or equivalence of one product versus a competing product that performs the same function".

However, this definition may not be broad enough. In the European context the Product Environmental Footprint (PEF) is an example of an LCA that will typically be subject to such review requirements. Even though a comparative assertion is not explicitly stated in the report, Environmental Product Declarations (EPD) and PEF aim to give data and information to be used in comparisons and they could therefore be regarded as a basis for comparative assertions. In fact, critical review by at least one independent and qualified external reviewer (or review team) is mandatory in the PEF methodology (European Commission 2013).

13.2.2 Chronology

A critical review can basically be performed in two alternative ways. The first is to review the LCA after the study is completed (a posteriori review). The second approach is an integrated/interactive review where the reviewer(s) follows the study from the definition of goal and scope, through data collection to the conclusion (concurrent review). In the a posteriori approach at least one iteration of review comments and associated modifications of the study are performed and the critical review report should reflect the entire review process. In the concurrent review scheme the reviewer(s) can be involved in several of the different steps throughout the conduction of the study, i.e. (ISO 2014):

(a) "the goal and scope definition;
(b) inventory analysis including data collection and modelling;
(c) impact assessment;
(d) life cycle interpretation;
(e) draft LCA report"

with the critical review statement being issued for the final version of the LCA report. ISO 14044/ISO TS 14071 do not specify any requirements or preferences to one or the other approach (neither does PEF), and they can hence always be freely chosen.

Most literature recommends the concurrent review (Weidema 1997; Klöpffer 2005, 2012; Hamilton and Ayer 2013; Schulz and Mersiowsky 2013). An a posteriori critical review involves a risk of delays in the final phase. The reviewer(s) has to comment on the draft final reports usually within a few weeks and there is a risk that serious flaws in methodology or data quality, or new aspects appear. Doing the necessary corrections may be hindered by budget and timing. Also, the review process requires communication between the practitioner and the reviewer(s) and in some cases the practitioner may not be available after the completion of the study (Klöpffer 2005).

The concurrent review approach has the benefits that potential problems can be corrected at an early stage of the study. There may be some extra time needed at the beginning of the study to guide it onto the right track, but this will likely be less time consuming than delays caused by new aspects surfacing at the end of the study or by the need to figure out how assumptions and calculations influence the results. This obviously also influences the timing of the study in itself since the practitioner has to wait for review comments at different milestones throughout the study. Typically, one month additional time should be expected (Schulz and Mersiowsky 2013). A minor concern raised by Curran and Young (2014) is the risk that reviewers may become vested in the study and thus lose their independence.

13.2.3 Requirements

According to ISO 14044 (2006a), "the critical review process shall ensure that:

- the methods used to carry out the LCA are consistent with the international standard;
- the methods used to carry out the LCA are scientifically and technically valid;
- the data used are appropriate and reasonable in relation to the goal of the study;
- the interpretations reflect the limitations identified and the goal of the study; and
- the study report is transparent and consistent".

The European PEF guide (European Commission 2013) outlines the same requirements although it additionally mentions that the data quality should meet requirements and that the study report shall be accurate. This list gives guidance for the reviewer(s) and may also serve as the structure for review reports (Klöpffer 2012). In most instances the reviewer is expected not only to do an "administrative ISO check", but to also be a discussion partner accompanying the LCA project (Curran and Young 2014).

Thus, as stated by ISO TS 14071 (2014) "The critical review should cover all aspects of an LCA, including data appropriateness and reasonability, calculation procedures, life cycle inventory, impact assessment methodologies, characterisation factors, calculated LCI and LCIA results, and interpretation". Regarding two aspects ISO TS 14071 leaves it optional whether or not the critical review includes them:

1. Assessment of the life cycle inventory (LCI) model,
2. Assessment of individual data sets.

Curran and Young (2014) note that in contrast to the usually comprehensive review of methods and assumptions, there is often a limited examination of data and quantitative results. This may be due to a combination of limitations of time (budget), weak transparency and/or poor accessibility of data sets. In order to perform the critical review it is important that reviewers are granted access to the data and inventory model by the commissioner and practitioner.

In LCA it is difficult to establish objective quality criteria, and specific criteria for whether or not a study is correct cannot be defined. Therefore, much of the critical review has to rely on professional judgement regarding the consistency between goal and scope, data and models used, interpretations applied and the robustness of the conclusions drawn. The previous chapters in this part of the book specify in detail the requirements for conducting an LCA and thus also the aspects that reviewers should be aware of when performing a critical review. Several authors discuss in further detail the specific considerations and questions to ask during the review process, and the interested reader is referred to those (Consoli et al. 1993; Klöpffer 1997; Weidema 1997).

13.2.4 Deliverables

The deliverables of the critical review are (ISO 2006a, 2014):

- Comments to specific intermediate steps of the LCA (goal and scope definition, LCI, LCIA, interpretation) for a concurrent review,
- Comments to the final draft LCA report,
- Review report,
- Review statement.

The review report documents how the critical review was conducted including all reviewer comments and recommendations given plus a response (to each comment/recommendation) from the practitioner that may indicate consequent changes applied to the study and/or the report or a justification of the respective issue in the study or the report in respect to the comment. Annex A of ISO TS 14071 (2014) contains an informative template for a critical review report. Hamilton and Ayer (2013) suggest that "In general, you should ensure that the following steps in the process are documented:

- Review panel comments to the study team
- Study team responses to the review panel
- [...]
- Correspondence between the panel and the study team".

The critical review statement is a short text that clearly states whether or not the study is conform to the requirements of ISO 14040 and 14044. It should also discuss "any particular strengths, limitations and remaining improvement potentials of the LCA study or the critical review process" (ISO 2014). ISO TS 14071 clearly states what has to be included in the critical review statement (ISO 2014):

- "Title of the study;
- The commissioner of the LCA study;
- The practitioner of the LCA study;
- The exact version of the report to which the critical review statement belongs;
- The reviewer(s) or, in the case of a panel review, the panel members, including the identification of the panel chairperson;
- A description of the review process, including information on:
  - Whether the review was performed based on ISO 14044:2006, 6.2 or 6.3;
  - Whether the review was performed in parallel or at the end of the study;
  - Whether the review included or excluded an assessment of the LCI model;
  - Whether the review included an analysis of individual data sets;
- A description of how comments were provided, discussed and implemented;
- A statement of the result of the critical review, i.e. whether the study was found to be in conformance with ISO 14040 and ISO 14044 or not".

Hamilton and Ayer (2013) also recommend that "It is important that the final critical review statement includes:

- The date of issuance [...] of the study
- [...]
- Documentation of any outstanding issues that were not resolved during the review
- A summary of the comments/responses from the review process".

The final LCA report has to mandatorily include the review statement and review report, as well as all comments and recommendations of the reviewer(s) and any responses by the practitioner to them. It is a requirement of the ISO 14044 standard that the review statement and review report must be included in the LCA report (typically as an appendix to the report). The critical review statement has to be signed by the chairperson and should also be signed by the other reviewers. This signature is strictly individual and personal and cannot be representing an institution nor be replaced by an institutional stamp or label. This means that the reviewer (s) publically (if the report is published) state the conformance or non-conformance of the study to ISO 14040 and 14044 with their names and signatures, which ensures that especially intentional manipulations (but also larger mistakes) that would affect the LCA's conformance to ISO should have been identified and corrected. This can be seen as a sort of quality insurance, making the reviewer(s) personally responsible for the review process and content.

13.3 Reviewer Qualifications, Tasks and Selection

Since the purpose of a critical review is to perform a critical expert judgement as to whether the ISO 14044 criteria are fulfilled, the expert(s) should of course be independent of the LCA, but not necessarily external to the company. In fact, the foremost requirement for any reviewer, internal or external, in the context of a critical review is **complete independence** from the study (but not necessarily its commissioner or practitioner), i.e. not involved in the commissioner's or the practitioner's project team, nor otherwise implicated in the definition of the scope or the conducting of the LCA. In the case of an **internal expert**, this person may, however, be full-time or part-time employee of either the commissioner or the practitioner of the study, or otherwise be related to either or both of them, while still being independent of the study. An **external expert** has no financial dependency on either the commissioner or the practitioner, nor any political or other interest in the study results.

The necessary qualifications for reviewers performing a critical review depend on a number of factors, such as the type of review scheme (a posteriori or concurrent) and the goal and scope of the LCA:

1. Critical review practice is essential for at least one reviewer who has to be well experienced with the process of a critical review according to ISO 14044 (2006a) and ISO TS 14071 (2014). For a panel-based critical review this will usually be the chairperson of the review panel.
2. LCA expertise: As a general rule, there has to be at least one expert on LCA methodology and practice as well as the ISO 14040/14044 requirements. For a panel-based critical review this will usually be the chairperson of the review panel.
3. Technical expertise mostly concerns the LCI phase and is required in order to ensure that the underlying product system model and data are representative and adequately modelled according to goal and scope of the LCA. Technological

experts do not necessarily have to be familiar with the LCA methodology. This may comprise specific expertise, such as on the

- Product, service or organisation,
- Process(es) and technology,
- Relevant practice(s) including national or regional specificities if needed.

4. Scientific expertise may be required to ensure adequate consideration of environmental and/or social issues and phenomena of relevance for a given goal and scope definition. This applies particularly to the LCIA phase, but may also be relevant for aspects of the other LCA phases, notably the interpretation.
5. Other expertise may in some cases be necessary depending on the goal and scope, e.g. legal issues, stakeholder concerns, NGOs, etc.

Proficiency in the language of the study is of course required from all reviewers. The number of reviewers or review panel members is then a function of the expertise required and the expertise each reviewer brings into the process. So far, there is no official accreditation or certification required (or available) and the expertise of reviewers will usually be evaluated via their curriculum vitae including a list of relevant references. ISO TS 14071 (2014) also proposes an example of a self-declaration statement that can be used. To the authors' knowledge, several organisations intend to establish critical reviewer databases, but until the finalisation of this book no database has reached formal recognition in the global LCA community.

The selection of reviewer(s) will typically be done by the practitioner and/or the commissioner of an LCA. In the case of a panel review, they appoint an external independent expert as chairperson, who then selects other independent experts for the review panel. All experts are contracted by the commissioner or practitioner. This contract normally involves adequate remuneration, a commitment to tasks and timing and a non-disclosure agreement to ensure confidentiality of information and data that need to be accessed by the experts in order to complete the critical review. The reviewers' contract cannot contain conditions that influence the result of the critical review process. ISO TS 14071 also explicitly states that reviewer tasks cannot be subcontracted or delegated and thus have to be performed by the contracted reviewers personally.

ISO TS 14071 lists the respective tasks of the two principal roles in a critical review process, the chairperson and the reviewer. The reviewer's role essentially involves:

- Commenting on the LCA report (or parts of it during a concurrent critical review);
- Contributing to the critical review report;
- Expressing agreement or disagreement concerning the critical review statement including a justification in case of disagreement.

The chairperson has the same role as a reviewer but with the additional tasks of running the critical review process via:

- Setting up of the review panel;
- Distribution of tasks relative to each panel member's competencies;
- Coordination of the review process including ensuring a common understanding of the required tasks among all reviewers and their relation to ISO 14040 and 14044;
- Recording and sharing each reviewer's comments within the panel and with the practitioner/commissioner;
- Resolve potential conflicting positions between reviewers, aiming at a consensual critical review statement or if that is not possible including a minority position in the statement;
- Enable and support a smooth communication among all panel members and with the practitioner and commissioner;
- Ensure the generation and panel approval of review report and statement.

Consequently, the workload and required experience level regarding the critical review process will be higher for the chairperson which should be reflected in the contractual conditions. Furthermore, this also means that besides technical qualifications, the chairperson should be particularly skilled in communication and project management.

How to become a critical reviewer is a frequently asked question and there are several ways once an interested candidate has acquired the necessary competencies and experience. The opportunity to participate in such a process may come via different channels, typically via colleagues who may have been asked first and refer to you, via mailing lists, or via a direct contact with an offer. In any case, it is advisable to first participate as expert in a reviewer panel a few times in order to get acquainted with the process and usual practice. Having participated in a few critical reviews, you could propose to take on the additional responsibility of acting as chairperson.

13.4 Conclusions

Critical review is an important element of an LCA study that helps ensuring conformance to the relevant LCA standards ISO 14040 and 14044 and thus building credibility and trust in its methodology, data, results, the robustness of its conclusions, and ultimately increasing its acceptance among stakeholders. In the authors' experience, critical review can trigger a tremendous improvement of an LCA's rigour, transparency, technical quality and robustness, especially if conducted concurrently to the study. It also helps bringing in external and independent views and experiences, which typically enriches the methodological aspects, such as modelling, data, and the interpretation of results in a study. It is, however, not a guarantee that the study is perfect or even as perfect as possible since there will always be aspects that could not be considered, that were overlooked, or that could not be addressed. Also, a critical review does not verify or validate the goals of an LCA or how its results will be used (ISO 2006b), which means that if the objective itself is problematic the LCA study may be stated as conform to ISO, while still supporting misleading conclusions or recommendations beyond the LCA report. Critical review is also not a validation process against objective evidence, such as measurements or other observations.

This chapter provides a broad overview of several complementary aspects related to the critical review, without discussing them all in detail. The authors recommend the cited references for further reading, particularly the ISO standards of course, along with the publications by Klöpffer (1997, 2005, 2012), Weidema (1997), Hamilton and Ayer (2013), and Curran and Young (2014), which will provide further depth and details, practical aspects, and experiences for the interested reader. They are certainly essential reads for aspiring critical reviewers and chairpersons, but also for practitioners of an LCA that will be exposed to a critical review.

References

Consoli, F., Allen, D., Boustead, I., Fava, J., Franklin, W., Jensen, A., de Oude, N., Parrish, R., Perriman, R., Postlethwaite, D., Quay, B., Séguin, J., Vigon, B.: Guidelines for Life Cycle Assessment: A "Code of Practice". SETAC, Pensacola, FL (1993)

Curran, M.A., Young, S.B.: Critical review: a summary of the current state-of-practice. Int. J. Life Cycle Assess. **19**, 1667–1673 (2014). doi:10.1007/s11367-014-0778-2

EC-JRC: International Reference Life Cycle Data System (ILCD) Handbook: Review schemes for Life Cycle Assessment (LCA). EUR 24710 EN (2010a). doi:10.2788/39791

EC-JRC: International Reference Life Cycle Data System (ILCD) Handbook: Reviewer qualifications for Life Cycle Inventory data sets. European Commission, Joint Research Centre, Institute for Environment and Sustainability, Ispra, Italy (2010b). doi:10.2788/95543

European Commission: Recommendation 2013/179/EU on the use of common methods to measure and communicate the life cycle environmental performance of products and organisations. Off J Eur Union 210 (2013). doi:10.3000/19770677.L_2013.124.eng

Gleick, P.H.: Hummer versus Prius "Dust to Dust" Report Misleads the Media and Public with Bad Science. Pacific Institute, Oakland, CA (2007)

Hamilton, M., Ayer, N.: Navigating the LCA Critical Review Process - EarthShift White Paper (2013)

ISO: Environmental Management—Life Cycle Assessment—Requirements and Guidelines (ISO 14044). ISO, the International Organization for Standardization, Geneva (2006a)

ISO: Environmental Management—Life Cycle Assessment—Principles and Framework (ISO 14040). ISO, the International Organization for Standardization, Geneva (2006b)

ISO: Environmental Management—Life Cycle Assessment—Critical Review Processes and Reviewer Competencies: Additional Requirements and Guidelines to ISO 14044:2006 (ISO 14071). ISO, the International Organization for Standardization, Geneva (2014)

Jolliet, O., Saade-Sbeih, M., Shaked, S., Jolliet, A., Crettaz, P.: Environmental Life Cycle Assessment. CRC Press Taylor & Francis, Boca Raton, FL (2015)

Klöpffer, W.: Peer (Expert) review in LCA according to SETAC and ISO 14040. Int. J. Life Cycle Assess. **2**, 183–184 (1997). doi:10.1007/BF02978411

Klöpffer, W.: The critical review process according to ISO 14040-43. Int. J. Life Cycle Assess. **10**, 98–102 (2005). doi:10.1065/lca2004.12.190

Klöpffer, W.: The critical review of life cycle assessment studies according to ISO 14040 and 14044. Int. J. Life Cycle Assess. **17**, 1087–1093 (2012). doi:10.1007/s11367-012-0426-7

Lang, B., Lupi, C., Omlin, M., Reinhardt, I.: Wo speist man ökologischer? Möglichkeiten und Grenzen von Ökobilanzen im Restaurant-Vergleich. GAIA—Ecol Perspect Sci Soc **3**, 108–115 (1994)

Schulz, M., Mersiowsky, I.: Critical review of LCA—essential for quality and understanding. In: 6th International Conference on Life Cycle Management. LCM 2013. Gothenburg, Sweden, p. 4 (2013)

Weidema, B.P.: Guidelines for critical review of product LCA. Belgium, Brussels (1997)

Author Biographies

Ralph K. Rosenbaum LCA expert and environmental modeller focusing on LCIA development since early 2000s. Contributed to several UNEP/SETAC working groups towards global harmonisation of LCA methodology. Interested in LCIA modelling of emissions and water/soil resource use, operationalisation of uncertainty management and spatial differentiation.

Stig Irving Olsen LCA expert both as researcher and as consultant. Involved in the development of LCA methodologies since mid 1990s. Contributed to UNEP/SETAC working groups on LCIA methodology. Main LCA interest is human toxicity impacts, emerging technologies, and decision making.

Chapter 14
Use of Input–Output Analysis in LCA

Tuomas J. Mattila

Abstract Input–output analysis can be used as a tool for complementing the traditionally process-based life cycle assessment (LCA) with macroeconomic data from the background systems. Properly used, it can result in faster and more accurate LCA. It also provides opportunities for streamlining the LCA inventory collection and focusing resources. This chapter reviews the main uses of input–output analysis (IO) to ensure consistent system boundaries, to evaluate the completeness of an LCA study and to form a basis for in-depth inventory collection. The use of IO as a data source for social and economic sustainability metrics is also discussed, as are the limitations of the approach. All aspects are demonstrated through examples and references both to recent scientific literature and publicly available datasets are provided. The aim of the chapter is to present the basic tools for applying IO in practical LCA studies.

Learning Objectives
After studying this chapter, the reader should be able to:

- Understand the historical background of input–output analysis and how it relates to LCA.
- Understand the basic equations of input–output analysis.
- Use input–output datasets to find background information on product systems and processes.
- Use hybrid input–output analysis to identify hotspots and the effect of cut-off in process-LCA.
- Use input–output analysis to improve process-LCA dataset.

T.J. Mattila (✉)
Finnish Environment Institute SYKE & Luonnonkoneisto
Engineering Co-operative, Helsinki, Finland
e-mail: tuomas.mattila@ymparisto.fi

M.Z. Hauschild et al. (eds.), *Life Cycle Assessment*,
DOI 10.1007/978-3-319-56475-3_14

- Use input–output analysis as a basis for collecting more detailed process-LCA data.
- Find social and economic data to supplement environmental LCA.
- Understand the strengths and the limitations of using input–output analysis as supplement to process-LCA.

14.1 Introduction

This chapter introduces how to use input–output analysis (IO) in life cycle assessment (LCA). IO was initially developed for macroeconomic systems analysis and planning, but it shares many approaches and methods with process-based LCA. After decades of separate methodological development, the recent trend is to combine the tools into environmentally extended input–output analysis (EEIO), hybrid IO-LCA and comprehensive sustainability assessment. The application of IO together with LCA is assisted by the fact, that it shares the same structure as attributional LCA, linking environmental impacts to economic demand through a product system.

An important problem in conventional process-based LCA is cut-off, or the omission of certain parts of the product system (see Chap. 9). LCA attempts to model every environmental, social and economic impact caused by a product throughout its life cycle from "cradle to grave", integrated over time and space. In practice, this is impossible, and certain simplification for the system boundaries have to be introduced. Everything outside those system boundaries is considered to be "cut-off" from the analysis. If this cut-off is allowed to be subjective, it ruins the idea of comparable and repeatable results. Therefore detailed cut-off criteria, product category rules, standards and handbooks have been developed for standardising and harmonising system boundary setting (EC-JRC 2010).

The product system of an LCA can be thought of as a branching tree. It starts from the functional unit and branches out to the first tier of inputs needed to supply the functionality. Each of these first tier inputs then branches out into second tier inputs and so forth (EC-JRC 2010). This branching out is repeated until all the identified inputs and outputs are either resources extracted or emissions emitted to the environment (i.e. "elementary flows"). In practice, only a part of this branching out is done in an individual study. In a typical study, primary data is collected for the foreground processes, which are closest to the final user (see more about foreground process in Sect. 8.2.3). The remaining inputs are connected to LCA databases, which include product systems from previous studies. This forms the background system. The result is a branching process diagram, which proceeds from an individual product towards more general background processes. In addition, there are processes and flows for which no data can be found, and they are considered cut-off. This dataset is then used to estimate, how much environmental impacts should be allocated to the product system in question. In comparison to this branching bottom-up approach, IO ends up with the same result from the top-down, starting from economy wide statistics and narrowing down to industries and product systems.

The IO-based sustainability assessment does not start from a product, but inventory data are collected at the whole economy level. Then the total environmental, social and economic results are allocated to specific industries. This will give a set of "satellite accounts", which describe how much direct impacts each sector causes during a year of production. Using economic allocation, these direct impacts are then combined into embodied impacts for each produced good or service (i.e. how much impact is caused by the whole upstream processing of a good or service). This results in a simultaneous IO-based LCA of all the products in the macroeconomic system. The embodied impact intensities for each product or service can then be used to calculate footprints for subsystems of the economy (e.g. countries, sectors, individual consumers).

A key assumption in IO is that the relationship between production and impacts is linear. This same assumption is shared by attributional LCA but not by consequential LCA. The attributional LCA proceeds by attributing a certain share of the global impacts to a product (e.g. "What fraction of airplane emissions is attributed to an air-freight package?"). Consequential LCA estimates the consequences of changing a part of the economy (e.g. "How much do global emissions change in response to one additional package? What if airfreight increases tenfold?") (see more about attributional and consequential LCA in Sect. 8.5.3). Thus far, attributional LCA has been used much more than consequential LCA. While the consequential approach may be more relevant for decision-making, it also produces nonlinear models which are challenging to integrate with linear models such as IO. As the focus of this chapter is on introducing IO and its applications, the following will include applications to only attributional LCA.

The IO-based approach has two main benefits: it is fast and it is comprehensive. Unlike a process-based LCA, which includes choices about system boundaries and is limited by the resources for inventory collection, an IO-based LCA has the whole economy as its system boundary. It shows indirect and feedback relationships among processes and sectors and is rapid and inexpensive to conduct. Therefore, it is a good screening level tool. In spite of these benefits, it also has several drawbacks. Because IO relies on readily available statistics, the resolution of products is limited by the availability of statistics. This results in aggregation errors when the footprint of "steel products" is used instead of the footprint of "an office chair, of specified make and manufacturer". In addition, the data is usually at least a few years old, as it takes time for the statistical office to collect and harmonise the data from individual companies. These problems are also present in process-LCA databases, but usually the product disaggregation and technology mixes are more diverse. A major drawback is also the limited coverage of environmental impact categories. Sector specific emissions for toxic substances especially are highly limited compared to the accuracy commonly found in process-LCA databases. Using process-LCA together with IO can utilise the benefits of IO and minimise the problems. In a hybrid-LCA, process-LCA data is used for foreground systems and for reliable process-LCA datasets. IO-LCA is then used to capture all the missing flows. Ideally, this results in a comprehensive system boundary and high data quality.

The structure of the chapter is to first give an outline of IO, starting from the background where it rose from. This gives perspective on the current applications. Then the three main uses of IO in LCA are discussed: filling gaps in process-LCA, providing a first draft template to identify hotspots for process-LCA data collection and using IO as a data source for economic and social sustainability assessment. The approach is practical more than theoretical. Each topic has a worked out example using real data to highlight the use of IO. A more mathematical description of IO and an application to the Finnish economy can be found from the dissertation of the author (Mattila 2013).

14.2 Introductory Examples to Environmentally Extended Input–Output Analysis (EEIO)

The origins of IO are in economic planning and the analysis of multiplier effects. These effects can be demonstrated with a very simple example.

"Assume that a farmer needs to supply 1000 kg of grain. Each 1000 kg of grain requires 30 kg of grain as seed. How much total grain has been produced to supply 1000 kg to a consumer?" This problem presents a loop: the outputs of the process are used as its inputs. This results in an infinite series of tiers in the supply chain. For producing 1000 kg of grain, 30 kg of grain is needed for seed (1st tier), the production of 30 kg of grain requires 0.9 kg of seed (2nd tier), for which 0.027 kg of seed (3rd tier) was needed, etc. As each tier is much smaller than the previous tier, the total amount can be approximated by calculating a few tiers and then adding the results. For an accurate answer, the solution can be found from the input–output relations. If the production of 1000 kg requires 30 kg of seed, the input–output ratio is 30/1000 = 0.03. The net output per unit of production is then $1 - 0.03 = 0.97$. The total amount of grain needed for a net output of 1000 kg is then $(1/0.97) \cdot 1000$ kg = 1030.928 kg. In this case, there is a very small multiplier effect (0.03 units of additional production for each unit of demand). In historical times when yields were lower and part of the grain was used as feed for the working animals, the input–output ratio was much higher and much of the production of grain was used to meet the inputs of producing that grain. In more general terms, the total amount of production $x = y/(1 - a)$, where x is the total amount of production, y is the final demand and a is the input coefficient.

These kinds of feedback loops are simple, when a process uses its own outputs as inputs. The problem becomes more challenging, when a process supplies outputs across the economy and uses inputs from several sources. The same feedback loops are present, but they can cycle through several tiers of production. These delayed feedback loops are very common in complex supply chains (or more accurately supply networks), and make economic planning difficult. The problems of planned economies were what made Wassily Leontief develop input–output analysis. He studied in the USSR and Germany, but later moved to the United States, where the wartime economy and subsequent restructuring of the economy provided a good testing ground and plenty of resources for applying the theory. His work with development of input–output analysis earned him a Nobel prize in economy in 1973.

In order to understand IO, let us look at an imaginary production system in a planned economy (Fig. 14.1). Assume that the goal is to build 1,000,000 trucks, and that needs inputs from four economic sectors: truck manufacture, metal manufacture, machine manufacture and ore mining. The sectors are deeply interconnected with trucks needing inputs from metals and machinery; metals needing metals, machinery and ores; machinery needing metals and machines; and ores needing machinery. In addition, each sector needs trucks to transport goods and raw materials. The system clearly has several feedback loops at different levels. It could be solved stepwise, following each loop until the additional production needed would be very small. In a sense, it reminds us of life cycle assessment and interconnected unit processes. A stepwise approach is feasible, if the system is quite small, but what if the system has thousands of sectors and millions of interactions

Fig. 14.1 The same product system described as a flowchart and an input coefficient matrix (**A**) and final demand vector (**y**)

(a)

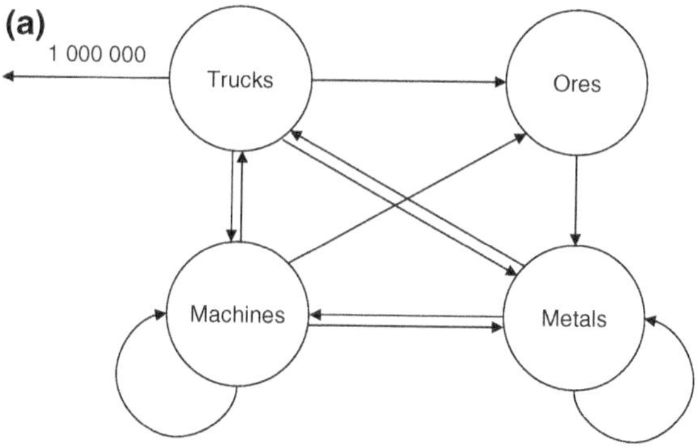

(b)

$$Y = \begin{array}{l} \text{Trucks} \\ \text{Metals} \\ \text{Machines} \\ \text{Ores} \end{array} \begin{bmatrix} 1\,000\,000 \\ 0 \\ 0 \\ 0 \end{bmatrix}$$

$$A = \begin{array}{l} \\ \text{Trucks} \\ \text{Metals} \\ \text{Machines} \\ \text{Ores} \end{array} \begin{array}{c} \begin{array}{cccc} \text{Trucks} & \text{Metals} & \text{Machines} & \text{Ores} \end{array} \\ \begin{bmatrix} 0 & 0.02 & 0.02 & 0.05 \\ 0.15 & 0.2 & 0.2 & 0 \\ 0.2 & 0.1 & 0.2 & 0.4 \\ 0 & 0.1 & 0 & 0 \end{bmatrix} \end{array}$$

like the world economy? Fortunately, the solution is almost as simple as in the case of the grain and seed, and almost all of IO can be summarised in a single equation.

The economic system can be described by using an input coefficient table (**A** in Fig. 14.1). Each column represents a sector, showing the inputs needed to produce one unit of output from that sector. For example, it takes 0.1 units of ores to make one unit of metals in the imaginary truck example. The outputs from the economic system are accounted separately in a final demand vector (**y**). It does not matter what the units are, although commonly a single unit of monetary value is used for each sector.

Now the total amount of produced goods (**x**) is the sum of final demand **y** and the amount of production needed for intermediate demand (i.e. for making all the intermediate products needed to supply the final product). The amount of intermediate production is in direct relation to the total production in each sector (**x**) and the amount of intermediate inputs each sector needs from other sectors. Written as an equation:

$$\boldsymbol{x} = \boldsymbol{y} + \boldsymbol{Ax} \tag{14.1}$$

If we have only a single sector, this results in the same solution as in the grain example: $x = y/(1 - a)$. When there are several sectors, the structure of the equation is the same but matrix inversion replaces scalar division. This gives the core equation of economic input–output analysis:

$$\boldsymbol{x} = (\boldsymbol{I} - \boldsymbol{A})^{-1}\boldsymbol{y} \tag{14.2}$$

where **I** is an identity matrix, which has ones on the diagonal and zeros elsewhere. In linear algebra, it has the same role as one in scalar algebra. $(\mathbf{I} - \mathbf{A})^{-1}$ is the inverse of (**I** − **A**), which can be thought of as the equivalent of division in matrix algebra. (The example can be followed in a spreadsheet program by using the functions MMULT() for matrix multiplication and MINVERSE() for inversion). This inverse of the input coefficient table is commonly known as the Leontief inverse, and it shows the system wide interconnections of each sector with other sectors in its supply chain.

Applying Eq. (14.2) to the system in Fig. 14.1 gives a solution to the truck problem (Fig. 14.2). In order to produce 1,000,000 trucks for final demand, 1,012,608 trucks need to be manufactured. The elements of $(\mathbf{I} - \mathbf{A})^{-1}$ describe the total production needed to provide one unit of final demand from the sector. These are often called indirect multipliers. For example, it takes 0.26 units of metals to produce a truck, while the direct input (**A** matrix in Fig. 14.1) is only 0.15. The indirect inputs take into account all the feedback loops in the system and are always bigger than the direct inputs.

However, how does this relate to sustainability assessment, since many of the "sustainability aspects" are externalities or outside the economic sectors? This has been solved through the introduction of "satellite accounts" and environmental extensions, thus resulting in an environmentally extended input–output table

$$\underbrace{\begin{bmatrix} 1\,012\,608 \\ 264/34 \\ 299\,480 \\ 26\,473 \end{bmatrix}}_{X} = \underbrace{\begin{array}{c|cccc} & \text{Trucks} & \text{Metals} & \text{Machines} & \text{Ores} \\ \hline \text{Trucks} & 1.01 & 0.04 & 0.03 & 0.06 \\ \text{Metals} & 0.26 & 1.32 & 0.34 & 0.15 \\ \text{Machines} & 0.3 & 0.24 & 1.32 & 0.54 \\ \text{Ores} & 0.03 & 0.13 & 0.03 & 1.01 \end{array}}_{(I-A)^{-1}} \underbrace{\begin{bmatrix} 1\,000\,000 \\ 0 \\ 0 \\ 0 \end{bmatrix}}_{Y}$$

Fig. 14.2 A linear algebra solution to the system in Fig. 14.1

(EEIO). The environmental extension describes how much emissions or resources are used for each unit of production on a sector. These "direct emission intensities" are often collected as part of national statistics, especially for greenhouse gases and energy consumption. While the extension may sound difficult, it makes only minor additions to Eq. (14.2):

$$g = Bx = B(I - A)^{-1}y \tag{14.3}$$

where **g** is a vector of embodied environmental impacts associated with final demand **y**, and **B** is a matrix of direct environmental impact multipliers for each sector.

If we were interested in land use and assume that the manufacturing sectors each require 0.01 m^2 of land area and mining requires 1.0 m^2 of land area (i.e. $B = [0.01\ 0.01\ 0.01\ 1.0]$), the total land area demand of the truck example is $g = 0.01 \cdot 1{,}012{,}608 + 0.01 \cdot 264{,}734 + 0.01 \cdot 299{,}480 + 1.0 \cdot 26{,}473 = 42{,}241\ m^2$, with 26,473 m^2 or 63% coming from the mining sector.

The same equation can also be written in a different form:

$$g = Bx = B(I - A)^{-1}y = Cy \tag{14.4}$$

where **C** is a matrix of embodied environmental impact intensity (impact/monetary unit) for all products in the system. It can be thought of as a life cycle inventory (LCI) dataset and is a very valuable in constructing hybrid LCAs and making first estimates for products and services for which process-LCA data is hard to find (e.g. insurance services).

This simple example contains all the basic elements of EEIO and IO, which are used in all common applications of input–output analysis ranging from product level to societal level. However, the example is deceptively simple, the actual usefulness of IO becomes more obvious when one uses a real world example.

Example 14.1 Compare Danish and Chinese steel industry inputs from WIOD datasets 2000 and 2008. Look at total volume of inputs, direct input coefficients and indirect input coefficients.

The WIOD (World Input–Output Database) is one of the publicly available multiple region input–output (MRIO) datasets. It is available from wiod.org. The dataset includes both the input–output tables as well as the socio-economic and environmental accounts. For this example, we will have a look at the monetary input–output table and derive the direct and indirect inputs for Danish and Chinese steel industries.

The WIOT (world input–output table) is arranged in a sector-by-sector format, with all the sectors for a given country in one unit (Fig. 14.3). For the excel file, the country code and sector codes for Danish steel are DNK 27t28 "Basic metals and fabricated metal". In the spreadsheet, the total output (**x** vector) is the last of the columns and was \$5753 M in year 2000 and \$13 141 M in 2008. For Chinese steel production (CHN 27t28), the output was \$211,880 M in 2000 and \$1,251,139 M in 2008. Therefore, the Chinese metal production is considerably larger than the Danish and is growing at a rapid pace. However, has the production technology changed as well?

The emissions of a sector can be considered from a production and life cycle perspective. For the production perspective, a key indicator is the direct emission intensity, which describes the fuel consumption of that sector per monetary unit of output. The direct emission intensity of Chinese metal production has decreased. In 2000, the emissions were 272 Mt CO_2 (1.28 kg CO_2/\$) and in 2008 they were 578 Mt CO_2 (0.46 kg CO_2/\$). For Danish metal industry, the corresponding figures were 0.4 Mt CO_2 (0.07 kg CO_2/\$) in 2000 and 0.4 Mt CO_2 (0.03 kg CO_2/\$) in 2008. Both industries had obtained considerable reductions in emission intensity, but was this at the cost of increased outsourcing and more embodied emissions in the inputs? For this, we need the life cycle perspective of a sector level carbon footprint.

A first step in calculating the carbon footprint is to convert the monetary flow data into input coefficients (i.e. how much inputs are needed to provide one unit of output; the **A** matrix). This is obtained by dividing each column j of the monetary flows by the corresponding total output (x_j). (In this case, the **x** contains zero elements reflecting that some of the sectors are not active in the country, which results in an error. This can be avoided by replacing the zeros with a very small number such as 1/1000,000,000.) The input coefficients can be used for a rough

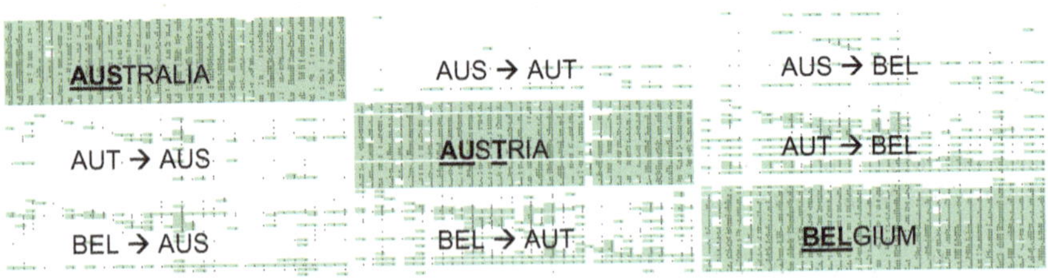

Fig. 14.3 A screenshot from a subset of the WIOT table from WIOD-database. Intra-country transactions are on the diagonal, while trade between countries is arranged on a grid. Each country has 35 sectors

comparison of the value added and input intensity of the sectors. The sum of all input coefficients for Danish steel in 2000 was 0.55 and for Chinese steel it was 0.77. This indicated that the Danish steel industry was able to produce more value added (i.e. less inputs needed for outputs produced) for each unit than Chinese. By 2008, the input coefficients for the industries had increased further to 0.62 for Denmark and 0.80 for China. This indicated that the industries had outsourced their input production to other sectors or countries and/or moved to lower refinement value products.

The trend to outsourcing can be seen from the highest input coefficients (Table 14.1). For both countries, the Basis Metals and Fabricated Metal sector has a considerable amount of inputs from companies within itself. In addition, China has its own mining operations and imports ores from Australia (input coefficient increased by 140% from 2000 to 2008). In contrast, the Danish steel industry has most of its purchases from retail and wholesaletrade services, and imports mainly processed metals from Germany. The largest change in the Danish steel industry has been the increase of recycling (input coefficient change of 100%).

The monetary inputs are interesting, but as we are interested in the carbon footprint, a few more stages are necessary. The first stage involves calculating the Leontief inverse $(\mathbf{I} - \mathbf{A})^{-1}$. The identity matrix **I** can be constructed in a spreadsheet by defining a table, where the elements are set to 1 if the row and the column have the same index [i.e. (1,1) or (2,2)] and 0 elsewhere. After this each element of **A** is subtracted from the corresponding element of **I** and the resulting matrix is inverted (MINVERSE() function in spreadsheet programs). For the WIOD, the inversion will take a lot of memory and some time on most desktop computers. Closing additional programs and copying **I** and **A** matrices to a new spreadsheet document will help conserve memory. After the inversion, it makes sense to copy the inverted matrix to a new spreadsheet to avoid the program from repeating the calculation every time the document is changed.

Table 14.1 A comparison of top 5 direct input coefficients of Chinese and Danish steel products in 2000 and 2008

Sector	Country	In 2000	In 2008	Change (%)
Input coefficient—China basic metals				
Basic metals and fabricated metal	CHN	0.33	0.33	1
Mining and quarrying	CHN	0.07	0.09	29
Electricity, gas and water supply	CHN	0.04	0.05	25
Machinery, nec	CHN	0.02	0.03	50
Mining and quarrying	AUS	0.01	0.02	140
Input coefficient—Denmark basic metals				
Wholesale trade and commission trade	DNK	0.06	0.08	33
Basic metals and fabricated metal	DNK	0.07	0.06	−14
Basic metals and fabricated metal	DEU	0.05	0.07	40
Retail trade	DNK	0.03	0.03	0
Manufacturing, nec; recycling	DNK	0.01	0.02	100

The second stage involves some manual work (or macro programming) in collecting the emissions for each sector in the economy. The CO_2 emissions for each sector in each country have to be collected into a column, which has the same ordering as the sectors and countries in the input–output table. The WIOD has this dataset arranged in separate files for each country, which means that the files need to be combined. After this step, the emissions are divided by the corresponding total output $\mathbf{x}$ to yield the $\mathbf{B}$ matrix (in this case we have only CO_2 emissions, so it is a vector instead of a matrix). After this the $\mathbf{B}$ matrix is arranged to be vertical (a row vector, using TRANSPOSE() function) and is multiplied with the $(\mathbf{I} - \mathbf{A})^{-1}$ matrix (MMULT() function). The result is a row vector, which contains the carbonfootprint of all the products in the world ($\mathbf{C}$ matrix). It is a very useful dataset for recalculation of the examples in this chapter and in other applications.

For Chinese metal products in 2008, the carbon footprint was 2.20 kg CO_2/\$, or almost five times the direct emission intensity. For Danish metal products, the carbon footprint was 0.32 kg CO_2/\$ or almost ten times the direct emission intensity. Both industries have most of their carbon emissions in the supply chain. A first step in locating those emissions is multiplying the input coefficients of the sectors (in $\mathbf{A}$ matrix) with the carbon footprint intensities to have a look, which inputs have the highest embodied emissions.

For Denmark, the emissions diverge globally at the first tier of the supply chain (Table 14.2). The top 5 embodied emissions include metal products from Germany, Russia, Denmark and Rest of the World (RoW; a statistical grouping of economies which were not included in the detailed country analysis of WIOD). If a top 20 listing of emissions had shown, it would have included several more countries in and outside Europe. In contrast, the Chinese metal production has its supply chain focused in China. Most of the inputs were energy, machinery and raw materials for metal production.

Although the metal product sectors of China and Denmark are so different that direct comparison is not meaningful, they provided an example of the use of EEIO to learn about global supply chains and their technological differences. This example also serves as a kind of a warning for using EEIO results in LCA without looking at the product mix in the sector. Using the Chinese industry average for a finished metal product would probably result in a major overestimation of the impact.

Table 14.2 The inputs with the highest share of the carbon footprint of the basic metals sectors in Denmark and China in 2008

	Denmark	kg CO_2e/\$		China	kg CO_2e/\$
Direct emission		0.03	Direct emission		0.46
Basic metals	DEU	0.04	Basic metals	CHN	0.74
Basic metals	RoW	0.02	Electricity	CHN	0.49
Basic metals	RUS	0.02	Mining	CHN	0.16
Basic metals	DNK	0.02	Non-metallic mineral	CHN	0.05
Electricity	DNK	0.01	Machinery	CHN	0.05
Total upstream		0.28	Total upstream		1.73

14.3 Avoiding Cut-Off Through Comprehensive System Boundaries

Most modern supply chains branch out globally, as was seen in the example of Danish metal products. Collecting process-LCA data on a supply chain, which rapidly spreads over several countries and continents, is difficult. Also wholesale and retail trade, which may cover 10–20% of all the inputs to a product manufacturing, often have no process-LCA datasets. The practical consequence of global supply chains and a shift to more services is an increase of cut-off in process-LCA.

Cut-off has always been unavoidable. Usually, it was assumed that the cut-off flows would be insignificant, but later studies have shown that the omission is often 30% or even much larger in some impact categories (Suh et al. 2004).

In principle, there are two sources of cut-off: the identified cut-off and the non-identified cut-off. The identified cut-off consists of flows that are identified during the process-LCA, but which have no LCI data available. The unidentified cut-off is flows which are omitted, since they are intangible (not related to energy or material flows) or simply overlooked. A real-life example of the latter would be ignoring maintenance services in a pulp and paper mill, although the maintenance services consume tools and specialty metals, with considerable impacts to metal depletion (Mattila 2013). Other classical examples would be ignoring insurance, facility rent, retail trade, marketing or software development. Although they may be below a specified cut-off limit at each stage, if these are omitted in all parts of the process-LCA product system, the complete omission will be significant. If economic or social indicators are considered, the omission will be even larger. In a case study of smartphone sustainability assessment, much of embodied child labour was in trade services and warehouse work in developing countries in the parts of the supply chain that supplied parts for smartphone assembly. This came as a surprise both to the analysts and the social responsibility people of the smartphone manufacturers, wholesale trade had previously been ignored in the inventory for child labour.

Fortunately, IO can be used to estimate both identified and non-identified cut-off flows. The first case is termed missing inventories and the second is termed checking for completeness. Both are applications of so-called hybrid-LCA. For a more detailed description of different ways of constructing a hybrid-LCA, see (Suh and Huppes 2005).

A critique for using the IO dataset to fill gaps is that it usually contains very few LCIA impact categories, most often climate impacts from fossil fuels. However, for some types of products and technologies there is a strong correlation between this category and many other LCIA impact categories (excluding toxic impacts and land use) (Laurent et al. 2012), so one approach is to use the ratio of process-LCA climate impact to cut-off impact as a "correction factor" or estimate of cut-off magnitude.

14.3.1 Estimating Missing Inventories from IO Data

Process-LCA has traditionally focused on physical processes and products. Consequently most LCA databases lack services. It is quite straightforward to complete these missing inventory items by using input–output results in a tiered analysis. The analysis consists of four stages:

1. Convert physical flows to monetary flows using price data or for example import statistics, which report both mass and price flows
2. Find an appropriate IO dataset (good geographical and year coverage, relevant environmental extensions included)
3. Convert consumer prices to producer prices (by removing value added tax as well astrade and transport margins)
4. Convert the monetary flow to the currency and year of the IO dataset using producer price indexes
5. Multiply the monetary flows with the corresponding LCI results from the IO dataset (matrix **C** in Eq. 14.4).

It is easiest to describe this process again through an example.

Example 14.2 Estimate the carbon footprint for a wedding trip planned to be from Denmark to San Francisco. The planned flight distance is 18,000 km, some estimated costs would be $40 for public transportation, $3000 for hotels and $1000 for restaurants and $100 for travel insurance.

Assuming that the emission intensity of airplane travel is 0.11 kg CO_2-eq/tkm (ecoinvent 2.2), the climate impact of the flight would be 3960 kg CO_2-eq. We will use the USEIO-LCA model for the economic flows (www.eiolca.net). The EIO-LCA model has a base year of 2002 both in producers and purchasers prices. For the purposes of this example, we will use the purchasers price model, which avoids translating the prices to producers prices (for now).

In order to use the model the prices have to be converted to year 2002 prices. This can be achieved through the detailed consumer price indexes (CPI), available from the US Bureau of Labor Statistics (www.bls.gov). Finding the right statistical category for each commodity requires some research and guesswork. For this example the CPI are presented in Table 14.3. Since prices have increased considerably from year 2002, the purchases of $4440 in 2014 would have been only $3265 in 2002.

Using the converted prices, the carbon intensities from the EIO-LCA can be used to calculate the carbon footprint from the monetary flows (Table 14.3). Based on the results the overall footprint associated with the monetary flows would be 1844 kg CO_2-eq, thus, compared to the emissions from the flight (3960 kg CO_2-eq) the emissions of the monetary flows would be considerable. The major contributor is the stay at the hotel, contributing 1367 kg CO_2-eq. The EIO-LCA presents a detailed description of the components for each of the carbon footprints. In the case of hotels, the main contribution is from the power generation and supply sector (59%), followed by direct emissions from hotel heating (14%).

Table 14.3 Commodity price indexes for 2014 and 2002 for the four goods in the example, their carbon intensities and the contribution to the overall carbon footprint (excluding the flight)

Commodity	CPI 2014	CPI 2002	Purchase in 2014	In 2002 prices	Carbon intensity (kg CO_2-eq/ $\$_{2002}$)	Carbon footprint (kg CO_2-eq)
Taxi	297	184	\$40	\$25	1.870	46
Hotels	308	251	\$3000	\$2445	0.559	1367
Restaurants	155	113	\$1000	\$729	0.580	423
Insurance	318	211	\$100	\$66	0.117	8
Total			\$4440	\$3265		1844

Based on the quick calculation, a good leverage point for reducing the emissions of the trip would be choosing a hotel with high energy efficiency and renewable energy. However, this example has two oversimplifications: first of all the process-based inventory for the flight probably has cut-off, so it represents an underestimation of the total impact; second, the emissions which occur high in the atmosphere have a larger radiative forcing than those close to the ground (therefore the contribution of the other purchases to the whole impact are probably less than the example indicates, and it would be best to avoid the flight altogether).

Example 14.3 The EIO-LCA dataset used in Example 14.2 is quite old (2002). How much would the results change if WIOD year 2008 data would be used instead?

Let us repeat the calculation, but with a different base year (2008) and with producer's prices, since WIOD is based on those. For the conversion from purchasers' to producers' prices, we will just remove the California sales tax (9%), by dividing the costs with 1.09. Since none of the purchases included transportation or retailtrade, we avoided the difficulty of finding the statistics for those.

The results are presented in Table 14.4. Based on the results, the carbonfootprint for the monetary flows would be 1291 kg CO_2-eq, much lower than with EIO-LCA but still significant. The main reasons for the difference are the reduced emission intensity from 2002 to 2008 and the aggregation errors introduced by the WIOD dataset. The EIO-LCA has 428 sectors, with a very detailed disaggregation.

Table 14.4 Commodity price indexes for 2014 and 2002, correction to producers' prices and the carbon footprint using WIOD 2008 data

Commodity	CPI 2014	CPI 2008	Purchase in 2014	In 2002 producers prices	Carbon intensity (kg CO_2-eq/$\$_{2002}$)	Carbon footprint (kg CO_2-eq)
Taxi	297	240	\$40	\$30	0.75	24
Hotels	308	301	\$3000	\$2690	0.33	968
Restaurants	155	135	\$1000	\$799	0.33	287
Insurance	318	271	\$100	\$78	0.14	12
Total			\$4440	\$3597		1291

In comparison, the WIOD only has 35 sectors for each country. Consequently, restaurant and hotel services are in the same category and have the same emission factor. Similar aggregation errors are common in the WIOD dataset in all supply chains, resulting in a more "blurry" image of the supply chain and its hotspots.

14.3.2 Estimating Completeness of the Process-LCA Dataset

Input–output can be useful for finding inventory data on flows that are commonly not found in process-LCA databases, such as insurance, financial services and hotels. However, it can also be used to estimate, how complete the process-LCA dataset is. This is based on estimating the input coefficient and value added in the process-LCA dataset. In Example 14.1, the Danish and Chinese basic metal industries were compared, and it was found that the Danish industry has a much lower sum of input coefficients (0.62) than the Chinese (0.8). It means that for each unit of production, the Danish industry produced value added for 0.38 units. If one calculates the input coefficients and value added for a process-LCA dataset and finds that the value added would be much higher (e.g. 0.9 units per unit of production) it either indicates a very profitable process, or much more likely an omission of some important costs (e.g. infrastructure rent, repairs, insurance and transport).

In constructing a process-LCA, it is straightforward to get financial data for the foreground processes, as one is collecting primary data from companies in any case. However, it may be much more difficult to collect financial data from the companies in the supply chain, since they are most likely not willing to reveal their production cost breakdown to a purchaser of their products. In this case, the input coefficients of the IO-table can be used as a template. The list of physical inputs from an LCA unit process database can be compared with the amounts found in the IO-table inputs, taking note of the main differences in inputs in the two datasets. The IO-table inputs can also be circulated to the companies providing the data with a questionnaire, so they can indicate if their inputs differ considerably from the industry average inputs (this can also be a benchmarking process for the participating companies, increasing their interest for participation).

A third approach for estimating the completeness of the process-LCA is to compare the carbon footprint composition between the process-LCA and the sector average carbon footprint. The formal tools for doing this are contribution analysis and structural path analysis (SPA). Contribution analysis maps out the location of direct emissions in the supply network, which contribute the most to the life cycle impacts. Structural path analysis converts the matrix representation of an IO-LCA into a description of process flows, which cause most of the impacts. The full details for these methods can be found in Heijungs and Suh (2002), but they are also incorporated into most LCA software. For the IO dataset, the EIO-LCA has a contribution analysis included in the toolbox and some IO datasets can be imported into LCA software. If neither case is applicable, one has to follow the approach presented in Example 14.1 (i.e. calculate the carbon footprint matrix **C** and multiply the elements of **A** with it to give a first tier breakdown of the supply chain). If the IO-LCA-based results show a significant carbon footprint from trade services, they probably should be included in the process-LCA inventory.

A problem in this straightforward approach is the lack of environmental extensions in IO-LCA. A given input might be highly significant for a single impact category (for example repair services for metal depletion), but if the impact category is not included in the IO dataset, it will not be identified as important. This problem will gradually be resolved as more impact categories are included in environmentally extended input–output (EEIO) models. The process is now underway in impacts related to land use and biodiversity, hopefully sometime soon global inventories for toxic emissions would be published.

14.3.3 Using Input–Output Analysis as a Template for LCA

Thus far, we have been discussing how to use IO-LCA to fill the gaps in process-LCA. However, the process may be reversed: start from IO-LCA and focus the process-LCI collection work on the parts of the IO-product system, which have the highest environmental impacts. This approach is known as the path exchange method (Lenzen and Crawford 2009). It is a highly effective way of collecting LCI inventories.

In practice, one performs a so-called Accumulative Structural Path Analysis (ASPA) (Suh and Heijungs 2007). The ASPA is conceptually simple: one multiplies all the direct inputs (**A** matrix) with the corresponding embodied impact intensities (**C** matrix). Then top ranking inputs are screened to the next step based on either a specified cut-off level (e.g. more than 1% of total impact) or a specified inclusion limit (together the included inputs must cover >90% of total impact). After the screening, the process is repeated for each of the selected inputs for the second tier. This results in a branching tree structure of the process system, which can be visualised with a Sankey diagram or a flow chart (Fig. 14.4). After the path analysis has extracted the most critical pathways, process-LCA is used to check how much the actual inputs in the foreground system differ from those assumed in the IO-table. Then the LCA proceeds by replacing the most critical inputs with process-LCA collected inventory data.

LCA software (such as SimaPro or OpenLCA) includes tools for drawing Sankey diagrams. If the IO dataset has been imported to the software, the path exchange method is straightforward (for import of IO data the reader is referred to instructional material for the respective LCA software). There is however a hidden risk in this simplicity. With the software, it is easy to overwrite the background IO data with the process-LCA which is collected. Because IO systems are so interconnected, this results in the change of every background process. For example, let us assume the studied product is in the basic chemicals sector, and electricity use is a critical input. If the process-LCI result for electricity consumption is much lower, we need to change the default from the IO. If we just replace the input coefficient in the identified process, we automatically change the amount of electricity needed in all the companies in the basic chemicals sector! This will influence all the inputs for the product system, for example, the packaging materials probably needed cardboard, which needed some basic chemicals to manufacture. Therefore, it is important to make copies of the identified processes before changing them.

It is possible to do the whole process manually in a spreadsheet (although using a mathematical programming language will make the work less tedious). The following example presents a simple iteration in carbon footprinting for a new product.

Example 14.4 Using IO to create a template LCA system boundary for an underwater exploration robot. The OpenROV is an open sourced underwater exploration robot kit.

The bill of materials and the estimated costs are found in the project web page (www.openrov.org). For the purposes of this example, the bill of materials of 35 items was aggregated to IO classifications (Table 14.5). From this onwards, the analysis proceeded by calculating the carbon footprint (using WIOD 2008) for each of the materials, ranking the results, choosing a new set of inputs for the second tier and repeating. In each tier, the input coefficients in the **A** matrix were multiplied with the monetary flow of the inputs from that sector. For example: $141 were from the rubber and plastics sector, which had an input coefficient of 0.22 for "Chemicals and chemical products". Therefore, the input coefficients for the chemicals sector were multiplied with $31.

Using a coarse cut-off limit of 5% of the total footprint, the following diagram was obtained in 30 min using spreadsheet software and drawing tools (Fig. 14.4). It highlights that from the bill of materials, the electronics, plastics and metals are the most relevant. Within the electronics supply chain, there are three components that should be investigated in detail: supply of basic metals, electricity and imported electronics. Within the plastic parts, inputs from chemical industry should be investigated, as should the electricity use. For metals, the direct emissions of metal manufacturing and the metal product inputs should be investigated. The only third tier input included (and it was just at the margin of 5% cut-off) was the direct emissions from the chemical manufacture needed for the plastic components.

Overall, the identified processes cover only 52% of the total footprint. Repeating the analysis with a lower cut-off limit (e.g. 1%) would result in a significantly higher number of highlighted processes.

Even with the coarse cut-off limit, the IO-based template seems reasonable. The main identified inputs were similar to what would have been identified using a

Table 14.5 A cost breakdown for the OpenROV 2.7 underwater exploration robot classified to WIOD IO-sector classes. For simplicity, it was assumed all purchases would be from USA

Input	Cost
Electronic and optical equipment	$313
Rubber and plastics	$141
Basic metals and fabricated metals	$56
Manufacturing, unspecified	$6

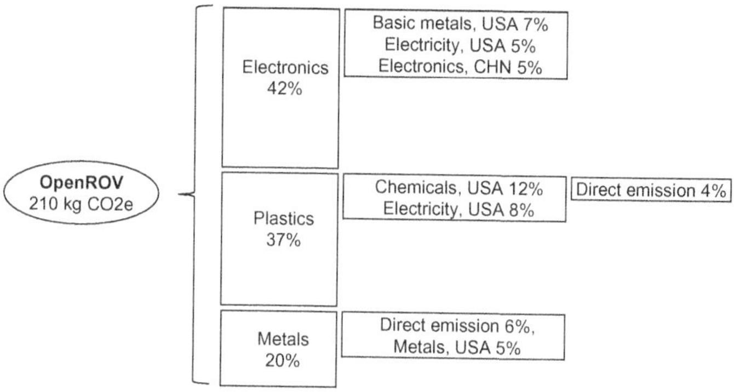

Fig. 14.4 A first estimate of the critical parts of the supply chain for an underwater exploration robot prototype using WIOD data and accumulative structural path analysis with a cut-off of 5%

process-LCA, but having a relative importance score added to them assists in priority setting for further inventory collection.

14.4 Using IO as a Source for Social and Economic Sustainability Assessment

While few IO datasets include many indicators on environment, almost all of them have detailed socio-economic accounts. This can be used as a comprehensive background dataset for social and economic sustainability assessment.

All national accounts include data on employment and value added. Some include the employment by worker category (gender, age and salary level). This can be used to find data for triple bottom line sustainability assessment (see Chap. 5), mapping out where economic activities are happening, where added value goes to and what kinds of salaries are paid to maintain and create the product system.

For example, the WIOD dataset includes the number of employees and the number of persons engaged, and the hours worked by these people and the amount of compensation paid. In addition, it includes a disaggregated dataset for high-medium and low-skilled labour (hours worked and compensation paid). This data can be used to map out, where in the product system work is being done, and the fairness of the compensation compared to the rest of the value added. Average pay in a given country or region is also straightforward to calculate from the data in order to facilitate interpretation.

The social hotspots database (SHDB, socialhotspot.org) has taken this analysis a step further. The database includes inventory and characterisation matrices for social issues (see more about Social LCA in Chap. 16). They are based on risks associated with worker conditions in a given country and sector. These are then used to multiply the hours worked in the supply chain in each sector and country to give an overall risk score for social sustainability, as well as individual indicator (137 indicators) and risk score results (134 risk scores). As more characterisation models become available for social life cycle assessment, there is increased opportunity to use them together with IO-LCA.

One of the benefits of using LCA and IO together is that the analytical tools created for LCA are also applicable to the IO datasets. It is as straightforward to do structural path analysis or a contribution diagram for employment or employee compensation as it is for climate impacts. For example, using the example of the underwater robot the work hour footprint is 13.27 h of work, with the majority of it being 8 h in the electronics supply chain. Of that embodied work, 2.5 h were in USA and 1.4 in the Chinese electronics sector. Approximately 45% of workers in the Chinese electronics sector were low-skilled and 8% were high skilled in 2008. The manufacture of components created some knowledge intensive work, which might be considered beneficial. The value added per hour worked in that sector was 4.7 $/h, of which 33% was wages (labour compensation), equalling 1.6 $/h wages. This is in line with the average manufacturing wages in 2008 in China (Bureau of Labor Statistics, USA), so the sector pays average wage. The calculation could be taken further, by using structural path analysis to map out the entire value tree and the hours worked and the wages paid. These could then be compared to the average wages in the country to evaluate whether the operation is increasing the average wages in the country.

The analysis presented above is based on average statistics. This is a limitation for companies that have a strong policy of social responsibility in the supply chain, as their suppliers might be very different from the overall average figure. For those cases, the benefit of this kind of analysis is to provide a checklist of potential hotspots and to make sure these are addressed in choosing suppliers and negotiating policies.

As the tools for social LCA become more widespread and sophisticated, the IO dataset provide a testing ground for using them. Relatively simple calculations can reveal valuable information about the amount and wages of workers. Complemented with other statistics collected for example by the United Nations International Labour Organization, the analysis can be taken deeper and more focused on issues such as work injuries or child labour.

14.5 Data Sources

14.5.1 Publicly Available EEIO Datasets

There are several publicly available EEIO datasets (Table 14.6), many of them are available for free through an academic license. The datasets however differ in the amount of regions they cover, their sector disaggregation and number of impact categories.

Table 14.6 A comparison of publicly available EEIO datasets

Database	Latest data year	Time series	Regions	Sectors	Impact categories
WIOD	2009	x	40	35	6
EIO-LCA	2002		1	428	20
EXIOBASE 2.0	2007		48	163	98
Waste input–output	2000		1	103	4
EORA	2011	x	190	Average 85	10
CEDA 4.0	2002		6	428	12

The WIOD (www.wiod.org) introduced in the examples of this chapter is a simple to use, relatively small and compact EEIO database. It has a resolution of 40 regions and 35 sectors, which makes it very aggregated and prone to aggregation errors. It also has very few impact categories (6).

In comparison, the EORA (www.worldmrio.com) has a much higher resolution for sectors (on average 85 but ranging from 25 to 428 depending on country) and regions (190 regions). In spite of extensive disaggregation of emission types and sources, the database includes only greenhouse gases, energy, ecological footprint, human appropriation of net primary production (HANPP) and some resource extraction impacts. The cost of a larger sector and country disaggregation is also that the full resolution multiple region input–output analysis (MRIO) cannot be processed with a spreadsheet, but has to be operated through a mathematical programming language (e.g. MATLAB or R). EORA however has a large amount of footprint results precalculated and it has time series of the data, improving analysis possibilities further.

EIO-LCA (www.eio-lca.net) contains some other datasets, but the core dataset is an input–output table of the US in 2002. The resolution is considerable with 428 sectors and the amount of impact categories (the LCIA method TRACI is used) is fairly high for an EEIO model. The web interface (www.eio-lca.net) makes using the tool relatively easy.

CEDA 4.0 (www.cedainformation.net) is based on the same data as EIO-LCA but is much more detailed on the environmental emissions. It has 14 pre-characterised impact categories and 2500 emission and resource depletion categories (LCI inventory level). Currently CEDA 4.0 is available for 6 countries, but the version 5.0 is planned to have global coverage.

The Waste Input–Output Table is a single country input–output table for Japan in 2000. For environmental impact assessment, it has only four impact categories, but the model has a unique approach to waste. Waste generation, processing and reuse have been modelled using separate sectors and technology specific coefficients. Although the data is not very useful in most analyses since it is old and focuses on a single country, the modelling approach is worth considering, especially if one is interested in circular economy research. Another dataset with

waste treatment for each sector. However, these are not common in EEIO datasets, and would have to be added through process-LCA in order to get a full cradle-to-grave assessment.

A related problem is that the IO dataset includes data for a given year. But what if the infrastructure needed has been built a long time ago and is no longer maintained? Moreover, what about the eventual demolition and recycling of the infrastructure? Although IO datasets are spatially complete, they are not temporally complete pictures of the life cycle. This is good to keep in mind, especially when comparing options that are very spread out over time. Mining and energy production systems are typical examples. Ignoring the impacts to future generations undermines the whole purpose of sustainability assessment.

As mentioned earlier, most EEIO datasets are based on a single year of production, while the emission intensities develop over time. For example, the carbon footprint of electricity production in China almost halved from 2002 to 2010 and the electricity production footprint in USA decreased by 37% (Fig. 14.5). Since the base year of EIO-LCA is 2002 and electricity generation is a major contributor to most of the carbon footprints, this means that many of the carbon footprints are now overestimated with the 2002 data. The rate of change is even more rapid in developing countries. This however is a problem which is common to both process and IO-LCA as background datasets are never up to date. A solution is to apply the path exchange method to update the emission intensities for the paths which are identified as important.

The aggregation of sectors is another problem in using the IO datasets for LCA. This can be outlined with an example. The WIOD dataset has 40 sectors and one metal product sector. We used WIOD to estimate the LCA for the underwater exploration robot in Example 14.4, where the main source of emissions was the electronics. The carbon footprint of "electronic and optical equipment" from USA was 0.28 kg CO_2-eq according to the WIOD dataset. The EORA dataset has a much more detailed classification, with 42 products listed under the category electronic

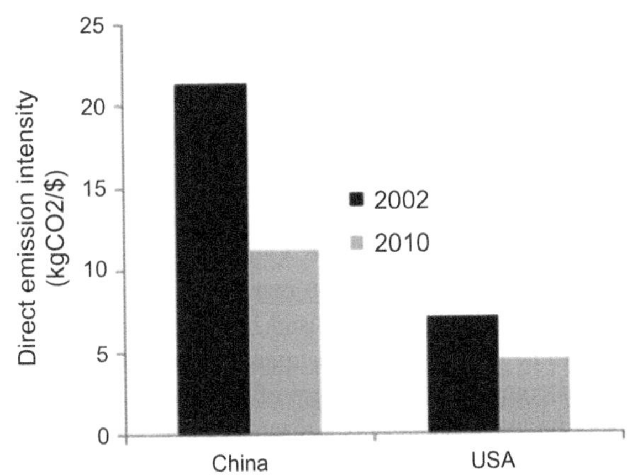

Fig. 14.5 Carbon footprint of electricity generation in China and USA in 2002 and 2010. *Source* EORA dataset factor multipliers

and measurement equipment. The carbon footprint of those products ranges from 0.38 kg CO_2-eq (electricity and signal testing) to 1.09 kg CO_2-eq (carbon and graphite products). The circuit boards and electronic components, which were most relevant for the example, would have a carbon footprint of approximately 0.58 kg CO_2-eq, which is almost twice the value obtained from the WIOD. The aggregation of expensive goods and cheaper products and components in one sector results in an underestimation of the impacts of the latter. While the aggregation has benefits in making the database easier to handle, it also results in loss of precision. The loss depends on the sector and the product, which is analysed, as well as the impact category considered. The effect is magnified, when the characterisation factors of emissions have a large spread and single substance emissions can dominate the whole result (as is the case for the toxicity-related impact categories). The more the product differs from the bulk of the sector's production, the larger the aggregation error. Fortunately, having access to a dataset like EORA means that we can double check the results from a more aggregated model against the disaggregated results, at least for the few impact categories which are included inEORA.

14.7 Summary

This chapter has outlined the application of IO in making better LCAs. The applications of IO have progressed from the research of late 1990s to applicability in case studies. The increased data availability in recent years has increased the possibilities for applying IO.

The main applications of IO in LCA are estimating inventories for flows, which are otherwise cut-off, evaluating the completeness of the LCA, highlighting potential hotspots for inventory collection and providing background data for social and economic sustainability assessment. In addition, the data sources in IO databases make it possible to evaluate the completeness and relevance of process-LCA datasets, by comparing the base year and country of the technology with the emission intensities recorded in the IO statistics.

IO-tables can be daunting at first, since they contain massive amounts of data. Once one gets used to them, they are a valuable addition to the toolbox of a LCA practitioner.

References

EC-JRC: European Commission—Joint Research Centre—Institute for Environment and Sustainability: International Reference Life Cycle Data System (ILCD) Handbook—General guide for Life Cycle Assessment—Detailed guidance. First edition March 2010. EUR 24708 EN. Publications Office of the European Union, Luxembourg (2010)

Heijungs, R., Suh, S.: The Computational Structure of Life Cycle Assessment. Kluwer academic publishers, Dortrecht (2002)

Laurent, A., Olsen, S.I., Hauschild, M.Z.: Limitations of carbon footprint as indicator of environmental sustainability. Environ. Sci. Technol. **46**(7), 4100–4108 (2012)

Lenzen, M., Crawford, R.: The path exchange method for hybrid LCA. Environ. Sci. Technol. **43**, 8251–8256 (2009)

Mattila TJ (2013) Input–output analysis of the networks of production, consumption and environmental destruction in Finland. Doctoral dissertation. Aalto University, Espoo, Finland. https://aaltodoc.aalto.fi/handle/123456789/10942

Suh, S., Heijungs, R.: Power series expansion and structural analysis for life cycle assessment. Int. J. Life Cycle Assess. **12**(6), 381–390 (2007)

Suh, S., Huppes, G.: Methods for life cycle inventory of a product. J. Clean. Prod. **13**(7), 687–697 (2005)

Suh, S., Lenzen, M., Treloar, G.J., Hondo, H., Horvath, A., Huppes, G., Jolliet, O., et al.: System boundary selection in life-cycle inventories using hybrid approaches. Environ. Sci. Technol. **38** (3), 657–664 (2004)

Author Biography

Tuomas J. Mattila Has worked on LCA development since late 2000s in two fields: impact assessment and environmentally extended input-output analysis. His background is in mathematical modeling and chemical engineering. In addition to his research work he also applies ecological engineering in consultancy and farming

Chapter 15
Life Cycle Costing: An Introduction

Jan-Markus Rödger, Louise Laumann Kjær and Aris Pagoropoulos

Abstract The chapter gives an introduction to life cycle costing (LCC) and how it can be used to support decision-making. It can form the economic pillar in a full life cycle sustainability assessment, but often system delimitations differ depending on the goal and scope of the study. To provide a profound understanding this chapter describes several approaches and terms, fundamental principles and different types of costs. A brief introduction is given to conventional LCC and societal LCC but the main focus is on environmental Life Cycle Costing (eLCC) as the LCC approach that is compatible with environmental Life Cycle Assessment (LCA) in terms of system delimitation. Differences are explained and addressed, and an overview is given of the main cost categories to consider from different user perspectives. As inventory data is often sensitive in financial analyses, a list of relevant databases is provided as well as guidance on how to collect data to overcome this hurdle. In an illustrative case study on window frames, the eLCC theory is applied and demonstrated with each step along the eLCC procedure described in detail. A final section about advanced LCC introduces how to monetarise externalities and how to do discounting.

All authors contributed equally

J.-M. Rödger (✉)
Division for Quantitative Sustainability Assessment,
Department of Management Engineering, Technical University of Denmark,
Kongens Lyngby, Denmark
e-mail: januw@dtu.dk

L.L. Kjær · A. Pagoropoulos
Section of Engineering Design and Product Development,
Department of Mechanical Engineering, Technical University of Denmark,
Kongens Lyngby, Denmark

M.Z. Hauschild et al. (eds.), *Life Cycle Assessment*,
DOI 10.1007/978-3-319-56475-3_15

Learning Objectives
After studying this chapter, the reader should be able to:

- Understand the fundamental principles of Life Cycle Costing.
- Know how to use Life Cycle Costing as a tool to make good decisions from different perspectives—as a product/service developer or someone who buys a product/service.
- Know historical and current application areas.
- Know different variants of Life Cycle Costing approaches and understand their differences e.g. in terms of system boundary approaches.
- Be familiar with the monetarisation of intangible elements, which approaches are available.
- Know how to deal with costs in the future (discounting).

15.1 Introduction

Life Cycle Costing (LCC) can form the economic pillar in a full life cycle sustainability assessment comprising the environmental, economic and social dimension (see Chap. 5). LCC is a versatile technique capable of being applied for a range of purposes and at different stages in the project or asset life cycle to support decision-making. It might be undertaken both as an absolute analysis (e.g. to support the process of budgeting) and as a relative analysis (e.g. in order to compare alternative technologies; Langdon 2007). Three variants of LCC can be distinguished. Conventional LCC, also termed financial LCC, is the original method, and in many ways synonymous with Total Cost of Ownership (TCO). Environmental LCC is aligned with LCA in terms of system boundaries, functional unit, and methodological steps. Lastly, Societal LCC includes monetarisation of other externalities, including both environmental impacts andsocial impacts.

For conventional LCC, standards from various government bodies and industry sectors have been developed, includingISO 15663,IEC 60300-3-3, BS 3843, AS/NZS 4536, ISO 15686. For environmental LCC the work of the scientific working group within SETAC on LCC resulted in the LCC methodology described in (Hunkeler et al. 2008), while societal LCC is still at an early stage of development, and more research work is required.

The three types of LCC will be explained in Sect. 15.2. As the approach that is most aligned with LCA, environmental LCC will be the type of LCC explained in depth and exemplified throughout this chapter. Section 15.3 presents the steps of an environmental LCC and provides some practical information for data gathering. A case study in Sect. 15.4 shows how to apply the approach. Section 15.5 elaborates on some advanced issues in LCC, including how to monetarise externalities and how to deal with discounting in LCC.

15.2 Fundamental Principles and Variants of LCC

LCC can be conducted for different purposes, and the methodological choices will depend on the goal and scope of the study. This section introduces the fundamental principles in LCC, including the different purposes and the influence of the target group. This is followed by a description of different types of cost and terminology, before the three variants of LCC are explained.

15.2.1 Fundamental Principles of LCC

As the name suggests, LCC is a technique that assesses costs over the life cycle of a product or a system. Literature features a multitude of terms synonymous to LCC to describe costing across the life cycle of a product, a system or a project, including Through-Life Costing (TLC), Whole-Life Costing (WLC) and Total Cost of Ownership (TCO). It should be noted that in the absence of any internationally recognised standard to describe these terms in detail, differences between them remain a subjective opinion based upon experience, field of study and economic standpoint (Boussabaine and Kirkham 2008).

Conducting an LCC can have different purposes. It may be used as a planning tool, an optimisation tool, a tool for hotspot identification, as part of a life cycle sustainability assessment of a specific product, or to evaluate investment decisions.

A primary consideration relates to the timing of the analysis, where two main types of LCC can be distinguished. *Ex ante* LCC is a prospective approach based on estimates, and is conducted at the early stages of decision-making. In contrast, *ex post* LCC is a retrospective approach based on actual results, usually conducted at the end of a project or a specific time period.

Another relevant consideration is the target group. The target group might be a single actor in a value chain such as a producer or a user or it might take the whole value chain into perspective. The choice of the target group during the goal and scope definition phase of the LCC has implication on the necessary level of detail.

Consider the life cycle cost of a passenger car (see Fig. 15.1). At first, taking the driver perspective (user in Fig. 15.1) for a passenger car, fuel and insurance costs, as well as taxes and potential maintenance are very relevant information in the operation phase, indicated by different grey shadings in the figure. In contrast, a manufacturer would be interested in a detailed analysis of the operational (OPEX) and capital expenditures (CAPEX) such as logistics, research & development (R&D), marketing and so on. Taking the view of a recycler, they would rather be interested in a detailed description of the constitution of the service fees and those expenditures related to recycling the product. A detailed description of costs to be considered from different perspectives and in the different life cycle stages is given in Sect. 15.3.

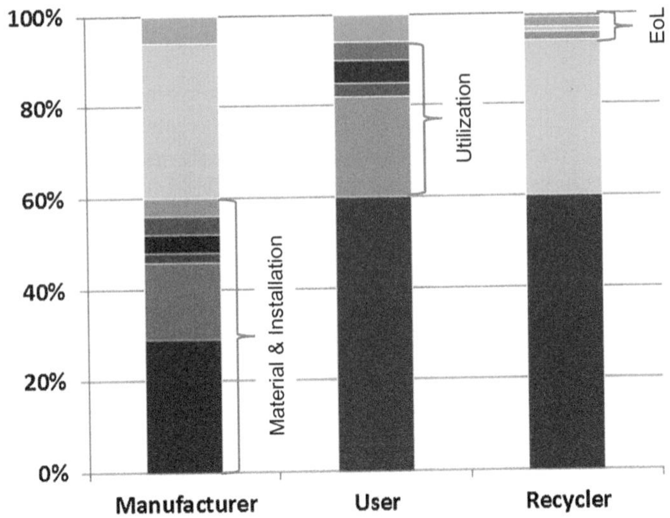

Fig. 15.1 Different level of details for different actors in life cycle costing of a passenger car

15.2.2 Different Types of Costs and Terminology

Costs, Revenues and Value Added

A cost is normally considered as being synonymous with a price of something—it is the monetary value that someone has to pay for something. In an LCC, costs are identified over the life cycle of the product.

LCC can also include revenues which are considered as negative costs. Hunkeler et al. (2008) argue that there are no fundamental problems involved in adding the revenues in the analysis, as long as it is clear how it is being carried out, although for practical reasons they are frequently left out. Depending on the context, inclusion of revenues may be required in order to effectively support decision-making. Consider an example where a window manufacturer uses LCC to compare the life cycle costs of two windows. The two windows are identical, except that one has an extra decorative feature. In this example, for LCC to be of practical use, it needs to evaluate thetrade-off between the extra costs of the feature versus the expected increase in sales. In cases where LCC covers multiple target groups—e.g. manufacturer and user in the passenger car example from before—adding revenues can be confusing, as the cost for one actor is often the revenue for another. In this case, it is important to clearly distinguish between costs and revenues for each target group. In environmental LCC, where multiple perspectives are common, only the value added for each life cycle stage is accumulated in LCC, in order not avoid double counting. See Sect. 15.3 for a detailed description of value added.

15.2.3 Temporal Distribution

Since in LCC, costs are accumulated over a lifespan, one needs to consider that the monetary flows occur at different times. This complicates the analysis for two reasons.

The first is that prices change due to the market dynamics. Looking at cars for example: all costs associated with a car, viz. steel, labour, fuel, plastics, taxes, are likely to change from year to year. In the long run there is a sustained increase in the general price of goods, which effectively alters the purchasing power of currency—a phenomenon known as inflation. In LCC one would like to compare costs based on a chosen reference year and therefore all costs needs to be adjusted to that year when doing the comparison. This is done by using inflation rates. Equation 15.1 shows how to calculate the price P of a product at time t (in years) assuming an inflation rate r, where $P(0)$ is the price at the reference year (t = 0).

$$P(t) = (1 + r)^t P(0) \tag{15.1}$$

The second complicating fact is that people are likely to have a time preference, and often prefer to spend money later rather than now. One solution to take these considerations into account in LCC when comparing future and present costs is discounting. Discounting essentially weights impacts by assigning a lower weight to costs in the future than present costs, and is discussed in greater detail in Sect. 15.4.

15.2.4 Internal Versus External Costs

Costs borne by actors directly involved in the life cycle of the product are termed *internal costs* (sometimes also referred to as ‘private costs’). However, a product or system may involve other costs, borne by other actors indirectly influenced by the product life cycle, e.g. as a result of pollution or othersocial impacts. These are termed *external costs*.

External costs (also termed externalities) are value changes caused by a business transaction, which are not included in its price, or value changes caused as side effects of the economic activity (Dodds and Galtung 1997; Hunkeler et al. 2008). For example, in the construction of a highway close to a residential area, one possible external cost that is not normally included in the life cycle costs of the highway is the value reduction of the houses close to the highway due to the increased noise levels. In conventional LCC, external costs are usually not included. If the external costs are already expressed in some monetary unit, they can be included in the environmental LCC. In societal LCC, externalities can be monetarised and included in the assessment. External costs and monetarisation in general is covered in Sect. 15.5 (Advanced LCC). Table 15.1 gives an overview of the most common terms used in LCC and their definitions.

Table 15.1 Definitions of terms used in LCC

Term	Definition
Price	The amount of money that will purchase a finite quantity, weight, or other measure of a good or service (Sullivan et al. 2006)
Revenue	The income generated from sale of goods or services, or any other use of capital or assets, associated with the main operations of an organisation before any costs or expenses are deducted
Internal cost	Costs borne by actors directly involved in the life cycle of the system under study
External costs	External costs (also termed externalities) are value changes caused by a business transaction, which are not included in its price, or which occur as side effects of economic activity (Dodds and Galtung 1997; Hunkeler et al. 2008)
Value added	Value added is the difference between the sales of products and the purchases of products or materials by a firm, covering its labour costs and capital costs as well as its profits (Hunkeler et al. 2008)
Life cycle costs	The sum of value added over the life cycle of a product or a system (Moreau and Weidema 2015)
Net Present Value (NPV)	NPV is the sum of all the discounted future cash flows that takes into account the time value of money over the entire life time (Park 2011)
Discounting	A method used to convert future costs or benefits to present values using a discount rate (Langdon 2007)
Inflation rate	A measure of the overall change in prices for goods and services over time
Exchange rate	Currency conversion between different currencies

15.2.5 Three Variants of LCC

To understand the differences between the three types of LCC, Fig. 15.2 shows how they relate to the three pillars of sustainability (people, planet, and profit) and which costs they include.

LCC can have various goals, depending on the needs and perspectives of the study commissioners. In an LCC of a private vehicle, the user might not be interested in the end-of-life stage of the car, while society might also include broader impacts, which are not borne by the user or the supply chain such as public health expenditures due to particulate matter emissions. To accommodate for these differences, three variants of LCC have been proposed (Hunkeler et al. 2008): Conventional LCC, Environmental LCC and Societal LCC. The differences between the three variants are summarised in Table 15.2.

Conventional LCC (cLCC)

Conventional LCC (also sometimes called financial LCC) was originally designed for procurement purposes in the U.S. Department of Defence (White and Ostwald 1976; Korpi and Ala-Risku 2008). LCC is mainly applied as a decision-making tool, to support acquisition of capital equipment and long-lasting products with high

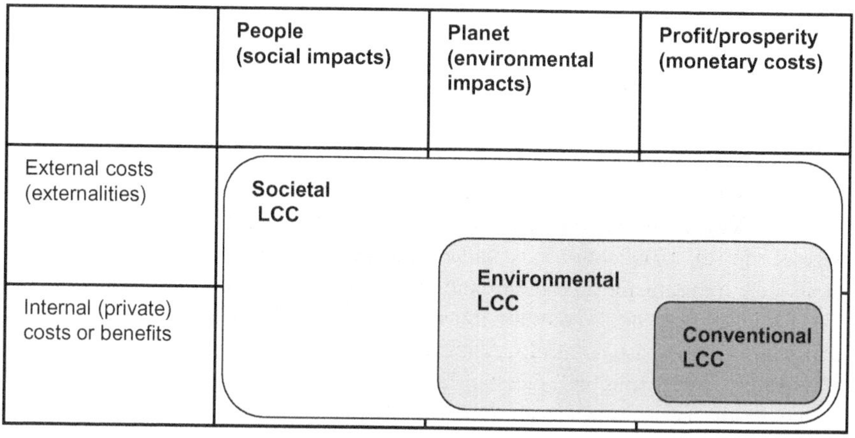

Fig. 15.2 Comparison of the three different types of Life Cycle Costing (Adapted from UNEP 2011 guideline)

investment costs (Hunkeler et al. 2008). Conventional LCC is done from the perspective of a single actor, often the user of a solution. An example would be the procurement of a car, where the driver evaluates different options from an economic viewpoint. In this case, focus is on acquisition costs, taxes, fuel costs and anticipated maintenance costs and might even considerend-of-life costs orrevenues (second-hand value) in the evaluation. Conventional LCC can also be done from the manufacturer point of view, breaking down the life cycle costs with specific focus on the production stages, and—if also borne by the manufacturer—end-of-life costs. In conventional LCC, only internal costs are considered, often ending up with one result for Total Cost of Ownership (TCO), or in cases of hotspot identification with a breakdown of activities, also known as Activity-Based-Costing (ABC). Discounting of the results is recommended. See more about discounting in Sect. 15.5.

Environmental LCC (eLCC)
Unlike the single actor perspective of the conventional LCC, environmental LCC (eLCC) is aligned with the ISO standard 14040 and 14044 on LCA in the sense that it takes the perspective of a functional unit and considers the whole life cycle, including all actors in thevalue chain or life cycle. Unlike the conventional LCC, which is industry driven, environmental LCC was rather developed to support LCA in the sense that it covers the economic dimension, and helps identify hot-spots in terms of both cost and environmental impacts. Besides the internal costs borne by actors in the life cycle, environmental LCC may also include external costs that are expected to be internalised in the near future. In the case of the car, this means that anticipated extra taxes on pollution from fuel combustion might be included in the operational cost. In principle, including external costs from environmental impacts that are quantified in the LCIA results in double counting, since the impacts are

Table 15.2 Comparison of the different variants of life cycle costing

	Conventional LCC	Environmental LCC	Societal LCC
Goal	The assessment of all life cycle costs that are directly covered by the main producer or user in the product life cycle	The assessment of all life cycle costs that are directly covered by all stakeholders connected to the product life cycle	The assessment of all life cycle costs that are covered by anyone in the society
Definition of the life cycle	Economic lifetime, often excluding end-of-life	Complete life cycle	Complete life cycle
Perspectives	Mainly one stakeholder, either manufacturer or user	One or more stakeholders connected to the life cycle	Anyone in the society, often governments
Reference unit	Product or project	Functional unit	functional unit
Types of costs	Internal costs of one stakeholder, focusing mainly on acquisition and ownership costs	Internal costs of stakeholders connected to the life cycle, plus external costs and benefits expected to be internalised such as CO_2 taxes	Internal costs of all actors plus external costs, i.e. impacts that production or consumption have on third parties
Adjustment to inflation	Yes	Yes	Yes
Discounting of results	Consistent, with discount factors ranging between 5 and 10%	No. Discounting the results of the LCC would make the analysis inconsistent with the steady-state assumption of LCA (see Sect. 15.5 on discounting)	Consistent but usually low discount factors (<3%)
Consistent with LCA?	No	Yes, but with a risk of double counting the monetarised environmental impacts	No, due to risk of double counting and inconsistencies with the quasi-dynamic approach in sLCC (see Hunkeler et al. (2008))
Standards	Multiple standards, including ISO 15663, IEC 60300-3-3, BS 3843, AS/NZS 4536, ISO 15686	None, but follows the LCA standards ISO 14040/14044	Currently no standards

accounted for in both analyses. This is not necessarily a problem as long as it transparently shown in the presentation of results and is done consistently for all alternatives being compared (Hunkeler et al. 2008). Like in LCA, the environmental LCC is a steady-state model, and therefore no discounting of the results is usually done. Section 15.3 explains the steps of an eLCC in detail.

Societal LCC (sLCC)
The aim of the societal LCC is to support decision-making on a societal level including governments and public authorities. It includes quantifying the environmental effects in monetary terms. As such, societal LCC (sLCC) includes selected external costs by assigning a monetary value on them. This process is called monetarisation of costs (or impacts). In practice, it is performed by translating the impact results from the LCA into monetary units, e.g. assessing damage costs (see Sect. 15.5 for different monetarisation methods). In this way, the sLCC incorporates the LCA results, and the LCA results should therefore be reported as a subset of the LCC to avoid double counting. An LCC that monetarises all environmental impacts from the LCA is in some cases termed full environmental LCC (Hoogmartens et al. 2014). A sLCC goes one step further and also monetarises social impacts such as: affected social well-being, job quality, etc. In this way, the LCC can be linked to Corporate Social Responsibility (CSR). An sLCC offers the possibility of presenting the result in one single monetary unit, essentially comprising all three pillars of sustainability in a combined Life Cycle Sustainability Assessment (LCSA) aimed at supporting e.g. policy decisions (see Chap. 5). However, this approach of combining all results in a single value is often criticised, mainly because of the uncertainties involved, stemming from both the fact that is difficult to ensure that all relevant external costs are taken into account and from the fact that the external costs are highly uncertain. Discounting is common in sLCC, see more in Sect. 15.5.

As a method for supplementing LCA with economic measures, the eLCC is recommended due to the consistency in the scope of the two analyses. The procedure and methodological considerations are presented in the following section, along with an application example using the case on window frames in Chap. 39 and also known from the other methodology chapters.

15.3 Environmental LCC (Aligned with LCA)

Environmental Life Cycle Costing (eLCC) is the only analysis comparable to the environmental Life Cycle Assessment (LCA) approach. This section gives guidance on how to conduct an eLCC in a consistent way and in parallel to an LCA. It covers three steps:

1. Goal and Scope definition
2. Data collection
3. Interpretation and sensitivity analysis

In general, the overall approach is very similar to the standardised LCA, but there are some important differences, which may both make the analysis easier and more laborious. One advantage is that characterisation or weighting of inventory data can be avoided in eLCC, since the aggregated cost data provide a direct measure of the financial impact and can be aggregated without further processing. On the other hand, the distribution of impacts over time is very important in LCC compared to LCA due to the use of discounting which depends on when the impact or cost occurs. If eLCC results are intended to be used in parallel to LCA, various assumptions must be aligned which will be described in detail below.

15.3.1 Goal and Scope Definition

The goal and scope definition in eLCC is similar to what is needed in LCA and henceISO 14040/44 should be used as a basis (see Chaps. 7 and 8). The goal and scope should be clearly defined but due to the iterative approach, the scope may be revised along the analysis. For instance, eLCC can be used both as a planning tool and as an accounting and reporting tool. Usually it is used for prospective, consequential, and change-oriented assessments to evaluate alternatives, in order to support the product or system design phase, which has utmost influence on prospective costs and emissions along the various life cycle stages.

Functional Unit
For an eLCC, the functional unit shall be defined in a similar manner as for an LCA (see Chap. 8). If the LCC is meant to be conducted in parallel to an LCA, the functional unit needs to be identical.

System Boundaries
System boundaries must be clearly defined and documented like in an LCA (see Chap. 8). If the eLCC is conducted in parallel to an LCA, system boundaries for both must be equivalent and assume the same user perspective. However, eLCC is coarser and it is not always necessary to break down all stages and collect all upstream processes. All real and anticipated money flows should be internalised in a systematic way. The inclusion of external costs in an eLCC is sometimes required while in other situations the system boundaries are negotiable. In general, the inclusion of external costs that are anticipated to be internalised in the decision-relevant future is required.

Cut-Off Criteria
There is an important difference between eLCC and LCA in terms of cut-off criteria. Especially for complex systems with more than a thousand processes, process-based LCA leaves out processes that are assumed to have a negligible contribution thus introducing cut-offs (see Chapt. 8). LCC on the other hand does not suffer from these truncation errors, as costs that occur upstream in the supply chain are assumed to be represented in the price of a product or a service. The cost of purchasing a car for example will include all costs associated with the production of the car, including raw materials, overheads, R&D, marketing, profits for the supply chain and so on. If this is not the case, someone in the supply chain would have to produce at a loss or zero profit, a situation that is clearly unsustainable in the long run.

Although reasonable, the hypothesis that upstream costs are always included in the price is not without exceptions. The price of most commodities is strongly determined by the laws of supply and demand. Market downturns often leave those suppliers with the highest production costs in dire straits, forcing them to either sell at a loss or lay up their assets for a period until the market recovers.

It should be highlighted that an eLCC can be used to inspire the system scoping of an LCA. When considering the life cycle of a product it might be easier for stakeholders to identify all monetary costs over the life cycle rather than environmental impacts and material uses. These costs can be used as a guidance to include all necessary processes (including services) needed to sustain a product or a system over its life cycle in the LCA. Services are often neglected in LCA and including them in the eLCC might inspire to also include them in the LCA. One example could be service and maintenance costs for a car, which is important for the life cycle costs, but might easily be forgotten in the LCA.

Lastly, for minor costs that are not likely to alter the result of the analysis, cut-off criteria need to be applied. For a single life cycle stage (e.g. raw material extraction, production or transport, etc.) a rule of thumb could be that costs that are likely to contribute to less than 1% of the total cost of that stage can be neglected (Hunkeler et al. 2008).

Allocation

Complex systems are subject to allocation to perform an eLCC. In LCA, it is recommended by the ISO 14040 to divide theunit process into sub-processes or to expand the system in order to avoid allocation (see Chap. 8). On the other hand, system expansion is not performed in eLCC. This is due to the fact the eLCC is solely an attributional indicator that can only describe costs, and does not trace the consequences of particular decisions. Special attention is required to ensure a consistent system definition.

The overhead costs are a good example where an allocation method is required. These costs describe all ongoing business expenses, which are not directly linked to production. For those categories, usually costs or revenues are de facto used as allocation keys. For example, in a refinery that refines crudeoil into a number of products, if 40% of the revenue comes from gasoline, then 40% of the overhead costs are allocated to gasoline production.

Inventory

In the inventory analysis, costs should be quantified in one currency (e.g. euro or US dollar) and be based on a common year. For example, if a previous version of a product is compared with the latest version, the costs of both must be aligned in terms of the actual value of the currency at those specific times. If two variants of the same product (e.g. a sedan and a station car) are compared it is necessary to include all relevant costs. Relevant means in particular if costs of the alternatives change (e.g. energy costs, material costs, transportation, etc.). In terms of absolute LCCs (stand-alone) all costs must be taken in account.

Simply adding costs of all actors in the life cycle would not yield any meaningful result. The cost of one actor is the revenue of another, and this process would end up aggregating the same costs multiple times. Instead, what should be considered in an eLCC is the value added at each stage of the life cycle. Value added is the difference between the sales of products and the purchases of products or materials by a firm, covering its labour costs and capital costs as well as its profits.

To get an overview of the hot-spots of the assessed product systems it is recommended to use cost categories on different aggregation levels (see Fig. 15.3). The 1st level consists of three life cycle stages (Manufacturing, Operation and End-of-Life) and external costs. For a manufacturer, the main objective is to analyse every cost in detail during manufacturing, thus the level of detail is higher compared to the other stages in the life cycle. For an operator or user the main focus is on the different costs during the use of the product or service. This affects the data collection strongly. To make the data collection more applicable it should be distinguished between the user perspective and the manufacturer perspective (see Sect. 15.2) thus each life cycle stage has several sub-categories at the second level. For example, if the life cycle costs from a user perspective is to be analysed,

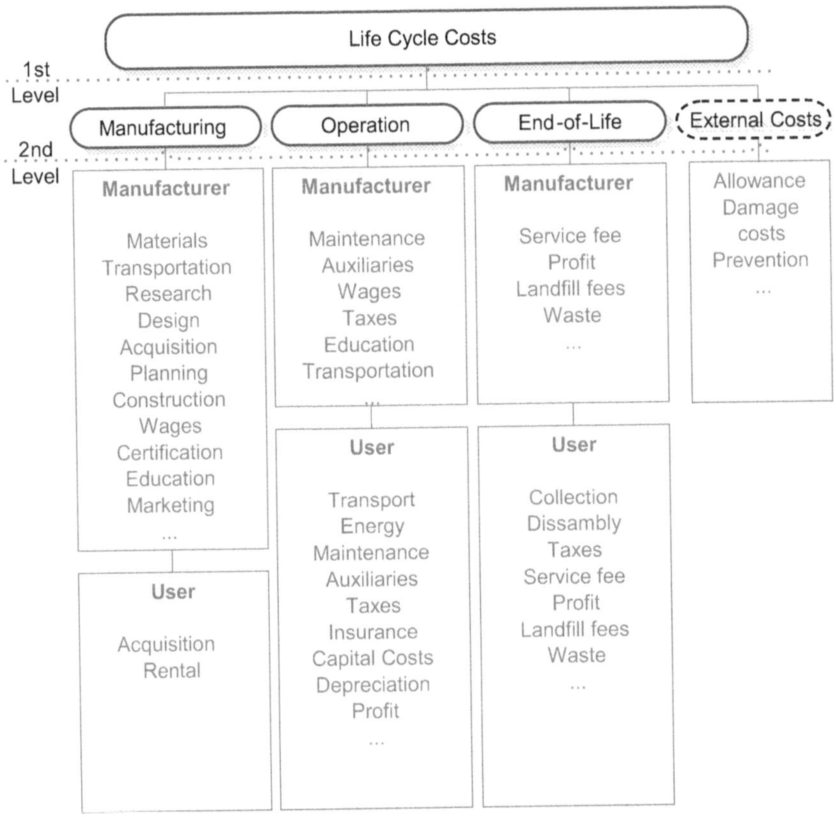

Fig. 15.3 Overview of cost categories distinguish between aggregation levels and between Manufacturer and User perspective

the level of detail for the manufacturing stage is much smaller and the acquisition costs can used as a decent proxy for every cost that occurs upstream. External costs as the fourth stage is another topic within the eLCC and consist mainly of costs for emission allowance (see Sect. 15.5), damage and prevention costs. If external costs are included in the assessment, care must be taken to avoid double counting as described in Sect. 15.2.

15.3.2 Collection of Data for Inventory Analysis

The availability of reliable cost data is crucial in order to perform a realistic life cycle cost analysis. Gathering financial data can be time-consuming and will depend on the collaboration with the involved companies and institutions. The following sections discuss issues in regards to information gathering, particularly for company-based data sources, independent data sources and indirectly derived data.

Company-Based Data Sources

When collecting data internally in a company, there are usually several existing data sources. Accessing them requires the collaboration and involvement of various departments within the company, a task which may be challenging due to unclear responsibilities, lack of resources in the departments and confidentiality issues as well as constraints against an additional economic assessment method.

Typical internal and external data sources in relation to each of the life cycle stages are:

- **Investment and Manufacturing stage**. *Internal*: R&D, Production, and Human Resource Departments
- **Use stage**. *Internal*: R&D, Product Development, accounting systems and Sales (e.g. consumption patterns and sale prices). *External:* Publicly available databases and industry statistics (e.g. fuel prices and taxation costs borne by the user).
- **End-of-life**: *Internal*: R&D or Product Development, who have dealt with the recycling or redistribution of product systems. *External:* EU-directives, international conventions etc.

Independent Data Sources

Financial data can be very sensitive, especially if the results are intended to be published. In these cases most of the data need to be gathered from other independent data sources and references.

Examples of public databases are shown in Table 15.3, giving an overview of different cost categories. These data are published at least annually. However, the scope of each database is different, and it is important to check each data source in terms of comprehensiveness, validity for different regions, currencies and time period to ensure that the data are comparable, while also taking the goal and scope definition into account.

Table 15.3 Public database for life cycle cost data

Type	Scope	Name	Link
Crude oil	Sectors, monthly, country	International Energy Agency	www.iea.org/statistics/topics/priceandtaxes
Plastics	Global, weekly	The Plastic Exchange	www.theplasticsexchange.com
Marine fuel oils	Sector, daily, global	Ship and Bunker's	www.shipandbunker.com/prices
Chemicals	Sector, daily, global	ICIS, Part of RELX Group	www.Icis.com/chemicals
Metals	Sector, daily, global	London Metal Exchanges	www.lme.com
Commodities	Sector, yearly, global	United Nations	www.comtrade.un.org/data
Inflation	Sector, country, monthly	World Bank	www.data.worldbank.org
Wages	Sector, country, yearly	International Labour Organization	www.ilo.org
Currency exchange rates	Yearly, monthly	World Bank	www.data.worldbank.org
Power, gas, coal, oil	Daily	European Stock Exchange	www.eex.com/en

CES EduPack (Granta Material Inspiration 2016) is a leading commercial database that comprises several material and process information. There is a specific Eco Design and Sustainable Development tool, which can analyse the product and design decisions in regards to costs. Another secondary data source for prices is economic input–output tables combined with mass flow analysis on industry sector level. If information on material amount is available (e.g. from the LCA) information on how much a sector pays for the amount can be extracted from comparing monetary supply–use tables and physical supply–use tables. See Chap. 14 for more details on input output tables and their application in LCA.

Indirectly Derived Data

If all these above mentioned data sources are not able to provide the necessary cost data, a cost estimation technique needs to be applied. Cost estimation techniques associate the cost of a product or activity to the available information at the time of the analysis (e.g. the cost of a window frame in regards to its size and bill of materials). The following techniques can be used to estimate costs:

- Surveys and interviews
- Expert opinions
- Cost estimation techniques (qualitative or quantitative).

Both surveys and interviews as well as expert opinions can provide useful cost estimates. While these methods are time-consuming, they can provide estimates in cases where data is either not available or hard to predict. Advanced survey methods such as the Delphi method (see Chap. 21) can help combine multiple expert opinions, with multiple rounds of questionnaires helping the group to converge towards a single number.

Cost estimation techniques can be broadly classified into qualitative and quantitative. *Qualitative techniques* identify similarities between products and are more appropriate to implement when time is limited. One example is case-based reasoning (Huang et al. 2012), which compares previous cases and finds the most suitable to adjust to the new case. An example would be the preliminary determination of the cost of a construction project based on similar construction projects that occurred in the past. *Quantitative techniques* on the other hand are more accurate as they take different product or resource parameters during a whole product life cycle into account (Niazi et al. 2006). The parametric approach for example assesses the characteristics of a product and determines mathematical relationships to describe its cost. It should be pointed out that in practice cost estimation is not necessarily linear, as twice the input does not necessarily produce twice the output. Reasons can include the associated economies (or diseconomies) of scale, the existence of large overheads and exponential relations between inputs. Use of advanced cost models such as simulation models or neural networks can take into account non linearity of costs and the dynamic behaviour of systems.

15.3.3 Interpretation and Sensitivity Analysis

Sensitivity analysis inLCC is very similar to LCA, and is covered in detail in Chap. 11. However, there are some differences. The main difference relates to the fact that—unlike environmental impacts and emissions—commodity prices are much more volatile due to the market mechanisms of supply and demand and the fact that commodities are traded in thestock exchange. Prices are very sensitive to cyclical effects such as financial circles, seasonality etc. Therefore in LCC, the timing of costs is very important, and costs with high price variability such as fuel costs should be subject to sensitivity checks.

15.4 Step-by-Step Application of LCC

In the following section, a case study of an environmental LCC is shown, elaborating the different steps to be conducted to identify the whole costs of a product. This procedure is explained by using the case study discussed in previous chapters (see Chap. 39 for a detailed description). The new window is expected by the company Nor-win to gain a market share of 20–30%. The new window differs from existing windows with respect toheat insulation properties, due to the combination of wood and a composite that is comprised of polyamide and glassfibre. Thus, the new window is expected to have a lower overall environmental impact compared to earlier products from the company. However, a quantitative, life cycle costing has not been conducted yet to identify the full economic profile for the manufacturer and customer. The analysis is based on the aforementioned steps of

- Goal and Scope definition
- Data collection for inventory analysis
- Interpretation and Sensitivity Analysis.

As the goal and scope and the inventory analysis are similar to the described LCA cases, this section mainly focuses on how to gather life cycle costing data and the interpretation of the results. Nonetheless a short summary introduces the goal and scope of the study.

15.4.1 Goal and Scope

The study aims to perform a stand-alone eLCC as guidance for the ongoing design of the new window. Economic hot-spots for the window will be identified. The manufacturer wishes to position itself as a proactive company in terms of sustainability, which entails life cycle costing as well. The target audiences are: (1) design departments at the manufacturer and (2) customers to provide transparency about their new product. The function that is analysed here provides the properties described in the LCA for a minimum of 20 years. The analysis includes all life cycle stages from manufacturing to operation and EoL. However, the level of detail differs compared to the LCA as the data for raw material extraction, primary and secondary material production and other upstream processes are reflected in the final material costs. No cut-off criteria were applied because all needed cost data could be found for the time span of 20 years.

15.4.2 Inventory Analysis

Based on the inventory analysis from the LCA, a list of the materials and production related energy consumption was extracted. But the view on production had to be expanded to cover all costs (CAPEX, and OPEX, indirect costs and immaterial costs).

Usually materials are traded hence these costs were found on specific market platforms or at thestock exchange (see Sect. 15.3). The market evaluation, the design and ramp-up of the production entail additional financial efforts and are covered by the R&D costs. Additionally entailed costs (e.g. labour, infrastructure, prototype production) were allocated by dividing the total costs with the total expected production volume. Usually the cost centre is broken down in very high detail to identify potential cut-downs to improve the return of investment. To improve this rate, the OPEX-related costs are very important as well as they reflect the direct product related costs (e.g. raw material, energy, labour, service and maintenance). The labour costs for assembling the window are dependent on the employer and the production site; therefore official wage levels from several statistical websites were used. The so-called overhead costs include production related costs, e.g. heating and ventilation as well as for lighting. Indirect costs entail site permits, regulatory requirements and prospective liabilities. Those were excluded in the case study. Another cost driver is the immaterial costs (e.g. marketing and competition). An essential part for a company is to promote their products on the market. These additional costs were allocated by dividing the total costs with the total expected production volume. Some products or services need additional certification to increase their market value or achieve competitive edge. These costs were allocated by dividing the total costs with the total expected production volume. Another offer for the customer can be transportation from the manufacturing site to the door sill. All these costs are shown for one window in Table 15.4. The main cost driver is the window pane with about three quarter of the total costs. The labour is second most important driver with about 15% of the overall manufacturing costs. All others are relatively small.

The operation costs are dominated by several drivers (see Table 15.5). Value added taxes (VAT), which are highly dependent on the market (e.g. Denmark 25%, Germany 19%), were added based on the price (costs plus profit for the manufacturer). A relatively low profit margin of 10% was assumed here. As some

Table 15.4 Calculation of the manufacturing costs of polyamide/glassfibre window

Manufacturing costs	Inventory value	Costs [€] per unit	Calculated costs [€]	Worst case [€]	Best case [€]
Materials					
Window frame (kg)	14.36	1.63	23.47	26.53	19.55
Window pane (kg)	61.46	5.90	362.39	435.05	289.95
Window packaging (kg)	1.20	0.75	0.90	1.08	0.72
Production					
Assembly (Electricity) (MJ)	165.00	0.02	3.69	4.05	3.33
R&D (h)	0.01	36.74	0.37	0.39	0.35
Marketing (h)	2.00	0.30	0.60	0.63	0.57
Certification (–)	1.00		0.00	0.00	0.00
Overhead (–)	1.00	2.00	2.00	2.10	1.90
Transportation (tkm)	52.83	0.16	8.23	9.88	6.59
Labour (h)	2.05	36.74	75.33	79.09	71.56
Total manufacturing costs			476.98	558.80	394.51

Table 15.5 Calculation of the operational costs of polyamide/glassfibre window

Operation costs	Inventory value	Costs [€] per unit	Costs [€]	Worst case [€]	Best case [€]
Acquisition costs					
Profit (%)	10	4.77	47.70	50.08	45.31
Taxes (VAT) (%)	25	5.25	131.17	152.22	109.96
Interests (%)	5	6.56	32.79	34.43	31.15
Transportation (tkm)	8	0.19	1.46	1.54	1.39
Installation (h)	4	36.74	146.96	154.31	139.61
Use (MJ)	12,400	0.03	334.76	351.50	267.81
Total operation costs			694.85	744.08	595.24

Table 15.6 Calculation of the EoL costs of polyamide/glassfibre window

EoL costs	Inventory value	Costs [€] per unit	Costs [€]	Worst case [€]	Best case [€]
Window frame (kg)	14.30	0.67	9.63	11.56	7.70
Window pane (kg)	61.40	0.11	7.02	8.43	5.62
Window packaging (kg)	1.20	15.35	18.43	22.11	14.74
Transportation (tkm)	3.89	0.28	1.10	1.32	0.88
Total EoL-costs			36.18	43.41	28.94

customers need a loan to purchase the window, interests were assumed here as well with an effective rate of 5% over one year. Usually the customer has to pay for the delivery of the windows as well, thus the transport performance from the inventory was used and multiplied with the specific costs (e.g. national statistical databases). The windows have to be installed as well and those costs entail the working hours as well as the travelling time of the craftsmen and are dependent on the country as well. However, the largest cost driver is the operation. The assumed 20 years, multiplied with the specific price for district heating including inflation adjustment rate (see Sect. 15.3), led to the result that almost half of the expenses are due to theheat losses.

The End-of-Life stage of a product is always uncertain and future costs must be predicted (see Table 15.6). Theaverage inflation rate of the previous years (e.g. 10 years) was assumed as a good estimate. Taking the actual market costs (e.g. from recycling plants) and adjusting them with the inflation rate (e.g. from national statistical offices) over time (e.g. 20 years) led to a valid estimate. Based on the information from the life cycle inventory, the window will be mainly incinerated and recycled. Those costs were available at incineration plants and recycling stations. Adjusting those with an average annual inflation rate of 2% the prospective costs are roughly 48% higher than those in 2015.

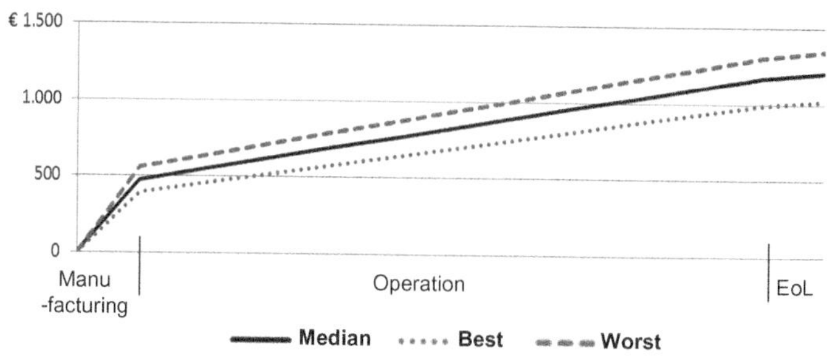

Fig. 15.4 LCC-results including best and worst case assumptions presented over its life cycle

It can be concluded, that the manufacturing and operation stages are most dominant for the new window (Fig. 15.4). Although the costs are based on real market prices, official national and international databases and realistic inflation rates were assumed, an LCC always includes uncertainties. Therefore in each cost centre a specific deviation was assumed and described in the tables above as worst and best case. Based on the assumptions it can be concluded that the life cycle costs have a range of about +11 and −16% compared with the median price of 1.208 € over the entire time span of 20 years.

15.5 Advanced LCC

This section covers some more advanced concepts in regards to LCC and monetarisation in life cycle assessments.

15.5.1 How to Monetarise?

A general problem with some goods and services is that they cannot be traded and it is therefore difficult to determine an objective price for them. Such cases of goods or services without a market can be grouped in the external costs category. Examples of external costs that may call for quantification in anLCC are: the societal cost of respiratory diseases due to air pollution from internal combustion engines, the societal benefits in regional biodiversity from an improved waste treatment plant or the individual benefits of reduced commuting time by using a private vehicle instead of public transport. In the absence of a market price for these values, it is necessary to use monetary valuation to determine their economic value.

15.5.2 Monetarisation of External Costs

Monetary valuation is important in both eLCC and sLCC. As mentioned before, eLCC expands the scope of the cLCC by including cash flows that are expected to be internalised, such as costs for waste disposal and emission taxes. sLCC goes even further, by adopting a societal perspective that includes both all internal costs and selected external costs, as defined in the goal and scope definition.

While monetarisation of external costs can be useful in LCC, especially for socio-economic assessments, it can also be valuable in relation to LCA, as it allows comparisons across impact categories, if the monetarisation is done onmidpoint impact category level. Several methods have been proposed for monetarising externalities, and an overview is given in Fig. 15.5, while Table 15.7 gives a short description of the different approaches.

Most methods try to determine the individuals'Willingness To Pay (WTP) for a particular benefit (or inversely, their willingness to accept a payment in return for a particular loss or disbenefit). An alternative principle determines the external costs in a more direct way by equating them to the costs that would have to be paid in order to avoid or counter balance the change. In order to determine the individuals' Willingness to Pay, different approaches exist, and within each approach multiple methods can be used—each with its own pros and cons. For a more detailed description, see (Boardman et al. 2010).

Table 15.8 shows examples of values for monetarisinggreenhouse gas emissions obtained with three different methods from Table 15.7 and illustrates how the result may be very different depending on the methodology used. The table shows the uncertainties involved in monetarised impacts and how it is important to understand the methodologies behind the analysis. Showing a financial value can easily be perceived as something very definite, with a risk of oversimplifying what are in fact complex issues. While actual market-based costs are factual, monetarised impacts always depend on perceptions and value judgements, which make the underlying assumptions critical for supporting interpretation and presentation of results of an LCC that includes external costs. Methodological choices should always reflect the goal and scope of the study, taking the target audience and the decision-making context into account.

15.5.3 Discounting

There are multiple reasons why costs that occur at different points in time are not directly comparable. From the perspective of behavioural economics, people have a time preference, and prefer to postpone payments as much as possible. A firm in the private sector will prefer to pay suppliers later, and in the meantime invest in expanding its own activities. To solve this problem, it is possible to give a higher weight to imminent costs andrevenues, and a lower weight to future payments.

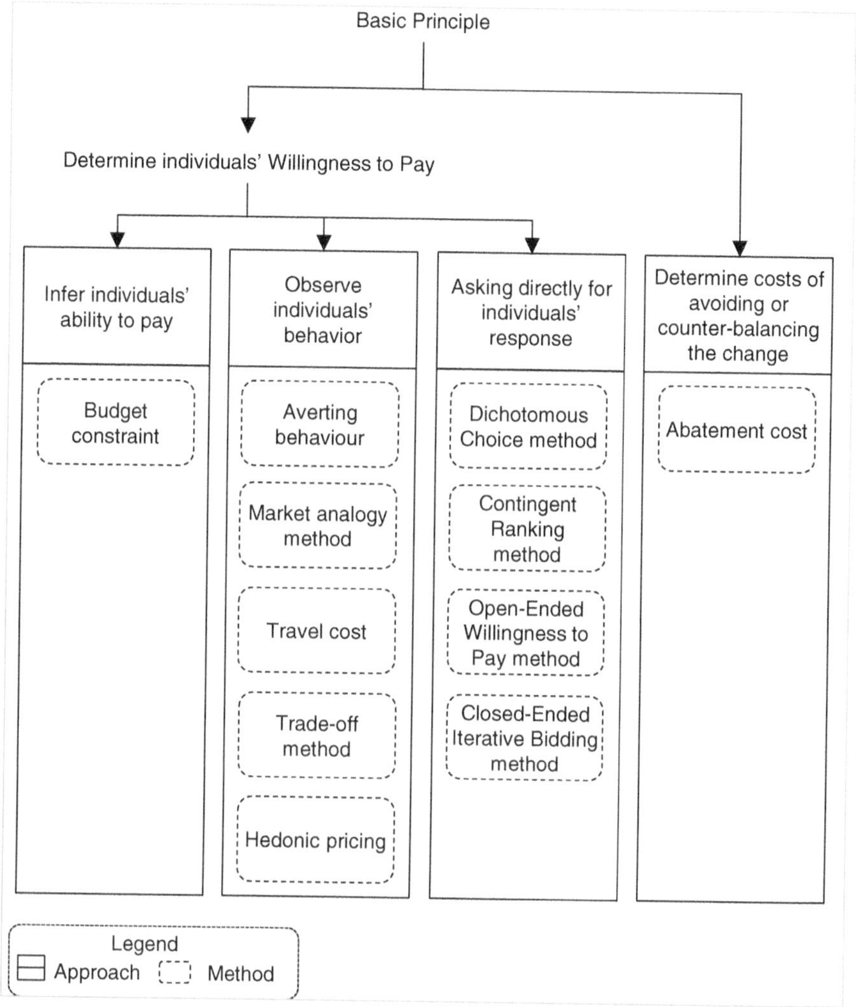

Fig. 15.5 Determining costs—approaches and methods. Based on (Boardman et al. 2010)

The way to determine those weights is through discounting. The weight $w(t)$ for payments occurring at time t is called the discount factor. The discount factor depends on the discount rate r, which is the rate by which the discount factor $w(t)$ decreases over time assuming a first order decrease. The discount factor is calculated as follows:

$$w(t) = \frac{1}{(1+r)^t} \tag{15.2}$$

Table 15.7 Determining costs—approaches' description (Sources: Pizzol et al. 2014; Boardman et al. 2010)

Approach	Description	Possible application area (examples)	Main weakness	Methods
Determine costs of avoiding or counter-balancing the change	The value of the external cost equals the cost of measures needed to mitigate it	Evaluate the cost of greenhouse gas emissions by assessing the costs for carbon sequestration	Does not value utility losses, and hence does not express individuals' attitudes, but rather external targets	Abatement cost
Asking directly for individuals' response	The goal is to elicit people's willingness to pay for changes in quantities or qualities of goods	Ask a number of individuals how much they are willing to pay for the preservation of a national park	Results are highly sensitive to potential sources of error in the survey, as for example the size and the representativeness of the sample of the respondents, and the wording of the questions	Contingent ranking method, Dichotomous choice method, Close-ended iterative bidding, Open-ended willingness to pay method
Observe individuals' behaviour	For cases where there may not be a market for the good or service of interest, its value may be reflected indirectly in the substitute market for a related good	Evaluate the benefit of newer catalysts in cars by evaluating its impact on healthcare costs for respiratory diseases	This approach assumes that people make decisions under full information, a situation that is not satisfied in practice	Market analogy method, Averting behaviour, Trade-off method, Travel cost, Hedonic pricing
Infer individuals' ability to pay	Determine willingness to pay for an additional Quality-Adjusted Life Year.	Evaluate the cost of a statistical life	The approach is only applicable specifically to the value of human well-being	Budget constraint

The sum of all the discounted costs and revenues is theNet Present Value (*NPV*) and is equal to:

$$\mathrm{NPV} = \sum \frac{P(t)}{(1+r)^t} = \frac{P(0)}{(1+r)^0} + \frac{P(1)}{(1+r)^1} + \frac{P(2)}{(1+r)^2} + \frac{P(3)}{(1+r)^3} + \cdots \frac{P(t_{\max})}{(1+r)^{t_{\max}}} \tag{15.3}$$

Table 15.8 Comparison of different approaches for monetarisation of greenhouse gasses in CO_2-eq

Example	Cost per tonne CO_2-eq (€)	Reference	Method
Emission trading scheme (ETS) system	8	(European Commission and Directorate-General Climate Action—B: European & International Carbon Markets 2010)	Market price
Carbon offset program	24	Carbon Offset Program Retrieved August, 2015, from http://www.myclimate.org	Abatement cost
LCIA method (Stepwise2006 v.1.2)	83 €	Weidema (2009)	Budget constraint

where $P(t)$ denotes the cash flows at time t, which is the time of the cash flow. Figure 15.6 shows a simplified flow chart for deciding on the correct discount rate, together with a most probable value. In cases where there is no single correct discount rate, the effect of different discount rates should be investigated through sensitivity analysis, in particular for studies with a long time duration.

The appropriate discount rate depends on the type of cost that is being discounted. For internal costs it is closely related to the cost of borrowing. For private companies a conservative discount factor might be anywhere between 5 and 15%, depending on the required return on investment (Hunkeler et al. 2008). After the financial crisis in 2008 and the worldwide public debt problem, a lower discount factor is more likely for private companies. In the public sector, national ministries of finance generally specify the discount rates to be used in the economic analysis of publicly funded projects. These typically fall into the range of 3–5% (Langdon 2007). In terms ofsocial impacts, British economist Frank P. Ramsey proposed a model in which society would attempt to maximise a social welfare function (Boardman et al. 2010). In that case, the discount rate would reflect on one hand impatience and on the other hand society's preference for smoothing consumption flows over time. Ramsey's formula for society's marginal rate of time preference gives:

$$r = d + g \cdot e \tag{15.4}$$

where r equals the pure rate of time preference (d), plus a term multiplying the long-run rate of growth in per capita consumption (g), by a constant (e). Ramsey's formula usually produces values in the range between 0.5 and 1.2% (Boardman et al. 2010). Finally for environmental impacts, the discount factor depends on the time horizon of the impacts under study. Toxicity from heavy metals can have a

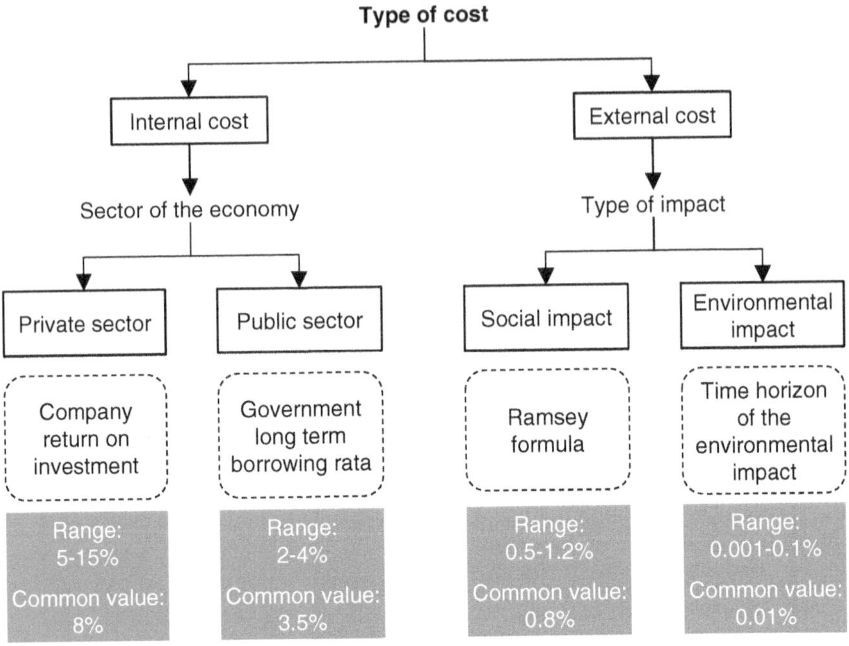

Fig. 15.6 Decision tree for choosing the correct discount factor

long time horizon, as heavy metals are often toxic to humans andecosystems and can remain available in the environment for thousands of years after emission. With that in mind, a discount factor of just 0.0001% that halves the importance of effects every 700,000 years might seem appropriate. On the other hand, toxicity from organic pesticides can have a much shorter time horizon. Diuron for example is a common pesticide with a half-life of approximately 2000 days, so by the same logic a much higher discount rate is appropriate. From the above example, it becomes clear that each environmental impact will require a different discount factor.

Table 15.9 shows the calculation of the NPV of the fuel costs for the average driver in the US, driving 24 miles a day to and from work, by means of a fuel efficient car of 25 miles/gallon with 240 working days per year between 1996 and 2002, where 1996 is used as the reference year. Figure 15.7 shows the discounted fuel costs and discount factors for a range of discount rates. The cost of fuel is shown as the baseline (discounting at 0%). Notice that in case of a high discount factor, fuel costs occurring earlier in time are significantly more important than future costs.

Table 15.9 Calculation of net present value for a discount factor equal to 5%, by applying Eqs. 15.2 and 15.3

Year	1996	1997	1998	1999	2000	2001	2002
Time (t)	0	1	2	6	4	5	6
Discount factor (r)	1	0.95	0.91	0.86	0.82	0.78	0.75
Gasoline price (USD/gallon)	1.20	1.30	1.00	0.97	1.50	1.49	1.25
Fuel cost (USD) [P(t)]	300	325	250	243	375	373	313
Present value (USD 1996)	300 * 1	325 * 0.95	250 * 0.91	243 * 0.86	375 * 0.82	373 * 0.78	313 * 0.75
Net present value (USD 1996)	1879						

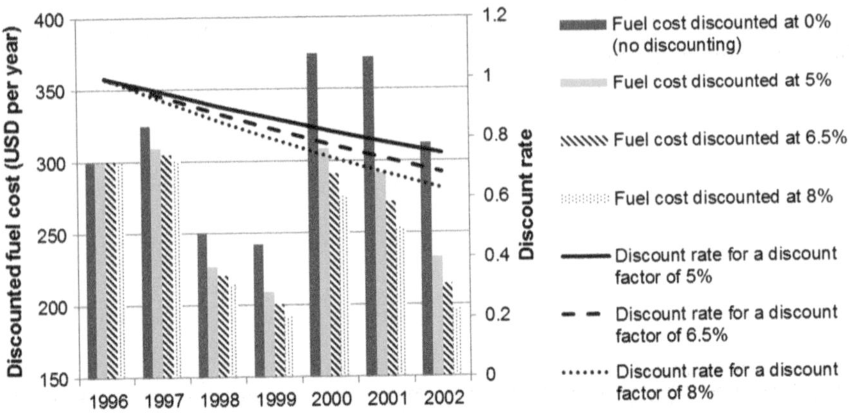

Fig. 15.7 Discounted fuel costs and discount factors for a range of discount rates for the average US driver

References

Boardman, A., Greenberg, D., Vining, A., Weimer, D.: Cost-Benefit Analysis, 4th edn. Prentice Hall, New Jersey (2010)

Boussabaine, A., Kirkham, R.: Whole Life-Cycle Costing: Risk and Risk Responses. Wiley, Hoboken (2008)

Dodds, K., Galtung, J.: Peace by peaceful means: peace and conflict. Int. Aff. R. Inst. Int. Aff., Development and Civilization (1997). doi:10.2307/2623565

European Commission, Directorate-General Climate Action—B: European and International Carbon Markets: Guidance on Interpretation of Annex I of the EU ETS Directive (excl. aviation activities), p. 26 (2010)

Granta Material Inspiration: CES EduPack. Granta Design, Cambridge (2016)

Hoogmartens, R., Van Passel, S., Van Acker, K., Dubois, M.: Bridging the gap between LCA, LCC and CBA as sustainability assessment tools. Environ. Impact Assess. Rev. **48**, 27–33 (2014). doi:10.1016/j.eiar.2014.05.001

Huang, X.X., Newnes, L.B., Parry, G.C.: The adaptation of product cost estimation techniques to estimate the cost of service. Int. J. Comput. Integr. Manuf. **25**, 417–431 (2012). doi:10.1080/0951192X.2011.596281

Hunkeler, D.D., Lichtenvort, K., Rebitzer, G., Ciroth, A.: Environmental life cycle costing. CRC Press, Pensacola (2008)

Korpi, E., Ala-Risku, T.: Life cycle costing: a review of published case studies. Manag. Audit. J. **23**, 240–261 (2008). doi:10.1108/02686900810857703

Langdon, D.: Life Cycle Costing (LCC) as a contribution to sustainable construction: A common methodology, Final Report. European Commission, Brussels (2007)

Moreau, V., Weidema, B.P.: The computational structure of environmental life cycle costing. Int. J. Life Cycle Assess. **20**, 1359–1363 (2015). doi:10.1007/s11367-015-0952-1

Niazi, A., Dai, J.S., Balabani, S., Seneviratne, L.: Product cost estimation: technique classification and methodology review. J. Manuf. Sci. Eng. **128**, 563 (2006). doi:10.1115/1.2137750

Park, C.S.: Contemporary Engineering Economics. Pearson Prentice Hall, New Jersey (2011)

Pizzol, M., Weidema, B., Brandão, M., Osset, P.: Monetary valuation in life cycle assessment: a review. J. Clean. Prod. **86**, 170–179 (2014). doi:10.1016/j.jclepro.2014.08.007

Sullivan, W.G., Wicks, E.M., Luxhoj, J.T.: Engineering Economy. Pearson Education, New Jersey (2006)

UNEP: Towards a Life Cycle Sustainability Assessment: Making Informed Choices on Products. United Nations Environment Programme, Nairobi (2011)

Weidema, B.P.: Using the budget constraint to monetarise impact assessment results. Ecol. Econ. **68**, 1591–1598 (2009). doi:10.1016/j.ecolecon.2008.01.019

White, G., Ostwald, P.F.: Life cycle costing. Manag. Account. **57**(7), 39–42 (1976)

Author Biographies

Jan-Markus Rödger Involved in LCA/LCC and LCM activities in various industries since 2010. Main activities are the application of science based assessment tools in actual product and production development processes. Interested in the integration of life cycle thinking in the manufacturing environment.

Louise Laumann Kjær Has been working with sustainability assessments and LCA within both the private and public sector as a consultant and researcher since 2009. Special interests include Environmental Input/Output LCA, the relationship between economic and environmental impacts of products and services, and assessing the sustainability of innovative business models such as circular business models and product/service-systems.

Aris Pagoropoulos Researcher of economic and environmental impacts of products and services, with a focus on the shipping and manufacturing industries. Main interests include LCC, data analysis and Environmental Input/Output LCA.

Chapter 16
Social Life Cycle Assessment: An Introduction

Andreas Moltesen, Alexandra Bonou, Arne Wangel and Kossara Petrova Bozhilova-Kisheva

Abstract An expansion of the LCA framework has been going on through the development of 'social life cycle assessment'—S-LCA. The methodology, still in its infancy, has the goal of assessing social impacts related to a product's life cycle. This chapter introduces S-LCA framework area and the related challenges. It outlines the main conceptual differences between LCA and S-LCA and discusses the barriers in terms of methodological development and potential application. Three case studies are presented applying S-LCA in different contexts and using varying methods. In the light of the outlined differences, perspectives for the future development of S-LCA are discussed.

Learning Objectives
After studying this chapter, the reader should be able to:

- Understand the methodological phases of S-LCA.
- Explain the main differences between LCA and S-LCA; the related challenges and implications.
- Explain how social impacts are often defined in the SLCA literature.
- Explain how social impacts depend on the conduct of the company rather than the nature of the process.
- Demonstrate an overview of S-LCA applications in different contexts and using different methods.
- Give examples where the use of SLCA for decision support may not benefit stakeholders in the product life cycle.
- Discuss the perspectives for the future development of S-LCA.

A. Moltesen (✉) · A. Bonou · A. Wangel · K.P. Bozhilova-Kisheva
Division for Quantitative Sustainability Assessment, Department of Management Engineering, Technical University of Denmark, Kgs. Lyngby, Denmark
e-mail: andreasmoltesen@gmail.com

M.Z. Hauschild et al. (eds.), *Life Cycle Assessment*,
DOI 10.1007/978-3-319-56475-3_16

16.1 Introduction

Since the 60s, there has been increasing awareness that constant growth in consumption and production within the limits of the finite planet is not viable for humans andecosystems. This realisation has led to a vision forsustainable development. The key term "sustainability" is defined in Chap. 5 as "the ability for meeting present human needs without compromising future generations" after the commonly referenced *Brundtland Report* from 1987 (WCED 1987). The chapter also discusses that the goal of sustainable development was one of the motivations behind the development of LCA, which aims to support environmental protection.

However, beyond the environmental concerns sustainability is also related to social aspects. The concerns on the social aspect of sustainability reflects in today'spolicy frameworks such as United Nations' Sustainable Development Goals, in various national and international initiatives focusing on sustainability of supply chains, and in standardisation frameworks of social nature such as the ISO 26000's Guidance on Social Responsibility (ISO 2010; UN 2010; UNDP 2015). In this context, to be able to give a more comprehensive assessment of a product's or system's contribution to sustainability, an expansion of the LCA framework to also include the impacts on social entities (e.g. workers, consumers, communities) has been going on since the early years of this millennium. This expansion of LCA is known as the 'social life cycle assessment'—S-LCA.

The ambition for S-LCA is to be a methodology, in other words a system of methods with corresponding procedural steps, which if followed will lead to an assessment of the social impacts of a product over its life cycle. The initial development of S-LCA was strongly influenced by LCA, with the scientific community assuming that S-LCA can assess social impacts in the same way that LCA can assess environmental ones. Its methodological phases are thus similar to the ones discussed in Chaps. 7–12:

- *Goal definition* addresses what is to be assessed and why the assessment is performed.
- *Scope definition* addresses the choices made in order to perform the assessment and the limitations of the assessment.
- *Inventory analysis* has the purpose of collecting the data outlined through the goal and scope definition.
- *Impact assessment* uses models to translate inventory data into impacts.
- *Interpretation* analyses the outcome of the previous phases in accordance with the goal of the study and tries to answer the question posed in the goal definition.

16.1.1 Status of S-LCA

As described in Chap. 5, (environmental) LCA has been standardised, e.g. in the ISO 14000 series standards and in theEuropean Commission'sILCD guideline (ISO 2006a, b; EC-JRC 2010), and is broadly acknowledged and applied in publicpolicymaking and private initiatives (see Part III of this book for examples of applications). In contrast, S-LCA is still in its infancy. The existing S-LCA literature thus presents a broad variety of approaches for the above methodological phases. Therefore, to characterise it as a consistent and consensual methodology will be misleading. Rather, one could probably speak of bits and pieces of methodological suggestions with the overall goal of assessingsocial impacts related to a product's life cycle.

To date the most important step towards thestandardisation of S-LCA has been the development of the "Guidelines to S-LCA" under the UNEP-SETACLife Cycle Initiative (Benoît and Mazijn 2009). This was the result of a consensus process involving researchers working on S-LCA, mainly fromEurope and North America. The process, which lasted several years, was the first step towards bridging the differences present in the S-LCA community at the time of publishing. Yet, since a limited amount of research had been published prior to the "Guidelines for S-LCA", this publication, rather than a definitive guide, can be considered as a first rough map, a skeleton for the future work on S-LCA. This was also emphasised by the main authors of the guidelines and has become evident in the later work on S-LCA where significant methodological problems have been revealed.

16.1.2 Focus of Chapter

The intention of this chapter is to give an introduction to the S-LCA area and the related challenges rather than to analyse its methodological aspects in detail or to give a stepwise description of how one *could* perform an S-LCA (for this we refer to the "Guidelines for S-LCA" which is more a "how to" guide).

We outline the main conceptual differences between LCA and S-LCA drawing on the background knowledge of the LCA framework that you will obtain by reading Chaps. 7–12. The chapter further discusses the barriers that these differences set in relation to using the methodological framework of LCA for assessing social impacts. By "barrier" is meant anything that could impede the *ease of use*, the *accuracy*, or the *meaningfulness* of the assessment. These observations are of key importance for the applicability and trustworthiness of S-LCA.

The chapter's structure follows the methodological phases outlined earlier in this section, however, as the interpretation of S-LCA does not differ from the LCA, this phase is not described. The methodological overview is followed by a summary, discussing the implications of the differences between S-LCA and LCA. After this, a short presentation of three case studies applying S-LCA in different contexts and using varying methods is given to illustrate real applications of S-LCA. Finally, in the light of the outlined differences, perspectives for the future development of S-LCA are discussed.

16.2 Overview of S-LCA Methodology

16.2.1 Goal Definition

S-LCA assesses "social impacts" rather than environmental impacts as done in the LCA. But what is meant by "assessing social impacts"? There is a general consensus in the S-LCA community that the ultimate purpose of an S-LCA is to assess how human well-being is affected by products or systems throughout their life cycle (Weidema 2006; Dreyer et al. 2006; Jørgensen et al. 2010b). Using the LCA terms, well-being can thereby be considered as the Area of Protection in S-LCA, i.e. the concept that S-LCA is most fundamentally attempting to assess impacts on in order to ensuresustainability. This also implies that S-LCA should provide a methodology not only for identifying the social changes caused by a product or system but also for characterising them and evaluating them in relation to how they contribute to some overall human well-being.

S-LCA is to assess impacts on well-being, but well-being of whom? In principle, any affected human is considered a stakeholder in S-LCA, implying that if the well-being of a person is affected by some activity in the product life cycle, it should be included in the assessment. Prevailing stakeholder groups (see also Table 16.1) considered in S-LCA are the workers across the life cycle (who have gained the largest attention in S-LCA research); the local or regional communities affected by the product life cycle stages; and the product users (Jørgensen et al. 2008). Additionally, S-LCA may consider other stakeholders who can affect or can be affected by decisions taken across the product life cycle, e.g. shareholders, company owners and other decision-makers (Benoît and Mazijn 2009).

16.2.2 Scope Definition

Impact Categories in S-LCA

The goal of S-LCA is to assess impacts from the product life cycle on stakeholders' well-being. However, before assessing how it is *affected* we first need to define what well-being *is*. Despite being at the foundation of S-LCA, "well-being" has been discussed to a rather limited extent by the S-LCA community (Jørgensen et al. 2010b). The concept goes beyond physical health, i.e. psychological aspects play a central role in its essence. Furthermore, well-being in S-LCA is a concept commonly related to a personal (and thus subjective) experience. Thus, objectively

Table 16.1 An overview of social impacts included in S-LCA approaches

Worker related issues
Non-discrimination
Freedom of association and collective bargaining
Child labour, including hazardous child labour
Forced and compulsory labour
Level and regularity of wages and benefits
Physical working conditions
Psychological working conditions
Training and education of employees
Society-related issues
Corruption
Development support and investments in society
Local community acceptance of company
Company commitment to sustainability issues
Product user-related issues
Integration of costumer health and safety concerns in product
Availability of product information to product users
Ethical guidelines for advertisements of product

observable living conditions, such as income, physical health, housing, etc. are necessary but not sufficient to gauge well-being.

In S-LCA, well-being is mainly understood in a descriptive way, meaning that S-LCA methodology developers have attempted to identify those social themes that contribute to human well-being and hence form the basis for the definition of impact categories for S-LCA. Indicatively in the "Guidelines for S-LCA", there are more than 30 themes. Table 16.1 summarises some of these per stakeholder group:

The social themes in Table 16.1 have been identified following three different approaches of which the first has been the dominant one.

(i) *Normative compliance*: Most of the themes related to employees and workers have been based on international conventions relating to working conditions, namely conventions from the International Labour Organisation (ILO 2016). This is a UN organisation working to establish a set of universal worker rights. Although ILO conventions have been adopted by most countries, their enforcement is often weak. Other less authoritative standards such as the ones made by the Global Reporting Initiative (GRI 2016) have also been used to identify relevant social aspects for S-LCA.

Normative requirements are undoubtedly useful for monitoringsocial impacts. Nonetheless, they should be perceived as the outcome of long political negotiations and compromises to reach international consensus rather than as scientifically valid instruments for assessing human well-being. Therefore, while the limits they set can be a reference for S-LCA, they are not absolute standards aiming to safeguard well-being and their direct adoption in S-LCA can be problematic.

(ii) *Social theory interpretation*: A second approach, less commonly used in the S-LCA literature, is to use social theories about human well-being and from these derive the social themes relevant to include in S-LCA (Jørgensen et al. 2010b). Yet, it remains a challenge to establish theoretically valid and to some extent mechanistic causal pathways (as also known from the environmental impact assessment in LCA) between various events in the product life cycle and well-being. Figure 16.1 shows an example of an impact pathway for child labour.

(iii) *Co-creation*: A third approach, which is more discussed in literature than actually carried out (Dreyer et al. 2006; Kruse et al. 2009), is to identify the social impacts relevant to include in the S-LCA through participatory processes involving the stakeholders that are affected. The principle is that the affected stakeholders know what influences their well-being and how, and therefore they should be the ones to define what is relevant to assess.

Even though it might seem preferable to base S-LCA on a combination of the two latter approaches, these introduce several challenges. One is that if the social impacts that affect well-being vary according to the perception of stakeholders, then aggregating impacts across the life cycle stages (which is a fundamental principle within the life cycle methodologies) might be problematic as different stakeholders along the life cycle will often have different perceptions. Another problem is related to the identification of relevant social themes. The aspects considered in the ILO conventions or standards have been publicly accepted as relevant and important to consider. This is not necessarily the case for the aspects identified through theoretical analysis of "well-being" or aspects defined by stakeholders themselves.

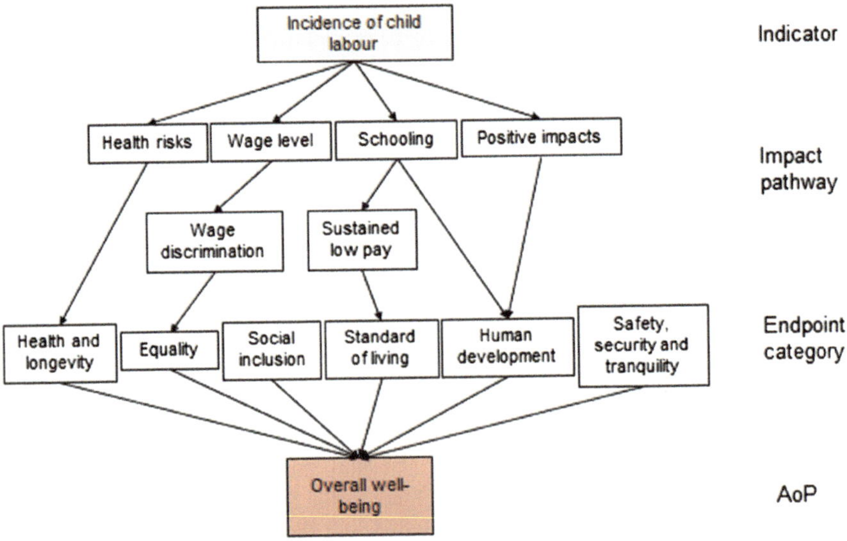

Fig. 16.1 Impact pathway for the impact category child labour (Jørgensen et al. 2010b)

These approaches may therefore be more difficult to relate to a decision-maker, let alone to be streamlined. As a compromise, it has been suggested to let the normatively based impact categories function as a core set of impact categories that should always be included in an S-LCA and then supplement by co-created impact categories according to the relevance in the specific study (Dreyer et al. 2006).

Setting the System Boundaries in S-LCA

System boundaries in S-LCA, like in LCA, define which parts of the life cycle and which processes belong to the analysed system, i.e. which processes are required for providing the function defined by the functional unit (see Sect. 8.4). A distinction is done here between attributional and consequential approaches (see Sect. 8.5). For attributional assessments, the system boundaries have not been discussed explicitly by the S-LCA community and most case studies to date use the same kind of system boundaries as an attributional LCA, i.e. following a general supply chain logic. However, in consequential assessments there is a difference between LCA and S-LCA. Consequential LCA modelling includes only the processes that change because of the decision assessed. This is based on the premise that it is from a change in these processes or product uses that environmental impacts arise. Therefore, if no process change occurs, no impact change occurs.

Social impacts on the other hand do not occur merely due to production processes or product uses. They occur in all of life's situations—also when not carrying out a process or using a product. Taking the example of a worker within production of footballs, he/she may experience impacts related to conducting the work (e.g. unsafe conditions). The worker also experiences other impacts that only partly (if at all) can be related to the work (e.g. access to education for the worker's children). This implies that when we are to assess the social impacts due to the change of a product or production process then we should account for both the direct and the indirect consequences, including those that would occur if the changes had not happened. In the example of the football worker, the social consequences of producing a number of footballs are that a number of labourers are needed, contributing to a certain employment rate in the community around the factory. A decision leading to a reduction of the production of footballs may lead to lowering the number of employed labourers. This means that less workers would be exposed to unsafe conditions, but on the other hand, more people would be unemployed. In other words, the change to be considered in a consequential S-LCA includes both the impacts associated with carrying out a process and those associated with not carrying it out in order to be able to judge the consequence of the change. Similar examples can be found for the product users (Jørgensen et al. 2010a). In a more schematic form, the life cycle stages in a consequential S-LCA include the following (Fig. 16.2):

This discussion about impacts of not producing may seem somewhat theoretical but consider the following real case: In 2006, the multinational footwear manufacturing company Nike discovered that one of their suppliers, Saga Sports in Pakistan, employed child labour. To avoid the risk of moral condemnation from their customers, Nike chose to cut their contract with the company. But since 70%

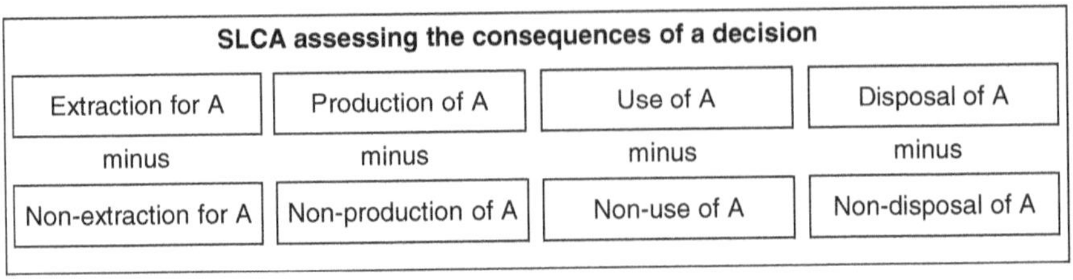

Fig. 16.2 The structure of an S-LCA for assessing the consequences of a decision to choose between product A and nothing reflecting that it must determine the difference between the induced activities and the status quo. In consequential (and attributional) LCA, all 'non' stages (representing the status quo) would normally be assumed to be zero

of Saga Sports' production went to Nike, many of the 4000 workers were fired, impacting not only the workers but also the local society, where an estimated 20,000 people depended on the income (Montero 2006). Assume now that an S-LCA was made to show the impacts related to producing a football at Saga Sports not including the impacts of not producing. The assessment would capture the impacts of child labour in the production, and show that if the balls were produced somewhere else where no children were employed, the child labour would (probably) be eliminated in the production, and all other being equal, this would create a socially better product. That would obviously not reflect the complete consequence of the situation outlined above where a large number of people were fired (and where the child workers may very well have entered into other forms of child labour, potentially under worse conditions). Given that the decision created negative social impacts in the local community, accounting for the impacts of not producing would give a more accurate picture. Including the impacts of both the production and the non-production/use/discarding is therefore essential in consequential S-LCA, and a distinct feature of S-LCA in comparison to LCA.

In Sect. 8.5.4 it was discussed whether LCA modelling should be based on a consequential or attributional approach depending on the decision context and the goal of the study (in accordance with the European Commission's LCA guidebook, theILCD handbook, EC-JRC 2010). Even though the international S-LCA community has not discussed the specifics of the modelling approach in detail, the same modelling principles as in LCA could be applied.

Identifying Causality Between Processes and Impact

The perhaps most important difference between S-LCA and LCA concerns the relationship between the product life cycle and the associated social or environmental impacts:

In LCA, generic life cycleunit process databases exist, that provide inventory data for various processes. A generic process accounts for certainelementary flows that lead to a certain assessment result. This result will be the same whenever the process is used. Although generic process data should only be used for the background processes (see the ILCD handbook and Sect. 9.3) they are generally considered representative of actual conditions with some accuracy. A good reason for this is found in the physicochemical properties of materials, processes and related emissions. Consider, for example, the process of melting iron. Factory parameters may influence the efficiency of the process, but in all cases, a certain minimum amount of energy will be required due to the physical properties of iron. A generic process could account for an amount of energy based on average global conditions. As for the type of energy, it could be based on average energy mix. The existence of generic processes leads consequently, to a causal relationship between the nature or type of process and the assessed impacts.

However, assessing social impacts is different. Even though no empirical studies have been conducted on the topic, there is a general consensus that the degree of causality between the type of process and social impacts is much weaker and less consistent compared to environmental impacts. To exemplify, as discussed previously, one of the issues very often considered in S-LCA is violations of ILO established labour rights. This includes workers' rights to organise in labour unions and abolishment of forced labour (anti-slavery). Consider now again the example of iron melting: there seems to be no causality between the actual process and the right of workers to organise in unions. Iron may be melted by workers who have the right to be organised or by workers who are denied this right. Rather than being related to the type of the process, it is therefore often stated in S-LCA literature that social impacts are related to the conduct of the company—i.e. it is how the company is managed that determines the social impacts that it creates, rather than what it produces.

The example of iron production illustrates well how the type of the process causes specific elementary lows leading to environmental impacts, but at the same time tells very little about the social impacts it creates. Note that there are other cases where a generic causality between a process and its social impacts is easier to establish. Consider for example different types of work-related injuries, which is another often-included impact category in S-LCA. For this type of impacts, it seems reasonable to expect a higher number of cuts and bruises for a technician compared to an office worker. This means that different job functions tend to be differently correlated to various impacts. Furthermore, when a job function can be closely related to a process, it seems reasonable to make the connection between the social impact and the nature of the process. Had anyone made an empirical investigation of the matter, we assume that that the general findings could be represented as in Fig. 16.3. This point has enormous implications for S-LCA, and we will return to this issue several times throughout this chapter.

The Issue of Impact Allocation

S-LCA is, like LCA, focussed on assessing impacts related to a functional unit. In order to provide the functional unit, a number of processes need to be operated throughout the product life cycle. But if it is the company's conduct rather than the operation of the process that causes the impacts, how should one allocate the impacts to each of the processes that the company performs and through that consistently to the life cycle of the product and the functional unit orreference flow? Several different approaches have been presented in literature. A frequent

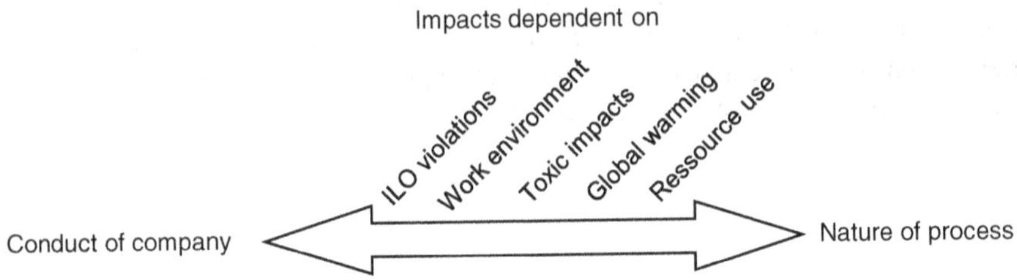

Fig. 16.3 The extent to which social and environmental impacts are controlled by the conduct of the company or the nature or type of process. In general, social can be considered much more dependent on the conduct of the company than environmental impacts

suggestion is to allocate social impacts of the company to the process, based on the working time required to perform it. Impact allocation can then be expressed by the following equation:

$$\text{Social impacts}_{\text{process}} = \text{Working time}_{\text{process}}/\text{Working time}_{\text{total in company}} \cdot \text{Social impacts}_{\text{total in company}} \quad (16.1)$$

Other allocation keys than working time are also suggested in literature. An example is to use value creation. In this case, the formula would be the same, except that "working time" would be substituted with "value creation". Although the goal of the study may indicate which approach is the right to use, it is in many cases up to the S-LCA practitioner to choose. This choice, if not arbitrary, will often depend on what information is available or on other motivations of the S-LCA practitioner. Consequently, two challenges arise. One, related to the freedom of choosing allocation key. This jeopardises the credibility of the method since the choice can heavily influence the S-LCA results. The second challenge is related to access to information. For a practitioner who is not deeply involved in the product life cycle (e.g. working in a lead company in thevalue chain) getting data on value creation and working time may be very difficult which may hinder the applicability or the ease of use of the assessment.

The goal of the assessment could specify what impacts to allocate to the process. Thus, here again, there is a difference between attributional and consequential approaches. If the goal of the study is to assess the consequences of a choice, calling for a consequential S-LCA then the allocation approach would be different than the one expressed in Eq. 16.1, since all social impacts that occur as a consequence of the decision should be included:

$$\begin{aligned}\text{Social impacts}_{\text{process}} = {} & \text{Social impacts}_{\text{total for world, process is performed}} \\ & - \text{Social impacts}_{\text{total for world, process is not performed}} \end{aligned} \quad (16.2)$$

In the football example in Sect. 16.2.2 above, we discussed that the assessment should include both the impacts that occur when the footballs are produced, and the impacts that occur when they are not. This means that identifying the consequence of a decision will necessarily include an estimation of a counterfactual. Such an assessment will be quite difficult in most cases and though it is a central point of S-LCA, it is still unclear how it can be done in practice.

16.2.3 Inventory Analysis

In both LCA and S-LCA, an inventory of data is made. In LCA, these data aim to capture environmental exchanges. Physical flows such as mass and energy to and from the processes are included in the assessment (as discussed in Chap. 9). Depending on the accuracy of the measurement techniques, these can often be determined with a very high degree of certainty. For assessing social impacts, the same "mass and energy balance approach" cannot be applied. Instead, we have to specify some interplay between the process and its social surrounding on which data should be collected.

Table 16.1 presents impact categories which we could include in an S-LCA. Nevertheless, identifying data that are both available and can capture the impact we are trying to assess is not straightforward. For example, as shown in Table 16.1, almost all S-LCA approaches consider discrimination towards workers as a relevant impact to include in an S-LCA. However, what data should be collected to assess its occurrence? Some have suggested using the ratio between male and female workers. Although corresponding data could be easy to collect, this does not seem to be a very accurate indicator for company induced discrimination. The reasons for a lower representation of a gender, e.g. women in the company, may for example be that the company gets more male than female applicants, which will lead to more male employees all other things being equal. A more accurate indicator, but by no means bulletproof, may relate to, e.g. workers' direct experiences of being discriminated due to gender, race, religion, etc. However, getting data on the actual experience of the worker can be quite difficult and time-consuming.

This case exemplifies the dilemma between the ease of use of the indicator (relating to access to information), and its accuracy (relating to how well the indicator captures the phenomena we are trying to assess). An underlying debate relates to the essence of well-being and to the extent to which the concept can be meaningfully described objectively or subjectively. In Sect. 16.2.2 we discussed both approaches in terms of choosing the social issues to be included in "well-being". A similar discussion is relevant regarding the indicators that can represent these issues.

Objective indicators relate to living or working conditions that can be identified without consulting the stakeholder about his or her perceptions. However, research on well-being indicates that there is a rather poor correlation between subjective experience, and objective living conditions. One is not necessarily happy when he/she is rich, healthy, has many friends, etc. Thus, in order to get an accurate measure of how a product life cycle changes the well-being of the affected stakeholders, subjective indicators are also needed. A subjective indicator may be an open question asked to the relevant person about, e.g. *how satisfied are you with X*. Existing S-LCA approaches have prioritised the development and application of objective indicators due to higher data availability and reproducibility at the expense of the limited accuracy in indicating actual changes in well-being.

Another methodological debate concerns whether one should use process—or result indicators, i.e. indicators that are related to the quality of a company's formal management system or to the company's measured social performance compared to the other companies in the product life cycle. The idea behind the first approach is that the occurrence of social impacts in a company will correlate with the initiatives in place to avoid them. For example, if a company has a strong system in place to ensure that discrimination in the hiring of employees is not occurring, then fewer cases of discrimination will occur. The second approach is about assessing the actual occurrence of social impacts based on reports or observations. The idea is simply that the reported incidences give an accurate picture of the impacts occurring.

Both approaches have pros and cons. The mere existence of a high-quality management system does not certify compliance and implementation in the everyday routines of the company. Likewise, a low reported or observed occurrence of impacts may be because the company (intentionally or unintentionally) or an external auditor does not report the incidences systematically. Which of the two approaches is most accurate, is difficult to tell. To date, the most common approach is to use performance indicators. For more information about the management indicators, the reader may refer to Dreyer et al. (2010).

The Data Collection Problem

While LCA may be performed at an acceptable level of accuracy using generic databases, the focus on company behaviour in S-LCA implies that site-specific data are indispensable. Specific information is needed not only for the company in question, but also for the context of national and regional regulatory frameworks, monitoring agencies, socio-economic conditions, etc. Obviously, this requirement for site-specific data imposes a tremendous burden in terms of costs and time spent. A second, but related, problem is the difficulty to identify the companies in the product chain and get relevant data. Often, only first-tier suppliers can be reached easily. Reasons for this may be that suppliers are unwilling to hand over information to the buyer about who their suppliers are in fear that the buyer would simply circumvent them. Another reason is that the goods might be bought on open markets with a large number of unidentified suppliers.

Three different approaches have been proposed to mitigate this data collection challenge:

One is to create databases of social impacts where one could find a specific company's performance. This would enable the S-LCA practitioner to circumvent the central problem of having to audit each implicated company. However, the strenuous task of company identification would still remain. Compiling such databases may seem very ambitious. Yet, the main challenge is not about collecting the data (many companies already undergo social audits which could potentially be used as data source in an S-LCA), but rather about making these data publicly available.

A second approach is to base S-LCA on indicators that are more closely related to the nature of the process. An example may be to relate the local value creation from a company in a product's life cycle to the increase in average lifetime of the population where the value creation results in increased income (Norris 2006). Then value creation, which is a relatively process-related phenomenon, could be used as an indicator for impacts on average lifetime in the affected population. However, whether this, or other more process-related indicators, will actually be able to capture the breadth of social impacts and well-being is questionable.

A third and probably the most feasible approach is to make databases of social impacts related to sectors and countries. These could provide a basis for the assessment and the S-LCA practitioner would only need to know where the various stages in the life cycle take place. An example is the Social Hotspot Database (SHDB 2016) presenting social impacts in a number of categories per working hour in different sectors and geographic regions. However, given that in many cases there will be significant differences in the social impacts within one sector in a country, the S-LCA based on this approach is generic and its representativeness for a specific product will be highly uncertain. Companies in the product's life cycle would risk being assigned an outright invalid score and this lack of accuracy makes this approach less useful for S-LCAs of specific products.

16.2.4 Impact Assessment

The impact assessment of an S-LCA, similar to LCA, consists of the elements classification, characterisation, normalisation and weighting (see Sect. 8.2.5). Of these, only classification and characterisation will be addressed below. Even though literature on the area is scarce, normalisation and weighting are considered to be performed like in LCA.

Classification

According to ISO 14044 (2006) classification is the element of the impact assessment, in which the inventory flows are assigned to different impact categories. Classification in LCA is central because of the nature of the inventory analysis. To capture the exchanges between a process and the environment, data collection is based on inputs and outputs of energy and mass. The same approach is not feasible in S-LCA, since there is no way to capture the total exchanges between a process (or a company) and the social world. Therefore, the inventory analysis in an S-LCA is designed to measure certain aspects of interest such as the ones shown in Table 16.1. It is thus known beforehand why this type of data is collected, and to what they contribute. Classification is in this way built into the indicators in S-LCA.

Characterisation

In LCA, hundreds of elementary flows may be included in the inventory. For a decision-maker to be able to evaluate this information there is a need for translating these flows into a number of meaningful environmental impact scores. This translation is essential, to indicate the importance of the flows. For example, emissions of benzene need to be translated into some measure of toxicity, which can be compared to and summarised with impacts from other toxic emissions, to give results that are meaningful for decision-makers.

In S-LCA, the situation is somewhat different. Similar to LCA, there is a list of impact categories. However, the number of social indicators (which are the equivalent for the elementary flows in LCA) is much smaller. In some cases, there is a one-to-one relationship between number of indicators and impact categories, e.g. when accounting for work-related diseases, ILO violations or the like. In this case, there will be no need for characterisation, i.e. the indicator results are directly meaningful for the decision-maker. In other cases several indicators are established for each impact category, e.g. in order to describe "decent working conditions". In the latter case, there will be a need for translating the data on these indicators into impacts. An example of such a translation is given in Spillemaeckers et al. (2004). Their approach is to collect data on certain conditions A, B, C and D. Then a certain impact is said to occur depending on the number and the extent to which the conditions are met. Another example can be seen in Dreyer et al. (2010).

A separate discussion, similar as in LCA (see Sect. 10.2.3), is whether the assessment should be done at a midpoint level in theimpact pathway, or whether the characterisation should aim to go all the way to an endpoint. An example for midpoint assessment is to establish impact groups such as "violations of ILO conventions", "non-lethal working accidents", etc. Whenever an incidence within each group occurs, then a score is assigned, e.g. if workers are not allowed to organise in unions (which is a violation of an ILO convention) in the product life cycle the "violation of ILO convention" impact group gets a score of 1. If there is also child labour (which is also a violation of an ILO convention), the "violation of ILO conventions" impact group gets a score of 1 more. In this way, social impacts can be grouped and characterised. However, the question here is, whether this is meaningful. What is the value for the decision-maker given that all kinds of nuances are disregarded through a more or less random grouping and scoring?

Earlier in this chapter, we discussed that the ultimate goal of SLCA is to assess the changes in human well-being. Consequently, S-LCA researchers have suggested that the midpoint-oriented impact categories should be further related to the Area of Protection in S-LCA, i.e. human well-being. Along these lines, Weidema (2006) established quantitative severity scores for various social impacts, whereby very different social impacts could be compared and summarised. More concretely, he suggested translating all impacts into loss of QALYs (Quality adjusted life years), according to the equation:

$$QALY = YLL + k^{*}YWL \qquad (16.3)$$

where YLL is years of life lost, YWL is years of well-being loss and k is a constant denoting the loss of life quality associated with the impact. When knowing what are the social impacts that affect life expectancy, how severe they are and their duration, the loss of QALYs can be calculated for each social impact. Then, impacts can simply be added to give a total score. The approach is similar to assessing DALYs in LCA (see Sect. 10.2.3). The advantage of a single score is that it supports an easy overview of the product performance. The weakness, however, is that one needs to assign severity scores to very different types of impacts, ranging from incidences of discrimination to cancer. This is a rather difficult and uncertain task, which might lack comprehensiveness and consistency.

16.3 Implications of the Problems Related to the S-LCA Methodology

As we have seen in this chapter, there are two main differences between LCA and S-LCA, which have a significant impact on the usability of S-LCA. The first relates to establishing a causality between processes and impacts. The environmental impacts depend on the nature of the process, whereas social impacts depend on multiple factors such as the conduct of the company and the culture in which it operates. This affects inventory analysis and data collection. In order to perform a reasonably accurate LCA we only need to know the types of the processes involved in the life cycle. However, this approach would drastically lower the accuracy of S-LCA, because of this low causal relationship between process and social impacts. Additional information about the company that operates the process is needed, which in most cases is going to be more difficult to get than simply getting an overview of the type of processes. The second difference is that when S-LCA is to be used for decision support, there is a need for assessing both the impacts of producing/using/discarding and of not producing/using/discarding the product. This adds complexity anduncertainty to data collection in comparison to LCA.

From an overall perspective, these differences indicate that the combined accuracy and ease of use of S-LCA is, and is likely to continue to be, poorer compared to LCA. Same accuracy would require detailed knowledge about the actual life cycle of the product and about the impacts of not producing. Same ease of use would require generic process data, which in most cases will give us assessments of very low accuracy. Existing initiatives, such as databases with social audit information about companies, partially address the issue. Yet, the challenge of identifying the companies that carry out each process, remain.

The third identified barrier is the meaningfulness of S-LCA results for providing decision support. For the case of LCA, *better decisions* are understood as decisions that lead to less environmental impacts. The LCA informs the decision-maker about the environmental impacts related to the entire life cycle of, e.g. two products with comparable functional units. The decision-maker can hereby choose the product that is associated with lower environmental impacts. The LCA hereby has an environmental effect if used in decision support by eliminating the 'bad' environmental choices, assuming that the LCA is carried out correctly.

One may think that the same argument is valid when it comes to S-LCA, with the only difference that it should improve social impacts when used for decision support. However, this may not be the case. The effect of using S-LCA in decision support may in fact be outright negative as the following example shows.

Assume that an S-LCA of a product shows that the workers in the product life cycles experience very poor working conditions. The decision-maker may on this basis choose not to buy or use the product. But how will this decision improve the working conditions? One way may be that the company with the poor working conditions will go out of business. This will eliminate the poor working conditions for the worker but will increase unemployment. Going unemployed will rarely help the worker despite the poor working conditions—remember that the worker took the job in the first place and probably only had worse alternatives. Another scenario could be that instead of going out of business, the company will become aware that the social conditions of the working place are a market parameter (measured through S-LCA). This realisation may lead to improving the working conditions at the working place. However, research on the topic indicates that creating improvements, which are not only improvements on paper but real experienced improvements for the workers, is very difficult, and will often require a change in working culture, which is not likely to happen as a result of living up to the standards set by the S-LCA (Barrientos and Smith 2007; Bezuidenhout and Jeppesen 2011). Further detail about the effect of using S-LCA is explored in Jørgensen et al. (2012) and it is outside the scope of this chapter to go into all details of the argument. Yet these examples indicate that the same logic, which is valid for LCA, may not be directly transferable to the SLCA area when it comes to the effect of using SLCA and LCA for decision support.

Whether these issues will deem S-LCA unusable is impossible to say—it will depend on the needs of the user. It seems though they may well prevent S-LCA from gaining the popularity and widespread use that is seen for LCA. Limitations for its usability can be exemplified for two main areas where LCA is used for decision support:

(i) Prospective assessments: in this case, LCA aims to assess the *expected* environmental impacts from new innovations. This assessment is only possible because we assume a causal link between process and environmental impact. Future environmental impacts can be estimated based on reference products and technologies. Thus, if there is no (or only a very weak) link between process and impact, as is the case for social impacts, this prospective assessment will have no or only a very limited accuracy.
(ii) Assessment of product families: Following a parallel argument as used above, it is possible to make a generic LCA for, e.g. vacuum cleaners, as they more or

less all include the same components and consume comparable amounts of electricity throughout their use. Again, this is possible because of the link between environmental impacts and process. In S-LCA, where there is no or only a very weak link this will impede the possibility for reaching an assessment of a product family with a reasonable degree of accuracy.

16.4 S-LCA Case Studies

While the S-LCA methodology is still immature, experiences from its application in product case studies are important drivers for its future development. This section will present three cases to illustrate how main challenges are addressed in current research.

16.4.1 Laptop Computer

The first case study by Ciroth and Franze (2011) concerns a lightweight laptop (ASUSTeK UL50Ag notebook for office use) and assesses environmental and social impacts in parallel. Thus, the goal is not a comparison of products, but (1) identification of social and environmental hotspots, (2) recommendations on company andpolicy level and (3) application of the UNEP/SETAC Guidelines for S-LCA on a complex product. Specifically regarding (4), the effectiveness of the EU Ecolabel (the Flower) criteria is discussed. The case study is very comprehensive and detailed; however, the use (and re-use) stage is not considered. Note that for this stage most S-LCA studies only account for the aspects included in the stakeholder group "consumers".

The case study points to human well-being as the ultimate goal of LCA and notes the pervasive significance of computer use in modern life. Nonetheless, it stops at the UNEP/SETAC Guidelines for S-LCA, which relates to company behaviour and to general behaviour within the specific industrial sector. Thus, the indicators proposed are found "not applicable to use phases as there are no companies or industries involved". The study is concerned with midpoint categories only, as "the use of endpoint implies the aggregation of results, which in turn reduces transparency and increases uncertainty".

The study acknowledges that interviews with directly affectedstakeholders are to be preferred to other data collection methods. However, it mentions that, with a few exceptions, the time needed for local and site-specific data collection is prohibitive. Although the study suggests a participatory approach in defining impact categories and indicators, there is no reflection on the assessment's validity, in relation to cultural differences between nations and regions.

Allocation is not applied. Instead, "each company is considered as one unit no matter which different products the company produces and which of these products are relevant for the study". Thus, if an impact is occurring in a company in the life cycle, all the company's products will be associated with this impact to the same extent regardless of, e.g. the working time used for producing each product. Also, an equalweighting factor for the companies included in the life cycle is used, meaning that regardless that one company contributes far more than another in terms of, e.g. the total working time, to the final product, all companies will 'count' the same in the final assessment.

The computer case study represents a thorough effort to test the UNEP/SETAC Guidelines for S-LCA and does substantiate a range of methodological problems as well as overall issues of relevance and comprehensiveness. Most significantly, it demonstrates that the S-LCA findings and conclusion bring no new insights beyond those that could already be expected prior to the study. Considering the costs and time involved in an S-LCA study like this, the question about what the UNEP/SETAC Guidelines for S-LCA have to offer compared to more simple audit tools remains unanswered.

16.4.2 Cut Roses from Ecuador and the Netherlands

The second case study, by the same authors (Franze and Ciroth 2011), compares the production in Ecuador and the Netherlands of a bouquet of cut roses with 20 flowers per bouquet, packaged and transported to the flower auction in Aalsmeer, the Netherlands. The main objective is to "try out" the UNEP/SETAC Guidelines for S-LCA. The study conducts in parallel an LCA and an S-LCA of the production system. It does recognise that social impacts are inter-related and may include many indirect effects. Nonetheless, the discrete impact categories associated with each stakeholder group and the wide range of sub-categories are considered satisfactory. Problems with quality of data from various sources, considering the motivation, structure of companies, NGOs and government institutions, are mentioned.

Not surprisingly, the study concludes that social impacts in the Netherlands are mainly positive, while environmental impacts, in particular during winter, are rather negative. Thus, from an environmental point of view, importing roses from Ecuador is to be preferred over producing them in the Netherlands. Yet, from a social perspective, the Netherlands is preferred over the production in Ecuador. Regarding social conditions, the study outlines a general scenario for improvement, but such an intention is beyond the scope of the UNEP/SETAC Guidelines for S-LCA. For the social impact assessment, a simple colour coding is used for scoring, and noweighting is performed. The use stage is only marginally considered in terms of health and safety of consumers.

16.4.3 Greenhouse Tomatoes

The third case study by Andrews et al. (2009) departs from the calculation of quantitative impacts based on the UNEP/SETAC Guidelines for S-LCA and asks the question "What percentage of my supply chain has attribute X". The X may represent an existing CSR indicator, and the basis for calculating the percentage is the total working hours within the chain. The case study points to the potential of life cycle attribute assessment (LCAA) "to piggyback off other initiatives" (ISO 14001, GRISustainability Reporting, SA 8000, FSC, and the US Green Building Council's LEED programme).

However, depending on different stakeholder interests, working hours may be substituted, e.g. by "forested acres" to check on the percentage of FSC certified acres. The study selects eight indicators, one of which is "wage levels", and asks the question whether wage levels have properties as an indicator in S-LCA that equal energy consumption in LCA which in many studies serve as "an important indicator that is closely related with results across many impact categories". All indicators are selected at a midpoint, i.e. regarded as means to an end. The study recognises that data quality declines as Input–Output tables at sector level are used instead of more detailed process flows. Therefore, primary data were collected through company interviews. The fact that the tomato company in this case dominates its own supply chain and that no supplies are produced overseas limits the data quality problem. The study manages to pinpoint the percentage of compliance with CSR criteria and the spots where more CSR activity is needed.

The three case studies respond to the call of the UNEP/SETAC Guidelines for S-LCA, except for the third, which adopts the holistic perspective of life cycle assessment and then aligns with CSR criteria. These selected case studies and other contributions to the S-LCA literature suggest solutions for a range of unresolved issues. However, establishing a methodological consensus and a base for comparative studies is still needed. In conclusion, the studies exemplify that S-LCA is not yet a mature methodology. Findings are often predictable, and the additional value of an S-LCA is not evident in comparison to other approaches, particularly when considering the heavy data requirements.

16.5 Future Development

The major driver for the S-LCA development has been to create a social assessment method that "mimics" as closely as possible the principles of LCA with a view for a possible integration of the two and also acknowledging that a life cycle perspective is relevant for social impacts as it is for environmental impacts. This is supported by a concept of sustainability, according to which societies are operating within environmental limits. Having elaborated LCA to some level of consensus and maturity, it is now time to tackle the social dimension of sustainability.

A fundamental problem in the social version of theLCA framework is that central differences between the environmental and social issues may be overlooked. One reason may be that natural scientists venture beyond their scope in the effort to establish S-LCA as a clone of LCA. Considering the well-established LCA paradigm and institutionalised LCA research community the risk of disregarding social sciences altogether cannot be excluded.

Seen in this light, it seems that future development of S-LCA might follow two paths. One is to continue the current trend and fully exhaust the 'LCA cloning' approach, which will call for more research within areas such as indicator development, characterisation modelling in S-LCIA, establishing and validatingimpact pathways, aggregation procedures, normalisation references and valuation methods. Another path, however, would be to more fully acknowledge existing social science research, which would raise fundamental questions about the foundations of the methodology. It would for example lead to reviewing recent concepts of human well-being in order to inspire a redefinition of an integrated set of social impact categories.

Regardless of whether S-LCA will succeed in integrating important lessons from the social sciences, S-LCA cannot escape its purpose of being a methodology that is (1) life cycle oriented and (2) aiming for social assessment. This conjunction will inevitably lead to significant data requirements for which there is no miracle cure. Without a solution to this issue, S-LCA studies will probably continue to be limited to one or a few companies. This will raise the question: "what makes S-LCA worthwhile to develop and use considering that assessments of social impacts in companies have long been developed?"

References

Andrews, E., Lesage, P., Benoît, C., et al.: Life cycle attribute assessment. J. Ind. Ecol. **13**, 565–578 (2009). doi:10.1111/j.1530-9290.2009.00142.x

Barrientos, S., Smith, S.: Do workers benefit from ethical trade? Assessing codes of labour practice in global production systems. Third World Q. **28**, 713–729 (2007)

Benoît, C., Mazijn, B.: Guidelines for Social Life Cycle Assessment of Products. UNEP/SETAC Life Cycle Initiative, Paris (2009)

Bezuidenhout, A., Jeppesen, S.: Between the state, market and society: labour codes of conduct in the Southern African garment industry. Dev. South Afr. **28**(5) (2011)

Ciroth, A., Franze, J.: LCA of an Ecolabeled Notebook, Consideration of Social and Environmental Impacts Along the Entire Life Cycle. Berlin (2011)

Dreyer, L., Hauschild, M., Schierbeck, J.: A Framework for social life cycle impact assessment (10 pp). Int. J. Life Cycle Assess. **11**, 88–97 (2006). doi:10.1065/lca2005.08.223

Dreyer, L.C., Hauschild, M.Z., Schierbeck, J.: Characterisation of social impacts in LCA: Part 1: development of indicators for labour rights. Int. J. Life Cycle Assess. **15**, 247–259 (2010). doi:10.1007/s11367-009-0148-7

EC-JRC: European Commission-Joint Research Centre-Institute for Environment and Sustainability: International Reference Life Cycle Data System (ILCD) Handbook—General Guide for Life Cycle Assessment—Detailed Guidance. First edition March 2010. EUR 24708 EN. Luxembourg. Publications Office of the European Union (2010)

Franze, J., Ciroth, A.: A comparison of cut roses from Ecuador and the Netherlands. Int. J. Life Cycle Assess. **16**, 366–379 (2011). doi:10.1007/s11367-011-0266-x

GRI: Global Reporting Initiative. https://www.globalreporting.org/Pages/default.aspx (2016). Accessed 29 June 2016

ILO: International Labour Organisation. http://www.ilo.org/global/lang–en/index.htm (2016). Accessed 29 June 2016

ISO: Environmental Management: Life Cycle Assessment—Principles and Framework (ISO 14040). ISO the International Organization for Standardization, Geneva (2006a)

ISO: Environmental Management: Life Cycle Assessment—Requirements and Guidelines (ISO 14044). ISO the International Organization for Standardization, Geneva (2006b)

ISO: Guidance on Social Responsibility (ISO 26000:2010). ISO the International Organization for Standardization, Geneva (2010)

Jørgensen, A., Le Bocq, A., Nazarkina, L., Hauschild, M.: Societal LCA methodologies for social life cycle assessment. Int. J. Life Cycle Assess. **13**, 96–103 (2008)

Jørgensen, A., Finkbeiner, M., Jørgensen, M.S., Hauschild, M.Z.: Defining the baseline in social life cycle assessment. Int. J. Life Cycle Assess. **15**, 376–384 (2010a). doi:10.1007/s11367-010-0176-3

Jørgensen, A., Lai, L.C.H., Hauschild, M.Z.: Assessing the validity of impact pathways for child labour and well-being in social life cycle assessment. Int. J. Life Cycle Assess. **15**, 5–16 (2010b). doi:10.1007/s11367-009-0131-3

Jørgensen, A., Dreyer, L.C., Wangel, A.: Addressing the effect of social life cycle assessments. Int. J. LCA **17**(6), 828–839 (2012)

Kruse, S., Flysjö, A., Kasperczyk, N., Scholz, A.J.: Socioeconomic indicators as a complement to life cycle assessment: an application to salmon production systems. Int. J. Life Cycle Assess. **14**, 8–18 (2009). doi:10.1007/s11367-008-0040-x

Montero, D.: Nike's dilemma: is doing the right thing wrong? Christ. Sci. Monit. (2006)

Norris, G.A.: Special issue honouring Helias A. Udo de Haes: broadening the scope of LCA social impacts in product life cycles towards life cycle attribute assessment. Int. J. Life Cycle Assess. **1**, 97–104 (2006). doi:10.1065/lca2006.04.017

SHDB: Social Hotspot Database. http://socialhotspot.org/ (2016). Accessed 29 June 2016

Spillemaeckers, S., Vanhoutte, G., Taverniers, L., et al.: Integrated product assessment—the development of the label "sustainable development" for products ecological, social and economical aspects of integrated product policy (2004)

UN: United Nations Global Compact Sustainable Supply Chains. http://supply-chain.unglobalcompact.org/site/index (2010). Accessed 20 Dec 2015

UNDP: Sustainable Development Goals. http://www.undp.org/content/undp/en/home/librarypage/corporate/sustainable-development-goals-booklet.html (2015)

WCED: World Commission on Environment and Development: Our Common Future. Oxford University Press, Oxford (1987)

Weidema, B.P.: The integration of economic and social aspects in life cycle impact assessment. Int. J. Life Cycle Assess. **11**, 89–96 (2006). doi:10.1065/lca2006.04.016

Author Biographies

Andreas Moltesen Has been working with LCA since 2006 with a particular focus on social life cycle assessment. He has later worked on life cycle assessments of biofuels and is currently particularly involved with life cycle assessments of transport systems.

Alexandra Bonou LCA expert and modeller focussing on ecodesign since 2009. Has worked on integrating life cycle thinking and environmental target setting in organisations of the private and public sector. Interested in life cycle management, product development, social life cycle assessment.

Arne Wangel Sociologist studying international technology transfer and capacity building in South countries. Since 2008 involved with LCA with main interests in application of LCA in global value chain analysis, strategies for building LCA capacities in South countries, and development of a methodology for Social LCA.

Kossara Petrova Bozhilova-Kisheva Experienced in LCSA of construction and demolition waste recycling into building materials and SLCA methodology development. Interested in LCSA, SLCA, health impact assessment, eco-design, product/technology development, built environment, energy systems and system analysis.

Part III
Applications

Chapter 17
Introduction to Part III: Application of LCA in Practice

Ralph K. Rosenbaum

Abstract While Part II of this book presents the theoretical foundation and methodology of LCA, Part III is dedicated to a comprehensive discussion of how this methodology has been adapted and applied in practice. The chapters of Part III provide an easily readable and accessible introduction to different fields of LCA application with their specific decision situations, user competences and stakeholder needs, and associated methodological challenges and adaptations.

While Part II of this book presents the theoretical foundation and methodology of LCA, Part III is dedicated to a comprehensive discussion of how this methodology has been adapted and applied in practice. The chapters of Part III provide an easily readable and accessible introduction to different fields of LCA application with their specific decision situations, user competences and stakeholder needs, and associated methodological challenges and adaptations. Chapters 18–25 deal with the role of LCA and life cycle thinking in various decision contexts and discuss the methodological adaptations for specific uses of LCA, such as policy support, organisational LCA, life cycle management, ecodesign, and ecolabelling and differences and synergies between LCA and the Cradle-to-Cradle concept and certification system:

18. Life Cycle Thinking and the use of LCA in policies around the world
19. Globalisation and mainstreaming of LCA
20. Organisational LCA
21. Future-oriented LCA
22. Life Cycle Management
23. Ecodesign implementation and LCA
24. Environmental labels and declarations
25. Cradle to cradle and LCA

R.K. Rosenbaum (✉)
IRSTEA, UMR ITAP, ELSA Research Group and ELSA-PACT – Industrial Chair for Environmental and Social Sustainability Assessment,
361 Rue Jean-François Breton, BP 5095, 34196 Montpellier, France
e-mail: ralph.rosenbaum@irstea.fr

M.Z. Hauschild et al. (eds.), *Life Cycle Assessment*,
DOI 10.1007/978-3-319-56475-3_17

The remaining chapters of Part III are all dedicated to the application of more 'classic' LCA to different technology domains comprising both some of the central sectors of society and some of the more specialised fields to give an introduction to the broad practical use of LCA in the assessment of products and technologies:

26. LCA of energy systems
27. LCA of electromobility
28. LCA of buildings and the built environment
29. LCA of food and agriculture
30. LCA of biofuels and biomaterials
31. LCA of chemicals and chemical products
32. LCA of nanomaterials
33. LCA of drinking water supply
34. LCA of wastewater treatment
35. LCA of solid waste management systems
36. LCA of remediation of soil and groundwater

Chapters 18–25 all have their individual structure and different learning objectives as stated in the beginning of each chapter, reflecting the diversity of the subjects that they cover. In contrast, Chaps. 26–36 on the use of LCA in different technology domains have a more harmonised structure covering the following aspects:

- **Introduction,** providing the context and a description of the sector or technology, a brief history of LCA application in this area, definitions of specific terminology relevant for the concerned application (e.g. first- and second-generation biofuels), and the main questions relative to the environment that LCA is used to answer (e.g. is organic agriculture better than intensive or extensive agriculture?).
  - **Literature review**, giving an overview over a selection of published case studies and essential further reading material focusing on
  - the main life cycle stages, drivers, and processes contributing to potential environmental impacts, and
  - what the main impacts are, including potential burden-shifting.
- **Specific methodological issues**, discussing what the literature survey has identified as the main methodological considerations and challenges when applying LCA to this technology field, structured into:
  - General issues (not specific to any of the following four methodological phases)
  - Goal and scope
  - Inventory and product system modelling
  - Impact Assessment
  - Interpretation.

- **Conclusions** of LCA studies for this sector/technology, summarising the main findings of LCA studies related to the respective technology field in terms of:
  - Main tendencies and shared conclusions among essential literature
  - Controversial conclusions among essential literature
  - Recent advances and achievements as well as remaining limitations
  - Perspectives and further research needs, i.e. the most important current methodological shortcomings for this technology/sector.

The learning objectives of the chapters on the use of LCA in different technology fields are also the same across all chapters and are thus presented only once, in the following. After studying one of the Chapters from 26 to 36, the reader should have acquired the following capabilities for the respective technology domain:

- Discuss the role, relevance and state of the art of LCA in the respective technology domain.
- Use the technology domain-specific terminology and definitions correctly.
- Explain the key methodological issues, challenges, limitations and good practice of applying LCA within the technology field and their implications for the results of an LCA.
- Contrast major environmental concerns and challenges related to the respective technology and its environmental performance against its benefits.
- Outline the main factors and hot spots influencing the environmental performance of a technology.
- Be aware of the main literature available for a technological application domain of LCA including its common findings as well as contrasted differences in conclusions drawn.

Enjoy the reading!

Chapter 18
Life Cycle Thinking and the Use of LCA in Policies Around the World

G. Sonnemann, E.D. Gemechu, S. Sala, E.M. Schau, K. Allacker, R. Pant, N. Adibi and S. Valdivia

Abstract The chapter explains what Sustainable Consumption and Production (SCP) is about, why it is about taking a life cycle approach and shows that SCP-related policies have been developed at the intergovernmental level and in different regions of the world. A key element at the international level is the 10-Year Framework of Programmes on SCP adopted in 2012 and the global agreements on the Sustainable Development Goals (SDGs) adopted in 2015. Life cycle thinking has become mature, moving from its academic origins and limited uses, primarily in-house in large companies, to more powerful approaches that can support the provision of more sustainable goods and services through efficient use in product development, external communications, in support of customer choice, and in public debates. Now governments can use LCA for SCP policies. For this purpose LCA databases are needed. LCA is in particular relevant for policies focusing on design for sustainability, sustainable consumer information, sustainable procurement and waste management, minimization and prevention as well as sector-specific policies like sustainable energy and food supply. Examples of life

---

G. Sonnemann (✉) · E.D. Gemechu
ISM-CyVi, UMR 5255, University of Bordeaux, Talence, France
e-mail: guido.sonnemann@u-bordeaux.fr

G. Sonnemann · E.D. Gemechu
ISM-CyVi, UMR 5255, CNRS, 33400 Talence, France

S. Sala · E.M. Schau · R. Pant
European Commission, Directorate D: Sustainable Resources, Bioeconomy Unit, Joint Research Centre, Ispra, Italy

K. Allacker
Department of Architecture, Faculty of Engineering Science, KU Leuven, Leuven, Belgium

N. Adibi
PLATEFORME [avniR]-cd2e, Loos-en-Gohelle, France

N. Adibi
WELOOP, 27 Rue du Stade, 62300 Lens, France

S. Valdivia
World Resources Forum, St. Gallen, Switzerland

M.Z. Hauschild et al. (eds.), *Life Cycle Assessment*,
DOI 10.1007/978-3-319-56475-3_18

cycle thinking and the use of LCA in policies are provided for numerous countries around the world but with a certain focus on the European Union. It can be expected that the use of LCA in policies for the sustainability assessment of products will further increase, also slowly covering more means of implementation such as incentives and legislative obligations.

Learning Objectives

After studying this chapter, the reader should be able to:

- Explain the basic principles of Sustainable Consumption and Production (SCP), covering exploration, extraction, product development, manufacturing, use and end of life options;
- Discuss which SCP policies have been developed at the intergovernmental level and in different regions of the world;
- Express the link to focus areas particularly relevant for LCA-based policies such as design for sustainability and sustainable consumer information;
- Value the examples of life cycle thinking and the use of LCA in policies provided for numerous countries around the world;
- Explain the opportunities of using LCA in policies for the sustainability assessment of products in the future aiming at showing the efficiency of the means used.

18.1 Introduction to Policies on Sustainable Consumption and Production

Over the last few decades, the need to transition into a more sustainable society has become more and more evident and pressing. In return, global efforts to address sustainability challenges have also significantly increased. To this end, especially 2015 was the year of sustainability, which included not only the global agreement on the Sustainable Development Goals (SDGs), but also the Paris Agreement. It was reached by the parties to the United Nations Framework Convention on Climate Change (UNFCCC) on 12 December 2015 in Paris and symbolizes a fundamentally new course in the two-decade-old global fight against climate change.

"The necessary shift tosustainable consumption and production (SCP) patterns will do much to improve the lives of some of the world's poorest people as well as protect the rich resources that nature provides. But we will not achieve this shift unless we have effective policies, social and technological innovation, public and private investment, and the engagement of governments, business, consumers, educators and the media. Each and every one of us will have a role to play …" (UNEP 2012a).

SCP is understood as the "The use of services and related products, which respond to basic needs and bring a better quality of life while minimizing the use of natural resources and toxic materials as well as the emissions of waste and pollutants over the life cycle of the service or product so as not to jeopardize the needs of future generations" (Norwegian Ministry of Environment, Oslo Symposium 1994). It means that SCP is a holistic approach that has at its core a life cycle perspective, which is the attitude of becoming mindful of how everyday life has an impact on the environment and society.

According to UNEP (2012a) SCP focuses on resource efficiency that is about ensuring that natural resources are efficiently produced and processed, and consumed in a more sustainable way, as well as about reducing the environmental impact from the consumption and production of products over their full life cycles. By producing more well-being with less material consumption, resource efficiency enhances the means to meet human needs while respecting the ecological carrying capacity of the earth. Such improvements can also increase the competitiveness of enterprises, turning solutions for sustainability challenges into business, employment and export opportunities. The fundamental objective of SCP is to decouple economic growth from environmental degradation.

SCP policies cover all the areas highlighted in Fig. 18.1. One of UNEP's six sub-programmes is on Resource Efficiency and SCP. The overarching aim of this sub-programme (UNEP 2013a) is to detach economic growth from unsustainable resource use and environmental degradation. In general, it can be observed that governments in support of a shift to sustainable consumption and production focus more on production in developing countries and on consumption in developed countries. SCP is covered under the SDG 12 on responsible consumption and production.

By applying the life cycle approach, priorities can be identified more transparently and inclusively and policies can be targeted more effectively so that the maximum environmental benefit is achieved relative to the effort expended (CEC 2005a).

The chapter focuses first on life cycle thinking and then on LCA in policies at the international level, with a particular focus on intergovernmental organizations including the European Union (EU) and selected countries around the world.

18.2 Policies Based on a Life Cycle Thinking at the International Level and Around the World

18.2.1 10-Year Framework of Programmes on SCP

On June 2012 established a landmark in the international recognition of SCP in policies with the adoption of "the 10-Year Framework of Programmes onSustainable Consumption and Production Patterns (10YFP)" by the Heads of State at the United Nations Conference on Sustainable Development (Rio+20)—as

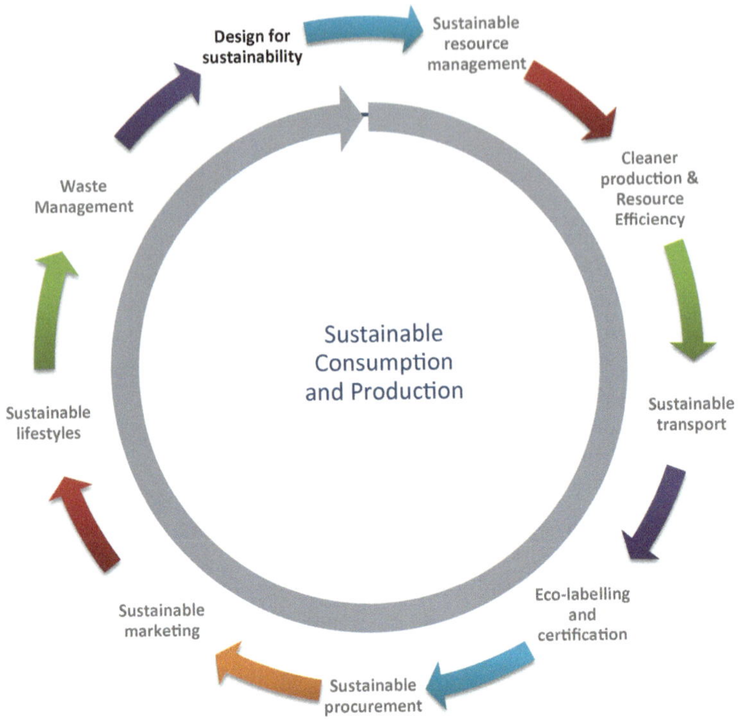

Fig. 18.1 SCP policies along the product life cycle (UNEP 2010a)

stated in paragraph 226 of the Rio+20 Outcome Document "The Future we Want" (UNCSD 2012a).

The 10YFP is a concrete and operational outcome of Rio+20. It is a global framework of action to enhance international cooperation to accelerate the shift towards SCP in both developed and developing countries. The framework will support capacity building and provide technical and financial assistance to developing countries for this shift. The 10YFP will develop, replicate and scale up SCP and resource efficiency initiatives, at national and regional levels, decoupling environmental degradation and resource use from economic growth, and thus increase the net contribution of economic activities to poverty eradication and social development. It responds to the 2002 Johannesburg Plan of Implementation, and builds on the 8-year work and experience of the Marrakech Process—a bottom-up multi-stakeholder process, launched in 2003 with strong and active involvement from all regions in the world. The 10YFP will also build on the work of the national cleaner production centres and other SCP best practices engaging a wide range of stakeholders (UNEP 2012b).

The adopted document in Rio+20—The Future we Want (UNCSD 2012b)—provides the vision, goals and values of the 10YFP as well as its functions, organizational structure, means of implementation, criteria for programmes design and

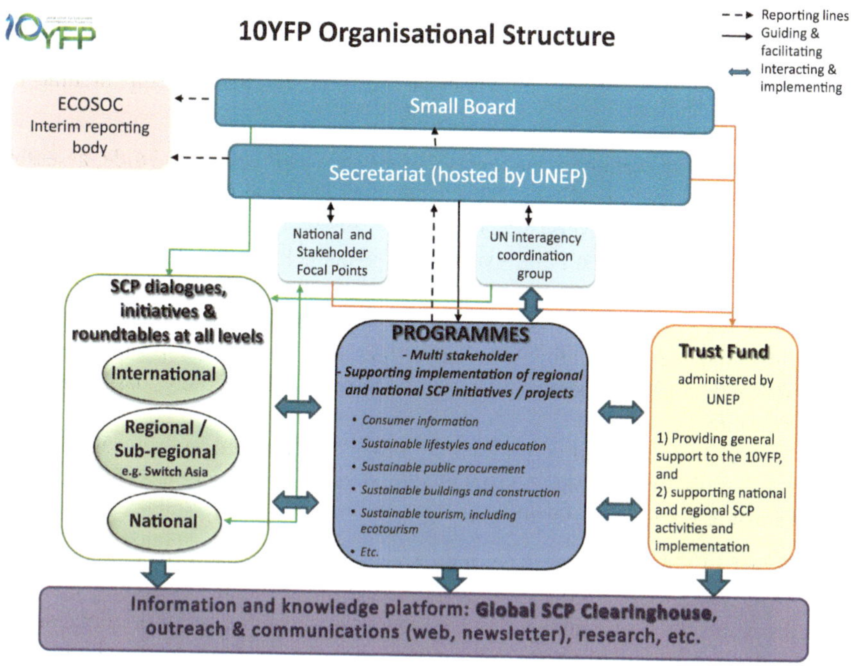

Fig. 18.2 10YFP operational structure (UNEP 2012b)

an initial, non-exhaustive list of five programmes. UNEP has been requested to serve as the 10YFP Secretariat and to establish and administer a Trust Fund to support SCP implementation in developing countries and countries with economies in transition. The 10YFP operational structure is summarized in Fig. 18.2.

The 10YFP is called to assist countries in reaching a common vision that promotes a life cycle perspective, among other aspects of SCP. This call for LCT in the development of SCP policies in countries is demanding life cycle based expertise, data, methodologies, skills andresources and the support of stakeholders and initiatives worldwide. SCP programmes need a solid scientific andpolicy knowledge base and the use of a mix of efficient instruments such as education, training and data collection.

Need of support from the LCA expert community for implementing the 10YFP could be on result-based indicators for the 10YFP, understanding how LCA could be used better for policymaking, guidance on product sustainability information (e.g. hot spot methodology), input for the global SCP Clearinghouse on Sustainable Consumption and Production platform[1] (including south–south cooperation), and the involvement of life cycle experts and regionalstakeholders in the development of the 10YFP programmes.

[1] www.scpclearinghouse.org/fr/.

18.2.2 UNIDO/UNEP Programme on Resource-Efficient Cleaner Production

Recognizing that resource efficiency requires Cleaner Production (CP) and vice versa, UNIDO and UNEP have moved towards Resource Efficient and Cleaner Production (RECP). RECP recognizes that CP methods and practices generate multiple benefits that are relevant to many of today's most pressing global challenges, including mitigation of GHG emissions and adapting to climate change; responding to increasing scarcity ofwater, fuels and other materials; providing decent jobs; and halting environmental degradation. RECP, therefore, builds upon CP in accelerating the application of preventive environmental strategies to processes, products and services to increase efficiency and reduce risks to humans and the environment.

UNIDO and UNEP launched in 1994 a joint programme to establish National Cleaner Production Centres/Programmes (NCPCs/NCPPs). They incorporated the lessons learned from the NCPCs in their joint RECP programme strategy. The strategy was approved in 2009 for implementation. It supports the global imperative to decouple economic development from further environmental degradation and resource depletion. The programme aims to improveresource efficiency and environmental performance of businesses and other organizations in developing and transition countries. The envisioned principal outcome is the widespread adaptation and adoption of RECP methods, practices, technologies and policies. The past decade has demonstrated that these are applicable and relevant. The challenge is now to scale-up their application so that they become common practice rather than isolated initiatives in a few selected enterprises (UNIDO/UNEP 2013).

18.2.3 OECD: Sustainable Materials Management and Green Claims

One of the policies of the Organisation for Economic Co-operation and Development (OECD) that is clearly stating a strong reference to LCT is Sustainable Materials Management (SMM). It is increasingly recognized as apolicy approach that can make a key contribution to green growth and the challenges that are posed by sustained global economic and demographic growth. One of the key challenges of the SMM approach is to effectively address the environmental impacts that can occur along the life cycle of materials, which frequently extends across borders and involves a multitude of different economic actors (OECD 2012).

The OECD Committee on Consumer Policy launched a project to examine ways to enhance the value and effectiveness of green claims in April 2009.

Environmental claims, also termed "green claims", are assertions made by firms about the environmentally beneficial qualities or characteristics of goods and services. They can refer to the manner in which products are produced, packaged, distributed, used, consumed and/or disposed of. In addition to the environmental aspects, these claims are sometimes defined to include the social responsible or ethical manner in which products are produced and distributed. The Committee's work underscores the complexity of the issues and challenges facing stakeholders in the field of environmental claims. There is agreement, however, on a number of basic principles that could enhance the value and effectiveness of claims (OECD 2010).

18.2.4 Asia/Pacific: Strategy of Green Growth and Circular Economy

In the Asia–Pacific region, SCP rides on the back of the economic growth and broader sustainable development agenda. The strategy of Environmentally Sustainable Economic Growth, or Green Growth, is an approach that is promoted by the UN Economic and Social Commission for Asia and the Pacific (UNESCAP), has been widely adopted by countries in the region. It was launched in 2005 at the Fifth Ministerial Conference on Environment and Development in Seoul, Republic of Korea, as a way to reconcile tensions between efforts to achieve two of the Millennium Development Goals, namely, poverty reduction and environmental sustainability. Green Growth promotes SCP, the development of sustainable infrastructure, and the introduction of green tax reform for reducing poverty, while UNESCAP has since provided capacity building to some national governments towards the development of Green Growth strategies (UNESCAP 2005).

An important role in Asia plays the Circular Economy promoted by the government in China and inspired by Japanese and German Recycling Economy Laws. China's rapid economic growth demands major supplies of all basic industrial commodities, in competition with other nations. China's emissions cross boundaries and oceans, impacting Korea, Japan, and North America. Its contribution to greenhouse gas emissions is rising rapidly. The Circular Economy approach to resource-use efficiency integrates cleaner production and industrial ecology in a broader system encompassing industrial firms, networks or chains of firms, eco-industrial parks and regional infrastructure to support resource optimization. State-owned and private enterprises, government and private infrastructure, and consumers all have a role in achieving the Circular Economy (Indigo Development 2009).

18.2.5 Latin America and the Caribbean (LAC): Regional Action Plan and Mercosur Policy on SCP

A Regional Council of Government Experts on SCP was set up in Latin America and the Caribbean in 2003 to support the implementation of the SCP regional strategy. The Regional Council has also provided inputs and advice to the LAC Forum of Ministers of the Environment. The LAC Forum of Minister of the Environment is the most representative and influential gathering of environmental policymakers in the region and endorsed important elements of the regional SCP strategy. In 2005 the Fifteenth Meeting of the Forum of Ministers of the Environment of Latin America and the Caribbean (Caracas, Venezuela) decided to foster the preparation of SCP policies, strategies and action plans. The Sixteenth Meeting of the Forum (in Dominican Republic in 2008) approved the regional Action Plan on SCP. The region has identified the following four priorities on SCP: National Policies and Strategies, Small and Medium-Sized Enterprises (SMEs), Sustainable Public Procurement and Sustainable Lifestyles (UNEP 2008).

Agreeing on the need for a common SCP policy with a focus on eco-efficiency and the reduction of hazards forhuman health and the environment, Mercosur member countries signed the Declaration on Cleaner Production Principles in October 2003. This led to the approval of the Mercosur Policy or Promotion and Cooperation onSustainable Consumption and Production in 2007 (Mercosur 2007). Signed by an important trade block of the world, this policy sets an important example for regional coordination on SCP. The policy contributed to the further development of national SCP action plans such as the Argentinian one (Decreto 1289, 2010).

18.2.6 Africa: African 10-Year Framework of Programmes on SCP

SCP activities in Africa started in the mid-1990s. The UNIDO and UNEP established National Cleaner Production Centres (NCPCs) in 1995, which have remained the major institutions for promoting SCP in the region. Since 2000, the African network of NCPCs started to convene biannual regional roundtables on SCP. In 2004, the NCPCs formed the African Roundtable on SCP (ARSCP) as a not-for-profit regional institution to promote SCP. The ARSCP is a multi-stakeholder forum and its activities include, but are not limited to, the organization of national and sub-regional SCP roundtables developing sub-regional and regional programmes and projects on SCP, and organizing trainings on selected SCP topics. The ARSCP pioneered the development of the African 10-Year Framework of Programmes (10YFP) on SCP adopted by the African Ministerial Conference on Environment (2005). Thestrategic focus of the 10YFP is linking SCP with the challenges of meeting basic needs in a more sustainable manner. The major achievements of the African 10YFP on SCP fall under five categories: (i) Mainstreaming, (ii) Energy, (iii) Water, (iv) Information-based instruments and (v) Sustainable Public Procurement (UNEP 2013b).

18.2.7 Reference to Life Cycle Thinking in Different SCP Policies Around the World and the Role of Trade

From the policy examples described above, we can conclude that the topic of addressing environmental impacts of products, materials and resources throughout their life cycles in an integrated way is mainly covered in the SCP policy framework of developed countries. In contrast, in general, the SCP policy programmes of the developing regions of the world focus on national policies, specificresources and business development.

The expansion of life cycle based environmental standards andregulations in industrialized countries could have significant impacts on market access of developing countries. Therefore, the fear in many developing countries is that stricter product standards in the markets of developed countries will act as trade barriers for their exports. Moreover, there is widespread suspicion that environmental restrictions are sometimes used as indirect means of protecting the industries in developed countries (Verbruggen et al. 1995).

18.3 LCA Promotion and Policies at the International Level

18.3.1 UNEP/SETAC Life Cycle Initiative

In 2002 UNEP jointly with the Society of Environmental Toxicology and Chemistry (SETAC) and partners from governments, academia, civil society, business and industry joined forces to promote life cycle approaches worldwide as a way to increase resource efficiency and to accelerate a transition towards more sustainable consumption and production patterns. After the publication of the ISO 14040 standard dealing with LCA, UNEP and SETAC, aware of the need for dissemination and implementation, jointly began to engage more partners to work on the articulation of science-based existing efforts around LCT and established the UNEP/SETAC Life Cycle Initiative (Toepfer 2002).

The Life Cycle Initiative's activities to date have been carried out in two phases, in which around 200 members of the global life cycle community have been actively involved. The first phase (2002–2007) focused on establishing the Life Cycle Initiative as a global focal point of life cycle-related knowledge and activities and on building an expert community of practitioners. Activities to move the Life

Cycle agenda forward concentrated on three important fields of work: (1) Life Cycle Management (LCM), (2) Life Cycle Inventory (LCI) and Life Cycle Impact Assessment (LCIA), as well as the cross-cutting area of social impacts along the life cycle. At the end of the first phase a process was started to help the creation of regional and national life cycle networks, in particular in developing countries, to support capability development. Phase 2 activities (2007–2012) saw the Life Cycle Initiative evolve to be more participative with regard to stakeholders, encouraging more involvement from key actors at the global level in order to achieve common understanding and agreement on tools and strategies being developed. The main outcomes of phase 2 were accomplished through close collaboration with crucial stakeholders in the field. In both phases, the Life Cycle Initiative was able to provide support in the application of sustainability-driven life cycle approaches based on lessons learned from leading organizations by its capacity of engaging with world-class experts and practitioners working in product policy, management and development (UNEP/SETAC 2012).

Building on the achievements from phases 1 and 2 and in particular the results of a stakeholder consultation process in 2011 and 2012, the vision for phase 3 (2012–2017) was coined as 'a world where life cycle approaches are mainstreamed'. Activities in phase 3 focus on creating the enabling conditions to (a) enhance the global consensus and relevance of existing and emerging life cycle methodologies and data management; (b) expand capabilities worldwide and make life cycle approaches operational for organizations; and (c) communicate current life cycle knowledge to influence and partner with stakeholders. Five flagship projects have been defined in the areas of (i) data and databases management, (ii) global guidance on environmental life cycle impact assessment indicators, (iii) product sustainability information 'meta' guidance, (iv) LCA for organizations and (v) global capability development and implementation. Moreover, a special effort on communication and stakeholder outreach has been initiated. These activities are expected to be implemented jointly with a number of other projects. Progress made in phase 3 are monitored every 2–3 years by key indicators and compared to a baseline survey carried out in 2012.

A crucial deliverable and ongoing activity of the UNEP/SETAC Life Cycle Initiative (Sonnemann et al. 2011) is to help overcoming the lack of consistent and high-quality LCA data worldwide and to support capacity building for developing countries is the Global Guidance Principles for Life Cycle Assessment databases published in 2011 by UNEP/SETAC (2011) and the follow-up activities in phase 3. These principles give guidance for proper gathering and management of data, which enable better, more reliable life cycle assessment results and improve their use for decision-making. Life cycle data availability had been recognized by UNEP as a strategic element for advancing SCP through the development and implementation of life cycle based tools and approaches that need these data.

18.3.2 International Resource Panel

The International Resource Panel was established in 2007 to provide independent, coherent and authoritative scientific assessment on the sustainable use of natural resources and the environmental impacts of resource use over the full life cycle and to contribute to a better understanding of how to "decouple" economic growth from environmental degradation (UNEP 2010a).

By providing up-to-date information and best science available information contained in the International Resource Panel's reports is intended to be policy relevant and support policy framing, policy and programme planning, and enable evaluation and monitoring of policy effectiveness (UNEP 2012c).

The broad scope of the Resource Panel requires a wide range ofsustainability experts who organize, review, validate, integrate and communicate findings from studies by appropriate scientists through activities at the working group level. The Resource Panel may limit its size to 30–50 members, while many more scientists are expected to be engaged in the various working groups.

Reports of the International Resource Panel include the following:

- Environmental Risks and Challenges of Anthropogenic Metals Flows and Cycles (2013);
- Measuring Water Use in a Green Economy (2012);
- Decoupling naturalresource use and environmental impacts from economic growth (2011);
- Priority products and materials: assessing the environmental impacts of consumption and production (2010);
- Assessing biofuels: towards sustainable production and use of resources (2009).

These reports take a life cycle perspective and often refer to LCA studies reviewed. The report on assessing the environmental impacts of consumption and production, for example, identifies priorities amongst global consumption activities, industrial sectors and materials from primary industries in terms of their environmental impacts and their resource use. This can play a role in directing environmental and resource policy to those areas that really matter. There is a significant opportunity to improve the basis for decision-making by assessing best available scientific information from a global perspective in order to direct the attention of decision-makers to the big problems first, while avoiding burden shifting in time, space and between environmental impacts (UNEP 2010b).

18.3.3 FAO Partnerships on Bioenergy and Livestock

Bioenergy

The Global Bioenergy Partnership (GBEP) was established to implement the commitments taken by the G8 in the 2005 Gleneagles Plan of Action to support "biomass and biofuels deployment, particularly in developing countries where biomass use is prevalent". Following a consultation process among developing and developed countries, international agencies and the private sector, the GBEP was launched at the 14th session of the Commission on Sustainable Development (CSD 14) in New York on 11 May 2006 (FAO 2009).

From 2007 to 2012 GBEP received a renewed mandate by the G8. The Camp David Summit declared to applaud the Global Bioenergy Partnership (GBEP) for finalizing a set of sustainability indicators for the production and use of modern bioenergy and for initiating capacity building activities through a Regional Forum in West Africa and to invite GBEP to continue implementing capacity building activities that promote modern bioenergy for sustainable development (The White House 2012).

In line with GBEP's Terms of Reference and the state of the international debate on bioenergy, a Task Force on GHG Methodologies was established under the leadership of the United States of America, co-chaired by United Nations Foundation, to analyse the full life cycle of transport biofuels and solid biomass, and to develop a common methodological framework for the use of policymakers and stakeholders when assessing GHG impacts by which the methodologies of GHG life cycle assessments could be compared on an equivalent and consistent basis (FAO 2013).

Livestock

The Partnership on the environmental benchmarking of livestock supply chains is looking to improve how the environmental impacts of the livestock industry are measured and assessed, a necessary first step in improving the sustainability of this important food production sector. At the Rio+20 sustainable development conference, governments agreed on the necessity of making agricultural production more sustainable, and stressed in particular the need to shift to more sustainable livestock production systems (FAO 2012).

FAO and governmental, private sector, and nongovernmental partners work together on a number of fronts to strengthen the science of environmental benchmarking of livestock supply chains. Activities planned for the initial 3-year phase of the project include the following (FAO 2012):

- Establishing science-based methods and guidelines on how to quantify livestock's carbon footprint, covering various types of livestock operations and rearing systems;
- Creating a database of greenhouse gas emission factors generated for the production of different kinds of animal feed–feed production and use offer significant opportunities for reducing livestock emissions;
- Developing a methodology for measuring other important environmental pressures, such aswater consumption and nutrient losses.

18.3.4 ITU Sustainability Standards for the Information and Communications Technology Industry

A number of global companies in the information and communications technology (ICT) sector are increasingly being asked by their customers, investors, governments and other stakeholders to report on their sustainability performance. In response to this growing demand, the Toolkit on environmental sustainability for ICT companies is an International Telecommunication Union (ITU) led initiative; that means it is carried out by ITU together with over 50 partners. The Toolkit provides plenty of detailed support on how ICT companies can build sustainability into the operations and management of their organizations, through the practical application of international standards and guidelines. The Toolkit provides a set of agreed upon sustainability requirements for ICT companies that allows for a more objective reporting of how sustainability is practiced in the ICT sector in these key areas: sustainable buildings, sustainable ICT in corporate organizations, sustainable products, end of life management, general specifications and Key Performance Indicators (KPIs), and an assessment framework for environmental impacts of the ICT sector. It puts international standards and guidelines into context and brings them to life with real-life examples, showing how ICT organizations around the world are dealing with their sustainability challenges (ITU 2012).

18.4 Examples of LCA Promotion and Use in Policies Around the World

18.4.1 Introduction

According to a recent survey done by theLife Cycle Initiative (UNEP/SETAC 2016), the main role of LCA in policies in the last years has been in environmental labelling and the formulation of regulations on product use and waste management mostly in developed countries and still in a very limited way in developing ones. Also, certain governments have been promoting life cycle based policies and encouraging the use of life cycle assessment, for example, to estimate GHG and other emissions of biofuels. Legislation and certification schemes for biofuels are currently emerging, like the global RSB[2]-certification, and mineral oil tax exemption for biofuels of national authorities (UK, Switzerland, Netherlands, Germany, California, etc.); these certification schemes include a range of life cycle impact assessment indicators (SQCB 2013). Policymakers in Denmark and Germany are using interpretations of LCA studies to distinguish between more or less environmentally friendly packaging systems and/or materials; and LCA has

[2]Roundtable on sustainable biofuels.

already been successfully used for the case of the Swedish waste incineration tax (Björklund and Finnveden 2007). However, there is a perceived risk that they are ignoring the uncertain and subjective nature of LCA assessments, which raises questions about the appropriateness of using an LCA-based estimate as a performance metric in public policy.

The examples show that there are high expectations of the future use of LCA in SCP policy areas—such as sustainable public procurement and eco-design directives as well as consumer information. However, there are still certain challenges to overcome such as the lack of good quality and available data, more capacity building and resources. International dialogue and consensus on those issues are required for advancing more life cycle based policies to influence the marketplace, in particular taking into consideration the special context of developing countries. The following examples from different regions of the world provide an overview of LCA promotion and use in policies with quite some detail for the European Union and for selected countries with a certain focus on databases.

18.4.2 Europe: EU and Switzerland

EU Policies Integrating Life Cycle Thinking and Life Cycle Assessment

In the EU at European and national level, over the last 20 years, there has been an increasing emphasis on integrated approaches in environmental policy. Policy has been focusing on linkages between environmental media (air, water, soil) and cross-cutting environmental themes (e.g. climate change, biodiversity etc.) that pay more attention to sustainable resource use. All of these policies aim at fostering the reduction of environmental impact and at further integrating resource use issues and the negative impacts associated to their use in a coordinated way (CEC 2005b). In a growing number of policies and business instruments, LCT and LCA have been recognized as useful approaches in policy support in terms of impact assessment, implementation measures and monitoring needs.

Since 1990, the European Council resolution of 7 May 1990 on waste policy invited the European Commission to submit as soon as possible a proposal for a European-wide eco-labelling scheme covering the environmental impact during the entire life cycle of the product. This resulted in the first EUregulation regarding Eco-label (CEC 1992) where the evaluation of the impact associated with product life cycle is the core of the label scheme. Hence, the first area integrating LCT and eco-design concepts was related to waste policy and to the need of informing consumers. Since then several policy initiatives integrated LCT.

It can be interpreted that the EU has already made significant steps, through various policies building from the Integrated Product Policy (IPP) (CEC 2003). In the IPP, the European Commission (EC) concluded that LCA provides the best framework for assessing the potential environmental impacts of products that are currently available. However, the need for more consistent data and consensus on LCA methodologies was underlined. The further integration of LCT and LCA within policies builds on achievements made in the context of the Thematic Strategy on the Prevention and Recycling of Waste (CEC 2005a), the Thematic Strategy on the Sustainable Use of Natural Resources (CEC 2005b), the Sustainable Consumption and Production and Sustainable Industrial Policy Action Plan of 2008 (CEC 2008a). In 2005, the IPP Communication was particularly strengthened by the EC's Thematic Strategy on the Sustainable Use of Natural Resources (CEC 2005b). It focuses on decoupling economic growth from environmental impacts. LCT is a core to this thematic strategy, being a foundation of the indicators that will be developed to monitor progress across the community. The global dimension is equally recognized through UNEP recommendation to establish the International Resource Panel. Among others, the Action Plans on Sustainable Consumption and Production and on Sustainable Industrial Policy (SCP/SIP) (CEC 2008a) helped to identify and overcome barriers for SCP. The plans built upon ongoing European initiatives and instruments, including the Eco-Management and Audit Scheme (EMAS), the Eco-Label Scheme, the Environmental Technology Action Plan (ETAP), Green Public Purchasing (GPP), the Eco-design of Energy-using Products (EuP) Directive as well as others. This was done in light of increasing coherence among the different related policy areas, while addressing gaps and supporting global interaction. In more recent updates more overarching policy documents such as the Resource Efficiency Flagship Initiative of the Europe 2020 Strategy (CEC 2011a), and another related Roadmap (CEC 2011b) that state, by 2050, the EU economy shall have developed in such a way as to accommodate resource constraints and planetary boundaries. In 2013, a landmark communication has been released: the Single Market for Green Products (CEC 2013a, b).

A brief description of the main initiatives over the last 10 years is given below, entailing initiatives of several EC Directorates General (DGs) such as DG Environment, DG Enterprise and Industry and DG Climate:

- ***Integrated Product Policy Communication Building on Environmental Life Cycle Thinking***—Integrated Product Policy (IPP) seeks to minimize environmental impacts by looking at all phases of a products' life cycle and taking action where it is most effective (CEC 2003).
- ***Stimulating technologies for sustainable development***—Assessments of technologies should verify the technological performance and the claimed performance from an economic and environmental viewpoint, taking into account the whole life cycle of the technology (CEC 2004).
- ***Thematic Strategy on the Prevention and Recycling of Waste***—In order to secure a higher level of environmental protection, the proposal is to modernize the existing legal framework—i.e. to introduce life cycle analysis in policy-making and to clarify, simplify and streamline EU waste law (CEC 2005a).
- ***Thematic Strategy on the Sustainable Use of Natural Resources***—To have a higher impact in reversing unsustainable trends, containing environment degradation and preserving the essential services that natural resources provide, environment policy needs to move beyond emissions and waste control (CEC 2005b).

- ***REACh Regulation on Chemicals***—Risk assessment and management of chemicals have integratinglife cycle thinking (EC 2006).
- ***Sustainable Consumption and Production and Sustainable industrial policy Action Plan***—The Action Plan aims to reduce the overall environmental impact and consumption of resources associated with the complete life cycle of goods and services (CEC 2008a).
- ***Public procurement for a better environment***—Procurement is described as a process whereby public authorities seek to procure goods, services and works with a reduced environmental impact throughout their life cycle when compared to goods, services and works with the same primary function that would otherwise be procured (CEC 2008b).
- ***Waste framework directive***—the directive aims at clarifying key concepts like the waste hierarchy; strengthening the measures that must be taken in regard to waste prevention; introducing an approach that takes into account the whole life cycle of products and materials and not only the waste phase (EC 2008).
- ***Eco-design directive***—The Eco-design Directive provides with consistent EU-wide rules for improving the environmental performance of energy-related products through eco-design (EC 2009a).
- ***Community Eco-Management and Audit Scheme***—The EMAS IIIregulation prescribes that for non-industrial organizations, such as local authorities or financial institutions, it is essential that they also consider the environmental aspects associated with their core business, these include, amongst others, product life cycle-related issues (EC 2009b).
- ***Eco-label***—The EU aims at establishing a voluntary eco-label award scheme intended to promote products with a reduced environmental impact during their entire life cycle (EC 2010a).
- ***Energy labelling directives***—In the directive text is stated that when the Commission reviews progress and reports on the implementation of the SCP/SIP Action Plan in 2012, it will in particular analyse whether further action to improve the energy and environmental performance of products is needed, including the products' environmental impact during their life cycle (EC 2010b).
- ***Resource efficiency flagship***—In the resource efficiency manifesto, one of the road map aims is to create better market conditions for goods and services that have lower impacts across their life cycles, (CEC 2011a, and the related road-map CEC 2011b).
- ***Building sector and Construction work regulation and Strategy for the sustainable competitiveness of the construction sector and its enterprises***—One of the basic requirements for construction works set in EU regulation (EC 2011b) state that the construction works must be designed and built in such a way that they will, throughout their life cycle. Moreover, in the strategy for thesustainability of the building sector (CEC 2012a), a coherent and mutually recognized interpretation of the performances through harmonized indicators is advocated.
- ***Proposal for a General Union Environmental programme to 2020***—Measures will also be taken to further improve the environmental performance of goods

and services on the EU market over their whole life cycle through measures to increase the supply of environmentally sustainable products and stimulate a significant shift in consumer demand for these products (CEC 2012b).

- ***Communication on Bioeconomy***—Actions are set towards the enhancement of bioeconomy markets, taking into account added value, sustainability, soil fertility and climate mitigation potential; supporting the future development of an agreed methodology for the calculation of environmental footprints, e.g. using LCA (CEC 2012c).
- ***Building the single market for green product***—The Single Market for Green Products initiative proposes a set of actions, establishing two methods to measure the environmental performance throughout the life cycle of products and organizations, theProduct Environmental Footprint (PEF) and the Organization Environmental Footprint (OEF); providing principles for communicating environmental performance, such as transparency, reliability, completeness, comparability and clarity; supporting international efforts towards more coordination in methodological development and data availability (CEC 2013a, b).

In the wide policy context presented before, there is an increasing need of life cycle based policy support activities. It is considered of the utmost relevance to develop a science-to-policy interface, due to the broad implications of the decisions supported by LCA. In this context, the Joint Research Centre of the European Commission (EC-JRC) is leading a "science-to-policy" process: gathering, capitalizing and evaluating existing knowledge in order to provide robust support to policy decision-making (Sala et al. 2012). The EC-JRC is working towards providing this policy support through a number of project and initiatives, such as follows:

- European Platform on LCA;
- Support for the product and organization environmental footprint;
- Development of life cycle based indicators for resources, products and waste (EC-JRC 2012);
- Definition of methods to include LCT in waste management (EC-JRC 2011a, b, c);
- Definition of methods to include resource efficiency criteria for energy using products (Ardente and Mathieux 2012);
- Use of LCA for building futurescenarios for policy evaluation (EC-JRC 2013).

The European Platform on LCA

In its Communication on Integrated Product Policy (CEC 2003), the European Commission concluded that LCA provides the best framework for assessing the potential environmental impacts of products currently available. In the document, the need for more consistent data and consensus LCA methodologies was underlined. It was therefore announced that the Commission will provide a platform to facilitate communication and exchanges on life cycle data and launch a coordination initiative involving both ongoing data collection efforts in the EU and existing harmonization initiatives. In 2005, DG Environment together with the Institute for Environment and Sustainability established the European Platform on Life Cycle Assessment (EPLCA). This Platform promotes the availability, exchange and use of quality-assured life cycle data and methods. The EPLCA aims to improve the credibility, acceptance and use of LCA in business and public authorities; to ensure greater coherence across LCA-based instruments and to provide robust decision support to a range of environmental policies and business instruments (see Fig. 18.3).

The main deliverables of the EPLCA are the International Reference Life Cycle Data System (ILCD), the European Reference Life Cycle Database (ELCD), the LCA Resources Directory and LCT Forum mailing list. The ILCD Handbook, launched in 2010, is a series of detailed technical documents, providing guidance for good practices in LCA in business and government, serving as a "parent" document for the development of sector- and product-specific guidance documents, criteria and simplified tools.

The European Commission's Environmental Footprint for Products and Organizations

To date, a company wishing to market its product as green in several EU Member State markets faces a confusing range of choices of methods and initiatives, and might find its needs to apply several of them in order to prove the product's green credentials. This is turning into a barrier for the circulation of green products in the Single Market. The recent EC initiative "Building the Single Market for Green Products—Facilitating better information on the environmental performance of products and organisations" is a step towards removing this ambiguity by improving the way how environmental performance of products and organizations is measured and communicated (CEC 2013a).

Within this initiative, the EC Environmental Footprint was adopted as a harmonized method for multi-criteria (i.e. multi-impact category) environmental LCA of products and organizations. Environmental Footprint (EF) is a new harmonized scheme for multi-criteria life cycle environmental assessment of products and organizations developed by the European Commission's JRC in close cooperation with Directorate-General for the Environment.

The two guidelines on Product EF (PEF) and Organization EF (OEF) provide specific and practical guidance for comprehensive, robust and consistent environmental assessment of products and organizations. To further support comparisons and comparative assertions within product groups and sectors, Product Environmental Footprint Category Rules (PEFCRs) and Organization Environmental Footprint Sector Rules (OEFSRs) are developed in a 3-year pilot phase starting in 2013. Chapter 24 offers more details on the PEF/OEF.

EU Policy Background and Rationale

The EC Environmental Footprint fits within the integrated product policy (IPP) and the SCP/SIP Communication of the European Union. More recently, in December 2010, the environmental ministers of the Member States of the EU met in the Environment Council and invited the European Commission to "develop a common

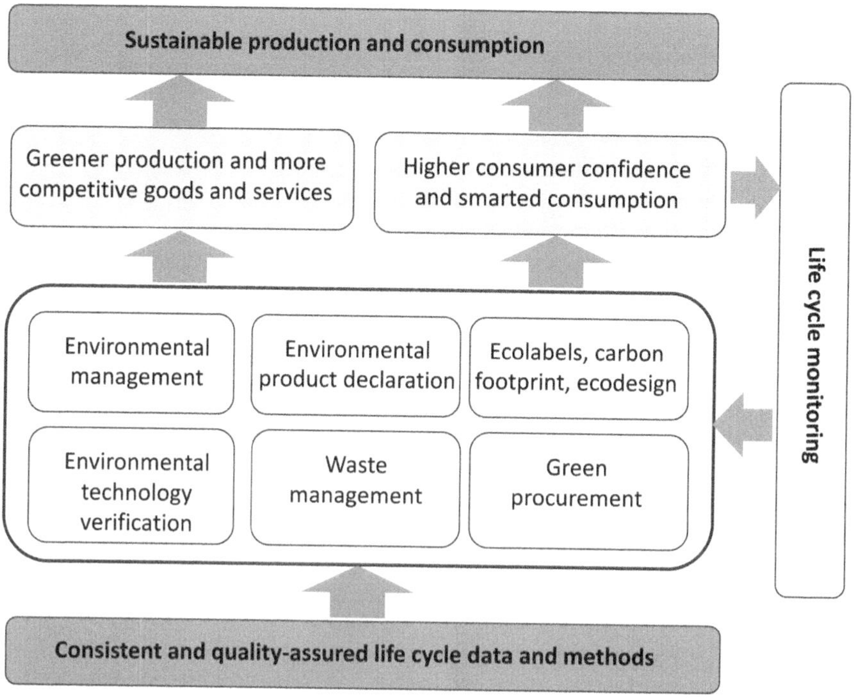

Fig. 18.3 Life cycle data and methods as the basis of tools and approaches for supporting sustainable production and consumption policies

methodology on the quantitative assessment of environmental impacts of products, throughout their life-cycle, in order to support the assessment and labelling of products" (Council of the European Union 2010).

The EC "Roadmap to a Resource Efficient Europe" (CEC 2011a) was an answer to this invitation and proposes ways to increase resource productivity and to decouple economic growth from both resource use and environmental impacts, taking a life cycle perspective. One of its objectives is to "Establish a common methodological approach to enable Member States and the private sector to assess, display and benchmark the environmental performance of products, services and companies based on a comprehensive assessment of environmental impacts over the life-cycle ('environmental footprint')".

In April 2013, the EC published the communication on "Building the Single Market for Green Products—Facilitating better information on the environmental performance of products and organisations" (CEC 2013a). The Commission Recommendation (CEC 2013b) that encourages EU Member States and the private sector to use the EC PEF and OEF methods to measure and communicate the environmental performance of products and organizations accompanies the communication. The PEF and OEF methods based on LCA are integral part of the

Recommendation. This is seen as an important step forward to ensure robust decision support for consumers, industry and policymakers.

France
France is among the European countries that are active in the transition towards sustainable production and consumption patterns. In recent years, many public and private initiatives were launched: "Grenelle environment" and "French Energy Transition" are among these initiatives.

Grenelle Environmental Labelling
Based on the Grenelle II law (French Ministry of Ecology, Sustainable Development and Energy 2010), in 2011–2012 France conducted a national experimentation on consumer product environmental information. The experimentation covered the quantification of environmental impacts and the communication of environmental footprints to the consumer. More than 160 companies participated. All sectors were represented, with about one-third from the food and beverage area. Several foreign companies—from Chile, Colombia, Sweden, etc.—were part of the experimentation.

The French governmental conclusion report now constitutes the roadmap for additional future developments including the development and consolidation of the technical tools (database, PCR, calculators). Furthermore, the French Government participate and contribute actively to the EU "PEF/OEF" Environmental Footprint pilot phase.

Energy Transition for Green Growth Act
The Energy Transition for Green Growth Act represents the French government's aim—linked to the Paris Agreement (COP21)—to reduce its greenhouse gas emissions, diversifying its energy model and increasing the deployment ofrenewable energy sources (French Ministry of Ecology, Sustainable Development and Energy 2015).

The actions focus on the buildings, transport, Circular Economy, renewable energy and nuclear energy. The initiative also aims to remove regulatory constraints. The government aims to implement the energy transition with the involvement of all stakeholders.

Even though the role of Life Cycle Thinking has been highlighted in these initiatives, the application of Life Cycle approaches in is yet far from being mainstream. Many LCA-related activities were conducted by private and public key actors including networks in France in recent years, either directly linked to or from beyond the aforementioned initiatives.

Switzerland
Federal Office for the Environment (FOEN) jointly with other Swiss Federal Offices backs ecoinvent, the Centre for Life Cycle Inventories, which is a Competence Centre of the Swiss Federal Institute of Technology Zürich (ETH Zurich) and Lausanne (EPF Lausanne), the Paul Scherrer Institute (PSI), the Swiss Federal Laboratories for Materials Testing and Research (EMPA) and the Swiss Federal Research Station Agroscope Reckenholz-Tänikon (ART). The ecoinvent centre

holds one of the world's leading LCA databases. The centre's mission is to establish and provide scientifically sound and transparent international life cycle assessment and life cycle management data and services to industry, consultancies, public authorities and research institutions. Switzerland provides also technical assistance to other countries aiming at building up LCA knowledge.

Switzerland supports an international approach through the use of the Global Guidance Principles delivered by the UNEP/SETAC Life Cycle Initiative, as well as the development of Product Category Rules and the environmental Life Cycle Impact Assessment Indicators under development. Another key issue is global interoperability; in this context FOEN supports activities to increase data availability, transparency, capabilities and the use of gate-to-gate unit process data in a "Lego bricks" approach to enhance the interoperability among databases (UNEP 2013b).

18.4.3 The North American Free Trade Agreement Countries: USA and Mexico

USA

The USA has developed LCA databases to support the work of the private and public sectors on sustainability. The most recent activity is the LCA Digital Commons Project at the UNDA National Agricultural Library. The goal is to develop a database and toolset intended to provide data for use in LCAs of food, biofuels, and a variety of other bio-products. Researchers at the University of Washington Design for Environment Laboratory have developed initial unit process data. OpenLCA provides core software for the Commons database. The development of visualization tools is underway with Earthster. Listed below are some of the organizations and resources data, which are contributing to the LCA Digital Commons project (UNDA National Agricultural Library 2013).
Organizations involved are:

- US EPA promotes the use of LCA to make more informed decisions through a better understanding of the human health and environmental impacts of products, processes, and activities and supported the development of the Global Guidance Principles on LCA Databases.
- National Renewable Energy Laboratory—created and maintains the U.S. Life Cycle Inventory (LCI) Database to help LCA practitioners answer questions about environmental impact.
- The Sustainability Consortium drives scientific research and the development of standards and IT tools, through a collaborative process, to enhance the ability to understand and address the environmental, social and economic implications of products.
- The OpenLCA Project creates modular software for life cycle analysis and sustainability assessments. The software is available as open source and is free.

- New Earth is a non-profit organization initiating, facilitating and implementing innovative strategies and tools to help achieve sustainable development on a global level such as Earthster.
- The American Centre for Life Cycle Assessment is a non-profit membership organization formed in 2001 to increase awareness of and to promote the adoption of Environmental LCA among industry, government and NGOs.
- The Innovation Centre for U.S. Dairy is working with the entire dairy industry to foster innovation and give consumers more of what they want, when and where they want it.

Data resources available include the following:

- Earthster—This website is the home of a new system that is web-based, free and open source (non-proprietary). It begins by inviting participation purely on the basis of providing zero-cost access to markets; this is something that any business can respond positively to. Next, buyers and consumers at one end of the system have the ability to send signals to producers about the desired environmental and social attributes or characteristics of products and their life cycles. Third, producers have the ability to download and use free software to rapidly benchmark themselves versus industry averages, and optionally to click-to-report environmental and social attributes of their processes and products to the marketplace. Fourth, LCA data providers, and developers of methodologies, scorecards, labelling systems, etc., all have the ability to process the publicly provided information using their own systems, providing decision-makers with customized reports with only the data of interest to the decision-maker.
- US Life Cycle Inventory Database provided by the National Renewable Energy Laboratory is publicly available and contains data modules for commonly used materials and processes, such as primary fuel production and combustion, electricity generation and transformation processes.
- The ecoinvent Database provided by the Swiss Centre for Life Cycle Inventories, which was and is supported by Swiss Federal Offices.
- GaBi Databases generated by PE International.
- ELCD Core Database provided by the Joint Research Centre of the European Commission.

Mexico

Major barriers and needs for LCA in Mexico consist in lack of information, regulations and capacities on LCA, as well as a lack of diagnosis of the market and its requirement for a good LCA and the right private–academic–public partnership. Conditions required to overcome these barriers are to establish the structure of global life cycle policies according to internal needs and to get the training for its development as well as open LCA to different sectors in order to promote partnership for the development of tools that can support the design, development and implementation of policies. Issues and deliverables needed for advancing LCA-based policies include the following:

- Diagnosis and assessments about how to introduce LCT and LCA in policies according to conditions in Mexico,
- Identification of which policies and sectors must be influenced to do so, as well as life cycle experts available in the country.

With regard to life cycle inventory datasets, from 2005 onwards data was collected from public and private sources. In 2010 the database format was designed in compliance with existing documentation formats. From 2010 the IT platform and data management have been implemented such as the identification of policies and sectors that need to be targeted (e.g. energy, transport). Since then data are available in the national database called the Mexican Life Cycle Inventory Mexicaniuh (UNEP 2013b).

In Mexico, a regulation for sustainable buildings (NMX 2013) on criteria and minimum environmental requirements needs the impact assessment of the whole life cycle of buildings (including the use phase). In case of the replacement of building materials, it is also demanded the use of third-party reviewed LCAs of alternative materials for comparative assertion purposes (Güereca et al. 2015).

18.4.4 Other OECD Countries: Japan and Australia

Japan

Since 2000 Type III-based Environmental Declarations have been developed in the so-called EcoLeaf programme. Moreover, since 2008 the Carbon Footprint Programme (CFP) has been developed. The business is aware of these programmes but their uptake is faced with challenges. It is not easy for consumers to understand the label so that they could be willing to buy labelled product. Therefore, it is needed to prepare the grounds for sales promotion with cost-efficiency. In addition, a carbon offset pilot programme based on the CFP where communities can collect credits has been developed, as were general guidelines on Supply Chain GHG Emission Accounting.

Two international LCA workshops were organized in Japan in February 2013, which focused on Future Utilization of Visualized Information on Environmental Impacts in Product Life Cycle and Corporate Value Chain as well as Sharing of experience and findings of world's major initiatives. Concerns were raised about the appropriate criteria for the selection of a large number of environmental impact categories and the corresponding data availability, in particular in developing countries. Moreover, before using the results for comparative assertion on uncertainly and accuracy of databases as well as the methodologies behind the data collected need to be discussed between database managers, developers and policymakers. Finally, the impact categories to be addressed will differ according to the product category. Hence, there is a need to get global consensus in developing Product Category Rules (PCRs) (UNEP 2013b).

Australia
The National AusLCI Initiative was set up to support national goals coming from the public and private sectors.

LCA database guidance has been developed with Australian industry and practitioners and is generally compliant with Ecoinvent guideline and UNEP-SETAC Global Guidance Principles on LCA Databases. The datasets are being presented in EcoSpold andILCD format. Unit process and system processes are both being provided. AusLCI next steps are to increase the coverage of building products by migrating data from the BPIC database into the AusLCI to allow access for more sectors, to further increase building and agriculture coverage and to begin publishing data in ecoinvent as part of a National Project Agreement. LCA has been successfully used in the Voluntary Green Building Rating. Challenges and opportunities cover the regional and international data harmonization and interoperability, acceptance of user-friendly decision support tools for different stakeholders, policy support lessons, including public procurement, as well as value and risk case studies as relevant areas (UNEP 2013b).

18.4.5 Emerging Economies in Asia: China and Thailand

China
The policies in China based on LCA are all very recent, from 2012: Technology assessment and implementation for energy conservation and emission reduction, evaluation and recommendation of energy efficiency products and eco-design of products. These policies are all supporting the Chinese policy approach aiming at the establishment of a circular economy.

Also encouraging is the policy on the Eco-design of Industrial Products Guidance of 2013 (MIIT/MEP/NRDC 2013) which is boosting mainstreaming of LCA in China by promoting its use in product design.

The Chinese Life Cycle Database (CLCD) has been developed since 2007 and published in 2010 by Sichuan University and IKE Technology. The goal is to have a fundamental LCA database representing Chinese technology and market average with more the 600unit processes of fundamental products in one core model.

More global agreements and practical guidelines are desirable, such as the UNEP/SETAC Global Guidance Principles for LCA Database that are a starting point and need to be actively disseminated. Overall in China, there is a huge need for capacity building and technical assistance, although a number of university, research centres and companies have identified the life cycle topic as a promising approach for the future and are catching up with regard to theinternational level (UNEP 2013b).

Thailand
Thai National Science and Technology Development Agency (NSTDA), which is part of the Ministry of Science and Technology, has a leading role in the development of a Thai LCA database. Progress has been made in LCT and LCA through the introduction of the Thai Green Label (Type 1), the green procurement activities, the promotion of biofuels (ethanol 2001, biodiesel 2005) and the National Green Growth Strategies (2013–2018). These topics are covered and need to further be implemented as LCA-related actions in the SCP and Green Economy Roadmap. Thailand is using life cycle inventory data to quantify the Green Gross Domestic Product (GDP) of its industrial sectors, as well as LCA and Life Cycle Costing to assess Phase 1 (2008–2011) of the Thai Green Public Procurement Plan (ORDER PRE/116/2008 2008) to decide whether and how to implement phase 2 (2014–2017) (Mungcharoen 2013).

As next steps, more capacity building activities, for example, on indicators for Life Cycle Impact Assessment in the field of agri-food, decoupling and Green GDP are needed as well as joint activities with other Asian countries, such as the set-up of the LCA Agri-Food Asia Network (UNEP 2013b).

18.4.6 Emerging Economies in Latin America and Africa

Brazil
At present LCA-based policies in Brazil include, for instance, the Brazilian Life Cycle Assessment Programme (2010), the National Solid Waste Policy (Federal Law No. 12.305, 2010, and Decree No. 7.404, 2010, MMA 2010) and the Brazilian Eco-label Type 1 Scheme. For example, the National Solid Waste Policy calls for shared responsibilities among relevant stakeholders along the life cycle of wastes and the use of LCA to promote products with fewer environmental impacts. Proposed future policies cover GHG inventory and Green Procurement.
The major needs and barriers for implementing LCA-based policies in Brazil and the conditions to overcome these barriers are as follows: (i) governmental funding to the National LCA Programme; (ii) private funding from industrial sectors; (iii) industrial awareness of main achievements, in terms of economic and environmental profits, using the methodology as a management tool; and (iv) public awareness and capacity building for policymakers. Focus areas the government needs to work on to advance the uptake of LCA are the following:

- Governmental funding for applied research and technological development projects during the next 10 years.
- Increasing participation of the private/industrial sectors in the development of Life Cycle Inventories, aiming to build a Brazilian primary data LCI database.
- Capacity building of policymakers and private sector through specific industrial associations support (similar approach used by EU with European associations).

The Brazilian government has made a major effort to develop a national LCA database (UNEP/SETAC 2011), working among others with UNEP and reviewers provided by the UNEP/SETAC Life Cycle Initiative to increase the quality of the datasets in this emerging database.

Needs for capacity building include the promotion of specific courses on LCA topics in the academy (undergraduate and graduate level) and in the industry educational system (professional level). Moreover, the organization of international seminars, in coordination with the European Commission and UNEP/SETAC, to disseminate successful case studies of industry in OECD countries on the utilization of LCA in their supply chain management and to train on international and global guidelines, data acquisition approaches and reviewing schemas are also required (UNEP 2013b).

South Africa

Carbon footprints are much debated and many private firms have theirs assessed, albeit most often not based on LCA and not captured in any national policy. There is some concern about trade barriers based on carbon footprints, and at sub-national government level it is especially the fruit- and wine-exporting Western Cape where this is a serious concern.

Some kind of 'life cycle thinking' is embodied in the Mineral Resources Development Act of 2002 through the 'planning for mine closure' regulations, but this is more of a temporal type of life cycle thinking, not one of shiftings of burdens to other players in the supply chain (e.g. through fuel switching) or between environmental compartments. The recently promulgated national waste management strategy is very strongly aligned to the waste hierarchy.

Environmental considerations were somehow considered in the establishment of the biofuels industrial strategy of 2007; the Energy Information Administration for a fuel ethanol plant has included a GHG balance. The umbrella environmental legislation, called the 'national environmental management act' (NEMA 1998) probably does include some loose reference to LCT and principles.

The ecoinvent project for developing a South African database is seen as a starting point for a possible national database. Participants of a roadmap discussion for a South African LCA Database organized in Cape Town on 3 and 4 Feb 2015 identified the potential key role of the National Cleaner Production Centre (NCPC) for access to the Department of Trade and Industry (DTI) and the Department of Environmental Affairs (DEA), as well as to its large industry network to take advantage of this seed project. In the longer term, it might make sense to create an independent national LCA database. In the short term, also capacity building on life cycle thinking and LCA is important to increase the maturity of those approaches in the country in order to provide the basis for the development of LCA-based policies.

Colombia

A SCP action plan explicitly incorporates LC thinking (MAVDT 2010). Also in the same year, a national public procurement policy (MinAm 2010) was issued which updates and integrates the national plan on green markets with the national SCP action plan. The national public procurement policy aims at providing life cycle based criteria for sustainable purchasing by public offices and at supporting the implementation of these criteria. No national database is under development in Colombia.

18.5 Conclusions

The present chapter explains what Sustainable Consumption and Production is about, why it is about taking a life cycle approach and shows that SCP-related policies have been developed at the intergovernmental level and in different regions of the world since the Johannesburg World Summit on Sustainable Development in 2002. A key element at the international level is the 10-Year Framework of Programmes on SCP that has been adopted in Rio+20 and provides multiple opportunities for promoting policies based on life cycle thinking and using LCA.

Life cycle thinking has been considered mature, moving from its academic origins and limited uses primarily in-house in large companies to more powerful approaches that can efficiently support the provision of more sustainable goods and services through efficient use in product development, external communications, in support of customer choice, and in public debates (Pennington et al. 2007).

Now governments can use LCA for SCP and resource efficiency policies. LCA is in particular relevant for policies focusing on design for sustainability, sustainable consumer information, sustainable procurement and waste management, minimization and prevention as well as sector-specific policies like sustainable energy and food supply. The execution of LCA studies is directly required in policies evaluating the environmental preference of biofuels in different countries. It can be expected that the use of LCA in policies for the sustainability assessment of products will further increase.

Conflicts between free trade and environmental requirements to products have become evident (e.g. ISO process on carbon footprinting) and need to be taken seriously. LCA as an analytical tool is not much the problem but its use in certification and trade relevant polices will continue to generate conflicts. Environmental standards and regulations have penitential impact on the market access of developing countries; hence, there is a fear in those countries that any further strictness on product standards in the developed countries' market could result in creating a significant trade barrier for their exports. In this context the International Organization for Standardization (ISO) has an important role to play. Under the World Trade Organization (WTO) Agreement (1994), Members (governments) are obliged to adopt international standards wherever feasible and this includes ISO standards. One consequence is that businesses (including governments as trading parties) can make adherence or certification/registration under ISO standards a term or condition of trade with a foreign business (UNEP 2013b). In particular, the European Commission is moving rather quickly in putting LCA into policy use.

18.6 Perspectives

Overall, we see that definitely some areas still need improvements to ensure better integration in policymaking, even in Europe. It is often considered a critical task to find the right balance between, e.g.

- Enhancing the comparability of LCAs by being prescriptive versus providing the required flexibility in order to apply LCA for many different types of applications in very diverse product groups or sectors.
- Allowing limited assessments on a few impact categories with a high degree of certainty versus pushing towards more comprehensive assessments including impact categories with a lower degree of certainty whilst being transparent about their need for improvement;
- Scientific robustness of available Life Cycle Impact Assessment models versus applicability and feasibility aspects;
- Cementing the status quo, towards "stability" of the recommendations over time, versus encouraging further improvements related to both LCIA method development and related;
- LCI data availability and quality;
- Ensuring sufficiently robust quality of LCA results, including the methods and underlying data used, via review and verification requirements versus applicability and feasibility aspects.

The above-mentioned aspects are crucial for any sustainability assessment methods and require actions by several stakeholders (from methods developed to policymaking) to ensure applicability and efficacy. From the policymaking side, there is a need to balance the stability of the recommendation (to be applied in a business and policy context) and the thriving scientific development, for example, in the field of impact assessment. Furthermore, finding the best solution to guarantee comparability among studies and being open to updated data, models and factors are of paramount importance.

Mainstreaming the use of LCA in such policies is currently hampered by the missing availability of high-quality data worldwide. There are concerns related to data availability for running the evaluations, ensuring robustness and representativeness of data. Knowledge mining and review of existing studies are extremely crucial for supporting policies.

Opportunities for the future of national databases are emerging through the Global Network of interoperable LCA Databases that is an initiative from the International Forum on LCA Cooperation International started by the EC and UNEP in 2012, and now supported by a number of national governments around the world. Its vision is to establish "a global network comprised of independently-operated and interoperable LCA databases that connects multiple data sources to support life cycle assessment in a way that facilitates sustainability-related decisions". More detailed objectives include to define and contribute to the availability of an electronic system and protocol to enable access by users to the majority of the LCA databases and other relevant sustainability data, meaning that the LCA datasets and other data therein can be easily accessed in an exchange format that allows to use them seamlessly in LCA software, with sufficient documentation of metadata that allows defining fitness for purpose by any user.

However, for this to happen it is not only necessary to contribute to the availability of an electronic system and protocol but also to foster capacity building in emerging economies and developing countries, in which more and more of the global production and consumption is taken place. For such capability development efforts to be successful they need to focus as much on the demand side by training on life cycle management in business and life cycle thinking in policies as on the technical aspects with regard to national LCA databases and regionalized life cycle impact assessment methods. These needs are addressed further in Chap. 19 on Globalisation and mainstreaming of LCA.

Finally, life cycle based policymaking in the future will have to address also the means to implement policies (incentives, legislative obligations and thresholds, etc.) based on life cycle assessment results. This implies that upcoming opportunities of using LCA in policies for the sustainability assessment of products need to be accompanied by ways to show the efficiency of the policies put in place.

SCP policies based on life cycle thinking are starting to be well developed at the international level and around the world, while there is a need for further promotion of LCA-based policies based on a widely accepted analysis of the benefits and limits of such polices. Our expectation, based on the past experience of using life cycle thinking and LCA for policy support, is that the use of life cycle methodologies and related methods and tools in policy support will continue to grow in influence in the foreseeable future.

Acknowledgements We would like to thank UNEP for the permission to use the material from the unpublished draft report (UNEP 2013b) on "Life cycle thinking and the use of life cycle assessment in sustainable consumption and production policies: Exchange of experiences and the need for quality data availability and capacity building". This report was initiated by Jodie Bricout from cd2e and includes the content of the presentations from the 2nd Intergovernmental Dialogue on LCA held in Paris, France, in April 2013. We would like to thank the delegations from Switzerland, USA, Mexico, Japan, Australia, China, Thailand and Brazil for these presentations, without which it would not have been possible to give a worldwide overview of LCA promotion in policies, covering all these countries. These delegations included well-known LCA experts such as Rolf Frischknecht from Switzerland, Nydia Suppen from Mexico, Atsushi Inaba from Japan, Greg Foliente from Australia, Hongtao Wang from China and Thumrongrut Mungcharoen from Thailand. Moreover, Harro von Blottnitz from the University of Cape Town provided valuable information on the situation in the South Africa. Finally, the authors would like to thank especially Llorenç Milà i Canals from UNEP and Bruce Vigon from SETAC for their support in moving the Life Cycle Initiative through Phase 3 with the vision of a world where life cycle approaches are mainstreamed. We hope to testimony with this chapter that the world is slowly moving in this direction.

References

African Ministerial Conference on Environment: African 10-Year Framework Programme on Sustainable Consumption and Production (2005)

Ardente, F.E., Mathieux, F.: Integration of Resource Efficiency and Waste Management Criteria in European Product Polices—Second Phase, Report No 3. Refined Methods and Guidance Documents for the Calculation of Indices Concerning Reusability/Recyclability/Recoverability, Recycled Content, Use of Priority Resources, Use of Hazardous Substances, Durability. JRC Technical Report 73188, European Union, Luxembourg (2012). ISBN: 978-92-79-25931-9

Björklund, A.E., Finnveden, G.: Life cycle assessment of a national policy proposal: the case of a Swedish waste incineration tax. Waste Manage. **27**(8), 1046–1058 (2007)

CEC: Integrated Product Policy—Building on Environmental Life-Cycle Thinking. Communication from the Commission to the Council and the European Parliament. COM (2003) 302 final (2003)

CEC: Stimulating Technologies for Sustainable Development: An Environmental Technologies Action Plan for the European Union. Communication from the Commission. COM (2004) 38 final (2004)

CEC: Thematic Strategy on the Prevention and Recycling of Waste. COM (2005) 666 (2005a)

CEC: Thematic Strategy on the Sustainable Use of Natural Resources. Communication from the Commission. COM (2005) 670 (2005b)

CEC: Sustainable Consumption and Production and Sustainable Industrial Policy Action Plan. Communication from the Commission. COM (2008) 397/3 (2008a)

CEC: Public Procurement for a Better Environment. Communication from the Commission. COM (2008) 400 (2008b)

CEC: A Resource-Efficient Europe—Flagship Initiative Under the Europe 2020 Strategy. Communication from the Commission. COM (2011) 21 final (2011a)

CEC: Roadmap to a Resource Efficient Europe. Communication from the Commission. COM (2011) 571 (2011b)

CEC: Strategy for the Sustainable Competitiveness of the Construction Sector and its Enterprises. Communication from the Commission. COM (2012) 433 (2012a)

CEC: Proposal for a decision of the European Parliament and of the Council on a General Union Environment Action Programme to 2020 "Living well, within the limits of our planet". COM (2012) 710 (2012b)

CEC: Innovating for Sustainable Growth: A Bio-Economy for Europe. Communication from the Commission. COM (2012) 60 (2012c)

CEC: Building the Single Market for Green Products—Facilitating Better Information on the Environmental Performance of Products and organisations. Communication from the Commission. COM (2013) 196 (2013a)

CEC: Commission recommendation of 9 April 2013 on the use of common methods to measure and communicate the life cycle environmental performance of products and organisations 2013/179/EU. Brussels (2013b)

Chomkhamsri, K., Pelletier, N.: Analysis of Existing Environmental Footprint Methodologies for Products and Organizations: Recommendations, Rationale, and Alignment. European Commission, Joint Research Centre, Institute for Environment and Sustainability, Ispra (2011)

Council of the European Union: Council conclusions on sustainable materials management and sustainable production and consumption: key contribution to a resource-efficient Europe 3061st ENVIRONMENT Council meeting, Brussels, 20 Dec 2010. Retrieved from http://www.consilium.europa.eu/uedocs/cms_data/docs/pressdata/en/envir/118642.pdf (2010)

Decreto 1289/10: Incorpórase al ordenamiento jurídico de la República Argentina la Decisión N° 26/07 del Consejo del Mercado Común del MERCOSUR. Argentina (2010)

EC: Regulation (EC) No1907/2006 of the European Parliament and of the Council of 18 December 2006 concerning the Registration, Evaluation, Authorisation and Restriction of

Chemicals (REACH), establishing a European Chemicals Agency, amending Directive 1999/45/EC and repealing Council Regulation (EEC) No 793/93 and Commission Regulation (EC) No 1488/94 as well as Council Directive 76/769/EEC and Commission Directives 91/155/EEC, 93/67/EEC, 93/105/EC and 2000/21/EC (2006)

EC: Directive 2008/98/EC of the European Parliament and of the Council of 19 November 2008 on waste and repealing certain Directives Text with EEA relevance (2008)

EC: Directive 2009/125/EC of the European Parliament and of the Council of 21 October 2009 establishing a framework for the setting of ecodesign requirements for energy-related products (2009a)

EC: Regulation (EC) No 1221/2009 of the European Parliament and of the Council of 25 November 2009 on the voluntary participation by organisations in a Community eco-management and audit scheme (EMAS), repealing Regulation (EC) No 761/2001 and Commission Decisions 2001/681/EC and 2006/193/EC (2009b)

EC: Regulation No 66/2010 of the European Parliament and of the Council of 25 November 2009 on the EU Ecolabel (2010a)

EC: Directive 2010/30/EU of the European Parliament and of the Council of 19 May 2010 on the indication by labelling and standard product information of the consumption of energy and other resources by energy-related products (2010b)

EC: Life Cycle Thinking and Assessment, Joint Research Centre, http://lct.jrc.ec.europa.eu/index_jrc (2011a). Accessed 18 Nov 2011

EC: Regulation No 305/2011 of the European Parliament and of the Council of 9 March 2011 laying down harmonised conditions for the marketing of construction products and repealing Council Directive 89/106/EEC (2011b)

EC-JRC (European Commission, Joint Research Centre, Institute for Environment and Sustainability): International Reference Life Cycle Data System (ILCD) handbook. Retrieved 26 Nov 2012, from http://lct.jrc.ec.europa.eu/assessment/publications (2010–2012)

EC-JRC (European Commission, Joint Research Centre, Institute for Environment and Sustainability): Supporting Environmentally Sound Decisions for Waste Management A technical guide to Life Cycle Thinking (LCT) and Life Cycle Assessment (LCA) for waste experts and LCA practitioners. European Union, Luxembourg, JRC Scientific and Technical Reports 65850; ISBN 978-92-79-21017-4 (PDF) (2011a)

EC-JRC: Supporting Environmentally Sound Decisions for Bio-Waste Management. A practical guide to Life Cycle Thinking (LCT) and Life Cycle Assessment (LCA). European Union, Luxembourg, JRC Scientific and Technical Reports 65851; ISBN 978-92-79-21019-8 (PDF) (2011b)

EC-JRC: Supporting Environmentally Sound Decisions for Construction and Demolition (C&D) Waste Management. A practical guide to Life Cycle Thinking (LCT) and Life Cycle Assessment (LCA). European Union, Luxembourg, JRC Scientific and Technical Reports 65852; ISBN 978-92-79-21021-1 (PDF) (2011c)

EC-JRC: Life Cycle Indicators Framework: Development of Life Cycle Based Macro-Level Monitoring Indicators For Resources, Products and Waste For the EU-27. European Commission, Joint Research Centre, Institute for Environment and Sustainability, European Union, Luxembourg, JRC Technical Report 73336 (2012). ISBN: 978-92-79-25937-1

EC-JRC (European Commission, Joint Research Centre, Institute for Environment and Sustainability): Sustainability assessment of future-oriented scenarios: a review of data modelling approaches in Life Cycle Assessment, draft, submitted to internal review (2013)

FAO: The Global Bioenergy Partnership Common Methodological Framework for GHG Lifecycle Analysis of Bioenergy. GBEP Secretariat, Rome, Italy (2009)

FAO: New effort to harmonize measurement of livestock's environmental impacts, News Article. http://www.fao.org/news/story/en/item/150555/ (2012). Accessed 1 Aug 2013

FAO: The Global Bioenergy Partnership. GBEP Secretariat, Rome, Italy (2013)

French Ministry of Ecology, Sustainable Development and Energy: Le Grenelle Environnement—Grenelle 2 law, Paris, France (2010)

French Ministry of Ecology, Sustainable Development and Energy: Energy Transition for Green Growth Act: User Guide for the Act and its Attendant Actions, Paris (2015)

Güereca, P.L., Ochoa, R., Gilbert, H., Suppen, N.: Life cycle assessment in Mexico: overview of development and implementation. Int. J. Life Cycle Assess. **20**(3), 311–331 (2015)

Indigo Development: China seeks to develop a circular economy. http://www.indigodev.com/Circular1.html#refs (2009). Accessed 29 July 2013

ISO (International Organization for Standardization): ISO 14040. Environmental Management—Life Cycle Assessment—Principles and Framework. ISO, Geneva, Switzerland (2006)

ITU: Toolkit on environmental sustainability for the ICT sector. http://www.itu.int/dms_pub/itu-t/oth/4B/01/T4B010000060001PDFE.pdf (2012). Accessed 17 Dec 2013

MAVDT: Política Nacional de Producción y Consumo Sostenible de Colombia. www.minambiente.gov.co/index.php/component/content/article?id=155:plantilla-asuntos-ambientales-y-sectorial-y-urbana-8#pol%C3%ADtica-y-criterios (2010). Accessed 25 Apr 2016

Mercosur: Política de Promocíon y Cooperacíon en Produccíon u Consumo Sostenibles en el Mercosur (2007)

MIIT/MEP/NRDC: Eco-design of industrial products guidance. Published in 2013 by the Ministry of Industry and Information Technology and the Ministry of Environmental Protection of the People's Republic of China, and the National Development of Reform Commission (NDRC). www.indigodev.com/sepa_eip_guideline.html (2013). Accessed 28 Feb 2015

MinAm: Guía Conceptual y Metodológica de Compras Públicas Sostenibles de Colombia. Ministerio de Ambiente y el Centro Nacional de Producción más Limpia de Colombia. www.minambiente.gov.co/index.php/component/content/article?id=155:plantilla-asuntos-ambientales-y-sectorial-y-urbana-8#gu%C3%ADa (2010). Accessed 25 Apr 2016

MMA: Politica Nacional de Resíduos Sólidos – Ministério do Meio Ambiente. Lei no 12.305/2010, Decreto No. 7.404/2010. www.abdi.com.br/ (2010). Accessed 28 Feb 2015

Mungcharoen, T.: LCA applications in Thailand. In: Proceedings of the 2014 Chinese Conference on Life Cycle Management (CLCM), Beijing. www.clcm.org.cn/ 2014 (2013)

NMX: Norma Mexicana NMX-AA-164-SCFI-2013: Edificacion Sustentable – Criterios y Requerimientos Ambientales Minimos (Sustainable building—minimum environmental criteria and requirements). http://biblioteca.semarnat.gob.mx/janium/Documentos/Ciga/agenda/DOFsr/DO3156.pdf (2013). Accessed 28 Feb 2015

Norwegian Ministry of Environment: Oslo Symposium (1994)

OECD: Environmental Claims—Findings and Conclusions of the, Committee on Consumer Policy, Paris (2010)

OECD: Sustainable Materials Management: Making Better Use of Resources. OECD, Paris (2012)

ORDER PRE/116/2008: [Thai] Green public procurement plan of the state general administration and its public bodies, and the managing bodies of the social security. Published on 21 January. www.measures-odyssee-mure.eu/public/mure_pdf/transport/SPA44.PDF (2008). Accessed 28 Feb 2015

Pennington, D., Wolf, M., Bersani, R., Pretato, U.: Overcoming barriers to the broader implementation of life cycle thinking in business and public administration. Int. J. Life Cycle Assess. **12**(7), 458–460 (2007). doi:10.1065/lca2007.07.355

Sala, S., Pant, R., Hauschild, M., Pennington, D.: Research needs and challenges from science to decision support. Lesson learnt from the development of the international reference life cycle data system (ILCD) recommendations for life cycle impact assessment. Sustainability **4**(7), 1412–1425 (2012)

Sonnemann, G., Vigon, B., Broadbent, C., Curran, M.A., Finkbeiner, M., Frischknecht, R., Inaba, A., Schanssema, A., Stevenson, M., Ugaya, C., Wang, H., Wolf, M.-A., Valdivia, S.: Process on "global guidance for LCA databases". Int. J. Life Cycle Assess. **16**(1), 95–97 (2011)

SQCB: Legislation and certification schemes for biofuels, sustainability quick check for biofuels. http://www.sqcb.org/ (2013). Accessed 25 Mar 2013

The White House: Camp David Declaration, G8 Summit (2012)

Toepfer, K.: Editorial for Int J LCA on the launch of the UNEP-SETAC life cycle initiative. Int. J. Life Cycle Assess. **7**(4), 191 (2002)

UNCSD: The future we want, Rio+20 outcome document, Rio de Janeiro (2012a)

UNCSD: 10-Year framework of programmes on sustainable consumption and production patterns, A/CONF.216/5 adopted document, Rio de Janeiro (2012b)

UNDA National Agricultural Library: LCA digital commons project. http://www.lcacommons.gov (2013). Accessed 31 July 2013

UNEP: Final Report of the Fourth Government Experts Meeting on Sustainable Consumption and Production for Latin America and the Caribbean, Sao Paulo (2008)

UNEP: final draft mission document. International panel for sustainable resource management—UNEP. http://www.unep.org/resourcepanel/Portals/24102/PDFs/Mission%20Document.pdf (2010a). Accessed 19 Dec 2013

UNEP: Assessing the environmental impacts of consumption and production: priority products and materials. In: Hertwich E., van der Voet E., Suh S., Tukker A., Huijbregts M., Kazmierczyk P., Lenzen M., McNeely J., Moriguchi Y.A. (eds.) Report of the Working Group on the Environmental Impacts of Products and Materials to the International Panel for Sustainable Resource Management. Paris (2010b)

UNEP: Global Outlook on SCP Policies: Taking Action Together. UNEP, Paris, France (2012a)

UNEP: The 10 Year Framework of Programmes on Sustainable Consumption and Production, 4p. UNEP, Paris, France (2012b)

UNEP: Responsible Resource Management for a Sustainable World: Findings from the International Resource Panel. UNEP, Paris, France (2012c). 36 p

UNEP: Resource efficiency and sustainable consumption and production. http://www.unep.org/roa/Programmes/resourceefficiency-old/tabid/7180/Default.aspx (2013a). Accessed 8 Mar 2013

UNEP: Life cycle thinking and the use of life cycle assessment in sustainable consumption and production policies: exchange of experiences and the need for quality data availability and capacity building. Unpublished draft report including presentations from the 2nd intergovernmental dialogue on LCA, Paris (2013b)

UNEP/SETAC: Global Guidance Principles for Life Cycle Assessment Databases—A Basis for Greener Processes and Products. UNEP/SETAC, Paris (2011)

UNEP/SETAC: Greening the Economy through Life Cycle Thinking—10 Years of the UNEP/SETAC Life Cycle Initiative. UNEP/SETAC, Paris, France (2012). 60 p

UNEP/SETAC: Opportunities for National Life Cycle Network Creation and Expansion Around the world with a Special Focus on Mainstreaming and LCA Database Development in Emerging Economies, Based on a Global Survey. UNEP/SETAC, Paris, France (2016)

UNESCAP: Report of the Ministerial Conference on Environment and Development in Asia and the Pacific, Bangkok, Thailand (2005)

UNIDO: Joint UNIDO-UNEP Programme on Resource Efficient and Cleaner Production in Developing and Transition Countries. http://www.unido.org/fileadmin/user_media/Services/Environmental_Management/Contacts/Contacts/RECP%20Programme%20Flyer%20April%202010%20(2).pdf (2010). Accessed 17 Dec 2013

UNIDO/UNEP: Joint UNIDO-UNEP Programme on Resource Efficient and Cleaner Production in Developing and Transition Countries. http://www.unep.fr/scp/cp/pdf/RECP%20Programme%20Flyer%20April%202010.pdf (2013). Accessed 10 Dec 2013

Verbruggen, H., Kuik, O., Bennis, M.: Environmental Regulations as Trade Barriers for Developing Countries: Eco-Labelling and the Dutch Cut Flower Industry, CREED Working Paper No 2, Institute for Environmental Studies, Amsterdam (1995)

Disclaimer

The designations employed and the presentation of the material in this publication does not imply the expression of any opinion whatsoever on the part of the UNEP/SETAC Life Cycle Initiative and the European Commission concerning the legal status of any country, territory, city or area or of its authorities, or concerning delimitation of its frontiers or boundaries. Moreover, the views expressed do not necessarily represent the decision or the stated policy of the European Commission, UNEP or SETAC, nor does citing of trade names or commercial processes constitute endorsement.

Author Biographies

Guido Sonnemann Involved in LCA development and promotion in the word since the mid-1990s. Has led the UNEP/SETAC Life Cycle Initiative Secretariat and publications such as the Global Guidance Principles on LCA Databases. Main LCA interests are chemistry and resources, integration with criticality and risk, plus LCM and public policy.

Eskinder Gemechu LCA expert since 2009 with a background in chemical, environmental and process engineering. Recent research focus was resource impact assessment modelling, working towards methodological integration of LCA and resource criticality assessment under the LCSA assessment framework.

Serenella Sala LCA expert focusing on environmental and ecological modelling since 2001. Main areas of research are related to models, methods and indicators for integrated impact assessment and sustainability assessment, including their role in the science-policy interface.

Erwin M. Schau Expert in LCA. Has a background in industrial economics and technology management. He has been involved in LCA for industry and academia since 2000, from 2012 stronger focus on the policy side with the European Union Environmental Footprint. Main interests are feed, food and fibre.

Karen Allacker LCA expert focusing on the construction sector since 2001. Research focus on quantitative sustainability assessment of the built environment from a life cycle thinking perspective. When working for the EC-JRC, she was involved in the development of the Environmental Footprint.

Rana Pant LCA expert with experience in industry and the Joint Research Centre (JRC) of the European Commission. Topics covered are applying Life Cycle Thinking to solid waste management, Life Cycle Impact Assessment (LCIA), the International Reference Life Cycle Data System (ILCD) Handbook and Environmental Footprint.

Naeem Adibi Currently working as Managing director of WELOOP in northern France. He has a multidisciplinary background in engineering, environmental sciences and management. He holds a PhD degree in civil engineering from French Ecole Centrale de Lille. He has been coordinating numerous LCM and LCA projects in building construction, automotive and recycling, as well as in a range of industrial sectors.

Sonia Valdivia Co-chair of the capacity building area of the UNEP/SETAC Life Cycle Initiative and Co-coordinator of the Social Life Cycle Alliance. As UNEP officer, she managed the UNEP/SETAC Life Cycle Initiative between 2005 and 2014. Her areas of interest include life cycle management, social LCA and LCA based policies.

Chapter 19
Globalisation and Mainstreaming of LCA

Arne Wangel

Abstract The chapter describes how a globalised economy exacerbates the need of a mainstreaming of LCA, in particular the emergence of long, complex and geographically highly dispersed global value chains (GVCs). In documenting the three phases of the UNEP-SETAC Life Cycle Initiative, a conventional roadmap for global mainstreaming of LCA is drawn. However, the questioning by some South governments of the rationale and a North methodological bias of LCA draws attention to the significance of national and local contexts in developing countries. The chapter argues a more elaborate concept for building capacity for LCA in developing countries and suggests how to strategize national LCA agendas.

Learning Objectives
After studying this chapter the reader should have a clear understanding of the importance of the context of globalisation for the development of LCA methodology, its dissemination of and the capacity building for LCA, in particular with regard to the adoption of LCA in developing and industrialising countries.

19.1 Introduction: The Global Challenge for LCA

The greenhouse effect spans the entire globe causing disruption of livelihood for millions of people across continents. The pattern of climate gas emissions mirrors the last century of human civilisation, when mass production and consumption emerged and became internationalised. Thus, mitigating as well as adapting to these environmental impacts constitutes a global challenge.

A. Wangel (✉)
Division for Quantitative Sustainability Assessment, Department of Management Engineering, Technical University of Denmark, Kongens Lyngby, Denmark
e-mail: arne.wangel@gmail.com

M.Z. Hauschild et al. (eds.), *Life Cycle Assessment*,
DOI 10.1007/978-3-319-56475-3_19

The organisational forms of internationalisation of production have changed from exchange of goods across borders by trading companies, to include foreign investment in overseas territories, and to establish subsidiaries of multinational corporations after the Second World War. Supported by improved information and transport technologies, selected segments have been outsourced or offshored to locations offering low tax regimes, low labour cost, same or better quality, delivery on-time and other advantages. More recently, some corporations are transforming from vertically integrated wholly owned companies into sliced up global supply chains organised by lead firms, e.g. branded manufacturers or major retailers. The lead firms cross national borders and combine value-added activities by a range of suppliers into global value chains (GVCs) for the manufacture of a final product.

The growth of emerging economies (first of all in BRIC countries: Brazil, Russia, India and China) is to a significant extent a result of this new form of internationalisation of production, as "GVCs began to concentrate in these giant countries that offered seemingly inexhaustible pools of low-wage workers, capable manufacturers, abundant raw materials and sizeable domestic markets. Thus, China became the 'factory of the world', India the world's 'back office', Brazil had a wealth of agricultural commodities, and Russia possessed enormous reserves of natural resources plus military technologies linked to its role as a Cold War Superpower" (Gereffi 2014).

Looking at trade statistics, the increasing importance of global production chains is reflected in the rising trade in intermediate inputs, which now represent more than half of the goods imported by OECD economies and close to three-fourths of the imports of large developing economies, such as China and Brazil. This phenomenon poses new challenges, because "imported inputs also account for a significant chunk of exports, blurring the line between exports and imports as well as between domestic products and imports. As part of global production chains, products at different stages of value added may be imported and re-exported multiple times, increasing the size of reported exports and imports relative to global and national value added" (World Economic Forum 2012).

The centre of gravity in world trade is shifting from West to East, as the financial and economic crisis in US and EU drags. The crisis has not "reversed globalization, but accelerated two long-term trends in the global economy: the consolidation of GVCs and the growing salience of markets in the South" (Cattaneo et al. 2010). However, inequalities among developing countries may increase.

Conventional trade statistics on gross trade flows between nations reflect the dispersed production process in quite imprecise terms. Thus, recently, an alternative measure, GVC income, which is defined as the income generated in a country by participating in global manufacturing production, has been suggested (Timmer et al. 2015). This will allow a detailed analysis at product level on the distribution of activities in a global value chain among suppliers and the amount of value added with each supplier. The World Input–Output Database (WIOD 2012) enables an analysis of the implications of production fragmented across borders, e.g. for shifting patterns in demand for skills in labour markets, or for local emissions of pollutants to the environment.

Life cycle assessment of GVCs is confronted with a similar data problem. Making use of generic databases developed in industrialised countries may produce imprecise or invalid results for processes and materials in developing countries, whose different properties and conditions of operation may cause a high degree of uncertainty.

19.2 Global Product Life Cycles

Thus, the length, complexity and geographical location of life cycle segments in GVCs, often in very diverse socio-economic contexts, present challenges for conducting LCA. The following three examples illustrate why an inventory of valid and precise site-specific data is crucial and why this objective may be hard to achieve.

The **farming of pangasius in Vietnam** and its processing into frozen fillets for export to industrialised countries have brought tremendous growth to the national economy. However, aquaculture involves serious environmental problems such as fish feed containing zinc, copper, cadmium and mercury; water pollution due to uneaten feed or faeces and improper discharge of wastewater from the ponds; loss of mangrove forest causing loss of biodiversity and natural barriers against tsunamis; and antibiotics residues in wild fish around farms.

Nonetheless, the strong competition in international markets, first of all from China, motivates a prime concern about compliance with health and food safety standards in Vietnamese aquaculture management. To perform a life cycle assessment of pangasius production is hampered by difficulties in getting site-specific data. A Dutch-Vietnamese research team conducted what they termed as a ‘stakeholder-based screening life cycle assessment up to the exit-gate of the fish farm’. The study identified two critical processes: (1) feed ingredient production, transport and milling; and (2) pond effluents in grow-out farming. For the first process, a generic inventory had to be used, as only five out of 30 feed producing companies in the Mekong delta provided data for the team. The team suspected that producers use secret formula, do not meet sanitary standards and are reluctant to provide information out of fear of government authorities (Bosma et al. 2011). Concerning the second critical process, which requires data to be collected from the farmers, the experience of food traceability systems shows that keeping records on inputs and outputs is an extra work task, for which no manpower can be spared during peak production (Yong 2008). Farmers are also concerned about data security, as some may use prohibited drugs in the ponds (own interviews 2009).

While feed producers' secrecy, out of fear of negative sanctions from public authorities and lack of capacity for record-keeping with primary producers, bar the access to local data in Vietnam, an example from Ghana shows that the final segments of the life cycle of some products may be completely hidden.

The **export of second-hand computers to Ghana**, some of which is illegal, extends the use phase of the life cycle of these products to meet needs of cheap computing for consumers in Africa. When the life time of the recycled product comes to a definitive end, final disposal is often performed under hazardous conditions, e.g. by open fire, causing severe problems for the environment and for the occupational health and safety of workers exposed. Only data collection on-site will be able to capture the processes in this and other informal sectors of the national economy.

In Malaysia, emerging capacity for assessing environmental impacts of a product having strategic importance for the national economy illustrates that the validity of foreign LCA studies based on generic data is being contested by local studies based on local data.

Palm oil production in Malaysia is a natural resource sector of strategic importance. The sector has developed over four decades and contributes 5–6% to GDP. In 2011, palm oil and palm oil-based products, ranked as the largest exports revenue earner with a total combined value of RM 80.30 billion, contributed 61.8% to total exports (MPOB 2011). Crude palm oil production accounts for app. 3.5% of the total environmental impacts in Malaysia (Yusoff and Hansen 2007). The country accounts for 39% of world production and 44% of world exports (MPOB 2014). Palm oil is sold on the world market in fierce competition with other vegetable oils and subjected to frequent price fluctuations.

The rationale of the strong R&D efforts is to safeguard the export revenue from this strategic commodity. This includes science-based arguments for the healthier properties—in terms of cholesterol content—of palm oil (MPOC 2016) as compared to soya bean oil. As a contested product, industry-driven research is directed to investigate the environmental concerns about palm oil production. LCA is adopted as a tool to drive back opposition from environmental, non-governmental organisations and from competitors and prove that palm oil has comparatively less environmental impact than other vegetable oils. In the scientific discussion, several Malaysian researchers claim that European databases are not representative for processes in Asia; thus in LCAs, according to them, palm oil appears to be worse than it really is.

These three examples point to the limitations involved when relying on generic databases only. Conditions in developing countries may be very different, when it comes to climate, habitats and natural resources characteristics and also with regard to the socio-economic and regulatory context. In the globalised economy, it is evident that a vibrant, strong and internationally well-connected LCA research community in any country is needed to produce relevant and valid LCAs of products and services. This is part of a general challenge of mainstreaming the use of LCA (Rebitzer and Schäfer 2009).

19.3 The UNEP/SETAC Life Cycle Initiative for Global Mainstreaming of LCA

The origin of capacity support specifically targeting LCA was the framework programme on Sustainable Consumption and Production established by UNEP DTI as part of its follow-up of the World Summit on Sustainable Development 2002. The 10-year framework programme focuses on SMEs as reliable suppliers, regional life cycle networks, and training programmes targeted at National Cleaner Production Centres (NCPCs) (De Leeuw 2006).

Joining with the Society of Environmental Toxicology and Chemistry (SETAC), the UNEP-SETAC Life Cycle Initiative focused on forming a focal point, i.e. a community of practitioners and stakeholders, and defined the following three objectives for its *first phase 2002–2006*:

1. Global representation in the various bodies of the initiative
2. Organisation of activities around the world
3. Organisation of capacity building material and of activities aiming at developing countries as well as small and medium enterprises (SMEs).

Thus, international outreach and capacity building was clearly targeted (Udo de Haes 2003). During this period, three regional networks were formed in Africa (Ramjeawon et al. 2005), Asia and Latin America; also an open forum with more than 1000 members from all over the world has been established. A range of awareness workshops, scientific conferences and outreach activities to cleaner production centres have been conducted. Practical tools in the form of training manuals and guides have been produced and disseminated covering Life Cycle Impact Assessment, Life Cycle Inventories, Social Life Cycle Assessment, Life Cycle Management and a Life Cycle Database Registry (Sonnemann 2003, 2004). The First Edition of the LCA Award 2006–2007 acknowledged pioneering works and individual commitment to Life Cycle Assessments in developing countries, e.g. research to assess the environmental impact of sugar production in South Africa, newsprint paper production in Zimbabwe and new approaches to assess impacts on biodiversity in Brazil (Sonnemann and Valdivia 2007).

The mission for the *second phase 2006–2010* of the Life Cycle Initiative was to bring science-based Life Cycle approaches into practice worldwide, thus explicitly setting capacity development on the agenda. The UNEP/SETAC Life Cycle Initiative served as an umbrella for a number of separate projects with different forms of affiliation to the Initiative. The Life Cycle Awards for projects using Life Cycle approaches in developing countries were being continued (Sonnemann and Valdivia 2007). A number of training and scientific events have been conducted. In the Asia Pacific Region, the National Institute of Advanced Industrial Science and Technology (AIST), Research Centre for Life Cycle Assessment, Tsukuba, Japan, was organising LCA workshops, e.g. focussing on food and waste chains in the region (Inaba et al. 2001, 205–206). Also, a survey has been completed, which compares levels of LCA implementation between nations in using indicators such as numbers of seminars, workshops, case studies, the establishment of a LCA Forum or Society, LCI Database development, LCIA methodology development, the extent of application by industries and worldwide technology transfer. A regular LCA event in Asia Pacific was reflected in the Seventh International Conference on EcoBalance.

The *third phase 2012–2016* targeted the mainstreaming of the use of life cycle approaches, including better accessibility to cost-effective, robust methodologies and tools based on reliable data, transfer of scientific knowledge to the wider society and improved global communication channels of the UNEP/SETAC Life Cycle Initiative via a number of flagship projects:

- Environmental life cycle impact assessment indicators
- LCA of Organisations
- Data and database management
- Global Principles and Practices for Hotspot Analysis
- Global capability development.

The flagship project on Global capability development (UNEP/SETAC Life Cycle Initiative 2016) had the aim to strengthen and consolidate the life cycle work in the regions, including documentation of local consultants and databases available. Focal points at Governmental offices (including national statistic offices for data management aspects) and chambers of commerce were identified and linked to the national networks. Some deliverables identified for this flagship include the following:

- Establishing a baseline on the level of Life Cycle Thinking worldwide, assessing the current capabilities on Life Cycle issues in non-OECD countries, with updates planned for every 3 years to trace the evolution.
- Life cycle tools (i.e. on life cycle management, life cycle based footprinting indicators and eco-design) spread across the emerging and rapidly growing economies via the Life Cycle Initiative's or local platforms.
- South–south (e.g. in Latin America) cooperation for increased implementation and North–South cooperation for methodologies' enhancement, data generation and exchange.
- Life cycle experts' and practitioners' network established in each region.
- Online tools, if possible, translated into several languages including English, Spanish, Chinese and Portuguese.

In 2001, LCA capacity constraints were documented for Argentina (Arena 2000). In 2006, UNEP DTI took stock of the situation in developing countries (Sonnemann and de Leeuw 2006). In 2007, the need for LCA of globalsupply chains of food products and transboundary movement of waste was highlighted (Inaba et al. 2007). In 2012, an analysis by Toolseeram Ramjeawon of the status of LCA in developing countries and of the need to build LCA capacities (Ramjeawon 2012) pointed to the lack of technical expertise and the absence of awareness as main barriers for improving beyond a very limited or non-existent level of implementation of LCA. A mapping based upon a six criterion definition, which resulted in a total of one hundred local, regional and global LCA networks around the world, confirms this situation. The survey received a response from only six networks in developing countries, primarily in South America and Southeast Asia, leaving Africa and Central Asia almost unrepresented.

The analysis by Ramjeawon (2012) suggested that joining global value chains provides one avenue for improvement with producers in developing countries, because the foreign lead firms will require LCA-based documentation of environmental performance from their suppliers to facilitate entry to markets in Europe and US (Ramjeawon 2012). Domestically, the key recommendation is for the government to create effective demand for LCA by launching national sustainable consumption and production action plans (Ramjeawon 2012). However, closing the enormous gap between levels of implementation in developing countries and industrialised countries calls for a wide range of capacity building activities to be adopted in developing countries. A roadmap was proposed with the following progressive steps (Ramjeawon 2012):

1. Introduction of life cycle topics in educational programmes and research activities;
2. Networking;
3. Setting up a national inventory database and development of tools to set up, maintain and disseminate data;
4. Development of national life cycle impact assessment methodologies;
5. Capacity development to apply LCA in industry and in public decision-making;
6. Promotion of LCA applications and creating a stock of success stories and dissemination;
7. Policy development.

In 2011, global guidance principles for LCA databases were launched (Sonnemann et al. 2013). In 2015, the status of life cycle management in emerging economies was assessed (Valdivia et al. 2015).

As one of few countries, Malaysia has launched a comprehensive plan for implementation of LCA. In 2006, the *National Initiative to Develop the Lifecycle Inventory Database for the Development of Eco-Friendly Products and Services* (SIRIM 2016) was initiated under the NinthMalaysia Plan. The initiative is hosted at SIRIM Berhad with support from the Japan International Cooperation Agency (JICA) for a number of the activities under the project. The main objective is to develop the National Life Cycle Inventory Database as the basis for LCA studies. This will support the National Eco-labelling Programme and facilitate compliance with environmental standards in international trade. The specific objectives are as follows:

- To develop the national life cycle inventory (LCI) database;
- To develop a critical mass of local LCA practitioners;
- To develop eco-labelling criteria documents for the National Eco-labelling Programme;

To create awareness among industry and consumer groups on the importance of LCA in today's manufacturing and procurement practice. Thus, the national LCA initiative intends to roll out basic resources for LCA practices by sourcing data on relevant processes in Malaysia, by supplying definitions of eco-labelling criteria, by initiating broad-based effort to create awareness the significance of LCA among stakeholders, and by supporting the creation of a pool of LCA resource persons.

In 2008, SIRIM completed a project under the EU Asia Pro Eco Programme *Sustainable Production and Consumption as the Long-term Solution to Reduce Urban Environmental Degradation—Developing a Reference Framework for Electrical and Electronic Products*, establishing a reference framework that links the roles and contribution of all stakeholders in the supply demand chain of electrical and electronic products, i.e. manufacturers, retailers and consumers.

All five major public universities, Universiti Sains Malaysia, Universiti Malaya, Universiti Teknologi Malaysia, Universiti Putra Malaysia and Universiti Kebangsaan Malaysia, have rather limited LCA research activities—currently there is no permanent research group specialising in LCA. Results are not communicated beyond those researchers, who are producing them, except as training modules produced for staff training within the Department of Environment (DOE).

Once completed, the national life cycles inventories (LCIs) will be made available to industry on a subscription basis. SIRIM has conducted an extensive outreach effort, in particular to SMEs; however, attendance to awareness and training workshops has been low. Only the plastic manufacturers have adopted LCA thinking and methods to highlight the environmental impact of the plastic bag product chain as compared to that of products with a similar function.

Environmental non-governmental organisations (NGOs) are not concerned with LCA thinking, except for the Business Council for Sustainable Development—Malaysia (BCSDM 2016). The Environmental Management and Research Association of Malaysia (ENSEARCH 2016), the membership of which is primarily drawn from environmental professionals in industry, is not introducing LCA thinking as such. Some years back, ENSEARCH widely publicised the concept of cleaner production. More recently, ENSEARCH supports the application of green technologies with a focus on energy efficiency and waste minimisation.

As one of the rapid industrialising countries in Asia, Malaysia benefits from transfer of knowledge on environmental management systems by transnational companies as a part of corporate policy in their overseas subsidiaries. However, the small- and medium-scale companies with local ownership, which constitute the majority of enterprises in most sectors, are unable to allocate resources or staff for improving environmental performance. Universities may be in a position to include research on life cycle assessment, possibly triggered by the availability of a foreign research grant, which is specific in scope and has a limited duration. The effort to develop a national LCA knowledge base, also servicing the private sector, may encounter financial and capacity constraints of the national research infrastructure.

19.4 LCA and South Policy Agendas

Global inequalities in resource distribution between the North and a number of developing countries in the South are the cause of serious capacity constraints also in the area of environmental management, in particular with regard to life cycle assessment. This has motivated some governments to assert the position that LCA is part of a 'green protectionism' agenda in trade policies of the North. Such agenda is seen as a push for industrial modernisation denying developing countries a growth potential, which countries in the North have enjoyed during a more than one-century long process of industrialisation.

Thus, stakeholders in developing countries originally adopted an altogether critical stand of confronting the rationale of LCA. In response to the influence of retail buyers, purchasing departments, product development teams, as they present their long 'arm' of environmental audits of suppliers in developing countries, delegates from the Third World claimed a 'one-sided focus on environment' at an LCA workshop during the World Summit on Sustainable Development, Johannesburg, September 2002. They argued that LCA tools are

(1) too complicated for practical use;
(2) too focused on environmental problems as defined by industrialised countries;
(3) one-sided in terms of ignoring costing and social issues such as work environment, human safety and employment. The delegates further observed that life cycle indicators will select against old-fashioned, polluting industries, thus not help to protect employment in developing countries (Udo de Haes 2004, 8).

In response, the need for simplification of LCA tools was recognised; an increased focus on soil erosion, water scarcity, and regional conditions was encouraged. Efforts to include consequences with regard to soil erosion, water scarcity, other regional conditions and land use into life cycle assessments have been recommended and in some cases practiced. Most prominently, land use consequences with regard to biofuel have been extensively analysed and discussed (Dallemand et al. 2010) and the enlargement of the scope of life cycle indicators to cover occupational health and safety, working conditions and other social issues was promoted.

Nonetheless, life cycle approaches aim to stimulate modernisation of industry favouring the development of modern, less-polluting industry. The environmental burdens in developing countries are typically higher than those in industrialised countries per functional unit (Udo de Haes 2004). This line of argument corresponds to the recommendation of *Aid for trade* as needed to effectively dissolve perceptions among developing countries that environmental and social standards are equivalent to green protectionism. Financial support to facilitate trade is considered to pave the way for long-term alliances between developing and developed countries linking trade and sustainable development. Udo de Haes outlined several options for combining environmental (or social) requirements as calculated by LCA with real financial support: "(1) the costs of such schemes could be funded by industrialised countries, because it is these countries that—justly!—ask for these schemes; (2) technical assistance could be given for the functioning of such schemes, including validation as to whether the respective criteria have been met; (3) funds may be provided for the transition into more modern, efficient technology" (op.cit., 10). A range of intergovernmental programmes to support the diffusion of climate mitigation are now in existence (de Coninck and Puig 2015).

The criticism that LCA methods are quite demanding in terms of time spent, data, software and analytical skills is raised when considering strategies to improve the livelihood of small producers (Riisgaard 2010, 10). Instead of conducting the time-consuming and costly exercise of a full LCA, several types of simplified LCA have been suggested (e.g. Hochschorner and Finnveden 2003; Hur et al. 2005; Xiaoming and Yi 2007). One approach is a stripped down version based on product categories ('product families', cf. Lenau et al. 2002). The effort is to focus on a few essential indicators, while maintaining the substance of environmental aspects. Another approach is to conduct a preliminary assessment in the format of a Life Cycle Check to identify 'hotspots' and the need for more detailed analysis (Wenzel et al. 2001).

An alternative, pragmatic approach combines LCA, risk analysis and scenario analysis into a systematic screening process prompting *go/no go* decisions (Klöpffer et al. 2007).

Basically, the environmental concerns in developing and industrialised countries have a different origin. While non-governmental organisations in industrialised countries have played a major role in alerting the public to the hazards of industrial pollution for human health and nature and pushing for regulation, trade policy conditionalities for entry into export markets currently provide the main motivation for complying with environmental standards in developing countries. This kind of regulation is driven by consumers in industrialised countries giving preference to, e.g. eco-labelled products. Also, many subsidiaries of multinational corporations operating in developing countries are directed to adopt corporate policies on environmental standards. However, for other companies, particularly those which are locally owned, efforts for improving eco-efficiency of products and services are driven as part of optimisation targeting cost savings to be gained. Furthermore, in many developing countries, the scope for civil society is restricted leaving environmental non-governmental organisations (NGOs) with few avenues for addressing policy agendas and public debate at large.

Thus, strategies for the mainstreaming of LCA worldwide need to consider how the LCA agenda of improving eco-efficiency of products and services relate to policy positions of the government, the private sector and public discourse.

19.5 Building Capacity for LCA in Developing Countries

The mere availability of relevant tools and trained professionals does not create transition. For a comprehensive strategy of mainstreaming LCA in developing countries, the concept of capacity development needs to be elaborated. In 2002, the United Nations Development Programme (UNDP) reviewed decades of technology transfer and observed "Donors can ship out four-wheel-drive vehicles, or textbooks, or computers; they can dispatch expatriate experts, whether on long-term secondment or on short-term consultancies. But they have not really appeared to transfer knowledge—or at least not in the catalytic way that might ignite a positive chain reaction throughout developing societies" (Fukuda-Parr et al. 2002, 3). Numerous cases of mismatch between ready-made technology packages and a different socio-economic and political context, into which the package was parachuted, have been documented in the history of technical assistance. On this basis, the UNDP report called for a complete change of paradigm, as "foreign experts … can run multiple seminars and courses that improve the individual skills of thousands of people. However, the capacity of local institutions and of countries as a whole has still not appeared adequate to meet the challenges of development" (ibid). The report identified three levels of capacity development: (1) the training of individuals which is only meaningful when jobs are available in relevant; (2) local institutions operating as well-functioning organisations and interacting with a conducive; and (3) enabling environment of related institutions, regulations and policies.

Accordingly, the new paradigm goes far beyond a simple identification of a relative absence of, e.g. a critical mass of LCA experts in developing countries: "Rather than starting from a mail-order catalogue of standard parts to be forced into likely looking slots, the challenge instead should be fully to understand the local situation and move forward from there—step by step" (op.cit., 13). The concept of capacity can be further elaborated to include five core capabilities:

1. The capability to self-organise and act. Actors are able to mobilise resources (financial, human and organisational); create space and autonomy for independent action; motivate unwilling or unresponsive partners; and plan, decide and engage collectively to exercise their other capabilities.
2. The capability to generate development results. Actors are able to produce substantive outputs and outcomes (e.g. health or education services, employment opportunities, justice and rule of law); sustain production over time; and add value for their clients, beneficiaries, citizens, etc.
3. The capability to establish supportive relationships. Actors can establish and manage linkages, alliances and/or partnerships with others to leverage resources and actions; build legitimacy in the eyes of key stakeholders; and deal effectively with competition, politics and power differentials.
4. The capability to adapt and self-renew. Actors are able to adapt and modify plans and operations based on monitoring of progress and outcomes; proactively anticipate change and new challenges; and cope with shocks and develop resilience.

The capability to achieve coherence. Actors can develop shared short- and long-term strategies and visions; balance control, flexibility and consistency; integrate and harmonise plans and actions in complex, multi-actor settings; and cope with cycles of stability and change (Morgan 2006, 8ff).

Ortiz and Taylor have further suggested assessing whether capacity development is supporting the development of 'standing capacities' that result in an organisational readiness to respond to new and unforeseen challenges. "Standing capacity' requires intangible qualities such as relationship leverage, programme design capabilities, innovative culture, autonomous self-motivation and agile, adaptive management response-ability" (Ortiz and Taylor 2008).

At the level of planning and implementing specific capacity development projects, several approaches have been proposed. One attempt is the *Results-oriented approach to capacity development and change (ROACH)*. It was launched at the request of the Danish International Development Agency (DANIDA) following their evaluations of capacity development components in existing projects. ROACH stresses these dimensions when embedding interventions into existing structures, enabling ownership and facilitating organisational learning:

- Both functional–rational and political aspects of change must be addressed. Moreover, they must be addressed inside and outside the organisation. Inside, capacity development in a functional–rational sense must ensure that 'the job is getting done', as this is supplemented by 'political' activities, e.g. to force change in internal power relations. Outside the organisation, the capacity development activities of the functional–rational kind will seek to create an 'enabling environment' for the organisation for getting its job done, while political activities will strive to get power relations right and accommodate the interests of stakeholders involved.
- The context of the target organisation is given full consideration, as both factors within the influence of the project and those beyond are identified and addressed (Boesen and Therkildsen 2005).

19.6 Outlook

Mainstreaming of LCA cannot be assumed to be completed as a straightforward and linear process. A conventional sequence of interventions, Transferring tools, building knowledge bases and training professionals, focusing on inputs (e.g. free software) to make the system work, will produce only limited results. *Handing down the torch* to national stakeholders with an ambition to build consensus and long-term knowledge networking needs to be followed up by local processes to integrate the inputs into the specific national and corporate context of environmental management.

In short, a national agenda must target 'home grown', demand-driven specific opportunities for the implementation of LCA according to existing capacities and enabling environment parameters, as determined by the authorities for civil society and the private sector.

Considering the basic North–South asymmetries, specific tasks include the following:

1. Adapting LCA methodology to conditions of developing economies both in terms of life cycle inventory data that are representative of the conditions of the country and impact assessment for regionally relevant impact categories and resource-based impact categories like land use and water use.
2. Strategizing the adoption of LCA in developing countries—at least these capacity constraints and opportunities must be explored: (1) causes of data insufficiency; (2) current constraints in organisational capacity of key stakeholders; (3) types of relationships between actors within the product chain; (4) gaps in institutional capacity of the enabling environment; and (5) strategy options through extensive dialogues with stakeholders.
3. Reconfiguring the relationship between 'sender' and 'recipient' in international development cooperation on LCA:

    a. In relation to government policy, LCA methodologies need to respond to the specific context of developing countries to fully incorporate socio-economic concerns of the private and public sector, and policy makers in developing countries. Also, programmes of action for LCA in developing countries must be strategized to integrate with the current level and scope of environmental management in a given country.
    b. In relation to company practices, research on simplified tools for small producers must be stepped up, and manuals for application must build upon examples relevant to production and services in developing countries. Data representing regional conditions must be made available for both inventory and impact assessment.
    c. In relation to the actors in domestic and export markets, the application of LCA in developing countries must produce immediate and tangible benefits as a contribution to transition towards national objectives of sustainable production and consumption, and as enabling steps to maintain or access positions in global value chains.

Examples of public and private sector actors pioneering such effort within the context of real economic flows of aid, trade and investment, are as follows:

- Joint capacity building at South partner universities facilitating LCA research, teaching, assessment of local products and contributions to a national life cycle inventory database, e.g. as facilitated by the Danish programme 'Building Stronger Universities' (BSU 2016).

- Transnational companies exploring business options for shared value projects targeting improved environmental performance and socio-economic benefits (Porter and Kramer 2011).

References

Arena, A.P.: Spreading life-cycle assessment to developing countries. J. Ind. Ecol. **4**(3) (2000)

BCSDM: BCSDM. http://www.wbcsd.org/regional-network/members-list/asia/bcsdmalaysia.aspx (2016). Accessed 8 June 2016

Boesen, N., Therkildsen, O.: A Results-Oriented Approach to Capacity Change. Ministry of Foreign Affairs, Danida, Copenhagen (2005)

Bosma, R., Anh, P.T., Potting, J.: Life cycle assessment of intensive striped catfish farming in the Mekong Delta for screening hotspots as input to environmental policy and research agenda. Int. J. Life Cycle Assess. **16**(9), 903–915. doi:10.1007/s11367-011-0324-4 (2011). Accessed 6 June 2016

BSU: Building Stronger Universities in developing countries. http://dfcentre.com/research/building-stronger-universities-bsu/ (2016). Accessed 8 June 2016

Cattaneo, O., Gereffi, G., Staritz, C.: Global value chains in a postcrisis world: resilience, consolidation, and shifting end markets. In: Cattaneo, O., Gereffi, G., Staritz, C. (eds.) Global Value Chains in a Postcrisis World: A Development Perspective, pp. 3–20. World Bank, Washington (2010)

Dallemand, J.F., et al.: Biomass for transport, heat and electricity: scientific challenges. Manag. Environ. Qual. Int. J. **21**(4), 523–547 (2010). http://www.emeraldinsight.com/doi/abs/10.1108/14777831011049142. Accessed 9 June 2016

De Coninck, H., Puig, D.: Assessing climate change mitigation technology interventions by international institutions. Clim. Change **131**(3), 417–433 (2015). doi:10.1007/s10584-015-1344-z

De Leeuw, B.: Life cycle management in developing countries: state of the art and outlook. Int. J. Life Cycle Assess. **11**(1), 123–126 (2006)

ENSEARCH: ENSEARCH. http://ensearch.org/ (2016). Accessed 8 June 2016

EU Asia: EU Asia Switch. http://www.switch-asia.eu/policy-support-components/rpsc/policy-dialogue/ (2016)

Fukuda-Parr, S., Lopes, C., Malik, K.: Capacity for Development: New Solutions to Old Problems. Earthscan Publications Ltd, London (2002)

Gereffi, G.: Global value chains in a post-Washington Consensus world. Rev. Int. Polit. Econ. **21** (1), 9–37 (2014)

Hochschorner, E., Finnveden, G.: Evaluation of two simplified life cycle assessment methods. Int. J. Life Cycle Assess. **8**(3), 119 (2003)

Hondo, H., et al.: Designing our future society using systems thinking. November 25–27, 2006, Tsukuba, Japan. Int. J. Life Cycle Assess. **12**(1), 66–69 (2007)

Hur, T., et al.: Simplified LCA and matrix methods in identifying the environmental aspects of a product system. J. Environ. Manag. **75**(3), 229–237. http://www.sciencedirect.com/science/article/B6WJ7-4FR8PTN-1/2/de380d30525f74b7da63b9ca0ab6e87a (2005)

Inaba, A., Grant, T., Moriguchi, Y.: Developing a research network for life-cycle assessment in the Asia Pacific Region. J. Ind. Ecol. **5**(3), 6–8 (2001)

Inaba, A., et al.: LCA of global supply chains—from production through to end of life management. Int. J. Life Cycle Assess. **12**(3), 205–206. http://www.springerlink.com/index/10.1065/lca2007.04.321 (2007). Accessed 26 July 2016

Klöpffer, W., et al.: In: Proceedings of the Workshop on Nanotechnology and Life Cycle Assessment (October 2006). ftp://ftp.cordis.europa.eu/pub/nanotechnology/docs (2007)
Lenau, T., et al.: Environmental Design in Product Families. Danish Environmental Production Agency, Copenhagen (2002). (in Danish)
Morgan, P.: The concept of capacity, Maastricht and Brussels, the Netherlands. The European Centre for Development Policy Management (ECDPM) (2006)
MPOB: Malaysian Palm Oil Board. http://www.palmoilworld.org/sustainability.html (2011). Accessed 8 June 2016
MPOB: Malaysian Palm Oil Board. http://www.mpoc.org.my/Malaysian_Palm_Oil_Industry.aspx (2014). Accessed 8 June 2016
MPOC: Malaysian Palm Oil Council. http://www.palmoilhealth.org/tag/cholesterol/ (2016). Accessed 8 June 2016
Ortiz, A., Taylor, P.: Emerging Patterns in the Capacity Development Puzzle. Why, What and When to Measure? Institute of Development Studies (IDS) (2008)
Porter, M.E., Kramer, M.R.: Creating Shared Value. How to Reinvent Capitalism—and Unleash a Wave of Innovation and Growth. Harvard Business Review (January–February). https://archive.harvardbusiness.org/cla/web/pl/product.seam?c=8062&i=8064&cs=1b64dfac8e4d2ef4da5976b5665c5540 (2011)
Ramjeawon, T., von Blottnitz, H., Kituyi, E., Mebratu, D.T.: LCA knowledge network in Africa (ALCANET). Int. J. Life Cycle Assess. **10**(6), 449 (2005)
Ramjeawon, T.: Building capacity for life cycle assessment in developing countries. In: Curran, M. A. (ed.) Life Cycle Assessment Handbook a Guide for Environmentally Sustainable Products. Wiley, Hoboken (2012)
Rebitzer, G., Schäfer, J.H.: The remaining challenge—mainstreaming the use of LCA. Int. J. Life Cycle Assess. **14**(S1), 101–102 (2009). http://link.springer.com/10.1007/s11367-009-0077-5. Accessed 26 July 2016
Riisgaard, L.: Integrating poverty and environmental concerns into value-chain analysis: a strategic framework and practical guide. Dev. Policy Rev. **28**(2), 195 (2010)
SIRIM: LCA Malaysia. http://lcamalaysia.sirim.my/ (2016). Accessed 8 June 2016
Sonnemann, G.: Int J LCA and the initiative collaborate to enhance global capacity of LCA and LCM. Int. J. Life Cycle Assess. **8**(5), 306–306 (2003). http://link.springer.com/10.1007/BF02978924. Accessed 26 July 2016
Sonnemann, G.: Corner: UNEP/SETAC life cycle initiative. Strengthening capacity building through regional networks. Int. J. Life Cycle Assess. **9**(5), 334 (2004)
Sonnemann, G., et al.: Global guidance principles for life cycle assessment databases: development of training material and other implementation activities on the publication. Int. J. Life Cycle Assess. **18**(5), 1169–1172 (2013)
Sonnemann, G., de Leeuw, B.: Life cycle management in developing countries: state of the art and outlook. Int. J. Life Cycle Assess. **11**(S1), 123–126 (2016). http://www.springerlink.com/index/10.1065/lca2006.04.020. Accessed 26 July 2016
Sonnemann, G., Valdivia, S.: Corner: UNEP/SETAC life cycle initiative. Int. J. Life Cycle Assess. **12**(7), 544 (2007)
Timmer, M.P., et al.: An illustrated user guide to the world input–output database: the case of global automotive production. Rev. Int. Econ. **23**(3), 575–605 (2015). doi:10.1111/roie.12178
Udo de Haes, H.A.: The UNEP/SETAC life cycle initiative—a personal view of the results after one year. Int. J. Life Cycle Assess. **8**(5), 307–309 (2003). http://link.springer.com/10.1007/BF02978925. Accessed 3 Aug 2016
Udo de Haes, H.A.: Life-cycle assessment and developing countries. J. Ind. Ecol. **8**(1–2), 8–10 (2004). http://doi.wiley.com/10.1162/1088198041269436. Accessed 8 June 2016
UNEP/SETAC Life Cycle Initiative: http://www.lifecycleinitiative.org/activities/phase-iii/global-capability-development/ (2016). Accessed 8 June 2016
Valdivia, S., et al.: Mainstreaming life cycle sustainability management in rapidly growing and emerging economies through capacity-building. In: Sonnemann, G., Margni, M. (eds.) Life Cycle Management. LCA Compendium-The Complete World of Life Cycle Assessment,

pp. 263–277, Springer, Dordrecht, Netherlands . doi:10.1007/978-94-017-7221-1. http://link.springer.com/10.1007/978-94-017-7221-1 (2015)

Wenzel, H., et al.: Product Life Cycle Check. A Guide (1st Draft). Institut for Produktudvikling, Danmarks Tekniske Universitet, Kgs. Lyngby (2001)

WIOD: World input–output database. http://www.wiod.org/new_site/home.htm (2012). Accessed 8 June 2016

World Economic Forum: The Shifting Geography of Global Value Chains: Implications for Developing Countries and Trade Policy. World Economic Forum (2012)

Xiaoming, X., Yi, D.: A simplified and practical life-cycle design system based on feature-based modeling. In: 2007 Chinese Control Conference, pp. 166–169 (2007)

Yong, A.: M-FIT Results (Malaysia Food Information and Traceability (M-FIT) Project). FoodReg Malaysia, Mimeo, Ho Chi Minh City (2008)

Yusoff, S., Hansen, S.B.: Feasibility study of performing an life cycle assessment on crude palm oil production in Malaysia. Int. J. Life Cycle Assess. **12**(1), 50–58 (2007)

Author Biography

Arne Wangel Sociologist studying international technology transfer and capacity building in South countries. Since 2008 involved with LCA with main interests in application of LCA in global value chain analysis, strategies for building LCA capacities in South countries, and development of a methodology for Social LCA.

Chapter 20
Organisational LCA

Julia Martínez-Blanco and Matthias Finkbeiner

Abstract The most applied and widespread approaches for environmental assessments at the organisation level have only recently extended their view beyond the factory gates. Even if they now consider the full value chain, they still mostly concentrate on a single environmental aspect like greenhouse gases (GHGs). While LCA was originally developed for products, its benefits and potential can be extended to the assessment of organisations. Organisational LCA is built on the principles, requirements and guidelines of ISO 14040 and ISO 14044, but requires some adaptations in the scope and inventory phases, when the unit of analysis and the system boundaries are defined. Also, the approach for data collection needs to be fixed. Organisational LCA is a compilation and evaluation of the inputs, outputs and potential environmental impacts of the activities associated with the organisation adopting a life cycle perspective. It includes not only the facilities of the organisation itself, but also the activities upstream and downstream the value chain. This methodology is capable of serving multiple goals at the same time, like identifying environmental hotspots throughout the value chain, tracking environmental performance over time, supporting strategic decisions, and informing corporate sustainability reporting. Several initiatives are on the way for the LCA of organisations: the UNEP/SETAC Life Cycle Initiative published the 'Guidance on organizational LCA', using ISO/TS 14072 as a backbone; moreover, the European Commission launched a guide for the organisation environmental footprint.

Learning Objectives
After studying this chapter, the reader should be able to,

- discuss the difference between product LCA and organisational LCA,
- explain the objective and utility of organisational LCA for an organisation,

J. Martínez-Blanco (✉) · M. Finkbeiner
Chair of Sustainable Engineering, Department of Environmental Technology,
Technische Universität Berlin, 10623 Berlin, Germany
e-mail: julia.martinezblanco@campus.tu-berlin.de; julia.martinez@uab.cat

M.Z. Hauschild et al. (eds.), *Life Cycle Assessment*,
DOI 10.1007/978-3-319-56475-3_20

- give a general overview of the existing methodological frameworks,
- implement the key elements of the organisational LCA methodology,
- understand the links between organisational LCA and other environmental tools.

20.1 Introduction

The life cycle approach was originally developed for products and is internationally accepted as the best tool available for assessing the environmental impacts of products (both goods and services). While product LCA can be used to support product decisions, e.g. as part of an ecodesign process (see Chap. 23), many environmental issues and aspects are rather managed on the organisational level, e.g. as part of an environmental management system. To support organisational management and decision-making, the application of LCA to another object, i.e. for the assessment of organisations, seems meaningful. The benefits and the potential of the LCA approach do definitely apply to organisational LCA as well. However, the evaluation of the environmental performance of an organisation can be even more demanding than that of products. The value chain of an organisation involves not only one chain of suppliers and other partners, but a network of them, which may be rather complex in big organisations.

The central element of this chapter is the so-called organisational LCA or LCA of organisations. This still relatively new member of the LCA family is defined as the compilation and evaluation of the inputs, outputs and potential environmental impacts (considering a multi-impact approach) of the activities associated with an organisation adopting a life cycle perspective.

The chapter provides an overview of the state of the art of organisational LCA. It discusses the need and features of the LCA of organisations, digs into the differences with product LCA, presents several leading initiatives that promote the concept, and finally provides some hints for the successful application of the methodology.

20.1.1 The Way Towards Organisational LCA

Organisations, including companies, corporations, firms, public institutions, non-governmental organisations (NGOs) etc. have a key responsibility to reduce their environmental impacts. A crucial step on the way to improve environmental performance is the quantification and the consideration of environmental aspects within the organisation's strategy and operation.

There are already some methodologies for quantifying the environmental performance of organisations and they are widespread among organisations

(Fig. 20.1). Most common approaches for the assessment at the organisational level are environmental management systems (EMS), which usually follow ISO 14001 or EMAS in the European context, schemes for carbon footprinting like ISO 14064 and ISO 14069 and the well-known GHG Protocol Initiative. The latter, together with product LCA, and derived approaches like Product Environmental Footprint (PEF) of the European Commission and Environmental Product Declarations (EPDs), constitute the set of tools which include the life cycle or value chain concept (see Chap. 24). Product LCA, PEF, EPDs and EMS include more than one environmental aspect, thus could be tagged as multi-impact tools.

Although existing approaches at the organisation level (like EMS and corporate carbon footprinting) consider the assessment of the value chain only optionally and/or focus mostly on a single aspect (like GHGs), their application and discussions have prompted and steered the future use of an LCA of organisations.

The first efforts in the life cycle community on organisational footprinting took place in the 1990s (Taylor and Postlethwaite 1996; Finkbeiner et al. 1998; Clift and Wright 2000). They were continued by combining input–output analysis with LCA (Huang et al. 2009). In the last years, several initiatives have further developed the concept (see Sect. 20.4).

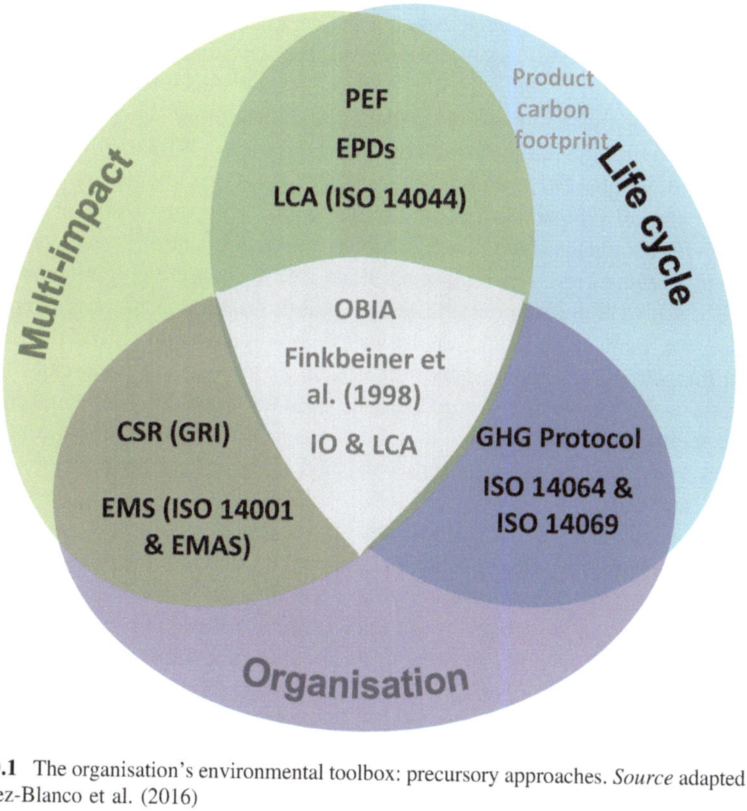

Fig. 20.1 The organisation's environmental toolbox: precursory approaches. *Source* adapted from Martínez-Blanco et al. (2016)

20.1.2 Underlining Organisational LCA Features

Organisational LCA is the first approach that analyses all the three dimensions shown in Fig. 20.1 in a comprehensive way. Therefore, unlike existing methods, organisational LCA:

- assesses the organisation, not a product (organisational dimension)
- assesses the value chain of an organisation, not only the own facilities (life cycle dimension)
- assesses a set of relevant environmental impacts and aspects, not just one (multi-impact dimension).

The management of organisations requires facts and figures for a better informed environmental decision-making and for setting effective improvement strategies. For this purpose, information and data are needed on the level at which the decisions are taken, i.e. on the organisational level. Organisational LCA draws a comprehensive picture of the environmental performance of an organisation and reveals the environmental hotspots (e.g. operations, facilities, suppliers, brands). This helps the organisation to prioritise targets, actions, programmes, efforts and budget in a more efficient and effective way than in the conventional gate-to-gate corporate environmental management view.

Previous studies and discussions about the environmental performance of organisations and their value chains, which focused mostly on GHG emissions, revealed that the upstream and downstream activities often significantly influence the overall environmental performance of organisations (WRI and WBCSD 2011; Downie and Stubbs 2013; Makower et al. 2014). As examples, only 2% of the corporate carbon footprint of the consumer goods company Unilever or 8% for the cosmetic company Natura originate in the manufacturing steps within the companies' boundaries. The remaining 98 and 92%, respectively, are emitted over the life cycle, i.e. throughout the value chains (Natura 2014; Unilever 2015). Because the management and decisions of an organisation affect the environmental impacts of its supplier network, as well as the use stage and end-of-life of the products in the portfolio, organisations need to focus environmental inquiries beyond their own facilities. Here it is proposed to do it by adopting the life cycle perspective for the organisation that will shed light on and support the management of the value chain.

Land, water and air are intricately influencing ecosystems and humans. Decisions made in the name of protecting one environmental 'medium' can result in the detriment of another, and even lead to consequences for human health (UNEP 2012). Therefore, as promoted by product LCA, a holistic approach is needed in organisational LCA in order to prevent trade-offs or the shifting of burdens.

20.2 Organisational LCA Versus Product LCA

As product LCA is the basis of organisational LCA, this section summarises the connections between the two methodologies. Furthermore, Sect. 20.5 provides an outline of the main elements in the scope and inventory phases. For further detail on the topic of this section, see Annex D in UNEP (2015) and Martínez-Blanco et al. (2015a).

20.2.1 Complementarities and Similarities

Organisational LCA follows the four-phase methodology stated by LCA standards (ISO 14040 and ISO 14044), including goal and scope definition, inventory analysis, impact assessment and interpretation. As a matter of fact, most of the principles, requirements and guidelines of the LCA standards apply also for organisational LCA (with minor terminology amendments). For instance, the requirements in ISO 14040/44 for impact assessment, reporting and review basically apply to both product and organisational LCA (Finkbeiner and König 2013).

Furthermore, product LCA and organisational LCA are complementary in three different ways:

- **Complementary levels of assessment**: they may accompany each other and provide different levels of information to the organisation. For instance, organisational LCA could be used to identify the environmental hotspots at the organisation and throughout the whole value chain, and product LCA could then provide further insights on the key products and activities identified.
- **Organisational LCA usually needs product LCA data**: in organisational LCA, the modelling of the environmental impacts of the products or services provided by suppliers very often involves the use of specific or generic product LCA datasets.
- **Transferability of the results**: a proxy organisational LCA could be calculated by the weighted summation of the product LCAs for the products in the organisation's portfolio (plus supporting activities). The other way around also works, generic product LCAs results could be generated from organisational LCA results based on specific allocation keys.

20.2.2 Main Differences

There is one fundamental difference between the two methods: organisational LCA is not designed for comparison between organisations, while one prominent aim of product LCA is to achieve comparability between different products providing the same service. As stated in ISO/TS 14072, "the results [of organisational LCA] are not intended to be used in comparative assertions intended to be disclosed to the public", as the results of product LCAs may be. In organisational LCA, the unit of comparison is not consistent between organisations, because the organisation and its product portfolio is unique and can differ strongly depending on the sector, the size, the location, and the overall business model.

The description of the scoping elements in product LCA aims to achieve comparability (apart from transparency and reproducibility) (see Chap. 8). The definition of 'reporting organisation', 'reporting flow' and 'system boundary' (see Table 20.1) in organisational LCA is motivated to guarantee a meaningful performance tracking. Performance tracking, a promising application for organisational LCA, is the regular assessment of the environmental performance of an organisation over time to measure its continuous improvement. This connects also perfectly with the overall goal of environmental management systems. Of course, this also considers a kind of comparison, but the subject compared over time is the same organisation, not a competing one.

Apart from the comparability issue, there is another obvious difference, i.e. the object of study: an individual product or an organisation, respectively.

As a consequence of these two main differences, some principles, requirements and guidelines from ISO 14040/44 do need to be adapted for organisational LCA. Main differences arise during the scoping and inventory phases. Table 20.1 summarises major discrepancies in the scope phase that are related to the definition of the unit of analysis and the boundaries. Due to the nature of the new object of study and the scoping changes, it is also necessary to bear in mind some new operational considerations during the inventory analysis (see Sect. 20.5).

20.3 Main Benefits and Applications

As mentioned above, comparisons between different organisations appear neither meaningful nor robust at this point in time. In accordance with ISO/TS 14072, an organisational LCA shall state that the results will not be used for studies envisaged to be used for comparative assertions between organisations intended to be disclosed to the public (e.g. ranking among organisations). However, there are a number of other benefits that an organisational LCA can generate. Some of these are listed in the Table 20.2.

Table 20.1 Differences between product LCA and organisational LCA (based on Martínez-Blanco et al. 2015a)

	Product LCA	Organisational LCA
Unit of analysis		
General	The object of study is the product, i.e. any good or service	The object of study is the organisation, i.e. a sole-trader, company, corporation, firm, enterprise, authority, partnership, charity or institution, or part or combination thereof, whether incorporated or not, public or private
	The main purpose of the unit of analysis is to find the relation between two or more products that provide the same function and are compared	The unit of analysis represents the comparability basis between years for environmental performance tracking of the organisation
Definition	**Functional unit** is the quantified performance of a product system for use as a reference unit	**Reporting organisation** defines the organisation under study to be used as a unit of analysis
	Functional unit is defined according to the main function(s) of the product	**Reporting organisation** includes: subject of study, consolidation method, and reference period
Quantification	**Reference flow** is a measure of the outputs from processes in a given product system required to fulfil the function expressed by the functional unit	**Reporting flow** is a measure of the outputs from the reporting organisation during the reference period
	Reference flow refers to the number of units needed to fulfil the functional unit	**Reporting flow** represents the quantification of the product portfolio of the reporting organisation: including type and quantity of products. It may be expressed per unit, weight or volume and by other measures, like number of employees or revenue
System boundary		
General	No distinction is done between direct and indirect impacts	System boundary includes direct and indirect activities
Defined by	The system boundary is derived from the type of product and it is not directly dependent on the functional unit considered	The definition of the reporting organisation is the determining issue for stating the system boundary, as the subject of study and the consolidation method are critical for the definition of the boundaries

(continued)

Table 20.1 (continued)

	Product LCA	Organisational LCA
Other		
Comparison	Comparison between products is expected and can be communicated, provided that the scope of the assessment is equivalent	External communication of comparative assertions between organisations is discouraged
Time	Generally, results of the study are relatively time-independent over a reasonable period	The environmental results of the organisation are reported for a given reference period
Supporting activities	Those activities that are not directly linked to the production (e.g. business travel, leased assets, heating, cleaning services, managerial offices) are usually disregarded	Those activities not directly related with the production process are included

Table 20.2 Main benefits of organisational LCA application

Analytical benefits	• Gain insight about the main actors and the impacts involved in internal operations and value chain • Identify environmental hotspots throughout the value chain for each of the environmental categories considered • Track the environmental performance of the organisation over time
Managerial benefits	• Get support to define which are the priority actions and targets at different levels • mprove organisational procedures, for instance in the collection and management of environmental data • Get the basis for voluntary or regulatory reporting and environmental communication with stakeholders • Show environmental awareness for marketing purposes
Societal benefits	• Reduce pressure on the environment and avoid future negative effects on the organisation • Incentivise suppliers in the value chain, consumers, and even competitors to adopt environmental friendly practices

20.3.1 Integration into Management and Decision Analysis Systems

LCA of organisations provides comprehensive information—along the value chain and for multiple impact categories—at the level at which decisions are taken and beyond the organisation's walls. Through its results, the organisation understands which are the risks and impact reduction opportunities and has strong arguments to elucidate which are the most effective actions to reduce the organisation's environmental impacts. This is also encouraged by the new version of EMS standard ISO 14001 (ISO 2015a), which stresses the relevance of life cycle thinking and the consideration of the supply chains.

Organisational LCA scheme allows the organisation to:

- **Run different scenarios**: to assess the effect of proposed actions or measures.
- **Set environmental targets**: to define quantified reductions for a certain impact category to be achieved in a target year on the basis of a reference year. Within the context of organisational LCA the organisation could define reduction targets for the whole organisation and for its parts, and at the short and long term.
- **Track performance**: to compare the environmental results of the organisation's own operations over time. The performance could be evaluated against the environmental targets.

20.3.2 Integration into the Environmental Toolbox of Organisations

Organisational LCA can complement the environmental toolbox of organisations (see Fig. 20.1) but also benefits from it. Existing experience applying other environmental or sustainability methodologies and the associated data collected will streamline the application of organisational LCA.

The second column of Table 20.3 summarises how the application and integration of organisational LCA is simplified depending on the type or class of tools already applied in the respective organisation. The third column presents some of the additional benefits and added value of applying organisational LCA in addition to the existing approaches and tools.

In general, the framework of interdepartmental work on environmental issues and communication channels with suppliers, established during the implementation of an environmental tool will facilitate the application of other methodologies, organisational LCA or others from the toolbox.

Furthermore, the results of an organisational LCA may be the backbone for environmental performance reporting with voluntary sustainability reporting schemes—like the Global Reporting Initiative (GRI), the Carbon Disclosure Project (CDP), and the United Nations Global Compact principles.

20.4 Existing Frameworks for Organisational LCA

Currently, there are three leading initiatives working on the development and agreement of approaches for the environmental multi-impact assessment of organisations and their value chains.

At the global level, ISO developed the technical specification "ISO/TS 14072: Environmental management—Life Cycle Assessment—Requirements and guidelines for Organisational Life Cycle Assessment" (ISO 2014), which adapts the

Table 20.3 Integration of organisational LCA at the organisation's environmental toolbox

Tool type	How does the tool streamline the application of organisational LCA?	How does organisational LCA complement the tool?
Organisational on-site assessment (e.g. EMS)	The organisational on-site assessment offers data for direct activities and guides the identification of the targeted suppliers	Organisational LCA complements and refreshes the EMS of organisations mainly by broadening the horizon from on-site to value chain improvements
LCA at product level (e.g., PEF, EPDs, product carbon footprint)	It may roughly identify some important hotspots in the value chain that should be further assessed. Organisational LCA may consist of the addition of the different LCAs weighted by the amount of products	It brings a more comprehensive understanding of the organisation environmental performance by including the whole portfolio and supports organisation-related decisions
Life cycle and single-indicator assessment for corporations (e.g. GHG Protocol, ISO 14064)	The overall analytical framework, the data collection procedures and tools developed for the single-indicator assessment may guide the scoping definition of the multi-impact approach	Organisational LCA will identify impacts beyond the specific single indicator, thus avoiding unintended consequences and burden shifting

requirements of product LCA to organisations and states some potential benefits that LCA can bring to organisations.

ISO/TS 14072 serves as the basic foundation of the 'Guidance on Organizational Life Cycle Assessment' (UNEP 2015), developed within a flagship project of the UNEP/SETAC Life Cycle Initiative (UNEP/SETAC 2016). This document builds on the ISO specification, but goes into greater detail with regard to the capabilities of organisational LCA and its methodological framework. Eleven case studies were included in the guidance document to illustrate some methodological aspects as well as the benefits that the methodology could bring to organisations.

At the regional level, the European Commission launched the so-called Organisation Environmental Footprint (OEF) Guide (European Commission 2013a). The document aims to increase reproducibility and comparability by underlining prescriptiveness over flexibility to ensure that the methodology is applied consistently (European Commission 2013a). At the European level, OEF together with an equivalent guide for product footprinting (i.e. PEF Guide) is supposed to achieve an important goal, the implementation of LCA into European environmental policy.

The three approaches are similar in many ways and share most of the principles, requirements and guidelines, but they are not totally interchangeable. The name of the methodology depicted by ISO/TS 14072, organisational LCA (OLCA), was adopted by the UNEP Guidance with the exception of the acronym, which includes a hyphen, i.e. O-LCA. The European Commission proposed the term Organisation Environmental Footprint (OEF).

Apart from the acronym, only few and minor differences exist between the ISO standard and the UNEP Guidance. However, the OEF Guide has some requirements that do not align with life cycle standard principles (Finkbeiner 2013; Galatola and Pant 2014; Martínez-Blanco et al. 2016). Examples include the recycling formula for end-of-life and the default set of impact categories and methods. In addition, the OEF Guide considers the option of comparative assertions intended to be disclosed to the public to be valid within the same sector and according to specific sectorial guides (which are under development).

20.4.1 *Frontrunners*

Although complete and rigorous applications of organisational LCA are not yet common practice, several examples already exist of frontrunners that have developed their own methodology, sometimes inspired by product LCA or corporate carbon footprinting. These examples encompass the application of organisational LCA. At this point, the foremost source of examples is the UNEP Guidance that includes 11 experiences from companies of different sectors, sizes and regions (UNEP 2015), like the U.S. food and beverage conglomerate Mondelēz, the hotel group Accor, the German car manufacturer Volkswagen, the retail group Colruyt, the Australian Inghams, the Japanese Shiseido, and the natural gas provider Storengy.

Furthermore, existing initiatives for organisational LCA road tested the methodological approaches and the application of their respective reference documents. The UNEP Guidance was road tested by twelve organisations that volunteered to take the lead in this process. The process is presented at the publication "Road testing Organizational Life Cycle Assessment around the world: Applications, experiences and lessons learned" (UN Environment 2017). The pilot phase of the OEF initiative focused on the development of two sectorial guides (retail and copper production) and aims to develop further ones for other sectors in the future.

20.5 Crash Course for Applying Organisational LCA

The summary of the main methodological issues of organisational LCA is presented here according to UNEP (2015) and ISO/TS 14072. Most of the requirements stated and the methodology presented here are also similar in the OEF Guide.

Organisational LCA follows the four-phase approach proposed in the ISO 14040/44. It is in the scoping phase that the major methodological differences with product LCA are found (see Table 20.1). The inventory analysis for organisational

LCA is governed by the same principles, requirements and guidelines as product LCA (see Chap. 9), however, it has its own particularities due to the higher complexity of the object assessed and the new scoping elements used.

As a consequence, this section will focus on the three key methodological issues of organisational LCA: two at the scoping level, i.e. the reporting unit and the boundaries, and finally one issue at the inventory level related to data collection.

The rest of the elements of the goal and scope phase (like allocation, data quality requirements, assumptions, value choices and optional elements, limitations, etc.) and the other two phases of an LCA, impact assessment and interpretation, are also included in a study of organisational LCA and should primarily follow ISO 14040/44.

20.5.1 Reporting Unit

As in product LCA, the scope defines the breadth, depth and detail of the study in accordance to the stated goals. The new elements that are specific for organisational LCA are the definition of the subject being assessed and how the system boundary is drawn. Both are presented in Fig. 20.2, along with a simplified example.

The unit of analysis in organisational LCA is the reporting unit that is defined as the quantified performance expression of the organisation under study to be used as a reference. It is broken down into two elements: the reporting organisation and the reporting flow.

The primary purpose of the **reporting organisation** is to describe the unit of analysis, i.e. what is to be understood by 'the organisation'. The definition includes three aspects summarised in Table 20.4.

The **reporting flow** represents the quantification of the unit of analysis. It is a measure of the outputs from the reporting organisation during the reference period. Reporting flow description shall use quantitative terms and is commonly based on

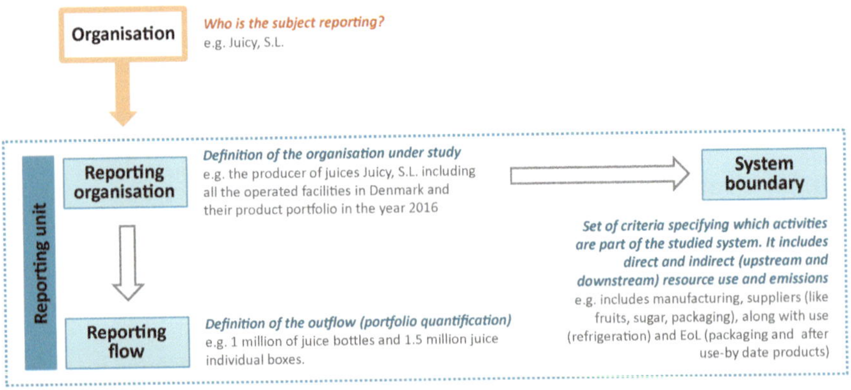

Fig. 20.2 Main elements in the scope phase of organisational LCA

Table 20.4 The three features to define the reporting organisation

Feature	Guidelines	Comments
Subject of study	The subject selected should represent a clear unit of operation, and shall be transparently justified and reported The assessment of the full organisation is recommended. If properly justified, the assessment may focus on segments or selected parts of an organisation (e.g. business divisions, brands, regions or facilities)	Assessing a segment of the organisation could be preferred if it is: • A pilot for a broad application in the future • An autonomous part pioneering the application
Consolidation method	It is a systematic approach that specifies which parts of the organisation should be considered in the study and which should not It is particularly important for big or complex organisations (i.e., including wholly owned operations, incorporated and non-incorporated joint ventures, subsidiaries, etc.) The consolidation approach should be chosen depending on the complexity of the organisation and what is to be prioritised: risk or effective tracking and implementation of management policies	Three methods exist: • Financial/operational control approaches: the organisation includes units over which it has financial/operational control • Equity share approach: the organisation includes units according to its share of equity interest
Reference period	Time period for which the organisation is reporting It is recommended to assess one operation cycle or fiscal year	For example 2016

the portfolio records. It is recommended to group the portfolio into clusters of similar products. The quantification could be based on:

- Physical terms: unit of goods or number of services or in terms of weight or volume.
- Non-physical terms: economic revenue or number of employees. To be used only in certain situations: e.g., big and very diverse portfolios or when each product in the portfolio is unique.

Ideally, apart from the type of products and the amounts produced for each of them, the reporting flow should also include information about the quality and the durability of the products, particularly when actions are taken that change any of those characteristics.

20.5.2 System Boundary: Direct and Indirect Activities

The other key element in Fig. 20.2 is the system boundary. The system boundary is the set of criteria specifying which activities are part of the studied system and which resource use and emissions associated with them are included in the study. It is in accordance to the reporting organisation previously selected. Figure 20.3 shows the list of proposed activities at the reporting organisation and through the value chain and examples of emissions and resources to include.

The system boundary includes and differentiates resource use and emissions and also linked activities that are:

- Direct: from sources that are owned or controlled by the reporting organisation, like, e.g. combustion or process emissions, natural resources consumption and leachates at facilities of the organisation.
- Indirect: occur at sources owned or controlled by another organisation or the consumer. They take place throughout the value chain (upstream or downstream) and are consequence of the activities of the reporting organisation. Examples include purchased electricity, employee commuting and use of sold products.

Unlike in corporate carbon footprinting, where only greenhouse gas emissions from the generation of electricity are mandatory apart from direct emissions, in organisational LCA direct resource use and emissions and all the relevant upstream resource use and emissions shall be included and it is also recommended to assess downstream burdens. Downstream activities should be included always, particularly if products use energy or generate emissions during their use phase.

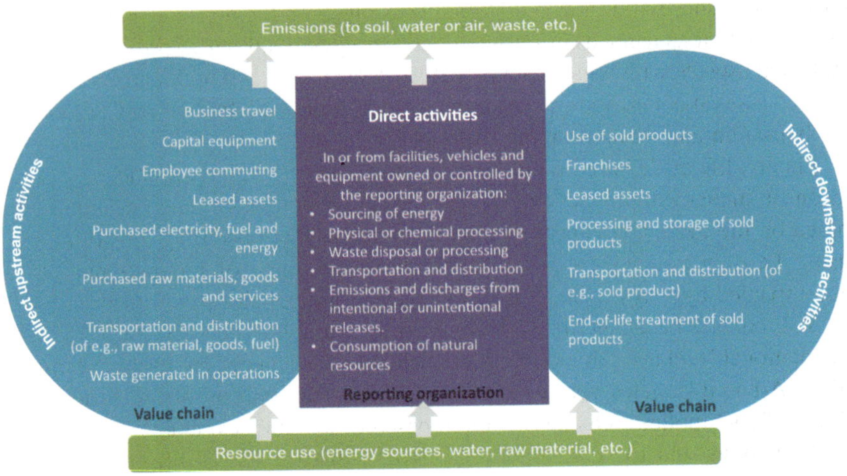

Fig. 20.3 Direct and indirect activities to be considered in an organisational LCA

20.5.3 *Inventory: Collecting Data from Whole Organisation*

As in product LCA, life cycle inventory is the phase that addresses data collection and modelling of the system and it should include the whole set of inputs and outputs from activities involved in the provision of the reporting flow and within the system boundary.

Two approaches can be used to collect the data of the inventory in organisational LCA:

- **Bottom-up approach**: it entails adding the different LCAs of the products in the portfolio of the reporting organisation, weighted by the amount of products that are produced during the reference period, together with the supporting activities. The organisation may define clusters or families of products and assess only representative or proxy products. See an example of clustering in Milà i Canals et al. (2011).
- **Top-down approach**: it considers the reporting organisation as a whole, and adds upstream (cradle-to-gate) models for all inputs of the organisation and downstream (gate-to-grave) models for all outputs.

Additionally, a hybrid approach or intermediate approach that is using both bottom-up and top-down data may be feasible.

As presented in Fig. 20.4, the data collected in the inventory should be differentiated between direct and indirect activities. The organisational LCA should include supporting activities. Those are activities of the organisation that are not directly involved in the production process (like heating, cleaning, canteen services, commuting of employees, research and marketing activities, etc.) but are part of the organisation's activities.

Different plans for the collection of the data may be designed for direct and indirect activities (see Fig. 20.4). In general, better quality and more specific data is expected for activities inside the reporting organisation, and for indirect activities identified as significant (on environmental terms, mass, economic or others). Specific data are also welcomed to model indirect activities, but higher use of assumptions, extrapolations and generic data is expected for them.

20.6 Final Remarks

Organisational LCA application may reveal environmental hotspots where the organisation should focus energies and intervention, throughout the value chain and among all the products and operations involved in the provision of the portfolio. Understanding risk and impact reduction opportunities gives a solid ground to strategic decisions at different levels, for instance, when making decisions on technologies, investments and new product lines. It may also serve as the

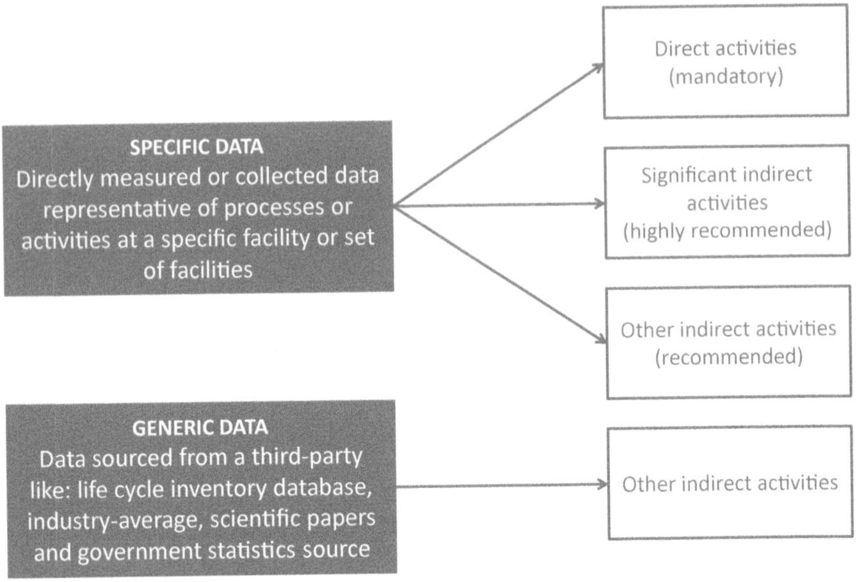

Fig. 20.4 Specific and generic data prioritisation

framework for tracking environmental performance over time and for informing corporate sustainability reporting.

Most of the principles, requirements and guidelines for product LCA apply also for organisational LCA with some minor terminology amendments. Major discrepancies between the product and organisational LCA are during the definition of the unit of analysis and the associated system boundary and for the completion of the inventory. Organisational LCA studies are not thought for comparative assertions intended to be disclosed to the public. Therefore, the need of consistency at the scope phase is for environmental performance tracking.

This still relatively new member of the LCA family has the potential to promote and spread life cycle approaches especially for those actors which do not apply them yet. Many organisations have implemented EMS over several years—in 2014 more than 300,000 organisations in about 170 countries had certified EMS according to ISO 14001 (ISO 2015b). Organisational LCA can complement and refresh the mature EMS of these organisations and pinpoint significant and more cost-effective improvement options upstream and downstream the gates of the organisation's sites.

Organisations of all sizes and sectors have a key responsibility in the efforts to reduce environmental impacts. Large corporations play a promising role, but the contribution of medium and small organisations is also important if addressed as a collective. In developing countries, almost 200,000 EMS were certified in 2013 (ISO 2015b), while there is still a need for checking and promoting the application of LCA. LCA of organisations may overcome some of the barriers for the implementation of LCA in SMEs and developing countries, since LCA of organisations provides an overall idea of the environmental performance without having to perform independent LCAs for many products. Furthermore, the threat of selecting against non-best available technologies is reduced because comparative assertions are discouraged in organisational LCA.

The existing initiatives for the LCA of organisations and consequently also this chapter focus so far on environmental impacts only. However, the organisational LCA approach can also be promising for assessing further sustainability dimensions, like social aspects. Social performance is determined by how an organisation conducts towards its stakeholders, while current Social LCA (S-LCA) rather focuses on the product level and has to face the challenges in relating the social impacts to the product when in fact they are mainly caused by the behaviour of organisations (see Chap. 16). Martínez-Blanco et al. (2015b) demonstrate and discuss how a social organisational LCA approach can overcome some of the methodological and practical challenges of the more product related S-LCA.

References

Clift, R., Wright, L.: Relationships between environmental impacts and added value along the supply chain. Technol. Forecast Soc. Change **65**, 281–295 (2000)

Downie, J., Stubbs, W.: Evaluation of Australian companies' scope 3 greenhouse gas emissions assessments. J. Clean. Prod. **56**, 156–163 (2013). doi:10.1016/j.jclepro.2011.09.010

European Commission: Commission recommendation of 9 April 2013 on the use of common methods to measure and communicate the life cycle environmental performance of products and organisations (2013/179/ EU). Off. J. Eur. Union (2013). http://eur-lex.europa.eu/legal-content/EN/TXT/?uri=CELEX%3A32013H0179. Accessed June 2016

Finkbeiner, M.: Product environmental footprint—breakthrough or breakdown for policy implementation of life cycle assessment? Int. J. Life Cycle Assess. **19**, 618 (2013). doi:10.1007/s11367-013-0678-x

Finkbeiner, M., König, P.: Carbon footprint and life cycle assessment of organizations. J. Environ. Account. Manag. **1**, 55–63 (2013). doi:10.5890/JEAM.2012.01.005

Finkbeiner, M., Wiedemann, M., Saur, K.: A comprehensive approach towards product and organisation related environmental management tools. Int. J. Life Cycle Assess. **3**, 169–178 (1998). doi:10.1007/BF02978825

Galatola, M., Pant, R.: Reply to the editorial "product environmental footprint—breakthrough or breakdown for policy implementation of life cycle assessment?" written by Prof. Finkbeiner (Int J Life Cycle Assess 19(2):266–271). Int. J. Life Cycle Assess. **19**, 1356–1360 (2014). doi:10.1007/s11367-014-0740-3

Huang, Y.A., Lenzen, M., Weber, C.L., et al.: The role of input–output analysis for the screening of corporate carbon footprints. Econ. Syst. Res. **21**, 217–242 (2009). doi:10.1080/09535310903541348

ISO: Environmental Management—Life Cycle Assessment—Requirements and Guidelines for Organizational Life Cycle Assessment (ISO/TS 14072). ISO, the International Organization for Standardization, Geneva (2014)

ISO: Environmental Management Systems—Requirements with Guidance for Use (ISO 14001: 2015). ISO, the International Organization for Standardization, Geneva (2015a)

ISO: ISO Survey 2014. ISO, the International Organization for Standardization, Geneva (2015b). http://www.iso.org/iso/iso_survey_executive-summary.pdf?v2014. Accessed June 2016

Makower, J., Mattison, R., Salo, J., Kelley, D.: State of Green Business 2014. GreenBiz Group and Trucost (2014). http://info.greenbiz.com/rs/greenbizgroup/images/state-green-business-2014.pdf. Accessed June 2016

Martínez-Blanco, J., Inaba, A., Finkbeiner, M.: Scoping organizational LCA—challenges and solutions. Int. J. Life Cycle Assess. **20**, 829–841 (2015a). doi:10.1007/s11367-015-0883-x

Martínez-Blanco, J., Lehmann, A., Chang, Y.-J., Finkbeiner, M.: Social organizational LCA (SO-LCA)—an organizational approach to strengthen the implementation of S-LCA. Int. J. Life Cycle Assess. **20**, 1586–1599 (2015b). doi:10.1007/s11367-015-0960-1

Martínez-Blanco, J., Inaba, A., Finkbeiner, M.: Life cycle assessment of organizations. In: Finkbeiner, M. (ed.) The Encyclopedia of Life Cycle Assessment. Springer, Dodrecht (2016). (Vol. 7, Special Forms of LCA)

Milà i Canals, L., Sim, S., García-Suárez, T., et al.: Estimating the greenhouse gas footprint of Knorr. Int. J. Life Cycle Assess. **16**, 50–58 (2011). doi:10.1007/s11367-010-0239-5

Natura: Natura report 2013 (2014). http://natu.infoinvest.com.br/ptb/4888/Relatrioanualemingls.pdf. Accessed June 2016

Taylor, A.P., Postlethwaite, D.: Overall business impact assessment (OBIA). In: Proceedings of the 4th SETAC LCA Case Study Symposium, Brussels, Belgium (1996)

UNEP: Greening the Economy Through Life Cycle Thinking: Ten Years of the UNEP/SETAC Life Cycle Initiative. Life Cycle Initiative, United Nations Environment Programme and Society for Environmental Toxicology and Chemistry, Paris (2012)

UNEP: Guidance on Organizational Life Cycle Assessment. Life-Cycle Initiative, United Nations Environment Programme and Society for Environmental Toxicology and Chemistry, Paris (2015). http://www.lifecycleinitiative.org/wp-content/uploads/2015/04/o-lca_24.4.15-web.pdf. Accessed June 2016

UNEP/SETAC: Life cycle initiative website (2016). www.lifecycleinitiative.org. Accessed June 2016

UN Environment: Road testing Organizational Life Cycle Assessment around the world: Applications, experiences and lessons learned. Life-Cycle Initiative, United Nations Environment Programme and Society for Environmental Toxicology and Chemistry, Paris (2017)

Unilever: Unilever sustainable living plan 2014: scaling for impact (2015). https://www.unilever.co.uk/Images/uslp-2014-uk-ire.compressed–3-_tcm1252-469820_1_en.pdf. Accessed June 2016

WRI and WBCSD: Corporate Value Chain (Scope 3) Accounting and Reporting Standard—Supplement to the GHG Protocol Corporate Accounting and Reporting Standard. World Resources Institute and World Business Council for Sustainable Development (2011). http://www.ghgprotocol.org/files/ghgp/Corporate%20Value%20Chain%20(Scope%203)%20Accounting%20and%20Reporting%20Standard.pdf. Accessed June 2016

Author Biographies

Julia Martínez-Blanco Involved in the use and development of life cycle thinking methodology since mid-2000s. Contributed to several UNEP/SETAC working groups. Main LCA interests and experience are the application to the organization level and to the agricultural and waste sectors.

Matthias Finkbeiner Developing and applying LCA, LCM, LCC, SLCA and footprints since the 1990s. Chair of the ISO-Committee for LCA, the LCM conference series and the UNEP/SETAC groups on carbon footprint and organizational LCA. Interested in the real world application of life cycle based sustainability assessment approaches in companies and policy.

Chapter 21
Future-Oriented LCA

Stig Irving Olsen, Mads Borup and Per Dannemand Andersen

Abstract LCA is often applied for decision-making that concerns actions reaching near or far into the future. However, traditional life cycle assessment methodology must be adjusted for the prospective and change-oriented purposes, but no standardised way of doing this has emerged yet. In this chapter some challenges are described and some learnings are derived. Many of the future-oriented LCAs published so far perform relatively short-term prediction of simple comparisons. But for more long-term time horizons foresight methods can be of help. Scenarios established by qualified experts about future technological and economic developments are indispensable in future technology assessments. The uncertainties in future-oriented LCAs are to a large extent qualitative and it is important to emphasise that LCA of future technologies will provide a set of answers and not 'the' answer.

Learning Objectives

After studying this chapter, the reader should be able to

- Explain the challenges of prospective LCA studies.
- Explain the differences between foresight and LCA.
- Provide an overview of some tools for performing prospective assessments.

S.I. Olsen (✉)
Division for Quantitative Sustainability Assessment, Department of Management Engineering, Technical University of Denmark, Kongens Lyngby, Denmark
e-mail: siol@dtu.dk

M. Borup · P.D. Andersen
Division for Technology and Innovation Management, Department of Management Engineering, Technical University of Denmark, Kongens Lyngby, Denmark

M.Z. Hauschild et al. (eds.), *Life Cycle Assessment*,
DOI 10.1007/978-3-319-56475-3_21

21.1 Introduction: The Need for Future-Oriented LCA

LCAs are often regarded as suitable for decision support also in decision-making that concerns the future, e.g. in technology and product development or strategic decision-making. Surveys on the use of LCA in Nordic countries and in European industry and business show that LCA is frequently used for strategic purposes (e.g. Roos et al. 2016; Frankl and Rubik 2000). Thus, LCA practitioners and researchers apply LCA as a tool in strategy processes and longer term planning. However, the basic economic, environmental and societal/social conditions change with time, and current conditions may therefore not be valid. The environmental interventions of future systems are a product of very complex interactions and dependencies of these basic conditions and thus uncertain. Additionally, scaling issues, i.e. that operational-scale technologies differ from laboratory or pilot scale, adds uncertainties (e.g. Frischknecht et al. 2009; Caduff et al. 2014). Distinctions between retrospective and prospective LCAs, and between state-oriented LCA with an accounting perspective and change-oriented LCA (attributional or consequential) are made (Rebitzer and Ekvall 2004) and traditional life cycle assessment methodology must be adjusted or changed for the prospective and change-oriented purposes. However, no standardised way of doing this has emerged.

As explained in Part II of this book, the core of process-based LCA has traditionally been a detailed analysis of the processes of the entire life cycle of the product. It is assumed that the product is known and fully specified and, in principle, all materials, resource consumptions, discharges and environmental impacts, including all subprocesses, should be assessed. In practice, there is always a limit to how much information is included and a tendency to focus more on some parts and processes than others.

These limitations can become a problem if the LCA is narrowly considered a question of obtaining the results and figures that come from the process. If instead the LCA is considered a process of learning in which the people and organisations involved in the process acquire knowledge about the product and its environmental impacts, the limitations will often appear less severe. One of the important outcomes of an LCA is the knowledge obtained about which parts of the life cycle process are uncertain and of which there exists no precise information. To make proper sense of the results, a detailed understanding of the background of the figures is needed.

Over the years, prospective elements and learning/decision elements have increasingly been introduced into LCA (e.g. Wender et al. 2014). The prospective elements of LCA concern both the production system or functional unit that is the object for the LCA, as well as the general societal conditions surrounding this system. In strategy management literature, the latter is often labelled the system's strategic environment; not to be confused with the system's physical environment which most often is in focus in life cycle assessments. In practice, it is often difficult to distinguish exactly between the product system and its societal condition. One practical way to distinguish this is that the characteristics of product systems can be affected (selection of materials and their flows) by its manufacturer whereas the societal conditions (such as cost of energy and commodities, availability of materials, cycles in national economies) cannot be affected. Second, future changes are not only affected by possible decisions. Future changes in both the product system and in its surrounding societal conditions are affected by four types of factors: (1) socio-cultural, (2) technological, (3) economic and (4) political/legal factors.

With this in mind we define prospective or future-oriented LCA as a systematic assessment of future events and developments in society, technology, economy and policy that in the long-term could considerably influence the product system (and/or functional unit) and its societal conditions and hereby the environmentally relevant flows.

It is argued that integration of long-term scenarios in LCA is needed for use of LCA in typical strategic planning and public policy planning. A more explicit and systematic handling of uncertain aspects are important dimensions in this and have been addressed, e.g. by Miller and Keoleian (2015).

A working group (WG) was established by the Society for Environmental Toxicology and Chemistry (SETAC) to address the issues of developing systematic scenarios as a basis for studies of future product systems (Weidema et al. 2004). Its report provides rather detailed recommendations for development of scenarios with a particular focus on the needs in LCA. Scenarios include three basic elements "the definition of alternative future circumstances, the path from the present to the future, and the inclusion of uncertainty about the future". In this chapter, the aim is to provide a broader understanding of the concepts and methods for future-oriented studies as well as linking up to literature on future-oriented LCA. Scenario analysis is therefore just one of the methods included here to address the uncertain future.

21.1.1 Foresight and Future-Oriented LCA

The practical processes to strategically deal with the future development of science, technology, economy and of the society has been studied since the 1940s (Jantsch 1967; Bell 2009). Many of the widely used foresight techniques were developed by American military planners during the Cold War period—aiming to 'think the unthinkable' and to prepare for it—and later-on adapted by large corporations to strengthen their intelligence capability. Over time, the conceptual and methodological development of prospective technology assessment (also known as and used interchangeably with the term 'foresight') has broadened, and produced a considerable and varied toolbox with ample applications in governmental policy-making and corporate strategizing. Today, it has become an important practice to deal with uncertainties in technology development and strategically guide decision-making, planning and actions.

Whereas LCA seeks a systematic and comprehensive analysis of environmental impacts over a product's life cycle, technology foresight usually does not focus particularly on environmental aspects of future technologies. On the contrary, technology foresight is often criticised for having a too optimistic and positive view on the future technologies disregarding the environmental impacts and risks that also are connected to the technologies. Furthermore, it is well documented that experts in general are over-optimistic in their assessment of the future potential of the technology in which they have their expertise (Tichy 2004). The principles of life cycle assessment are thus very different from the principles of technology foresight. On many points the two approaches are quite opposite, as shown in Table 21.1. But how can technological foresight then be relevant to LCAs that are usually made for present-time products and systems? The key elements of life cycle assessment are the identification and analysis of all the different processes in the entire life cycle of an industrial product or a technological system, including careful accounting of the materials and energy flows associated with the processes. Technology foresight studies normally address partial elements of the (future) technological systems in the sense of the main technical functionality, while a life cycle assessment gives a more comprehensive picture of the actual technological system in use and its different components. In its traditional form, LCA is a very detailed method, focusing on certainties and the most precise data available, while a foresight study is sketchier, process-oriented and—at least to some extent—trying to deal with the uncertain aspects of future developments.

The next section explains some of the methods applied in foresight studies a bit more in detail.

Table 21.1 Characteristics of the methodological approaches of LCA and technology foresight—in a traditional and archetypical form

Issue	Life cycle assessment	Technology foresight
Concept of time	Present and near future	Future, long-term perspective
Procedural focus	Analysis, assessment, interpretation	Synthesis, reflection, interpretation, elucidation
Data sources	Data, information, records, etc.	Information, opinions, questionnaires, statements, etc.
Analysis object	Products or services with the same functionality	Technical functionality of technologies and systems
Delimitation of analysis	Life cycle perspective, environmental impacts	Many issues and themes
Key results	Objectified results, aiming to provide precise answers	Sketches, scenarios, strategies and uncertainties

Adapted from Rasmussen et al. (2005), see also text

21.2 Prospective Methods

In this section the aim is to provide a broader view of different methods that can be applicable in future-oriented LCAs. Prospective technology analyses (or technology foresight) apply a variety of methods and approaches that have been adapted from standard social-science research methodologies. One group is the quantitative methods such as S-curve analyses, analogies, experience curves, and different sorts of extrapolations of time series. LCA studies most often apply this type of methods when there is need to predict the future and they are mainly applicable for the short-term analyses (see also Table 21.2). Quantitative methods are often difficult to apply in prospective analyses where uncertainty is high, time horizons long and changes in technology or market situation can be large. We are then left with qualitative methods (or judgmental methods) such as literature reviews, expert panels, scenarios, futures workshops and Delphi surveys.

An overview of the use of technology foresight methods in policy making can be found in Popper (2008). An overview of their use in corporate contexts can be found in Daheim and Uers (2008).

Over the last 50 years, systematic methods of analysing expected futures have been developed. Foresight has appeared as a common name designating these methods. The purpose of technology foresight is not prediction of the future or exclusively to identify data about a technology in the long-term future. The purpose is rather to establish a fuller understanding of the possible technology futures and the forces shaping the future developments. The goal is to support current strategic discussion and decision-making as well as possible rather than predicting precisely. Foresight develops a well-informed context for current decisions.

21.2.1 Diffusion Modelling, S-Curves and Analogies

Diffusion modelling, S-curves and analogies are all quantitative methods for prospective technology analyses. S-curves or 'Utterback-curves' are based on the assumption—or empirical observation—that technologies usually go through a distinct technological maturity life cycle on the market (Utterback 1996). Related models are diffusion models or Fisher-Pry substitution models (Fisher and Pry 1971). The S-curve hypothesis states that the market growth is slow in the early phase (infant or ascent phase). In the next phase market growth rapidly increases (growth phase), and in a third phase (market maturing) growth flattens out or even becomes negative, see Fig. 21.1. Hereafter, new technologies fulfilling the same needs or several needs in a better or cheaper way overtake the market. It is often useful to check scenarios for the future diffusion of the technology with traditional S-curve methodologies. Traditional market forecasts based on extrapolation of historical data in the early phases of a technology's market presence tend to underestimate the growth rates of the market, whereas the later forecasts overestimate the growth rate. The reason for this is that the technology has reached a more mature phase resulting in a less steep market growth. Anticipating an analogue technological maturity life cycle can be helpful, but the overall problem is to estimate when the market peak occurs.

21.2.2 Experience Curve (Learning Curve)

The concept of experience curves is based on learning curves first introduced by T.P. Wright reporting on a study of cost reductions in airplane production in America in the 1920s and 1930s (Wright 1936). The experience curve in contrast to learning curves applies not only to labour-intensive situations, but also to process-oriented ones. During the 1960s and 1970s experience curves were increasingly used in industrial forecasting and marketing strategy (Fusfeld 1970; Boston Consulting Group 1972). The experience curve describes how cost reductions appear in line with accumulated production. Accumulated production is used as a substitute for accumulated experience in the learning system. The mathematical expression of the experience curve can be written in the following way:

$$C_t = C_0 \cdot Q_t^{-k} \quad t = 1, \ldots, T, \tag{21.1}$$

where observations of the unit cost at time t, C_t are calculated as average over the observed variables. C_0 is the cost of the first unit produced. Q_t is the accumulated production at time t, and, finally, k is the learning factor.

In future-oriented LCA learning curves can be used for estimating future efficiencies of new technologies, and it has been shown that the learning factor (k) is fairly stable for each specific technology and is typically between 0.9 and 0.75 (Weidema et al. 2004). It has also been shown that emission coefficients are closely related to the cumulative investment and it is suggested that conservative learning factors (0.85–0.95) can be used as proxies for the physical efficiency improvements in flows (Weidema et al. 2004).

As examples, the performance and efficiency of operational-scale technologies will differ from those of laboratory-scale or pilot-scale equipment in terms of performance and efficiency figures gained with process modelling. Gavankar et al. (2014) suggest that LCAs based on immature data should be interpreted in conjunction with their technology and manufacturing readiness level. This can in practice relate to the learning curves.

21.2.3 Delphi Studies

Delphi is characterised by Linstone and Turoff (1975) "as a method for structuring a group communication process so that the process is effective in allowing a group of

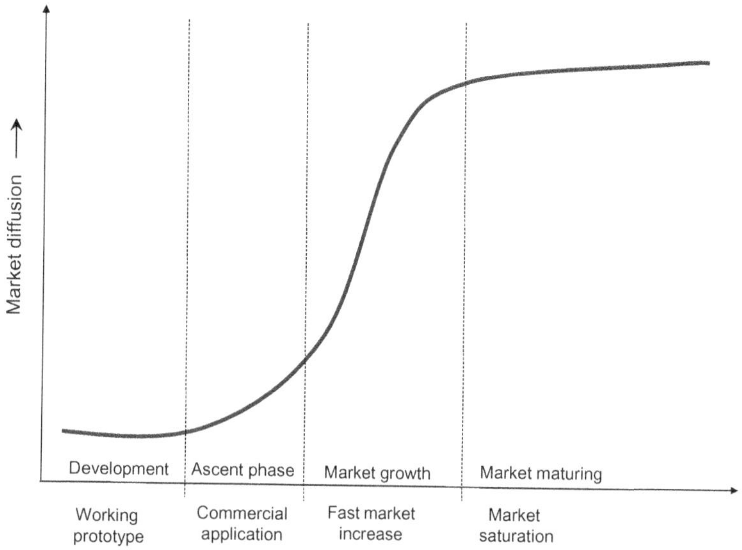

Fig. 21.1 Example of an S-curve

individuals, as a whole, to deal with a complex problem". In practice, Delphi studies are often based on questionnaires sent to a selected panel of experts. The foresight period may range from few years to 20 or 30 years. A Delphi process includes typically two rounds where the results from each round are communicated to the participants in order to achieve a consensus or reveal divergent (bipolar) viewpoints among the participants. Technology foresight studies based on the Delphi method have been carried out on a national level in Japan since the 1970s. Delphi surveys are often based on the formulation of 'statements' about the future societal, technological, economic or political development. An example could be 'More than 75% of all new wind turbines are without gear-boxes'. Statements can be formulated as a desk-study or using iterative processes at workshops or via questionnaires. The statements are then exposed to a number of questions such as: 'Period in which the statement will have first occurred', 'If realised what will be the impact on the environment?', etc. Lists of such statements and questions are sent in two rounds to selected experts. When filling in the questionnaire the experts are typically also asked to state their level of expertise on each of the statements. See the example in Table 21.2. For statement 1 some kind of normal distribution can be observed on the period in which the statement will first occur. For statement 4 the results indicate disagreement among the respondents.

Table 21.2 Example of Delphi statements and affiliated questions related to future environmental impacts of wind turbines

		Your level of expertise on the field of the statement			Period in which the statement will have first occurred						Impact on wind power's cost competitiveness					Environmental effects due to manufacturing and decommissioning of wind technology				
Statement No.	Statements about future wind power technology	Own field of work	Knowledgeable	No knowledge	Before 2005	2006 – 2010	2011 – 2015	2016 – 2020	After 2021	Never	Highly beneficial	Beneficial	Neutral	Harmful	Highly harmful	Highly beneficial	Beneficial	Neutral	Harmful	Highly harmful
1	10% of Europe's electricity from wind power	14	28	2		8	28	11	5		15	19	4			7	15	12	3	
2	More than half of all new turbines in Europe are placed offshore	11	29	4		6	10	10	8	6	5	17	8	7		4	11	15	5	
3	40% cost reduction of wind produced electricity relative to 2001	13	24	7	1	10	7	12	3	4	26	5	1			7	9	11	4	
4	Global implementation of Kyoto targets	5	30	9	1	8	9	3	10	5	13	18	2			10	19	12	1	
5	50% increase in EU and European national expenditure on wind power related research	13	21	12	7	8	9			11	7	17	6	3		6	13	9	5	1
6	Other renewable source of energy (other than hydro) becomes fully competitive with wind	7	20	8	1	7	8	9	10	2	3	11	11	6		4	10	15	1	
7	Competitive concept for storage of wind energy (e.g. based on hydrogen)	5	26	15		10	5	9	9		11	14	3			6	9	10	4	

Source Andersen and Bjerregaard (2001)

21.2.4 Scenarios

Scenarios can be defined as stories describing different but plausible futures. They are developed using techniques that systematise the perceptions of alternative futures (Schwartz 1998). Scenarios are basically tools for taking a long-term view in a world of great uncertainty. Up until the 1970s, most future studies aimed at predicting the future using various computer-based forecasting techniques. The rise of scenario analysis in strategic planning activities has largely been ascribed to the inability to provide credible forecasts and the perceived need for introducing tools for imagining, analysing, discussing, suggesting and preparing for sets of equally 'plausible' futures and running scenarios is essential when handling prospective assessments.

21.2.5 Technology Roadmaps

Technology roadmapping is a forward-looking approach developed and widely used to support strategic long-term planning within organisations like industrial companies (e.g. Phaal et al. 2004). As the name indicates, road map studies analyse and discuss the road ahead for the development of a specific industrial product or a specific technology. Roadmaps seek to capture the surrounding conditions, threats and opportunities for a particular group of stakeholders in a technology area or in an area of technology application.

Technology roadmaps can take on different forms. Usually, they include a graphical representation of the future developments as a central element. These often appear as multi-layered charts describing connections between different sub elements and different expected trends and developments. The connections between the different developments also indicate how and in what time period different actors are meant to contribute. Figure 21.2 shows an example of such graphical representations of roadmaps. Horizontally, it goes from the past to a future vision and vertically it goes from identification of, e.g. specific skills through development of a technology based on these skills, a product delivered by the technology and to the market for that product. Usually, a graphical representation of the future development and the interplay between different sub-elements is a central element in a roadmapping exercise.

Through describing and discussing the possible road ahead, including the problems and risks that can be expected, the roadmap perspective is built up.

The technology roadmapping approach is increasingly applied in foresight studies, especially in those exercises that are focused upon particular industrial sectors like, e.g. the energy sector. Traditional technology roadmapping describes a specific, partial perspective, e.g. the perspective of an industrial company or interest organisation with a clearly defined goal. The approach is thus explicitly subjective and normative. Within the limitations of the subjective perspective, the approach of technology roadmapping can lead to a comprehensive and multi-facetted understanding of a desirable development path for a technology and of the interplay between different kinds of activities (e.g. market, scientific, or industrial activities), different drivers of change, etc.

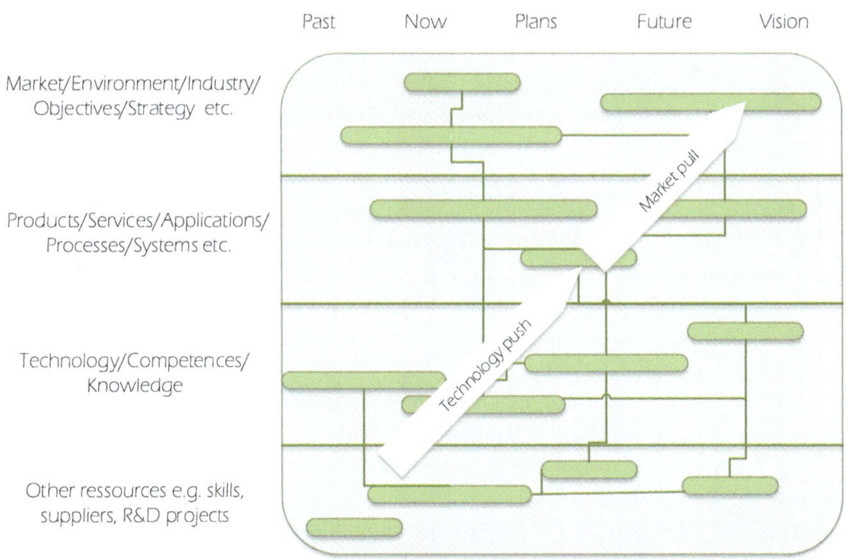

Fig. 21.2 Example of the architecture of a technology roadmap project. After Phaal et al. (2004)

21.3 Key Concepts in Future-Oriented LCA

21.3.1 Dealing with the Future

Future orientation is often categorised into short-term (1–5 years), medium-term (3–15 years), long-term (more than 20 years), and very long-term (more than 50 or 100 years). Several types of approaches to the future can be distinguished depending on the purpose of the study.

- *Predictive approaches*, which answer the question "What will happen?" are meant to be defined for simple objects and short-term studies. Predictive studies are forecasts (the likely scenario) and what-if (conditioned to some specific events). Predictive approaches aim at describing the most likely futures and generally involve forecasting current trends into the future creating 'surprise-free' or 'business as usual' like images of the future.
- *Explorative approaches*, which answer the question "What can happen?" aim at describing a number of plausible futures, which may be possible, desirable/feared and/or realisable, and start out from present trends leading to equally likely futures. They are external (related to exogenous conditions) and strategic (conditioned to some actions completed in a certain way). Cornerstone scenarios are also defined as explorative scenarios and are meant for complex objects (e.g. energy systems) and long-term time horizons.
- *Anticipative or normative approaches*, which answer the question "How can a specific target be reached?" are created on the basis of desirable or feared visions of the future. Anticipative approaches involve working backwards from a future state to find possible pathways to that particular future. This methodology is often termed 'back-casting'.

In practice, scenarios in future-oriented LCAs are often based on a mix of prediction, exploration and anticipation.

The SETAC WG on scenarios in LCA defined all methods for dealing with the future in future-oriented LCA and divided them into six groups of methods: extrapolation methods, exploratory methods, dynamic modelling, cornerstone scenarios, participatory methods and normative methods, and put them into an LCA application setting as illustrated in Fig. 21.3. When choosing and applying a future-oriented method (e.g. a scenario approach) it is important to keep in mind that the time horizon must be consistent with the goal of the study (Weidema et al. 2004).

An important distinction is made between 'what-if' scenarios and 'cornerstone 'scenarios. 'What-if' scenarios are used to compare two or more well-known situations. They are the most widely used and frequently applied in the sensitivity analysis as discussed in Chaps. 11 and 12. 'Cornerstone 'scenarios are more uncertain and do not necessarily provide quantitative results. They point out a potential direction of future development and have a more long-term perspective. Future-oriented technology assessment in most cases deals with cornerstone

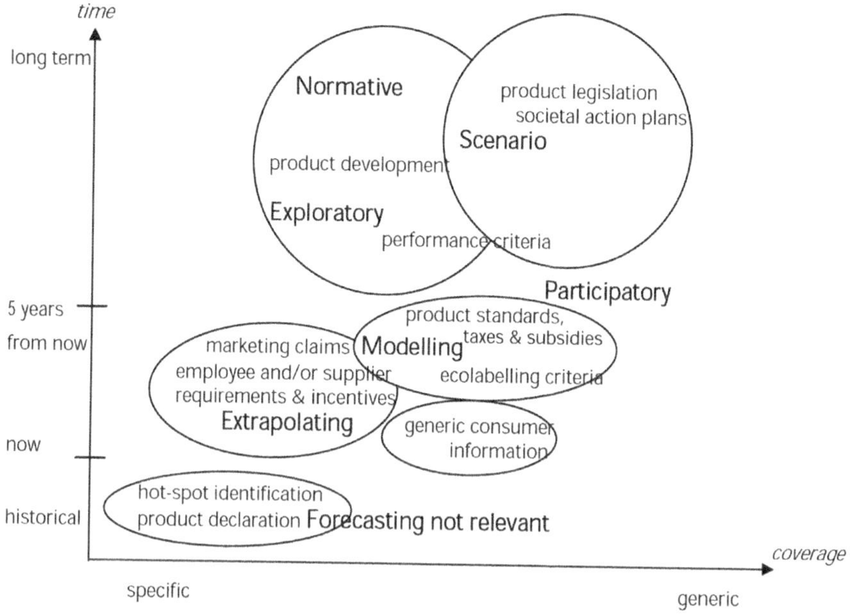

Fig. 21.3 Relevance of different future-oriented approaches in relation to applications of LCA (Weidema 2003)

scenarios and focus of the remainder of this chapter will be on approaches in this category, since these are less developed in the LCA literature.

21.3.2 Dealing with Time Horizons

Time horizons in a future-oriented LCA is a central issue. Choice of time horizon depends on the focus area and the goal of the LCA. For example, foresight for areas that are difficult to change, such as roads and energy supply infrastructure, needs to have a different time horizon than foresight for areas that change quickly, such as information and communication technology. In practice, future-oriented LCA studies often look towards the future in the long-term or medium-term, but set up possibilities for action and recommendations in the short-term.

21.3.3 Dealing with Uncertainty

In this context the term uncertainty does not refer to statistical uncertainty or uncertainty on measurements or data values of given parameters, but uncertainty in a more qualitative sense. It is uncertainty about what the characteristics of the technology are, how widespread the technology is going to be used, and other open questions and issues that are yet unknown. Therefore, uncertainty in future-oriented LCA studies will necessarily need to be addressed in a qualitative manner, at least in part of the analysis, and cannot be reduced to stochastic, systemic or methodological uncertainties only.

Traditionally, many scientific areas aim at identifying and describing certainties and neglect or underestimate the uncertain aspects and risks connected to the field and to the new knowledge produced. In connection with technology development and techno-scientific activities, this has been called the tradition of objectification or purification (Latour 1993). For example, questions of how the technology will be produced, which use-context it implies, which support technologies, infrastructures and support systems it requires, etc. can be hidden or ignored. Moreover, analyses within the history of science and technology and within the sociology of knowledge, show that a considerable amount of new uncertainties and risks are generated in connection with development of new technology and new knowledge (e.g. Beck 1992).

As a response to this, new forms of analytical practices that focus more explicitly on the uncertainties have been developed (e.g. Harremoës 2003). When dealing with sustainability and environmental aspects of new knowledge and technologies, the way uncertainties and risks are addressed by the different knowledge communities becomes of completely central importance (Funtowicz et al. 1999; Hisschemöller et al. 2001; EEA 2001; Lemons 1995).

An illustration of the traditional focus on certainties is a figure with a circle. Inside the circle is the known. Outside the circle is the unknown, called 'no-know' in Fig. 21.4. In between is a large, fuzzy area of partially known but uncertain issues. The fully known area is in fact very small and therefore, dealing with technology development and transition processes to new technological systems is identical to working with the uncertain issues. To be capable of analysis and assessment of the uncertainties is of central importance for the development of new technology areas since they include 'positive' opportunities as well as 'negative' effects and risks of the new developments.

Two main types of uncertainty are usually pointed out in the uncertainty literature: Epistemic uncertainty (lack of knowledge) and variability uncertainty (ontological uncertainty—due to inherent variability and indeterminacy). See Chap. 11 for further details on both. In connection with decision support through analysis and modelling, a distinction between three levels of uncertainty in between fully determined and total ignorance and indeterminacy have been pointed out (Walker et al. 2003):

1. Statistical uncertainty;
2. Scenario uncertainty;
3. Recognised ignorance.

Scenario uncertainty refers to assessment of possible, plausible futures and the making of—to some degree unverifiable—assumptions in connection with this.

Fig. 21.4 Understanding of 'known' and unknown ('no-know') (Harremöes 2003)

The uncertainties of future-oriented LCA activities belong to a large extent to level 2 and 3, rather than level 1.

From a technology foresight perspective, the differences between technology foresight and LCA can in practice be used productively in the design of strategic, future-oriented studies with sensitivity to environmental aspects (Rasmussen et al. 2005). For example, a combination of technology foresight and LCA were employed in a project in the wind energy area, making it possible to keep a strategic environmental perspective throughout the project (Andersen et al. 2007). From the technology foresight methods, trends mapping, a Delphi questionnaire, scenarios and a number of different expert panels were employed. In the first phase, a full present-time LCA was carried out. A later step in the process was a simplified LCA 'scanning' of selected aspects of the future wind power technology. Figure 21.5 shows the different phases of the project.

On the other hand, the need to forecast future product systems in prospective LCA can also draw a lot upon the principles of foresight, e.g. in the attempt to build scenarios. Some suggestions for the inclusion of forecasting methods are provided by Weidema et al. (2004) in a systematic form as presented in Table 21.3 and in Fig. 21.3.

There has been a number of studies applying prospective LCA in practice with some of them aiming more at the methodological aspects, in particular for assessing emerging technologies (Wender et al. 2014; Frischknecht et al. 2009). When performing LCA for emerging technology cases, practitioners have responded with a number of strategies to be prospective, including:

- Developing structured scenarios within LCA models (Pesonen et al. 2000; Hospido et al. 2010)
- Statistical time-resolved data (Zimmermann et al. 2015)
- Thermodynamic process modelling (Grubb and Bakshi 2011)

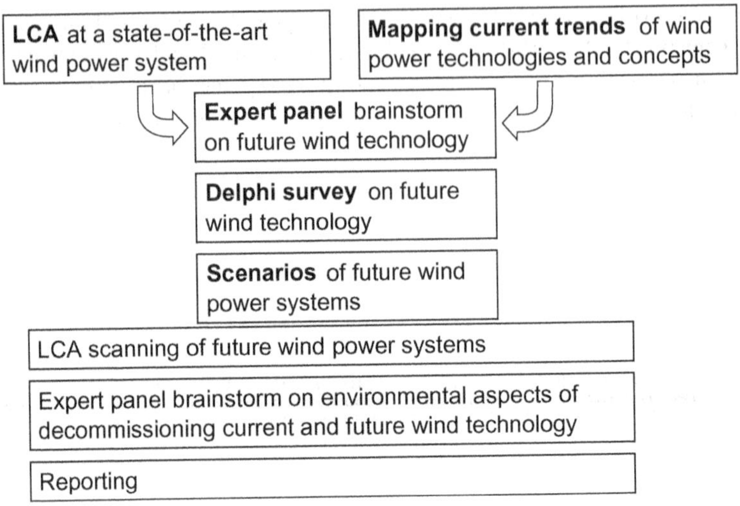

Fig. 21.5 Example of methodological design of a technology foresight-LCA project in the wind energy area. Modified from Andersen et al. (2007)

Table 21.3 Suggested forecasting methods for studies with different time frames and complexity

Term	Complexity	
	Specific and predictable	Less predictable and more complex
Short (1–5 years)–medium (3–20 years)	Extrapolation methods	Dynamic modelling and participatory methods
Long (>20 years)	Dynamic modelling, exploratory and normative methods	Cornerstone scenarios methods

Reproduced from Weidema et al. (2004)

- Consideration of experience curves from analogous industries to identify potential future improvements in efficiency (Wender and Seager 2014)
- Dimensional analysis to explore scaling effects (Caduff et al. 2014; Gavankar et al. 2014)
- Exploring market-driven impacts through consequential LCA (Weidema 2003), and
- Uncertainty bounding analyses to provide upper and lower limits to environmental impact (Eckelmann et al. 2012).

These advances allow the development of life cycle inventories descriptive of future technological developments and accounting for parameter and scenario uncertainty in exploring how the life cycle inventory may change with future developments and alternative process configurations (Wender et al. 2014). However, many of the advances do not address the complex long-term forecasting (Table 21.3).

21.4 Concluding Remarks

LCA is to a large extent being applied to support decisions that extend into the future. There are a number of challenges related to such future-oriented LCAs to which foresight methods may provide important inputs even though both methods are very different. LCAs aim to provide quantitative results, but much of our knowledge of the future is only qualitative. Therefore, LCA of future technologies will provide a set of answers and not 'the' answer. In addition to the steps already taken in any LCA, which to a large extent relate to different extrapolation methods, future-oriented LCA need to establish scenarios and to relate to uncertainties that are not just stochastic but rather linked to scenario uncertainty and recognised ignorance. Scenarios should be established through the help of qualified experts about future technological and economic developments, which are indispensable in future technology assessments. Different types of scenarios are relevant to different situations, e.g. what-if scenarios are relevant in comparison of well-known situations in the short-term and in specific cases, whereas cornerstone scenarios aim to point out a potential direction in the future development with a long-term perspective.

A number of prospective or foresight methods, from extrapolation to technology roadmaps, were presented which all can play a role when performing future-oriented LCA. However, providing more specific guidance in performing future-oriented LCA is difficult due to different requirements and conditions for each specific case in terms of, e.g. time horizon and complexity. Nonetheless, a set of questions about the future technology development in a given field is a helpful tool developed to systematically address the different kinds of driving forces shaping the future technology development in this field. The questions help getting, at first, an overview of the driving forces and barriers and, later, maintaining this overview in the further discussion of the driving forces. The set of questions is presented in Appendix.

Appendix: Questions

The questions are formulated as standard questions which can be specified further in the work in the different technology areas according to the needs. The questions ensure that different types of development mechanisms can be addressed in a systematic manner. It is not expected that answers can be found to all questions in all cases. There will probably be questions which cannot be answered or where only vague guesses can be suggested. One's first answers to the set of questions can be taken up again later in the analysis process, whereby some of the answers and the understanding of the development dynamics can be refined.

The set of questions below is an example of questions that can be used for illuminating future developments in an area (the area in the example is energy technology). The questions can either be used in a questionnaire survey, for dialogue with individual experts or for reflection internally among the LCA analysts. It is structured in three sections:

- Basics—what technology are we talking about.
- Drivers for technology change.
- Changes resulting from the drivers.

Both 'positive' and 'negative' drivers (barriers) are meant to be included. The types of driving forces addressed are:

- Technical and technological issues.
- Science and knowledge developments.
- Energy systems—infrastructures.
- Use of the technology—e.g. what role in the electricity systems?
- Where, on which markets, is the technology used—how wide spread is the use?
- Regional and geographical aspects.
- Industrial production of the technology.
- Innovation networks and innovation communities of the technology.
- Public regulation and public support.
- Societal and political concerns.
- Environmental challenges and possible risks.

Each question can be asked for (a) the near future; (b) the midterm future and (c) the long-term future.

1. **Basics (in brief)**
   - 1.1 What technology is addressed?
   - 1.2 What different basic technology concepts are available or seen as possible alternatives in the future? By technology concepts we mean for example, thin-film PV, silicon PV, etc.
   - 1.3 What are the main elements (sub-technologies) of these technology concepts? e.g. tower, blades, foundation, net connection, etc. of off-shore wind farms, etc.
2. **Drivers for technology change**
   By 'drivers' is both meant 'positive' and 'negative' (limiting) factors influencing the development of the technology.
   - 2.1 For each main element: What are the relevant developments in techniques and technological knowledge connected to this component?
   - 2.2 For each main element: What relevant influences from generic techno-scientific areas as material research, nanotechnology, biotechnology, biochemistry and information and communication technology can be identified as drivers for change? For example, functional surfaces, biochemical processes, corrosion knowledge, material techniques, sensor technology, microbiologic processes, etc.

2.3 What are the relevant developments in integration of the technology in the energy systems and infrastructures? For example, integration technologies, institutional and organisational arrangements, development of fuel supply chains, regulatory procedures, etc.
2.4 What are the relevant developments in the use of the technology: what role will it have in electricity systems? For example, central/decentral production; general purpose or specific purpose, niche markets, etc.
2.5 What are the relevant developments in dissemination of the technology—how widespread will the use be; on what specific markets?
2.6 What regional/national/geographical aspects can be identified as drivers for technology change? For example, specific conditions in some regional electricity systems, climatic aspects, etc.
2.7 What relevant developments in industrial production of the technology can be identified as drivers for technology change?
2.8 What relevant developments in the knowledge community and the network of innovators of the technology can be identified as drivers for technological change? For example, developments in the 'industrial sector' of the technology, industrial innovators/manufacturers, research programmes, other support institutions, etc.
2.9 What public regulation and public support can be identified as drivers for technology change? For example, market support, development programmes, etc.
2.10 What public, societal and political concerns can be identified as drivers for technology change? e.g. security of supply, employment, safety issues, emission restrictions, etc.
2.11 What developments in environmental challenges and risks can be identified and become drivers for technology change?

3. **Resulting changes from the drivers**

This section concludes from section 2, sketching the picture of the technology in the short-term future, medium-term future and long-term future and pointing out relevant LCI issues.

3.1 Taken into account the questions in section 2—What main development path can be identified for the technology?
3.2 Taken into account the questions in section 2—What relevant alternative/extreme development paths can be identified?
3.3 Direct changes: Technology change.
Picture of the future technology: What will, in brief, be the characteristics of the technology, its design, costs, use and life cycle?

- Total design and selection of technology concept (also covering material use).
- Design and the main sub-technologies/main parts.
- Production processes.

- Installation, e.g. system/support structure, foundation, site preparation, power conditioning equipment, land requirement and storage requirement.
- Operating and maintenance.
- Dismantling and waste handling.

3.4 Expected impacts on LCI issues
How will these changes lead to changes in LCI issues (material/resource consumptions, environmental impacts, etc.)

- Total design and selection of technology concept (also covering material use)
- Design and the main sub-technologies/main parts
- Production processes
- Installation, e.g. system/support structure, foundation, site preparation, power conditioning equipment, land requirement and storage requirement.
- Operating and maintenance
- Dismantling and waste handling

References

Andersen, P.D., Bjerregaard, E.: Prospective life-cycle assessment on wind power technology 2020. In: TA-Datenbank-Nachrichten, nr. 4, vol. 10 (2001)

Andersen, P.D., Borup, M., Krogh, T.: Managing long-term environmental aspects of wind turbines: a prospective case study. Int. J. Technol. Policy Manag. **7**(4), 339–354 (2007)

Beck, U.: Risk Society. Towards a New Modernity, 1st edn. SAGE Publications Ltd., Thousand Oaks, CA (1992). (13: 978-0803983465)

Bell, W.: Foundations of Futures Studies, vol. 1. Transaction Publishers, New Brunswick. IBSN: 978-0-7658-0539-3 (2009)

Boston Consulting Group: Perspectives on Experience, Boston, MA (1972)

Caduff, M., Huijbregts, M.A.J., Koehler, A., Althaus, H.-J., Hellweg, S.: Scaling relationships in life cycle assessment. J. Ind. Ecol. **18**(3), 393–406 (2014). doi:10.1111/jiec.12122

Daheim, C., Uerz, G.: Corporate foresight in Europe: from trend based on logics to open foresight. Technol. Anal. Strateg. Manag. **20**(3), 321–336 (2008)

Eckelman, M.J., Mauter, M.S., Isaacs, J.A., Elimelech, M.: New perspectives on nanomaterial aquatic ecotoxicity: production impacts exceed direct exposure impacts for carbon nanotoubes. Environ. Sci. Technol. **46**(5), 2902–2910 (2012). doi:10.1021/es203409a

EEA: Late Lessons From Early Warning: The Precautionary Principle 1896–2000. Environmental Issues Report no. 22. European Environment Agency, Copenhagen (2001)

Fisher, J.C., Pry, R.H.: A simple substitution model of technological change. J. Forecast. Soc. Change **3**, 75–88 (1971)

Frankl, P., Rubik, F. (eds.): Life Cycle Assessment in Industry and Business: Adoption Patterns, Applications and Implications. Springer, Berlin (2000)

Frischknecht, R., Büsser, S., Krewitt, W.: Environmental assessment of future technologies: how to trim LCA to fit this goal? Int. J. Life Cycle Assess. **14**(6), 584–588 (2009). doi:10.1007/s11367-009-0120-6

Funtowicz, S., O'Connor, M., Ravetz, J.: Scientific communication, international cooperation and capacity building for sustainable development. Int. J. Sustain. Dev. **2**(3), 363–367 (1999)

Fusfeld, A.R.: The technology progress function: a new technique for forecasting. Technol. Forecast. **1**, 301–312 (1970)

Gavankar, S., Suh, S., Keller, A.A.: The role of scale and technology maturity in life cycle assessment of emerging technologies a case study on carbon nanotubes. J. Ind. Ecol. **00**, 1–10 (2014). doi:10.1111/jiec.12175

Grubb, G.F., Bakshi, B.R.: Life cycle of titanium dioxide nanoparticle production. J. Ind. Ecol. **15** (1), 81–95 (2011)

Harremöes, P.: The need to account for uncertainty in public decision making related to technological change. Integr. Assess. **4**(1), 18–25 (2003)

Hisschemöller, M., Hoppe, R., Dunn, W.N., Ravetz, J.R.: Knowledge, Power, and Participation in Environmental Policy Analysis. Transaction Publishers, London (2001)

Hospido, A., Davis, J., Berlin, J., Sonesson, U.: A review of methodological issues affecting LCA of novel food products. Int. J. Life Cycle Assess. **15**(1), 44–52 (2010). doi:10.1007/s11367-009-0130-4

Jantsch, E.: Technological Forecasting in Perspective. OECD, Paris (1967)

Latour, B.: We have never been modern. Harvester Wheatsheaf, Birmingham (1993)

Lemons, J. (ed.): Scientific Uncertainty and Environmental Problem Solving. Blackwell Science, Oxon (1995)

Linstone, H., Turoff, M. (eds.): The Delphi Method: Techniques and Applications. Addison Wesley, Reading, MA (1975)

Miller, S.A., Keoleian, G.A.: Framework for analyzing transformative technologies in life cycle assessment. Environ. Sci. Technol. **49**(5), 3067–3075 (2015). doi:10.1021/es505217a

Millet, S.M., Honton, E.J.: A Manager's Guide to Technology Forecasting and Strategy Analysis Methods. Batelle Press, Columbus (1991)

Pesonen, H.-L., Ekvall, T., Fleischer, G., Huppes, G., Jahn, C., Klos, Z.S., et al.: Framework for scenario development in LCA. Int. J. Life Cycle Assess. **5**(1), 21–30 (2000). doi:10.1007/BF02978555

Phaal, R., Farrukh, C.J.P., Probert, D.R.: Technology roadmapping—a planning framework for evolution and revolution. In: Technology Forecasting and Social Change. Elsevier **71**(1–2), 5–26 (2004)

Popper, R.: How are foresight methods selected? Foresight **10**(6), 62–89 (2008)

Rasmussen, B., Borup, M., Andersen, P.D., Borch, K.: Prospective technology studies with a life cycle perspective. Int. J. Technol. Policy Manag. **5**(3), 227–239 (2005)

Ravetz, J.: What is post-normal science. Futures **31**, 647–654 (1999)

Rebitzer, G., Ekvall, T. (eds.): Scenarios in Life-Cycle Assessment. SETAC Press, Pensacola (2004)

Roos, S., Zamani, B., Sandin, G., Peters, G.M., Svanström, M.: A life cycle assessment (LCA)-based approach to guiding an industry sector towards sustainability: the case of the Swedish apparel sector. J. Clean. Prod. **133**, 691–700 (2016). doi:10.1016/j.jclepro.2016.05.146

Schwartz, P.: The art of the long view. Planning for the future in an uncertain world, p. 272. Wiley, New York (1998)

Tichy, G.: The over-optimism among experts in assessment and foresight. Technol. Forecast. Soc. Chang. **71**, 341–363 (2004)

Utterback, J.M.: Mastering the Dynamics of Innovation. Harvard Business School Press, Harvard (1996)

Walker, W.E., Harremoës, P., Rotmans, J., van der Sluijs, J.P., van Asselt, M.B.A., Janssen, P., Krayer von Krauss, M.P.: Defining uncertainty. A conceptual basis for uncertainty management in model-based decision support. Integr. Assess. **4**(1), 5–17 (2003)

Weidema, B., Ekvall, T., Pesonen, H., Rebitzer, G., Sonnemann, G., Spielmann, M.: Scenarios in Life-Cycle Assessment. Society of Environmental Toxicology and Chemistry (SETAC), Pensacola, FL (2004)

Weidema, B.: Market Information in Life Cycle Assessment. Environmental Project 863. Danish Environmental Protection Agency (2003). www.mst.dk

Wender, B.A., Seager, T.P.: Anticipatory life-cycle assessment of SWCNT-enabled lithium ion batteries. In: David R. (ed.) Nanotechnology for sustainable manufacturing, CRC Press pp. 247–262 (2014)

Wender, B.A., Foley, R.W., Prado-Lopez, V., Ravikumar, D., Eisenberg, D.A., Hottle, T.A., et al.: Illustrating anticipatory life cycle assessment for emerging photovoltaic technologies. Environ. Sci. Technol. **48**(18), 10531–10538 (2014). doi:10.1021/es5016923

Wright, T.P.: Factors affecting the cost of airplanes. J. Aeronaut. Sci. **3**, 122–128 (1936)

Zimmermann, B.M., Dura, H., Baumann, M.J., Weil, M.R.: Prospective time-resolved LCA of fully electric supercap vehicles in Germany. Integr. Environ. Assess. Manag. **11**(3), 425–434 (2015). doi:10.1002/ieam.1646

Author Biographies

Stig Irving Olsen LCA expert both as researcher and as consultant. Involved in the development of LCA methodologies since mid 1990s. Contributed to UNEP/SETAC working groups on LCIA methodology. Main LCA interest is human toxicity impacts, emerging technologies, and decision making.

Mads Borup Analysed energy technology development and energy sector change towards sustainability since early 2000s. Works with analyses of innovation system dynamics, transition processes, sustainable consumption and production system, and green technology foresight. Interested in socio-technical, systemic perspectives on technology innovation.

Per Dannemand Andersen Since mid-1990s research interests were technology foresight, strategy and strategy processes in industrial sectors, and innovation and research strategy in energy technologies. Besides technology foresight practice, interests included methodology development for foresight and the position of foresight exercises within policy development.

Chapter 22
Life Cycle Management

Niki Bey

Abstract This chapter gives an overview of Life Cycle Management (LCM)—a discipline that deals with the managerial tasks related to practicing sustainable development in an organisation. Just as Life Cycle Assessment, LCM advocates the life cycle perspective, and it applies this perspective in decision-making processes. The chapter shows that LCA can play a key role in LCM since LCA provides quantitative performance measurements. It also explains, which stakeholders need to be considered, how LCA and LCM relate, how LCA can be used to develop Key Performance Indicators, and addresses how LCM can be integrated into an organisation.

Learning Objectives
After studying this chapter, the reader should be able to

- Define Life Cycle Management (LCM) and describe its links to related approaches and terms
- Give an overview of the central elements of LCM and how LCM relates to Life Cycle Assessment (LCA)
- Identify central stakeholders and their areas of influence on decisions
- Describe factors needed to make LCM work in organisations
- Describe how to develop Key Performance Indicators for use in LCM practice
- Practice LCM activities and decision-making contexts by means of a case study

In general, this chapter takes the viewpoint of a Life Cycle Management practitioner.

N. Bey (✉)
Division for Quantitative Sustainability Assessment, Department of Management Engineering, Technical University of Denmark, Kongens Lyngby, Denmark
e-mail: niki@dtu.dk

M.Z. Hauschild et al. (eds.), *Life Cycle Assessment*,
DOI 10.1007/978-3-319-56475-3_22

22.1 Introduction—What Is LCM?

22.1.1 Definition of Life Cycle Management—And Distinction Against Related Terms

Private individuals and managers in organisations constantly make choices in the context of all kinds of activities; be it shopping in supermarkets, choosing a holiday destination, selecting means for business travel or refurbishing a family home or equipping an entire office building. When working towards sustainable development, any such choice will imply the question, which option may be the preferable one in a holistic perspective. The way advocated and described in this book is to answer this question through taking a view on the life cycles of the involved products and systems. Practicing the above in managerial decision-making is Life Cycle Management (LCM); i.e. the making of managerial decisions with a sustainability-oriented, holistic view on the life cycles of the products, service activities and systems that the managerial decisions are dealing with.

Once adopted, Life Cycle Management means taking a constant view on many life cycles of products and systems, and practicing LCM thus influences decision-making processes in many contexts, in fact in more or less all activities an individual or an organisation pursues. Examples of decision-making in an organisation may be selecting engineering materials for products (e.g. for window frames), choosing suppliers providing materials and services, selecting logistics solutions (e.g. rail transport vs. air transport), designing or redesigning factories—or schools or residential houses—and even selecting among different options for public transportation systems to be established by a municipality—and all this potentially in different geographical regions.

LCM can, with full legitimacy and good reasoning, be practiced without a focus on environment and sustainability, e.g. for pure cost optimisations. However, it will always involve a view on entire life cycles. LCM in this chapter—and in common understanding—does build on the principles of sustainable development and Life Cycle Thinking (LCT). And LCM applies LCT in a generic and flexible way to business management, potentially employing any life cycle-based approaches and methods, such as:

- Life Cycle Assessment (LCA), [sometimes also referred to as 'Environmental LCA' (E-LCA)]
- Social LCA (S-LCA or SLCA) (Chap. 16)
- Life Cycle Costing (LCC) (Chap. 15);
- Ecodesign/Life Cycle Design (LCD)/Sustainable Product Design (SPD), Design for Recycling/Circularity and others (Chap. 23).

This also covers combinations and/or simplifications of such approaches as well as classic business tools, such as stakeholder analysis, SWOT (strengths, weaknesses, opportunities, and threats) analysis (with sustainability focus), etc. In the remainder of this chapter, the term 'tool' also stands synonymously for 'method' and 'instrument'.

It is very important to acknowledge that it is always up to the individual organisation to decide which tools their particular Life Cycle Management approach should comprise—there is no 'one-size-fits-all' LCM approach. Rather, LCM draws upon the above outlined tools and approaches—something that can be called a toolbox (UNEP/SETAC 2007)—from which the organisation picks its individual set leading to their tailor-made LCM approach. This is because a large number of factors such as product type, business model, market presence, organisational maturity, level of ambition, value chain position, regulatory frameworks and others can play a decisive role for what a meaningful and feasible approach would be for a given organisation.

A prominent, already existing suggestion towards combining environmental, economic and social assessment is Life Cycle Sustainability Assessment (LCSA). The two principally possible methodological options to generate LCSA results are either to present three sets of impact category indicators next to each other (from LCA, LCC, and SLCA) or to develop methods for integrated inventory analysis and integrated impact assessment to reach integrated impact scores (Kloepffer 2008; Guinée et al. 2011). However, although principally targeted towards LCM, none of the two options seems immediately feasible for decision-support in Life Cycle Management for the following reasons: In the combined option, the resulting relatively large, combined number of indicators e.g. 20…30, may rather confuse than guide the decision-maker. The integrated approach, however, suffers from several methodological challenges, one of them being how to deal with location-specificity, which is crucial in both LCA and SLCA but not (yet) addressed to the extent needed. Thus, further research is required, potentially also integrating insights from, e.g. Multi-Criteria Analysis, Multi-Criteria Decision-Making, etc. (e.g. Linkov and Seager 2011; Prado et al. 2012).

Life Cycle Management is applicable in businesses and organisations of all kinds, sizes, markets and supply chain/value chain positions—be it manufacturers, retailers, service providers or other organisations—be they for-profit or not-for-profit, i.e. even a sports club can practise LCM. This chapter therefore uses the term 'organisation' as a placeholder for 'business', 'company' or any other product producing company and/or service-providing entity practicing LCM.

Since this is a Life Cycle Assessment textbook, one can appropriately say 'LCM puts LCA into *practice*' and also 'LCA is a key tool in the LCM toolbox'. However, unlike Life Cycle Assessment, LCM operates *without* being a methodology consisting of a number of well-described distinct phases or steps (and LCM is not ISO standardised either). In addition, also contrary to LCA, LCM has no deterministic character, i.e. from a given starting point and given constraints it does not necessarily lead to the same conclusions, for instance the same recommendations. Rather, LCM is *a management concept* with an underlying life cycle-sustainability-oriented mind-set, and it comprises a number of different tools that can be applied in combination or separately, by different departments of an organisation or any stakeholder, and in a variety of decision-making contexts, such as choice of manufacturing processes in production development or supplier choice in the purchase department. LCM thus *pulls* Life Cycle Assessment and other quantitative or qualitative decision-support tools into concrete *decision-making contexts*. This is explained in Example Box 22.1.

Example Box 22.1
A concrete LCM decision-making context is, for instance, whether or not to choose a certain new material that product developers—due to some technical properties—may find advantageous for a product type (e.g. for window frames). That material may also have adverse environmental properties, have a higher price per unit than alternative traditional materials, and may be more difficult to source and process in manufacturing than alternatives. In order to find out more on specifically the environmental properties of the material, either LCAs or other quantitative or qualitative assessments or data sources can be reviewed (in case they are at hand for the practitioner) or be actively commissioned. This would provide a basis for the material decision—and other factors, such as manufacturability, will have to be included in this decision context too. If it is decided to conduct an LCA, still some options exist as to how this LCA should be performed, e.g. whether a life cycle screening could suffice versus a full formal LCA, and whether to perform the assessment by own staff (that might even have to be hired/trained first) or by consultants (which may be expensive). Both choices influence the required time for the LCA, the incurred costs, and potentially other factors, e.g. the quality of the obtained LCA result. All this establishes the basis for the environment-oriented part of the decision of what material to select, and thus ultimately, all this may have an influence on the life cycle sustainability profile of the product, which the material is part of. In Life Cycle Management practice, the mechanical/technical, economic and manufacturing-related properties (as well as potentially additional ones) would have to be assessed as well—and communicated appropriately to individuals inside the organisation with different professional backgrounds, since the final decision is usually not made by one person alone.

The above example indicates the breadth of contexts that the LCM practitioner may have to take into account, incl. the variety of tools, like LCA, that can be employed in support of making life cycle-spanning decisions. Qualitative tools such as guidelines and checklists may be employed as well.

When looking at the LCA framework with its four phases and the Direct applications described in ISO 14040 (see Fig. 22.1), Life Cycle Management covers all of these Direct applications, such as Product development and improvement. LCM could thus be called the initiator/trigger or 0th phase for making LCAs, but also the result-user/-executor or 5th phase of the LCA, which is indicated by the double arrow in Fig. 22.1.

Life Cycle Management advocates taking a life cycle-wide view on business activities, and through this it can make the practitioner aware of two key

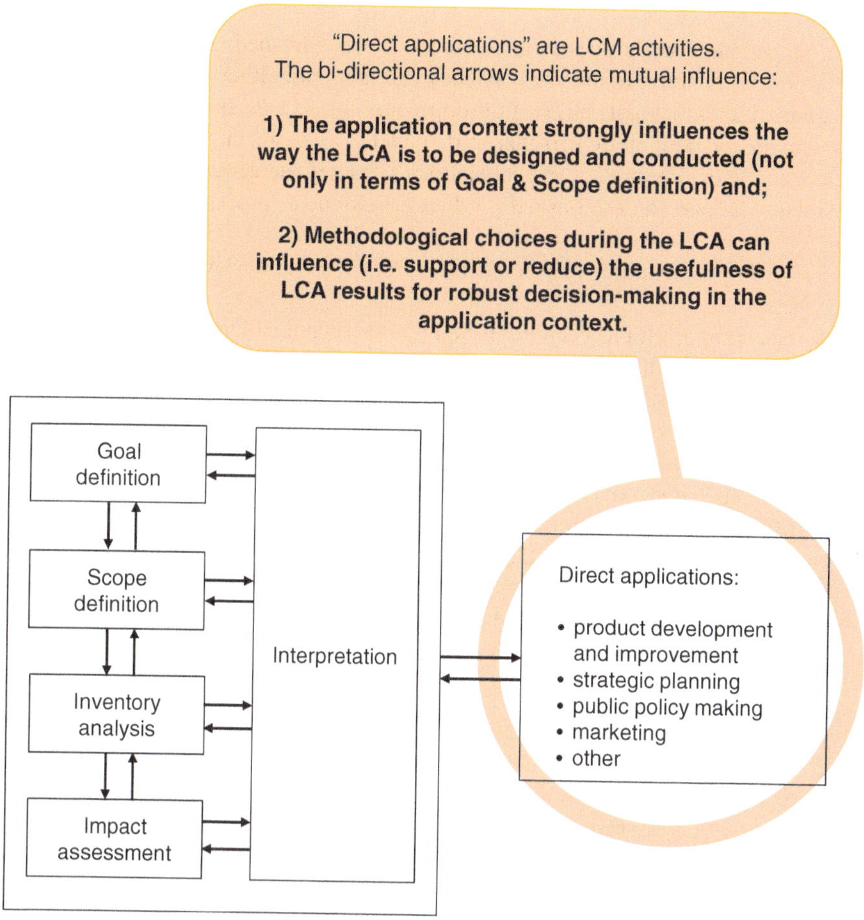

Fig. 22.1 The LCA framework of ISO 14040 augmented from an LCM perspective: key for successful support of LCM by LCA is that the LCA relates to the decision context of the direct application. Therefore, the direct application context could be called the initial 0th and/or 5th phase of an LCA (based on ISO 14040:2006)

circumstances: First, no life cycle stage can be neglected in the decision-making, since activities in any life cycle stage of a product or system (and thus decisions on whether or not to conduct those activities) can influence the sustainability profile of that product or system. Second, managing with a life cycle perspective initiates and sustains an organisational learning process of identifying what these influences may be, whether they are desired or non-desired ones, and which indicators to focus on. LCM thus enables identifying options and prioritising actions to reach more sustainable ways of running businesses.

In essence, Life Cycle Management deals with the *managerial tasks* related to practicing sustainable development in an organisation. It could thus also be referred to as Sustainability Management—and is in some organisations also referred to as such, or as Responsibility Management, CSR management (CSR = Corporate Social Responsibility) or similar. However, the term 'life cycle' in the title underlines Life Cycle Management's origins in the life cycle thinking (compare Sonnemann and Margni 2015, Chap. 2).

Sonnemann and Margni (2015) also provide the below, all-embracing definition:

> *"Life Cycle Management (LCM) is a management concept applied in industrial and service sectors to improve products and services while enhancing the overall sustainability performance of business and its value chains. In this regard, Life Cycle Management is an opportunity to differentiate through sustainability performance on the market place, working with all departments of a company such as research and development, procurement, and marketing, and enhance the collaboration with stakeholders along a company's value chain. LCM is used beyond short-term business success and aims at long-term achievements minimizing environmental and socioeconomic burden while maximizing economic and social value."*

Since it is an overarching, cross-functional discipline, Life Cycle Management shares aim and focus with a number of other approaches and concepts, coming from the sustainability field, the managerial field and other fields. In practical application in the organisation, all such concepts are usually combined rather than pursued separately and exclusive to each other. Thus, applied Life Cycle Management often adds the aspect of sustainability to other more established management disciplines. An example is Supply Chain Management (SCM): Originally, this activity is about securing a required input flow of supply of raw materials and pre-manufactured goods, etc. into a manufacturing company with focus on the supplies being provided in time, and at required quality and cost. LCM adds the sustainability angle on this activity, for instance in the form of making sure that in addition to the other requirements also specified work conditions at suppliers comply with company standards and/or in the form that emissions from in-flow logistics do not exceed certain emission values.

Management approaches and tools, which LCM relates to in such a way, include e.g.

- Product (life cycle) Data Management (PDM);
- Sustainable Supply Chain Management (S-SCM);
- Corporate Social Responsibility (CSR) and Corporate Responsibility (CR);
- Environmental management (potentially according to ISO 14001, EMAS, etc.);
- Environmental Health and Safety (EHS);
- Compliance management;
- Corporate governance;
- Risk management.

In order to avoid possible confusion, it is pointed out below in what ways LCM in this chapter's context differs from the above and other quite similar terms and approaches:

- Product Life Cycle Management (PLM) is often used as covering product-related data management, as is Information Life Cycle Management and Product Data Management (PDM). LCM in the context of this chapter has a broader scope than such PLM/PDM, since it does not only focus on data management but on all life cycle activities in an organisation.
- In some companies, Life Cycle Management is referred to as an activity starting after launch of a product, i.e. subsequently to product development. In this chapter however, LCM covers product portfolio management, product development, and after-launch/after-sales activities until end-of-life of individual products and entire product types on the market.
- LCM in this chapter's context is not an activity that relates exclusively to the marketing life cycle (from product launch over market maturity until it is taken off the market). LCM rather relates to the physical life cycle(s) (cradle-to-grave) of a given palette of products that the organisation has a certain influence upon.
- Life Cycle Management is also often referred to as an activity of Asset Management, e.g. for large infrastructural installations with relatively long life times. Again, LCM in the context of this chapter has a broader scope than just this management of use/after-sales and end-of-life.
- Life Cycle Management also differs from environmental management systems and schemes, such as the European EMAS, the British BS7750 and the international ISO 14001, since these are designed for production (site) management. ISO 14001—according to ISO, one of the most widely applied standards for environmental management—in its 2015 revision now requires a view on life cycles of the produced products as well—in addition to the production focus. This, in turn, exemplifies a tendency towards integrated approaches rather than to pursue several single-issue-focussed management systems in parallel.

Instead of focusing on a certain part of the organisation or of the product life cycle, Life Cycle Management takes as a starting point an overarching perspective, and it thus influences—and is influenced by—many parts of the organisation (Fig. 22.2).

It is important to note already here that there are *internal* and *external* factors influencing the accomplishment of the managerial tasks. This will be elaborated later in this chapter (Sect. 22.2).

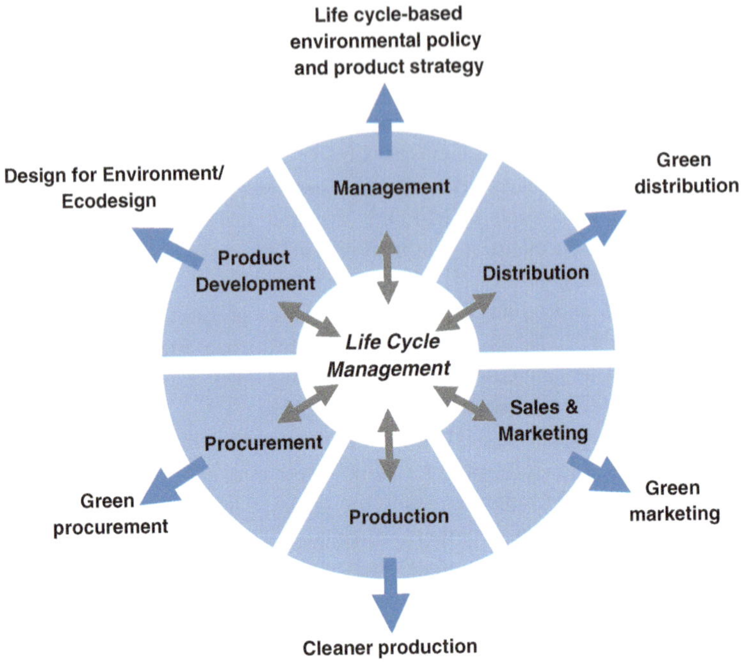

Fig. 22.2 Life Cycle Management as the central influencer and connector of different departments in an organisation and (shown on the circumference) examples of the outcome of its influence in each of those departments (based on UNEP 2007)

22.1.2 *Brief History of LCM—Why Did It Arise? How Does It Evolve?*

Life Cycle Management is a relatively young concept, which came up only about three decades ago. This timing can be explained as a logic consequence of two factors: First, the need of practitioners to operationalise the, then, new concept of Sustainable Development. Since the life cycle concept had been known and applied already since the mid-1970s (e.g. for military asset procurement; see also Chap. 15 on Life Cycle Costing), dealing with sustainability matters in a life cycle perspective was an obvious path to go. Second, a number of environmental incidents and catastrophes happened during the 1970s and 1980s. Immense disasters for instance with chemicals in the cities of Bhopal and Seveso, with oil tankers such as Amoco Cadiz and Exxon Valdez, with nuclear power plants such as Three Mile Island and Chernobyl as well as the controversial end-of-life treatment of the off-shore oil storage buoy Brent Spar all filled much in the public debate as well as in public and corporate consciousness in those years—and they were increasingly communicated through rising global news networks and environmental NGOs

(Non-Governmental Organisations). Therefore, the question of dealing with these issues in a strategically feasible and desired way—i.e. sustainably—became evident for governments, businesses, and the upcoming environmental NGOs alike.

A key initiator and driver for LCM discussion and development since 2002 is the UNEP-SETAC Life Cycle Initiative, which has issued a large number of publications, e.g. the LCM Guide (UNEP/SETAC 2007) and online material such as the LCM Navigator for SMEs (UNEP/SETAC 2008). The UNEP-SETAC Life Cycle Initiative has also been patron and sponsor for the bi-annual conference series on LCM that took place for the first time in 2001 in Copenhagen, Denmark. The LCM field is also promoted and elaborated by particular companies and, to a certain extent, also by business organisations and other NGOs, e.g. the World Business Council for Sustainable Development (WBCSD), through e.g. guidelines and models developed in their regime.

Life Cycle Management is today an established discipline, albeit pursued under many different names (as explained in Sect. 22.1). Current trends in LCM include:

1. Interest in mainstreaming LCM, i.e. getting it better integrated in standard procedure of organisations in many contexts (Sonnemann and Margni 2015);
2. Using LCM in organisational capability and maturity development support (e.g. Swarr et al. 2015; Pigosso et al. 2013);
3. Improving data flow integration and exploitation, e.g. in relation to socio-technical trends such as Internet-of-Things (i.e. the interconnectedness of various products, other than computers, through the Internet), and Big Data (i.e. the tracking, storing and making-available of large amounts of data; in fact relating to 'PLM/PDM');
4. Implementing LCM coherently and in the long term in ever-changing organisations. The 17 Sustainable Development Goals (SDGs) agreed upon in the UN General Assembly in late 2015, are increasingly understood as managerial targets that also drive future developments [e.g. in the SDG Compass (GRI/UN Global Compact/WBCSD 2015)];
5. Developing approaches to ensure companies' long-term survival, e.g. if their business offerings rely on non-renewable resources.

22.1.3 LCM as Integral Part of a Management System

To manage means, in general, to have control of, to take care of and to make decisions about. The goal of Life Cycle Management is to contribute to sustainable development, through operationalising it in organisations. LCM does this through (1) requiring the practitioner to identify those sustainability matters that are relevant for the particular organisation (e.g. via conduction of LCAs and other assessments), and (2) connecting these with possible managerial actions (i.e. for instance advising on who shall do what and when). The subsequent managerial actions of monitoring/checking and potentially initiating corrective or alternative actions are part of LCM in the same way as they are part of managerial procedure in general.

Similar to any other management concept, LCM can be integrated into a management system that the organisation may be using, and LCM can herein serve as a tool, e.g. for Sustainability Strategy Development. In its support of continuous improvement, LCM fits seamlessly together with widely applied management systems developed by ISO, such as ISO 9001 for Quality management and ISO 14001 for Environmental management, the latter having undergone a substantial revision in 2015, now emphasising its relevance not only in production contexts but also for product life cycle improvements, see ISO (2015).

The particular managerial tasks of Life Cycle Management, mentioned in Sect. 22.1.1, can be distinguished into four types:

1. Setting (measurable) targets—for the entire organisation or for parts of it, as part of, for instance, a strategy development process, as well as creating the basis for execution and planning, so that targets can be reached;
2. Executing the plan;
3. Tracking execution and performance; and
4. Taking corrective actions or setting new targets—depending on the performance.

In shorter terms, this sequence is described as Plan, Do, Check, Act or just PDCA (UNEP/SETAC 2007; ISO 2015), and in the sense of continuous improvement, the four phases will be repeated cyclically for an infinite number of times (see Fig. 22.3), each time at a slightly higher level of sophistication. Such a standard managerial activity cycle is practiced in many organisations and is a backbone concept in many ISO standards. Life Cycle Management follows those same four phases but distinguishes itself particularly in what the first and second phase deal with, namely target-setting for above sustainability matters and creating structures so that sustainability targets can be reached.

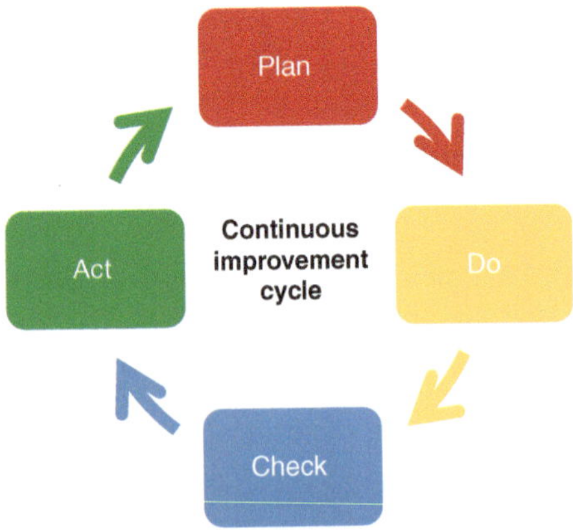

Fig. 22.3 A generic PDCA cycle of continuous improvement, consisting of the four phases plan, do, check, and act, which are run through continuously in management concepts such as LCM (based on ISO work, e.g. ISO 2015)

As a graphic representation of the different managerial factors' interconnectedness, Herrmann (2010) suggests an integrated model of LCM, the Total Life Cycle Management framework. It is developed to serve as a model defining relations with other management disciplines and as consistent frame-of-reference for LCM work and distinguishes two disciplines within Life Cycle Management: life cycle stage *spanning* management disciplines and life cycle stage *related* ones.

Disciplines spanning over several life cycle stages in this framework are:

- Social Life Cycle Evaluation;
- Economic Life Cycle Evaluation;
- Ecological Life Cycle Evaluation (i.e. Life Cycle Assessment);
- Information and Knowledge Management;
- Process Management.

Disciplines focusing on a certain life cycle stage are

- Product Management;
- Production Management;
- After-Sales Management;
- End-of-life Management.

The integrated model of Total Life Cycle Management is shown in Fig. 22.4.

As pointed out earlier, Life Cycle Management is no step-by-step methodology, but rather a management concept with an underlying mindset of thinking in life cycles and holistic contexts. Integrating this mindset into everyday practice of the organisation is key for a successful implementation of Life Cycle Management. This integration involves many different stakeholders as elaborated in the next section.

22.2 Who is Involved in LCM? Stakeholders and Their LCM Activities

The archetypical organisation applying Life Cycle Management is probably a company doing business with products of some kind (Note: The term 'products' in this chapter is generally understood as 'goods and/or services', but 'services' is sometimes mentioned for reasons of practical clarity). Such a company may or may not have, for instance, own product development activities, own production facilities, own logistics operations, etc. However, it may just as well be a different type of organisation such as an NGO, e.g. a consumer organisation or environmental activists group, or it can be a governmental organisation (e.g. an Environmental Protection Agency). The key common characteristic is that the organisation exerts influence on life cycles of products and/or their surrounding systems through its decisions. Using an LCA term, this could be called influence on the 'processes' taking place during the life cycle. For any such organisation, and

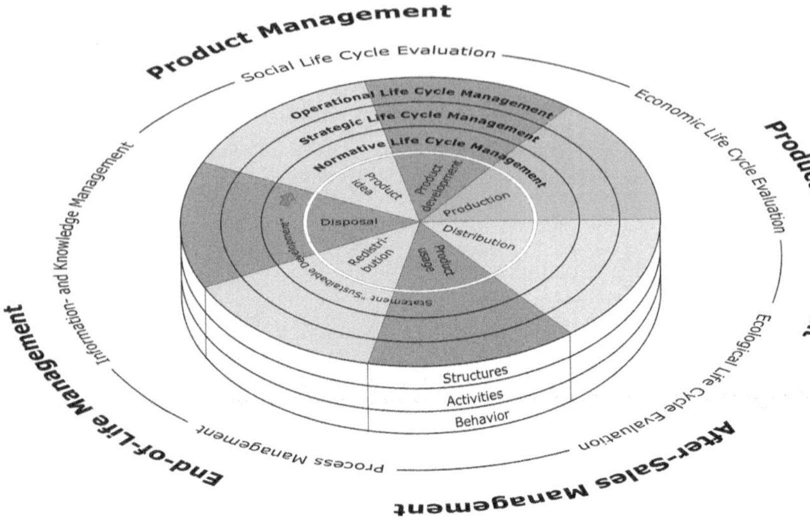

Fig. 22.4 A framework model for Life Cycle Management suggested by Herrmann (2010). This total Life Cycle Management model shows how sustainability as management philosophy can be unfolded in a consistent way in relation to management on different levels (normative, strategic, and operational level) and to the management objects (structures, activities, behaviour of/in the organisation). The model also shows, how the different levels and objects relate to life cycle stage-focused management disciplines (product, production, after-sales, end-of-life) and life cycle-spanning disciplines being information/knowledge management, social evaluation, economic evaluation, process management and—environmental life cycle assessment (reproduced with permission from the author)

that can be many, it may be meaningful to include Life Cycle Management in their management approach in order to exert the influence in a more structured, coordinated and informed way.

As an example, for a company producing a certain type of windows and seeking the environmentally preferable solutions, the decision context can, for instance, consist in having to follow EU and national regulation on particularly their type of windows, or that for certain markets specific end-of-life regulation may be given which may even be contrary to end-of-life regulation on other markets.

Life Cycle Management thus involves many stakeholders inside and outside the organisation with the organisation itself being a stakeholder too. All stakeholders have individual areas of influence; each can be a catalyst but also a potential unsurmountable obstacle for sustainability efforts, and each stakeholder may thus be able to substantially influence the success of such efforts.

The separation of external and internal stakeholders in the following two sections is only generic and exemplary. Depending on the size of the organisation, the function of some stakeholders may be internal or external. In small companies, the function of a legal department may, for instance, be fulfilled by an external law firm, whereas in larger companies, this function will be represented by an in-house department. However, the point is that a stakeholder dealing with legal issues needs to be taken into account, since, e.g. legal issues may hinder the implementation of any concrete optional solution. Thus, the LCM manager has to integrate this factor when selecting optional paths for solutions.

22.2.1 External Stakeholders

Seen from the perspective of an industrial company, there is a variety of external stakeholders. Three types of such external stakeholders can be separated as having a more or less direct influence on the way a company runs parts of its business and thus as having a more or less direct influence on activities and decisions made in the company:

- Customers/Consumers—being the key targets of company activities;
- Governmental bodies and authorities in general—setting legislative frames and regulatory requirements around the activities;
- NGOs of various kinds characterised by not being formal governmental bodies or authorities and not being declared representatives of industry. This includes environmental organisations, standardisation bodies, think tanks and academia —setting societal agendas, providing scientific insight and influencing company activities through de-facto standards (e.g. a certain ISO standard may be a voluntary instrument, but can be a de-facto requirement for certain products and/or markets).

Competitors could be considered a fourth one, but are here part of the stakeholder type Industry, incl supply chain since this is where competitors exert their indirect influence on activities of a given company. Similarly, other industrial companies could collectively be considered a fifth stakeholder, e.g. as in branch organisations, but they would not exert influence on the activities other than as covered by already mentioned types of stakeholders. Last but not least, shareholders are not described as distinct stakeholder because they do not exert direct or indirect influence on a company's business offering.

Figure 22.5 shows these three stakeholders as well as a company/industry offering the product or service. The figure also indicates each stakeholder's direct or indirect influence on the product (or service), and thus their way of influencing the decision space of the company. Customer feedback on product performance, etc. is a very typical and direct source of influence on activities at companies, e.g. related to product design improvements. Authorities have a similar direct influence, e.g. via product-type-specific regulation (e.g. the European Directives and international trade requirements). NGOs, in contrast, have typically only indirect influence on products/designs but still potentially directly on the company itself (see note B in figure caption).

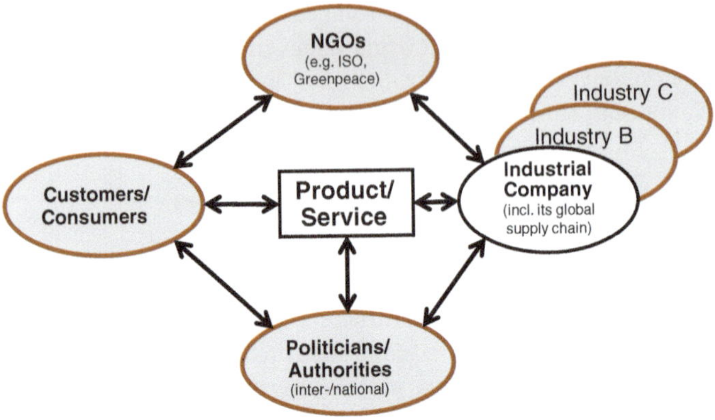

Fig. 22.5 The three generic types of external stakeholders influencing (directly or indirectly) the product/service business solution offered by a company: customers, politicians/authorities, and NGOs, as well as industry (incl. competitors and supply chain). Mutual relations indicated by *arrows*. *Note A* Investors and potential shareholders are not shown since they normally do not exert influence on the product/service solution. *Note B* Not all relations are indicated by *arrows* in order to keep simplicity in illustration, e.g. politicians/authorities' relation with NGOs

If the company for instance offers windows, a specific direct influence comes from international and national laws and regulations for buildings and from local authorities that the windows must comply with, and an incentive might come from potentially existing Green Public Procurement (GPP) schemes, as they exist, e.g. in Europe for some product types. The influences will differ depending on the different markets, i.e. regions, where the windows are intended to be sold. If the product shall be sold globally, obviously the number of influences (e.g. legal compliance requirements) increases highly. This means for Life Cycle Management work, that country/market-specific approaches may make sense, rather than trying to develop and apply one global approach.

22.2.2 *Internal Stakeholders—Departments in an Organisation*

Seen from a company perspective there are also a number of internal stakeholders, being the departments with their individual agendas and targets. Internal stakeholders can exert a great push and/or pull on the LCM function and the product/service offering of the company; and this can relate to all, both top management as well as workers in the production shop. A generic set of internal stakeholders includes the following ones (below and Fig. 22.6):

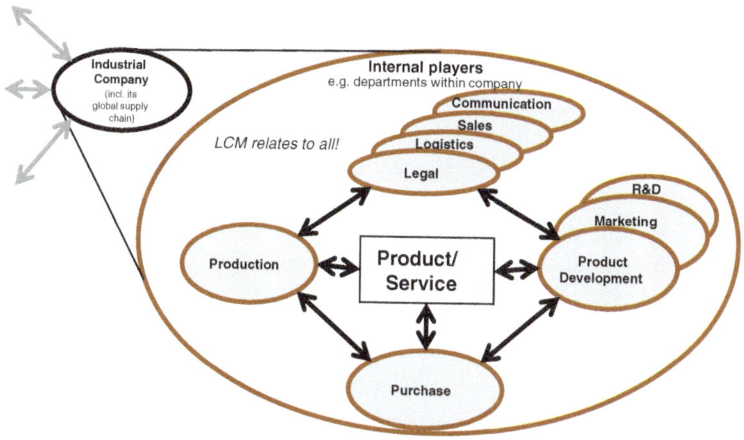

Fig. 22.6 The generic types of internal stakeholders of industrial companies, mutual relations between them and their (direct or indirect) influence on how the product/service is designed and marketed. LCM relates to all of them

- Marketing
- Research & Development (R&D)
- Product development
- Purchase
- Production
- Sales
- Legal
- Communication (to internal and external stakeholders)

Typically, *decision-making contexts* have many elements, and the LCM manager can take the role of pulling these different elements together in order to create a broad basis for decision-making. Looking at the window case, it may be that the company identified a particular material combination as environmentally preferable. However, if the production department has, e.g. difficulties in producing larger quantities at same quality levels, and or the purchase department would not be able to secure large enough quantities of it, the material combination would not be a feasible solution anyway. The same goes without saying for cost issues, although these also depend on the applied cost calculation model and business model. Only when having such a broad, combined view of influencing factors, truly holistic business decisions can be made. The LCM function closely interacts with such classic functions, like purchase, and advocates sustainability-related matters in that function's decision context, e.g. a ban of sourcing of conflict minerals (i.e. minerals mined in conflict zones) or e.g. a design for easier disassembly.

22.3 How do LCA and LCM Relate to Each Other?—The Key Role of LCA in LCM

Augmenting the LCA framework of ISO 14040, Sect. 22.1 already introduced a relation between Life Cycle Assessment and Life Cycle Management in a graphic way. The present section elaborates this relation further and points out the key role that LCA can play for practicing Life Cycle Management.

On a framing note, it can be stated that Life Cycle Assessment and Life Cycle Management have a lot in common, but also have some clear differences. Both build upon life cycle thinking, both are a basis for decision-making, both only make sense within defined system borders, and both consider activities today or in the near future and address potential effects that may occur in the far future (which require the practitioner to make assumptions and build scenarios). However, contrary to Life Cycle Assessment, LCM (usually) neither establishes concrete cause–effect chains—and not at all in a quantitative way using algorithms and computer models—nor does LCM employ the concept of impact categories, and midpoints and endpoints.

Although, LCM can be practiced without employing LCA, and although LCA cannot support all decision-making in LCM (e.g. cost-related issues), the section shows that LCA offers support on two managerial key areas: status determination and target-setting.

22.3.1 LCA Provides Environmental Quantifications Needed for LCM

A classic saying goes: "What you cannot measure, you cannot manage!" Environmental Life Cycle Assessment is a prominent tool that enables such measuring and performance tracking for managerial purposes. Using Life Cycle Assessment, a practitioner can, for example:

- Determine in which environmental condition or state a given product system is, e.g. a product on the market (shown as its contributions to a number of environmental impact categories)
- Identify environmental hot spots in that system
- Compare potential alternative solutions (e.g. new design suggestions)
- Make scenario analyses and establish a ranking of such alternative scenarios.

Conducting Life Cycle Assessments requires both thinking backwards in time and thinking forwards/ahead—i.e. life cycle thinking. In addition, LCAs create outcomes in the form of insights and results, but also in the form of questions that arise along the way. The latter are dealt with through making assumptions on past and/or future conditions in the life cycle. However, LCA cannot provide suggestions for managerial actions that put such outcomes into practice within a decision context. For instance, an LCA-based overview of impact potentials of different options may show one option as being preferable. However, by definition, the LCA cannot show, whether an environmentally preferable option, e.g. a certain engineering material for a window frame, would still be preferable, if other aspects were taken into account, e.g. the capability of a potential supplier to deliver that preferable material in the required quantity, quality, price and time frame. Exactly here, Life Cycle Management comes into play, since LCM can be used to integrate several aspects and related tools, including LCA, into one holistic, sustainability-oriented set of decision criteria (see Sect. 22.1.1).

22.3.2 Developing Key Performance Indicators for Application in LCM

Once an organisation has determined its mission, identified all its stakeholders, and defined its goals, it needs a way to measure progress towards those goals. Life Cycle Assessment is a crucial tool to conduct and support such progress measurements, since results from LCAs can be used to define performance indicators, which are a standard means used in management processes. Indicators are typically aggregated *indications* of the state of a given system at a certain point in time and/or of the performance of a system over a given period of time. One can distinguish several types of indicators; e.g. leading (target-setting) versus lagging (monitoring) indicators, result indicators versus performance indicators, and key result indicators (KRIs) as well as key performance indicators (KPIs), see, e.g. Parmenter (2015).

Irrespective of what type of indicator the Life Cycle Manager decides to use, LCA results can often crucially enrich such indicators or even entirely populate them, since they are quantitative by nature, thus LCA is a key tool in setting targets in LCM. An example of an indicator in LCM may be energy requirement of a manufacturing process—or of an entire production site (then often further specified as energy requirement per product produced, per year, per production line, etc.). This indicator can be populated using different units, for instance Mega Joules [MJ], kilowatt-hours [kWh] or other physical SI units. However, it could also be quantified in LCA-based units such as Global Warming Potential [kg CO_2-eq]. This could then support a comparison with and/or addition to other activities of the organisation. Transport and logistics activities of the organisation would, for instance, typically be monitored in terms of kilometres [km] (and as with production indicators, typically be specified further, e.g. per product). In the LCA context, one would additionally track employed means of transport (i.e. road, air, rail, ship) and potentially a particular vehicle type (e.g. five ton-truck, container vessel, bulk freight train, etc.). Using LCA, the transport activities would be quantifiable in kilograms of CO_2-eq as well, and a comparison with the above energy requirements of the production site or the entire organisation could be done—made possible through LCA.

Key aspects of managerial indicators are summarised below (compare, e.g. Gries and Restrepo 2011):

- They help an organisation define and measure progress toward organisational goals
- They are quantifiable metrics, agreed to beforehand, that reflect the critical success factors of an organisation
- They differ depending on the organisation, e.g.
  - A company may have as one of its KPIs the percentage of its income that comes from return customers, or revenue from eco-labelled products
  - A school may focus its KPIs on graduation rates of its students
- They must reflect the organisation's goals, must be key to its success/activities, and must be quantifiable/measurable (!)
- They need to be defined, incl.:
  - instructions on how to calculate them, => different individuals must always reach the same KPI value when calculating a particular KPI with the same background data
  - a definition that does not change from year to year (i.e. long-term definition needed);
- They can be used to set targets

When developing KPIs for any application context, the following needs to be observed:

- The landscape of data sources may be non-harmonised (e.g. not all source data may cover the same year, same location, etc.). This means: Observe 'data landscape' and availability of input data
- KPI inflation—Aim is *not* to measure what *can* be measured, but to measure what *should* be measured. This means: The fewer KPIs the better
- Pseudo accuracy: since 'Bad data in' leads to 'bad data out', this means: Interpret the KPI values related to their underlying data quality

In order to support the reader in developing KPIs genuinely meaningful for LCM work, and in reflection of current paradigm-setting developments in business and science, the remainder of this section deals with very concrete characteristics of environmental sustainability-relevant KPIs, points out problems and describes ways to address these.

Based on studying company reporting, one can distinguish two types of environmental sustainability-related KPIs seen today in many companies' communications: Intensity-based KPIs and company-wide KPIs, and e.g. (Bjørn et al. 2016a) pointed out problems that both types have:

- Intensity-based KPIs:
  - Examples: *kg CO_2-eq emitted per product* or *Emissions per \$ of revenue*

Main problems:

  - Performance depends largely on the chosen reference (e.g. last year's product model? Market average? Competitor's product?)
  - Total impact may still increase, if sales increase (which usually is a company target—depending on business model)

- Company-wide KPIs:
  - Example: *kg CO_2-eq emitted per year*
    Main problem:
    - Misleading, for example if company is outsourcing and does not account for the outsourced activities

One can argue that an additional problem of both KPI types is, that they do not indicate how much impact reduction—or which concrete indicator value—would be enough in order to reach sustainable levels. Organisations such as the WBCSD and the Science-Based Targets initiative (SBT)—a collaboration established in 2014 between UN Global Compact, Carbon Disclosure Project (CDP), World Resources Institute (WRI), and the World Wide Fund For Nature (WWF)—have begun to address this missing link in suggesting methods to determine concrete targets for application in the company context. The SBTs relate to greenhouse gas emissions only, i.e. they only relate to this part of environmental sustainability, see e.g. www.sciencebasedtargets.org and Krabbe et al. (2015).

An example for such a KPI is given below, incl. related problems:

- Absolute-target-based KPIs

- "We as company x will by year y reduce our global emissions to z tons CO_2-eq per year".

Problems with this absolute type of KPI are, however

1. When communicating such targets, the company is bound to them, since they are clear and trackable. This kind of lock-in is by many companies considered a risk and thus not pursued by many companies, even if they recognise Earth as a finite system in their reporting, compare, e.g. (Bjørn et al. 2016a).
2. The challenge of defining targets for other environmental impact categories than Global Warming (i.e. for impact categories where there is no international agreement and/or where targets are per se more difficult to determine, e.g. for regional and local impact categories).

The development of absolute KPIs is accelerated by the Two-degree-target agreed upon at the UN Climate Change summit COP21 in late 2015, where the global community committed to keeping greenhouse gas emissions at levels, so that average global temperature rise by the year 2050 stays well below 2 degrees centigrade above pre-industrial levels. It has also been shown, that such absolute indicators can be integrated into existing LCA methodology for other global impact categories than Global Warming Potential, e.g. for Terrestrial Acidification (Bjørn and Hauschild 2015; Bjørn et al. 2016b). Ways of consistently breaking down high-level targets into targets for lower levels of decision-making are being suggested, e.g. by Rödger et al. (2016). In conclusion, the development of internationally agreed absolute indicator limits and their subsequent broad application cannot be seen to happen in the near future, but individual companies may adopt the principle and develop their own absolute indicators, as some companies already do.

The above examples and issues show that KPI development requires overview, understanding and utmost care by the practitioner in order to produce meaningful KPIs, which can trustfully be used as the central management instrument they are. This is even more important when applied within LCM, since it can be considered an even more complex field than many other management fields.

22.4 How is LCM Applied in Practice? A Brief LCM Case Study

Application of Life Cycle Management fundamentally incorporates that the practitioner makes choices *before* and *during* the application itself, rather than that she or he follows a predetermined procedure. The reason for this is that LCM is a management concept—as explained earlier in this chapter—and not a deterministic method or algorithm, with concrete steps or rules, that would lead to concrete, repeatable outputs, if triggered by same inputs. LCM application can be described as consisting of two elements:

1. Individually selecting one or several tools from the collection that may be called the 'LCM toolbox' (UNEP/SETAC 2007)—and;
2. Doing this in the concrete context of the given organisation—i.e. in collaboration with different departments and under recognition of a variety of factors, such as the organisation's position in supply chains and the intended application field

Thus, there is never a one-size-fits-all way of applying LCM. Rather, each organisation needs to determine for itself and for the concrete product or activity, which combination of tools they consider appropriate for managing life cycle matters of that product or activity.

A typical starting point for applying LCM is product innovation and product development processes. From here, LCM activities often radiate back and forth to and from production and operations and to and from marketing and communications, especially sustainability reporting (see e.g. McAloone and Bey (2009)).

Taking the example of deciding among alternative materials for a new type of façade window frames, an LCM practitioner may come into play in the following, diverse decision contexts: Starting point would probably be that the window company has decided which markets, i.e. global regions and/or countries, the new type of windows (and frames) shall be sold in, and in which target quantities and at which target sales price. In parallel, the design engineering/R&D (Research & Development) department has probably determined a number of technically feasible materials (e.g. a metal, a type of wood, and a composite material—plus maybe combinations of them). Although usually not practiced, the company might already here set a target environmental impact for the windows as LCM activity, and quantify this target in terms of a certain maximum Global Warming Potential the windows should not exceed (potentially further described, e.g. for modules of the window or for an annual production of all production sites or for one selected site, etc.). The target figure may be related to the window company's sustainability strategy and sustainability targets, which they may have published in their sustainability reporting. Supportive methods for such detailed target-setting are suggested, e.g. by Rödger et al. (2016).

Taking into account the different generic external and internal stakeholder types of customers, competitors, NGOs, governmental organisations as well as departments within the organisation (see Sect. 22.2), the company would check legal requirements and de-facto market requirements on the selected markets. This could refer to, e.g. legally banned materials or substances on that market for that type of product, but also to certain environmental labels and certifications that key competitors on that market have certified their products with. Or it could mean to apply for newly introduced, not yet widely applied labels that may represent a competitive edge on that market. Both, legislation and de-facto market requirements, are often quite different from country to country and especially from region to region, e.g. between Asia and North America. Obviously, if the chosen market is global this would require further decisions, e.g. whether or not to prioritise one label over others or to work towards certification against all labels on the global market, which of course has an impact on the cost of the product. Also, sustainability-related campaigns by NGOs, e.g. against specific labour practices or certain technologies, would be mapped as far as possible and analysed in terms of potential threats to sales of the product on the selected market(s). If a stakeholder campaign or similar activity is identified, decision options are either to ignore, to fight or to accept the respective stakeholder demand. An example of the latter is the following one from the paper and pulp industry: the global tissue paper producer Kimberly-Clark Corp. and Greenpeace agreed in 2009 on a new fibre policy and on a regular review process, ending a several-years Greenpeace campaign—an agreement to the benefit of all parties.

Back to the window example: In parallel to related information becoming available for the LCM decision-maker, the window producer's R&D department may practice ecodesign in making life-spanning scenarios for the product and taking design actions with focus on selected life cycle stages. The developers may, for instance, design the windows for easy disassembly (focus: end-of-life stage of the windows' life cycle) or design them in a modular way, allowing to easily attach and exchange components such as blinds, motoric actuators, etc. or optimise insulation capabilities—all this for the different materials under consideration (i.e. focus: use stage). At the same time, purchasers and designers could be supported in choosing the exact material composition, e.g. the alloy of a metal and the supplier(s) selected to deliver raw materials and semi-manufactured goods (focus: materials stage and transportation stage). Furthermore, the production department as well as logistics and also service/after-sales departments could be drawn into the materials decision in order to, e.g. ensure manufacturability and secure availability, etc. (focus: manufacturing stage). Early in this process, the respective departments would also identify legal requirements on the target markets, such as safety norms, flammability norms and insulation capacity as well as potentially important voluntary requirements, which represent de-facto market requirements, such as certain environmental labels (e.g. certified wood, or eco-labels for the entire product).

Last, but not least, a core department responsible for business model options—i.e. the board of management—would be strongly involved, if a material choice should represent a strategic change, as for instance seen in the shift to aluminium used for car bodies which some automotive companies and suppliers have made. In the window case, this could be the shift to composite materials, requiring, e.g. entirely new production technologies and design options and constraints. In conjunction with potential regulative developments, e.g. as currently seen in Europe towards Circular Economy and the released action plan (EU Commission 2015), the window producer could consider a business model that incorporates take-back of their windows at their end-of-life—an option which then would have to be analysed for its environmental and economic viability.

22.5 What Does LCM Require and Yield in an Organisation? Application and Integration of LCM

Life Cycle Management costs time, requires learning curves at the different departments and—in essence—necessitates new thinking by all involved individuals. Such costs and obstacles often surface in many places and on many occasions, e.g. in the course of time-consuming data collection, life cycle modelling, communications, or when trying to determine, what in the particular company's context actually is meant with the term 'sustainability'. However, although very difficult to pin-point and quantify, LCM also yields gains, e.g. in the form of increased knowledge about the organisation's own processes and life cycle chains, incl. better insight into conditions at suppliers, due to the 'total' overview. LCM may also reduce risk and increase opportunities, as well as improve the ability to respond early to new legislation and market trends in the field. In that sense, LCM generally increases the resilience of the organisation that practices LCM, since it encourages taking both short-term, detailed views and long-term, helicopter-perspective views on company activities.

Sustainability matters are manifold and so are the organisations that work with them. LCM can, of course, be practiced in the multi-national corporation, in the small or medium-sized enterprise, in the family-run company and in the one-person firm—and this irrespective of where on the Planet they are located. Since LCM is a translator of sustainability matters into business practice, LCM itself needs to be manifold and highly adaptable (a bit comparable to a Swiss Army knife which integrates many tools and makes them available in one pocket-sized product), so that pursuing LCM can produce improvements of the sustainability matters for any organisation and in any department of that organisation. Due to this necessarily multifaceted character, Life Cycle Management may well be perceived as diffuse. However, LCM is *the* means to put sustainability approaches into practice, since no measurement and assessment tool (such as LCA) can tell by itself what its results are to be used for or, e.g. what ranges of these results the organisation should deem desirable (beyond legal threshold values, if applicable, and also this context is not known to the tool).

Thus, having tools available in an organisation is not enough (even if these tools should in fact be easy and effective to use, which is not always the case): It requires an overall management concept, such as LCM, to reach a coherent, holistic approach, practiced in the relevant parts of the organisation. Last, but not least, the long-term survival—i.e. the long-term successful application—is key for any sustainability-supporting approach. Thus, the integration and anchoring of LCM in the organisation are just as important as the applicability of the tools it comprises. Addressing this key aspect, approaches to map and quantify organisational maturity and to build adapted capability have been developed and are increasingly in focus (as introduced in Sect. 22.1.2).

LCM can be integrated in an organisation in two principle ways: via projects and/or via functions (plus via combinations of both). The way of choice depends on the type of organisation and on preferences of the organisation. In project-based organisations, typically companies that design, produce and sell one-of-a-kind products or small series, most prominently construction companies, LCM can be integrated into the Stage/Gate process that the projects usually are managed by. Project managers and portfolio managers, in this case, need in to have a good understanding of the LCM concept and act as life cycle managers, unless a dedicated LCM specialist is included in the project team. This depends on the size of the project team and of the organisation behind, leading to the circumstance that often only larger teams and organisations have a dedicated LCM role in project teams—and this role can even be taken on by a consultant. In function-based organisations, typically mass-producing or batch-producing manufacturing companies (which our window company probably would be), and in combined structures, such as matrix organisations, Life Cycle Management can be integrated by establishing dedicated departments and/or appointing individuals and adapting set organisational processes, so that LCM aspects relevant in the particular company ultimately become integrated into day-to-day procedures.

In larger organisations, Life Cycle Management is typically not one outspoken activity or position, but it is in the vast majority of such companies embedded in and dealt with by several departments and under many titles, often two to three departments (typically a corporate one plus the production-environmental department plus, maybe, a product-related one, see Sect. 22.1.1), rather than by one department alone. Ultimate goal in both principal integration situations is, that all project members and all departments are aligned and that, eventually, the LCM function becomes obsolete and a self-running part of all other functions. In either way of integration, the definition of KPIs is key in order to be able to set targets and track performance (see Sect. 22.3.2), e.g. at departments.

As explained earlier, any LCM approach is tailor-made in the course of finding the individually best-suited combination of tools that together represent the company-specific approach. Figure 22.7 shows an example of such a combined approach towards sustainability/LCM work at a company, integrating several elements.

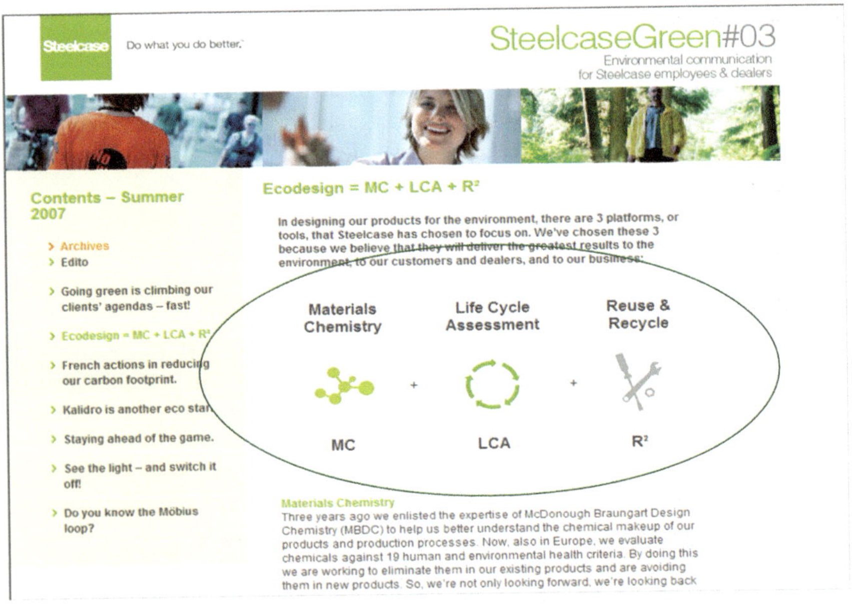

Fig. 22.7 An example of an integrated approach (*screen shot of an intranet page*): The company Steelcase, globally providing office furniture and office space solutions, combined three elements/tools in their approach. Due to product type, material types, life cycles, markets and other factors, they chose to use both LCA, Materials Chemistry (i.e. cradle-to-cradle) and Reuse and recycling—already in 2007—and communicated this to internal and external stakeholders

22.6 Conclusions

This chapter described Life Cycle Management and the different aspects it involves—both from its contents and from its application and integration in organisations. Many examples relate to industrial companies but the principles are the same in municipalities and other organisations. In all contexts, Life Cycle Management has the below key characteristics:

- LCM is no step-by-step methodology or one-size-fits-all approach but a management concept that can and needs to be adapted to any organisation's context, i.e. LCM is always tailored to the specific organisation.
- LCM deals with the managerial tasks related to practicing sustainable development in an organisation—it could thus also be referred to as Sustainability Management.
- LCM requires collaboration between several departments within the organisation and with external stakeholders, such as supply chain partners.
- As with any other management discipline, practicing LCM requires setting targets and tracking performance. Such targets and related performance indicators can be set based on relative terms but may also be set as absolute terms.

Practicing LCM in a consistent way aims at ensuring that improvements in system performance can be achieved while also making sure that this does not lead to either sub-optimisations or burden shifting in the company/organisation as a whole and/or in the relevant life cycles.

References

Bjørn, A., Hauschild, M.Z.: Introducing carrying capacity based normalization in LCA: framework and development of references at midpoint level. Int. J. Life Cycle Assess. (2015). doi:10.1007/s11367-015-0899-2

Bjørn, A., Bey, N., Georg, S., et al.: Is earth recognized as a finite system in corporate responsibility reporting? J. Clean. Prod. (2016a). doi:10.1016/j.jclepro.2015.12.095

Bjørn, A., Margni, M., Roy, P.O., et al.: A proposal to measure absolute environmental sustainability in life cycle assessment. Ecol. Indic. (2016b). doi:10.1016/j.ecolind.2015.11.046

EU Commission.: COM(2015) 614 final, ANNEX to the COMMUNICATION FROM THE COMMISSION Closing the loop—An EU action plan for the Circular Economy (2015)

GRI/UN Global Compact/WBCSD.: SDG Compass—UN Sustainable Development Goals Compass. www.sdgcompass.org (2015)

Gries, B., Restrepo, J.: KPI measurement in engineering design—a case study. In: International Conference on Engineering Design ICED11 (2011)

Guinée, J.B., Heijungs, R., Huppes, G., et al.: Life cycle assessment: past, present, and future. Environ. Sci. Technol. **45**, 90–96 (2011). doi:10.1021/es101316v

Herrmann, C.: Ganzheitliches life cycle management. (2010). doi:10.1007/978-3-642-01421-5

ISO.: Environmental Management systems—Requirements with Guidance for Use (ISO 14001:2015), 3rd ed. ISO, The International Organization for Standardization, Geneva (2015)

Kloepffer, W.: State-of-the-art in life cycle sustainability assessment (LCSA) life cycle sustainability assessment of products. Int. J. LCA **13**, 89–95 (2008). doi:10.1065/lca2008.02.376

Krabbe, O., Linthorst, G., Blok, K., et al.: Aligning corporate greenhouse-gas emissions targets with climate goals. Nat. Clim. Change **5**, 1057–1060 (2015). doi:10.1038/nclimate2770

Linkov, I., Seager, T.P.: Coupling multi-criteria decision analysis, life-cycle assessment, and risk assessment for emerging threats. Environ. Sci. Technol. **45**, 5068–5074 (2011). doi:10.1021/es100959q

McAloone, T.C., Bey, N.: Environmental improvement through product development—a guide (DTU's eco-design webpages www.ecodesign.dtu.dk) (2009)

Parmenter, D.: Key Performance Indicators: Developing, Implementing, and Using Winning KPIs, 3rd edn. Wiley, Hoboken (2015)

Pigosso, D.C.A., Rozenfeld, H., McAloone, T.C.: Ecodesign maturity model: a management framework to support ecodesign implementation into manufacturing companies. J. Clean. Prod. **59**, 160–173 (2013). doi:10.1016/j.jclepro.2013.06.040

Prado, V,, Rogers, K,, Seager, T.P.: Integration of MCDA tools in valuation of comparative life cycle assessment. In: Life Cycle Assessment Handbook: A Guide for Environmentally Sustainable Products. (2012). doi:10.1002/9781118528372.ch19

Rödger, J.-M., Bey, N., Alting, L.: The Sustainability Cone—a holistic framework to integrate sustainability thinking into manufacturing. CIRP Ann. Manuf. Technol. (2016). doi:10.1016/j.cirp.2016.04.033

Sonnemann, G., Margni, M., et al.: Life cycle management. LCA Compend—Complete World Life Cycle Assess. (2015). doi:10.1007/978-94-017-7221-1

Swarr, T.E., Asselin, A.-C., i Canals, L.M., et al.: Building organizational capability for life cycle management. In: Sonnemann, G., Margni, M. (eds.) Life Cycle Management, pp. 239–256 (2015)

UNEP/SETAC: Life cycle management: a business guide to sustainability. UNEP/SETAC Life Cycle Initiative, Paris (France) (2007)

UNEP/SETAC: LCM navigator for SMEs. www.lifecycleinitiative.org/LCM_navigator/index_c.html (2008)

Author Biography

Niki Bey mechanical engineer and expert in applied life cycle thinking in organisations. Works since the late 1990s with Ecodesign, Sustainable manufacturing, Product/Service Systems, Life Cycle Management and Sustainability strategy development in consultancy and academia across many industrial sectors with focus on integration of approaches in real life applications.

Chapter 23
Ecodesign Implementation and LCA

Tim C. McAloone and Daniela C.A. Pigosso

Abstract Ecodesign is a proactive product development approach that integrates environmental considerations into the early stages of the product development process so to improve the environmental performance of products. In this chapter, the ecodesign concept will be discussed, in terms of its implementation into manufacturing companies. Existing methods and tools for ecodesign implementation will be described, focusing on a multifaceted approach to environmental improvement through product development. Additionally, the use of LCA in an ecodesign implementation context will be further described in terms of the challenges and opportunities, together with the discussion of a selection of simplified LCA tools. Finally, a seven-step approach for ecodesign implementation which has been applied by several companies will be described.

Learning Objectives

After studying this chapter, the reader should be able to:

- Define ecodesign and understand its importance in the context of sustainability.
- Understand the extensive variety of ecodesign methods and tools.
- Understand the main challenges of Life Cycle Assessment (LCA) implementation in the context of ecodesign.
- Understand how to communicate LCA results within an ecodesign activity.
- Explain simplified LCA approaches for implementation into ecodesign programmes.
- Understand how to measure progress and set goals for ecodesign implementation.
- Carry out a seven-step approach for ecodesign implementation into companies.

T.C. McAloone (✉) · D.C.A. Pigosso
Section of Engineering Design and Product Development,
Department of Mechanical Engineering, Technical University of Denmark,
Kongens Lyngby, Denmark
e-mail: tmca@dtu.dk

M.Z. Hauschild et al. (eds.), *Life Cycle Assessment*,
DOI 10.1007/978-3-319-56475-3_23

23.1 Introduction to Ecodesign

There are many reasons for the environmental problems we experience in the world. Massive population growth and an increase in relative wealth (and thus growing consumerism) on a global level are two significant contributors to the strain on our fragile ecosystem. Manufactured products are essential for the wealth of society and for our desired quality of life. However, our growing consumption of products lies directly or indirectly at the root of a great deal of the pollution and depletion of resources that the consumerist society causes (Commission of European Communities 2001). Environmental impacts are caused by every product in some way or another, from the extraction of raw materials, through their production and use, to the management and final disposal of waste (Baumann et al. 2002).

Regardless of the nature, size and time of occurrence of environmental impacts for a product, the vast majority of environmental impacts are actually decided already in the very early phases of product development. In fact it is estimated that approximately 80% of a product's environmental performance[1] is fixed during the early phases of the product development process (McAloone and Bey 2011). It is during product development that materials, technologies and the product's lifetime are decided. The product developer has thus a great influence on the product's life cycle and therefore also on the later occurring environmental impacts and on the environmental performance of the products. For this reason, it is important that the product developer integrates environmental considerations carefully and systematically into the product development activity (McAloone and Bey 2011). This integration of environmental considerations into product development is called *ecodesign*.

Ecodesign is a proactive approach to environmental management during product development, with the aim of integrating environmental considerations into the product development process. The goal is to minimise environmental impacts throughout the product's life cycle, without compromising other essential criteria such as performance, functionality, aesthetics, quality and cost (Johansson 2002; van Weenen 1995). Ecodesign requires a balanced view of the whole product life cycle, focusing attention on the reduction of the major environmental impacts of the product, throughout its lifetime.

Ecodesign calls on the knowledge and competencies of many disciplines in the product development process, as considerations about materials, processes, logistics, recyclability—and many more—are likely to arise as potential contributors to an improved environmental profile of the product design in hand (Brones and Carvalho 2015). The involvement of many functions and professions in this process gives rise to multiple viewpoints and increases the likelihood for optimal solutions (McAloone and Bey 2011).

[1]The environmental performance of a product can be determined by the sum of all the environmental impacts it causes during its lifetime (Nielsen and Wenzel 2002).

Taking a systematic approach to understanding where and why a product has environmental impacts in its lifetime can lead to competitive advantages for the company. It has been demonstrated, for instance, that environmental thinking in product development leads to efficient products, which are both economically viable to produce, cheaper to operate and maintain, and more robust during their lifetimes (de Caluwe 2004; Eagan and Finster 2001).

Designing products with improved environmental performance is a necessary action for industries to ensure both competitive and environmental advantages (Bey et al. 2013). The systematic incorporation of environmental considerations during the product development process (i.e. ecodesign implementation) is not an easy task, especially in the early stages, which are characterised by greater degrees of design freedom, but also limited information about the product and its pending manufacturing processes.

Over the past couple of decades, several approaches and methodologies have been developed to support manufacturing companies to integrate ecodesign into their product development processes (Baumann et al. 2002; Pigosso et al. 2014). For ecodesign to be successful, activity at three main levels of the company is required:

- *Strategic (or managerial)*, to set the goals and expectations throughout the whole organisation;
- *Tactical,* to schedule and prioritise the good intentions of management; and
- *Operational,* to deploy ecodesign methods and tools directly, within product development projects.

Strategic approaches are related to the integration of ecodesign into the strategic decision-making and business processes. Some examples of activities carried out in the context of a strategic ecodesign implementation include: definition of environmental targets for the product portfolio; deployment of responsibilities across different hierarchical levels; development of a communication strategy to customers and stakeholders; development of strategic competences in the organisation; etc. In other words, strategic-level ecodesign implementation creates the foundation, the goals and the resources in the organisation for the ecodesign process to be a success (Pigosso et al. 2013a, b).

At the *tactical* level of the company, the task is to ensure that the goals, strategies, and visions of the strategic management group are prioritised and organised in a way that they can be integrated into the product development process. This activity is very important, as without it, no decisions can be made to ensure a systematic approach to ecodesign. It is at this level that: (1) candidate ecodesign projects are chosen, (2) methods and tools are prioritised and integrated into the product development process and (3) the product development process is generally updated to include environmental considerations. It is also here that ecodesign implementation roadmaps are made and deployed (Pigosso et al. 2013a, b).

The actual development of products with improved environmental performance takes place during the so-called *operational* ecodesign implementation, which starts in the early phases of the product development process, where the greatest improvement opportunities lie. Identifying the desired environmental performance of products, the environmental hotspots (environmental aspects and life cycle stages that have the highest environmental impact), developing alternative product concepts based on ecodesign guidelines, selecting concepts to be further developed based on their environmental performance, etc., are some examples of activities carried out in the operational implementation of ecodesign.

The ambition for ecodesign implementation is often correlated to the main internal and external drivers of the company for ecodesign implementation. Usually, companies that are applying ecodesign due to a legislative compliance driver (e.g. compliance with European directives) have a lower ambition when compared to companies that are implementing ecodesign due to an internal strategic driver (e.g. to a sustainability strategy or organisational values). Companies that are implementing ecodesign due to customer requirements, for example, usually begin the ecodesign implementation with a limited ambition, which is subsequently expanded as the companies learn the other business benefits linked to ecodesign implementation.

23.2 Ecodesign Methods and Tools

Since the establishment of ecodesign as a product development practice and as a research object for scientists, several ecodesign methods and tools have been developed—both by academics and in industry. Currently, more than 150 ecodesign methods and tools exist (Pigosso et al. 2014), and the number continues to grow.

The methods and tools that exist have various goals and focuses, such as "evaluate environmental impacts", "reveal potential trade-offs" (Byggeth and Hochschorner 2006) and "facilitate the choice between different aspects" (Baumann et al. 2002; Byggeth and Hochschorner 2006). In this section, we provide an overview of a selection of existing ecodesign methods and tools.

Ecodesign methods and tools are defined as any systematic means for the management and implementation of ecodesign at an operational level. Ecodesign methods and tools are usually applied in the early phases of the product development process (Fig. 23.1), where the largest improvement opportunities lie.

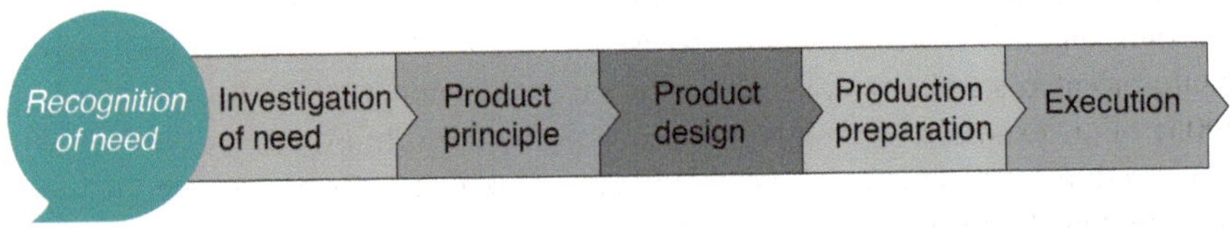

Fig. 23.1 One way to depict the product development process (Andreasen and Hein 1987)

As mentioned earlier, a large amount of the environmental impacts of a product's life cycle are determined in the early phases of its development (McAloone and Bey 2011).

According to their main purpose, the ecodesign methods and tools can be classified into three main groups (Pigosso et al. 2011a, b):

- Prescriptive: Present generic guidelines (from a pre-established set of best practices to minimise the environmental impacts);
- Comparative: Compare the performance of different products, concepts or design alternatives for a given product;
- Analytical: Identify improvement potentials by means of an assessment of the most relevant environmental aspects.

A few examples of ecodesign methods and tools, classified according to the product development phase (Fig. 23.1) and type of tool (prescriptive, comparative and analytical), are presented in Table 23.1.

Most of the existing ecodesign methods and tools are focused on the early stages of product development, mainly in the "product principle" and "product design" phases. Furthermore, there is a tendency to the use of analytical tools rather than to comparative and prescriptive ones.

The ecodesign methods and tools presented in Table 23.1 are further described in Table 23.2, with an indication of references, where more information can be obtained.

23.3 LCA and Ecodesign

In this section we explore in some more depth, how LCA can be used in an ecodesign context, to support the development of products with improved environmental performance.

Table 23.1 Examples of ecodesign methods and tools (all are cited in Tables 23.2 and 23.3)

	Analytical	Comparative	Prescriptive
Recognition of need	Eco-QFD STRETCH (Strategic environmental challenge)		
Investigation of need	Eco-roadmap	EcoBenchmarking	
Product principle	Eco-function matrix	LiDs wheel	Ten golden rules
Product design	LCA Environmental effect analysis (EEA)	DfE matrix MECO matrix	EcoDesign pilot
Production preparation	–	–	–
Execution	EcoValue	Eco communication matrix	

Table 23.2 Description of presented ecodesign methods and tools

Method/tool	Description
Eco-QFD (quality function deployment) (Ernzer et al. 2005)	Supports the transfer of market insights to the product's requirements list. In this method, the environmental demands are acquired from an expert team instead of directly from the customer. It allows the product developer to focus on the internal (product properties) and not the external environmental parameters (e.g. generating energy, waste treatment)
STRETCH (strategic environmental challenge) (Cramer and Stevels 1997; Stevels 2007)	Focuses on assessing improvements in the most promising environmental opportunities throughout the product life cycle. Considers product business units and market strategies, in addition to potential changes in environmental pressure exerted by external stakeholders. It consists of five steps: (1) Identify the major forces that influence business strategy, (2) Develop scenarios that the company can adopt based on core strengths and develop a list of potential marketing strategies for the product. (3) Specify opportunities for environmental improvements for each scenario. (4) Select the environmental challenges that will lead to substantial improvements in the environmental performance of products. (5) Address the selected environmental challenges
Eco-roadmap (Donnelly et al. 2006a, b)	Concise graphical tool that captures short- and long-term environmental drivers (legislation [enacted and future] and customer requirements) in one document. The eco-roadmap contains the actual product-relevant legislation and customer requirements in the scope of sustainable and environmentally compliant product design. The eco-roadmap also highlights draft legislation, emerging customer requirements, and industry trends for future sustainable and environmentally compliant product features
EcoBenchmarking (Boks and Diehl 2005; Boks and Stevels 2003; Wever et al. 2005, 2007)	Supports organisations to understand and develop a critical attitude towards their own products, to create awareness about environmental issues in and outside a company, and thus to find environmental improvement options for their products that are feasible for implementation. The method is based on 10 steps with specific goals and questions to be answered and can be adjusted in two ways (light vs. extended; information vs. physical) depending on the context and needs
Eco-function matrix (Lagerstedt et al. 2003)	A communication platform for functional priorities and environmental impacts is established by combining the Environmental Profile and Functional Profile of the product. The functional profile describes and evaluates properties, areas and activities that are associated with the functionality of the product and its commercial viability. The Environmental Profile identifies the characteristics of the products that are correlated with the generation of environmental impacts. It can be applied at different stages of product development, according to the degree of specification and detailing of Environmental and Functional Profile

(continued)

Table 23.2 (continued)

Method/tool	Description
EcoStrategy wheel (Knight and Jenkins 2009)	Contains eight strategies (function optimisation, reducing the impact on the use stage, reducing the use of materials, choosing the right materials, optimisation of lifetime, production optimisation, optimisation of treatment of end of life, improved distribution) to environmentally improve products and is based on knowledge of the team members. Strategies are used as a checklist and are sources of inspiration to support meetings using brainstorming techniques
Ten golden rules (Luttropp and Lagerstedt 2006)	Consists of a summary of several guidelines and manuals used by companies from various sectors, with recommendations of environmental strategies. It can be used to improve the environmental performance of the concept of a product or to compare various alternative concepts. In order to be used by a particular company, it must first be transformed and customised according to the characteristics of the company and its developed products
Life cycle assessment (LCA) (Bhander et al. 2003)	Quantifies environmental impacts of the whole life cycle of goods and services. It involves all successive stages of a product system, ranging from extraction of raw materials and energy required in manufacturing, use and distribution until the final disposition of the product, which may include recycling of materials and components, and other ways of post-consumption treatments
Environmental effect analysis (EEA) (Lindahl 1999, 2000)	Qualitative dialogue process with a starting point in the use of available experiences and environmental requirements from stakeholders. The tool was developed to assist product development teams in quick and effective assessment of environmental issues, clarifying their goals and objectives, and toward fulfilling them in real product development efforts. The basic principle is to list all activities considered to have significant environmental influence, and for each activity judge the quantity and seriousness of each aspect, as well as to suggest ways for making improvements that will reduce the impacts of the proposed product
DfE (design for environment) matrix (Eagan and Finster 2001)	The matrix raises questions about the environmental impacts of a product through 100 issues that allocate a wide range of environmental and design issues and provides a semi-quantitative analysis of the product design alternatives. The totals for each life cycle stage (pre-manufacturing, manufacturing, packaging and distribution, use and maintenance, end-of-life) and environmental impacts indicate improvement areas in terms of the environmental attributes of a product throughout its life cycle. The total score of the matrix is a relative measure of environmental product attributes and complements the economic parameters of customer value and manufacturability that should also be evaluated

(continued)

Table 23.2 (continued)

Method/tool	Description
MECO matrix (Hochschorner and Finnveden 2003; Wenzel et al. 1997)	Estimates the environmental impact of each life cycle stage (raw material, manufacturing, use, disposal and transport) and is performed by estimating the amount of material (M), energy (E), chemical (C) and other materials (O) used in the product life cycle. All input and output flows must be considered for a category in relation to a time base according to the functional unit of product and the stage of the life cycle chosen
EcoDesign pilot (Wimmer et al. 2005)	Once the environmental strategies are selected for the improvement of an existing product, different design rules and guidelines guide the designer during the design process. To improve the environmental performance, each product requires specific measures depending on its environmental impact at different stages of its life cycle (extraction of raw materials, manufacturing, transportation, use and disposal). A good set of measures of the design rules and guidelines can be found in the EcoDesign PILOT software
EcoValue (Gheorghe and Ishii 2007; Jones et al. 2001; Kengo et al. 2001; Pascual and Stevels 2005, 2006)	Ecovalue is defined as the ratio between a monetary amount (price) and the environmental load over the life cycle of the product/service concerned. Ecovalue acknowledges market diversity, where consumers value different attributes in products. For this purpose, the criteria used to set priorities rely on market composition, consumption power, and a product's environmental load. Units used include retail price in monetary units and a product's environmental load, expressed in millipoints (mPt)
Eco communication matrix (Stevels 2001)	Supports the development of the marketing and communication strategies. Most of the data necessary for completing the matrix are derived from the earlier phases of the project (benchmarking, ecodesign matrix, etc.). Comparisons can be made between products of different generations and/or of competing products. The rows of the matrix correspond to energy consumption, materials, packaging and transportation, substances, durability/recyclability, manufacturing and life cycle perspective. The columns of the matrix correspond to the company internal benefits, benefit to clients/customers and benefits to stakeholders and society. Each of these benefits are divided into tangible, intangible and emotion perception

LCA is one of the most well-known methods that can support ecodesign implementation (Brezet et al. 1999; Cappelli and Delogu 2006; Hunkeler and Vanakari 2000; Munoz et al. 2006). It provides a quantification of environmental aspects and impacts across the product life cycle and supports the between concepts and design options. LCA involves all successive stages of a product life cycle, ranging from extraction of raw materials through the environmental impacts of manufacturing, distribution and use of the product, all the way until its final disposal, which may include subsequent activities such as recycling of materials and components, plus other ways of treating post-consumption (Azapagic and Clift 1999).

LCA has gained broad acceptance in industry as a trustworthy method to quantify the environmental aspects and potential impacts of the life cycle of products. The LCA methodological framework is defined by ISO 14040 and 14044 standards (ISO 2006a, b), which describe the minimum requirements for its correct use and performance. The holistic systems perspective, which is applied in LCA, enables the company to disclose the 'problem shifting' which occurs when solutions to environmental problems at one place in a product's life cycle create new problems elsewhere in the life cycle (Jeswiet and Hauschild 2005).

23.3.1 LCA Challenges

It is important to understand the challenges and limitations of LCA in the context of ecodesign, in order to be able to use the two approaches to product environmental improvement. Many authors have written about the challenges of LCA within ecodesign (Alting et al. 2007; Keoleian et al. 1994; Portney 1993), which can be expressed in five main areas, as described in the following.

The first challenge relates to the dilemma of opportunity (and cost) versus knowledge (and numbers). In the early phases of the product development process, the cost of making a design decision is very low and the window of opportunity to affect ecodesign improvements and integrate environmentally enhancing features into the product is the largest (Bhander et al. 2003; Keoleian et al. 1994). However, by nature of this early phase of the project, it is here where we know the absolute least about our product, thus rendering it very difficult to quantify the contents of related processes for the manufacture of a product that has not yet even been fully conceptualised (Bhamra et al. 1999; Tchertchian et al. 2013). The later in the product development process one waits, the more quantitative the data one has to model in an LCA, yet the smaller the window of opportunity to affect any changes—and the higher the cost of doing so. A number of tools and guidelines exist, to bridge this opportunity-knowledge gap, but it remains a limitation. One obvious action is to perform an LCA on a previous product or a competitor's product, as there is almost always some product on the market with similar functionality and ingredients to the product we are designing.

Challenge number two relates to the required knowledge and competencies of the product developer (Portney 1993). LCA is in itself a detailed and highly specialised approach, belonging to a strong scientific knowledge domain (Ny et al. 2006). Ecodesign, within the context of product development, is usually dominated by well-trained, highly skilled and well-practiced designers and engineers, with competencies in the systematic design and development of products and systems (Diehl 2005; Hesselbach and Herrmann 2003). In other words, LCA is a highly analytical (natural science dominated) activity and ecodesign is a highly synthesis-oriented (technical, engineering) activity. LCA has two important places in a manufacturing company: (i) the environmental, health & safety (EH&S) function of the organisation, where reporting and high-level (maybe product family) assessments are carried out; and (ii) in the product development department, where the knowledge of products, processes, ecosystems and use scenarios provide important guidance for ecodesign. The challenge is always, how to ensure the right level of LCA knowledge in the mind of the product developer and/or how to compensate for the lack of LCA knowledge through a combination of bridging tools (Poudelet et al. 2012; Tchertchian et al. 2013), plus the pairing of LCA specialists with product development specialists. The solution to this challenge lies in strategic management recognition, paired with a tactical management prioritisation.

The third challenge with LCA in the ecodesign activity relates to completely new products, where we cannot rely on previous product releases or competitors' products, to create some form of benchmark for the ecodesign effort in especially the early phases of product development (Trappey et al. 2011). Completely new products are more of a marginal case, when compared to incrementally innovated products, but there are still examples that need environmental attention (e.g., electronic products, many clean-tech products, nano-based products, etc.). In such product development cases, LCA tends to play a lagging, rather than a leading role (Poudelet et al. 2012).

Challenge number four relates mostly to the misconception of the scope and merits of LCA. LCA is an analysis method—a very well developed and accepted one at that. Challenges arise when the company expects great product development and ecodesign advances, based alone on the results of an LCA (Brezet et al. 1999; Russo et al. 2014). It is not possible to "analyse oneself to a better product", i.e. to improve a product solely based on an analysis of its environmental performance, and therefore LCA should not be deployed as the only method to ecodesign improvement for a project. Instead, one should pair the analysis activity with the task of synthesis (product development), as the methods and tools exist in abundance, to support the good ecodesign process afterwards.

The fifth and final noteworthy challenge that LCA as a scientific field faces lies in the claims from "competing approaches". A popular claim within the current Cradle to Cradle (C2C) approach is that LCA belongs to the realm of eco-efficiency, which is a reductionist and limiting agenda, whereas C2C belongs to an agenda of positivism, growth and innovation (Hauschild 2015; McDonough and Braungart 2010; Rossi et al. 2006; see Chap. 25). Whilst the mental model created by this claim is compelling and easy to understand, it is not necessarily entirely useful. Ecodesign, LCA, C2C and a number of other approaches to enhancing environmental performance of products, can easily be used interchangeably and must respect each other's basic philosophies and scientific bases, if they really are to be deployed with successful environmental improvement to follow (Bakker et al. 2010; Bjørn and Hauschild 2011; Reay et al. 2011).

23.3.2 Using and Communicating LCA for Ecodesign in the Organisation

Referring again to the two key areas of a company in which LCA plays a role, namely the EH&S department and the product development department, it may be that the company has LCA information, data and results available, but the question is, how to access these data and what to use them for (Miettinen and Hämäläinen 1997). If a full LCA has been carried out, the final *improvement assessment* phase ought to point to areas of ecodesign priority for a particular product. However, it may be that only the inventory analysis and impact assessments have been carried out, which will pose difficulties for the ecodesign activity within the company, to directly use the data (Poudelet et al. 2012). The age and the scope of existing LCA studies in the company will also dictate their usefulness for the ecodesign process, but nevertheless, key focus areas should be possible to derive from the LCA activity (Chang et al. 2014).

Importantly also, the existence of LCA studies inside the company indicates that some previous attention has been given to the environmental improvement of products and processes in the organisation. Tracing the people, the projects and the types of data gathered and analysed will give a good idea of the intentions of the company with respect to ecodesign improvements and give good starting points for future activities. Especially understanding the goals and scope for existing LCA studies in the company is important, as it will uncover some details regarding the existing (or maybe earlier) environmental strategy of the company.

Identifying existing LCA results is one important task, and communicating these across the organisation is another. Understanding how to interpret and communicate LCA results is very important, in order to ensure that both management and product development professionals understand how to make further improvements based on the results calculated (Tingström and Karlsson 2006). There are a number of ways in which an LCA can be reported, and for the sake of ecodesign, it is important to choose a presentation of LCA studies in a way that supports the product development task. In other words, a presentation of environmental impact categories (global warming potential, air pollution, solid waste, etc.) is probably less useful for a product developer than presenting a more detailed model of the actual product or system (down to component level), showing comparative analyses, maybe with aggregated calculations (person equivalents, eco-points, energy, etc.). It may also be sufficient to carry out a faster but less detailed screening LCA to support the ecodesign task, as opposed to a much more trustworthy but also relatively time-consuming full product LCA (Simon et al. 2000).

23.3.3 Simplified Approaches Aimed At Integrating LCA Into Ecodesign

The practical use of environmental LCA methods and software tools in industry has revealed the need for simplification for product development projects. Hence, streamlined life cycle assessment methods have been derived from experience with the complex full methods (Jeswiet and Hauschild 2005).

Simplified LCA, also known as Streamlined LCA, has emerged over the years, as an efficient way to evaluate the environmental attributes of a product, process, or service life cycle. The aim of simplifying LCA is to provide essentially the same type of results as a detailed LCA, i.e. covering the whole life cycle, but in a superficial way (e.g. using qualitative and/or quantitative generic data), followed by a simplified assessment, thus reducing significantly the expenses and time expended.

Simplified LCA should still include all relevant aspects, but good explanations (e.g. company guidelines, materials negative lists or materials black lists, preferred mode of transportation) can to some extent replace resource-demanding data collection and treatment. The assessment should focus on the most important environmental aspects and/or potential environmental impacts and/or stages of the life cycle and/or phases of the LCA and give a thorough assessment of the reliability of the results (Zackrisson et al. 2008).

Full-scale LCA is traditionally quantitative. However, it is recognised that where quantification is not possible (for reasons of time, cost or data availability, for example), qualitative aspects can—and should—be taken into account (Heijungs et al. 2010). Simplified-LCA (S-LCA) is not meant to be a rigorous quantitative determination, but rather a tool for identifying environmental "hot spots" and highlighting key opportunities for effecting environmental improvements.

It is not complicated to apply quantitative and detailed LCAs to simple products, such as packaging, since they consist of few components or types of material, where information on most of the commonly used materials is available (and, if necessary, it is quick and easy to collect). For more complicated products, such as, e.g. televisions, a complete LCA may prove to be very resource-demanding and at the same time somewhat imprecise, due to the number of possible processes, materials, suppliers, etc., being very high and varied. Furthermore, the database on "not so common" materials is limited so for these cases, S-LCAs are more helpful, especially in the early stages of product development. In the case of improvements in already existing product systems, the use of (full) LCA may become easier, once data from a reference system can be used (with a well-known life cycle).

Streamlined approaches and other ecodesign methods and tools, such as design checklists and matrices, are essential to support ecodesign implementation in the early design phases. The practical use of these tools in product development depends on the nature and complexity of the product system (e.g. new vs. established), the product development cycle (time-to-market constraints), availability of technical and financial resources, and the design approach (integrated vs. serial).

These factors influence the role and scope of LCA in an ecodesign process. Effective communication and evaluation of environmental information and the integration of this information with cost, performance, cultural and legal criteria will also be critical to the success of design initiatives based on the life cycle framework.

Some examples of simplified LCA methods and tools are presented in Table 23.3. These methods and tools present a life cycle perspective and provide an analysis or comparison of the environmental impacts associated to a product, using or providing qualitative or semi-quantitative data. In order to avoid repetition, the simplified LCA methods and tools presented in Table 23.2 [STRETCH (Strategic Environmental Challenge), and MECO matrix)] are not replicated in Table 23.3.

Table 23.3 Simplified LCA methods/tools (Pigosso et al. 2011a, b)

Ecodesign method/tool	Summary	Criteria for assessment	Approach
Design abacus (Bhamra and Lofthouse 2007)	Used to rate a product on social, economic and environmental areas, in both the analysis and planning of a design. It helps you identify design goals, compare many design variables and compare different product designs across the product life cycle	Defined by the user (example: energy, material, usability, cost, life span, end of life)	Qualitative
Eco-compass technique (Sun et al. 2003)	Used to evaluate the environmental impact of an existing product. Combining the cost and benefit, a product's life locus tree can be built up and the environmental impact of a product is assessed on the performance of process and life stages of a product using these eight indices	Mass intensity, energy intensity, health and environmental potential risk, revalorisation, resource conversation, and service extension	Semi-quantitative
ECODESIGN checklist method (ECM) (Wimmer 1999)	Points out purposefully redesign tasks in order to increase the environmental performance of a product. Based on a holistic view of the product in three analysis levels (part-, function-,	Usability of product (customer's needs oriented), low energy consuming	Semi-quantitative

(continued)

Table 23.3 (continued)

Ecodesign method/tool	Summary	Criteria for assessment	Approach
	and product level) the method shows clearly, where the weak points of a product are and how to realize reuse, recycling of parts, where to integrate, omit or create functions and where to reduce consumption or increase efficiency, usability of the whole product	product (use stage), low resource consumption and avoiding waste (manufacturing stage),durable product, reuse of product-parts, recycling of product-materials	
Ecodesign web (Bhamra and Lofthouse 2007)	Provides a quick way of helping designers to identify which areas of the product should be focused on to improve its environmental performance. It works by comparing seven design areas with each other to identify a "better than"/"worse than" output	Materials selection, materials usage, distribution, product use, optimal life, end of life	Qualitative
Environmental design strategy matrix (EDSM) (Lagerstedt et al. 2003)	Identifies some design strategies based on characteristics of products at the different life cycle stages	Life cycle length, energy consumption, resource consumption, material requirement, configuration and disposal route	Qualitative
Green design advisor (GDA) (Ferrendier et al. 2002)	Provides a direction of improvement, as well as the design features with the highest improvement potential and shows the weak points, as well as good design features. Additional design guidelines exist; however, there are no automatically generated design alternatives	Number of materials, mass, recycled content, recyclability, toxicity, energy use, time for disassembly, disposal cost	Semi-quantitative
Green design tool (Kassahun et al. 1995)	Based on analysing "top level greenness attributes" of a product, providing to the designer an overview of the environmental status of product design. It can be applied using the basic concept of the product	Reusability, label, internal joints, material variety, material identification, recycled content, chemical usage, additives, surface	Semi-quantitative

(continued)

Table 23.3 (continued)

Ecodesign method/tool	Summary	Criteria for assessment	Approach
		finishes, external joints and hazards level of material	
MET matrix (Byggeth and Hochschorner 2006b)	Aims to find the most important environmental problems during the life cycle of a product, which can be used to define different strategies for improvement. The environmental problems should be classified into the categories	Material cycle, energy use, toxic emissions	Qualitative or quantitative
The environmentally responsible product assessment matrix (ERPA) (Hochschorner and Finnveden 2003)	The central feature of ERPA assessment is a 5 × 5 matrix. One dimension is the life cycle stages and the other is environmental concerns. The method can be used to evaluate products, processes, facilities, services or infrastructure. Each element of the matrix is assigned a rating from 0 (highest impact) to 4 (lowest impact), according to a checklist. The rating is based on the seriousness but also on whether possibilities of reducing impacts have been utilized or not	Materials choice, energy use, solid residues, liquid residues and gaseous residues	Semi-quantitative

23.4 Creating Goals and Measuring Progress with Ecodesign

Whether it be LCA-driven/supported or not, any ecodesign process is best supported if a set of measurable goals and performance indicators are established for the activity. These are a fundamental element of any successful ecodesign activity, as they can provide an early warning to prevent environmental damage (Issa et al. 2013).

The use of environmental performance indicators (EPIs) to monitor product performance is often identified as one of the successful factors for effective ecodesign implementation, as such indicators can help to pinpoint improvement opportunities and prevent environmental damage through the product under development (Fiksel et al. 1998; Herva et al. 2011).

Most methods and tools to measure the environmental performance of products, such as Life Cycle Assessment (LCA), still present high complexity and large data requirements to the ecodesign activity (Hur et al. 2005), providing results that can be classified as *lagging EPIs*. Lagging EPIs measure the product's impacts on the environment, as a final result of a process.

In contrast to lagging EPIs, *leading EPIs* aim to produce simpler measures of environmental aspects that can inspire effective actions towards improving products' environmental performance. Environmental aspects are defined as elements of the organisation's activities, products or services that interact with the environment. In this context, the use of leading product-related EPIs can be seen as a simpler and faster quantitative approach to ensure performance measurement and improvement during the product development process (Bovea and Perez-Belis 2012).

Databases of leading product-related EPIs exist, in some cases including accompanying guides, with the aim of supporting companies to select the most relevant EPIs, based on the developed products and their strategies (see e.g. Issa et al. 2015).

23.5 Carrying Out Ecodesign, Step by Step

Many approaches and processes are advocated and published, proposing a process of ecodesign (Bhamra et al. 1999; Brones and Carvalho 2015; Herva et al. 2012; Kengpol and Boonkanit 2011; Luttropp and Lagerstedt 2006; Pigosso et al. 2013a, b; Poole et al. 1999; Rio et al. 2013), and in at least a large company setting, the absolute most important route to success is to integrate ecodesign decisions and considerations into the established product development process for the company (Pigosso et al. 2013a, b). However, a generic process of decision-making and ecodesign implementation can be derived from the various ways and processes suggested.

One such approach has been published by the Technical University of Denmark (DTU), as the result of a sponsored campaign by the Danish Environmental Protection Agency and the Confederation of Danish Industry. The approach, named *Environmental Improvement through Product Development: A Guide* (McAloone and Bey 2011), describes seven generic steps towards ecodesign implementation, and is created based on a detailed analysis of other existing approaches, plus a number of trial implementations in industry. The following gives a summary of this approach, which steps the user through an analytical point of departure, through a creative-synthesis ecodesign approach, before considering how to implement the proposed ecodesign changes in the organisation on a more permanent basis. Examples of the application of each one of the proposed steps can be found in the guide.

23.5.1 *Environmental Improvement Through Product Development: A Seven-Step Approach*

The following seven steps guide the user through a solution-oriented process, towards environmental improvement. The seven generic steps are meant to provide a simple and inspiring way of approaching ecodesign, by isolating the task from the ordinary product development tasks in the company, the idea being to gain focus in the product development team, about the "ideal ecodesign approach" in a workshop setting. The approach attempts to create space for innovation by focusing solely on environmental issues. At the end of the approach, it is the intention that the practitioner considers how to integrate the seven steps into their own organisation's product development process.

The approach is constructed with a focus on:

- Gaining an *overview* of a current product's environmental problems;
- Providing *insight* into important details concerning the product's environmental impacts, its use and its users;
- Creating *solutions and concepts* that lead to environmental improvements; and
- Creating *foresighted proposals* for the creation of an environmental strategy for product development.

The approach is designed as a chain of exercises that ought to be completed from start to finish, in the order that the steps are presented. The approach charts an eco-redesign process, so as to ensure that there is an established benchmark product beforehand; it therefore requires that a product is chosen in advance as the object for environmental improvement. The product can be either an already marketed product, which will serve as a reference product, or a product that is currently under development. The first case is the simplest, as it is easier to identify data about the product's life cycle.

The first six steps of the seven-step approach isolate the environmental task and focus on identifying environmental problems. Subsequently improvement proposals are created. Step 7 provides a framework for an action plan and the basis for systematic integration of the proposed environmental improvements into the product development process. See Fig. 23.2.

Step 1: Describe the use context

As the very first exercise, it is important to reach a common understanding of the product and its value contribution under use. This provides a common starting point for discussions about the environmental product improvement possibilities for the product, for use later in the process, when the creation of product alternatives is in focus. It is important that the alternatives created meet the same requirements for the customer. Redundant product attributes should be considered as waste, both from an environmental and a customer perspective. Step 1 is intended to reach a

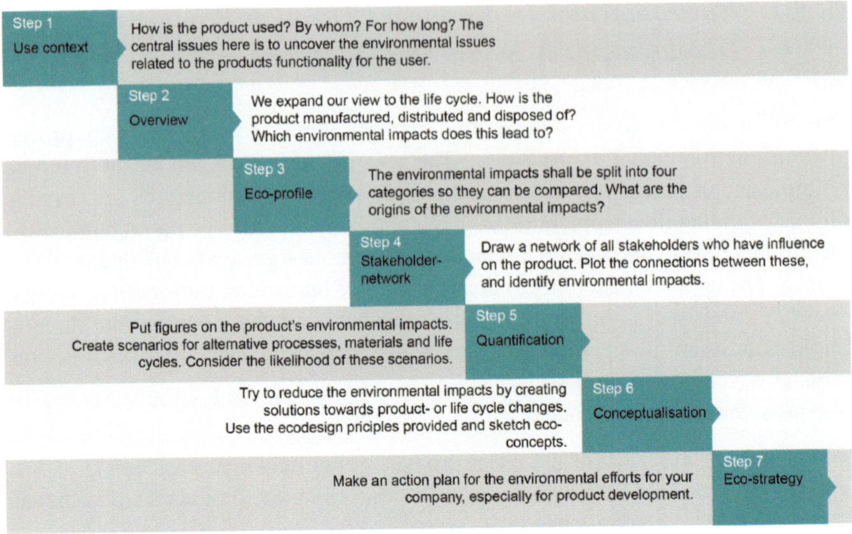

Fig. 23.2 Seven-step approach to ecodesign, as described by McAloone and Bey (2011)

description of the product's functionality to the user. This description provides the benchmark for all subsequent decisions and can also be used when, for example, alternative concepts shall be compared.

The description of the use context is achieved by answering the following questions:

- "*What should the product be used for*?" which leads to a description of the basic task that the product must carry out for the user.
- "*What does the product do*?" which allows for a description of the product's functionality, including the technological principle and the features that the product must possess in order to deliver the service to the user.
  - "... *For whom*?" leads to a description of the main user or user group.
  - "... *How long*?" and "... *how often*?" lead to a definition of the time frames in which the product must operate.
  - "... *Where in the world*?" leads to a definition of the geographical area in which the product must operate.

The responses to the above questions lead to a clear description of the product in the form of the *value contribution that the product delivers to the user*, or in other words, the product's *functional unit*.

Step 2: Create an overview of the environmental impacts

In this step, the aim is to create an overview of the product's life cycle and all significant environmental impacts that may occur throughout the life cycle of the product. A product life cycle typically consists of five main stages:

- *Materials* covers materials extraction and manufacturing (e.g. the manufacture of plastic granules from crude oil) and semi-finished products (e.g. steel profiles from iron ore), etc.
- *Manufacture* includes the purchase of components, plus the manufacturing and assembly processes, both at suppliers and in in-house production facilities.
- *Transport* covers the entire logistics chain, from suppliers to the end-user and beyond, including distribution activities by ships, trains, planes, trucks, vans and cars.
- *Use* includes the actual usage and possible ancillary products that are necessary for the product to perform its function (e.g. paper filters for a coffee maker). The use stage also includes installation and possible maintenance activities.
- *Disposal* includes reuse/recycling, incineration and landfill. The actual distribution of these disposal options depends on many factors, including regulatory requirements where the product is disposed of, who disposes of the product (an individual or a company), etc. It is obviously difficult to predict how the product will be disposed of, as this stage is typically far in the future.

Depending on preference, one may choose to integrate the transport life cycle stage into all of the other product life stages, as transport is in itself the "glue" between each life cycle stage. However, many choose to specify and gather all transport activities into one stage for itself, so as to (i) remember to pay attention to transport's environmental impacts and (ii) highlight the transport activity's contribution to the overall environmental footprint of the product.

Figure 23.3 shows a picture of how one could organise the overview of environmental impacts by means of adding sticky notes to a schematic of the product life cycle. The advantage of this approach is that it is a simple way of granting access to all team members in the product development team, to come with their own proposals of environmental impacts.

Step 3: Create your environmental profile and find root causes

Having created an overview of the product's main life cycle stages and environmental impacts in Step 2, the idea of Step 3 is to begin to categorise the identified environmental impacts according to their type. Subsequently the possible causes for the environmental impacts' emergence should be noted, before beginning to gather data on the environmental impacts that can be quantified.

The idea with this step is to create a more nuanced picture of the physical relationships that underpin each environmental focus area, than was created in Step

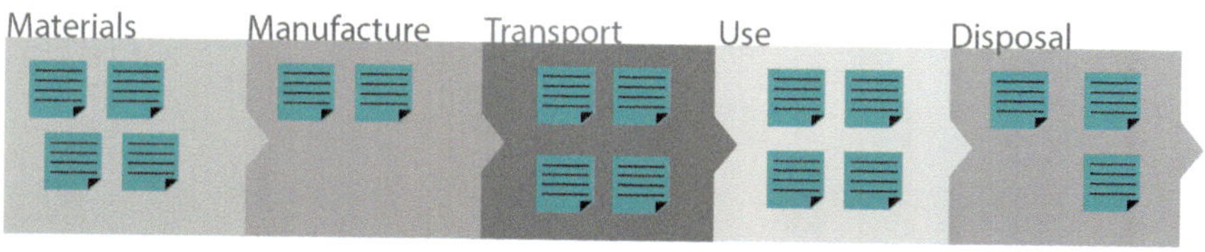

Fig. 23.3 Identifying environmental impacts throughout the product's life cycle

2. A number of focus areas can then be prioritised, based on the team's consideration of the need for action.

The already identified environmental impacts will now be organised into one of four categories: *M*aterials, *E*nergy, *C*hemicals or *O*ther (Field et al. 1993; Wenzel et al. 1997; Hochschorner and Finnveden 2003):

- *Materials:* This includes resource and disposal aspects of each life cycle stage, i.e. whether a material is based on a scarce resource, whether it can be easily recycled, or whether it must be landfilled, etc. Remember also to consider whether ancillary materials are used, particularly in the use stage, e.g. paper filters for coffee makers.
- *Energy:* This includes energy sources and energy aspects in the product life cycle stages. There can, for example, be large differences in energy consumption for material processing, depending on whether one takes new or recycled raw materials into consideration. Remember also to consider component suppliers. The transport and use-related energy consumption is also recorded under this category.
- *Chemicals*: This includes chemical consumption and chemical-related emissions of each life cycle stage, such as toxic chemicals used in manufacture.
- *Other:* In this category all other aspects are noted, that one has chosen to consider. For example, health and safety in own (or suppliers') manufacturing plants, aspects related to Corporate Social Responsibility (CSR), or general economic concerns.

This categorisation of the environmental impacts is created in a so-called *MECO matrix* (Fig. 23.4).

The reason for first carrying out Step 2 before this categorisation in Step 3, is that the MECO matrix can be limiting for the product developer, who feels compelled to fill in the matrix in a systematic manner, from the top left to the bottom right. By inserting a Step 2 and simply identifying the environmental impacts at a more abstract life cycle stage level, the environmental impacts are in focus, rather than

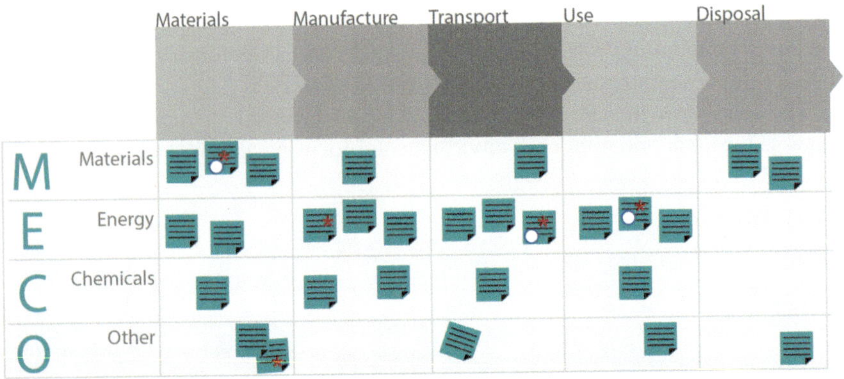

Fig. 23.4 MECO-matrix and product life résumé with environmental impacts categorised

their MECO categories. Having carried out Step 2, the sticky notes created can be simply moved down from the life cycle arrows to the respective MECO categories. One will then see that it is fine to have empty MECO cells, which may not actually have environmental impacts for the particular product under consideration.

Step 4: Sketch the stakeholder-network

Classic environmental efforts in companies take their point of departure in a product or a technology, directing special attention towards the improvement of the product's life cycle performance. This approach (which is represented in Steps 1–3) is a useful way of identifying environmental focus areas for the product itself, if one assumes that the product is used in a specific way, and in a specific context.

A weakness of taking the product-technology approach in isolation, however, is that it is built on a large series of assumptions, about the use, the user, the product development activity, the supply chain and many other stakeholders connected directly or indirectly to the product.

Step 4 of this approach therefore proposes that the product development team identifies the various stakeholders who are connected to a particular set of activities, within which the product plays a role. It is these stakeholders who experience "value" and "goodness" from the product. Environmental impacts often occur in the exchanges between stakeholders, e.g. in negotiations along the supply chain and/or as a result of lack of overview of the roles and responsibilities in the product's so-called *stakeholder-network*.

A stakeholder-network consists of several types of stakeholders, for example the manufacturing company, a component supplier, an external designer, a freight forwarder, the authorities, customers, users, a disposal company and so on.

Sketches of the stakeholder-network give an insight into which stakeholders are affected by certain environmental impacts. To clarify the relationships between stakeholders and the impacts that occur, one can outline *information exchanges, material flows* and the resulting *environmental impacts*.

Step 5: Quantify the environmental impacts

Many decisions about the product's environmental profile can be taken on the basis of experience, dialogue and scenario-building. At some stage, however, some of the judgments and choices in product development must be based on hard numbers and quantitative assessments.

Step 5 in the seven-step approach is therefore focused on the quantification activity, of the environmental impacts, with the help of a quantitative life cycle assessment technique. The figures created in this exercise are used to carry out an internal comparison of product alternatives and visualisation of the orders of magnitude between the impacts of the product across their life cycle stages and alternative life cycles. A detailed overview of LCA tools and a discussion of the different types of LCA (full LCA, simplified LCA) have been given earlier in this chapter. It is here where the LCA tools fit well into the ecodesign process, in order to provide overview and priority for ecodesign improvement action.

As previously mentioned and exemplified, there are many possible methods and tools to choose from, when quantifying the environmental impacts of a product at this stage of the ecodesign process. Ultimately, the choice of method will depend on:

- Who will apply the method: a product developer, an industrial designer or an environmental specialist;
- How much one knows about the product at the time of the use of the method;
- Whether one wishes to use a fully tailored computer tool, or whether a spreadsheet or pocket calculator would suffice.

Common for all methods is that one must define the important environmental impacts in the product's life cycle, and model these within the framework of the chosen method. Some methods include data on materials and processes, which ease the quantification task, especially if the method is software-based. Figure 23.5 shows an example of the outcome of the quantification task.

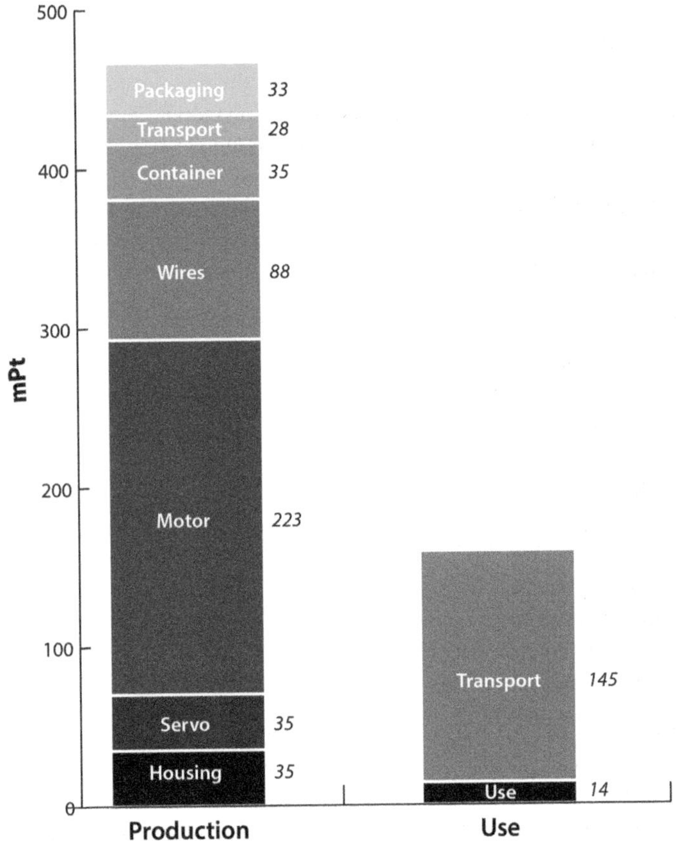

Fig. 23.5 Example of screening LCA exercise, to highlight key environmental focus areas during ecodesign

Step 6: Create environmental concepts

Step 6 of the ecodesign process described here is concerned with creating environmentally superior alternatives to those identified in Step 5, through the process of conceptualisation. This step is probably the closest to the normal product development and conceptualisation activity of all the seven steps described here. There are various tools available to aid this task:

- Approaches which are generic for any conceptualisation process, such as *brainstorming, brainwriting, sketching,* etc.
- The company may have *checklists* and *negative lists* (of materials, processes, chemicals), which prompt environmental thinking.
- Creation of *future scenarios* (e.g. "Outline the world's least energy-consuming house, which can be realised in the year 2020") in order to make a leap forward and perhaps find radical environmental concepts to back-cast from.
- The deployment of relevant *environmental principles* that can inspire and guide the environmental conceptualisation process.

There are hundreds of environmental principles to choose from when carrying out ecodesign (Pigosso et al. 2014; Vezzoli and Manzini 2008). For the sake of providing a generic list of exemplar principles, the seven-step ecodesign guide proposes ten environmental principles. The principles are meant as a way of viewing the "ideal" ecodesigned solution that the product developer could be expected to produce, if there were no other constraints in the product development process. The idea with these ten principles is to push the solution space for ecodesign to a large and creative space. It is clear that not all ten principles will ever be possible to fulfil for any one design; for this reason the concepts that may arise from following any one of the principles could be called "ideal concepts". The subsequent task for the ecodesigner would be to reconcile the ideal concepts arising from following each of the ten principles in turn, into a set of consolidated ecodesign concepts that consider as many as possible of the ten principles. The ten environmental principles are as follows:

- Reduce the *material intensity* of the product or service
  "By reducing the amount of material in the product, fewer material resources are required for manufacturing, the product requires less transport work, and there is less material to belandfilled or recycled. Attempt also to reduce the indirect material requirements, which are related to, e.g. the extraction of raw materials".
- Reduce the *energy intensity* of the product or service
  "As energy supply today is not based on 100% renewable sources, and as fuels are often fossil, the consumption of energy typically leads to environmental loads that can be reduced by changing the design".
- Reduce the *dispersion of harmful substances* through the product
  "Substances which in themselves are harmful, but which are used to achieve certain product characteristics—e.g. brominated flame retardants—can seep

from the product out into nature and into the food chain, for instance by evaporation".

- Increase the amount of *recycled and recyclable materials* in the product
  "It is a good idea to improve the possibility of recycling, for example by producing the product with few materials and by making them easily separable. At the same time, it is essential to apply increasing amounts of recycled materials in the product, since this will increase market demand for these materials".
- Optimise the product's *durability*
  "Unless the product has a very high environmental impact in the use stage, it is a good idea to make products that last for a long time, as this makes the production of new products for the same purpose unnecessary. At the same time, it is not useful to invest too much in the durability of products which are known to have a short use stage due to, for example, rapid technology obsolescence".
- Incorporate *environmental features* into the product
  "Make sure that the product is designed to reduce environmental loads, for instance by using standby functions, low-energy features or duplex features on printers".
- *Signal the product's environmental features* through the physical design
  "Make the product's environmental features visible to the user by, for example, placing the standby button on the front of the product or by setting the duplex mode as default in the printer driver settings".
- Maximise the use of *sustainable resources and supply chains*
  "Is there a link between recyclability of the product and use of recycled materials in the production? Do we know the origins of the materials and resources we use (with respect to both environmental and ethical standards)? Have we considered alternative materials on the merits of their environmental performance?"
- Optimise the product's *performance*
  "It is environmentally advantageous to combine several complementary functions in one product and to focus on the effectiveness of the product as a whole. The customer will evaluate the product's value, based on both its usability and its ability to efficiently meet a specific demand/desire. High perception of utility often leads to efficient use and increased durability of the product".
- *Design the life cycle first* and then the product
  "By thinking through all stages of the life cycle, one can achieve a very good understanding of the environmentally relevant properties the product must have, and these are then taken into account in the development process. Products which are developed on the basis of thorough knowledge of users' activities and needs have a better chance of achieving optimised life cycles and environmental profiles".

A simple example of the use of the above ten environmental principles would be when trying to produce an ecodesigned office chair. Following the first environmental principle (*reduce material intensity*), the ecodesigner may arrive at an ideal product concept, where the chair had very few materials, and maybe even the same materials throughout. By following a subsequent environmental principle, (e.g. *optimise the product's durability*), the ecodesigner may arrive at an ideal concept, where the chair's dimensions and materials choices were much more voluminous and hardwaring than for the first principle. The task of product development in general is dependent on creating sub-concepts that some way or other come into conflict with each other, in their ideal states, before applying compromises and design decisions, to end up at a final solution. The same idea is intended here for the environmental principles, where the creation of the final solution should be a product of the best possible consolidated solution to the principles posed. The office chair, in this example, would probably combine light-weight materials and smart structural designs, together with an easy maintenance (easy-wipe, sturdy, high scratch resistant) material choices, for as much of the chair's components as possible.

Step 7: Develop an environmental strategy

Steps 1–6 above describe an approach that isolates a product development team in an intensive ecodesign workshop-type activity. For environmental efforts, ideas and requirements to become rooted in the organisation, a strategy and prioritisation of efforts is required. Step 7 therefore prompts the product developer to reflect on the achievements gained in this activity, in order to consider which of the possible improvement proposals are worthy not just of the product currently being ecodesigned, but of being integrated generally into the company's product development process.

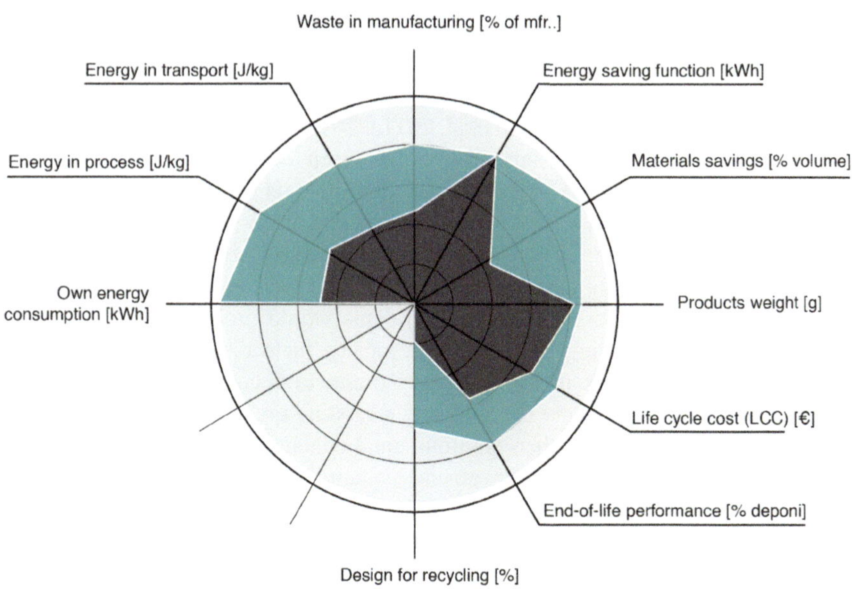

Fig. 23.6 First attempt at ecodesign strategy

The last task in this approach is therefore to decide on an environmental strategy, by using the product that has been the case of consideration for the first six steps in the approach, to attempt a generalisation of the environmental product development effort for the whole company. In the first instance this will have to be sketched out; later on it can be refined and made concrete, so that it can become a part of the company's strategic foundation and action plan.

Figure 23.6 shows an example of an initial sketch of an ecodesign strategy for a company. The next task is to begin to put real numbers and improvement goals onto the strategy wheel, in order to create a set of concrete ecodesign improvement goals for the company for an agreed period. The grey area represents the current profile, and the green area represents the targets for the future.

23.6 Final Remarks

This chapter has focused on the process of ecodesign in a company context. It has paid particular attention to the role that LCA plays in the ecodesign activity, and where the interplay between LCA and ecodesign can be optimised, to create efficient and effective improvements to products, processes and systems. It can be seen that a careful and systematic approach to integrated analysis (LCA) and synthesis (ecodesign) is optimal in order to achieve environmentally enhanced solutions through the ecodesign activity.

The chapter has discussed just a few examples of methods and tools, but has demonstrated through categorisation, that there are ways to support the product developer to choose the most suitable methods and tools according to the type of product, stage of the product development process and ambition of the product development team. Although, we have only scratched the surface of the many methods and tools that exist, there are several categorisations and collections of tools to be studied in literature (Pigosso et al. 2011a, 2014; Bovea and Pérez-Belis 2012).

Finally, a generic ecodesign process has been described. This process was chosen for this chapter due to its simplicity, to its balancing of analysis and synthesis, and because it has been tested in industry. It is important to remember, however, that it is seldom in an established manufacturing company that one would have the luxury of being able to carry out such a dedicated ecodesign activity for every single product development project. This would probably not be the most efficient way of prioritising efforts in the company. Instead, the task of the strategic and tactical practitioners in any company ought to focus on the *integration* of ecodesign considerations into the company's existing product development process (Pigosso et al. 2013a, b). When environmental enhancements are expected on the same level as cost saving achievements, quality enhancing design efforts and manufacturability considerations, we can be sure that ecodesign is truly integrated into the organisation's way of thinking about product development, business creation, and general purpose.

This chapter has had its primary focus on the *activity* of ecodesign, and not so much on the *purpose*. From one company to the next, the purpose/ambition for carrying out ecodesign may differ, whether it be to achieve a certain ecolabel, to adhere to a given legislation, to out-compete a competitor, or to achieve the ambitious goal of circular economy. Regardless of the purpose, the activity at the level of ecodesign remains the same. For simplicity—and to be loyal to the field, this chapter has stuck to the *discrete product* as the object of ecodesign. In recent years, the ecodesign activity has broadened, from single products, through the ecodesign of more complex systems (Cluzel et al. 2012), through services or combined product/service-systems (Lindahl et al. 2014), and to ecodesign of communities and societies. For all of these newer ecodesign focus areas, the basic principles remain the same, namely focus on life cycle; attention to a healthy balance of analysis and synthesis; choice of most suitable methods and tools; and application of common product development good practice, in order to achieve promising results. The good news is that the vast majority of methods, tools and guidelines for ecodesign now exist. The large and exciting task remains in mastering how to deploy an ecodesign process that results in environmentally excellent improvements, which are also great successes on the marketplace!

References

Alting, L., Hauschild, M., Wenzel, H.: Life Cycle Engineering and Management. In: Seliger, G. (ed.) Sustainability in Manufacturing: Recovery of Resources in Product and Material Cycles, pp. 31–67. Springer, Berlin (2007)

Andreasen, M.M., Hein, L.: Integrated Product Development. Springer, Berlin (1987)

Azapagic, A., Clift, R.: Life cycle assessment as a tool for improving process performance: a case study on boron products. Int. J. Life Cycle Assess. **4**, 133–142 (1999)

Bakker, C.A., Wever, R., Teoh, C., De Clercq, S.: Designing cradle-to-cradle products: a reality check. Int. J. Sustain. Eng. **3**, 2–8 (2010)

Baumann, H., Boons, F., Bragd, A.: Mapping the green product development field: engineering, policy and business perspectives. J. Clean. Prod. **10**, 409–425 (2002). doi:10.1016/S0959-6526(02)00015-X

Bey, N., Hauschild, M.Z., McAloone, T.C.: Drivers and barriers for implementation of environmental strategies in manufacturing companies. CIRP Ann. Manuf. Technol. **62**, 43–46 (2013). doi:10.1016/j.cirp.2013.03.001

Bhamra, T., Lofthouse, V.: Design for Sustainability: A Practical Approach, 1st edn. Gower Publishing Limited, Hampshire (2007)

Bhamra, T., Evans, S., McAloone T.: Integrating environmental decisions into the product development process. I. The early stages. In: EcoDesign'99: First 1–5 (1999)

Bhander, G.S., Hauschild, M., McAloone, T.: Implementing life cycle assessment in product development. Environ. Prog. **22**, 255–267 (2003). doi:10.1002/ep.670220414

Bjørn, A., Hauschild, M.Z.: Cradle to cradle and LCA—is there a conflict? In: Glocalized Solutions for Sustainability in Manufacturing—Proceedings of the 18th CIRP International Conference on Life Cycle Engineering, pp. 599–604 (2011)

Boks, C., Diehl, J.C.: EcoBenchmarking for all. In: 4th International Symposium on Environmentally Conscious Design and Inverse Manufacturing, 2005. EcoDesign'05, pp. 792–798 (2005)

Boks, C., Stevels, A.: Theory and practice of environmental benchmarking in a major consumer electronics company. Benchmarking Int. J. **10**, 120–135 (2003)

Bovea, M.D., Perez-Belis, V.: A taxonomy of ecodesign tools for integrating environmental requirements into the product design process. J. Clean. Prod. **20**, 61–71 (2012). doi:10.1016/j.jclepro.2011.07.012

Brezet, H., Stevels, A., Rombouts, J.: LCA for EcoDesign: the Dutch experience. Ab Stevels design for sustainability program, Delft University of Technology. In: First International Symposium on Environmentally Conscious Design and Inverse Manufacturing, 1999. Proceedings. EcoDesign'99, pp. 36–40 (1999)

Brones, F., de Carvalho, M.M.: From 50 to 1: integrating literature toward a systemic ecodesign model. J. Clean. Prod. **96**, 44–57 (2015). doi:10.1016/j.jclepro.2014.07.036

Byggeth, S., Hochschorner, E.: Handling trade-offs in Ecodesign tools for sustainable product development and procurement. J. Clean. Prod. **14**, 1420–1430 (2006). doi:10.1016/j.jclepro.2005.03.024

Cappelli, F., Delogu, M.: Integration of LCA and EcoDesign guideline in a virtual CAD framework. In: Proceedings of Life Cycle Engineering, pp. 185–188 (2006)

Chang, D., Lee, C.K.M., Chen, C.-H.: Review of life cycle assessment towards sustainable product development. J. Clean. Prod. **83**, 48–60 (2014)

Cluzel, F., Yannou, B., Leroy, Y., Millet, D.: Proposition for an adapted management process to evolve from an unsupervised life cycle assessment of complex industrial systems towards an eco-designing organisation. Concurr. Eng. **20**, 111–126 (2012). doi:10.1177/1063293X12446663

Communities C of the E: Green paper on integrated product policy. Office for Official Publications of the European Communities, Brussels (2001)

Cramer, J., Stevels, A.: Strategic environmental product planning within Philips sound & vision. Environ. Qual. Manag. **7**, 91–102 (1997)

de Caluwe, N.: Business benefits from applied EcoDesign. IEEE Trans. Electron. Packag. Manuf. **27**, 215–220 (2004). doi:10.1109/AGEC.2004.1290904

Diehl, J.C.: Ecodesign knowledge transfer: how to take the economical and cultural context of the receiver into consideration. In: Fourth International Symposium on Environmentally Conscious Design and Inverse Manufacturing, 2005. Eco Design 2005, pp. 86–92. IEEE (2005)

Donnelly, K., Beckett-furnell, Z., Traeger, S., et al.: Eco-design implemented through a product-based environmental management system. J. Clean. Prod. **14**, 1357–1367 (2006a). doi:10.1016/j.jclepro.2005.11.029

Donnelly, K., Olds, R., Blechinger, F., et al.: ISO 14001—effective management of sustainable design. J. Sustain. Prod. Des. **4**, 43–54 (2006b). doi:10.1007/s10970-006-0004-9

Eagan, P., Finster, M.: Creating Business Value by Linking Industrial Ecology with Business Strategy and Product Design (IEEE Keynote) (2001). doi:10.1109/ECODIM.2001.992478

Ernzer, M., Oberender, C., Birkhofer, H.: Methods to support EcoDesign in the product development process. In: 3rd International Congress of Design Research, Rio de Janeiro, p. 13 (2005)

Ferrendier, S., Mathieux, F., Rebitzer, G., et al.: Eco-design Guide: Environmentally Improved Product Design Case Studies of the European Electrical and Electronics Industry, ECOLIFE Thematic Network, pp. 21–34. (2002)

Field, F., Isaacs, J.A., Clark, J.P.: Life cycle design and its role in product and process development. Int. J. Environ. Conscious Des. Manuf. **2**(2), 13–20 (1993)

Fiksel, J., Mcdaniel, J., Spitzley, D.: Measuring product sustainability. J. Sustain. Prod. Des. **6**, 7–18 (1998)

Gheorghe, R., Ishii, K.: Eco-design value alignment-keys to success. In: ASME 2007 International Mechanical Engineering Congress and Exposition (IMECE2007), pp. 267–277 (2007)

Hauschild, M.Z.: Better—but is it good enough? On the need to consider both eco-efficiency and eco-effectiveness to gauge industrial sustainability. Proc. CIRP **29**, 1–7 (2015)

Heijungs, R., Huppes, G., Guinée, J.B.: Life cycle assessment and sustainability analysis of products, materials and technologies. Toward a scientific framework for sustainability life cycle analysis. Polym. Degrad. Stab. **95**, 422–428 (2010). doi:10.1016/j.polymdegradstab.2009.11.010

Herva, M., Franco, A., Carrasco, E.F., Roca, E.: Review of corporate environmental indicators. J. Clean. Prod. **19**, 1687–1699 (2011). doi:10.1016/j.jclepro.2011.05.019

Herva, M., Franco-Uria, A., Carrasco, E.F., Roc, E.: Application of fuzzy logic for the integration of environmental criteria in ecodesign. Expert Syst. Appl. **39**, 4427–4431 (2012). doi:10.1016/j.eswa.2011.09.148

Hesselbach, J., Herrmann, C.: Knowledge management as a support of an efficient Ecodesign. In: EcoDesign'03 2003 3rd edn., pp. 1–4 (2003)

Hochschorner, E., Finnveden, G.: Evaluation of two simplified life cycle assessment methods. Int. J. Life Cycle Assess. **8**, 119–128 (2003)

Hunkeler, D., Vanakari, E.: EcoDesign and LCA: Survey of current uses of environmental attributes in product and process development. Int. J. Life Cycle Assess. **5**, 145–151 (2000)

Hur, T., Lee, J., Ryu, J., Kwon, E.: Simplified LCA and matrix methods in identifying the environmental aspects of a product system. J. Environ. Manag. **75**, 229–237 (2005). doi:10.1016/j.jenvman.2004.11.014

ISO: Environmental Management—Life Cycle Assessment—Principles and Framework (ISO 14040). ISO, The International Organization for Standardization, Geneva (2006a)

ISO: Environmental Management—Life Cycle Assessment—Requirements and Guidelines (ISO 14044). ISO, The International Organization for Standardization, Geneva (2006b)

Issa, I.I., Pigosso, D.C.A., Mcaloone, T.C., Rozenfeld, H.: Product-related environmental performance indicators : a systematic literature review. In: Ecodesign 2013, Jeju, South Korea (2013)

Issa, I.I., Pigosso, D.C.A., McAloone, T.C., Rozenfeld, H.: Leading product-related environmental performance indicators: a selection guide and database. J. Clean. Prod. **108**, 321–330 (2015). doi:10.1016/j.jclepro.2015.06.088

Jeswiet, J., Hauschild, M.: EcoDesign and future environmental impacts. Mater. Des. **26**, 629–634 (2005). doi:10.1016/j.matdes.2004.08.016

Johansson, G.: Success factors for integration of ecodesign in product development: a review of state of the art. Environ. Manag. Health **13**, 98–107 (2002). doi:10.1108/09566160210417868

Jones, E., Stanton, N.A., Harisson, D.: Applying structured methods to eco-innovation. An evaluation of the Product Ideas Tree diagram. Des. Stud. **22**, 519–542 (2001). doi:10.1016/S0142-694X(01)00007-2

Kassahun, B., Saminathan, M., Sekutowski, J.C.: Green design tool. In: Proceedings of the 1995 IEEE International Symposium on Electronics and the Environment ISEE (Cat. No. 95CH35718), pp. 118–125. IEEE (1995)

Kengo, K., Rei, S., Kenji, Y., Satoko, W.: Eco-value as an indicator for sustainable design. In: Proceedings Second International Symposium on Environmentally Conscious Design and Inverse Manufacturing, pp. 1106–1109 (2001). doi:10.1109/ECODIM.2001.992532

Kengpol, A., Boonkanit, P.: The decision support framework for developing Ecodesign at conceptual phase based upon ISO/TR 14062. Int. J. Prod. Econ. **131**, 4–14 (2011). doi:10.1016/j.ijpe.2010.10.006

Keoleian, G., Glantschnig, W.J., McCann, W:. Life cycle design: AT&T demonstration project. In: Proceedings of the International Symposium on Electronics and the Environment, San Francisco, USA (1994)

Knight, P., Jenkins, J.: Adopting and applying eco-design techniques: a practitioners perspective. J. Clean. Prod. **17**, 549–558 (2009). doi:10.1016/j.jclepro.2008.10.002

Lagerstedt, J., Luttropp, C., Lindfors, L.-G.: Functional priorities in LCA and design for environment. Int. J. Life Cycle Assess. **8**, 160–166 (2003)

Lindahl, M.: E-FMEA—a new promising tool for efficient design for environment. In: Proceedings First International Symposium on Environmentally Conscious Design and Inverse Manufacturing, Tokyo, pp 734–740 (1999)

Lindahl, M.: Environmental Effect Analysis (EEA)—An Approach to Design for Environment Licentiate Thesis in Industrial Ecology, Department of Chemical Engineering and Technology, Royal Institute of Technology, Stockholm, Sweden, ISBN 9 1-973906-2-3 (2000)

Lindahl, M., Sundin, E., Sakao, T.: Environmental and economic benefits of Integrated Product Service Offerings quantified with real business cases. J. Clean. Prod. **64**, 288–296 (2014). doi:10.1016/j.jclepro.2013.07.047

Luttropp, C., Lagerstedt, J.: EcoDesign and the ten golden rules: generic advice for merging environmental aspects into product development. J. Clean. Prod. **14**, 1396–1408 (2006). doi:10.1016/j.jclepro.2005.11.022

Mcaloone, T.C., Bey, N.: Environmental improvement through product development—a guide, 1st edn. Denmark, Copenhagen (2011)

McDonough, W., Braungart, M.: Cradle to Cradle: Remaking the Way We Make Things. Macmillan, New York (2010)

Miettinen, P., Hämäläinen, R.P.: How to benefit from decision analysis in environmental life cycle assessment (LCA). Eur. J. Oper. Res. **102**, 279–294 (1997)

Munoz, I., Rieradevall, J., Domenech, X., Gazulla, C.: Using LCA to assess eco-design in the automotive sector—Case study of a polyolefinic door panel. Int. J. Life Cycle Assess. **11**, 323–334 (2006). doi:10.1065/lca.2005.05.207

Nielsen, P.H., Wenzel, H.: Integration of environmental aspects in product development: a stepwise procedure based on quantitative life cycle assessment. J. Clean. Prod. **10**, 247–257 (2002)

Ny, H., MacDonald, J.P., Broman, G., et al.: Sustainability constraints as system boundaries—an approach to making life-cycle management strategic. J. Ind. Ecol. **10**, 61–77 (2006). doi:10.1162/108819806775545349

Pascual, O., Stevels, A.: Ecodesign operationalization and company performance in electronics industry. In: Fourth International Symposium on Environmentally Conscious Design and Inverse Manufacturing, 2005. Eco Design 2005, pp. 807–813 (2005)

Pascual, O.A., Stevels, A.: Maximizing profitability with ecovalue. In: Proceedings of Eco Design 2006 Asia Pacific Symposium. NPO EcoDesign Promotion, Network, Tokyo (2006)

Pigosso, D.A., Grandi, C., Rozenfeld, H.: Strategic implementation of design for environment at Embraer. In: 8th International Symposium on Environmentally Conscious Design and Inverse Manufacturing, Jeju, South Korea, pp. 2–5 (2013a)

Pigosso, D.C.A., McAloone, T.C., Rozenfeld, H.: Systematization of best practices for ecodesign implementation. In: International Design Conference—DESIGN 2014, Dubrovnik, Croatia, pp. 1651–1662 (2014)

Pigosso, D.C.A., Rozenfeld, H., Seligerl, G.: EcoDesign Maturity Model: criteria for methods and tools classification. In: Seliger, G., Khraisheh, M.M.K., Jawahir, I.S. (eds.) Advances in Sustainable Manufacturing, 1st edn., pp. 239–243. Springer, Berlin (2011a)

Pigosso, D.C.A., Souza, S.R., Ometto, A.R., Rozenfeld, H.: Life Cycle Assessment (LCA): discussion on full-scale and simplified assessments to support the product development process. In: 3rd International Workshop—Advances in Cleaner Production, São Paulo (2011b)

Pigosso, D.C.A., Rozenfeld, H., McAloone, T.C.: Ecodesign maturity model: a management framework to support ecodesign implementation into manufacturing companies. J. Clean. Prod. **59**, 160–173 (2013). doi:10.1016/j.jclepro.2013.06.040

Poole, S., Simon, M., Sweatman, A, et al.: Integrating environmental decisions into the product development process: part 2 the later stages. In: Proceedings. EcoDesign'99: First International Symposium on Environmentally Conscious Design and Inverse Manufacturing, 1999, pp. 334–337. IEEE (1999)

Portney, P.R.: The price is right making use of life cycle analysis. Issues Sci. Technol. **10**(2), 69–75 (1993)

Poudelet, V., Chayer, J.-A., Margni, M., et al.: A process-based approach to operationalize life cycle assessment through the development of an eco-design decision-support system. J. Clean. Prod. **33**, 192–201 (2012). doi:10.1016/j.jclepro.2012.04.005

Reay, S.D., Mccool, J.P., Withell, A.: Exploring the feasibility of cradle to cradle (product) design: perspectives from New Zealand scientists. J. Sustain. Dev. **4**, 36–44 (2011)

Rio, M., Reyes, T., Roucoules, L.: Toward proactive (eco)design process: modeling information transformations among designers activities. J. Clean. Prod. **39**, 105–116 (2013). doi:10.1016/j.jclepro.2012.07.061

Rossi, M., Charon, S., Wing, G., Ewell, J.: Design for the next generation incorporating cradle-to-cradle design into Herman Miller products. J. Ind. Ecol. **10**, 193–210 (2006). doi:10.1162/jiec.2006.10.4.193

Russo, D., Rizzi, C., Montelisciani, G.: Inventive guidelines for a TRIZ-based eco-design matrix. J. Clean. Prod. **76**, 95–105 (2014). doi:10.1016/j.jclepro.2014.04.057

Simon, M., Poole, S., Sweatman, A., et al.: Environmental priorities in strategic product development. Bus. Strategy Environ. **9**(6), 367–377 (2000)

Stevels, A.: Adventures in ecodesign of electronic products (1993–2007). Ipskamp, Enschede (2007)

Stevels, A.: Application of EcoDesign: ten years of dynamic development. In: Proceedings of the Second International Symposium on Environmentally Conscious Design and Inverse Manufacturing, pp. 905–915 (2001)

Sun, J., Han, B., Ekwaro-Osire, S., Zhang, H.-C.: Design for environment: methodologies, tools, and implementation. J. Integr. Des. Process. Sci. Manuf. Des. **7**, 59–75 (2003)

Tchertchian, N., Yvars, P.-A., Millet, D.: Benefits and limits of a constraint satisfaction problem/life cycle assessment approach for the ecodesign of complex systems: a case applied to a hybrid passenger ferry. J. Clean. Prod. **42**, 1–18 (2013). doi:10.1016/j.jclepro.2012.10.048

Tingström, J., Karlsson, R.: The relationship between environmental analyses and the dialogue process in product development. J. Clean. Prod. **14**, 1409–1419 (2006). doi:10.1016/j.jclepro.2005.11.012

Trappey, A.J.C., Ou, J.J.R., Lin, G.Y.P., Chen, M.-Y.: An eco- and inno-product design system applying integrated and intelligent qfde and triz methodology. J. Syst. Sci. Syst. Eng. **20**, 443–459 (2011). doi:10.1007/s11518-011-5176-8

van Weenen, J.: Towards sustainable product development. J. Clean. Prod. **3**, 95–100 (1995)

Vezzoli, C., Manzini, E.: Design for Environmental Sustainability, 1st edn. Springer, London (2008)

Wenzel, H., Hauschild, M., Alting, L.: Environmental Assessment of Products. Kluwer Academic Publishers, Dordrecht (1997)

Wever, R., Boks, C., van Es, H., Stevels, A.: Multiple environmental benchmarking data analysis and its implications for design: a case study on packaging. In: Fourth International Symposium on Environmentally Conscious Design and Inverse Manufacturing, 2005. Eco Design 2005, pp. 799–806 (2005)

Wever, R., Boks, C., Marinelli, T., Stevels, A.: Increasing the benefits of product-level benchmarking for strategic eco-efficient decision making. Benchmarking Int. J. **14**, 711–727 (2007). doi:10.1108/14635770710834509

Wimmer, W., Pamminger, R., Stachura, M., Grab, R.: ECODESIGN in the electronics industry—achieving legal compliance with the EU-directives and environmentally improving products by using the new EEE-PILOT. In: Fourth International Symposium on Environmentally Conscious Design and Inverse Manufacturing, 2005. Eco Design 2005, pp. 671–677 (2005)

Wimmer, W.: The ECODESIGN Checklist Method : A Redesign Tool for Environmental Product Improvements. In: First International Symposium On Environmentally Conscious Design and Inverse Manufacturing, 1999. Proceedings. EcoDesign'99, pp. 685–688 (1999)

Zackrisson, M., Rocha, C., Christiansen, K., Jarnehammar, A.: Stepwise environmental product declarations: ten SME case studies. J. Clean. Prod. **16**, 1872–1886 (2008). doi:10.1016/j.jclepro.2008.01.001

Author Biographies

Tim C. McAloone actively working into the fields of ecodesign, eco-innovation, product/service-systems and product development since the late 1990s, with a particular focus on innovation and sustainability strategies. Has led a series of research projects and innovation networks including industrialists, academics and other organisations.

Daniela C.A. Pigosso has been working with ecodesign, eco-innovation sustainable design and product/service-systems since the late 2000s. Developed a maturity-based framework for ecodesign and sustainable design. Main interests are on the managerial integration of sustainability best practices into business and innovation processes.

Chapter 24
Environmental Labels and Declarations

Jeppe Frydendal, Lisbeth Engel Hansen and Alexandra Bonou

Abstract Based on the terminology and structure developed by the International Organization for Standardization, a description is given on the types of ecolabels that build on life cycle assessments. Focus is on type I labels that point out products and services with an overall environmental preferability within a specific product category. Type I labels include official labels set up by government and international institutions. Examples are given on operation of labelling schemes, development and focus area for criteria that must be met to obtain a label, effects on environment and legislation of labelling, the use of ecolabels in marketing, and the way ecolabels help build a market for "greener products". Type III labels—or Environmental Product Declarations—are also briefly described with indicative examples from the building sector, a declaration for office furniture, and an introduction is given to the European Commission's programme for product—and organisational environmental footprints.

Learning Objectives
After studying this chapter, the reader should be able to:

- Explain the process of ecolabelling design and development
- Know the main types of environmental labelling as they are standardised by ISO
- Understand the purpose of ecolabels in communicating sustainability efforts
- Explain the main challenges in using ecolabelling for decision support
- Understand the extensive variety of environmental labelling schemes
- Understand the purpose and the importance of harmonising ecolabelling schemes.

J. Frydendal (✉) · L.E. Hansen
Danish Standards Foundation, Göteborg Plads 1, Nordhavn, Denmark
e-mail: jf@ecolabel.dk

A. Bonou
Department of Management Engineering, Technical University of Denmark, Kongens Lyngby, Denmark

M.Z. Hauschild et al. (eds.), *Life Cycle Assessment*,
DOI 10.1007/978-3-319-56475-3_24

24.1 Introduction

Environmental labels and declarations are a form of sustainability performance measure. They provide information about certain environmental aspects of a product or service to non-environmental experts. The intention is to inform and influence consumer and professional purchasers to take into account such concerns when choosing between products and services. By changing their consumption patterns in favour of environmentally friendlier options consumers and professional purchasers can thus contribute to more sustainable consumption and support the goal of sustainable development (see Chap. 5).

Environmental labels have been developed since the 70s (first of them being the Blue Angel developed in 1978 by the German Federal Ministry of Interiors) due to a growing concern for the environment, both on government, business and consumer levels. But they evolved mostly after the late 80s and particularly after the '92 UN conference on environment and development when the adoption of the Agenda 21 put the target of sustainable development in the political dialogue globally (UN 1992; UNOPS 2009).

From a producer perspective, the growing environmental concern opened a new market opportunity and therefore the so-called "green marketing" emerged. This aimed to enhance company reputation to the consumers and show a responsible code of conduct by promoting products of presumable environmental superiority. To capitalise on environmentally friendlier practices companies used green claims, in many cases expressed in the form of some kind of product logo, declaration and labelling.

In the beginning, there were no standards or guidelines for developing and using such labels, which means that there was no credibility or validation of the claims. Thus, there was a risk for distorting the market and confusing the consumers if not misleading them, which is commonly referred to as "greenwashing". To tackle this challenge and provide an objective basis for verifying a company's claims about its product's performance, standardisation initiatives took place. Most significant is the development of the ISO 14000 family, which started already in 1991. The family contains more than 20 standards which are designed to guide a voluntary environmental management system (IISD 1996; ISO 2009). Additionally, since the 1990s governments in many countries have been introducing national or regional programmes or schemes for environmental labels. Labelling programmes have also been initiated by private companies and non-governmental organisations. The goal is to obtain environmental improvement by using market forces, and giving the consumers credible labels that point out products with a lower environmental burden.

Despite the existing efforts, at the UN summit in 2002, 10 years after the adoption of Agenda 21, a need was identified on a global scale to "develop and adopt, where appropriate, on a voluntary basis, effective, transparent, verifiable, non-misleading and non-discriminatory consumer information tools to provide information relating to sustainable consumption and production" (UN 2002). Therefore, the purpose of this chapter is to outline different types of such tools and to present the benefits from and the barriers for their implementation.

24.2 Types of Labels

The private initiative "the Ecolabel Index" (www.ecolabelindex.com) has listed more than 400 different labels from 199 countries and 25 industry sectors. Some of these deal with single environmental issues, for example in relation to carbon emissions or water consumption. Single-issue labels can be used by a variety of sectors, e.g. a forest management label can be used for various wood-containing products. Single issues can also be sector specific like the organic cotton labels used in the textile industry, window energy rating schemes used in the building sector, or the dolphin-safe label specifically for tuna products. Other labels deal with multiple environmental issues. Such is the case of the EU and the Nordic ecolabels. These can be attributed to a broad range of products (cosmetics, white goods, windows, etc.). Multiple-issue labels can also be sector specific as seen in the case of electronic or ecotourism specific labels.

Regarding the organisations that administer the various labels there is regional variation and overlapping initiatives also occur. Taking the example of organic food, there are countries with comprehensive legislation and corresponding labelling such as the US, France, Canada, Denmark and Japan. EU has established related legislation and a label additionally to national schemes. In countries where a corresponding law is not in place, labelling is based on nonprofit organizations, private companies and others.

Beyond the environmental considerations there are also labels dealing with other dimensions of sustainability (see Chap. 5) touching upon ethics and social issues. For example the Coalition for Consumer Information on Cosmetics deals with animal testing while other labels certify wild life protection and animal welfare. Regarding human relations there are labels certifying fair trade conduct, fair labour practices, abolishment of child labour, socially responsible investing, just to name some examples.

While most labels are voluntary, in some cases, they can be *mandatory*. An example is the EU Energy label (regulated by the EU's Energy Labelling Directive) that shows the energy consumption efficiency of energy-related products such as white goods. Similar mandatory energy rating labels are found around the globe, e.g. in Australia and Singapore.

Figure 24.1 from Rubik and Frankl (2005) gives an overview of the different labelling types. It also indicates that consumers are exposed to a large number of different labels that follow different sustainability principles and meet different criteria. This variety in characterising products can be difficult to understand in the purchase situation. This chapter refers only to labels that deal with environmental issues and that are voluntary.

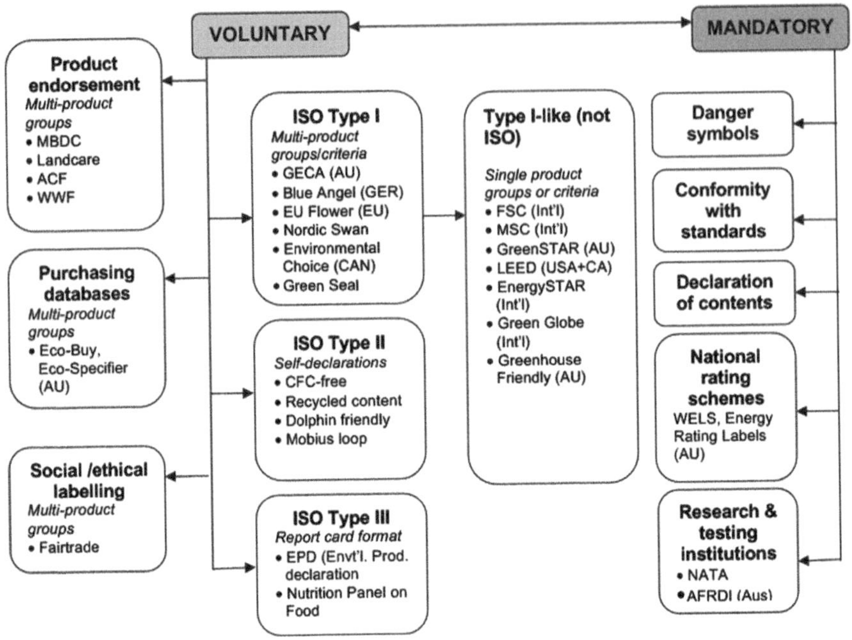

Fig. 24.1 Taxonomy of labels for communication of environmentally relevant information about products, adapted from Rubik and Frankl (2005)

24.2.1 ISO-Definitions

As discussed, to support voluntary initiatives the International Organization for Standardization (ISO) started a process of developing a set of rules and guidelines for different types of environmental labels. As part of the ISO 14000 series of environmental management standards, the ISO 14020 series defines three broad types of labels/declarations. Two of these ('Type I' and 'Type III') are life cycle based as part of the definition, whereas the last ('Type II') is not. Apart from these three types, ISO has also developed a general set of principles for labels that are neither of these types.

Type I Environmental Labelling (ISO 14024)

Type I environmental labels are defined as voluntary, multi-criteria-based and third party-verified labels that indicate an overall environmental preference in a life cycle perspective of a product or service within a specific product category (ISO 1999). This type of environmental label, or Ecolabel, is a tool to help the market stimulate continuous environmental improvements. All the label examples in Fig. 24.2 are Type I ecolabels.

European Flower

1992- European Commission/National competent bodies

Nordic Swan

1989-Nordic Council of Ministers

Blue Angel

1978- Government of Germany/independent bodies

China Environmental Labelling

1993- Government of China

Ecomark

1991- Government of India

Austrian ecolabel

1991- Government of Austria

EcoMark Japan

1989- Non-profit/Government of Japan

Good Environmental Choice Australia (GECA)

2000- Private non profit

ECOLOGO (North America)

1988- for profit

Fig. 24.2 Examples of type I ecolabels (name, year of establishment and management body)

Type II Self-declared Environmental Claims (ISO 14021)
The overall objective of the ISO standard 14021 is to harmonise the use of self-declared environmental claims and thereby try to reduce the number of inaccurate and misleading claims (ISO 2016). However, it is important to keep in mind that many self-declared environmental claims do not follow the standard of ISO 14021 and some might also be in conflict with the marketing regulation. As all other marketing claims—environmental claims are also regulated by e.g. the marketing legislation. Since they lack a foundation in LCA, the self-declared environmental claims will not be discussed further in this chapter.

Type III Environmental Declarations (ISO 14025)
Type III environmental declarations are quantified environmental data based on LCA (life cycle inventory data or impact assessment results) primarily intended for business-to-business communication for the purchaser to be able to compare the environmental performance of different products fulfilling the same function (ISO 2006a). This aim requires consistency between the studies underlying the declaration of the compared products, and the standard encourages harmonisation between different declaration schemes.

24.3 Ecolabels

To clarify the terminology, ecolabels are thus a subset of environmental labels that identify environmental preferability of a given product or service compared to other products in the same product group. It is the purpose of the ISO 14024 standard to ensure more consistent consumer information and credibility by setting a number of minimum requirements that a Type I ecolabelling scheme has to fulfil:

- It is based on the life cycle perspective;
- It is multi-criteria based (not only looking at single environmental issues, such as climate change)
- Environmental criteria are based on sound scientific and engineering principles. To ensure objectivity, a broad range of stakeholders is involved in the selection of criteria (e.g. government, consumers, industries, etc.);
- It includes functional requirements (fitness for use) to ensure a sufficient quality of labelled products and services;
- Criteria are time-limited and revised if the situation has changed (e.g. if new and better technologies have been introduced) to ensure that the criteria continuously support identification of the products that have an overall environmental preferability;
- There is transparency in all stages of its operation and development, which, e.g. includes, but is not limited to, the following aspects:
  - Publicly available criteria,
  - Public hearing or hearing among interested parties of criteria,
  - Information about the funding sources for the programme development,
  - Public listing of all certified products and services;
- It is accessible to all potential applicants;
- It involves third-party certification;
- There is compliance monitoring after the licence is awarded.

Out of the more than 400 labels mentioned in the introduction, which include all sustainability related labels, the Global Ecolabelling Network (GEN) (www.globalecolabelling.net) has less than 30 members worldwide. GEN is a non-profit membership association for Type I ecolabelling organisations (including governmental, non-governmental, non-profit, etc.). Members include the organisations that award the most-used Type I labels in the world such as the ecolabels of the EU, Germany, China, India, North America, Brazil and Australia. A more detailed example for the Nordic ecolabel is given in Box 24.1.

Box 24.1 The Nordic Ecolabel—The Official Ecolabel in the Nordic Countries

The Nordic Ecolabel is the official Ecolabel of the Nordic countries but can be used and is recognised globally.

It was initiated in 1989 by the Nordic Council of Ministers with the purpose of providing an environmental labelling scheme that would contribute to a sustainable consumption and production.

The Nordic Ecolabel is a voluntary, positive ecolabelling of products and services and was also initiated as a practical tool for consumers to help them actively choose environmentally-sound products. The Nordic Ecolabel is an ISO 14024 type I ecolabel and Nordic Ecolabelling is a third-party verification body.

The Nordic Ecolabel is well established and internationally recognised. Annual consumer surveys show that 90–95% of the consumers in the Nordic Countries recognise the Nordic Ecolabel.

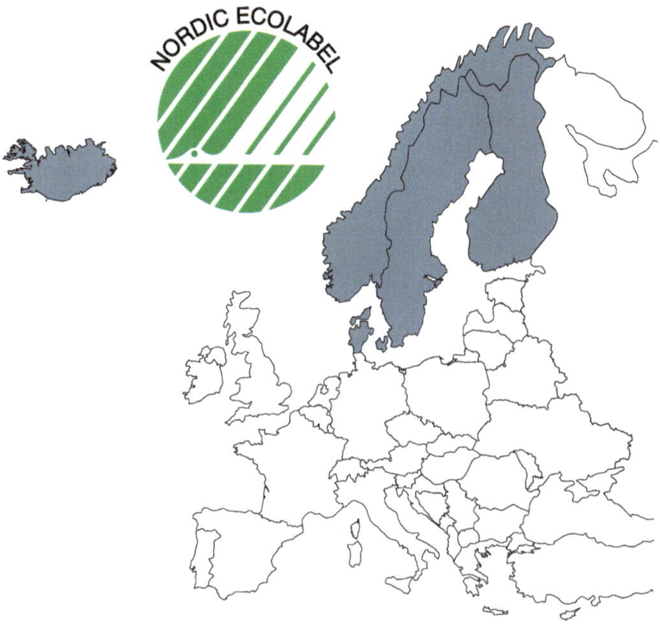

24.3.1 Criteria for Type I Ecolabels

Ecolabels ensure consistency with the two main principles of LCA presented in the introduction to this book (see Chap. 2): (i) the life cycle perspective and (ii) the multiple environmental issues.

Life Cycle Perspective

The publicly available criteria document should list all the requirements that a product or service has to fulfil to be awarded an ecolabel and these criteria should cover the life cycle. For the example of printed matter products, Box 24.2 shows a comparison between the criteria for a Type I ecolabelling scheme and a labelling scheme dealing with a single environmental issue—sustainability of forestry.

Box 24.2 Two Types of Product Labelling for Printed Matter

There is a large number of labels in the graphic industry used by companies to promote the products as environmentally friendly. Type I ecolabels and raw material labels claiming sustainable forestry are among the most common. The figure below illustrates the difference between the Nordic Ecolabel for printed matter products—a Type I ecolabel and a raw material label focusing on sustainable forestry.

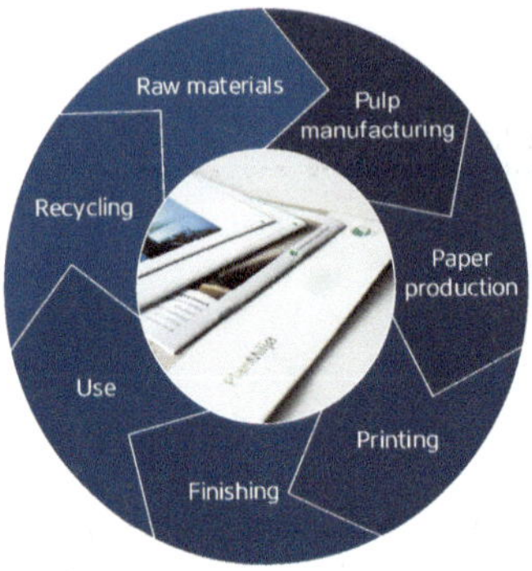

Life cycle stage of printed matter	Type I ecolabel	Raw material label requirements
Raw materials	• Sustainable forestry and/or recycled fibres in a high proportion of the paper used	• Sustainable forestry and/or recycled fibres
Pulp manufacturing	• Chemical requirements • Consumption of energy	
Paper production	• Chemical requirements • Consumption of energy	
Printing	• Printing inks and other chemicals • %-waste/shredded paper • energy consumption	
Finishing	• Glue and coating	
Use	• None directly, but the chemical requirements of the other stages will influence the chemical exposure of the user	
Disposal/recycling	• Design for recycling	

For a type I ecolabel, requirements are set for the environmental performance in all relevant stages of the life cycle whereas the raw material label only sets requirements for the forestry.

Since the purpose of type I ecolabel, is to drive improvement of the environmental performance of products, the criteria should focus on points where the manufacturer's decision affects the environmental performance. This means that requirements will not have to be set for all stages of the life cycle as it may not be possible for the manufacturer to influence the whole life cycle. For example, it is not relevant to set requirements for consumer behaviour of an ecolabelled product, but it is possible to set requirements for the product that allow and encourage the consumer to ensure low impacts in the use and disposal stages. To take the example of an ecolabelled laundry machine, such requirements could include:

- Low-temperature washing as standard programme;
- High spinning as standard (relevant especially in countries where tumble drying is common);
- Design for disassembly and recycling.

Multiple Environmental Impact Categories

Criteria for Type I ecolabelling also have to consider multiple environmental impact categories to prevent burden shifting from one impact to another which is a risk when there is single focus on one environmental impact category—such as climate change (see Chap. 2). Different environmental impact categories are considered depending on their relevance for each product group. Taking the example of the European ecolabel for hand dishwashing detergents, the criteria aim at promoting products that have a reduced impact on aquatic ecosystems while for the case of paints and varnishes, there is special focus on volatile and semi-volatile organic compounds (EC 2011a, 2014a). Note that different ecolabelling schemes use different classification schemes for product groups. If corresponding criteria have not been developed for a product group, the products within the group cannot be eligible for labelling or environmental claims (see also Sect. 24.3.4). For the EU ecolabel, every non-food and non-medical product marketed in the European economic market is entitled and various actors across the supply chain can apply (e.g. producers, manufacturers, retailers, wholesalers, importers). Labelling schemes can also be sector specific such as the Type I ecolabelling scheme operated by the nonprofit organization natureplus (www.natureplus.org), targeting building and accommodation with validity across the whole of Europe according to uniform criteria.

The specifics of environmental criteria as well as technical and quality requirements also differ across ecolabelling schemes. For the example of hand-washing detergents, the EU ecolabel provides formulas for calculating the toxicity to aquatic organisms using as an indicator the Critical Dilution Volume for all substances while the Australian Standard Good Environment Choice for Australia (GECA) does not. The requirements and upper limits of substances allowed to be used can also differ. For the quoted example, EU thus excludes the use of formaldehyde while GECA has an upper limit of 0.1% by weight. Additionally, the substances allowed to be used in the products can be regulated by different regulations (national and international standards), e.g. GECA requires that colourants in cleaning products are either compliant with certain EU regulation or are approved for use in foods by the Australian government standards (EC 2011a; GECA 2015).

24.3.2 Setting and Revising the Criteria

When setting new criteria or revising existing criteria, a public consultation with key experts and relevant stakeholders is compulsory, which may take substantial amounts of time and resources. In the case of the European ecolabel, the preparatory work, which is the first step of the process, includes feasibility, environmental and market studies, improvement analysis and revision of existing life cycle assessment or implementation of new studies where necessary. Depending on the results, the criteria are drafted and iterated by the European Union Eco-labelling board. The outcome is circulated for approval among the relevant European Commission services. The process is then brought to a member state level where a vote is taken by national regulatory authorities (EC 2016a). Note that these decision steps are subject to change and continuous improvement. Currently, from the start until the adoption of criteria through a Commission Decision, the process takes 2 years on average.

For all ecolabelling schemes, the criteria have to be evaluated from time to time and if necessary revised (setting stricter requirements) to make sure that the requirements continuously favour the products on the market that have the best environmental performance and thus ensuring a continuous improvement incentive. Box 24.3 gives a comprehensive example of the criteria development cycle for a product group within Nordic Ecolabelling.

Box 24.3 The Criteria Development Cycle for the Nordic Ecolabel

The process of developing new criteria starts with a *feasibility study* in which the potential (see below) of labelling a product area is investigated looking at both environmental aspects and the market situation. During *criteria development* a set of criteria are developed and input from stakeholders (including a 2 months public consultation) are obtained. When the criteria have been adopted, companies can obtain a licence to use the ecolabel if they demonstrate compliance with the criteria. Before the criteria expire, an evaluation is performed and based on that the criteria are revised in a process similar to criteria development.

When looking into the life cycle aspects of a given product group during the development of criteria, some of the tool used by Nordic Ecolabelling are the MECO-matrix (Wenzel et al. 1997), existing LCA knowledge, literature studies plus information and data collected from the industry.

Nordic Ecolabelling has developed a tool called RPS that is used when evaluating the potential in developing ecolabel criteria for new product groups and when setting up the specific requirements for a product group. RPS is an abbreviation for Relevance, Potential and Steerability.

Relevance. As the overall goal of ecolabelling is to have a positive effect on sustainable consumption and production it makes sense that the product groups that are selected for ecolabelling have a high environmental relevance. Likewise, the specific requirements for a product group also have to have a high environmental relevance or a positive effect on product quality or health impacts.

Potential. When selecting a new product group for ecolabelling, or when setting a specific requirement for a given product group, it is very important that it is not only environmentally relevant, but also that the products on the market differ so that the ecolabel can point to the better products. If all products on the market have the same environmental performance, there would be no potential for the ecolabel to make a positive impact.

Steerability. Relevance and potential is not enough when selecting a product group for ecolabelling. There also has to be an interest between the stakeholders of the product group to use ecolabels to increase the supply or demand for ecolabelled products. If no manufacturer is interested, the development of the criteria might be a waste of time as there will be no direct impact from ecolabelling if there are no ecolabelled products on the market. However, there might be some indirect effects as shown later in this chapter. The specific requirements within a product group also have to be steerable meaning that, e.g. it has to be possible for the licence holders to influence and to document and control the fulfilment of the criteria.

When revising the criteria an overlap will be ensured between criteria generations so that the existing licence holders have time to implement the changes needed to fulfil the new and stricter requirements and to go through the recertification process.

After the public hearing, the criteria proposal is finalised and sent to the Nordic Ecolabelling Board who will decide whether to adopt the criteria. The Nordic Ecolabelling Board consists of the chairmen of the national ecolabelling boards that collaborate on the scheme and represent a wide range of stakeholders including NGOs, authorities, retailers and industry. In Denmark the national ecolabelling board is appointed by the Danish Minister of Environment.

24.3.3 The Certification Process

The certification for a Type I ecolabel—also referred to as licensing—has to be carried out by an independent third party. In order to be awarded a licence the applicant must be in compliance with the general rules of the programme and the product must meet all product environmental criteria and functional requirements as defined in the publicly available criteria document.

In general, the process is that the applicant will collect all necessary documentation including documentation from suppliers and test laboratories and send it together with an application to the ecolabelling body running the programme that has to do the verification. After being awarded the licence, the licence holder is responsible for ensuring that the products continuously meet the ecolabel requirements and the ecolabelling body has to be informed about any changes in the product or manufacturing process that might influence the compliance. Typically, the licence lasts for 3–4 years.

The licensing specifics depend on the ecolabelling scheme. Taking some examples from Fig. 24.1: For the EU ecolabel, licensing is managed in each country by a dedicated national independent organisation called 'Competent Body', typically anchored in the ministry of environment. The competent body assists with the application process, evaluates the application and decides on the award of the label (EC 2016b). In Australia GECA, a nonprofit organization, is responsible for the management of the scheme and setting the rules that detail the requirements and procedures for products to be certified. The fulfilment of requirements is verified through audits by third party conformity assessment bodies (CABs) which are required to comply with corresponding standards such as ISO/IEC 17065 (Conformity assessment—Requirements for bodies certifying products, processes and services) and ISO/IEC Guide 28 (Guidance on a third-party certification system for products) (ISO 2004, 2012). The CABs need to be accredited by the government appointed accreditation body for Australia and New Zealand (GECA 2015). In China, it is a governmental body, the State Environmental Protection Agency, which issues the guide and requirements for accrediting the labelling and supervises the management and certification. The documentation is reviewed by a certification centre and the licensing process includes onsite inspection and sampling (CEPACEC 2016).

Regarding the monitoring of continued compliance with the ecolabel criteria, testing is done through verified bodies. For example, in China testing is done by the adequate agency. In the EU it is done by qualified laboratories, preferably accredited under ISO 17025 (ISO 2005), or equivalent, that should be approved by the competent body. As for the monitoring, in EU it includes sampling from time to time, factory inspections and product tests. A file of the test results and all relevant documentation needs to be kept and be available at all times. In China, annual inspections take place until the label expires.

Box 24.4 describes the certification process in the Nordic Ecolabelling system.

Box 24.4 Certification at Nordic Ecolabelling The Nordic Council of Ministers has set up requirements that Nordic Ecolabelling, besides the requirements of ISO 14024 for type I ecolabelling schemes, also needs to fulfil the ISO 17065 standard when it comes to the certification process. The standard includes, among other, requirements related to independence and the quality of the certification process.

Further, Ecolabelling Denmark is by law subject to the rules of the public administration act when it comes to the certification process.

Before awarding a licence, Nordic Ecolabelling will be checking all submitted documentation for compliance with the requirements of the criteria. An evaluator will go through the documentation and make sure that all relevant documentation is present and shows compliance. A double check is performed by another staff member, that has not been involved in the checking before, and who also goes through the case to verify compliance. Before the licence is awarded, Nordic Ecolabelling will do an inspection visit at the production site to ensure that the situation is in line with the documentation sent. Inspection visits are performed all over the world where the production of ecolabelled products takes place.

After certification, Nordic Ecolabelling has procedures for monitoring the continued compliance, e.g. by follow-up inspection visits, spot checks of ecolabelled products on the market and other types of follow-up evaluation during the validity of the licence.

During the recertification/renewal process, the same certification process applies as for a new application—including inspection visit at the production site.

Inspection visit at a textile production site in Bangladesh

24.3.4 Ecolabels as a Marketing Tool

Ecolabels are market-driven tools. By using the label in their marketing, manufacturers or suppliers can provide credible information showing that the product or service has a good environmental performance. Indeed, such an attribute is positively evaluated by consumers. Results from a pan-European survey showed that 95% of respondents considered environmental change to be an important issue while the majority is willing to pay more for environmentally friendly products/services (EC 2014b). Ecolabels can be a useful tools for promoting environmental policies aiming to reward the environmentally best performing products and companies, such as those established by the European Commission (EC 2008, 2016c). Ecolabelling is further promoted as a marketing tool in emerging economies, e.g. for African products (UNEP 2016). Labels can be useful towards reaching the UN Sustainable Development Goals, particularly Goal 12. Which is to 'ensure sustainable consumption and production patterns' (UN 2016).

However, the proliferation of environmental claims and labels also has adverse marketing effects. Indicatively in the aforementioned European surveys, a decline in confidence in environmental claims was observed, from 52% in 2007 to 47% in 2011. This has led to guidelines for green claims developed by national authorities, self-regulatory bodies, and the private sector. An example for Denmark is the Guidance from the Danish Consumer Ombudsman on the use of environmental and ethical claims in marketing (Danish Consumer Ombudsman 2016). This stipulates that it is not allowed to promote a product with general environmental claims such as "environmentally friendly", "environmentally correct", "gentle on the environment", "green", "blue", "more environmentally friendly", "smaller environmental footprint", "better for the environment" etc., unless being substantiated by a proven, significantly lower environmental burden compared to similar products. The Consumer Ombudsman further states that this would normally require that a complete product life cycle assessment has been carried out. However, the use of such general environmental claims is justified if a product is awarded a license to use the ecolabel of an official ecolabel scheme, such as 'the Swan' from the Nordic Ecolabel or 'the Flower' from the European Union Ecolabel. The latter point indicates that official ecolabels have a unique status in the use as a marketing tool as they are assumed credible and not misleading.

Criticism

Even if ecolabels are a powerful marketing tool, there is also criticism from parts of the business sector. One objection is the fact that in most schemes a fee has to be paid for the verification process and the subsequent use of the label. It has been argued that this puts a financial burden on companies that are actually doing something beneficial for the environment. However, the rationale behind the scheme is that consumers are actually willing to pay a little bit more for ecolabelled products, and in most cases the fee will not affect the final price for the consumer significantly.

Another criticism from the business sector is that the criteria are set up in a way that obstructs innovation, because they are based on analysis of existing technologies focusing on their strengths and weaknesses. The criteria are, however, not prescribing the use of a specific technology but aim at minimizing consumption of resources and emissions to the environment. In cases where totally new technologies emerge during the validity period of a set of criteria, these might not fit totally to that new technology, but in cases like that a number of schemes have the possibility to change or adjust the criteria.

24.3.5 Increasing the Demand for Green Products Using Ecolabels

To ensure a maximum positive impact of an ecolabelling scheme, it is important to not only target suppliers of products, but also work with increasing the demand for ecolabelled products to create a pull effect in the market. Here there are three main target audiences—all with a great potential for increasing the demand and thereby yielding environmental benefits from changing the consumption to more sustainable products and services:

- Consumers;
- Public procurement;
- Procurement in private companies and organisations.

Until recently, it has been difficult to realise the potential in public procurement at the European market because of EU regulation that did not allow the public authorities to set ecolabels as a requirement for procurement in tenders. However, in the context of sustainable consumption and production, since 2008 EU has supported green public procurement initiatives, "a process whereby public authorities seek to procure goods, services and works with a reduced environmental impact throughout their life cycle when compared to goods, services and works with the same primary function that would otherwise be procured" (EC 2016d). A challenge for the public sector has been that purchasers might not have sufficient skills to set up environmental requirements in a tender. Even when they do, they might lack resources for defining relevant requirements, and for verifying the fulfilment by the tenderers. In this context, EU passed a new directive in 2014 on the coordination of procedures for the award of public works contracts, public supply contracts and public service contracts, which explicitly allows the use of ecolabels as a requirement (EC 2014c). Thus, the possibility to use official ecolabels in the tenders facilitates the implementation of relevant environmental requirements while saving resources. Additionally, the implementation of the directive in the national legislation of the EU member states may have a positive influence on the demand for ecolabelled products and services in the public sector in Europe.

The Private sector does not have similar regulatory restrictions in demanding ecolabelled products. Yet, if companies (particularly the ones with large market share and considerable purchasing power) start demanding ecolabelled products and services, a significant difference can be made. Box 24.5 explains the example of a procurement network in the Nordic countries that successfully focuses on increasing demand for ecolabelled products and services in the private sector.

Box 24.5 Increasing the Demand for Ecolabelled Products Through a Professional Procurement Network

Ecolabelling Denmark, like its sister organisations in the other Nordic countries, has established a procurement network to help increase the demand for ecolabelled products and services in the Danish market and thereby increase the environmental benefits of the official ecolabelling schemes using a market pull effect.

The current members include—among other—financial companies and banks, insurance companies and consultants. The network has, for example achieved an increased supply of ecolabelled cleaning services by demanding ecolabelling from their suppliers and thereby established a more competitive market for ecolabelled cleaning services. Apart from cooperating on increased supply of ecolabelled goods and services, the members share experience with procurement of green products.

Members of the network have to fulfil the following requirements:

1. Have a publicly available purchasing policy in which it is clearly stated how environmental requirements are proactively used in tenders and other procurement situations. The use of the official ecolabels in procurement has to form an important part of the policy.
2. Comply with the principles of UN Global Compact and banks and investment companies further have to comply with the UN Principles for Responsible Investment.
3. Commit to report the purchase of ecolabelled products and services every year and with the reporting be able to document a significant improvement every year.

The network has been very efficient in motivating the suppliers of the members in adapting to the ecolabelling requirements and applying for an ecolabel licence.

24.3.6 *Positive Side Effects of Ecolabels*

Besides the direct effects, ecolabelling has also shown to have indirect effects with a positive influence on the environment (see Box 24.6 for the so-called rippling effect and Box 24.7 for the link to ecodesign and regulation). Some of these positive side effects are:

- Some companies use ecolabelling criteria for benchmarking to get an idea on environmental improvement options—and even though they do not necessarily take active part by applying for an ecolabel, they still might implement some improvements based on this.

- An increased demand by single consumers or companies can have a much wider impact than the 'few' products that they buy. When experiencing an increasing demand for ecolabelled products a manufacturer will often in response implement improvements in their standard products that is also supplied to the consumers not actively demanding ecolabels, and in that way their consumption inadvertently becomes more sustainable.
- Ecolabelling criteria and ecolabelled products on the market can demonstrate that it is possible to provide greener products on the market, and ecolabelling can thereby be a driver for stricter environmental requirements in legislation.

Box 24.6 The Rippling Effect of Ecolabelling

Ecolabelled goods or services have a broad influence that goes beyond the reduction of environmental burdens due to a single purchase. This is because implementing ecolabelling requirements might cause a rippling effect. For example, a demand for ecolabelled hotel services can lead to a demand for corresponding ecolabelled products and services such as laundering service. Consequently, there will be a demand for laundry detergents fulfilling the ecolabel requirements in every stage of the life cycle. This can bring environmentally improved products on the market and thus make them available not only for those who actively demand for them, but for a broader audience because the required product improvements are often implemented in the standard products. Further, when e.g. the leading company in a branch starts ecolabelling their products and services—then other companies will follow and a wider range of the products on the market would be developed to meet the ecolabelling requirements. Therefore, what initially might start as a demand for ecolabelled hotel services could spread like ripples in the water with a higher influence on sustainable production and consumption.

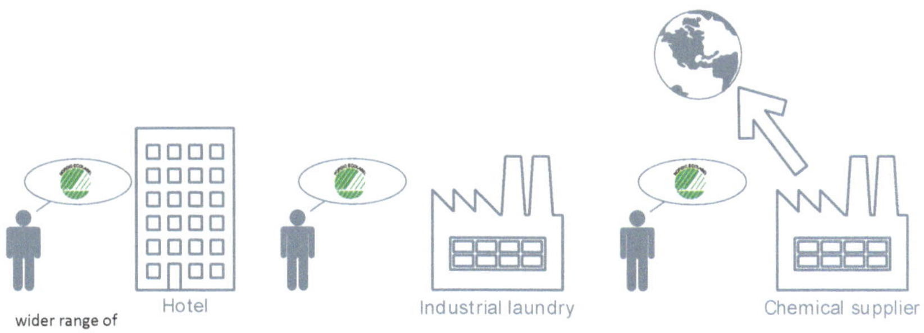

Box 24.7 Ecolabels as a Driver for Ecodesign and Stricter Mandatory Regulation

Since 2008, according to the Danish legislation all fireplaces sold in Denmark cannot emit more than 10 g of particulate matter per kg of wood burned. In 2015, the limit value was reduced to 5 g/kg and from 2017 a maximum of 4 g/kg is allowed.

Since the first criteria for fireplaces were adopted in 2001 and since the first licence was awarded in 2004, Nordic Ecolabelling has shown that it is possible to produce fireplaces with lower emissions than the legal limit and the continuously stricter ecolabelling requirements have been a driver for the development of cleaner fireplaces and demonstrated to the authorities that it is possible to set up a tighter regulation. Even with a stricter legislation enforced by the authorities, the Nordic Ecolabel will still be driving the development of fireplaces with lower emissions as the requirements of the ecolabel is also becoming stricter. Hence, in 2015 new ecolabel requirements for emissions of particulate matter is 3 g/kg and by 2017 only 2 g/kg.

On top of the strict requirements on emissions of particulate matter, the Nordic Ecolabel certainly also has numerous other requirements to ecolabelled fireplaces—e.g. the efficiency of the fireplace and requirements related to the production process and raw material sourcing.

24.4 Environmental Product Declarations (EPDs)

ISO Type III Environmental declarations, also referred to as "environmental product declarations", are documents that transparently communicate environmental information and that can be used to compare the environmental performance of different products fulfilling the same function.

Worldwide, there is a number of different programme operators of Type III declaration schemes. The Global Environmental Declarations Network (GEDnet) is an international organisation of Type III environmental declaration bodies and practitioners with about 10 members (www.gednet.org)—most of them responsible for different national declaration schemes. Nevertheless, there are more operators.

One sector worth mentioning, where the use of environmental product declarations has especially increased globally is the one of building products. In Europe the reason is that EU regulation from 2011 includes a reference to the use of environmental declarations, which has increased the demand for them, stating that: "For the assessment of the sustainable use of resources and of the impact of construction works on the environment Environmental Product Declarations should be used when available" (EC 2011b).

As a response to the legislation EPD programme operators have come to cooperate and harmonise approaches. Indeed, the ECO Platform (www.eco-platform.org) (see Box 24.8) was initiated by a group of EPD programme operators, LCA practitioners and European building sector branch organisations and aims to harmonise the different declaration schemes so that the resulting environmental declarations can be used in all European countries. This is in accordance with the original intention of the ISO 14025 standard for type III declaration schemes. The harmonisation within the sector is further supported by a European standard EN 15804 that provides core product category rules for type III environmental declarations for any construction product or service. Similarly in the US, where the LEED certification developed by the non-profit organisation US Green Building Council (USGBC), prevails. This includes a set of rating systems for the design, construction, operation, and maintenance of green buildings, homes and neighbourhoods, and it is one of the most popular green building certification programmes used worldwide (EC-DG Energy 2014; LEED 2016).

Box 24.8 ECO Platform—Harmonising Environmental Declarations in the Building Sector

"The objective of ECO Platform is the development of verified environmental information of construction products, in particular type III declarations called EPD (Environmental Product Declarations). The added value of EPD under the ECO Platform framework is the possibility to use these declarations in all European but also international markets.

ECO PLATFORM

ECO Platform is not a programme operator. It is a group of them together with LCA practitioners, industrial associations and other stakeholders working to guarantee a coherent framework for EPD".

24.4.1 *Product Category Rules (PCR)*

According to ISO 14025, an environmental declaration scheme has to develop PCRs for each product group where the scheme operates. These are meant to enable transparency and comparability between EPDs. The PCR will normally establish for example:

- functional unit for the product area,
- allocation rules,
- system boundaries,
- LCIA methods,
- data sources.

The development of PCR of a Type III environmental declaration scheme has three recommended steps according to the standard ISO 14025 (ISO 2006a):

1. *Define the product category*:
2. *Collect and/or produce appropriate LCA*;
3. *PCR: Specify common goal and all relevant rules for product category LCA, predetermined parameters, rules on additional environmental information, requirements for reporting. Write instructions on how to produce the data required for the declaration.*

The development of a PCR requires consultation involving interested parties. Yet there is no global consensus for developing sound PCRs. Box 24.9 gives an example of a PCR developed by the Norwegian EPD Foundation.

Box 24.9 Product Category Rules for Seating—The Norwegian EPD Foundation

The product category rules for seating for the Norwegian EDP system, e.g. defines the functional unit to use, what stages to include, the system boundary, data quality requirements and the calculation rules and impact categories that have to be applied.

Functional unit:

"*Production of one unit of seating provided and maintained for a period of 15 years*".

The PCR indicates that secondary materials are included only as recycling processes and that electricity is included as national grid mix of either the country or the region where main energy-consuming processes take place. This is very different from the overall rules of the Danish pilot EPD scheme (www.MVD.dk) where a consequential approach is applied, stipulating the use of data for the marginal suppliers (see Chap. 8 for a discussion of the difference between consequential and attributional perspectives in LCA). Such differences in methodology can lead to large variations in the results of the declaration.

In the Norwegian EPD system, a panel of stakeholders is involved in the development of the PCRs and the documents are sent out for consultation to ensure acceptance and transparency. After the consultation period, the technical committee (TC) of the programme operator decides on the final PCR for the product group. The TC of the Norwegian declaration scheme shall consist of no less than 5 LCA/EPD experts.

Verification of Environmental Declarations

Even though it is not an explicitly stated requirement of the ISO standard, most type III environmental declarations make use of third party verification to make sure that the LCA follows the product category rules and that the overall requirements of the declaration scheme is fulfilled.

24.4.2 Benefits and Drawbacks of Environmental Declarations

The intention of Type III environmental declarations is to provide reliable, detailed information about the environmental performance in the life cycle of a given product or service to the decision maker in a purchasing situation. In this way, the purchaser can use it to choose products with a lower environmental burden. Thereby, as marketing tools, the environmental declarations can influence the shift to more sustainable products and yield a positive effect in the environment. Additionally, even if they are driven by marketing purposes, Type III declarations can, like Type I ecolabels, inspire product development changes with positive effects on the environment. For example, the information and knowledge gained from performing a life cycle assessment can spark new ideas for incremental product development to improve the environmental performance, leading to ecodesign initiatives (see Chap. 23). Suppliers of raw materials and intermediates can also supply their customers with cradle-to-gate data in the form of an environmental declaration, which makes it possible for manufacturers to enhance the accuracy of their LCAs.

Environmental declarations are seen by some companies as a possibility to provide much more detailed information about product improvements than the Type I label and it also supports further distinction between products that all have a Type I label. However, since it can be produced for all products—no matter the relative environmental burden of the product within that product group, an environmental declaration, contrary to the ecolabel, is not an indicator of the overall environmental preferability. This means that a purchaser, whose expertise in interpreting environmental information is often limited, needs to make dubious choices, e.g. when buying a product with low climate change but high ecotoxicity impact potential. Also, in case of product comparisons, to make a meaningful decision, a purchaser would have to compare environmental declarations for two different products with the same functional unit covering the whole life cycle and the same impact categories. However, declarations are only available for a limited number of products and are not always consistent in their descriptions of technical and environmental information. Given the need for validation of environmental claims, there is thus a requirement for clear product rules. This requirement is set in various standards, i.e. except for the PCRs in ISO 14025 there are also the Product Rules in the GHG Protocol Product Life Cycle Accounting and Reporting Standard, and the Supplementary Requirements in PAS 2050. Other standards, such as BP X30 (France), SMRS (Sustainability Consortium), TS 0100 (Japan) and ISO 14067 on carbon footprint, also require the use of PCRs. The result is that various EPD programmes have been using any of the above standards to develop PCRs independently and without cross recognising other programmes. As Box 24.10 exemplifies, these reasons combined, make it difficult to use EPDs in practice. Consequentially, as discussed in Sect. 24.5 there is a need for harmonisation in order to improve the validity of EPDs.

Box 24.10 Environmental Declaration of Office Chairs—Two Examples

Steelcase was the first contract furniture company to offer environmental declarations to its customers on the international market. HÅG, which today is part of Scandinavian Business Seating, has also been one of the pioneers within the use of environmental declarations. Both companies have chosen two different approaches as Steelcase is not following an established scheme of a programme operator, but instead uses external experts to review the LCA and declarations, whereas HÅG uses the Norwegian EPD system.

Even though both declarations seem to be for the same product type and for a 15-year use, it is not possible for a purchaser to compare the environmental performance based on the declarations, as they do not have identical impact categories. Furthermore, the allocation, calculation and data modelling rules differ which can have a significant influence on the result. Even for LCA experts this can be complicated, let alone the average procurer with limited LCA competence.

24.5 The Need for Harmonisation

The standardisation efforts have tackled the issue of legitimacy and credibility of environmental claims. Yet, the rising number of methods for measuring the environmental performance of products, services and organisations has been overwhelming. By 2010, for carbon emissions only, there were 62 leading initiatives and methods on product carbon footprinting and 80 on carbon reporting (EC 2016e). This trend seems to be continuing in the future since governments are more and more using LCA in voluntary or mandatory public policy initiatives (see also Chap. 18). Apart from the potential confusion on the market discussed in the introduction of this chapter, this proliferation of methods often leads to additional costs for cross border trading since companies might have to comply with requirements of schemes that have divergent methodological choices some of which are even left to the user to decide. Comparability is not established either because the methods are different or due to their inherent flexibilities. The situation is even more blurred when trying to measure environmental performance with time, since consistency of the corresponding methodological choices (e.g. year by year) would need to be guaranteed.

There is therefore a need for aligning and harmonising the different EPD schemes to avoid distortion in the market, to ensure transparency, increase comparability and avoid unfair competition. Benefits from doing so can be identified in relation to different stakeholders:

- *Consumers*: Will gain by improved comparability and more informed decisions;
- *Companies*: By ensuring both credibility and harmonisation, companies can make legitimate environmental claims and benchmark themselves within a certain sector or product category. This allows to also monitor their performance, to compare to their peers and therefore to better focus any improvement efforts. Aside the incremental changes, this can potentially lead to more radical innovations for strong improvements that may shift the reference environmental performance of the whole sector/product category.
- *Small Medium Enterprises (SME)s*: SMEs increasingly have to provide inventory data as part of their communication about their products in global supply chains. They also have to comply with diverging schemes in order to compete with their multinational counterparts. Having a common methodological reference will thus facilitate them to reduce complexity and cost.
- *Investors*: Along the same line, investors can better target their decisions by having a consistent reference to the sector and a common ground for assessing how companies perform.
- *Policy makers*: These benefits also relate to governmental actors and policy makers. By knowing the environmental performance of stakeholder groups they can better identify gaps, allocate resources and incentivize consumption of reliably greener alternatives. They can apply this information in policies, e.g. by setting environmental limits for products or by linking economic instruments to environmental performance so that in the long term they can support the sustainable development goals.

In practice, it is very difficult to use EPDs for comparisons of products, because a meaningful comparison requires that the declarations include all stages of the life cycle, use the same environmental indicators, and are based on LCAs with the same scope, methodology and data quality. This means that they should be based on the same Type III declaration scheme where a specific set of product category rules has been defined. Despite the challenges, there are more and more initiatives for criteria harmonisation. Such are the aforementioned for the building sector and the broad EU initiative discussed in Sect. 24.5.1. Across countries, there are several recognition arrangements occurring not only in Europe (see Sect. 24.3.4) but also in North America, Asia and elsewhere (DigitalEurope 2015).

24.5.1 EU's Product and Organisational Environmental Footprints (PEF/OEF)

In 2011, the European Commission started working on methodology harmonisation for Product Environmental Footprints (PEFs) and Organisational Environmental

Footprints (OEFs) which are the EU Commission's term for the impact profiles resulting from LCAs, i.e. similar to Type III based environmental declarations for products or organisations (EC 2016f).

Rather than suggesting a new approach, the idea was to build upon well-established, verified and broadly used methods, standards and guidelines, such as ISO 14040-44, ISO 14064, PAS 2050 and WRI/WBCSD GHG protocol (ISO 2006b, c; EC-JRC 2010; BSI 2011; WRI/WBCSD 2016). Two years later the Product Environmental Footprint (PEF) and Organizational Environmental Footprint (OEF) Guides where released under the "Single Market for Green Products Initiative". The aim of these voluntary initiatives is to provide a common methodological basis and support a single metric for a single market. PEF relates to single products (i.e. goods or services) while OEF refers to an organisation comprising a well-defined portfolio of products and/or services. The latter can be calculated using aggregated data (thus, it is not required to have individual PEFs and sum them up to get the OEF). Both apply common rules, which further allow to explore synergies between the organizational and product levels. PEF/OEF are life cycle based and cover 15 impact categories in the ILCD method (see Chap. 10). Still, in line with what was discussed in Sect. 24.3.1 the choice of impact categories to communicate could further be tailored to each product category and sector so that consumers can more easily grasp the information (EC 2016e).

The release of the PEF/OEF guides was followed by a 3-year period (2013–2016) of testing and refining of the method on the basis of 26 selected pilot case studies within different sectors. An additional intention is to complement them with product category and sector specific rules to facilitate streamlining of the LCA and ensure comparability between similar products and sectors. The pilot studies are also investigating strategies for communicating life cycle environmental performance to business partners, consumers and other company stakeholders.

From a policy point of view, PEF/OEF are considered to strengthen existing product instruments such as Ecolabel, Green Public Procurement and Ecodesign. ISO Type I labels can also benefit since PEFs can consistently inform on the most relevant environmental impacts and life cycle stages. The possibility for benchmarking will additionally allow to refine existing labels based on market performance. PEF/OEF also have a place in the bigger framework of circular economy, which is the strategic vision for European economic development (EC 2015). Overall, regardless its European focus, the initiative has attracted global attention given the global nature of the economy, the importance of the European market and the UN Sustainable Development Goals.

Yet, such a broad harmonisation attempt does not come without challenges. Critics undermine the comprehensiveness of the approach, i.e. within the PEF framework one could still get different results for the same product, which jeopardises the principles of consistency and credibility. They also question the suitability of PEF for consumers, who would still lack the competences that would enable them to weigh different environmental impacts and their trade-offs. Additionally, the attempt to make the methodology more feasible may lead to lower accuracy and meaningfulness, e.g. PEF includes only selected parts of the spectra of problems, which is not always sufficient. New developments take place both on product system modelling level, e.g. with consequential LCA, and on a LCIA level, e.g. with endpoint modelling. Once consensus is reached on such methodological choices, the PEF/OEF methodology will constantly need to be updated in order to keep up with methodological developments that improve the reliability of LCA results. Despite the drawbacks indicated, the PEF/OEF address the need for harmonisation and take a step towards it. Rather than redundant tools that add complexity, PEF and OEF are conceived and further refined as tools that act in synergy with existing schemes.

References

BSI: PAS 2050:2011 Specification for the assessment of the life cycle greenhouse gas emissions of goods and services (2011)

CEPACEC: Certification scope and procedure for the environmental labelling products in China. In: China Environ. United Certif. Cent. http://www.sepacec.com/cecen/ (2016). Accessed 30 Jun 2016

Danish Consumer Ombudsman: Environmental and ethical marketing claims. http://www.consumerombudsman.dk/Regulatory-framework/dcoguides/Environmental-and-ethical-marketing (2016). Accessed 30 Jun 2016

DigitalEurope: Ecolabels—criteria harmonisation and recognition arrangements. http://www.digitaleurope.org/DesktopModules/Bring2mind/DMX/Download.aspx?Command=Core_Download&EntryId=951&PortalId=0&TabId=353 (2015)

EC: Communication from the commission to the European parliament, the council, The European economic and social committee and the committee of the regions on the sustainable consumption and production and sustainable industrial policy action plan. Brussels (2008)

EC: 2011/263/EU: Commission Decision of 28 April 2011 on establishing the ecological criteria for the award of the EU Ecolabel to detergents for dishwashers (notified under document C (2011) 2806). Off. J. Eur. Union **111**, 22–33 (2011a)

EC: Regulation (EU) no 305/2011 of the European parliament and of the council of 9 March 2011 laying down harmonised conditions for the marketing of construction products and repealing Council Directive 89/106/EEC. Off. J. Eur. Union **88**, 5–43 (2011b)

EC: 2014/312/EU: Commission Decision of 28 May 2014 establishing the ecological criteria for the award of the EU Ecolabel for indoor and outdoor paints and varnishes (notified under document C(2014) 3429). Off. J. Eur. Union **164**, 45–73 (2014a)

EC: Consumer Market Study on Environmental Claims for Non-Food Products. Brussels (2014b)

EC: Directive 2014/24/EU of the European parliament and of the council of 26 February 2014 on public procurement and repealing directive 2004/18/EC. Off. J. Eur. Union **94**:65–242 (2014c)

EC: Circular Economy Strategy. http://ec.europa.eu/environment/circular-economy/index_en.htm (2015)

EC: EU Ecolabel criteria development and revision. In: EC-Environment. http://ec.europa.eu/environment/ecolabel/criteria-development-and-revision.html (2016a). Accessed 30 Jun 2016

EC: How to apply for EU Ecolabel. In: EC-Environment. http://ec.europa.eu/environment/ecolabel/how-to-apply-for-eu-ecolabel.html (2016b). Accessed 30 Jun 2016

EC: Single market for green products initiative. In: EC-Environment. http://ec.europa.eu/environment/eussd/smgp/index.htm (2016c). Accessed 30 Jun 2016

EC: Green public procurement. In: EC-Environment. http://ec.europa.eu/environment/gpp/index_en.htm (2016d). Accessed 30 Jun 2016

EC: Questions about the product environmental footprint (Pef) and organisation environmental footprint (Oef) methods. In: EC-Environment. http://ec.europa.eu/environment/eussd/smgp/pdf/q_a.pdf (2016e). Accessed 30 Jun 2016
EC: Single market for green products-policy background. In: EC-Environment. http://ec.europa.eu/environment/eussd/smgp/policy_footprint.htm (2016f). Accessed 30 Jun 2016
EC-DG Energy: ENER/C3/2012-436—market study for a voluntary common European Union certification scheme for the energy performance of non-residential buildings (2014)
EC-JRC: International Reference Life Cycle Data System (ILCD) Handbook—General guide for Life Cycle Assessment—Detailed guidance, 1st edn. Publications Office of the European Union, Luxembourg (2010)
GECA: Good Environmental Choice Australia-Certification Scheme Rules (version 15). New South Wales (2015)
IISD: Global Green Standards-ISO 14000 and Sustainable Development. Manitoba (1996)
ISO: ISO 14024:1999 Environmental Labels and Declarations—Type I Environmental Labelling—Principles and Procedures. ISO, the International Organization for Standardization, Geneva (1999)
ISO: ISO/IEC Guide 28:2004 Conformity Assessment—Guidance on a Third-Party Certification System for Products. ISO, the International Organization for Standardization, Geneva (2004)
ISO: ISO/IEC 17025:2005 General Requirements for the Competence of Testing and Calibration Laboratories. ISO, the International Organization for Standardization, Geneva (2005)
ISO: ISO 14025:2006 Environmental Labels and Declarations—Type III Environmental Declarations—Principles and Procedures. ISO, the International Organization for Standardization, Geneva (2006a)
ISO: Environmental Management—Life Cycle Assessment—Principles and Framework (ISO 14040). ISO, the International Organization for Standardization, Geneva (2006b)
ISO: Environmental Management—Life Cycle Assessment—Requirements and Guidelines (ISO 14044). ISO, the International Organization for Standardization, Geneva (2006c)
ISO: Environmental Management—The ISO 14000 Family of International Standards. ISO, the International Organization for Standardization, Geneva (2009)
ISO: ISO/IEC 17065:2012 Conformity Assessment—Requirements for Bodies Certifying Products, Processes and Services. ISO, the International Organization for Standardization, Geneva (2012)
ISO: ISO 14021:2016 Environmental Labels and Declarations—Self-Declared Environmental Claims (Type II Environmental Labelling). ISO, the International Organization for Standardization, Geneva (2016)
LEED: LEED. http://leed.usgbc.org/leed.html (2016). Accessed 30 Jun 2016
Rubik, F., Frankl, P.: The Future of Eco-labelling: Making Environmental Product Information Systems Effective. Greenleaf Pub, Sheffield (2005)
UN: Agenda 21: programme of action for sustainable development. In: United Nations Conference on Environment and Development (UNCED), p. 343. United Nations, Rio de Janerio (1992)
UN: Johannesburg declaration on sustainable development and plan of implementation of the world summit on sustainable development: the final text of agreements negotiated by governments at the world summit on sustainable development. In: World Summit on Sustainable Development, and United Nations. United Nations Department of Public Information, Johannesburg (2002)
UN: Goal 12: ensure sustainable consumption and production patterns. In: Sustain. Dev. Goals. http://www.un.org/sustainabledevelopment/sustainable-consumption-production/ (2016). Accessed 30 Jun 2016
UNEP: Ecolabelling—as a potential marketing tool for African products (an overview of opportunities and challenges). http://www.unep.org/roa/docs/pdf/Eco-labelling-Brochure.pdf (2016)
UNOPS: A guide to environmental labels for procurement practitioners of the United Nations system (2009)

Wenzel, H., Hauschild, M.Z., Alting, L.: Environmental Assessment of Products, vol. 1. Kluwer Academic Publishers, Berlin (1997)
WRI/WBCSD: Greenhouse gas protocol. http://www.ghgprotocol.org/ (2016). Accessed 30 Jun 2016

Author Biographies

Jeppe Frydendal Working with LCA/LCM since late 1990s in industry, university, consultancy and ecolabelling. The focus and interest has always been to disseminate life cycle thinking to companies ensuring that they convert LCA knowledge to real life environmental improvements. Co-author of the UNEP Business Guide on Life Cycle Management.

Lisbeth Engel Hansen Participated in some of the first Danish assessments of materials in a life cycle perspective. Has since the early 1990s been involved in development of life cycle based ecolabels both methodological and in practice, among other things as a board member of GEN. Participate in the development of the ISO standards on ecolabelling.

Alexandra Bonou LCA expert and modeller focussing on ecodesign since 2009. Has worked on integrating life cycle thinking and environmental target setting in organisations of the private and public sector. Interested in life cycle management, product development, social life cycle assessment.

Chapter 25
Cradle to Cradle and LCA

Anders Bjørn and Michael Z. Hauschild

Abstract Cradle to Cradle (C2C) offers a positive vision of a future, where products are radically redesigned to be beneficial to humans and the environment. The idea is not to reduce negative impacts (as in LCA), but to increase positive impacts. This chapter presents the C2C concept and its relationship with the circular economy, the C2C certification and examples of C2C certified or inspired products and systems. This is followed by a comparison of C2C with eco-efficiency and LCA. Because of their important differences, we conclude that care should be taken when combing C2C and LCA, e.g. using LCA to evaluate products inspired by C2C. We then provide an in-depth analysis of the conflicts between C2C and LCA and offer solutions. Finally, we reflect upon how LCA practitioners can learn from C2C in terms of providing a vision of a sustainable future, creating a sense of urgency for change and communicating results in an inspiring way.

Learning Objectives
After reading this chapter, one should be able to

- Explain the Cradle to Cradle® (C2C®) concept and its three key principles,[1]
- Outline the C2C certification scheme,
- Provide examples of C2C inspired or certified products,
- Discuss similarities and differences, complementarities and conflicts between C2C and LCA.

[1]Cradle to Cradle® and C2C® are trademarks of MBDC, LLC. Used with permission.

A. Bjørn (✉) · M.Z. Hauschild
Division for Quantitative Sustainability Assessment, Department of Management Engineering, Technical University of Denmark, Kongens Lyngby, Denmark
e-mail: anders.bjoern@polymtl.ca

A. Bjørn
CIRAIG, Polytechnique Montréal, 3333 Chemin Queen-Mary, Montréal, QC, Canada

M.Z. Hauschild et al. (eds.), *Life Cycle Assessment*,
DOI 10.1007/978-3-319-56475-3_25

25.1 Background

> Imagine a world in which industry, yes, every factory and every building is as wasteful and as useful as a cherry tree in full bloom. A world, in which buildings – just like trees – use the sun's energy, produce nutrients and oxygen, provide living space for other creatures, cleanse water, purify the air and even change to adapt to the seasons. A world without environmental pollution or waste, where only products with materials that are beneficial to both man and the environment are manufactured. A world, in which materials are of such high value that they flow in specially designed material cycles.
>
> A world, in which humans can actually be pleased about the benefits their consumption has on the environment. A world, in which humans are freed from and no longer have to live under the restraints and limitations placed on them by always having to save, reduce and cut down on certain things for the sake of the environment. That is exactly the kind of world that the Cradle to Cradle® design concept opens up to all of us.
>
> – Excerpt from Michael Braungart's Cradle to Cradle Vision (EPEA 2013)

The words above do not resemble those of an ordinary engineering discipline. But then again, the Cradle to Cradle concept is by no means an ordinary engineering discipline.

25.1.1 History

Cradle to Cradle® (C2C) is based upon the idea of imitating nature in the approach to sustainable product and system design. The first use of the term Cradle to Cradle is generally attributed to the Swiss architect Walter R. Stahel in the end of the 1980s (PLI 2013). The term originated as a reaction to the newly emerged idea of companies being responsible for their products and systems from "cradle to grave". Stahel argued that the "cradle to grave" perspective was merely reinforcing the existing linear economical model and relied on end-of-pipe solutions. He argued that the really sustainable solution was to use durable goods in a loop from "cradle back to cradle" in a circular economy (PLI 2013). At the turn of the century, the German chemist Michael Braungart and US architect and designer William McDonough further developed the C2C concept and provided examples and guidance for its use in the design of products and system. In 2002 this work was compiled in the book "C2C—Remaking the way we make things" (McDonough and Braungart 2002). A decade later "The Upcycle" was published (McDonough and Braungart 2013). This sequel to the 2002 book clarifies the theoretical basis of C2C, addresses some common misconception and provides additional examples of how the concept can be applied to the design of products and systems at various scales.

25.1.2 *Influence*

The C2C books have broad scopes and address a variety of topics, such as environmental science, product and system design, organizational theory and philosophy. The books are written in a visionary, provocative and popular style and have consequently reached an audience beyond the sole fields of civil engineering and product design. Apart from attracting a large numbers of readers, the C2C concept has had a number of concrete impacts on business, civil society and policy.

The trademark C2C is owned by McDonough Braungart Design Chemistry (MBDC) located in North America and the Environmental Protection Encouragement Agency (EPEA), located in Hamburg has the license to use it. The C2C Product Innovation Institute (California) has the license to certify products according to the Cradle to Cradle Certified™ Product standard, i.e. the C2C certification scheme. These institutions have moreover trained consultancies around the world to assist companies in going in a C2C direction and/or complying with the certification requirements. This has resulted in a number of products and system designs based on the concept and C2C certified products (see Sect. 25.4).

Some environmental NGOs are also praising the concept for its positive agenda and a few are even entirely dedicated to moving the world in a C2C direction, such as the Danish CradlePeople. Also, Cradle to Cradle has inspired the financially secure Ellen MacArthur Foundation (EMF) to promote the idea of a Circular Economy, defined as a restorative or regenerative industrial system by intention and design (EMF 2012). This charity has since 2012 published a series of comprehensive reports and books with the overall aim of convincing business and policy-makers that a strategic transformation from a linear to a circular economy is not only possible, but also in their enlightened self-interest, as it can increase the wealth and resilience of companies and societies (e.g. EMF 2012, 2015a, b).

Direct influences of Cradle to Cradle and Circular Economy on policy have initially been modest, but in 2015 the European Commission adopted an "EU action plan for the Circular Economy" (EC 2015a). The declared aims of this action plan is to "contribute to 'closing the loop' of product lifecycles through greater recycling and reuse, and bring benefits for both the environment and the economy" (EC 2015b). The plan includes among other things common EU targets for recycling, economic incentives for producers of recyclable products and measures to stimulate industrial symbiosis.

25.2 Key Principles of C2C

C2C aims for positive impacts on the environment, including humans. To achieve this, three key design principles must be followed: Waste Equals Food, Use Current Solar Income and Celebrate Diversity.

25.2.1 Key Principle 1: Waste Equals Food

According to the C2C concept, humans are not inherently harmful to the environment. Rather than considering environmental impacts as unwanted, but inevitable, consequences of human activities, they should be seen as symptoms of design failures. Waste, understood as a physical flow with no use for anyone and therefore no economic value, is such a symptom of design failure. Waste is a phenomenon unknown to nature in which materials are continuously cycled between different ecosystem species. The waste of one organism becomes food for another organism. Waste equals food. The very concept of waste should therefore be eliminated and the focus should instead be to design "healthy emissions", meaning that the emissions that inevitably result from industrial processes should be engineered as resources to be taken up by other industrial processes or ecosystems. The focus should thus shift from trying to reduce the amounts of emissions to designing emissions with beneficial effects (either for organisms in nature or for other industrial processes). This applies to emissions occurring throughout the life cycle of a product and also to the product itself when it reaches its disposal stage. To ensure that such emissions can undergo recycling in continuous loops without loss of quality, materials should either be defined as technical or biological nutrients, see Fig. 25.1.

The (short) definition of a biological nutrient is "*a product usable by defined living organisms to carry on life processes such as growth, cell division, synthesis of carbohydrates, energy management, and other complex functions*" (PII 2016). Analogously, a technical nutrient can (in short) be defined as "a product capable of "feeding" technical systems" (PII 2016). Feeding may be in the form of dismantle and reuse, physical transformation (e.g. plastic remoulding) and chemical transformation (e.g. plastic depolymerisation). It should be noted that materials in the technical cycle are therefore free to take part in any product, as long as they

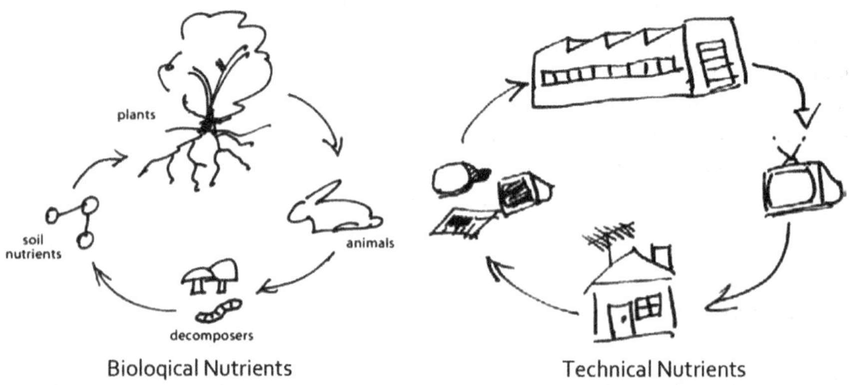

Fig. 25.1 The technical and biological nutrient cycles. Image Copyright© MBDC, LLC. Used with permission

maintain their value. It is even encouraged to *increase* material value as they cycle from product to product. This process is termed ‘upcycling’, which is also the title of the recent C2C book (McDonough and Braungart 2013). Materials harmful to the environment or humans are accepted as technological nutrients in C2C if they fulfil the above definition and as long as they do not end up in the environment and as long as humans are not exposed to them.

Some materials may qualify as biological nutrients, but also be able to take part in the technical nutrient cycle until a critical point, where their quality is too low to be of any value in the technical cycle. At this point, they should enter the biological cycle. For example, paper produced from wood can undergo recycling by pulping. This process, however, leads to the rejection of the fraction of the wood fibres that have become too short to be of use. When a wood fibre becomes too short due to pulping and no other industrial processes in the technical cycle can use it, it can no longer be characterised as a technical nutrient. Because wood fibres also qualify as biological nutrients the rejected wood fibres can therefore enter the biological cycle through, e.g. anaerobic digestion, composting or spreading of ashes following incineration, provided that they are not contaminated (see below) (MIE 2011).

Products composed of biological nutrients, such as wood fibres in paper, are inherently degradable. They should therefore naturally be ‘consumed’ by the consumer, who may choose to nourish his or her garden soil with the biological nutrients contained in the worn out product (e.g. a piece of textile).

By contrast, products composed of technical nutrients, are per definition not ‘consumable’. Their value to the user is the function they provide, not the materials they are composed of, which means that they may, potentially, be leased rather than sold to the users. Products composed of technical nutrients should thus take part in Product Service Systems (PSS). In PSS, users pay for services (such as the ability to watch television 500 h per year for 5 years) rather than products (a television). For the consumer this has the advantage that they are guaranteed the quality of the service (if the television malfunctions or breaks down, the supplier is obliged to fix it) and that they do not have to bother with disposing the product when it is no longer useful (e.g. taking an obsolete or broken television to the recycling centre). For the company engaging in PSS has the advantage that they maintain control over the materials embedded in the products. Products may thus be designed with durability in mind and for easy replacement of individual parts. Modular designs may also be advantageous since parts of redundant products may be fed into new products as resources. The potentials of PSS as driver of environmental improvements are obvious: It can ensure cleaner fractions of used products, to a higher extend designed for disassembly and ensure a higher recovery fraction than for non-PSS goods, where the continuous cycling of technical nutrients to a large extent relies on the good will of consumers.

The distinction between the biological cycle and the technical cycle is also at the core of the circular economy, and so is the distinction between consumers and users.

A key message in the C2C concept is that biological and technical nutrients should not be mixed beyond easy separability, as this creates a risk of a product that neither fits the biological nor the technical nutrient cycle. Such a product is termed a *monstrous hybrid* and can never truly be recycled and the result is a "downcycled" product of lower quality and value (McDonough and Braungart 2002). Moreover, the separation of biological and technical nutrients in the recycling process is technically complicated and often requires high inputs of energy and chemicals, which may cause damage if emitted to the environment. The authors use ordinary cellulose-based office paper as an example of a monstrous hybrid, since it is composed of biological materials (cellulose fibres) as well as technical materials (e.g. coating agents, dyes and inks). Therefore, it can neither be part of the biological cycle (the chemicals are persistent and possibly toxic to the environment) nor the technical cycle, at least not indefinitely (cellulose fibres are shortened during the pulping process). Attempts to recycle ordinary office paper thus inevitably results in downcycling. Recycled office paper has a lower functionality than virgin office paper. To compensate for the mix of biological and technical nutrients its recycling process requires additional input of virgin wood fibres and chemicals for bleaching and de-inking the pulp, which are emitted from the recycling plant causing damage to the environment.

25.2.2 Key Principle 2: Use Current Solar Income

The second key principle dictates that the energy required to fuel a continuous-loop C2C society must all originate from "current solar income," defined as photovoltaic, geothermal, wind, hydro, and biomass. All these energy sources are effects of solar radiation on Earth's surface (except geothermal energy, which originates from nuclear processes in the core of the Earth). The solar income must be 'current'. Otherwise, fossil fuels would be permitted since they are 'old' solar income. This key principle is inspired by nature since all processes occurring in nature are fuelled by current solar income.

It is important to note that there are no quantitative restrictions on the amount of energy used throughout the life cycle of a C2C product or system. The quantity of energy used is considered irrelevant as long as the energy quality (i.e. energy source) meets the requirements of current solar income (McDonough and Braungart 2002).

25.2.3 Key Principle 3: Celebrate Diversity

Avoiding one-size-fits-all designs is the main point of the last key principle. Products and systems should be designed with respect for local cultures, economies, and environments. Ecosystems differ with respect to structure, processes, function and the services they may offer to humans, depending on varying natural conditions (e.g. climatic and geologic) across the globe. Similarly, the organisms populating the ecosystems each fulfil a specific task and biologists generally agree that a high diversity of species (biodiversity) lead to more robust or resilient systems. Consequently, products and systems designed by humans should: "*draw information from and ultimately "fit" within local natural systems ... express an understanding of ecological relationships and enhance the local landscape where possible ... draw on local energy and material flows ... take into account both the distant effects of local actions and the local effects of distant actions*" (McDonough and Braungart 2002).

The third key principle also encourages that one should 'become native' and realize one's role as a species among other species. Members of ecosystems are dependent on receiving, transforming and passing on nutrients to each other. Therefore the aim should not be to reduce impacts on the environment, as suggested by the eco-efficiency concept, as this would result in isolation from other species (McDonough and Braungart 2002). Instead we should exchange nutrients with the environment and integrate it into our system designs (e.g. through the use of green roofs for storm water management and constructed wetlands for waste water treatment). On a more philosophical note, the authors advocate the abandonment of the mental image of 'mother nature', which we feel guilty about hurting. Instead nature should be perceived as a companion that we may learn from and support (Tobias 2010).

25.2.4 The Cherry Tree Metaphor

C2C proponents often use the metaphor of a cherry tree to sum up the three key principles: "*Thousands of blossoms create fruit for birds, humans, and other animals, in order that one pit might eventually fall to the ground, take root, and grow...although the tree actually makes more of its "product" than it needs for its own success in an ecosystem, this abundance has evolved...to serve rich and varied purposes. In fact, the tree's fecundity nourishes just about everything around it*".

Rather than being eco-efficient the cherry tree is being eco-effective. C2C translates this into *doing the right thing* rather than *doing the thing right*. Thus, eco-effectiveness is focused on achieving the right goal, as opposed to eco-efficiency, which is focused on optimizing the means to achieving some predefined goal.

25.3 C2C Certification Program

Since 2005, companies have been able to apply for a product-level C2C certification. The certification program was initially administered by MBDC in the US and by EPEA in Europe. Since 2010, the non-profit California based Cradle to

Cradle Products Innovation Institute™ has also been licensed to carry out certifications. The institute trains and certifies consultants around the world, who assist companies in complying with the certification requirements.

25.3.1 Certification Criteria

The certification can be awarded to products at five levels; Basic, Bronze, Silver, Gold and Platinum, of which the criteria for Platinum are the most strict (PII 2016). It should be noted that not even a platinum awarded product guarantees a 'true' C2C product, i.e. one that fulfils all three key principles for all aspects. Rather, products awarded a certification should be seen as 'on the path' to C2C. The progressive nature of the certification also urges for stepwise product improvement from a lower to a higher certification level. The certification criteria cover five categories. For each category, a number of criteria must be met, depending on the level of certification. The overall certification level of the product is determined by the category with the lowest achievement level (PII 2017a), see example in Table 25.1.

Below an illustrative selection of these criteria for version 3.1 of the certification program (PII 2016) is presented.

Material Health

The applicant must obtain an overview of all homogenous materials in the product and materials or substances present at a concentration of 100 ppm (parts per million) or higher must be reported (Bronze and above). Additionally, banned chemicals (e.g. some metals, flame retardants and phthalates) must be reported at any level. After receiving the material list the certifying body evaluates all materials according to their human and environmental risk potentials and to the recyclability of each material and award a colour score (red, yellow, green, grey and banned) for each material following their classification methodology. The higher the certification level, the lower the allowed red and grey (unknown) material content in the product. In addition, Gold- and Platinum level applicants must also demonstrate compliance with Cradle to Cradle emissions standards. These define maximum

Table 25.1 Example of C2C certification scorecard

Certification criterion	Basic	Bronze	Silver	Gold	Platinum
Material health				X	
Material reutilization			X		
Renewable energy and carbon management		X			
Water stewardship			X		
Social fairness				X	
Overall certification level		X			

Based on PII (2017b)

values for the off-gassing of problematic volatile organic compounds (VOCs) from the product in its use stage. At platinum level process chemicals must also be assessed, among which no red chemicals are allowed.

Material Reutilization

The applicant must demonstrate that the product has successfully been designed as either a technical or a biological nutrient (or both if the materials are easily separable, thus avoiding the creating of a "monstrous hybrid"). Furthermore, applicants must be in the process of developing a plan for end of life product recovery. At Gold-level a "Well-defined nutrient management strategy" must be in place and this plan must be implemented at Platinum level. A minimum nutrient reutilization score is required within each certification level. The score is a weighted average of the percentage of the product considered recyclable/compostable (weight of 2) and the percentage of recycled/rapidly renewable content (weight of 1). A material may be classified as recyclable based on its inherent qualities, independently of the existence of an infrastructure for its recovery.

Renewable Energy and Carbon Management

Applicants must supply information on the quantity and sources of electricity and on-site emissions used in the "final manufacturing stage" of the product. The industrial processes covered by the final manufacturing stage in the certification program varies across product categories, see PII (2015). For all levels, except Basic, an applicant is required to present a strategy for supplying the energy needed for the final manufacturing stage of a product through current solar income (photovoltaic, geothermal, wind, hydro, and biomass). At Gold-level at least 50% of the energy required in final manufacturing must come from current solar income (5% for Silver and 100% for Platinum). Finally, for the Platinum level the embodied energy associated with the product from Cradle to Gate must come from current solar income. To increase the share of energy from current solar income, it is allowed to purchase specified renewable energy certificates documenting that the electricity used comes from renewable sources.

Water Stewardship

For all levels except Basic, a facility wide water-audit must be conducted, meaning that all water flows associated with product final manufacturing are fully characterized. For Silver and Gold levels, there are furthermore requirements to characterize and optimize product related chemicals in effluent to develop a strategy for supply-chain water issues for tier 1 suppliers. At Platinum level, all water leaving the final manufacturing facility must meet drinking water quality standards.

Social Fairness

For all levels, applications must perform a streamlined self-audit related to fundamental human rights and any identified issues must be addressed by management procedures. At Bronze level this self-audit must be more thorough and for Silver level and higher applicants must fulfil criteria related to social conditions at suppliers (e.g. through purchasing fair trade materials or FSC certified wood) or initiate local social projects. At platinum level a facility level audit must be completed by a third party following internationally recognized standards (such as SA8000 or BCorp).

A study of the application of the five certification criteria to an aluminium can, belonging to the technical cycle, found that the main learnings for the company manufacturing the cans were (Niero et al. 2016):

- *Material health*: substances even at ppm level, often originating from additives giving the desired functional properties to the base material, have an impact on recyclability.
- *Material reutilization*: ensuring recyclability, e.g. through a change in lacquer, is a prerequisite for achieving the volumes of recycled aluminium required for a high recycled content of new aluminium cans.
- *Renewable energy and carbon management, water stewardship and social fairness*: these certification criteria require interventions at the company's (and to some extend suppliers') processes, rather than in the design of the aluminium can.

25.3.2 Certified Products

As of March 2017, 482 certifications were in place (PII 2017b). These cover more than 482 individual products, since more than one of a company's products may be covered by a single certification, if they are materially very similar (e.g. a series of textile products only differing in patterns and colours). The distribution between certification levels was: 3% Basic, 44% Bronze, 35% Silver, 18% Gold and 0.2% (one product) Platinum. The five largest product categories were Building Supply and Materials (177), Interior Design and Furniture (158) Home and Office Supply (44), Packaging and Paper (27) and Fashion and Textiles (21). It can be seen that the certifications cover both products of a business-to-business and business-to-consumer nature.

25.3.3 Comparison of Certification Program with Type 1 Eco-labels

The term eco-label may cover a wide variety of schemes (see more in Chap. 24). In Europe the most recognized and widespread labels belong to what ISO-14024 classifies as Type 1 environmental labelling (ISO 1999): 'a voluntary, multiple-criteria based, third party program that awards a license which authorizes the use of environmental labels on products indicating overall environmental preferability of a product within a particular product category based on life cycle considerations.' The Nordic swan and the EU flower are examples of Type 1 labels (NEL 2016).

When comparing the C2C certification scheme with Type 1 labels, an obvious difference is that the C2C scheme applies uniform criteria across product categories, whereas Type 1 labels define specific criteria for each product category. The Nordic swan (Type 1) currently contains 56 product categories (e.g. copy and printing paper, cleaning products and computers) (NEL 2016). From a consumer perspective applying uniform criteria means that the same criteria apply for all C2C products (e.g. using 50% current solar income in final manufacturing for all Gold certified products), which makes the criteria easier to understand. However, it also means that products in product groups composed mainly of homogenous materials may relatively easily be granted a C2C certification and that the opposite may be true for other product groups. A C2C label guarantees that the company behind it has gone through the trouble of becoming certified (including the often demanding task of identifying all substances present above 100 ppm), but it does not guarantee that a product has a better environmental performance than its competitors within a given product group. In fact, many criteria, especially those at Basic level, are related to intentions rather than the actual performance of an existing product. By comparison, Type 1 labels are based on the LCA concept of the functional unit, which is used to define and distinguish between the many product categories. Therefore the performance of existing products is all that matters for Type 1 labels. Criteria for each Type 1 label product category are defined and continuously redefined so only a fraction (typically around one third) of the market is able to fulfil them.

Another important difference is that Type 1 labels are based on a life cycle perspective with eco-efficiency as its underlining concept. The C2C certification is also based on a life cycle perspective, but with a focus on the processes within the applying company and the end of life stage. Criteria related to upstream processes at suppliers are included only at the highest certification level (i.e. platinum). In addition, the C2C scheme has a strong focus on the chemical content of the product itself and how this may affect the users and the recyclability of the product. In comparison, Type 1 labels are typically more focused on the inputs and outputs of processes in the life cycle. It should be noted that eco-efficiency and C2C concepts differs considerably with respect to their approach to a sustainable production (see Sect. 25.5). However, as noted in Sect. 25.3.1 C2C certified products are by no means ideal C2C products and in fact the scheme has many similarities with eco-efficiency based Type 1 labels, such as the Nordic swan, meaning that it is just as focused on doing "less bad" as it is on doing "more good".

25.4 Examples of C2C Products and Systems

The C2C concepts and certification has inspired a number of products, systems and initiatives. This section presents six examples: three certified products, a C2C inspired system, a C2C-based product concept and a C2C inspired company network. Note that the three certified product examples all have been on the market, but may not be so anymore (or may not have renewed certifications) at the time you read this, due to the dynamic nature of consumer products.

25.4.1 C2C Certified Technical Product: Returnity© Upholstery Textile

Returnity is an example of a C2C Gold certified series of upholstery products designed to fit into the technological cycle (PII 2017b). The textiles are manufactured by Austria based Backhausen and are reportedly 100% recyclable through a process of chemical hydrolysis, where the nylon polymers are broken down into monomers, which are then fed into the production of new nylon products. In this process downcycling is avoided as the resulting nylon monomers are neither reduced in quantity nor quality. Backhausen has organized a take back system, so when the upholstery has reached the end of its useful life, the user may report it by email, after which the product is being transported from the user to the recycling facility.

25.4.2 C2C Certified Biological Product: Trigema© Clothing

Trigema is an example of a series of C2C Silver certified clothing products (t-shirts, trousers, pyjamas, etc.) designed to fit into the biological cycle (PII 2017b). Trigema is a German-based company, who has designed their certified series to be "completely safe for humans and the environment". Trigema is made of organic cotton. Little information is provided on how the product should be disposed of, but composting, anaerobic digestion and incineration with recovery of nutrients from ash are all in theory viable options that ensure that the biological nutrients (the cotton, dye, etc.) of the product returns to the soil.

25.4.3 C2C Certified Technical Product: TerraSkin© Mineral Paper

TerraSkin is an example of a series of C2C Silver certified mineral paper products designed to fit into the technological cycle (PII 2017b). The TerraSkin company mainly operates in the US and their factory is located in Asia. TerraSkin is composed of 75% $CaCO_3$ powder from limestone and 25% high-density polyethylene (HDPE) binder. It is applied both as graphical paper and packaging paper and has some qualitative advantages over cellulose based paper since it has a very smooth surface and is water and tear-resistant. Currently TerraSkin does not fulfil the first key principles of C2C. This is because no dedicated post-consumer waste management system exists. TerraSkin may be disposed of through plastic municipal collection schemes covering high-density polyethylene (HDPE), where it will be transformed to plastic lumber and consequently take part in, e.g. benches, roadside curps or playground equipment. This represents downcycling since the HDPE content in TerraSkin has acquired reduced functionality, quality and thus economic value.

25.4.4 C2C Inspired System: Ford Motor Company

The C2C inspired initiatives at the Ford manufacturing site in Detroit, US, offer an example of how the C2C concept may be applied to large scale systems (MBDC 2011). After having operated for almost a century, the soils at the factory were heavily polluted and Ford was concerned about how to comply with stricter stormwater regulation related to the Rouge River that received water leaching from the soils. The solution was to use green spaces (lawns and green roofs) in combination with replacing impervious paving with porous paving for improved storm water filtration and retention. To reduce the problem of contaminated soil Ford has initiated a number of small scale phyto-remediation experiments, in which polyaromatic hydrocarbons (PAH) are taken up and broken down by plants. Finally, the installation of photovoltaic cells to cover part of the factory's electricity consumption has been proposed. Other than solving their initial purpose (stormwater retention and filtration, absorption and neutralization of pollutants) the initiatives have reportedly also lead to increased biodiversity at the factory grounds.

25.4.5 C2C Product Concept: Running Shoes for Rent Concept

The last example is not an existing product, but a product concept presented in (McDonough and Braungart 2002). The concept illustrates how biological and technical nutrients may be integrated in the same product and provide different functions. In the running shoe concept, the sole should be composed of biological nutrients, with the aim of releasing beneficial nutrients to the environment with each footstep through abrasion. In contrast, the upper part of the shoe would be composed of technical nutrients. This opens up for the possibilities of replacing the sole of the shoe as it wears down and even to lease the upper part of the shoes out in product service systems.

25.4.6 C2C Inspired Network of Companies: Carlsberg Circular Community

The Carlsberg Circular Community is a cooperation platform launched in January 2014 featuring the brewing company Carlsberg and a selection of global partners (e.g. packaging suppliers) aiming to rethink the design and production of traditional packaging material. Inspired by the C2C design framework, the stated objectives of the Carlsberg Circular Community is: "to rethink the design and production of packaging material and develop the next generation of packaging products that are optimized for recycling and reuse while retaining or improving their quality and value" (Carlsberg Group 2015). The initiative aims both at modifying standard packaging to fulfil C2C certification criteria and at developing new packaging types, such as the Green Fiber Bottle, which is an idea of creating a beer bottle made from sustainably sourced wood fibres.

25.5 C2C Compared to Eco-efficiency and LCA

As alluded to above, C2C has a very different approach to environmental sustainability than eco-efficiency and the related measurement tool LCA, although many similarities also exist. Table 25.2 summarizes the main characteristics of C2C, eco-efficiency and LCA.

Table 25.2 shows that C2C, eco-efficiency and LCA have quite different characteristics. In a position paper on the 'Usability of Life Cycle Assessment for Cradle to Cradle purposes' the problem of combining the use of C2C and LCA is summarized as: 'Measuring a qualitative plan for creating a beneficial footprint by using a quantitative instrument designed to measure an existing environmentally damaging footprint' (MIE 2011). Consequently, the position paper recommends that LCA is not used as a sole means to assess environmental impacts of C2C Certified products. Below we elaborate on some conflicts, and possible solutions, between C2C and LCA that one should be aware of when trying to combine the concepts.

Table 25.2 Main characteristics of C2C, eco-efficiency and LCA

Characteristic	C2C	Eco-efficiency	LCA
Aim	To inspire the creation of products with a positive impact on humans and the environment	To reduce negative environmental impacts per product, service or unit of GDP	To systematically quantify all environmental impacts from life cycles of products and systems
Targets	Mainly industrial designers and CEOs	Business (eco-efficiency was publicized by the World Business Counsel for Sustainable Development) (WBCSD 2000)	Depend on the goal and scope definition. Often CSR department at companies
Vision of sustainable society	Follow the three key principles	Large reductions in negative impacts per unit of GDP compared to today "to a level at least in line with the Earth's estimated carrying capacity" (WBCSD 2000)	No explicit vision of society as a whole, since reducing impacts from the delivery of single functions is the basis of the method
Absolute guidance for product development?	Yes. Always follow the three key principles to the extent possible	No. Achieving increases in eco-efficiency requires case-specific analysis	No. Achieving increases in eco-efficiency requires case-specific analysis
Life cycle approach and coverage of environmental issues	The designer must carefully choose the material composition of a product to obey the three key principles and plan for the fate of the materials through multiply life cycles ("cascades"). No environmental issues are covered explicitly because no negative impacts are assumed to happen when the three key principles are followed	A life cycle perspective is encouraged. The types of environmental impacts covered depend on the assessment tool and the case	Designed to quantify impacts of product life cycles. The LCIA step aims to cover a comprehensive list of environmental issues (see Chap. 10). The goal and scope definition of an LCA determines which issues to include (see Chaps. 7 and 8)
Includes the idea of positive environmental impacts?	Yes	Not explicitly. Reducing negative impacts is the focus	Potentially (see Sect. 25.6.3)

(continued)

Table 25.2 (continued)

Characteristic	C2C	Eco-efficiency	LCA
Quantitative?	Partly. However, qualitative aspects precede quantitative aspects. E.g. quantity of energy used is irrelevant when the right quality is obtained (current solar income). Quantities of emissions are also irrelevant as long as they are the right emissions ('the right material, at the right place at the right time')	Yes	Yes
Driver of innovation?	The stated goal of C2C is to stimulate radical innovation ("remaking the way we make things"). However, not all C2C certified or inspired products appear to be radically innovative, perhaps because it is very difficult to follow the three key principles completely in today's world	The idea of doing more with less environmental impact has been driving energy efficiency improvements for decades (e.g. LED lights)	Depends on the goal and scope definition of an assessment. However, LCA is often applied to existing products on the market, since the LCI of a product not yet developed is often difficult to predict during the product's development
Values intentions?	Yes. Since C2C calls for radical redesign, defining a plan for such a process is valued, which is reflected in the certification system	No. Only actual increases in eco-efficiency are valued	Intentions are not valued by themselves, but a scenario analysis can be used to quantify the projected outcomes of intentions

25.6 Conflicts and Solutions When Combining C2C and LCA

25.6.1 C2C Prioritizes Qualitative Aspects—LCA Measures Quantitative Aspects

Eco-effectiveness is often presented as "doing the right thing", whereas eco-efficiency is "doing the thing right". In pursuing the right thing, C2C is primarily using a qualitative approach. Values and principles come before quantitative parameters.

For example, the quantity of energy used throughout the life cycle of a product or system is irrelevant from a C2C perspective as long as the quality is right (current solar income). This approach is reflected in the C2C certification scheme where there is no restriction on the quantity of energy used throughout the life cycle. However, it should be noted that, despite of the second key principle, the certification allows for a share of the energy use originating from sources other than current solar income. Also, the energy related certification criteria only cover the "final manufacturing stage" (except for the platinum level, for which upstream processes are covered), which will often only account for a modest share of the total life cycle energy consumption, especially for products consuming energy in the use stage. This allowable fraction ranges from 0% at Platinum level to 95–100% at Silver, Bronze and Basic levels (of which 82% of all products currently certified belongs to). Moreover, from a consequential LCA-perspective (see Sect. 8.5.3) purchasing renewable energy certificates (which is credited as using current solar income in the certification scheme) will only have a positive impact on the LCA results if the purchasing leads to an increase in the installation of renewable energy capacity (e.g. more solar cells and wind turbines). If this is not the case, the marginal energy sources are the same as if no certificate had been purchased and the purchasing of certificates should have no impacts on the LCA results. A C2C certified product may therefore, in theory, score relatively badly compared to a reference product in an LCA, because the certification scheme does not encourage energy efficiency.

Another conflict between C2C and LCA arising due to the former's focus on quality is that the quantities of materials are not considered, as long as they are designed as technical or biological nutrients. This means that C2C does not (actively) support the idea of dematerialization, which may lead to relatively material-intensive products. This is not a problem per se, but if the energy to produce the products is not current solar income (see above) or if the infrastructure to ensure the recycling of the product (be that in the technical or biological cycle) is not in place (see Sect. 25.6.2) it may result in a relatively bad LCA performance.

Finally, it can be argued that a focus on quantity of emissions and resource use is crucial in a world with a finite resource base and emission absorption capacity and a growing demand for goods and service. Section 25.6.4 will further explore these issues.

25.6.2 C2C Values Intentions—LCA Values Measurable Outcomes

When scrutinizing the C2C certification criteria it becomes clear that many of them, especially at the lower levels, relate exclusively to intentions. This is in line with

C2C being characterized as a qualitative plan (see Sect. 25.5). Hence, the Basic certification level can largely be achieved solely by a combination of documenting the existing characteristics of the product and stating intentions for improving it in the future. LCA does not value intentions. Therefore, a product that has obtained a C2C certification without having undergone any change in elementary flows of its life cycle in the process scores exactly the same in an LCA as the product assessed before being awarded the certification.

On a more general note, designers wanting to design a product to fit into the technical cycle often face a paradox: When introducing their novel products to the market they cannot be sure that it will remain in the technical cycle due to the lack of recycling infrastructure. However, if they decide to wait for such an infrastructure to be established before introducing their product, they may have to wait for a long time, because the construction of the infrastructure is often driven by demand in the form of products in need of new recycling technology. Consequently, the designer may choose to market the product, in spite of the lack of infrastructure, in hope that such an infrastructure will eventually be feasible to establish because of the volumes of the product that have entered the market and possibly other products fitting into the same infrastructure. In such cases, intentions of recycling the product may in time lead to its recycling. This can be captured in an LCA, depending on its goal and scope (Bjørn and Hauschild 2013), but LCA results based on assumptions of a future recycling system are inevitably more uncertain than results based on modelling an existing system.

25.6.3 C2C Operates with Positive Environmental Impacts—LCA Operates with Negative Environmental Impacts

A central perception in C2C is that the environment can be "benefitted" from the addition of nutrients. This is used as an argument to integrate nature into human designs instead of aiming to conserve nature through the separation of human and natural processes. LCA, on the other hand, separates the environment from the economy abstractly via its ecosphere/technosphere distinction. The ecosphere is composed of a number of environmental compartments that respond with a negative impact when concentrations of chemicals, or other indicators of the ecosystem state, are increased marginally as a result of the environmental exchanges of a product system. The technosphere may contain organisms of natural origin, but they are highly manipulated and hence not considered part of the environment. A pot plant is an example of this. The positive impacts on the quality of the pot plant soil from the addition of biological nutrients are not classified as benefits to nature in LCA, because the pot plant belongs to the technosphere. Instead, LCA can account for the negative environmental impacts that have been avoided if the biological nutrients added to the pot plant displace the use of synthetic fertilizer. This accounting practice falls within the general scope of LCA of quantifying negative impacts on the environment (whether caused or avoided) with the aim of reducing these.

One could argue that positive impacts on the ecosphere ought to be captured in LCA by spatially resolved characterization factors, which take into account that nutrients that end up in ecosystems that are naturally nutrient poor can increase the number of species and their populations within that ecosystem. However, because LCA aims to conserve natural conditions as much as possible, an increase in the number of species and their populations is generally not considered positive, at least not if it happens at the cost of a reduced number of original species, adapted to nutrient-poor conditions, and their populations. For example, the endpoint LCIA model of freshwater eutrophication adopted in the ReCiPe LCIA methodology (Huijbregts et al. 2015) derives the marginal damage from an increased phosphorous concentration from a set of monitoring data that excludes data from nutrient-poor locations (less than 0.1 mg phosphorous per litre freshwater). This exclusion is intentional as the authors "do not account for a potential overall increase in species richness due to an increase in P for oligotrophic situations" (Struijs et al. 2010). Because of the conservation ethics underlying LCA any change in natural conditions, even if they result in more species and higher populations, are considered negative, because changes essentially lead to cultural landscapes. The conservation ethics can also be defended from a biodiversity perspective, because an increase in local species (and genomic) diversity through the addition of nutrients may come at the expense of a reduction in regional or global species (and genomic) diversity. The species thriving in a nutrient-poor environment may become threatened or go extinct, if the practice of adding nutrients happens at a large scale, which is the case in many regions of the world due to modern agriculture's reliance on synthetic fertilizer (Steffen et al. 2015).

In some cases, LCA can actually handle genuine positive environmental impacts (not just reduced or avoided negative impacts). This applies when the function of the studied system(s) is to restore a previously manipulated piece of land or aquatic system (technosphere) to a natural state (ecosphere). For such a study, an example of a functional unit is "reintroducing 0.5 m of top soil and natural vegetation during no more than 5 year to 1 ha of eroded land". In this case, the result of the LCIA will still be expressed as negative impacts for a number of impact categories. It would then be up to the decision-makers to decide whether the ecological benefit of the functional unit outweighs the negative impacts quantified by the LCA of delivering the functional unit (e.g. from treating potential pollution at the site and from extracting and transporting top soil from another location).

The above points mean that one must be very specific about the type of environment to be benefitted when trying to evaluate a C2C product using LCA. What exactly is meant by "benefit"? Is the benefit intended on a manipulated environment that in LCA would be characterised as part of the technosphere? Is the benefit supposed to increase species diversity in naturally nutrient-poor environments? Can the benefit be considered an obligatory or positioning property, meaning that it may be captured by the functional unit of an LCA study (see Sect. 8.4)?

25.6.4 C2C is Positive Towards Material Growth: LCA is Not

In line with C2C proponents not supporting the idea of dematerialization at the product scale, they also do not support it at the economic scale. Rather than decoupling negative environmental impacts from GDP, C2C advocate a "positive recoupling" of the relationship between economy and ecology, which is argued to support future economic growth (Braungart et al. 2007). This stand can also be inferred from the 2002 book: "The key is not to make human industries and systems smaller, as efficiency advocates propound, but to design them to get bigger and better in a way that replenishes, restores and nourishes the rest of the world" (McDonough and Braungart 2002). The C2C argument is that as long as resources are circulated within continuous loops, then the amounts of resources circulating and their rate of circulation does not need to be restricted. Also, when accepting the premise that the introduction of biological nutrients to the biological cycle is beneficial to the environment (see Sect. 25.6.3), then the introduction of even more biological nutrients are seen as even more beneficial.

This obviously conflicts with the perspective in LCA that impact is proportional to the quantity of elementary flows, which increases with the increase of product flows and other flows that are internal to the technosphere. Currently this argument holds for all products, inspired by C2C or not, because societies generally do not respect the three key principles in C2C. In other words, in today's world an increase in products inevitably results in an increase in consumption of renewable and non-renewable resources, CO_2 emissions, and other emissions that occur due to the lack of closed material loops.

Would increasing consumption, as argued by C2C proponents, then be unproblematic in a perfect C2C world, i.e. one with no violations of the three key principles? Probably not, when considering that economies having an increasing consumption historically have coincided with the accumulation of materials in societal stocks, such as buildings and infrastructure. This conclusion was also reached by the European Environmental Agency (EEA 2011) for the European economy as a whole: "even maximum recycling cannot cover all EU demand for resources. This is due to the accumulation of goods in a growing EU economy, for example in the construction sector, which acts as a long-term store for materials, making them unavailable for recycling for many years". So even if 100% recycling, i.e. no waste generation, is achieved in a materially growing economy, the accumulation of materials must be compensated by a supply of virgin resources, as illustrated in Fig. 25.2 (when disregarding the theoretical option of obtaining resource from other economies that are materially shrinking).

This situation is fundamentally unsustainable because it will eventually lead to resource scarcity. This scarcity is further driven by the increasing need for minerals going into the renewable energy infrastructure (windmills, photovoltaic cells, transmission cables, etc.) needed to fuel increasing material throughputs (Kleijn and van der Voet 2010).

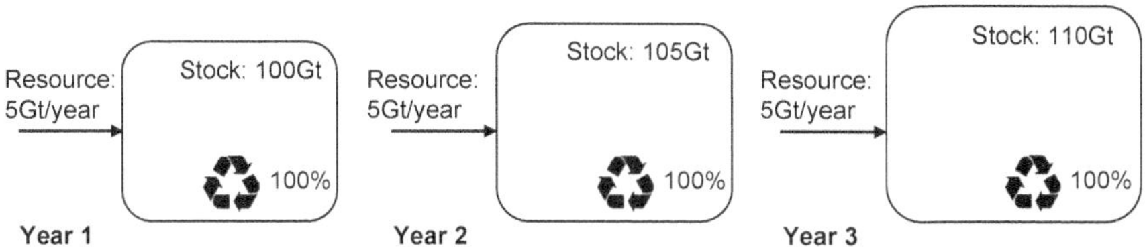

Fig. 25.2 Illustration of how resource input is required in a materially growing economy that recycles 100% of its resources

A growing economy may also cause scarcity of renewable resources because the planet has a limited (short term and long term) capacity to produce biomaterials through photosynthesis (Running 2012). This became evident when increased use of biofuels from especially corn caused food prices to increase dramatically in 2007 thus resulting in food scarcity for the poorest (Partzsch and Hughes 2010). Moreover, biodiversity is generally lower in managed biological systems (agriculture, plantations, etc.) than in natural systems. An economy with a growing intake of biomaterials, therefore, threatens to reduce biodiversity.

25.7 Learnings from C2C to LCA Experts and Practitioners

So far, mostly C2C's shortcomings have been addressed by scrutinizing the concept from the perspective of LCA. However, C2C does point out some thought-provoking weaknesses of LCA and the underlining concept of eco-efficiency. These weaknesses and their potential solutions should, in the authors' opinions, be addressed by LCA experts and practitioners in their pursuit of a sustainable world.

25.7.1 Avoid Perfectly Optimized Wrong Systems

First of all, eco-efficiency and LCA are more focused on the means to achieve sustainability than on how a sustainable future might actually look. The guiding principle is reductions of environmental impacts per functional unit, but the complex dynamics between the different functions making up a society is poorly captured in LCA and to a large extent excluded from the system boundaries of an assessment for pragmatic reasons.

This may lead to what C2C proponents sarcastically terms "highly optimized wrong systems". Such a situation arises when in the pursuit of eco-efficiency systems or technologies that are inherently unsustainable, are optimized. LCA generally tells you what is *greener* in today's non-ideal world, while C2C tells you what is *green* in an ideal future. Often, both things cannot be pursued simultaneously. This is because an ideal future may never be realized if products and systems are optimized according to what is greener in today's non-ideal world, since this optimization of "wrong systems" could serve to lock societies into an unsustainable trajectory.

Solid waste incineration is a good candidate of a "highly optimized wrong systems". When optimizing the incineration process the efficiency of the flue gas cleaning and the conversion of the chemically stored energy in the waste into heat and electricity are increased. This increases the eco-efficiency of the system. However, for a large part of the solid waste fractions currently incinerated, the practice is inherently unsustainable. Non-renewable materials are lost (for instance petroleum-based plastics), decreased in quality due to contamination (for instance metals), or chemically transformed into a form where they can hardly be utilized (for instance the decrease in bioavailability of phosphorus (Thygesen et al. 2011)). Furthermore, investments in capacity for waste incineration necessitates that plants operate for several decades into the future. This decreases the incentive to install capacity for recycling, clearly conflicting with the aim within the circular economy concept of closing material loops. Moreover, state-of-the-art recycling technologies may perform worse in an LCA than state-of-the-art incineration, because more efforts have gone into the optimization of the latter than into the former. However, as noted above, recycling technologies may have the potential to become the environmentally preferable waste management option if given the chance to develop and mature. The scales may also tip to the favour of recycling due to the tendency of decreasing average environmental impacts per unit of energy consumption as a consequence of increasing shares of renewable energy in energy mixes of most nations. This means that, in time, environmental impacts associated with energy consumption in the recycling processes becomes smaller and, in parallel, the environmental impacts associated with average grid-electricity production and district heating that are superseded by energy production from waste incineration (in an attributional LCA) becomes smaller and less advantageous to replace.

An idea for how to avoid the risk of perfectly optimized wrong systems is to embed a vision of a sustainable society, whether inspired by C2C or not, into the background system of a LCI model. This would mean that background processes would reflect the vision of a sustainable society and thus reward product systems that are designed to fit such a vision. For example, the electricity mix could be composed of 100% renewable sources. Challenges of such a strategy are the choice among competing visions of sustainable societies and the translation of such visions to future models of the thousands of unit processes commonly made available in commercial LCI databases. For example, what modes of transport (private cars, busses, trains?) will exist in a sustainable future and how will these be fuelled (biofuels or electricity or hydrogen produced from renewable sources?)? These challenges are discussed in Chap. 21 on future-oriented LCA. In other cases it may be more appropriate to carry out a straightforward qualitative assessment, where a product system is evaluated based on the extent to which it fulfils a range of sustainability conditions (such as "being completely based on renewable materials" or "not relying on fossil fuels in use stage"), as proposed by Ny et al. (2006) and de Pauw et al. (2015).

25.7.2 Improve LCI Modelling of Material Quality and Multiple Life Cycles

In a life cycle inventory model, the fate of products and their material components in the technosphere is usually treated as an issue of multifunctionality (see Sect. 8.5). For example, a recycling facility provides the co-function (in addition to managing waste) of delivering recycled materials and this co-function is typically handled by performing system expansion or allocation (see Sect. 8.5.2). However, in doing so, it is difficult to capture the difference in quality between the original material and the material leaving the recycling facility (that may have been "downcycled"). Typically, this quality issue is handled quantitatively in LCA by applying substitution factors, that take into account that a certain amount of virgin material is needed to make up for a loss in quality (e.g. a factor of 0.7 for paper recycling). The problem with this approach is that the value of a recycled material depends on which new product the material will take part in. For example, a certain alloy composition of recycled aluminium may be seen as an advantage in one product and a disadvantage in another product. Also, current modelling practice of recycling processes typically does not account for subtle differences in material inputs to recycling or recycled materials (e.g. differences in alloy elements present and their concentrations), which can be decisive for the value of recycled materials in a new life cycle.

To overcome these weaknesses, LCI modelling practice needs to improve the representation of a material's constituents, considering (1) intended heterogeneity (e.g. in the case of coating or metal alloys to improve functionality) and unintended heterogeneity (e.g. a packaging material becoming "polluted" with food during the use stage), (2) how heterogeneity affects material recyclability (for example, some plastic recycling processes cannot operate if impurities exceed a certain threshold), (3) the composition of recycled materials. It may also be feasible to consider multiple life cycles in the LCI modelling, which could be aligned with the C2C idea of planning for a "cascade" for a material (see Table 25.2). If doing so, the functional unit should cover all the functions that a material deliver to the planned life cycles, rather than the function on the first life cycle, as is currently common practice (Niero and Olsen 2016).

25.7.3 Avoid False Sense of Sustainability Progress

LCA results are generally interpreted in a relative manner (see Chap. 2). That is, calculated impacts are compared with impacts from other human activities, e.g. different products in a comparative assessment and/or background impacts (per person) in a reference year through the normalization stage of LCIA. Therefore, LCA results do not provide any information on the seriousness of an impact, understood as its size compared to the carrying capacity of affected ecosystems, i.e.

their capacity to generate resources and absorb pollution (Sayre 2008). If carrying capacities are exceeded, the structure and functions of ecosystems may fundamentally change, leading to the disappearance of species and loss of ecosystem services. Past increases in global consumption of products and services have resulted in the exceeding of several carrying capacities at various scales and this trend is projected to continue (WRI 2005; Steffen et al. 2015), as discussed in Chapt 5. This means that "better" at the product level is not necessarily "good enough". In other words, presenting individual products and services as "green" or "sustainable" because they have an incrementally lower environmental impact than reference products and services may create a false sense of sustainability progress, and in some situations qualify as "greenwashing" (UL 2010).

C2C does not contain the concept of carrying capacity, because C2C aims for positive impacts, rather than limiting negative impacts to levels that nature can cope with, and because calls for the integration of nature and societies. Yet, C2C's ambitious absolute guidance (the three key principles) encourages radical changes in the ways we produce and consume, creates a sense of urgency, and offers a positive vision of a sustainable future to aim for. This may serve as a wakeup call for some societal actors who have not previously been engaged in environmental issues.

LCA practitioners can learn from the absolute perspective of C2C in the interpretation stage of many types of LCAs. In a comparative LCA it is important to note that "better" is not necessarily "good enough" and to put the environmental impacts of the studied product system(s) into a wider sustainability framing, whether such a frame is based on C2C or the concept of carrying capacity, as integrated in normalization references (Bjørn and Hauschild 2015).

25.7.4 Emphasize Positive Impacts and Opportunities When Communicating Results

LCA informs decisions on how to reduce negative environmental impacts. As argued by C2C proponents the idea of doing less of something bad (two negative terms combined) may be psychologically demotivating. Doing more of something good sounds more appealing, which undoubtedly contributes to the wide interest in C2C. The idea of making an effort to create a "delightful diverse, safe, healthy and just world, with clean air, water, soil and power—economically, equitably, ecologically and elegantly enjoyed" is simply more inspiring than the old mantra of reducing negative ecological impacts, accompanied by terrifying images of drowning polar bears, suffocated fish and birds covered in oil.

Can LCA practitioners adopt a positive framing of the need for sustainable development, while still being honest about the many unsustainable trends facing the world, the enormity of the challenge and the insufficiency of incremental improvements? We would have liked to answer this question by "yes, of course!", but at this point we can only come up with a couple of fuzzy suggestions.

Firstly, a closer connection between the functional unit of LCA and happiness can be created, as proposed by Hofstetter et al (2006). This can serve to motivate the task of maintaining or creating (the conditions for) happiness, while minimizing or completely abolishing the "unhappiness" imposed on other living creatures (human or not) by the provisioning of product systems. Perhaps more importantly, a closer connection between what is assessed in LCA and happiness can make stakeholders reflect upon whether a product system is really needed in the pursuit of happiness. Maybe another product system or activity fulfilling a completely different function and having a radically lower environmental impact can be just as good, or better, at creating the conditions for happiness? In the illustrative example of Hofstetter et al. (2006) it is proposed (hypothetically) that the activities of gardening and going to a weekend house on the country side might be equally good at creating conditions for happiness, even though they provide much different functions. The former activity obviously has a much lower environmental impact.

Secondly, LCA may be framed within a larger struggle against business as usual. LCA can inform this struggle since it can show the gap between the required reductions in environmental impacts (e.g. to avoid exceeding carrying capacities) and the actual reductions taking place. The struggle against business as usual is also about social justice. During the last three decades political and economic systems have failed at reducing environmental impacts (most impacts have increased) and have increased income inequality within and between nations. Therefore, it could be motivating for LCA practitioners and their audience to position themselves as agents of change in the struggle for an environmentally sustainable and just world. In the words of Naomi Klein: "What if global warming isn't only a crisis? What if it's the best chance we are ever gonna get to build a better world?" (Klein 2014).

References

Bjørn, A., Hauschild, M.Z.: Absolute versus relative environmental sustainability. J. Ind. Ecol. **17**, 321–332 (2013). doi:10.1111/j.1530-9290.2012.00520.x

Bjørn, A., Hauschild, M.Z.: Introducing carrying capacity based normalization in LCA: framework and development of references at midpoint level. Int. J. Life Cycle Assess. **20**, 1005–1018 (2015)

Braungart, M., McDonough, W., Bollinger, A.: Cradle-to-cradle design: creating healthy emissions—a strategy for eco-effective product and system design. J. Clean Prod. **15**, 1337–1348 (2007). doi:10.1016/j.jclepro.2006.08.003

Carlsberg Group: Carlsber Group CSR Report 2014: Growing Responsibility Together (2015)

de Pauw, I.C., Kandachar, P., Karana, E.: Assessing sustainability in nature-inspired design. Int. J. Sustain. Eng. **8**, 5–13 (2015). doi:10.1080/19397038.2014.977373

EC: Closing the loop—an EU action plan for the Circular Economy. COM (2015) 614 final. European Commission (2015a)

EC: Implementation of the Circular Economy Action Plan. European Commission (2015b). http://ec.europa.eu/environment/circular-economy/index_en.htm

EEA: Earnings, jobs and innovation: The role of recycling in a green economy. EEA Report 8/2011. European Environment Agency. ISSN: 1725-9177. Copenhagen, Denmark (2011)

EMF: Towards the Circular Economy Vol. 1: An Economic and Business Rationale for an Accelerated Transition. Ellen MacArthur Foundation, Cowes (2012)
EMF: Delivering the Circular Economy: a Toolkit for Policymakers. Ellen MacArthur Foundation, Cowes (2015a)
EMF: Towards a Circular Economy: Business Rationale for an Accelerated Transition. Ellen MacArthur Foundation, Cowes (2015b)
Hofstetter, P., Madjar, M., Ozawa, T.: Happiness and sustainable consumption—psychological and physical rebound effects at work in a tool for sustainable design. Int. J. Life Cycle Assess. **11**, 105–115 (2006)
Huijbregts, M.A.J., Steinmann, Z.J.N., Elshout, P.M.F., et al.: ReCiPe2015: a life cycle impact assessment method at midpoint and endpoint level report I : characterisation factors (2015)
ISO: ISO 14024:1999. Environmental Labels and Declarations—Type I Environmental Labelling—Principles and Procedures. International Organization for Standardization, Geneva (1999)
Kleijn, R., van der Voet, E.: Resource constraints in a hydrogen economy based on renewable energy sources: an exploration. Renew. Sustain. Energy Rev. **14**, 2784–2795 (2010). doi:10.1016/j.rser.2010.07.066
Klein, N.: This Changes Everything. Simon & Schuste, New York (2014)
MBDC: The Making of the Next Industrial Revolution at Ford Motor Company. McDonough Braungart Design Chemistry, Charlottesville (2011)
McDonough, W., Braungart, M.: Cradle to Cradle—Remaking the Way We Make Things, 1st edn. North Point Press, New York (2002)
McDonough, W., Braungart, M.: The Upcycle: Beyond Sustainability-Designing for Abundance. North Point Press, New York (2013)
MIE: Usability of Life Cycle Assessment for Cradle to Cradle Purposes—Position Paper. Ministry of Infrastructure and the Environment, Utrecht (2011)
NEL: Product Groups. Nordic Ecolabelling (2016). http://www.nordic-ecolabel.org/criteria/product-groups/?p=1
Niero, M., Olsen, S.I.: Circular economy: to be or not to be in a closed product loop? A Life Cycle Assessment of aluminium cans with inclusion of alloying elements. Resour. Conserv. Recycl. **114**, 18–31 (2016). doi:10.1016/j.resconrec.2016.06.023
Niero, M., Hauschild, M.Z., Hoffmeyer, S.B., Olsen, S.I.: Combining eco-efficiency and eco-effectiveness for continuous loop beverage packaging systems: learnings from the Carlsberg Circular Community. J. Ind. Ecol. (in review) (2016)
Ny, H., Macdonald, J.P., Broman, G., et al.: Sustainability constraints as system boundaries—an approach to making life-cycle management strategic. J. Ind. Ecol. **10**, 61–77 (2006)
Partzsch, L., Hughes, S.: Food versus fuel: governance potential for water rivalry. Food Ethics (2010). doi:10.1007/978-1-4419-5765-8_10
PII: Guidance for Applying the Final Manufacturing Stage Requirements in the Cradle to Cradle CertifiedTM Product Standard, Versions 3.0 and 3.1 (2015)
PII: Cradle to Cradle Certified(TM) Product Standard Version 3.1. Cradle to Cradle Products Innovation Institute (2016)
PII: Overall Product Scoring. In: Get Cradle to Cradle Certified™. Cradle to Cradle Prod. Innov. Inst. (2017a). http://www.c2ccertified.org/get-certified/levels
PII: Cradle to Cradle Certified Products Registry. In: Cradle to Cradle Prod. Innov. Inst. (2017b). http://www.c2ccertified.org/products/registry
PLI: Cradle to Cradle. In: Prod. Inst. (2013). http://www.product-life.org/en/cradle-to-cradle
Running, S.W.: A measurable planetary boundary for the biosphere. Science **337**, 1458–1459 (2012). doi:10.1126/science.1227620
Sayre, N.F.: The genesis, history, and limits of carrying capacity. Ann. Assoc. Am. Geogr. **98**, 120–134 (2008). doi:10.1080/00045600701734356
Steffen, W., Richardson, K., Rockström, J., et al.: Planetary boundaries: guiding human development on a changing planet. Science (Washington D C) **347**, 736 (2015). doi:10.1126/science.1259855

Struijs, J., Beusen, A., Zwart, D., Huijbregts, M.: Characterization factors for inland water eutrophication at the damage level in life cycle impact assessment. Int. J. Life Cycle Assess. **16**, 59–64 (2010). doi:10.1007/s11367-010-0232-z

Thygesen, A.M., Wernberg, O., Skou, E., Sommer, S.G.: Effect of incineration temperature on phosphorus availability in bio-ash from manure. Environ. Technol. **32**, 633–638 (2011). doi:10.1080/09593330.2010.509355

Tobias, C.: Exclusive interview with Dr. Michael Braungart: Businesses should do more good, rather than less bad. In: eco-business.com—Asia Pacific's Environ. Bus. community (2010). http://www.eco-business.com/news/2010/oct/25/businesses-should-do-more-good-rather-less-bad/

UL: The Sins of Greenwashing: Home and Family Edition. Underwriters Laboratories (2010)

WBCSD: Eco-efficiency—creating more value with less impact. World Business Council for Sustainable Development (2000)

WRI: Millennium Ecosystem Assessment: Ecosystems and Human Well-Being. Synthesis. Island Press, World Resources Institute, Washington, DC (2005)

Author Biographies

Anders Bjørn Part of the LCA community since the early 2010s. Main focus is interpretations of sustainability and integration of sustainability targets in LCA to enable absolute sustainability assessments.

Michael Z. Hauschild Involved in development of LCIA methodology since the early 1990s. Has led several SETAC and UNEP/SETAC working groups and participated in the development of the ISO standards and the ILCD methodological guidelines. Main LCA interests are chemical impacts, spatial differentiation and science-based boundaries in LCIA.

Chapter 26
LCA of Energy Systems

Alexis Laurent, Nieves Espinosa and Michael Z. Hauschild

Abstract Energy systems are essential in the support of modern societies' activities, and can span a wide spectrum of electricity and heat generation systems and cooling systems. Along with their central role and large diversity, these systems have been demonstrated to cause serious impacts on human health, ecosystems and natural resources. Over the past two decades, energy systems have thus been the focus of more than 1000 LCA studies, with the aim to identify and reduce these impacts. This chapter addresses LCA applications to energy systems for generation of electricity and heat. The chapter gives insight into the LCA practice related to such systems, offering a critical review of (i) central methodological aspects, including the definition of the goals and scopes of the studies, their coverage of the system life cycle and the environmental impacts, and (ii) key findings of the studies, particularly aimed at identifying environmental hotspots and impact patterns across different energy sources. Based on this literature review recommendations and guidelines are issued to LCA practitioners on key methodological aspects that are important for a proper conduct of LCA studies of energy systems and thus ensuring the reliability of the LCA results provided to decision- and policy-makers.

A. Laurent (✉) · M.Z. Hauschild
Division for Quantitative Sustainability Assessment, Department for Management Engineering, Technical University of Denmark, Produktionstorvet 424, 2800 Kongens Lyngby, Denmark
e-mail: alau@dtu.dk

M.Z. Hauschild
e-mail: mzha@dtu.dk

N. Espinosa
Department of Energy Conversion and Storage, Technical University of Denmark, Frederiksborgvej, 399, 4000 Roskilde, Denmark
e-mail: nieves.espinosa@ec.europa.eu

N. Espinosa
Product Policy Bureau, European Commission Joint Research Centre, Inca Garcilaso, 3, 41092 Sevilla, Spain

M.Z. Hauschild et al. (eds.), *Life Cycle Assessment*,
DOI 10.1007/978-3-319-56475-3_26

26.1 Introduction

Over the past decades, energy systems have increasingly received attention from stakeholders, including from high policy-makers, due to the combination of four major factors. Although different trends can be observed across countries, energy demand is expected to keep increasing worldwide, hence putting an increasing pressure on the supply side. The total primary energy supply, which amounted to 560 EJ globally in 2012, is thus expected to increase by 20–35% by 2040 (IEA 2015a). Conventional fossil resources are still anticipated to absorb that increase although depletion issues, in particular of conventional oil resources, have been widely acknowledged. As a result, initiatives to find alternative resources to fulfil the services that are currently relying on petroleum products have emerged (e.g. electric transportation to replace fossil-fuelled ones; see Chap. 27). In parallel, the increasing risk of disruptions of oil and natural gas supplies have led nations to define strategies to ensure secured energy supply, including establishing of emergency oil stocks for short-term disruptions and/or long-term planning to transition to more renewable and local sources (IEA 2014). Finally, energy systems are the primary source of anthropogenic greenhouse gas (GHG) emissions responsible for climate change. Electricity and heat production alone were thus responsible for 25% of the total GHG emissions in the world in 2010 while transportation was reported to account for 14% (IPCC 2014). In that setting, the key role of energy systems as support for entire economies combined with the triple issues of fossil resource depletion, climate change and energy security has put them at the centre of the sustainability debate. The development and dissemination of renewable energy technologies, deployment of carbon capture and storage systems, fuel switching, continued use of nuclear power and gains in energy efficiency are mechanisms, which can help mitigate these issues and have therefore become the focus of most energy policies (IEA 2015b).

Energy systems embody a wide range of systems and technologies and can be regarded as a “supporting sector”, i.e. a sector that feeds into all other application sectors, e.g. transportation, building sectors, industrial sectors, etc. In relation to life cycle assessment (LCA), it therefore means that energy systems can be considered relevant to nearly all LCA studies ever done until now. According to Chen et al. (2014) and Hou et al. (2015), between 1998 and 2013, approximately 7500 scientific articles and proceedings papers were published in the field of life cycle assessment and 1067 of them could be categorised within the subject “Energy and Fuels” . Simply taking the keywords “energy systems”, “energy technologies”, “power systems”, “power plants”, “electricity systems”, “heat systems”, combined with LCA leads to the non-exhaustive identification in Web of Science of 674 scientific articles published up to 2015, see Fig. 26.1. Matching the pattern observed by Chen et al. (2014) for all LCA-related publications, an exponential trend can be observed.

Energy systems and technologies considered in this chapter are limited to the energy supply systems and can be categorised in two major groups: electricity and heat production systems and fuels for transportation. Further differentiation can be done depending on energy sources (e.g. coal, wind, nuclear power, etc.),

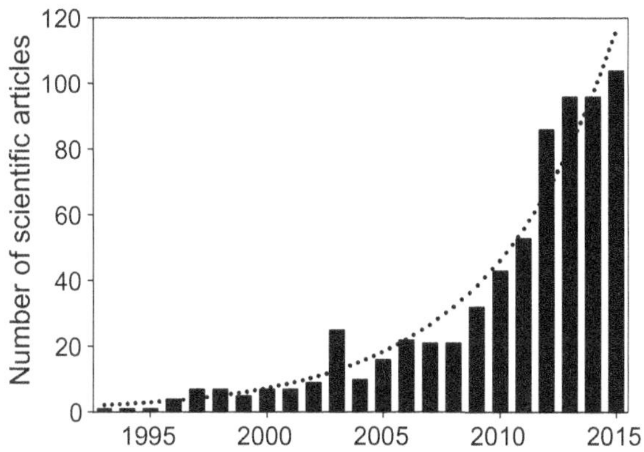

Fig. 26.1 Number of scientific articles addressing LCA and energy systems (non-exhaustive; total retrieved of 674 papers). Search made in ISI Web of Science using the keyword LCA combined with either "energy systems", "energy technologies", "power systems", "power plants", "electricity systems" or "heat systems" (Thomson Reuters, New York, NY). Exponential trend displayed in dotted line ($r^2 = 0.95$)

technology types (e.g. concentrated solar power and photovoltaics for solar power), application types (e.g. electricity or heat only, or combined heat and power plants), or fuel types (e.g. trains running from electricity or diesel in railway transportation). Overall, these sub-categories and differentiations are not addressed exhaustively in this chapter, which is intended to remain overarching and generic to all energy systems. In addition, the present chapter is limited to only addressing LCA in relation to electricity and heat production systems and therefore does not cover fuels for transportation. For the latter, the reader is referred to Chap. 27, which addresses e-mobility and touches upon that topic in relation to road transportation, and Chap. 30, which specifically addresses biofuels.

26.2 Literature Review

This section is intended to provide a non-exhaustive overview of research in the field of LCA applied to electricity and heat systems. It aims to provide an analysis of the key points of published LCA studies, addressing both methodological aspects and main findings.

26.2.1 Goal and Scope of the Studies

LCA studies on electricity and heat systems can roughly be divided into two main categories, which differ by the scoping/scaling, complexity and overarching goals of the study:

1. Studies assessing a specific energy technology/source/system at a power plant level (with possible inclusion of transport and distribution system) or sub-power plant level (e.g. specific component of the system). The goals of the study typically include weak-point analyses for eco-design, reporting/documentation of environmental performances of a newly developed technology, benchmarking against other technologies using the same or other energy sources (renewables and/or non-renewables).
2. Studies assessing energy systems in a context perspective, typically at meso- and large-scale. These studies relate the supply systems to context-dependent parameters, including the energy demand, types/settings of the application of the system, etc. They are primarily associated with goals oriented towards policy analysis or decision- and policy-making at urban, national or regional scales. They include retrospective and foresight studies looking into national energy scenarios, penetration of renewables into electricity grid mixes, installation and deployment of micro-grids for buildings, etc.

Most LCA studies made on energy systems are Category 1 studies, while the conduct of Category 2 studies is typically post-2010. Over the years, Category 1 studies have been commissioned and/or performed by electricity suppliers and researchers in academia for individual technologies, energy sources and national or regional grid mixes. The accumulated large pool of data can now be found in LCI databases, such as ecoinvent (Weidema et al. 2013), where hundreds of single processes, differentiated by energy sources, technologies and locations and typically defined as the supply of 1 KWh or 1 MJ, are available to LCA practitioners. A non-exhaustive glimpse of Category 1 studies is provided in Sect. 26.3.1 and in Table 26.4 (placed in Appendix); an overview of Category 2 studies is given in Table 26.1.

The definitions of the scope of the studies vary significantly between the two categories of studies as well as within a same category. Most of the choices with regard to the scope definition are not harmonised and are often made by the LCA practitioners based on previous studies and/or reference guidelines, such as the ISO standards or the ILCD Handbook. An example is the choice of the LCI modelling framework, with studies relying on attributional modelling with use of allocation while others use consequential modelling (see Sect. 8.5). These choices are not always clearly justified in studies, in particular with respect to the goal of the study.

Although not always transparently reported in the past studies, an important step in the scope definition is the elaboration of a properly defined functional unit (FU). Because Category 2 studies look at the energy system in relation to its context while Category 1 studies do not, different functional units can be observed. Two major types of functional units can be found in Category 1 studies: (i) FUs defined as the generation of 1 kWh or MJ of electricity/heat at power plant/heat unit, and (ii) FUs defined as the supply of xx kWh of electricity to the grid (thus including the energy transport and distribution systems). These definitions are by far the most common and relate to studies looking at the output of the energy production system. Other types of functional units, with more focus on the fuel inputs to the system, can also

Table 26.1 Examples of Category 2 studies i.e. systemic, context-driven studies

Scale	Functional unit	Short description (incl. modelling)	Reference
Macro-scale (global); prospective	(*Not explicitly defined*) Interpreted as the supply of electricity to match the global demand up to 2050 (demand fixed by different scenarios)	– Assessment of environmental impacts associated with the BLUE Map scenario compared to the business as usual scenario, as defined by the International Energy Agency over the period 2007–2050 – Use of a hybrid LCA model combining multi-regional input–output model and process LCIs. Inclusion of a dynamic perspective (e.g. evolution of grid mixes over time, etc.)	Hertwich et al. (2015)
Macro-scale (global, regional, national); retrospective	(*Not explicitly defined*) (i) Supply of electricity matching demand in each country in a given year (demand fixed by statistics for each country in each year); (ii) 1 kWh of electricity consumed in a given country in a given year	– Retrospective assessment of environmental impacts from electricity generated in each country/region over the period 1980–2011 – Use of process LCI and historical statistics on electricity produced from different energy sources and technologies in each country/region	Laurent and Espinosa (2015)
Macro-scale (EU); prospective	(*Not explicitly defined*) Interpreted as the supply of electricity matching the demand in the EU for the period 2005–2010	– Assessment of environmental impacts caused by each of two policy scenarios over 2005–2010: bioenergy policy and business as usual policy – Use of consequential LCA to capture impacts of the policy implementation, e.g. increase in biomass demand in non-EU countries	Dandres et al. (2011)
Macro-scale (EU); prospective	(*Not explicitly defined*) Interpreted as the supply of electricity matching the demand in the EU in the year 2050	– Assessment of environmental impacts associated with 44 scenarios electricity supply in the EU in 2050 – Use of a hybrid LCA model combining multi-regional input–output model and process LCIs, incl. requirements for accommodating the variability of wind and solar power (e.g. storage) and changes in grid mixes for production processes	Berril et al. (2016)

(continued)

Table 26.1 (continued)

Scale	Functional unit	Short description (incl. modelling)	Reference
Macro-scale (Mexico); present perspective	Total annual amount of electricity generated by public sector in 2006, i.e. 225,079 GWh	– Assessment of environmental impacts of electricity generation in Mexico in 2006 – Process LCI data used	
	Santoyo-Castelazo et al. (2011)		
Macro-scale (Estonia); prospective	1 MWh of grid electricity consumed in Estonia	– Comparative assessment of 3 scenarios for 2020 (i.e. nuclear, oil shale, natural gas scenarios) compared to "current" situation in 2002 – Correction of process LCI to adapt future scenarios	Koskela et al. (2007)
Macro-scale (Denmark); prospective	1 kWh of electricity consumed in Denmark	– Comparative assessment of 2 scenarios for 2030 (2030-Green and business as usual) in Denmark compared to "current" situation in 2010 – Consequential LCA to model possible future Danish power systems (future changes for power generation technologies included)	Turconi et al. (2014)
Macro-scale (United Arab Emirates); prospective	Supply of 1 kWh of net electricity	– Comparative assessment of a number of scenarios for 2020, 2030 and 2050 (planned policies, planned policies with carbon capture and storage systems after 2030, nuclear scenario, renewables scenario), also compared to "current" situation in 2010 – Technologies foreseen in use in 2030 based on literature sources. Combination with process LCI.	Treyer and Bauer (2016)
Meso-scale (Island of Koh Jig); present/ prospective	Supply of 265 kWh of electricity per day to Koh Jig Island for 20 years (i.e. 1934.5 MWh)	– Three alternative microgrid systems of electrification for the entire island of Koh Jig (1.2 km^2) – Interpreted as attributional model with system expansion used for recovered materials only (not energy)	Smith et al. (2015)
Meso-scale (house); present	Total power generation in one year	– Comparisons of 9 different power generation systems to sustain energy requirements of a standalone mobile house in Turkey	Sevencan and Ciftcioglu (2013)

be found in literature, e.g. studies assessing different fuel inputs to a power plant and focusing on their different energy contents.

With regard to Category 2 studies, the functional unit is often defined as the supply of an amount of energy based on the demand of the country, region or entity supported by the energy systems under study in a temporal perspective, i.e. past, present or future-oriented (see Table 26.1) illustrating the variety of Category 2 studies. As reported in Table 26.1, two main types of functional units are often used. They differ by the amount of energy, which defines the "quantity" aspect of the functional unit. This quantity may either match the total energy demand/consumption defined by the scenario(s) considered (e.g. Hertwich et al. 2015; Berril et al. 2016) or be normalised to the consumption of one kWh for all scenarios (e.g. Turconi et al. 2014). In the former, some practical challenges may arise. In studies encompassing a wide scoping with several scenarios and sub-systems, the quantification of the functional unit may thus become difficult. For example, in Laurent and Espinosa (2015), the environmental impacts associated with the electricity generated in each country in the world for each year within the period 1980–2011 were assessed. It means that for national assessments, as many functional units as numbers of countries and numbers of years included in the study need to be quantitatively defined although the primary functions are the same, i.e. the supply/generation of electricity matching the demand in each country and each year. Similar issues can be observed in future-oriented studies, for example in Hertwich et al. (2015), where the potential environmental impacts of the BLUE map scenario (IEA 2015b) are compared against those of the business as usual scenario: each scenario entails different energy demands, which are accounted for in the analysis of the results to demonstrate the benefits of renewables in electricity supply systems. As indicated above, other studies, which have assessed future energy scenarios, have defined their functional units as one kWh of electricity consumed/generated (e.g. Turconi et al. 2014; Treyer and Bauer 2016).

26.2.2 *Life Cycle Coverage*

One of the strengths of LCA is the adoption of a life cycle perspective (see Chap. 2). Including all the life cycle stages, from the raw materials extraction to the final disposal stage, is important to prevent environmental burden-shifting from one life cycle stage to another. For example, renewable energy technologies are often improperly flagged as "green" in different media. However, this denomination often only holds when they are considered solely in their use stage and mainly in relation to climate change impacts (see also Sect. 26.2.3). Renewables have important environmental impacts outside their use/operation stage, e.g. production (see Sect. 26.3.1). Therefore, when taking the whole life cycle of renewables-based energy systems, one may demonstrate that they are "greener" than fossil-based energy systems, but they are not free of any environmental impacts.

In LCA studies of energy systems, the life cycle has often been truncated, in particular with the disregard of the disposal stage and, to a lesser extent, of the use stage. Arvesen and Hertwich (2012) thus showed in a review of LCA studies of wind power (44 reviewed studies) that the manufacturing stage was the only life cycle stage common to all studies. Most studies were reported to consider the operation and maintenance of the wind power plants even though different assumptions we made. The end-of-life was either omitted or modelled using assumptions for the decommissioning and recovery of materials/energy. Likewise, in their review of LCA studies of thin-film photovoltaics (PV) systems, Chatzisideris et al. (2016) found that out of 46 studies, all addressed the production stage (incl. raw materials extraction) while only 29 (i.e. 63% of studies) and 11 (i.e. 24%) studies encompassed the use and disposal stages, respectively.

As indicated in Sects. 26.2.3 and 26.3.1, environmental impacts of renewable energy sources stem from the production of the different materials, infrastructure and equipment supporting the systems, e.g. PV modules and supporting infrastructure for photovoltaics (e.g. Espinosa et al. 2015), or components of wind turbines (e.g. Arvesen and Hertwich 2012). Important positive effects can arise in the total environmental burden of the systems when materials are recycled at the end-of-life of the systems, thus substituting the production of virgin materials, or when energy recovery accompanies incineration of materials, thus substituting the generation of heat and electricity from conventional, often fossil-based energy sources. Although inconsistencies and lack of transparency have been observed across studies addressing the disposal stage, most studies point out the high relevance of the disposal stage in the total environmental burden of the energy systems (e.g. Arvesen and Hertwich 2012; Espinosa et al. 2015). In addition, for some energy sources, specific environmental impacts are largest during their use/operations, e.g. water use impacts for hydropower (Pfister et al. 2011). These observations thus highlight the great risk of truncating the life cycle of the energy systems and only limiting it to the materials and production stages. Important biases may be associated with results of such narrowly scoped studies, for example if a study points out high impacts during the production of materials while overlooking that these materials end up being recycled with high efficiency in the disposal stage, thus reducing considerably their respective environmental impacts.

26.2.3 Impact Coverage

Because of the strong focus of energy policies on mitigating climate change and maximising energy efficiency, a large majority of the LCA studies focusing on energy systems have limited their impact assessment to the sole quantification of life cycle GHG emissions (expressed in mass unit of CO_2 equivalents) and energy demand (e.g. use of cumulative energy demand indicator, energy payback time). A number of reviews focusing on specific energy technologies or systems have identified and reported such patterns. Examples of such reviews include Schreiber et al. (2012), who focused on electricity generation with use of carbon capture and storage (CCS) systems (15 studies), Arvesen and Hertwich (2012), who reviewed LCA studies of wind power (44 studies), and Chatzisideris et al. (2016), who assessed the body of LCA studies on thin-film photovoltaics (33 studies). In some situations, this simplification is a conscious choice made by the authors of the studies, who sometimes acknowledge the limitations of the study and recommend that other environmental impacts be considered (e.g. Burkhardt et al. 2012). Other situations show ambiguity as to whether the authors are aware that GHG emission accountings and energy demand assessments do not necessarily represent the total environment burden. Such authors often use the terms "environmental impacts", "life cycle assessment", "environmental LCA" to refer to assessments or studies that only deal with life cycle GHG emission and/or energy demand accountings, and, more importantly, without making clear to the reader the distinction between them and the possible limitations to their conclusions (e.g. Sherwani et al. 2010; Chua et al. 2014).

The inclusion of a limited number of environmental impacts may invalidate the support provided to decision-makers if one aims to assess the total environmental burden of a system or technology. Such situations can be the result of environmental burden-shifting from one impact category to another, i.e. if decisions directed to reducing one impact inadvertently lead to increase in others, which are overlooked in the assessment. Figure 26.2 illustrates this phenomenon at the level of individual energy sources by considering the switch from fossil fuels to renewables per kWh of electricity produced (updated from Laurent et al. 2012).

When moving from fossil-based energy sources (in brown dots in Fig. 26.2) to renewables (coloured dots), reductions of 1–2 orders of magnitude in the climate change impacts (x-axes) are observed for a same electricity output. What is interesting is that other environmental impacts such as acidification and particulate matter (Fig. 26.2a, b) are being decreased at the same time because these stem from the same emission sources as for GHG emissions. However, for other environmental impacts, notably the toxicity-related impacts (Fig. 26.2c) or non-renewable resource depletion (Fig. 26.2d), such trend may not be observed and while the climate change impacts are being reduced, these impacts may remain at the same level or even increase. This is for example suggested for wind power or solar power in Fig. 26.2c, d, where the human toxicity impacts and resource depletion impacts are, respectively, comparable and increased compared to those of electricity produced from natural gas or hard coal (due to larger emissions of heavy metals and use of rare metals through the life cycle of the energy systems).[1]

Therefore, a study only assessing climate change runs the risk to overlook these trends in other environmental impacts and provide recommendations to policy- and decision-makers that could either be further optimised, or worse, lead to

[1]Note that these results may also be sensitive to the selected LCIA methods (particularly for resource depletion, for which no widely accepted indicator exists) and to the LCI data present in ecoinvent database (differences in system boundaries of technologies; disregard of evolving technological level in renewable energy sources).

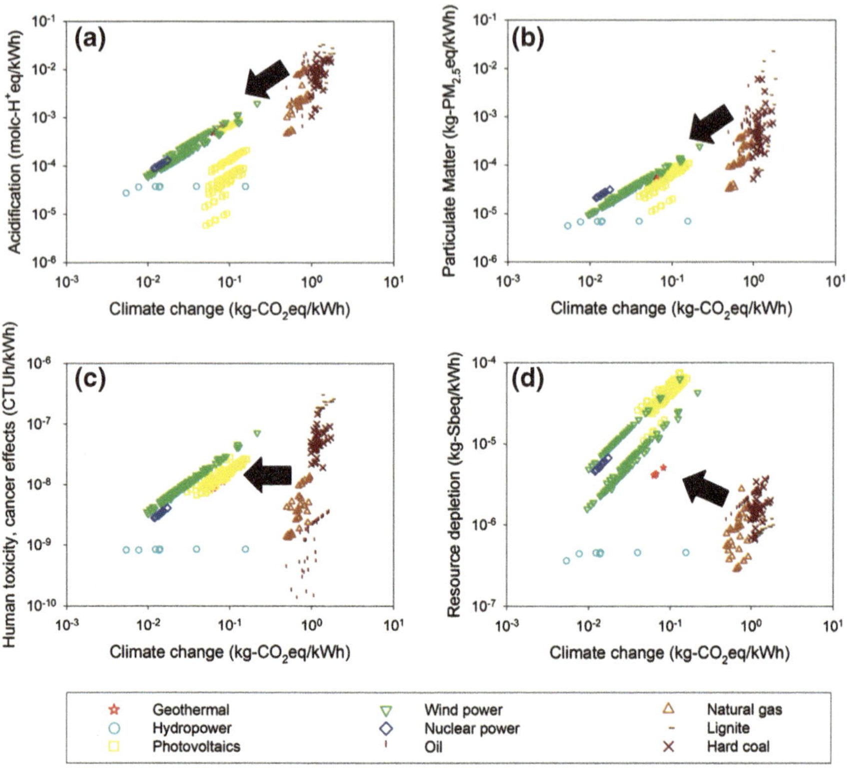

Fig. 26.2 Selected environmental impacts for electricity production plotted against climate change impacts: **a** Acidification, **b** particulate matter, **c** human toxicity—cancer effects, **d** resource depletion (updated from Laurent et al. 2012). *Black arrows* reflect the trends when switching from fossils to renewable energy sources; they are marked for indicative purpose and disregard variations across energy sources. Logarithmic scales are used on both axes. Study performed using ecoinvent 3.1 LCI database and ILCD LCIA methodology in SimaPro LCA software

unsustainable pathways (Laurent et al. 2012). At the level of national electricity mixes, occurrences of environmental burden-shifting have been observed in the past. A prime example is the French grid mix, for which the switch from fossils to nuclear power after the oil crisis in the 70s has contributed to decrease the climate change impacts from the electricity sector by more than 60% between 1980 and 2011 (in spite of increased electricity demand) whereas other environmental impacts have increased in the same period, e.g. ca. 50% for freshwater ecotoxicity impacts and ca. 600% for ionising radiation (Laurent and Espinosa 2015). This calls for covering the whole spectrum of environmental impacts when performing life cycle assessments of energy systems.

26.3 Main Findings of Published LCA Studies

26.3.1 Analysis of Environmental Hotspots

The life cycle of heat and electricity generation systems can be regarded as the inter-section of two life cycles: (i) the life cycle of the power plant unit, including the transmission and transport infrastructure and the equipment at the plant; and (ii) the life cycle of the fuels (see Fig. 26.3). The latter is irrelevant for systems relying on wind power, solar power, hydropower and geothermal power, for which the energy source is assumed directly available without additional processes than those already encompassed in the life cycle of the power plant itself. These are also energy sources for which no fuel combustion takes place.

LCA studies have demonstrated that two different patterns exist in the localization of the largest environmental impacts in the life cycles of heat and electricity generation systems, with a major distinction between systems based on fossils, biomass and nuclear power (i.e. where there is fuel combustion) and those relying on wind power, solar power, hydropower and geothermal power (i.e. where no fuel combustion occurs).

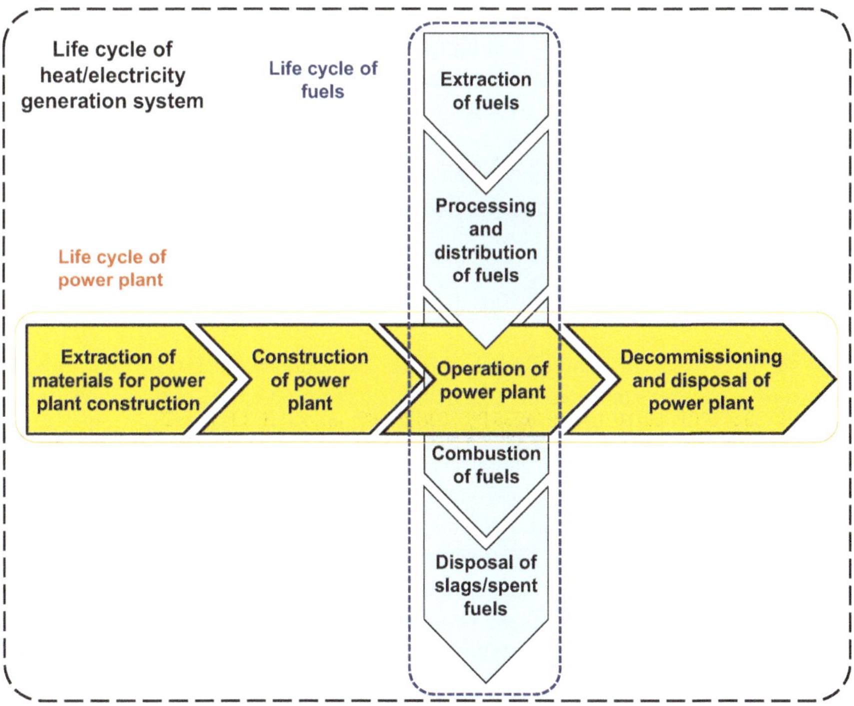

Fig. 26.3 Life cycle of heat and electricity generation systems with intersecting life cycles of the power plant and the fuel required for the operations

In that regard, the capital goods (e.g. power plant facilities, turbines, machineries, etc.) are a relevant part of the systems to address, in particular the extent to which they contribute to the overall environmental burden and their ability to be disregarded or not by practitioners. While capital goods are the main drivers of the impacts for hydropower, wind, solar and geothermal power (no fuel), and hence should not be disregarded for those systems, their contributions in other systems, e.g. fossil-based, is less obvious. Frischknecht et al. (2007) have thus demonstrated a dependency on the type of impact categories considered in the assessment. Generally, the non-toxicity impact categories, such as climate change, are negligibly affected by capital goods whereas toxicity-related impact categories and resource use and depletion impacts (e.g. metal depletion) are more sensitive to the inclusion of capital goods. Capital goods may thus contribute to 94% and 85% to metal/mineral depletion and land use for coal-fired power plant systems, respectively (Frischknecht et al. 2007). For natural gas power plants, other impact categories may also be significantly affected by capital goods, e.g. if the natural gas supply in the assessed region relies on long-distance gas transport (Frischknecht et al. 2007). Assuming a full coverage of environmental impacts, these results therefore call for the systematic inclusion of capital goods when assessing energy systems. Note that these are included by default in many process-based LCI databases, e.g. ecoinvent 3 (Weidema et al. 2013).

Other distinctions can also be observed within the two aforementioned categories of systems, but they are often limited to specific impact categories (e.g. water use or land use between wind and geothermal power) and are technology-dependent (e.g. reservoir-based *vs.* run-of-river hydropower). Table 26.2 provides an overview of the environmental hotspots per impact category and major energy source based on the generation of a kWh-unit of electricity. A summary per group of energy source is provided in the following subsections.

26.3.1.1 Coal-, Gas- and Oil-Based Systems

With the exception of metal/mineral resource depletion indicators, which indicate distribution of impacts between the materials requirements for the power plant construction and those of the infrastructure for the mining activities, all impacts stem predominantly from the operation in the use stage of the power plants, in particular from the life cycle of the coal, gas or oil fuels.

Three major environmental hotspots can thus be identified: (i) the mining activities, which contribute to freshwater eutrophication and toxicity-related impacts through the resulting spoils, to water use, land use and fossils depletion categories through the use of these resources, and to metal depletion (i.e. mining infrastructure); (ii) the fuel combustion, which is a major contributor to all airborne-emission-driven impacts such as climate change, acidification, terrestrial and marine eutrophication, particulate matter or toxicity-related impacts; and (iii) the disposal of the heavy metals contained in the combustion slag and bottom ashes, which primarily contribute to toxicity-related impact categories.

Table 26.2 Location of environmental hotspots in heat and electricity generation systems per impact category for each energy source (colour coding differentiating the patterns)

Impact categories	Coal	Nat. gas	Oil	Nuclear power	Wind power	Solar power	Hydro-power	Geo-thermal	Biomass
Climate change	U (U)	U (U)	U (U)	U (RP)	RP	RP	RP	RP	U (RP/U)
Stratospheric ozone depletion	U (RP)	U (RP)	U (RP)	U (RP)	RP	RP	RP	U	U (RP)
Acidification	U (U)	U (U)	U (U)	U (RP)	RP	RP	RP	RP	U (RP/U)
Terrestrial eutrophication	U (U)	U (U)	U (U)	U (RP)	RP	RP	RP	RP	U (RP/U)
Freshwater eutrophication	U (RP)	U (RP)	U (RP)	U (RP)	RP	RP	RP	RP	U (RP/U)
Marine eutrophication	U (U)	U (U)	U (U)	U (RP)	RP	RP	RP	RP	U (RP/U)
Photochemical ozone formation	U (U)	U (U)	U (U)	U (RP)	RP	RP	RP	RP	U (RP/U)
Ionising radiation (human health)	U (RP)	U (RP)	U (RP)	U (D)	RP	RP	RP	RP	U (RP)
Particulate matter	U (U)	U (U)	U (U)	U (RP)	RP	RP	RP	RP	U (RP/U)
Human toxicity	U (RP/U/D)	U (RP/U/D)	U (RP/U/D)	U (RP/D)	RP/D	RP/D	RP/D	RP/D	RP/U (RP/U/D) /D
Ecotoxicity	U (RP/U/D)	U (RP/U/D)	U (RP/U/D)	U (RP/D)	RP/D	RP/D	RP/D	RP/D	RP/U (RP/U/D) /D
Water use	U (RP)	U (RP/U)	U (RP/U)	U (U)	RP	RP	RP/U	U	U (RP)
Land use	U (RP)	U (RP)	U (RP)	U(RP)	RP/U	RP/U	RP/U	RP/U	U (RP/D)
Fossils depletion	U (RP)	U (RP)	U (RP)	U (RP)	RP	RP	RP	RP	U (RP)
Metal/mineral resource depletion	RP/U (RP)	RP/U (RP)	RP/U (RP)	U (RP/D)	RP/D	RP/D	RP/D	RP/D	RP/D

Based on assessments of ecoinvent 3.1 energy production processes using ReCiPe and ILCD LCIA methodologies (Weidema et al. 2013; Huijbregts et al. 2015; Hauschild et al. 2013). Sensitivity to long-term emissions for freshwater eutrophication and toxicity-related impacts was included in the identification of the hotspots (addition of life cycle stage hotspots when inclusion, if different picture from exclusion)

For fossil-based, bio-based and nuclear power, the life cycle of the fuels is considered part of the use/operation stage of the power plants (see Fig. 26.3). The first letter code therefore indicates the position of the hotspots within the life cycle of the power plants; the letter code in the brackets further specifies the hotspots when stemming from the operations of the power plant by giving their positions within the life cycle of the fuel. Same designations are used to represent the different life cycle stages. For the power plants: *RP* raw materials extraction and construction of power plants; *U* use/operation stage of the heat/electricity generation plant; *D* decommissioning/disposal of the plant. For the fuels: *RP* mining operations and/or resource production (e.g. biomass), refining and distribution, *U* fuel combustion; *D* slag or spent fuel disposal

26.3.1.2 Nuclear Power Systems

All impacts are concentrated in the life cycle of the nuclear fuel (i.e. operation of the power plant). A large number of impacts, including climate change, stratospheric ozone depletion, acidification, eutrophication, photochemical ozone formation, particulate matter, fossil depletion and land use primarily stem from the extraction and processing of the uranium, for which important energy supplies are needed (e.g. diesel for machineries, electricity/heat). The extraction of uranium also contributes to uranium resource depletion, typically accounted for in the metal depletion impact category. Toxicity-related impacts (dominated by long-term emissions of heavy metals) and freshwater eutrophication also arise from this process due to the disposal of the tailings and spoils from the mining activities. The disposal of the spent nuclear fuel is a second important source of impacts, in particular for ionising radiation, for which it is the primary source, and for toxicity-related impacts and metal depletion, both resulting from the requirements of steel for the fuel conditioning (e.g. steel canisters, etc.). A third hotspot stems from the significant water requirements during the operations of the nuclear power plant, which dominate the water use impacts.

26.3.1.3 Biomass-Based Systems

Environmental impacts of bioenergy systems are largely influenced by the type of fuel used, hence a majority of the impacts stemming from the operations of the plant and more specifically from the life cycle of the fuel. Impacts such as climate change, acidification, eutrophication, photochemical ozone formation, fossils depletion and particulate matter may stem from either the biofuel or biogas combustion itself or from the biomass production, i.e. from growing and harvesting (e.g. first generation biofuels; see Chap. 30 on biofuels and bioproducts). If the energy source is a bio-waste or residue not utilised elsewhere, processes associated with this waste stream should not be accounted for in the assessment, thus shifting the environmental impacts for these categories solely to the combustion processes (e.g. incineration, biogas plants).

Because of the large variability across fuels, toxicity-related impacts can stem from any place in the life cycles of the power plants and the fuels. For example, if bio-waste is used as fuel and has no content of toxic elements, the hotspots will arise from the life cycle of the power plant itself, while the hotspots will lie in the production of the fuel if the fuel production is considered and requires high energy requirements and/or is associated with important direct emissions of toxic substances (e.g. pesticides in farming practices).

Likewise, for water use and land use, different fuels will have different hotspots. Water use impacts would typically be concentrated in the production of the biomass (if any is considered and if irrigation is applied). Land use impacts will also stem from the production of the fuels, which may also entail indirect land use impacts (see Weiss et al. 2012). For further details on LCA applied to biomass systems, the reader is referred to Chap. 30.

26.3.1.4 Wind, Solar, Geothermal and Hydropower Systems

All impacts but land use and water use impacts stem from the production of the power plant unit (incl. raw materials extraction). The exact sources of the impacts vary from one energy source to another as well as across technologies within a same energy source. The production of the raw materials and components of the power plant unit, such as PV modules (e.g. Si wafers), wind turbines (steel, composite materials) or dams (reinforced steel), are the primary causes to most impact categories including climate change, acidification, photochemical ozone formation, eutrophication, particulate matter, ionising radiation, water use and fossils depletion. These contributions are largely explained by the large energy requirements in these manufacturing processes, e.g. steel production. With respect to freshwater eutrophication and toxicity-related impact categories, the sulfidic tailings and spoils from mining activities contribute significantly to the impacts due to emissions of heavy metals and phosphorous compounds. For human toxicity and ecotoxicity, the disposal of the scrap metals (e.g. steel, copper) is also an important contributor, notably for renewable technologies like solar power or wind power. These disposal processes, along with the metal extraction processes at the beginning of the life cycle, contribute to metal depletion, which can be influenced by the presence of recycling. Water use impacts show different hotspots depending on the energy source and technology in use. Water requirements in the production of the components for wind and solar power plants as well as for run-of-river hydropower plants drive the impacts for these energy sources, while reservoir-based hydropower and geothermal power plants concentrate the water use impacts during their operations. Same dependencies can also be observed for land use, which typically can stem from either the mining operations (e.g. photovoltaics, run-of-river hydropower) or the installation sites and the associated distribution network (e.g. wind farms, reservoir-based hydropower, geothermal power).

26.3.2 Key Findings

Because of the large number of LCA studies on energy systems, providing a comprehensive analysis of their findings can easily become a laborious exercise. Instead, Table 26.4 in Appendix provides an overview of main findings of LCA studies assessing different energy technologies, including environmental performances, environmental hotspots, etc. For further details, the reader is referred to the references provided in Table 26.4; several of them are reviews performed on LCAs of specific energy sources or technologies. Figure 26.4 additionally provides an illustration of the variations of results for selected impact categories across fossil-based and renewable energy sources and technologies.

26.3.2.1 Technology Dependence

As reflected in Fig. 26.4, there is a strong dependence of the impact results on the type of technologies, even within a same energy source. Two parameters are particularly important in the differentiation of the technologies and their resulting impacts: the existence of cleaning technologies and the conversion efficiencies of the plant (Turconi et al. 2013). Existence of cleaning technologies has been shown to potentially yield significant reductions in impacts, e.g. use of carbon capture and storage (CCS) systems for reducing climate change for coal and natural gas power plants (see Fig. 26.4) or cleaning of the coal prior to combustion to reduce downstream emissions and associated environmental impacts (see Ryberg et al. 2015). However, it is important to note that these cleaning technologies often target one or few specific impact categories (e.g. CCS systems to reduce climate change impacts), and may thus lead to burden-shifting from those targeted impact categories to other environmental problems. See for example the changes in the impact results for particulate matter between systems with and systems without CSS systems for coal and natural gas in Fig. 26.4. While climate change impacts are significantly reduced by the implementation of CCS systems, these impacts tend to increase. This reinforces the need to encompass a full impact coverage.

The power plant conversion efficiencies are another influential source of differentiated impact results across technologies. They are calculated as the ratio between the useful energy output (as electricity and/or heat) and the energy input. Power plant efficiencies typically range within 30–45% for coal and natural gas (conventional), 90% for hydropower, 30–50% for wind turbines, 5–20% for solar cells (large variations between technologies), etc. These efficiencies are constrained by theoretical maximums determined by thermodynamics laws (i.e. Carnot's efficiency law). However, improvement of these efficiencies, particularly for thermal power sources (coal, gas, oil), can be made by introducing energy recovery systems that will increase that theoretical maximum. This is for example the case when implementing combined cycles, where the waste heat from the first cycle is used through additional cycles to recover more energy (e.g. in gas power plants). Co-generation of heat and power can also significantly increase these efficiencies, for example in utilising the waste heat from the power plants to district heating purposes. Such co-generation systems can then result in efficiencies above 90%.

26.3.2.2 Performances Across Energy Sources

Most studies include comparisons of heat or electricity produced from different energy sources, e.g. to position the analysed system(s) relative to the currently applied system with respect to environmental impacts. Trends vary considerably

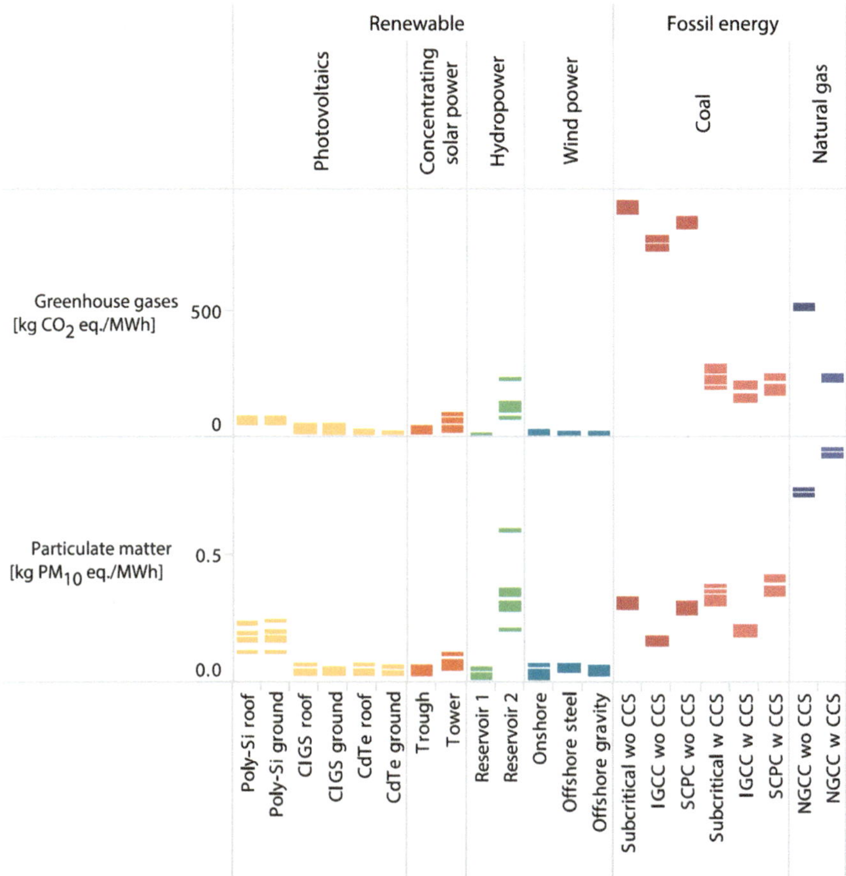

Fig. 26.4 Ranges of impact results for climate change and particulate matter impacts for different energy sources and technologies (extracted from Hertwich et al. 2015). *CCS* CO_2 capture and storage, *CdTe* cadmium telluride, *CIGS*: copper indium gallium selenide, *IGCC* integrated gasification combined cycle coal-fired power plant, *NGCC* natural gas combined cycle power plant, *Offshore gravity* offshore wind power with gravity-based foundation, *Offshore steel* offshore wind power with steel-based foundation, *Reservoir 2* type of hydropower reservoirs used as a higher estimate, *SCPC* supercritical pulverized coal-fired power plant

depending on the technology assessed (see above) and the assumptions made in the assessment (e.g. modelling and coverage of life cycle and impacts; see Sects. 26.2.2 and 26.2.3). Higher shares of renewables and nuclear power in energy systems are typically associated with lower environmental impacts for several impact categories, including climate change and eutrophication (e.g. Hertwich et al. 2015; Laurent and Espinosa 2015). Other impact categories show less conclusive results, e.g. toxicity-related impacts, land use impacts and water use impacts, e.g. land use and water use impacts from hydropower reported as larger than those of fossils-based power generation (e.g. Hellweg and Mila ì Canals 2014; Hertwich et al.

2015). Metal depletion is often reported to be an impact category where renewables perform worse than fossil-based systems (Hertwich et al. 2015; Berril et al. 2016; Laurent et al. 2012). Overall, at a global scale, two patterns seem to characterise the use of electricity generation technologies in current electricity supply systems, with developing economies having relied on energy policies ineffectively targeting environmental problems, thus resulting in "dirtier grid mixes", while developed economies, which progressively integrate higher shares of renewables, move towards "cleaner grid mixes" (Laurent and Espinosa 2015). With respect to renewables, wind power often emerges as the renewable technology with the lowest overall environmental impact (Hertwich et al. 2015; Berril et al. 2016; Astrubali et al. 2015). For example, although including a large variability in the impact results, solar power is reported to lead to higher impacts than wind power per unit of electricity produced due to large impacts stemming from material production and a lower ability to generate electricity over the same period of time (Hertwich et al. 2015; Berril et al. 2016).

In the assessment of renewables, two alternative, noteworthy indicators have often been used as criteria for assessing system performances: the energy payback time (EPBT) and the energy return on investment (EROI). The EPBT is defined as the time (typically in years) for a system to compensate for the use of energy for its production, installation and end-of-life, and start producing more energy than what has been invested through its life cycle. For example, if a system has a lifetime of 20 years and its EPBT was found to be 3 years, it means that "free energy" is produced for 17 years. The EROI is defined as the amount of usable energy supplied by a system in its lifetime over the energy required to produce, implement and dispose of it, which is equal to the EPBT, and is dimensionless (e.g. 20:3 in the example above). EROI ratios below one are not considered viable technologies on the market. PV technologies are currently associated with EPBT of 1–4.1 years, with cadmium telluride (CdTe) and copper indium gallium diselenide (CIGS) technologies showing lowest EPBTs, and EROI of 8.7–34.2 (Bhandari et al. 2015). Albeit outdated to some extent, wind power technologies typically show EPBT of few months to 1–2 years with typical ERIO of 8–40 among recent studies (Davidsson et al. 2012). Such figures are comparable to the performances of some fossil-based energy sources, such as natural gas and oil, for which ERIO are decreasing due to lower availability of the resources (increasing amount of energy spent to recover oil or natural gas).

26.4 Specific Methodological Issues

From the published LCA studies of energy systems a number of issues can be identified. They relate to either influential methodological choices or assumptions on which no consensus currently exists, or to inconsistencies or malpractice observed in studies (most of them being noted in Sect. 26.2). This section, therefore, builds on Sect. 26.2 (and contains several cross-references to it) to focus on key issues that are central to the consistency and reliability of the assessment results, and it provides guidelines and recommendations to address them when performing LCA of energy systems.

26.4.1 General Issues

The review performed in Sect. 26.2 highlighted a latent problem of transparency in the LCA studies on energy systems. Important methodological aspects and assumptions are often not sufficiently documented. In addition to compromising the reproducibility principle that each study should fulfill, it makes the results difficult to interpret and compare across studies. Examples of poorly reported aspects include the handling of multifunctional processes, e.g. use of system expansion, the data sourcing, the use of electricity mixes, which are not always specified, the accounting of energy used and produced, for which different methods can be used, the coverage of the life cycle inventories (e.g. materials required), the potentially missing impact pathways (e.g. no accounting of rare earth metals), or the assumptions made to model the disposal stage. Such lack of transparency is not a problem specific to LCA studies applied to energy systems (e.g. LCA of waste management systems, see Chap. 35).

To remediate this issue, some review studies have provided guidance to ensure a better reporting and harmonisation in the LCA practice (e.g. Davidsson et al. 2012, for wind power; Frischknecht et al. 2016 for PV power systems). In general, LCA practitioners are strongly recommended to use Appendices (for reports) or Supporting Information (for scientific publications) to document clearly and transparently their data, methodological assumptions and modelling (see overall guidance in Methodological Chaps. 8 and 9 of this textbook).

26.4.2 Goal and Scope Definition

Building on the review presented in Sect. 26.2, four key aspects are addressed below for the goal and scope definition: (i) the definition of the functional unit, (ii) the scoping of the system boundaries, (iii) the selection of the impact categories and (iv) the LCI modelling framework and handling of multifunctional processes.

26.4.2.1 Functional Unit

The functional unit must be defined as the primary service provided by the system, i.e. its “raison d’être”. The role of the energy systems assessed in LCA studies typically consists in supplying electricity or heat to allow other activities to operate. As a consequence, for studies under Categories 1 and 2 (see Sect. 26.2.1), the functional unit needs to be defined based on an energy output (whether it meets a known demand or not). An example of malpractice is the definition of functional units based on a specific area of PV modules in comparative studies of PV technologies. Such definitions prevent to account for different efficiencies of the compared PV module alternatives, and hence for their different electricity amounts generated from a same PV module area (see Box 8.1 in Chap. 8; case 1). It is therefore important to relate to the main function of the system when defining the functional unit.

Defining an appropriate functional unit also contributes to ensure a comparability of alternatives or scenarios in the performed LCA studies. In studies with a demand-driven context that compare base-load with intermittent energy technologies, such as wind power or PV power systems, this can however be challenging due to the different "reliability of supply" of the two systems. This can usually be eluded by modelling the intermittent source with a storage system (Gagnon et al. 2002) or by adding a compensating source whenever the intermittent source cannot supply electricity. Similar challenges arise when comparing electricity supply systems matching base-load electricity demand with those matching peak-load electricity demand (Turconi et al. 2013).

In line with the review presented in Sect. 26.2, two categories of studies were identified from the published LCA studies applied to electricity and heat systems: (i) studies assessing specific energy technologies/sources/systems at a power plant or sub-power plant level, and (ii) studies, typically at meso- and large-scale, assessing energy systems in a context perspective (see details in Sect. 26.2.1). These call for different definitions of functional units, which are gathered from LCA practice; they are provided in Table 26.3, which provides recommendations for practitioners undertaking LCA of energy supply systems.

Table 26.3 Recommendations for defining functional units of energy systems (non-exhaustive list of situations)

Type of situations/goal of studies	Recommendations for FU definition
Category 1 studies (power plant or sub-power plant level)	
Focus on fuel input comparisons (with disregard of energy output)	Provision of xx MJ of fuel energy content (or primary energy) to power plant z
Focus on supply of electricity and/or heat	-"Generation of 1 kWh or MJ of electricity/heat at power plant/heat unit in country x" (without transport and distribution system) -"Supply of 1 kWh of electricity to the grid in country x" (with transport and distribution system)
Category 2 studies (context perspective; meso- and large-scale assessments)	
Investigation of how the environmental impacts of the grid mix will change/evolve	Supply or consumption of 1 kWh of net electricity in country or region x
Investigation of environmental impacts from whole electricity supply system over time (with consideration of demand)	Supply of electricity to match the global demand in country or region x in year y (quantified demand fixed by different scenarios)

As reflected in Table 26.3, a simplistic functional unit defined as the supply or consumption of a unit of electricity output (e.g. 1 kWh) is often appropriate to the case study. One important exception is, however, the assessment of meso- or large-scale systems taken in their context and with consideration of energy demands modelled as scenario analyses. With a simplistic definition of the functional unit as indicated above, such studies become limited to only address the question of how the environmental impacts of the grid mix will change. As indicated in Laurent and Espinosa (2015) with assessments of national electricity supply systems, a scenario A may show lower environmental impacts than a scenario B on a 1 kWh-basis (grid mix level) but a reversed tendency may be observed when accounting for the total demand. The total demand may indeed differ between Scenario A and Scenario B because of different energy policies, which could for example influence the consumers' behaviour and their overall demand in different ways, lead to different efficiencies in the smart grid for matching the demand with the supply, integrate different measures for energy efficiencies, etc. A total demand, which ends up higher in Scenario A than in Scenario B, may therefore compensate the better performances of the grid mix in Scenario A, thus resulting in Scenario B being the most environmentally preferable. These observations can be linked to the differences between eco-efficiency (here: the grid mix having lower environmental impacts, but with no guarantee of overall reduction of environmental impacts at the societal level) and eco-effectiveness (here: the whole electricity supply system to support the total demand having lower impacts, thus ensuring lower environmental impacts at societal level). Chapter 5 addresses these concepts in a more generic and detailed way. Consequently, in studies supporting policy analysis or policy-making, and where scenarios need to be assessed, it is important that the whole perspective, including not only the changes on the electricity grid mix but also the changes in the demand, be encompassed in the analysis of the results. And this is why the functional unit may have different quantities relating to the energy demands in its definition (since these vary from one system to another), while still maintaining comparability of the energy systems/scenario under study.

26.4.2.2 System Boundaries

The life cycle of the energy systems should include both the life cycle of the power plants and that of the fuels, the latter being relevant for all energy sources but wind, solar, geothermal and hydropower sources due to the absence of fuel per se. For hydropower and geothermal power sources, the use of water (which could be regarded as the fuel to some extent) and the associated impacts should, however, be carefully evaluated. For fossil-based and biomass-based systems, the life cycle of the fuel is important to include, as it is the main source of impacts (see Table 26.2).

As a general rule, to avoid overlooking any potentially large impacts and possible burden-shifting, the practitioners are recommended to include the entire life cycle of electricity and heat generation systems. In practice, this can sometimes be challenging, for example in the inclusion of the power plant life cycle. Based on the analysis in Table 26.2, practitioners are invited to consider the following guidance to scope the system boundaries of their electricity and heat generation systems, including as a minimum:

1. The life cycle of the fuels for all fossil-based, nuclear-based and biomass-based systems. The life cycle of power plants (excluding the operation stage, thus mainly consisting of the plant construction and decommissioning) typically shows minor contributions to most environmental impacts associated with the supply of heat and electricity (see Sect. 26.3.1).
2. The life cycle of the power plants and equipment for renewable energy systems. Environmental impacts typically stem from the production stage and possible crediting can be gained through the disposal stage, which thus should not be dismissed.

Note that these rules are general, non-exhaustive and are not technology-specific: the practitioner shall still adopt a case-by-case approach before ruling out part of the energy system life cycles. Although Table 26.2, which shows environmental hotspots per life cycle stage and per energy technology, may be used as a screening step, the practitioners should assess any possible exceptions to these patterns in relation to their systems under study. For example, in situations (1), for biomass-based energy systems relying on waste, little impacts may be credited to the waste generation itself (e.g. zero-burden assumption), possibly making the life cycle of the power plants non-negligible in the total environmental burden: in such cases, the life cycle of the power plants should be comprehensively covered. The addition of carbon capture and storage system to fossil-fuelled power plants is another example, where the practitioners should also look into the life cycle of the power plants.

26.4.2.3 Selection of Impact Coverage

As reflected in the review of the impact coverage in Sects. 26.2.3 and 26.3.1, to avoid burden-shifting from one impact category to another, all impact categories are relevant for inclusion when assessing electricity and heat generation systems. In particular, practitioners should put emphasis on consistently including toxicity-related and resource-use-based impact categories in addition to the non-toxicity-related impact categories, such as climate change, acidification or eutrophication. Toxicity-related impacts associated with renewables-based electricity production have been shown to potentially remain at the same level as those related to fossil-based electricity production. A sole focus on climate change can thus be deceiving if one aims to assess the total environmental burden. Resource use indicators often turn out to be highly relevant for renewable energy sources, e.g. water use for hydropower, metal depletion for wind and solar power, land use for bioenergy systems, hydropower and wind power, etc.

26.4.2.4 LCI Modelling Framework and Handling of Multifunctional Processes

The ILCD guidelines being only recently available, a limited number of studies have performed LCAs on energy systems while attempting to follow these guidelines. The LCI modelling framework and handling of multifunctional processes have thus often been limited to choosing between the attributional and consequential modelling and in the selection of materials and energy mixes used in system expansion.

In practice, energy systems do not differ from other systems when it comes to define the LCI modelling framework and the respective handling of multifunctional processes. Examples of multifunctionality in energy systems typically include the co-generation of heat and electricity or the recycling of materials, which can affect the production stage (recycled materials used for construction/production of power plants, e.g. wind turbines) and the disposal stage (materials sent to recycling, e.g. PV module components, batteries, etc.). To address those, the detailed methodological guidance provided in Chap. 8 is therefore sufficient; the steps can be summarised as follows:

- In line with the identified decision context situations (i.e. A, B, C1, C2) in the goal definition, decide which of consequential or attributional modelling framework should be adopted.
- Characterise the multifunctional processes, for which subdivision, system expansion or allocation is required.
- In cases of system expansion: identify which processes should be used.
- In cases of allocation: identify, determine and describe the allocation key(s) used.

The detailed documentation of the processes used for system expansion or of the allocation key(s) should be reported in the LCI analysis section. Procedures and guidelines to do so are given in Chap. 9, to which the reader is referred for details. In the following Section, details are specifically provided to address allocation of energy co-generation processes and marginal energy mixes in system expansion cases, with a particular focus on the marginal technologies, respectively.

26.4.3 *Inventory Analysis and System Modelling*

The data collection and the building of the modelling generally do not differ from that of other systems, and guidance from Chap. 9 can thus be followed for performing the LCA phase. Our aspects are, however, emphasised in below sections, as they require attention from LCA practitioners in specific situations: (i) the LCI data availability to match the temporal, technological and spatial representativeness; (ii) the allocation principles for electricity and heat co-generation processes; (iii) the identification and modelling of marginal energy technologies; and (iv) the comprehensive scoping of the sensitivity analyses. Other aspects of relevance are the use of IO modelling, which are increasingly used for large-scale assessments of energy systems, and the modelling of indirect land use change, particularly relevant for bio-based systems; the reader is referred to Chaps. 14 and 30, respectively, which specifically address these issues.

26.4.3.1 LCI Data Availability with Temporal, Technological and Spatial Requirements

As indicated in Chaps. 8 and 9, the data collected in the LCI phase should match to the best possible extent the required data representativeness indicated in the scope definition. This aspect, which can be relatively simple for some product systems, can be challenging for some energy systems, for example when performing future-oriented studies or when assessing emerging technologies. Below are a number of points that should be considered along with guidance to address them, wherever applicable:

- *Systems with a time-oriented perspective*: the temporal and technological representativeness must be carefully addressed, e.g. in studies comparing different scenarios and different technologies in the future. Besides the definition of scenarios, LCA practitioners should ensure that the collected LCI data integrate a prospective dimension, e.g. including future technology developments, future evolutions of the market and future practices (e.g. in waste management). A typical example is the consideration of electricity mixes in consistency with the time period imposed by the scenarios analysed in the study. Evolutions of these mixes over time should thus be considered. Even more relevant in future-oriented modelling than in conventional case studies, it is important to document any assumptions or choices that make the modelling diverge from the data representativeness requirements as such discrepancies often can significantly influence the conclusions (and should thus be tested in sensitivity analysis). Future-oriented LCA is further discussed in Chap. 21.
- *Systems with spatial variation*: the spatial representativeness should be addressed, e.g. in studies with specific locations. Energy systems are strongly country- or region-specific, e.g. electricity grid mixes can vary considerably from one country to another. The modelling of energy systems should capture these geographical specificities with sufficient accuracy. LCI processes for electricity grid mixes are typically the best covered in available LCI databases, e.g. 50 countries differentiated in ecoinvent v.3 database (Treyer and Bauer 2013, 2014). If LCI processes are not readily available, LCA practitioners should either create processes or adapt existing ones to match the local or regional conditions (e.g. adapting the electricity grid mix in an ecoinvent process for a given country).

As indicated in Chap. 8, the geographical, temporal and technological representativeness are intertwined and it is likely that the two above aspects/sets of recommendations will apply to the same study, e.g. studies assessing the future deployment of a new energy technology on the market, including a comparison with existing ones.

26.4.3.2 Allocation of Electricity and Heat Co-generation Processes

In the case of allocation, energy indicators could be needed to perform allocation of co-generation processes. Three approaches may be selected: (i) allocation of all impacts to one of the output, electricity or heat, assuming that it is the main purpose of the process, (ii) allocation based on the energy content, assuming that a MJ of electricity is equal to a MJ of heat, thus using the respective electricity and heat outputs to derive the allocation key, and (iii) allocation based on the energy quality, recognising the higher quality of electricity over heat, for example in using exergy of the electricity and heat outputs as a basis for the allocation key (Fruergaard et al. 2009). Exergy indicates the extent of the energy that can be converted to work: while electricity has an exergy factor of 1, heat has a variable exergy factor typically around 0.15–0.20 depending on the temperature of the delivered heat and the temperature of the surroundings (Fruergaard et al. 2009). Approach (i) is rare and requires to be well argued by the practitioner if used. Approaches (ii) and (iii) are the most commonly applied approaches for allocation of energy processes. Note that allocation based on energy quality will associate most of the burden to electricity, while allocation based on energy content will shift most of the burden to heat production.

26.4.3.3 Modelling of Marginal Energy Technologies

By definition, marginal data represent the technology or process actually affected by the changes (Weidema et al. 1999). The time perspective is important to consider when identifying that technology or process. For example, if an increase in electricity demand in a country like Denmark that relies heavily on wind turbines for electricity generation occurs in an hour or day when wind blows, the marginal technology for electricity supply at that moment could be wind power (and may change later on if wind stops). This type of very short-term/instantaneous marginal is however not relevant in LCA studies, where aggregation over time is performed. Averages of marginal technologies would be more relevant to use, for example estimating that wind is the marginal technology for a cumulative two months of the year and other sources are marginal technologies for the remaining cumulative 10 months. This leads to the creation of mixes of marginal technologies. Such examples only consider short-term marginal technologies, i.e. existing technologies capable to respond to a change in demand (no impact on capital investments). They should be distinguished from long-term marginal technologies, i.e. technologies for which the production capacities are impacted in a long-term perspective (e.g. >10 years), like the closure of old coal-fired power plants or the installation of new wind turbines.

For studies in decision context situation B, a mixed consequential/attributional modelling is required, with the use of system expansion for solving process multifunctionality. The processes impacted by structural changes in the background system should be modelled using mixes of long-term marginal processes while the others are modelled using short-term marginal or average processes (see Chap. 8).

Difficulties arise in identifying and determining the mixes of long-term marginal processes, and important differences in the results might arise depending on what marginal technologies are assumed (e.g. renewables versus fossils-based energy sources). Although Chap. 9 provides some practical guidance to support LCA practitioners in that effort, to which the reader is referred, no consensus currently exists on ways to identify these mixes of long-term marginal technologies. This results in important uncertainties for processes that are included in nearly all LCAs. With respect to energy processes, if these are decisive for the outcome of the study, the use of explorative scenarios is typically recommended to model several possible mixes of long-term marginal technologies (e.g. Schmidt et al. 2011; Münster et al. 2013). LCA practitioners should include these as part of their sensitivity analyses, which will thus enable them to assess and understand the range of potential environmental consequences associated with the implementation of their analysed systems.

26.4.3.4 Importance of Sensitivity Analysis

As part of the LCI analysis phase, practitioners need to prepare the basis for uncertainty and sensitivity analyses (see Chap. 9). This can be regarded as a scoping and identification of key parameters that need to be varied in the assessments. This identification is an iterative process, e.g. going back and forth with the LCIA phase and the results obtained to pinpoint the processes and associated key parameters that are influential on the results.

With respect to energy systems, there is a case dependency on which parameters to include. As for any LCA studies, the identification of major modelling assumptions, such as the identification of mixes of long-term marginal technologies or the inclusion of indirect land use change effects, should systematically lead to sensitivity analyses. Additional sensitivity analyses may also stem from the large application of LCA to emerging technologies and/or to systems taken in a prospective dimension (e.g. future-oriented assessments). These types of studies are associated with large uncertainties due to the use of scenarios and the inadequacy of data (e.g. lab-scale data for an emerging energy technology to represent a fully deployed system in the future) or, worse, the lack of it (data not yet generated). Such situations call for sensitivity analyses to address the temporal dimension and inherent uncertainties in the modelling. Practitioners are therefore recommended to develop explorative scenarios based on all key parameters pertaining to the evolution of the technologies or systems in time. Examples of such parameters include the efficiencies of the plants, the lifetime of the infrastructure, the type and performances of disposal routes (e.g. recycling), the emission factors, etc.

26.5 Conclusions

This chapter provides a glimpse at how LCA has been applied to energy systems and technologies in the past two decades and what learnings can be gained from the large body of LCA studies. The review provided herein is not intended to be exhaustive because of the large extent and diversity of energy systems. Nevertheless, it brings sufficient insights to realise that the application of some key methodological steps could be improved. For example, a comprehensive coverage of the system life cycle (e.g. including the often-overlooked disposal or decommissioning stage) and of all relevant environmental impacts (e.g. not just addressing climate change or energy-related questions) should be better ensured in future studies.

Life cycle assessment is still a relatively young field and the methodology is constantly being improved. In that respect, several methodological aspects relevant to assessments of energy systems need to be further developed and accepted within the LCA community. Some of them relate to the LCI or system modelling, e.g. the inclusion of indirect land use change for bio-based systems or methodologies to consistently identify mixes of long-term marginal technologies. Others relate to LCIA and are not necessarily specific to energy systems, like for example the assessment of climate change impacts in a dynamic perspective (e.g. relevant to use of carbon capture and storage systems). The inclusion of the temporal perspective in LCA studies of energy systems is particularly relevant as many policy makers currently define and/or fine-tune energy pathways for the future decades (e.g. IEA 2014), and require foresight assessments that can anticipate the impacts in the future from current and forthcoming energy technologies. Frameworks for consistently conducting such foresight LCA studies still need to be developed (Laurent and Espinosa 2015). This development can also be expected to run in parallel to a continued increase in the application of LCA to large-scale energy systems, such as electricity supply systems at urban, national or regional scales, and thus efficiently and effectively support high-level energy policy-makers.

Appendix

See Table 26.4.

Table 26.4 Non-exhaustive overview of LCA studies assessing the different energy technologies/sources

Energy sources/technology focus	Main findings
Hard coal and lignite Direct combustion, gasification combustion, flue gas cleaning, physical/chemical cleaning process, carbon capture and storage technologies	– Emissions in combustion processes and supporting processes (e.g. demineralization of fuels) drive the impacts, with process efficiency and types of technology (e.g. direct combustion vs. gasification combustion, inclusion of cleaning processes, etc.) as key factors (e.g. Gagnon et al. 2002; Turconi et al. 2013; May and Brennan 2003; Ryberg et al. 2015; Masanet et al. 2013). Fuel supply chain contributes to a lesser extent, with contributions depending on plant settings, e.g. existence of flue gas cleaning system, and fuel type, e.g. sulphur content, metal content (Dones et al. 2005) – Important emission reductions can be achieved (e.g. on old power plants) via cleaning of the fuels although environmental trade-offs exist between the added impacts from cleaning processes and the resulting saved impacts (e.g. Nomura et al. 2001; Ryberg et al. 2015). This also induces shift of impacts from the combustion processes to the cleaning processes or to other parts of the life cycle (e.g. fuel supply chain) – The review/meta-analysis by Schreiber et al. (2012) showed that the three carbon capture and storage technologies (post-combustion, oxyfuel, pre-combustion) lead to the expected reductions in climate change impacts but to increases in many other environmental impacts regardless of capture technology, time horizon and fuel type (coal, lignite or natural gas). Three influential parameters are the (i) power plant efficiency and the added energy requirements from capture process, (ii) the CO_2 capture efficiency and purity, (iii) the fossil fuel origin and composition (Schreiber et al. 2012)
Natural gas Single-cycle or combined cycle turbines	– The majority of the literature has a narrow scope on CO_2 emissions and NO_x and SO_2, to a lesser extent – Emissions of greenhouse gases dominated by use stage, but with significant contributions also from the fuel provision due to fugitive emissions of methane and energy requirements in gas extraction and transportation (Gagnon et al. 2002; Dones et al. 2005; Masanet et al. 2013) – For use of carbon capture and storage technologies, see above row on hard coal and lignite (Schreiber et al. 2012)
Shale gas Combined cycle gas turbines	– Albeit with large uncertainties and variability in the impact results, depending on assumptions and types of technology in use, impact results from various studies suggest that shale gas seems comparable to conventional gas for climate change, with ranges of with a range of 416–730 g CO_2-eq/kWh. Taking the study by Stamford and Azapagic (2014) on electricity in UK from shale gas produced by fracking, it ranges within 412–1102 g CO_2-eq/kWh, thus in the lower end of fossil fuels but in the higher end of the renewables (see also Fig. 26.4). Toxicity impacts are found to be higher than for conventional gas (Stamford and Azapagic 2014) – Differences of impacts with other energy sources are due to assumptions regarding fugitive emissions during shale gas extraction and due to differences in the recoverable resources (Stamford and Azapagic 2014)

(continued)

Table 26.4 (continued)

Energy sources/technology focus	Main findings
Oil Fuel oil cycle	– Emissions (e.g. GHG, NO_x) mainly stem from the use/operations of the power plant. Like-for coal, fuel supply chain contributes to a lesser extent, with contributions depending on plant settings, e.g. existence of flue gas cleaning system, and fuel type, e.g. sulphur content, metal content (Dones et al. 2005) – Energy utilisation efficiencies are an additional influential parameter, inducing variations in the impact results, e.g. base-load power plants with efficiencies up to 58% (ca. 530 g CO_2-eq/kWh) versus peak-load power plants with efficiencies of 30–40% (ca. 750–900 g CO_2-eq/kWh) (Turconi et al. 2013)
Nuclear energy Closed/open fuel cycles, pressurized/boiling water reactors	– Large focus on GHG and little focus on other environmental impacts (Turconi et al. 2013; Poinssot et al. 2015a, b) – Nuclear energy production is the most important activity source, which contributes to the impact category ionising radiation in LCA studies – Uranium extraction and enrichment processes are the drivers for most environmental impacts, i.e. >70% (Poinssot et al. 2015a, b; Masanet et al. 2013) – Large variability in emission factors (and resulting impacts) depending on the type of technology and the assessment approaches, incl. assumptions on uranium extraction and enrichment processes and handling of nuclear waste (Warner et al. 2012; Turconi et al. 2013). For climate change, such variations of up to one order of magnitude were observed in studies (Turconi et al. 2013; Masanet et al. 2013) – Ecodesign initiatives should focus on optimising nuclear fuel cycle, reduce impacts at mining of uranium, and improving the recycling of uranium and plutonium from spent fuel (Poinssot et al. 2015a, b; Masanet et al. 2013)
Wind power On-shore, off-shore (less common	– Important focus on emissions and/or energy accounting, e.g. EPBT (e.g. Wang and Sun 2012; Dolan and Heath 2012; Schleisner 2000) – Manufacturing of the wind turbines is the only life cycle stage that is common to all LCA studies (Arvesen and Hertwich 2012). Other stages are omitted in some studies – On-shore and offshore turbines can have similar emission factors because larger emissions during the construction phase of offshore turbines can be compensated by their higher efficiency during use (Turconi et al. 2013) – The environmental impact of wind technologies is concentrated mainly in the manufacturing stage and to a smaller extent in the disposal stage but is at a minimum in the operational/use stage. Impact per generated power is strongly influenced by the operating lifetime, quality of wind resource, conversion efficiency and size of the wind turbines (Masanet et al. 2013; Caduff et al. 2012)

(continued)

Table 26.4 (continued)

Energy sources/technology focus	Main findings
	– Three major components mainly contribute to the overall environmental impacts: rotor (due to fibre glass), tower (incl. foundation; due to reinforced steel) and nacelle (due to energy-requiring fibre glass and use of copper in different electrical components) (Martinez et al. 2009; DONG Energy 2008) – Disposal stage can contribute to significant decreases of impacts when recycling of valuable materials replaces production of virgin materials (Martinez et al. 2009; Weinzettel et al. 2009)
Solar power—Photovoltaics technologies e.g. crystalline silicon (Si) photovoltaics (PV), cadmium telluride PV, copper indium gallium diselenide PV, organic PV, perovskites, etc.	– Large body of studies assessing PV technologies (see reviews—total of over 400 studies—by Hsu et al. 2012; Kim et al. 2012; Gerbinet et al. 2014; Sherwani et al. 2010; Chatzisideris et al. 2016). Large focus on climate change and energy indicators, e.g. EPBT and EROI – Large variability in the results due to differences in methodological choices (e.g. system boundaries) and to types/specificities of systems assessed, e.g. source of electricity used in manufacturing, solar panel typology, climatic conditions, irradiation (Sherwani et al. 2010; Masanet et al. 2013) – For Si-PV technology, impacts stem mainly from the manufacturing stage due to the energy requirements in the upstream processes, e.g. production of Si and PV wafers. Gains in energy efficiencies are foreseen, e.g. as already observed with Si ingot growth by the Czochralsky process (Frankl et al. 2006) – Balance of system (BOS) components, e.g. inverters, insulators, supporting structure, have largely been omitted in past studies and only recently have started to be included in LCA studies (Gerbinet et al. 2014; Espinosa et al. 2015). Results suggest that their impacts are not negligible (e.g. Espinosa et al. 2015; Gerbinet et al. 2014) – In organic photovoltaics, silver used as electrode is overall the largest source of impacts (e.g. toxicity-related impacts, metal depletion), thus calling for establishing efficient recovery systems in the disposal stage (Espinosa et al. 2015) – Disposal stage/end-of-life of the materials has been largely omitted in past LCA studies (e.g. see reviews of Gerbinet et al. 2014; Chatzisideris et al. 2016). Recycling the material in PV modules is already economically viable, mainly for concentrated and large-scale applications. Projections are that between 80 and 96% of the glass, ethylene vinyl acetate, and metals (tellurium, selenium, lead) will be recycled. Other metals, such as cadmium, tellurium (Te), tin, nickel, aluminium and copper, should be saved or they can be recycled by other methods (IPCC 2012). Recycling of materials has been demonstrated to be influential on the impact results in some LCA studies (e.g. Espinosa et al. 2015)
Solar power—Concentrated solar power technologies	– Studies have primarily focused on parabolic trough and tower technologies, and to a lesser extent on parabolic dish – Overall, studies have a strong focus on climate change and energy indicators, with little consideration of other environmental impacts (e.g. land use)

(continued)

Table 26.4 (continued)

Energy sources/technology focus	Main findings
e.g. solar towers, parabolic trough, dish Stirling (i.e. parabolic dish), linear Fresnel reflectors	– For parabolic trough plants, the solar field was reported to be a main driver of impacts, primarily stemming from the manufacturing stage due to the steel, molten salt and synthetic oil requirements (e.g. Burkhardt et al. 2011; Ehtiwesh et al. 2016). The storage system was reported to have a relatively lower contribution to environmental impacts. While Burkhardt et al. (2011) report a significant contribution from the power plant unit, Ehtiwesh et al. (2016) found a negligible influence. Differences in the methodologies considered and in the assumptions made may explain these discrepancies, and more comprehensive studies are needed – Studies have suggested that the electrical efficiency is important for the environmental performances of the systems, e.g. fewer mirrors for a same electrical output yielding lower impacts (e.g. land use, etc.) (e.g. Viebahn et al. 2008)
Hydropower Reservoir-type, run-of-river (both small and large scale)	– Materials manufacturing and construction of plants are important drivers of the impacts, e.g. cement, reinforced steel, electric equipment and energy used in construction activities. Remoteness of hydropower plants increasing contribution from construction and transportation stages and decreasing transmission efficiencies (Suwanit and Gheewala 2011; De Miranda and da Silva 2010; Masanet et al. 2013) – Differentiation of impacts between run-of-river and reservoir technologies due to important aspects, e.g. CH_4 emissions from anaerobic degradation of biological materials in reservoirs for climate change; evaporative losses of water for water use in reservoirs; large stagnant water areas in reservoirs causing eutrophication, land use and human toxicity impacts (Masanet et al. 2013). Large influence of these aspects on the impact results, thus leading to large variations in the impact results in the literature as a consequence of different assumptions (Masanet et al. 2013; Suwanit and Gheewala 2011) – Strong focus on climate change and energy indicators (CED, EPBT). Other impacts addressed to a lesser extent, with less consideration for land use, water use and toxicity-related impacts – None of the retrieved studies made quantitative considerations regarding the secondary utilisation of water (drinking, irrigation or navigation purposes)
Geothermal power Hydrothermal resources producing hot water and/or steam	– Few LCA studies on geothermal power plants (Bayer et al. 2013), coupled with high degree of technology- and site-specificity of the case studies, e.g. Italy (Buonocore et al. 2015; Bravi and Basosi 2014) or Germany (Frick et al. 2010; Lacirignola and Blanc 2013), make it difficult to generalise (Bayer et al. 2013) – The assessment of toxicity-related impacts, e.g. from heavy metals, have not been sufficiently addressed in past studies to discuss their potential relevance in geothermal power generation (Bayer et al. 2013) – Strong influence of quality of geothermal source, maturity of the plant, conversion efficiency, etc.
Bioenergy	See Sect. 26.3.1, Cherubini and Strømman (2011) and the extensive discussion in Chap. 30

IEA: World Energy Outlook 2014. International Energy Agency, Paris (2014)
IEA: Key World Energy Statistics 2015. International Energy Agency, Paris (2015a)
IEA: Energy Technology Perspectives 2015. International Energy Agency, Paris (2015b)
IPCC: Renewable energy sources and climate change mitigation. In: Edenhofer, O., Pichs-Madruga, R., Sokona, Y., Seyboth, K., Matschoss, P., Kadner, S., Zwickel, T., Eickemeier, P., Hansen, G., Schlömer, S., von Stechow, C. (eds.) Special Report of the Intergovernmental Panel on Climate Change. Cambridge University Press, Cambridge (2012)
IPCC: Summary for policymakers. In: Edenhofer, O., Pichs-Madruga, R., Sokona, Y., Farahani, E., Kadner, S., Seyboth, K., Adler, A., Baum, I., Brunner, S., Eickemeier, P., Kriemann, B., Savolainen, J., Schlömer, S., von Stechow, C., Zwickel, T., Minx, J.C. (eds.) Climate Change 2014: Mitigation of Climate Change. Contribution of Working Group III to the Fifth Assessment Report of the Intergovernmental Panel on Climate Change. Cambridge University Press, Cambridge (2014)
Kim, H.C., Fthenakis, V., Choi, J.K., Turney, D.E.: Life cycle greenhouse gas emissions of thin-film photovoltaic electricity generation: systematic review and harmonization. J. Ind. Ecol. **16**(S1), S110–S121 (2012). doi:10.1111/j.1530-9290.2011.00423.x
Koskela, S., Seppälä, J., Lipp, A., Hiltunen, M.R., Pold, E., Talve, S.: Estonian electricity supply scenarios for 2020 and their environmental performance. Energy Policy **35**(7), 3571–3582 (2007). doi:10.1016/j.enpol.2007.01.001
Lacirignola, M., Blanc, I.: Environmental analysis of practical design options for enhanced geothermal systems (EGS) through life-cycle assessment. Renew. Energy **50**, 901–914 (2013). doi:10.1016/j.renene.2012.08.005
Laurent, A., Espinosa, N.: Environmental impacts of electricity generation at global, regional and national scales in 1980–2011: What can we learn for future energy planning? Energy Environ. Sci. **8**(3), 689–701 (2015). doi:10.1039/C4EE03832K
Laurent, A., Olsen, S.I., Hauschild, M.Z.: Limitations of carbon footprint as indicator of environmental sustainability. Environ. Sci. Technol. **46**(7), 4100–4108 (2012). doi:10.1021/es204163f
Martínez, E., Sanz, F., Pellegrini, S., Jiménez, E., Blanco, J.: Life cycle assessment of a multi-megawatt wind turbine. Renew. Energy **34**(3), 667–673 (2009). doi:10.1016/j.renene.2008.05.020
Masanet, E., Chang, Y., Gopal, A.R., Larsen, P., Morrow, W.R., Sathre, R., Shehabi, A., Zhai, P.: Life-cycle assessment of electric power systems. Annu. Rev. Environ. Resour. **38**(1), 107–136 (2013). doi:10.1146/annurev-environ-010710-100408
May, J.R., Brennan, D.J.: Life cycle assessment of australian fossil energy options. Process Saf. Environ. Prot. **81**(5), 317–330 (2003). doi:10.1205/095758203770224351
Münster, M., Finnveden, G., Wenzel, H.: Future waste treatment and energy systems: examples of joint scenarios. Waste Manag **33**(11), 2457–2464 (2013). doi:10.1016/j.wasman.2013.07.013
Nomura, N., Inaba, A., Tonooka, Y., Akai, M.: Life-cycle emission of oxidic gases from power-generation systems. Appl. Energy **68**(2), 215–227 (2001). doi:10.1016/S0306-2619(00)00046-5
Pfister, S., Saner, D., Koehler, A.: The environmental relevance of freshwater consumption in global power production. Int. J. Life Cycle Assess. **16**, 580–591 (2011)
Poinssot C, Boullis B, Bourg S: Role of recycling in advanced nuclear fuel cycles. In: Reprocessing and Recycling of Spent Nuclear Fuel. Woodhead Publishing Oxford. http://www.sciencedirect.com/science/article/pii/B9781782422129000022 (2015a). Accessed 05 2016
Poinssot, C., Bourg, S., Boullis, B.: Improving the nuclear energy sustainability by decreasing its environmental footprint. Guidelines from life cycle assessment simulations. Prog Nuclear Energy (2015). doi:10.1016/j.pnucene.2015.10.012
De Miranda, R.F., da Silva, G.A.: Life-cycle inventory for hydroelectric generation: a Brazilian case study. J. Clean. Prod. **18**(1), 44–54 (2010). doi:10.1016/j.jclepro.2009.09.006

Ryberg, M.W., Owsianiak, M., Laurent, A., Hauschild, M.Z.: Power generation from chemically cleaned coals: Do environmental benefits of firing cleaner coal outweigh environmental burden of cleaning? Energy Environ. Sci. **8**(8), 2435–2447 (2015). doi:10.1039/C5EE01799H

Santoyo-Castelazo, E., Gujba, H., Azapagic, A.: Life cycle assessment of electricity generation in Mexico. Energy **36**(3), 1488–1499 (2011). doi:10.1016/j.energy.2011.01.018

Schleisner, L.: Life cycle assessment of a wind farm and related externalities. Renew. Energy **20** (3), 279–288 (2000). doi:10.1016/S0960-1481(99)00123-8

Schmidt, J.H., Merciai, S., Thrane, M., Dalgaard, R.: Inventory of country specific electricity in LCA - Consequential and attributional scenarios. Methodology report v2. Inventory Report v2, 26. http://lca-net.com/p/212 (2011). Accessed 05 2016

Schreiber, A., Zapp, P., Marx, J.: Meta-analysis of life cycle assessment studies on electricity generation with carbon capture and storage. J. Ind. Ecol. **16**, S155–S168 (2012). doi:10.1111/j.1530-9290.2011.00435.x

Sevencan, S., Ciftcioglu, G.A.: Life cycle assessment of power generation alternatives for a stand-alone mobile house. Int. J. Hydrogen Energy **38**(34), 14369–14379 (2013). doi:10.1016/j.ijhydene.2013.09.029

Sherwani, A.F., Usmani, J.A., Varun, : Life cycle assessment of solar PV based electricity generation systems: a review. Renew. Sustain. Energy Rev. **14**(1), 540–544 (2010). doi:10.1016/j.rser.2009.08.003

Smith, C., Burrows, J., Scheier, E., Young, A., Smith, J., Young, T., Gheewala, S.H.: Comparative life cycle assessment of a Thai Island's diesel/PV/wind hybrid microgrid. Renew. Energy **80**, 85–100 (2015). doi:10.1016/j.renene.2015.01.003

Stamford, L., Azapagic, A.: Life cycle environmental impacts of UK shale gas. Appl. Energy **134**, 506–518 (2014). doi:10.1016/j.apenergy.2014.08.063

Suwanit, W., Gheewala, S.H.: Life cycle assessment of mini-hydropower plants in Thailand. Int. J. Life Cycle Assess. **16**(9), 849–858 (2011). doi:10.1007/s11367-011-0311-9

Treyer, K., Bauer, C.: Life cycle inventories of electricity generation and power supply in version 3 of the ecoinvent database—part I: electricity generation. Int. J. Life Cycle Assess. (2013). doi:10.1007/s11367-013-0665-2

Treyer, K., Bauer, C.: Life cycle inventories of electricity generation and power supply in version 3 of the ecoinvent database—part II: electricity markets. J. Life Cycle Assess, Int (2014). doi:10.1007/s11367-013-0694-x

Treyer, K., Bauer, C.: The environmental footprint of UAE's electricity sector: combining life cycle assessment and scenario modeling. Renew. Sustain. Energy Rev. **55**, 1234–1247 (2016). doi:10.1016/j.rser.2015.04.016

Turconi, R., Boldrin, A., Astrup, T.: Life cycle assessment (LCA) of electricity generation technologies: overview, comparability and limitations. Renew. Sustain. Energy Rev. **28**, 555–565 (2013). doi:10.1016/j.rser.2013.08.013

Turconi, R., Tonini, D., Nielsen, C.F.B., Simonsen, C.G., Astrup, T.: Environmental impacts of future low-carbon electricity systems: detailed life cycle assessment of a Danish case study. Appl. Energy **132**, 66–73 (2014). doi:10.1016/j.apenergy.2014.06.078

Viebahn, P., Kronshage, S., Trieb, F., Lechon, Y.: Final Report on Technical Data, Costs, and Life Cycle Inventories of Solar Thermal Power Plants. Deliverable no. 12.2—RS 1a for EU FP6 project "New Energy Externalities Developments for Sustainability" (NEEDS). Project no. 502687 (2008)

Wang, Y., Sun, T.: Life cycle assessment of CO_2 emissions from wind power plants: methodology and case studies. Renew. Energy **43**, 30–36 (2012). doi:10.1016/j.renene.2011.12.017

Weidema, B., Frees, N., Nielsen, A.-M.: Marginal production technologies for life cycle inventories. Int. J. Life Cycle Assess. **4**(1), 48–56 (1999). doi:10.1007/BF02979395

Weidema, B.P., Bauer, C., Hischier, R., Mutel. C., Nemecek. T., Reinhard. J., Vadenbo. C.O., Wernet. G.: Overview and methodology. Data Quality Guideline for the Ecoinvent Database Version 3. Ecoinvent Report 1(v3). The ecoinvent Centre, St. Gallen, CH (2013)

Weinzettel, J., Reenaas, M., Solli, C., Hertwich, E.G.: Life cycle assessment of a floating offshore wind turbine. Renew. Energy **34**(3), 742–747 (2009). doi:10.1016/j.renene.2008.04.004

Weiss, M., Haufe, J., Carus, M., Brandão, M., Bringezu, S., Hermann, B., Patel, M.K.: A review of the environmental impacts of biobased materials. J. Ind. Ecol. **16**, S169–S181 (2012). doi:10.1111/j.1530-9290.2012.00468.x

Author Biographies

Alexis Laurent Working with LCA since 2010 with a strong LCIA focus, particularly on normalisation aspects. Main LCA interests include development of LCIA methods, LCA applications and footprinting of large-scale systems for policy-making (e.g. nations, sectors), and LCA applied to various technology domains, including energy systems.

Nieves Espinosa Background in industrial engineering with experience on LCA since 2008, with a focus on ecodesign of selected energy technologies such as organic photovoltaics, and other energy related products and processes. Special attention is given to the implementation and integration of these technologies with a policy oriented perspective.

Michael Z. Hauschild Involved in development of LCIA methodology since the early 1990s. Has led several SETAC and UNEP/SETAC working groups and participated in the development of the ISO standards and the ILCD methodological guidelines. Main LCA interests are chemical impacts, spatial differentiation and science-based boundaries in LCIA.

Chapter 27
LCA of Electromobility

Felipe Cerdas, Patricia Egede and Christoph Herrmann

Abstract Private transportation is increasingly responsible for a significant share of GHG emissions. In this context, electric vehicles (EVs) are considered to be a key technology to reduce the environmental impact caused by the mobility sector. While EVs do offer an opportunity to decrease the production of greenhouse gases radically by avoiding the generation of tailpipe emissions, different technological challenges must be overcome. On the one side, the production of the battery system is of significant importance as it is reckoned to be responsible for around 40–50% of the total CO_2-eq. emissions of the vehicle's manufacturing stage. Moreover, the additional requirements for metals like copper and aluminium for the battery system as well as rare earth metals for the production of electric motors might lead to shifting the problem to other life cycle stages or areas of impact. On the other side, the source of the energy used to power an EV has an ultimate influence on the environmental impact caused during the vehicle's use stage. The life cycle assessment methodology is normally used to measure the environmental impact of electric vehicles and to identify potential problem shifting. In this chapter, we present an overview of the application of the methodology within the electric mobility sector.

27.1 Introduction

27.1.1 Current Context

Transportation poses great challenges for the sustainability agendas of countries worldwide. As reported in the latest climate change report by the IPCC, by 2010 direct anthropogenic greenhouse gas (GHG) emissions from the transportation

F. Cerdas (✉) · P. Egede · C. Herrmann
Chair of Sustainable Manufacturing and Life Cycle Engineering,
Institute for Machine Tools and Production Technology, Technische Universität Braunschweig, Langer Kamp 19b, Brunswick, Germany
e-mail: juan.cerdas-marin@tu-braunschweig.de

M.Z. Hauschild et al. (eds.), *Life Cycle Assessment*,
DOI 10.1007/978-3-319-56475-3_27

sector increased from 2.8 Gt CO_2-eq. in 1970 to 7.08 Gt CO_2-eq. (Sims et al. 2014). In that year, this sector contributed to up to 23% of the total energy related CO_2 emissions (Sims et al. 2014). In addition, individual transportation ownership is projected to grow up to 2 billion vehicles by 2050 (IEA 2009) reaching rates of 12 Gt CO_2-eq./year (Sims et al. 2014). The transition towards a more sustainable individual transportation urges for radical changes regarding transportation means, fuel consumption and resulting GHG emissions. In this regard, different strategies have emerged worldwide, including:

- The development of alternative fuels (e.g. biofuels see Chap. 30).
- The reduction of driving distances by facilitating other modes of commuting, improving public transportation or its accessibility, promoting car sharing programs, among many others.
- The optimisation of existing technologies (or the development of new ones) to increase the vehicle's energy efficiency. For instance, by improving the efficiency of the drive train.
- The reduction of the vehicle's weight by substituting materials and implementing new design concepts or modern production technologies (e.g. 3D printing).
- The development of alternative powertrains such as electric vehicles together with increased production of renewable energy.

Regarding the latter, electric vehicles (EVs[1]) are seen by many countries as a promising technology to achieve significant reductions of GHG emissions. In Germany, for example, the national electric mobility plan disclosed a set of ambitious objectives which among others include the mass scale production of lithium-ion batteries, as well as the production of battery and plug-in hybrid electric vehicles (BEV and PHEV) to ultimately place 1 million electric vehicles on the country's roads by 2020 (German Federal Government 2009). Moreover, in countries like Sweden and the United States, market sales shares of EVs reached in 2014 over 1%[2] (OECD/IEA 2015).

The successful penetration of Electric Vehicles (EVs) in the automotive market depends on three key factors, i.e. costs, customer satisfaction and engineering performance. In this regard, the environmental impact of an EV [or its potential reduction when compared to a conventional vehicle (CV[3])] plays an important role towards its market acceptance and is one of the main reasons for their development in the first place.

As they do not produce tailpipe emissions, EVs are believed to radically decrease greenhouse gas emissions. Yet, tailpipe emissions are only one aspect of

[1]We use the term EV to refer to battery electric vehicles (BEV), hybrid electric vehicles (HEV) and plug-in hybrid electric vehicles (PHEV).

[2]In Norway, the EV market share represents more than 12% and in the Netherlands more than 3%.

[3]In this chapter, we refer as conventional vehicles to vehicles powered with an internal combustion engine.

the analysis. The question of whether driving an EV is better than driving a conventional one from an environmental stand point, demands a more comprehensive analysis.

27.1.2 Technical Context of Electric Vehicles

The term electric vehicle EV refers to a vehicle that is fully or partially powered by electricity supplied by an electrochemical or electrostatic energy storage system and fully or partially propelled by an electric motor (Guzzella and Sciarretta 2005). In this chapter, we classify electric vehicles into battery electric vehicles (BEV) and hybrid electric vehicles[4] (HEV).

According to the degree of hybridisation, HEVs can be classified as mild and full HEVs. Figure 27.1 shows the configuration of the aforementioned types of EV and the respective flow and type of energy. EVs mostly use lithium-ion batteries the capacity of which varies depending on its size (see Fig. 27.1) and permanent magnet synchronous motors as traction motors. BEVs rely completely on electricity from the grid.

Broadly speaking, this electrical energy is converted by an electric motor into mechanical energy that is ultimately transmitted to the wheels. While its electrical range is larger than that of the other types of EVs, BEV depends almost completely on an on-board battery whose charging can take hours and thereby restrict its autonomy. Vehicles like the Nissan Leaf, the BMWi3 and the Tesla model S are some examples of mass produced BEV.

HEVs, in turn, depend on two different energy sources: fuel and electricity generated during regenerative breaking. Three types of HEV are distinguished according to how the energy flows between the vehicle's components. These include parallel HEV[5] (represented in Fig. 27.1), series HEV[6] and combined series–parallel HEV. Roughly, in an HEV the internal combustion engine (ICE) converts the fuel's chemical energy into mechanical energy, whereas the battery is charged internally through the energy produced by the electric motor/generator. Examples of mass produced hybrid vehicles are the Toyota Prius, the Ford Fusion and the VW Jetta hybrid.

Plug-in electric vehicles (PHEVs) are a type of HEV that can be plugged-in to be recharged from the electricity grid, providing the vehicle with an all-electric range

---

[4]The term "hybrid vehicle" is used to distinguish a car that combines an engine and an electric motor/generator (Guzzella and Sciarretta 2005).

[5]Parallel HEV: The energy flow follows two parallel routes: (i) The fuel tank feeds the internal combustion engine (ICE) which delivers mechanical energy to the wheels, (ii) the battery delivers electrical energy to an electric motor which delivers mechanical energy to the wheels.

[6]Series HEV: The energy flow follows a single route. The fuel tank feeds an ICE that is couple to a generator. The generator charges the battery which provides electrical energy to an electric motor. The electric motor delivers mechanical energy to the wheels.

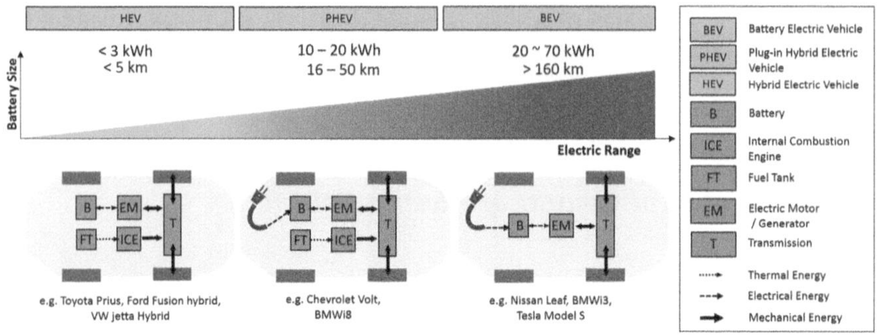

Fig. 27.1 Technical characteristics of electric vehicles

that is usually enough for a daily urban utilisation.[7] In the case of a PHEV, the ICE works as a range extender. Examples of mass produced PHEVs are the Chevrolet Volt and the BMW i8. A more detailed description of the EV's main components that should be considered for an LCA is presented in Sect. 27.2.2. For more general information regarding technical characteristics of electric vehicles, their components and their well to wheel (WTW—see Fig. 27.2) energy efficiency, we refer the reader to the work of Helmers and Marx (2012) and Yong et al. (2015).

27.1.3 Role of LCA in the (Electric) Mobility Sector

Imagine for example that we are interested in comparing the environmental impact of driving 100 km at a fixed speed, with a Land Rover Discovery V6 against a Suzuki Alto 1.1 both conventional gasoline-engine-driven vehicles. This example is naturally very trivial one could easily argue that the heavier the car is, the more energy is required to move it, and therefore the larger its fuel consumption and thus its carbon footprint. The environmental impact of a CV is driven by how much fuel is used and how efficiently it is combusted during its operation. In this sense, a common practice is to compare the potential environmental benefits of two or more fuel combustion technologies by framing the study on a well to wheel (WTW) approach. As represented in Fig. 27.2, the WTW analysis is focused on assessing the life cycle environmental impact of the energy carrier used to power a vehicle (e.g. liquid fuel, natural gas, hydrogen and electricity).

This approach is usually divided into well to tank (WTT) and tank to wheel (TTW) analysis. WTT analysis examines the upstream supply chain of an energy carrier, namely all the different conversion and distribution steps necessary to deliver energy to the vehicle. A TTW approach, contemplates exclusively the

[7] In the eLCAr guidelines, daily commuting in an urban environment is characterised by 40 km of driving range and a max speed of 160 km/h.

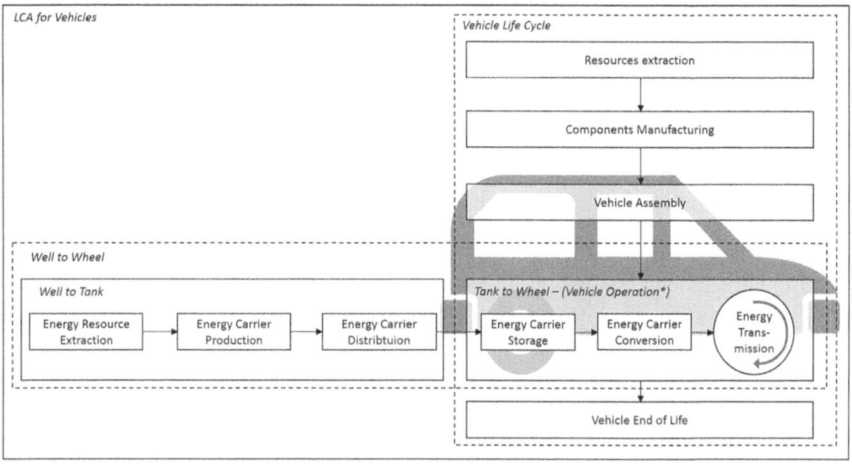

Fig. 27.2 Different perspectives for the environmental burden of vehicles. Adapted from Nordelöf et al. (2014)

efficiency with which the energy is converted by the engine and transmitted to the wheels. Here, the results are expressed in terms of the amount of tailpipe emissions per delivered traction or distance driven.

The example becomes less trivial if the analysis focuses on evaluating the impact associated with those strategies developed to reduce fuel consumption (e.g. through reducing the weight of the car) and tailpipe emissions (e.g. through developing alternative drive trains or non-fossil fuels). While the adoption of alternative lighter materials for the production of automotive components leads to a reduction of the overall vehicle's weight, the environmental consequences of such a measure require an analysis made with both eyes opened.

Compared to aluminium, a steering wheel made of magnesium can present GHG emission savings in the production stage if the magnesium is produced through an electrolysis process. When compared to an average magnesium Pidgeon process, the GHG emissions can be up to four times higher which leads to increase in GHG life cycle emissions (Ehrenberger 2013). Accordingly, if we consider the life cycle of the vehicle in our example, a measure to reduce weight through integrating magnesium intensively in the vehicle components may lead to an increase of the life cycle emissions.

In the case of an EV, the environmental trade-offs between materials, components, vehicle characteristics and specific factors influencing each of the life cycle stages are more complex to identify and to analyse.

To illustrate this case, consider the generic break-even analysis represented in Fig. 27.3. The graphic is divided in three parts including EV production, EV use stage and EV end-of-life as shown. Suppose now that we want to compare the environmental burden of driving a battery electric vehicle (BEV) against a CV. Disregarding the nature of the impact represented in the figure, before having driven the first kilometre both cars are already responsible of a certain amount of impact U (U_{ev} for the EV and U_{cv} for the conventional vehicle). The environmental burden of the production stage of an EV is estimated to be larger than that of the conventional vehicle due mainly to the impact caused during the production of the battery pack.

Generally, a battery pack contains large amounts of aluminium and copper required as current collectors in the cells, as well as large quantities of other important metals that are necessary for the production of the electrode active materials (e.g. nickel, cobalt, manganese and lithium among many others) whose upstream supply chain in some cases includes significant mining processes characterised by being very energy- and SO_2 emission-intensive and in some cases producing significant amounts of toxic emissions. Furthermore, for the specific case of a BEV, the battery can represent up to 30% of the total vehicle's weight, a fact that has repercussions in the energy consumption of the vehicle in the use stage.

As both vehicles enter operation (see EV use stage in Fig. 27.3), their environmental impact is driven mainly by the fuel or energy demand. Given fixed conditions for the comparison (e.g. acceleration, speed, number of passengers, vehicle characteristics and driver behaviour), both vehicles are subject to basically the same nature of forces acting on them. Broadly speaking, these forces lead to a mechanical energy demand, mostly driven by the vehicle's weight that must be supplied to the vehicle's wheels.

The slope of the curves m represents the impact produced per kilometre driven. It is determined by (a) the total energy consumed by the vehicle including energy losses due to powertrain inefficiencies and the energy consumed by non-propulsion-related components; and (b) the impact produced due to maintenance and service. On a TTW perspective, not only is the conversion of energy and its transmission to the wheels much more efficient in a BEV, but also this process produces no tailpipe emissions at all. Nevertheless, from a WTW perspective, the environmental burden of both vehicle types will be different as the impact depends on the source (or mix of sources) of the energy carrier delivered to power the vehicle. In this regard, Fig. 27.3 presents three hypothetic scenarios for the vehicles under comparison: (BEV-1) EV powered with a moderate fossil energy mix, (BEV-2) EV powered with a large share of fossils within the energy mix and (BEV-3) EV powered with a low CO_2-eq. intensity energy mix. Important to notice here is that the sooner the EV breaks even, the larger is the potential reduction of its environmental burden in a life cycle perspective.

The right side of Fig. 27.3 (EV EoL) represents the end-of-life stage. The potential future reuse and/or remanufacture of components as well as the production of secondary material from spent automotive parts, implies the possibility of earning environmental credits which ultimately help reducing the overall life cycle impact.

As seen, the environmental impact of an EV can be influenced by many different factors making the range of results varying greatly from one scenario to another. LCA offers in this regard a straightforward methodology that not only enables a fair comparison between technologies, but also helps identify the many different

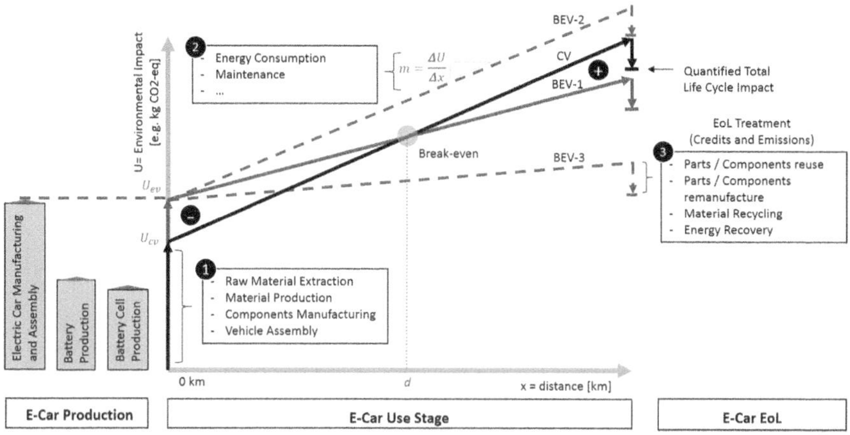

Fig. 27.3 Representation of a break-even analysis for the environmental burden of EVs (BEV-1, BEV-2 and BEV-3 represent electricity grid mixes causing moderate, high and low emissions of GHGs, respectively)

environmental trade-offs that can rise out of design strategies and transport policies. In short, LCA in the field of (electric)-mobility can be used to give well-grounded answers to the challenges:

- Comparisons between different types of EVs and CVs;
- The effect of the energy mix used to power vehicles;
- The effect of driving behaviour and local climate conditions (ambient temperature);
- The evaluation of weight reduction strategies;
- The analysis of the contribution of the traction battery to the overall environmental impact of an EV;
- The analysis of end-of-life scenarios mainly regarding the treatment of main components, especially batteries, electric motors and car body.

27.2 LCA of Electric Vehicles: Specific Methodological Issues

In this section, we introduce the reader to the application of LCA within the (electric)—mobility sector. To begin, a general overview of the application of the LCA methodology in the (electric)-mobility field is presented. Particularly, this section discusses issues of goal definition and problem scoping (see more about goal and scope definition in Chaps. 7 and 8) and gives a brief overview on data collection for the three major life cycle stages.

27.2.1 General Methodological Issues

EVs exist in several segments and configurations, each of which is composed by a specific set of components. These components are all responsible of a share of the total vehicle's environmental impact as they: (i) need to be produced, assembled and eventually disposed or recovered, (ii) may add additional weight to the vehicle which ultimately means energy consumption; and (iii) may affect the energy conversion efficiency thus increasing energy demand. In addition, due to the physical and technical interdependencies between components, a modification in one of them can lead to important changes in terms of design and/or performance in other components of the vehicle. This situation is particularly important to consider when defining the boundaries of the system and the modelling approach.

27.2.1.1 Definition of the Goal for the LCA of EVs

A comprehensive statement of the goal of the study may prevent an increase in the complexity of the analysis and reduce the degree of variability of the results (Nordelöf et al. 2014). Defining clearly and unambiguously the intended application of the expected results and the purpose of the study helps in general to identify and to describe the system that will best represent the product system.

For example, if the study is intended to make assertive comparisons about the environmental impact caused by driving an EV a certain distance using different types of batteries, the scope and inventory analysis of the study should explicitly contain information regarding battery characteristics, its energy density, weight and the potential interdependencies between components.

The reason for carrying out the study and the decision context also play an important role for the further definition of the product system and its boundaries. If an LCA aims to provide support during a decision making process, it is essential to consider all the potential effects of that specific decision. In this regard, framing the study within a decision context is important as it helps to define methodological and quality needs.

In line with the eLCAr guidelines (Del Duce et al. 2013) and the ILCD (European Commission—Joint Research Centre—Institute for Environment and Sustainability 2010), an LCA in the field of (electric)-mobility can be set in a situation context A (micro-level, product or process-related decision support) or B (meso-level and macro-level, strategic ("policy") decision support). Both contexts address potential consequences of a certain decision; however, the extent and nature are very different from each other (see more about decision context in Sect. 8.5.4).

These differences should be taken into account as they imply drastic changes regarding the way in which the involved supply chains are modelled. In general, an LCA intended to analyse short-term effects would most probably be best represented by situation A. This is for example the case where an LCA is performed on an operational level (e.g. comparison of two different brands of electric vehicles, introduction of a new vehicle model or technology) and its modelling is based on current supply chains.

As there is a possibility that the market share of EVs strongly increases in the mid-term, large-scale consequences might occur on structures of adjacent systems linked to the product system being assessed. Some examples that can be identified in this regard are: i. the required infrastructure, ii. the production of electricity to meet the additional demand, iii. the rise of new material supply such as rare earth metals, lithium, copper and aluminium among other. An LCA that aims to address these circumstances is to be set in a situation context B. Comprehensive examples of goal and scope definition for the specific case of an electric vehicle are presented in the eLCAr guidelines.

27.2.1.2 Product System and System Boundary

In this section, we focus the analysis on a situation type A. The principal activities to be considered for an LCA of an EV are shown in Fig. 27.4. These can be grouped in four different stages: (i) The production of components which in turn comprises background activities located in the upstream supply chain (e.g. mining processes, production of materials and transportation among others). Although the battery is technically part of the drive train, we consider the battery system as an extra component. In that way, the component production activities are distributed in: (a) production of drive train, (b) battery and (c) car body; (ii) the vehicle assembly stage including, for example, the respective energy consumption, materials waste, painting processes and their emissions; (iii) the vehicle's use stage including charging patterns and driving behaviours, effect of local climate, production and distribution of energy, maintenance and service activities, and charging and road infrastructure; and (iv) the vehicle's end-of-life stage including credits for material and energy recovery from recycling, reuse and remanufacturing activities.

Notice that the boundaries are presented in a very general form. The actual system boundary should be described by tracking down the supply chain of the product system defined.

27.2.1.3 Functional Unit and Reference Flow

Finding an equivalent functionality for the comparison of EVs and conventional vehicles with an LCA can be very challenging. While the function can be expressed as a unit of service provided by the respective vehicle (e.g. a journey from point *A* to point *B*), it might be necessary to take a more descriptive approach so as to enable a fairer comparison and an easier interpretation of the results.

A more comprehensive picture of the systems being compared can be characterised by addressing questions such as how often, for how long, how well (or efficiently), under which conditions, among others.

As previously discussed, all the components of an EV have an influence on the energy demand during the use stage, which is mainly due to the weight added to the

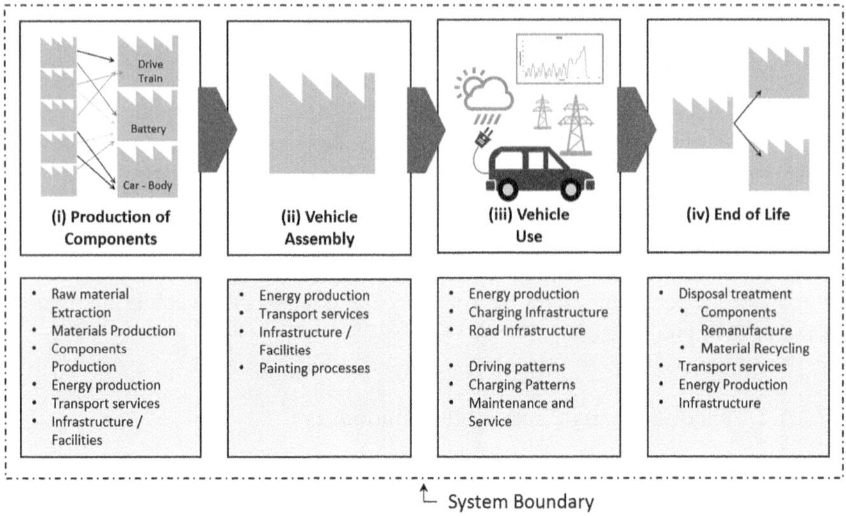

Fig. 27.4 Generic representation of the product system for the LCA of an EV

vehicle, but could also be related to issues of mechanical, electrical, thermal and electrochemical efficiency. This interdependency forces the LCA practitioner to consider a whole-vehicle perspective to approximate a description of an equivalent functionality.

EVs increase this complexity. On the one side, its efficiency is governed by properties such as ratio of discharge, energy density, number of cycles and by how the battery is used, but to add complexity, these properties differ from one battery system to another depending on both materials and production processes. On the other side, the mass added by the battery is responsible for a significant portion of the vehicle's energy demand during the use stage.

After overcoming this challenge, the vehicles under comparison are comprehensively defined in size and technical properties, and then a description of the conditions under which the vehicle is used should be given. This is particularly important for the case in which an LCA intends to compare an EV against a CV. As EVs have limited autonomy in terms of driving range and charging times compared to CVs, the functional unit should make reference to the type of application. These autonomy limitations are not relevant if the study aims to compare daily urban transportation as the technical requirements can easily be satisfied by the technologies being analysed. If instead the focus lies on analysing an average vehicle use, these mentioned limitations should be considered by for instance including a rented car or the use of public transportation to model the (long-)distance that the EV is not able to complete (Del Duce et al. 2013).

Another important aspect to consider is the variation of the overall vehicle efficiency from one to another driving cycle, which makes the LCA only valid for the driving cycle (or its mix) considered for the comparison. In brief, as described in the eLCAr guidelines, the definition of the functional unit for an LCA in the field of (electric)-mobility can be formulated by describing parameters such as:

- Properties of the vehicle and/or its key components: vehicle class, weight, range per charge, lifetime, number of passengers per ride, maximum number of passengers among others.
- Interaction between the vehicle and its components and how this influences the unit of service.
- Location, time, geography, driving cycles, weather, etc.

27.2.2 Life Cycle Inventory Analysis in LCA of EV

While the definition of the goal and the respective decision context selected might limit the system to be studied, the analysis could take different focuses. For example, we may be interested in comparing the impact of two different battery configurations for a specific BEV. Although the battery of an EV could be modelled to a high extent independently from the vehicle, its weight might influence design considerations in other components that might end up adding more weight to the vehicle. This situation can lead to an increase of the upstream material supply chain and of course to more energy demand during the use stage of the vehicle. In such a system, disregarding the fact that the focus of the analysis lies on the battery system, the peripheral components affected shall be considered in the foreground system.

The interdependency matrix presented in the eLCAr guidelines[8] provides an overview of the most common interactions between components that might be taken into account. In this regard, the eLCAr guidelines distinguish two frequent situations in which an LCA may be focused. On the one hand, an LCA may be set to analyse one component. If the interaction between the component and the rest of the vehicle does not have a significant effect, the LCA can be restricted to the life cycle of that specific component placing it in the foreground system. If the component has an interaction with another vehicle component, then the foreground system of the LCA should include the entire vehicle, however, the level of detail of the analysis is not necessarily the same as for the component under study. On the other hand, the focus of the LCA may be on the complete vehicle, and in this case the foreground system should always include the whole vehicle.

A third situation can occur when, for example, the focus of the study lies in one of the product's life cycle stages. For instance, if the LCA aims to analyse the driving behaviours, geographic effects (i.e. climate, energy mixes, etc.) or business

[8]The interdependency matrix introduced in the eLCAr guidelines includes information on how one specific component might influence other components in an EV. These interdependencies are based on different assumptions including vehicle and driving characteristics. The interdependency matrix is available online.

models (e.g. car sharing or fleet application for example), the study might place the EV as being part of the background system. Once the foreground system is defined, the analysis focuses on identifying the data to be collected. We restrict this section to the description of the data collection process for the vehicle production stage, the estimation of the energy consumption during the use stage and the identification of unit processes to be modelled in the end-of-life.

27.2.2.1 Data Collection for the Vehicle Production Stage

A generic data collection plan for the vehicle production stage is presented in Fig. 27.5. The figure presents a simplification of the components to be considered as well as the processes and materials associated. Notice that the foreground system for the production stage of an EV is divided into production of components and vehicle assembly.

As recommended by the eLCAr guidelines, the production of each of the components in the foreground system should be subdivided into unit processes, which can be described as an independent operation and ultimately characterised in terms of their exchanges with the background system and the environment. In this regard, Fig. 27.5 includes a list of the most representative manufacturing processes used within the automotive industry as well as the main components that are recommended to be included.

Three groups of components are in focus: the glider, the drive train and the battery. Data collection in this field can be very exhaustive and time consuming. The eLCAr guidelines provide a rather general set of recommendations and only

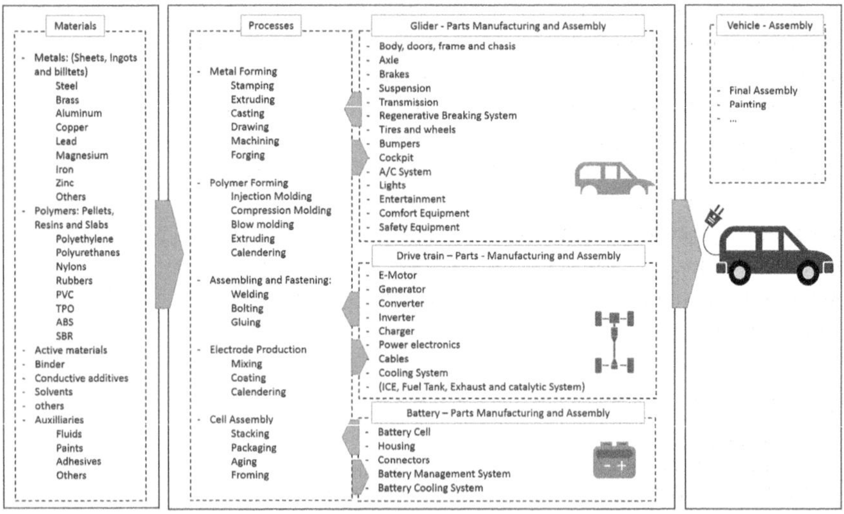

Fig. 27.5 Data collection plan for the production stage of EV

aim at pointing out the main issues to be addressed. In general, many LCAs done in the automotive industry are based on data modelled with the GREET model (Greenhouse gases, Regulated Emissions and Energy use in Transportation model) (Burnham et al. 2006).

The GREET model is an open source software originally created to evaluate vehicles on a WTW perspective. The current version includes a vehicle-cycle model, which contains information about the production and end-of-life stages of automotive components including alternative drive trains.

Furthermore, a common practice when modelling life cycle inventories of electric vehicles is to adapt or convert data from conventional vehicles (e.g. by replacing fuel tank with battery, combustion and exhaust systems with charging and power electronic systems). In this regard, the LCI of the VW Golf (Schweimer and Levin 2000) has often been adapted and extrapolated for research purposes. Moreover, detailed LCI for electric vehicles can be found in the research of Zackrisson et al. (2010), Hawkins et al. (2013) and Notter et al. (2010).

In addition, a few studies have published LCIs for the production of specific components. For example, the research by Majeau-Bettez et al. (2011), Ellingsen et al. (2014), Zackrisson et al. (2010) and Dunn et al. (2015) present detailed inventories for the production of traction batteries and research from Sullivan et al. (2013) offer models for the estimation of materials and energy used during the vehicles manufacturing and assembly stage.

27.2.2.2 Energy Consumption During the EV's Use Stage

The environmental impact of EVs during their use stage is influenced by different elements. Although the consideration of factors such as the charging and road infrastructure is also recommended, this section focuses on the estimation of the energy demand of the vehicle during its use (i.e. TTW in Fig. 27.2). For more information regarding the potential impact due to maintenance services and the generation of non-exhaust emissions, we refer the reader to the works of Del Duce et al. (2014) and Simons (2013).

The energy demanded by an EV to be included in an LCA can be measured or estimated. It can be divided into mechanical energy demand (i.e. the energy required to drive the vehicle from A to B), consumption of auxiliary devices, consumption due to air conditioning and heating requirements inside the vehicle, energy losses in the battery when the vehicle stays on standstill and the extra consumption of the battery due to charging losses.

The mechanical energy demand[9] of driving a vehicle over a certain distance is defined by the specific speed profile (v) of the trajectory and the power at the wheel

[9]The development of this section is based on the work done by Hofer (2014) and Guzzella and Sciarretta (2005).

(P_w) at every instant of the journey. P_w can be calculated by estimating the conservative and dissipative forces acting on a vehicle (Fig. 27.6).

The sum of all these forces is called traction force or force at the wheel (F_w) and is defined as:

$$F_w = F_a + F_r + F_g + F_k \tag{27.1}$$

Following, P_w can be defined as:

$$P_w = F_w \cdot v \tag{27.2}$$

Notice that P_w can be positive (i.e. the drive train needs to deliver torque to the wheels to propel the vehicle), negative (during braking) or zero (e.g. during coasting or when the vehicle is stopped). The mechanical energy demand can therefore be calculated by integrating P_w along a specific driving cycle.[10] The research from Hofer (2014) includes an estimation of the contribution of specific forces to the total mechanical energy demand of a vehicle for different driving cycles. Notice in Fig. 27.7 how for the same vehicle not only the total mechanical energy demand varies along the different driving cycles, but also the contribution of the different forces.

The electric passenger car transport and vehicle dataset developed for the ecoinvent v3 dataset (Del Duce et al. 2014) followed a similar approach. The energy consumption in the dataset is calculated based on the NEDC driving cycle with a power train efficiency of 70% and by increasing the efficiency by 5% for the parts where an urban-driving condition was modelled (i.e. therefore considering regenerative braking). The consumption of auxiliaries was estimated by Del Duce et al. (2014) assuming an average speed of 50 km/h and nominal powers of 3 kW

[10]A driving cycle is a description of a vehicle journey. It is usually represented by the variation of the speed against time on a specific road topography. Driving cycles are developed by executive/legislative bodies from different countries in order to standardise the measurement of the emissions produced and the consumption of fuel of a vehicle after a determined distance. Depending on the nature of speed changes (i.e. abrupt, gradual), driving cycles are divided into transient and steady state driving cycles. Transient driving cycles are those in which the speed changes constantly during the cycle. Examples of transient driving cycles are the New European Driving Cycle (NEDC) and the Worldwide harmonised Light vehicles Test Procedures (WLTP). The NEDC is supposed to represent the average usage of a passenger car in Europe and is composed of four urban cycles and one highway cycle. The WLTP is still under development but is based on statistical driving conditions from the EU, Japan, Korea, India and the US and is expected to replace the NEDC in the near future. Steady-state driving cycles basically represent a constant sequence of speed over time. For more information on driving cycles refer to Mock et al. (2013) and Barlow et al. (2009).

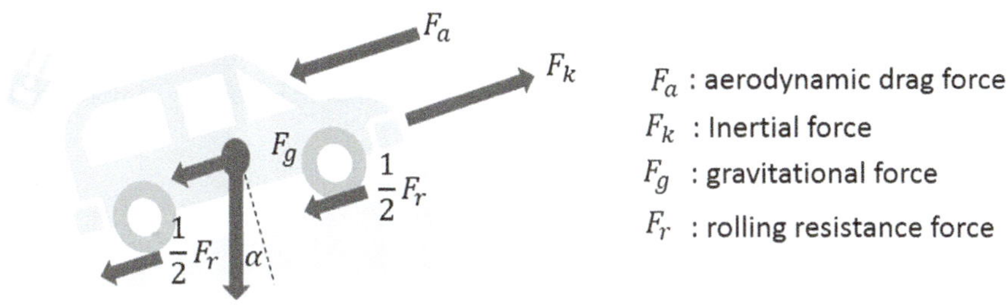

Fig. 27.6 Forces acting on a vehicle in movement. Based on Guzzella and Sciarretta (2005)

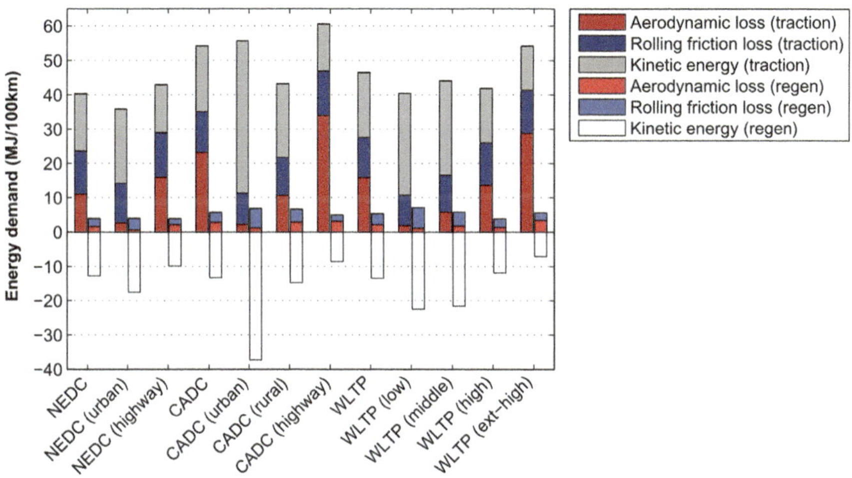

Fig. 27.7 Contributions to P_w for the NEDC and WLTP driving cycles. From Hofer (2014)

for heating, 0.6 kW for air conditioning and 0.5 kW for auxiliaries (i.e. radio, lights and so on). For every 100 km, a total of 2 kWh for the heating system and 0.4 kWh for the cooling system[11] were calculated.

From a WTT perspective, the energy mix dataset to be selected for the analysis will have a strong impact and shall therefore be carefully considered. Depending on the energy chosen mix, its contribution to the final results will vary among impact categories ranging from being negligible (e.g. when supplying energy from photovoltaic panels) to being the dominant hotspot (e.g. for the case of energy mixes with high shares of energy from coal power plants). Moreover, this mix can change in composition during the day or between seasons. In addition, as some regions are

[11]A more detailed approach for the estimation of the energy demand from auxiliaries and air conditioning systems can be found on the research from Egede (2016). More detailed calculations regarding energy losses on the battery system due to charging inefficiencies and to the discharging mechanisms while idling can be found in Del Duce et al. (2013).

interconnected, selecting energy mix datasets for specific places (e.g. countries) might mislead the interpretation of results. As recommended in the eLCAr guidelines, to enhance comparability the studies within the European Union should include in the analysis the European mix (EU-27).

27.2.2.3 Processes to Be Considered During the End-of-Life Stage of an EV

As recommended by the eLCAr guidelines, all the relevant processing steps involved after the use stage of an EV should be considered. First, the vehicle must be partly disassembled. This involves the treatment of specific hazardous materials (e.g. fluids, airbags, etc.) according to local regulations. Further components are recovered to be either recycled, remanufactured or disposed.

Of particular importance is the processing of the E-Motor and the battery system. Depending on the condition of the E-Motor, the motor can be remanufactured and therefore brought to a like-new condition to be reused. In this case, its life cycle should also be considered as it may be shorter than the motor's first cycle. If remanufacturing is not possible, the E-motor should be disassembled and its most important parts should be reused. Major processes for the treatment of the electric motor include the recycling of metals and the recovery of the permanent magnets and rare earth metals.

The end-of-life treatment of the traction battery should be considered within the system boundary. Modelling the recycling processes of a traction battery is difficult as they, being an emerging technology, are currently strongly under development. In other words, the LCA practitioner will have to deal with modelling a process that, partially or completely, does not exist. Nevertheless, as the topic of battery recycling becomes increasingly relevant due to environmental, economic and political reasons, future recovery of important metals such as aluminium, copper, nickel and cobalt is expected to contribute to the overall life cycle impact of an EV.

Current available research on battery recycling has identified several possible processing steps such as: battery dismantling down to the cell level and further processing of the cells. Since the battery dismantling process is mainly composed of mechanical steps, identifying the subprocesses and the material to be recovered is simple. However, the treatment of the cells might take several directions (e.g. hydrometallurgical and pyro-metallurgical processes) each of which involves technology-specific processes and thus differing from each other in terms of nature and amount of material recovered, processing costs (important to considered as it could indicate future market trends) and environmental impact.

27.3 Environmental Impact of EVs

While this field is still young, there is a relatively large amount of research trying to describe specifically the what, how and where of the interaction of electric vehicles and the environment. In the recent years, numerous LCA studies have been published not only in academic journals but also as environmental certifications from car manufacturers. Yet, results vary strongly between publications and benchmarking the results is a challenging exercise as most of the research available failed to express unambiguously the scope of the study and there are often inconsistencies in the application of the methodology. Moreover, lack of data and use of rough assumptions regarding energy consumption during the use stage, life time of the battery and inconsistencies on the selection of electricity mixes are common issues within the current research on LCA for EVs. Nevertheless, a few studies have provided very transparent life cycle inventories providing a more comprehensive understanding of the potential environmental impacts of electric vehicles.[12]

27.3.1 EVs Versus Conventional Vehicles

In this section, we make reference to the study done by Hawkins et al. (2013). Although the inventory is mostly based on secondary data sources, it is perhaps (and to our knowledge) the most complete and transparent inventory of an EV that is currently publicly available.

Their research compares the environmental impact produced by driving 1 km in a conventional vehicle (diesel or gasoline) against an EV. It includes the modelling of a generic glider adapted to meet the specific configuration of each technology under study. Their inventory considers the production of the glider, drive train and the battery and comprises around 140 vehicle subcomponents.[13]

For the case of the EV, the researchers compared the impact of two different battery chemistries, namely Li-$FePO_4$ and Li-NCM, and analysed the use of three different energy mixes. The energy consumption during the use stage was estimated based on the NEDC driving cycle. Based on data from manufacturers, they assumed a consumption of 0.63 MJ/km for the EV and 68 ml/km and 53.5 ml/km for the gasoline and the diesel vehicle, respectively. For this study, the authors assumed a vehicle lifetime of 150,000 km.

The global warming potential (GWP) impact for each specific case was spread over the total life cycle and the results are shown in Fig. 27.8. In general, the largest contributor to the environmental impact of the six scenarios was found to be the use stage.

[12]For a comprehensive state of the research, we refer the reader to the literature reviews done by Hawkins et al. (2012) and Nordelöf et al. (2014).

[13]The complete LCI can be accessed online.

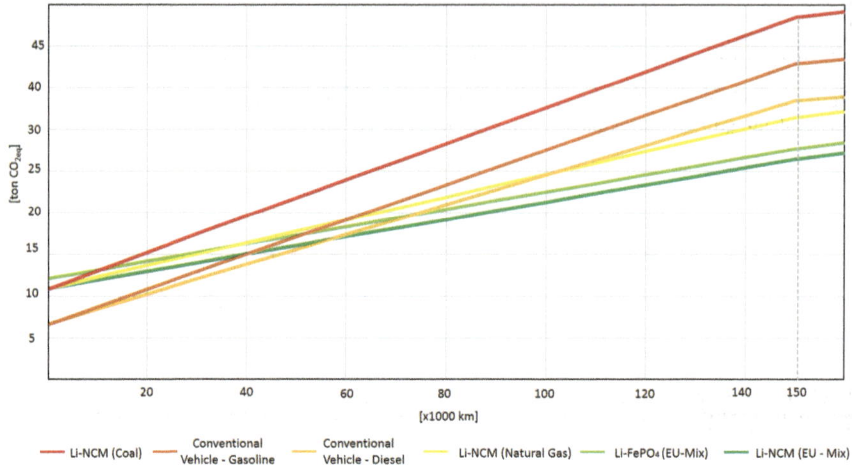

Fig. 27.8 Comparison of the life cycle GWP of a CV and an EV as function of the distance driven

For the case in which the European energy mix is used there is a general reduction in the life cycle CO_2-eq. emissions from EVs in comparison to the conventional ones. Under this condition, the difference on the GWP impact between the different battery cell chemistries is due only to the production stage. As seen in the figure, when EVs are powered with coal-based electricity, their life cycle CO_2-eq. emissions present increments of around 17–27% compared to gasoline and diesel, respectively.

The production stage also presents several interesting differences. The production of a conventionalvehicle was found to emit around half of the GHG emissions that are emitted during the production of an EV. The study reported that a GWP impact of around 13 tonne CO_2-eq. is produced during the production of an EV, 35–41% of which is caused by the production of the traction battery.[14] In addition, the cooling system required by the battery system is identified to contribute around 18% of the total CO_2-eq. from the production of an EV. Results for the other environmental impact categories are shown normalised to the scenario with the highest impact for each impact category in Fig. 27.9.

Several points are worth noticing from these results. First, the impact categories terrestrial acidification (TAP) and particulate matter formation (PMFP) behave very similarly in the EV and the conventional vehicle. As argued, the portion of hard coal and lignite used for the generation of the European electricity mix prevents the EVs to perform better than their conventional counterpart and therefore reduction in

[14]We refer the reader to the research done by Ellingsen et al. (2014) as it includes a very comprehensive LCI for a battery system similar to the battery in the study from Hawkins et al. (2013). As estimated by the author, for a lifetime of 150 000 km the battery reaches a GWP impact of approximately 4.7 tonnes CO_2-eq.

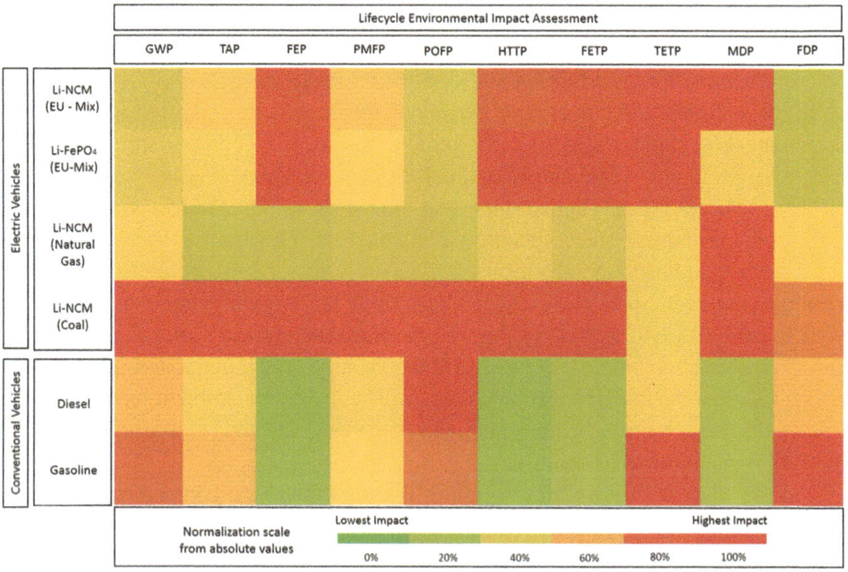

Fig. 27.9 Environmental impact of conventional and EV for different Impact Categories. Normalised from the absolute results

the life cycle TAP impacts can most probably be reached by powering EVs with electricity with lower sources of sulphur (e.g. natural gas).

As expected, the category photochemical oxidation formation (POFP) impact of EVs is significantly lower than for conventional vehicles as this impact is predominantly related to combustion processes. Human toxicity potential (HTP) is identified as a potential problem shifting by the authors. In this category, EV presents increments of up to 290% compared to CV which is mainly due to the increase in the use of metals like copper and nickel. These materials are usually produced through mining activities which are characterised by producing important amounts of toxic refuses. Finally, fossil resource depletion potential (FDP) potentially decreases if the EV is powered with average European electricity, but as shown, the advantages are not determinant if the coal-intensive energy mixes are used instead.

27.3.2 The Environmental Impact of a Lightweight Electric Vehicle

Using lightweight materials is one common strategy to reduce the weight of CVs. Lightweight materials used in CVs aim to reduce the energy consumption and thereby the environmental impact in the use stage of the vehicle. The savings in the use stage have to outweigh the higher environmental impacts of the lightweight material in comparison to the reference material, which usually occur in the raw material acquisition, production and end-of-life stage.

In EVs, lightweight materials can have the same effect. Different parameters determine how lightweight electric vehicles perform in comparison to reference EVs and CVs and when and if a break-even is reached. For the comparison of electric vehicles—both lightweight and non-lightweight electric vehicles—with CV, regional and use stage specific parameters are relevant.

In both vehicle types, electricity or fuel is necessary to operate the air conditioning. However, in CV the excess heat from the engine is used to heat up the vehicle cabin. In EV, this is not possible because the electric motor is very efficient and generates almost no excess heat. Hence, heat has to be generated when needed, which requires the use of additional energy. Therefore, the ambient temperature has a strong influence on the energy demand of the EV and only a minor influence on the fuel consumption of the CV.

For the comparison of a lightweight EV with a reference EV, the electricity mix must be considered as well as material properties (e.g. the lightweight factor of the lightweight material in comparison to the reference material) and vehicle properties (e.g. the energy saving per kilometre for each reduced kilogramme). To cover all influencing parameters, the comparison of lightweight electric vehicles with CV and reference vehicles requires a systematic approach. First, a detailed system description of the vehicles and their use is necessary. Examples are the description of the daily and seasonal use pattern. Then, the modelling of the interdependencies of the parameters like: the energy mix, the ambient temperature, the use pattern and the properties of the lightweight materials is required.

Finally, an adequate visualisation of results simplifies the interpretation of results for both LCA and non-LCA experts. The visualisation of LCA results in form of a map is useful because the results of the comparative assertion of (lightweight) electric vehicles and conventional vehicles depend on the regional parameters electricity mix and ambient temperature.

Figure 27.10 shows the LCA world map of the comparison of a gasoline CV and an EV with a lithium iron-phosphate (Li-$FePO_4$) battery for the impact category climate change. The use pattern represents a commuter using the vehicle in the morning and the afternoon (daily use) evenly throughout the year (seasonal use). Blue and green colours indicate that the CV is advantageous. Red, orange and yellow colours indicate that the EV performs better than the CV. Due to the consideration of the ambient temperature, the results vary within one country. For some countries like Spain, Argentina or Mexico no clear decision for or against one vehicle type can be given.

When lightweight materials are used, the question arises if a break-even point x is reached during the lifetime of the vehicle. Do the savings in the use stage outweigh the higher environmental impact of the lightweight material in the production and end-of-life stage ($i_{P,lw}$ and $i_{E,lw}$) which exists in comparison to the environmental impact of the reference material ($i_{P,ref}$ and $i_{E,ref}$)? This also depends on the lightweight factor of the lightweight material in comparison the reference

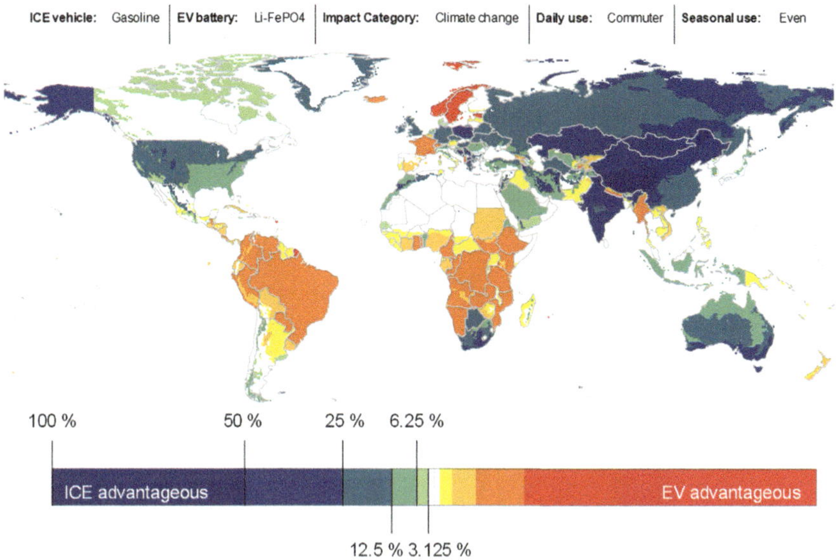

Fig. 27.10 LCA world map for the comparison of gasoline vehicle and electric vehicle with Li-$FePO_4$ battery for the impact category climate change, world map created with R and r world map. Taken from Egede (2016)

material (a_{lw}), the environmental impact of the electricity mix (i_e) and the saving in energy consumption of the vehicle per kilometre for each reduced kilogramme (c_{erv}). The following equation shows the calculation of this break-even point, X (distance driven in the vehicle):

$$x = \frac{a_{lw}(i_{P,lw} + i_{E,lw}) - (i_{P,ref} + i_{E,ref})}{c_{erv} * i_e(1 - a_{lw})} \tag{27.3}$$

Displaying the break-even point in relation to the energy mix leads to the chart in Fig. 27.11. The chart immediately allows to see when the use of a specific lightweight material leads to a break-even for a given specific energy mix, identified by its GHG emission intensity (the example of Germany in indicated in the figure). It is important to note that often ranges are given for the environmental impact of a material. This leads to ranges in the results for the break-even point. In the given example, the range of material 1 is more narrow than the range of material 2.

As a result, in Germany a break-even for material 1 is achieved for a total driving distance between 100,000 and 240,000 km whereas the break-even for material 2 is reached between 110,000 and 340,000 km. Further information on the environmental assessment of lightweight electric vehicles can be found in Egede (2016).

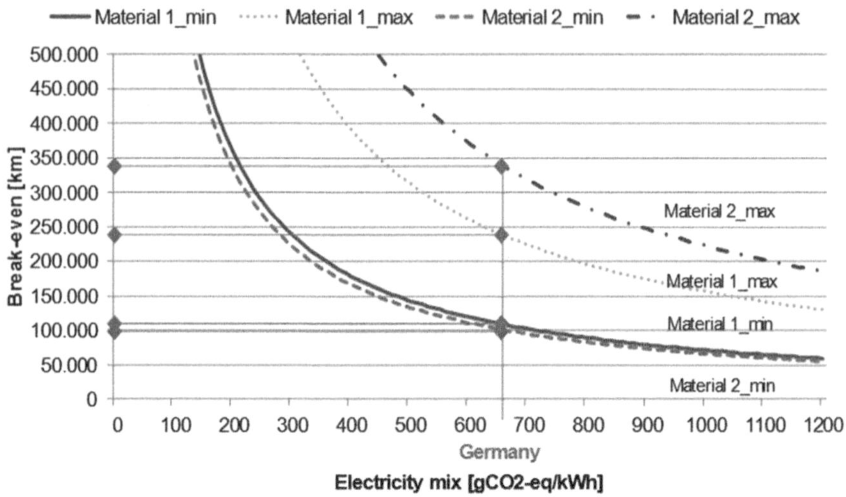

	Reference	Material 1_min	Material 1_max	Material 2_min	Material 2_max
$i_P + i_E$ (kgCO2-eq/kg)	2.38	2.7	2.9	2.8	3.3
α_{lw}(-)	-	0.94	0.94	0.92	0.92
erv (Wh/km/kg)	0.0369				

Fig. 27.11 Break-even analysis of two different materials. Taken from Egede (2016)

27.4 Concluding Remarks and Perspectives

The move towards a sustainable private transportation is challenging. There are many different strategies being conceived to achieve reduction in energy consumption and production of GHG emissions in the sector specifically.

While it is true that the development and promotion of EVs could lead to cutting tailpipe GHG emissions, the actual environmental effect of such a measure needs a more comprehensive analysis. The LCA methodology, if applied transparently and unambiguously, offers the possibility of broadening the understanding of the consequences of a potential electrification of personal transportation.

To close this chapter, we address issues of concern related to the technological sector analysed and the application of the methodology for its evaluation. The lack of methodological harmony is a central issue in the discussion.

One of the largest challenges to overcome towards understanding the environmental implications of EVs is the high level of inconsistency among undergoing research. While the eLCAr guidelines aimed at harmonising the application of the LCA methodology to obtain more accurate information, much of the research available still fails to give a proper definition of the system being analysed and its scope.

Accordingly, benchmarking results among different studies reported in the literature is difficult due to aspects such as ambiguous setting of boundaries, variations in the product lifetime analysed and lack (or inexistence) of a functional equivalency to enable comparability between two or more vehicles analysed. In spite of the lack of consensus regarding the application the LCA methodology in this field, a common shared conclusion is that the source of the energy used to power the vehicle to a great extent defines its environmental impact. In other words, substantial improvements in several impact categories can be reached if EVs are powered with low impact energy sources and therefore, promoting the market penetration of EVs in regions where electricity comes mostly from fossil sources can mislead to an increase in the global GHG emissions from transportation.

The production stage of an EV is estimated to be up to two times more environmentally intensive than a CV. In particular, the battery system poses the biggest challenges. On the one side, current massively produced battery systems for mobility applications contain large amounts of metals whose mining processes are usually characterised by being very harmful on a local/regional level.

In this regard, the analysis of significant changes in the material supply chain of batteries is needed. Considering the recycling of battery systems and the potential recovery of materials is important as this is estimated to minimise local environmental impacts and reduce the overall energy use and emissions of its production (Dunn et al. 2015).

On the other side, there is a raising concern regarding the materials intended to be used for batteries. As these differ abruptly regarding their state of development, most of these materials are difficult to screen and evaluate from an environmental standpoint as their behaviour on an industrial scale (massively produced) might be unpredictable. A more detailed consideration of the vehicle's production, and especially of the battery system, within the application of the LCA is essential as this could lead to identifying potential problem shifting issues.

References

Barlow, T., Latham, S., Mccrae, I., Boulter, P.: A reference book of driving cycles for use in the measurement of road vehicle emissions. **280** (2009)

Burnham, A., Wang, M., Wu, Y.: Development and applications of GREET 2.7—the transportation vehicle-cycle model. Energy. **124** (2006). doi:10.2172/898530

Del Duce, A., Egede, P., Öhlschläger, G., Dettmer, T., Althaus, H.-J., Bütler, T.S.E.: eLCAr—guidelines for the LCA of electric vehicles (2013)

Del Duce, A., Gauch, M., Althaus, H.-J.: Electric passenger car transport and passenger car life cycle inventories in ecoinvent version 3. Int. J. Life Cycle Assess. (2014). doi:10.1007/s11367-014-0792-4

Dunn, J.B., Gaines, L., Kelly, J.C., et al.: The significance of Li-ion batteries in electric vehicle life-cycle energy and emissions and recycling's role in its reduction. Energy Environ. Sci. **8**, 158–168 (2015). doi:10.1039/C4EE03029J

Egede, P.: Environmental Assessment of Lightweight Electric Vehicles, 1st edn. Springer International Publishing, Berlin (2016)

Ehrenberger, S.: Life Cycle Assessment of Magnesium Components in Vehicle Construction (2013)

Ellingsen, L.A.-W.W., Majeau-Bettez, G., Singh, B., et al.: Life cycle assessment of a lithium-ion battery vehicle pack. J. Ind. Ecol. **18**, 113–124 (2014). doi:10.1111/jiec.12072

European Commission—Joint Research Centre—Institute for Environment and Sustainability: International Reference Life Cycle Data System (ILCD) Handbook: Review schemes for Life Cycle Assessment (LCA). EUR 24710 EN (2010)

German Federal Government: German Federal Government's National Electromobility Development Plan (2009)

Guzzella, L., Sciarretta, A.: Vehicle Propulsion Systems, 2nd edn. Springer, Berlin (2005)

Hawkins, T.R., Gausen, O.M., Strømman, A.H.: Environmental impacts of hybrid and electric vehicles—a review. Int. J. Life Cycle Assess. **17**, 997–1014 (2012). doi:10.1007/s11367-012-0440-9

Hawkins, T.R., Singh, B., Majeau-Bettez, G., Strømman, A.H.: Comparative environmental life cycle assessment of conventional and electric vehicles. J. Ind. Ecol. **17**, 53–64 (2013). doi:10.1111/j.1530-9290.2012.00532.x

Helmers, E., Marx, P.: Electric cars: technical characteristics and environmental impacts. Environ. Sci. Eur. **24**, 14 (2012). doi:10.1186/2190-4715-24-14

Hofer, J.: Sustainability Assessment of Passenger Vehicles: Analysis of Past Trends and Future Impacts of Electric Powertrains. ETH Zurich (2014)

IEA: Transport, Energy and CO_2: Moving Toward Sustainability. Paris, France (2009)

Majeau-Bettez, G., Hawkins, T.R., Strømman, A.H.: Life cycle environmental assessment of Li-Ion and nickel metal hydride batteries for plug-in hybrid and battery electric vehicles. Supporting information. Zh. Eksp. Teor. Fiz. 1–51 (2011). doi:10.1021/es103607c

Mock, P., German, J., Bandivadekar, A. et al.: From laboratory to road. **77** (2013)

Nordelöf, A., Messagie, M., Tillman, A.M., et al.: Environmental impacts of hybrid, plug-in hybrid, and battery electric vehicles-what can we learn from life cycle assessment? Int. J. Life Cycle Assess. (2014). doi:10.1007/s11367-014-0788-0

Notter, D.A., Gauch, M., Widmer, R., et al.: Contribution of li-ion batteries to the environmental impact of electric vehicles. Environ. Sci. Technol. **44**, 6550–6556 (2010). doi:10.1021/es903729a

OECD/IEA: Global EV Outlook 2015 (2015)

Schweimer, G.W., Levin, M.: Life cycle inventory for the Golf A4. Environ. Res. 1–40 (2000)

Simons, A.: Road transport: new life cycle inventories for fossil-fuelled passenger cars and non-exhaust emissions in ecoinvent v3. Int. J. Life Cycle Assess. (2013). doi:10.1007/s11367-013-0642-9.1

Sims, R., Schaeffer, R., Creutzig, F., et al.: Transport. In: Edenhofer, O., Pichs-Madruga, R., Sokona, Y., Farahani, E., Kadner, S., Seyboth, K., Adler, A., Baum, I., Brunner, S., Eickemeier, P., Kriemann, B., Savolainen, J., Schlömer, S., von Stechow, C., Zwickel, T., Minx, J.C. (eds.) Climate Change 2014: Mitigation of Climate Change: Contribution of Working Group III to the Fifth Assessment Report of the Intergovernmental Panel on Climate Change, pp. 599–670. Cambridge University Press, Cambridge (2014)

Sullivan, J.L., Burnham, A., Wang, M.Q.: Model for the part manufacturing and vehicle assembly component of the vehicle life cycle inventory. J. Ind. Ecol. **17**, 143–153 (2013). doi:10.1111/j.1530-9290.2012.00515.x

Yong, J.Y., Ramachandaramurthy, V.K., Tan, K.M., Mithulananthan, N.: A review on the state-of-the-art technologies of electric vehicle, its impacts and prospects. Renew. Sustain. Energy Rev. **49**, 365–385 (2015). doi:10.1016/j.rser.2015.04.130

Zackrisson, M., Avellán, L., Orlenius, J.: Life cycle assessment of lithium-ion batteries for plug-in hybrid electric vehicles—critical issues. J. Clean. Prod. **18**, 1519–1529 (2010). doi:10.1016/j.jclepro.2010.06.004

Author Biographies

Felipe Cerdas mechanical and environmental engineer involved in the application of the LCA in industry since 2009. His work has been focused on the lifecycle environmental and economic assessment of energy storage systems for electric vehicles. Interested in the application of LCA for the assessment of emerging technologies in early stages.

Patricia Egede focus on the LCA of electric vehicles, particularly in fleet operations or with lightweight materials. Main interest in the differences of environmental impacts related to regional influencing factors and resulting visualization challenges, member of the European eLCAr project to adapt ILCD Handbook to electric vehicles.

Christoph Herrmann has led national and international research projects in the context of sustainable manufacturing and life cycle engineering since 1997. Main interest is the further development of LCA as an engineering tool that drives innovation.

Chapter 28
LCA of Buildings and the Built Environment

Benjamin Goldstein and Freja Nygaard Rasmussen

Abstract How we design human settlements has a profound influence on society's environmental pressures. This chapter explores the current state of LCA applied to two scales of human settlements; individual buildings and the built environment, where the built environment is understood as a collection of autonomous buildings along with the infrastructure and human activity between those buildings. The application of LCA to buildings has seen growing interest in recent years, partly as a result of the increased application of environmental certification to buildings. General findings are that the use stage of the building tends to dominate environmental impacts, though as buildings become increasingly energy efficient, life cycle impacts shift towards other stages. LCA of built environments has been a useful supplement to mass-based urban environmental assessments, highlighting the importance of embodied environmental impacts in imported goods and showing interesting trade-offs between dense urban living and the greater purchasing power of wealthy urbanites. LCAs of human settlements also face difficult challenges; the long use stage (often decades) introduces high uncertainty regarding the end-of-life stage; evolving electrical mixes throughout the use stage; gaps in consumption data at the city level. This chapter endeavours to elucidate the strengths, research needs and methodological shortcomings of LCA as applied to buildings and the built environment, showing that they can act as complimentary tools to help society's shift towards a sustainable future.

B. Goldstein (✉)
Division for Quantitative Sustainability Assessment,
Department of Management Engineering,
Technical University of Denmark, Kongens Lyngby, Denmark
e-mail: bgol@dtu.dk

F.N. Rasmussen
Faculty of Engineering and Science, Danish Building Research Institute,
Aalborg University, Copenhagen, Denmark

M.Z. Hauschild et al. (eds.), *Life Cycle Assessment*,
DOI 10.1007/978-3-319-56475-3_28

28.1 Introduction

Settlements are comprised of buildings along with the spaces and infrastructure between them, the design of which strongly influences the environmental performance of the overall system. A settlement is no mere assemblage of buildings, but an interplay of buildings, infrastructure, space, environment and institutions that help shape the behaviour of residents and visitors alike, and by proxy, their consumptive regimes and environmental impacts. In understanding the environmental pressures of how we construct the places we inhabit, research has focused on two units of analysis: the individual *building* and the *built environment* (assessments at the nation-state and planetary level notwithstanding). The building is an independent structure that provides shelter from the elements to facilitate one or multiple human activities (living, manufacturing, trading, etc.) The built environment is an umbrella term for the buildings, infrastructure and the human activity between buildings (e.g. mobility, leisure, etc.), ranging from the rural to the urban, the latter of which will be the focus of discussion in this chapter since cities now house more than half of humanity and a much larger share of economic activity (Kennedy et al. 2015). Figure 28.1 illustrates the difference between the two systems.

In terms of the scale of this resource use and environmental degradation, the contributions of buildings and the built environment to global totals are significant. According to UNEP's Sustainable Buildings and Climate Initiative, buildings account for 40% of global energy use, 38% of global greenhouse gas emissions and 40% of the solid waste streams in developed countries (UNEP 2012). When moving up to the city, the impacts are larger: an estimated 70% of greenhouse gas emissions and over 66% of global electricity use eminate from urban activites (Fragkias et al. 2013). Cities are also the drivers of global material consumption, typically in a linear fashion, that pulls resources from their hinterlands and beyond for use within the city and then disposal outside the city, disrupting bio-geochemcial ecological cycles. Nutrient use is a salient example of this, whereby the nutrients incorporated in food are exhaust to local waterways through human waste, which has become the single largest source of nutrient emissions to surface waters globally since the 1940s (Morée et al. 2013).

28.1.1 Buildings and the Built Environment: Crucial Differences

Of concern is that the environmental impacts from buildings and the built environment show little sign of abating: gross global energy and material consumption continue to grow for both the building sector and cities in lockstep with urbanisation and economic development. LCA has a role to play in informing future designs of buildings and urban environments during the transition towards a sustainable future, helping ensure that the benefits of economic growth do not

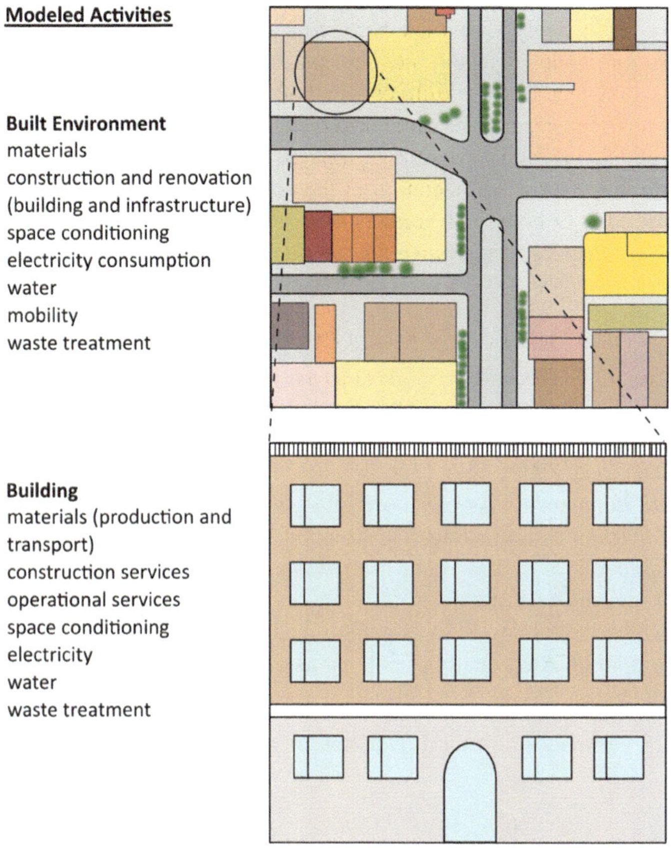

Fig. 28.1 Juxtaposition of the built environment (*top*) and a single building (*bottom*) incorporated within the dense urban fabric

undermine global ecosystems functioning. LCAs of buildings and the built environment are not identical in methodology and have distinct framings of the systems they assess:

1. Scale: Building LCAs focus on a single building or building type and attempt to model this to a *high* degree of accuracy. Built environment LCAs model an agglomeration of buildings (neighbourhood, city, conurbation) and attempt to model this to a *reasonable* degree of accuracy.
2. Temporal Scope: Building LCAs focus on the entire lifetime of a building, typically decadal. LCAs of the built environment take a snapshot of the material, energy demands and waste generation of the study system over a short period, typically a calendar year.
3. LCI Method and Data: Building LCA strives for accuracy and concerns itself with minutia (exact masses and lifetimes of building components, precise construction techniques, etc.) preferably with buildings specific data.

Table 28.1 Rough outline of the differences between the application of LCA to buildings and the built environment

Assessment type	Scale assessed	Temporal scope	LCI components	LCI data
Building	Single building or building type	Building service life (decades)	Materials, energy consumption patterns, water usage, construction methods, disposal technologies	Site tailored
Built environment	Neighbourhood, city, conurbation	Single year	Major categories of consumption: construction aggregates, metals, plastics, food, wood, fuels (transport and heating), water, electricity, waste generation (solid, liquid and gas)	Expenditure surveys, census data, waste statistics, industry reports

Built environment LCAs are more interested in capturing general trends in a city's environmental loading (construction aggregates, metals, transport fuels, etc.) based off of coarser data sets (waste statistics, household consumption surveys, census data, etc.)

4. LCA Method: Building LCA is predominantly done using process-based LCA. Input–output LCA is equally as popular as process-based LCA in assessing built environments.

Table 28.1 provides a general overview of these methodological disparities.

28.1.2 Complimentary Methods to Inform the Design of Human Settlements

The differences between the two applications of LCA do not stop at methods but also their strengths. A simple example illustrates this clearly. Imagine a new neighbourhood comprised of extremely energy efficient homes built at a great distance from areas of recreation, work and shopping, and that this neighborhood lacks viable public transit options, necessitating personal vehicle use for most errands. Now imagine a dense city with reliable transit and nearby amenities that negate the need for significant automobile use, but that the building stock is comprised predominantly of old and energy-inefficient buildings in a continental clime (hot summers, cold winters). Which system would you guess is sustainable? An LCA of the buildings in both situations would identify the first situation as superior. But, if we scaled our assessment up to the built environment (both neighbourhoods), we would find that neither is preferable since both hypotheticals rely on large energy imports to the system that are very likely fossil fuel based (transport for the first, the latter for space conditioning).

This does not mean that the neighbourhood scale LCA is superior, since such an LCA would only be able to identify the major drivers of the impacts (transport and building energy), but is far too coarse to propose specific design interventions to rectify these sub-optimisations. Informing the design of sustainable human settlements requires a multifaceted LCA approach, leveraging both the detail-oriented building perspective and the broader built environment viewpoint. The goal of the chapter is to show how LCA can be applied to these complementary scales of the human settlement in order to help aid in the societal shift towards a sustainable future. Both *buildings* and the *built environment* will be discussed in sequence to convey the methodological considerations when performing LCA on these systems, summarise major findings in the use of LCA on buildings and built environments, and finally impart the reader with the skills to differentiate between more and less rigorous applications of LCA to these systems.

28.2 LCA of Buildings

Since the oil crises in the 1970s, a major concern within building design and operation has been to limit the need for operational energy and hence the need for oil-based heating and electricity. Increasing regulatory requirements on the energy performance of buildings has taken the building design to ever more complex levels where additional materials and technologies are used in order to reduce the energy consumed in operating the building and in servicing the needs of the users. This development of buildings towards increasingly complicated products coupled with the attribute of relatively long product service lives makes LCA an obvious part of the environmental evaluation of buildings.

LCAs within the application area of buildings are mainly used to compare different choices of shape, design or material at a single building level. Either the comparison is made with the potential impacts of alternative design solutions or the results are evaluated against a benchmark performance of the specific type of building and use. A more holistic LCA methodology applied to buildings has received increasing interest over the past decade, also following an increasing focus on life cycle thinking, development of building sustainability certification systems (e.g. BREEAM, DGNB) and the parallel development of standards and LCA methodology in general. For instance, the ISO/TC 59 SC 17 and the European CEN/TC 350 standards series on sustainability assessment of buildings and constructions provide harmonised approaches for structuring and evaluating environmental impacts of a building's life cycle.

Even though harmonised approaches to structuring and calculating building LCAs exist, horizontal comparison of the environmental impacts of one building with those of another building is difficult. This is due to the uniqueness of the service provided by each assessed building, reflecting a vast range of specific requirements including:

- Building type (e.g. office, multifamily residential)
- Site and location specific requirements (e.g. relation to surrounding built environment)
- Technical requirements (e.g. thermal transmittance of building envelope)
- User/owner specific requirements (e.g. low maintenance, adaptability of design, aesthetics).

Although most building LCAs are performed in the later stages of the design or even as the building is finished, there is a general agreement within the sector of the need for developing measures to include LCA-based decisions in the earlier stages of the building design. As opposed to the as-build accounting of impacts, intervention in the early design stage can change the actual physical design of the building in order to improve the environmental efficiency of the building. However, regardless of the temporal focal point for assessing the environmental sustainability of buildings, a range of subjects related to system boundaries and study set-up are still not harmonised in the building LCA practice. This is further explained in the following sections.

28.2.1 The Building Life Cycle Stages

The life cycle stages of buildings are generally divided into three main stages of pre-use, use and after-use. Within these three main stages, additional substages as illustrated in Table 28.2 are often specified depending on the study.

Table 28.2 Main life cycle stages and substages of these seen in building LCA studies

Main life cycle stages	Substages seen in building LCA studies
Pre-use	Extraction of raw materials Transport to manufacturing Manufacturing Transport to retailer Transport to building site Construction site activities Construction worker's transport
Use	Use (e.g. emissions from installed materials) Maintenance Repair Replacement Refurbishment Energy demand for building operation Use specific energy demand Water consumption
After-use	Demolition Waste processing Disposal Next product system/recycling potential

The use stage of the building spans the expected service life of the building, i.e. the assumed number of years in operation. In practice, but often without further justifications, 50–80 years is habitually used as reference study period in assessments, even though the physical structure of an average building has the potential to last longer. Still, 50–80 years is a substantial amount of time in which annual impacts from energy and material consumption are added to the total results of a building's LCA. Thus, the use stage has traditionally contributed considerably to the calculated life cycle impacts, for instance, by 95% primary energy consumption in a 2003 study of a Michigan university campus with a service life of 75 years (Scheuer et al. 2003). Correspondingly, in a 1996 study of a generic office building with a service life of 50 years, the use stage contributed with 80 and 90% of the life cycle energy in the locations Vancouver and Toronto, respectively (Cole and Kernan 1996).

However, the continuous effort in reducing the operating energy in buildings has led to a change in the role of life cycle stages, subsystems and materials in LCAs of recent low-energy buildings, where embodied impacts then are gaining importance (Ramesh et al. 2010). An important consideration in the analyses of use stage impacts from energy consumption versus embodied impacts from building materials lies in the system boundaries set for each study. Specifically, whether the user-related electric requirement—the plug load—(for cooking, cleaning, entertainment etc.) is included or excluded. This share of electricity consumption may cause impacts of the same magnitude as the impacts from a low-energy building's operational energy consumption (heating, ventilation, etc.) and can thus be a prominent contributor to the overall potential impacts from a building.

28.2.2 System Boundaries of the Building Life Cycle

The system boundaries of a building life cycle are important to the assessment at two distinct levels:

- The primary level is the boundary of the life cycle stages and substages included or excluded from the assessment, for instance; is the maintenance of the building components included?
- The secondary level is the boundary of the life cycle inventories included or excluded within each assessed life cycle stage, for instance; is the detergent for the window cleaning included in the maintenance stage or is it only the biennial layer of paint?

Although the boundaries at both levels should be established in accordance with the goal and scope of the assessment, simplifications without further explications can be seen in many case studies.

28.2.2.1 Next Product System: An Additional Life Cycle Stage?

The influence on results of an additional life cycle stage within the scope can be seen in a study by Thormark (2002), where an—at this point—additional life cycle stage, the recycling potential, is evaluated in the context of low-energy row-houses with assumed service lives of 50 years. The recycling potential expresses how much of the embodied energy and natural resources used in a building or a building element could, through reuse or recycling, be made usable in the next product system after demolition of the building in which the materials were originally installed. What can be made usable in the next product system is then deducted from the impacts of the building system under scope. Results showed that 37–42% of the embodied energy could be recovered through recycling and that the recycling potential was about 15% of the total energy use during an assumed lifetime of 50 years (Thormark 2002).

Calculating the recycling potential is but one approach of several to deal with recyclable materials in the building system. Other approaches referred to in building LCA studies include the 'second allocation method', whereby the recycling of materials benefits the building profile only by a reduction in waste generation (Scheuer et al. 2003) or the 'cut-off approach' (see Frischknecht 2010) whereby the product system is cut off at the point in time where the recyclable items cease to be waste, i.e. when it regains a market value. Thus, the life cycle stages at the building after-life, the waste processing and the recycling of materials, are potentially influential to the LCA results obtained, although to a very varying degree depending on the allocation approaches used in the specific study (see more about allocation in Chap. 8).

28.2.2.2 Simplifications of Life Cycle Stages and LCI Input

A Swiss study by Kellenberger and Althaus (2009) explored the potential impacts of different building components and at different levels of simplifications often seen in building LCAs, both regarding life cycle stages and the inventoried materials used for the construction. Results showed that for all studied components the additional materials play an important role with up to 30% of the total impacts from a component. For heavier materials, the transport process had quite a high impact (>10% of total impact from component), but the installation process for the components and the cutting waste could be neglected as they influenced results to a minor degree.

The above-mentioned study on simplifications furthermore confirms what is highlighted in several studies; that with the contemporary low-energy buildings there is no single element or life cycle stage certain to dominate the impact results of a building LCA. On the contrary, different life cycle stages and material scopes can prove important at different variations of building design and geographical preconditions. Specific goal and scope definitions of a study can justify simplifications of life cycle stages and inventory, but it is important to be aware of potentially misleading results if simplifications are conducted without further reasons in a building LCA study.

28.2.2.3 Scenario Evaluation

The very long service life of buildings sets the studies apart from many other LCA applications. Because the use stage of the building is so long, the life cycle impacts from the use stage and the after-use stage are very much depending on the defined scenarios, e.g. for maintenance frequencies or the annual heating demand in the building. However, applying scenario testing as part of sensitivity analyses of studies is not that common within the field; the exception being scenarios for the technologies behind the energy provided for the use stage. In this regard, several studies can be found that evaluate the sensitivity of results to the geographical and technological scope of the electricity production. For instance, assessing a Norwegian building, the generated impact results will prove much different depending on whether the national mix (primarily hydro power), the Nordic mix (where nuclear- and coal-based CHP technologies influence to a larger degree) or the European mix (dominated by fossil fuel-based technologies) is used.

A few studies also evaluate the temporal scope of the energy production, i.e. how will the technologies change in the course of the building service life? This is relevant because the annual energy used is assumed constant for the building service life of 50–80 years, but the technologies providing this annual input of energy cannot be static as the energy system in fact does change. Depending on the purpose of the building LCA study, there is thus reason in evaluating the dynamics of the system.

28.2.2.4 Impact Categories Assessed

An additional aspect of comprehensiveness lies within the scope of the impact categories assessed in the building LCAs (see more about impact assessment in Chap. 10). The prevailing focus on energy within the building sector is reflected in the early generation of environmental assessments of buildings, where the (primary) energy consumption is the single most used indicator (Khasree et al. 2009). The inherent connection between the materials used in the building construction and the capability of the installed materials to reduce the energy consumed, means that energy balances of buildings remains a prevalent topic of the sectoral LCAs. The exclusive focus on energy performance does not capture the full extent of resource uses and problematic emissions also generated by the building sector. Hence, a more complete set of indicators must be applied to ensure comprehensiveness of the assessment. Furthermore, as the energy consumed in the use stage of new building diminishes due to improved building envelopes, the embodied impacts of the buildings become apparent. Table 28.3 sums up a general picture of the assessed impacts categories found in building LCA studies.

Table 28.3 Impact categories seen in building LCA studies

Often included categories	Less frequently included categories	Rarely included categories
Primary energy demand Global warming	Acidification Eutrophication Photochemical ozone creation Ozone depletion Resource depletions	Ionising radiation Toxicity (human/ecosystem) Land transformation Land occupations

A study by Heinonen et al. (2016) also points to this relevant issue of lack of comprehensiveness in assessed impact categories. However, on a European scale at least, explanations can be found on this current state of the art of assessed impacts, namely in the fact that within the framework of the European standards for Environmental Product Declarations (EPDs) of building materials (EN 15804), a predefined set of indicators is established. The set corresponds with the CML methodology plus additional resource use and waste generation categories. As building material EPDs form the basis of many building specific LCAs, this scope of impact categories from a material level is transferred to the building level. In this sense, the sectoral application and standards development affects the practice of conducted building LCAs, also at the scientific level.

28.2.3 Notable Studies

Having outlined the general areas of application related and methodological attention points of the building scale LCA, this section and Table 28.4 briefly introduce a range of notable studies highlighting selected aspects of relevance in the practice and development of building LCA.

Methodological issues of building LCA highlighted in the studies in Table 28.4 concern the use of dynamic modelling of the important use stage of the building (see Collinge et al. 2013) as well as the previously mentioned significance of simplifications at system level, input level and indicator level. The two different modelling approaches of input–output-based LCA (see Chap. 14) and process-based LCA modelling seem in general to be applied at the different levels of national building sector scale and single building scale, respectively. Nässén et al. (2007) discusses difference in results from applying the two different approaches to the production stage of buildings.

Future application of building scale LCA may well continue its importance in the post-construction evaluation of certified buildings although the application to early stage design (see Basbagill et al. 2012) remains an important area of development in order to identify environmentally preferable design solutions before construction takes place. Furthermore, incorporation of the financial and social aspects of building construction alongside the environmental assessment (Ostermeyer et al. 2013) will have a profound relevance to the decision takers in the

Table 28.4 Selected building LCA studies highlighting different aspects of methodological and application issues

Study	Aspect	Highlights
Nässén et al. (2007)	IO-LCA of buildings versus process-based LCA	Energy use (GJ/m^2) of relevant sectoral processes such as transport, construction activities and service sectors may be grossly underestimated in process-based LCA of buildings
Kellenberger and Althaus (2009)	Relevance of simplifications in LCA building components	How typical simplifications of LCI and life cycle stages may have significant relevance to LCA results depending on component type
Blengini and Di Carlo (2010)	Significance of impacts from life cycle stages in a low-energy building	How embodied impacts from the pre-use and maintenance of the building supersedes the operating energy in majority of assessed mid-point impact categories in a current low-energy building
Basbagill et al. (2012)	Application of LCA to early building design	Introduces a method enabling designers to understand the relative global warming potential implications of building component decisions
Collinge et al. (2013)	Dynamic modelling	Use stage scenario testing by dynamic modelling of characterisation factors and electricity mixes
Ostermeyer et al. (2013)	LCA coupled with life cycle costing (LCC) and social LCA (SLCA) in a refurbishment project	The study introduces a Pareto-optimisation approach to refurbishment activities of residential buildings and highlights the need for further development of SLCA to be included as evaluation of the sustainability of building activities
Heinonen et al. (2016)	Simplifications of LCIA categories included in assessment	Based on a case study: how only eight of 17 mid-point categories of the ReCiPe methodology correlates to the GWP which is oftentimes used in studies as the single environmental indicator

construction process. This may hold true especially for the vast body of western post-war buildings ripe for refurbishment actions, because these existing buildings are already deeply defined within a site-specific social and economic context that needs to be taken into account when a change in design and functionality is regarded.

28.3 LCA of Built Environment

The quest to understand the environmental impacts of cities was initially an inwardly focused effort to improve conditions for the working poor that had amassed in recently industrialised cities around the world in the nineteenth century. John Snow's study of an 1854 cholera outbreak in London linked the infectious disease to contaminated drinking water, providing partial impetuous for the study of water and waste flows in the city and the development of the city's modern sewage system and drinking water network (McLeod 2000). This type of urban self-assessment was championed by the reform-urbanism movement at the turn of the twentieth century, which fought the pernicious effects of poor air, water and waste management in cities, eventually formalising into the sanitisation standards and modern land-use planning enshrined in modern cities.

It was not until the 1960s that the attention shifted from a public health focus to an environmental focus. In addition to the question, 'how is the environment in the city affecting the inhabitants?', researchers began to ask, 'how is the city, as a whole, interfering with the functioning of the environment?' Widely acknowledged as the first researcher to explicitly address this question was Abel Wolman's study 'The metabolism of cities' (1965). Wolman's study estimated the fluxes of materials and energy consumed by and the air pollution from an 'average' US city of one million inhabitants. This 'urban metabolism', measured as the material, energy and waste treatment demands of a city, is exactly synonymous with the LCI components considered as part of an LCA of the built environment that were introduced in Sect. 28.1.1. This part of the chapter will show how quantifying the primarily mass-based urban metabolism can be used as starting point of a full LCA of the built environment, contrast this with methodologies of LCAs of the built environment and highlight some of the recent developments in this field.

28.3.1 *From Urban Metabolism to LCA of the Built Environment*

Kennedy et al. (2007) suggest a definition of urban metabolism as: 'the sum total of the technical and socioeconomic processes that occur in cities, resulting in growth, production of energy, and elimination of waste'. Since Wolman's (1965) paper, the number of studies of various cities' urban metabolisms is in the dozens and could very well over 100 (Decker et al. 2000; Kennedy et al. 2007, 2010; Zhang 2013; Stewart et al. 2014). The majority of these studies have been undertaken in the past two decades and have been performed by researchers in the field of industrial ecology. Industrial ecology is itself a diverse research field of the environmental implications of socio-technical systems in general, and this wide scope is mirrored in different methodologies that have been employed to assess urban metabolism: ecological footprint, eMergy (energy memory), carbon footprint, material flow analysis (MFA), etc.

It is the MFA of urban metabolism that is most relevant to our discussion, since it yields results in terms of mass flows through the study area, which can be interpreted as an LCI of the built environment and readily fed into the LCA framework for environmental profiling. MFA of an urban metabolism is akin to a material balance of an urban system, with the exception that non-mass flows such as electricity are often included as well, with accounting almost invariably performed over the period of a year. Kennedy (2012) provided a formalised mathematical description of how MFA would ideally be applied to a built environment, accounting for fluxes of materials through the study area, material additions to stock and waste generation. Though no consensus exists regarding what types of flows need to be included in an MFA of a built environment to adequately characterise a city's environmental footprint, Table 28.5 outlines the material and energy flows to be accounted if one is to make a comprehensive assessment of the metabolic, as adopted from Kennedy and Hoornweg (2012). These material and energy flows seek parsimony between the need to be sensitive to current data limitations, whilst capturing important activities driving environmental impacts and resource consumption.

Table 28.5 Comprehensive list of material and energy flows that a holistic LCA of a built environment would account for

Inflows	*Outflows*
Biomass (t and J)	Waste emissions (t)
Food	Gases
Wood	Solid
Fossil fuel (t and J)	Wastewater
Transport	Heat (J)
Space conditioning/industrial	Substances (t)
Electricity (kWh)	Produced goods (t)
Natural energy (J)	
Water (t)	*Stocks (inflows that do not exit system within assessment period)*
Drinking (surface and groundwater)	Infrastructure/buildings (t)
Precipitation	Construction aggregates
Substances (t)	Metals
E.g. salts, degreasers, etc.	Wood
Produced goods (t)	Other materials
	Other (machinery, durable) (t)
Production (inflows to technosphere generated within urban territory)	Metals
Biomass (t and J)	Plastics
Minerals (t)	Other materials
Energy (J)	Substances (t)

Typical units of measure are indicated in square brackets. Adopted from (Kennedy and Hoornweg 2012)

It is often the case that there is little information about the material flows in and out of a city, and it is thus rare to find a study that has this level of completeness in accounting a city's urban metabolism. It is far more likely for studies to cobble together disparate data sources to build an MFA inventory. This is typically done in two ways; bottom-up (or 'activity-based') using basic data on economic activity and demographics to estimate flows (e.g. number of housing construction starts in a year times the average amount of concrete in a house to estimate concrete demands of a city's construction sector) (Kennedy et al. 2007) or top-down using regional trade data to balance production, imports and exports to gauge a city's demand (Rosado et al. 2014). Moreover, many of the flows of key materials in terms of embodied environmental impacts and future resource scarcity (plastics, rare earth metals) are embedded within electronics and other consumer goods, for which data is scarce.

In assessing the sustainability of the built environment, MFA has proven to be invaluable in exposing the extent to which cities continue to rely on unsustainable, non-renewable resource regimes to fuel their daily metabolism and growth. MFA studies have also illuminated one of the most pernicious aspects of modern cities; their linear metabolism that uses the urban hinterland and beyond as a source for both essential imports and waste assimilation, whilst recycling only marginal amounts of total inputs (Kalmykova et al.'s 2012). Nonetheless, MFA is not without its shortcomings. Most salient is the limitation of using mass as a proxy for environmental impacts. An obvious example of this is the way that high 'biomass' flows in MFA studies are almost always considered sustainable, despite the fact that the food flows lumped into this biomass category are some of the largest drivers of land-use change, greenhouse gas emissions and threats to biodiversity at the global level. Next is that most MFA studies only estimate direct mass and energy demands of a city, eschewing the 'ecological rucksack' of indirect mass and embodied energy, which can account for more than 50% of a system's burdens (Goldstein et al. 2013), as shown in Fig. 28.2 which highlights the embodied mass and energy aspects of five different cities. Lastly, when MFAs consider pollutant loading in terms of GHGs, they tend to include scope 1 (direct combustion within city limits) and scope 2 (imported electricity) impacts, ignoring the significant impacts embodied within imported goods.

28.3.2 Linking MFA to LCA

Coupling MFA with LCA is a natural solution to the methodological shortcomings of MFA focused studies in order to provide a more holistic assessment of the environmental performance of built environments. LCA of the built environment only requires a change of perception, whereby the MFA is viewed as the LCI for the *use stage* of a city's annual demands (see Chap. 9). In this line of thinking, the MFA morphs into a crucial part of the process. With the MFA viewed in this

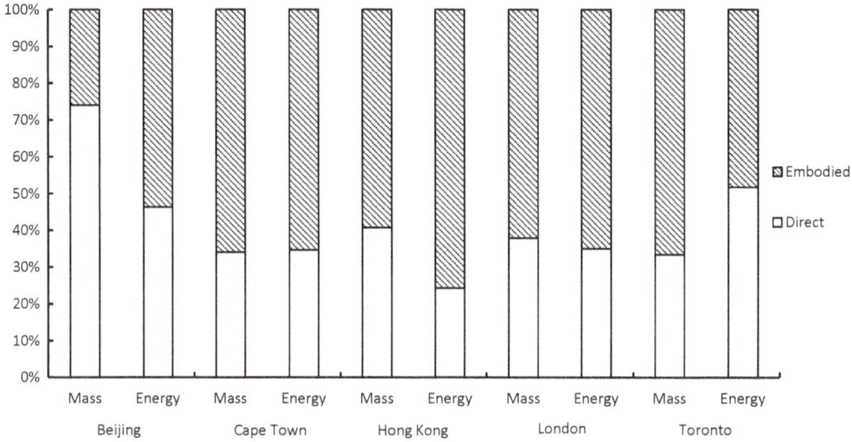

Fig. 28.2 Relative importance of embodied mass and energy impacts in LCAs of five cities. The direct mass and energy represent what is traditionally captured in MFA studies. Note Beijing's lower embodied mass, a result of the frenetic construction activity in the city and the resulting concrete and aggregate flows. *Source* Goldstein et al. (2013)

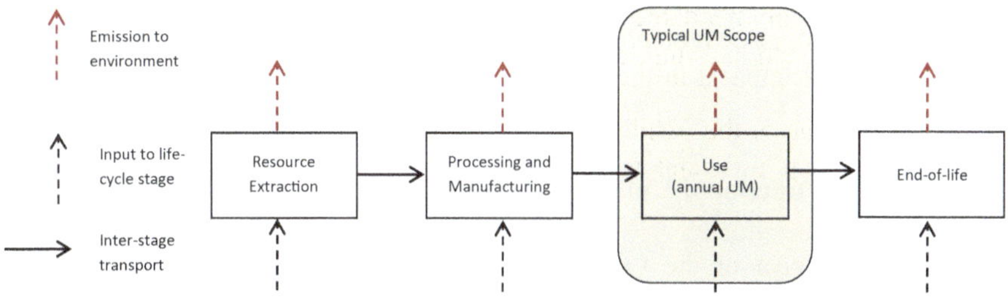

Fig. 28.3 Fusing MFA with LCA: urban metabolism (UM) is taken as the use stage of the metabolic flows, with supporting upstream production and downstream waste management activities built around this. *Source* Goldstein et al. (2013)

complimentary manner, it is easy to imagine building the other life cycle stages up- and downstream of the use stage characterised by the MFA (LCI) of the built environment, which can then be modelled using traditional LCA methods. Impacts from consumption occurring within the city, such as fuel combustion or electricity, can also be modelled using the same principle; where the assessor's job is to identify the most representative processes and align them with the demands of the study region. Figure 28.3 illustrates this.

Due to the durability of many of the goods consumed by cities, it is challenging to accurately model the end-of-life phase of these goods. Rosado et al. (2014) developed a method to account for the lifetimes of goods imported into the Lisbon region, which could theoretically be employed in the framework described here, but the proposed method can obviate the uncertainties in future recycling rates and evolving waste treatment technologies. To sidestep this issue, one could ignore the disposal phase of the durable goods used in the built environment and only include the impacts from waste treatment performed in the study region over the assessment period, though this has the double issue of underestimating impacts due to waste treatment in the future, whilst simultaneously ignoring the fact that many goods with substantial embodied impacts may be recycled at end of life (structural steel, aluminium, glass and maybe someday, precious metals in electronics). There is no right or wrong method here, but transparency in communicating the method chosen and its shortcomings is paramount.

Another methodological consideration is that the level of detail of the MFA is normally very coarse. The LCI for the use stage typically consists of bulk categories of goods such as 'steel', 'aluminium' and 'wood', which the assessor then has to turn into a life cycle. This means that generic material production processes in an LCA database usually employed in modeling ignoring more detailed processes related to manufacturing of specific goods (forming, assembly, etc.). Moreover, transport is also difficult to come by, meaning that one is either left excluding it from the assessment, or using international trade data to make informed judgements about plausible sources of materials imported into a city. Considering the uncertainties regarding specific types of goods consumed and sources of those goods, it might be defensible to omit these from assessments of the built environment, since assembly and transport are often not the dominant sources of environmental impacts in the life cycles of many products

28.3.2.1 The Functional Unit: What are We Actually Assessing?

One essential aspect of the LCA is the functional unit (see Chap. 8 on scope definition for more information about the functional unit). In an LCA of the built environment, what are we assessing exactly? Because built environments do not perform any one function, and many of the functions performed are impossible to quantify (foster community, facilitate cultural exchange, build institutional capacity), developing a functional unit moves from a methodological necessity to an esoteric philosophical exercise in semantics, and has been largely avoided in urban scale LCAs. Notwithstanding, since LCA is ostensibly to be used either for benchmarking or to decision-support between urban design alternatives, a common basis of assessment should be used to facilitate appraisals. This typically takes the form of taking gross impact potentials for the built environment and normalising these to the per-capita level, since this will at least show the environmental intensity of providing for the metabolic activities of the average denizen

28.3.3 Process-Based and Input–Output LCA of the Built Environment

Both process and input–output (IO) LCA have been applied to the built environment, with IO methods figuring most prominently in urban LCA literature. For a detailed description of IO-LCA, the reader can refer to Chap. 14 of this book.

Section 28.3.1 largely outlined what could be considered a process-based approach to assessing the built environment with LCA: determine the direct material demands of the study region and include the embodied impacts up- and downstream, as well as relevant impacts during the use stage. The reason that the process-based method has seen less application is that the general tenor amongst urban sustainability researchers has been that in order for LCAs of the built environments to be as complete as possible, an IO-LCA or hybrid-LCA approach should be employed (Chester et al. 2012). This rationale is absolutely justified; if data for final demand is available, inputting this into the IO-LCA framework yields a more complete inventory and accounting of environmental impacts. Moreover, if multiregion-IO methodology is used, then trade interdependencies between the study region and other economies can be ascertained. Another advantage of the IO-LCA method is that it provides a true demand-side analysis of a built environment, accounting for the environmental burdens of the study area's final demands. Process-based assessment has the shortcoming of not allocating fuels and electricity used in the production of goods manufactured in the study region but ultimately exported to the final consumer, meaning that some of the burdens for these exported goods are incorrectly ascribed to the producing city.

IO-LCA of the built environment faces numerous methodological challenges. Much like the process-based approach, data availability is a challenge, this time in terms of getting final consumption data (in terms of final expenditures) at the subnational level, which means that IO-LCAs of built environments are often built from scaled national-level data (consumer household expenditure surveys have been used in a US context to scale down to the sub-urban level). Moreover, the IO tables are also at the national level, ignoring the regional industry interdependencies, though advances in multi-scale IO models may be able to overcome this (Bachmann et al. 2014). Though comprehensive in terms of value-chain coverage, the number of substances (~ 100) covered by IO-LCA are meagre in comparison to the process approach (over 1000), meaning that until now IO-LCA on the built environment has primarily focused on accounting GHGs (arguably meaning that these were not full multi-criteria LCAs as per ISO standards). Moreover, the IO-LCA method is not compatible with existing LCIA methods, missing out on the indicator sets and communicative power of these tools. IO-LCA is caught in a permanent present tense, whereby it models the impacts of present final-demands, ignoring life cycle stages beyond production and use.

The story is that process-based and IO-LCAs of the built environment are not incompatible, but have different strengths that the assessor should leverage depending on available data and study aims. Process-based LCA is best applied when the benefits of the LCIA methodologies are wanted and/or detail beyond the level of the economic sector is wanted. IO is optimised for inventory completeness.

28.3.4 Opportunities and Challenges

The application of LCA to the built environment is new and evolving, but there are already a number of exciting envisioned uses for this tool. Most obvious is the use of LCA as an environmental screening tool to identify weak-points in the environmental performance of a neighbourhood, city or conurbation, and as a benchmarking method to assess longitudinal environmental performance related to policy changes or growth in a study area. LCAs of the built environment have been able to identify important characteristics of relating the environmental impacts of a study region to the urban form, economic development, population dynamics and local climate. Table 28.6 shows the effects of various urban attributes on the environmental performance of the built environments, pinpointing where policy interventions might be best applied.

LCA of the built environment can also be used for scenario analysis, testing out the environmental efficacy of urban design interventions (i.e. how would the environmental performance of a city change if large-scale food production were employed or if a certain type of built form was pursued?) or policies (how would GHG emissions change with implementation of a congestion charge?).

LCA of the built environment also provides exciting opportunities to explore other aspects of their environmental performance. *Nexus interactions* are one such area, whereby single metabolic activities that drive environmental impacts on multiple fronts, and therefore, that driver acts at the *nexus* of the drivers. Increase in private motor vehicle usage is one such metabolic driver that sits at the nexus of

Table 28.6 Findings of LCA applied to the built environment

Study region attribute	Typical effect on urban environmental performance
Low population density	Impact potentials from mobility take on increased importance (Heinonen et al. 2011)
Climate variability (hot summers, cold winters)	Impacts from space conditioning take on increased importance (Goldstein et al. 2013)
High population growth rate or economic development	Impacts from capital formation (building and infrastructure construction) take on increased importance (Goldstein et al. 2013)
Low population growth rate	Impacts from household consumption takes on increased importance (Goldstein et al. 2013)
High disposable income	Impacts from household consumption takes on increased importance (Heinonen et al. 2011)
Compromised waste management system	Local impacts take on increased relevance (Goldstein et al. 2013)

non-renewable resource use (fossil fuels and metals), GHG emissions (combustion) and the embodied impacts of construction aggregate use (new road construction). LCA can play a role in quantifying the collinearity of these impacts through urban design. Another interesting prospective use of LCA at the scale of the built environment is to quantify *boundary effects* at the border of a city or region, such as impacts from waste expelled into neighbouring geographic regions and fluxes across the system boundary (e.g. through traffic). LCA could predict the severity of environmental disruption from these types of boundary effects and highlight both the benefits and burdens of adjacent human settlements.

Challenges also abound in the application of LCA to the built environment. Data shortcomings cannot be overstressed. Often environmental assessments at the scale of the built environment rely on data from other sources that are used as a proxy for material and energy demands at the neighbourhood, city or regional level. The use of 'big data' to develop more representative consumption profiles could increase the robustness of LCAs of the built environment, and provide finer scaled assessments in terms, both spatially and demographically. Another area of improvement would be the current 'black box' nature of LCA-based assessments, which ignores the way that interactions between subsystems within the built environment generate the study system's emergent metabolism. The 'black-box' perspective results in static models unable to capture non-linear behavior, reducing their predictive power. Combining the environmental auditing power of LCA with other dynamic modelling tools such as system dynamics (Tam et al. 2014) in the urban realm would enhance the applicability of LCA in the urban realm and justify its place at the table when considering urban design or policy interventions.

An emerging approach to the LCA of the built environment is 'territorial LCA'; the application of LCA framework to mixed rural–urban systems (Loiseau et al. 2014). The method is closely aligned with those of this chapter with the noteworthy divergence that the territorial LCA method focuses on land uses and programmes as a method for describing functional units, providing a new perspective to overcome the ambiguity or lack of functional unit in previous LCAs of the built environment. Loiseau et al. have already applied this method to regions along the coastline of Southern France (2013, 2014), highlighting the method's potential applicability.

28.3.5 Notable Studies

We will close the dedicated discussion of LCAs of the built environment with a concise list of studies that exemplify the methods we have discussed, explored ways to overcome current methodological challenges or pushed LCA of the built environment into new exciting directions. The list is not restricted strictly to LCAs, including MFA and carbon footprint studies as well since these can be readily incorporated within the LCA framework. Table 28.7 provides an overview of selected studies. In general, LCAs and MFAs of cities have shown strong links between urbanisation and increasing per-capita material flows (Kennedy et al.

2007), and as mentioned above, assessments of the built environment have related the metabolic profiles of cities to levels of economic development, city morphology, local climate and population dynamics (Goldstein et al. 2013).

In terms of important methodological developments, Lenzen and Peters (2010) showed how multiregional IO-LCA can be applied at the scale of the urban household. Using a simpler, single region IO-LCA of Helsinki, Finland, Heinonen et al. (2011) demonstrated that dense living may reduce transport emission within the city, but increase consumption in other areas by affluent residents erased these benefits, highlighting the tension between urban morphology and wealth.

Goldstein et al.'s (2013) assessment of five cities was the first to apply a process-based approach and use the full suite of indicators available to LCA practitioners. Figure 28.4 shows how local environmental improvement with economic development (particulate matter formation) can be juxtaposed with increasing environmental pressure in private consumption (agricultural land

Table 28.7 Details of notable studies of urban environmental performance

Study	Method	Highlights
Lenzen and Peters (2010)	Multiregion IO-LCA	Applied a multiregion IO of GHGs to Sydney, Australia
Hillman and Ramaswami (2010)	Process-based LCA of GHGs of 8 US cities	Identified 6 key cross urban boundary activities that can be used for expedited, yet complete, GHG accounting of cities and explore boundary effects (i.e. air travel)
Heinonen et al. (2011)	IO-LCA of GHGs of Helsinki, Finland neighbourhoods	Dense urban living reduces per capita transport emissions, but increased wealth of inhabitants ultimately results in higher overall consumption and carbon footprint
Chen and Chen (2012)	GHG accounting of Vienna, Austria	Combined network analysis with GHG accounting to show interconnections between urban subsystems
Goldstein et al. (2013)	Process-based LCA of five cities using multiple indicators	Application of hitherto unexplored indicators at the urban level (eutrophication, ecotoxicity), linking economic development and local environmental performance
Tam et al. (2014)	System dynamics MFA model of Shenzhen, China construction sector	System dynamics approach to predict future C&D waste (legal and illegal) generation for different policy scenarios
Rosado et al. (2014)	MFA of Lisbon, Portugal	Comprehensive trade statistics used to account for over 10,000 goods used in the city and included embodied mass within goods
Kennedy et al. (2015)	MFA of global megacities	Identified links between urban morphology and electricity demands, and showed the super-linear scaling of urban metabolism and population

formation). Highlighting the utility of MFA as a means of LCI development, Rosado et al.'s (2014) study of Lisbon, captured over 10,000 goods entering and exiting the urban system using trade statistic, while concurrently accounting for indirect mass flows embodied within those goods.

Hillman and Ramaswami's (2010) GHG accounting of eight US cities laid down a solid methodology to account for the majority of GHG impacts and showed and explored the boundary effects of transport across municipal borders. Chen and Chen's (2012) paper illustrated how network analysis can be used to open up the urban 'black box', elucidating how material and energy fluxes between urban subsystems influence GHG emissions in Vienna. Tam et al. (2014) combined MFA with system dynamics to assess policy scenarios on C&D waste in Shenzhen, illustrating that focusing on a single aspect of a city's metabolism can yield detailed and relevant results. Lastly, Kennedy et al.'s (2015) MFA of the worlds megacities (population > 10^7) showed how environmental impacts scale super linearly with urbanisation, serving as a prescient reminder of the need to improve the environmental performance of the built environment as urbanization continues globally.

28.4 Methodological Challenges and Best Practice

Having outlined the application of LCA to buildings and the built environment, it is prudent to end with a discussion of current methodological challenges related to this type of assessment. Data, LCI components, LCA method and other noteworthy methodological aspects are discussed in sequence, and are summarised in Table 28.8.

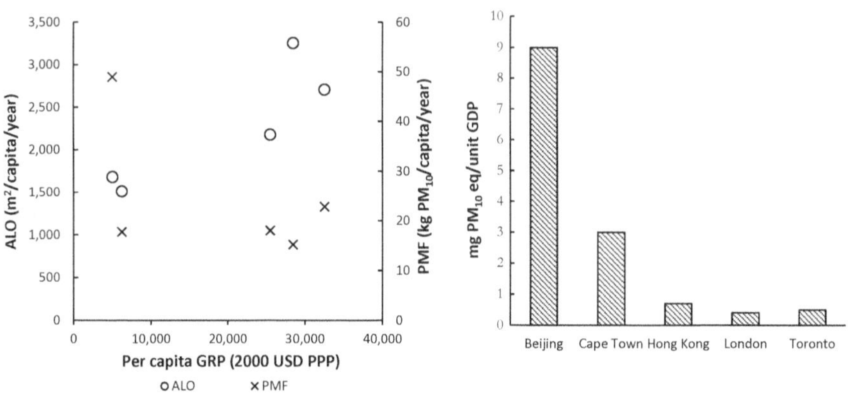

Fig. 28.4 Per-capita agricultural land occupation (ALO) as m^2 $year^{-1}$ and PMF as kg particulates <10 μm $year^{-1}$ are shown on the *left*. Local air pollution issues in Beijing and Cape Town are disproportionately high relative to the amount of economic activity, particularly when the predicted impact per unit of economic activity is taken into account as shown on the *right*. Wealthier cities show a tendency to minimise air pollution while the exported environmental pressure of ALO increases with the wealth of the residents. *Source* Goldstein et al. (2013)

28.4.1 General Challenges

Sensitivity testing of the scenarios applied in building LCAs remains an important issue in order to evaluate the potential impacts of a building. It is especially important due to the long service life of the building for a great deal of societal and technological changes can happen within the time frame of 50–80 years. Thus, the scenarios for the use stage and the end-of-life stage should preferably be tested to ensure some sort of likelihood of the obtained results. Such a particularly long time frame also comes with a difficulty of correctly interpreting the impact scores obtained. Due to the aggregation of emissions and their characterisation with time-integrated characterisation factors, emissions and resource extractions occurring at different moments in time during 50–80 years or more are represented as if they would take place simultaneously, and affecting the same (human) generation. For most product LCA applications, which will typically have time frames in the range of 10–30 years, this is not of particular importance, but needs to be considered in this context. An additional challenge to the practice of building LCAs is for the practitioners to harmonise, if not the inclusion then at least the description of the life cycle stages accounted for. For now, it often presents a challenge to compare results of different studies because of the opacity in the explanations of included life cycle stages and processes.

Other methodological notes for LCA of the built environment include the static nature of the assessment and the black box perspective on the study system. Integration of LCA within the system dynamics framework or network analysis should be able to help overcome this, though work in this direction remains cursory. Lastly, built environments are socio-technical systems, and it should always be kept in mind that the predominantly environmental assessment provided by LCA, though linked to the system's economic and social attributes, should always be balanced against the more detailed work of social scientists, economists and political scientists who look at equity, living conditions, crime, institutions, employment, artistic and cultural practices, psycho-geography, public health and a myriad of other important aspects of life in human settlements. This last sentiment is probably truer for LCA of the built environment than any other application of LCA.

28.4.2 Goal and Scope

Like any LCA, looking at buildings and the built environment with this tool requires a decision between process and IO-LCA. Both process and IO have their strengths; detail and completeness, respectively. Neither is preferred, though the process-based approach does have the advantage of better compatibility with the various LCIA methodologies and related indicator suites. Relatedly, as a rule, LCAs of the built environment should use multi-indicator assessments to minimise the risk of burden shifting between environmental issues, a practice that is difficult for

Table 28.8 Overview of the challenges of applying LCA to buildings and built environments

	Building		Built environment	
	Shortcoming(s)	Best practice	Shortcoming(s)	Best practice
LCI data	Lack of transparency in EPDs of building materials	Applying generic data to assessments of early design and specific data for footprinting and accounting	Local data typically absent for many metabolic activities	Use local data or adjust proxy data to local conditions
LCI components	Simplifications are frequently performed in studies without estimations of the impacts they might be responsible of	Including all inputs relevant to the purpose of the study	Only selected metabolic activities covered Coarsely aggregated fluxes Embodied impacts in imported goods ignored	Major environmental drivers (construction, mobility, building energy, food) represented Fluxes should also be as disaggregated as possible Imported goods should be included to avoid underestimations
Goal and scope	Energy use and global warming potential often the only evaluated categories	Inclusion of a comprehensive set of impact categories to avoid burden shifting	Single indicator assessment Defining functional unit	Multi-indicator toolsets should be used to avoid burden shifting Transparency in functional unit definition
General challenges	Scenarios for use/EoL stage rarely evaluated as part of sensitivity testing Lack of transparency and lack of consistency in life cycle stages included in assessments Lack of transparency in EPDs of building materials	When relevant according to the purpose, testing of the energy, maintenance/replacements and EoL scenarios Addressing all relevant life cycle stages in relation to the purpose. Applying estimates for the stages where there is a lack of data. Describing thoroughly the processes involved in the included life cycle stages Applying generic data to assessments of early design and specific data for footprinting and accounting	Static and black box models Social issues largely eschewed	Combine LCA with system dynamics and/or network analysis Compliment assessment with work from social scientists, public health specialists and economists

Current best practices are highlighted to give the reader a guideline for performing LCAs and appraise the work of others

IO-LCA practitioners due to database limitations. In the future, best practice in this application area might consist of hybrid-IO-LCA that marries the best features of process and IO-LCA, though this has yet to see application. Lastly, there have been a number of GHG studies on the built environment that have focused on scope 1 (direct) and scope 2 (imported electricity) emissions (see some of the background data for the carbon disclosure project; www.cdp.net). This practice ignores GHGs embodied in imported goods and should be avoided since it can vastly underestimate the true carbon footprint related to a built environment's metabolism.

28.4.3 Inventory and Product System Modelling

28.4.3.1 Data

Data used on building LCAs is seen at different levels of specificity, generic data describing average production impacts and product specific data relating to products from specific manufacturers (see Chap. 9 for more general information on Inventory and Product System Modelling in LCA). Generic data on building materials is found in most general LCA databases, but are limited in the sense that they cover mainly industrial products; thus a range of, e.g. biomaterials and innovative materials cannot be found. A whole category of technical components is also underrepresented in current databases. The product specific data is widely promoted as generating the most correct results when assessing a specific building. However, as the product specific data is mainly marketed in the format of EPDs, there is often a lack of consistency and transparency across the different national EPD programmes and the product category rules they each use. This affects the system boundaries and allocation methods applied in the LCA calculations. For instance, different national EPD programmes will account differently the biogenic carbon storage possible in wood products. An ongoing effort in harmonising EPD programmes and assessment methods (e.g. the European Product Environmental Footprint—PEF) will increase the possibilities of using the EPDs for studies, although the format of the EPDs continues to be 'black box' oriented, meaning that only LCIA data and not LCI data is presented in order to protect property rights of the manufacturers.

At the built environment level, a succinct lack of consumptive data at the sub-regional scale is a recurring theme, since trade statistics of goods/materials, food balances and household consumption surveys are normally performed at the national or regional level. Waste statistics, though normally available at the local level, are notoriously coarse in terms of disaggregating constituent flows within waste streams. Activities, such as private automobile use are also normally procured through surveys that are at best generalisations of travel patterns. This does not mean that good data is not available at the level of the built environment, but many studies rely on data from larger regions as proxies for urban metabolic activities and use coarse local level data. Jones and Kammen (2014) have shown how publicly available consumer expenditure data can be used to map carbon footprints at the neighborhood level, while commercial geo-demographic data has also been employed to move more granular assessments (Minx et al. 2013). Moreover, with the proliferation of so-called 'big data', rich data sets of real urban metabolic activities should become available to researchers in the future. Ideally, high quality, locally contextual data should be used (see Rosado et al. 2014), and this data should be disaggregated enough to capture important goods within metabolic activities (e.g. 'biomass' should be broken down into food and non-food, with the former preferably disaggregated further to meat, fruits, etc.). Where national or regional data are used as proxies, these should be made locally contextual using economic data to account for the disparity in purchasing power between the area to be assessed by LCA and the region covered by the data (this holds for process and IO-LCAs).

28.4.3.2 LCI Components

There is great disparity in the way building LCAs delimit the input flows of the system under scope; the main reason for this disparity probably being the different purposes of studies. If the purpose of the study is to assess the building within a specific certification framework, the completeness of the LCI may not be warranted as long as it aligns with the guidelines of the certification scheme. In this sense, it is important to keep in mind that certification-related assessment may not strive for the absolute accuracy of LCA results, rather it aims to place the building performance within a relative benchmark system developed for the certification scheme in question. For studies detached from these or similar relative performances, the basic LCA principle of comprehensiveness naturally applies (see more about inventory analysis in Chap. 9). Transparency and exact descriptions of the materials and components included are paramount in the reporting, because even materials seemingly irrelevant to the results, for instance the fittings and fixtures, can affect certain impact categories. Thus, as no harmonised approach exists in the reporting of the LCI components, it can be difficult to interpret whether fittings may be included in the respective elements where they are installed, e.g. a wall, or if the fittings are not included in the study at all.

LCI of the built environment is usually driven by data availability. There are numerous studies that have had to reduce their scope due to lack of quality data. Best practice would be to include all of the urban metabolic activities listed in Sect. 28.3.1, but this is an optimistic assertion. One aspect to be cautious about is to ensure that the LCA broadly covers the major environmental drivers; construction, building energy, mobility and household goods (and sometimes water) to a reasonable extent, unless the assessment is explicitly focused on a particular aspect of a city's environmental performance (nutrients, mobility, etc.). As mentioned above, biomass should always be disaggregated. To simply lump all biomass together as a renewable resource since it is produced by the planets annual solar budget ignores the reality that the consumption of some food items, namely, meat and dairy, are some of the largest drivers of global environmental pressure.

28.5 Conclusion

To perform an LCA of a human settlement is a complex exercise. What is a clear is that parsimony between the reductionist perspective of the autonomous building and the generality of the territory should be struck in order to address the multi-scalar sustainability challenges of the buildings and their agglomerations into the built environment. At the building scale, the environmental impacts during the use stage, as a rule, dominate, though environmental burdens shift to other life cycle stages with increased building operational efficiency. For the built environment, it is typically the mobility, space conditioning and nutritional needs of the residents that drive the majority of environmental impacts, with antagonism existing between lowering building energy use in new construction and increased transport energy from dispersed nature of these developments. Recent advances in LCA at the built environment have seen the application of IO-LCA, often at a multiregional level, and models that include dynamics of the subsystems that constitute a city. Next steps in the assessment of built environments will include harnessing novel data sources provided by ubiquitous computing ('internet of things') to ameliorate data gaps and provide real time monitoring and feedback of urban environmental performance, and the use of network science and systems thinking to shed light on the inner workings of the urban 'black box'. At building level, recent advances include the standardisation of LCA calculation procedures (Khasreen et al. 2009); although the harmonised studies following this have yet to be seen on a larger scale. Further development of building LCA may be found within the area of early design stage interventions as well as more holistic approaches to evaluating the life cycle of not only new constructions but also existing constructions in relation to the social and economic preconditions.

References

Bachmann, C., Roorda, M., Kennedy, C.: Developing a multi-scale multi-region input–output model. Econ. Syst. Res. **27**, 172–193 (2014)

Basbagill, J., Flager, F., Lepech, M., Fischer, M.: Application of life cycle assessment to early stage building design for reduced embodied environmental impacts. Build. Environ. **60**, 81–92 (2012). doi:10.1016/j.buildenv.2012.11.009

Blengini, G.A., Di Carlo, T.: The changing role of life cycle phases, subsystems and materials in the LCA of low energy buildings. Energy Build. **42**, 869–880 (2010). doi:10.1016/j.enbuild.2009.12.009

Chen, S., Chen, B.: Network environ perspective for urban metabolism and carbon emissions: a case study of Vienna, Austria. Environ. Sci. Technol. **46**, 4498–4506 (2012). doi:10.1021/es204662k

Chester, M., Pincetl, S., Allenby, B.: Avoiding unintended tradeoffs by integrating life-cycle impact assessment with urban metabolism. Curr. Opin. Environ. Sustain. **4**, 451–457 (2012). doi:10.1016/j.cosust.2012.08.004

Cole, R.J., Kernan, P.C.: Life-cycle energy use in office buildings.pdf. Build. Environ. **31**, 307–317 (1996). doi:10.1016/0360-1323(96)00017-0

Collinge, W.O., Landis, A.E., Jones, A.K., et al.: Dynamic life cycle assessment: framework and application to an institutional building. Int. J. Life Cycle Assess. **18**, 538–552 (2013). doi:10.1007/s11367-012-0528-2

Decker, E.H., Elliott, S., Smith, F.A., et al.: Energy and material flow through the urban ecosystem. Annu. Rev. Energy Environ. **25**, 685–740 (2000). doi:10.1146/annurev.energy.25.1.685

Fragkias, M., Lobo, J., Strumsky, D., Seto, K.C.: Does size matter? Scaling of CO_2 emissions and U.S. Urban Areas. PLoS One **8**, 1–8 (2013). doi:10.1371/journal.pone.0064727

Frischknecht, R.: LCI modelling approaches applied on recycling of materials in view of environmental sustainability, risk perception and eco-efficiency. Int. J. Life Cycle Assess. **15**, 666–671 (2010). doi:10.1007/s11367-010-0201-6

Goldstein, B., Birkved, M., Quitzau, M.-B., Hauschild, M.: Quantification of urban metabolism through coupling with the life cycle assessment framework: concept development and case study. Environ. Res. Lett. **8**, 1–14 (2013). doi:10.1088/1748-9326/8/3/035024

Heinonen, J., Kyrö, R., Junnila, S.: Dense downtown living more carbon intense due to higher consumption: a case study of Helsinki. Environ. Res. Lett. **6**, 034034 (2011). doi:10.1088/1748-9326/6/3/034034

Heinonen, J., Säynäjoki, A., Junnonen, J.-M., et al.: Pre-use phase LCA of a multi-story residential building: can greenhouse gas emissions be used as a more general environmental performance indicator? Build. Environ. **95**, 116–125 (2016). doi:10.1016/j.buildenv.2015.09.006

Hillman, T., Ramaswami, A.: Greenhouse gas emissions footprints and energy use benchmarks for eight U.S. cities. Environ. Sci. Technol. **44**, 1902–1910 (2010). doi:10.1021/es9024194

Jones, C., Kammen, D.M.: Spatial distribution of U.S. household carbon footprints reveals suburbanization undermines greenhouse gas benefits of urban population density. Environ. Sci. Technol. **48**, 895–902 (2014). doi:10.1021/es4034364

Kalmykova, Y., Harder, R., Borgestedt, H., Svanäng, I.: Pathways and management of phosphorus in urban areas. J. Ind. Ecol. **16**, 928–939 (2012). doi:10.1111/j.1530-9290.2012.00541.x

Kellenberger, D., Althaus, H.-J.: Relevance of simplifications in LCA of building components. Build. Environ. **44**, 818–825 (2009). doi:10.1016/j.buildenv.2008.06.002

Kennedy, C.: A mathematical description of urban metabolism. In: Michael, P., Weinstein, R., Eugene, T. (eds.) Sustainability Science: The Emerging Paradigm and the Urban Environment, pp. 275–291. Springer, New York, Dordrecht, Heidelberg, London (2012)

Kennedy, C., Hoornweg, D.: Mainstreaming urban metabolism. J Ind Ecol. **16**, 780–782 (2012). doi:10.1111/j.1530-9290.2012.00548.x

Kennedy, C., Cuddihy, J., Engel-yan, J.: The changing metabolism of cities. J. Ind. Ecol. **11**, 43–59 (2007). doi:10.1162/jie.2007.1107

Kennedy, C., Pincetl, S., Bunje, P.: The study of urban metabolism and its applications to urban planning and design. Environ. Pollut. (2010). doi:10.1016/j.envpol.2010.10.022

Kennedy, C.A., Stewart, I., Facchini, A., et al.: Energy and material flows of megacities. Proc. Natl. Acad. Sci. (2015). doi:10.1073/pnas.1504315112

Khasreen, M.M., Banfill, P.F.G., Menzies, G.F.: Life-cycle assessment and the environmental impact of buildings: a review. Sustainability **1**, 674–701 (2009). doi:10.3390/su1030674

Lenzen, M., Peters, G.M.: How city dwellers affect their resource Hinterland. J. Ind. Ecol. **14**, 73–90 (2010). doi:10.1111/j.1530-9290.2009.00190.x

Loiseau, E., Roux, P., Junqua, G., et al.: Adapting the LCA framework to environmental assessment in land planning. Int. J. Life Cycle Assess. **18**, 1533–1548 (2013). doi:10.1007/s11367-013-0588-y

Loiseau, E., Roux, P., Junqua, G., et al.: Implementation of an adapted LCA framework to environmental assessment of a territory: important learning points from a French Mediterranean case study. J. Clean. Prod. **80**, 17–29 (2014). doi:10.1016/j.jclepro.2014.05.059

Minx, J., Baiocchi, G., Wiedmann, T., Barrett, J., Creutzig, F., Feng, K., Förster, M., Pichler, P.-P., Weisz, H., Hubacek, K.: Carbon footprints of cities and other human settlements in the UK. Environ. Res. Lett. **8**, 35039 (2013) doi:10.1088/1748-9326/8/3/035039

McLeod, K.S.: Our sense of snow: the myth of John Snow in medical geography. Soc. Sci. Med. **50**, 923–935 (2000). doi:10.1016/S0277-9536(99)00345-7

Morée, A.L., Beusen, A.H.W., Bouwman, A.F., Willems, W.J.: Exploring global nitrogen and phosphorus flows in urban wastes during the twentieth century. Glob. Biogeochem. Cycles **27**, 836–846 (2013). doi:10.1002/gbc.20072

Nässén, J., Holmberg, J., Wadeskog, A., Nyman, M.: Direct and indirect energy use and carbon emissions in the production phase of buildings: an input–output analysis. Energy **32**, 1593–1602 (2007). doi:10.1016/j.energy.2007.01.002

Ostermeyer, Y., Wallbaum, H., Reuter, F.: Multidimensional Pareto optimization as an approach for site-specific building refurbishment solutions applicable for life cycle sustainability assessment. Int. J. Life Cycle Assess. **18**, 1762–1779 (2013). doi:10.1007/s11367-013-0548-6

Ramesh, T., Prakash, R., Shukla, K.K.: Life cycle energy analysis of buildings: an overview. Energy Build. **42**, 1592–1600 (2010). doi:10.1016/j.enbuild.2010.05.007

Rosado, L., Niza, S., Ferrão, P.: A material flow accounting case study of the Lisbon metropolitan area using the urban metabolism analyst model. J. Ind. Ecol. (2014). doi:10.1111/jiec.12083

Scheuer, C., Keoleian, G.A., Reppe, P.: Life cycle energy and environmental performance of a new university building: modeling challenges and design implications. Energy Build. **35**, 1049–1064 (2003). doi:10.1016/S0378-7788(03)00066-5

Stewart, I., Kennedy, C., Facchini, A.: Metabolism of megacities: a review and synthesis of the literature (2014)

Tam, V.W., Li, J., Cai, H.: System dynamic modeling on construction waste management in Shenzhen, China. Waste Manag. Res. (2014). doi:10.1177/0734242X14527636

Thormark, C.: A low energy building in a life cycle—its embodied energy, energy need for operation and recycling potential. Build. Environ. **37**, 429–435 (2002). doi:10.1016/S0360-1323(01)00033-6

United Nations Environment Programme (UNEP): Building design and construction: Forging resource efficiency and sustainable development. Sustainable buildings and climate initiative (2012)

Wolman, A.: The metabolism of cities. Sci. Am. **213**(3), 179–190 (1965)

Zhang, Y.: Urban metabolism: a review of research methodologies. Environ. Pollut. **178**, 463–473 (2013). doi:10.1016/j.envpol.2013.03.052

Author Biographies

Benjamin Goldstein urban systems researcher with a focus on the environmental assessment of built environments. Interests include industrial ecology, political ecology and urban metabolism.

Freja Nygaard Rasmussen works within the field of LCA methodology applied to building design and building materials. Has contributed to national development of LCA tools and methodology for building certification systems.

Chapter 29
LCA of Food and Agriculture

Teunis J. Dijkman, Claudine Basset-Mens, Assumpció Antón and Montserrat Núñez

Abstract This chapter deals with the application of Life Cycle Assessment to evaluate the environmental sustainability of agriculture and food processing. The life cycle of a food product is split into six stages: production and transportation of inputs to the farm, cultivation, processing, distribution, consumption and waste management. A large number of LCA studies focus on the two first stages in cradle-to-farm gate studies, as they are the stages where most impacts typically occur, due to animal husbandry and manure handling, production and use of fertilisers and the consumption of fuel to operate farm machinery. In the processing step, the raw agricultural product leaving the farm gate is converted to a food item that can be consumed by the user. Distribution includes transportation of the food product before and after processing. In the consumption stage, environmental impacts arise due to storage, preparation and waste of the food. In the waste management stage, food waste can be handled using a number of technologies, such as landfilling, incineration, composting or digestion. A number of case studies are looked at here where the life cycles of typical food products (meat, cheese, bread, tomatoes, etc.), and an entire diet are discussed. Other case studies deal with what LCA can conclude on the differences between conventional and organic farming, and the perceived advantages of local food items. Finally, methodological issues in agricultural LCA are discussed: the choice of functional unit, setting the boundary

T.J. Dijkman (✉)
Division for Quantitative Sustainability Assessment, Department of Management Engineering, Technical University of Denmark, Kongens Lyngby, Denmark
e-mail: tedi@dtu.dk

C. Basset-Mens
CIRAD, UPR Hortsys, ELSA—Research Group, Campus International de Baillarguet, TA B-103/C, 34398 Montpellier Cedex 5, France

A. Antón
IRTA, Food and Agricultural Research Institute, Centre de Cabrils, Barcelona, Spain

M. Núñez
Irstea, UMR ITAP, ELSA-LCA Research Group, 361 Rue Jean François Breton, 34196 Montpellier, France

M.Z. Hauschild et al. (eds.), *Life Cycle Assessment*,
DOI 10.1007/978-3-319-56475-3_29

between technosphere and ecosphere, modelling flows of nutrients and pesticides, and the generally limited number of impact categories included in LCA studies.

29.1 Introduction

Agricultural production systems have changed dramatically over the course of the past century. The introduction of the combustion engine around 1900 started the replacement of human labour with fossil energy. In 1910, the Haber–Bosch process to bind nitrogen from air was commercialised. Synthetic fertilisers could now be produced on an industrial scale. In the decades after the Second World War synthetic pesticides became widely used, which together with improved plant breeding led to the so-called Green Revolution: a spectacular increase in yields worldwide, but especially in Asia and Latin America. More recently, techniques to alter the genetic material of crops have been applied to develop new plant varieties, such as herbicide-resistant maize. Especially in Europe this development has led to controversy.

These developments do not mean that there are no challenges remaining for agriculture: the world population keeps growing. The global population increased from 1.65 billion in 1900 to 6 billion in 2000. In 50 more years, an additional 2 to 4.5 billion people will be added to the population (FAO 2012). Part of this growing population is becoming increasingly affluent. More affluence results in a higher demand for food in general—a year-round supply of fruits and vegetables either imported off-season from distant countries or produced in artificialised production systems such as heated greenhouses—for meat in particular. Part of this population, especially in poor countries, is also increasingly living in cities (50% in 2007 at the world scale, Kulikowski 2007), implying the import of food from rural areas and the development of urban farming. Urban farming is showing a great potential for securing food supply, creating jobs and alleviating poverty in the Global South, but it can be accompanied by environmental and human health risks due to intensified and not always well-controlled practices.

Today, food production is associated with major environmental problems. Rachel Carson's famous 1962 book Silent Spring addressed an important environmental problem: the effects of pesticide use on non-target animals, including humans. Later, focus shifted to issues such as eutrophication due to nutrient runoff. Greenhouse gas (GHG) emissions caused by agriculture are high on the agenda today. For example, the United Nations Food and Agricultural Organization (FAO) estimates that livestock alone contributes 18% to the global GHG emissions and that 50% of the methane emitted into the atmosphere by human activity is due to crop and livestock production (FAO 2013). Another environmental problem is the disruption of the nitrogen cycle. Human extraction of nitrogen from air, mainly for fertiliser production, is larger than all natural processes extracting nitrogen from air. Subsequent application of the produced fertiliser results in extensive nitrogen emissions to surface water. Last, but not least, agriculture is the economic sector with the largest requirements of water and land and the main driver of land use change, e.g. through the conversion of forests into agricultural land (FAO 2013).

In order to gain insight into the environmental performance of agriculture and agricultural products, a considerable number of Life Cycle Assessments (LCA) of these products have been carried out in recent years. Also, several initiatives encourage the environmental life cycle-based assessment of food products, such as the Envi Food protocol (Food SCP RT 2013), the Product Environmental Footprint (PEF) pilots (European Commission 2016), and the well-established international LCA Food conference.

This chapter focuses on the LCA of food products. Foods are here defined as products of plant or animal origin that provide macro and micro nutrients and energy to the human body. They can be either produced by a form of land-based agriculture and aquaculture or collected in the environment, such as seafood or fungi. The structure of the chapter is as follows. First, the life cycle of food products is presented and discussed. After that, a selection of LCA studies is reviewed, illustrating typical LCAs of food products and describing some relevant cases. Finally, the main results and methodological issues found in the case studies are summarised.

29.2 The Life Cycle of Agricultural Food Products

Six stages can be distinguished in the life cycle of agricultural products, which are somewhat different from the usual stages in LCA (see Chap. 6). A large part of the agricultural LCAs carried out are cradle-to-farm gate studies (see Fig. 29.1), because of the importance of agricultural production, and because the agricultural stage often bears the largest environmental impacts.

29.2.1 Production and Transportation of Inputs

In most modern food systems, the first stage of the life cycle is the production and transportation of inputs, such as agrichemicals, machines, building elements, seeds and energy carriers such as fuel and electricity to the farm. The production of these inputs has a vast geographic scope and requires substantial transportation.

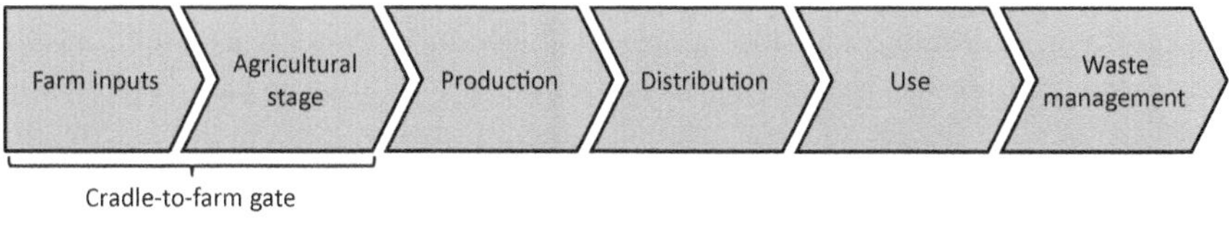

Fig. 29.1 The six stages in LCA of agricultural products

Agrichemicals include pesticides and fertilisers. Pesticide is a general term covering all chemicals (insecticides, fungicides, herbicides or others) used to protect the farm product from different pests, diseases or undesired plant growth. Both the production of pesticides and of fertilisers contribute to environmental impact potentials on the site of their production. The manufacturing of both nitrogenous and phosphorous fertilisers requires substantial energy inputs: according to a 2004 estimate (Swaminathan and Sukalac 2004 as cited in IPCC 2007), the fertiliser industry consumes 1.2% of the total annual energy use, and contributes similarly to GHGs emissions, mainly carbon dioxide and nitrous oxides. Moreover, phosphorus is derived from phosphate rock, which is a non-renewable, overexploited resource whose reserves may be depleted in 50–100 years (Cordell et al. 2009). In animal husbandry, the production and transport of animal feed is associated with emissions of N_2O and CO_2, mainly caused by fertiliser production and fuel use.

The production of capital goods, such as machinery, buildings for animals, greenhouses or glasshouses can have contrasted contributions to overall food impacts. Usually, farm equipment is used for a longer period of time, whilst an LCA study typically considers the production of a given mass of product, or the production from 1 ha of land, so the impacts of the production and disposal of the farm equipment would have to be allocated over different product systems. Consequently, the impacts of the equipment are often relatively small. In some cases though, for example growing tomatoes in tunnel greenhouses, the production of the greenhouse may be an important source of environmental impacts as reported by Torrellas et al. (2012a). Previous studies have shown the importance of including agricultural capital goods in environmental assessments. In particular, for protected crops, structural components of unheated greenhouses may account for nearly 30% of the total impacts in environmental impact categories such as resource depletion and global warming (Antón et al. 2014). In accordance with ISO standards (ISO 2006a, b), in order to be accurate when assessing the environmental impact of products, infrastructure must be taken into account as capital goods are explicitly part of the production system. Most guides recommend including capital goods in the assessment when they contribute more than 5% of the total impacts per impact category (EU-JRC 2010).

29.2.2 Agricultural Stage

The second stage in the life cycle of food products is the agricultural stage. This stage starts at the origin of the food product, for example seeds, fertilisers, pesticides, water and energy in case of crops, and breeding of animals in case of meat or dairy products. In the agricultural stage, all these inputs are used to produce the food product. In this section, the most important processes are described. However, some LCA studies may include other processes than the ones listed here.

29.2.2.1 Application of Agrichemicals

Aside from their beneficial effects on the crop growth and yield, application of agrichemicals results in emissions from the field and from the buildings in which animals are kept. Pesticides are mainly considered because of their toxic impacts on non-target species during or after application and bioaccumulation in harvested parts of the crops, potentially contributing to ecotoxicity and human toxicity impacts. Because pesticides are distributed over the agricultural field when used, the disposal stage is usually not included in LCAs. However, pesticide residues are present in packaging materials, in the sprayer and in the water used to clean the sprayer after application (van Zelm et al. 2014). Ideally, the disposal and subsequent fate of these residues should be included in LCA studies.

Fertilisers are applied to supply nutrients to crops, mainly nitrogen (N), phosphorous (P) and potassium (K). Fertiliser consumption typically contributes to potential impacts due to field emissions into all environmental compartments: air, water and soil. More specifically, on-farm use of fertilisers results in emissions of ammonia (NH_3), nitrous oxide (N_2O) and nitrogen oxides (NO_x) to air, contributing to impact categories such as acidification, climate change and eutrophication. In addition, emissions of N in the form of nitrate (NO_3^-), and of P in the form of phosphate (PO_4^{3-}) and particulate P through erosion, result in eutrophication of nearby water bodies and ultimately of the ocean and the sea. In LCIA practice, N emissions are considered to result in marine eutrophication, and P emissions result in eutrophication of freshwater bodies (see Chap. 10).

Farming practice influences the emissions of pesticides and fertilisers to the environment. When performing an LCA study, the choice of agricultural data should reflect the goal and scope of the LCA study. The study might focus on either average practices, i.e. the agricultural practice that is most common among farmers, even if these are non-optimal practices, or innovative and alternative agricultural practices. Worst, best or alternative practices should only be modelled when primary data is present to document this behaviour, or when the LCA study aims to compare different practices (van Zelm et al 2014).

29.2.2.2 Animal Husbandry and Manure Management

Apart from application of agrichemicals for forage production specified in the section above, enteric fermentation and manure handling are important contributors to impacts such as climate change, eutrophication and acidification. Enteric fermentation in the digestive tract of ruminants (cattle, sheep, goat) produces methane. This process is one of the major contributors to GHG emissions in LCAs of products from these animals. In a typical cradle-to-farm gate study, it will contribute approximately one-third of total climate change impacts. Excretion and manure handling from both ruminants and monogastric animals (pigs, chickens and other birds) result in direct emissions of CH_4 and N_2O. Manure can be applied as fertiliser, thus replacing synthetic fertilisers. Prior to application on the field, manure can be treated mechanically, biologically or chemically. Ten Hoeve (2015) showed that all manure treatment technologies have inherent environmental advantages and disadvantages, and hence, the choice of technology depends on local policy preferences, costs and practicality.

29.2.2.3 Farming Operations

Fuel is consumed during a number of on-farm operations. Ploughing is an especially energy-intensive activity because of the large volume of soil that needs to be moved. Other fuel-consuming activities are the application of fertiliser and pesticides, roughage production, harvesting, heating of greenhouses, transport of the product, etc. Because the fuel used on farms is normally from a fossil origin, these actions induce non-renewable resource use and climate change impacts.

The use of machinery in farm operations negatively affects soil quality through compaction and erosion. Degraded soils are less productive and require extra inputs of fertilisers to maintain food production steady on the short term. In addition, soil ecological functions, such as buffering and filtering of toxic chemicals, water retention and soil-biota, are also affected. Hence, at the long term, unsustainable farming practices may lead to irreversible soil degradation. To maintain food production, land cover is transformed to accommodate new cropland areas, many times at the expense of natural vegetation. Today, croplands and pastures have become one of the largest biomes on the planet, occupying around 40% of the land surface (FAO 2011) and still expanding to feed the growing population.

29.2.2.4 Irrigation

Irrigated crops sustain 40% of the global food production (Abdullah 2006) and are responsible for around 70% of global water withdrawal taken from surface and ground water bodies (FAO 2014). Especially in arid countries (e.g. Australia, India, Spain) and for water-intensive-crops (e.g. almonds, rice) impacts derived from water consumption might be one of the main contributors to overall food production environmental impacts. Water leaving the farm is a vector of salts, toxic and nutrient-rich pollutants, potentially affecting aquifers and surface water bodies downstream. In many situations, water withdrawal is also responsible for a large use of non-renewable energy resources, with important associated environmental impacts, in order to transport the water to the field.

29.2.3 Processing

Processing is any step to convert the raw farm product into a (packaged) food item that will be preceded and followed by logistic phases further described below.

There is a wide variety of processing steps that can be performed, some of which have been included in LCA studies. Both the types and importance of potential environmental impacts from food processing vary a lot in function of the food item considered, as showed in the selected case studies commented in Sect. 29.3.

29.2.4 Distribution

All food life cycles include distribution stages dealing with (often refrigerated) transportation to the warehouse and to the retailer, sorting fruits, conditioning, packaging and cool storage for good maintenance of food properties. Together these processes can contribute a large share to the potential impacts of food products. These impacts are mostly related to the use of non-renewable energy, such as fuel use in transportation and electricity in cool storage. These distribution phases are particularly important in fresh products' life cycles such as fruits and vegetables. In animal products' life cycle, where the contribution of the agricultural stage is large, the impacts from distribution will appear minor. In contrast, in fruit and vegetable products, the stage of refrigerated transport and storage can have a major contribution to the total impacts. This is particularly true for fruits and vegetables transported over long distances, especially when these are air-freighted (see Table 29.1). For instance, Sim et al. (2007) revealed that the transportation of French beans by airplane from Kenya to England constituted 95% of their overall impacts. The mode of transportation more than the distance itself will play a role.

29.2.5 Consumption

The use stage of food mostly includes food transport from the retailer to the point of consumption as well as energy use for cooking and storing. This stage might include, depending on the LCA case study, private households or restaurants and institutional kitchens (Sonesson et al. 2003). Most studies that have included the use stage conclude that its environmental impacts are related to food storage or preparation. In the complete life cycle of a food product; however, these steps usually are minor contributors to overall impacts in most categories (Schau and Fet 2008).

Table 29.1 CO_2-eq emissions per tkm (1 tonne transported over 1 km) for different modes of transport (Cristea et al. 2013)

Mode of transport	Emissions per tkm (g CO_2-eq)
Road	120
Rail	23
Ship	5–12
Air	475–1000

An increasing part of meals are consumed in restaurants, institutional kitchens, caterers, etc. According to some studies, the variations in energy use for cooking between restaurants and institutional kitchens are large as are the variations between different dishes. Dishwashing, which is a direct effect of home cooking and eating, uses large and varying amounts of either cold or hot water. Hot water is an important contributor to a household's total energy use.

An aspect that is currently finding its way into LCA practice is exposure of humans to pesticide residues in food during consumption, leading to human toxicity impacts. Even though most countries have regulations in place to limit human exposure to pesticides to levels considered safe, LCA practice aims to quantify any effects on humans, no matter how small these may be. With the release of the dynamiCROP model (Fantke et al. 2011) and USEtox 2.0 in 2015 (USEtox 2015), this pathway is covered in LCIA practice.

29.2.6 Waste Management

The disposal stage of a food product consists of both handling of food waste generated along the entire life cycle, and the treatment of human excretion resulting from food intake. An alternative not considered here is to allocate the impacts of waste handling to the life cycle stage where the waste arises.

The food sector is wasteful. About one-third of all food produced in Europe for human consumption is lost or wasted before people consume it. For fruits and vegetables, this number may reach ~45%. In general, 20% of food produced is wasted along the supply chain, from agricultural production (9%) to post-harvest handling and storage (4%), processing and packaging (5%) and distribution (3%). The consumer discards between 15 and 33% (Williams and Wikström 2011). Reduction of food waste in the use stage has been shown to be an effective way to reduce food environmental impacts. Avoidance or at least reduction of waste must be the priority. Once waste is produced it is also a challenge to close loops of nutrients and other materials. The potential to extract valuable bio-chemicals or recover energy and nutrients from various waste streams is significant, namely the recovery of energy and nutrients through digestion and composting is one of the most common methods in the food sector (Ellen MacArthur Foundation 2015). A number of studies have dealt with food waste handling, considering options such as landfilling, incineration, centralised and decentralised composting, digestion to produce biogas and conversion to animal feed. The results of the studies do not uniformly point in one direction and there also appear to be trade-offs between different impact categories.

For some impact categories, food waste and human excretion can result in substantial contributions to the impacts found. For example, various GHGs are emitted from wastewater treatment and subsequent sludge disposal. Depending on the wastewater treatment facility, emissions of N and P to surface water may contribute to the eutrophication potential (see more on LCA of wastewater treatment in Chap. 34).

29.3 Selected LCA Studies on the Food Sector

This section will start by describing a number of contrasted examples of LCAs of food products, followed by a number of case studies about various developments intended to lower the environmental cost of supplying food. This selection of LCA case studies does not aim at being exhaustive, but rather at illustrating the diversity of case studies.

29.3.1 Examples of Food Product LCAs

Below, we describe a number of LCAs of different types of food products: meat, cheese, bread and tomato. The studies have been chosen based on representativeness of their outcomes among LCA studies for similar products, and inclusion of processes beyond the farm gate. This chapter closes with an LCA study of a full diet to give an overview of the relative magnitudes of impacts of different food items.

29.3.1.1 Meat

Dalgaard et al. (2007) conducted an LCA of Danish pork exported to the United Kingdom to determine the environmental hotspots. The functional unit of the study was 1 kg of Danish pork (carcass weight) delivered at the port of Harwich.

The system boundaries in this study included the pig farm, the slaughterhouse and the use and maintenance of transport infrastructure required to transport the pork to the UK. Because of the chosen consequential-LCA approach (see Chaps. 8 and 9), the feed products considered were limited to grain and soybean meal, which is the most competitive feedstock. Grain is mixed with soybean meal to achieve the optimal protein content in the feed.

Manure and other by-products are considered as co-products of pork meat. The manure is used as natural fertiliser or is anaerobically digested into biogas for district heating and electricity production. The animal by-products are used as feed or for energy purposes. For these by-products, a system expansion was done in which environmental impacts of synthetic fertiliser, fossil fuels, grain and soy for feed production were subtracted as avoided impacts.

The study was limited to three environmental impact potentials: global warming, eutrophication and acidification. The impact assessment method used was EDIP97 with updated global warming characterisation factors for methane and nitrous oxide. Normalisation and weighting steps, which are optional in LCA, were not done in this study.

The results, summarised in Table 29.2, showed that the environmental hotspots were in the farm's input stage (i.e. production of grain and soybean meal used as forage). Approximately two-thirds of the global warming impact, which was 3.6 kg CO_2-eq over the entire life cycle, was attributed to the production of farm inputs out of which 2 kg CO_2-eq came from the production of grain. The authors did not specify where in the production of grain these impacts arise. The agricultural stage (mainly pig housing) contributes a further 0.9 kg CO_2-eq due to methane emissions from manure. Meat processing in the slaughterhouse contributed about 5% of the global warming impacts, while transport to the UK contributed less than that. The use of manure as natural fertiliser resulted in a negative impact: although methane was emitted from manure, the avoided impacts from not producing synthetic fertiliser were greater.

More than 99% of the eutrophication impacts were associated with grain production, manure application and ammonia emissions from pig housing: about 122 and 47 g NO_3^--eq, respectively, out of 232 g NO_3^--eq in total. The acidification impacts are also highest in the agricultural stage: ammonia emissions from pig housing contributed 53% and grain production contributed 38% to the total impact of 45 g SO_2-eq.

Based on their findings, the authors proposed to further reduce the protein consumption in pig feed by shifting from soy meal to grain. Reducing the protein content in feed will reduce nitrogen excretion and emissions from pig manure in the pig housing or when spread out on the field as manure. Moreover, as soy meal has a

Table 29.2 Contribution of the life cycle stages to the overall impacts of 1 kg of pork, at UK port

Process	Global warming potential	Eutrophication potential	Acidification potential	Photochem. ozone formation	Ozone depletion potential
Soybean meal	8	1	5		
Grain	61	53	39		
Pig housing	26	20	56		
Energy use in pig housing	4	1	1		
Manure application to field	-6	27	1		
Slaughterhouse	5	-1	1		
Transport after slaughterhouse	1	1	1		
Total impact	3.6 kg CO_2-eq	232 g NO_3^--eq	45 g SO_2-eq	1.3 g C_2H_4-eq	0.7 mg CFC11-eq

The contributions are expressed as percentages of total impact (%) (Dalgaard et al. 2007). The total percentage does not sum up to 100% because contributions between 0 and 1 or −1% are all referred to as, respectively, 1 or −1%

higher global warming impact per kg than barley, the global warming impact will decrease as well when shifting to grain.

Now, one can ask whether pork production is a good representative of meat production. Or, how would the picture look like when looking at LCAs of poultry and beef? De Vries and De Boer (2010) reviewed 16 cradle-to-farm gate LCA studies on livestock production in nonorganic farming systems in OECD-countries. They recalculated the results found in the reviewed papers to fit three functional units: kg of meat, kg of protein and kg daily intake. Here we will focus on their findings for the first functional unit. All impacts found were fully allocated to the edible part of the products. The impact potentials considered were fossil energy use, global warming, acidification, eutrophication and land use.

Among all meat products, beef showed the greatest fossil energy use. Energy use results for pork and chicken production were in the same range (see Table 29.3). For global warming, the reviewers found the highest impacts for beef, followed by pork, then chicken. Global warming, energy use and land use impacts from beef were considerably higher than impacts from pork and chicken. Ruminants (beef) emit high amounts of methane that contribute to global warming impacts. Energy consumption is also greater for beef production than for pork and chicken production. In terms of land use, beef has greater feed requirements (i.e. feed conversion ratio, kg feed per kg meat), which means that a larger extension of land is needed, both as direct land use (pasture), and as indirect land use for forage production. Moreover, cows live longer and because most cows only have one calf per year a larger breeding stock is needed. Regarding acidification and eutrophication, the variation within each type of meat was larger than the variation between the three types of meat. These variations were mainly attributed to differences in emissions of NH_3 caused by different agricultural practices and climatic conditions.

Concluding, chicken and pork meat at the farm gate show comparable impacts, while beef meat generally causes higher impacts. The internal variability of the impacts within each meat product was high, reflecting the variability of practices and environmental conditions.

Table 29.3 Comparison of minimum and maximum environmental impact potentials per kg of chicken, pork and beef (De Vries and De Boer 2010)

Livestock animal	Energy use (MJ)	Global warming (kg CO_2-eq)	Acidification (g SO_2-eq)	Eutrophication (g PO_4^{3-}-eq)	Land use (m^2)
Chicken	15–29	3.7–6.9	0.062–0.29	0.004–0.079	8.1–9.9
Pork	18–34	3.9–10	0.043–0.74	0.032–0.17	8.9–12.1
Beef	34–52	14–32	0.11–0.90	0.063–0.33	27–49

29.3.1.2 Cheese

A full LCA of a Swedish cheese was carried out by Berlin (2002). Though now relatively old, this study yielded conclusions that were confirmed by later studies.

As a functional unit "1 kg of Ängsgården semi-hard cheese wrapped in plastic" was chosen. The product system included milk production at farm level, cheese production at dairy factory, retailing, household consumption and waste management. Data from 1 farm and 1 dairy, both in Southwest Sweden were used. In order to produce cheese, apart from milk, several other ingredients were used: rennet (enzymes from calves' stomach), calcium chloride, saltpetre, salt and water. Packaging material was made of plastic and cardboard. A number of cleaning agents in the dairy was also included. Capital goods such as buildings and equipment were excluded from the study.

The impacts from the farm were allocated over the milk and meat produced. The dairy factory produced four different kinds of cheese as well as a range of other products. An economic criterion was used to allocate impact over the different co-products of the factory. The following impact potentials and flow indicators were considered: global warming potential, acidification, eutrophication, photochemical ozone formation, material and energy use. Ozone depletion and ecotoxicity were dealt with in a qualitative way. For this reason, these 2 impact categories will not be extensively discussed here. Impact assessment was stopped at the characterisation step. Hence no normalisation or weighting was done. Characterisation factors from different methods and guidelines were used to calculate impacts. The results are summarised in Table 29.4.

Almost 95% of global warming impacts were attributed to farming, with methane emissions from fermentation in the cow rumen being the most important impact source. N_2O emissions from soil processes and fertiliser production also played a key role. The remaining 5% of the impacts were attributed to cheese

Table 29.4 Results of the LCA of 1000 kg of Swedish semi-hard cheese (Berlin 2002)

Life cycle stage	Global warming (kg CO_2-eq)	Acidification (kg SO_2-eq)	Eutrophication (kg O_2-eq)	Photochemical ozone formation (kg C_2H_4-eq)
Farm inputs and agricultural stage	8300	135	2120	2.4
Other inputs	67	<1	3.3	<0.1
Processing	369	<1	10	<0.1
Retail	48	<0.1	<1	<0.1
Use	12	<0.1	<1	<0.1
Waste management	−2	<−0.1	<−1	<−0.1
Total	8794	136	2134	2.5

production and retailing. Here, CO_2 emissions from natural gas and fuel use caused the largest share of the global warming impact. Impacts from electricity use were low, because the Swedish electricity grid mix was used for modelling the product system. Since Sweden's electricity mix is largely made up of hydropower and nuclear power, GHG emissions from electricity production are low.

Of the photochemical ozone formation impacts, in which only volatile organic compounds (VOC) were taken into account, 93% of the impact was attributed to farm emissions. As was the case for global warming impacts, the cow-produced methane was the main source of VOCs.

The farming step in the life cycle of cheese was the source of more than 99% of the eutrophication and acidification impacts. These impacts were attributed to ammonia volatilisation from manure for both impact categories. Nitrate leaching from the soil was another important contributor to eutrophication.

We have seen that, in this study, the major contribution to the impact categories considered arose from the farm inputs and agricultural stage (i.e. the process of milk production). The cheese making process accounted for most of the remaining impacts. The study was published in 2002, but most of the data are from the mid to late 1990s. Despite the age of the study, the conclusion that most impacts of dairy products arise at the farm input stage and agricultural stage was also found in more recent LCAs. However, compared with newer studies, the share of these stages (93 to >99%) as found by Berlin (2002) is at the high end. For example, in their LCA study of the production of cheddar and mozzarella in the USA, Kim et al. (2013) found that the farm inputs and agricultural stages (feed production and on-farm emissions) contributed to more than 60% in seven out of nine impact categories. For cumulative energy demand, marine and freshwater eutrophication and human toxicity, other steps such as manufacturing, retail and consumption were identified as hotspots. In other studies dealing with cheese production in the Netherlands (Van Middelaar et al. 2011), New Zealand (Basset-Mens et al. 2007), Spain (González-García et al. 2013) and Serbia (Djekic et al. 2014), the agricultural stage was confirmed to be the most important environmental hotspot. In the distributive step the mode of transport more than the distance itself plays a role. In their study of New Zealand cheese exported to England, Basset-Mens et al. (2007) found that the contribution of ship transport over more than 20,000 km from New Zealand port to England was lower than the contribution of truck and consumers' car transport together in all impact categories except acidification (Basset-Mens et al. 2007). These authors also found that the average sewage treatment in England for human excretion was the second main contributor for eutrophication (31%) after farm production.

Comparing the Berlin (2002) study with newer studies raises two relevant points. Firstly, the limited number of impact potentials that were quantified shows that the field of LCA has seen a considerable methodological development in the last 15 years. Secondly, the lack of data that led to the ozone depletion potential being discussed qualitatively in the study illustrates the data availability issue that LCA practitioners traditionally have to deal with. For example, the authors of the study described the locations where cooling equipment was used, as well as the refrigerants used, but could not quantify the resulting environmental impacts because data about refrigerant leaking was not available at that time.

29.3.1.3 Bread

One of the first studies of a whole food product was an LCA of bread production carried out by Andersson and Ohlsson (1999). The main aim of this study was to compare different scales of baking. A secondary goal was to identify environmental hotspots. This study was selected for this chapter to illustrate some typical pitfalls of LCA application, which often arise due to the complexity of the modelled systems in real life, data unavailability and lack of transparency of the modelling choices made.

In cooperation with industrial partners from the bakery industry that provided data, the authors compared the environmental impacts of 1 kg of white bread ready for consumption at home produced in four different Swedish scenarios. Two of these were industrial bakeries. The first produced bread distributed through the entire country, the second operated at a regional scale, the third a small local bakery, whilst the fourth corresponded to the baking of bread at home.

Included in the system boundaries were the farm inputs, wheat cultivation, wheat milling and baking, production of packaging material, the household stage (freezing the bread), waste handling and transport. Capital goods were excluded and so were all other ingredients of the bread other than wheat. These ingredients were excluded because little variation was observed between the recipes. Unless the impacts of these ingredients were minor, which has not been tested, excluding them makes it hard to conclude on the environmental hotspots: an unknown part of the impacts are not quantified. The wholesale and retail steps were also excluded because they were expected to contribute little to the overall impacts.

Impacts from wheat cultivation and milling were allocated to wheat and flour, respectively. Allocation of the impacts of the bakeries was done differently. For the two industrial bakeries, allocation was done on basis of the mass of the products produced there. In contrast, economic allocation was used in the local bakery scenario. This option was chosen because data for the masses produced were not available.

The impact categories and indicators included in the study were energy use, land use, global warming potential, eutrophication potential, acidification potential and photo-oxidant formation potential. No normalisation or weighting step was done. The results are presented in Table 29.5. We will not discuss the details of the results here, but summarise the main findings of the hotspot analysis. A hotspot was defined as a sub-system that contributes more than 20% to the impact in a given impact category or flow indicator. With the exception of primary energy use and photo-oxidant formation, the farm inputs and agriculture stages were hotspots in all of the impact categories for all scenarios. Eutrophication impacts were dominated by wheat cultivation. Transport was a large contributor to global warming and acidification potentials in the industrial bakery scenarios. In the bakeries using natural gas or oil in their ovens, the baking process also was a hotspot for the global warming impacts. Food processing, i.e. ethanol released when baking bread, was the main hotspot for the photo-oxidant formation impacts, even though transport was also a hotspot in this impact category.

Summarising, we see that in contrast to what was observed in the previous chapters, the agricultural stage is not the single dominating stage in this case study.

The main aim of the authors was to compare different scales of bread production. It is, however, doubtful how useful the results were to study differences of scale, because of a number of methodological inconsistencies that a practitioner should try to avoid.

First, the system boundaries were set in a way to include the consumers' transport to buy the bread, or the raw materials to bake the bread at home, except in the local bakery scenario. The authors claim that the local bakery is visited on foot or by car on the way home. As the transport by car hence does not add extra kilometres, the impacts were not accounted for, while the impacts due to the transportation could also have been allocated to both the travel back from work and the bread production. It is not clear from the study why this would not be the case when purchasing industrially produced bread at supermarket. Lack of transparency in modelling choices makes it difficult to identify if the systems compared are equal.

Second, the allocation procedure is not done consistently: the impacts from the industrial bakery were allocated by mass, those from the local bakery economically. The authors justify the approach by mentioning that the fractions of impacts allocated to the bread were found to be similar for the industrial bakeries on one hand and the local bakery on the other hand. However, this does not validate the approach: the similarity in outcomes may be accidental.

A factor that further complicates the interpretation of the comparison relates to the differences between bakeries. One of the industrial bakeries used a natural gas oven while the local bakery had an oil-fuelled oven. In the second industrial bakery and the home baking scenarios, electric ovens were used. It is not clear from the

Table 29.5 Summary of the results of the LCA of bread production

Studied systems	Primary energy (MJ)	Global warming (kg CO_2-eq)	Acidification (mol H^+)	Eutrophication (g O_2)	Photochem. oxidant formation (g C_2H_4-eq)
Industrial bakery 1 (national)	22	940	0.15	160	5.4
Industrial bakery 2 (regional)	14	630	0.1	99	3.2
Local bakery	12	660	0.1	120	2.6
Home baking (heat from electricity)	18	620	0.078	88	2.4
Home baking (heat from oil)	17	630	0.09	89	2.6

Functional unit 1 kg of white bread ready for consumption at home. Electricity production from waste incineration is assumed to replace electricity produced from oil combustion (Andersson and Ohlsson 1999)

study whether or not the use of different fuels for the oven was related to the different scales of the bakeries. If not, this would mean that the systems compared were not equivalent regarding the goal set, i.e. compare different scales of baking. However, if the aim of the study had been to compare most common practices of baking at a given scale, the same definition of systems would have been fair to answer the question aimed. Summarising, when conducting an LCA it is important to act consistently and to design a model of a product system that suits the aim of the study.

29.3.1.4 Tomato

As an example of an LCA of a vegetable product, we will discuss here the study on tomatoes produced in a greenhouse in Spain, carried out by Torrellas et al. (2012a). The study was carried out to investigate how the environmental performance of greenhouse tomatoes could be improved.

The functional unit of the study was 1 tonne of loose tomatoes at the farm gate, indicating that this study is a cradle-to-farm gate study. Included in the system boundaries were the manufacturing of greenhouse components, auxiliary equipment such as the irrigation and water collection systems and the substrates in which tomato plants are grown, products needed for greenhouse management such as fertilisers, pesticides, water and electricity, waste handling including transport to waste management from the site of the greenhouse.

The environmental impacts were calculated for six categories and flow indicators: cumulative energy demand, abiotic depletion, acidification, eutrophication, global warming and photochemical oxidation. If we use the same definition of a hotspot here as in the last paragraph, i.e. a process or group of processes to which more than 20% of an impact can be attributed, then the results (see Fig. 29.2) showed that the structure of the greenhouse was a hotspot in all of the categories. The steel in the frame and the plastic of the cover were the main contributors. Auxiliary equipment represented a hotspot in all impact categories apart from eutrophication. For acidification, eutrophication and global warming, fertiliser production and application constituted a hotspot as well. A large part of the environmental impacts was associated with the greenhouse itself: for the impact categories abiotic depletion, acidification, photochemical oxidation and cumulative energy demand the sum of greenhouse structure and auxiliary equipment accounted for more than 75% of the total impacts. For eutrophication and global warming the percentages were 46 and 66%, respectively. Therefore, the impacts of the greenhouse were relatively high. Because the studied greenhouse was located in Spain, no heating was needed. Another study that assessed different geographical greenhouse tomato scenarios showed that the most important contributor was the greenhouse heating, highlighting the need to reduce energy consumption and use renewable energy sources where greenhouse heating is necessitated (Torrellas et al. 2012b).

Since the study by Torrellas et al. (2012a) was a cradle-to-farm gate LCA, the role of transport in the life cycle of vegetables and fruit is not highlighted. Other work, for example the aforementioned study by Sim et al. (2007) on French beans flown from Kenya to England, showed that transportation may be determining for the environmental impacts of vegetables and fruit. The same was concluded by Mithraratne et al. (2010), who analysed the carbon footprint of a tray of kiwifruit produced in New Zealand and consumed in Europe. Here, shipping contributed to 44% of the footprint.

29.3.1.5 A Full Diet

In the previous chapters various food items were discussed in isolation, with functional units of 1 kg (or 1 tonne) of the food item. Although this gives a good overview of the impacts of single food products, it provides little information about where environmental impacts arise in a diet. After all, humans usually do not consume identical amounts of different foods. Therefore, we finish this chapter by reviewing a study by Muñoz et al. (2010), who did a full LCA of an average Spanish diet. The functional unit of their study was given as 'the supply of food for a Spanish citizen in the year 2005'. The authors included all processes needed to provide food to the consumer in the study: agriculture stage, processing of food, distribution, retailing, storage at home and preparation. Furthermore, the end of life of food was considered via food waste management and wastewater treatment of human excretion.

Impact assessment was limited to global warming, acidification, eutrophication and primary energy use. The results, presented in Fig. 29.3, showed that food production, including agricultural and processing stages, represented the largest source of impacts in all categories. Meat and dairy alone made up 54% of the global

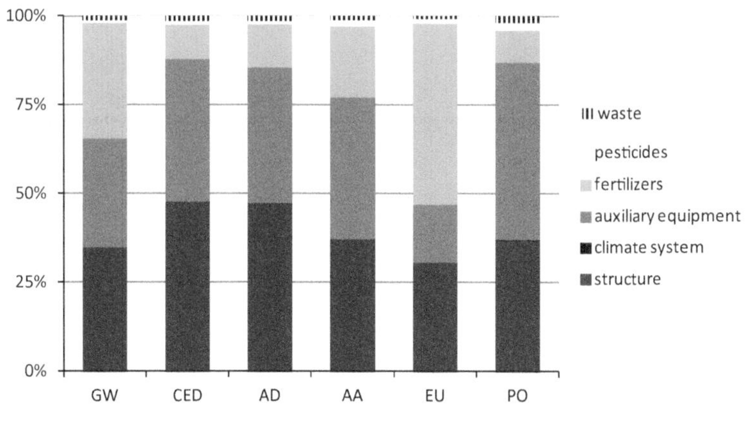

Fig. 29.2 Results of the LCA of the production of 1 tonne cold greenhouse tomatoes. Adapted from Torrellas et al. (2012a)

warming potential. Wastewater treatment contributed 17% of the impact potential when including emission of biogenic CO_2 while it was only 3% when excluding these emissions. These CO_2 emissions can be excluded from the impact assessment, as they are formed from carbon that was taken up by the plant during its growth.

Meat, dairy and beverages represented 60% of the eutrophication potential. Here, wastewater treatment was the second most important stage, contributing 17% to the total impact. Home storage and cooking, which was considered as one process, was the second contributor to the acidification potential, with a contribution of 12%. This process was also the second largest user of primary energy (22%).

With regard to the percentages above, the authors mentioned that there was a fair level of uncertainty involved in the study. As an example, for some food products data were missing. For other products the data used were collected for Danish rather than Spanish circumstances, following a consequential approach, instead of the attributional approach used by the authors.

From the results it is clear that for a full diet, the agricultural stage and the production its inputs are the stages contributing most to the impacts. Among the different products, meat production was associated with the largest impacts in all of the four impact categories studied. Dairy and beverages were other considerable contributors to environmental impacts.

The authors of this study did not discuss options to reduce the environmental impacts of diets. Other authors recommend exploring a reduction of food waste and

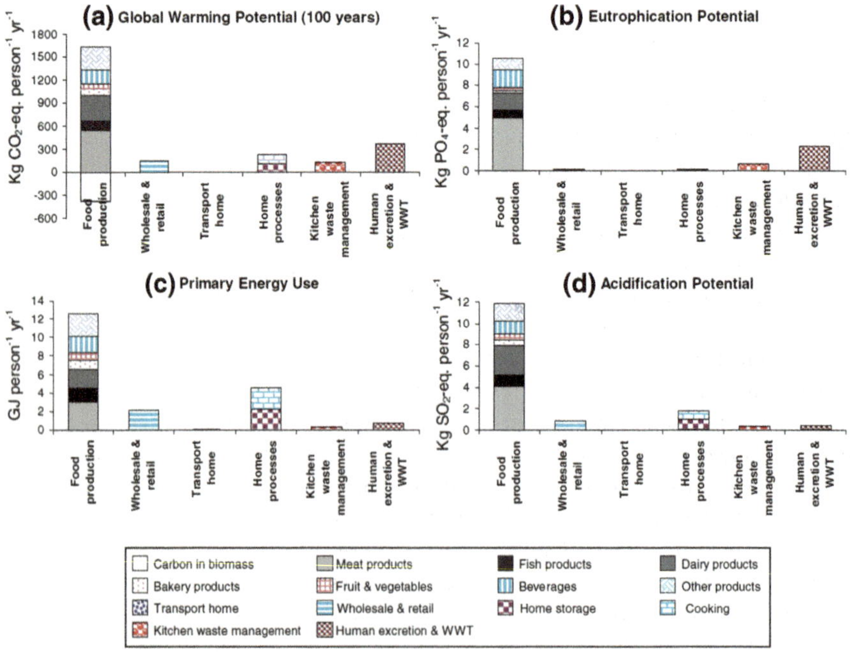

Fig. 29.3 results for the LCA study of the average Spanish diet. Taken from Muñoz et al. (2010)

switching to a (partly) vegetarian diet. In the study of the Spanish diet, the authors found that 23% of food purchased was discarded. This waste consisted of a smaller fraction of inedible parts of the food products, but the largest waste fraction were edible parts not consumed for one reason or another.

Regarding switching to a diet containing less meat, Saxe and Jensen (2014) compared the environmental consequences of switching from an average Danish diet (ADD) to a New Nordic Diet (NND). In the NND, focus is on local ingredients, produced organically. In addition, the diet contains less meat, but more fish, wholegrain products, nuts, fruits, berries and vegetables than the average Danish diet. The NND was shown to reduce environmental impacts, but at the expense of increased cost for the consumer.

29.3.2 Case Study: Conventional, Integrated and Organic Farming

Most agricultural practices in Western countries are chemically intensive, aiming at maximising the production by using external inputs. This production mode is associated with environmental problems mostly due to nutrients and pesticides emissions to environmental compartments and decreased near farms.

In order to reduce the environmental burdens of agriculture, new production methods have been introduced. Organic agriculture aims at producing while sustaining the health of soils and people and preserving biodiversity. In practice, this form of agriculture differs from conventional agriculture in the sense that it avoids the use of synthetic-agrochemical pesticides and mainly uses manure and compost as fertiliser. In organic milk farming, as discussed below, cows spend most of their time outside in order to stimulate their natural behaviour. Feed should consist of 60% of roughage, produced organically, preferentially on-farm. Another alternative is integrated farming. Based on the principles of integrated pest management, it aims to achieve optimal long-term results from both an environmental and economic point of view. Furthermore, pesticide application has to be targeted and limited, the soil has to be protected in winter and the crop rotation needs to be diversified.

In this chapter, two comparative cases are discussed: organic and integrated farming in Switzerland, and organic and conventional milk production in the Netherlands.

29.3.2.1 Crop Production

Nemecek et al. (2011) compared the environmental impacts of conventional/integrated and organic farming in Switzerland. Environmental impacts were reported as yearly averages calculated from 7-year crop rotations. The authors identified three functions of agriculture, and defined functional units accordingly: land management, providing income to the farmer and production of food. The focus in this chapter will be on the first and third functions. In the land management function, the impacts are expressed per hectare per year. In the food production function, they are reported per kg dry matter or per MJ net energy content, depending on the food or feed product.

The environmental impacts were assessed using the Swiss Agriculture Life Cycle Assessment (SALCA) framework. In this framework, the usual impact categories and flow indicators energy resources, global warming potential, ozone formation, eutrophication, acidification, terrestrial and aquatic ecotoxicity and human toxicity are considered along with biodiversity and soil quality.

The results found for normal fertilisation levels are shown in Fig. 29.4.

The results for biodiversity, not given in Fig. 29.4, showed that biodiversity was higher in an organic farming system, mainly as a consequence of banning synthetic pesticides. The soil quality indicators did not vary much between the three systems.

The results presented in Fig. 29.4 show that, when looking at the land management function, the potential impacts of organic farming were considerably less than those caused by conventional/integrated farming. However, the land required to produce the same amount of product was 25–30% higher for organic farming. For that reason, the differences between organic and conventional/integrated farming were smaller when looking at the results from the production function. Still, the authors showed a significantly lower impact for organic farming for most of the impact categories when looking at the impacts per kg dry matter, especially for toxicity impacts. The exceptions here were land use, which was higher for organic farming, and ozone formation, acidification and eutrophication, where the differences were not statistically significant. The large difference in toxicity-related impact categories was explained by the reduced use of pesticides in the organic farming systems. Not taken into account in the method applied was the use of copper as a fungicide in organic farming. Therefore, the toxicity impacts of organic farming have been underestimated. The reduction in energy demand and global warming potential in the organic farming systems was due to reduced use of mineral fertiliser in this type of farming.

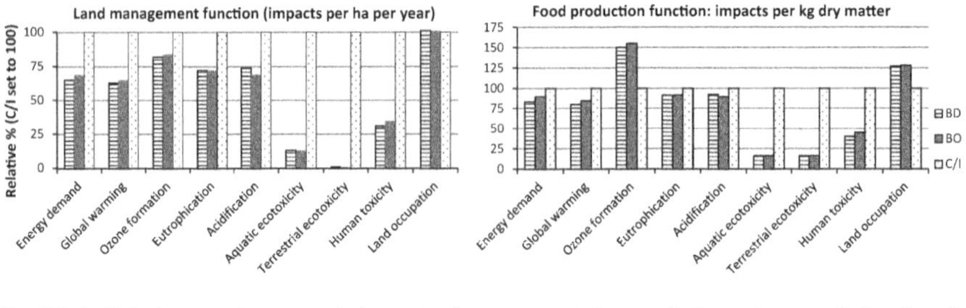

Fig. 29.4 Relative environmental impacts for crop rotations of three types of farming in Switzerland. Impacts for conventional/integrated farming set to 100% (reference system). *Abbreviations BD* bio-dynamic, *BO* bio-organic, *C/I* conventional/integrated

In addition to calculating the impacts associated with crop rotations, the authors also looked at the impacts of individual crops in organic and integrated crop production. Here it was found that some organic products had higher environmental impacts in some categories than their conventional/integrated counterparts. This was partially explained by lower yields in organic farming. Based on this the authors stressed the importance of looking at product systems instead of looking at products in isolation.

Summarising, depending on the selected function, the outcome is either that organic farming is the environmentally favourable option, or that the results are not conclusive to decide on the most environmentally friendly option for agriculture.

29.3.2.2 Milk Production

Based on Dutch data from 2003, Thomassen et al. (2008) performed a comparative LCA of conventional and organic milk production. This cradle-to-farm gate study used 1 kg of fat and protein corrected milk leaving the farm gate as the functional unit.

The studied system included the on-farm processes and the inputs required on the farm: breeding of animals, production of feed concentrates, roughage and bedding material, transport of animal manure used as fertiliser. In addition, the conventional milk production also required pesticides and artificial fertilisers for feed and roughage production.

The studied systems had a number of multifunctional processes, most notably the cow which produced not only milk, but also meat, calves and hides. In order to deal with these co-products, economic allocation was applied.

The study considered five impact categories and flow indicators: land use, energy use, climate change, acidification and eutrophication. Stratospheric ozone depletion was excluded because previous studies had shown that milk production does not produce significant ozone depletion impacts. Furthermore, human toxicity, terrestrial and aquatic ecotoxicity were excluded. The reason for this was the absence of data on pesticides and heavy metals used. The impacts obtained are summarised in Table 29.6.

The difference in land use observed in Table 29.6 was mainly explained by the lower yields for feed production and a lower intensity of animals per hectare in the organic system. The difference in energy use is related to the absence of fertiliser and pesticide production in the organic system and the smaller use of concentrate in the feed mix. Likewise, the production and use of concentrates and artificial fertiliser explained the higher eutrophication impacts in the conventional farming system. The small difference in acidification impacts may seem counterintuitive: the conventional system emitted more ammonia from manure storage and application, and fertiliser application per hectare. However, the animal intensity was much higher in the conventional system, leading to lower overall impacts when looking at mass of milk produced. The climate change impacts were similar for both systems. The organic system had higher on-farm impacts because of the higher number of

Table 29.6 Comparison of environmental impacts of conventional and organic milk production per kg of fat and protein corrected milk leaving the farm gate (Thomassen et al. 2008)

Impact category	Conventional system	Organic system
Land use (m^2)	1.3	1.8
Energy use (MJ)	5.0	3.1
Eutrophication (kg NO_3^--eq)	0.11	0.07
Acidification (g SO_2-eq)	9.5	10.8
Climate change (kg CO_2-eq)	1.4	1.5

cows needed to produce the same volume of milk. In contrast, the off-farm emissions were higher in the conventional system, mainly due to the purchase of feed concentrates. Concentrates are feed products that contain a high density of digestible nutrients and are usually low in fibre content (FAO 1995).

This case shows the importance of the perspective of the study, as reflected in the functional unit, on the outcome of the study. For example, taking the perspective of a farmer with a fixed land area at his disposal who wants to produce milk in a more sustainable way, switching from conventional to organic farming would probably result in lower environmental impacts. In this case, the functional unit will be area-related (e.g. hectare per year). However, the main motivation of a farmer usually is to maximise income for the agricultural production. Looking at how to minimise environmental impacts per currency unit might be more realistic from the farmer viewpoint. In that case, the functional unit will be income-related (e.g. €). As can be seen from Table 29.6, using the perspective of producing food, the functional unit chosen was kg of milk. The picture of which option is most favourable to the environment is not clear.

In both the crop and the milk examples, results showed that neither conventional farming nor organic farming was a more environmentally favourable practice in all respects. LCA was powerful at identifying environmental hotspots and margins of improvement of the studied systems. Agriculture is multifunctional and LCA outcomes depend a lot on the agricultural function studied, especially when comparing systems with contrasted intensification levels. This conclusion is also found in studies conducted by other authors, such as Cederberg and Mattsson (2000), De Backer et al. (2009) and the review by Foster et al. (2006). In addition, based on a review of 34 agricultural LCAs, Meier et al. (2015) argue that many LCA studies do not sufficiently capture the differences between two production systems, for a number of reasons. The goal and scope definition does not differentiate the characteristics of the farming systems. The inventory data used for N and C flows from the field are based on models, often developed for modelling conventional farming, and do not represent actual circumstances. Finally, LCA studies apply LCIA indicators of all readily available impact categories. Other important categories for agriculture, such as land use impacts on biodiversity and soil quality, water use and (terrestrial) toxicity are currently the object of fast scientific development and will be normal practice in a few years.

29.3.3 Case Study: 'Local' Food

Consumers are increasingly aware of environmental problems connected to food production. Many food products are transported over long distances: air-freighted French beans from Kenya, apples from Chile or peas from Egypt can be found in European and North American supermarkets. The "food miles" concept (defined in 1994 by Paxton as the distance food travels from producer to consumer) has become popular in the UK and USA and has led to increased interest in local food, which has been produced in close proximity of where it is consumed. Apart from questionable environmental benefits, local food is also associated with other values: taste, naturalness, local economy, to mention a few examples given by Edwards-Jones et al. (2008). These authors have analysed the interest in local food and food miles from an environmental and ethical point of view. Here we will focus on the environmental aspects that have been analysed using LCA.

Transporting food products over large distances may appear as a waste of resources and an unnecessary cause of GHG emissions. Especially air transport causes high emissions per tonne kilometre. In order to have a measure of the distance over which food is transported the concept of 'food miles' is used. However, when looking from an LCA perspective, transport is only one of the stages in the life cycle of a food product where GHG and acidifying emissions occur. Apart from the distance, the transportation mode is of the utmost importance for the environmental impacts of the transportation phase. Moreover, for many field-grown crops the production of fertiliser has a large global warming impact potential, whilst in crops grown in greenhouses the use of electricity for heating and lighting can cause considerable impacts. Therefore, only looking at the transported distance is not sufficient to conclude on the environmental benefit of local food.

To underline this, the authors give an example of an assessment of global warming potential for apples consumed in the UK, taking a life cycle perspective. Researchers from the UK had found that local production results in the lowest GHG emissions, whilst researchers from New Zealand found the opposite: apple production in New Zealand followed by transport to the UK results in the lowest global warming impacts. Therefore, there are two contrasting conclusions. Which one is right? The answer is, surprisingly, both. The two studies used different system boundaries. When looking at a full calendar year, and including the cold storage that is required to store the apples between harvest and consumption, it can be shown that apples from the UK are favourable in most parts of the year. However, when these apples have to be stored for a long time, importing freshly harvested apples from New Zealand is environmentally favourable. Similarly, it was found that, in terms of energy, it is more efficient to import off-season tomatoes from Spain to the UK, rather than growing the tomatoes locally in heated greenhouses (Smith et al. 2005). Unfortunately, in terms of water use impacts, the imported tomatoes have much greater impacts than their local counterpart (Payen et al. 2015).

In addition, in the study about Danish pork meat exported to the UK, which was described in Sect. 29.3.1 the 'food miles' concept was criticised. The authors pointed out that the different transport steps contribute less than 1% to the overall global warming impacts. Hence looking at reducing these impacts, transport is not the place where significant reductions can be obtained. The authors of the pork study call the concept of food miles misleading.

Food-miles can be concluded to be a simple social representation of a complex system that help people engage but it is not reliable as an indicator of the environmental impacts of a product system.

29.4 Methodological Issues

In the previous sections, a number of Life Cycle Assessments of food and food products has been discussed. These studies showed that in the full life cycle of a food product, it is often the farm inputs and agricultural stages where most environmental impacts arise. Within these agricultural stages, a few trends can be observed. Firstly, global warming impacts can be attributed to animal husbandry or fertiliser production. Secondly, acidification and eutrophication impacts are associated with the production and use of fertiliser or animal manure. Finally, toxic impacts are related to pesticide use and are still seldom considered.

The two papers described in the section about organic, integrated and conventional farming not only gave an indication of the differences in environmental impacts between these farming systems, they also illustrated the importance of considering land use in LCAs. The case studies on Swiss farming practices and local foods stressed the need to consider systems as a whole. Studying crops in isolation or only at a certain moment in time might lead to deceptive conclusions.

Besides those, a number of other methodological issues relating to the environmental assessment of food and food production remain.

First of all, the choice of the functional unit of a study has to reflect a product's function. As discussed in Sect. 29.3.2, some authors ascribe three functions to agriculture: land management, providing an income for farmers and production of food. The choice of the functional unit will depend on the goal and scope of the LCA study (see Chaps. 8 and 9). It will often be relevant to express the results by different functional units within the same study to give a fair picture of the compared systems, especially in the case of highly contrasted intensification levels. From a consumer's point of view, the primary function of food products is to provide sufficient energy and nutrients to the human body. Beyond the mass of food produced, research is therefore looking for more qualitative functional units, taking into account the food's nutritional value in a harmonised way. As an example, increasing yields of wheat may negatively affect the nutritional quality of the grains. It is therefore recommendable to clearly define the nutritional quality of a food product in the functional unit, especially in comparative LCA studies (Schau and Fet 2008).

A second issue is the definition of the system boundaries (see Chap. 9). This discussion has two important aspects. The first aspect is setting the border between technosphere (the product system) and ecosphere (the natural environment). This border is essential in LCA, as only material and energy flows crossing it are considered inputs or outputs. Especially in LCAs of agricultural and food products this boundary might be hard to identify clearly, because in food production the technosphere is closely linked to the ecosphere. An example can be found in the emissions of pesticides to agricultural soil. It can be argued that these should be marked as emissions to the ecosphere because the pesticides might affect various forms of life in the soil: worms, beetles, which are not necessarily the target organisms for the pesticide. On the other hand, one can reason that the soil of a field is part of the technosphere, because it is manipulated by humans to an extent where it is incomparable to natural soils. Setting the system boundaries is dependent on the goal and scope of the LCA study: it is not possible to objectively define one correct boundary setting that works for all agricultural LCAs (Dijkman 2014; Rosenbaum et al. 2015). For this reason, it is important to explicitly define the system boundaries in the goal and scope definition. Ideally, the boundaries between technosphere and ecosphere, and thus of LCI and LCIA for modelling the inflows and outflows, should be defined uniformly in order to produce a consistent LCA study. The second aspect of setting the system boundaries relates to the processes that are included in the study. Often, a cradle-to-farm gate study is done because it is assumed that most impacts arise at the agricultural stage or because the post-farm gate processes are identical. This might result in overlooking product losses during processing or consumption, while the reduction of food waste can contribute to lowering environmental impacts elsewhere in the life cycle.

A third issue in the LCA of food products is the inclusion of the diversity of production systems. Most LCA studies in the past have relied on a very small number of farms, while agricultural production systems are generally very diverse due to the interaction between the farmers' skills and practices and their environment. In studies where the variability of systems has been explored, the variability of LCA results is larger within one production group than between the studied alternatives. This also leads to the question of uncertainty of LCA results, which is generally not evaluated (see Chap. 11).

As a fourth issue, the modelling of flows from the agricultural field can be improved. Meier et al. (2015) argued that the modelling of nutrients needs to be improved in LCA practice, because especially nitrogen flows are responsible for many environmental impacts from on-field processes. Many studies reviewed by Meier et al. (2015) did not calculate the N balance from the field. When the N balance was calculated, differences in the N surplus (defined by Meier et al. (2015) as the nitrogen potentially emitted to the environment via different pathways) between conventional and organic farming systems were not always reflected in differences in the eutrophication potentials calculated for both farming systems. However, because the N surplus is a measure of the amount of N that is available for leaching to surface water or other environmental compartments, differences in the eutrophication potential should be related to differences in the N surplus. Moreover, Meier et al. (2015) found that the N emissions exceeded the N surplus in four processes representing Swiss agriculture in the EcoInvent database.

In addition, N emissions from manure, which are dependent on the excretion of N by livestock animals, are seldom adjusted for differences in the animals' dietary composition. Because the N balance of a field is dependent on many factors, such as the chemical or organic fertilisers used, the uptake into the crop and local soil and climate conditions, a simple model of the N balance of a field for the use in LCA is not available. The authors also stated the need for better modelling of flows of carbon from the field and manure. Apart from nutrient flows, modelling of pesticide flows can be improved. Databases with inventory data, such as ecoinvent (Ecoinvent Centre 2007) and the US LCI database (NREL 2003), often assume fixed emissions to one or two environmental compartments, independent of the pesticide applied, the application technology used, the climatic circumstances and the crop or soil onto which the pesticide is applied. Here, models such as PestLCI 2.0 (Dijkman et al. 2012) and the forthcoming results of ongoing pesticide consensus work can be used to better represent the influence of chemical properties and local circumstances on pesticide emissions (Rosenbaum et al. 2015).

A final issue in agricultural LCAs is the limited number of impact categories and flow indicators that are usually included: global warming, acidification, eutrophication, ozone depletion and energy use. Land use-related impacts are increasingly considered. This is an important impact category, because land is a scarce resource. Most land that is suitable for agriculture, is currently already in use as such. Moreover, considering land use may also help to illustrate the trade-offs of, e.g. organic farming: more land is required per unit of product, resulting in an expanded use of land for that product when switching to organic farming (under the assumption that the demand for the product remains unchanged). This direct land use change in turn results in indirect land use change (iLUC) because the organic product displaces another product, which ultimately results in the conversion of grasslands or forest into agricultural land. At the same time, chemical pesticide use is avoided, and the nutritional quality of grains may be at optimum when grown below maximum yield levels. So, even though all farmers switching to organic farming would considerably reduce certain environmental impacts of agriculture such as toxicity impacts, the amount of food available may be reduced as well on the short term.

Another impact category that is usually omitted in LCAs is toxicity (Meier et al. 2015). Because of neglecting toxicity, the effects of pesticide use are not well reflected in LCA results. Historically, toxicity was excluded because of the unavailability of emission data and impact assessment methods for these categories. On the LCI side, models such as PestLCI 2.0 can now be applied to calculate pesticide emissions to air, surface water and groundwater (Dijkman et al. 2012). Likewise, in impact assessment, models such as USEtox (Rosenbaum et al. 2008) can be used to calculate toxicity impacts. However, LCI and LCIA models do not necessarily apply the same boundaries between ecosphere and technosphere, both in terms of time and space. In order to overcome this inconsistency, and to provide guidance to LCA practitioners about modelling pesticide emissions to the environment and their impacts in LCA, a series of international workshops has been held in an effort to establish a global consensus. Rosenbaum et al. (2015) report the objectives of the effort and the recommendations of the first workshop.

Finally, water use-related impacts are frequently omitted in LCA studies. Given the importance of water resources for economic activities in general and for food production in particular as well as for human and ecosystems health, this impact category has rapidly evolved in the last few years. Nowadays, operational methods based on regional water stress are available (e.g. Berger et al. 2014, Pfister and Bayer 2014) and the new AWARE consensus impact assessment method (WULCA 2016) to asses water deprivation impacts (see Sect. 10.15). Despite these developments, much work to improve environmental relevance of methods addressing the consequences of water use on the environment is ahead. Water use is also associated with other long lasting problems such as salinisation (Payen et al. 2016) and desertification (Núñez et al. 2010), included in a number of LCA pilot studies.

Therefore, even though quite some steps have been made in the field of LCA of food products during the last 25 years, a number of challenges and methodological issues remain to be improved.

29.5 Applying LCA to Food Products in Southern Countries

In a context of demographic increase, especially in cities, southern and generally poor countries are facing immense challenges in terms of food security, poverty reduction, food safety and environmental protection. In such contexts, global assessment tools such as LCA can help stakeholders in food supply chains focus, improve and develop the most promising technical alternatives and support the eco-design of livestock and cropping systems. However, the application of LCA to food products in southern countries is recent and associated to difficult and numerous challenges. All the previously mentioned methodological challenges are relevant, but in an even more critical way due to the extreme diversity of production systems (often including perennial crops) and practice, the data scarcity on production systems, the lack of knowledge and appropriate models for estimating the field fluxes and finally the predominance of environmental impacts for which no consensual methodology exists. This concerns all environmental impacts associated to the use of land and water, including aspects of biodiversity, water deprivation, soil quality and fertility, carbon balance and GHGs from soils, salinisation impacts, etc. This also concerns all toxicity impact categories which are complex and for which available inventory approaches are either not valid or difficult to implement due to data scarcity. Particularly important in these situations are also the social impacts associated to food supply chains for which ambitious research programmes are starting. Finally, caring for the environment in such subsistence economies where the priority is to feed the people, often appears as a luxury for wealthy people. Therefore, awareness-raising and education about eco-friendly practices, human health risks and environmental sustainability are important priorities in association to more methodological studies to implement LCA in southern countries. Some research teams have started to implement LCA to food products in southern countries with examples on peri-urban tomato in the South of Benin (Perrin 2013), palm oil in Indonesia (Bessou et al. 2016), clementine in Morocco and mango in Brazil (Basset-Mens et al. 2016).

29.6 Conclusions

We have seen that LCAs of food products can be divided into six stages: inputs production and transportation, agricultural stage, processing, distribution, use and waste management. A large number of LCA studies are cradle-to-farm gate studies, and include only the two first stages. As a consequence of production and flows of nutrients and pesticides from the field, as well as from livestock and manure handling, the agricultural stage is often found to be the major contributor to many impact categories. This was illustrated in a number of case studies. Different impacts can arise in the food production stage. Environmental impacts from the use stage are often related to energy consumed in food storage and preparation. In the waste management stage, impacts arise due to food waste handling and treatment of human excretion. Throughout the life cycle of a food product, food waste is a major problem.

The case study about conventional and organic farming showed that, depending on the choice of functional unit and the impact categories included in the assessment, LCA can be used to conclude in favour of both conventional and organic farming practices. The case study about local food showed that local food is not by definition more sustainable. Transport is not often the decisive factor when it comes to environmental impacts of food products, so the circumstances during production weigh more heavily in determining where and how locally produced food is more sustainable. Overall, the key strength of LCA lies in the identification of the hot-spots and margins of improvement of each system.

Despite methodological improvements, a number of challenges remain for agricultural LCA. Firstly, the functional unit is currently often defined on basis of mass produced, or production per unit of area, without considering the nutritional quality of the product. Secondly, setting the system boundary between technosphere and the ecosphere is difficult in those production systems showing a direct interface with nature. In the life cycle inventory, modelling of flows of nutrients, water, salt and pesticides can be improved in many studies. A challenge in the LCIA phase remains the limited number of impact categories included in most LCA studies. Moreover, some impact categories relevant for agriculture, such as land use-related impacts including soil quality aspects and biodiversity damage, remain to be further developed and operationalised.

References

Abdullah, K.: Use of water and land for food security and environmental sustainability. Irrig. Drain. **55**(3), 219–222 (2006). doi:10.1002/ird.254

Andersson, K., Ohlsson, T.: Life Cycle Assessment of bread produced on different scales. Int. J. Life Cycle Assess. **4**(1), 25–40 (1999). doi:10.1007/bf02979392

Antón, A., Torrellas, M., Raya, V., Montero, J.I.: Modelling the amount of materials to improve inventory datasets of greenhouse infrastructures. Int. J. Life Cycle Assess. **19**(1), 29–41 (2014). doi:10.1007/s11367-013-0607-z

Basset-Mens, C., McLaren, S., Ledgard, S.: Exploring a comparative advantage for New Zealand cheese in terms of environmental performance. Presented at 5th international conference LCA in foods, Gothenburg, Sweden, 25–26 Apr 2007 (2007)

Basset-Mens, C., Vannière, H., Grasselly, D., Heitz, H., Braun, A., Payen, S., Koch, P., Biard, Y.: Environmental impacts of imported and locally-grown fruits for the French market: a cradle-to-farm-gate LCA study. Fruits **71**(2), 93–104 (2016). doi:10.1051/fruits/2015050

Berger, M., Van der Ent, R., Eisner, S., Bach, V., Finkbeiner, M.: Water accounting and vulnerability evaluation (WAVE): considering atmospheric evaporation recycling and the risk of freshwater depletion in water footprinting. Environ. Sci. Technol. **48**(8), 4521–4528 (2014). doi:10.1021/es404994t

Berlin, J.: Environmental Life Cycle Assessment (LCA) of Swedish semi-hard cheese. Int. Dairy J. **12**(11), 939–953 (2002). doi:10.1016/S0958-6946(02)00112-7

Bessou, C., Basset-Mens, C., Latunussa, C., Vélu, A., Heitz, H., Vannière, H., Caliman, J.-P.: Partial modelling of the perennial crop cycle misleads LCA results in two contrasted case studies. Int. J. Life Cycle Assess. **21**(3), 297–310 (2016). doi:10.1007/s11367-016-1030-z

Cederberg, C., Mattsson, B.: Life Cycle Assessment of milk production: a comparison of conventional and organic farming. J. Clean. Prod. **8**(1), 49–60 (2000). doi:10.1016/S0959-6526(99)00311-X

Cordell, D., Drangert, J.O., White, S.: The story of phosphorus: global food security and food for thought. Glob. Environ. Change **19**(2), 292–305 (2009). doi:10.1016/j.gloenvcha.2008.10.009

Cristea, A., Hummels, D., Puzzello, L., Avetisyan, M.: Trade and the greenhouse gas emissions from international freight transport. J. Environ. Econ. Manag. **65**(1), 153–173 (2013). doi:10.1016/j.jeem.2012.06.002

Dalgaard, R., Halberg, N., Hermansen, J.E.: Danish Pork Production: An Environmental Assessment. DJF Animal Science No. 82. University of Aarhus, Tjele (2007)

De Backer, E., Aartsens, J., Vergucht, S., Steurbaut, W.: Assessing the ecological soundness of organic and conventional agriculture by means of Life Cycle Assessment (LCA): a case of leekproduction. Br. Food J. **111**(10), 1028–1061 (2009). doi:10.1108/00070700910992916

De Vries, M., De Boer, I.J.M.: Comparing environmental impacts for livestock products: a review of Life Cycle Assessments. Livest. Sci. **128**(1–3), 1–11 (2010). doi:10.1016/j.livsci.2009.11.007

Dijkman, T.J., Birkved, M., Hauschild, M.Z.: PestLCI 2.0: a second generation model for estimating emissions of pesticides from arable land in LCA. Int. J. Life Cycle Assess. **17**(8), 973–986 (2012). doi:10.1007/s11367-012-0439-2

Dijkman, T.J.: Modelling of pesticide emissions for life cycle inventory analysis: model development, applications and implications. Ph.D. thesis. Technical University of Denmark, Kongens Lyngby (2014)

Djekic, I., Miocinovic, J., Tomasevic, N., Smigic, N., Tomic, I.: Environmental life-cycle assessment of various dairy products. J. Clean. Prod. **68**, 64–72 (2014). doi:10.1016/j.jclepro.2013.12.054

Ecoinvent Centre: Ecoinvent Reports No. 1–25. Swiss Centre for Life Cycle Inventories, Dübendorf (2007)

EU-JRC: International Reference Life Cycle Data System (ILCD) Handbook. General Guide for Life Cycle Assessment—Detailed Guidance. Publications Office of the European Commission, Luxembourg. Retrieved 20 Apr 2016 from http://eplca.jrc.ec.europa.eu/?page_id=86 (2010)

Edwards-Jones, G., Milà i Canals, L.L., Hounsome, N., Truninger, M., Koerber, G., Hounsome, B., Cross, P., York, E.H., Hospido, A., Plassmann, K., Harris, I.M., Edwards, R.T., Day, G.A. S., Tomos, A.D., Cowell, S.J., Jones, D.L.: Testing the assertion that 'local food is best': the challenges of an evidence-based approach. Trends Food Sci. Technol. **19**(5), 265–274 (2008). doi:10.1016/j.tifs.2008.01.008

Ellen MacArthur Foundation: Growth Within: A Circular Economy. Vision for a Competitive Europe. Retrieved 20 Apr 2016 from https://www.mckinsey.de/sites/mck_files/files/growth_within_report_circular_economy_in_europe.pdf (2015)

European Commission: The Environmental Footprint Pilots. Retrieved on 21 Apr 2016 from http://ec.europa.eu/environment/eussd/smgp/pef_pilots.htm (2016)

Fantke, P., Juraske, R., Antón, A., Friedrich, R., Jolliet, O.: Dynamic multicrop model to characterize impacts of pesticides in food. Environ. Sci. Technol. **45**(20), 8842–8849 (2011). doi:10.1021/es201989d

FAO: Interactions between Livestock Production Systems and the Environment. Retrieved on 3 May 2016 from http://www.fao.org/wairdocs/lead/x6123e/x6123e00.htm#Contents (1995)

FAO: FAOSTAT Land Use Data. Retrieved on 25 Apr 2016 from http://faostat3.fao.org/download/E/EL/E (2011)

FAO: World Agriculture Towards 2030/2050: The 2012 Revision. ESA working paper no. 12-03. Retrieved on 8 June 2016 from http://www.fao.org/docrep/016/ap106e/ap106e.pdf (2012)

FAO: FAO Statistical Yearbook 2013—World Food and Agriculture—Part 4: Sustainability Dimensions. Food and Agriculture Organisation of the United Nations, Rome (2013)

FAO: Water Withdrawal by Sector, Around 2007. Retrieved on 20 Apr 2016 from http://www.fao.org/nr/water/aquastat/tables/WorldData-Withdrawal_eng.pdf (2014)

Food, S.C.P.R.T.: ENVIFOOD Protocol, Environmental Assessment of Food and Drink Protocol, European Food Sustainable Consumption and Production Round Table (SCP RT), Working Group 1. Food SCP Round Table, Brussels (2013)

Foster, C., Green, K., Bleda, M., Dewick, P., Evans, B., Flynn, A., Mylan, J.: Environmental Impacts of Food Production and Consumption: Final Report to the Department for Environment, Food and Rural Affairs. Manchester Business School, Manchester (2006)

González-García, S., Hospido, A., Moreira, M.T.: Environmental Life Cycle Assessment of a Galician cheese: San Simon da Costa. J. Clean. Prod. **52**, 253–262 (2013). doi:10.1016/j.jclepro.2013.03.006

IPCC: Contribution of working group III to the fourth assessment report of the intergovernmental panel on climate change. Cambridge University Press, Cambridge (2007)

ISO: Environmental Management—Life Cycle Assessment—Principles and Framework (ISO 14040). ISO, The International Organization for Standardization, Geneva (2006a)

ISO: Environmental Management—Life Cycle Assessment—Requirements and Guidelines (ISO 14044). ISO, The International Organization for Standardization, Geneva (2006b)

Kim, D., Thoma, G., Nutter, D., Milani, F., Ulrich, R., Norris, G.: Life Cycle Assessment of cheese and whey production in the USA. Int. J. Life Cycle Assess. **18**(5), 1019–1035 (2013). doi:10.1007/s11367-013-0553-9

Kulikowski, M.: Mayday 23: World Population Becomes More Urban than Rural. North Carolina State University Newsroom. Retrieved on 26 Apr 2016 from http://news.ncsu.edu/releases/2007/may/104.html (2007)

Meier, M.S., Stoessel, F., Jungbluth, N., Juraske, R., Schader, C., Stolze, M.: Environmental impacts of organic and conventional agricultural products: are the differences captured by life cycleassessment? J. Environ. Manag. **149**, 193–208 (2015). doi:10.1016/j.jenvman.2014.10.006

Mithraratne, N., Barber, A., McLaren, S.J.: Carbon footprinting for the kiwifruit supply chain: report on methodology and scoping study. Final report. Landcare Research, Auckland (2010)

Muñoz, I., Milà i Canals, L.L., Fernández-Alba, A.R.: Life Cycle Assessment of the average Spanish diet including human excretion. Int. J. Life Cycle Assess. **15**(8), 794–805 (2010). doi:10.1007/s11367-010-0188-z

Nemecek, T., Debois, D., Huguenin-Elie, O., Gaillard, G.: Life Cycle Assessment of Swiss farming systems: I. Integrated and organic farming. Agric. Syst. **104**(3), 217–232 (2011). doi:10.1016/j.agsy.2010.10.002

NREL: US LCI Database. Retrieved on 6 Sept 2013 from www.nrel.gov/ (2003)

Núñez, M., Civit, B., Muñoz, P., Pablo Arena, A., Rieradevall, J., Anton, A.: Assessing potential desertification environmental impact in Life Cycle Assessment. Part 1: methodological aspects. Int. J. Life Cycle Assess. **15**(1), 67–78 (2010). doi:10.1007/s11367-009-0126-0

Paxton, A.: The food miles report: the dangers of long distance food transport. The SAFE Alliance, London (1994)

Payen, S., Basset-Mens, C., Perret, S.R.: LCA of local and imported tomato: an energy and water trade-off. J. Clean. Prod. **87**(15), 139–148 (2015). doi:10.1016/j.jclepro.2014.10.007

Payen, S., Basset-Mens, C., Núñez, M., Follain, S., Grünberger, O., Marlet, S., Perret, S.R., Roux, P.: Salinisation impacts in Life Cycle Assessment: a review of challenges and options towards their consistent integration. Int. J. Life Cycle Assess. **21**(4), 577–594 (2016). doi:10.1007/s11367-016-1040-x

Perrin, A.: Evaluation environnementale des systèmes agricoles urbains en Afrique de l'Ouest: implications de la diversité des pratiques et de la variabilité des émissions d'azote dans l'Analyse du Cycle de Vie de la tomate au Bénin. Ph.D. thesis. AgroParisTech- Sciences agronomiques et écologiques, Paris (2013)

Pfister, S., Bayer, P.: Monthly water stress: spatially and temporally explicit consumptive water footprint of global crop production. J. Clean. Prod. **73**, 52–62 (2014). doi:10.1016/j.jclepro.2013.11.031

Rosenbaum, R.K., Bachmann, T.M., Gold, L.S., Huijbregts, M.A.J., Jolliet, O., Juraske, R., Koehler, A., Larsen, H.F., MacLeod, M., Margni, M., McKone, T.E., Payet, J., Schuhmacher, M., Van de Meent, D., Hauschild, M.Z.: USEtox: the UNEP-SETAC toxicity model: recommended characterisation factors for human toxicity and freshwater ecotoxicity in life cycle impact assessment. Int. J. Life Cycle Assess. **13**(7), 532–546 (2008). doi:10.1007/s11367-008-0038-4

Rosenbaum, R.K., Antón, A., Bengoa, X., Bjørn, A., Brain, R., Bulle, C., Cosme, N., Dijkman, T. J., Fantke, P., Felix, M., Geoghega, T.S., Gottesbüren, B., Hammer, C., Humbert, S., Jolliet, O., Juraske, R., Lewis, F., Maxime, D., Nemecek, T., Payet, J., Räsänen, K., Roux, P., Schau, E.M., Sourisseau, S., Van Zelm, R., Von Streit, B., Wallman, M.: The Glasgow consensus on the delineation between pesticide emission inventory and impact assessment for LCA. Int. J. Life Cycle Assess. **20**(6), 765–776 (2015). doi:10.1007/s11367-015-0871-1

Saxe, H., Jensen, J.D.: Does the environmental gain of switching to the healthy New Nordic Diet outweigh the increased consumer cost? J. Food Sci. Eng. **4**(6), 291–300 (2014). doi:10.17265/2159-5828/2014.06.004

Schau, E.M., Fet, A.M.: LCA studies of food products as background for environmental product declarations. Int. J. Life Cycle Assess. **13**(3), 255–264 (2008). doi:10.1065/ica2007.12.372

Sim, S., Barry, M., Clift, R., Cowell, S.J.: The relative importance of transport in determining an appropriate sustainability strategy for food sourcing. Int. J. Life Cycle Assess. **12**(6), 422–431 (2007). doi:10.1007/s11367-006-0259-3

Smith, A., Watkiss, P., Tweddle, G., McKinnon, A., Browne, M., Hunt, A., Treleven, C., Nash, C., Cross, S.: The validity of food miles as an indicator of Sustainable development. Final report prepared by AEA Technology Environment for DEFRA. DEFRA, London (2005)

Sonesson, U., Janestad, H., Raaholt, B.: Energy for preparation and storing of food—models for calculation of energy use for cooking and cold storage in households. SIK-Rapport 709. The Swedish Institute for Food and Biotechnology, Gothenburg (2003)

Swaminathan, B., Sukalac, K.E.: Technology transfer and mitigation of climate change: the fertilizer industry perspective. Presented at the IPCC Expert Meeting on Industrial Technology Development, Transfer and Diffusion, Tokyo, Japan, 21–23 Sept 2004 (2004)

Ten Hoeve, M.: Environmental consequences of pig slurry treatment technologies: a life cycle perspective. Ph.D. thesis. University of Copenhagen, Copenhagen

Thomassen, M.A., Van Calker, K.J., Smits, M.C.J., Iepema, G.L., De Boer, I.J.M.: Life cycle assessment of conventional and organic milk production in the Netherlands. Agric. Syst. **96**(1–3), 95–107 (2008). doi:10.1016/j.agsy.2007.06.001

Torrellas, M., Antón, A., López, J.C., Jose Baeza, E., Perez Parra, J., Pere, M., Ignacio Montero, J.: LCA of a tomato crop in a multi-Tunnel greenhouse in Almeria. Int. J. Life Cycle Assess. **17** (7), 863–875 (2012a). doi:10.1007/s11367-012-0409-8

Torrellas, M., Antón, A., Ruijs, M., García Victoria, N., Stanghellini, C., Montero, J.I.: Environmental and economic assessment of protected crops in four European scenarios. J. Clean. Prod. **28**, 45–55 (2012b). doi:10.1016/j.jclepro.2011.11.012

USEtox: USEtox 2.0 Official Launch. Retrieved on 12 Jan 2016 from http://www.usetox.org/events/usetox-20-official-launch (2015)

Van Middelaar, C.E., Berentsen, P.B.M., Dolman, M.A., De Boer, I.J.M.: Eco-efficiency in the production chain of Dutch semi-hard cheese. Livest. Sci. **139**(1–2), 91–99 (2011). doi:10.1016/j.livsci.2011.03.013

Van Zelm, R., Larrey-Lassalle, P., Roux, P.: Bridging the gap between life cycle inventory and impact assessment for toxicological assessments of pesticides used in crop production. Chemosphere **100**, 175–181 (2014). doi:10.1016/j.chemosphere.2013.11.037

Williams, H., Wikström, F.: Environmental impact of packaging and food losses in a life cycle perspective: a comparative analysis of five food items. J. Clean. Prod. **19**(1), 43–48 (2011). doi:10.1016/j.jclepro.2010.08.008

WULCA: The AWARE method. Retrieved on 21 Apr 2016 from http://wulca-waterlca.org/project.html (2016)

Author Biographies

Teunis J. Dijkman has worked with method development in LCA since 2010, focusing on Life Cycle Inventory modelling. Interested in the boundary setting between technosphere and ecosphere for agrochemicals. Contributed to consensus building and model development for global pesticide emission quantification.

Claudine Basset-Mens agronomist and LCA scientist since early 2000s, adapting LCA to agri-food chains (animal and crop production) in a wide range of situations and countries. Her main research interests relate to inclusion of agricultural systems diversity, modelling of field emissions, modelling of perennial crops within LCA and the inclusion of uncertainty.

Assumpció Antón expert in the application of LCA methodology in agriculture since early 2000s. Final goal of her work focus on the environmental improvement of agricultural systems. Main interests are development of agricultural LCI and LCIA methods related to agricultural processes (agrochemical emissions, land use and water consumption).

Montse Núñez environmental scientist with interest in LCA, agriculture, and modelling of environmental impacts from resource use in agricultural activities. Involved in development of LCIA methods to assess impacts of land use and water use since 2007.

Chapter 30
LCA of Biofuels and Biomaterials

Susanne Vedel Hjuler and Sune Balle Hansen

Abstract Biofuels and biomaterials can today substitute many commodities produced from fossil resources, and the bio-based production is increasing worldwide. As fossil resources are limited, and the use of such resources is a major contributor to global warming and other environmental impacts, the potential of bio-products as substitutes for fossil-based products is receiving much attention. According to many LCA studies, bio-products are environmentally superior to fossil products in some life cycle impact categories, while the picture is often opposite in others. Bio-products is a highly diverse group of products and the environmental profile of bio-products relative to their fossil counterparts is case specific and to a high degree depending on the feedstock used. This illustrates the importance of conducting case specific LCAs for determining the environmental profile of bio-products relative to fossil ones, and emphasises the importance of including all relevant impact categories, in order to avoid problem shifting.

30.1 Introduction

Many conventional petrochemical products, such as chemicals, polymers and fuels, can today be produced from biomass. There are multiple drivers for substitution of petrochemical products. One driver is that fossil resources are limited. Another driver is that fossil fuel consumption is a major contributor to global warming and to other important environmental impacts as well. Furthermore, biofuels and biomaterials (hereafter referred to together as bio-products) provide an option for basing production on more local feedstock, creating jobs and promoting reduced

S.V. Hjuler · S.B. Hansen
Division of Quantitative Sustainability Assessment, Department for Management Engineering, Technical University of Denmark, Produktionstorvet Building 424, 2800 Kongens Lyngby, Denmark

S.V. Hjuler (✉)
Novozymes A/S, Krogshøjvej 36, 2880 Bagsværd, Denmark
e-mail: hjuler.susanne.vedel@gmail.com

M.Z. Hauschild et al. (eds.), *Life Cycle Assessment*,
DOI 10.1007/978-3-319-56475-3_30

dependence on fossil resources. Following this, the potential of bio-products as substitutes for fossil-based ones is high on the global political agenda, and the production worldwide is increasing. In 2003, 8–10% of the feedstock for the European chemical industry was biomass (Rothermel 2008), whereas the global share of bio-based chemicals was 2% in 2008 and is expected to be at least 22% by 2025, according to USDA (2008). For the global polymer production, the bio-based share has reached more than 8% (Carus et al. 2013). Approximately 2% of the global liquid fuel consumption was covered by biofuels in 2010 (EIA 2012). Direct substitution of some of the same fuels, chemicals and materials which are conventionally produced based on fossil resources is possible, but also the development of new materials and products with different properties is pursued.

Along with the growing demand for bio-products, there is an equally increasing need for awareness of the environmental performance of the bio-products relative to their petrochemical counterparts. The aim of this chapter is to provide a review of the state-of-the-art within LCAs on bio-products, as well as an introduction to methodology and methodological challenges and uncertainties in the field. Definitions and system boundary choices are introduced initially to illustrate the systems in question, after which general LCA results in the field are presented, followed by discussions of specific methodological issues and their potential implication for the assessment results.

An important portion of this chapter is based on, and for some parts taken from, Jørgensen (2014) and further details on many of the discussed issues can be found there.

30.1.1 Definition of Bio-products

Bio-products are products made from biomass feedstocks. The definition of biomass can vary among literature sources but one of the most comprehensive definitions can be found in directives from the European Union, e.g. the Renewable Energy Directive, EU-RED, (European Parliament 2009), stating that "biomass means the biodegradable fraction of products, waste and residues from biological origin from agriculture (including vegetal and animal substances), forestry and related industries including fisheries and aquaculture, as well as the biodegradable fraction of industrial and municipal waste".

Bio-products can also be defined in opposition to fossil products, made from e.g. oil or natural gas. Both fossil- and bio-based products come from biomass but the key difference is that in the case of fossil products, this biomass went through fossilisation. Fossil resources have been formed from ancient biomass during millions of years of geological processing and storage. The carbon present in these resources was thus removed from the atmosphere through photosynthesis many millions of years ago and is no longer a part of the present day natural carbon cycle balance. This carbon is termed 'fossil carbon' (European Commission 2010).

On the contrary, bio-products are based either completely or partly on biomass feedstock, in the form of plants or biogenic residues/waste (Weiss et al. 2012). As plants take up and store CO_2 from the atmosphere during their growth through photosynthesis, carbon exchanges between biomass, including bio-products, and the atmosphere are part of the present-day natural carbon cycle. The carbon pool that is constituted by carbon in biomass and bio-products is termed 'biogenic carbon'. Whether this biogenic carbon can be considered 'CO_2 neutral' or 'carbon neutral' is a specific key issue for bio-products and their relevance for climate change, and it is discussed in Sect. 30.4.4.

There are also issues related to the timing of carbon sequestration and release which affects the 'carbon neutrality' aspect, as temporary release or storage of carbon may also play a role in terms of climate change, especially on short-term targets (e.g. Cherubini et al. 2012b; Jørgensen et al. 2015). The issue of the potential role of temporary carbon storage in relation to climate change is addressed in Sect. 30.3.1.

While the feedstock of bio-products consists of biogenic carbon, the production of the feedstock as well as other processes in the product life cycle may include fossil fuel consumption, or in other ways contribute to greenhouse gas (GHG) emissions. Thus, it is crucial to conduct an LCA in order to determine the actual climate change impacts of bio-products, along with other environmental impacts.

There is no standard classification of bio-products. However, a proposal is given in Fig. 30.1 providing an overview of existing bio-products, considering bio-product application and type. The same distinction into biofuels and biomaterials was used throughout this chapter. This chapter focuses on bio-products that serve for the substitution of conventional petrochemical products, such as fuels, chemicals and polymers, rather than bio-products based on traditional biomaterials such as wood, paper and textiles. However, many of the aspects discussed here apply for traditional biomaterials as well.

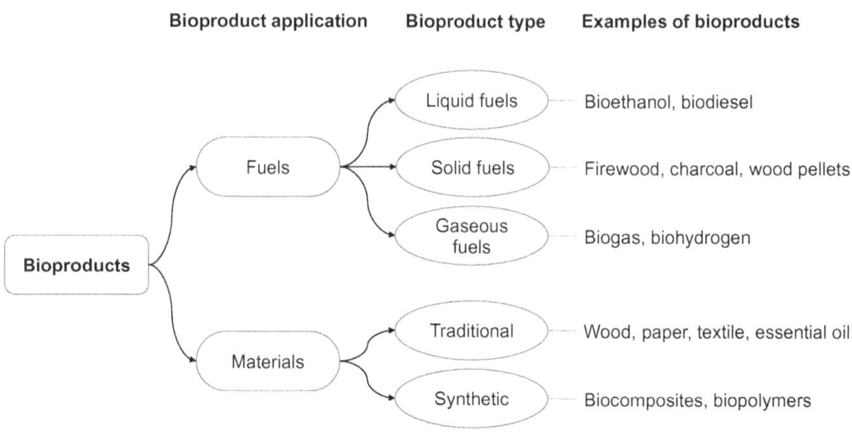

Fig. 30.1 Overview of existing bio-products. Image Copyright © Anthony Benoist. Used with permission

Generally, bio-products are distinguished into different 'generations': *First generation* (1G) covers bio-products that are produced by conventional methods, based on sugar, starch, vegetable oil or animal fats. Most bio-products today are 1G, but this feedstock entails certain competition with food production and there is a major focus on moving on to the *Second generation* (2G). The use of the term 2G is not completely consistent in the literature. Predominantly, it refers to the change from the conventional technology to production of bio-products using lignocellulosic feedstock, e.g. agricultural residues or energy crops. However, some use the term to e.g. refer to a conversion route or end product, rather than feedstock (Cherubini 2010). Here, the first meaning of the terminology is used. Today, 2G biofuel production is reaching commercial scale. However, use of lignocellulosic feedstock also entails competition with its prior use, such as soil carbon replenishment for lignocellulosic residues, and land use competition for energy crops. Further, a so-called *Third generation* (3G) of bio-products is discussed in the literature, but like for 2G, the use of the term varies. The term 3G is often used to characterise the use of certain feedstocks like algae (e.g. Sander and Murthy 2010; Carus and Dammer 2013), or microalgae (Posten and Schaub 2009), due to its potential to address many of the concerns about 1G and 2G feedstock (Sander and Murthy 2010), its high yield potentials (Posten and Schaub 2009; Sander and Murthy 2010) and the different growing conditions, compared to terrestrial plants (Posten and Schaub 2009). However, others use the term 3G to refer to a follow up on 2G, and do not refer to a shift in feedstock in the same way as the shift between 1G and 2G (Bessou et al. 2011). Bio-products beyond 1G are sometimes termed as 'next generation' or 'advanced' bio-products. As outlined, the use of terminology in this field is not consistent in the literature so caution should be taken when reading the terms.

Biofuels can be classified into three categories (Fig. 30.1):

1. Liquid biofuels: the most common are bioethanol and biodiesel
2. Gaseous biofuels: Biogas, syngas and biohydrogen
3. Solid biofuels: Pellets, lignin, biochar and wood

Among the biofuels, this chapter will concentrate on the liquid biofuels biodiesel and bioethanol. These are among the focus areas in the EU-RED (European Parliament 2009) in Europe and the Renewable Fuel Standard, RFS (Schnepf and Yacobucci 2012) in the USA, both of which set the official (political) standards for permitted life cycle emissions of a biofuel to be labelled 'renewable'. In both documents, the renewability of a biofuel is judged solely on GHG emissions. This in turn steers the industrial and to a large extend the academic focus onto GHG emissions, which could lead to potential problem shifting towards other impact categories.

Biomaterials can be classified in various ways—here two main distinctions are made (Fig. 30.1):

1. Synthetic biomaterials (e.g. biopolymers or biocomposites) substituting conventional materials, such as various types of petrochemical plastic, cement or chemicals.
2. Traditional biomaterials, e.g. wood, paper and textiles.

The biomaterials in focus here are those of the first group.

An example of such a biomaterial is polylactic acid (PLA), a widespread biopolymer for example used for biomedical applications, disposable cups and cutlery, food containers and biodegradable bags for compostable waste.

30.1.2 System Boundaries

Bio-products are most often created to substitute fossil resource-based products. As such, it is of high importance that the system boundaries are set in a manner to make direct impact comparisons with fossil resource-based products possible. In assessments of bio-products, the most controversial system boundary decisions pertain to the assessment of land use change (LUC) (direct and indirect), the modelling of waste products, use stage and the 'end-of-life handling' (for biomaterials). Often, these aspects have simply been sidestepped or ignored. A number of studies have included these aspects and found that they are in most cases very important (e.g. Majer et al. 2009). As such, they should never be omitted from the assessment without proper documentation that they are negligible.

Other types of system boundaries are pertaining to inclusion of impact categories and time horizons. These are dealt with in Sect. 30.2 Sustainability of Biofuels and Biomaterials and Sect. 30.3 Specific Issues for Biomaterials.

It must be kept in mind that the system boundaries should be drawn to include what is important in relation to the goal of the study (see Chap. 8). Inclusion of aspects like LUCs and in particular indirect land use change (ILUC) may lead to increased representativeness of the study, but also increased uncertainties. There can thus be great variations in LUC emissions depending on land uses, soil type, geography, climate, etc.; and it can be difficult to determine which products are ultimately substituted. The needed data for a specific assessment are often difficult to come by and simplifications/assumptions may be necessary. It is important that these issues, along with simplifications and assumptions, are dealt with and described.

Conventionally, attributional system boundaries (see Chap. 8) have been used for bio-products as with most other products. In recent years, consequential system boundaries (see Chaps. 8 and 9) have been introduced; either for the entire system or partially, to describe (in)direct LUCs (see Sect. 30.4) and for the evaluation of the residue use and by-products, which cross the system boundaries.

30.2 Sustainability of Biofuels and Biomaterials

More LCA results can be found on biofuels than on biomaterials (e.g. Patel et al. 2005), and most are based on first generation (1G), where fewer LCA results are available for the next generations of bio-products.

For the environmental sustainability of bio-products relative to their fossil-based counterparts, conclusions cover a wide range. This is further complicated by LCA not having consensus on a standardised impact assessment methodology resulting in various impact categories being used in studies on bio-products (Singh et al. 2010). However, generally, LCA studies conclude the following:

- With respect to fossil fuel consumption and climate change impacts, bio-products generally perform better than their petrochemical counterparts (Weiss et al. 2007, 2012; Cherubini and Strømman 2011; Wang 2010; Patel et al. 2005). This picture can, however, change if the feedstock is planted on previously high carbon stock land (Kim et al. 2009) or when including GHG emissions from ILUC, which can be substantial (Weiss et al. 2012). The issues of LUC and ILUC are covered in Sect. 30.4.
- When it comes to other impact categories, bio-products do not necessarily perform better than their conventional counterparts. Many reviews conclude that bio-products often have a higher impact than conventional products in the case of eutrophication (e.g. Weiss et al. 2007, 2012; Cherubini and Strømman 2011) and stratospheric ozone depletion (e.g. Weiss et al. 2007, 2012).
- For acidification, some reviews conclude that bio-products generally have a higher impact than their petrochemical counterparts (e.g. Cherubini and Strømman 2011; Tabone et al. 2010; Luo et al. 2009; Weiss et al. 2007). Other studies, however, find inconclusive results for this impact category, which may indicate that the results vary between different types of bio-products (Weiss et al. 2012).
- Some studies suggest that biomaterials have a lower impact in terms of human toxicity, terrestrial ecotoxicity, as well as carcinogenic potential, but a higher aquatic ecotoxicity, compared to conventional materials (Weiss et al. 2012), while other studies conclude that bio-product systems lead to increased human toxicity and ecotoxicity in most cases (Cherubini and Strømman 2011). However, those categories are often not included in LCA studies and results are based on few studies (Weiss et al. 2012).

In addition, biomass feedstock production use land and thus include a number of impacts related to that land use (see more on impact assessment of land use in Sect. 10.14). These are, however, in many cases not consistently included in LCA, as mentioned in Sect. 30.4.3.

As mentioned, most current conclusions on LCA results of bio-products compared to their fossil-based counterparts are primarily based on 1G bio-products. For 2G bio-products, these results are expected to improve, as this feedstock is generally expected to have a better environmental performance than 1G. This is further discussed in Sect. 30.5.

These results make the current overall environmental performance of bio-products compared to conventional products inconclusive from a general point of view, and illustrate the importance of including all relevant impact categories when considering bio-based product sustainability, in order to avoid problem shifting. Further, it illustrates that environmental sustainability of bio-products relative to fossil counter products is rather case specific, emphasising the need for LCAs on a case level.

30.2.1 Biofuels or Biomaterials?

Most LCA studies on bio-products focus on their environmental performance relative to petrochemical counter products. Comparing whether biofuels or biomaterials are preferable, in terms of optimal biomass use from an environmental perspective, has been less studied. However, Patel et al. (2005) concludes that when comparing use in biofuels or biomaterials from the perspective of energy savings and reduced GHG emissions, biomaterials currently seem to be preferable. This conclusion is supported by Brehmer et al. (2012) stating that dedicated biochemical production can outperform biofuels in terms of fossil fuel replacement potential. As innovation continues in those technologies, these are preliminary results and further investigation is needed. Also in terms of to what extent the issue between biofuels and biomaterials is competition and to what extent it is complementary (Patel et al. 2005).

The potential complementarity between biofuels and biomaterials is to some extent pursued in the biorefinery concept—a bio-based analogue to the existing petrochemical refinery, producing both fuel and materials in an integrated process. However, even though some commercial and pilot scale biorefinery plants exist, the technology is still rather new and only few LCA studies are available. Cherubini and Jungmeier (2010) conclude that the biomass energy and material recovery is maximised if applying the biorefinery concept, joining a variety of technical processes. Apart from the potentially increased efficiency, LCA results resemble the general LCA results for bio-products, as the biorefinery concept is simply an integrated production of various bio-products.

30.2.2 Major Impact Processes in the Life Cycle

The environmental impacts from bio-products originate from a number of activities, but some contribute more than others. Here, some of the major influencing aspects are highlighted:

- *LUCs* can potentially contribute very significantly to especially the GHG emissions in a bio-based product system. LUCs are discussed in Sect. 30.4.

- A substantial share of environmental impacts from bio-products originate from *industrial farming practices*, e.g. from the use of fertilisers, pesticides and water. Further, solid and liquid wastes can cause significant impacts if not treated properly, but they also have the potential to be used directly as solid fuel or be used as feedstock for second generation bio-products and thereby improve the environmental performance of the system.
- The *production stage* for bio-products can also have a large contribution to the overall environmental impact, e.g. for bioethanol (Tabone et al. 2010; Wang 2010).

30.2.3 Reliability of Results

Conclusions of environmental impacts of bio-products compared to conventional petrochemical products from different LCA studies often differ substantially, as previously mentioned. Here, we distinguish between:

- *Variations* due to real differences in studied systems, e.g. due to difference in products, feedstock, production routes or spatial variability due to different regional scopes
- *Delimitations* due to lack of the entire LCA perspective in studies, either by including only one or a few impact categories, or using a scope that does not cover the entire life cycle (e.g. using only 'cradle-to-gate' perspective, see Sect. 8.6)
- *Uncertainty* due to limitations in data, insufficient knowledge on which to base reliable assumptions etc.

Variations and Delimitations

When discussing bio-products in general, there is bound to be variations due to the different types of bio-products. However, many LCA studies on bio-products also differ in, e.g. modelling choices of system boundaries, functional units, allocation procedure and life cycle scenarios (Weiss et al. 2012; Malca and Freire 2011). Often, there are considerable variations in the LCAs on bio-products due to difference in product systems; e.g. in terms of feedstock type, production technology or yields (Weiss et al. 2012; Wang 2010; Dornburg et al. 2003). Furthermore, results vary with regional scope, as aspects such as electricity generation and end-of-life handling of products differ substantially between countries (Patel et al. 2005).

Many LCA studies on bio-products are also delimited in their life cycle approach, either in terms of impact categories, scope or coverage of life cycle stages. For one thing, a substantial part of LCA studies on bio-products focuses only on fossil fuel consumption and GHG emissions (Von Blottnitz and Curran 2007). While this may be relevant in terms of specific political targets, it entails a risk of problem shifting. In later years, it seems that the inclusion of more impact categories within this field has increased (Weiss et al. 2012). Some studies only include a cradle-to-gate perspective, rather than the full LCA perspective cradle-to-grave, or leave out important aspects in some life cycle stages, with the limitations and potential problem shifting that entails.

Other points to be aware of are that there is an overweight of LCAs on European products, and that some results are based on pilot scale production while others are based on large industrial scale (Weiss et al. 2012) and others again are based largely on generic data.

Such variations and delimitations make comparison of results difficult and meta-analysis has become a popular tool for reducing some of the differences. In meta-analysis, available studies are made comparable by, e.g. altering the functional unit and recalculating results accordingly (Weiss et al. 2012).

Example 30.1 The interest in 3G biofuels has seen an important increase in the last decade. Many theoretical process chains have been proposed by coupling biomass production facilities (as open ponds or photobioreactors) and concentration and transformation processes (centrifugation, chemical separation, wet or dry extraction of compounds of interest, etc.). Since the first microalgal biodiesel LCA (Lardon et al., 2009), around 300 studies dealing with this topic have been published in scientific literature. All these studies try to capture the environmental drawbacks of these emerging systems for which real industrial data do not (yet) exist. The variability of LCIs combined with unclear system boundaries (as well-to-wheel or well-to-tank (see Sect. 27.1.3), co-product management rules, etc.) explains large variations in results. To illustrate this variability, Collet et al. (2013) reviewed 13 microalgal biofuel LCA publications; they found a range of climate change impact from −75 to 531 g CO_2-eq per MJ produced. Collet et al. (2015) list 24 main issues driving variation in LCA results due to differences in goal and scope, LCI and LCIA steps, which should be assessed to allow result comparisons among 3G systems.

Uncertainty

Uncertainties in inventory data may be rather large and often vary substantially between bio-product studies (Weiss et al. 2012; Patel et al. 2005). One of the main reasons for this is the substantial use of assumptions (Davis et al. 2009), which is partially due to the fact that especially production of biomaterials is still relying on rather new technologies and LCA modelling is often based on small scale data (Weiss et al. 2007). In addition, when conducting an LCA on bio-products it must be kept in mind that especially biofuel is a highly politicised topic and that existing data and assumptions may be biased. A vast number of academic papers and reports have been published using a wide array of methodologies and assumptions, making comparison of studies difficult if not impossible in many cases. Thus, for the purpose of conducting an LCA and collecting data, as well as choosing existing studies for comparison, strict review of the assumptions and data foundation in existing studies is important.

The potentials for production of bio-products in developing and emerging countries are rapidly increasing. Specific data in these countries is often sparse, non-verifiable and not publicly available and it cannot be assumed that data from industrial countries apply. Note that Intergovernmental Panel on Climate Change (IPCC) guidelines (e.g. IPCC 2006) often do not apply to tropical conditions as the background data are collected from industrialised countries. Special focus on data integrity and clearly stated assumptions is thus necessary.

30.2.4 Areas Getting Increased Attention

Current LCA results generally do not include ILUC impacts, but significant work is dedicated to tackling this issue in LCA. Further, several land use related impacts are receiving increased attention, such as water consumption and quality downgrading, changes in biodiversity, losses of soil carbon and erosion of soil, as well as changes in surface albedo and other biogeophysical impacts. However, many such land use related impacts have not yet been included in general LCAs (e.g. Weiss et al. 2012; Wang 2010; Patel et al. 2005). ILUC is further discussed in Sect. 30.4.1 and land use related impacts are further discussed in Sect. 30.4.2.

N_2O emissions from agricultural land using industrial and/or organic fertilisers is another issue, which has been neglected in many studies as data is sparse (Cherubini et al. 2009). However, some studies have highlighted the potentially very significant impacts from this strong greenhouse gas (e.g. Crutzen et al. 2008) and recent studies have quantified some potential emissions, though generic data from IPCC (2006) is often used rather than site specific data (e.g. Majer et al. 2009).

30.3 Specific Issues for Biomaterials

In many aspects, biofuels and biomaterials are comparable as they are based on the same feedstock types and thus have similar opportunities and challenges with respect to feedstock issues. For that reason, they can to a large extend be addressed together, which is what has been done in the previous sections in this chapter. However, there are some issues that differ between biofuels and biomaterials, due to one main difference, which is the product lifetime. Whereas the use stage for fuels is also the disposal stage and the fuel is not expected to have a long lifetime, this is different for some biomaterials. This difference leads to a number of issues that need to be considered specifically for biomaterials.

30.3.1 Temporary Carbon Storage

When biomaterials are produced, they store the biogenic carbon from their biomass origin, thus keeping it out of the atmosphere until, the carbon is released again, which e.g. happens if the product is incinerated at its end of life. There is an ongoing discussion on how, and even if, such temporary carbon storage contributes to reducing global warming issues, and many suggestions for the handling of temporary carbon storage exist (see, e.g. Brandão et al. 2013; Jørgensen et al. 2015).

The ILCD handbook handles temporary carbon storage as follows (European Commission 2010):

Temporary removal of CO_2 from the air by e.g. storage in long-lived bio-products is accounted for in the inventory, but generally not included in the total impact calculation of the LCA, due to the general infinite time horizon in LCA. It should only be considered in the total impact calculation if the short-term perspective rather than the normal infinite perspective is considered. Generally, all emissions occurring during the first 100 years of the analysis are inventoried as normal elementary flows, whereas emissions occurring after those 100 years are inventoried as 'long-term emissions' and are not included in the short-term emissions. Thus, accounting for temporary carbon storage is done by introducing a 'correction elementary flow', using IPCC GWP100 factors and including the duration of the temporary carbon storage within the first 100 years of the analysis. Using this approach, the derived GWP100 impact factor is −0.01 kg CO_2-eq for 1 kg CO_2 stored 1 year.[1]

This way of handling temporary carbon storage is analogous to the handling of delayed emissions, with the exception that fossil delayed emissions do not have the benefit of prior uptake of CO_2 from the atmosphere.

Example 30.2 If a stock of biopolymer products has a mass of carbon equivalent to 1 tonne CO_2 ($\sim$273 kg carbon) and is stored for 5 years, the correction flow will be:

$$1000\ \text{kg}\ \ CO_2 * 5\ \text{years} * -0.01\ \text{kg}\ CO_2\text{-eq/kg}\ CO_2/\text{year} = -50\ \text{kg}\ CO_2\text{-eq} \quad (30.1)$$

If the study requires inclusion of the value of temporary carbon storage in the impact assessment, the correction flow of −50 kg CO_2-eq means that 50 kg CO_2-eq are subtracted from the total GHG impact of the product stock over its lifetime, thus lowering the carbon footprint of the products.

[1]0.01 kg CO_2-eq per kg CO_2 stored 1 year corresponds to the GWP100 for 1 kg CO_2 of 1 kg CO_2-eq over 100 years, if assuming linearity, and considering a short-term perspective, rather than the normal infinite one, as explained in the text.

Time Horizons

A central issue in the discussion of the handling of temporary carbon storage in LCA is the use of time horizons. When assessing climate change impacts in LCA, a time horizon of 100 years is often used for the characterisation factor, reflecting the time horizon adopted in the Kyoto Protocol. In approaches for assessing the value of temporary carbon storage, the use of such a time horizon is by some interpreted as implying that impacts occurring after this time should not be included in the assessment (e.g. Moura-Costa 2002; Clift and Brandão 2008). To distinguish between these interpretations, this latter interpretation, of disregarding all impacts occurring after the time horizon, is here referred to as 'accounting period'. It is important to understand the major difference between the time horizon used in the Kyoto Protocol and such an accounting period. The 100-year time horizon adopted in the Kyoto Protocol includes impacts of the first 100 years of every greenhouse gas emission, regardless of when the emission occurs, cutting off only the so-called 'tails' of the emissions that are left 100 years after the emission. For CO_2 emissions, for example, approximately 20% stays in the atmosphere for many thousand years (Archer et al. 1997). The approach is illustrated in Fig. 30.2:

In contrast to this, the accounting period, suggested by some for temporary carbon storage crediting, starts counting from the time the carbon is stored and cuts off every impact occurring after the accounting period, regardless of when along the accounting period the carbon is released again. This means that some of the impacts from an emission will be 'hidden' or completely disregarded due to the accounting period, which is a very different result than using the time horizon as adopted in the Kyoto Protocol (Jørgensen and Hauschild 2013; Brandão et al. 2013). For example, if carbon is stored for 30 years before being released, using the 100-year time horizon will still give a 100 year impact inclusion whereas the 100-year accounting period will only give an inclusion of 70 years impact. The difference in the two ways of interpreting time horizons is illustrated in Fig. 30.3.

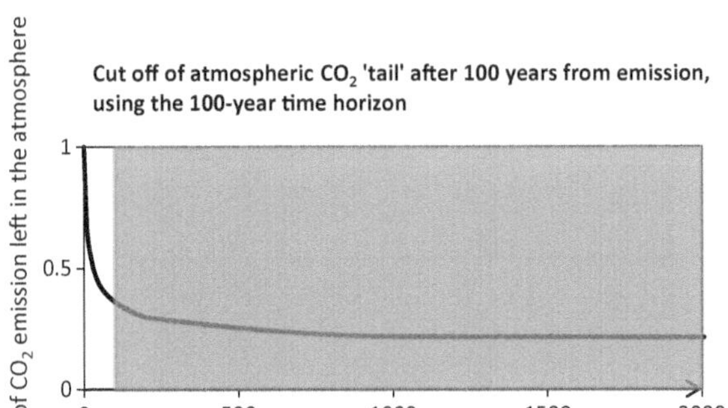

Fig. 30.2 Implication of the 100-year time horizon. The *black line* indicates the fraction of a CO_2 emission left in the atmosphere as a function of time after emission, whereas the *grey area* represents the 'tail' of the emission, which is cut off due to the use of the 100-year time horizon (modified from: Jørgensen and Hauschild 2013)

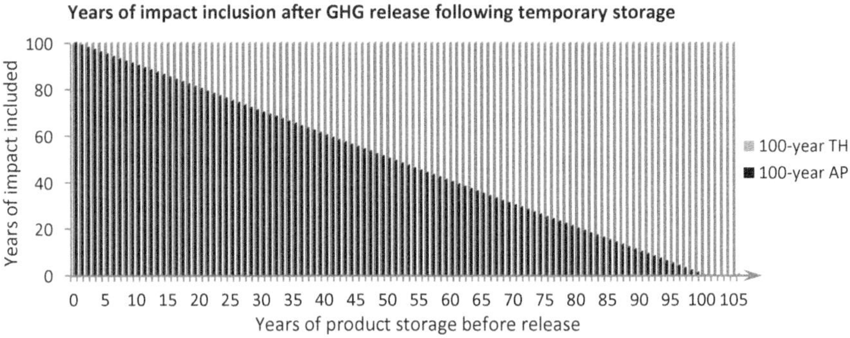

Fig. 30.3 Difference in using a 100-year time horizon (TH) and a 100-year accounting period (AP). Inclusion of impacts after a GHG emission, following a certain time of temporary carbon storage in bio-products depending on whether a 100-year TH or AP is used

The use of a 100-year accounting period can thus not be justified with reference to the normal use of a 100-year time horizon in LCIA of climate change impacts. As the choice is essential for the outcome, it should be ensured that it properly reflects the real climate change mitigation value of temporarily storing carbon (e.g. Jørgensen and Hauschild 2013; Brandão et al. 2013).

Another time-related issue is the lifetime of the product, which determines the length of temporary carbon storage and thus influences the potential value of the storage relative to the product carbon footprint. Since product lifetimes are generally very short compared to the duration of impacts from GHG emissions, temporary carbon storage does not change much in terms of long-term climate impacts (Jørgensen and Hauschild 2013). Note that terminology for above-discussed types of time horizons differ in the literature, and may often not be clearly distinguished.

Tipping Points

The issue of long-term climate change implications is, however, not the only relevant issue when considering the potential role of carbon storage in mitigating climate change. Due to increased global warming, there is also an issue of urgency, with the risk of passing critical climate change levels, so-called climatic 'tipping points'. A tipping point is a point where a structural change occurs in a system, which starts a 'chain reaction', meaning that it is no longer external forcing, but rather internal mechanisms in the system, which drives the process of change (IPCC 2007). Crossing such a tipping point is expected to lead to dramatic climate system changes which may be virtually irreversible (IPCC 2007), meaning that it is not realistic to return to the situation as it was before crossing the tipping point.

Mitigation of the rise in atmospheric GHG concentration is urgently required if the passing of expected climatic tipping points should be avoided. Thus, even short-term storage of carbon may have value in terms of climate change mitigation, if it can add to staying below such climatic tipping points, until more permanent solutions are reached (Jørgensen et al. 2015). It is therefore suggested to distinguish between long-term and short-term impacts when addressing the climate change mitigation value of temporary carbon storage (Jørgensen and Hauschild 2013; Jørgensen et al. 2015).

30.3.2 Biodegradability

There is a widespread misconception that all biomaterials are biodegradable, making them more sustainable than, and distinguishing them from, petrochemical materials. There are several mistakes in this belief.

In the first instance, biomaterials are not necessarily biodegradable, while some petrochemical materials can be biodegradable. Biodegradability is a material property, which depends on the molecular structure of the material, not the feedstock (PlasticsEurope 2013). In fact, materials can be identical once produced, regardless of whether they were made from a biomass or petrochemical feedstock.

Second, biodegradability does not inherently equal sustainability. Even if biodegradability does increase environmental sustainability of a certain material in the disposal stage, there may be other impacts in the life cycle of that biodegradable product that cancel out or overshadow the benefit. However, in some cases, being biodegradable may also in itself add no benefit, or may even have a damaging effect, relative to the environmental sustainability profile. In some countries, plastic is incinerated to produce heat and power, thereby replacing alternative fuels that are often fossil. The concept of using biomass first for products and subsequently for energy recovery through incineration is referred to as 'carbon cascading' (Weiss et al. 2012). If instead composting biodegradable materials, the benefits from the composting need to be counterbalanced against what is lost from the heat and power recovery from waste incineration. A study of the biopolymer polylactic acid (PLA) has shown that product-specific GHG emissions may vary by 20% depending on whether or not energy recovery from incineration is included (Hermann et al. 2011). In case biodegradable materials end up being incinerated, which is likely under the current waste management situation in a country like Denmark, it does not matter whether they are biodegradable.

If biodegradation takes place under anaerobic conditions, the carbon in the material will be converted into a mixture of methane and CO_2 by digestion, rather than just converted to CO_2 as under aerobic conditions in e.g. composting. If anaerobic digestion takes place in a biogas plant or a landfill with capture of landfill gases, the methane can be used to substitute fossil energy. However, if taking place in landfills where the methane is not captured, or in poorly managed home composting systems leading to anaerobic conditions (Song et al. 2009), as applicable in many places in the world, the methane production may lead to higher overall climate change impacts from biodegradable materials than for non-biodegradable ones, due to the relative high characterisation factor of methane compared to CO_2

(e.g. Patel et al. 2005). When conducting studies on biodegradable polymers, composting is often considered as the waste handling method, whereas the option of digestion is rarely included (Patel et al. 2005). Some studies suggest that composting biodegradable materials can be more beneficial than incineration, if the compost is used for agricultural carbon soil replenishment (e.g. Weiss et al. 2012; Hermann et al. 2011).

For the issue of potential storage of carbon in the composted material, however, it is most likely that between 80% and all of the initially sequestered carbon will be released during composting of biodegradable polymers, due to their ability to rapidly decompose (Patel et al. 2005) in which case there will not be much left for soil replenishment and storage in the soil. Furthermore, composting biomaterials only have a nutritional value for the soil if the biomaterial in question includes nutrients such as nitrogen, phosphorous or potassium, which is normally not the case. If, on the other hand the biomaterial does include nitrogen, then composting includes the risk of potential emissions of N_2O, which due to its large global warming potential can reverse any benefits from the compost (IPCC 2006). Finally, for using biodegradable products in compost and soil replenishment, all substances in the bio-products have to comply with requirements for this use, such as heavy metal thresholds (e.g. Song et al. 2009).

Biodegradability is therefore a property like many others, which may or may not have a positive impact on the environmental profile of a product when considering all life cycle impacts under the relevant circumstances.

30.4 Land Use Change, LUC

Three major drivers for land use change (LUC) exist globally: (1) Timber harvesting, (2) Infrastructure development, (3) Agricultural/horticultural expansion (Kim et al. 2009). Often, the LUC is due to a combination of the above. Biofuels and biomaterials are by default directly linked to the agri- and horticultural sectors and direct and indirect land use change will in most cases contribute significantly to the life cycle impacts of bio-based products. In some cases, LUC impacts can be higher than the impacts of the rest of the life cycle impacts combined. Potential LUCs related to bio-product crops are:

1. Forest or other virgin land to bio-product crop
2. Agricultural crop to bio-product crop
3. Fallow land to bio-product crop
4. Urban land to bio-product crop (happens only on a very small scale and is not dealt with here)

30.4.1 Direct and Indirect Land Use Change

Direct land use change (DLUC) is the man-made conversion of a piece of land from a previous land use (like another crop type or forest) to a bio-product crop.

Indirect land use change (ILUC) occurs when land carrying a crop 'A' is converted to a bio-product crop and market demand for product(s) that are derived from the earlier crop 'A' (directly or through other indirect crop displacements) drives conversion of marginal land elsewhere to meet that demand. This process is shown in Fig. 30.4. DLUC is often followed by ILUC if the DLUC takes place on previously cultivated land. Marginal land describes the land ultimately most likely to be converted. Unless there is an unlikely decline in demand for agricultural land in the region, the marginal land will be virgin land like forest or abandoned land like secondary forest or fallow.

Example 30.3 If rapeseed oil in Europe is used for biodiesel production, less vegetable oil will be available on the market. Palm oil, which is produced mainly in Malaysia and Indonesia, is currently the cheapest and fastest growing vegetable oil crop in the world and as such it is the marginal oil crop (i.e. the oil most likely to be produced if vegetable oil demand increases). Thus, the supply deficit in vegetable oil when rapeseed oil is used as fuel will most likely be covered by additional production of palm oil in Malaysia and Indonesia. The marginal land in Malaysia and Indonesia is to a large extent forest, so the production of biodiesel from rapeseed in Europe will likely result in deforestation in South East Asia. Note that in this example no direct land use change occurs when producing fuel from rapeseed oil as only the use of the crop changes.

As ILUC is highly dependent on market forces, and it creates scenarios that are beyond the influence of farmers and bio-product producers, it becomes highly uncertain to predict or model, and a formal consensus has not been reached for a methodology to include ILUC in LCA.

30.4.2 Impacts of Land Use Change

Historical and current LCA literature on bio-products focuses mainly on carbon emissions from LUC, but other impacts like biodiversity loss, water use impacts and biogeophysical impacts like albedo changes are starting to attract more interest.

Biodiversity

The loss of biodiversity is a major environmental concern, and habitat loss and (local) species extinction due to a changed land use is expected to be the main driver of biodiversity changes in terrestrial ecosystems (Sala et al. 2000). However,

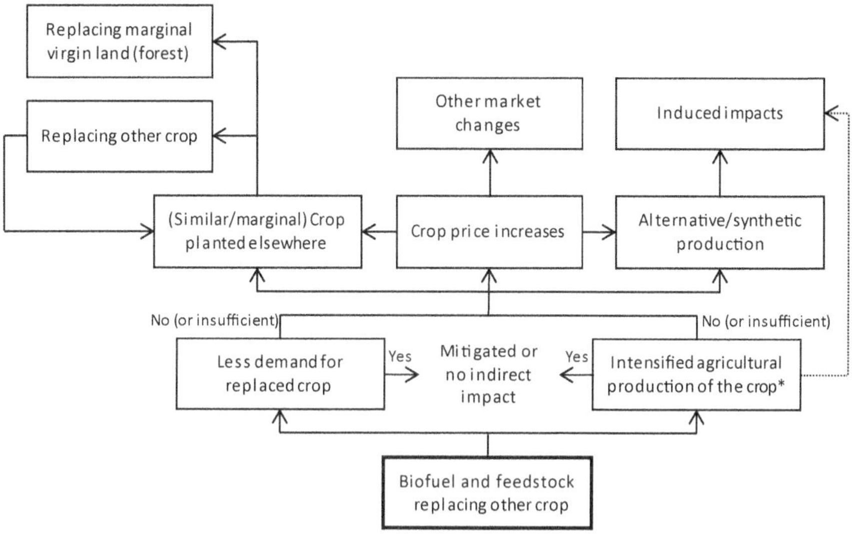

Fig. 30.4 Potential ILUC effects can be very complex making accurate predictions very difficult. The figure provides a simplified overview of the potential consequences of LUC. ILUC can be avoided or mitigated by less demand for—or intensified agricultural production of—the replaced crop. *Asterisk* Intensified agricultural production can induce impacts through e.g. increased use of chemicals and soil degradation

approaches on how to handle the consideration of biodiversity impacts in LCA are still on a rather preliminary level (e.g. Chaudhary et al. 2015; de Baan et al. 2013; Koellner et al. 2013; Michelsen 2008). The current lack of consensus on an assessment method (which over time may be solved by the UNEP-SETAC Life Cycle Initiative's effort to establish a global consensus on LCIA models and methods) and the lack of geographical specific data are main obstacles for obtaining quantitative and reliable results of biodiversity impacts in LCA. However, due to the importance of this aspect, obtaining qualitative or rough quantitative results of best available approaches is considered better than disregarding biodiversity impacts.

Albedo

Albedo is a measure of how much of the incoming solar radiation is reflected by a surface. Albedo values vary with seasons and differ a lot depending on the type of land cover. Surfaces covered in snow and ice thus have much higher albedo than darker surfaces, for instance areas covered with forest (e.g. Cherubini et al 2012a). The effect of increased albedo is cooling, whereas decreased albedo of a surface leads to warming. As LUC can lead to substantial changes in the albedo of an area, which can play an important role in terms of climate impacts, albedo changes is increasingly included in climate impact assessments of biofuel systems (e.g. Bright et al. 2012a, b; Cherubini et al. 2012a). Reliability of results depends on availability of relevant geographical data for similar land uses. Other biogeophysical effects exist as well, but on a global scale the albedo effect is the dominating direct climate forcing of these, especially in locations with seasonal snow cover (Claussen et al. 2001; Randerson et al. 2006; Bala et al. 2007).

30.4.3 Carbon Pools

Through the conversion of land a number of carbon pools are affected (Germer and Sauerborn 2008):

- Above ground biomass: All living plants above ground
- Below ground biomass: All living biomass in the soil; e.g. roots
- Litter and deadwood: Fallen trees, leaves and branches
- Soil carbon: Organic carbon residues left in the soil after degradation of biomass

The various carbon pools vary greatly depending on climate, vegetation and soil type. The net carbon balance for LUC is the difference in the four carbon pools between the previous land use and the bio-product crop land use. LUC can thus result in net emission as well as a net sequestration of carbon depending on the carbon stored in the previous land use and the bio-product crop land use. Note that carbon sequestration credits are the difference in carbon stock at the beginning and the end of the assessment time horizon. If the land is cleared in the last crop cycle of the assessment time horizon then no permanent carbon sequestration should be credited. Thus, unless there is documented reason to believe that the land will be left with a certain carbon pool at the end of the assessment time horizon then no permanent carbon sequestration should be credited. In the case of plantations, which are replanted, e.g. every 30 years, it can be argued that there is temporary carbon storage for these 30 years. However, as the carbon is sequestered and stored approximately linearly over the 30 year period, the average retention of the carbon is only half of that, i.e. 15 years. If the goal of the study requires inclusion of temporary carbon storage, then the temporary carbon storage model from ILCD (European Commission 2010) could be applied (see Sect. 30.3.1). Applying this, the temporary carbon storage in the plantations should be credited 15% of the full carbon storage potential (see application in e.g. Hansen et al. 2014). It must be clearly stated that the storage is biogenic. If organic residues of the bio-product crop (e.g. trunks) are treated and stored in a product thereby preventing the release of the fixed carbon, then this temporary carbon storage should not be inventoried under LUC, but rather under the production stage of the bio-product, in which the residue is generated.

Although biomass is often considered CO_2 neutral as the carbon has been captured recently, this is not the case if the biomass is from virgin sources, such as virgin forest, where there has been carbon equilibrium for thousands or millions of years, or if such land has been cleared for enabling the production of the biomass. So even though the individual trees/plants may not be old, the carbon stored in the land must be considered a permanent carbon pool, which is lost if the forest is converted to agriculture. All virgin forest conversion emissions should thus be inventoried as fossil carbon dioxide (European Commission 2010). Before a forest is converted to agricultural land it is most often logged for valuable timber. The extraction of trees suitable for timber has a market value and can in attributional LCA be considered as belonging to a separate product system. The carbon stored in the timber trees should thus not be counted in the LUC emissions in the bio-product system. In a consequential approach it is necessary to consider whether the timber is substituting timber harvesting somewhere else, in which case the carbon stored in a similar quantity of marginal timber can be subtracted from the total carbon loss from the LUC including the timber. Special local conditions, which are not covered specifically in the ILCD Handbook (European Commission 2010), can occur as per following example:

Example 30.4 In the case of palm oil derived biodiesel from Malaysia, the Malaysian government has selected areas still under forest cover for timber extraction and future development, e.g. oil palm plantations. These areas have already been or will be harvested for timber by selective logging. Clear cutting is only used if a plantation is established immediately after the logging. The extracted timber would thus be cut whether an oil palm plantation is established or not and thus felling of that timber in the virgin forest should in this case not be allocated to the palm oil biodiesel, but to the timber production. The emissions from a logged-over forest would thus be more suitable to use in the assessment of the palm oil production. However, if the logged-over forest was left idle it would likely recover and the actual loss of carbon when converting to oil palm is thus that of a recovered logged forest. ILCD recommends that the emissions from land use change should be allocated to the first 20 years of agricultural use or in the case of plantations the first planting cycle, which is often 20–30 years (more details on timing issues can be found in Sect. 30.4.4). Applying this amortization period, the emissions from conversion of forest (which is harvested for timber before conversion) should be those of a logged-over forest after 20–30 years of recovery. The emissions from the logged-over forest should still be counted as fossil emissions as per above whereas the carbon sequestered during recovery is biogenic. Note that with a few changes in details or assumptions (e.g. that the logged forest dies rather than recovers), this example could turn out differently.

ILCD (European Commission 2010) labels the emissions from conversion of secondary forests, crops and plantations to bio-products crops as biogenic carbon dioxide.

The IPCC (2006) provides generic carbon data for various land uses under various climatic conditions. It is recommended to use IPCC data when no site or regional specific data is available (European Commission 2010). However, it must be stressed that—as also highlighted in IPCC (2006)—carbon stocks even in forests of similar type and geography vary significantly and using generic/mean values would result in large uncertainties.

30.4.4 Timing issues and Payback Time

Immediate Emissions

Immediate LUC emissions are emissions taking place during the conversion of the land. Above and below ground biomass as well as litter and deadwood are included. In accordance with the ILCD (European Commission 2010) guidelines, LUC emissions for conversion to sub-annual, annual and bi-annual crops should be attributed to the first 20 years of the new land use, unless it is known that the new land use period will be shorter than 20 years. Thus, for any given year the production at that piece of land should be allocated 1/20 of the total immediate LUC emissions. For a plantation with trees or palms, it is recommended that the emissions are allocated to the first cycle of trees/palms, i.e. 20–30 years.

Long-Term Emissions

The degradation or sequestration of soil carbon takes place over a number of years until a new equilibrium is established. As for the immediate emissions, it is recommended to allocate the emissions/sequestration over 20 years unless the actual period, when 90% of the emissions/sequestration have occurred is known, in which case that number is used as n in Eq. (30.2) (European Commission 2010). As opposed to the immediate emissions, the long-term emissions cannot be allocated linearly over the allocation period. Instead, larger fractions of the total emissions are allocated to the earlier years:

$$x = (100 * 2)/(n + 1) * (n - i)/n, \tag{30.2}$$

where

- x is the % of the total carbon inventory allocated to year i
- n is the amortization period for the total carbon allocation, e.g. 20 years
- i is the number of years after the LUC ($i \leq n$)

Example 30.5 Peat soils are soils with very high organic contents. They are created through biomass accumulation under water logged, anoxic conditions, which slows the organic degradation and makes the accumulation rate higher than the degradation rate. When peat land is converted to agricultural land the peat is drained, thus exposing the peat to oxygen and increasing the rate of decomposition. Without proper land management, the oxidation of the peat can continue for several decades and emit carbon dioxide in quantities several times larger than the immediate emissions from land clearing depending on the site specific soil conditions. Special considerations must thus be given to the management practices as well as site specific soil conditions when conducting an assessment of a bio-product crop planted on peat soil. Whereas the largely homogenous peat in temperate climate regions has been studied intensively, the large localised variations in tropical peat composition as well as effects of diverse management practices of cultivated peat land have not yet been fully understood in terms of potential and actual carbon emissions. The large ranges of emissions from cultivated tropical peat are evident from IPCC (2014) and these ranges may increase further as more data is produced from the numerous peat composition and management practices scenarios. Note that peat land conversion to palm oil plantation is not uncommon in, e.g. Indonesia. As palm oil is the marginal vegetable oil on the world market, this can have impacts on all other biodiesel feedstocks through ILUC, as explained in Sect. 30.4.1.

For both immediate emissions and long-term emissions no allocation to individual years is needed if the assessment uses an average of the emissions from the bio-product crop of a period equal to or longer than the amortization period. In that case, the total LUC and ILUC emissions are simply added to the rest of the life cycle emissions (European Commission 2010).

Payback Time

Payback time is often used to describe the time it takes a bio-product to pay back the carbon emissions related to LUC; the so-called carbon debt. Almost all bio-products save carbon emissions compared to fossil products when (indirect) LUCs are not taken into consideration, but the picture can change when including these. By dividing the total LUC carbon emissions for a hectare of land converted to bio-product crop production with the carbon savings from replacing the fossil products with bio-products produced from that hectare for a year (not including LUC emissions) it is calculated how long it will take before the bio-product system actually starts saving carbon emissions.

Example 30.6 A land use change has resulted in a total emission of 50 tonne CO_2-eq/ha. 1 tonne biodiesel is produced per ha/year and the carbon emissions from the biodiesel production are 1.7 tonne CO_2-eq/tonne biodiesel. The emissions for extraction, refining and combustion of fossil diesel are 2.7 tonne CO_2-eq/tonne biodiesel equivalent (biodiesel has a lower heating value than fossil diesel, so less than 1 tonne diesel is needed per tonne biodiesel equivalent). The net saving for the biodiesel production is thus 1 tonne CO_2-eq/tonne biodiesel. Paying back the land use change emissions will thus take 50 years. Only then will the biofuel system actually start saving CO_2 emissions. (Note that values used in the example are generally realistic values but can vary from case to case.)

30.5 Concluding Remarks and Perspectives

- *Competition for Biomass*

Due to limitations in land availability, only the most optimistic outlooks expect availability of enough sustainable biomass to cover future replacement of all replaceable fossil fuels and products with bio-products, while still covering demands for food and feed of a growing population (see e.g. Jørgensen 2014). Therefore, the biomass use needs to be based on prioritisation.

- *Integrated Biorefineries*

As mentioned in Sect. 30.2, the biorefinery concept is already to some extend existing today and breakthroughs within this area are expected in the coming years (Weiss et al. 2012). Biorefineries are expected to provide the optimal utilisation of biomass for production of both biofuels and biomaterials, resulting in a maximised utilisation of the biomass feedstock. However, there are today still significant challenges in reducing production cost.

- *Maturity of Competing Technologies*

While the petrochemical industry has been optimised through a long time period and is based on mature technology, production of many bio-products is still relying on rather new technologies. Some types of biofuels can to a certain extent make use of conventional processes, but this is often not the case for biomaterials (Weiss et al. 2007). Furthermore, many bio-products are not yet available in commercial scale. Thus, there is a rather large potential for optimising efficiency, while substantial improvements are also expected for the integrated production of biofuels and biomaterials in biorefineries (Patel et al. 2005).

Major reduction potentials in environmental impacts from biomass feedstock production have also been identified to be obtainable if changing agricultural management practices. For example, changing to no-tillage farming, including cultivation of winter crops or changing to extensive farming, as well as utilising agricultural residues for biofuel or biomaterial production, except the fraction that is essential for preserving soil organic carbon (Weiss et al. 2012). However, it should be kept in mind that some changes in farming practices can also lead to lower productivity and thus need for additional land use.

- *Future Bio-product Feedstocks*

As mentioned, most current general conclusions on environmental sustainability of bio-products relative to fossil products relate to 1G, whereas it is expected that the environmental performance of 2G will have substantial potential to improve. The potentials lie both in reduced GHG emissions and non-renewable energy use when substituting fossil products, compared to 1G (e.g. Dornburg et al. 2008) and in reduced impacts in terms of water use and water quality, unless the feedstock crops are from irrigated plantations (Gnansounou 2010). Compared to feedstocks for 1G bio-products, feedstocks for 2G bio-products generally have lower need for agricultural input such as fertiliser, pesticides and irrigation (e.g. Dohleman et al. 2010) have higher soil organic carbon sequestration potentials (e.g. Anderson-Teixeira et al. 2009) and can reduce erosion (e.g. Somerville et al. 2010). Furthermore, feedstock for 2G bio-products can decrease the land use compared to that for 1G bio-products, and especially the use of agricultural land. A major part of the reason for improvement expectations for 2G is that it can utilise biomass feedstock types which can grow on land not suited for agricultural production, as well as agricultural and forest residues. Such feedstocks are not competing with global food production, and in the case of residues, this feedstock is in some cases even seen as a waste product. However, currently much of such residues are left in the fields/forests, adding to soil carbon replenishment. The aspect of soil carbon replenishment should thus be considered when assessing the sustainably available amount of crop residues, e.g. by considering a minimum amount of residues to either be left in the field or be earmarked for biochar production through pyrolysis, which also produces bio-oil. Biochar is mostly non-degradable carbon, which will potentially stay in the soil for millennia to avoid soil degradation and act as carbon sequestration (Lehmann 2007).

Bio-products based on algae biomass (3G) also present some interesting perspectives. Promising features of algae biomass as feedstock for bio-products are that they do not compete for agricultural land, there is a large production potential, and CO_2 capture and biomass yields are high. Furthermore, a range of products are available from microalgae, including hydrogen, biogas, biodiesel, bioethanol and other starch based bio-products (Posten and Schaub 2009). However, there are also challenges, e.g. in terms of power requirements and production cost, and improvements are needed in order to make algae as feedstock for bio-based production a sustainable and commercially viable reality (e.g. Sander and Murthy 2010). Also, algae feedstock may still show higher impacts in most environmental impact categories, including fossil energy use and GHG emissions, compared to conventional crops (Clarens et al. 2010). As it is a rather new technology, it is hard to predict the actual potential.

- *Reference Fossil Product Feedstock*

Conventional petrochemical reference products are typically oil-based. However, as conventional crude oil resources are moving towards depletion, other petrochemical options will take over in the future. Shale gas and oil sands have received increased attention as oil prices have made these options feasible to exploit. These feedstocks, however, seem to have much larger impacts on the environment than conventional oil. Production of synthetic crude oil from oil sands is generally reported to have substantially higher GHG emission impacts per barrel compared to production of conventional crude oil, however large variations in results exist (e.g. National Energy Technology Laboratory 2008; Charpentier et al. 2009), and the life cycle water consumption for shale gas and oil sands is approximately double that of conventional oil (King and Webber 2008), while also affecting quality and availability of land (Jordaan 2012). Thus, if these are the fossil resources that will become the future conventional feedstock against which to compare the bio-products, the latter will become more competitive in terms of environmental performance.

Acknowledgements The authors gratefully acknowledge the helpful inputs from Arnaud Hélias (Montpellier SupAgro/INRA) and Anthony Benoist (CIRAD).

References

Anderson-Teixeira, K.J., Davis, S.C., Masters, M.D., Delucia, E.H.: Changes in soil organic carbon under biofuel crops. GCB Bioenergy **1**, 75–96 (2009)

Archer, D., Kheshgi, H., Maier-Reimer, E.: Multiple timescales for neutralization of fossil fuel CO_2. Geophys. Res. Lett. **24**(4), 405–408 (1997)

Bala, G., Caldeira, K., Wickett, M., Phillips, T.J., Lobell, D.B., Delire, C., Mirin, A.: Combined climate and carbon cycle-cycle effects of large-scale deforestation. PNAS **104**, 6550–6555 (2007)

Bessou, C., Ferchaud, F., Gabrielle, B., Mary, B.: Biofuels, greenhouse gases and climate change. In: Lichtfouse, E., et al. (eds.) Sustainable Agriculture, vol. 2 (2011). Springer, New York—EDP Sciences 2011. doi: 10.1007/978-94-007-0394-0_20

Brandão, M., Levasseur, A., Kirschbaum, M.U.F., Weidema, B.P., Cowie, A.L., Jørgensen, S.V., Hauschild, M.Z., Pennington, D.W., Chomkhamsri, K.: Key issues and options in accounting for carbon sequestration and temporary storage in life cycle assessment and carbon footprinting. Int. J. Life Cycle Assess. **18**, 230–240 (2013)

Brehmer, B., Boom, R.M., Sanders, J.: Maximum fossil fuel feedstock replacement potential of petrochemicals via biorefineries. Chem. Eng. Res. Des. **87**(9), 1103–1119 (2009)

Bright, R.M., Cherubini, F., Strømman, A.H.: Climate impacts of bioenergy: Inclusion of carbon cycle and albedo dynamics in life cycle impact assessment. Environ. Impact Assess. Rev. **37**, 2–11 (2012a)

Bright, R.M., Cherubini, F., Astrup, R., Bird, N., Cowie, A.L., Ducey, M.J., Marland, G., Pingoud, K., Savolainen, I., Strømman, A.H.: A comment to "Large-scale bioenergy from additional harvest of forest biomass is neither sustainable nor greenhouse gas neutral": important insights beyond greenhouse gas accounting. GCB Bioenergy **4**(6), 617–619 (2012b)

Carus, M., Dammer, L.: Food or non-food: which agricultural feedstocks are best for industrial uses? Nova-Institut GmbH, nova paper#2 on bio-based economy 2013-07 (2013)

Carus, M., Baltus, W., Carrez, D., Kaeb, H., Ravenstijn, J., Zepnik, S.: Market study on bio-based polymers in the world, capacities, production and applications: Status Quo and Trends towards 2020 (leaflet). Nova-Institut GmbH, Version 2013-07 (2013)

Charpentier, A.D., Bergerson, J.A., MacLean, H.A.: Understanding the Canadian oil sands industry's greenhouse gas emissions. Environ. Res. Lett. 4 (2009)

Chaudhary, A., Verones, F., de Baan, L., Hellweg, S.: Quantifying land use impacts on biodiversity: combining species-area models and vulnerability indicators. Environ. Sci. Technol. **49**, 9987–9995 (2015)

Cherubini, F.: The biorefinery concept: using biomass instead of oil for producing energy and chemicals. Energy Convers. Manage. **51**, 1412–1421 (2010)

Cherubini, F., Jungmeier, G.: LCA of a biorefinery concept producing bioethanol, bioenergy, and chemicals from switchgrass. Int. J. LCA **15**(1), 53–66 (2010)

Cherubini, F., Strømman, A.H.: Life cycle assessment of bioenergy systems: state of the art and future challenges. Bioresour. Technol. **102**, 437–451 (2011)

Cherubini, F., Bird, N.D., Cowie, A., Jungmeier, G., Schlamadinger, B., Woess-Gallasch, S.: Energy and greenhouse gas-based LCA of biofuel and bioenergy systems: key issues, ranges and recommensations. Resour. Conserv. Recycl. **53**, 434–447 (2009)

Cherubini, F., Bright, R.M., Strømman, A.H.: Site-specific global warming potentials of biogenic CO_2 for bioenergy: contributions from carbon fluxes and albedo dynamics. Environ. Res. Lett. **7**, 045902 (2012a)

Cherubini, F., Guest, G., Strømman, A.H.: Application of probability distributions to the modeling of biogenic CO_2 fluxes in life cycle assessment. GCB Bioenergy **4**, 784–798 (2012b)

Clarens, A.F., Resurreccion, E.P., White, M.A., Colosi, L.M.: Environmental life cycle comparison of algae to other bioenergy feedstocks. Environ. Sci. Technol. **44**, 1813–1819 (2010)

Claussen, M., Brovkin, V., Ganopolski, A.: Biogeophysical versus biogeochemical feedbacks of large-scale land cover change. Geophys. Res. Lett. **28**, 1011–1014 (2001)

Clift, R., Brandão, M.: Carbon Storage and Timing of Emissions. Centre for Environmental Strategy, University of Surrey, GU2 7XH (2008). ISSN: 1464-8083

Collet, P., Hélias, A., Lardon, L., Steyer, J.-P., Bernard, O.: Recommendations for life cycle assessment of algal fuels. Appl. Energy **154**, 1089–1102 (2015)

Collet, P., Spinelli, D., Lardon, L., Hélias, A., Steyer, J.P., Bernard, O.: Life-Cycle Assessment of Microalgal-Based Biofuels. In: Pandey, A., Lee, D.-J., Chisti, Y., Soccol, C.R. (eds.) Biofuels from Algae, pp. 287–312. Elsevier, Amsterdam (2013)

Crutzen, P.J., Mosier, A.R., Smith, K.A., Winiwarter, W.: N_2O release from agro-biofuel production negates global warming reduction by replacing fossil fuels. Atmos. Chem. Phys. **8**, 389–395 (2008)

Davis, S.C., Anderson-Teixeira, K.J., DeLucia, E.H.: Life-cycle analysis and the ecology of biofuels. Trends Plant Sci. **14**(3), 140–146 (2009)

de Baan, L., Alkemade, R., Koellner, T.: Land use impacts on biodiversity in LCA: a global approach. Int. J. Life Cycle Assess. **18**, 1216–1230 (2013)

Dohleman, F.G., Heaton, E.A., Long, S.P.: Perennial Grasses as second-generation sustainable feedstocks without conflict with food production. In: Khanna, M., et al. (eds.) Handbook of Bioenergy Economics and Policy Natural Resource Management and Policy, vol. 3. Springer, New York (2010). doi:10.1007/978-1-4419-0369-3_3

Dornburg, V., Lewandowski, I., Patel, M.: Comparing the land requirements, energy savings, and greenhouse gas emissions reduction of biobased polymers and bioenergy: An analysis and system extension of life-cycle assessment studies. J. Ind. Ecol. **7**(3–4), 93–116 (2003)

Dornburg, et al.: Scenario projections for future market potentials of biobased bulk chemicals. Environ. Sci. Technol. **42**, 2261–2267 (2008)

EIA.: US Energy Information Administration (2012) as presented by Index Mundi on www.indexmundi.com/energy.aspx. Accessed on 23 July 2012

European Commission.: International Reference Life Cycle Data System (ILCD) Handbook—General guide for Life Cycle Assessment—Detailed guidance, 1st edn. March 2010. EUR 24708 EN. European Commission—Joint Research Centre—Institute for Environment and Sustainability. Publications Office of the European Union, Luxembourg (2010)

European Parliament.: Directive 2009/28/EC of the European Parliament and of the Council. Off. J. Eur. Union L 140/16–62 (2009)

Germer, J., Sauerborn, J.: Estimation of the impact of oil palm plantation establishment on greenhouse gas balance. Environ. Dev. Sustain. **10**, 697–716 (2008)

Gnansounou, E.: Production and use of lignocellulosic bioethanol in Europe: Current situation and perspectives. Bioresour. Technol. **101**, 4842–4850 (2010)

Hansen, S.B., Olsen, S.I., Ujang, Z.: Carbon balance impacts of land use changes related to the life cycle of Malaysian palm oil-derived biodiesel. Int. J. Life Cycle Assess. **19**, 558–566 (2014)

Hermann, B.G., Debeer, L., deWilde, B., Blok, K., Patel, M.K.: To compost or not to compost: LCA of biodegradable materials' waste treatment. Polym. Degrad. Stab. **96**(6), 1159–1171 (2011)

IPCC.: 2006 IPCC Guidelines for National Greenhouse Gas Inventories. National Greenhouse Gas Emissions Programme, IGES (2006)

IPCC.: Climate Change 2007—the physical science basis. In: Solomon, S., Qin, D., Manning, M., Chen, Z., Marquis, M., Averyt, K.B., Signor, M., Miller, H.L. (eds.) Contribution of Working Group I to the Fourth Assessment Report of the Intergovernmental Panel on Climate Change (IPCC). Cambridge University Press, Cambridge, New York (2007)

IPCC.: 2013 Supplement to the 2006 IPCC Guidelines for National Greenhouse Gas Inventories: Wetlands. In: Hiraishi, T., Krug, T., Tanabe, K., Srivastava, N., Baasansuren, J., Fukuda, M., Troxler, T.G. (eds.). IPCC, Switzerland (2014)

Jordaan, S.M.: Land and water impacts of oil sands production in Alberta. Environ. Sci. Technol. **46**, 3611–3617 (2012)

Jørgensen, S.V., Hauschild, M.Z.: Need for relevant timescales when crediting temporary carbon storage. Int. J. Life Cycle Assess. **18**, 747–754 (2013)

Jørgensen, S.V.: Environmental assessment of biomass based materials: with special focus on the climate effect of temporary carbon storage. Ph.D. thesis, Department of Management Engineering, Technical University of Denmark (2014)

Jørgensen, S.V., Hauschild, M.Z., Nielsen, P.H.: The potential contribution to climate change mitigation from temporary carbon storage in biomaterials. Int. J. Life Cycle Assess. **20**, 451–462 (2015)

Kim, H., Kim, S., Dale, B.E.: Biofuels, land use change, and greenhouse gas emissions: some unexplored variables. Environ. Sci. Technol. **43**, 961–967 (2009)

King, C., Webber, M.: Water intensity of transportation. Environ. Sci. Technol. **42**(21), 866–7872 (2008)

Koellner, T., de Baan, L., Beck, T., Brandão, M., Civit, B., Margni, M., Milà i Canals, L., Saad, R., de Souza, D.M., Müller-Wenk, R.: UNEP-SETAC guideline on global land use impact assessment on biodiversity and ecosystem services in LCA. Int. J. Life Cycle Assess. **18**, 1188–1202 (2013)

Lardon, L., Hélias, A., Sialve, B., Steyer, J.-P., Bernard, O.: Life-cycle assessment of biodiesel production from microalgae. Environ. Sci. Technol. **43**, 6475–6481 (2009)

Lehmann, J.: A handful of carbon. Nature **447**, 143–144 (2007)

Luo, L., Voet, E., Huppes, G., Udo de Haes, H.A.: Allocation issues in LCA methodology: a case study of corn stover-based fuel ethanol. Int. J. Life Cycle Assess. **14**, 529–539 (2009)

Majer, S., Mueller-Langer, F., Zeller, V., Kaltschmitt, M.: Implications of biodiesel production and utilization on global climate—a literature review. Eur. J. Lipid Sci. Technol. **111**, 747–762 (2009)

Malca, J., Freire, F.: Life-cycle studies of biodiesel in Europe: a review addressing the variability ofresults and modeling issues. Renew. Sustain. Energy Rev. **15**, 338–351 (2011)

Mascia, P.N., et al. (eds.) Plant Biotechnology for Sustainable Production of Energy and Co-products. Biotechnology in Agriculture and Forestry 66, Springer, Berlin (2010). doi:10.1007/978-3-642-13440-1_15

Michelsen, O.: Assessment of Land use impact on biodiversity. Proposal of a new methodology exemplified with forestry operations in Norway. Int. J. Life Cycle Assess. **13**, 22–31 (2008)

Moura-Costa, P.: Carbon accounting, trading and the temporary nature of carbon storage. The Nature Conservancy U.S. (2002)

National Energy Technology Laboratory (NETL):. Development of Baseline Data and Analysis of Life Cycle Greenhouse Gas Emissions of Petroleum-Based Fuels. DOE/NETL-2009/1346 (2008)

Patel, M., Bastioli, C., Marini, L., Wuerdinger, E.: Life cycle assessment of bio-based polymers and natural fiber composites. Biopolymers (2005, online)

PlasticsEurope.: The plastics portal, Q&As (2013). Accessible at http://www.plasticseurope.org/what-is-plastic/types-of-plastics-11148/bio-based-plastics/qas.aspx. Accessed Feb 2016

Posten, C., Schaub, G.: Microalgae and terrestrial biomass as source for fuels—a process view. J. Biotechnol. **142**, 64–69 (2009)

Randerson, J.T., Liu, H., Flanner, M.G., Chambers, S.D., Jin, Y., Hess, P.G., Pfister, G., Mack, M.C., Treseder, K.K., Welp, L.R., Chapin, F.S., Harden, J.W., Goulden, M.L., Lyons, E., Neff, J.C., Schuur, E.A.G., Zender, C.S.: The impact of boreal forest fire on climate warming. Science **314**, 1130–1132 (2006)

Rothermel, J.: Raw material change in the chemical industry—The general picture. Presentation at HLG Chemicals—Working Group Feedstock, Energy & Logistics, February 7, 2008 Brussels (2008). Available at: http://ec.europa.eu/enterprise/sectors/chemicals/files/wg_7_8fer08/01rothermel_raw_material_change_en.pdf. Accessed Jan. 2014

Sala, O.E., Chapin III, F.S., Armesto, J.J., Berlow, E., Bloomfield, J., Dirzo, R., Huber-Sanwald, E., Huenneke, L.F., Jackson, R.B., Kinzig, A., Leemans, R., Lodge, D.M., Mooney, H.A., Oesterheld, M., Poff, N.L., Sykes, M.T., Walker, B.H., Walker, M., Wall, D.H.: Global biodiversity scenarios for the year 2100. Science **287**, 1770–1774 (2000)

Sander, K., Murthy, G.S.: Life cycle analysis of algae biodiesel. Int. J. Life Cycle Assess. **15**, 704–714 (2010)

Schnepf, R., Yacobucci, B.D.: Renewable fuel standard: overview and issues. CRS Report for Congress, Congressional Research Service (2012)

Singh, A., Pant, D., Korres, N.E., Nizami, A.S., Prasad, S., Murphy, J.D.: Key issues in life cycle assessment of ethanol production from lignocellulosic biomass: challenges and perspectives. Bioresour. Technol. **101**, 5003–5012 (2010)

Somerville, C., Youngs, H., Taylor, C., Davis, S.C., Long, S.P.: Feedstocks for lignocellulosic biofuels. Science **329**, 790–792 (2010)

Song, J.H., Murphy, R.J., Narayan, R., Davies, G.B.H.: Biodegradable and compostable alternatives to conventional plastics. Philos. Trans. R. Soc. B **364**, 2127–2139 (2009)

Tabone, M.D., Gregg, J.J., Beckman, E.J., Landis, A.E.: Sustainability metrics: life cycle assessment and green design in polymers. Environ. Sci. Technol. **44**, 8264–8269 (2010)

USDA: S Biobased Products Market Potential and Projections Through 2025, OCE-2008-01, US Department of Agriculture (2008)

Von Blottnitz, H., Curran, M.A.: A review of assessments conducted on bio-ethanol as a transportation fuel from a net energy, greenhouse gas, and environmental life cycle perspective. J. Clean. Prod. **15**(7), 607–619 (2007)

Wang, M.: Life-cycle analysis of biofuels. Biotechnol. Agric. For. **66**, 385–408

Weiss, M., Patel, M.K., Heilmeier, H., Bringezu, S.: Applying distance-to-target weighing methodology to evaluate the environmental performance of bio-based energy, fuels, and materials. Resour. Conserv. Recycl. **50**(3), 260–281 (2007)

Weiss, M., Haufe, J., Carus, M., Brandão, M., Bringezu, S., Hermann, B., Patel, M.K.: A review of the environmental impacts of biobased materials. J. Ind. Ecol. **16**, S169–S181 (2012)

Author Biographies

Susanne Vedel Hjuler has been working with LCA since the early 2010s. Particularly experienced within LCA of biomaterials, climate change related LCIA method development, as well as experienced LCA practitioner. Participated in the UNEP/SETAC task force on global warming LCIA. Main LCA interests are climate change, biomass use, as well as use of LCA in practice.

Sune Balle Hansen LCI expert working intermittently on applied LCA of biofuels and edible oils since 2004 with focus on palm oil and palm oil derived biodiesel. Special LCA research attention has been given to the inclusion of residue use and land use change aspects.

Chapter 31
LCA of Chemicals and Chemical Products

Peter Fantke and Alexi Ernstoff

Abstract This chapter focuses on the application of Life Cycle Assessment (LCA) to evaluate the environmental performance of chemicals as well as of products and processes where chemicals play a key role. The life cycle stages of chemical products, such as pharmaceuticals drugs or plant protection products, are discussed and differentiated into extraction of abiotic and biotic raw materials, chemical synthesis and processing, material processing, product manufacturing, professional or consumer product use, and finally end-of-life. LCA is discussed in relation to other chemicals management frameworks and concepts including risk assessment, green and sustainable chemistry, and chemical alternatives assessment. A large number of LCA studies focus on contrasting different feedstocks or chemical synthesis processes, thereby often conducting a cradle to (factory) gate assessment. While typically a large share of potential environmental impacts occurs during the early product life cycle stages, potential impacts related to chemicals that are found as ingredients or residues directly in products can be dominated by the product use stage. Finally, methodological challenges in LCA studies in relation to chemicals are discussed including the choice of functional unit, defining the system boundaries, quantifying emissions for many thousands of marketed chemicals, characterising emissions in terms of toxicity and other impacts, and finally interpreting LCA results. The chapter is relevant for LCA students and practitioners who wish to gain basic understanding of LCA studies of products or processes with chemicals as a key aspect.

P. Fantke (✉) · A. Ernstoff
Division for Quantitative Sustainability Assessment, Department of Management Engineering, Technical University of Denmark, 2800 Kongens Lyngby, Denmark
e-mail: pefan@dtu.dk

A. Ernstoff
EPFL Innovation Park, Quantis International, Bât. D, 1015 Lausanne, Switzerland

M.Z. Hauschild et al. (eds.), *Life Cycle Assessment*,
DOI 10.1007/978-3-319-56475-3_31

31.1 LCA and Chemicals: Introduction and Context

31.1.1 Chemicals and Their Relevance in Society

Chemicals are everywhere. Almost every second a new entry is added to the list of more than 100 million unique chemical substances registered in the Chemical Abstracts Service (CAS; www.cas.org), the world's authority on chemical information. Since industrialisation, the welfare of modern society largely builds on extensively mining minerals and fossil fuels including coal, petroleum and natural gas to produce large quantities of synthetic chemicals ('synthetic' simply means man-made and should not be confused with 'artificial', which implies that a chemical does not occur naturally). Consequently, the enormity and diversity of the chemical industry is astounding and poses various challenges for the management of environmental and human health impacts related to chemicals production and use. In this chapter, we outline important aspects to know about chemicals in the context of LCA.

Fundamentally, chemicals are substances composed of one or more atoms, and make up every material thing on earth—including our bodies. The atomic composition of chemicals classifies them essentially as 'organic' (chemicals with molecules built on a skeleton of interlinked carbon atoms and primarily consisting of carbon, oxygen, and hydrogen) and 'inorganic' (chemicals with molecules generally lacking carbon-to-carbon bonds, but instead based on the rest of the elements, including metals). In this sense, 'organic' has nothing to do with 'organic food' or 'organic farming' or 'organic lifestyle' as these terms generally refer to promoting sustainability. The atomic composition, molecular structure and ionisation (positive/negative charge) all influence chemical reactivity and behaviour in the environment as well as in living organisms. Because of this, chemical behaviours can be predicted and tested, and chemicals can be designed by industries to fulfil biological (e.g. medical) and physical (e.g. solvent) functions.

Chemicals may also be classified according to functional groups (e.g. alcohols, amines, acids and bases), structural groups (e.g. polycyclic aromatic hydrocarbons), physical structure (e.g. nanotubes), feedstock sources (e.g. petrochemicals derived from fossil fuels, biochemicals derived from starch- and sugar-based feedstocks), physicochemical properties (e.g. volatile, lipophilic), use function (e.g. surfactants, warfare agents), means of creation (e.g. reaction intermediates, metabolites), main economic sector (e.g. cosmetics, agrochemicals), toxicity endpoints (e.g. carcinogens, neurotoxins, endocrine disruptors), and other aspects.

Established nomenclatures or patents can be used to name chemicals. Most chemicals have an assigned CAS Registry Number except some metabolites of natural processes or grouped chemicals, such as polychlorinated dibenzofurans. CAS numbers are the most discriminant method for chemical reference. Of the chemicals registered by CAS, more than ten thousand are currently in commercial use, some with annual production volumes of millions of tonnes, while most chemicals are produced at levels of less than thousand tonnes per year. Worldwide, the production of chemicals has risen to several hundred million tonnes per year and sales were valued in 2013 at 3156 billion Euro with an average annual growth of 10.3% between 2003 and 2012 (CEFIC 2014). China dominates world chemical sales with a share of 33.2% followed by the European Union (16.7%), USA (14.8%), and Japan (4.8%) in 2013. Databases, such as the European Chemicals Agency (ECHA) Registered Substances database (http://echa.europa.eu/information-on-chemicals), the Household Product Database (http://householdproducts.nlm.nih.gov), the Hazardous Substances Data Bank through ToxNet (http://toxnet.nlm.nih.gov), and the Chemical and Product Categories Database (http://actor.epa.gov/cpcat) attempt to keep track of chemicals, their uses, properties and/or toxicity, but large data gaps still remain.

Several major environmental and health concerns associated with chemicals have led to various shifts in the global chemicals market. As an example, potentially toxic and highly persistent polychlorinated biphenyls (PCBs) have been replaced by chlorinated paraffins in various applications. While PCBs have been primarily produced in USA and Europe with a total historical production volume of 1.3 million tonnes between 1930 and 1995, chlorinated paraffins are currently almost exclusively produced in China and reach production volumes of more than one million tonnes per year (Fantke et al. 2015).

The chemical and pharmaceutical industries are a major driver of the welfare of modern society and scientific progress. These industries rely on the extraction, purification and synthesis of both naturally occurring and artificial chemicals and are among the largest and most influential economic sectors at the global scale. Main production segments are petrochemicals (e.g. benzene, styrene), consumer chemicals (e.g. detergents, fragrances and flavours), speciality chemicals (chemicals used for providing a special performance or effect, e.g. paints, dyes, adhesives), basic inorganics (fertilisers, industrial gases like nitrogen and oxygen), and polymers (e.g. plastics, synthetic rubbers and fibres). One of the largest segments is the production of organic chemicals with, e.g. formaldehyde, aromatics, acids, alcohols and esters providing the building blocks for drugs, agrochemicals, cosmetics and many other applications.

Along with societal advantages, the rise of chemical industries has also caused various undesirable consequences. Health impacts associated with air pollution are increasing worldwide and there is currently insufficient information to fully assess the impacts of chemicals on humans and the environment. Rachel Carson's book *Silent Spring* published in 1962 documented the detrimental impacts of chemicals on wildlife and humans, especially related to using synthetic organic pesticides, and marked a major change in public awareness that eventually inspired regulation of industry and for example the creation of the United States Environmental Protection Agency. Since that time, a remarkable amount of research correlates and demonstrates impacts on human and ecosystem health as well as the environment (e.g. the ozone layer) caused by intentional and unintentional chemical releases both indoors and outdoors. Some reported impacts are directly related to the chemical industry, whereas other impacts are related to the use or disposal of chemicals by other industries. In the following sections, we overview strategies for chemical management, focusing particularly on life-cycle assessments of chemicals production processes and chemical products.

31.1.2 Chemicals Management in Relation to LCA

Depletion of the ozone layer by chlorofluorocarbons used as refrigerants and solvents, soil and water pollution with heavy metals from ore mining and processing, pesticide emissions and residues in food, the formation of dioxins by incomplete combustion processes, and leaching of fertilisers into groundwater are just examples of the many problems associated with chemical releases to the environment. Hence, managing human and environmental risks posed by chemicals that are potentially toxic or may lead to other impacts is a major concern of regulators, industries, consumers and other stakeholders. As a consequence, the chemicals industry is one of the most regulated industries with main focus on regulating chemicals in consumer products and minimising chemical emissions to the indoor (workplace, public buildings and household) and outdoor environments along product life cycles. In the context of chemicals management, risk is defined as the probability of a chemical to cause an adverse effect (hazard) occurring as a result of a given contact between the chemical and humans or ecosystems (exposure). Risks associated with chemical emissions from a given product or process can arise at a later time after a chemical emission or exposure has occurred and depend on chemical background concentrations due to all release sources. In LCA, information on emission location and time as well as information on background concentrations, e.g. from sources outside the considered product system, is usually not available. Hence, modelled impacts in LCA are not interpreted in terms of actual risk, i.e. real environmental effects, but in terms of 'potential impacts' (Chap. 10) used as environmental performance indicators for comparing and optimising products or systems with respect to a defined functional unit (Hauschild 2005). However, models applied in LCA can also be applied in other fields of research and can be advanced and adapted to consider background concentrations as well as spatiotemporal resolution (e.g. daily or seasonal changes), and in such cases estimated potential impacts can be interpreted as estimates of actual risk.

Chemicals management occurs from local to global scale, from specific product–chemical combinations to entire industries and from raw material acquisition to waste handling, depending on the intended purpose. The Montreal Protocol on Substances that Deplete the Ozone Layer (http://ozone.unep.org) and the Stockholm Convention on Persistent Organic Pollutants (POPs; www.pops.int) are examples of global chemicals management treaties, whereas the Registration, Evaluation, Authorisation, and Restriction of Chemicals (REACH) is a recent example of an international legislative framework for managing industrial chemicals in the European Union. At all levels and scopes, effective chemicals management relies on assessment tools and guiding principles to ensure consistency and the achievability of defined goals. There are many examples of chemicals assessment tools and guidance, such as risk assessment, green and sustainable chemistry, chemical alternatives assessment, life cycle assessment, and a market for entrepreneurs to create industry-specific interfaces and applications. In the following sections, risk assessment (Sect. 31.1.3), green and sustainable chemistry (Sect. 31.1.4), and

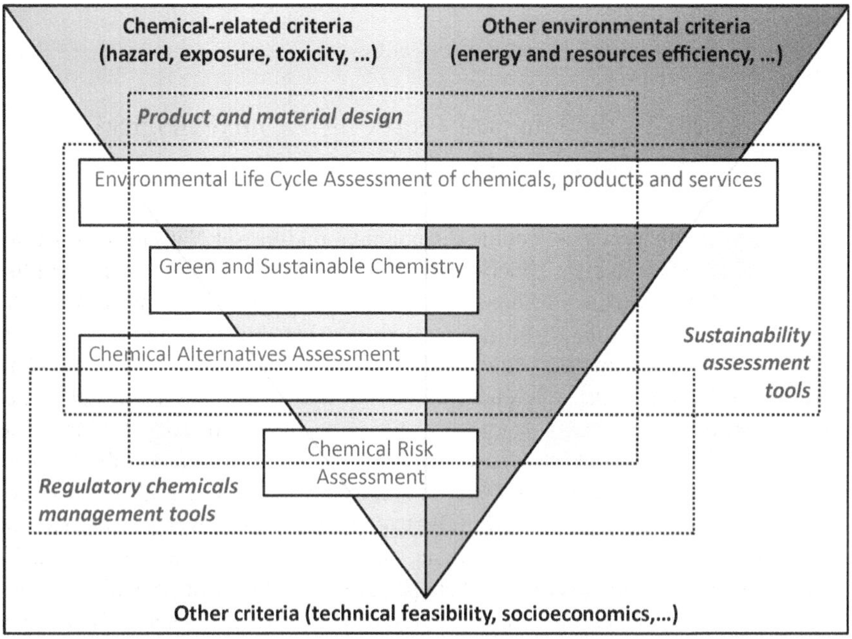

Fig. 31.1 Conceptual relationships of main chemical management tools

chemical alternatives assessment (Sect. 31.1.5) are discussed as commonly used chemical management tools that have both complementary and overlapping aspects with LCA as illustrated in Fig. 31.1.

31.1.3 Risk Assessment and Safety

Chemical risk assessment or chemical safety assessment is implemented in various regulatory frameworks and is one of the most widely used chemicals management tools. Risk assessment (‘*How risky is a situation*?’) as an integral part of risk management (‘*What shall we do about it, if a situation is risky*?’) essentially emerged at the start of the nineteenth century from studying hazards and risks associated with different occupations. Risk assessment mainly consists of hazard identification, dose-response assessment, exposure assessment and risk characterisation. Depending on the context, ‘risk’ and ‘safety’ have different meanings with regulatory policy commonly seeking to minimise risk while optimising safety. In this context, risk is generally defined as the probability of harm, whereas safety is described as the absence of harm (Embry et al. 2014).

Chemical ‘safety’ is defined by legislators or regulators and can vary from country to country, and evolves over time as science progresses. In this sense, ‘safe’ is not synonymous with 'natural' as it is often perceived. In fact, using the word 'natural' is misleading in the context of chemical safety (and LCA) and there are many naturally occurring chemicals that have very harmful properties like arsenic, nicotine or radon. As a consequence, we need to acknowledge that it is not always the 'natural' chemicals or solutions that are most 'environmentally friendly'—a common misconception in different science-policy fields and among consumers. Defining safety *thresholds*, e.g. chemical concentrations in different environmental media (e.g. ambient air, soil, water) or in food, is a common strategy in chemical risk assessment, and generally refers to levels below which a situation is considered 'safe' by a risk manager, meaning that any risk below threshold is regarded as 'acceptable'. As an example, chemical exposure resulting in one additional cancer case or less over lifetime in a population of one million people is regarded as an acceptable risk, i.e. safe, in the U.S. (van Leeuwen and Vermeire 2007). Using units like 'part per million' (ppm) as in one cancer case in a million or 'part per billion' is common for assessing (very small) amounts of chemicals in the environment. To get an impression of how much one ppm actually is, we can use 1 teaspoon of salt (5.5 g) in 5.5 tonnes of potato chips corresponding to one part of salt per one million parts of potato chips.

Thresholds are also applied when managing environmental systems and for developing chemical pollution control strategies, such as allowable nutrient releases from wastewater treatment plants or setting greenhouse gas emission targets, or in the context of 'planetary boundaries' in an attempt to assess if the pressure from chemical pollution (analogous to the amount of receiving environment required to dilute pollution below a threshold level) exceeds a planetary boundary (analogous to the amount of receiving environment available) for a 'safe operating space' for human activities (MacLeod et al. 2014). Chemical pollution levels have recently been expressed as 'chemical footprints' that can be compared with respective planetary or other boundaries for chemical pollution (Posthuma et al. 2014) to assess how companies or nations perform with respect to different chemicals management issues.

Risk assessment approaches take a receptor perspective (Fig. 31.2, right-side box), where thresholds are set in order to protect specific receptors, i.e. exposed humans or ecosystem species. In a receptor perspective, all relevant sources of a chemical or target chemicals are typically considered. In contrast, impact assessment tools in LCA are generally not receptor-oriented or threshold-based. This is because LCA takes a 'producer' (or 'emitter') perspective (Fig. 31.2, left-side box) by comparing potential impacts relative to each other across compared products and life cycle stages, aiming at minimising impacts considering various receptors (entire human populations, freshwater ecosystems, marine ecosystems, etc.). Differences and commonalities of risk assessment and LCA have been contrasted elsewhere (e.g. Bare 2006; Pennington et al. 2006), and there are several attempts combining, blending or integrating both concepts (Harder et al. 2015).

An increasing number of chemicals is approved for use in commerce, e.g. in food contact materials, but many chemicals lack adequate information to characterise risks (Neltner et al. 2013). In response, high-throughput screening ('first tier'

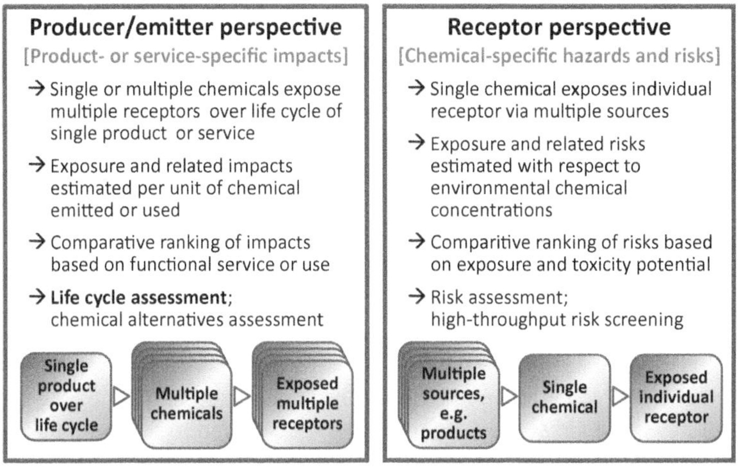

Fig. 31.2 Examples and underlying characteristics of dichotomous perspectives followed in different chemicals management approaches

assessments) of chemical risks has emerged as a strategy for prioritising and ranking chemicals for more in-depth study ('higher-tier' assessments). First-tier screening usually relies on ranking chemicals with respect to hazard (e.g. chemical toxicity) combined with estimates of exposure. 'High-throughput' refers to processing dozens to thousands of chemicals via resource efficient methodologies, such as robotic in vitro bioassays (instead of animal in vivo experiments) and low-tier computational models relying on databases (instead of data-intensive complex and time-consuming modelling). LCA impact assessment models have been used in high-throughput risk screening offering dual purpose and a promising area of interdisciplinary overlap to manage chemical risks (e.g. Shin et al. 2015).

31.1.4 Green and Sustainable Chemistry

'Green chemistry' is a concept that was coined by the U.S. Environmental Protection Agency in the early 1990s in response to the Pollution Prevention Act and increasing attention to chemical pollution. This concept builds upon a set of 12 Principles of Green Chemistry defined by Anastas and Warner (1998) aiming at reducing or eliminating hazardous substances in the design, manufacture and application of chemical products. Thereby, 'green' refers to more environmentally benign (less hazardous) chemicals. The concept of 'sustainable chemistry' is broader than the scope of green chemistry and strives towards 'eco-efficiency'. In addition to chemical hazards, sustainable chemistry centrally focuses on optimising the use of finite resources, while reducing environmental impacts of chemical production (OECD 2012). Sustainable chemistry—sometimes also referred to as sustainable chemistry and engineering—is rooted in the concept of Sustainable Development established in Rio de Janeiro in 1992 at the United Nations Conference on Environment and Development and is guided by 9 Principles of Green Engineering postulated at the Sandestin conference (Abraham and Nguyen 2003).

Green and sustainable chemistry are concepts focusing on the technological approaches aiming at the reduction of resource consumption and pollution prevention in chemical production processes rather than focussing on the assessment of chemicals in the environment. Hence, green and sustainable chemistry—often relying on comparing qualitative or semi-quantitative indicator results—are primarily applicable in the design phase of products to guide innovation and to support sustainable production goals.

Green chemistry in relation to LCA has been discussed in more detail elsewhere (e.g. Anastas and Lankey 2000). In summary, compared to green and sustainable chemistry, LCA aims at fully quantifying potential impacts associated with a chemical product or production system over its entire life cycle. Using LCA in early stages of chemical product and process design of various sectors including emerging technologies (e.g. bio- and nanotechnologies) has provided insight into the relationship between chemical and process parameter selection and related impacts on humans and the environment (Kralisch et al. 2015). LCA results have moreover demonstrated that quantitative methods are needed to assess the environmental performance of 'green' chemicals (Tufvesson et al. 2013). This is especially relevant as green chemistry usually focuses on optimisation of (production) processes, including some specific end-of-life problems related to chemicals, which may still risk sub-optimisation when a full life cycle perspective is lacking.

Using LCA in early product development stages, for example before a product has been created and marketed, comes with methodological and practical challenges, such as low data availability, uncertainty related to future product applications, and unclear scale of production for a changing market. Therefore, LCA has mostly been applied to chemical products and processes that are already well established and operational at the market scale, which often leads to LCA supported decision making being reactive instead of proactive.

31.1.5 Chemical Alternatives Assessment

Chemical alternatives assessment (CAA) aims to identify, compare and select safer alternatives to substitute (replace) harmful chemicals in materials, processes and products on the basis of their hazards, performance and economic viability (Hester and Harrison 2013). CAA emerged from the U.S. Environmental Protection Agency's Design for Environment (DfE) program in the late 1990s to promote less hazardous chemicals in various products and applications, and to avoid unintended consequences of harmful alternatives resulting in incremental improvements or even 'regrettable substitution' situations (Fantke et al. 2015). Ideally, CAA tools would evaluate hazard, exposure, life cycle and social impacts, economic feasibility and technical performance of alternative solutions, and consider chemicals, materials, products or technologies, and behavioural changes as potentially viable solution options. In reality, however, most CAA tools focus only on comparisons of hazard scores and exclusively consider chemicals as potential solutions. Several existing CAA tools have been compiled into the OECD Substitution and Alternatives Assessment Toolbox (www.oecdsaatoolbox.org).

The concept of 'acceptable risk' (as applied in risk assessment) is usually avoided in CAA in order to support selecting *relatively* less hazardous chemicals and materials in products (Whittaker 2015). Despite the current focus on assessing chemical hazard, including exposure, life cycle, and social considerations are lately also gaining more attention (Jacobs et al. 2016), focusing the CAA discussion around using more quantitative and chemical function-based methods and tools (Tickner et al. 2015). However, the need for rapid screening of numerous viable alternative solutions prevents CAA from simply adopting the use of LCA tools due to high complexity and data demand.

CAA is mainly used to identify and evaluate solutions to hazardous chemicals in products that have been targeted for market phase-out, and to inform early product development to minimise reliance on hazardous chemicals. With that, CAA takes the 'producer' perspective similarly to LCA (Fig. 31.2, left-side box), focusing on the impact of chemicals and their alternatives on various receptors. The main difference between CAA and LCA is that while CAA focuses on seeking for viable alternatives to harmful chemicals, LCA considers the life cycle of whole products or processes not focusing specifically on the content of one or more chemicals that might be considered 'hazardous', but instead evaluating the overall product or process environmental performance.

31.2 LCA Applied to Chemicals

Chemicals play a central role in the LCA framework for different reasons. Hundreds of chemical emission (inventory) flows typically occur along the life cycle of products or systems (Fig. 31.3) and are quantified as part of the Life Cycle Inventory (LCI; see Chap. 9) phase. Chemicals are also often precursors of product materials, and input for manufacturing and disposal processes. Chemical emissions associated with energy conversion during manufacturing, transport of goods and end-of-life treatment processes often dominate overall emission profiles for many product categories that can be characterised in the Life Cycle Impact Assessment (LCIA; see Chap. 10) phase.

Chemicals contribute to nearly all LCIA impact categories affecting human health and ecosystem quality as two main areas of protection in LCA (Hauschild et al. 2013). In LCIA, chemicals contribute to global warming, stratospheric ozone depletion, formation of photochemical ozone in the troposphere, air pollution (via respiratory particles and precursors), aquatic and terrestrial acidification and eutrophication, and last but not least human toxicity and aquatic and terrestrial ecotoxicity. Only a handful

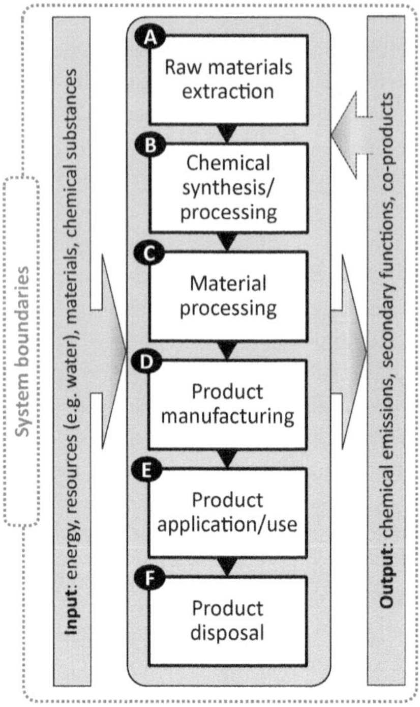

Relevant questions for different life cycle stages...

A: Fossil or bio-based resources; virgin or recycled materials; alternative use of resource or material (e.g. food) or waste (e.g. shell or husk); location of extraction (e.g. rain forest, ocean, conflict zone); secondary functions (e.g. algae waste treatment)?

B: Reactor configuration (e.g. batch or continuous); number of reaction steps or unit operations; solvents; catalysts (e.g. enzymes, rare metals)?

C: Machine lubricants; chemical contents and concentrations?

D: Machine lubricants; solvents; adhesives?

E: Product longevity; secondary functions; exposure during use; nature of use (e.g. amount used; additional requirements such as hot water); societal function (e.g. to cure a disease)?

F: Special handling (e.g. pharmaceuticals); disposal location; process (e.g. incineration or landfil); reuse opportunity; circularity of waste streams?

A-D and **F:** Waste handling (e.g. municipal or industrial; type); co-products (e.g. materials; energy from incineration); existing or new infrastructure; efficiency; worker exploitation or occupational exposure?

Fig. 31.3 Generic life cycle stages and system boundaries for chemical products or materials and LCA-related questions. In some cases, chemical processing may be followed by material production (e.g. polymers) before manufacturing a product (e.g. plastic bottles), while in other cases, chemicals (e.g. solvents) may be directly added to products or product manufacturing processes. *Underlined topics* are mostly lacking methods or not included in environmental LCA studies

of chemicals are associated with the majority of abovementioned impact categories, such as carbon dioxide, methane and other greenhouse gases contributing to global warming impacts or ammonia, nitrogen oxides, phosphate and some other nitrogen and phosphorus containing chemicals contributing to aquatic eutrophication. In contrast, thousands of chemicals can be characterised as potentially toxic to humans and/or ecosystems (Rosenbaum et al. 2008).

The generic life cycle stages shown in Fig. 31.3 are applicable to a chemical product (e.g. pharmaceutical or dye) or material (e.g. polymer), from raw materials extraction to product disposal, often referred to as ‘cradle to grave’ (Fig. 31.3, stages A–F). A ‘cradle to grave’ LCA study can provide valuable insight regarding which stages dominate the impacts throughout a product life cycle. Some of these life cycle stages, however, may not be relevant or may be assumed to be equal in two compared systems depending on the goal (Chap. 7) and scope (Chap. 8), and the product system under study. For example, the ‘material processing’ stage may not be relevant in cases where a chemical is directly added into a product as an ingredient, such as fragrances in cosmetics or detergents in cleaning products. As another example, the ‘product application/use’ or ‘product disposal’ stages may not be relevant for comparing the environmental performance of chemical synthesis or production processes as long as the compared processes do not influence the chemical amount used in a product or for product disposal.

An LCA study from raw material extraction to chemical product manufacturing, i.e. without considering product use and disposal stages, is referred to as ‘cradle to gate’ (Fig. 31.3, stages A–D), where ‘gate’ refers to the manufacturing or production facility (which could be the ‘gate’ of a chemical or product ‘factory’, depending on the focus of the study). In Table 31.1, different assessment scopes for LCA studies focusing on chemicals in materials, products and processes are contrasted and associated with relevant chemicals management questions.

LCA can help identify a variety of impacts associated with chemical production, use, and disposal, that are either intrinsic to a chemical (e.g. toxicity potential) or related to supporting industrial chemical processes (e.g. water consumption, greenhouse gas emissions). The main uses of LCA for managing chemicals and chemical processes are to compare impacts between products or services, or to identify ‘hot spots’ within a life cycle that contributes greatly to the impacts of a product or service. With respect to chemicals, LCA can be applied to various combinations of the generic life cycle stages in Fig. 31.3 depending on the LCA study goal and chosen system boundaries. In some cases, individual life cycle stages and associated inputs or outputs may be found to be irrelevant for the system considered or question asked. The chemical industry developed a guidance document to support the assessment of the environmental performance of chemical products based on attributional LCA, i.e. referring to process-based modelling and excluding market-mediated effects (WBCSD 2014).

In the following sections, an overview is given of how LCA has been applied to consider these various life cycle stages and the general lessons learnt from these studies. Thereby, LCA can be used to compare impacts at the level of chemicals in materials, products and formulations or at the level of chemical synthesis and production processes.

31.2.1 Chemicals in Materials, Products, and Formulations

A subset of materials, products, formulations (combination or mixture of chemicals) and processes are intrinsically reliant on the functionality of key chemical ingredients. In this section, main trends in the application of LCA- or LCA-based methodologies are summarised. This may include also partial LCA studies, e.g. methods only considering a subset of life cycle stages (i.e. cradle to gate or gate to gate), with focus on chemicals in materials, products and formulations.

Table 31.1 Relevant life cycle assessment scopes and life cycle stages for selected chemicals management questions and example studies

Chemicals management questions	Assessment scopes and considered life cycle stages	Example studies
What is the environmental performance of different products with respect to chemical emissions?	• Cradle to grave • Stages A–F (Fig. 31.3) • Focus on chemicals consumption and emissions	• Cleaning products (Van Lieshout et al. 2015) • Textiles (Roos et al. 2015)
What are the environmental profiles of the production of different chemicals?	• Cradle to (factory or consumer) gate • Stages A–D or a subset of these stages (Fig. 31.3) • Focus on chemical manufacturing	• Pharmaceuticals (Wernet et al. 2010)
Which stage of a chemical product life cycle contributes most to environmental impacts?	• Cradle to grave • Hotspot analysis including stages A–F (Fig. 31.3) • Focus on chemicals as products	• Plant protection products (Geisler et al. 2005)
Which chemical synthesis and/or manufacturing processes contribute most to environmental impacts?	• Cradle to (factory) gate • Hot-spot analysis including stages A–B or A–C (Fig. 31.3) • Focus on chemical manufacturing	• Pharmaceuticals (De Soete et al. 2014) • Nano-materials (Pati et al. 2014)
Which life cycle stage of a chemical in a product contributes most to human exposure?	• Cradle to grave • Partial LCA (only human e.g. exposure estimates) including stages A–F (Fig. 31.3) • Focus on chemicals in products	• Cosmetics (Ernstoff et al. 2016a)
Which feedstock provides the most environmentally friendly substrate for biochemical synthesis?	• Cradle to (factory) gate • Stages A or A–B (Fig. 31.3) • Focus on chemicals and raw materials consumption	• Acrolein (Cespi et al. 2015) • PET (Akanuma et al. 2014)

LCA studies have focused on pharmaceuticals (e.g. De Soete et al. 2014), cleaning products (e.g. Van Lieshout et al. 2015) and pesticide formulations (e.g. Geisler et al. 2005) as examples of products where chemicals provide the main product functions. Other LCA studies on chemicals with in-product functions include studies focusing on flame retardants in electronics (Jonkers et al. 2016), nano-materials used in bandages and cosmetics (Botta et al. 2011), and polymers used in food packaging (Hottle et al. 2013). Chemicals required for industrial processes have also been assessed in LCA studies, including industrial solvents (Zhang et al. 2008) and chemicals used for the production of treated water, oil and gas, printing paper and dyed textiles (e.g. Alvarez-Gaitan et al. 2013; Parisi et al. 2015).

When analysing LCA studies on chemical-based functions, a few generalisations emerge. For example, it is important to consider life cycle thinking early on in the design phase of products and processes whenever possible and it has been shown that simplified tools may help in this process (e.g. De Soete et al. 2014). Furthermore, it has been demonstrated that hybridised LCA tools or metrics can be useful to improve communication and management for specific stakeholders (e.g. Alvarez-Gaitan et al. 2013).

Several LCA studies indicate that being sceptical of services deemed 'green' or 'sustainable' is crucial, especially when an LCA has not yet been performed. Case studies on, e.g. 'green' solvents (Zhang et al. 2008) or 'sustainable' bio-based chemicals and materials (e.g. Hottle et al. 2013) demonstrate that materials and products guided by principles of 'sustainability', 'eco-friendliness' or 'green chemistry' can have significant, but often disregarded or unassessed, environmental impacts. An example is given in Fig. 31.4, where environmental life cycle impacts of petro- and bio-based polymers are contrasted based on data from Hottle et al. (2013).

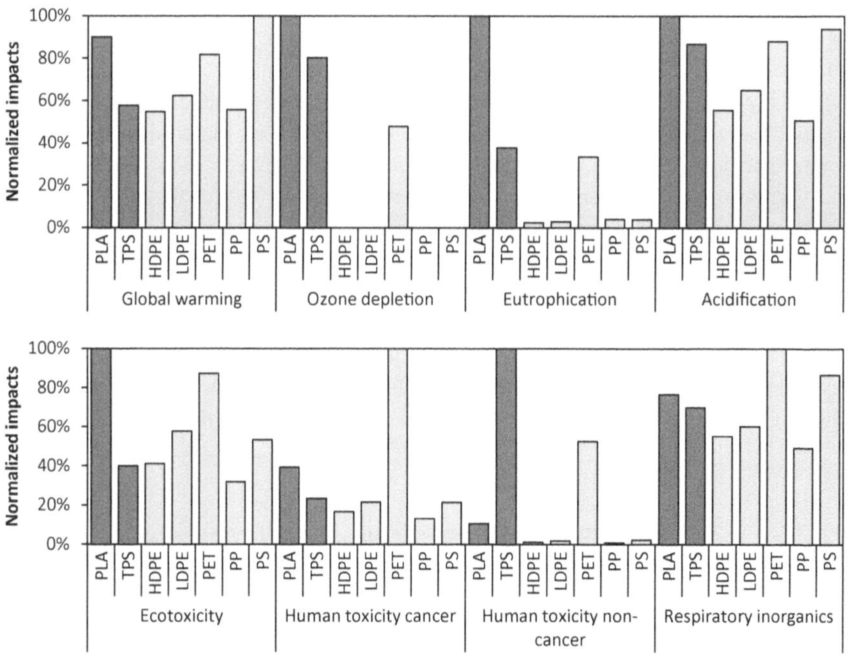

Fig. 31.4 Impact scores for LCAs of two bio-based polymers (*dark bars*; *PLA* Polylactic acid, *TPS* Thermoplastic starch) compared to petroleum-based polymers (*light bars*; *HDPE* High-density polyethylene, *LDPE* Low-density polyethylene, *PET* Polyethylene terephthalate, *PP* Polypropylene, *PS* Polystyrene) per kg of produced granule, normalised for each category to the polymer with highest impacts (based on data from Hottle et al. 2013)

According to this study, bio-based polymers lead to higher impacts than petro-based polymers for several impact categories, which contradicts assumptions that bio-based automatically implies 'green' or 'sustainable' (see also Chap. 30). Higher impacts for bio-based polymers are mainly associated with feedstock-related agricultural emissions of fertilisers (eutrophication) and pesticides (human toxicity and ecotoxicity), as well as deforestation (impacts related to changes in land use). However, the relative importance (i.e. contribution to overall environmental impacts) of the different impact categories also needs to be considered when evaluating the overall environmental performance of different polymers or other chemical products and processes.

Often products are referred to as 'green' or 'sustainable' based on a single environmental issue (e.g. reducing greenhouse gas emissions), or based on following the principles of green chemistry in chemical design only. However, chemical products that are claimed to be 'green' or 'sustainable' may in fact lead to greater impacts on the environment or humans than the conventional alternatives. For example, 'eco-friendly' food packaging made of plant fibres may increase exposure and environmental emissions of highly hazardous fluorinated chemicals (Yuan et al. 2016), and 'green' solvents can have higher impacts across many impact categories when compared to conventional solvents (Zhang et al. 2008). Furthermore, the production of bio-based raw materials (such as corn, sugar cane, or soy for feedstock) may or may not be associated with lesser greenhouse gas emissions and consumptions of fossil resources, but may have *equal or greater impacts* in other impact categories (e.g. land use, toxicity related to using pesticides, eutrophication related to using fertilisers) than fossil-based materials (see Chap. 30 for further details). These phenomena are commonly referred to as burden shifting (e.g. between environmental issues or compartments). Identifying these is a fundamental application principle unique to LCA.

LCA is a tool that can be useful for comparing products and processes for identifying such burden shifting and how to minimise impacts across a variety of impact categories. However, it is important always to ensure as a practitioner that all relevant chemical emissions are inventoried and all impact pathways are characterised. These general cautions are also relevant for LCA studies focusing on chemical synthesis and production processes as discussed in the following section.

31.2.2 Chemical Synthesis and Production Processes

LCA is a useful tool for improving existing processes and designing new processes for the synthesis and production (Fig. 31.3, stages A–D) as well as for the end-of-life treatment (Fig. 31.3, stage F) of chemicals and chemical products, to inform process systems engineering decisions (Jacquemin et al. 2012). In this section, LCA case studies focusing on chemical synthesis and production processes across various economic sectors are discussed.

A major issue illustrated by several LCA studies is that management decisions based on single indicators or criteria can lead to increasing other impacts (that were not considered in the decisions), thereby indicating the strength of LCA as an approach to assess multiple indicators and related trade-offs. An example is the development and application of new plant protection products (pesticides) designed with the intention to reduce human toxicity and ecotoxicity potentials associated with emissions after application in agricultural crop protection or elsewhere (e.g. household pesticides). A related LCA study revealed that the production of a new and more effective plant growth regulating pesticide with less intrinsic toxicity (preferable from a risk perspective) than a functionally equivalent earlier marketed pesticide comes at the expense of increased impacts associated with pesticide synthesis and production processes (Geisler et al. 2005). The higher impacts for the new pesticide are mostly explained by the high complexity of its molecular structure requiring more synthesis and processing steps. In general, impacts related to the production of chemicals have been attributed to energy consumption which tends to increase with increasing complexity of a chemical molecule. Highly specialised chemicals, such as pharmaceuticals, can thereby be associated with higher energy consumption and related impacts from synthesis and production processes than other chemicals (Wernet et al. 2010).

Not only complexity of chemical synthesis and production processes, but also the difference in raw materials used drives environmental performance profiles of chemicals and chemical products. This is shown in another set of LCA studies contrasting chemical production from fossil fuel-based versus renewable (bio-based) resources. Synthesising and producing chemicals from biomass (e.g. sugar cane) instead from fossil fuels (e.g. petroleum or natural gas) has been proposed as a 'sustainable technology' with respect to reducing reliance on fossil resources and greenhouse gas emissions. However, a full sustainability analysis has typically not been conducted, which is why several LCA studies have focused on this claim.

As an example, a simplified overview of the different chemical synthesis and processing steps involved in polyethylene terephthalate (PET) polymer production is given in Fig. 31.5. While terephthalic acid used in the production of the chemically identical PET and bio-PET is in both cases derived from petroleum, ethylene glycol can be derived from natural gas as a fossil resource (for PET) or from sugar cane as a biomass feedstock (for bio-PET). The process of natural gas refinement to create ethylene glycol alone consists of several steps including cracking (breaking down) into ethylene and other chemicals, ethylene separation and purification involving several distillation processes (not shown in Fig. 31.5). Accordingly, LCA studies have found that bio-based chemical production usually can lead to less greenhouse gas emissions than fossil-based chemical production, mainly because less refinement of fossil fuels is required. However, growing, harvesting and processing bio-based feedstocks may lead to other impacts related to agriculture production systems, e.g. land use (see Chaps. 29 and 30), which are highly variable with respect to the type of biomass used (Tabone et al. 2010; Akanuma et al. 2014). Furthermore, the type of biomass used can influence the energy required, and

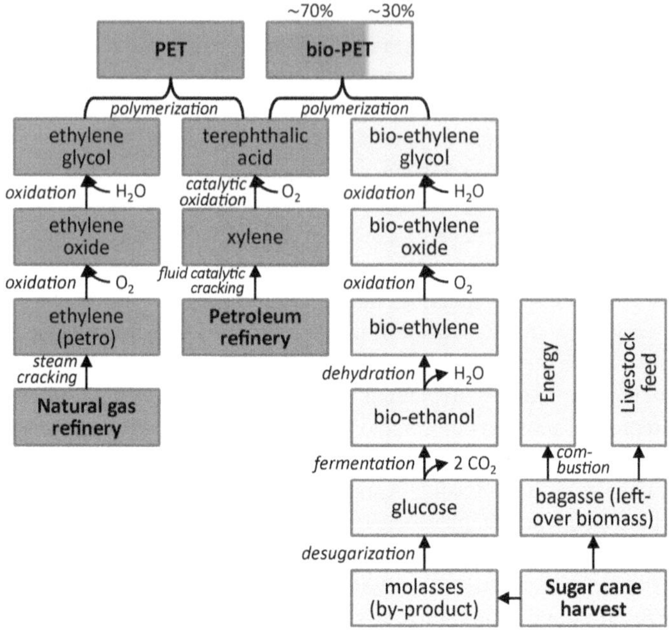

Fig. 31.5 Production process steps for chemical synthesis of polyethylene terephthalate (PET) derived from fossil fuels and bio-PET (partly) derived from bio-resources (modified from Tabone et al. 2010)

post-processing of bio-based products and residues greatly influence the overall related environmental performance.

Other LCA studies have focused on specific aspects of chemical synthesis and processing, such as comparing continuous and batch reactor types (e.g. Wang et al. 2013) or different catalysis and fermentation processes (e.g. Pati et al. 2014). It is further important to consider which catalysts are used in other processing steps that petro- and bio-based materials like PET have in common, such as antimony trioxide found at concentrations of 200–300 ppm in PET or other, metal-free catalysts used in the polycondensation process as part of polymerisation. Several studies have concluded that processes with higher yields have a lower impact per chemical production unit. The use of solvents has additionally been identified as an important component influencing environmental performance of chemical products (De Soete et al. 2014). Generally, and specifically for chemical synthesis and processing, it is important to be sceptical of processes and products labelled or deemed 'green' or 'sustainable' without performing a full LCA as shown, e.g. for 'green' nano-materials synthesis (Pati et al. 2014). An overview of aspects that are relevant for assessing 'green' chemical synthesis and production processes is given by Kralisch et al. (2015).

31.3 Specific and General Methodological Issues for LCA of Chemicals

Applying LCA, specifically in the chemical and pharmaceutical industry, and in other sectors where chemicals play a central role, comes with several methodological and practical challenges. Generally, gathering chemical inventory data, quantifying impacts, and interpreting results constitute challenges for LCA studies across sectors. In the following sections, some of the most relevant challenges focusing on chemicals in LCA are discussed in relation to the definition of the goal and scope of an LCA study, product system modelling and quantification of life cycle chemical emissions in the inventory analysis, characterisation modelling in the impact assessment, and finally interpretation of LCA results in different contexts.

31.3.1 Goal and Scope Definition

Consistently defining the goal and scope for chemical products or processes (e.g. functional unit of the considered product or service system and related reference flow(s) and system boundaries) is not trivial and needs to be critically considered by a practitioner. Examples of relevant issues when defining functional unit, reference flow(s), and system boundaries are discussed in the following.

Functional Unit (FU)

LCA (and other types of assessments) can be designed to compare functionally equivalent chemicals and chemical products as classified by *chemical function* (e.g. solvents, catalysts), *material function* (e.g. nanotubes, polymers) or *product function* (e.g. herbicides). It is hence important to define the level of 'functionality' based on which a study will be conducted. This functionality must be captured in the definition of the FU of an LCA study as basis for comparing products or systems.

Performing an LCA study is useful for providing valuable insight into which alternative, functionally equivalent chemicals or products provides the function with the lowest overall environmental impact profile, thereby focusing on avoiding burden shifting between different types of environmental impacts. To screen multiple alternatives to harmful chemicals in a particular product application, in contrast, the focus often is not mainly on environmental performance, but on a combination of regulatory compliance, economic and technical feasibility, along with considering hazard and human, environmental and social impacts. In such cases, a chemical alternatives assessment (CAA) might be the preferred approach to identify the most viable solution(s).

Chemicals and chemical products can also fulfil more than a single function and, hence, a partial definition of the functional unit could lead to inconsistent comparisons if the appropriate product systems are not considered as demonstrated in Example 31.1.

Example 31.1 **Functional Unit (FU)** Take a cosmetic product like shampoo, where different chemical ingredients provide different functions as part of the final shampoo product, e.g. to provide clean, shiny and fragrant hair for one person over 24 h. If the **FU is defined** with respect to a single shampoo product (one-product system) that **cleans the hair of one person** (by containing detergents) **and makes it shiny** (by containing siloxanes) **for 24 h**, a functionally equivalent service could be also provided by applying two distinct products (two-product system), one being a shampoo that only cleans hair (and does not make it shiny) and another being a conditioner that makes the hair shiny (and does not clean). However, both the one-product and two-product systems should not provide fragrance in order to be consistently compared via the same FU (bold text above) that excludes fragrance.

Likewise, if the FU is defined to just clean hair for one person over 24 h, comparing LCA results of a shampoo that only provides clean hair to a shampoo that provides clean, shiny and fragrant hair could yield the misleading outcome that the former shampoo 'performs better', because the production and related impacts of additional chemicals of the latter shampoo (containing siloxanes for making the hair shiny and terpenes for making the hair fragrant) are related to functions not fulfilled by the shampoo that only cleans hair. Hence, the comparison would be biased by comparing products fulfilling distinct functions.

Defining an appropriate FU for multi-functionality (see Chap. 8) is also important. For example, water and propylene glycol are both effective chemical solvents and, thus, both would fulfil an FU defined with respect to providing the function of a solvent in, e.g. a shampoo product. Propylene glycol, however, provides other functions that water does not provide (e.g. stabiliser, humectant, emulsifier). Therefore, a comparison of propylene glycol and water in an LCA study based on a solvent-based FU would not capture the multi-functionality of propylene glycol. Defining the FU with respect to all functionalities and then providing system expansion when necessary (e.g. water plus a stabiliser plus a humectant plus an emulsifier is functionally equivalent to propylene glycol in shampoo) can be an important consideration in any LCA on chemicals or product systems. It is, hence, important to ensure the product(s) or chemical(s) investigated in an LCA study are functionally equivalent and the FU captures this equivalency appropriately.

Reference Flow

The reference flows (Chap. 8) in an LCA study reflect the overall amount of goods and/or services that are required to fulfil the defined FU. Taking a no-wash (dry) shampoo versus a conventional (liquid) shampoo as examples, the reference flows to fulfil an FU of cleaning the hair of one person for one day could be 10 g of the liquid shampoo product plus the (hot) water used to wash the hair. The reference flow for the dry, leave-in, no-wash shampoo could be simply 5 g of powdered product (with no wash-water needed). Furthermore, if functionally equivalent products or chemicals provide different efficiencies to fulfil a defined FU, the different efficiencies need to be accounted for in the reference flow. This issue also points to a problem for cradle to gate LCA studies on chemicals, where it is possible that a chemical could have greater cradle to gate impacts than another chemical per unit mass emitted, but far less of the former chemical is required to fulfil the same FU. Here, pesticides with different efficiencies towards the same pest offer a typical example.

System Boundaries

The system boundaries (Chap. 8) of any defined chemical product or service systems in an LCA study need to capture all relevant processes for the systems being compared. For example, if the purpose of an LCA study is to compare bio- with fossil-based chemical synthesis, the system boundaries must include and differentiate all raw material acquisition processes, namely all refining processes for the fossil-based chemical and the crop production and processing steps for the bio-based chemical (see also Fig. 31.5). However, for these systems, it may not be relevant to include chemical use and disposal stages in the study, whenever these life cycle stages are equivalent in both cases. Such systems are referred to as 'cradle to (factory) gate' systems and are common in LCA studies on chemical synthesis and other chemical production processes (Jimenez-Gonzalez and Overcash 2014). In contrast, if the purpose of the study is to compare two distinct fossil-based materials fulfilling the same function, the disposal stage could be a relevant driver of the difference between the compared product systems.

For several chemical products and production processes, consistently defining system boundaries is challenging. An example is the application of plant protection products containing chemical pesticide active ingredients (e.g. carbamate insecticides) applied in agricultural crop production, where the FU could be defined to provide a specified amount of crop in a season. Allocating field buffer strips (i.e. non-agricultural areas that are among other functions introduced to reduce the impact of applied pesticides on non-treated areas), which may be required by law, to the technosphere would apparently influence the crop yield per hectare and amount of pesticide used compared with an equivalent system, where the buffer strips are defined as part of the environment (Rosenbaum et al. 2015). Including buffer strips in the considered technosphere system or not will, hence, influence the related impacts and also defines the scope of the environmental distribution processes of pesticides in the LCI and LCIA phases. As a consequence, the definition of the system boundaries needs to be aligned with the selected pesticide inventory data and characterisation models to avoid overlaps, double counting of processes and potential gaps along the pesticide impact pathways.

31.3.2 Product System Modelling and Inventory Analysis

There are several obstacles that need to be considered in the product system modelling and inventory analysis phase (Chap. 9), after the goal and scope of an LCA study have been defined.

Data Availability and Quality
All relevant chemical elementary flows from a given product system to the environment need to be quantified in the LCI phase. When using LCA software, emission quantities are often available through an LCI database, for example for processes occurring in Europe or the 'rest of the world.' LCI databases generally rely on typical or average emission inventories or an inventory taken by one industry for a given unit process, which may be outdated or tied to, e.g. a specified electricity mix. Thus, it is always preferred to gather primary data, especially for the foreground system modelling (Chap. 9), of the specific LCA case under study. This poses a particular challenge to LCA practitioners, who may or may not have access to company-specific data to resolve the nuances of a particular supply chain. While in some cases, a particular commissioner of an LCA study might provide such data, while in other cases such data have to be collected from different parties. An example is the application of plant protection products, where pesticide manufacturers will know the concentration of a pesticide active ingredient in a formulation product, but where the different farmers might know the effectively used amount that is applied on agricultural fields and this usually depends on pest-, climate-, soil- and application-specific conditions.

Emission Estimation and Modelling
Most chemical synthesis and material or product manufacturing processes involve several steps, which can yield usable by-products that have to be considered in an LCA study (see Chap. 9 for further details). As an example, harvesting sugar cane yields refined sugar, but also molasses (sugar refining by-product) and bagasse (dry leftover biomass after extracting the juice from the sugar cane). While molasses can be further used to produce biochemicals, bagasse is usually burned (for energy conversion) or used as livestock feed (Fig. 31.5). In an LCA study, usually only one of these products (sugar, biochemical, energy, livestock feed) is in focus and the other products must be accounted for through subdivision or system expansion or, if it cannot be avoided, through different types of allocation (see Chap. 9).

When building a product system model, different tools and software packages are available. Specifically for simulating material and energy balances of chemical production and processing, there exist several (open-source and commercial) chemical process simulators, such as Aspen HYSYS for oil and gas process simulation and Aspen Plus for chemical process optimisation (www.aspentech.com), BatchReactor and BatchColumn for chemical reactor and batch distillation columns simulation, respectively (www.prosim.net), or the CHEMCAD software suite for chemical process simulation and optimisation including batch operations (http://www.chemstations.com). Such software packages may include proprietary data from the chemical and other industries that are otherwise not accessible and may intrinsically use different allocation systems. The responsibility of ensuring transparency and consistency when building a product system including proper consideration of co-products and by-products lies with the LCA practitioner. However, several documents exist for LCA practitioners to seek guidance, and working examples of co-product considerations for the chemical industry can be found elsewhere (e.g. Weidema 2000; Karka et al. 2015).

In most LCA studies, an inventory covers hundreds of processes and emission flows but not all chemical emissions are usually able to be considered. Often missing from LCI databases are, e.g. emissions to the occupational and consumer environments, and the ingredients (e.g. chemicals) in a product, which can be emitted indoors during product use or outdoors during post-use as demonstrated in Example 31.2.

***Example 31.2* LCI Emission Pathways** When a **consumer product** (e.g. perfume) or industrial product (e.g. agricultural pesticide) is used, the chemicals within the product can follow different transport and fate mechanisms that can lead to exposures of humans and ecosystems. Consider that a colleague at work **applies an air freshener or perfume in the office**. Perhaps you smell or even taste it in the first minutes after application (indication of exposure), maybe the scent remains in the office for some days (indicating sorption and desorption to and from indoor walls and other surfaces), and maybe you can even smell it just outside the office building (indicating transport outdoors via ventilation).

In some cases, a large proportion of chemicals within products can be taken in by humans during and after product use, which is a major concern amongst regulators and researchers. In LCA, such considerations are currently largely missing, but first efforts were made to include indoor fate and exposure pathways (referred to as ‘near-field’) into the toxicity characterisation model USEtox 2.0 (http://usetox.org). Without accounting for near-field fate and exposure pathways, LCA studies typically may assume a fixed fraction like 100% of product ingredients being emitted to the environment. In general, assuming such emission distributions could lead to an underestimation of resulting human toxicity potentials and in some cases also to an overestimation of environmental or ecosystem impacts. This is illustrated in Fig. 31.6 for d-limonene as commonly found chemical in a shampoo product, where assuming 100% of the used product being washed-off (left-side pathway in figure) instead of modelling a more complex yet more realistic distribution (right-side pathway in figure) yields a difference of more than three orders of magnitude for freshwater ecotoxicity impacts, which is beyond the uncertainty range for this impact category.

Emissions can also occur from chemical residues in products that are related to cross-contamination, i.e. such chemicals are not purposefully added to a product and enter a product from using, e.g. recycled material where not all chemical ingredients are known. Often, inventory data related to cross-contamination

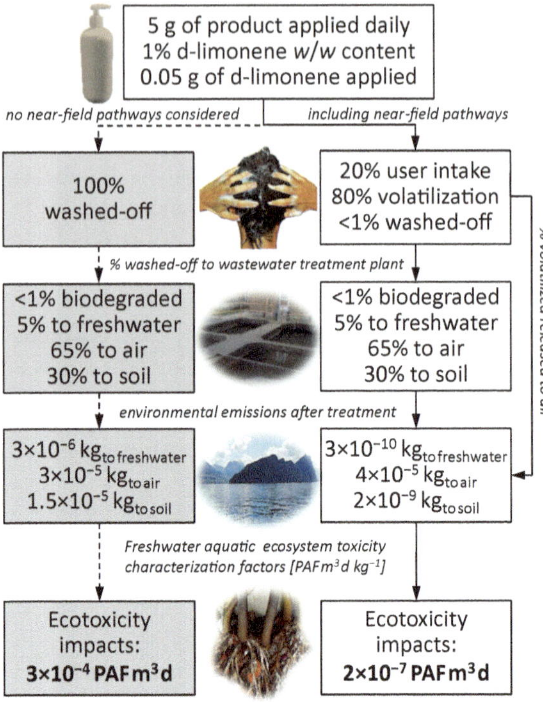

Fig. 31.6 Illustrative example of assumptions for emission distributions of chemicals in consumer products in life cycle assessment showing a substantial decrease in the estimated potentially affected fraction (PAF) of freshwater species from the chemical ingredient d-limonene (CAS: 5989-27-5) in shampoo when accounting for indoor fate and exposure of cosmetic products. Adapted from results from Ernstoff et al. (2016a) combined with freshwater ecotoxicity characterisation factors from USEtox 2.0 (http://usetox.org). Air emissions were assumed to be to urban air, water emissions to continental freshwater, and soil emissions to continental natural soil

pathways are very limited if at all available. Using similar processes or pathways as proxy might be a possibility to address this limitation, but also introduces additional uncertainty in the emission estimates.

Spatiotemporal Variability in Emissions

Time (e.g. year, season or duration) and location can influence variations in emissions, referred to as 'spatiotemporal' variability. In many cases, LCI results do not capture the time of emissions from systems, e.g. agricultural practices (e.g. harvesting, applying fertilisers) can occur according to daily or seasonal cycles according to the geographic location of the farm. Likewise, emissions of landfill leachate are influenced by changes in environmental conditions (e.g. acidity and temperature) which can change through time and according to location (Bakas et al. 2015).

Incomplete Emission Inventories

It is important to be aware of the incompleteness of some emission inventories. For example, energy conversion processes generally are well detailed in LCI which can result in high toxicity related impacts resulting from energy conversion, but other processes, e.g. related to chemical synthesis may have less complete inventories and, hence, related toxicity impacts might be underestimated (Laurent et al. 2012).

31.3.3 Impact Assessment

Characterising chemical emission flows resulting from the LCI in terms of their impacts on humans and the environment requires a careful consideration of study context (e.g. spatial region), number and relevance of chemicals to be characterised (in many cases, most chemicals contribute marginally to overall impacts, while only few chemicals dominate overall impact profiles). In the following, challenges and pitfalls in the impact assessment of chemical products and processes are discussed with focus on toxicity-related impacts, where special challenges exist mainly due to the countless chemicals to be characterised and the complexity of related impact pathways.

Limited Substance Coverage

USEtox, a scientific consensus model for characterising human and ecotoxicological impacts of chemicals, presently provides characterisation factors for more than 3000 chemicals, which constitutes the largest list currently available in LCIA. However, with tens of thousands of chemicals on the market, inventoried chemical emissions either documented in an LCI database or by a practitioner investigating a specific system or process, may in many cases not have existing characterisation factors or the data required to develop new characterisation factors (e.g. toxicity dose-response information). This limitation to substance coverage in LCIA is important when interpreting results, because a lack of data for many chemicals does not preclude their possible impacts.

Chemical Degradation Products

When a chemical does not degrade, or degrades very slowly, it is considered ‘persistent.’ Persistent chemicals thereby can be linked to greater impacts because they are not or very slowly removed from the system through degradation. In current LCIA methodologies, abiotic (e.g. where a chemical is transformed via interactions with sunlight) and biotic (e.g. when a chemical is metabolised by soil bacteria) degradation essentially ‘removes’ organic chemicals from the system and no further impacts are characterised. In reality, degradation processes transform a chemical into one or more degradation or transformation ‘products’, including other chemicals or gases, which can also impact the environment. Degradation products can have greater or lesser impacts than their parent compounds; for example aminomethylphosphonic acid (AMPA), which is the main degradation product of the broad-spectrum herbicide glyphosate, is more persistent and more toxic than the glyphosate parent compound. As an example, not including AMPA in an LCA study that considers agricultural processes where this herbicide is used could underestimate the impacts of using glyphosate. Therefore, an LCA practitioner should include estimates of persistent degradation products and appropriate characterisation factors (in this case for AMPA, not glyphosate) to better capture the impacts of chemicals. While this approach will not be feasible for all chemicals (due to data limitations), it should be performed when the issue is known and data are available.

Impacts from Chemical Mixtures

Impacts towards humans or different ecosystems, related to chemical emissions, are a function of the simultaneous prevalence of other chemicals, which might have synergistic (enhancing) or antagonistic (counteracting) properties with respect to the effect of a considered chemical. Since information on the site-specific mixture of chemicals in any environmental medium or compartment is not usually available, and the impacts of such a mixture on humans or the environment are not known, synergistic or antagonistic effects are usually not considered, and instead additivity of exposure and related effects is assumed. This means that the effects of all chemicals contributing to the same impact category, e.g. freshwater aquatic ecosystem toxicity or ozone depletion, are summed up to arrive at an overall product system-related impact score. If for any LCA study the emission location and time is known and related background levels are available for all relevant chemicals, this assumption could be evaluated by identifying and quantifying the synergistic or antagonistic effect potentials. However, the potentially added accuracy in an LCA context is most likely not relevant given existing uncertainty attributable to other aspects in the characterisation of chemical emissions. Besides, the large number of chemicals present and emitted into the environment yields an almost limitless amount of possible mixtures, rendering it impossible to quantify the specific effect potentials for each mixture.

Missing Fate and Exposure Processes and Pathways

In order to reduce the demand put on LCA practitioners, to streamline workflows, and to allow for science-based and consensus-driven solutions, LCIA often relies on predefined methodologies. While hundreds or even thousands of chemicals might be inventoried for various processes in an LCA study, characterisation factors or a LCIA method for a given impact at mid-point or end-point level might in some cases be missing, especially for toxicity-related impacts (see Chap. 10). Moreover, certain exposure settings (occupational, consumer) or routes (e.g. dermal exposure) or target organisms (e.g. exposures of bees) may be missing from an LCIA model. Effect factors may also be missing or inconsistent, e.g. in the case of human toxicity, effects of allergy or endocrine disruption (i.e. interaction with the hormone system) are often not included, but may be highly relevant for chemicals in consumer products. Finally, many of the methodological gaps in LCIA are also due to the reliance on simplifying assumptions. The LCA practitioner, who is constrained by resources (time, money, data access), is responsible for compiling the necessary data and for ensuring that the LCIA methodology chosen (or developed) is suitable for the defined goal and scope of an LCA study. Specifically, to characterise a chemical's impact, several assessment factors are required and must be sufficiently scrutinised within the chosen LCIA method, such as the chemical environmental fate, ecosystem and/or human exposure if relevant, and subsequent effects with respect to given impact categories. Each of the related data requirements poses its own challenges. To avoid the misleading conclusion that missing aspects of the chosen LCIA method do not cause impacts because they were not assessed, it is important to be familiar with which processes (e.g. biotransformation), environmental compartments (e.g. indoor air), exposure pathways (e.g. dermal uptake), and effects (e.g. endocrine disruption), may be missing from the selected characterisation methods but are relevant for the system under study. In some cases such missing aspects can be addressed by the practitioner by developing new methods or by adapting existing methods; if not, it is important to be aware of how this could influence results.

Spatiotemporal Variability of Impacts
LCIA methods are generally based on regional or global averages for various chemical, environmental and pathway data and processes, e.g. how long it takes chemicals emitted to freshwater to reach the sea (i.e. residence time), or how many persons live in an urban area (i.e. population density). Studies have shown, intuitively, using a continental average instead of 'spatially differentiated' regionalised models can yield large uncertainty in the estimated impacts (e.g. Kounina et al. 2014). Thus, if the location of the emissions (e.g. from a specific factory) in an LCA study is known, using a model with characterisation factors specific for that region can reduce uncertainty of model results. If emission locations are not known (as is the case for most chemicals in typical LCA studies), characterisation results for regions can be applied that are parameterised, i.e. averaged for the characteristics of a particular region. The same rule applies for temporal aspects, where in LCA mostly steady-state conditions and continuous emissions are assumed, which might not be true for, e.g. agricultural pesticides that are applied on specific days only (i.e. pulse emissions). In such cases, accounting for the dynamics of the chemicals in the modelled environmental system may reduce uncertainty in characterisation results (e.g. Fantke et al. 2012), but whenever temporal information on emission patterns is not available, parameterised characterisation results can be applied that account for the most important temporal aspects of a modelled system.

Impacts Versus Benefits
Life cycle *impact* assessment inherently focuses on quantifying 'negative' impacts on humans and the environment. A stakeholder could in some cases argue that their product or service offers a benefit to society that is not accounted for, meaning that an LCA yields misleading results. When facing such an argument as an LCA practitioner, it is important to go back to the fundamentals of LCA. The impact assessment phase of LCA is designed to assess environmental 'benefits' in the form of 'avoiding environmental impacts.' For example, a wastewater treatment plant design that also decreases environmental pollution compared to another design offers a 'benefit' that is quantifiable in an LCA context (see Chap. 34 on LCA of wastewater treatment). Furthermore, when comparing functionally equivalent products or services, their benefits (e.g. restoring a wetland to yield a level of biodiversity, or designing a car with a certain safety rating) is often captured in the functional unit of an LCA study, which defines a unit of the (beneficial) service being provided. There are special cases where considering societal benefits that are not captured in the functional unit or by the assessment methods can be extremely important when guiding decision-making. In some cases, LCA may not be the appropriate tool to assess such benefits; however, developing LCA-compatible methods to quantify societal benefits (specifically positive human health outcomes)

is a topic of high interest when assessing human nutrition and dietary patterns, where two functionally equivalent diets can have very different health impacts or benefits (Nemecek et al. 2016).

31.3.4 Interpretation

The interpretation of results is fundamental for the findings and reliability of every LCA study and subsequent guidance provided to stakeholders, and to LCA in general (see Chap. 12). Robust and transparent interpretation of results from an LCA study can offer sound council for the stakeholders and when aggregating with other LCA studies can elucidate generalisable findings important for sustainable development. As an example of nuances of interpretation, the 'New Plastics Economy' report (WEF 2016) cites interpretation of several LCA studies and implies that a major shortcoming of LCA is its inability to identify and support 'target states', such as moving towards increased production and use of bio-based plastics. Indeed, as previously discussed, LCA studies on bio-based versus fossil-based plastics have demonstrated similar, if not greater impacts (e.g. on land use and toxicity potentials) for bio-based plastics due to agricultural practices (see Chap. 30), which is a finding that may be unintuitive or undesirable to some (e.g. stakeholders in the bioplastics industry). When interpreting such LCA results, it is important to distinguish what an LCA says about 'here and now' versus what it could mean for future sustainability goals or targets of stakeholders. For example, LCA results showing bio-based plastics have 'greater impacts' than fossil-based plastics do not discredit bio-based plastics as a sustainability *goal*, but they do indicate that bio-based plastics face sustainability challenges given current agricultural practices, which thus must be addressed to avoid impact trade-offs. Furthermore, LCA results can help indicate which feedstock is the most eco-efficient (less impacts per kilogram) to work towards a bio-based 'target'. In practice, LCA may not be able to easily identify target states often elucidated according to societal values (which may include socioeconomic or political factors) or intuitive/consensual sustainability goals, but LCA can be instrumental in reaching goals and target states in a holistically sustainable manner and shedding light on challenges faced when working towards such goals. In the following, some additional challenges in interpreting LCA results are outlined.

Contribution to Impact Results

Especially for LCAs on chemical products or processes, it is important to transparently report and document the contribution of different chemicals to impacts related to product life cycle stages and individual processes. This can help identify potential problems in the processing of LCI or LCIA results (e.g. if one chemical dominates results). Interpreting LCA results might be particularly challenging if it is not clear whether toxicity-related impacts are associated with chemical *emissions* occurring along the product life cycle or, in contrast, with chemicals that are product *ingredients* (Roos and Peters 2015). As an example, glass used as food packaging can show higher potential toxicity impacts compared with plastic packaging due to transport-related emissions of toxic chemicals from fossil fuel burning (Humbert et al. 2009), which is linked to the fact that glass is usually heavier than plastic. However, plastic food packaging can likely lead to greater exposures to various chemicals through food than glass, but this aspect is not (yet) considered in current LCIA toxicity models (Ernstoff et al. 2016b). Further, it might be unclear whether worker and/or consumer exposure pathways are included as these are currently beyond the scope of LCA studies focusing primarily on environmental emissions. The covered pathways and exposed populations should always be clarified in an LCA study to avoid possible misinterpretation of results. This is of specific relevance for the comparison of chemicals and chemical products and processes, where such ambiguities can cause confusion regarding the contribution of chemicals and related impact pathways and life cycle stages.

Identification of Considered Chemicals

In any of the aforementioned contexts, it must be acknowledged that most chemicals have various common names (lindane, CAS RN: 58-89-9, is for example also commonly known as HCH, hexachlorobenzene, or cyclohexane, etc.). Hence, it is important to ensure that names for chemicals in the different phases of an LCA study (e.g. inventory analysis and impact assessment) are consistently chosen based on using CAS registry numbers or similar as chemical identifier to, e.g. avoid double counting or neglecting chemicals with ambiguous names. This exercise can prove to be challenging as LCA software packages often report chemical inventories by chemical name and not by CAS number.

Quality and Uncertainty

Quality checks across the large number of inventoried chemicals is usually difficult, but inventory results should nevertheless be verified by, e.g. checking the mass balance of only those chemicals that drive overall impact results, for examples heavy metals that often dominate toxicity impact profiles. Furthermore, it is essential to report and discuss uncertainties of LCA data and results with respect to each impact category as integral part of the analysis, and consider such uncertainties in the interpretation of results and guidance provided to decision makers (see Chap. 11).

Particularly uncertainty associated with toxicity characterisation results is high compared with other impact categories and results can furthermore differ between toxicity characterisation methods, which can in some cases influence the overall ranking of compared product systems. Uncertainty (lack of data or understanding) and variability (data heterogeneity) are distinct concepts, but are sometimes (incorrectly) aggregated. For example, often 'high uncertainty' is perceived negatively or seen to discredit a particular LCIA method. However, such 'uncertainty' can be a direct reflection of reality and variabilities in temporal and spatial chemical fate and organism disease responses (see Chap. 11 for further details). Likewise, if an impact category has low or no associated uncertainty, this is perceived as positive but should in fact be a warning sign that there may be a lack of understanding of what uncertainties and/or variabilities exist or that the environmental relevance (or representativeness) of an indicator may be low (which introduces an uncertainty in the interpretation phase, but this is usually not quantified). To begin transparently addressing this issue, impacts should ideally be cross-compared using different LCIA characterisation methods with particular focus on identifying which chemicals contribute the most to impacts in each LCIA method (which are often not the same). Moreover, uncertainty ranges for toxicity-related impacts should be reported in logarithmic scale to put average uncertainties of two to three orders of magnitude into perspective of more than 15 orders of magnitude in the variability across chemicals. This is shown in Fig. 31.7 for 786 chemicals with available measured ecotoxicity effective concentrations for 50% of the exposed species (EC50; mg/L) for aquatic ecosystems. EC50 values are used to calculate effect factors as part of toxicity characterisation in LCIA (see Sect. 10.11). The relation between uncertainty and cross-chemical variability is not much different for toxicity impacts than for other impacts, where uncertainty in characterisation results (of usually only a handful of contributing chemicals) and related variability across contributing chemicals are both less broad. However, uncertainty ranges vary widely between chemicals, but chemical-specific uncertainty around characterisation factors is usually not available in LCA, except for specific pathways, e.g. exposure to pesticide residues in food crops (Fantke and Jolliet 2016), where also the underlying method to quantify chemical-specific uncertainty is outlined.

Comparison with Results from Other Methods
Comparing results from an LCA study with results from a different method can help identify methodological inconsistencies that require further inspection. As an example, it might be desired to compare the ranking of chemicals in terms of their potential toxicity impacts on humans and/or ecosystems in an LCA study with the ranking of chemicals based on persistence, bioaccumulation and toxicity or other criteria used, e.g. by risk regulators. In this context, it is important to acknowledge that inconsistencies can result from the primary data used in an LCA versus another

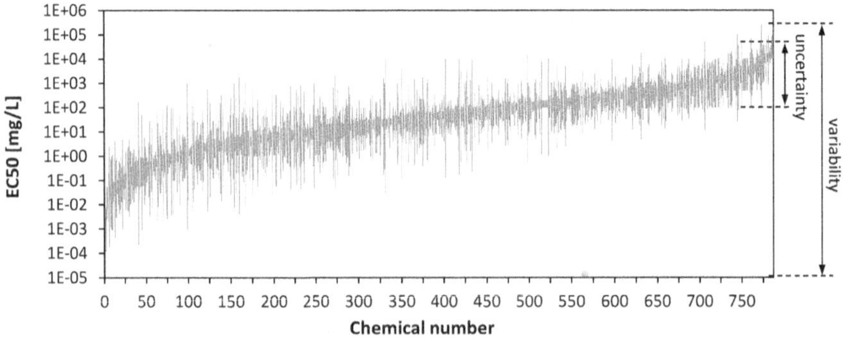

Fig. 31.7 Ranges of measured chemical-specific ecotoxicity effective concentrations (50% of exposed species affected), EC50, for aquatic ecosystems collected and indicated as reliable for 786 chemicals based on REACH (echa.europa.eu/regulations/reach)

method, or assumptions and cut-offs may be based on different criteria, e.g. worst case versus best estimate or most sensitive species versus average ecosystem sensitivity (Harder et al. 2015). This might lead to problems when comparing chemical rankings based on different assessment methods and data sources. Chemical toxicity results may furthermore differ between regions, countries or assessment methods, and thereby the consideration of chemicals as, e.g. 'non-carcinogenic' in LCA toxicity characterisation models may not be consistent with a specific regulatory context, such as the Registration, Evaluation, Authorisation, and Restriction of Chemicals (REACH) legislation framework of the European Union.

31.4 Conclusions

Stakeholders commissioning an LCA study can drive the goal and scope, the selection of inventory processes, and the selection of impact categories. In many cases, this can lead to an assessment that is restricted, for example to greenhouse gas emissions and a focus on climate change. The limited scope of such studies must be considered in the interpretation and application of their results, whenever other important impact pathways for chemical production, use, and disposal are not assessed. It is always important to be critical towards LCA outcomes and understand their limitations and scope, and respect that no tool (including LCA) can answer all questions related to chemicals and sustainability.

Not only can the scope of an LCA study be intentionally restricted according to its goal and scope, but there are several remaining challenges that also limit LCA, such as partial coverage of chemical inventory data, fate modelling (e.g. regional variation), exposure pathways (e.g. dermal exposure of consumers), and characterisation of potential human and ecosystem toxicity impacts. Given that there are tens of thousands of commercially used chemicals, and often little data on their properties or effects, the challenge of addressing chemical risks and impacts is not unique to LCA. Generally, the various methods for characterising risk and impacts of chemicals face similar challenges of data availability, but they also face methodological challenges and intentional differences. For example, results of an LCA addressing several impact categories and hundreds of chemicals, where often the exact emission location and timing is unknown, are difficult to cross-compare with results of a toxicity-focused risk assessment considering specific (e.g. worst-case) conditions and only one or several chemicals of concern (Harder et al. 2015).

Attempts of combining LCA with principles of green and sustainable chemistry, combining LCA- and risk-based approaches, and including life cycle impacts in chemical alternatives assessment frameworks demonstrate the growing complementarity and relevance of the life cycle approach in other science-policy fields (Jimenez-Gonzalez and Overcash 2014; Harder et al. 2015; Jacobs et al. 2016). Overall, the number of LCA studies focusing on chemicals or chemical products or processes is growing; thus, increasing discourse and trust in LCA methods as well as improving existing inventory and impact characterisation approaches.

Over the past years of research, LCA has developed into a powerful tool to identify and assess trade-offs and burden shifting between different environmental issues, identify hotspots and minimise overall environmental impacts of chemicals emitted along the life cycles of products and processes. With rising interest in creating 'environmentally friendly' chemicals and products, LCA is particularly important to help avoiding 'green washing' and unsupported sustainability claims. A common example is the comparison of products that can be developed purely from petrochemicals and also from a combination of petro- and biochemicals. Larger potential greenhouse gas emissions in the petrochemical production are confronted with often larger land use and pesticide-related toxicity impacts from agricultural crop production when serving as feedstock for biochemical production (Tabone et al. 2010; Cespi et al. 2015). Only comparing climate change impacts in this context would lead to false conclusions (i.e. that bio-based chemicals are always 'greener') and does not help identify how to optimise production processes and resource use when moving from petrochemicals to biochemicals in, e.g. plastics production. This is especially relevant when assessing emerging technologies, where there is a high level of optimisation potential in the years to come for upscaling lab-level processes to a commercial level.

Future research related to chemicals and LCA should focus on identifying and resolving areas of high uncertainty (such as changes through space and time), filling data gaps (for example with high-throughput exposure and toxicity modelling approaches), and addressing issues of high concern such as consumer and occupational exposure and other toxicity endpoints (e.g. toxicity to bees). Furthermore, applying LCA in case studies and analyses to address issues of existing and emerging technologies can help pinpoint and corroborate solutions towards more sustainable production and consumption of synthetic and naturally occurring chemicals.

References

Abraham, M.A., Nguyen, N.: "Green engineering: defining the principles"—results from the Sandestin conference. Environ. Prog. **22**, 233–236 (2003)

Akanuma, Y., Selke, S.E.M., Auras, R.: A preliminary LCA case study: comparison of different pathways to produce purified terephthalic acid suitable for synthesis of 100% bio-based PET. Int. J. Life Cycle Assess. **19**, 1238–1246 (2014)

Alvarez-Gaitan, J.P., Peters, G.M., Rowley, H.V., Moore, S., Short, M.D.: A hybrid life cycle assessment of water treatment chemicals: an Australian experience. Int. J. Life Cycle Assess. **18**, 1291–1301 (2013)

Anastas, P.T., Lankey, R.L.: Life cycle assessment and green chemistry: the yin and yang of industrial ecology. Green Chem. **2**, 289–295 (2000)

Anastas, P., Warner, J.: Green Chemistry: Theory and Practice. Oxford University Press, New York (1998)

Bakas, I., Hauschild, M.Z., Astrup, T.F., Rosenbaum, R.K.: Preparing the ground for an operational handling of long-term emissions in LCA. Int. J. Life Cycle Assess. **20**, 1444–1455 (2015)

Bare, J.C.: Risk assessment and Life-cycle impact assessment (LCIA) for human health cancerous and noncancerous emissions: Integrated and complementary with consistency within the USEPA. Hum. Ecol. Risk Assess. **12**, 493–509 (2006)

Botta, C., Labille, J., Auffan, M., Borschneck, D., Miche, H., Cabié, M., Masion, A., Rose, J., Bottero, J.-Y.: TiO_2-based nanoparticles released in water from commercialized sunscreens in a life-cycle perspective: structures and quantities. Environ. Pollut. **159**, 1543–1550 (2011)

CEFIC: The European Chemical Industry Facts and Figures 2014, p. 54. European Chemical Industry Council, Brussels (2014)

Cespi, D., Passarini, F., Mastragostino, G., Vassura, I., Larocca, S., Iaconi, A., Chieregato, A., Dubois, J.L., Cavani, F.: Glycerol as feedstock in the synthesis of chemicals: a life cycle analysis for acrolein production. Green Chem. **17**, 343–355 (2015)

De Soete, W., Debaveye, S., De Meester, S., Van der Vorst, G., Aelterman, W., Heirman, B., Cappuyns, P., Dewulf, J.: Environmental sustainability assessments of pharmaceuticals: an emerging need for simplification in life cycle assessments. Environ. Sci. Technol. **48**, 12247–12255 (2014)

Embry, M.R., Bachman, A.N., Bell, D.R., Boobis, A.R., Cohen, S.M., Dellarco, M., Dewhurst, I. C., Doerrer, N.G., Hines, R.N., Morett, A., Pastoor, T.P., Phillips, R.D., Rowlands, J.C., Tanir, J.Y., Wol, D.C., Doe, J.E.: Risk assessment in the 21st century: roadmap and matrix. Crit. Rev. Toxicol. **44**, 6–16 (2014)

Ernstoff, A.S., Fantke, P., Csiszar, S.A., Henderson, A.D., Chung, S., Jolliet, O.: Multi-pathway exposure modelling of chemicals in cosmetics with application to shampoo. Environ. Int. **92–93**, 87–96 (2016a)

Ernstoff A, Muncke J, Trier X, Niero M, Fantke P (2016b) Exposure to chemicals in food packaging as a sustainability trade-off in LCA. In: 10th International Conference on LCA of Food, 19–21 Oct 2016, Dublin, Ireland, pp. 336–343

Fantke, P., Jolliet, O.: Life cycle human health impacts of 875 pesticides. Int. J. Life Cycle Assess. **21**, 722–733 (2016)

Fantke, P., Wieland, P., Juraske, R., Shaddick, G., Itoiz, E.S., Friedrich, R., Jolliet, O.: Parameterization models for pesticide exposure via crop consumption. Environ. Sci. Technol. **46**, 12864–12872 (2012)

Fantke, P., Weber, R., Scheringer, M.: From incremental to fundamental substitution in chemical alternatives assessment. Sustain. Chem. Pharm. **1**, 1–8 (2015)

Geisler, G., Hellweg, S., Hofstetter, T.B., Hungerbühler, K.: Life-cycle assessment in pesticide product development: methods and case study on two plant-growth regulators from different product generations. Environ. Sci. Technol. **39**, 2406–2413 (2005)

Harder, R., Holmquist, H., Molander, S., Svanström, M., Peters, G.M.: Review of environmental assessment case studies blending elements of risk assessment and life cycle assessment. Environ. Sci. Technol. **49**, 13083–13093 (2015)

Hauschild, M.Z.: Assessing environmental impacts in a life-cycle perspective. Environ. Sci. Technol. **39**, 81A–88A (2005)

Hauschild, M.Z., Goedkoop, M., Guinée, J., Heijungs, R., Huijbregts, M., Jolliet, O., Margni, M., De Schryver, A., Humbert, S., Laurent, A., Sala, S., Pant, R.: Identifying best existing practice for characterization modeling in life cycle impact assessment. Int. J. Life Cycle Assess. **18**, 683–697 (2013)

Hester, R.E., Harrison, R.M.: Chemical Alternatives Assessments. RSC Publishing, Royal Society of Chemistry, Cambridge (2013)

Hottle, T.A., Bilec, M.M., Landis, A.E.: Sustainability assessments of bio-based polymers. Polym. Degrad. Stab. **98**, 1898–1907 (2013)

Humbert, S., Rossi, V., Margni, M., Jolliet, O., Loerincik, Y.: Life cycle assessment of two baby food packaging alternatives: glass jars vs. plastic pots. Int. J. Life Cycle Assess. **14**, 95–106 (2009)

Jacobs, M.M., Malloy, T.F., Tickner, J.A., Edwards, S.: Alternatives assessment frameworks: research needs for the informed substitution of hazardous chemicals. Environ. Health Perspect. **124**, 265–280 (2016)

Jacquemin, L., Pontalier, P.-Y., Sablayrolles, C.: Life cycle assessment (LCA) applied to the process industry: a review. Int. J. Life Cycle Assess. **17**, 1028–1041 (2012)

Jimenez-Gonzalez, C., Overcash, M.R.: The evolution of life cycle assessment in pharmaceutical and chemical applications—a perspective. Green Chem. **16**, 3392–3400 (2014)

Jonkers, N., Krop, H., Ewijk, H., Leonards, P.E.G.: Life cycle assessment of flame retardants in an electronics application. Int. J. Life Cycle Assess. **21**, 146–161 (2016)

Karka, P., Papadokonstantakis, S., Hungerbühler, K., Kokossis, A.: Life cycle assessment of biorefinery products based on different allocation approaches. Comput. Aided Chem. Eng. **37**, 2573–2578 (2015)

Kounina, A., Margni, M., Shaked, S., Bulle, C., Jolliet, O.: Spatial analysis of toxic emissions in LCA: a sub-continental nested USEtox model with freshwater archetypes. Environ. Int. **69**, 67–89 (2014)

Kralisch, D., Ott, D., Gericke, D.: Rules and benefits of life cycle assessment in green chemical process and synthesis design: a tutorial review. Green Chem. **17**, 123–145 (2015)

Laurent, A., Olsen, S.I., Hauschild, M.Z.: Limitations of carbon footprint as indicator of environmental sustainability. Environ. Sci. Technol. **46**, 4100–4108 (2012)

MacLeod, M., Breitholtz, M., Cousins, I.T., de Wit, C.A., Persson, L.M., Rudén, C., McLachlan, M.S.: Identifying chemicals that are planetary boundary threats. Environ. Sci. Technol. **48**, 11057–11063 (2014)

Neltner, T.G., Alger, H.M., Leonard, J.E., Maffini, M.V.: Data gaps in toxicity testing of chemicals allowed in food in the United States. Reprod. Toxicol. **42**, 85–94 (2013)

Nemecek, T., Jungbluth, N., Canals, L.M., Schenck, R.: Environmental impacts of food consumption and nutrition: where are we and what is next? Int. J. Life Cycle Assess. **21**, 607–620 (2016)

OECD (2012) The Role of Government Policy in Supporting the Adoption of Green/Sustainable Chemistry Innovations. Organisation for Economic Co-operation and Development, Paris, p. 50

Parisi, M.L., Fatarella, E., Spinelli, D., Pogni, R., Basosi, R.: Environmental impact assessment of an eco-efficient production for coloured textiles. J. Clean. Prod. **108**, 514–524 (2015)

Pati, P., McGinnis, S., Vikesland, P.J.: Life cycle assessment of "green" nanoparticle synthesis methods. Environ. Eng. Sci. **31**, 410–420 (2014)

Pennington, D.W., Margni, M., Payet, J., Jolliet, O.: Risk and regulatory hazard-based toxicological effect indicators in life-cycle assessment. Hum. Ecol. Risk Assess. **12**, 450–475 (2006)

Posthuma, L., Bjørn, A., Zijp, M.C., Birkved, M., Diamond, M.L., Hauschild, M.Z., Huijbregts, M.A.J., Mulder, C., van de Meent, D.: Beyond safe operating space: finding chemical footprinting feasible. Environ. Sci. Technol. **48**, 6057–6059 (2014)

Roos, S., Peters, G.M.: Three methods for strategic product toxicity assessment—the case of the cotton T-shirt. Int. J. Life Cycle Assess. (2015). doi:10.1007/s11367-11015-10895-11366

Roos, S., Posner, S., Jönsson, C., Peters, G.M.: Is unbleached cotton better than bleached? Exploring the limits of life-cycle assessment in the textile sector. Cloth. Text. Res. J. **33**, 231–247 (2015)

Rosenbaum, R.K., Bachmann, T.M., Gold, L.S., Huijbregts, M.A.J., Jolliet, O., Juraske, R., Koehler, A., Larsen, H.F., MacLeod, M., Margni, M.D., McKone, T.E., Payet, J., Schuhmacher, M., van de Meent, D., Hauschild, M.Z.: USEtox—the UNEP-SETAC toxicity model: recommended characterisation factors for human toxicity and freshwater ecotoxicity in life cycle impact assessment. Int. J. Life Cycle Assess. **13**, 532–546 (2008)

Rosenbaum, R.K., Anton, A., Bengoa, X., Bjørn, A., Brain, R., Bulle, C., Cosme, N., Dijkman, T. J., Fantke, P., Felix, M., et al.: The Glasgow consensus on the delineation between pesticide emission inventory and impact assessment for LCA. Int. J. Life Cycle Assess. **20**, 765–776 (2015)

Shin, H.-M., Ernstoff, A.S., Arnot, J.A., Wetmore, B., Csiszar, S.A., Fantke, P., Zhang, X., McKone, T.E., Jolliet, O., Bennett, D.H.: Risk-based high-throughput chemical screening and prioritization using exposure models and in vitro bioactivity assays. Environ. Sci. Technol. **49**, 6760–6771 (2015)

Tabone, M.D., Cregg, J.J., Beckman, E.J., Landis, A.E.: Sustainability metrics: life cycle assessment and green design in polymers. Environ. Sci. Technol. **44**, 8264–8269 (2010)

Tickner, J.A., Schifano, J.N., Blake, A., Rudisill, C., Mulvihill, M.J.: Advancing safer alternatives through functional substitution. Environ. Sci. Technol. **49**, 742–749 (2015)

Tufvesson, L.M., Tufvesson, P., Woodley, J.M., Börjesson, P.: Life cycle assessment in green chemistry: overview of key parameters and methodological concerns. Int. J. Life Cycle Assess. **18**, 431–444 (2013)

van Leeuwen, C.J., Vermeire, T.G.: Risk Assessment of Chemicals: An Introduction, 2nd edn. Springer Press, Dordrecht (2007)

Van Lieshout, K.G., Bayley, C., Akinlabi, S.O., von Rabenau, L., Dornfeld, D.: Leveraging life cycle assessment to evaluate environmental impacts of green cleaning products. Procedia CIRP **29**, 372–377 (2015)

Wang, Q., Vural Gürsel, I., Shang, M., Hessel, V.: Life cycle assessment for the direct synthesis of adipic acid in microreactors and benchmarking to the commercial process. Chem. Eng. J. **234**, 300–311 (2013)

WBCSD: Life Cycle Metrics for Chemical Products, p. 120. World Business Council for Sustainable Development, Geneva (2014)

WEF: The New Plastics Economy: Rethinking the Future of Plastics, p. 36. World Economic Forum, Geneva (2016)

Weidema, B.: Avoiding co-product allocation in life-cycle assessment. J. Ind. Ecol. **4**, 11–33 (2000)

Wernet, G., Conradt, S., Isenring, H.P., Jiménez-González, C., Hungerbühler, K.: Life cycle assessment of fine chemical production: a case study of pharmaceutical synthesis. Int. J. Life Cycle Assess. **15**, 294–303 (2010)

Whittaker, M.H.: Risk assessment and alternatives assessment: comparing two methodologies. Risk Anal. **35**, 2129–2136 (2015)

Yuan, G., Peng, H., Huang, C., Hu, J.: Ubiquitous occurrence of fluorotelomer alcohols in eco-friendly paper-made food-contact materials and their implication for human exposure. Environ. Sci. Technol. (2016). doi:10.1021/acs.est.1025b03806

Zhang, Y., Bakshi, B.R., Demessie, E.S.: Life cycle assessment of an ionic liquid versus molecular solvents and their applications. Environ. Sci. Technol. **42**, 1724–1730 (2008)

Author Biographies

Peter Fantke develops methods for LCIA, health impact assessment and chemical alternatives assessment since 2006. Has contributed to UNEP/SETAC LCIA working groups and is USEtox Manager. He is mainly interested in quantifying and characterising chemical emissions, uncertainty analysis, consumer exposure, chemical substitution and model parameterisation.

Alexi Ernstoff Studied various aspects of chemical fate, transport, and exposure since 2007. Recent focus is modelling human exposure to chemicals in products for LCIA. Main interest is ensuring human health impacts, mediated by using consumer and food products, is consistently considered in quantitative sustainability assessments.

Chapter 32
LCA of Nanomaterials

Mirko Miseljic and Stig Irving Olsen

Abstract Application of nanomaterials in products has led to an increase in number of nanoproducts introduced to the consumer market. However, along with new and improved products, there is a concern about the potential life cycle environmental impacts. Life cycle assessment is able to include a wide range of environmental impacts but, due to data limitations, it is commonly applied with focus on the cradle-to-gate part of the nanoproducts life cycle, neglecting use and disposal of the products. These studies conclude that nanomaterials are more energy demanding and have an inferior environmental profile than conventionally used materials, but functional units of these comparisons need to consider the use stage benefits attained through nanomaterials. A particular assessment challenge is the lack of understanding of the toxicological mechanisms related to potential release, fate and effects of nanomaterials when penetrating into living organisms. This is especially relevant for the freshwater compartment, as it is a common recipient.

32.1 Introduction

The basis of the nano-technology terminology is the nanometre, which is one billionth of a metre (10^{-9} m). Nano-scale is defined as the range from 0.1 to 1000 nm, nanomaterial as a material with at least one dimension within 1–100 nm, and nanoparticle as a particle with all three dimensions within the 1–100 nm range (ISO 2008; SCENIHR 2007). Nanomaterials and nanoparticles can be naturally or unintentionally produced, and they have always been present in the human surroundings. Examples of these are, e.g., soil (dust) and salt particles in air, to which humans are commonly exposed. These naturally occurring nanomaterials are in general known to cause little harm to humans (Buzea et al. 2007). Other sources of naturally occurring nanoparticles are, e.g. forest fires or volcanos, the particles from

M. Miseljic (✉) · S.I. Olsen
Division for Quantitative Sustainability Assessment, Department of Management Engineering, Technical University of Denmark, Kgs. Lyngby, Denmark
e-mail: mimi@dti.dk

M.Z. Hauschild et al. (eds.), *Life Cycle Assessment*,
DOI 10.1007/978-3-319-56475-3_32

which may have the same impacts as the anthropogenic, unintentionally produced particles arising from combustion activities. The other path to production of nanomaterials is by manufacturing activities through either downscaling (from bulk) or upscaling (from atomic or molecular) materials. These activities may also cause unintentional emissions of nanoparticles. The manufacturing of engineered nanomaterials (ENMs) is linked to application of nanomaterials in/on consumer products, as alteration of materials from bulk to nano-size leads to an increase in surface area and improved functionality. As an example, gold becomes reactive at nano-size and may be used to increase oxidation in car catalytic converters so emission of pollutants is reduced. Further, ENMs may also provide products with, e.g. improved material properties like hydrophobicity (lotus effect), strength and/or electrical conductivity.

For quite some time particles from combustion have been known to cause harmful human impacts, even though the exact mechanisms are still being researched. ENMs and nanoparticles (ENPs) also cause concerns, but their behaviour in the environment and potential impacts to environment and humans are to a large extent still unknown (Miseljic and Olsen 2014; Jolliet et al. 2013). In order to embrace the entire life cycle of ENMs, and to avoid burden shifting, the LCA approach is favourable in order to quantify potential environmental impacts—not only as a single approach but also as a framework to be applied along with other methods (Som et al. 2010).

32.1.1 Nanoproducts and Environmental Assessment

Nano-technology and the application of ENMs in products have developed much in recent years. The reason for this is that ENMs are able to improve properties and functionalities of different materials, and thereby the consumer products. This has meant that companies have developed new and smart products, resulting in more nanoproducts being introduced to the consumer market (see Figs. 32.1 and 32.2). The more common nanoproducts are within the product category of health and fitness, with TiO_2-enhanced sunblock and Ag-enhanced clothing as prominent examples. The home and garden product category contains the second most widely used nanoproducts (mainly as sealants and coatings), as seen in Fig. 32.1. The most commonly used nanomaterials are based on Ag, TiO_2, and carbon as shown in Fig. 32.2.

In line with the rapid introduction of nanoproducts, and thereby ENMs, concerns are raised in terms of the potential environmental impacts these may have along their life cycle. Currently, chemical risk assessment (RA) is commonly performed on ENMs, but this approach has a different scope compared to LCA. RA is a procedure applied in order to estimate if a toxicological risk occurs from a substance, thus calculating, measuring or modelling the existence of risk derived from chemical exposure (European commission 2007; ECHA 2010, see also Sect. 31.1.3). LCA on

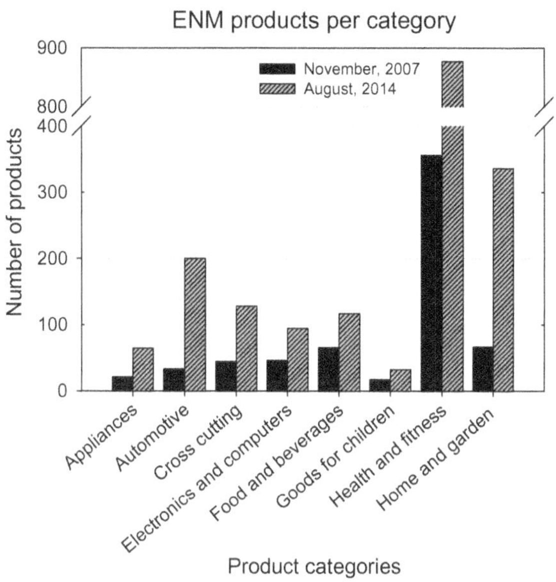

Fig. 32.1 Number of ENM products per product category (PoEN 2014; Hansen 2009; Miseljic 2014)

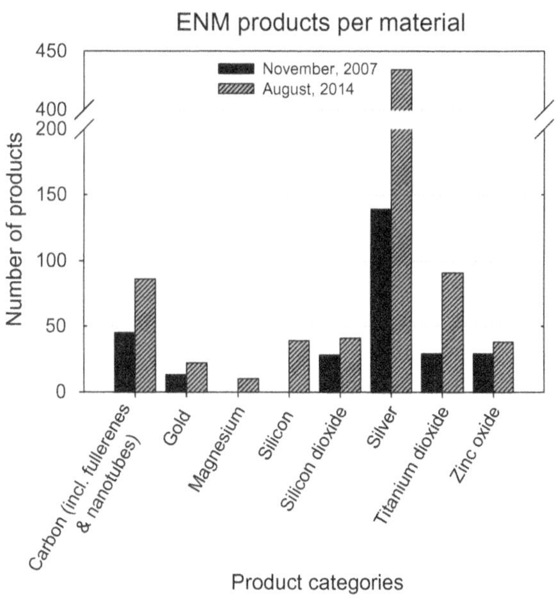

Fig. 32.2 ENM products per most frequently used material, in year 2007 and 2014 (PoEN 2014; Hansen 2009; Miseljic 2014)

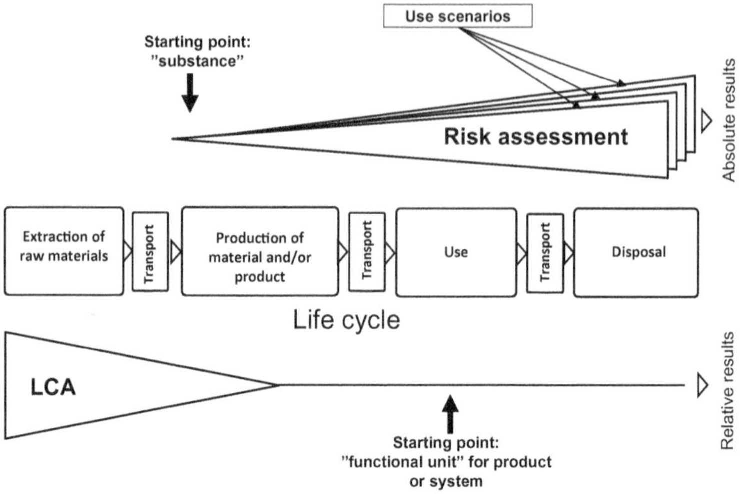

Fig. 32.3 RA and LCA coverage and starting point when addressing a product or system (Grieger et al. 2012)

the other hand is a relative environmental impact assessment method that also considers a wide range of environmental impact categories (ISO 2006). In LCA, the starting point of the assessment is the functional unit, where products or systems are studied, whereas RA is commonly substance oriented, see also Fig. 32.3.

There are benefits and challenges to applying both RA and LCA, due to these two approaches being developed with different initial purposes. The benefit of applying RA for ENMs is that it targets specific emission, exposure and dose-response conditions, but the downside is that the uncertainty is still high due to lack of data and lack of proper regulation. A refinement of regulation is needed, e.g. for the characterisation of ENMs applied in laboratory testing and thereby the use of appropriate metrics for expressing hazard and exposure. In addition to ENM mass, Aitken et al. (2011) and Hankin et al. (2011) proposed particle number and surface area as additional characteristic metrics.

LCA conveys an assessment of a wider range of environmental impacts, but as for RA there are also challenges to this approach. One of these is, as for RA, the lack of needed data and another is the lack of consideration of impact from ENM/ENP released to the environment.

Sweet and Strohm (2006) and Som et al. (2010) outlined that RA should consider more life cycle concepts and LCA should be more risk-based, when dealing with future environmental impact assessment of ENMs. This underlines the general approach that LCA should be applied as a complimentary framework along with other environmental impact assessment approaches (Grieger et al. 2012).

32.2 Literature Review

In order to evaluate the current assessment status of ENMs, and the application of LCA in this context, published scientific articles were reviewed. Only 29 LCA case studies were found and reviewed according to strengths and weaknesses, and the challenges the assessments represent (Miseljic and Olsen 2014). Jolliet et al. (2013) identified 43 studies but use a broader definition of LCA. Some studies did not perform LCA according to usual guidelines (e.g. ISO 2006), but claimed to include life cycle thinking and commonly focused on energy consumption of ENM manufacturing.

32.2.1 Impact Hotspots

Commonly, among the reviewed studies in Miseljic and Olsen (2014) and Jolliet et al. (2013), it can be concluded that, on a same produced mass basis, the manufacture of nanomaterials is found to have significantly higher energy requirements compared to the production of conventional materials such as aluminum or steel (shown e.g. in Khanna et al. 2007, 2008). Even though there are discrepancies of up to several orders of magnitude between the energy requirements for the same nanomaterials reported by different studies this often results in a less favourable cradle-to-gate environmental impact profile of ENM products compared to comparable conventional products. The high energy demand for production emphasise that one should be careful with normally accepted cut-off criteria when making inventories of nanoproducts since the product generally contains only few percent of ENMs. In terms of impact categories, those that are dominating for ENM production are nonrenewable resource depletion, global warming, acidification and impacts caused by airborne inorganics. However, due to the improved product functionality, the use stage of ENM products is often more environmentally friendly than for conventional products. This is exemplified In Lloyd and Lave (2003) and Lloyd et al. (2005), where ENM enhanced clay-polypropylene ENMs in car body panels and platinum-group metal particles in car catalysts have an environmentally friendlier use stage. However, potential release of ENMs/ENPs to the environment is not considered in these studies.

Most of the reviewed LCA studies focused on the manufacturing of nanomaterials, thus limiting the assessment scope to the energy consumption of the cradle-to-gate part. The use and disposal stages are commonly not considered, with few exceptions, e.g. Walser et al. (2011) who also consider toxicity from ENMs/ENPs from nanoproducts. However, the tendency is also that these studies rely on generic data, as seen in e.g. Grubb and Bakshi (2010), Osterwalder et al. (2006), Tibbetts et al. (1994), Hwang et al. (2005), Healy (2006), and Healy et al. (2008).

32.2.2 *Overall Findings*

The tendencies across the identified LCAs on ENMs can be summed up and also reflect the LCA state of knowledge regarding ENMs (Miseljic and Olsen 2014):

- Usually the studies consider cradle-to-gate or manufacturing system boundary.
- Use and disposal stages of the life cycle are poorly covered.
- Functional unit does not always consider functional benefits of ENMs.
- Common use of generic LCI data and assumptions, e.g. for upscaling of laboratory data.
- Almost no consideration of ENM release (e.g. in use or disposal stages) and the potential toxic impacts from these (fate, exposure and effect consideration). Walser et al. (2011) and Meyer et al. (2010) are exceptions.
- Cradle-to-gate LCA comparison of counterpart products (with ENMs and without) show that ENM products are more energy demanding and therefore have a worse cradle-to-gate environmental profile, e.g. in polymer nanocomposites versus steel and socks with and without nano (Moign et al. 2010; Meyer et al. 2010).
- Cradle-to-grave LCA comparison of counterpart products (with ENMs and without) show that the use stage is better for ENM products as usually an improved functionality is achieved, e.g. comparing clay-propylene nanocomposites with steel or aluminium in light duty vehicles (Osterwalder et al. 2006). End-of-life performance of ENM products is rarely considered.

32.3 General Methodological Issues

32.3.1 *Goal and Scope Definition*

Generally, it is found that the *system boundaries* for a fair comparison to products containing ENMs should include both the use stage where the beneficial functionalities of the ENMs are expressed, and the disposal stage. Certain reviewed studies consider the use and disposal stages, but their coverage is rather incomplete (Lloyd and Lave 2003; Lloyd et al. 2005; Babaizadeh and Hassan 2012; Manda et al. 2012; Roes et al. 2007; Steinfeldt et al. 2010). In the quoted example by Lloyd and Lave (2003) on clay-polypropylene ENMs in car body panels, the use stage is assessed by solely including the resource savings (fuel consumption) when driving with the lighter ENM enhanced panels. Other supplementary materials are not considered, meaning that release of other agents is not considered in the use stage (Som et al. 2010). In addition, the disposal stage is commonly not dealt with due to lack of knowledge in end-of-life treatment of ENMs and also which disposal processes they will be subject to (landfilling, incineration, or recycling).

In order to perform a comparable LCA the *functional unit* is central, as a comparable functionality of products or systems needs to be applied. Certain ENM studies tend to apply a simplified mass based functional unit, e.g. relating to 1 kg of an ENM product (Joshi 2008; Kushnir and Sandén 2008; Grubb and Bakshi 2010). However, a mass-based functional unit does not make sense when comparing ENMs with conventional products, as functionality is not proportional with weight (Hischier and Walser 2012). The improved material functionality, when using ENMs in products, needs to be considered in the functional unit. This means that higher resource and energy use in ENM production, compared to conventional additives, may be justified in the use stage by leading to less environmental impacts.

As an example, Roes et al. (2007) include elasticity (Young modulus) and tensile strength in the functional unit when comparing polypropylene (PP)/layered silicate nanocomposites with conventional PP, reason being that the nanocomposite and PP mix obtains the needed material properties at a lower weight. In this approach Roes et al. (2007) scale the functional unit, but in general it may be difficult to identify the most important properties that are to be applied in a functional unit of a fair LCA comparison.

32.3.2 Inventory Data

The inventory modelling of a nanoproduct typically involves background processes that are not specific to nano and thus not of specific interest here, while foreground process (e.g. data for production of ENM; and direct nanoparticle emissions) should be specific of the considered process (Jolliet et al. 2013). Commonly, as identified in the 29 LCA on ENMs studies from Miseljic and Olsen (2014), the studies rely on generic data for production of ENMs. Primary process data, on the other hand, is often difficult to acquire from the ENM industry. This can be due to several factors, but mainly to the relative novelty of the scientific field and the competition within the technology domain. This ongoing tendency leads to a higher level of uncertainty, as generic data and estimations need to be applied (e.g. in Bauer et al. 2008; Joshi 2008; Merugula et al. 2010; Isaacs et al. 2006). The immaturity of the field also lies behind the incomplete life cycle coverage in the performed LCAs with the very frequent omission of use and disposal stages from the product system modelling.

Life cycle inventory (LCI) modelling is gradually improving for ENMs, studies such as Geranio et al. (2009), Köhler et al. (2008), Künninger et al. (2010), Som et al. (2011), Suppen et al. (2005), Durucan et al. (2006) and Gutowski et al. (2010) do not perform LCAs per se, but are providing valuable LCI data on specific processes as well as some estimates of the direct release of nanoparticles. Geranio et al. (2009) quantifies the release of Ag ENPs during textile washing and Künniger et al. (2010) the release of nano-Ag from facades due to weathering.

Concerning the direct emissions of nanoparticles, the physical and chemical characteristics of the emitted nanoparticles are also of interest since these characteristics are needed to link up with the impact assessment.

32.3.3 *Impact Assessment*

Generally, the impact from background and foreground process is no different from other LCAs. Therefore, the main aspect to consider is the potential impacts from direct emissions of nanoparticles. Here the potential toxic effects have caused highest concern and the next chapter will deal exclusively with this.

32.3.4 *Interpretation*

It is evident from the previous text that LCAs of nanoproducts are rather uncertain and that an understanding of the main uncertainties is important (Jolliet et al. 2013). It is suggested to use tools for uncertainty assessment and sensitivity analysis as explained in Chap. 11.

32.4 Specific Methodological Issues for Ecotoxicological Impact Characterisation of ENMs/ENPs

As concluded in the previous section, the potential release of ENMs throughout the life cycle of ENM products is generally not considered. A sensitivity analysis of the importance of potential freshwater ecotoxicity impacts from Ag and TiO_2 ENM release from products underlined the need to consider potential impacts of such releases throughout a product's entire life cycle (Miseljic and Olsen 2014). The analysis also illustrates the differences between the impacts of different ENPs (i.e. higher freshwater ecotoxicity from Ag ENMs, compared to TiO_2 ENMs). The assessment of impacts of ENM release from cradle to grave improve the LCA of ENMs, and address an increasing environmental concern (Buzea et al. 2007; Bauer et al. 2008; Oberdörster et al. 2007; Jolliet et al. 2013). However, current lack of understanding of the mechanisms leading to toxicity, more precisely the ENM release, fate and potential effects when penetrating into living organisms, pose challenges for the assessment while being highly relevant; especially for the freshwater compartment, as it is a common recipient (Quik et al. 2011; Lowry et al. 2012; Som et al. 2010). According to Jones and Grainger (2009), the main hurdle is to predict the actual fate of the released ENMs. In the following, the most important aspects related to such an assessment are discussed as an example of the challenges specific for LCIA of nano-technology.

32.4.1 Particle Appearance

The primary ENP appearance differs over time and according to environmental conditions, as ENPs tend to agglomerate (coagulate), aggregate (fuse) or a combination thereof. The interaction and bonding happens in order to reduce the high-surface energy. Interaction between two ENPs, e.g. in liquid and air, can in theory be described by forces of van der Walls attractions and electrostatic repulsions (Rupasinghe R-A-TP 2011). The appearance of ENPs, along with the forces causing this, influences the toxicity of ENMs in e.g. water (Oberdörster et al. 2007).

32.4.2 Transformation

After release to freshwater, which is considered a common recipient, the ENMs are either subject to biotic (interaction with plants, water flea, fish etc.) or abiotic (interaction with water, sand, light, etc.) transformation. These can alter the shape, size, surface chemistry, and ultimately the fate of the released ENMs. Thus, physico-chemical properties of ENMs are important for the differentiated behaviour of ENMs in water and the water-phase processes considered important are:

- Aggregation/agglomeration
- Dissolution
- Sedimentation (and resuspension)
- Change in surface structure of ENMs/ENPs.

Commonly, freshwater fate of ENMs tends to be dominated by sorption to high-surface-area colloids with subsequent sedimentation (Klaine et al. 2008). Within the sediment the ENMs can be transported, and also re-suspended to the water phase, see Fig. 32.4.

32.4.3 Transport

Transport of ENMs is partially controlled by aggregation/agglomeration, which subsequently is followed by sedimentation. The aggregation will depend on parameters such as (Lowry and Casman 2009; Lowry et al. 2012):

- Hydrophobicity
- Chemical bonding between nanoparticles
- Ionic strength
- Ionic composition.

Ionic strength, pH and the presence of divalent cations such as Ca^{2+} and Mg^{2+} will influence the rate and extent of aggregation/agglomeration (e.g. a higher ionic

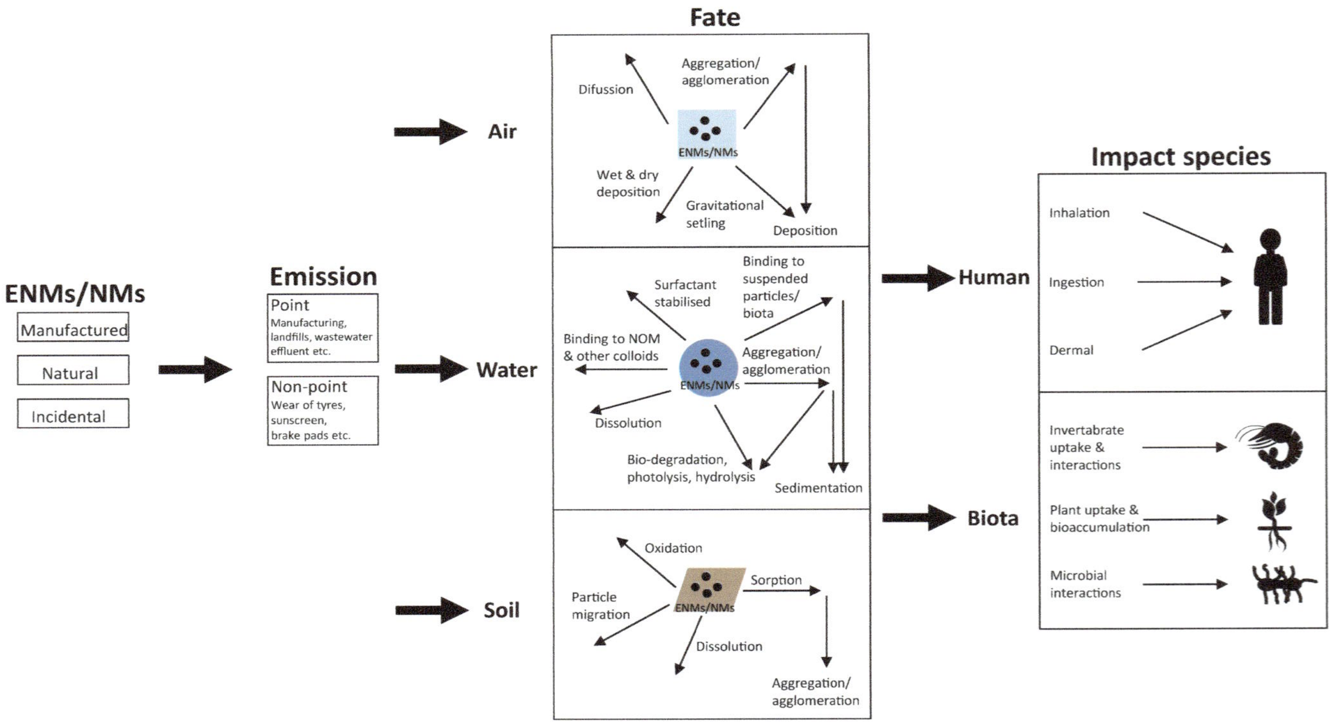

Fig. 32.4 Potential environmental pathways of released engineered and naturally occurring nanomaterials (ENMs and NMs) in air, water, and soil and related to common impact organisms (Miseljic and Olsen 2014)

strength will lead to more aggregation/agglomeration, as is the case for marine waters compared to freshwater). Brant et al. (2005) exemplify that if the ENM C_{60} fullerene is released into water with an ionic strength higher than 0.001 M the formed aggregates/agglomerates will sorb to other particles and media and eventually become immobilised. This happens in particular when pH is close to the isoelectric point (i.e. pH at which a molecule has no net electrical charge), since the particle charge is then lower and a change in repulsive forces is able to promote aggregation/agglomeration (Franklin et al. 2007; Illés and Tombácz 2006).

In water, the gravitational forces cause the aggregates/agglomerates to sediment, thus becoming less available to certain aquatic organisms, but more to the benthic organisms (Klaine et al. 2008; Lowry et al. 2012). Subsequently, turbulent motion in benthos and bio-turbulation in the sediment can cause the ENM to be re-suspended and become more available again in the water phase.

32.4.4 Important Physico-Chemical Characteristics

Considering the developing understanding of the fate of ENMs in freshwater, the following physico-chemical ENM properties are important to consider (Batley and McLaughlin 2010; Klaine et al. 2008):

- Chemical composition
- Mass
- Particle number and concentration
- Surface area concentration
- Size distribution
- Specific surface area
- Surface charge/zeta potential
- Surface contamination and the nature of any shell and capping
- Solubility
- Crystal structure.

In addition to the ENM specific properties, the natural conditions of the surrounding environment are also important when dealing with ENM fate. However, aggregation/agglomeration and dissolution alongside other co-related mechanisms are neither fully understood nor well represented with characterisation data, in relation to mechanisms shown in Fig. 32.4 (Farré et al. 2011).

32.4.5 Toxicity

The toxic effects from ENPs and ENMs depend on several parameters, e.g. size, surface/crystal structure, dissolution and aggregation/agglomeration. Size of ENPs is proven to have an influence on the level of toxicity, e.g. 48 h testing on Daphnia magna showed, according to Zhu et al. (2009), a 143 mg/L LC_{50} for <25 nm TiO_2 (20% rutile and 80% anatase crystal structure), while Heinlaan et al. (2008) observed a 20,000 mg/L LC_{50} for 25–75 nm TiO_2 (crystal structure not reported). These two studies along with several others (e.g. Kashiwada 2006; Hussain et al. 2009) indicate large variations between different nominal sizes of ENPs, but it needs to be underlined that tests are difficult to compare due to variations in test conditions. In addition to size, crystal structure also has an influence on toxicity, where, e.g. TiO_2 in anatase crystal structure is known to be more toxic to organisms than in rutile form.

Toxicity is dependent on the intrinsic toxicity potential of the ENMs, influenced by, e.g. size, and ions formed through oxidative dissolution (Scheringer et al. 2010). The high toxicity potential of free Ag ions in natural waters may be disrupted by the presence of complexing ligands, as they reduce the Ag ion concentration and thus the bioavailability (Scheringer et al. 2010). In addition, toxicological effects also depend on the surface structure of ENPs, where surface structure can be removed/altered, e.g. by natural and anthropogenic chemicals in the environment. Change in surface structure may result in enhanced mobility, bioavailability, aggregation (mainly hydrophobic surfaces), sedimentation, dissolution and dispersion (mainly hydrophilic surfaces), and consequently the actual exposure and toxicity may increase (Vonk et al. 2009; Lowry and Casman 2009; Ratte 1999). Further, the pH and presence of adsorbing molecules and ions have an influence on ENM fate and eventually the toxicity.

The correlation of various mechanisms in ENM behaviour and impacts, in contrast to single-chemical behaviour, means that single-chemical impact models are not suitable to be applied for ENMs (Lowry and Casman 2009).

32.5 Conclusion: What to Consider When Performing LCA on ENMs

Currently, LCA of ENMs is deficient in several areas. First, novelty of the nano-technology is limiting the availability of LCI data. Second, the potential release of ENMs/ENPs is difficult to include in LCA at this point. This is due to both lack of LCI data and an incomplete understanding of fate, exposure and effects (eco- and human toxicity characterisation factors). Figure 32.5 illustrates the current state of LCA for nano-technological products.

Based on already performed LCAs on ENMs there should be awareness of the following when aiming at performing an LCA on ENMs:

- Goal and scope: If possible consider the whole life cycle and be realistic when setting goals (see Fig. 32.5). The functional unit needs to take into account the potential differences in functionality of the product when using ENMs.
- LCI: Data is difficult to acquire, so either collaborate with the industry or base the data on already published studies and generic processes from databases as

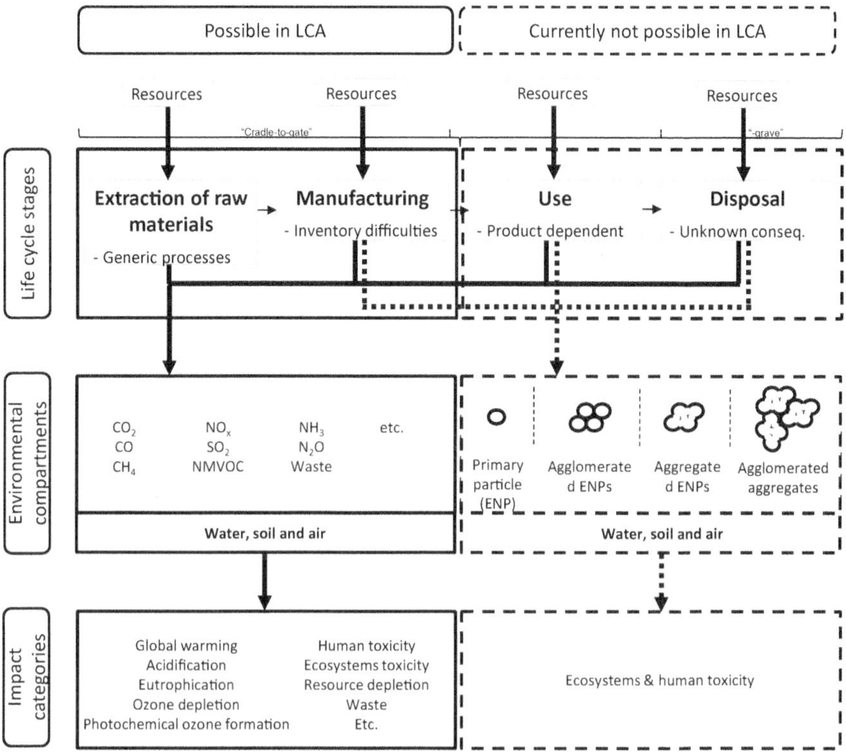

Fig. 32.5 Possibilities and limitations of LCA: what currently can be assessed in LCA (Miseljic and Olsen 2014). *Full lines* illustrate what is currently done (and is possible at the stage of current research) while *dotted lines* illustrate the challenges, especially related to the assessment of release of nanoparticles

e.g. ecoinvent. Be aware that lab-scale production data can be misleading and need to be scaled up.

- LCIA: A completely holistic impact assessment cannot yet be performed, mainly due to the challenges related to quantification of ENM/ENP release and the related impacts (Som et al. 2010; Bauer et al. 2008):
  - How much ENM/ENP is released to environmental compartments (e.g. water) and technosphere (e.g. waste water treatment)?
  - Which exposures to ENM/ENP occur in the environment and what are the effects on biota and humans?
  - At different times, what appearance (size, shape and composition) do the ENMs/ENPs take in the environment (primary particles (ENPs), agglomerated ENPs, aggregated ENPs, agglomerated aggregates)?
  - What are the environmental consequences from different end-of-life treatment of ENM products?

So far, the LCAs performed on ENMs have been used to assess the accountable production-related emissions. Future LCAs should seek to develop the areas that are currently poorly covered, so that impact burden shifting is avoided. This may be done by including other approaches (e.g. RA) using LCA as a framework for gathering the best developed approaches in order to perform a holistic environmental impact assessment.

References

Aitken, R.A., Bassan, A., Friedrichs, S., et al.: Specific Advice on Exposure Assessment and Hazard/Risk Characterisation for Nanomaterials Under REACH (RIP-oN 3). Final Project Report, RNC/RIP-oN3/FPR/1/FINAL. http://ec.europa.eu/environment/chemicals/nanotech/pdf/report_ripon3.pdf (2011). Accessed 9 Nov 2011

Babaizadeh, H., Hassan, M.: Life cycle assessment of nano-sized titanium dioxide coating on residential windows. Constr. Build. Mater. **40**(March), 314–321 (2012)

Batley, E.G., McLaughlin, J.M.: Fate of Manufactured Nanomaterials in the Australian Environment. CSIRO Niche Manufacturing Flagship Report. Department of the Environment, Water, Heritage and the Arts. http://www.environment.gov.au/system/files/pages/371475a0-2195-496d-91b2-0a33f9342a6d/files/manufactured-nanomaterials.pdf (2010). Accessed 5 Oct 2013

Bauer, C., Burchgeister, J., Hischier, R., Poanietz, W.R., Schebek, L., Warsen, J.: Towards a framework for life cycle thinking in the assessment of nanotechnology. J. Clean. Prod. **16**(8–9), 910–926 (2008)

Brant, J., Lecoanet, H., Wiesner, M.: Aggregation and deposition characteristics of fullerene nanoparticles in aqueous systems. J. Nanopart. Res. **7**(4–5), 545–553 (2005)

Buzea, C., Blandino, P. II, Robbie, K.: Nanomaterials and Nanoparticles: Sources and Toxicity. Department of Physics, Gastrointestinal Diseases Research Unit & Department of Physiology, Queens University at Kingston General Hospital, Kingston, ON, Canada (2007)

Durucan, S., Korre, A., Munoz-Melendez, G.: Mining life cycle modelling: a cradle-to-gate approach to environmental management in the industry. J. Clean. Prod. **14**(12–13), 1057–1070 (2006)

European Chemicals Agency (ECHA): Guidance on information requirements and chemical safety assessment. http://guidance.echa.europa.eu/docs/guidance_document/information_requirements_en.htm?time=1289468158 (2010). Accessed 9 Nov 2011

European Communities (2007) Corrigendum to Regulation (EC) No1907/2006 of the European Parliament and of the Council of 18 December 2006 concerning the Registration, Evaluation, Authorisation and Restriction of Chemicals (REACH), establishing a European Chemicals Agency, amending Directive 1999/45/EC and repealing Council Regulation (EEC) No793/93 and Commission Regulation (EC) No1488/94 as well as Council Directive 76/769/EEC and Commission Directives 91/155/EEC, 93/67/EEC, 93/105/EC and 2000/21/EC (OJ L 396, 30.12.2006). Off. J. Eur. Union. L136:3–280

Farré, M., Sanchís, J., Barceló, D.: The fate and the behavior of nanomaterials in the environment, analysis and assessment of the occurrence. TrAC Anal. Chem. **30**(3), 517–527 (2011)

Franklin, N.M., Rogers, N.J., Apte, S.C., Batley, G.E., Gadd, G.E., Casey, P.S.: Comparative toxicity of nanoparticulate ZnO, Bulk ZnO, and ZnCl2 to a freshwater microalga (*Pseudokirchneriella subcapitata*): the importance of particle solubility. Environ. Sci. Technol. **41**(24), 8484–8490 (2007)

Geranio, L., Heuberger, M., Nowack, M.: The behaviour of silver nanotextiles during washing. J. Environ. Sci. Technol. **43**(21), 8113–8118 (2009)

Grieger, K.D., Laurent, A., Miseljic, M., Christensen, F., Baun, A., Olsen, S.I.: Analysis of current research addressing complementary use of life-cycle assessment and risk assessment for engineered nanomaterials: Have lessons been learned from previous experience with chemicals? J. Nanopart. Res. **14**, 958 (2012)

Grubb, F.G., Bakshi, R.B.: Life cycle of titanium dioxide nanoparticle production. J. Ind. Ecol. **15** (1), 81–95 (2010)

Gutowski, T.G., Liow, J.Y.H., Sekulic, D.P.: Minimum energy requirements for the manufacturing of carbon nanotubes. In: 2010 IEEE International Symposium on Sustainable Systems and Technology (ISSST), vol. 1(6), pp. 17–19 (2010)

Hankin, S.M., Peters, S.A.K., Poland, C.A., et al.: Specific Advice on Fulfilling Information Requirements for Nanomaterials Under REACH (RIP-No. 2). Final Project Report, RNC/RIP-oN2/FPR/1/FINAL. http://ec.europa.eu/environment/chemicals/nanotech/pdf/report_ripon2.pdf. Accessed 9 Nov 2011

Hansen, S.F.: Regulation and risk assessment of nanomaterials—too little, too late? Ph.D. Thesis. Department of Environmental Engineering. Technical University of Denmark (2009)

Healy, M.: Environmental and economic comparison of single-wall carbon nanotube production alternatives. Master's thesis. Northeastern University, Boston. United States of America (2006)

Healy, M.L., Dahlben, L.J., Isaacs, J.A.: Environmental assessment of single-walled carbon nanotube processes. J. Ind. Ecol. **12**(3), 376–393 (2008)

Heinlaan, M., Ivask, A., Blinova, I., Dubourguier, H.C., Kahru, A.: Toxicity of nanosized and bulk ZnO, CuO and TiO_2 to bacteria Vibrio fischeri and crustaceans Daphnia magna and *Thamnocephalus platyurus*. Chemosphere **71**(7), 1308–1316 (2008)

Hischier, R., Walser, T.: Life cycle assessment of engineered nanomaterials: state of the art and strategies to overcome existing gaps. Sci. Total Environ. **425**(15), 271–282 (2012)

Hussain, S.M., Braydich-Stolle, L.K., Schrand, A.M., Murdock, R.C., Yu, K.O., Mattie, D.M., Schlager, J.J., Terrones, M.: Toxicity evaluation for safe use of nanomaterials: recent achievements and technical challenges. Adv. Mater. **21**(16), 1549–1559 (2009)

Hwang, C.-L., Ting, J., Chiang, J.-S., Chuang, C.: Process of direct growth of carbon nanotubes on a substrate at low temperature. U.S. Patent 6,855,376, Industrial Technology Research Institute, Chutung (Taiwan). http://www.freepatentsonline.com/6855376.html (2005). Accessed 20 Nov 2012

Illés, E., Tombácz, E.: The effect of humic acid adsorption pH-dependent surface charging and aggregation of magnetite nanoparticles. J. Colloid Interface Sci. **295**(1), 115–123 (2006)

Isaacs, A.J., Tanwani, A., Healy, L.M.: Environmental assessment of SWNT production. In: Proceedings of the 2006 IEEE International Symposium on Electronics and the Environment, pp. 38–41. ISEE (2006)

ISO: Environmental management—life cycle assessment—principles and framework (ISO 14040). ISO, The International Organization for Standardization, Geneva (2006)

ISO: Technical Specifications ISO/TS 27687:2008 (E): Nanotechnologies—Terminology and Definitions for Nano-objects–Nanoparticle, Nanofibre and Nanoplate. ISO, The International Organization for Standardization, Geneva (2008)

Jolliet, O., Rosenbaum, R.K., Laurent, A.: Life cycle risks and impacts of nanotechnologies. In: Malsch, I., Edmond, C. (eds.) Nanotechnology and Human Health, pp. 213–277. Taylor & Francis, London (2013)

Jones, C.F., Grainger, D.W.: In vitro assessments of nanomaterial toxicity. Adv. Drug Deliv. Rev. **61**(6), 438–456 (2009)

Joshi, S.: Can nanotechnology improve the sustainability of biobased products? J. Ind. Ecol. **12**(3), 474–489 (2008)

Kashiwada, S.: Distribution of nanoparticles in the see-through medeka (Oryzias latipes). Environ. Health Perspect. **114**(11), 1697–1702 (2006)

Khanna, V., Bakshi, B.R., Lee, L.J.: Life cycle energy analysis and environmental life cycle assessment of carbon nanofibers production. In: Proceedings of the 2007 IEEE International Symposium on Electronics & the Environment, 7–10 May 2007, pp. 128–133 (2007)

Khanna, V., Bakshi, B.R., Lee, L.J.: Assessing life cycle environmental implications of polymer nanocomposites. In: Proceedings of the 2008 IEEE International Symposium on Electronics and the Environment, pp. 1–6. IEEE Computer Society, Washington, DC, USA (2008)

Klaine, S.J., Alvarez, P.J.J., Batley, G.E., Fernandes, T.F., Handy, R.D., Lyon, D.Y., Mahendra, S., McLaughlin, M.J., Lead, J.R.: Nanomaterials in the environment: behavior, fate, bioavailability, and effects. Environ. Toxicol. Chem. **27**(9), 1825–1851 (2008)

Köhler, R.A., Som, C., Helland, A., Gottshalk, F.: Studying the potential release of carbon nanotubes throughout the application life cycle. J. Clean. Prod. **16**(8–9), 927–937 (2008)

Künninger, T., Fischer, A., Gerecke, A., Heeb, M., Kunz, P., Ulrich, A., Vonbank, R.: Release of Conventional and Nano-Sized Biocides from Coated Wooden Façades during Weathering: Consequences for Functionality and Aquatic Environment Proceedings of the International Convention of Society of Wood Science and Technology and United Nations Economic Commission for Europe—Timber Committee October 11–14, 2010, Geneva, Switzerland (2010)

Kushnir, D., Sandén, B.A.: Energy requirements of carbon nanoparticle production. J. Ind. Ecol. **12**(3), 360–375 (2008)

Lloyd, M.S., Lave, B.L.: Life cycle economic and environmental implications of using nanocomposites in automobiles. Environ. Sci. Technol. **37**(15), 3458–3466 (2003)

Lloyd, M.S., Lave, B.L., Matthews, H.S.: Life cycle benefits of using nanotechnology to stabilize platinum-group metal particles in automotive catalysts. Environ. Sci. Technol. **39**(5), 1384–1392 (2005)

Lowry, G.V., Casman, E.A.: Nanomaterial transport, transformation, and fate in the environment. NATO Science for Peace and Security Series. Nanomaterials: Risks and Benefits, pp. 125–137 (2009)

Lowry, G.V., Gregory, K.B., Apte, S.C., Lead, J.R.: Transformations of nanomaterials in the environment. Environ. Sci. Technol. **46**(13), 6893–6899 (2012)

Manda, B.M., Blok, K., Patel, M.K.: Innovations in papermaking: an LCA of printing and writing paper from conventional and high yield pulp. Sci. Total Environ. **15**(439), 307–320 (2012)

Merugula, A.L., Khanna, V., Bakshi, R.B.: Comparative life cycle assessment: reinforcing wind turbine blades with carbon nanofibers. In: Proceedings of the 2010 IEEE International Symposium on Sustainable Systems and Technology (ISSST 2010). 5507724 (2010)

Meyer, E.D., Curran, M.A., Gonzalez, M.A.: An examination of silver nanoparticles in socks using screening-level life cycle assessment. J. Nanopart. Res. **13**(1), 147–156 (2010)

Miseljic, M.: Improvement of methodological and data background for life cycle assessment of nano-metaloxides, Ph.D. thesis, Handed in November 2014 & Defended in February 2015, Technical University of Denmark (2014)

Miseljic, M., Olsen, S.I.: Life-cycle assessment of engineered nanomaterials: a literature review of assessment status. J. Nanopart. Res. **16**(6), 2427 (2014)

Moign, A., Vardelle, A., Themelis, N.J., Legoux, J.G.: Life cycle assessment of using powder and liquid precursors in plasma spraying: the case of yttria-stabilized zirconia. Surf. Coat. Technol. **205**(2), 668–673 (2010)

Oberdörster, G., Stone, V., Donaldson, K.: Toxicology of nanoparticles: a historical perspective. Nanotoxicology **1**(1), 2–25 (2007)

Osterwalder, N., Capello, C., Hungerbühler, K., Stark, J.W.: Energy consumption during nanoparticle production: How economic is dry synthesis? J. Nanopart. Res. **8**(1), 1–9 (2006)

PoEN (Project on Emerging Nanotechnologies): Consumer products inventory. http://www.nanotechproject.org/cpi (2014). Accessed 5 Aug 2014

Quik, T.K.J., Vonk, A.J., Hansen, F.S., Baun, A., Van De Meent, D.: How to assess exposure of aquatic organisms to manufactured nanoparticles? Environ. Int. **37**(2011), 1068–1077 (2011)

Ratte, H.T.: Bioaccumulation and toxicity of silver compounds: a review. Environ. Toxicol. Chem. **18**, 89–108 (1999)

Roes, A.L., Marsili, E., Nieuwlaar, E., Patel, M.K.: Environmental and cost assessment of a polypropylene nanocomposite. J. Polym. Environ. **15**(3), 212–226 (2007)

Rupasinghe, R.-A.-T.P.: Dissolution and aggregation of zinc oxide nanoparticles at circumneutral pH; a study of size effects in the presence and absence of citric acid. Master thesis, University of Iowa (2011)

SCENIHR: The existing and proposed definitions relating to products of nanotechnologies. Scientific Committee on Emerging and Newly Identified Health Risks. http://ec.europa.eu/health/ph_risk/committees/04_scenihr/docs/scenihr_o_012.pdf (2007). Accessed 20 Sept 2012

Scheringer, M., Macleod, M., Behra, R., Sigg, L., Hungerbühler, K.: Environmental risk associated with nanoparticulate silver used as biocide. H and PC Compendium on Detergency, vol. 6(2), April/June 2011, pp. 27–29 (2010)

Som, C., Berges, M., Chaudry, Q., Dusinska, M., Fernandes, F.T., Olsen, S.I., Nowack, B.: The importance of life cycle concepts for the development of safe nanoproducts. Toxicology **269** (2–3), 160–169 (2010)

Som, C., Wick, P., Krug, H., Nowack, B.: Environmental and health effects of nanomaterials in nanotextiles and façade coatings. Environ. Int. **37**, 1131–1142 (2011)

Steinfeldt, M., von Gleich, A., Henkle, J.L.L., Endo, M., Morimoto, S., Momosaki, E.: Environmental relief effects of nanotechnologies by the example of CNT composite materials and films. In: International Conference; 9th, Ecobalance; Towards and Beyond 2020 (2010)

Suppen, N., Carranza, M., Huerta, M., Hernández, A.M.: Environmental management and life cycle approaches in the Mexican mining industry. J. Clean. Prod. **14**(12–13), 1101–1115 (2005)

Sweet, L., Strohm, B.: Nanotechnology—life-cycle risk management. Hum. Ecol. Risk Assess. **12**, 528–551 (2006)

Tibbetts, G.G., Bernardo, C.A., Gorkiewicz, D.W., Alig, R.A.: Role of sulfur in the production of carbon fibers in the vapor phase. Carbon **32**(4), 569–576 (1994)

Vonk, J.A., Struijs, J., van de Meent, D., Peijnenburg, W.J.G.M.: Nanomaterials in the Aquatic Environment: Toxicity. Exposure and Risk Assessment. RIVM Report 607794001/2009, RIVM Bilthoven, Nederlands. http://www.rivm.nl/bibliotheek/rapporten/607794001.pdf (2009). Accessed 10 Oct 2012

Walser, T., Demou, E., Lang, J.D., Hellweg, S.: Prospective environmental life cycle assessment of nanosilver t-shirts. Environ. Sci. Technol. **45**(10), 4570–4578 (2011)

Zhu, X., Chang, Y., Chen, Y.: Toxicity and bioaccumulation of TiO_2 nanoparticle aggregates in *Daphnia magna*. Chemosphere **78**, 209–215 (2009)

Author Biographies

Mirko Miseljic Background in development and application of LC(I)A methodology for different technology domains. Contributed with improvement of methodological and data background for LCA of nano-metaloxides, and LCA work within e.g. nanomaterials, waste, water treatment, food. Interested in LCIA modelling, spatial differentiation and uncertainty.

Stig Irving Olsen LCA expert both as researcher and as consultant. Involved in the development of LCA methodologies since mid 1990s. Contributed to UNEP/SETAC working groups on LCIA methodology. Main LCA interest is human toxicity impacts, emerging technologies, and decision making.

Chapter 33
LCA of Drinking Water Supply

Berit Godskesen, Noa Meron and Martin Rygaard

Abstract Water supplies around the globe are growing complex and include more intense treatment methods than just decades ago. Now, desalination of seawater and wastewater reuse for both non-potable and potable water supply have become common practice in many places. LCA has been used to assess the potentials and reveal hotspots among the possible technologies and scenarios for water supplies of the future. LCA studies have been used to support decisions in the planning of urban water systems and some important findings include documentation of reduced environmental impact from desalination of brackish water over sea water, the significant impacts from changed drinking water quality and reduced environmental burden from wastewater reuse instead of desalination. Some of the main challenges in conducting LCAs of water supply systems are their complexity and diversity, requiring very large data collection efforts, with multiple sources of information, many of them not public and requiring cooperation. Important for product and system LCAs with substantial water use, it is emphasized that standard life cycle inventory databases do not reflect the significant variance in environmental impacts of water supply across locations and technologies.

B. Godskesen (✉)
HOFOR (Greater Copenhagen Utility), Ørestads Boulevard 35,
2300 Kbh S, Copenhagen, Denmark
e-mail: berg@env.dtu.dk

N. Meron
The Porter School of Environmental Studies, Tel-Aviv University,
69978 Tel-Aviv, Israel

M. Rygaard
Department of Environmental Engineering, Technical University of Denmark,
Bygningstorvet Building 115, 2800 Kongens Lyngby, Denmark

M.Z. Hauschild et al. (eds.), *Life Cycle Assessment*,
DOI 10.1007/978-3-319-56475-3_33

33.1 Introduction

33.1.1 Water Consumption and Water Treatment Technologies

Water supply systems are unique for every region around the globe. They are based on a variety of water resources and technologies. Importantly, some of them have been based on the same traditional technologies for more than 100 years, while others are rapidly changing to cope with the urban development. The differences are also large between neighbouring countries. For example, while Denmark is 100% based on groundwater abstraction for its water supply, the neighbouring countries Sweden and Germany were sourcing just 22 and 61%, respectively of their water supply from groundwater in 2010. Instead of groundwater these countries use a variety of surface water, spring water and artificially recharged groundwater (IWA 2014). Other countries are now heavily reliant on water reuse and desalination, for example Spain, USA, Israel, Singapore and Saudi Arabia (IWA 2014; Tal 2006; GWI 2010). Desalination and reuse are increasingly used and the rapid development is underlined by the rapid growth in desalination capacity around the globe (Fig. 33.1).

33.1.2 Water Systems Growing Complex

There is a wide variety of water systems, which may interact with many processes and systems (Fig. 33.2). Although, water systems normally include: production (abstraction and treatment or desalination), transmission and distribution of water to various users, each process may apply various technologies, and the systems may

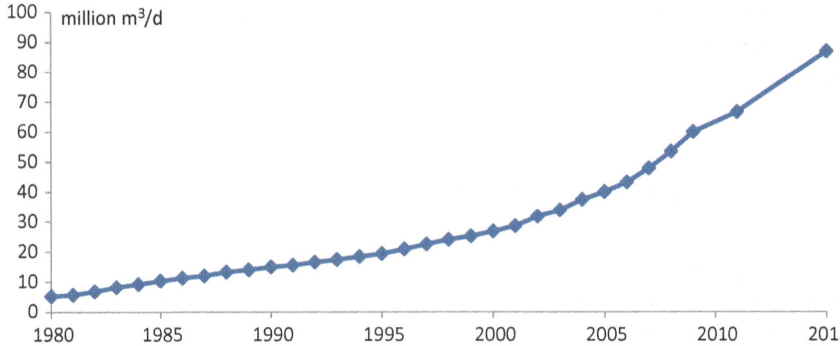

Fig. 33.1 Installed capacity of desalination plants registered by International Desalination Association. *Source* Pankratz (2010) and http://idadesal.org/desalination-101/desalination-by-the-numbers/

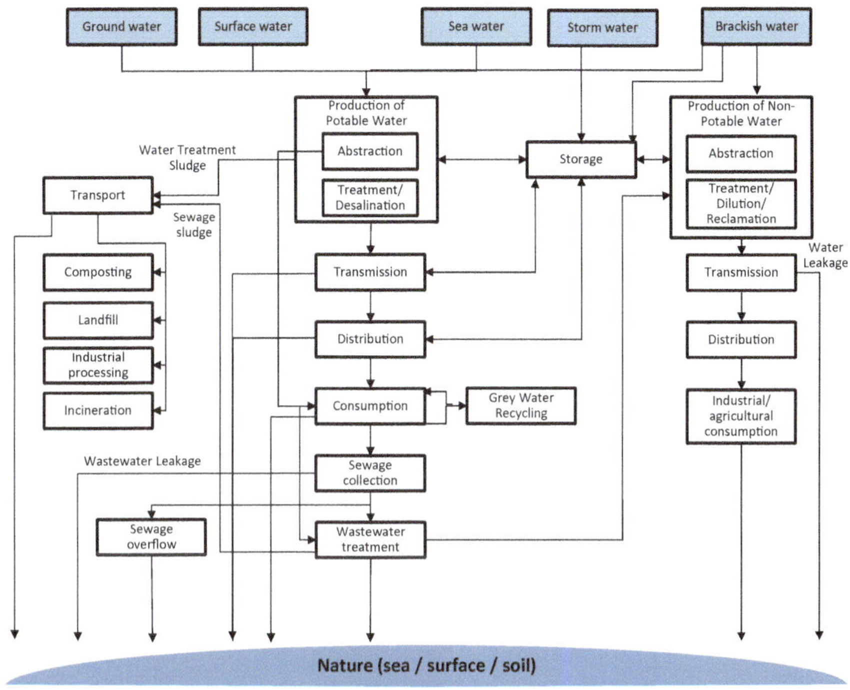

Fig. 33.2 General water supply chain. Adapted from Meron et al. (2016)

use multiple resources and multiple output water qualities for different users. In order to account for the different environmental profiles of water production pathways, Hospido et al. (2013) proposed the concept of water supply mix for the example of irrigation in Spain and inspired by the concept of the electricity mix. Water losses throughout the supply system also vary considerably and can range between 5 and 55% (Ratnayaka et al. 2009), thus may have an important effect on systems' impact. The environmental performance of water systems can therefore greatly vary.

From a systems perspective, the new sources of water and combinations of new and traditional water treatment technologies makes planning decisions difficult. Multiple water resources and differences in the direct and indirect impacts on the environment from each process in the system are complex aspects to consider in the process of finding the best solution for a particular situation. To complicate things, the drivers of change in water systems also push in the direction of having systems with multiple water qualities, treatment technologies and resources in use at the same time. For example, until now, Danish water utilities have managed urban water systems based on groundwater abstraction, simple low-intensive treatment, distribution of one water quality (drinking water). After use, a one-stringed sewer system would divert wastewater to central treatment plants before discharging the treated wastewater to the recipients. Now, water utility managers foresee a

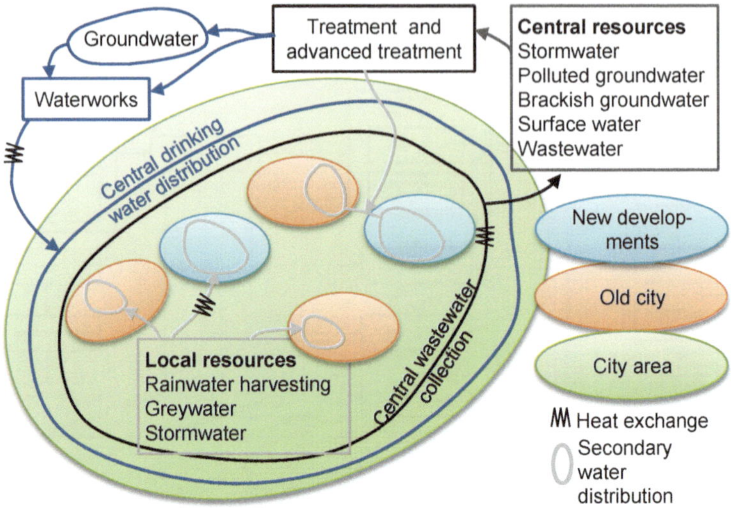

Fig. 33.3 Visions for the future urban water systems in larger cities in Denmark as developed by the two biggest Danish water utilities Aarhus Vand and HOFOR. The water systems are foreseen to become increasingly complex and cover multiple water sources, a mix of central and decentralized systems and vary between old and new parts of the city (Rygaard et al. 2012)

diversification of their water systems that include distribution of multiple water qualities, decentralized handling of wastewater and reclamation of wastewater for distribution for non-potable purposes (Fig. 33.3). Also decision support in much more water scarce areas around the world such as China and Australia have dealt with the difficult choice between a vulnerable, but simple water system and a more robust but significantly more complex diversified water system (Kenway et al. 2011; Lane et al. 2015; Li et al. 2016). In such cases LCA provides insight to the environmental trade-offs between future water supply scenarios.

33.1.3 Water Supply Technologies: Traditional and New Possibilities

With the development in especially membrane filtration processes, it has become common practice for many water supplies to treat water that just a couple of decades ago was considered economically infeasible to use. Back then, most water suppliers based their production on simple treatment techniques such as aeration, flocculation and filtration with activated carbon and sand filters, and disinfection by chlorine. These techniques are adequate to remove many common unwanted substances, for example methane and hydrogen sulphide, iron, organic pollutants, and to deactivate or remove pathogens. With newer treatment techniques like membrane filtration and advanced oxidation methods it is possible to treat wastewater and remove salt from ocean water to make it suitable for drinking. Potable wastewater reuse, seawater and brackish desalination can now be established at total costs around 0.5–2 US$ per produced m^3 water, which is similar to production costs of more simple and traditional water treatment (Greenlee et al. 2009; Rygaard et al. 2011; Wols and Hofman-Caris 2012). However, aeration and simple filtration requires little energy and few other resources in the operation, while removing salts from water using reverse osmosis requires advanced membranes, high pressure pumps and chemicals to keep the membranes clean. Typical simple groundwater treatment requires around 0.3 kWh/m^3 while state-of-the-art desalination of ocean water requires around 2.5–7 kWh/m^3 produced (Rygaard et al. 2011; Plappally and Lienhard 2012). Shifting from traditional and simple treatment technologies to more advanced treatment gives access to huge additional water resources in the wastewater stream and seawater. On the other hand, it can have significant impact on the material and energy use in the production of water.

Membrane-based treatment technologies can produce very clean product water with essentially no pollutants and minerals in it. This may be particularly beneficial for some industry processes demanding ultraclean water. The option of remineralizing the demineralized water makes it possible to optimize water quality for specific uses, e.g. drinking water with a certain mineral content (Birnhack et al. 2008; Rygaard et al. 2011a).

33.2 Literature Review

In urban water management, LCA is found to be the most dominant and appropriate tool to assess the environmental impacts (Godskesen et al. 2013). Other tools such as carbon and water footprint are also being used but they are not as comprehensive as they only focus on one or two environmental aspects and might not cover the entire life cycle from cradle to grave.

LCA has been applied in the water sector for years and numerous LCA studies of water processes and subprocesses have been reported. Publications have been made on the abstraction, production, transport and distribution, and on entire urban or regional water systems. These studies lead to various conclusions. A meta-analysis of water supply systems and subsystems has confirmed that there is a large variation in the impacts of water supply systems. For example, global warming potential ranges between 0.16 and 3.4 kg CO_2-eq per m^3 of supplied water (Meron et al. 2016).

33.2.1 LCA to Identify Hot Spots in Water Supply

Several studies have shown that water production has the highest contribution to the impacts of the entire water supply system (Friedrich et al. 2009; Godskesen et al.

2013; Lemos et al. 2013; Tarantini and Ferri 2001; Uche et al. 2013), and desalination in particular (Del Borghi et al. 2013). Only a few publications specifically report the contribution of the abstraction and transport and distribution subsystems to the impact of entire water supply systems, although in some cases they may also have significant contribution. For example, in Tarragona, Spain (Amores et al. 2013), and in Lasi, Romania (Barjoveanu et al. 2014), the distribution subsystem has the highest impacts in all categories except eutrophication due to pumping electricity consumption.

Activities affecting water used at households were also shown to be important. For example, water boiling to improve water quality after its deterioration through the distribution subsystem has the highest contribution to the impacts of the supply system in Hanoi, Vietnam (Homäki et al. 2003). The impacts of activated carbon-based filter to improve water quality at domestic level are also considerably higher scores than the impacts of the centralized water supply in Milan (Nessi et al. 2012). Environmental impacts of household activities can be reduced if water is softened at the treatment stage (Godskesen et al. 2012).

The importance of electricity consumption in LCAs of water supply systems has been reported in many studies (Lemos et al. 2013; Lundie et al. 2004; Tarantini and Ferri 2001; Lane et al. 2015). Energy is also found to be a significant factor in water supply subsystems: in abstraction (Buckley et al. 2011), in production through treatment of freshwater (Buckley et al. 2011; Igos et al. 2014; Lyons et al. 2009; Racoviceanu et al. 2007), in desalination (Lyons et al. 2009; Raluy et al. 2005a; Stokes and Horvath 2006; Tarnacki et al. 2012; Uche et al. 2013), in pumping (Amores et al. 2013) and in landfilling sludge in the case of wastewater reclamation (Li et al. 2016).

Energy also has an important contribution to the environmental impacts of water through background processes. The production of treatment chemicals (Bonton et al. 2012; Buckley et al. 2011) and materials for construction of decentralized water supply systems (Godskesen et al. 2013) are reported to have significant impacts.

Production technologies were also studied and compared. Large variation was reported in the impacts of water production. For example, the average Global Warming Potential (GWP) of thermal desalination have been found to be about ten times the average of reverse osmosis desalination's GWP, and about 100 times higher than freshwater technologies' GWP (Meron et al. 2016). Raluy et al. (2005a, b) studied several desalination technologies and import of water from a distant river to a local water body. The paper underlines that even though desalination has high energy requirements it has become competitive and transfer of water is not always the best solution dependent on energy needs for long distance transport.

Within the process of freshwater treatment, chemicals production may play an important role (e.g. Barrios et al. 2008). Yet in some cases chemicals had low contribution (Arpke and Hutzler 2006; Tarantini and Ferri 2001; Jeong et al. 2015).

The contribution of the materials and construction of the distribution infrastructure may be significant and up to 60% of the overall impact of distribution, while only up to 15% of the abstraction impacts and up to 20% of the production impacts (Meron et al. 2016). Infrastructure construction is shown to have only limited contribution to the total impact of sea water desalination (Raluy et al. 2005b; Uche et al. 2013) and of groundwater treatment plants (Igos et al. 2014; Uche et al. 2013), but infrastructure does represent a significant contribution to the impact of water distribution (Barjoveanu et al. 2014; Slagstad and Brattebø 2014; Uche et al. 2013; Jeong et al. 2015). A study of the impact of pipes compared different materials used in the water transport and distribution network, using several impact categories. The study showed that the installation stage is especially relevant for constructive solutions with smaller pipe diameters (e.g. 90 mm diameter HDPE), whereas the production of the pipe becomes more relevant with larger pipe diameters (e.g. 200 mm diameter HDPE). The reduction of environmental impacts involves the optimisation of the trench dimensions and the process of installation as well as the selection of pipe materials with lower environmental impacts in the production stage (Sanjuan-Delmás et al. 2014).

Several LCAs studied the impacts of urban rainwater capture as an option to supplement or replace water demand from centralized water supply systems. These studies show that rain harvesting can reduce environmental impacts in some cases (Godskesen et al. 2013) whereas in some locations rain water harvesting is not the best choice (Mithraratne and Vale 2007; de Haas et al. 2011). Rain tank impacts are mainly due to electricity consumption (de Haas et al. 2011; Mithraratne and Vale 2007; Angrill et al. 2012) and in some cases infrastructure, depending on which materials are used (e.g. concrete or plastic) (Mithraratne and Vale 2007; Angrill et al. 2012).

33.2.2 LCA of Water Reuse

In several studies LCA was applied to study the environmental impacts of water reuse. Tertiary treatment has a relatively low impact compared to the impact of the entire wastewater treatment plant, as well as compared to desalination (Pasqualino et al. 2011). Production from freshwater sources has also been shown to have similar impacts to tertiary treatment (Meneses et al. 2010). Reclaimed wastewater that replaces freshwater resources used for irrigation may reduce the environmental burden of the water system, compared to systems without reuse (Fang et al. 2016). Wastewater reclamation, water transfer, and desalination were compared in different locations including California (Stokes and Horvath 2006), Arizona (Lyons et al. 2009) and northern China (Li et al. 2016). In all these studies desalination has the highest environmental impacts in all of the impact categories, except in the freshwater withdrawal impact. In summary, LCA has been used to show the reduced environmental burden from water systems turning towards water reuse, instead of expanding surface water treatment or turning to desalination.

LCA studies have also highlighted the need for looking beyond standard impact categories, like carbon footprint, and include toxicity impacts in the decision-making. For example, a study compared eco-toxicity of four alternatives for use of wastewater after secondary treatment: no-reuse, direct use, and use after two different tertiary treatment technologies based on ozonation. From the eco-toxicity perspective, use after tertiary treatments is the best choice. The study emphasized also that LCAs of wastewater reusing systems assessing toxicity should include wastewater pollutants such as heavy metals, pharmaceuticals and personal care products, which can contribute above 90% of the toxicity impact (Muñoz et al. 2009).

33.2.3 LCA as a Tool in Water Supply Management

The previous sections have shown how LCA have revealed the environmental burden of various water systems and included processes and technologies. LCA is also used in cases where there is a lack of water and need for strategic choices in the planning of future water supplies. One good example of the typical application of LCA in water supply management is the comparison of possible solutions to water scarcity, where two or more water production scenarios are considered. Muñoz and Fernández-Alba (2008) showed that a shift from ocean water to brackish groundwater could significantly reduce the environmental impact from water supply operations. They found that desalinating groundwater with salt content of 15 g/L reduces environmental impacts to nearly half of a seawater-based desalination plant treating water with salt content of 36 g/L. The difference is mainly explained by the electricity consumption in both cases. Similarly, several other cases have used life cycle thinking approaches in decision support before changing the water systems with the aim to obtain a better environmental performance in the utilization of water resources, water treatment technologies, etc. (Rygaard et al. 2014). Another example is an LCA of the Sydney water planning aiming to evaluate several initiatives and to bring down the environmental impacts. The study included several scenarios for changing water supply and wastewater systems and the outcome is a decision support tool for future planning of the complex water system (Lundie et al. 2004).

33.2.4 Tap Versus Bottled Water

Under some circumstances consumers prefer to buy bottled water instead of drinking water from the tap, but what is the environmental perspective on this choice? A comparison of an LCA on centralized drinking water production for Copenhagen, reported in Godskesen et al. (2013) with studies of CO_2-emissions for bottled water in Niccoluci et al. (2010) or Jungbluth (2006) reveals the environmental benefits of centralized drinking water supply in terms of carbon footprint. In the mentioned studies bottled water production emits between 0.14 and 0.18 kg CO_2-eq/L when including water intake, production of the bottle and transport. Water supply based on groundwater as in Copenhagen from source to tap emits

Table 33.1 Climate change impacts from production and distribution of 1L of water from bottled water or centralized drinking water supply (Godskesen et al. 2013)

References	System	Country of study	kg CO_2-eq/L
Jungbluth (2006)	Bottled water in non-returnable bottle	Switzerland	0.18
Niccoluci et al. (2010)	Bottled water in non-returnable bottle	Italy	0.14
Godskesen et al. (2013)	Centralized groundwater based drinking water supply	Denmark	0.00019
Godskesen et al. (2013)	Centralized drinking water supply, desalination of seawater	Denmark, hypothetical	0.0013

740–920 times less kg CO_2-eq/L, and even the hypothetical case of desalinated water in Copenhagen emits 110–140 times less CO_2-eq/L (Table 33.1). Similarly, Botto et al. (2011) found that tap water had ecological and carbon footprints 300 times less than bottled water. This comparison emphasizes that when it comes to carbon footprint, centralized water supply is strongly preferable, especially when the water source is groundwater but also when it is seawater even though desalination processes require much more electricity even in the Copenhagen case for 2013, which is a system relying heavily on fossil fuels.

33.2.5 Case Study of Four Technologies for Drinking Water Supply in Copenhagen, Denmark

Water supply in Denmark is based on groundwater abstracted from well fields located on primarily rural or agricultural land. This is also the case for the capital of Denmark, Copenhagen, where water is abstracted from groundwater sources located outside the city limits and transported to the waterworks where it is treated by aeration and sand filtration before distribution. The basic structure of Copenhagen's water supply was established more than 150 years ago, and the structure remained largely unchanged until now apart from additional well fields and waterworks.

The European Water Framework Directive (EU-WFD) is implemented in the EU-Member States through River Basin Management Plans which among other parameters regulate the water flow requirements for rivers and streams and the utilizable amount of water in each freshwater (ground- and surface water) compartment (European Commission 2012). The implementation has revealed that groundwater is not an abundant resource when the requirements to the quality of the freshwater environment have to be met as stipulated in the EU-WFD, and the water utility in Copenhagen has been forced to seek new water resources or new approaches to sustain the water withdrawal permissions in order to supply the city with sufficient water for urban purposes. One could say that the Copenhagen water scarcity is more political than physically founded when compared to other more water stressed areas in the Mediterranean, region, Northern Africa, or India where freshwater resources are far more scarce and often overexploited (Smakhtin et al. 2004; Gleeson et al. 2012) putting pressure on water supplies to shift to other water sources than freshwater. As an example the water service providers in Melbourne built a desalination plant due to a severe drought and an increase in the number of inhabitants. The building of the desalination plant was finalized in 2012. In a case study for the Copenhagen region, we identified four relevant options for water supply which fulfil the EU-WFD and which can either alone or as a mix constitute the future water supply. We performed an environmental evaluation using LCA on the four options since environmental performance is a well-established criterion and should per se be included in any optimization of future supply options in search for the optimal water supply solution.

In this case study system boundaries need to be placed so the LCA also includes effects of changed water quality in the households which is relevant when evaluating water systems delivering water of different water hardness (Godskesen et al. 2012). Also, some of the proposed alternatives are located in areas with combined sewers which means that rain and wastewater are transported in the same sewer system. Therefore, system boundaries should reflect this difference among the alternatives. Finally, the impacts of the water supply system on freshwater resources can be very important in relation to water supply and a method for this was further developed with local specificity for the Copenhagen region (Godskesen et al. 2013).

33.2.5.1 Cases

The four cases were: A1 rain- and stormwater harvesting, A2 compensating actions, A3 new well fields and A4 desalination. The existing system was also included as the base case A0, enabling us to compare the environmental impacts of the four alternatives with today's water production. We defined the functional unit as: Replacing 1 m^3 of potable drinking water as of today in a way that fulfils the EU-WFD's water flow requirements. Schematic diagrams of the options and their location in relation to the urban area are shown in Fig. 33.4.

The LCA was performed according to the ISO 14044 standard procedure (ISO 2006) also including a weighting step. The systems were modelled with the GaBi 4.4 software delivered by PE International and environmental impacts were assessed using EDIP 1997 (Wenzel et al. 1997).

A0 Base Case

In 2009 the city of Copenhagen (population of 0.52 million) used a total volume of 29.8 million m^3 drinking water. The water was abstracted from groundwater

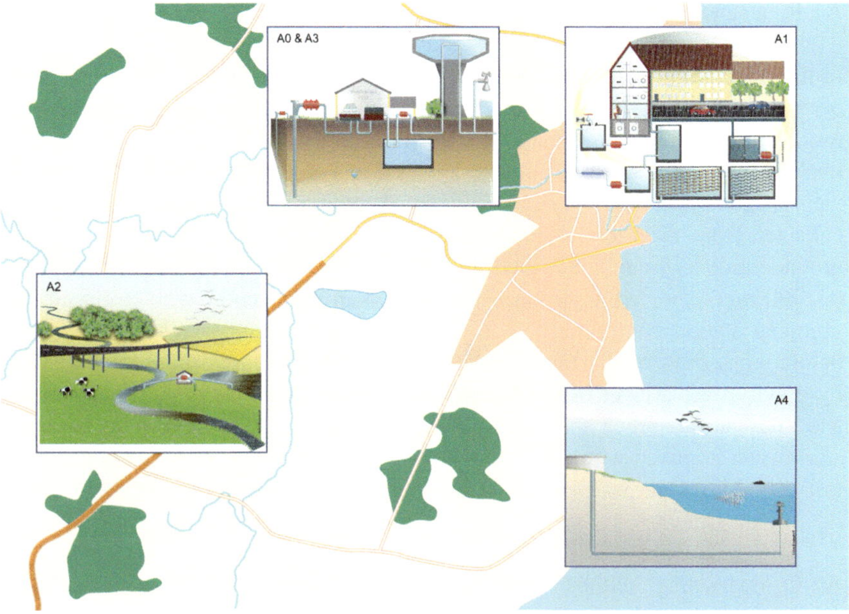

Fig. 33.4 The four options included in the case study: *A0* base case relying on groundwater abstraction; *A1* rain- and stormwater harvesting from several blocks; *A2* compensating actions consisting of water transfer in the affected catchment areas; *A3* establishing well fields 20 km further away from the waterworks; *A4* desalination of seawater from Øresund. The background is a hypothetical map but it emphasizes where the alternatives are located in relation to the urban area (*dark orange*) (Godskesen et al. 2013)

sources located outside the city, requiring only simple treatment at the waterworks in terms of aeration and sand filtration before distribution. During aeration CH_4 and H_2S are emitted and these emissions are also included in the LCA. The water abstraction, treatment and distribution consume only 0.27 kWh per m^3 drinking water. Since the groundwater originates from chalk aquifers the hardness of the water is 362 mg/L as $CaCO_3$ and it is categorized as very hard drinking water (USGS 2012). Primary data from the Copenhagen water supply on use of materials and auxiliaries for water supply was used in the assessments. After use drinking water is considered as wastewater and is transported via combined sewers to the wastewater treatment plants where it is treated before discharged to the Sea (Øresund). Electricity consumption for wastewater transportation is based on average consumption in the period 2007–2009 and modelling of the processes at wastewater treatment plants is based on registered data for consumptions from 2005 to 2009 (Danva 2010).

A1 Rain- and Stormwater Harvesting

In the A1 case rain- and stormwater is considered harvested from an urban area of 68,500 m^2 (roof area 20,200 m^2; main road area 8500 m^2) populated by 1000 residents and 200 employees. The water is of non-potable quality and is used for flushing toilets and washing clothes. The case is hypothetical as it does not exist but was designed and dimensioned as a potential option (Petersen 2011). Rainwater is collected from the roofs of residential and office buildings and led to an underground basin (750 m^3). Stormwater from main roads is collected in large pipes (Ø1000 mm) and led to a basin established in connection with a clarifier and pumping station controlling the flow. The clarifier separates oils from the water before it passes through a dual porosity filter. In the dual filtration, stormwater floats by gravity over a layer of $CaCO_3$ particles where suspended solids, heavy metals and PAHs are adsorbed and thereby removed (Jensen 2009). Afterwards the treated stormwater is mixed with rainwater and stored in a basin. Prior to distribution back to the buildings the water is UV-treated.

A2 Compensating Actions

Compensating actions included transfer of water from lakes and groundwater compartments with surplus of groundwater to water courses where the water flow is reduced due to the water utility's groundwater abstraction. Also included was reestablishment of wetlands from agricultural land. Besides these compensating actions A2 included all processes in the base case (A0).

A3 New Well Fields

Assuming that it would be possible for the water utility to find well fields with a surplus of available groundwater according to the EU-WFD within an additional distance of 20 km, the new well fields case (A3) is equivalent to the base case A0 but with addition of a 20 km longer pipeline from well fields to the waterworks. In comparison, in A0, water is transported 5 km. The longer distance means increased electricity consumption for pumping of abstracted groundwater.

A4 Desalination

Copenhagen is situated at the entrance to the Baltic Sea (Øresund) with brackish water, and desalination of seawater is thus an option. The treatment plant is considered to be located 5 km south of the city. First, water is filtrated mechanically (150 μm) to remove large particles, a coagulant is added and pH adjusted and the water is ultra-filtrated whereby 10% of the water is lost and returned to Øresund after separation of dry matter. Anti-scaling chemicals are added before the water passes through a two-step reverse osmosis membrane. Finally, calcium hydroxide is added and the water UV treated. The water has a hardness of 108 mg/L as $CaCO_3$ when distributed as drinking water.

33.2.5.2 Methodological Challenges

This case study gives examples where system boundaries must be defined with great care to make the comparisons based on the results from the LCA trustworthy.

33.2.5.3 Water Hardness

Although, central softening at waterworks uses energy and chemicals, the case study showed that these negative effects are more than compensated for by positive effects of reduced water hardness encountered in the households (Fig. 33.5).

The negative environmental effects in the study originate from the softening processes of chemical precipitation of $CaCO_3$ in a pellet reactor at the waterworks. The positive effects located in the households are, e.g. prolonged service life of household equipment like washing machine, dishwasher, coffee maker and kettle; and reduced consumption of energy, cleaning agents, laundry detergent, soap and shampoo. Thus, from an environmental viewpoint it is preferable to reduce the water hardness at the waterworks of very hard water supplies. Decentralized softening of water was not included in our study. The study emphasizes the importance of including effects of changed water hardness in the LCA scoping, when the choice of water supply technologies produces different hardness and therefore causes effects of importance for the overall environmental assessment.

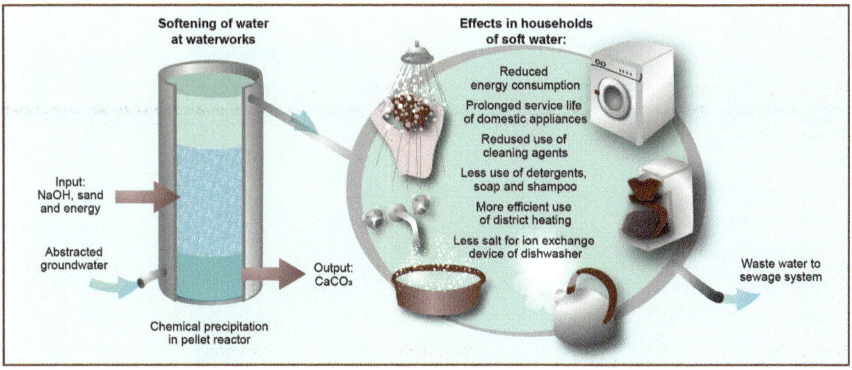

Fig. 33.5 Processes occurring at waterworks and in the households when central softening of drinking water is introduced (Godskesen et al. 2012)

In the case study of alternative technologies for water supply the cases A1 (rain- and stormwater based supply) and A4 (desalinated seawater) deliver water of a lower hardness (171 and 108 mg/L as $CaCO_3$ respectively) compared to the drinking water in the base case (A0) (362 mg/L as $CaCO_3$). For desalination of seawater (A4) consideration of the beneficial effects of the lower water hardness reduces the environmental impacts by approximately 40% while the rain- and stormwater case (A1) shows a reduction of environmental impacts by 35%—results not shown here but can be found in Godskesen et al. (2013). However, desalination (A4) is still the technology with the highest impact though not as severe when the effects of reduced hardness are included (Table 33.2).

Table 33.2 Normalized impact scores per 1 m^3 water delivered by the four options to replace 1 m^3 of potable water, grouped after Environmental impacts, Toxicity impacts and Resource consumption (Godskesen et al. 2013)

	A0 Base case	A1 Rain- and storm-water	A2 Compensating actions	A3 New well fields	A4 Desalination
Environmental impacts, μPET (person equivalent targeted, weighted result)					
Total environmental imp.	124	82	124	138	205
Global warming	82.5	65.5	82.8	91.9	151.4
Acidification	24.6	10.3	24.7	27.5	36.3
Nutrient enrichment	14.5	7.6	14.5	16.2	23.6
Photochem. ozone form	1.9	−1.5	1.9	2.2	−6.5
Toxicity impacts, μPET (person equivalent targeted, weighted result)					
Total toxicity imp.	176	126	180	194	181
Ecotoxicity water chronic	63.7	24.9	64.8	70.1	85.7
Human toxicity soil	69.9	69.8	70.3	78.7	58.8
Human toxicity water	42.4	31.0	45.2	44.9	36.1
Resource consumption, μPR (person reserve)					
Chromium	17.3	−34.1	17.4	17.3	−38.3
Copper	0.05	−3.0	0.057	0.063	−5.3
Hard coal	2.6	1.2	2.6	2.9	5.1
Natural gas	1.7	1.1	1.7	1.9	2.4

33.2.5.4 Combined Sewers

When conducting the LCA of the case study it was found that the combined sewers in the city which transport the discharge (rain and wastewater) to the wastewater treatment plants where it is treated also have an effect on the system boundaries.

The decoupling of the rain- and stormwater from the sewer system is a significant environmental advantage of A1 as electricity consumption for transport and treatment of sewage water is reduced. Therefore, the system boundaries had to be defined so that this difference is taken into account. Hence, this work (Table 33.2) shows that rain- and stormwater harvesting in areas with combined sewers is environmentally beneficial while other authors have found that rainwater harvesting in areas with separate sewer systems (rain and wastewater is handled in separate sewer systems) has a higher environmental impact than, e.g. import of freshwater (Crettaz et al. 1999).

When the modelled system is expanded to include the wastewater system, the strongest environmental impacts originate from wastewater treatment mainly due to the high electricity consumption for treating wastewater (Godskesen et al. 2011; Lundie et al. 2004). Therefore, we found that it is important to include the wastewater system when collecting rain- and stormwater in areas with combined sewers.

33.2.5.5 Results of the Case Study of Alternative Technologies for Water Supply

The results for the alternatives differ markedly for the different impact categories (Table 33.2) and show that the rain- and stormwater harvesting option (A1) has the lowest total aggregated environmental impact (82 μPET/m^3). The cases relying on groundwater abstraction (A0, A2 and A3) have environmental impacts of 124–138 μPET/m^3. A1 has a low environmental impact mainly due to the role of combined sewers and the positive effects of reduced water hardness in the households. Desalination has the highest total environmental impact score (205 μPET/m^3), primarily due to the high electricity demand of this technology.

The environmental impact category with the highest importance is global warming potential (Table 33.2). The contribution from water treatment is higher for A4 compared to the others (Fig. 33.6). The alternatives relying on groundwater abstraction (A0, A2, A3) show very similar patterns with little contribution from water production and more than 50% from wastewater transport and treatment in the global warming impact category and total environmental impact. If wastewater treatment had not been included, these three options would have had the lowest impact, but then they would not have been comparable since the rain- and stormwater harvesting reduces the amount of wastewater to be treated. This emphasizes the importance of a thorough assessment of proper system boundaries (in this case by including the combined sewers and wastewater treatment processes), functional unit, etc. in the preparation of an LCA (ISO 2006).

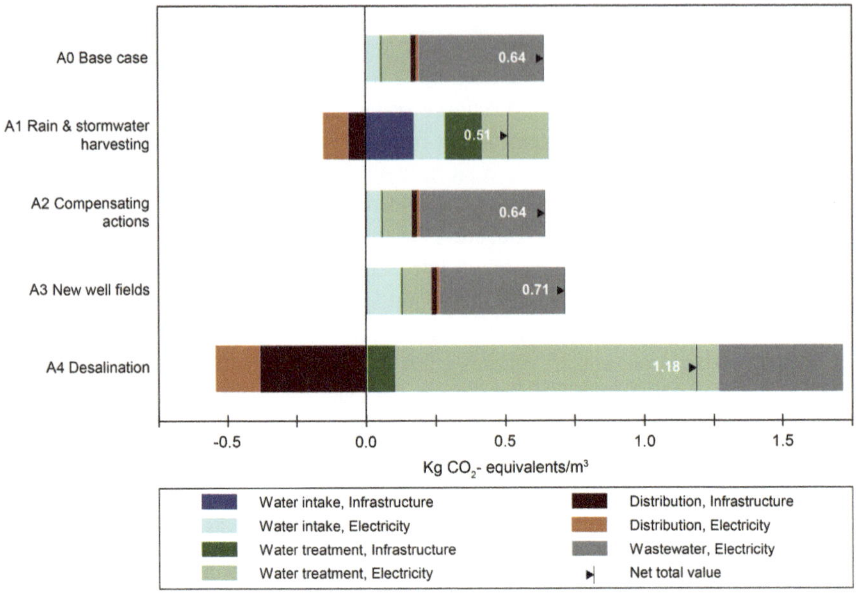

Fig. 33.6 Distribution over the life cycle of contributions to global warming potential for the base case and the four alternative options for water supply (Godskesen et al. 2013)

In conclusion, the LCA showed that the rain- and stormwater harvesting scenario (A1) has the lowest environmental impact (82 μPET/m^3) followed by the options relying on groundwater abstraction (124–138 μPET/m^3), and that A4 Desalination (205 μPET/m^3) has a noteworthy increase in total environmental impacts. If the rain- and stormwater is not harvested it is led to combined sewers in the city which makes it environmentally beneficial to prevent it from discharging into the sewers, e.g. by harvesting and recycling for non-potable purposes. Figure 33.6 shows by the reduction in environmental impacts (negative numbers which reduce the environmental impacts of A1 and A4) that it is essential to include the beneficial effects of reduced water hardness in households when comparing the environmental impacts of water supply cases leading to water of different hardness.

33.3 Specific Methodological Issues for the Application of LCA to Water Supply

33.3.1 General

Following the ISO standard 14044 (ISO 2006) most LCA studies report objectives as part of the goal and scope description, as well as functional unit and system boundaries definitions. However, a widely accepted standard of a uniform set of indicators to describe the water supply system under study is still missing. Each study therefore describes the studied system in a different way, in many cases missing important descriptors. As a minimum, it is recommended to include the following descriptors:

- Analysed region population and its area
- Total length of pipes
- Distances between abstraction and production, and average distance between production and consumers
- Average difference in height from production to consumers
- Water losses
- Energy mix of electricity
- Details of the water sources and their respective share of contribution
- Production technologies in place, their capacity, and actual supply
- Product water quality

33.3.2 Goal and Scope

Published case studies of LCA of water supply indicate that the goal is often to compare different technologies for water supply in order to identify the most environmentally sound technology for water production or water supply system (Lundie et al. 2004; Lassaux et al. 2007; Klaversma et al. 2013; Godskesen et al. 2013; Stokes and Horvath 2006; Lyons et al. 2009). The goal may also be to identify hot spots in the system allowing for optimizing the environmental performance.

Studies use different functional units (FU). The most common is one m^3 of water produced at the period of time for which the analysis is valid, at the user or at the end of the process. For example, the most common FU for the entire water supply systems is one m^3 of potable water at the consumer tap. Another example of the definition of a FU would be the annual consumption of water at the end user.

When infrastructure is the focus of the LCA FUs may take another form. For example, a study on reverse osmosis membranes used one membrane module as FU (Lawler et al. 2015) and a study on pipes used a metre of pipe network (Herz and Lipkow 2002; Sanjuan-Delmás et al. 2014). Such definitions may serve adequately the specific systems, in which they were used, but they are not applicable for comparison between systems or case studies, in which case the specific FU should be linked to the commonly used FUs.

Quality of source and product water can vary substantially and should therefore be reported. Owens (2001) proposed water quantity and quality indicators to make LCAs compatible with environmental management and reporting systems.

When carrying out the case study of water technologies for Copenhagen it was found that the service life of different components in the system differ and therefore it is important to carefully go through each component and gather data or best available estimates on the expected service lives. This applies to components such as pumps, different types of pipes (polyethylene, polypropylene, concrete, cast iron, etc.), materials used in building of waterworks (bricks, concrete, steel), etc.

33.3.3 Inventory and Product System Modelling

The system boundaries of studies of water supply systems vary by the life cycle stages included in each analysis: infrastructure construction, operation, maintenance and demolition. Studies also vary in the included activities: material production, material transportation, equipment use and energy production (Meron et al. 2016).

Some studies of recycling systems include raw wastewater treatment (e.g. Tangsubkul et al. 2005; Pasqualino et al. 2011 Lyons et al. 2009; Li et al. 2016) while some studies start only with the secondary effluent entering tertiary treatment (Muñoz et al. 2009; Meneses et al. 2010; Stokes and Horvath 2006). A few studies do not report LCA results of water supply systems and the wastewater collection and treatment separately, due to availability of aggregated data only. For example, the study of Aveiro (2008) in Portugal (Lemos et al. 2013), and of the city of Atlanta (2005–2009) in the USA (Jeong et al. 2015) report the distribution of water and collection of wastewater together because electricity monitoring could not be separated. Having a more detailed monitoring system of electricity consumption as well as other operational data can be an important recommendation to water systems managers, which enable more accurate LCAs in the future.

A review of studies analysing rainwater harvesting tanks showed considerable difference between ex-ante theoretical calculations and measured data from established systems. Where the theoretical calculation had a median electricity consumption of 0.2 kWh/m^3, the median measured data was 1.4 kWh/m^3 (Vieira et al. 2014). This shows the importance of establishing better data inventories based on actual measured data rather than generic estimations. Therefore, it is recommended to collect actual data from the water supply system for the LCA modelling. If data is unavailable literature values and estimates may be used. Upstream and downstream background data is usually available in the LCI databases such as GaBi, Ecoinvent, etc.

In the case study of water technologies for supplying the city of Copenhagen the system boundaries where placed to reflect equal effects of the water hardness of the drinking water. This had a significant effect on the results of the LCA especially in the global warming potential impact category. The case study shows that system boundaries must be defined so the alternatives compared in the LCA are equal also when it comes to product water quality.

33.3.4 Impact Assessment

Various LCIA methods were used in different studies, including CML 2000 (e.g. Barjoveanu et al. 2014), CML 2001 (e.g. Li et al. 2016), and CML-IA (e.g. Muñoz and Fernández-Alba 2008; Meneses et al. 2010; Amores et al. 2013), ReCiPe (e.g. de Haas et al. 2011; Jeong et al. 2015; Slagstad and Brattebø 2014), Eco-indicator 95 (Mohapatra et al. 2002), Eco-indicator 99 (e.g. Uche et al. 2013), IPCC GWP 2007a (e.g. Uche et al. 2013), IMPACT 2002+ (e.g. Bonton et al. 2012), USES-LCA (e.g. Muñoz et al. 2009; Tarantini and Ferri 2001), USEtox (e.g. Li et al. 2016) or EPD 2013 (e.g. Del Borghi et al. 2013). Most papers use a variety of units to present the impacts, whereas some papers transform the results to a single unit such as "eco-point" (e.g. Raluy et al. 2005a; El-Sayed et al. 2010; Uche et al. 2013) and EDIP's "person equivalent" (Godskesen et al. 2011).

Several studies have pointed out the uncertainties resulting from using early stage impact models (Jeong et al. 2015; Lane et al. 2015; Zhou et al. 2011). A comprehensive LCIA harmonization for GWP-100 showed that the maximum difference between GWP scores obtained with different LCIA methods is 7%, while in some impact categories, such as human toxicity and marine ecotoxicity, variability between different LCIA methods is very high and scores are incomparable (Meron et al. 2016). Selection of impact models is important and future research is required in order to generate an agreed set of models.

Impact categories to describe water depletion have been the subject of many studies. Many have expressed the volume of freshwater withdrawn for water supply, (Sharma et al. 2009; Lundie et al. 2004; Lane et al. 2015; Jeong et al. 2015; Li et al. 2016) e.g. by water foot-printing (Hoekstra et al. 2011) where water is considered a resource for man rather than an environmental media with environmental impacts when withdrawn. More recent methods have been suggested to integrate freshwater use into the LCA methodology by treating freshwater consumption as an environmental impact category with an impact on the freshwater environment (Núñez et al. 2016) and human health (Boulay et al. 2015b). The relative Available WAter REmaining (AWaRe) indicator was developed by the Water Use in LCA (WULCA) working group of the UNEP-SETAC Life Cycle Initiative as a (proxy-) midpoint indicator to assess the environmental performance regarding freshwater consumption (Boulay et al. 2015a). The indicator aims to represent the potential of water deprivation, to humans or ecosystems, based on the assumption that the less water remaining available per area, the more likely another user will be deprived (Boulay et al. accepted). Further details on water consumption LCIA can be found in Sect. 10.15.

Reporting impacts of water supply systems in the water use category is important because water supply systems are the major source of direct impacts in this category and without it any impact assessment of products that use water will be incomplete.

To compare the significance of various impact categories of water supply systems, a normalization analysis of 10 supply system models and 15 production systems has been carried out using ReCiPe (H) V1.12/World (Meron et al. 2016). The highest value is associated with marine ecotoxicity, with consistently highest values in nine of the ten models of water supply systems. Normalized values of freshwater ecotoxicity, freshwater eutrophication, fossil depletion, human toxicity, and GWP follow. Other impact categories have considerably lower scores.

33.3.5 Interpretation

It is recommended to include a sensitivity and uncertainty analysis. In the study of drinking water technology for Copenhagen a sensitivity analysis on the future prediction of the Danish electricity mix for the years 2020 and 2050 evaluated the changes in the global warming potential impact category and showed that global warming potential values will decrease and other impact categories will be higher compared to the others (Godskesen et al. 2013). Similarly to the electricity mix, the scarcity of freshwater resources will change in the future due to population increase (demand) and climate change (local availability) as demonstrated for Spain by Núñez et al. (2015), which may be of significance for studies with longer time horizons. Therefore, a sensitivity and uncertainty analysis may change the outcome of the results and also affect the interpretation of the LCA.

33.4 Concluding Remarks and Outlook

LCAs of water supply systems are of growing importance as an increasing number of regions in the world rely less on nature (rain) and shift to more treatment intensive water resources, e.g. desalination of sea or brackish water. It has been shown that the differences among water supply systems result in significant variation in environmental impacts. However, site-specific LCAs of regional supply systems have been carried out only in a limited number of regions, mostly in Europe.

Studies of systems in water-stressed regions are therefore needed. In particular, it is important to carry out studies of supply systems in regions where desalination is heavily used. LCAs of water supply systems in rapidly developing countries (e.g. India, Brazil, Indonesia, Turkey) are also needed as the impacts of their water supply systems may be significantly different from the impacts identified in available studies.

Using standard LCI databases is a common practice although impacts of water supply systems can vary significantly (Meron et al. 2016). In the assessment of products that consume large amounts of water, using datasets from other regions may result in misleading conclusions. Relying on correct selection from available studies can serve as a basis for receiving more accurate results than straightforward use of standard datasets.

Our case study and the work of others show that LCA is useful for assessing the environmental impacts of water supply technologies and it provides a platform for integrating environmental considerations in the decision-making process and planning of future water systems. When conducting LCA of water supply it is important that:

- the system boundaries are defined carefully so that compared alternatives are fully comparable, e.g. shares the same product water quality
- a typical functional unit could be the annual consumption or supply of one m^3 of water at the end user
- a hot spot analysis is performed to better understand where and what processes in the water supply system contribute most to the environmental impacts;
- impacts of freshwater use are considered. This is especially relevant to include when working with water supply systems because of the intrinsically large use of water
- an uncertainty and sensitivity analysis is carried out that accounts for data availability, estimations versus measured data and also considers the future predictions of electricity mix and water scarcity.

LCA can also be used for integrating environmental aspects in the decision-making process within other areas of water systems. In Copenhagen the water utility is using LCA to reach better overall environmental performance of the water utility through:

- Evaluation of alternative options for supplying a new neighbourhood under development in Copenhagen with non-potable water
- Evaluation of stormwater management solutions
- Water supply strategy development, e.g. the choice of establishing new well fields and waterworks within the city limits or extending the water import from well fields and waterworks located 30–50 km outside the city.

LCA is not only relevant for the analysis of future urban water management. As for most other production activities, water utilities are also met with requirements or intentions to declare environmental impacts, carbon and water footprint, green accounting, etc. Therefore, it is in the water utility's interest to evaluate their production and transport of water, as well as handling of wastewater to provide transparent efficiency measures, decision support for daily operations that thoroughly covers environmental aspects.

To reach a full sustainability assessment, LCA can be combined with an economic and social evaluation, such as multi-criteria decision analysis (Sombekke et al. 1997; Lundie et al. 2006; Lai et al. 2008; Godskesen 2012). The combination of these criteria completes the three-dimensional sustainability approach as suggested by the first political definitions of sustainability (WCED 1987; UNEP 1992).

References

Amores, M.J., Meneses, M., Pasqualino, J., Antón, A., Castells, F.: Environmental assessment of urban water cycle on Mediterranean conditions by LCA approach. J. Clean. Prod. **43**, 84–92 (2013)

Angrill, S., Farreny, R., Gasol, C.M., Gabarrell, X., Vinolas, B., Josa, A., Rieradevall, J.: Environmental analysis of rainwater harvesting infrastructures in diffuse and compact urban models of Mediterranean climate. Int. J. Life Cycle Assess. **17**(1), 25–42 (2012)

Arpke, A., Hutzler, N.: Domestic water use in the United States: a life-cycle approach. J. Ind. Ecol. **10**(1–2), 169–184 (2006)

Barjoveanu, G., Comandaru, I.M., Rodriguez-Garcia, G., Hospido, A., Teodosiu, C.: Evaluation of water services system through LCA. A case study for Iasi City, Romania. Int. J. Life Cycle Assess. **19**, 449–462 (2014)

Barrios R, Siebel M, van der Helm A, Bosklopper K, Gijzen H (2008) Environmental and financial life cycle impactassessment of drinking water production at Waternet. J Clean Prod 16:471–476

Birnhack, L., Penn, R., Lahav, O.: Quality criteria for desalinated water and introduction of a novel, cost effective and advantageous post treatment process. Desalination **221**, 70–83 (2008)

Bonton, A., Bouchard, C., Barbeau, B., Jedrzejak, S.: Comparative life cycle assessment of water treatment plants. Desalination **284**, 42–54 (2012)

Botto, S., Niccolucci, V., Rugani, B., Nicolardi, V., Bastianoni, S., Gaggi, C.: Towards lower carbon footprint patterns of consumption: the case of drinking water in Italy. Environ. Sci. Policy **14**, 388–395 (2011)

Boulay, A.-M., Bare, J., De Camillis, C., Döll, P., Gassert, F., Gerten, D., Humbert, S., Inaba, A., Itsubo, N., Lemoine, Y., Margni, M., Motoshita, M., Núñez, M., Pastor, A.V., Ridoutt, B., Schencker, U., Shirakawa, N., Vionnet, S., Worbe, S., Yoshikawa, S., Pfister, S.: Consensus building on the development of a stress-based indicator for LCA-based impact assessment of water consumption: outcome of the expert workshops. Int. J. Life Cycle Assess. **20**, 577–583 (2015a)

Boulay, A.-M., Motoshita, M., Pfister, S., Bulle, C., Muñoz, I., Franceschini, H., Margni, M.: Analysis of water use impact assessment methods (part A): evaluation of modeling choices based on a quantitative comparison of scarcity and human health indicators. Int. J. Life Cycle Assess. **20**, 139–160 (2015b)

Buckley, C., Friedrich, E., Von Blottnitz, H.: Life-cycle assessments in the South African water sector: a review and future challenges. Water SA **37**, 719–726 (2011)

Crettaz, P., Jolliet, O., Cuanillon, J.-M., Orlando, S.: Life cycle assessment of drinking water and rain water for toilets flushing. Aqua (Oxford) **48**(3), 73–83 (1999)

Danva: Udviklingen i spildevandsselskabernes energiforbrug 2005–2009 (eng.: Development of Energy consumption at the wastewater treatment plants 2005–2009). Danva, Dansk vand-og spildevandsforening (2010)

de Haas, D., Lane, J., Lant, P.: Life cycle assessment of the Gold Coast urban water system. Water **38**, 57–64 (2011)

Del Borghi, A., Strazza, C., Gallo, M., Messineo, S., Naso, M.: Water supply and sustainability: life cycle assessment of water collection, treatment and distribution service. Int. J. Life Cycle Assess. **18**, 1158–1168 (2013)

GWI: Water Market Middle East 2010, Global water intelligence (2010)

El-Sayed Mohamed Mahgoub, M., van der Steen, N.P., Abu-Zeid, K., Vairavamoorthy, K.: Towards sustainability in urban water: a life cycle analysis of the urban water system of Alexandria City, Egypt. J. Clean Prod. **18**, 1100–1106 (2010)

European Commission: The EU water framework directive—integrated river basin management for Europe [WWW Document] (2012). http://ec.europa.eu/environment/water/water-framework/index_en.html. Accessed 14 Jan 2013

Fang, L.L., Valverde Perez, B., Damgaard, A., Plósz, B.G., Rygaard, M.: Life cycle assessment as development and decision support tool for wastewater resource recovery technology. Water Res. **88**, 538–549 (2016). doi:10.1016/j.watres.2015.10.016

Friedrich, E., Pillay, S., Buckley, C.: Carbon footprint analysis for increasing water supply and sanitation in South Africa: a case study. J. Clean. Prod. **17**, 1–12 (2009)

Gleeson, T., Wada, Y., Bierkens, M.F.P., van Beek, L.P.H.: Water balance of global aquifers revealed by groundwater footprint. Nature **488**, 197–200 (2012)

Godskesen, B.: Sustainability evaluation of water supply technologies—by using life-cycle and freshwater withdrawal impact assessment and multi-criteria analysis. PhD Thesis. DTU Environment (2012)

Godskesen, B., Zambrano, K.C., Trautner, A., Johansen, N.B., Godskesen, B., Rygaard, M., Albrechtsen, H.J., Johansen, N.B., Thiesson, L., Andersen, L., Clauson-Kaas, J., Neidel, T.L., Andersen, L., Clauson-Kaas, J., Neidel, T.L., Kløverpris, N.H.: Life cycle assessment of three water systems in Copenhagen—a management tool of the future. Water Sci. Technol. **63**(3), 565–572 (2011)

Godskesen, B., Hauschild, M.Z., Rygaard, M., Zambrano, K.C., Albrechtsen, H.J.: Life cycle assessment of central softening of very hard drinking water. J. Environ. Manage. **105**, 83–89 (2012)

Godskesen, B., Hauschild, M.Z., Rygaard, M., Zambrano, K.C., Albrechtsen, H.J.: Life-cycle and freshwater withdrawal impact assessment of water supply technologies. Water Res. **47**(7), 2363 (2013)

Greenlee, L.F., Lawler, D.F., Freeman, B.D., Marrot, B., Moulin, P.: Reverse osmosis desalination: Water sources, technology, and today's challenges. Water Res. **43**, 2317–2348 (2009)

Herz, R., Lipkow, A.: Life cycle assessment of water mains and sewers. Water Supply **2**, 51–58 (2002)

Hoekstra, A.Y., Chapagain, A.K., Aldaya, M.M., Mekonnen, M.M.: The Water Footprint Assessment Manual: Setting the Global Standard. Earthscan (2011). ISBN: 978-1-84971-279-8

Homäki, K., Nielsen, P.H., Sathasivan, A., Bohe, E.L.: Life cycle assessment and environmental improvement of residential and drinking water supply systems in Hanoi, Vietnam. Int. J. Sust. Dev. World **10**, 27–42 (2003)

Hospido, A., Núñez, M., Antón, A.: Irrigation mix: how to include water sources when assessing freshwater consumption impacts associated to crops. Int. J. Life Cycle Assess. **18**, 881–890 (2013)

Igos, E., Dalle, A., Tiruta-Barna, L., Benetto, E., Baudin, I., Mery, Y.: Life cycle assessment of water treatment: what is the contribution of infrastructure and operation at unit process level? J. Clean. Prod. **65**, 424–431 (2014)

ISO: Environmental Management—Life Cycle Assessment—Requirements and Guidelines (ISO 14044). ISO, the International Organization for Standardization, Geneva (2006)

IWA: International Statistics for Water Services. International Water Association, London (2014)

Jensen, M.B.: Basisrapport: Dobbeltporøs filtrering. Pilotafprøvning—Rensning af vejvand i Ørestad. Hændelse 1–25, januar–juli 2007 (eng.: Double Porous Filter—Pilot project). KU Life, Denmark (2009)

Jeong, H., Minne, E., Crittenden, J.C.: Life cycle assessment of the City of Atlanta, Georgia's centralized water system. Int. J. Life Cycle Assess. **20**, 880–891 (2015)

Jungbluth, N.: Comparison of the environmental impact of drinking water vs. bottled mineral water. Manuscript for the SGWA information bulleting and GWA. Uster (2006)

Kenway, S., Gregory, A., McMahon, J.: Urban water mass balance analysis. J. Ind. Ecol. **15**(5), 693–706. doi:10.1111/j.1530-9290.2011.00357.x (2011)

Klaversma, E., van der Helm, A.W.C., Kappelhof, J.W.N.M.: The use of life cycle assessment for evaluating the sustainability of the Amsterdam water cycle. J. Water Climate Change **04**(2), 103–109 (2013)

Lai, E., Lundie, S., Ashbolt, N.J.: Review of multi-criteria decision aid for integrated sustainability assessment of urban water systems. Urban Water J. **5**(4), 315–327 (2008)

Lane, J.L., de Haas, D.W., Lant, P.A.: The diverse environmental burden of city-scale. Water Res. **81**, 398–415 (2015)

Lassaux, S., Renzoni, R., Germain, A.: Life cycle assessment of water from the pumping station to the wastewater treatment plant. Int. J. Life Cycle Assess. **12**(2), 118–126 (2007)

Lawler, W., Alvarez-Gaitan, J., Leslie, G., Le-Clech, P.: Comparative life cycle assessment of end-of-life options for reverse osmosis membranes. Desalination **357**, 45–54 (2015)

Lemos, D., Dias, A.C., Gabarrell, X., Arroja, L.: Environmental assessment of an urban water system. J. Clean. Prod. **54**, 157–165 (2013)

Li, Y., Xiong, W., Zhang, W., Wang, C., Wang, P.: Life cycle assessment of water supply alternatives in water-receiving areas of the South-to-North water diversion project in China. Water Res. **89**, 9–19 (2016)

Lundie, S., Peters, G., Ashbolt, N., Lai, E., Livingston, D.: Water resources—a sustainability framework for the Australian water industry—the best of current strategic planning and sustainability assessments. Water **33**(7), 3465–3473 (2006)

Lundie, S., Peters, G.M., Beavis, P.C.: Life cycle assessment for sustainable metropolitan water systems planning. Environ. Sci. Technol. **38**(13), 3465–3473 (2004)

Lyons, E., Zhang, P., Benn, T., Sharif, F., Costanza, M., Li, K., Crittenden, J., Chen, Y.S.: Life cycle assessment of three water supply systems: importation, reclamation and desalination. Water Sci. Technol. Water Supply. **9**(4), 439–448 (2009)

Meneses, M., Pasqualino, J.C., Céspedes-Sánchez, R., Castells, F.: Alternatives for reducing the environmental impact of the main residue from a desalination plant. J. Ind. Ecol. **14**, 512–527 (2010)

Meron, N., Blass, V., Garb, Y., Kahane, Y., Thoma, G.: Why going beyond standard LCI databases is important: lessons from a meta-analysis of potable water. Int. J. Life Cycle Assess. **21**, 1134–1147 (2016)

Mithraratne, N., Vale, R.: Conventional and alternative water supply systems: a life cycle study. Int. J. Environ. Sustain. Dev. **6**(2), 136–146 (2007)

Mohapatra, P., Siebel, M., Gijzen, H., Van der Hoek, J., Groot, C.: Improving eco-efficiency of Amsterdam water supply: a LCA approach. Aqua J. Water Supply Res. Technol. **51**, 217–228 (2002)

Muñoz, I., Fernández-Alba, A.R.: Reducing the environmental impacts of reverse osmosis desalination by using brackish groundwater resources. Water Res. **42**(3), 801–811 (2008)

Muñoz, I., Rodríguez, A., Rosal, R., Fernández-Alba, A.R.: Life cycle assessment of urban wastewater reuse with ozonation as tertiary treatment: a focus on toxicity-related impacts. Sci. Total Environ. **407**(4), 1245–1256 (2009)

Nessi, S., Rigamonti, L., Grosso, M.: LCA of waste prevention activities: a case study for drinking water in Italy. J. Environ. Manage. **108**, 73–83 (2012)

Niccoluci, V., Rugani, B., Botto, S., Gaggi, C.: An integrated footprint based approach for environmental labelling of products: the case of drinking bottled water. Int. J. Design Nature Ecodyn. **5**(1), 68–75 (2010)

Núñez, M., Pfister, S., Vargas, M., Antón, A.: Spatial and temporal specific characterisation factors for water use impact assessment in Spain. Int. J. Life Cycle Assess. **20**, 128–138 (2015)

Núñez, M., Bouchard, C., Boulay, A.M., Bulle, C., Margni, M.: Critical analysis of life cycle impact assessment methods addressing consequences of freshwater use on ecosystems and recommendations for future method development. Int. J. Life Cycle Assess. (2016). doi:10.1007/s11367-016-1127-4

Owens, J.W.: Water resources in life-cycle impact assessment: considerations in choosing category indicators. J. Ind. Ecol. **5**(2), 37–54 (2001)

Pankratz, T.: IDA Desalination Yearbook 2009–2010. Media Analytics Ltd, Oxford (2010)

Pasqualino, J.C., Meneses, M., Castells, F.: Life cycle assessment of urban wastewater reclamation and reuse alternatives. J. Ind. Ecol. **15**, 49–63 (2011)

Petersen, S.: Central Sekundavandsopsamling (eng. Centralized rainwater harvesting for non-potable purposes). Student project carried out in cooperation between HOFOR and DTU (2011)

Plappally, A.K., Lienhard, V.J.H.: Energy requirements for water production, treatment, end use, reclamation, and disposal. Renew. Sustain. Energy Rev. **16**(7), 4818–4848 (2012)

Racoviceanu, A.I., Karney, B.W., Kennedy, C.A., Colombo, A.F.: Life-cycle energy use and greenhouse gas emissions inventory for water treatment systems. J. Infrastruct. Syst. **13**, 261–270 (2007)

Raluy, R.G., Serra, L., Uche, J.: Life cycle assessment of water production technologies—part 1: life cycle assessment of different commercial desalination technologies (MSF, MED, RO) (9 pp). Int. J. Life Cycle Assess. **10**(4), 285–293 (2005a)

Raluy, R.G., Serra, L., Uche, J.: Life cycle assessment of water production technologies—part 2: reverse osmosis desalination versus the Ebro river water transfer. Int. J. Life Cycle Assess. **10** (5), 436–454 (2005b)

Ratnayaka, D.D., Brandt, M.J., Johnson, M.: Twort's water supply, 5th edn. Butterworth-Heinemann, London (2009)

Rygaard, M., Albrechtsen, H.-J., Jensen, I., Pedersen, C., Sørensen, S., Zambrano, K.: Catalogue of development projects—identified at three workshops with AarhusVand, Københavns Energi and DTU (in Danish: Udviklingskatalog—Opsamling på 3 workshops afholdt i samarbejdet Fremtidens vandhåndtering i Storbyer). DTU, Miljøvej, Bygning 113, DK-2800 Kgs. Lyngby (2012)

Rygaard, M., Arvin, E., Bath, A., Binning, P.J.: Designing water supplies: optimizing drinking water composition for maximum economic benefit. Water Res. **45**(12), 3712–3722 (2011)

Rygaard, M., Godskesen, B., Jørgensen, C., Hoffmann, B.: Holistic assessment of a secondary water supply for a new development in Copenhagen, Denmark. Sci. Total Environ. **497–498**, 430–439 (2014). doi:10.1016/j.scitotenv.2014.07.078

Sanjuan-Delmás, D., Petit-Boix, A., Gasol, C.M., Villalba, G., Suárez-Ojeda, M.E., Gabarrell, X., Josa, A., Rieradevall, J.: Environmental assessment of different pipelines for drinking water transport and distribution network in small to medium cities: a case from Betanzos, Spain. J. Clean. Prod. **66**, 588–598 (2014)

Sharma, A.K., Grant, A.L., Grant, T., Pamminger, F., Opray, L.: Environmental and economic assessment of urban water services for a greenfield development. Environ. Eng. Sci. **26**(5), 921–934 (2009)

Slagstad, H., Brattebø, H.: Life cycle assessment of the water and wastewater system in Trondheim, Norway—a case study: case study. Urban Water J. **11**, 323–334 (2014)

Smakhtin, V., Revenga, C., Doll, P.: A pilot global assessment of environmental water requirements and scarcity. Water Int. **29**(3), 307–317 (2004)

Sombekke, H.D.M., Voorhoeve, D.K., Hiemstra, P.: Environmental impact assessment of groundwater treatment with nanofiltration. Desalination **113**(2–3), 293–296 (1997)

Stokes, J., Horvath, A.: Life cycle energy assessment of alternative water supply systems. Int. J. Life Cycle Assess. **11**(5), 335–343 (2006)

Tal, A.: Seeking sustainability: Israel's evolving water management strategy. Science **80**(313), 1081–1084 (2006)

Tangsubkul, N., Beavis, P., Moore, S., Lundie, S., Waite, T.: Life cycle assessment of water recycling technology. Water Resour. Manage. **19**, 521–553 (2005)

Tarantini, M., Ferri, F.: LCA of drinking and wastewater treatment systems of Bologna city: final results. In: Proceedings of the 4th Inter-Regional Conference on Environment-Water, 27–30 Aug 2001, Fortaleza (2001)

Tarnacki, K., Meneses, M., Melin, T., van Medevoort, J., Jansen, A.: Environmental assessment of desalination processes: reverse osmosis and Memstill®. Desalination **296**, 69–80 (2012)

Uche, J., Martínez, A., Castellano, C., Subiela, V.: Life cycle analysis of urban water cycle in two Spanish areas: Inland city and island area. Desalin. Water Treat **51**, 280–291 (2013)

UNEP: Rio declaration on environment and development. In: United Nations Publication (1992)

USGS: Water Hardness and Alkalinity (2012). http://water.usgs.gov/owq/hardness-alkalinity.html

Vieira, A.S., Beal, C.D., Ghisi, E., Stewart, R.A.: Energy intensity of rainwater harvesting systems: a review. Renew. Sustain. Energy Rev. **34**(June), 225–242 (2014). doi:10.1016/j.rser.2014.03.012

WCED: Report of the World Commission on Environment and Development: Our Common Future. Oxford University Press, New York (1987)

Wenzel, H., Hauschild, M.Z., Alting, L.: Environmental Assessment of Products 1: Methodology, Tools and Case Studies in Product Development. Chapman & Hall, London (1997)

Wols, B.A., Hofman-Caris, C.H.M.: Review of photochemical reaction constants of organic micropollutants required for UV advanced oxidation processes in water. Water Res. **46**, 2815–2827 (2012)

Boulay, A.-M., Bare, J., Benini, L., Berger, M., Lathuillière, M., Manzardo, A., Margni, M., Motoshita, M., Núñez, M., Pastor, A.V., Ridoutt, B., Oki, T., Worbe, S., Pfister, S. (accepted): The WULCA consensus characterization model for water scarcity footprints: assessing impacts of water consumption based on available water remaining (AWARE). Environ. Sci. Technol.

Zhou, J., Chang, V.W.C., Fane, A.G.: Environmental life cycle assessment of reverse osmosis desalination: the influence of different life cycle impact assessment methods on the characterization results. Desalination **283**, 227–236 (2011)

Author Biographies

Berit Godskesen LCA practitioner with focus on sustainability assessment of urban water systems where LCA is combined with other decision support tools. Main LCA interests are assessing environmental impacts of potential system changes and incorporating results in the decision making process.

Noa Meron Electronics engineer by background and involved in environmental sciences since 2008. Developed a best proxy methodology for LCAs where detailed LCI databases do not exist. Main LCA interests are in mathematical modelling and water supply systems.

Martin Rygaard An expert on urban water management systems and how these interact with the environment, humans and economy. Has a special interest in effects of changing water quality, e.g. through desalination. Another focus is combined use of LCA and economic assessments to provide holistic evaluations of water systems and technologies.

Chapter 34
LCA of Wastewater Treatment

Henrik Fred Larsen

Abstract The main purpose of wastewater treatment is to protect humans against waterborne diseases and to safeguard aquatic bio-resources like fish. The dominating environmental concerns within this domain are indeed still potential aquatic eutrophication/oxygen depletion due to nutrient/organic matter emissions and potential health impacts due to spreading of pathogens. Anyway, the use of treatment for micro-pollutants is increasing and a paradigm shift is ongoing—wastewater is more and more considered as a resource of, e.g. energy, nutrients and even polymers, in the innovations going on. The focus of LCA studies addressing wastewater treatment have from the very first published cases, been on energy and resource consumption. In recent time, the use of characterisation has increased and besides global warming potential, especially eutrophication is in focus. Even the toxicity-related impact categories are nowadays included more often. Application of LCA for comparing avoided against induced impacts, and hereby identifying trade-offs when introducing new technology, is increasingly used. A typical functional unit is the treatment of one cubic metre of wastewater which should be well defined regarding composition. Depending on the goal and scope of the study, all life cycle stages have the potential of being significant, though disposal of infrastructure seems to be the least important for the impact profile in many cases. No inventory data and none of the conventional impact categories (except stratospheric ozone depletion if emission of N_2O is excluded) should be ruled out; but eutrophication and ecotoxicity are in many cases among the dominating ones.

34.1 Introduction

The history and the present status of wastewater treatment including the overall use of LCA within this technology domain are briefly described below.

H.F. Larsen (✉)
Danish Road Directorate, 2640 Hedehusene, Denmark
e-mail: hfjlarsen@gmail.com

M.Z. Hauschild et al. (eds.), *Life Cycle Assessment*,
DOI 10.1007/978-3-319-56475-3_34

34.1.1 History

For thousands of years water has been used for sanitary purposes with the resulting wastewater (WW) being emitted to the environment. However, due to an accelerating global population with increased sanitary demands combined with the industrial revolution, the pollution potential has reached a severe level in modern time. In order to protect the aquatic bio-resources (e.g. fish and crustaceans), human health and the aquatic ecosystems, wastewater treatment (WWT) is now widespread and becomes more and more advanced. Starting with simple systems for sedimentation (mechanical or primary treatment), more advanced processes for removing organic matter and nutrients (ammonia/nitrate), like activated sludge treatment (secondary treatment), have now been used for some decades in industrialised and densely populated countries. In recent years, the focus is more on tertiary treatment (removing phosphorus) and processes for removing micro-pollutants including ozonation and activated carbon treatment. A paradigm shift is ongoing—wastewater is more and more considered as a resource of, e.g. energy (biogas from anaerobic digestion of sludge), nutrients (especially phosphorus) and polymers (sludge). Innovations addressing these issues are ongoing these years.

The focus of LCA case studies within this area has consistently been on energy and in some cases combined with resource consumption. More recently, characterisation has increasingly been included and besides global warming potential (including direct emissions of CH_4 and N_2O from WWT) especially eutrophication is in focus. With the enhanced awareness of micro-pollutants in effluent and sludge also the toxicity-related impact categories are nowadays included more often. Using LCA for identifying trade-offs and comparing the relative sustainability of alternative treatment systems has also become widespread.

34.1.2 Present Status

Today, the number of wastewater treatment technologies is quite large with optimization and new technologies currently being introduced. This process is mainly driven by legislation like the EU Water Framework Directive for Europe (EC 2000) or the Australian Guidelines for Water Recycling being part of the National Water Quality Management Strategy for Australia (Australian Government 2015). The wastewater treatment technologies or systems may be divided into at least three main groups with some overlap:

- Treatment systems for removal of organic matter (e.g. sedimentation, activated sludge).
- Treatment systems for removal of nutrients (e.g. nitrification/denitrification, P-precipitation).
- Treatment systems for removal of micro-pollutants (e.g. ozonation, activated carbon).

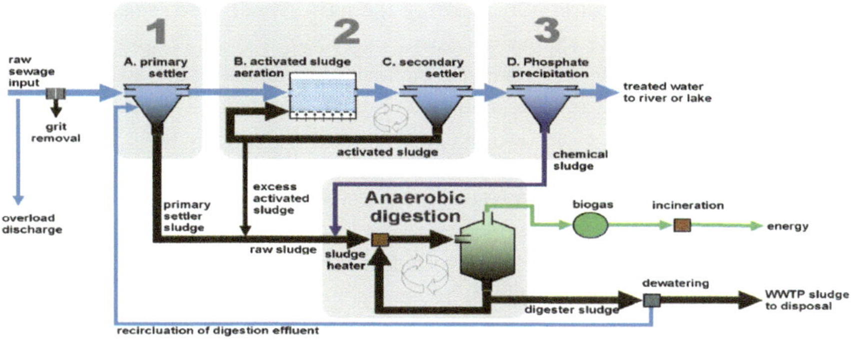

Fig. 34.1 Conventional wastewater treatment plant (Doka 2007, with permission)

Among new technologies that presently have only reached lab or pilot scale are, e.g. microbial fuel cells and advanced oxidation processes like manganese oxidation. Less advanced technologies/systems like source separation (e.g. separating toilet water containing faeces from urine containing water and bathing water) are also part of the innovation going on.

The traditional aim of sludge treatment is to reduce volume and mass in order to save disposal costs. However, as sludge has been and still is used as fertiliser on agricultural land (or, e.g. woods), removing pathogens is also a focus. Therefore, different physical, mechanical and biological technologies like dewatering, digestion, incineration and, e.g. heating for hygienic treatment are widespread. The increased focus on resource recovery/recycling in recent years has led to enhanced use of, e.g. anaerobic digestion for energy recovery (biogas/CH_4). Furthermore, new technologies like sludge inertisation (wet oxidation, pyrolysis) and sludge triage (separating primary and secondary sludge before treatment) for improved exploitation/recovery of, e.g. phosphorus, are part of the innovation going on.

The main environmental concerns within the wastewater treatment domain are still potential aquatic eutrophication/oxygen depletion due to nutrient/organic matter emissions and potential health impacts due to spreading of pathogens. Anyway, the focus on the potential ecotoxic effect of organic micro-pollutants (e.g. pharmaceuticals) and metals (e.g. mercury) is increasing together with efforts to improve the energy balance (e.g. optimise biogas production) and resource recovery (e.g. phosphorus).

The process steps in a typical conventional wastewater treatment plant (WWTP) are shown in Fig. 34.1.

34.2 Review of Existing LCA Case Studies on WWT

At least more than 60 LCA studies on wastewater treatment have been performed since the mid 1990s, with the paper by Emmerson et al. (1995) being among the first ones. Today, several review papers exist with Larsen et al. (2007), Corominas et al. (2013) and Zang et al. (2015) being the most recent. The review by Larsen et al. (2007) includes 22 studies and focus on eco-toxicity-related impacts though conventional impact categories are also included. Corominas et al. (2013) reviewed 45 studies but excluded specific studies on sludge treatment. This is also the case for the review by Zang et al. (2015) that includes 53 studies (among these several Asian ones) on different technologies but focus on activated sludge plants. This chapter is mainly based on the review by Larsen et al. (2007) supplemented by the two other, more recent reviews (Corominas et al. 2013; Zang et al. 2015) together with some of the most comprehensive studies including Larsen et al. (2010). The results of the Larsen et al. (2007) review are briefly shown in Table 34.1 in Appendix. The reviewed studies include to varying degrees life cycle stages, LCA impact categories, micro-pollutants, and more, and present LCA profiles for wastewater treatment. The results are presented and discussed in the following sections on the importance of different life cycle stages for the impact profile, the relevance of different impact categories for this application domain, and the degree to which wastewater specific issues like micro-pollutants and pathogens are included. Finally, spatial differentiation, normalisation and weighting are addressed.

34.2.1 Importance of Life Cycle Stages

The life cycle of the service of wastewater treatment comprises different stages, i.e. material stage (production of raw materials, e.g. oil) including the construction of the plant, use stage (running the plant), transport "stage" (in some cases an integrated part of the other stages) and finally disposal, waste or reuse/recycling stage (e.g. landfill). These stages are dealt with in the subsections below.

Material and Construction Stage

Some of the LCA studies included in Table 34.1 in Appendix, like Emmerson et al. (1995) and Tillman et al. (1998), have included the construction of the wastewater treatment plant(s) in a detailed way. In the case Emmerson et al. (1995), the results show that although the energy consumption is overall dominated by the operation stage at one of the WWTPs analysed, it is of the same order of magnitude in both the construction and the operation stages at the two other WWTPs included. Also the studies by Tangsubkul et al. (2005), Vlasopoulos (2004) and Vlasopoulos et al. (2006) point at the possible importance of infrastructure for several different processes, e.g. constructed wet lands and sand filters. Newer studies, not included in Table 34.1 in Appendix, like Larsen et al. (2010) confirm that infrastructure/capital goods may play a significant role when dealing with newer and upcoming technologies like ozonation and sludge inertisation. That infrastructure needs to be addressed in all cases, either by including or arguing for excluding, is also stated in the review by Corominas et al. (2013).

Use Stage

The use stage (or plant operation stage) plays an often dominating role is documented in almost all studies. The main reason is the typical use of electricity, fuels and especially the emission of pollutants from the wastewater to air, with effluents and sludge.

Transport

Transport may or may not play a significant role (but typically not dominating) in the LCA profile of a Wast Water Treatment Technology (WWTT) depending on the created scenario and its scoping. An example of significant importance of transport is the Australian study by Beavis and Lundie (2003) focusing on energetically efficient distance to place of application of biosolids (based on sludge) used for fertilisation of agricultural land. In their specific cases threshold transport distance of 172 km (aerobic digested sludge) and 143 km (anaerobic digested sludge) could be estimated. Another example of the importance of distance to place of application is described in the paper by Houillon and Jolliet (2005) showing by sensitivity analysis that doubling the distance results in a 23% increase in the overall energy consumption. In the study by Dixon et al. (2003) on small-scale WWTPs, the transport in the case of reed bed contributed with 30% of the total energy consumption. Also, transportation of the wastewater may be important in scenarios where it is collected in tanks and transported to the treatment plant over long distances.

Disposal Stage

The importance of including the disposal of waste (in some cases as a resource for reuse or recycling) in LCA studies on wastewater is documented in several studies. One example is the disposal of sludge for agricultural application. Including the substitution of fertiliser production and the potential impact from especially the metal content of the sludge is very important (Beavis and Lundie 2003; Tangsubkul et al. 2005; Hospido et al. 2005), which has also been shown in more recent studies like Larsen et al. (2010) not included in Table 34.1 in Appendix. Another example is whether or not the methane production from anaerobic digestion is utilised (substituting fossil energy) or is emitted to air and hereby contributing significantly to the global warming potential (Tillman et al. 1998).

34.2.2 Relevance of Different Impact Categories

The environmental impact categories are here divided into the typical energy-related ones and typical toxicity (or chemical)-related ones. This is because a typical challenge in wastewater treatment is the achievement of higher effluent water quality at the expense of higher energy consumption. The energy-related categories comprise global warming, acidification and photochemical ozone formation, in a wastewater treatment system all primarily attributable to the combustion of fossil fuels in stationary or mobile processes. The toxicity-related impact categories include ecotoxicity and human toxicity. Eutrophication which in many other cases is primarily energy related is here looked upon separately due to its high relevance for wastewater effluent. Resource consumption, stratospheric ozone depletion, land use, photochemical ozone formation, and waste generation are also treated separately.

Energy-Related Impact Categories

The typically high importance of the energy-related impact categories are documented in most of the studies reviewed. For example in the study by Clauson-Kaas et al. (2006), the induced potential impact (global warming, acidification, indirect eutrophication) related to the energy consumption from running two of the investigated treatment technologies (MBR and ozonation) is at least in the main scenario higher than the avoided potential impact (aquatic ecotoxicity) achieved by cleaning the water (normalised or weighted impact potentials). In the study by Beavis and Lundie (2003) focusing on disinfection technologies for effluents and digestion of sludge, the potential impacts related to energy consumption also plays a dominating role. In the review by Corominas et al. (2013), global warming, acidification and eutrophication is evaluated in 38, 27 and 28 of the 45 studies included, respectively. Newer impact categories like ionising radiation and particulate matter formation are also important as they are typically related to energy production.

Toxicity-Related Impact Categories

The importance of the toxicity-related impact categories, i.e. ecotoxicity and human toxicity, when doing LCA on wastewater treatment—especially if the chemical/toxic emission from the WWTP is actually included—is documented in several studies. In, for example, the Dutch study by Roeleveld et al. (1997) focusing on municipal wastewater treatment, the normalised results show aquatic ecotoxicity to be the second most important impact category only exceeded by eutrophication. Main contributors to the ecotoxicity of the effluent are metals (about 90%; Hg, Cd) whereas the included non-specified organic micro-pollutants account for the rest. That other micro-pollutants than just metals can play an important role for aquatic ecotoxicity in the LCA comparison of different wastewater treatment options is documented in the study by Clauson-Kaas et al. (2006) including endocrine disruptors and other organics. Terrestrial ecotoxicity may also in some cases play an important role. This is seen especially in cases involving agricultural application of sludge containing metals. One example is the study by Hospido et al. (2005) comparing anaerobic digestion of sludge with different thermal alternatives. In this case, the anaerobic digestion scenario includes agricultural application and gets the overall highest normalised impact score on terrestrial ecotoxicity due to the content of metals in the sludge. In the same study and same scenario, the impact category on human toxicity gets the second highest normalised impact score (human exposure to metals via food chains) showing that at least in a few cases human toxicity may play an important role in an LCA study of wastewater treatment technologies. That also human toxicity related to air emission from energy production may play an at least not negligible role in this context is shown in, for example, two Danish studies (Clauson-Kaas et al. 2001, 2006). More recent studies including pharmaceuticals and more, like Larsen et al. (2010), confirm the overall results of this review. The importance of toxicity-related impact categories is also reflected in the reviews by Corominas et al. (2013) and Zang et al. (2015).

Eutrophication and Oxygen Depletion Due to Emission of Organic Matter
Reduction in emission of organic matter (COD, BOD) and nutrients (N, P) has always been a key challenge for municipal WWTPs. That it is also important in LCAs of wastewater treatment is documented in many studies. For example in the paper by Roeleveld et al. (1997) focusing on municipal wastewater treatment in The Netherlands, the impact share of eutrophication is clearly the highest with 4.4%, whereas the second highest, aquatic ecotoxicity, only amounts to 2.4% and energy consumption only 0.6% (normalised on basis of the total potential impact of all Dutch societal activities). Another example is the study by Hospido et al. (2004) on a Spanish municipal wastewater plant showing that eutrophication is the dominating impact category after normalisation with a share of about 65%. The typical dominance of eutrophication when wastewater effluent is included is confirmed by newer studies as described in the reviews by Corominas et al. (2013) and Zang et al. (2015). Distinguishing between emissions to freshwater (typically P-deficient) and marine water (in many cases N-deficient), and if possible include spatial (and temporal) differentiation is important for this impact category.

Stratospheric Ozone Depletion
The impact category stratospheric ozone depletion is included in 6 out of the 22 reviewed studies. It may play some (minor) role in ranking different alternative wastewater treatment technologies as shown for advanced oxidation processes by, e.g. Muñoz et al. (2005, 2006) and García-Montaño et al. (2006). However, after normalisation, the importance is typically negligible as regards WWTPs (Roeleveld et al. 1997; Hospido et al. 2004, 2005). This insignificant importance is confirmed by the Corominas et al. (2013) review but it should be noted that emission of N_2O, which is in focus regarding global warming potential related to WWTPs, is considered to be today's dominant ozone layer depleting emission (UNEP 2013).

Photochemical Ozone Formation
That the impact category on (tropospheric) photochemical ozone formation (POF) in some cases may play at least a minor role is shown in several studies. In the study by Vlasopoulos et al. (2006) comparing 20 different technologies for cleaning petroleum process waters, the POF is showing a normalised contribution, that is, at the same level as the one for eutrophication. Another example is the study by Tangsubkul et al. (2006) analysing microfiltration processes where the POF plays a relative important role (due to its relation to energy production, in this case electricity production) and is shown to be microfiltration flux dependent. That the emission of volatile organic compounds (VOC) from fossil fuel combustion (transport vehicles, machines etc.) can make the impact category for photochemical ozone formation significant in the comparison of different sludge treatment scenarios is shown by Suh and Rousseaux (2002). However, in the study on a municipal WWTP by Hospido et al. (2005) the normalised contribution from POF was found to be negligible.

Waste Generation
The disposal of "waste", for example, sludge produced during the wastewater treatment, is in a number of cases characterised by the use of the other impact categories. For example, the disposal of sludge on agricultural land is in some cases characterised by the use of the impact categories for terrestrial ecotoxicity, human toxicity and more (e.g. Hospido et al. 2005). The importance of addressing waste generation and its disposal is documented in several studies (e.g. Beavis and Lundie 2003; Tangsubkul et al. 2005). However, in many impact assessment cases and studies all or some of the waste is "only" included as, e.g. "hazardous waste", "slag and ashes", "solid waste" etc. (e.g. Clauson-Kaas et al. 2001; Tillman et al. 1998) or not at all (e.g. Dixon et al. 2003). One should always aim for characterising all waste disposals by the well-established impact categories including emissions to the biosphere (like freshwater ecotoxicity and human toxicity) and not just different waste categories.

Land Use
Only three studies have included land use in the LCA and only as occupied square metres or square metres times years of occupation. In the case of Muñoz et al. (2006), the land use is associated with the construction of the plant and reflects the large area needed for the solar field. The results of Dixon et al. (2003) reflect the difference between the land use for small conventional plants and a constructed wetlands with the same capacity, i.e. the included wetlands require a factor of 17–40 times larger area than the corresponding conventional plants. Mels et al. (1999) analysed three different large (100,000 p.e.) wastewater treatment plants (one reference and two alternatives) and come up with an area need of 8000–10,000 m^2 depending on the plant. A general exclusion of land use can therefore not be recommended as it may play a role especially if the LCA includes constructed wetlands, high space demanding energy production or the like. That only a few studies have actually included land use until now is confirmed by the most recent review study by Zang et al. (2015).

Resource Consumption
8 out of the 22 LCA studies reviewed include an impact category for resource consumption/depletion. However, in more cases resource consumption data is included in the inventory data presented. That resource depletion may play an important role in the impact assessment of wastewater treatment and that, in many cases, it is associated with consumption of fossil fuels is shown by Roeleveld et al. (1997), Gasafi et al. (2004) and Suh and Rousseaux (2002). Later studies like Larsen et al. (2010) confirm this. Water consumption/use as a separate category has also been included in a few recent studies as described in Zang et al. (2015).

34.2.3 Micro-Pollutants and Pathogens in Effluent and Sludge

The reviewed papers (Table 34.1 in Appendix) only include micro-pollutants to a limited degree and pathogens are not included at all in an impact relevant manner.

Regarding inorganic micro-pollutants, the evaluation of potential toxic impact from metals in effluent or sludge is included in 8 out of the 22 studies reviewed. Two studies include metals only in the assessment of the wastewater effluent (Clauson-Kaas et al. 2001, 2006), three studies apparently include metals in both effluent and sludge (Roeleveld et al. 1997; Beavis and Lundie 2003; Tangsubkul et al. 2005), and the other three studies only include metals in sludge (Suh and Rousseaux 2002; Hospido et al. 2004, 2005).

Organic micro-pollutants in general are only dealt with in two studies and only specified as single substances (not groups) in one case (i.e. Clauson-Kaas et al. 2006). The Clauson-Kaas study includes linear alkyl benzene sulphonate (LAS), diethylhexyl phthalate (DEHP) and polycyclic aromatic hydrocarbons (PAHs), i.e. benzo(*a*)pyrene, benzo(*b*)fluoranthene, benzo(*g*,*h*,*i*)perylene, benzo(*k*)fluoranthene and indeno(1,2,3-*cd*)pyrene for effluent emissions. A newer study by Larsen et al. (2010), not included in Table 34.1 in Appendix, is one of the most comprehensive ones regarding organic micro-pollutants and includes 22 pharmaceuticals/ metabolites for effluent emissions, and LAS, nonylphenol, DEHP and benzo(*a*) pyrene for sludge applied on agricultural land.

Potential impacts of pathogens are not included in any of the 22 LCA studies. Reduction of pathogens by WWT is, however, included in two studies (Clauson-Kaas et al. 2006; Beavis and Lundie 2003) and pointed out as an important issue for sludge used for agricultural application (Hospido et al. 2004). Further, the lack of including human health risk caused by the presence of pathogens in wastewater is pointed out as a limitation "that can affect the use of LCA in decision support in water recycling planning" (Tangsubkul et al. 2005). A preliminary method on how to include pathogens in LCA has been developed by Larsen et al. (2009) and most recently this issue has been addressed regarding sewage sludge management (Harder et al. 2016).

34.2.4 Spatial Differentiation

Site dependency with regard to aquatic ecotoxicity is only included on a general level as a differentiation between fresh water aquatic environment and marine (saltwater) aquatic environment and only in six of the reviewed studies. However, several studies include site-dependent inventory data when specific existing wastewater treatment works are looked upon (e.g. Tillman et al. 1998; Emmerson et al. 1995; Muñoz et al. 2006). For WWT spatial differentiation seems especially relevant for impacts related to aquatic ecotoxicity and eutrophication.

34.2.5 Normalisation and Weighting

Twelve of the reviewed studies use normalisation with five of them supplementing with a weighting based on value choices. The normalisation is typically done on basis of the total societal (land, region or global) potential impact per citizen within a reference year and the normalised results, for example, expressed in percentages of the total societal impact in each impact category (see also Sect. 10.3 on normalisation). By introducing value choices weighting factors may be estimated for each impact category or anticipated weighting factors (e.g. 0.5 and 1) may be used in sensitivity analysis as in the study by Suh and Rousseaux (2002). In the study by Clauson-Kaas et al. (2006), weighting factors (1.0–1.7) based on distance to political reduction targets, i.e. governmental and international conventions on reduction targets (actually the same as a normalisation reference for a future scenario) are used. In the case of Tillman et al. (1998) and Svanström et al. (2004), the "monetary" principle "willingness to pay", i.e. the willingness of society to pay for restoration of impacts on "areas of protection" is used. In the recent review by Corominas et al. (2013), the use of the hierarchist perspective (archetypes) for weighting is found in WWT LCA cases. The strength of using normalisation and weighting is that it makes comparison between different WWT alternatives more simple and creates the opportunity to aggregate all the impact potentials into one common impact score. On the other hand, the weakness is that weighting is based on value choices and not natural science and therefore debatable. Probably due to this and a very stubborn (site-specific) risk-based approach on how to do environmental assessment within this applicationdomain, the LCA approach has had a hard time gaining a foothold. Using normalisation references at different scales (catchment, region, nation etc.) and different weighting principles may therefore be a good idea in trying to test the robustness of a result and gain acceptance.

34.3 Methodological Issues

When modelling LCA cases on wastewater the issue in focus is typically the service of treating one volume unit (i.e. m^3) of more or less contaminated water. The processes leading to the contamination (e.g. sanitation and consumption) are only included in a limited way and in most cases not at all.

Goal and Scope

The most commonly used functional unit (see Sect. 8.4.2) is one cubic metre of (ingoing) wastewater. If comparison among technologies is the aim, it is highly important to define the wastewater composition strictly (P content, COD content, etc.) in order to avoid introducing a bias in the comparison. Depending on the goal and scope "population equivalents" (e.g. based on BOD5) or nutrient content (kg phosphorus content) may also be used as functional unit. Defining the life time of the technologies in question is also important and may play a significant role. Scoping according to the goal is essential and may include the whole water cycle if the goal is mapping hot spots in a region's sanitary system or only specific process parameters if the aim is assessing the environmental performance of different technical process optimisations. It may, for example, be of high importance for the impact contribution from infrastructure whether or not the sewer system is included. Including all (relevant) emissions and resource consumption/recovery like the handling/treatment and final disposal of sludge are very important as this emission route in many cases contains the major part of the pollutants (heavy metals, eutrophying substances). The direct emission of greenhouse gases (CO_2, CH_4 and N_2O) from the sewer system and/or from processes like activated sludge treatment (including nitrification/denitrification), sludge processing (e.g. anaerobic digestion), special treatments like ANaerobic AMMonium OXidation (ANAMMOX) and even emissions after disposal of sludge to agricultural land or landfill, may be important depending on the goal of the study. In this context, one should be aware of the content of fossil-based carbon in the sewage water, which may be at a level of up to 25% according to the review by Zang et al. (2015). For more information on the definition of goal and scope, see Chaps. 7 and 8, respectively.

Inventory

Foreground data is typically based on real plant measurement, laboratory/pilot tests or a combination with literature values and estimates. Background data (both upstream and downstream) are in most cases based on LCI databases like ecoinvent and GaBi. Transparency is always important in order to secure the possibility of a third part reproducing the study. For more detailed information on inventory in LCA, see Chap. 9.

Impact Assessment

In order to achieve as robust an impact assessment as possible, the use of more than one impact assessment method is recommended (see Chap. 10 on life cycle impact assessment). Depending on the goal and scope, presentation of results at all relevant levels, i.e. inventory, impact potentials, normalised and weighted (single score) results should be done. Both midpoint and endpoint (damage) results should be included (see Chap. 10). This may be achieved using, for example, the ILCD recommended methods (EU 2013) and the ReCiPe method (Goedkoop et al. 2013). As described above, the typically important impact categories for LCIA on wastewater include the toxicity related ones (human toxicity, freshwater ecotoxicity, marine ecotoxicity, terrestrial ecotoxicity and particulate matter formation), eutrophication (marine and freshwater), global warming, acidification, ionising radiation and (in some cases) water use, land use and stratospheric ozone depletion. Attempts to include special impact categories like pathogens and acute toxicity (due to, e.g. water emission of ammonia), and spatial and temporal differentiation (e.g. regarding eutrophication) when doing site-specific assessments may be relevant depending on the goal and scope of the study.

Interpretation

With the aim of optimising the reliability and robustness of the result the use of sensitivity analysis but also uncertainty estimations if possible is highly recommended (see Chap. 12 on interpretation).

An Alternative Approach on How to Do LCA on Wastewater Treatment

The approach used to reach the goal of an LCA on wastewater treatment is typically based on the general approaches like hot spot identification in the life cycle of a specific technology or comparing the impact profile of different technologies performing the same service, e.g. phosphorus removal down to a specific level. An alternative is the "avoided against induced impacts" approach, where the impacts avoided by, e.g. introducing a new technology are compared to the impacts induced by this technology. This approach is illustrated in Figs. 34.2 and 34.3 and reflects a typical challenge in wastewater treatment, i.e. the achievement of higher effluent water quality at the expense of higher energy consumption or higher consumption of, e.g. precipitation chemicals.

This approach has now been used in several studies and was introduced by Wenzel et al. (2008) and in a more comprehensive way by Larsen et al. (2007, 2010). The approach puts special demands on the toxicity related impact categories and the eutrophication potential.

34.4 Concluding Remarks

When performing LCA on wastewater treatment, one should be aware of the following more or less domain specific issues:

- A typical functional unit is the treatment of one cubic metre of wastewater. Defining its composition/characteristics and/or using limit values for effluent is crucial for the reliability of the study in case of comparative studies
- All life cycle stages have the potential of being significant, and therefore need to be considered even though decommissioning and disposal of infrastructure in many cases seems to be the least important for the impact profile
- It is not generally possible in advance to consider some inventory data as unnecessary, but depending on the goal and scope large parts of the product system may be omitted (e.g. parts equal among alternatives)
- None of the conventional impact categories (except stratospheric ozone depletion if emission of N_2O is not included) should be excluded but eutrophication and ecotoxicity are in many cases among the dominating ones

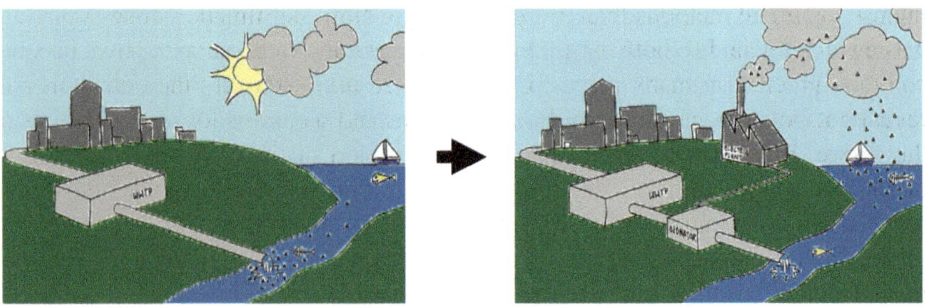

Fig. 34.2 By avoiding an obvious problem in one place we may induce a bigger problem somewhere else (sub-optimisation) (Larsen et al. 2010)

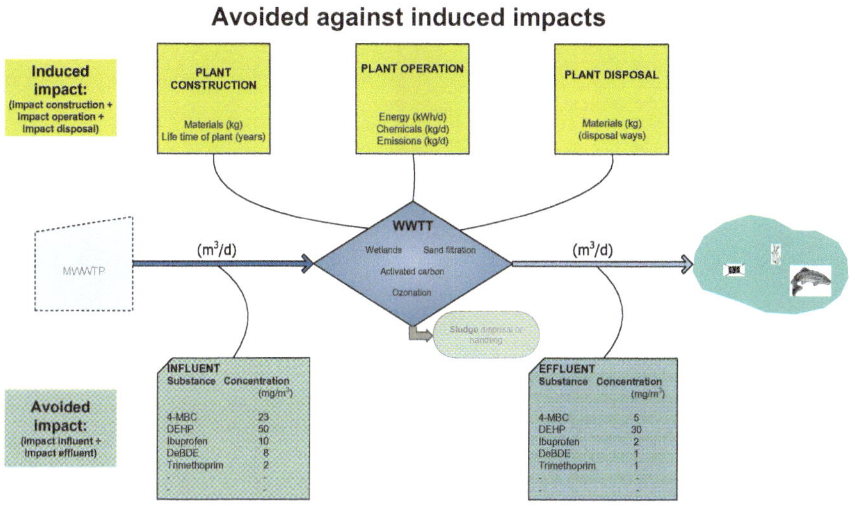

Fig. 34.3 The principle of avoided against induced impact illustrated for micro-pollutant polishing, by different wastewater treatment technologies (WWTT), e.g. ozonation of wastewater from a municipal wastewater treatment plant (MWWTP) (Larsen et al. 2007)

Furthermore, if the aim is to perform an environmental product declaration (EPD), one should consult the already existing product category rules (PCR) for "Wastewater collection and treatment services" (Environdec 2014).

It should be noted that a future change in the relative importance of the different impact categories is likely due to coming improved inventory data and enhanced LCIA methodology including an increase in the number of characterisation factors or the use of, e.g. whole effluent toxicity (Larsen et al. 2009) regarding aquatic ecotoxicity. As discussed by Larsen et al. (2010), existing methodologies only cover a minor part of the possible toxicity impact of pollutants in wastewater due to lack of (good) characterisation factors. Including the specific toxic modes of action of, e.g. endocrine disrupters in a proper way, have the potential of increasing the importance of aquatic ecotoxicity significantly as shown in Larsen et al. (2009, 2010). Another example is the achievement of better inventory data on N_2O and CH_4 emissions from WWT which might change the importance of global warming drastically (Zang et al. 2015).

Appendix

See Table 34.1.

Table 34.1 Results of review on 22 LCA WWT cases (Larsen et al. 2007)

Study on	Impact categories included for LCIA	Scoping, functional unit (FU) and life time	Potential impact of effluent quantified by [micro-pollutants]	References
Industrial wastewater: process water from extraction of oil and gas Sand filtration, ozonation and 20 other technologies	GW, DAR, AC, NE/ET, POF	+ Material stage + Construction + Use stage ÷ Transport ÷ Waste treatment FU: Cleaning of 10,000 m^3 wastewater to certain water quality levels[a] Life time: 15 years	No quantification [none]	Vlasopoulos et al. (2006) Vlasopoulos (2004)
Municipal wastewater, related to the WFD Sand filtration Membranbioreactor Ozonation	GW, AC, NE/ET, POF, CHTS, CHTW, AHTA, CETS, CETF, AETF, CETSW	+ Material stage + Construction + Use stage ÷ Transport + Waste treatment[b] FU: Further treatment of 1 m^3 wastewater treated conventionally, i.e. MBNC Life time: 20 years	CETF, CETSW [Cd, Pb, Ni, NPE, LAS, DEHP, EE2, E2, PAH, (Zn, Cu, Hg, Cr)]	Clauson-Kaas et al. (2006)
Industrial wastewater: kraft mill bleaching wastewater Heterogeneous photocatalysis (PhC)	GW, DAR, AC, NE/ET, POF, HT, CETF, OLD	+ Material stage ÷ Construction[d] + Use stage + transport ÷ Waste treatment FU: Removal of 15% DOC from 1 m^3 kraft pulp mill wastewater Life time: ? (laboratory experiment in Pyrex cells)	No quantification [none]	Muñoz et al. (2005)

(continued)

Table 34.1 (continued)

Study on	Impact categories included for LCIA	Scoping, functional unit (FU) and life time	Potential impact of effluent quantified by [micro-pollutants]	References
Industrial wastewater (synthetic)[c] Heterogeneous photocatalysis (PhC)	GW, AC, NE/ET, POF, HT, CETF, OLD, LU, EC	+ Material stage + Construction[d] + Use stage + Transport + Waste treatment[e] FU: Treatment of 1 m^3 synthetic α-methyl-phenyl-glycine (MPG) solution (500 mg/L) in order to obtain an inherent biodegradable effluent Life time: 15 years	Only for DOC, COD, *N*-ammonia and *N*-nitrate [none]	Muñoz et al. (2006)
Industrial wastewater (synthetic)[f] Photo-Fenton (PhF) process	GW, DAR, AC, NE/ET, POF, HT, CETF, CETS, CETSW, OLD	+ Material stage ÷ Construction[d] + Use stage + Transport + Waste treatment[g] FU: Removal of 80% DOC from 1.2 L of 250 mg/L Cibacron RED FN-R synthetic wastewater from simulated batch dyeing Life time: ? (laboratory experiment in Pyrex cells)	Only for DOC, COD, *N*-ammonia and *N*-nitrate [none]	García-Montaño et al. (2006)
Municipal wastewater in the Netherlands (total sustainability in society)	GW, DAR, AC, NE/ET, POF, HT, CETF, CETS, OLD, (discharge of	+ Material stage + Construction + Use stage	COD, NE/ET, CETF	Roeleveld et al. (1997)

(continued)

Table 34.1 (continued)

Study on	Impact categories included for LCIA	Scoping, functional unit (FU) and life time	Potential impact of effluent quantified by [micro-pollutants]	References
Activated sludge $FeCl_3$ (phosphor removal) Activated carbon	COD), (production of normal, toxic and nuclear waste)	+ Transport (÷ Waste treatment) FU: 24,000,000 p.e. (not defined) Life time: Not defined	[Hg, Cu, Cd, Zn, organic phosphorus containing compounds]	
Municipal wastewater in Australia (focus on disinfection and nutrients) UV Chlorination/dechlorination by chlorine or hypochlorite Dissolved air flotation	GW, AC, NE/ET (freshwater and marine), POF, HT, CETS, CETF, CETSW	+ Material stage ÷ Construction + Use stage + Transport (÷ Waste treatment) Case study I: FU: Disinfection (measured by CFU) of 1000 m^3 tertiary treated wastewater (chlorine residual less than 0.01 mg/L) Case study II: fu: 1000 m^3 wastewater yielding 714 kg biosolids (25% solids, anaerobic digestion) or 1250 kg biosolids (18% solids, aerobic digestion) Life time: Not defined	Case study I: CFU-measurements Case study II: At least CETF Sludge: (At least CETS, but not specified) [common concentrations of metals, pesticides and chlordane assumed]	Beavis and Lundie (2003)
Municipal wastewater in Australia (focus on recycling for irrigation application)	GW, NE/ET, HT, CETS, CETF, CETSW, salinisation	+ Material stage + Construction + Use stage + Transport	Effluent as irrigation water, i.e. emission to soil: NE/ET, HT, CETS, CETF, CETSW, salinisation	Tangsubkul et al. (2005)

(continued)

Table 34.1 (continued)

Study on	Impact categories included for LCIA	Scoping, functional unit (FU) and life time	Potential impact of effluent quantified by [micro-pollutants]	References
Ozonation + continuous microfiltration (CMF) Membrane bioreactor (MBR) + reverse osmosis (RO)		(÷ Waste treatment) FU: 1000 m^3 recycled wastewater meeting national threshold limits for irrigation of sensitive crops (incoming raw sewage of "medium" strength, i.e. 500 mg/L, 40 mg/L tot-N, etc.) Life time: Not defined	Sludge as biosolids, i.e. emission to soil: NE/ET, HT, CETS, CETF, CETSW [besides mentioning metals no specification]	
Municipal wastewater in Spain (focus on environmental performance, i.e. removal of organic matter and hotspots) Only primary and secondary treatment	GW, DAR, AC, NE/ET, POF, HT, CETS, OLD	+ Material stage ÷ Construction + Use stage + Transport + Waste treatment FU (1): 53,349 m^3/day (humid season) FU (2): 49,214 m^3/day (dry season) Life time: Not defined	NE/ET, (HT) Sludge: CETS, HT (GW, DAR, AC, NE/ET, POF, OLD) [Cd, Cr, Cu, Hg, Ni, Pb and Zn]	Hospido et al. (2004)
Municipal wastewater (focusing on energy consumption and reuse potential) Constructed wetlands, UV, Sequencing batch reactor	"No impact categories", i.e. only comparison of removal efficiency (BOD, Tot-N and Tot-P), nutrient recycling and energy consumption	÷ Material stage ÷ Construction + Use stage ÷ Transport (+ waste treatment) FU: 1 m^3 Life time: Not defined	No quantification [none]	Brix (1999)

(continued)

Table 34.1 (continued)

Study on	Impact categories included for LCIA	Scoping, functional unit (FU) and life time	Potential impact of effluent quantified by [micro-pollutants]	References
Municipal wastewater (focusing on small-scale treatment) Constructed wetland	Besides LU (included as m^2 occupied by plant) impact categories are only included as CO_2 emitted, energy consumption and solid emission (i.e. waste)	+ Material stage + Construction + Use stage + Transport ÷ Waste treatment FU: 1 p.e. (0.2 m^3/day) treated to acceptable discharge standards, i.e. 10 mg/L BOD, 25 mg/L SS and 5 mg/L ammonia: Scales: 12, 60 and 200 p.e. Life time: 10 years	No quantification [none]	Dixon et al. (2003)
Sludge form municipal wastewater plant (focus on hotspots and process design) Supercritical water gasification (SCWG)	GW, DAR, AC, NE/ET	+ Material stage ÷ Construction + Use stage + Transport (+ Waste treatment) FU: 1 ton DM undigested sewage sludge (3% DM equals 33 ton wet sludge) Life time: Not defined	Effluent from SCWG plant via WWTP: NE/ET [none]	Gasafi et al. (2004)
Industrial wastewater, brewery (focus on operation conditions for membrane filtration)	GW, NE/ET, HT, CETS, CETF, CETSW, POF	+ Material stage + Construction + Use stage + Transport (÷ Waste treatment)	No quantification (assumed to be equal for all scenarios) [none]	Tangsubkul et al. (2006)

(continued)

Table 34.1 (continued)

Study on	Impact categories included for LCIA	Scoping, functional unit (FU) and life time	Potential impact of effluent quantified by [micro-pollutants]	References
		FU: 1000 m^3/day (produced from incoming settled effluent water from a sequencing batch reactor (SBR), 40 mg/L TSS and permeate turbidity's of <1 NTU) Life time: 20 years (membrane 5 years)		
Municipal wastewater (focus on physical-chemical pre-treatment) activated sludge, *Pre-precipitation, Flotation*	No characterization only inventory (energy balance, final sludge production, effluent quality, use of chemicals, space requirements, i.e. LU)	÷ Material stage ÷ Construction + Use stage (÷ Transport) (÷ Waste treatment) FU: 7,120,000 m^3/year (19,500 m^3/day) (treated to acceptable discharge standards, i.e. <10 mg/L BOD, <50 mg/L COD, <10 mg/L Tot-N, <1 mg/L Tot-P, <10 mg/L SS) Life time: Not defined	No quantification (assumed to be equal for all scenarios) [none]	Mels et al. (1999)
Sludge form municipal wastewater plant (focus on final disposal), Incineration Agricultural land application	GW, NE/ET, HT, CETS, CETF, CETSW, POF, DAR	+ Material stage ÷ Construction + Use stage + Transport (+ Waste treatment)	No quantification For sludge: GW, NE/ET, HT, CETS, CETF, CETSW, POF, DAR, AC	Suh and Rousseaux (2002)

(continued)

Table 34.1 (continued)

Study on	Impact categories included for LCIA	Scoping, functional unit (FU) and life time	Potential impact of effluent quantified by [micro-pollutants]	References
Landfill		FU: 1 ton DM mixed sewage sludge Life time: Not specified but more than 30 years mentioned in argumentation for leaving out construction	[Substances and concentrations according to threshold limits in French regulation on leachate from landfills and content of sludge for agricultural land application. Only metals included for land application. No further specification]	
Predominantly municipal wastewater sludge (focusing on hot spots and energy consumption) Supercritical water oxidation	GW, POF, DAR For EPS2000: Human health, biological diversity, ecosystem production, resources and aesthetic values) For EcoIndicator99: Human health, ecosystem health and resources	+ Material stage ÷ Construction + Use stage + Transport (÷ Waste treatment) FU: 1000 kg wet sludge (7% TS) treated at specific plant. Water effluent and gases assumed to have no adverse impacts Life time: Not defined	No quantification (assumed to have no adverse impact for both water effluent and wet solid effluent) [none]	Svanström et al. (2004)
Sludge form municipal wastewater plant (focus on energy and global warming) Agricultural land application Incineration Wet oxidation	GW	+ Material stage ÷ Construction + Use stage + Transport (+ Waste treatment) FU: 1000 kg DM sludge disposed	No quantification [none]	Houillon and Jolliet (2005)

(continued)

Table 34.1 (continued)

Study on	Impact categories included for LCIA	Scoping, functional unit (FU) and life time	Potential impact of effluent quantified by [micro-pollutants]	References
Pyrolysis Incineration in cement kilns Landfill		(wet sludge: 0.3% dry solid content) Life time: Not defined		
Municipal wastewater treatment plant in Denmark (focus on hot spots) Activated sludge: nitrification/denitrification, phosphor removal, sludge incineration	GW, AC, NE/ET, Persistent toxicity ((CETS + CETF + CHTS + CHTW)/4), AHTA, AETF, slag and ashes	+ Material stage ÷ Construction + Use stage ? Transport (÷ Waste treatment) FU: 29,800,000 m^3 wastewater treated (i.e. 1 year, 1998) Life time: Not defined	NE/ET, Persistent toxicity, AETF [Pb, Hg, Cu, Zn, Cr, Ni, Cd, Se, As, dioxin]	Clauson-Kaas et al. (2001)
Sludge form municipal wastewater plant (focus on post-treatments) Anaerobic digestion + agricultural land application Incineration Pyrolysis	GW, DAR, AC, NE/ET, POF, HT, CETS, OLD	+ Material stage ÷ Construction + Use stage + Transport (+ Waste treatment) FU: 1000 kg DM sludge managed (thickened mixed: 1% dry matter) Life time: Not defined	No quantification For sludge: GW, DAR, AC, NE/ET, POF, HT, CETS, OLD [Cd, Cr, Cu, Hg, Ni, Pb and Zn]	Hospido et al. (2005)

(continued)

Table 34.1 (continued)

Study on	Impact categories included for LCIA	Scoping, functional unit (FU) and life time	Potential impact of effluent quantified by [micro-pollutants]	References
Municipal wastewater (focus on energy consumption and sludge production) Activated sludge: nitrification/denitrification, biological P-removal	No characterization only inventory (oxygen requirements, sludge production and auxiliaries requirement/production (methanol, $FeCl_3$ and methane)	÷ Material stage ÷ Construction + Use stage ÷ Transport ÷ Waste treatment FU: 2750 m^3/day wastewater treated (fixed discharge levels: <15 mg/L BOD_5, <15 mg/L TSS, <0.5 mg N/l ammonia, <10 mg/L Tot-N, <1 mg/L Tot-P Life time: Not defined	No quantification (assumed to be equal for all scenarios) [none]	Bagley (2000)
Municipal wastewater (focus on small-scale WWTPs) Activated sludge	No (quantitative) characterization only inventory (energy consumption, material use, waste, air emission from energy production—especially CO_2 (but also SO_2, CO and more)	+ Material stage + Construction + Use stage + Transport (+ Waste treatment) FU: 15 years of functioning, i.e. 1,095,000 m^3 (compliance of effluent and sludge (agricultural application) with regulatory framework) Life time: 15 years	No quantification (assumed to be equal for all scenarios [none]	Emmerson et al. (1995)

(continued)

Table 34.1 (continued)

Study on	Impact categories included for LCIA	Scoping, functional unit (FU) and life time	Potential impact of effluent quantified by [micro-pollutants]	References
Municipal wastewater (focus on change from conventional central WWTPs to local systems), sand filter, filter bed, urine separation	No specification, only inventory data used in detailed impact assessment, i.e. energy consumption; air emission of CO_2, methane, SO_2 etc.; water emission of N-tot, P-tot, BOD etc.; waste such as sludge, hazardous waste etc.	+ Material stage + Construction + Use stage + Transport (+ Waste treatment) FU: The treatment of wastewater from 1 p.e. during 1 year. p.e. not defined in paper Life time: Stated that this is taken into account (full technical life time for each component) but not specified in paper	No direct specification in paper of impact categories but at least assessed on basis of COD, BOD, N-tot and P-tot. [No specification]	Tillman et al. (1998)

MWWTP Municipal wastewater treatment plant; *WWT* Wastewater treatment; *ST* Sludge treatment; *WFD* Water framework directive (EC 2000, 2001); *MBNC* Mechanical (settlement)/biological/nitrification-denitrification/chemical phosphate removal; *GW* Global warming; *POF* Photochemical ozone formation; *AC* Acidification; *NE/ET* Nutrient enrichment/eutrophication; *DAR* Depletion of abiotic resources (sometimes divided into mineral resources and fossil energy resources); *OLD* Ozone depletion; *LU* Land use; *EC* (non-renewable) Energy consumption; *HT* Human toxicity; *CHTS* Chronic human toxicity soil; *CHTW* Chronic human toxicity water; *AHTA* Acute human toxicity air; *CETS* Chronic ecotoxicity soil; *CETF* Chronic ecotoxicity fresh water; *AETF* Acute ecotoxicity fresh water; *CETSW* Chronic ecotoxicity salt water; *DOC* Dissolved organic carbon p.e.: "wastewater" person equivalents; *CFU* Colony forming units; *NTU* Nephelometric turbidity units; + included; ÷ not included

[a]Related to end-use categories (irrigation of wheat, cotton, barley, alfalfa, sorghum, rhodes, citrus and for industrial use 'cooling system feed' and 'boiler feed')

[b]Treatment of waste from decommissioning of equipment included. For sludge only sludge incineration as disposal included

[c]Synthetic solution of α-methyl-phenyl-glycine (a pharmaceutical precursor)

[d]Same type of UVA lamp used in all cases—only differences in running time for achieving 15% reduction in DOC

[e]Transport and land filling of used catalyst included. For sludge only sludge incineration as disposal included

[f]Synthetic solution of the reactive azo dye Cibacron Red FN-R (C.I. Reactive Red 238)

[g]For sludge: dewatering, thickening and finally deposited at a landfill. Leachate (COD, BOD_7, NO_3^- and NH_4^+) and gas emission (CO_2, CH_4, NO_x and NH_3) estimated by ORWARE. 50% capture of gas (burned) and 90% capture of leachate (treated in WWTP) assumed

References

Australian Government: National water quality management strategy. Australian Government, Department of Environment. http://www.environment.gov.au/water/quality/national-water-quality-management-strategy (2015). Accessed Sept 2015

Bagley, D.M.: Life cycle analysis of municipal waste water treatment. In: 2000 Annual Conference Abstracts, vol. 2, pp. 437–443. Canadian Society for Civil Engineering (2000)

Beavis, P., Lundie, S.: Integrated environmental assessment of tertiary and residuals treatment—LCA in the wastewater industry. Water Sci. Technol. **47**(7–8), 109–116 (2003)

Brix, H.: How "green" are aquaculture, constructed wetlands and conventional wastewater treatment systems? Water Sci. Technol. **40**(3), 45–50 (1999)

Clauson-Kaas, J., Poulsen, T.S., Jacobsen, B.N., Guildal, T., Wenzel, H.: Environmental accounting—a decision support tool in WWTP operation and management. Water Sci. Technol. **44**(2–3), 25–30 (2001)

Clauson-Kaas, J., Dahl, F., Dalgaard, O., Høibye, L., Kjølholt, J., Wenzel, H., Larsen, H.F.: Videregående renseteknologier for kommunalt spildevand (Advanced treatment technologies for municipal waste water). Book in Danish with English summary. DANVA, Miljøstyrelsen, Lynettefællesskabet og Spildevandscenter Avedøre. DANVA Forsknings- og Udredningsprojekt nr. 2. DANVA, Denmark (2006)

Corominas, L.l., Foley, J., Guest, J.S., Hospido, A., Larsen, H.F., Morera, S., Shaw, A.: Life cycle assessment applied to wastewater treatment: state of the art. Water Res **47**:5480–5492 (2013)

Dixon, A., Simon, M., Burkitt, T.: Assessing the environmental impact of two options for small-scale wastewater treatment: comparing reedbed and an aerated biological filter using a life cycle approach. Ecol. Eng. **20**, 297–308 (2003)

Doka, G.: Part IV: life cycle inventory of wastewater treatment. Life cycle inventories of waste treatment services—EcoInvent report No. 13. Swiss Center for Life Cycle Inventories, Dübendorf, Switzerland (2007)

EC: Directive 2000/60/EC of the European Parliament and of the Council of 23 October 2000. Establishing a framework for community action in the field of water policy. http://ec.europa.eu/environment/water/water-framework/index_en.html (2000)

EC: Decision No. 2455/2001/EC of the European Parliament and of the Council of 20 November 2001. Establishing the list of priority substances in the field of water policy and amending Directive 2000/60/EC (2001)

Emmerson, R.H.C., Morse, G.K., Lester, J.N., Edge, D.R.: The life-cycle analysis of small-scale sewage-treatment processes. Water Environ. J. **9**, 317–325 (1995)

Environdec: Waste water collection and treatment services. Product group: UN CPC 9411 & 9423. Version 1.01. Product category rules according to ISO 14025:2006. Date: 26 Feb 2014. The International EPD System. www.environdec.com (2014). Accessed 2 Oct 2015

EU: Commission recommendation of 9 April 2013 on the use of common methods to measure and communicate the life cycle environmental performance of products and organisations. Official J Eur Union. L 124 (2013/179/EU) (2013)

García-Montaño, J., Ruiz, N., Muñoz, I., Domènech, X., García-Hortal, J.A., Torrades, F., Peral, J.: Environmental assessment of different photo-Fenton approaches for commercial reactive dye removal. J. Hazard. Mater. **A138**(2006), 218–225 (2006)

Gasafi, E., Meyer, L., Schebek, L.: Using life-cycle assessment in process design. supercritical water gasification of organic feedstocks. J. Ind. Ecol. **7**(3–4), 75–91 (2004)

Goedkoop, H.R, Huijbregts, M.A.J., De Schryver, A., Struijs, J., van Zelm, R.: ReCiPe 2008. A life cycle impact assessment method which comprises harmonised category indicators at the midpoint and the endpoint level. Version 1.08. http://www.lcia-recipe.net/ (2013)

Harder, R., Peters, G.M., Molander, S., Ashbolt, N.J., Svanström, M.: Including pathogen risk in life cycle assessment: the effect of modelling choices in the context of sewage sludge management. Int. J. LCA **21**(1), 60–69 (2016)

Hospido, A., Moreira, M.T., Fernández-Couto, M., Feijoo, G.: Environmental performance of a municipal wastewater treatment plant. Int. J. LCA **9**(4), 261–271 (2004)

Hospido, A., Moreira, M.T., Martin, M., Rigola, M., Feijoo, G.: Environmental evaluation of different treatment processes for sludge from urban wastewater treatments: anaerobic digestion versus thermal processes. Int. J. LCA **10**(5), 336–345 (2005)

Houillon, G., Jolliet, O.: Life cycle assessment of processes for the treatment of wastewater urban sludge: energy and global warming analysis. J. Clean. Prod. **13**, 287–299 (2005)

Larsen, H.F., Hauschild, M., Wenzel, H., Almemark, M.: Homogeneous LCA methodology agreed by NEPTUNE and INNOWATECH. Deliverable 4.1. EC Project "NEPTUNE", Contract No.: 036845. www.eu-neptune.org (2007)

Larsen, H.F., Olsen, S.I., Hauschild, M., Laurent, A.: Methodology for including specific biological effects and pathogen aspects into LCA. Deliverable 4.2. EC Project "NEPTUNE", Contract No.: 036845. www.eu-neptune.org (2009)

Larsen, H.F., Hansen, P.A., Boyer-Souchet, F.: Decision support guideline based on LCA and cost/efficiency assessment. Deliverable 4.3. EC Project "NEPTUNE", Contract No.: 036845. www.eu-neptune.org (2010)

Mels, A.R., van Nieuwenhuijzen, A.F., van Graaf, J.H.J.M., Klapwijk, B., de Koning, J., Rulkens, W.H.: Sustainability criteria as a tool in the development of new sewage treatment methods. Water Sci. Technol. **39**(5), 243–250 (1999)

Muñoz, I., Rieradevall, J., Torrades, F., Peral, J., Domènech, X.: Environmental assessment of different solar driven advanced oxidation processes. Sol. Energy **79**(2005), 369–375 (2005)

Muñoz, I., Peral, J., Ayllón, J.A., Malato, S., Passarinho, P., Domènech, X.: Life cycle assessment of a coupled solar photocatalytic-biological process for wastewater treatment. Water Res. **40** (2006), 3533–3540 (2006)

Roeleveld, P.J., Klapwijk, A., Eggels, P.G., Rulkens, W.H., van Starkenburg, W.: Sustainability of municipal wastewater treatment. Water Sci. Technol. **35**(10), 221–228 (1997)

Suh, Y.-J., Rousseaux, P.: An LCA of alternative wastewater sludge treatment scenarios. Resour. Conserv. Recycl. **35**, 191–200 (2002)

Svanström, M., Fröling, M., Modell, M., Peters, W.A., Tester, J.: Environmental assessment of supercritical water oxidation of sewage sludge. Resour. Conserv. Recycl. **41**, 321–338 (2004)

Tangsubkul, N., Beavis, P., Moore, S.J., Lundie, S., Waite, T.D.: Life cycle assessment of water recycling technology. Water Resour. Manag. **19**, 521–537 (2005)

Tangsubkul, N., Parameshwaran, K., Lundie, S., Fane, A.G., Waite, T.D.: Environmental life cycle assessment of the microfiltration process. J. Membr. Sci. **284**, 214–226 (2006)

Tillman, A.-M., Svingby, M., Lundström, H.: Life cycle assessment of municipal waste water systems. Int. J. LCA **3**(3), 145–157 (1998)

UNEP: Drawing down N_2O to protect climate and the ozone layer. A UNEP Synthesis Report. United Nations Environment Programme (UNEP), Nairobi, Kenya. www.unep.org/pdf/UNEPN2Oreport.pdf (2013)

Vlasopoulos, N.: Life cycle assessment of wastewater treatment technolgies. M.Sc. Dissertation. Imperial College, London, UK (2004)

Vlasopoulos, N., Memon, F.A., Butler, D., Murphy, R.: Life cycle assessment of wastewater treatment technologies treating petroleum process waters. Sci. Total Environ. **367**(2006), 58–70 (2006)

Wenzel, H., Larsen, H.F., Clauson-Kaas, J., Høibye, L., Jacobsen, B.N.: Weighing environmental advantages and disadvantages of advanced wastewater treatment of micro-pollutants using environmental life cycle assessment. Water Sci. Technol. **57**(1), 27–32 (2008)

Zang, Y., Li, Y., Wang, C., Zhang, W., Xiong, W.: Towards more accurate life cycle assessment of biological wastewater treatment plants: a review. J. Clean. Prod. (2015). doi:10.1016/j.jclepro.2015.05.060

Author Biography

Henrik Fred Larsen long term experience as both a researcher and a consultant with environmental sustainability assessment and environmental impact assessment (RA of chemicals, cleaner technology assessments). Contributed to UNEP/SETAC working groups on LCA methodology. Main LCIA interest is ecotoxicity impacts.

Chapter 35
LCA of Solid Waste Management Systems

Ioannis Bakas, Alexis Laurent, Julie Clavreul, Anna Bernstad Saraiva, Monia Niero, Emmanuel Gentil and Michael Z. Hauschild

Abstract The chapter explores the application of LCA to solid waste management systems through the review of published studies on the subject. The environmental implications of choices involved in the modelling setup of waste management systems are increasingly in the spotlight, due to public health concerns and new legislation addressing the impacts from managing our waste. The application of LCA to solid waste management systems, sometimes called "waste LCA", is distinctive in that system boundaries are rigorously defined to exclude all life cycle stages except from the end-of-life. Moreover, specific methodological challenges arise when investigating waste systems, such as the allocation of impacts and the consideration of long-term emissions. The complexity of waste LCAs is mainly derived from the variability of the object under study (waste) which is made of different materials that may require different treatments. This chapter attempts to address these challenges by identifying common misconceptions and by providing methodological guidance for alleviating the associated uncertainty. Readers are also provided with the list of studies reviewed and key sources for reference to implement LCA on solid waste systems.

I. Bakas (✉) · A. Laurent · M. Niero · M.Z. Hauschild
Division for Quantitative Sustainability Assessment,
Department of Management Engineering, Technical University
of Denmark, Kongens Lyngby, Denmark
e-mail: ioannis.bakas@bobagroup.dk

J. Clavreul
Residual Resources Engineering, Department of Environmental Engineering,
Technical University of Denmark, 2800 Kongens Lyngby, Denmark

A.B. Saraiva
SAGE/COPPE, Federal University of Rio de Janeiro, Rio de Janeiro, Brazil

E. Gentil
Copenhagen Resource Institute, Copenhagen K, Denmark

M.Z. Hauschild et al. (eds.), *Life Cycle Assessment*,
DOI 10.1007/978-3-319-56475-3_35

35.1 Introduction

Over the past century, both material use and waste generation have been constantly increasing in quantity and complexity at an unsustainable pace. Globally, waste generation from all sources amounts to around 17 billion tonnes, and is expected to reach 27 billion tonnes by 2050 (Karak et al. 2012). Municipal waste generation has also been increasing and, in Europe, only a few examples exist of decoupling municipal waste generation from economic growth, although recently efforts towards waste prevention are undertaken (EEA 2014). Currently, it is estimated that about 1.3 billion tonnes of municipal solid waste is generated worldwide and trends show that this number will increase in the future due to population increase, urbanisation and socioeconomic development of low-income populations (Hoornweg and Bhada-Tata 2012).

The organised and systematic collection and central treatment of waste originally begun for reasons pertaining to public health and safety, e.g. for combating diseases or reducing odours in public space. Only in recent years has waste been associated with environmental concerns, such as climate change, toxicity to humans and ecosystems or resource depletion. The links between waste management activities and emissions that cause specific environmental impacts have now been proven, e.g.: methane emissions from landfills contribute to climate change, halocarbons in discarded cooling systems or in-use foams contribute to stratospheric ozone depletion, while insufficient or inefficient recycling leads to increased resource depletion.

In light of these environmental concerns, pieces of legislation around the world have attempted to regulate waste management activities and to promote more sustainable systems for waste handling (e.g. Directive 2008/98/EC). The regulations may address technical issues, such as quality standards for recyclables or management issues, such as the promotion of recycling and the reduction of landfilling. In recent years, the role of waste as a pool for material resources extraction has been acknowledged and waste is now more and more viewed as a valuable resource instead of unwanted materials. Along these lines, new legislation and initiatives attempt to integrate waste management into a new vision of a circular economy, with increased quantity and quality of recycling.

In order to conform with legislation, but also to tackle significant environmental considerations, and motivated by issues around the effectiveness and cost of waste treatment, public authorities have started designing integrated management systems that comprehensively address waste generation and that are differentiated according to waste source or waste material (fraction). Although, there are relatively few options to consider regarding waste treatment (the three main ones being recycling —or biological treatment for organic waste, incineration and landfilling), their combinations for each waste type (defined by source of waste) and waste fraction are numerous. Therefore, the complexity of integrated waste management systems has become significant, highlighting the need to adopt systems approaches.

It is thus necessary to use appropriate tools that address activities related to waste management in a systematic and comprehensive manner. LCA can credibly assess the full environmental consequences of waste management, in particular accounting for the interlinks between the waste sector and other sectors of the economy. For example, the energy produced in an incineration plant or processed scrap metals feed respectively into the energy and metal manufacturing sectors. Life cycle thinking helps map all exchanges with other sectors and estimate environmental impacts accurately.

LCA applied to waste management systems is often termed "waste LCA" as it includes only the End-of-Life phase of a product. Waste LCAs are mostly of a comparative nature (e.g. assessing different treatment options for a material or a waste type) and thus, the previous life cycle stages of a material/product in question can be omitted. This is also called the "zero-burden assumption" (Ekval et al. 2007). In this respect, waste LCAs use different system boundaries assumptions than product LCAs.

Another particularity of waste LCAs is that waste treatment in many cases happens locally, close to the waste source. This fact facilitates the collection of site-specific data and thus increases the geographical resolution of the assessment.

35.1.1 Definition and Scope

A straightforward and descriptive definition of waste is as follows:

"Waste is a left-over, a redundant product or material of no or marginal value for the owner and which the owner wants to discard" (Christensen 2011).

In this chapter, only management of solid waste is addressed. Although, several definitions of solid waste exist, it is defined here as waste, which is neither water (wastewater) nor airborne (flue gases) (Christensen 2011). For application of LCA to wastewater management systems, see Chap. 34.

Towards their end-of-life, most goods and commodities eventually become discarded and typically enter solid waste management systems. The waste product thus goes through a number of activities, which can be divided in four main phases: (1) generation, (2) collection and transport, (3) treatment, and (4) recycling, utilisation or landfilling, illustrated in Fig. 35.1 (Christensen 2011).

Within the domain of waste management, the primary use of LCA is to inform about the environmentally preferable option when decision-making or policy-making communities evaluate different alternatives of solid waste management in a specific region. For instance when assessing the impact of integrating recycling in an existing municipal waste management system based on landfilling and incineration. The applications of LCA encompassed in this chapter are therefore service-oriented, focusing on assessments of processes, technologies and systems handling solid waste, and do not consider upstream activities prior to waste generation. The uses of specific types of waste as feedstock for manufacturing products are only discussed when describing the environmental offsets from

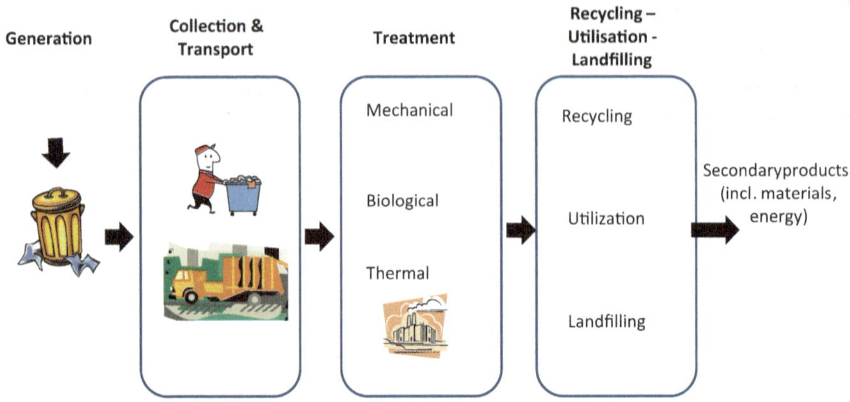

Fig. 35.1 The four phases of solid waste management systems (based on Christensen 2011)

material recycling and energy recovery, which can be credited to the assessed waste management systems (see Sect. 8.6.1).

35.1.2 Solid Waste Management Technologies and Practices

The technologies and practices involved in a solid waste management system can be arranged into distinct system stages.

The *collection and transport* stages refer to the collection of waste from its source of generation, which may include a large number of fractions (e.g. households) or fewer (e.g. industrial waste) depending on the waste type. Waste is then transported to central facilities for processing and/or treatment. The collection of waste may take place either as mixed waste or by targeting specific fractions that are separated at source (e.g. paper and cardboard destined to recycling). The type of collection system usually depends on the further treatment, e.g. for recycling waste is usually separated at the source in order to increase the homogeneity of the collected material.

The *treatment* stage refers to the processing of waste in order to modify its physical or chemical properties. Physical treatment may involve shredding and compacting of waste in order to reduce its volume. On the other hand, mechanical and biological treatment (MBT) facilities and thermal plants (such as incineration) mainly affect the chemical properties of waste, aiming at reducing its volume and environmental hazardousness.

The *Recycling-Utilisation-Landfilling* stage includes all final treatment options that follow the waste processing stage. The state and composition of waste determines the suitability of each of the alternative final options. Homogenous materials are suitable for recycling or utilisation (e.g. composting of organic waste), while

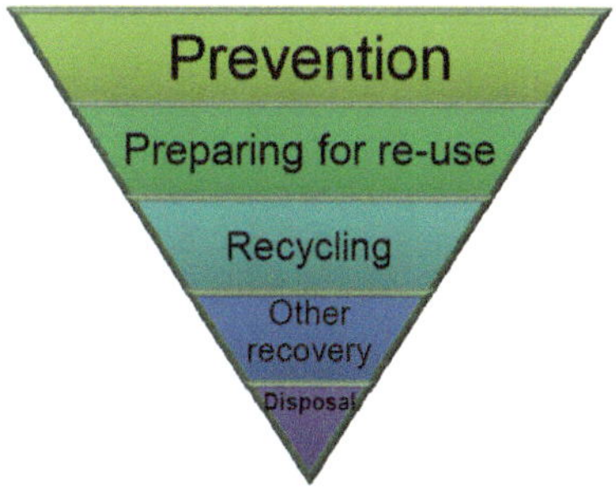

Fig. 35.2 Waste hierarchy indicating a scale of environmental preference for the five main treatment alternatives (EU-JRC 2011)

mixed waste usually ends up in landfills. This stage includes a large variety of technologies for recycling waste, composting (central composting or anaerobic digestion) and landfilling (engineered landfills, collection of landfill gas and leachate).

The selection of a treatment technology (but also the design of an integrated waste management system in general) has direct implications for the environmental impacts caused. Since 1980s many countries around the world have established a waste hierarchy to prevent or limit the impacts of waste management operations on the natural resources, ecosystems and human health. The hierarchy (see Fig. 35.2) has been included in the EU legislation as a legally binding framework for designing or improving a waste management system (see Directive 2008/98/EC or the EU Waste Framework Directive). The hierarchy is based on a "rule of thumb" regarding the environmental ranking of waste treatment options. Deviations from the hierarchy, according to the legislation are accepted if justified by means of appropriate tools such as LCA.

Waste prevention is mentioned as the first priority in the waste hierarchy. The assessment of prevention in LCA terms is fundamentally different compared to the other steps of the hierarchy as it involves upstream processes of a waste material, thus extending the system boundaries.

35.1.3 Main Environmental Concerns

The recent shift in the perception of waste as a resource is reflected in the waste hierarchy. Re-use and recycling are the highest ranking treatment options. The ambition of the legislators is to integrate waste management in a circular economy structure, where waste activities deliver recovered resources and close loops in material cycles. The reduction of the depletion of natural resources, such as fossil fuels, metals, as well as nitrogen and phosphorus is, therefore, a priority for waste management operations.

The main environmental concerns related to waste management, besides resource efficiency, are:

- Climate change and the related energy security issue. Greenhouse gases are emitted from various processes in waste management such as transport or landfilling. A major opportunity for climate change mitigation lies in the avoided emissions through waste materials recycling in a system expansion approach (see Sect. 8.6.1). Climate benefits also arise through waste incineration, where electricity and/or heat is produced locally, substituting other (usually fossil) sources of energy.
- Toxic emissions (to ecosystems and to humans) related to (1) processing waste (e.g. incineration) and (2) the eventual disposal of waste (e.g. in landfills). Toxic emissions also have a temporal aspect, since they may be released at a very slow rate (e.g. from landfills) and create problems at a much later time than when the waste deposition takes place.

Different impacts are related to different waste technologies:

- *Landfilling*: leachate created by water infiltrating the waste mass and not collected may pollute the surrounding soil and groundwater with organic and inorganic (metals) pollutants. Landfill gas created by the anaerobic degradation of organic matter contains methane, a strong greenhouse gas.
- *Incineration*: airborne emissions can affect local ecosystems. CO_2 is emitted from the incineration of carbon (e.g. fossil carbon in waste plastics). Bottom ash contains significant concentrations of toxic heavy metals that can potentially leach after its deposition. Benefits from incineration depend strongly on the local energy mix, substituted by the energy delivered by incineration.
- *Recycling*: recycling operations are linked mainly to energy use for processing the waste and chemicals used for recovery operations (e.g. de-inking of paper).

The complexity and variety of environmental issues arising from waste management calls for a comprehensive approach such as LCA that addresses all potential environmental impacts (e.g. JRC 2011).

35.2 LCA Applied to Solid Waste Management Systems (SWMS)

LCA is increasingly used to assess solid waste management systems. Two comprehensive review articles were published in 2014 on how LCA is applied on waste systems and which issues require special attention due to particularities in the field of waste LCA:

1. Laurent A, Bakas I, Clavreul J, Bernstad A, Niero M, Gentil E, Hauschild MZ, Christensen TH (2014) Review of LCA studies of solid waste management systems—Part I: Lessons learned and perspectives. Waste Management 34 (2014) 573–588

2. Laurent A, Clavreul J, Bernstad A, Bakas I, Niero M, Gentil E, Christensen TH, Hauschild MZ (2014) Review of LCA studies of solid waste management systems—Part II: Methodological guidance for a better practice. Waste Management 34 (2014) 589–606

The review and its results provide a useful background for describing the operational aspects of LCA applied to solid waste management systems. The authors analysed a wide range of peer-reviewed scientific articles and reports for their methodological approach to the LCA and contrasted it with the guidelines of the ILCD Handbook (JRC 2011). Key findings of the review are presented in this section, but the reader is referred to the original papers for a more in-depth analysis.

35.2.1 Review Process and Focus Areas

The selection of the studies was performed by choosing studies in English, peer-reviewed and referring to solid waste, excluding sewage sludge. The results of the review were summarised into a table, a version of which is presented in Table 35.3. The original version of the table included all elements of the review, namely:

1. References/sources
2. Type of LCA studies (public report, scientific article…)
3. Standard compliance (e.g. None, ISO, ILCD, …)
4. Goals (intended use/users of study)
5. Context situation (situation A, B, C1, C2)
6. Object(s) of study considered/compared
7. Type of waste (e.g. only organic waste included in study) (goal/scope)
8. Defined functional unit
9. System boundaries: Included/Excluded processes; included phases within the MSW based on intro definition (e.g. collection, incineration…).
10. Impact coverage (e.g. GW only)
11. Geographical coverage
12. Time scope (for validity of LCA results)
13. Date(s) of the collected primary (specific) data and secondary data (database) (to identify data representativeness in time, e.g. old/up-to-date)
14. Handling of multifunctional processes: approaches (e.g. allocation or syst. expansion) and type of data used to solve them (e.g. marginal/average data)
15. LCA software/Databases used for secondary data
16. LCIA method used
17. Use of normalisation and/or weighting (incl. method description for weighting if any)
18. Use of sensitivity analysis: what key parameters were identified and changed?

19. Main findings (e.g. significant impacts, comparative performances of 2 alternatives) ⇒ also in relation to goals is possible
20. Identified method shortcomings (*incl. "solutions" or "actions taken" for each identified problem*)
21. Identified modelling shortcomings (*incl. "solutions" or "actions taken" for each identified problem*)
22. Identified data uncertainties (e.g. which data are difficult to collect…) (*incl. "solutions" or "actions taken" for each identified problem*)

Table 35.3 in "Appendix" only includes some descriptive elements of the review, in order to inform about the external characteristics and variations of the reviewed studies. The more elaborate LCA elements of the review are included in Sects. 35.2.2–35.2.6 and 35.3. In these subsequent sections, the most important review findings are listed and methodological considerations are analysed in order to outline how a credible LCA should be applied on solid waste systems.

The majority of the studies reviewed (around 94%) were scientific articles published in peer-reviewed journals. Most of the studies claimed compliance with the ISO standards, namely that the methodology proposed by ISO was followed by the studies. In fact, many studies did not actually comply with the ISO provisions, despite their claim to do so. The review revealed that only about one out of five studies actually complied with the ISO standards. This is because of a number of elements (or combinations of elements) missing that are essential in the ISO standards provisions. In the following sections, these omissions that lead to deviations from the standards are described more in detail.

35.2.2 Goal

According to the ISO standards, the goal definition refers to the intended uses of the LCA case study and its potential users (ISO 2006). Moreover, the ILCD Handbook, launched by EU's Joint Research Centre, specifies the definition even further into six aspects, namely the intended applications, limitations in using the results, drivers for performing the study, target audience, disclosure to the public and the commissioner of the study (EC 2010a, b) (see also Chap. 7).

Since the majority of the studies reviewed come from scientific journals, they are rarely commissioned directly by an entity intending to use the results for decision support (some articles are based on larger reports that might be more complete and support decisions). This means that the goal definition is often out of focus for the study authors, which do not refer to potential users as the ISO standards require. Many of the studies only describe the intended use of the study, while many others do not have a specific purpose except for analysing methodological aspects of waste LCAs or tackling specific issues. Omitting the adequate description of the goal has a profound effect on the interpretation of the studies by the readers. The absence of context when considering the results might lead to overlooking the weaknesses of the study (such as the inclusion of only a small part of impact categories) or unjustified generalisations (e.g. generalising the environmental superiority of a treatment option over another when the results of the study refer to specific local conditions). These identified shortcomings might also apply to LCA studies from other technology fields published as scientific articles.

Another consolidated reference to the intended use of the study and the size of the consequences of the study's results is the decision context situation. The ILCD Handbook describes four cases of decision contexts: A (micro-level decision support), B (meso/macro-level decision support), C1 and C2 (accounting with no decision support) (see also Sect. 7.4). Most of the reviewed studies belong to situation B, followed by A. This was expected as normally the investigation of waste management systems happens on a larger scale (national, regional or municipal geographical units). Situation A refers to studies mainly assessing specific technologies or comparing them to others. However, although the classification of the study into a decision context situation helps put the results in perspective, none of the reviewed studies explicitly referred to a context situation.

35.2.3 Scope Definition

The *object* of the reviewed studies varies greatly among the studies. Different waste systems or parts of systems were assessed, while referring to various types of waste. Figure 35.3 shows the amount of studies investigating each distinct aspect of a solid waste management system. The traditional treatment options, as well as collection and transport feature as the most popular topics for investigation, while emerging technologies such as thermal and some forms of biological treatment are starting to gather attention. Due to the difficulty in framing LCAs on waste prevention in appropriate system boundaries or the lack of focus on prevention by decision-makers, only two studies were found that deal with this topic.

The *functional unit* in waste LCAs is expressed mainly according to four types: (1) unitary, (2) generation-based, (3) input-based and (4) output-based (see Box 35.1. with examples).

Box 35.1. Examples of Functional Units Used in the Reviewed Studies

1. Unitary: "management of 1 tonne of municipal solid waste"
2. Generation-based: "Management of the waste generated in Copenhagen municipality"
3. Input-based: "100 tonnes of waste entering a waste incineration plant"
4. Output-based: "Production of 500 kWh from a dedicated incineration plant for industrial waste"

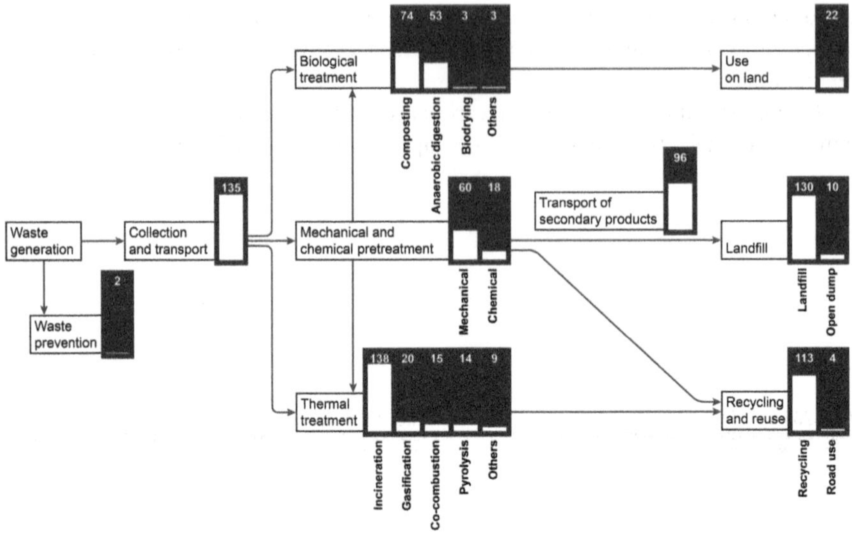

Fig. 35.3 Waste management technologies assessed in the studies. Many studies investigate more than one aspect of the system

Although, the definition of a functional unit is relatively straightforward, in many cases LCA practitioners neglect to specify an adequate functional unit, as shown in Fig. 35.4. It also seems that a unitary functional unit is by far preferred in the reviewed studies reflecting a more theoretical or methodological goal for the LCA or potentially a confusion of the functional unit with the reference flow of the system. The functional unit refers to a quantified description of the primary function of the system under study, while the reference flow refers to the physical flow required for the system to fulfil its function (see also Sect. 8.4). The use of a reference flow in the cases, where the functional unit is defined as unitary neglects the appropriate description of the functional unit in several aspects, such as the composition of the waste that is treated.

As already mentioned, the *system boundaries* in waste LCAs are set in order to include only the end-of-life stage of the products' life cycles. This is justified as waste LCAs are normally of comparative nature and it is therefore assumed that for the waste in question in each case, the previous life cycle stages are identical for the systems compared and therefore can be omitted.

Similar to all types of LCAs, a central issue in defining system boundaries is the inclusion of capital goods, i.e. the construction and use of infrastructure, plant facilities and equipment used in the assessed system. 62% of the studies reviewed did not mention the capital goods at all, 12% included them and 26% of the studies excluded them with justification.

Collection and transport processes are also occasionally excluded from the system boundaries, due to their minor contribution to the overall impact categories'

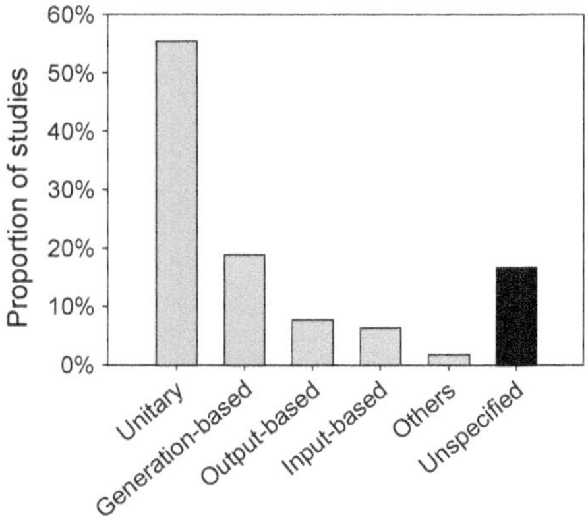

Fig. 35.4 Proportions of studies for each class of functional unit. A number of studies are classified into more than one category, including studies, for which the functional unit was only implicitly mentioned

results. 16% of studies chose to exclude such processes due to reasons such as their lack of relevance or identical contribution to all scenarios assessed.

Another set of processes that is often excluded from the system boundaries refers to secondary products (i.e. valuable outputs such as digestate from anaerobic digestion) and secondary waste stemming from waste treatment processes (e.g. air pollution control ashes from incineration). Secondary products and secondary waste are only included in 44 and 53% of the reviewed studies respectively.

Regardless of the impact of including/excluding such processes from the system boundaries, it is always recommended to address the issue transparently and in a case-by-case manner, as different systems with varying characteristics may justify opposite decisions regarding the definition of system boundaries.

The literature review also analysed LCA practitioners' preferences with respect to *included impact categories.* Figure 35.5 demonstrates their preference for already established, traditional impact categories. Almost all reviewed studies included, at least partially, non-toxic impact categories, while toxicity was considered by more than half of the studies. Resource impact categories such as land and water use were underrepresented, but as research in these fields advances, it is likely that they will become more central in future waste LCA evaluations. However, it should be underlined that incomplete assessments, as in the majority of the reviewed studies, may reduce credibility of the results: maybe the burden is shifted to one of the impact categories that is not assessed.

In order to better understand the characteristics of the reviewed studies, their distribution over space and time offers valuable insights. Figure 35.6 shows the geographical distribution of the studies with European countries and the US dominating the map. China, Australia and Japan follow in number of studies assessing waste produced in these countries. It is evident from the map that Africa and large parts of Asia are underrepresented in the waste LCA applications.

The time evolution of the studies as well as their distribution among the main scientific journals are shown in Fig. 35.7. As expected, the number of LCAs performed increases with time, alongside with the popularity of the tool and its

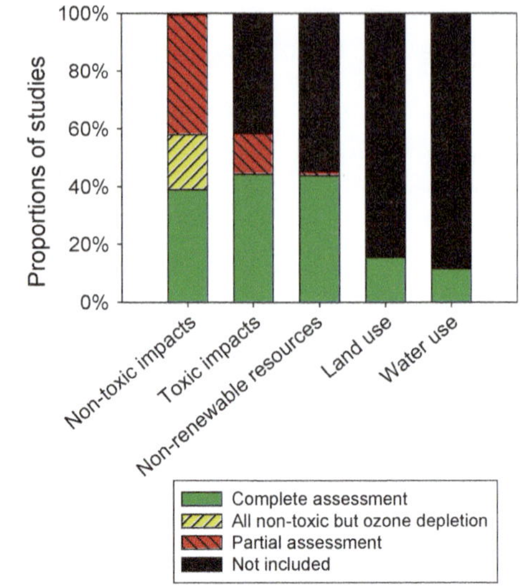

Fig. 35.5 Proportions of impacts covered in the assessments. Non-toxic impacts include climate change, stratospheric ozone depletion, acidification, photochemical ozone formation, eutrophication. Toxic impacts include aquatic and terrestrial ecotoxicity, human toxicity and impacts from particulate matters

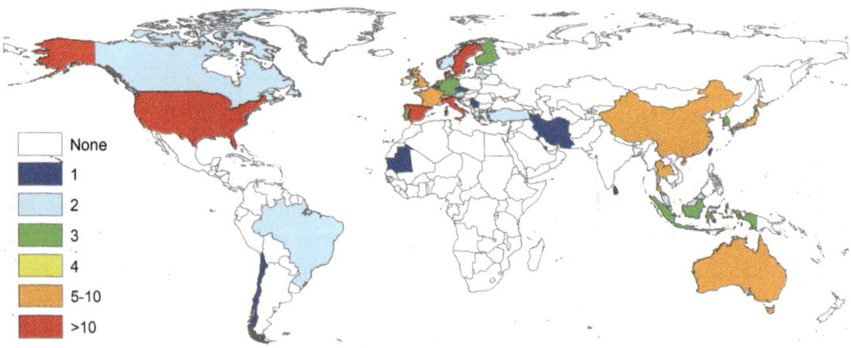

Fig. 35.6 Geographical distribution of case studies based on locations of waste management systems under study. Generic cases and European cases as well as technical reports were excluded from the figure

establishment as a mainstream evaluation method for waste management systems. The figure also shows the adoption of important European legislation to illustrate the influence of legislative measures on the intensity of the research on waste management's environmental impacts.

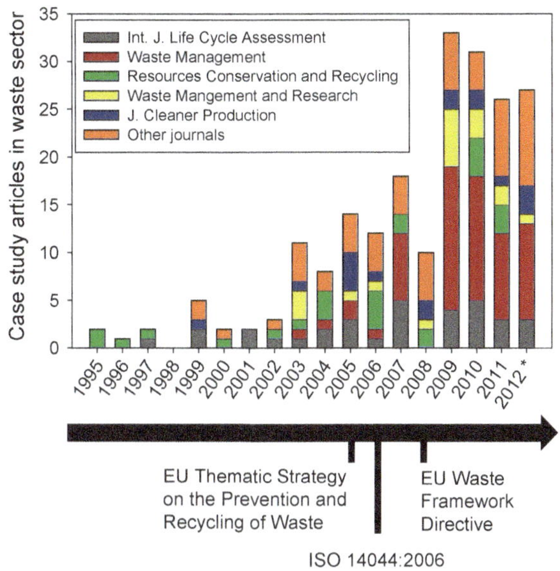

Fig. 35.7 Distribution of publications describing LCAs of solid WMS across peer-reviewed journals. For multiple articles describing the same case study, only the main article supporting the case was accounted for. Literature search was stopped mid-2012, hence 2012 is an incomplete year (indicated by *asterisk*)

35.2.4 Inventory Modelling

The inventory part of waste LCAs is given particular attention by practitioners due to its role in increasing the results' credibility. The accurate representation of the studied system with appropriate data is decided in the inventory preparations.

In the reviewed studies of LCA applied to solid waste management systems, the authors tried in general to use site-specific information and when not possible, search for data in literature and databases. Around 70% of the reviewed studies included at least partly primary data, as Fig. 35.8 shows, but in most of the studies, practitioners had to supplement their inventory analysis with literature information and/or generic data. Although, as mentioned before, waste LCAs typically refer to a specific geographical region, the quest for primary data is rarely fruitful. The reason is that data collection for LCIs is a difficult and time consuming process, leading many researchers to use generic data from widespread LCA databases, such as ecoinvent. These databases aim mainly at modelling the background system, but in the absence of relevant information they are often used as data sources for the foreground system as well. This solution also has the drawback that time representativeness is not followed. Databases are updated irregularly and usually after many years. Therefore, the use of generic database information in general reduces the relevance and representativeness of the study.

The use of LCA databases and the compilation of the inventory data are facilitated by the inclusion of LCA databases in LCA software. Both generic and waste-customised LCA tools are used, the latter ones enabling the specific modelling of different waste fractions through processes that can be parameterised (see

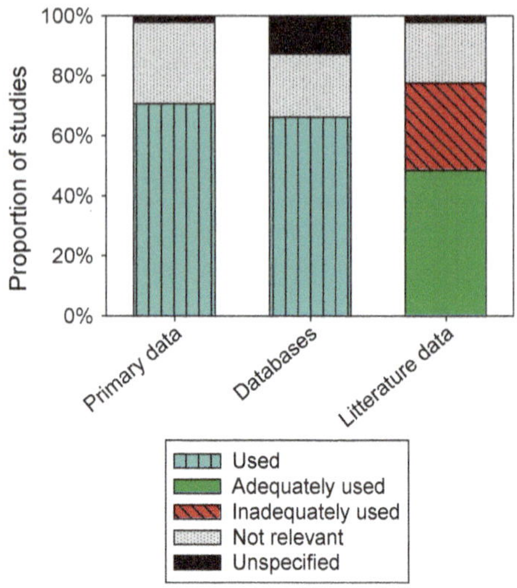

Fig. 35.8 Types of data sources used in studies. Primary data refers to site-specific data from, e.g. field investigations. Databases mainly refer to LCA software-embedded databases. Adequacy of literature data was based on a rough consistency check with the temporal and geographical scopes of the study

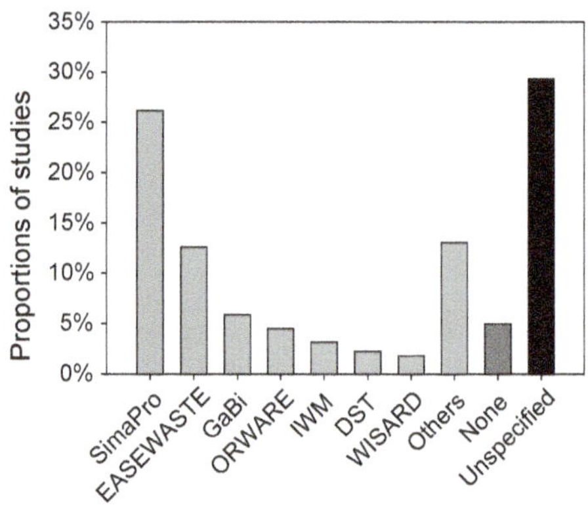

Fig. 35.9 LCA software used in studies. The category "Others" includes TEAM, UMBERTO, GEMIS, WRATE, LCAiT, JEMAI-LCA, EIME, WAMPS software

e.g. Clavreul et al 2014 for an example of such tools). The most popular LCA software among the reviewed studies was the generic LCA tool SimaPro (Fig. 35.9).

One important aspect in waste LCAs is the long-term emissions associated with waste landfilling. The issue around handling long-term emissions has been a subject of strong debate within the LCA community (Hischier et al. 2010). LCA in principle integrates emissions regardless of when they occur. This principle works well when considering relatively short time spans. But the time integration of emissions occurring in low concentrations over very long time spans (such as metal emissions leaching from landfills) leads to an estimation of very high and unrealistic impacts in toxic impact categories which are linked to toxic metal emissions when landfilling waste (Bakas et al. 2015). A suggestion has been to cut off all long-term emissions beyond the arbitrary threshold of 100 years from the waste deposition and either discard them (Hischier et al. 2010) or treat them in a separate impact category (Hauschild et al. 2008). These suggestions along with other proposals, all have inherent problems that do not allow them to become operational and widely accepted in the LCA community (Bakas et al. 2015). The lack of consensus and the absence of an adequate method to account for long-term emissions in LCA has led many practitioners in the reviewed studies to omit or assign less credibility to toxicity-related impact categories.

Within the LCI phase, reference needs to be made to the handling of multi-functional processes (see Sects. 8.5 and 9.2.2). This choice is particularly relevant as waste operations often lead to the production of secondary products such as secondary materials (recycling) and energy (incineration, landfill gas extraction). According to the ISO standards, system expansion is the preferable option for dealing with such processes and, as Fig. 35.10 shows, practitioners follow this recommendation to a great extent (around 75%), with a few cases of studies reported to resort to allocation. The allocation key varied among the studies with mass, heat value, waste volume, exergy and economic value all used by practitioners.

On the other hand, the choice between marginal and average data for crediting a waste system delivering secondary products is more evenly divided. In many cases, also, the choice is not sufficiently justified, which could be attributed to the difficulties of practitioners in identifying the proper approach and the lack of adequate framing of the goal and scope of their study. The choice between marginal and average data depends on the goal of the study and the context situation it belongs to (see Chap. 7).

System expansion is often very crucial for estimating the final results, as it strongly influences the benefits of one waste treatment option over another. Thus, the lack of transparency in how this is performed may substantially reduce the

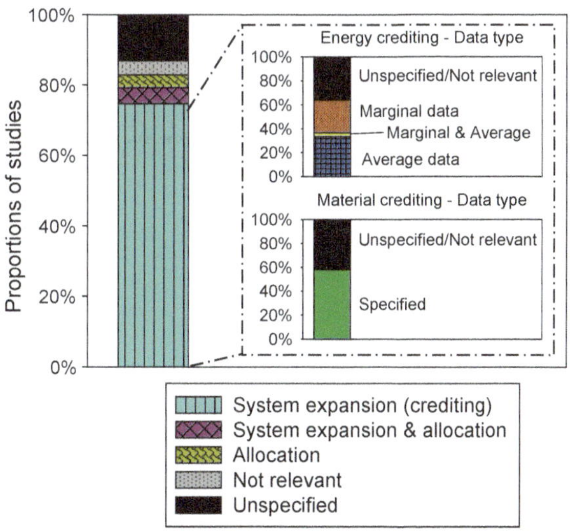

Fig. 35.10 Handling of multifunctional processes. Energy includes both heat and electricity. The differentiation between marginal and average data for material crediting is largely omitted in the studies, and is hence not reported in the figure

credibility of the final results and conclusions. A systematic framework for consistent modelling of recycling, co-production and energy recovery has recently been developed by Schrijvers et al. (2016), describing the relation between the LCA goals and the attributional/consequential approach. Most of the reviewed studies assume a 1:1 substitution ratio between primary and secondary material production and/or quality similar to the substituted product. However, an overestimated substitution ratio or grade of the recovered materials can significantly impact the benefits gained from recycling and alternative methods based on the average market consumption mixes of primary and secondary materials have been proposed for calculating the environmental credits of end-of-life material recovery in attributional LCA (Gala et al. 2015). One of the main challenges for LCA in the circular economy is to address the continuous loop of materials and account for the benefits from recycling in a consistent way (Niero et al. 2016).

35.2.5 *Life Cycle Impact Assessment*

The constant updating of existing and development of new impact assessment methods (see also Chaps. 10 and 40) makes it difficult to accurately map the popularity of specific LCIA methods among LCA practitioners in solid waste management. Figure 35.11 attempts to map the use of LCIA methods among the researchers and practitioners of the reviewed studies. CML is strongly preferred, followed by EDIP and Ecoindicator 99. Interestingly, around 20% of the studies failed to report on the LCIA method choice.

This mapping reveals information on the selection criteria applied by practitioners, and also the perception of credibility of LCIA methods by LCA practice. Additionally, the time of conducting the study is important as newly developed methods (such as ReCiPe; Goedkoop et al. 2009) are absent from Fig. 35.11 showing historical data, although they might be more widely used today. This information also needs to be put to perspective regarding the impact coverage

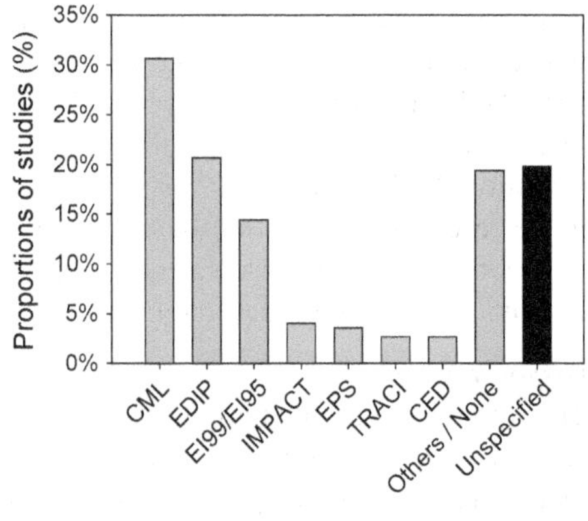

Fig. 35.11 LCIA methods used. Some studies used more than one LCIA method; all have been included here. Category "Others" includes the use of specific models, which are not considered as whole LCIA methods, e.g. IPCC (2007)

analysed in the previous chapter. The selection of a complete LCIA method does not guarantee its proper implementation as many cases demonstrate that some of the impact categories in the methods were omitted. This implies in most cases a reduced credibility of the LCA results (unless the omission is well justified and in line with the goal and scope of the study), despite the use of a well-established LCIA method.

Within the LCIA phase, *normalisation and weighting* steps are an option when appropriate. Performing these steps or not, is strongly associated with the choice of LCIA method and the normalisation and weighting frameworks these recommend. Most of the reviewed studies are concluded at the characterisation step, while 46% perform normalisation and 26% weighting. The majority of these cases perform weighting because of the choice of the Ecoindicator 99 LCIA method which is a damage-oriented method, offering its own weighting scheme (Goedkoop and Spriensma 2001).

With respect to weighting, a particular case arises when examining the impacts of long-term emissions from landfills. As mentioned before, this case poses particular challenges in an LCA framework when trying to characterise this type of emissions. Another aspect of this case is related to weighting, as some impacts from landfilling might occur in many millennia from waste deposition and might be weighted differently by some stakeholders. So far, there is no widespread weighting method for addressing time-differentiated impacts. Thus, this point was not addressed adequately by any of the reviewed studies.

In general, there are some specific *methodological considerations* during the execution of an LCIA on solid waste management systems that should be given particular attention when performing a waste LCA. The first consideration refers to the handling of the biogenic carbon contained in waste material and its contribution to global warming potential. Biogenic carbon can be considered either neutral or as contributing to climate change, depending on the approach, but this choice needs to be consistent throughout the study. Criteria for assigning global warming emission factors to biogenic carbon have been developed in the literature (Christensen et al. 2009).

Another particular consideration refers to the already mentioned issue of long-term emissions. The LCA community has not reached a consensus in the proposed impact assessment method (Bakas et al. 2015) and this causes significant confusion among practitioners. A new approach has recently been published that applies time differentiation on long-term emissions, estimating toxicity separately for distinct future time periods (Bakas et al. 2017).

35.2.6 Interpretation of Results/Conclusions

The interpretation phase of an LCA should present the results of the study in the context of the defined goal and scope, according to the ILCD Handbook (see also Chap. 12). Therefore, practitioners of waste LCAs should reflect on the goal and scope of the study and put their results in this perspective. The majority of the reviewed studies did not include an adequate interpretation section: instead the results were often presented out of context and only a fragmented commenting was included.

Many of the LCAs performed on solid waste management systems are comparative assertions on treatment technologies for a specific waste stream or material. The review of the LCA cases revealed some trends regarding the environmental superiority of some treatment options compared to other. Based on studies selected because of their higher quality, some generic statements of the superiority of different treatment options are presented in Fig. 35.12.

A central part of the interpretation is the *sensitivity analysis*, often accompanied by uncertainty analysis (see Chaps. 11 and 12). Sensitivity analysis is used for evaluating the dependence of the LCA results on input data, modelling choices and hypothesis made. Although, there are many methods for performing sensitivity analysis, in LCAs applied to solid waste systems, a scenario analysis is often used. Scenario analyses are based on constructing an alternative scenario to the main one, which includes a different assumption or data input. In the reviewed studies, many

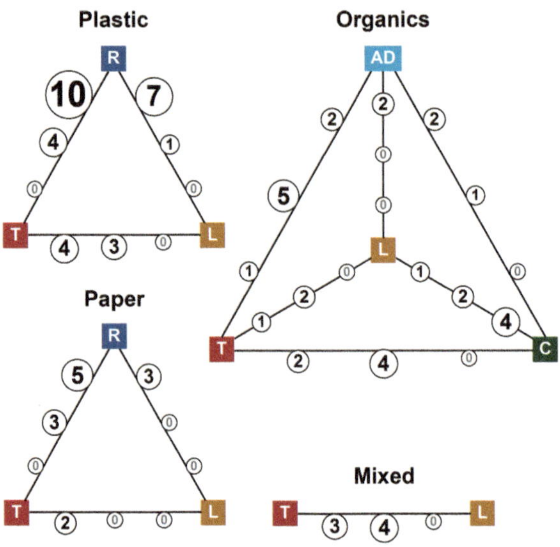

Fig. 35.12 Comparative analysis of key findings for selected waste treatment technologies applied to paper, plastic, organic and mixed waste fractions (total of 34 studies). The nodes "R" stand for recycling, "L" for landfilling, "T" for thermal treatment, "C" for composting, "AD" for anaerobic digestion. For each pair comparison, *three circled numbers* are indicated, representing the number of studies concluding on the better environmental performance (i.e. lower overall environmental impact) of one waste treatment technology over another (numbers closer to each of the two nodes), or reaching either inconclusive results or results with similar environmental burden (*numbers in the middle*). The size of the circles is proportional to the number of studies

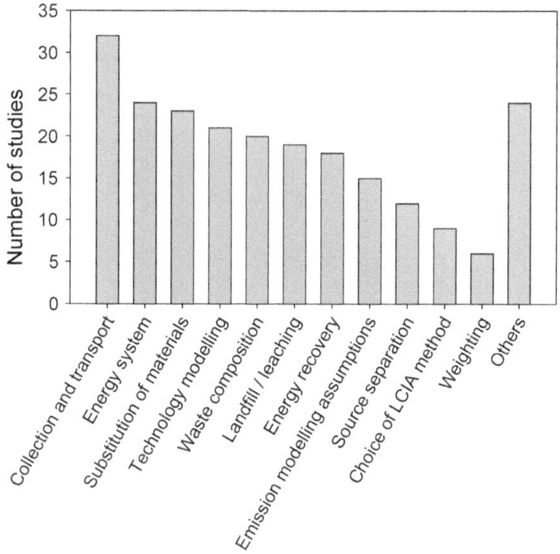

Fig. 35.13 Issues covered in sensitivity analysis (total of 101 studies). The category "Others" (25 studies) includes carbon accounting methods (6), inclusion of secondary materials (4), allocation rules (4), accounting of waste containers (3), time horizon in impact assessment (3), testing of other treatments (2), choice of databases (2) and normalisation (1)

different aspects of a waste system were processed in sensitivity analyses, as Fig. 35.13 shows. Preferred elements to include in sensitivity analyses are collection and transport.

35.3 Central Issues to Consider When Performing or Using Data from LCA Studies on Waste Management Systems

The analysis above provides the necessary background for identifying methodological issues of particular importance when conducting an LCA of waste management options. These issues are identified due to their importance in ensuring credibility of the LCA results and also the frequency by which researchers fail to address them properly. As a general recommendation, in Table 35.1 the main methodological issues are presented along with proposed solutions and recommendations.

The specific methodological challenges for waste LCAs comprise aspects like the differentiation in system boundaries, the zero-burden convention and specific capital goods. The particularities of waste LCAs also include the product system itself, which typically consists of more local installations and smaller geographical dispersion. Specific modelling issues also arise in waste LCAs: the inclusion of biogenic carbon in the modelling, which arises in many waste streams; also, the inclusion of long-term emissions when landfilling waste, which modifies the perception of temporal boundaries one needs to consider in the LCA.

Table 35.1 Key methodological issues and proposed solutions for application of LCA on solid waste management systems

	Methodological and consistency issues	Proposed solutions/recommendations
Goal	Absence of intended use, target audience and limitations of use	Follow ISO recommendations
	Consistency among goal elements	Check consistency among goal elements iteratively
Scope	Elements of functional unit definition missing	Define the functional unit comprehensively. Functional unit not to be confused with reference flow
	Lack of transparency in choices around the LCI	Ensure transparency and assess the choices in terms of uncertainty
	Fragmented description of system boundaries, especially in relation to capital goods and waste transportation	Document assumptions pertaining the definition of system boundaries
	Impact coverage lacking comprehensiveness and representativeness	Follow the selected LCIA method's recommendation for a comprehensive set of impact categories. If an impact category is excluded, justification should be provided
	Insufficient justification of modelling choices (e.g. allocation) and assumptions (e.g. data types used)	Key choices and assumptions, vital for the LCA results, should be transparently documented
	Consistency with the defined goal	Define scope elements within the context of the goal. Revise goal if necessary to ensure consistency
LCI (incl. modelling)	Lack of geographical and temporal data representativeness	Further data and information need to be collected to ensure a sufficient data representativeness
	Lack of data representing areas other than Europe and North America	More efforts for data collection from other parts of the world than Europe and North America
	Lack of documentation of data collection processes	Explain thoroughly how and why data sources are used (literature, databases, etc.)
	Distinction between fore- and background data sources missing and sources misused	Describe and assess the consequences of using background data for the foreground system if necessary
	Lack of data on long-term emissions	Consensus on how to deal with long-term emissions needed
	Use of non waste-specific LCA software	The use of waste-specific LCA software facilitates the more accurate waste system's modelling

(continued)

Table 35.1 (continued)

	Methodological and consistency issues	Proposed solutions/recommendations
LCIA	Missing impact categories (e.g. occupational)	The exclusion of the results in specific impact categories should be avoided or, if unavoidable, documented
	Modelling of impacts from long-term emissions missing	Assess the consequences of omitting the effects of long-term emissions
	Normalisation and weighting	Describe and justify where modelling stops (characterisation, normalisation, weighting)
Interpretation	Interpretation step often missing altogether	Interpretation of results is vital to putting results in context and should always be addressed
	Superficial analysis of impact potentials	Refer to contribution analyses at substance and process level, identify hotspots and recommend improvement potential
	Frequent absence of sensitivity analysis and sensitivity checks	Conduct sensitivity checks on most relevant processes and use the results in interpretation
	Negative impacts obtained from the disposal stage can mislead interpretations of LCA studies	Relate results to goal and system boundaries selection

35.4 Sources/Links to Access Information on LCA Applied to Solid Waste Management Systems

Table 35.2 presents a non-exhaustive list of sources for obtaining data, software tools and methodological guidance on LCA applied on solid waste management systems.

35.5 Concluding Remarks

This chapter attempts to provide guidance and recommendations in specific issues that differentiate waste LCAs from normal product LCAs. Due to these particularities, waste LCA has developed into its own sub-field encompassing its own sub-definitions of LCA elements, dedicated databases and software.

Legislators, through for example the official endorsement of the waste hierarchy in Europe, have acknowledged the importance of LCA in operating as a reliable tool for providing credible information to decision-makers. Waste generation is increasing globally, while new emerging waste streams appear for the first time (e.g. nanomaterials or composite plastics). The assessment of the environmental

Table 35.2 Key sources for information and tools addressing LCA applied on solid waste management systems

Sources	Short description	Used for
EU-JRC (2011)	Guidance document on application of LCA on SWMS	Guidance in conducting waste LCAs
Cleary (2009)	Review of methodological issues from applying LCA on SWMS	Better understanding of methodological challenges in LCA and waste management
Christensen (2011)	Description of technologies used in SWMS	Obtain knowledge on SWMS
http://www.wrate.co.uk/	Presentation of a waste LCA dedicated software	Modelling SWMS in an LCA context
https://www.epa.gov/warm	Presentation of the US EPA waste software, including a life cycle approach to greenhouse gas emission estimation	Modelling SWMS, combined with LCA elements
Gentil et al. (2010)	Review of nine software tools applying LCA on SMWS	Collecting information on dedicated waste LCA software
Doka (2009)	Information on the ecoinvent inventories for SWMS	Waste LCA inventories

implications of the management of new waste streams or of emerging treatment technologies, will remain an important topic in the future.

The new challenges bring also new methodological challenges to waste LCA practitioners with respect to environmentally sound treatment of new waste materials or new technologies. On the other hand, old debates still remain unresolved, such as the proper allocation procedure and the handling of long-term emissions. In any case, practitioners are encouraged to address all methodological challenges, present in waste LCAs, by following best practice examples and applying transparency.

Appendix: Reviewed Studies

See Table 35.3.

Table 35.3 Essential elements of reviewed studies of LCA applied on solid waste management systems

References/sources (short; details to be provided in spreadsheet 'LCA studies list')	Technology assessed	Type of waste (e.g. only organic waste included in study)	Impact coverage (e.g. GW only) (*use glossary in first spreadsheet*)	Geographical coverage	LCIA method used
Abduli et al. (2011)	L, C	MSW	GW, ODP, ET, O	IR	Eco-indicator 99
Abeliotis et al. (2012)	L	MSW	RD, GW, AC, EU, HT, PO	GR	CML 2
Al Maaded et al. (2012)	L, R	Mixed plastic waste	GW, AP, ADP, HTP	QA	NS
Al-Salem et al. (2009)	L, I, B	MSW	GW, AP, EU	KW	NS
Andersen et al. (2012)	C, I, L	Organic household waste	GW, PO, EU, AC, AET, TET, HT, O	DK	EDIP 1997
Arena et al. (2003)	I, L	Residual waste	GW, AC, EN, LW, WA	IT	NA
Arena et al. (2003)	L, I, R	PE/PET containers	GW, RD, LW, O	IT	NA
Assamoi et al. (2012)	L, I	Residual waste	GW, AC, EU, O	CA	NA
Assefa et al. (2005a)	I, B, T	Biodegradable waste	GW, EN	SE	NS
Assefa et al. (2005b)	T, I, L	Industrial waste	GW, AC, EU, EN, EI, PO	SE	NS
Aye and Widjaya (2006)	Int	Market waste	GW, AC, EU, PO	ID	NS
Banar et al. (2009)	Int	MSW	RD, GW, AC, EU, HT, PO, EI	TR	CML 2000
Beccali et al. (2001)	Int	MSW	GW, AC, EU, HT, TET, AET, EN, LW	IT	NA
Beigl and Salhofer (2004)	Int	MSW	GW, AC, EN	AT	NA
Bergsdal et al. (2005)	I	Municipal, commercial and special waste	GW, AC, EU, PO, TET, AET, EI, O	NO	CML 2
Bernstad et al. (2011)	I, C, B	OFMSW	GW, AP, EP, ODP, POP	SE	EDIP 1997
Bernstad et al. (2011)	R	MSW	GW, AP, EP, ODP, POP	SE	EDIP 1997
Bernstad et al. (2012)	COl	Food waste	GW, AC, EU, EN	SE	EDIP2003

(continued)

Table 35.3 (continued)

References/sources (short; details to be provided in spreadsheet 'LCA studies list')	Technology assessed	Type of waste (e.g. only organic waste included in study)	Impact coverage (e.g. GW only) *(use glossary in first spreadsheet)*	Geographical coverage	LCIA method used
Bientinesi and Petarca (2009)	Int	WEEE	HT, GW, OD, ET, AC, O	NA	Ecoindicator 99, Impact 2002+
Bigum et al. (2012)	R	high-grade WEEE	GW, AC, NE, POF, HT, ET	NR	EDIP 1997
Birgisdottir et al. (2007)	L, R	MSWI bottom ash	GW, PO, EU, EC, HT, AET, TET, RD, O	DK	EDIP 1997
Bjorklund and Finnveden (2007)	Int	All waste	EN, GW, OD, PO, AC, EU, AET, TET, HT	SE	CML 2000
Björklund et al. (1999)	Int	Non-hazardous MSW, park and yard waste, sewage sludge and industrial biodegradable wastes	GW, AC, EU, PO, O	SE	NA
Blenghini et al. (2012)	Int	MSW	GW, EN, RD	IT	IPCC
Blenghini et al. (2012)	Int	Residual waste	GW, EN, RD	IT	IPCC
Blengini and Garbarino (2010)	R	C&DW	HT, IR, OD, PO, AET, TET, AC, EU, GW, EN, LU, O	IT	IMPACT 2002+, Ecoindicator 99
Blengini et al. (2008b)	B, C, L	OFMSW	GW, AC, NE, POF, EU	IT	NS
Blengini et al. (2008a)	C, L	OFMSW	GW, OD, AC, PO, EU, EN	IT	SEMC 2000, EcoIndicator 99
Boldrin et al. (2010)	L, C	Organic kitchen waste	GW, AC, EU, PO, AET, TET, HT, O	DK	EDIP 1997

(continued)

Table 35.3 (continued)

References/sources (short; details to be provided in spreadsheet 'LCA studies list')	Technology assessed	Type of waste (e.g. only organic waste included in study)	Impact coverage (e.g. GW only) (*use glossary in first spreadsheet*)	Geographical coverage	LCIA method used
Boldrin et al. (2011)	I, C	Garden waste	GW, PO, EU, AC, AET, TET, HT	DK	EDIP 1997
Boughton et al. (2006)	L, I, R	Shredder residue	GW, AC, EU, PO, HT, AET, TET	US	CML
Bovea et al. (2010)	Int	NA	AC, EU, OD, GW, PO	ES	CML
Bovea and Powell (2006)	Int	MSW	GW, ADP, ODP, POF, AP, EP	ES	CML 2001
Brambilla-Pisoni et al. (2009)	Int	MSW	GW, OD, AC, EU, AET, TET, HT, PM, RD, LU	IT	Ecoindicator 99, CML2, EPS2000
Briffaerts et al. (2009)	R	Waste batteries	GW, AC, EU, AET, TET, HT, PM, RD, LU	BE	Ecoindicator 99
Buttol et al. (2007)	Int	MSW	GW, AC, EU, RD, TET, ET, AET, HT	IT	IPCC, CML, USES 2.0
Cabaraban et al. (2008)	C, L	MSW	GW, AC, EU, PO, AET, HT, EN	US	Ecoindicator 95, IPCC
Cadena et al. (2009)	C	OFMSW	GW, OD, AC, EU, PO, HT	ES	NA
Carballa et al. (2011)	B	Kitchen waste and sewage sludge	ADP, GW, EP, HTP, ETP	NR	NS
Carlsson-Reich (2005)	Int	MSW	GW, AC, EU, PO, EI, O	SE	NS
Chaya and Gheewala (2006)	I, L, B	MSW	GW, AC, EU, OD, PO, EN, O	TH	Ecoindicator 95
Chen et al. (2011a)	R, I	Plastics waste	GW, O	CN	IPCC (2006)

(continued)

Table 35.3 (continued)

References/sources (short; details to be provided in spreadsheet 'LCA studies list')	Technology assessed	Type of waste (e.g. only organic waste included in study)	Impact coverage (e.g. GW only) (*use glossary in first spreadsheet*)	Geographical coverage	LCIA method used
Chen and Christensen (2010)	I	MSW	GW, AC, PO, EU, HT, AET, LW, O	CN	EDIP 1997
Chen et al. (2011b)	R	Iron and steel production wastes	GW, EU	CN	NS
Cherubini et al. (2008)	L, I, B	MSW	EN, GW, AC, EU, O	IT	NS
Cherubini et al. (2009)	L, B, I, R	MSW	EN, NR	IT	NA
Chevalier et al. (2003)	I	NA	GW, AC, HT, AET, RD, O	FR	IPCC 1995, CML 2000
Ciacci et al. (2010)	L, R, I	Automobile shredder residue (ASR)	GW, OD, AC, EU, AET, TET, HT, PM, RD, LU	IT	EcoIndicator 99
Ciroth et al. (2002a, b)	I	MSW	GW, AC, EU, PO, OD, HT, O	DE,CH,ES	EcoIndicator 95
Clauzade et al. (2010) and Lecouls et al. (2010)	I, R, L	Tyres	GW, AC, PO, EU, RD, WA, EN, LW	FR	NA
Coelho and de Brito (2012)	L, R	C&DW	GW, AC, O	PT	NA
Colón et al. (2012)	Int	OFMSW	GW, AC, PO, EU, HT, ADP, OD.	ES	CML 2001
Colon et al. (2010)	C	Left-overs of raw fruit and vegetables and pruning wastes	RD, AC, EU, GW, OD, PO, EN	ES	CML 2001
Consonni et al. (2005)	I	Residual MSW	GW, HT, AC, PO, EI, O	IT	CML 2001
Cook et al. (2012)	L, I	Unused pharmaceuticals (hazardous waste)	GW, OD, AC, EU, PO, HT, AET, TET, PM	US	TRACI
Corti and Lombardi (2004)	Int	Tyres	GW, OD, AC, EU, PO, EN, O	IT	Ecoindicator 95

(continued)

Table 35.3 (continued)

References/sources (short; details to be provided in spreadsheet 'LCA studies list')	Technology assessed	Type of waste (e.g. only organic waste included in study)	Impact coverage (e.g. GW only) (*use glossary in first spreadsheet*)	Geographical coverage	LCIA method used
Craighill and Powell (1996)	R, L	Household waste	GW, AC, O	UK	IPCC, NA
Dahlbo et al. (2007)	R, L, T, I	Discarded newspaper	GW, OD, AC, EU, AET, TET, HT, PM, RD, LU, WA, EI, O	FI	DAIA, Ecoindicator 99, EPS2000
Damgaard et al. (2011)	L	Household waste	GW, PO, OD, AC, EU, HT, TET, AET, O	DK	EDIP 1997
Damgaard et al. (2010)	I	MSW	GW, PO, AC, EU, HT, TET, AET	EU	NA
de Feo and Malvano (2009)	Int	MSW	EN, GW, AC, EU, O	IT	NA
DEFRA (2006)	Int	Portable batteries	GW, AC, EU, OD, HT, AET, TET, RD	NA	CML 2001
Del Borghi et al. (2007)	L	Urban waste	GW, OD, AC, EU, PO, RD, LU, WA, LW	IT	NA
Delgado et al. (2007)	R, I	Plastics	GW, EN	EU	NA
Di Maria and Fantozzi (2004)	L, T	MSW	GW, ETP, AP, EP, LU, ODP, O	IT	Ecoindicator 95
Diggelman and Ham (2003)	C, I, L	Food waste	GW, O	US	NA
Dodbiba et al. (2008)	I, R	Plastic wastes from discarded TV	GW, RD, AC, PO, EU, HT	JP	Guinée 2002
Eggels et al. (2001)	Int	plastic packaging	GW, OD, AC, EU, RD, EN, HT, AET, PO, LW, O	EU	CML
Emery et al. (2007)	L, R, C	MSW	GW, OD, AC, EU, EN	UK	CML, WMO, IPCC
Eriksson et al. (2005)	I, L, B, R	MSW	GW, AP, EU, PO, O	SE	NS

(continued)

Table 35.3 (continued)

References/sources (short; details to be provided in spreadsheet 'LCA studies list')	Technology assessed	Type of waste (e.g. only organic waste included in study)	Impact coverage (e.g. GW only) (*use glossary in first spreadsheet*)	Geographical coverage	LCIA method used
Fallaha et al. (2009)	Int	NA	GW, OD, AET, TET, LU, AC, EU, EN, O	NA	IPCC, IMPACT2002 +
Finnveden et al. (2000)	I, L, R, B, C	Food waste, newspaper, corrugated board, mixed cardboard, PP, PE, PS, PET, PVC	EN, NEN, RD, LW, GW, OD, PO, HT, AET, TET, AC, EU, O	SE	EDIP, USES-LCA
Fruergaard and Astrup (2011)	I, T, B	SRF and organic waste	GW, AC, EU, PO, HT, AET, TET	DK	EDIP 1997
Fruergaard et al. (2010)	L, T	APC	GW, AC, NE, POF, HT, ET	DK	EDIP 1997
Gamberini et al. (2010)	Col	WEEE	GW, OD, AC, PO, HT, ET, IR, PM, RD, LU	IT	Ecoindicator 99
Gentil et al. (2010)	P	Food waste, unsolicited mail, beverage packaging (MSW)	GW, AC, AET, HT, EU, PO	EU	EDIP 1997
Giugliano et al. (2011)	Int	MSW	GM, AC, HT, PO	IT	CML 2001
Grant et al. (2005)	R	C&DW	GW, EN, RD, PO, EU, HT, LU, WA	AU	NA
Grant et al. (2001)	L, R	All commonly recycled materials (newspaper + packaging)	GW, PO, EN, WA, O	AU	IPCC 2000
Grant et al. (2003)	Int	MSW	GW, RD, HT, AET, TET, PO, EU	AU	NA
Guereca et al. (2006)	L, C	Biowaste	GW, OD, AC, EU, PO, HT, ET, LU, WD, RD	ES	NA
Gunamantha et al. (2012)	L, I, B, T	MSW	GW, AC, EU, PO	ID	NA

(continued)

Table 35.3 (continued)

References/sources (short; details to be provided in spreadsheet 'LCA studies list')	Technology assessed	Type of waste (e.g. only organic waste included in study)	Impact coverage (e.g. GW only) (*use glossary in first spreadsheet*)	Geographical coverage	LCIA method used
Güereca et al. (2007)	C, B, I, L	Biowaste	GW, OD, AC, PO, EU, AET, TET, HT, RD, WA, LU	ES	TRACI
Hanandeh and El-Zein (2010)	Int	MSW	GW, AC, O	AU	NA
Hassan et al. (1999)	L, I, C	Urban waste	GW, EI, LW	MY	IPCC
Hellweg et al. (2005)	L, I, T	MSW	O	CH	CML2001, Ecoindicator 99, Ecological Scarcity
Hischier et al. (2005)	Int	WEEE	ADP, GW, EP, HTP, ETP, AP	CH	CML
Hong and Li (2012)	R	Paper waste	Midpoint: OD, AET, TET, AC, EU, GW, EN, O. Endpoint: Human health, ecosystems quality, climate change, and abiotic resources	CN	IMPACT 2002+
Hong et al. (2010)	L, I, C	MSW	Midpoint: OD, AET, TET, AC, EU, GW, EN, O. Endpoint: Human health, ecosystems quality, climate change, and abiotic resources	CN	IMPACT 2002+
Hong et al. (2006)	C, L, I	MSW	GW, AC, EU	CN	NA
Hunt (1995)	L, I, C	Paper and plastic wastes	GW, O	NA	NA
Iriarte et al. (2009)	Col	Organic, glass, paper, packaging and rest	RD, GW, OD, HT, AET, TET, PO, AC, EU, EN	ES	CML 2000

(continued)

Table 35.3 (continued)

References/sources (short; details to be provided in spreadsheet 'LCA studies list')	Technology assessed	Type of waste (e.g. only organic waste included in study)	Impact coverage (e.g. GW only) (*use glossary in first spreadsheet*)	Geographical coverage	LCIA method used
Jenseit et al. (2003)	R	Plastic parts of end-of-life vehicles	EN, RD, GW, AC, PO, EI, O	EU	Jensen 1996, IPCC, UBA, own method
Johansson and Björklund (2009)	R, I	NA	GW, ADP	SE	CML 2001
Kaplan et al. (2008)	Int	MSW	GW, EU, O	US	NS
Khoo et al. (2009)	B, C, I	Food waste	GW, AC, EU, PO, EN	SG	EDIP 2003
Khoo (2009)	T	MSW, wood, organic waste, scrap tyres	GW, AC, EU, PO	SG	EDIP 2003
Khoo and Tan (2010)	C, I, L	Plastic bags	GW, AC, PO	SG	EDIP 2003
Kiatkittipong et al. (2009)	L, B, I	Bagasse waste	GW, AC, EU, PO	TH	EDIP, UMIP 97
Kim et al. (2009)	R	Waste home appliances	GW, RD, OD, PO, AC, EU, EC, HT, EI	KR	Ecoindicator 95
Kirkeby et al. (2006)	B, I	Household waste	GW, AC, PO, EU, HT, AET, TET, OD	DK	EDIP 1997
Klang et al. (2008)	L, C, I, B	MSW	GW, AP, EP	SE	NA
Koci and Trecakova (2011)	Int	Mixed municipal waste	RD, AC, EU, AET, GW, HT, OD, PO, TET	CZ	CML 2001

(continued)

Table 35.3 (continued)

References/sources (short; details to be provided in spreadsheet 'LCA studies list')	Technology assessed	Type of waste (e.g. only organic waste included in study)	Impact coverage (e.g. GW only) (*use glossary in first spreadsheet*)	Geographical coverage	LCIA method used
Koneczny and Pennington (2007)	L, R, I, C, B	Wet biodegradable wastes, paper and cardboard, plastic, glass, iron and steel, aluminium, other	Midpoint: AC, AET, TET, EU, GW, HT, EN, OD, PO, O. Endpoint: Ecosystem impacts, human well-being, economic production	PL	IMPACT 2002+, EDIP2003
Koroneos and Nanaki (2012)	L, R, B	MSW	GW, RD, EN, O	GR	Ecoindicator 99
Kreißig et al. (2003)	R, I, L	Cable waste (PVC)	GW, EN, AC, EI, O	EU	CML
Larsen et al. (2010)	R	Household waste	GW, AC, PO, EU, HT, RD, O	DK	EDIP 1997
Larsen et al. (2009)	Int	Residual household waste, paper, glass, bulky waste	GW, AC, PO, EU	DK	EDIP 1997
Le Borgne and Feillard (2001)	L, R	Bumper skin (car equipment)	EN, GW, RD, AC, EU, OD	NA	CML, EPS, critical volumes
Lee et al. (2007)	L, C, B	Food waste	GW, HT, AC, EU, ET	KR	CML 2002
Lelah et al. (2011)	Col	Waste glass packaging	GW, EN, ADP, O	FR	NS
Li et al. (2010)	R, T	Tyres	GW, O	CN	Ecoindicator 99
Liamsanguan and Gheewala (2007)	I	MSW	GW, PO, AC, EU	TH	IPCC, EDIP 1997
Liamsanguan and Gheewala (2007)	L, I	MSW	GW, EU	TH	NS

(continued)

Table 35.3 (continued)

References/sources (short; details to be provided in spreadsheet 'LCA studies list')	Technology assessed	Type of waste (e.g. only organic waste included in study)	Impact coverage (e.g. GW only) (*use glossary in first spreadsheet*)	Geographical coverage	LCIA method used
Lundie and Peters (2005)	C, L	Household food waste	EN, GW, WA, HT, AET, TET, AC, EU	AU	Heijungs et al., 1992
Manfredi and Christensen (2009a)	L	Household waste	GW, EU, OD, AC, PO, HT, O	NA	NA
Manfredi et al. (2009b)	L	Household, industrial, healthcare, public administration, education, trade and commercial	GW, EU, OD, AC, PO, HT, O	FI	NA
Manfredi et al. (2010)	L	Mixed waste	GW, OD, AC, EU, PO, AET, TET, HT, O	NA	EDIP 1997
Manfredi et al. (2011)	L, I, R	Organics, recyclable paper, recyclable plastic, aluminium and glass	GW, OD, AC, EU, PO, HT, AET, TET	DK	NA
Marinkovic et al. (2010)	R	Concrete wastes	GW, AC, EU, PO, EN, RD, O	RS	CML 2001
Martínez-Blanco et al. (2010)	C	OFMSW	RD, AC, EU, GW, OD, PO, EN	ES	CML 2001
Menard et al. (2004)	L	MSW	GW, OD, AC, EU, PO, HT, AET, TET, RD, LW	NA	EDIP 1997
Mendes et al. (2003)	L, C, B	MSW	GW, AC, EU	BR	EDIP 1997
Mendes et al. (2004)	L, I	MSW	GW, AC, EU	BR	EDIP 1997
Mercante et al. (2012)	Int	C&DW	GW, OD, PO, AC, EU	ES	CML
Merrild et al. (2012)	I, R	Household glass, plastics, paper, cardboard, steel and aluminium	GW, AC, EU, PO	DK	EDIP 1997
Miliute and Staniskis (2009)	Int	MSW	GW, AC, EU, PO	LT	NA
Molgaard (1995)	R, T, I, L	Plastic fraction in MSW	GW, OD, AC, EU, PO, LW	DK	NA

(continued)

Table 35.3 (continued)

References/sources (short; details to be provided in spreadsheet 'LCA studies list')	Technology assessed	Type of waste (e.g. only organic waste included in study)	Impact coverage (e.g. GW only) (*use glossary in first spreadsheet*)	Geographical coverage	LCIA method used
Morris (2005)	R, L	MSW	GW, AC, EU, HT, ET	US	TRACI
Morris (2005)	R, L	MSW	GW, AC, EU, HT, ET	US	TRACI
Morselli et al. (2005)	I	MSW	AC, EU, GW, RD, AET, TET, HT,, PO, OD, O	IT	NA
Morselli et al. (2008)	Int	MSW	GW, AP, EP, HTP, OD, PO	IT	Ecoindicator 95
Morselli et al. (2007)	I	MSW	GW, OD, AC, PO, HT, ET, IR, PM, RD, LU	IT	EcoIndicator 99
Muller (2011)	B, C	biogas digestate	GW, AC, EN	DE	CML
Munoz et al. (2004)	Int	Household, industrial-commercial waste	GW, AC, OD, HT, EU, PO, RD, WA, EN, LW	ES	NR
Munoz and Navia (2011)	R, L	Industrial waste	GW, AC, EU, PO, TET, AET, HT, LU, RD, PM	CL	Ecoindicator 99, CML 2000
Møller et al. (2011)	I	MSW	AC, ET	DK	EDIP 1997
Nakatani et al. (2010)	R	PET bottles	GW, O	JP/CN	NA
Navia et al. (2006)	R	Contaminated volcalic soil	GW, ODP, ET, AP, EP, LU, O	ES	NS
Nishijima et al. (2012)	R	Plastic wastes	GW, O	JP	NS
Niskanen et al. (2009)	L	All waste	GW, AC, EU, PO, OD, HT, TET, AET, O	FI	EDIP 1997
Noon et al. (2011)	L, I, R	Computer monitors	GW, EN, O	US	NA
Ortiz et al. (2010)	L, R, I	Construction waste	GW, AC, EU, AET, TET, HT, EN, WA, O	ES	CML 2001

(continued)

Table 35.3 (continued)

References/sources (short; details to be provided in spreadsheet 'LCA studies list')	Technology assessed	Type of waste (e.g. only organic waste included in study)	Impact coverage (e.g. GW only) (*use glossary in first spreadsheet*)	Geographical coverage	LCIA method used
Passarini et al. (2012)	Int	Automotive shredder residue	HT, GW, ET, AC, EU, LU, RD	IT	Ecoindicator 99
Perugini et al. (2005)	R, I	Plastic packaging	GW, WA, O	IT	NS
Pires et al. (2011)	Int	MSW	GW, AC, RD, EU, HT, PO	PT	CML 2000
Pires and Martinho (2012)	Int	Waste lubricant oil (WLO)	GW, AD, AP, EP, HT, ET, ET, POP	PT	NA
Pisoni (2009)	Int	MSW	GW, OD, ET, AC, EU, ET, O	IT	Ecoindicator 99
Riber et al. (2008)	I	MSW	GW, AC, EU, PO, OD, HT, AET	DK	EDIP
Rieradevall et al. (1997)	L	Household waste	GW, AC, EU, HT, RD	ES	NS
Rigamonti et al. (2009a)	R, I	MSW	GW, HT, AC, PO	IT	CED, CML
Rigamonti et al. (2010)	Int	Packaging waste, municipal biowaste and MSW	CED, GW, HT, AC, PO	IT	CED, CML 2001
Rigamonti et al. (2009b)	Int	MSW	GW, HT, AC, PO	IT	CED, CML
Rivela et al. (2006)	R, I	Wood waste	GW, HT, RO, RI, ODP, ET, EU, AC, LU, O	ES	NS
Rives et al. (2010)	Int	MSW	GW, AC, EU, OD, PO, HT, TET, EN, O	ES	CML
Rodríguez-Iglesias et al. (2003)	Int	Residual MSW	GW, AC, EU, O	ES	IPCC, Ecoindicador 95
Ross and Evans (2003)	R	Plastics	GW, PO, RD	AU	NA

(continued)

Table 35.3 (continued)

References/sources (short; details to be provided in spreadsheet 'LCA studies list')	Technology assessed	Type of waste (e.g. only organic waste included in study)	Impact coverage (e.g. GW only) (*use glossary in first spreadsheet*)	Geographical coverage	LCIA method used
Saft (2007)	T	Paint waste	RD, GW, OD, PO, AET, TET, HT, AC, EU, O	NL	CML
Salhofer et al. (2007)	L, I, R	Refrigerators, PE films, EPS, paper	GW, OD, AC, EU, PO, HT, EN	AT	CML
Santos Vieira and Horvath (2008)	R	Concrete wastes	GW, EN	US	NS
Scharnhorst et al. (2006)	I, L, R	Antenna rack	TET, AET, OD, PO, NEN, GW, O	EU	IMPACT 2002+
Scipioni et al. (2009)	I	MSW	GW, OD, AC, PO, HT, ET, IR, PM, RD, LU	IT	Ecoindicator 99
Shen et al. (2010)	R	PET	GW, AC, EU, PO, HT, AET, TET, RD, EN	EU	CML, IPCC
Sonesson et al. (1997)	C, L	Organic	GW, AC, EU, PO,OD, EN	SE	CML
Sonesson et al. (2000)	Int	Organic	GW, AC, EU, PO, RD, EN, EI	SE	NA
Stichnote and Schuchardt (2010)	C, B	Empty fruit bunches (EFB) and palm oil mill effluent (POME)	RD, AC, EU, AET, GW, HT, OD, PO, TET	ID	CML 2001
Tabata et al. (2010)	Col	MSW	GW, AC, O	JP	IPCC, NEDO
Tan et al. (2006)	Int	MSW	GW, AC, ET, RD	SG	NA
Tarantini et al. (2009)	Int	MSW	GW, AP, EP, ET, HT, POP, EN	IT	NA
Toller et al. (2009)	R, L	Wood ash, MSWI bottom ash	NR	SE	NR
Tonini et al. (2012)	I, R, T, B	Residual MSW	GW, AC, EU, AET, HT	DK	EDIP 1997
Tukker (1999)	I, T	Hazardous waste	HT, GW, OD, PO, AC, EU, LW, EN, O	NL	CML
Tunesi (2011)	I, T	MSW	GW, AC, AD	UK	NA

(continued)

Table 35.3 (continued)

References/sources (short; details to be provided in spreadsheet 'LCA studies list')	Technology assessed	Type of waste (e.g. only organic waste included in study)	Impact coverage (e.g. GW only) (*use glossary in first spreadsheet*)	Geographical coverage	LCIA method used
Turconi et al. (2011)	I	Residual MSW	GW, AC, EU, PO, ET, HT	IT/DK	EDIP 1997
Van-Haaren et al. (2010)	C	Yard wastes	GW, AC, EU, O	US	Ecoindicator 99
Welsh Assembly Gvt (2003)	Int	C&D, industrial, commercial, municipal waste	GW, OD, AC, EU, HT, RD	UK	CML, IPCC, WMO
Wenisch et al. (2004)	I	Household waste	GW, AC, HT, RD, WA, LW	FR	NA
Werner et al. (2007)	R, I	Creosote-treated beech-wood railway sleeper	GW, AC, OD, PO, EU, HT, TET, AET, RD	CH	CML
Winkler et al. (2007)	L, I, R	Household waste	GW, HT, AC	DE	CML
Wollny et al. (2002)	I, R, T	Mixed plastic waste packaging	GW, AC, EN, EI	DE	NA
WRAP (2008)	L, I, R	Mixed waste plastic	GW, OD, AC, EU, PO, HT, RD, LW, EN	UK	CML 2002
Wäger et al. (2011)	R, I, L	WEEE	GW, AC, EU, PO, OD, ARD, ET, HT	CH	CML 2002, Ecoindicator 99
Yi et al. (2011)	L, R, I, B	MSW	Midpoint: GW, OD, RD, HT, AC, EU, PO, O Endpoint: human health, social assets, biodiversity and primary production	KR	NA
Zaman (2010)	I, L, T	MSW	GW, AP, EP, ABD, ODP, HT, ET, POP	SE	CML

(continued)

Table 35.3 (continued)

References/sources (short; details to be provided in spreadsheet 'LCA studies list')	Technology assessed	Type of waste (e.g. only organic waste included in study)	Impact coverage (e.g. GW only) (*use glossary in first spreadsheet*)	Geographical coverage	LCIA method used
Zhang et al. (1999)	I, R	PCB and plastics contained in dismantled non-nuclear components	GW, AC, OD, EU, PO, HT, AET, TET, EI, O	US	EDIP 1997
Zhao et al. (2009a)	Int	MSW	GW, AC, EU, PO, OD	CN	EDIP 1997
Zhao et al. (2012)	Int	MSW	GW, AC, EU, PO, OD	CN	EDIP 1997
Zhao et al. (2009b)	I	Medical waste	RD, GW, OD, HT, AET, TET, PO, AC, EU	NA	CML 1999
Özeler et al. (2006)	Int	MSW	GW, AP, EP, HTP, CED	TR	NS

The studies on sewage sludges are excluded from the table as the focus is on solid waste management systems

References

Bakas, I., Hauschild, M.Z., Astrup, T.F., Rosenbaum, R.K.: Preparing the ground for an operational handling of long-term emissions in LCA. Int. J. Life Cycle Assess. **20**, 1444–1455 (2015)

Bakas, I., Hauschild, M.Z., Rosenbaum, R.K., Astrup, T.F., McKone, T.E.: Assessing Terrestrial Ecotoxicity Impacts of Long-Term Metal Emissions from Landfills in LCA—An Operational Methodology. Submitted to IJLCA (2017)

Christensen, T.H., Gentil, E., Boldrin, A., Larsen, A., Weidema, B.: Hauschild M (2009) C balance, carbon dioxide emissions and global warming potentials in LCA-modelling of waste management systems. Waste Manag. Res. **27**, 707–715 (2009)

Christensen, T.H.: Solid Waste Technology & Management. Blackwell Publishing Ltd., Chichester (2011)

Clavreul, J., Baumeister, H., Christensen, T.H., Damgaard, A.: An environmental assessment system for environmental technologies. Environ. Model Softw. **60**(2014), 18–30 (2014)

Cleary, J.: Life cycle assessments of municipal solid waste management systems: a comparative analysis of selected peer-reviewed literature. Environ. Int. **35**, 1256–1266 (2009) (Coelho, A., de Brito, J., 2)

Doka, G.: Life Cycle Inventories of Waste Management Treatment Services. Ecoinvent report No. 13. Swiss Centre for Life Cycle Inventories, St. Gallen (2009)

Ekvall, T., Assefa, G., Bjorklund, A., Eriksson, O., Finnveden, G.: What life-cycle assessment does and does not do in assessments of waste management. Waste Manage. **27**, 989–996 (2007)

EC-JRC: European Commission—Joint Research Centre—Institute for Environment and Sustainability: International Reference Life Cycle Data System (ILCD) Handbook - General guide for Life Cycle Assessment—Detailed guidance. First edition March 2010. EUR 24708 EN. Publications Office of the European Union, Luxembourg (2010a)

EC-JRC: European Commission—Joint Research Centre—Institute for Environment and Sustainability: International Reference Life Cycle Data System (ILCD) Handbook—Framework and requirements for LCIA models and indicators. First edition March 2010. EUR 24586 EN. Publications Office of the European Union, Luxembourg (2010b)

EC-JRC: European Commission—Joint Research Centre—Institute for Environment and Sustainability: Supporting Environmentally Sound Decisions for Waste Management—A Technical Guide to Life Cycle Thinking (LCT) and Life Cycle Assessment (LCA) for Waste Experts and LCA Practitioners. EUR 24916 EN. Publications Office of the European Union, Luxembourg (2011)

European Environment Agency: Waste Prevention in Europe—The Status in 2013. Publications Office of the European Union, Luxembourg (2014)

Gala, A.B., Raugei, M., Fullana-i-Palmer, P.: Introducing a new method for calculating the environmental credits of end-of-life material recovery in attributional LCA. Int. J. Life Cycle Assess. **20**, 645–654 (2015). doi:10.1007/s11367-015-0861-3

Gentil, E., Damgaard, A., Hauschild, M., Finnveden, G., Barlaz, M., Thorneloe, S., Kaplan, P.O., Eriksson, O., Matsui, Y., Ii, R., Thomas, B., Jones, R., Christensen, T.H.: Models for waste life cycle assessment: review of technical assumptions. Waste Manag. **30**, 2636–2648 (2010)

Goedkoop, M., Heijungs, R., Huijbregts, M., De Schryver, A., Struijs, J., van Zelm, R.: ReCiPe 2008; a life cycle impact assessment method which comprises harmonised category indicators at the midpoint and the endpoint level—report I: characterisation. Minist. van VROM, Den Haag (2009)

Goedkoop, M., Spriensma, R.: The Eco-indicator 99; a damage oriented method for Life Cycle Impact Assessment—Methodology Report. PRé Consult. Amersfoort, Netherlands (2001)

Hauschild, M., Olsen, S.I., Hansen, E., Schmidt, A.: Gone…but not away—addressing the problem of long-term impacts from landfills in LCA. Int. J. Life Cycle Assess. **13**, 547–554 (2008). doi:10.1007/s11367-008-0039-3

Hischier, R., Weidema, B., Althaus, H.-J., Bauer, C., Doka, G., Dones, R., Frischknecht, R., Hellweg, S., Humbert, S., Jungbluth, N., Köllner, T., Loerincik, Y., Margni, M., Nemecek, T.: Implementation of Life Cycle Impact Assessment Methods; Ecoinvent Report No. 3, v2.2. Swiss Cent. Life Cycle Invent, Dübend (2010)

Hoornweg, D., Bhada-Tata, P.: What a Waste—A Global Review of Solid Waste Management. World Bank, Washington DC (2012)

ISO: Environmental Management—Life Cycle Assessment—Requirements and Guidelines (ISO 14044). ISO, The International Organization for Standardization, Geneva (2006)

Karak, T., Bhagat, R.M., Bhattacharyya, P.: Municipal solid waste generation, composition, and management: the World Scenario. Crit. Rev. Environ. Sci. Technol. **42**, 1509–1630 (2012)

Laurent, A., Bakas, I., Clavreul, J., Bernstad, A., Niero, M., Gentil, E., Hauschild, M.Z., Christensen, T.H.: Review of LCA studies of solid waste management systems—part I: lessons learned and perspectives. Waste Manag. **34**(2014), 573–588 (2014a)

Laurent, A., Clavreul, J., Bernstad, A., Bakas, I., Niero, M., Gentil, E., Christensen, T.H., Hauschild, M.Z.: Review of LCA studies of solid waste management systems—part II: methodological guidance for a better practice. Waste Manag. **34**(2014), 589–606 (2014b)

Niero, M., Negrelli, A.J., Boas, S.H., Olsen, S.I., Birkved, M.: Closing the loop for aluminium cans: life cycle assessment of progression in Cradle-to-Cradle certification levels. J. Clean. Prod. **126**, 352–362 (2016)

Schrijvers, D.L., Loubet, P., Sonnemann, G.: Developing a systematic framework for consistent allocation in LCA. Int. J. Life Cycle Assess. (2016). doi:10.1007/s11367-016-1063-3

Author Biographies

Ioannis Bakas LCA expert focusing on environmental impacts from solid waste management systems. Developed method for accounting long-term emissions from landfilling. Interested in LCA modelling of soil emissions, integrated waste systems' evaluation and policy assessments.

Alexis Laurent working with LCA since 2010 with a strong LCIA focus, particularly on normalisation aspects. Main LCA interests include development of LCIA methods, LCA applications and footprinting of large-scale systems for policy-making (e.g. nations, sectors), and LCA applied to various technology domains, including energy systems.

Julie Clavreul LCA expert since the early 2010s. Her background is with a focus on the modelling and life cycle assessment of waste management systems and the analysis of uncertainties. Interest in the EU environmental footprint methodology.

Anna Bernstad Saraiva has been working with environmental engineering since 2008. Has done Post-doctoral studies on LCA of biorefinery processes and biomaterials. Research focus is on solid waste management, co-treatment of household waste flows, sustainable materials and chemicals and sustainable urban development.

Monia Niero background as chemical engineer, LCA expert since late 2000s. Involved in several LCA industrial projects. Main areas of expertise are LCI modelling and LCM in waste management, packaging and agri-food sectors. Interested in developing decision support tools for circular economy implementation at product and organization levels.

Emmanuel Gentil has been working in life cycle thinking, related to waste management systems since 1999, where he was involved in the development of a life cycle assessment software.

Research focus and consultancy services on life cycle assessment for waste management policy for the European Commission. Teaching LCA for American undergraduates studying abroad in Denmark.

Michael Z. Hauschild involved in development of LCIA methodology since the early 1990s. Has led several SETAC and UNEP/SETAC working groups and participated in the development of the ISO standards and the ILCD methodological guidelines. Main LCA interests are chemical impacts, spatial differentiation and science-based boundaries in LCIA.

Chapter 36
LCA of Soil and Groundwater Remediation

Gitte Lemming Søndergaard and Mikołaj Owsianiak

Abstract Today, there is increasing interest in applying LCA to support decision-makers in contaminated site management. In this chapter, we introduce remediation technologies and associated environmental impacts, present an overview of literature findings on LCA applied to remediation technologies and present methodological issues to consider when conducting LCAs within the area. Within the field of contaminated site remediation, a terminology distinguishing three types of environmental impacts: primary, secondary and tertiary, is often applied. Primary impacts are the site-related impacts due to the contamination in the ground, secondary impacts are the impacts related to clean-up of the site, and tertiary impacts are the impacts associated with the future use of the site. The major methodological issues to consider when conducting LCA are: (i) defining a functional unit that considers time frame and efficiency of remediation, which are important for assessment or primary impacts; (ii) robust assessment of primary impacts using site-specific fate and exposure models; (iii) weighting of primary and secondary (or tertiary) impacts to evaluate trade-offs between life cycle impacts from remediation and reduced pressure locally; and (iv) comparison with a no action scenario to determine whether there is a net environmental benefit from remediation. Overall, LCA is an important tool for the assessment of the secondary environmental impacts of remediation, and occasionally it has also been used to assess primary and tertiary impacts. In order to obtain robust decisions for the management of contaminated sites, the combination of LCA with other tools is necessary, including multi-criteria decision analysis tools, site-specific fate and exposure models and consideration of stakeholders' views.

G.L. Søndergaard (✉)
Department of Environmental Engineering, Technical University of Denmark, 2800 Kongens Lyngby, Denmark
e-mail: gile@env.dtu.dk

M. Owsianiak
Division for Quantitative Sustainability Assessment, Department of Management Engineering, Technical University of Denmark, 2800 Kongens Lyngby, Denmark

M.Z. Hauschild et al. (eds.), *Life Cycle Assessment*,
DOI 10.1007/978-3-319-56475-3_36

36.1 Introduction

Many sites around the world require clean-up due to the risk they pose to humans and ecosystems. It was estimated that in the US, a total of 300,000 sites will need clean-up during the next 35 years (US EPA 2004). In the EU member states, out of nearly 3 million sites with potentially polluting activities approximately 250,000 sites require clean-up (EEA 2007). Remediation technologies have been and are continuously being developed, but the main focus has thus far been put on technological aspects, such as design and optimisation of a technology for the contaminant of interest. Environmental impacts associated with remediation of contaminated sites activities are rarely evaluated, and only recently life cycle based approaches were considered as tools to support decision on the choice of a remediation technology among available alternatives.

In this chapter, we introduce remediation technologies and associated environmental impacts, present an overview of literature findings on LCA applied to remediation technologies and present methodological issues to consider when conducting LCAs. We focus mainly on environmental life cycle impacts from remediation of contaminated soil and groundwater, but due to the similarity in the nature of the problems, we also include studies concerning remediation of contaminated sediments, sludge and wetlands.

36.1.1 Remediation Technologies

Remediation technologies can be divided into in situ and ex situ technologies respectively, depending on where the remediation takes place (Collaran 1997; US EPA 2000). In situ remediation technologies target the contamination in the subsurface, i.e. without extracting or excavating contaminated soil or groundwater. Examples of technologies that can be used in situ are bioremediation, chemical degradation (oxidation or reduction), phytoremediation, thermally enhanced remediation and permeable reactive barriers. Some in situ remediation methods (thermal remediation and soil vapour extraction) require a treatment system for extracted vapours and are therefore not strict in situ methods. Ex situ remediation technologies involve the excavation of contaminated soil or extraction of contaminated groundwater followed by a treatment either on-site, e.g. in biopiles or an on-site groundwater treatment unit, or the soil can be transported to an off-site treatment facility. Ex situ remediation technologies cover biological, chemical and thermal remediation methods. Moreover, containment methods exist, which are remediation methods that seek to establish a barrier to immobilise contaminants and cut off the exposure routes instead of removing the contamination source. Surface capping and barrier installations, e.g. sheet piling, as well as placement in secured landfills are examples of containment methods. While ex situ remediation technologies or containment methods were traditionally applied at contaminated sites, in situ remediation technologies were developed as alternatives that required less intrusion at the site and less disturbance of site residents and neighbours. However, in situ remediation causes a lower level of nuisance during the remediation it may have a longer timeframe and a higher uncertainty related to reaching the remedial clean-up target (Caliman et al. 2011). Out of all ca. 2400 remediation projects applied in the US Superfund programme in years 1982–2008 (US EPA 2010), most (nearly 800) technologies was applied to ex situ groundwater treatment (pump-and-treat). There were nearly 600 projects on ex situ source control (dominated by solidification/stabilisation, with smaller contribution of incineration, thermal desorption and bioremediation). In situ source control technologies (ca. 540 projects) were dominated by soil vapour extraction, followed by a smaller number of projects that used bioremediation. Bioremediation, however, was the most frequently used methods for in situ groundwater treatment (in total 350 projects) followed by air sparging and chemical treatment.

36.1.2 Environmental Impacts from Remediation Technologies

Within the field of contaminated site remediation, a terminology distinguishing three types of environmental impacts, primary, secondary and tertiary, is often applied (Lesage et al. 2007a; Sparrevik et al. 2011). The primary impacts are the environmental impacts caused by the on-site contamination and cover human toxicity and ecotoxicity impacts due to the potential exposure via soil, groundwater and air. The primary impacts will most often be local in nature and may be difficult to assess with existing generic LCIA methodologies because these impacts are strictly site-specific (Lesage et al. 2007a; Lemming et al. 2010c). Secondary impacts are the environmental impacts associated with the intervention at the site, i.e. the remediation technology. The remediation of the site may for example include heavy machinery work on-site such as excavation and drilling, the use of materials for installations, e.g. polymers, steel and concrete as well as activated carbon for water or air treatment. In addition, electricity is often applied at the site for pumping, heating or injection. Moreover, transportation of soil, equipment and personnel can be a significant activity. All of these remediation activities cause environmental impacts both at the local, regional and global scale due to the emissions taking place in many geographical locations. Lesage et al. (2007a, b) took a consequential LCA approach (see Sect. 8.5.3) and introduced the term 'tertiary impacts' to account for the environmental impacts related to the fate of a brownfield site after the treatment. The future use of a site will depend on the state in which it is left after the remediation. If the exposure risk is not sufficiently reduced, it may not be possible later to use the site for residential/commercial purposes. Consequently, suburban greenfield may need to be developed depending on the local situation and demand for new land. The environmental impacts related to the development of greenfield is covered in the tertiary impacts in the study by Lesage et al. (2007a, b).

36.2 Literature Review

Table 36.2 in the Appendix gives an overview of existing LCA studies regarding remediation of contaminated sites, including the covered technologies and contaminants, types of impacts assessed (primary, secondary, tertiary), definition of the functional unit, time boundary for remediation and a main result or conclusion. Among the main findings from earlier reviews on this topic by Lemming et al. (2010a) and Morais and Delerue-Matos (2010), covering all studies published until 2009, were that the majority of the LCA studies dealt with ex situ remediation methods and soil contaminants (mainly metals and hydrocarbons), while only few studies dealt with in situ remediation of groundwater and common contaminants (such as chlorinated ethenes). Furthermore, their reviews identified a need for more site-specific assessment of primary impacts as the existing LCIA models do not take the groundwater compartment into account and will only provide a crude assessment of primary impacts due to the generic nature of the models. Here, we present recent trends in application of LCA to remediation technologies, corroborating the two earlier reviews and an earlier viewpoint article on this topic published by Owsianiak et al. (2013), by including more studies, followed by recommendation for better and consistent use of LCA-based methods in the future.

36.2.1 Contaminants and Technologies Assessed in LCA Studies

The majority of studies on LCA applied to remediation technologies for contaminated sites deal with soil remediation, with excavation combined with treatment and/or disposal the most investigated technology, followed by bioremediation. A significant number of studies deal with groundwater remediation, where mainly pump-and-treat or permeable reactive barriers (PRB) technologies are in scope. Interestingly, two studies that focus on remediation of contaminated sediment and wetland have recently been published, whereas one study compared bioremediation methods for contaminated sludge at a lab-scale. Most studies cover technologies that are well established and have already found field-scale applications. An exception is the recent paper by Lubrecht (2012), who analysed the performance of horizontal directional drilling (HDD), a relatively new drilling technique that can be used as an alternative to traditional vertical wells used for groundwater treatment.

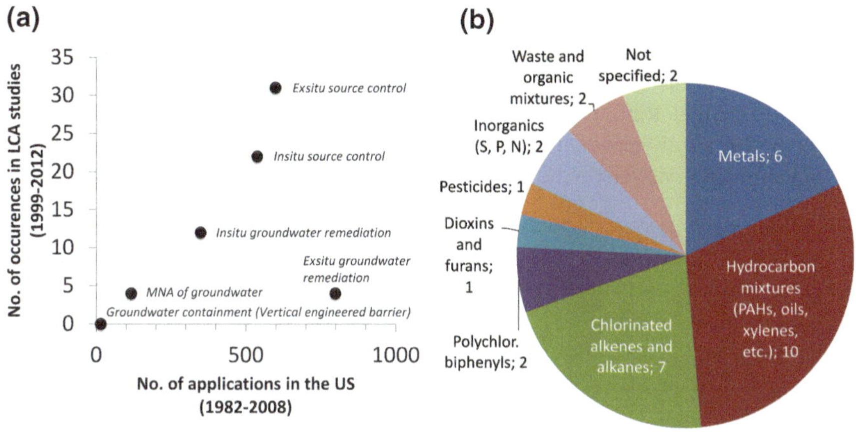

Fig. 36.1 Correlation between number of applications of remediation technologies and number of remediation technologies assessed in reviewed LCA studies (**a**); and percentage contribution of types of contaminants assessed in LCA studies in years 1999–2012 (**b**)

On the other hand, some of the relatively widely applied technologies have only been assessed a few times in a life cycle perspective, if at all (Fig. 36.1a). In addition, most studies focus on recognised contaminants, such as metals, polycyclic aromatic hydrocarbons (PAHs), hydrocarbon mixtures or polychlorinated biphenyls. However, there is an increase in number of studies dealing with chlorinated hydrocarbons. Pesticides or nutrients were rarely a scope of the study, while no studies deal with emerging threats such as pharmaceuticals or nanomaterials. These trends may reflect limited availability of data needed to carry out an LCA for emerging technologies and emerging contaminants, both on the life cycle inventory and impact assessment sides. This is supported by the fact that heavy metals, petroleum hydrocarbons and PAHs are the main contaminants found in soil, whereas petroleum hydrocarbons and chlorinated solvents are the main groundwater contaminants (EEA 2007), and these are also the most frequently studied compounds in LCA literature (Fig. 36.1b). The lag between the occurrence of pollutants in the environment and their occurrence in case studies indicates that achieving environmental sustainability goals for remediation of emerging contaminants and technologies can be challenged. Even if inventories are made for the relevant remediation processes, characterisation factors (CFs) for some pollutants may not be available (see Chap. 10).

36.2.2 Impact Assessment Methods Employed in LCA Studies

Out of 32 studies, 15 evaluated primary impacts (in addition to secondary impacts), while only two studies (Lesage et al. 2007a, b; Hou et al. 2014b) evaluated all three types of impacts. While earlier studies often covered both primary and secondary impacts, the recent ones, with few exceptions, focus solely on secondary impacts. This can be considered as a drawback that may bias a comparison, as it has been demonstrated that primary impacts can contribute substantially to human health impact categories (Lemming et al. 2010c, 2012) and their inclusion in the assessment can change the comparative ranking among alternatives (Sparrevik et al. 2011).

Comparing to the earlier reviews by Lemming et al. (2010a) and Morais and Delerue-Matos (2010), a few observations can be made with respect to methodology employed in the assessment. While papers published in the 90s and in the following decade were expected to report the use of earlier impact assessment methods, such as EDIP97, TRACI or IMPACT2002+, these methodologies are still being widely employed today. More recent methodologies such as ReCiPe or USEtox are occasionally used. Overall, there is an increasing number of studies which use LCA-based assessment tools, or link full LCAs with other assessment tools to support decisions. For example, Evans and Wilkie et al. (2010) assessed the performance of nutrient remediation, focusing on balance between bioenergy produced and consumed, and fossil fuel energy expenditures avoided and consumed by the remediation process (the so called net energy balance ratio, NEBR). They showed net energy benefits when biomass was used for biogas production or compost production. In addition, both utilisation pathways showed monetary benefits, as calculated using a benefit–cost ratio (BCR) life cycle-based analysis. Cappuyns et al. (2011) used LCA-based risk reduction, environmental merit and costs (REC) and demonstrated that in situ thermal treatment of soil has lower global environmental impacts than soil excavation and off-site treatment. They also concluded that non-LCA-based assessment tools such as best available technology not entailing excessive costs (BATNEEC) analysis could be used as a primary screening among remediation alternatives, followed by a full LCA for selected technologies. Inoue and Katayama (2011) compared the performance of life cycle costing (LCC) (see Chap. 15) and economic input–output LCA (see Chap. 14) with a rescue number for soil (RN_{SOIL}) representing the risk reduction obtained using each remediation technology, and demonstrated that different rankings appear between the remediation options depending on whether the ranking was based on the risk–cost, the risk–energy consumption or the risk–CO_2 emission scale. Sparrevik et al. (2012) have developed a method that integrates risk assessment (RA), LCA, and multi-criteria decision analysis (MCDA) into a common framework for decision support, and demonstrated its applicability to selection of the best sediment management alternative. Hou et al. (2014a) showed how life cycle impacts of remediation depend on site conditions and proposed a framework to select the most environmentally sustainable technology under various site conditions. A recent study (Beames et al. 2015) included land use in their impact assessment. In addition to land use related to production, they added the land use occupation as a consequence of remediation reflecting the duration of the different remediation techniques. Hybrid LCA (see Chap. 14) was used in one of the studies (Hou et al. 2014b). The hybrid LCA was built on the UK 123 sector national input–output table, which includes environmental and socio-economic data. The study claims that the use of hybrid LCA reduced the truncation errors and gave more complete system boundary than a process-based LCA.

In some cases the assessment is limited to energy demand and/or CO_2 emissions (Witters et al. 2012). For example, Lubrecht (2012), employed the SiteWise method (based on summing up GHG, NO_x, SO_x and PM_{10} emissions, arriving at four types of impact scores) to compare two groundwater remediation alternatives. The comparison, showed only modest reduction in included substances, which was sufficient for the author to claim the technology was sustainable, even though no definition of the functional unit was provided (the comparison was based on target depth of the plume only), system boundaries were not clearly defined, and only few impact categories were considered. We note that analyses of this kind have little in common with LCA, at least according to requirements presented in ISO standards or other authoritative guidelines (ISO 2006a, b; ILCD 2010), and should not be used to draw any conclusions about better or worse environmental performance of one remediation technology over other.

36.2.3 Main Drivers of Environmental Impacts

Our results tend to confirm earlier observations (Lemming et al. 2010a) that the main processes contributing to secondary impacts are on-site electricity/energy use for thermal remediation, as well as steel use for on-site installations. For ex situ remediation involving excavation and transport of the contaminated soil, the transportation processes also contribute significantly to secondary impacts, especially in cases where a remotely located site is remediated (Sanscartier et al. 2010).

Beames et al. (2015) focused on land use impacts in their study of in situ versus ex situ methods. The study demonstrated that there is a trade-off between energy use and land use in remediation. Energy-intensive methods such as excavation will use a high amount of energy, but will occupy the contaminated land for a shorter period, resulting in lower impacts on land use. On the other hand, less energy-intensive methods will have longer timeframes and therefore higher impact on land use.

Agents used in chemical oxidation typically generate high impacts due to the large requirements for production and transportation of the chemical, which is applied in very large quantities (Cadotte et al. 2007; Lemming et al. 2012). On the other hand, bioremediation technologies are the most favourable among remediation alternatives. However, Lemming et al. (2010c, 2012) demonstrated that enhanced bioremediation of chlorinated ethenes can have very long duration (several decades) in low-permeability media and that the higher toxicity of chlorinated intermediates can potentially lead to a larger primary toxic impact to the groundwater aquifer than if no remediation was initiated (Lemming et al. 2012). These findings highlight the need for employing site-specific fate and exposure models for assessment of primary impacts.

Phytoremediation of soil or wetlands appears to be an attractive remediation alternative, particularly when combined with energy recovery combustion of the biomass (Evans and Wilkie 2010; Witters et al. 2012; Vigil et al. 2015). In these cases, the environmental benefits were mainly associated with reduced CO_2 emissions. Similarly, the use of biomass-derived activated carbon for sediment capping was shown to perform better than other capping materials especially due to the CO_2 sequestration effect when this material is added to the seafloor (Sparrevik et al. 2011). Some of the recent studies that compared bioremediation alternatives focus on use of different substrates or electron donors. For example, lab-scale analysis has shown that methanol was the best alternative among electron donors in anaerobic dechlorination of pentachloroaniline (PCA), and that the environmental burden was reduced when the concentration of electron donors were reduced (Hong and Li 2012). Extrapolation of these results to field-scale applications has not been addressed by the authors, but Lemming et al. (2010c) have also noted that substrate demand for enhanced reductive dechlorination was an important contributor to global warming.

36.3 Specific Methodological Issues

Our earlier analysis of studies on LCA applied to remediation technologies (including all the studies published until 2012 presented in Appendix, Table 36.2), showed an increasing frequency of examples with serious methodological problems compared to requirements in ISO standards or authoritative guidelines (ISO 2006a, b; JRC 2010; Owsianiak et al. 2013). Here, we present methodological issues that need to be considered when conducting LCA-based comparison of remediation alternatives.

36.3.1 Issues in Goal and Scope Definition

In comparative assessments, remediation alternatives must be compared based on a function they provide, that is clean-up of a contaminated site. Even for the same site, there are many means by which the function can be fulfilled, depending on the type of the technology. It is often the case that the two performance parameters, i.e. remediation efficiency and remediation time frame, differ considerably between technologies. For example, monitored natural attenuation often takes more time and sometimes does not allow reaching clean-up levels that can be obtained using faster, invasive methods. Notably, these two parameters are often used by remediation practitioners as arguments for choosing a technology to clean-up a site (if there are no economic constraints). LCA practitioners however, are met with challenges because the functional unit should consider both the efficiency and time horizon aspects. The inclusion of performance parameters in the definition of the functional unit is important because they directly decide on the magnitude of primary impacts. We, thus, recommend defining a functional unit that only includes remediation efficiency, and subsequently determine time frames for each compared option. Alternatively, a functional unit can only include time frame for remediation and then option-specific efficiencies can be determined. Only in very rare cases when both performance parameters are the same, the function is the same and the alternatives can be compared without specifying efficiency and time horizon in the functional unit definition. Appendix Table 36.2 shows that there is little consideration of time and efficiency of remediation when defining the functional unit. In many studies, the functional unit (clearly defined in 26 out of all 32 studies), is based on the treatment of a certain volume or mass, without considering the performance aspects. This is an incorrect definition and a source of potential bias. Its magnitude will depend on the contribution of primary impacts to total impacts in the remediation life cycle. To qualify the timeframe prediction for long-term remediation scenarios (and for better assessment of primary impacts, as will be discussed later) site-specific, remediation performance models should be employed. If these models are not available and little is known about performance of a technology from a practical side, remediation alternatives can still be compared if they all provide an acceptable minimum level of remediation efficiency; for example: (i) reduction of contaminant mass to 99% of its initial value, or (ii) reduction of contaminant concentration to the level posing no risk. Note however, that even a small difference in remediation efficiency between two techniques that both satisfy the no risk level criterion can cause a difference in primary impacts that may influence the comparison. This aspect is particularly important when the difference in remediation time frames between two alternatives is large. If such is the case, uncertainty scenarios about remediation efficiency and/or time frame should be considered when conducting an LCA and interpreting results.

36.3.2 Issues in Life Cycle Inventory

Life Cycle Inventory (LCI) compiles all relevant environmental exchanges to and from the assessed remediation system during its lifecycle, i.e. energy and material inputs and outputs (Chap. 9). The majority of the reviewed studies made use of commercially available LCI databases, such as ecoinvent (Frischknecht et al. 2007) and the US LCI databases (NREL et al. 2004) to model inventories, depending on the location of the project. The use of these generic databases was sometimes combined with additional data collection for remediation-specific processes not included in the databases, e.g. production of activated carbon, chemical analyses and production of specific remedial soil amendments. These technology-specific processes must be included in the assessment if they are shown or expected to be important contributors to secondary impacts.

According to the ILCD requirements (JRC 2010), attributional LCA is recommended for micro-level decision support (decision context-situation A) related to specific products and for accounting purposes, i.e. descriptive documentation of a system, assuming average technological conditions (Sect. 8.5). Situation A assumes that remediation activity will not result in structural changes in the analysed system. Consequential LCA is recommended by ILCD for decision support on the meso or macro scale related to the strategic level, e.g. raw material strategies, technology scenarios, and policy options (decision context-situation B) (see Sect. 7.5). Remediation projects fall somewhere between the micro scale and the meso scale, depending on the size of the project. Both attributional and consequential approaches can, therefore, be argued for. Our review shows that all studies, except Lesage et al. (2007a, b) who conducted a consequential LCA, take an attributional approach to inventory modelling. Lesage et al. demonstrated how the consequential approach and inclusion of tertiary impacts influences LCA results; a brownfield rehabilitation scenario was favoured over the risk minimisation scenario due to the need for the development of suburban residential sites if the brownfield site was not remediated sufficiently for rehabilitation. The main driver of tertiary impacts from the development of suburban sites was the increased person car transport from the suburban areas during the 40-year timeframe of the analysis.

36.3.3 Issues in Life Cycle Impact Assessment

In the Life Cycle Impact Assessment (LCIA) phase the emissions and resource consumption collected in the LCI are translated to environmental impacts using LCIA models (Chap. 10). Below, methodological issues regarding assessment of primary, secondary and tertiary impacts are discussed. Table 36.1 shows the applied impact assessment methods.

While a large part of the earlier studies (before 2010) included the assessment of both primary and secondary impacts of site remediation, the newer studies generally focus on the secondary impacts with a few exceptions (Lemming et al. 2010b, c, 2012; Sparrevik et al. 2011; Hou et al. 2014b). This is a drawback as primary impacts can contribute substantially to human health impact categories and their inclusion in the assessment can change the comparative ranking among remediation alternatives. The reason for not including primary impacts may be that their quantification is not straight forward. Assessing the primary impacts most often requires site-specific fate and exposure models, while existing toxicity models, which are well suited to generic LCIA, employ generalised box models and exposure scenarios that in many cases are not representative of the specific conditions at the contaminated site. Furthermore, deep soil layers and the groundwater compartment are usually disregarded in the generic models, which make it questionable to use them for sites where groundwater contamination is the main concern. Earlier studies used the generic LCIA toxicity models for the evaluation of primary impacts, whereas newer studies (Lemming et al. 2010c, 2012; Sparrevik

Table 36.1 LCIA methodologies used in LCA studies applied to remediation technologies

Reference	Life cycle impact assessment		Comment
	Midpoint	Endpoint	
Diamond et al. (1999)	X		
Page et al. (1999)	x		
Volkwein et al. (1999)	X		
ScanRail Consult et al. (2000)	x (EDIP97)		
Vignes (2001)			LCIA based on pollution factors for each emission relative to the limit value for each emission
Ribbenhed et al. (2002)	x (USES-LCA + other)		
Blanc et al. (2004)			Intentionally terminated at inventory level
Godin et al. (2004)	x (EDIP97)		
Toffoletto et al. (2005)	x (EDIP97)		
Bayer and Finkel (2006)	X		
Cadotte et al. (2007)	x (TRACI)		
Lesage et al. (2007a, b)		x (IMPACT2002+)	
Higgins and Olson (2009)	x (TRACI)		
Lemming et al. (2010b)	x (EDIP97)		
Lemming et al. (2010c)	x (EDIP2003 + USEtox)		
Evans and Wilkie (2010)			Study uses only on net energy balance ratio (NEBR) and benefit–cost ratio (BCR)
Sanscartier et al. (2010)		x (IMPACT2002+)	
Cappuyns et al. (2011)			LCA-based REC (risk reduction, environmental merit and costs)

(continued)

Table 36.1 (continued)

Reference	Life cycle impact assessment		Comment
	Midpoint	Endpoint	
Hu et al. (2010)	x (IMPACT 2002 + + IPCC + Ecoindicator 99)	x (IMPACT 2002 +)	
Inoue and Katayama (2011)			Energy consumption and CO_2 emissions determined applying economic input–output LCA
Mak and Lo (2011)	x (TRACI)		
Sparrevik et al. (2011)		x (ReCiPe)	
Suer and Andersson-Sköld (2011)	x (EPD)	x (ReCiPe)	
Busset et al. (2012)	x (CML2001)		
Hong and Li (2012)	x (IMPACT 2002+)	x (IMPACT 2002 +)	
Lubrecht (2012)			SiteWise method—emission summed up and expressed as mass (GHG, NO_x, SO_x, PM_{10})
Witters et al. (2012)			"Analysis is limited to the Global Warming Potential (GWP) of CO_2"
Lemming et al. (2012)	X (EDIP2003 + USEtox)		
Hou et al. (2014a)	x (TRACI)		
Hou et al. (2014b)		x (ReCiPe)	
Beames et al. (2015)	x (ReCiPe)		Includes modification to the land use impact
Vigil et al. (2015)		x (ReCiPe)	

et al. 2011) included site-specific fate and exposure models. As none of the existing LCIA toxicity models takes formation of metabolites into account, these should also be addressed using site-specific models. Alternatively, metabolites can be included in life cycle inventory and characterised in the LCIA phase. Again, attention must be paid to representativeness of the (generic) CF to site conditions which are often far from the average situation conditions, for which CF are developed. Overall, we recommend using site-specific models to quantify primary impacts.

Of the reviewed LCA studies of site remediation published before 2010, all except Lesage et al. (2007b) applied midpoint characterisation models (Chap. 10). Within the recent 5-year period, endpoint characterisation has gained a larger application and was used in seven studies published after 2010. Midpoint characterisation models are probably sufficient to assess environmental impacts, given that no endpoint characterisation approaches are mature enough to be recommended by ILCD (Hauschild et al. 2013). A number of studies applied simplified methodologies, focusing either solely on the global warming impact (Inoue and Katayama 2011; Witters et al. 2012) and/or the energy use (Inoue and Katayama 2011; Evans and Wilkie 2010) or using one of the simplified screening tools REC (Cappuyns et al. 2011) or Sitewise (Lubrecht 2012). The REC tool (Beinat et al 1997) focuses mainly on energy-related impacts combined with some local impacts (water use, soil quality), but exclude material manufacturing from the inventory. The SiteWise tool (NAVFAC et al. 2011) only employs an actual impact assessment for the global warming impact, whereas emissions of NO_x, SO_x and PM_{10} are not characterised, but only presented as the summed mass of each emission type. A major drawback of these simplified screening tools is the incompleteness of the assessment and the limited impact focus. Such assessments may bias the comparison of remediation alternatives or result in burden-shifting due to the exclusion of potentially relevant impact categories and incomplete characterisation of emissions. In this context, it is important to note here that LCA applied to remediation technologies are subjected to similar constraints as LCAs applied to other systems; in all cases normative choices with respect to the applied LCIA methodology can influence the results and their interpretation (Dreyer et al. 2003). Care must be taken to base decisions on results without quantifying the associated uncertainties, particularly if some, potentially important, impact categories are not assessed. To complement site-specific assessment of primary impacts, we thus recommend including all life cycle impact categories in assessment of secondary impacts. In addition, we also advocate including a no action scenario as a reference point to which any remediation option should be compared to illustrate if there is net-benefit from remediation.

Assessment of tertiary impacts was done in the study be Lesage et al. (2007a, b) as a part of the consequential LCA taking the future use of the site into account and also done in the study by Hou et al. (2014b). The assessment of tertiary impacts also is not straight forward as it requires assumptions to be made for the unknown future. The assessment needs to answer questions such as "if this site is not cleaned up, which land (if any) will then be used instead?" The answer to this question may depend on many factors such as the local land market and demand as well as political decisions on urban development. One solution for dealing with this uncertainty is to include different future scenarios for use of the site and the possible new sites developed as a consequence of not remediating the contaminated site. In the study by Hou et al. (2014b), the assessment of the tertiary impact shifted the overall net-benefit of the remediation project to being positive. Hou et al. (2014b) therefore concludes that it is important to include all three types of impacts in order to assess the overall benefit of a remedial measure.

36.4 Conclusions

Today, there is increasing interest in applying LCA to support decision-makers in contaminated site management (Holland et al. 2009, 2011; Holland 2011). Indeed, the extent of soil and groundwater pollution globally suggests that even if a fraction of all sites is cleaned up in the future, a selection of less polluting technologies can potentially lead to reduction of environmental impacts from remediation. The most important question in this context are: (i) Is there a net-benefit from remediation? (ii) Do primary, secondary and tertiary impacts have equal weight? The limited number of studies without serious methodological problems (see Owsianiak et al. 2013 for statistics) and only occasional comparisons with no action scenario suggest that the answer to the first question cannot yet be given with certainty. The answer to the second question will depend on stakeholder views and perspectives, and may vary depending on the site and its future application (e.g. providing access to clean water may be more important than providing clean land for housing). In conclusion, LCA is an important tool for the assessment of the secondary environmental impacts of remediation. However, in order to obtain robust decisions for the management of contaminated sites, the combination of LCA with other tools may be necessary (e.g. multi-criteria decision analysis tools, use of site-specific fate and exposure models and consideration of stakeholder views).

Appendix

See Table 36.2.

Table 36.2 Overview of LCA case studies of contaminated site remediation and the impacts included

Reference	Technologies included		Contaminants	Impacts included			Functional unit	Time boundary for LCI	Result or conclusion
				Pri	Sec	Ter			
1. Diamond et al.	Ex situ	Excavation and disposal (S) Soil washing (S)	Not specified	X	X		Production of an equivalent amount of treated soil and groundwater (mass/volume)	25 years	"The framework facilitates a methodic investigation of activities associated with site remediation, and guides the analysis of potential environmental, human health, and resource depletion impacts. The framework allows for the consideration of a wide range of potential impacts by expanding consideration beyond a contaminated site itself, and the temporal boundary of on-site activities"
	In situ	No action Soil vapour extraction (S) In situ bioremediation (G)							
	Containment	Encapsulation (S)							
2. Page et al. (1999)	Ex situ	Excavation and disposal (S)	Lead	X	X		Production of an equivalent amount of treated soil and groundwater (mass/volume)	25 years	"The effects of the excavation and disposal remediation option extend beyond the contaminated site itself, and only become evident when analyzed from a life-cycle perspective"
3. Volkwein et al. (1999)	Ex situ	Excavation and on-site secured disposal (S) Excavation and decontamination (S)	PAHs, mineral oil, chromium		X		The ensemble of activities to achieve a certain risk level	Not specified	"The tool supplements the environmental information gained by a risk assessment with information about secondary impacts"
	Containment	Surface sealing with asphalt (S)							

(continued)

Table 36.2 (continued)

Reference	Technologies included		Contaminants	Impacts included			Functional unit	Time boundary for LCI	Result or conclusion
				Pri	Sec	Ter			
4. ScanRail Consult et al. (2000)	Ex situ	Excavation and external biological treatment (S)	Chlorinated ethenes, hydrocarbons	X	X		Not specified	Not specified	"The environmental costs of excavation, biosparging, and bioventing were generally of the same order of magnitude, whereas the environmental costs of the reactive wall were higher and the environmental costs of the biological wall lower compared to the other techniques"
	In situ	Biosparging (G) Bioventilation (S) Permeable reactive barrier (G) Biological barrier (G)							
5. Vignes (2001)	Ex situ	Pump-and-treat (on-site vacuum steam stripping) (G) Pump-and-treat (on-site activated carbon treatment) (G) Excavation and thermal treatment (S)	1,2,3-Trichloropropane (TCP) and total xylenes (G) Mix of organic contaminants (S)	X	x		Not specified	Not specified	"Doing nothing and excavation with thermal treatment are by far the worst options both locally and globally"
	In situ	No action Aerobic bioremediation (S) Anaerobic bioremediation (S)							
	Containment	Cap and contain (S)							
6. Ribbenhed et al. (2002)	Ex situ	Thermal treatment (S) Bioslurry (S) Soil washing (S)	PAHs, mercury, cadmium	x	x		1000 kg of dry sediment into treatment	Not specified	"The largest environmental impact is caused by the energy and/or electricity consumption and by the transportation of the material in the case where treatment of the material takes place off site"

(continued)

Table 36.2 (continued)

Reference	Technologies included		Contaminants	Impacts included			Functional unit	Time boundary for LCI	Result or conclusion
				Pri	Sec	Ter			
7. Blanc et al. (2004)	Ex situ	Excavation and off-site landfilling (S) Excavation and on-site containment (S) Excavation and liming stabilization (S) Excavation and bio-leaching (S)	Sulphur		X		A treatment of the site that allows environmental risks to be reduced to an acceptable level over the short term	Short term	"On-site containment appears to be the most environmentally-friendly technique, whereas bio-leaching and off-site landfilling result in the most important environmental burdens"
8. Godin et al. (2004)	In situ	No action	Spent potlining landfill	X	X		The management of 460,000 m^3 of wastemix and 200,000 m^3 of contaminated soil for a period of 50 years	50 years	"The LCA identified no action as having the least environmental impacts. However […] contaminant concentrations 50 years from the present could be approximately 30–40 times the regulatory criteria if this option is retained"
	Ex situ	Excavation and on-site secured disposal (S) Excavation and treatment (S) Excavation and incineration (S)							
9. Toffoletto et al. (2005)	Ex situ	Excavation with on-site biopiles (S)	Diesel oil	X	X		Remediation during 2-year period of 8000 m^3 of diesel-contaminated soil to the Quebec B criterion	2 years	"One major observation was the fact that the soil itself is responsible for an important fraction of the system's total impact, suggesting that it is beneficial to reach the highest level of remediation"

(continued)

Table 36.2 (continued)

Reference	Technologies included		Contaminants	Impacts included			Functional unit	Time boundary for LCI	Result or conclusion
				Pri	Sec	Ter			
10. Bayer and Finkel (2006)	Ex situ	Pump-and-treat (G)	PAHs, Tar		X		Control of a certain contaminated aquifer zone by complying with a certain concentration level	30 years	"A crucial finding that can be applied to any other site is the central role of steel, which particularly derogates the valuation of the funnel-and-gate system due to the associated emissions that are harmful to human health"
	In situ	Permeable barrier (G) (Funnel-and-gate)							
11. Cadotte et al. (2007)	Ex situ	Pump-and-treat (G) Excavation with on-site biopiles (S)	Diesel oil	X	x		Remediation of a 375 m^3 diesel-contaminated site to the Quebec B criterion in soil (700 mg kg^{-1}) and to the detectable limit of C_{10}–C_{50} for potable groundwater and surface water (0.1 mg L^{-1})	2–300 years depending on technology	"The all biological in situ scenario showed the least secondary and primary impacts […]. On the other hand, the in situ biological + chemical scenario produced the most secondary impacts and quickly removed primary impacts"
	In situ	Natural attenuation (S) Bioventing (S) Chemical oxidation (G) Biosparging (G) Oil removal (NAPL) Bioslurping (NAPL)							
12. Lesage et al. (2007a, b)	Ex situ	Excavation and off-site disposal (S) (brownfield rehabilitation)	Metals, PAHs, hydrocarbons	X	x	x	Legal and appropriate intervention on 1 ha of the tracked brownfield	4 years[a] 44 years[b]	"The ALCA[a] results show no clear preference for either intervention option […]. The CLCA[b] unequivocally supports rehabilitation if it is followed by residential use, as long as the development of suburban sites is avoided"
	Containment	Covering with 30 cm of clean soil (S) (risk minimization)							

(continued)

Table 36.2 (continued)

Reference	Technologies included		Contaminants	Impacts included			Functional unit	Time boundary for LCI	Result or conclusion
				Pri	Sec	Ter			
13. Higgins and Olson (2009)	Ex situ	Pump-and-treat (PTS)(G)	Chlorinated ethenes and ethanes		x		The system-specific requirements (energy, materials) needed to provide effective capture of the contaminant plume and treatment for 30 years	30 years	"Potential impacts of the model PRB are driven by the ZVI reactive medium and the energy usage during construction, while for PTS they are driven by the operational energy demand"
	In situ	Permeable reactive barrier with zero-valent iron (PRB) (G)							
14. Lemming et al. (2010b)	Ex situ	Excavation, off-site aeration and disposal (S)	Perchloroethene (PCE)	X	x		Treatment of the 7500 m^3 of contaminated soil within a 30 year timeframe	30 years/100 years	"The most favorable remediation technique was soil vapor extraction when a time boundary of 30 years was used. […] If a more realistic time frame (100 years) is used, the soil vapor extraction method becomes less favorable and the thermal method is the preferred option"
	In situ	Soil vapour extraction (S) In situ thermal desorption (S)							
15. Lemming et al. (2010c)	Ex situ	Excavation, off-site treatment and disposal (S)	Trichloroethene (TCE)	X	x		Treatment of 700 m^3 of contaminated soil resulting in a 98% removal of the contaminant mass within this volume	1200 years	"The primary human toxic impacts were high for ERD due to the formation and leaching of chlorinated degradation products […]. However, the secondary human toxic impacts of ISTD and excavation are likely to be even higher, particularly due to upstream impacts from steel production"
	In situ	Natural attenuation (S) Enhanced reductive dechlorination (ERD) (S) In situ thermal desorption (ISTD) (S)							

(continued)

Table 36.2 (continued)

Reference	Technologies included		Contaminants	Impacts included			Functional unit	Time boundary for LCI	Result or conclusion
				Pri	Sec	Ter			
16. Evans and Wilkie (2010)	In situ	Nutrient remediation by the aquatic plant hydrilla combined with production of biogas and compost (*Hydrilla verticillata*) (W)	Phosphorus and nitrogen		x		1 ha of aquatic plant harvesting	Not specified	"Net energy and economic gains were found using moderate data assumptions, which suggests that plant harvest may be an attractive management strategy for many lakes affected by hydrilla" "The respective energy and economic value outputs are largely decoupled, as energy output is dominated by biogas and fertiliser output, while economic output is dominated by the value of removing nutrients from aquatic systems and avoiding the use of herbicides"
17. Sanscartier et al. (2010)	Ex situ	Excavation, bioremediation on-site (biopiles) and disposal in an unlined landfill (S) Excavation, mechanical mixing off-site, and disposal in an unlined landfill (S)	Diesel oil		x		Treatment of 112 m^3 of soil contaminated with diesel to an acceptable risk level over the short term	Not specified	"On-site ex situ bioremediation in a temporary facility, followed by disposal in an unlined landfill, was found to have environmental impacts similar to in situ treatment, but far less than those for off-site treatment"
	In situ	Paving with asphalt and bioventing (S)							

(continued)

Table 36.2 (continued)

Reference	Technologies included		Contaminants	Impacts included			Functional unit	Time boundary for LCI	Result or conclusion
				Pri	Sec	Ter			
18. Cappuyns et al. (2011)	Ex situ	Excavation and off-site cleaning (S)	Exxsol, an aliphatic hydrocarbon (C-13)		x		Not clearly defined, but the study aims at "remediation values" equal 1500 mg/kg and 500 μg/L of Exxsol in soil and groundwater, respectively	Not specified	"According to the REC analysis, in situ thermal treatment showed a lower global environmental impact than soil excavation and off-site treatment, mainly because there were fewer emissions from the transport of contaminated soil" "Within the environmental aspects group of the BATNEEC method, soil excavation performed better than thermal soil remediation because it obtained a better score to meet the legal objectives for soil and groundwater quality"
	In situ	Remediation based on thermal conduction (S)							
19. Hu et al. (2010)	Ex situ	High Temperature Incineration, IHTI (S) Base catalysed decomposition, BCD (S)	Polychlorynated biphenyl (PCB)		x		Treatment of 10,000 tons of PCB contaminated soil from 800 to 1000 ppm to less than 5 ppm	Not specified	"BCD potentially has a lower environmental impact than IHTI technology in the PCB contaminated soil remediation process" "The major environmental impacts through the whole lifecycle arose from energy consumption in both IHTI and BCD processes. For IHTI, primary and secondary combustion

(continued)

Table 36.2 (continued)

Reference	Technologies included		Contaminants	Impacts included			Functional unit	Time boundary for LCI	Result or conclusion
				Pri	Sec	Ter			
									subsystem contributes more than 50% of midpoint impacts concerning with carcinogens, respiratory inorganics, respiratory organics, terrestrial ecotoxity, terrestrial acidification/eutrophication and global warming"
20. Inoue and Katayama (2011)	In situ (hypothetical case)	Landfarming (S)	Dieldrin		x		To reduce a contaminant concentration below the soil criterion for 1000 m^3 soil of agricultural field	15 days–60 months	"Energy consumption and CO_2 emission were determined from a life cycle inventory analysis using monetary-based intensity based on an input–output table. The values of RN_{SOIL} based on risk–cost, risk–energy consumption and risk–CO_2 emission were calculated, and then rankings of the candidates were compiled according to RN_{SOIL} values. A comparison between three rankings showed the different ranking orders"
	Ex situ (hypothetical case)	Disposal (S) High temperature thermal desorption, HTTD (S) Excavation and biopile (S)							
21. Mak and Lo (2011)	In situ	Permeable reactive barrier (PRB). Construction methods, materials of reactive media, and groundwater	Chromium (VI) or chromium (VI) and arsenic (V)		x		Successful treatment of 20 000 m^3 of the contaminated groundwater to the treatment goal, while the temporal boundary is 30 years	30 years	"Trench-based construction method can reduce the environmental impacts of the remediation remarkably compared to the

(continued)

Table 36.2 (continued)

Reference	Technologies included		Contaminants	Impacts included			Functional unit	Time boundary for LCI	Result or conclusion
				Pri	Sec	Ter			
		constituents are compared (G)							caisson-based method due to less construction material consumption by the funnel"
22. Sparrevik et al. (2011)	In situ	Natural recovery Capping using clay (SED) Capping using crushed limestone (SED) Capping using fossil anthracite coal-based active carbon (SED) Capping using biomass-based active carbon (SED)	Polychlorinated dibenzo-*p*-dioxins and —furans (PCDD/Fs)	X	x		Remediation of an area of sediments the same size as to the whole inner fjord (23.4 km^2), conservatively assessed for a 90-year time period	90 years	"Use of biomass-derived activated carbon, where carbon dioxide is sequestered during the production process, reduces the overall environmental impact to that of natural recovery"
23. Suer and Andersson-Sköld (2011)	Ex situ	Excavation, landfilling and refilling with pristine material (S)	Mineral oil		x		Not specified	20 years and 40 days	"The biofuel remediation had great environmental advantages compared to the ex situ excavation remediation. With the ReCiPe impact assessment method, which included biodiversity, the net environmental effect was even positive, in spite of the fact that the wood harvest was not utilised for biofuel production, but left on the contaminated site"
	In situ	Phytoremediation using willow (*Salix viminalis*) (S)							

(continued)

Table 36.2 (continued)

Reference	Technologies included		Contaminants	Impacts included			Functional unit	Time boundary for LCI	Result or conclusion
				Pri	Sec	Ter			
24. Busset et al. (2012)	Ex situ	Bioremediation with mechanical aeration (S) Bioremediation with electric aeration (S) Incineration with natural gas (S)	Polychlorinated biphenyl (PCB)		x		Treating 600 t of PCB contaminated moist soil (20% moisture) to reduce its PCB concentration from 200 to 50 mg/kg of soil	Not specified	"In most compared categories, the bioremediation processes are favourable. Of the bioremediation options, the lowest environmental footprint was observed for electric aeration. Irrespective of the aeration option, bioremediation was better than incineration"
25. Hong and Li (2012)	Lab-scale	Biodechlorination under methanogenic conditions using acetate as electron donor (SLUDGE) Biodechlorination under methanogenic conditions using lactate as electron donor (SLUDGE) Biodechlorination under acidogenic/methanogenic conditions using	Pentachloroaniline (PCA)		x		Dechlorination of 1 μM PCA in sludge	Not specified	"Optimizing the concentration of amended electron donors and increasing the population size of dechlorinating microorganisms are highly important in reducing the environmental burden by PCA bioremediation" "Results showed that the methanol scenario was the

(continued)

Table 36.2 (continued)

Reference	Technologies included		Contaminants	Impacts included			Functional unit	Time boundary for LCI	Result or conclusion
				Pri	Sec	Ter			
		methanol as electron donor (SLUDGE) Biodechlorination under acidogenic/methanogenic conditions using methanol + glucose as electron donor (SLUDGE)							most suitable option determined in this research"
26. Lubrecht (2012)	In situ	Horizontal directional drilling (HDD) crossing a groundwater plume (G) Auger drilling (wells spread across a plume) (G)	Not specified groundwater contaminant		x		Not specified. The basis for comparison is target depth of the plume. No remediation targets specified	Not specified	"Although both auger and horizontal drilling are considered to be high-emissions activities, comparing the two operations with the SiteWise application confirms that air emissions can be reduced during well construction by using HDD"
27. Witters et al. (2012)	In situ	Phytoremediation using willow (*Salix* spp.) combined with digestion and combustion (S)	Cadmium		x		Not clearly defined, but the study focuses on "energy production and CO_2 abatement per hectare per year	1 year	"Taking into account the marginal impact of the metals in the biomass on the energy conversion efficiency and on the potential use of

(continued)

Table 36.2 (continued)

Reference	Technologies included		Contaminants	Impacts included			Functional unit	Time boundary for LCI	Result or conclusion
				Pri	Sec	Ter			
		Phytoremediation using silage maize (*Zea mays* L.) combined with digestion and combustion (S) Phytoremediation using rapeseed (*Brassica napus* L.) combined with digestion and combustion (S)							the biomass and its rest products after conversion, digestion of silage maize with combustion of the contaminated digestate shows the best energetic and CO_2 abating perspectives"
28. Lemming et al. (2012)	In situ	Enhanced reductive dechlorination (ERD) In situ chemical oxidation (ISCO) Long-term monitoring Long-term monitoring with activated carbon treatment at waterworks	Trichloroethene (TCE)	X	x		The management of the target treatment zone leading to a 99% removal of the contaminant mass	830 years	"The aggregated environmental impact generated on the global, regional and local scales was greater than the local environmental impact removed in all the assessed scenarios. Overall, long-term monitoring and ERD were found to be the preferable management options as they resulted in the lowest secondary environmental impacts" […] "ISCO generates especially high levels of secondary impacts due to the applied permanganate"

(continued)

Table 36.2 (continued)

Reference	Technologies included		Contaminants	Impacts included			Functional unit	Time boundary for LCI	Result or conclusion
				Pri	Sec	Ter			
29. Hou et al. (2014a)	In situ	Enhanced in situ bioremediation (EIB), permeable reactive barrier (PRB), and in situ chemical reduction (ISCR)	Chlorinated ethylene		x		To to reduce the levels of PCE and its daughter products (i.e. TCE, DCE and vinyl chloride) within the treatment zone of the GSCM to levels below the California Maximum Contaminant Level (Cal-MCL) for drinking water with a 30-year timeframe. The Cal-MCL is 5 mg/L for PCE, 5 mg/L for TCE, 6 mg/L for 1.1-DCE and cis-1.2-DCE, and 0.5 mg/L for vinyl chloride	30 years	"In general source zone treatment technologies (i.e. EIB and ISCR) tend to have less life cycle impact than containment technologies (i.e. P&T and PRB)" "Site-specific parameters can have profound effects on the secondary life cycle impact of remediation alternatives at chlorinated solvent sites. In evaluating four remediation methods, P&T, PRB, EIB and ISCR, plume dimension parameters and hydrogeological parameters were found to have the most extensive and significant effects"
	Ex situ	Pump–and-treat (P&T)							
30. Hou et al. (2014b)	In situ	No action	Contaminated sediment containing heavy metals	x	x	x	For the studied case, when evaluating the different planning options, the functional unit was managing approximately 2500 m of waterways adjacent to the London Olympic Park for 100 years; when evaluating the different treatment methods, the functional unit was dredging approximately 30,000 m^3 of sediment and disposing the sediment in accordance with all applicable regulations	100 years	"The hybrid LCA offers a more complete system boundary than traditional process-based LCA" "In comparing soil washing with landfilling, the present study found that soil washing was superior to landfilling, in most social, economic, and environmental impact categories"
	Ex situ	Soil washing Landfilling							

(continued)

Table 36.2 (continued)

Reference	Technologies included		Contaminants	Impacts included			Functional unit	Time boundary for LCI	Result or conclusion
				Pri	Sec	Ter			
31. Beames et al. (2015)	In situ	In situ Multiphase Extraction (MPE)	Mix of benzene, toluene, ethylbenzene, xylene, mineral oil, chlorinated volatile organic compounds, polychlorinated biphenyls, chlorophenols, phenols and cresol		x		The functional unit in this study is the removal of approximately 80% of the estimated 500 metric tons of contaminant mass from a subsurface soil volume of approximately 40,000 m^3	6 years/15 years	"The results show that there is a trade-off between greenhouse gas emissions and land availability […]. Excavation leads to greater impacts in all the standard ReCiPe impact categories. The proposed impact assessment amendments show that Excavation yields the benefit of the site itself as a resource being available sooner"
	Ex situ	Excavation and ex situ treatment of soil							
32. Vigil et al. (2015)	In situ	Phytoremediation and biogas production Phytoremediation and landfill	Heavy metals (lead)		x		One hectare of decontaminated land	40 days/32 years	"Phytoremediation can provide a net sustainable benefit when used to recover soils polluted with heavy metals. The production of SNG (synthetic natural gas) is a key factor to the sustainability of a phytoremediation project because while contaminated land is being remediated,
	Ex situ	Excavation and landfill							

(continued)

Table 36.2 (continued)

Reference	Technologies included	Contaminants	Impacts included			Functional unit	Time boundary for LCI	Result or conclusion
			Pri	Sec	Ter			
								fossil fuel depletion is avoided and the metal-rich biomass is efficiently managed. A comparison of the same scheme, but with the biomass going to landfill, emphasises that phytoremediation sustainability is compromised when the biomass produced is not valorized in some way"

Pri Primary impacts, *Sec* Secondary impacts and *Ter* Tertiary impacts. *S* soil, *G* groundwater, *SED* sediment, *SLUDGE* sludge, *W* = wetland

[a]Attributional LCA

[b]Consequential LCA

References

Bayer, P., Finkel, M.: Life cycle assessment of active and passive groundwater remediation technologies. J. Contam. Hydrol. **83**, 171–199 (2006). doi:10.1016/j.jconhyd.2005.11.005

Beames, A., Broekx, S., Heijungs, R., Lookman, R., Boonen, K., Van Geert, Y., Dendoncker, K., Seuntjens, P.: Accounting for land-use efficiency and temporal variations between brownfield remediation alternatives in life-cycle assessment. J. Clean. Prod. **101**, 109–117 (2015). doi:10.1016/j.jclepro.2015.03.073

Beinat, E., van Drunen, M.A., Janssen, R., Nijboer, M.H., Koolenbrander, J.G.M., Okx, J.P., Schütte, A.R.: REC: a methodology for comparing soil remedial alternatives based on the criteria of risk reduction, environmental merit and costs. CUR/Nobis report 95-10-3, Gouda, The Netherlands (1997)

Blanc, A., Metivier-Pignon, H., Gourdon, R., Rousseaux, P.: Life cycle assessment as a tool for controlling the development of technical activities: application to the remediation of a site contaminated by sulfur. Adv. Environ. Res. **8**, 613–627 (2004). doi:10.1016/s1093-0191(03)00034-0

Busset, G., Sangely, M., Montrejaud-Vignoles, M., et al.: Life cycle assessment of polychlorinated biphenyl contaminated soil remediation processes. Int. J. Life Cycle Assess. **17**, 325–336 (2012). doi:10.1007/s11367-011-0366-7

Cappuyns, V., Bouckenooghe, D., van Breuseghem, L., van Herreweghe, S.: Can thermal soil remediation be sustainable? A case study of the environmental merit of the remediation of a site contaminated by a light non-aqueous phase liquid (LNAPL). J. Integr. Environ. Sci. **8**, 103–121 (2011). doi:10.1080/1943815x.2011.564187

Cadotte, M., Deschenes, L., Samson, R.: Selection of a remediation scenario for a diesel-contaminated site using LCA. Int. J. Life Cycle Assess. **12**, 239–251 (2007). doi:10.1065/lca2007.05.328

Caliman, F.A., Robu, B.M., Smaranda, C., et al.: Soil and groundwater cleanup: benefits and limits of emerging technologies. Clean Technol. Environ. Policy **13**, 241–268 (2011). doi:10.1007/s10098-010-0319-z

Collaran, E. (Ed.).: Methods in Biotechnology, Vol. 2: Bioremediation Protocols. Human Press, Totowa (1997)

Diamond, M.L., Page, C.A., Campbell, M., McKenna, S., Lall, R.: Life-cycle framework for assessment of site remediation options: method and generic survey. Environ. Toxicol. Chem. **18**, 788–800 (1999). doi:10.1002/etc.5620180427

Dreyer, L.C., Niemann, A.L., Hauschild, M.Z.: Comparison of three different LCIA methods: EDIP97, CML2001 and Eco-indicator 99—does it matter which one you choose? Int. J. Life Cycle Assess. **8**, 191–200 (2003). doi:10.1065/lca2003.06.115

EEA: Progress in management of contaminated sites. Report CSI 015. European Environment Agency, Copenhagen (2007)

Evans, J.M., Wilkie, A.C.: Life cycle assessment of nutrient remediation and bioenergy production potential from the harvest of hydrilla (*Hydrilla verticillata*). J. Environ. Manag. **91**, 2626–2631 (2010). doi:10.1016/j.jenvman.2010.07.040

Frischknecht, R., Jungbluth, N., Althaus, H.J., Doka, G., Heck, T., Hellweg, S., Hischier, R., Nemecek, T., Rebitzer, G., Spielmann, M., Wernet, G.: Overview and methodology. Ecoinvent Report No. 1. Swiss Centre for Life-Cycle Inventories, Dübendorf (2007)

Godin, J., Menard, J.F., Hains, S., et al.: Combined use of life cycle assessment and groundwater transport modeling to support contaminated site management. Hum. Ecol. Risk Assess. **10**, 1099–1116 (2004). doi:10.1080/10807030490887159

Hauschild, M.Z., Goedkoop, M., Guinee, J., et al.: Identifying best existing practice for characterization modeling in life cycle impact assessment. Int. J. Life Cycle Assess. **18**, 683–697 (2013). doi:10.1007/s11367-012-0489-5

Higgins, M.R., Olson, T.M.: Life-cycle case study comparison of permeable reactive barrier versus pump-and-treat remediation. Environ. Sci. Technol. **43**, 9432–9438 (2009). doi:10.1021/es9015537

Holland, K.S.: A framework for sustainable remediation. Environ. Sci. Technol. **45**, 7116–7117 (2011). doi:10.1021/es202595w

Holland, K.S., Favara, P.: The sustainable remediation forum. Remediat J **19**, 115–119 (2009). doi:10.1002/rem.20211

Holland, K.S., Lewis, R.E., Tipton, K., et al.: Framework for integrating sustainability into remediation projects. Remediat. J. **21**, 7–38 (2011). doi:10.1002/rem.20288

Hong, J.L., Li, X.Z.: Life cycle assessment comparison of substrates for the bioremediation of pentachloroaniline under acidogenic/methanogenic conditions. Int. J. Life Cycle Assess. **17**, 79–88 (2012). doi:10.1007/s11367-011-0338-y

Hou, D., Al-Tabbaa, A., Luo, J.: Assessing effects of site characteristics on remediation secondary life cycle impact with a generalised framework. J. Environ. Plan. Manag. 1–18 (2014a). doi:10.1080/09640568.2013.863754

Hou, D., Al-Tabbaa, A., Guthrie, P., Hellings, J., Gu, Q.: Using a hybrid LCA method to evaluate the sustainability of sediment remediation at the London Olympic Park. J. Clean. Prod. **83**, 87–95 (2014b). doi:10.1016/j.jclepro.2014.07.062

Hu, X.T., Zhu, J.X., Ding, Q.O.: Application of life cycle assessment on technology selection for POPs contaminated sites management and remediation. In: Selected Proceedings of the Fifth International Conference on Waste Management and Technology (ICWMT 5), pp. 509–513 (2010)

Inoue, Y., Katayama, A.: Two-scale evaluation of remediation technologies for a contaminated site by applying economic input–output life cycle assessment: risk-cost, risk–energy consumption and risk–CO_2 emission. J. Hazard. Mater. **192**, 1234–1242 (2011). doi:10.1016/j.jhazmat.2011.06.029

ISO.: Environmental management—life cycle assessment—principles and framework (ISO 14040). ISO, the International Organization for Standardization, Geneva (2006a)

ISO.: Environmental management—life cycle assessment—requirements and guidelines (ISO 14044). ISO, the International Organization for Standardization, Geneva (2006b)

JRC.: ILCD Handbook: General Guide for Life Cycle Assessment—Provisions and Action Steps, 1st edn. European Commission, Joint Research Centre, Institute for Environment and Sustainability, Ispra, Italy (2010)

Lemming, G., Hauschild, M.Z., Bjerg, P.L.: Life cycle assessment of soil and groundwater remediation technologies: literature review. Int. J. Life Cycle Assess. **15**, 115–127 (2010a). doi:10.1007/s11367-009-0129-x

Lemming, G., Friis-Hansen, P., Bjerg, P.L.: Risk-based economic decision analysis of remediation options at a PCE-contaminated site. J. Environ. Manag. **91**, 1169–1182 (2010b)

Lemming, G., Hauschild, M.Z., Chambon, J., et al.: Environmental impacts of remediation of a trichloroethene-contaminated site: life cycle assessment of remediation alternatives. Environ. Sci. Technol. **44**, 9163–9169 (2010c). doi:10.1021/es102007s

Lemming, G., Chambon, J.C., Binning, P.J., Bjerg, P.L.: Is there an environmental benefit from remediation of a contaminated site? Combined assessments of the risk reduction and life cycle impact of remediation. J. Environ. Manage. **112**, 392–403 (2012). doi:10.1016/j.jenvman.2012.08.002

Lesage, P., Ekvall, T., Deschenes, L., Samson, R.: Environmental assessment of Brownfield rehabilitation using two different life cycle inventory models. Int. J. Life Cycle Assess. **12**, 391–398 (2007a). doi:10.1065/lca2006.10.279.1

Lesage, P., Ekvall, T., Deschenes, L., Samson, R.: Environmental assessment of Brownfield rehabilitation using two different life cycle inventory models—part 2: case study. Int. J. Life Cycle Assess. **12**, 497–513 (2007b). doi:10.1065/lca2006.10.279.2

Lubrecht, M.D.: Horizontal directional drilling: a green and sustainable technology for site remediation. Environ. Sci. Technol. **46**, 2484–2489 (2012). doi:10.1021/es203765q

Mak, M.S.H., Lo, I.M.C.: Environmental life cycle assessment of permeable reactive barriers: effects of construction methods, reactive materials and groundwater constituents. Environ. Sci. Technol. **45**, 10148–10154 (2011). doi:10.1021/es202016d

Morais, S.A., Delerue-Matos, C.: A perspective on LCA application in site remediation services: critical review of challenges. J. Hazard. Mater. **175**, 12–22 (2010). doi:10.1016/j.jhazmat.2009.10.041

NAVFAC, USACE, Batelle: SiteWiseTM version 2. User guide. UG-2092-ENV. Developed by US Navy—Naval Facilities Engineering command (NAVFAC), United States Army Corps of Engineers (USACE), and Battelle (2011)

NREL, Athena Institute, Franklin Associates, Sylvatica: U.S. LCI Database Project—user's guide. National Renewable Energy Laboratory, USA, Athena Sustainable Materials Institute, Canada, Franklin Associates, Ltd., USA, Sylvatica, USA, NREL/BK-35854 (2004)

Owsianiak, M., Lemming, G., Hauschild, M.Z., Bjerg, P.L.: Assessing environmental sustainability of remediation technologies in a life cycle perspective is not so easy. Environ. Sci. Technol. **47**, 1182–1183 (2013). doi:10.1021/es305279t

Page, C.A., Diamond, M.L., Campbell, M., McKenna, S.: Life-cycle framework for assessment of site remediation options: case study. Environ. Toxicol. Chem. **18**, 801–810 (1999). doi:10.1897/1551-5028(1999)018<0801:lcffao>2.3.co;2

Ribbenhed, M., Wolf-Watz, C., Almemark, M., Palm, A., Sternbeck, J.: Livscykelanalys av marksaneringstekniker fàr fàrorenad jord och sediment [Life cycle assessment of remediation technologies for contaminated soil and sediment]. IVL Rapport/report B1476. IVL Swedish Environmental Research Institute Ltd., Stockholm, Sweden (2002)

Sanscartier, D., Margni, M., Reimer, K., Zeeb, B.: Comparison of the secondary environmental impacts of three remediation alternatives for a diesel-contaminated site in Northern Canada. Soil Sediment Contam. **19**, 338–355 (2010). doi:10.1080/15320381003695256

ScanRail Consult, HOH Water Technology A/S, NIRAS Consulting Engineers and Planners A/S, Revisorsamvirket/Pannell Kerr Forster: Environmental/economic evaluation and optimising of contaminated sites remediation. EU LIFE Project No. 96ENV/DK/0016. Copenhagen, Denmark (2000)

Sparrevik, M., Saloranta, T., Cornelissen, G., et al.: Use of life cycle assessments to evaluate the environmental footprint of contaminated sediment remediation. Environ. Sci. Technol. **45**, 4235–4241 (2011). doi:10.1021/es103925u

Sparrevik, M., Barton, D.N., Bates, M.E., Linkov, I.: Use of stochastic multi-criteria decision analysis to support sustainable management of contaminated sediments. Environ. Sci. Technol. **46**, 1326–1334 (2012). doi:10.1021/es202225x

Suer, P., Andersson-Skold, Y.: Biofuel or excavation? Life cycle assessment (LCA) of soil remediation options. Biomass Bioenergy **35**, 969–981 (2011). doi:10.1016/j.biombioe.2010.11.022

Toffoletto, L., Deschenes, L., Samson, R.: LCA of ex situ bioremediation of diesel-contaminated soil. Int. J. Life Cycle Assess. **10**, 406–416 (2005). doi:10.1065/lca2004.09.180.12

US EPA.: Abstracts of remediation case studies, Vol. 4, Federal remediation technologies roundtable, EPA 542-R-00-006 (2000)

US EPA.: Cleanup of the nation's waste sites: markets and technology trends. EPA 542-R-04-015. U.S. Environmental Protection Agency, U.S. Government Printing Office, Washington, DC. U. S. Environmental Protection Agency, U.S. Government Printing Office, Washington (2004)

US EPA: Superfund Remedy Report, Thirteenth edn. United States Environmental Protection Agency, Office of Solid Waste and Emergency Response, Washington (2010)

Vigil, M., Marey-Pérez, M.F., Martinez Huerta, G., Álvarez Cabal, V.: Is phytoremediation without biomass valorization sustainable? Comparative LCA of landfilling vs. anaerobic co-digestion. Sci. Total Environ. **505**, 844–850 (2015). doi:10.1016/j.scitotenv.2014.10.047

Vignes, R.P.: Use limited life-cycle analysis for environmental decision-making. Chem. Eng. Prog. **97**, 40–54 (2001)

Volkwein, S., Hurtig, H.-W., Klöpffer, W.: Life cycle assessment of contaminated sites remediation. Int. J. Life Cycle Assess. **4**, 263–274 (1999). doi:10.1007/bf02979178

Witters, N., Mendelsohn, R.O., Van Slycken, S., et al.: Phytoremediation, a sustainable remediation technology? Conclusions from a case study. I: energy production and carbon dioxide abatement. Biomass Bioenergy **39**, 454–469 (2012). doi:10.1016/j.biombioe.2011.08.016

Author Biographies

Gitte Lemming Søndergaard LCA expert with a focus on remediation of contaminated sites. Main LCA interests are combining LCA of contaminated site remediation with remediation performance modelling, and integrating LCA into multi-criteria decision support models for contaminated site management.

Mikołaj Owsianiak involved in development and application of life cycle impact assessment methods in sustainability assessment of technologies. Has worked on issues associated with: soils (remediation), metals (toxic impact assessment), biodiesel (fate in the environment), and carbonaceous materials (biochar and hydrochar).

Part IV
LCA Cookbook

Chapter 37
LCA Cookbook

Michael Z. Hauschild and Anders Bjørn

Abstract The LCA cookbook presents the provisions and actions from the ILCD Handbook that are central in the performance of an LCA. The selection is intended to cover all those activities that an LCA practitioner needs to undertake in a typical process-LCA, and the presentation follows the normal progression of the LCA work according to the ISO framework. For explanation of the reasoning behind the actions, the reader is referred to the presentation of the methodological elements in Part 2 of the book.

37.1 Introduction

This chapter is a cookbook with recipes on how to perform an LCA. It is intended to guide the LCA practitioner through the many steps, activities and decisions ("actions") that are needed to perform an LCA according to the ILCD Handbook (EC-JRC 2010). The cookbook follows the main structure of the ISO 14044 standard and gives detailed and concrete instructions on the main steps and activities that are relevant for most LCAs. The instructions are based on provisions and actions in the ILCD Handbook that are needed in order to perform an LCA. We have chosen to base the cookbook on the ILCD Handbook since it is the most recent detailed LCA guideline, based on the body of existing LCA methods, developed through an extensive public-, expert- and stakeholder consultation process. The ILCD Handbook offers very detailed guidance and requirements to the LCA practitioner, so detailed that the important provisions sometimes drown in the detail that is offered in the documentation of less important provisions. For the cookbook, we have therefore performed a selection of those provisions and

M.Z. Hauschild (✉) · A. Bjørn
Division for Quantitative Sustainability Assessment, Department of Management Engineering, Technical University of Denmark, Kongens Lyngby, Denmark
e-mail: mzha@dtu.dk

A. Bjørn
CIRAIG, Polytechnique Montréal, 3333 Chemin Queen-Mary, Montreal, QC, Canada

M.Z. Hauschild et al. (eds.), *Life Cycle Assessment*,
DOI 10.1007/978-3-319-56475-3_37

recommended actions under each provision that are most important and that need to be considered in the majority of LCA studies.

We also find that the ILCD Handbook is useful as a reference for documentation of the applied methodology in a way that increases the transparency and reproducibility of an LCA study. There may be reasons why an LCA practitioner wishes to deviate from the guidelines of the ILCD Handbook, just as the chapters of Part II of this book sometimes do. In such cases the guidelines are still useful for the documentation of the methodology applied in the LCA study to be able to state where provisions have been followed, where deviations were made and what actions were taken instead.

The focus of the cookbook is on answering the "what" and "how" questions to carrying out an LCA study. The reader is referred to the presentation and discussion of the methodological elements of LCA in Chaps. 7–13, for answer to any "why" questions that may arise during the use of the cookbook, and indeed, it is advisable to read these chapters prior to attempting to use the recipes of the cookbook. With respect to the reporting of an LCA study, the reader is referred to Chap. 38, which offers a template for an LCA report based on the reporting provisions of the ILCD Handbook.

The intended use of this chapter is thus to serve as a quick reference for the practitioner who is already familiar with the rationale behind the different elements of the LCA methodology and simply needs guidance on which steps to undertake and how to do them in order to perform an LCA. It can also be used as a checklist to ensure that all needed activities have been performed. As illustrated in Fig. 37.1 and explained in Sect. 6.3, the performance of an LCA involves several iterations where earlier phases are revisited and refined based on the insights gained in later phases and the use of sensitivity and uncertainty analyses. This iterative approach should be followed, also when you use the cookbook.

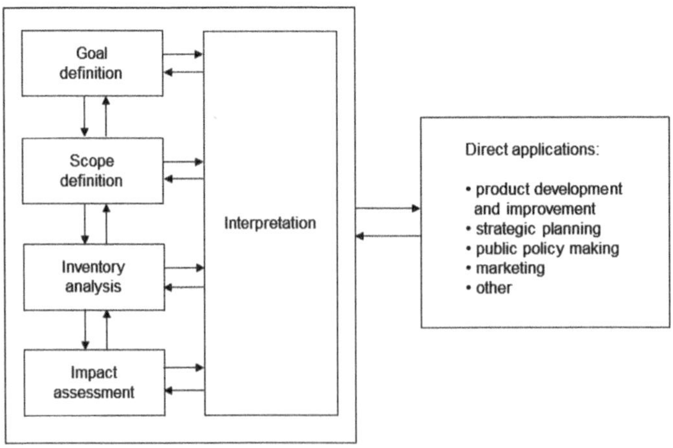

Fig. 37.1 Framework of LCA based with the main phases (modified from the ISO 14040 standard)

Inspired by the ISO standard, the ILCD Handbook operates with three degrees of strength in the requirements: "Shall", "Should" and "May".

- "Shall" means that the provision is a mandatory requirement that must be followed
- "Should" is a weaker requirement where the provision must be followed but deviations are allowed if they are clearly justified in the report. A justification can be that the provision or parts of it are not applicable, or that another solution can be demonstrated to be more appropriate
- "May" means that the provision is to be seen as a recommendation only (sometimes the term "recommended" is used instead of "may")

The guidelines in the ILCD Handbook follow the ISO standard but in some provisions, the ILCD requirement is stronger than the requirement in the ISO standard. There are also cases where the ILCD provision addresses an aspect that is not covered by the ISO standard. These deviations from the standard are marked with **[ISO!]** and **[ISO+],** respectively.

Each set of provisions has a number that refers to the section of the ILCD Handbook where it is discussed. This reference has been kept to allow the reader the possibility to consult the Handbook for further in-depth explanation of the requirements.

37.2 Goal Definition

The goal definition is discussed in Chap. 7 of this book. It is the first phase of the LCA, where the purpose of the study is defined and described. The goal definition is decisive for all following phases of the LCA. It guides the details of the scope definition, which sets the scene for both the inventory analysis and the impact assessment and determines the quality and level of precision that is needed from these phases. Finally, it frames the interpretation phase where the questions posed in the goal definition are attempted answered based on the outcome of the other methodological phases.

37.2.1 The Seven Aspects of Goal Definition

The central provisions and actions from the ILCD guideline on the seven[1] aspects of goal definition are the following:

[1]In the introduction to goal definition in Chap. 7, only six aspects are discussed because aspects IV and V are combined under one—"Target audience".

Seven aspects of goal definition (Provisions 5.2)

1. SHALL—**Intended applications**: Unambiguously identify the intended applications of the deliverable of the LCI or LCA study.
2. SHALL—**Limitations of study**: Unambiguously identify and detail any initially set limitations for the use of the LCI/LCA study. These can be caused by the following:

2.1. **Impact coverage limitations** such as in Carbon footprint calculations

2.2. **Methodological limitations** of LCA in general or of specific method approaches applied

2.3. **Assumption limitations**: Specific or uncommon assumptions/scenarios modelled for the analysed system [ISO+]

> Note that the initially identified limitations may need to be adjusted during the later LCA phases when all the related details are clear.
>
> Other possible limitations due to lack of achieved LCI data quality may also restrict the applicability; these are identified in the later interpretation phase of the study.

3. SHALL—**Reasons for study**: Unambiguously identify the internal or external reason(s) for carrying out the study and the specific decisions to be supported by its outcome, if applicable.
4. SHALL—**Target audience of study**: Unambiguously identify the audience(s) to whom the results of the study are foreseen to be communicated.
5. SHALL—**Type of audience**: Classify the targeted audience(s) as being "internal", "restricted external" (e.g. specific business-to-business customers) or "public". Differentiate also between "technical" and "non-technical" audience. [ISO+]
6. SHALL—**Comparisons involved?** Unambiguously state whether the study involves comparisons or comparative assertions across systems (e.g. products) and whether these are foreseen to be disclosed to the public. [ISO!]
7. SHALL—**Commissioner**: Identify the commissioner of the study and all other influential actors such as co-financiers, LCA experts involved, etc.

Each of the seven aspects must be considered when performing an LCA. While aspects 1 and 3 are central for *doing* an LCA because they have pervasive influence on decisions made in later LCA phases, aspects 2, 4, 5 and 6 mainly relate to *communicating the results* of an LCA, and aspect 7 addresses the organisational setup of the study.

37.2.2 Determining the Decision Context

The decision context of the LCA influences some of the later actions of the LCA, in particular related to the handling of multifunctional processes and modelling of the background system during the inventory analysis. The decision context is determined according to the following provision:

Classifying the decision context (Provisions 5.3)

Table 37.1 gives an overview of the resulting, practically relevant three archetype goal situations that will be referred to throughout this document to provide the required, differentiated methodological guidance. This relates to the subsequent provisions on classifying the decision context of the LCA study:

1. SHALL—**Identify applicable goal situation**: Identify the type of decision context of the LCI/LCA study, i.e. to which of the archetype goal situations A, B, C1 or C2 the study belongs. Draw on the goal aspects "intended applications" and "specific decisions to be supported"), as follows: [ISO!]

1.1. **Situation A—"Micro-level decision support"**: Decision support, typically at the level of products, but also single process steps, sites/companies and other systems, with no or exclusively small-scale consequences in the background system or on other systems, i.e. the consequences of the analysed decision alone are too small to overcome thresholds and trigger structural changes of installed capacity elsewhere via market mechanisms. Situation A covers among others the LCA applications listed below; any deviating assignment to another goal situation than A shall be justified and be in line with the above provisions (see also the specific provisions below for differentiating between Situation A and B, and between Situation C and A/B):

- Identification of Key Environmental Performance Indicators (KEPI) of a product group for Ecodesign/simplified LCA
- Weak point analysis of a specific product
- Detailed Ecodesign/Design-for-recycling

Table 37.1 Combination of two main aspects of the decision context: decision orientation and kind of consequences in background system or other systems

<table>
<tr><td rowspan="4">Decision support?</td><td rowspan="3">Yes</td><td colspan="2">Kind of process-changes in background system/other systems</td></tr>
<tr><td>None or small-scale</td><td>Large-scale</td></tr>
<tr><td>Situation A "Micro-level decision support"</td><td>Situation B "Meso/macro-level decision support"</td></tr>
<tr><td>No</td><td colspan="2">Situation C "Accounting"
(with C1: including interactions with other systems, C2: excluding interactions with other systems)</td></tr>
</table>

- Perform simplified KEPI-type LCA/Ecodesign study
- Comparison of specific goods or services
- Benchmarking of specific products against the product group's average
- Green Public or Private Procurement (GPP)
- Development of life cycle-based Type I Ecolabel criteria
- Development of Product Category Rules (PCR) or a similar specific guide for a product group
- Development of a life cycle-based Type III environmental declaration (e.g. Environmental Product Declaration (EPD)) for a specific good or service
- Development of the "Carbon footprint", "Primary energy consumption" or similar indicator for a specific product
- Greening the supply chain
- Providing quantitative life cycle data as annex to an Environmental Technology Verification (ETV) for comparative use
- Clean Development Mechanism (CDM) and Joint Implementation (JI)
- Development of specific, average or generic unit process or LCI results data sets for use in Situation A

1.2. **Situation B—"Meso/macro-level decision support": Decision support for strategies with large-scale consequences in the background system or other systems. The analysed decision alone is large enough to result via market mechanisms in structural changes of installed capacity in at least one process outside the foreground system of the analysed system. Situation B covers among others the LCA applications listed below; any deviating assignment to a goal situation other than B shall be justified and be in line with the above provisions (see also the specific provisions below for differentiating between Situation A and B and between Situation C and A/B)**:

- Policy development: Forecasting and analysis of the environmental impact of pervasive technologies, raw material strategies and related policy development
- Policy information: Identifying product groups with the largest environmental improvement potential
- Development of specific, average or generic unit process or LCI results data sets for use in Situation B

> It is important to note that the LCI modelling provisions for Situation B refer exclusively to those processes that are affected by these large-scale consequences. The other parts of the background system of the life cycle model will later be modelled as "Situation A", i.e. typically all the processes with a smaller contribution to the overall results.

1.3. **Situation C—"Accounting"**: From a decision-making point of view, a retrospective accounting/documentation of what has happened (or will happen based on extrapolating forecasting), with no interest in any additional consequences that the analysed system may have in the background system or on other systems. Situation C has two sub-types: C1 and C2. C1 describes an existing system but accounts for interactions it has with other systems (e.g. crediting existing avoided burdens from recycling). C2 describes an existing system in isolation without accounting for the interaction with other systems. This may cover the LCA applications listed below; any deviating assignment to a goal situation other than C1 or C2 shall be justified and be in line with the above provisions. See also the specific provision below for differentiating between Situation C and A/B:

1.3.1. **Situation C1—"Accounting with interactions"**:

- Monitoring environmental impacts of a nation, industry sector, product group or product
- Policy information: Basket-of-products (or -product groups) type studies
- Policy information: Identifying product groups with the largest environmental impact
- Corporate or site environmental reporting including indirect effects under Environmental Management Systems (EMS)
- Certified supply type studies or parts of the analysed system with fixed guarantees along the supply chain
- Development of specific, average or generic unit process or LCI results data sets for use in Situation C1

1.3.2. **Situation C2—"Accounting without interactions"**:

- Accounting studies that according to their goal definition do not include any interaction with other systems
- Development of specific, average or generic unit process or LCI results data sets for use in Situation C2

> Note that any decision support that would be derived needs to employ the methods under Situation A or B, with Situation C having a preparatory role only. Note, however, that due to the simplified provisions of this document, the modelling of Situation A studies (micro-level decision support) is identical to that of Situation C1 studies, but not vice versa.

2. SHALL—**Situation A or B**: Where a study cannot initially be clearly assigned to either Situation A or B, for example when analysing major strategies of market-dominating companies or product-related questions of market-dominating products. In this situation, the guiding criteria shall be whether the consequences of the analysed decision alone are big enough to overcome related thresholds and/or other constraints and result in large-scale consequences in the installed

production capacity outside the foreground system. Then: Situation B. If not: Situation A. Large-scale consequences shall generally be assumed if the annual additional demand or supply, triggered by the analysed decision, exceeds the capacity of the annually replaced installed capacity of the additionally demanded or supplied process, product or broader function, as applicable. If that percentage is bigger than 5%, 5% should be used instead. [ISO!]

3. SHALL—**Situation C1 or A/B**: In the case a study cannot initially be clearly assigned to either Situation C1 or A/B, for example when it is a monitoring study but involves a comparative decision support. In this situation, the guiding criteria shall be whether a comparative decision support is to be given by the LCI/LCA study, i.e. whether the study shall be used to support decisions on alternatives with better or worse environmental performance. Then Situation A or B applies, depending on small-scale or large-scale consequences; see related provisions. If not, i.e. the study is only retrospectively informing about better performance in the past, then Situation C applies. [ISO!]

Table 37.2 presents the classification to goal situation A, B or C for a wide range of LCA applications.

37.3 Scope Definition

The scope definition is discussed in Chap. 8 of this book. It is the phase where the LCA is scoped in accordance with the goal and intended application as formulated in the goal definition. Together with the goal definition it determines how the other LCA phases should be performed (Inventory analysis, Impact assessment and Interpretation, including uncertainty and sensitivity analysis) and how the reporting of the LCA should be done. It is an overarching aim of the scope definition to ensure the consistency of the applied methods, assumptions, and data and to strengthen the reproducibility of the study.

A scope definition encompasses the following nine scope items[2]:

1. Deliverables
2. Object of the assessment
3. LCI modelling framework and handling of multifunctional processes
4. System boundaries and completeness requirements
5. Representativeness of LCI data
6. Preparation of the basis for the impact assessment
7. Special requirements for system comparisons
8. Needs for critical review

[2]The ILCD Handbook operates with 10 scope items but here the aspect of data quality requirements, which the handbook proposes as a separate scope item, is considered under scope item 4 and 5.

Table 37.2 Most common types of LCI/LCA study deliverables required for specific LCA applications (indicative overview)

Application areas/purposes	LCA applications (from perspective of life cycle information user or provider)	LCI/LCA type of deliverable and/or application required as direct input for the "LCA application"[1, 2, 3]	Applicable goal situation	Related ISO standard (next to 14040 and 14044:2006)
Product improvement	Identification of Key Environmental Performance Indicators (KEPI) of a product group for Ecodesign/simplified LCA	d or e or i; and f	A	
	Weak point analysis of a specific product	f and d	A	ISO/TR 14062 (2002)
	Detailed Ecodesign/Design-for-recycling	f	A	ISO/TR 14062 (2002)
	Perform simplified KEPI-type LCA/Ecodesign study	i	A	
Product comparisons and procurement	Comparison of specific goods or services	e, h, or j	A	
	Benchmarking of specific products against the product group's average	e	A	
	Green Public or Private Procurement (GPP)	e, h, or j	A	ISO 14015 (2001)
Communication	Development of life cycle-based Type I Ecolabel criteria	d, e, g, or i	A	ISO 14024 (1999)
	Development of Product Category Rules (PCR) or a similar specific guide for a product group	e or d; and f	A	ISO 14025 (2006)
	Development of a life cycle-based Type III environmental declaration (e.g. Environmental Product Declaration (EPD)) for a specific good or service	d or g; and f	A	ISO 14025 (2006)
	Development of the "Carbon footprint", "Primary energy consumption" or similar indicator for a specific product	d, g, or f	A	ISO 14025 (2006)
	Calculation of indirect effects in Environmental Management Systems (EMS)	b or d	C1	ISO 14001 (2015)
	Greening the supply chain	h, j, or e	A	ISO 14015 (2001)
	Providing quantitative life cycle data as annex to an Environmental Technology Verification (ETV) for comparative use	h, d, or g	A	

(continued)

Table 37.2 (continued)

Application areas/purposes	LCA applications (from perspective of life cycle information user or provider)	LCI/LCA type of deliverable and/or application required as direct input for the "LCA application"[1, 2, 3]	Applicable goal situation	Related ISO standard (next to 14040 and 14044:2006)
Across several areas	Development of specific, average or generic unit process or LCI results data sets for use in different applications	a or b	A, B, C1, or C2	
	Clean Development Mechanism (CDM) and Joint Implementation (JI)	d, h, g, or f	A	
Strategic decision support	Policy development: Forecasting & analysis of the environmental impact of pervasive technologies, raw material strategies, etc. and related policy development	<u>e</u>	B	
	Policy information: Identifying product groups with the largest environmental improvement potential	<u>e</u>	B	
Accounting	Monitoring environmental impacts of a nation, industry sector, product group or product	d or b	C1	
	Policy information: Basket-of-products (or -product groups) type of studies	e	C1	
	Policy information: Identifying product groups with the largest environmental impact	e	C1	
	Certified supply type studies or parts of the analysed system with fixed guarantees along the supply chain	b, d, e, or h	C1	
	Corporate or site environmental reporting	d	C1	ISO 14015 (2001), ISO 14031 (2013)
	Accounting studies that according to their goal definition do not include any interaction with other systems	d	C2	

[1]Basic type as input for LCA application: *a* Unit process data set; *b* LCI results data set; *c* LCIA results data set; *d* LCA study, non-comparative; *e* Comparative LCA study; *f* Detailed LCI model of system. Application as input for other LCA applications: *g* KEPIs-based tool; *h* EPD; *i* Criteria set for life cycle-based Type I Ecolabel; *j* Life cycle-based Type I Ecolabel of the system

[2]Several LCA applications typically use at least alternatively the outcome of other LCA applications as input, e.g. Green Procurement often works with KEPI or Type I Ecolabel criteria. This is additionally indicated in table

[3]Note that LCA studies (*d* and *e*) as basic form of application can already directly provide the required LCA application, e.g. a weak point analysis of the specific product or the comparison of products in support of procurement. In that case, the letters *d* and *e* are underlined

9. Planning reporting of results

Items 2–6 are central for *doing* an LCA because these have pervasive influence on decisions made in later LCA phases. Aspects 1, 7, 8 and 9 mainly relate to *reporting and communicating* an LCA study.

37.3.1 Deliverables

The intended deliverable depends on the intended application of the results of the LCA study. Table 37.2 gives an overview of a broad range of applications and the deliverable that is needed from the LCA to support the application. The table also links the application to the goal situation as discussed in Sect. 37.2.2.

The following guidance is given on the determination of the type of LCA deliverable and intended application based on the overview given in Table 37.2.

Types of LCA deliverables and intended applications (Provisions 6.3)

1. SHOULD—**Types of deliverables**: Derive from the intended application(s) identified in the goal definition and any potential pre-settings, the appropriate type(s) of deliverable(s) that the LCI/LCA study should provide. Table 37.2 gives an overview. The following types are most common, listed in order of increasing comprehensiveness and/or complexity: [ISO!]

1.1. Life Cycle Inventory ("LCI") study and/or data set, in the following variants:

1.1.1. Unit process study and/or data set, with two sub-types:

1.1.1.1. Single operation unit process (variants: fixed or parameterised)

1.1.1.2. Black box unit process (variants: fixed or parameterised)

1.1.2. Partly terminated system data set (variants: fixed or parameterised)

1.1.3. Life Cycle Inventory results ("LCI results") study and/or data set

1.2. Life Cycle Impact Assessment results ("LCIA results") study and/or data set

1.3. Non-comparative Life Cycle Assessment study ("LCA study"), i.e. including impact assessment and interpretation

1.4. Comparative Life Cycle Assessment study ("Comparative LCA study"), in the following variants:

1.4.1. Non-assertive comparative Life Cycle Assessment study ("Non-assertive comparative LCA study")

1.4.2. Comparative assertion Life Cycle Assessment study ("Comparative assertion LCA study"), with superiority, inferiority or equality of any compared alternatives are explicitly concluded

1.5. Detailed LCI model of the analysed system

37.3.2 *Object of the Assessment*

The object of the assessment must be defined in a functional unit to ensure comparability of the studied alternatives in the case of a comparative LCA. If the purpose is to develop unit process LCI data or an environmental product declaration, the object of the assessment is defined as a reference flow (typically of one unit of the product, material or service).

Function, functional unit and reference flow (Provisions 6.4)

1. SHALL—**Identify system or process**: Identify in line with the goal and with the other scope settings the to-be-analysed system(s) or process(es) (e.g. good, service, technology, strategy, country, etc.) and describe it/them in an unambiguous way.
2. MAY—**Photos, specifications**: Provide photos, and/or technical specifications, and/or descriptions of the system(s), if and as appropriate for the addressees. [ISO+]
3. SHALL—**Identify function(s) and functional unit(s)**: One or more function(s) and quantitative, measurable functional unit(s) of each of the system(s) shall be clearly identified, if applicable and appropriate for the type of system.
4. SHALL—**Functional unit, details**: The functional unit(s) shall be identified and specified in detail across all the following aspects:

4.1. Function provided (what),
4.2. in which quantity (how much),

> Note that, even though the "how long" information is important, the use intensity and resulting overall quantity of the performed function is key to valid comparisons.

4.3. for what duration (how long), and
4.4. to what quality (in what way and how well is the function provided).
4.5. Changes in the functional performance over time (e.g. due to ageing of the product) shall be explicitly considered and quantified, as far as possible. [ISO+]

5. MAY—**Obligatory and positioning properties**: If product systems are analysed, it is recommended to use obligatory and positioning properties for the quantitative and qualitative aspects of their function, respectively. [ISO+]
6. SHALL—**Measurement methods**: ISO or national harmonised standards shall be used as measurement methods, as far as possible and wherever available and appropriate for use in an LCA context. Own measurement methods should only be used in case of unavailable or inappropriate harmonised standards only. They shall be clearly specified and documented and later be subject to critical review.

7. SHOULD—**Alternatives and complements to the functional unit**: It is noted that a functional unit cannot always be given or is not appropriate/useful. In such cases, it should be replaced or complemented by another clearly defined, quantitative and measurable item as outlined below; deviations shall be concisely justified: [ISO!]

7.1. **Materials and other application-unspecific products**: A functional unit cannot generally be given. Only the reference flow that includes the main technical specification of the product should be provided. In this case, the reference flow is also the declared unit, but not the functional unit.

7.2. **Multifunctional processes**: For each function one functional unit and/or reference flow should be given, as appropriate, depending on the kind of co-function/co-product (see other items in this sub-list). Otherwise the technical specification of the process and functions should be provided in the accompanying documentation.

7.3. **Monofunctional systems**: For systems (e.g. products) with only one relevant function or combination of functions, the functional unit(s) should be specified. In addition, one reference flow with a clear and detailed system name should be provided. The functionally relevant technical specification should be provided as part of the reference flow name and/or in the accompanying documentation.

7.4. **Multifunctional systems**: For multifunctional systems with multiple, parallel functions, the detailed technical specification should be provided. The corresponding functional units should be given in addition and when appropriate to the given case. One reference flow with a clear and detailed system name should be provided. (This one reference flow can be split up into each one reference flow for each function in case the data set is directly used in comparative studies. This to allow substitution of single functions to achieve equivalence of compared alternatives).

7.5. **Systems with alternative functions**: For systems with alternative functions, the most relevant alternative functions and functional units should be specified. In addition, one reference flow with a clear and detailed system name shall be provided. The functionally relevant technical specification should be provided as part of the reference flow name and/or in the accompanying documentation.

8. SHOULD—**Highly variable functions**: For highly variable functions of processes and systems, the way that the variable and parameters relate to the system's performance and to its inventory should be documented. This should be in form of mathematical relations or in another suitable form. The use of parameterised data sets is recommended to support appropriate documentation and efficient use.

9. SHALL—**Comparative studies**: For comparative studies, see Provisions 6.10 in Sect. 37.3.7. Among others, they shall be compared based on their reference flow.

37.3.3 LCI Modelling Framework and Handling of Multifunctional Processes

The choice of LCI modelling framework (attributional or consequential) must be made in accordance with the classification into goal situation A, B or C as described in Sect. 37.2.1. The handling of multifunctional processes must be in accordance with the chosen framework as described in the following provisions.

LCI modelling provisions for Situations A, B and C (Provisions 6.5.4)

1. SHALL—**LCI modelling provisions to be applied**: A specific combination of LCI modelling framework (attributional or consequential) and LCI method approaches (allocation or system expansion/substitution) is identified for each of the goal situations A, B, C1 and C2. The provisions cover scenario and uncertainty calculation. The provisions shall be applied as follows: [ISO!]

1.1. **Situation A—"Micro-level decision support" (6.5.4.2)**:

1.1.1. **Life cycle model**: The life cycle model of the analysed system(s) shall be modelled as an attributional model, i.e. depicting the existing supply-chain processes).

1.1.2. **Subdivision and virtual subdivision for black box unit processes and multifunctionality**: It shall be aimed at avoiding black box unit processes and solving multifunctionality by subdivision or virtual subdivision, as far as possible. The following applies for cases of system–system relationships and cases of multifunctionality, if subdivision/virtual subdivision is not possible or not feasible.

1.1.3. **Cases of system–system relationship**: if the analysed system's secondary function acts within a context system, where it only affects the existing processes operation, system expansion shall be performed via substitution with the short-term marginal.

Note that the analysed system may also have influenced the installed capacity of the context system, if it had been considered when planning the context system. For example, the heat generated by office equipment may have been considered when dimensioning the heating and cooling system of an office building.

Part-system relationships require no specific modelling provision, but the correct identification of the processes within the system boundary.

1.1.4. **Cases of multifunctionality—general**:

1.1.4.1. **Substitution of market mix of specific alternatives**: (Simplification compared to full consequential model): If for the not required[3] specific co-function, functionally equivalent alternative processes/systems are operated/produced to a sufficient[4] extent: the not required co-function shall, as far as possible, be substituted with the average market[5] consumption mix of the processes or systems that it supersedes, excluding the to-be-substituted function from this mix. If the to-be-substituted function has a small share in the overall environmental impact of the market mix, the market mix can be used instead, if the results are not relevantly changed.

1.1.4.2. **Substitution of market mix of general, wider alternatives**: If such alternative processes/systems do not exist[6] or are not operated to a sufficient extent, alternative processes/systems of the not required co-function in a wider sense should be used for substitution,[7] applying the same provisions as set out in the preceding sub-provision.

1.1.4.3. **Situation B?** If also such alternative processes/systems for the wider function do not exist or do not meet the named requirements, the study is

[3]i.e. in contrast to the one that is analysed or within the system boundary in the background system.

[4]"Sufficient" means that the not required co-function can quantitatively be absorbed by the market. That shall be assumed to be the case, if the annually available amount of the to-be-substituted co-function is not more than the annual amount produced by the annually replaced installed capacity of the superseded alternative process(es) or system(s). Note that this refers to the amount of co-function provided by the analysed process. E.g. if the study refers to a specific producer that contributes only a small share to the total production of the co-function, only this small amount counts, i.e. it is very likely that it can be absorbed by the market. If the study refers to the total production of a certain product that has the not required co-products, there is the chance that this much larger amount of co-products cannot be absorbed by the market.

[5]This "market" is the market where the secondary function is provided. E.g. for products produced from end-of-life and waste management this is the market of the primary production at the time and the location (e.g. country, region or global etc. market) where the end-of-life product or waste is known or forecasted to undergo recycling, reuse, or energy recovery. If this market cannot be clearly determined, the most likely market shall be assumed and well justified; this most likely market shall be on a continental scale or at least cover a group of countries/markets.

[6]As is the case e.g. for wheat grain and straw production, many oil refinery products, etc.

[7]E.g. for NaOH, as co-product of Chlorine production, apart from NaCl electrolysis no alternative route is operated to the sufficient extent. However, NaOH provides in a wider sense the function of neutralising agent (next to some other, quantitatively less relevant functions) and hence other, technically equivalent and competing neutralising agents such as KOH, $Ca(OH)_2$, Na_2CO_3, etc. can be assumed to be superseded; their mix would be used to substitute the not required NaOH. For the example of a wheat grain study and the not required co-product straw: instead of straw, other dry biomass (e.g. Miscanthus grass, wood for heating, etc.) provides equivalent functions and its market mix can be assumed to be superseded.

in fact a Situation B type study, as this implies large-scale consequences on other systems.

1.1.4.4. **Allocation**: (Simplification compared to full consequential model): if modelling of substitution is not feasible[8] and generic data is not sufficiently accurate to represent the superseded processes/systems: the two-step allocation procedure of Provisions 7.9.3 in Sect. 37.4.4 can be applied instead. Allocation shall, however, not be performed if it would relevantly favour the analysed process/system. This fact shall be argued or approximated. If allocation is performed, the resulting lack of accuracy shall be reported and explicitly be considered later in the result's interpretation. For multifunctional products and the alternative second step in allocation, Quality Function Deployment (QFD) is the preferred alternative to market price allocation.

1.1.4.5. **No substitution of main function(s)**: (Simplification compared to full consequential model): The determining co-function(s) shall not be substituted. In the case the determining and dependent co-functions cannot be clearly identified, the determining co-function(s) should be assumed to be those that jointly contribute more than 50% to the combined market value of all co-functions of the analysed multifunctional process or system.[9] (The market value is for this purpose the value of the co-functions as provided by the multifunctional process, i.e. without any further processing). In this case, the two-step allocation procedure shall be applied (see Provisions 7.9.3 in Sect. 37.4.4).

1.1.4.6. **Considering functional differences**: Differences in functionality between substituted and superseded function shall be considered either preferably by substituting the actually superseded amounts, or by substituting the market value corrected amount of the function.

1.1.5. **Cases of multifunctionality—waste and end-of-life treatment**: (note the simplifications given here for Situation A):

1.1.5.1. **Recyclability substitution of primary route market mix**: (Simplification compared to full consequential model): For waste and end-of-life treatment as cases of multifunctionality: system expansion shall be performed in accordance with the provisions for the cases of general multifunctionality. The avoided primary production of the reused part, recycled good, or recovered energy shall be substituted. This shall apply the recyclability

[8]"not feasible" refers to cases where many alternative processes/systems or alternatives for the function in a wider sense exist e.g. where over 10 alternative processes/systems make up over 80% of the market for the to-be-substituted function, and/or where the superseded processes/systems themselves have a number of co-functions.

[9]The reasoning is that in that case it is likely that the determining co-functions would be substituted.

substitution approach, with the simplification of substituting the average primary route consumption mix of the market where the secondary good is produced.

1.1.5.2. **Recyclability substitution of general, wider alternatives**: For "open loop—different primary route" cases, the market consumption mix of alternative goods in a wider sense should be used for substitution, along the same provisions as set out in the preceding sub-provision.

1.1.5.3. **Situation B?** Especially for the case of "open loop—different primary route" and for secondary goods with relevantly changed/downcycled properties, in addition, verification is needed on whether for the reused part, recycled material, or recovered energy, functionally equivalent, alternative processes or systems, or functional equivalents in a wider sense exist. If this is the case, it needs additional verification whether these are operated to a sufficient extent. Otherwise, the study is in fact a Situation B type study, as this implies large-scale consequences on other systems.

1.1.5.4. **Allocation**: (Simplification compared to full consequential model): if modelling the substitution is not feasible and generic data is not sufficiently accurate to represent the superseded processes/systems, then the two-step allocation procedure applied to waste/end-of-life according to Provisions 7.9.3 in Sect. 37.4.4 can be applied instead. This shall not be done if it would relevantly favour the analysed process/system; this fact shall be argued or approximated. If allocation is performed, the resulting lack of accuracy shall be reported and explicitly be considered later in the results interpretation.

1.1.5.5. **Considering functional differences**: Differences in functionality between substituted and superseded function shall be considered either and preferably by substituting the actually superseded amounts. As second priority and if the superseded amounts are not known, market value correction of the amount of the substituted function shall be performed. Note that this applies to all cases of waste and end-of-life treatment that generate any valuable secondary good, i.e. "closed loop", "open loop—same primary route" and "open loop—different primary route").

1.1.6. **Comparative studies, scenarios, uncertainty calculation**:

1.1.6.1. If among the to-be-compared systems, one or more systems have additional functional units, comparability shall be achieved by system expansion.

1.1.6.2. For comparative studies of Situation A, the main model for each of the compared alternatives shall each be complemented with assumption scenarios of reasonably best and reasonably worst cases. Optionally further assumption scenarios can be defined. Uncertainty calculation shall be performed, unless it has already been used to derive the reasonably best and worst-case scenarios. These scenarios serve to later perform the sensitivity check. The interested parties shall be involved towards a best attainable consensus on the definition of the reasonably best and reasonably worst-case assumption scenarios (and uncertainty calculation) that can in principle vary all data and method provisions and assumptions for Situation A **except for** the "shall" provisions and assumptions/conventions. It is recommended to also perform and report such assumption scenarios and uncertainty calculations for non-comparative LCI and LCA studies.

> Note that for LCI data sets that are intended to support comparative studies, the reasonably best and worst-case scenarios may be included within these data sets or be provided as complement.

1.2. **Situation B "Meso/macro-level decision support" (6.5.4.3)**:

1.2.1. **Provisions as for Situation A with two differences**: The above provisions for Situation A shall also be applied for Situation B, with two differences:

1.2.1.1. **Large-scale consequences**: Processes that have been identified as being affected by "big"[10] large-scale changes as a consequence of the analysed decision shall be modelled as the expected mix of the long-term marginal processes.

1.2.1.2. **Comparative studies, scenarios, uncertainty calculation**: (Additional flexibility for assumption scenarios), for comparative studies of Situation B: The assumption scenarios and uncertainty calculation can in principle vary all data and method provisions and assumptions for Situation B **including** the "shall" provisions and assumptions/conventions of the ILCD Handbook, while not those of ISO 14040 and 14044.[11]

> Note that comparative Situation B studies often include a "zero" option, i.e. include a scenario of "no action" (e.g. "no change in existing policy Y", or "no strategic measure on raw material X security of supply").

1.3. **Situation C—"Accounting" (6.5.4.4)**:

1.3.1. **Provisions as for Situation A with two differences**: The provisions for Situation A shall also be applied for Situation C. With two differences:

1.3.2. **Remaining cases of multifunctionality**: These shall be solved as follows:

1.3.3. **Situation C1**: Multifunctionality of processes and systems shall be solved with substitution via system expansion, as in Situation A, but independently

[10] Large-scale ("big") consequences shall generally be assumed if the annual additional demand or supply that is triggered by the analysed decision exceeds the capacity of the annually replaced installed capacity of the additionally demanded or supplied process, product, or broader function, as applicable.

[11] i.e. these scenarios and uncertainty calculation allow to apply the full range of method and modelling options of ISO 14044.

of the absolute amount of the not required co-function(s) that will be substituted.[12] The other provisions apply analogously.

1.3.4. **Situation C2**: General cases of multifunctionality of processes and systems shall be solved with allocation (i.e. applying the two-step allocation procedure; for details see Provisions 7.9.3 in Sect. 37.4.4). Cases of waste and end-of-life treatment shall be solved via allocation.

1.3.5. **Comparative studies**: Note the restrictions for direct comparative decision support of accounting data.

Note that Situation C1 is thereby modelled identically to Situation A, while independently of the size of the system or processes.

Note that substitution can lead to negative elementary flows or in rare cases even negative overall environmental impacts of the analysed systems. This must be explicitly addressed in reporting, explaining all implications and helping to avoid misinterpretation and misleading conclusions.

37.3.4 System Boundaries and Completeness Requirements

The scoping of the system with the setting of the system boundaries and the decision of which processes to include and which not to include needs to be done in a way which is in accordance with the goal definition and in particular the completeness requirements that follows from the intended application.

Deriving system boundaries and cut-off criteria (completeness) (Provisions 6.6)

1. SHALL—**Scope of LCA**: The following shall be covered by the LCI or LCA study:

1.1. potential impacts on the three areas of protection Human health, Natural environment and Natural resources,

1.2. that are caused by interventions between Technosphere and Ecosphere, and this

1.3. during normal and abnormal operation, but excluding accidents, spills and similar.[13]

[12]The reasoning is that the effect of superseding alternative processes/systems is existing, other than in Situation A where an additional amount of co-function is pushed into the market, i.e. in Situation C1, the check whether alternative processes/systems are operated or produced to a sufficient extent is unnecessary, as the superseding factually already occurs.

[13]i.e. excluding accidents, indoor and workplace exposure, as well as impacts related to direct application or ingestion of products to humans.

1.4. other kinds of impacts outside the scope of LCA that are found relevant for the analysed or compared system(s) may be identified and their relevance be justified. [ISO+]

2. SHALL—**Processes within the system boundary**: The final system boundary/ies of the analysed system(s) shall as far as possible include all relevant life cycle stages and processes that

2.1. are operated within the technosphere, and

2.2. that need to be included along the provisions of identifying to-be-included processes under attributional or consequential modelling, but with the specific provisions and simplifications for the applicable Situation A, B, or C.

2.3. any relevant deviation/omission from the above shall be clearly documented and in case of LCA studies later be considered in the interpretation.

3. SHALL—**Flows across the system boundary**: Next to the reference flow(s) that provide the functional unit(s) and permissible waste flows, no relevant other flows shall cross the boundary between the analysed system(s) and the rest of the technosphere, as far as possible. Only elementary flows (including permissible measurement indicators and flow groups) should cross the boundary between the analysed system(s) and the ecosphere. Any relevant deviation/omission from the above shall be reported and in case of LCA studies later be considered in the interpretation. [ISO!]

4. SHALL—**System boundary diagram**: The extent of the system model shall be identified and a schematic system boundary diagram be prepared.[14] Next to the included life cycle stages, the following shall be provided for the different types of deliverables: [ISO!]

4.1. **For single operation unit processes**: the process step to be represented.

4.2. **For black box unit processes**: the to-be-represented, e.g. process-chain, plant, site, etc. and the first and last process step included.

4.3. **For LCI results, LCIA results and non-comparative LCA studies**: the included life cycle stages.
Finally, the first and/or last process step included shall be given, unless the life cycle starts or ends with the cradle or grave, respectively.

4.4. **For comparative LCA studies**: for each of the compared options the included life cycle stages. In addition, for each of the options the first and/or last process steps included shall be given, unless the respective life cycle starts or ends with the cradle or grave, respectively.

4.5 **Flow chart**: Especially for the foreground system, it is recommended to already prepare technical flow charts on the main process steps.

---

[14]Other systems that become part of the analysed system in case system expansion is applied should not be shown in this diagram, but the quantitatively most relevant cases of multifunctional processes (as identified in the sensitivity analysis) shall be listed. This includes the quantitatively relevant cases of part-system relationships, which only exceptionally require an expanded system boundary diagram (e.g. if the analysed product would be the "part" of a part-system relationship such shall be provided).

5. SHALL—**List of exclusions**: Prepare an initial list of any types of activities, specific processes, product and waste flows, elementary flows or other parts that would be foreseen to be excluded from the analysed system, if any. [ISO+]

Note that this initial list is to be (iteratively) updated to reflect the situation at the end of the study.

Note that any final exclusion will need to be justified referring to the cut-off criteria and may limit the applicability of the resulting data set or the conclusions that can be drawn from a comparative study.

6. SHALL—**Part-system and system–system relationships**: For studies on parts that have a part-system relationship and on systems that have a system–system relationship, obtain data on the effects on the related systems and their data, as far as this is necessary in line with the goal and scope of the study. [ISO!]
7. SHALL—**System-external off-setting**: Off-set emissions (e.g. due to carbon off-setting by the Clean Development Mechanism, system-external carbon credits), and other, similar measures outside the analysed system shall not be included in the system boundaries, as far as they are relevant for the results. The related (reduced) emissions shall not be integrated into the inventory or used in LCA results interpretation. [ISO+]
8. SHALL—**Quantitative cut-off criteria**: Define the cut-off % value to be applied for the analysed system's product, waste and elementary flows that cross the system boundary, but that are not quantitatively[15] included in the inventory,[16] as follows:

8.1. **Overall environmental impact**: The cut-off % value shall generally relate to the quantitative degree of coverage of the approximated overall environmental impact of the system.[17] For comparative studies, the cut-off shall additionally also always relate to mass and energy. Two alternative options exist how to address the overall environmental impact: [ISO!]

[15]The respective flows shall, however, be foreseen to be identified and stay in the inventory, but without stating an amount and being marked as "missing relevant" or "missing irrelevant", as applicable.

[16]Note that co-functions are initially part of the inventory and only later removed via allocation or addressed with system expansion/substitution.

[17]While the true absolute overall impact (i.e. the "100% completeness") cannot be known in LCA and other such models, it can be approximated in practice in an iterative manner and with sufficient precision to serve as practical guidance and use for cut-off.

8.1.1. apply the cut-off individually for each of the to-be-included[18] impact categories. This requires that the LCIA methods have been identified at that point.

8.1.2. apply the cut-off for the normalised and weighted overall environmental impact. This requires that the LCIA methods, normalisation basis and the weighting set have been identified at that point.

8.2. **Identify the aimed-at% cut-off**: The aimed at quantitative cut-off/completeness percentage shall be identified as follows:

8.2.1. **For unit processes, LCI results and LCIA results**: the cut-off value has either already been defined in the goal phase (e.g. "Development of a single operation unit process data set of 95% completeness") or is to be derived from the respective completeness need of the intended application in the iterative scope steps.

8.2.2. **For non-comparative LCA studies**: the cut-off value has been identified depending on the detail of interest when analysing the system for key contributing processes and elementary flows; this has been defined typically in the goal of the study

8.2.3. **For comparative LCA studies**: the cut-off value is set depending on how much precision, accuracy and completeness is needed to show significant differences between the compared systems. This is done in the iterations of the LCA work after at least an initial LCI model has been modelled and analysed.

Note that, unless it was initially defined, the cut-off can only roughly be approximated in the initial scope phase and has to be adjusted iteratively.

Note that later deviations from the initially set cut-off criteria, e.g. due to lack of data, are to be identified in the subsequent LCI data collection and modelling and are to be documented at the end of the LCI/LCA study. The finally achieved cut-off (and any possible deviations) shall be reported and have to be fully reflected in the interpretation phase, in case of an LCA study. Both may lead to a revision of the supported intended applications of the LCI/LCA study. These issues are to be checked in the respective phase of the LCA work.

37.3.5 Representativeness of LCI Data

The representativeness of process data that is collected in the LCI relative to the processes that it is intended to represent the product system is addressed in three dimensions—technological-, geographical—and time-related representativeness.

[18]For studies with limited impact coverage (e.g. Carbon footprint), only these categories are to be considered, accordingly.

Technological representativeness (Provisions 6.8.2)

1. SHALL—**Good technological representativeness**: The overall inventory data shall have an as good as required technological representativeness, meeting the goal requirements of the study (note that technological, geographical and time-related representativeness are closely interrelated). For both analysed processes and systems, this includes all quantitative and qualitative aspects of the functional unit(s) and/or reference flow(s), and/or technical specification(s). This applies especially for those aspects that matter in terms of leading to relevant differences in the LCI data.
2. SHALL—**Specific way or mode of process?** Identify along the goal of the study and especially the intended applications whether the data needs to represent a specific way or mode of operating the technology/technique (e.g. a specific load factor for transport, or a specific start, closure, cycle step of a process, etc.), if this differs from the average, typical or integrated operation. [ISO+]
3. SHALL—**Different technologies for attributional and consequential modelling**: Note that attributional and consequential modelling often require very different processes (and to some degree also systems) for the background system. But see the simplifications set for all Situations, except for the processes that face "big" changes in Situation B: [ISO!]

3.1. **Attributional modelling**: The following should be used:

3.1.1. **Foreground system**: Technology-specific primary data for the foreground system and for the specifications of the products and wastes that connect the foreground system with the background system. Secondary data of the actual suppliers/downstream actors should be preferred to other (third-party) secondary data. Technology-specific, generic or average data from third parties should be used in those parts of the foreground system where this for the given case is of higher quality (i.e. more accurate, precise, complete) than available technology-specific primary or secondary data from suppliers/downstream actors.

3.1.2. **Background system**: Average technology as market consumption[19] mix data.

3.2. **Consequential modelling**: The following should be used:

3.2.1. **Foreground system**: The same applies as described above for attributional modelling. Here this includes the suppliers'/downstream actors' technology-specific secondary data of the contractually fixed or planned supply chain.

3.2.2. **Background system**: The short-term or long-term marginal technology mixes should be used, as appropriate for the applicable Situation A, B, C1

[19]This also applies if a market production mix data set is developed: the fact that the data set is to represent the production mix would be achieved by combining the representative mix of producing technologies of that market according to their production share. For the data in the background system of the individual routes, nevertheless the respective consumption mix data are to be used.

and C2. Among these, the named long-term technology mix only applies to those processes under Situation B that face "big" changes in consequence of the analysed decision, and—optionally—to the assumption scenarios. The technology mix of marginal processes should be identified, depending among others on the market conditions and the cost-competitiveness of the potential marginal processes.

3.3. **Using not fully representative data**: For both attributional and consequential modelling, not fully technologically representative data can be used only along the following conditions:

3.3.1. **For LCI and LCIA data sets/non-comparative LCI/LCA studies**: The use of not fully technologically representative data is justifiable only if this is not relevantly changing the overall LCIA results compared to using fully representative data; otherwise, the lower achieved representativeness shall be documented in the data set/report. For data provided for a competitor's product, lower representativeness shall not lead to higher overall environmental impacts of the LCIA results calculated for that product. For data provided for own products or for products without any competition situation (e.g. generic data from consultants or research projects for general background use), lower representativeness shall not lead to lower impacts of the overall LCIA results calculated for that product.

3.3.2. **For comparative LCA studies**: The conclusions or recommendations of the study should not be affected, as far as possible. Otherwise the lower achieved technological representativeness shall explicitly be considered when drawing conclusions and giving recommendations. Especially shall the use of less representative data not relatively disfavour any competitors' products to a relevant degree.

Note that this can be implemented only in the subsequent iterative steps of the LCA work.

4. SHALL—**Non-scalable supplies**: For the life cycle model of Situation A, B and C1, the following shall be applied: if the supply of a specific required function (e.g. product) cannot relevantly be increased in the analysed market and due to inherent constraints (e.g. as for hydropower in many countries) the market consumption mix of the specific function that the product provides (e.g. electricity in the above example) shall be used as far as possible, and not the data for the specific supplier/product. To not contradict the provisions on solving multifunctionality, this provision does not apply to required co-functions. [ISO!]

Geographical representativeness Provisions: 6.8.3)

For LCI results, LCIA results, LCA studies: be aware that the declared geographical scope of all later to be used inventory data needs to enable a correct impact assessment. This is to be checked especially carefully if a non-generic impact assessment (e.g. with differentiated characterisation factors by country, region or even site) is applied.

1. SHALL—**Good geographical representativeness**: The overall inventory data shall have an as good as required geographical representativeness, according to the goal of the study. This applies especially, where this matters in terms of relevant differences in the LCI data of different geographical scope.
2. SHALL—**Different geographical scope for attributional and consequential modelling**: Note that attributional and consequential modelling may require processes/products of a different geographical scope in the background system. But see the simplifications set for all Situations, except for the processes that face "big" changes in Situation B: [ISO!]

2.1. **Attributional modelling**: The following should be used:

2.1.1. **Foreground system**: Site or producer/provider specific data for the foreground system, supplier-specific data for the products that connect the foreground with the background system. Generic data of geographical mixes can be used also in parts of the foreground system if for the given case justified as being more accurate, precise and complete than available specific data (especially for processes operated at suppliers).

2.1.2. **Background system**: Average market consumption mix data for the background system.

2.2. **Consequential modelling**: The following should be used:

2.2.1. **Foreground system**: Site or producer/provider specific data for the directly controlled processes of the foreground system, suppliers' site specific data of the contractually fixed or planned supply chain of the foreground system plus for the products and wastes that connect the foreground with the background system. Generic data of geographical mixes can be used also in parts of the foreground system if for the given case justified as being more accurate, precise and complete than available specific data (especially for processes operated at suppliers).

2.2.2. **Background system**: The short-term or long-term marginal geographical mixes should be used for the background system, as appropriate for the applicable Situation A, B, C1, and C2. The geographical mix of the marginal processes should be identified, depending among others on the market conditions and cost-competitiveness of the potential marginal processes.

2.3. **Using not fully representative data**: For both attributional and consequential modelling, not fully geographically representative data can be used only along the following conditions:

2.3.1. **For LCI and LCIA data sets/non-comparative LCI/LCA studies**: The use of not fully geographically representative data is justifiable only if this is not relevantly changing the overall LCIA results compared to using fully representative data; otherwise the lower achieved representativeness shall be documented in the data set/report.

2.3.2. **For comparative LCA studies**: The conclusions or recommendations of the study should not be affected; otherwise, the lower achieved geographical representativeness shall explicitly be considered when drawing conclusions and giving recommendations. Especially shall the use of less representative data not relatively disfavour any competitors' products in a relevant degree.

Time-related representativeness (Provisions 6.8.4)

1. SHALL—**Good time-related representativeness**: The overall inventory data shall have an as good as required time-related representativeness, according to the goal of the study. This applies especially, where this matters in terms of relevant differences in the LCI data that represent a different time.

> Note that the represented year of a process or system shall refer to the actually represented year and not the year when the data set was calculated or the year of publication of used secondary data sources.

2. SHALL—**Specific seasonal or diurnal situation?** Check along the goal of the study and the intended applications whether the data needs to represent a specific seasonal or diurnal situation, if this differs from the average annual data. [ISO+]

3. SHOULD—**Time-related representativeness of future processes**: For processes that run more than 5 years in the future or past from the time of study (e.g. of the use and end-of-life stage of long-living products or in case of backward looking analysis), fully time-representative future/past scenario data should be used, if possible. If this is not possible: [ISO!]

3.1. **BAT and recent data**: For both attributional and consequential modelling, Best Available Technology (BAT) mix data should be used as second option, if BAT data can be argued to be sufficiently representative for the required time. The most recent data are the third option.

3.2. **Using not fully representative data**: Not fully time-representative data can be used only along the following conditions:

3.2.1. **For LCI and LCIA data sets/non-comparative LCI/LCA studies**: The use of not fully time-representative data is justifiable only if this is not relevantly changing the overall LCIA results compared to using fully time- representative data; otherwise the lower achieved time-representativeness shall be documented in the data set/report.

3.2.2. **For comparative LCA studies**: The conclusions or recommendations of the study should not be affected; otherwise, the lower achieved time-representativeness shall explicitly be considered when drawing conclusions and giving recommendations. Especially shall the use of less time-representative data not relatively disfavour any competitors' products in a relevant degree.

37.3.6 Preparation of the Basis for the Impact Assessment

The preparation of the basis of the later impact assessment phase the scope definition serves two main purposes: One is to ensure that the impact assessment is done in accordance with the goal definition and the intended application of the LCA. The other is to prepare the basis to ensure that the inventory analysis compiles the relevant data on elementary flows from the product system to support the assessment of the relevant impact scores. The ILCD guidance document presents the following provisions for the preparation of the basis of the impact assessment:

Preparing the basis for the impact assessment (Provisions 6.7)

Note that an impact assessment is required for all types of LCI/LCA studies at least for systematically assessing and improving the overall data quality, including applying the cut-off rules.

Impact categories and LCIA methods:

1. SHALL—**Goal-conform selection of impact categories and LCIA methods**: Select the impact categories to be included and the corresponding LCIA methods in accordance with the goal of the study. [ISO!]
2. SHOULD—**Requirements for impact categories**:

2.1. All impact categories that are environmentally relevant[20] for the LCI/LCA study shall be included, as far as possible and unless the goal definition would explicitly foresee exclusions (e.g. for Carbon footprint studies). Further ones can be included optionally.

[20]As this can be judged only in view of the LCIA results, i.e. after LCI data collection, modelling, etc., it is recommended to initially foresee the inclusion of all of the default impact categories (see next action). If the impact assessment later shows irrelevance of one of more impact categories, they can be left out; see also further provisions. For principally restricted assessments (e.g. Carbon footprint), see the respective action below.

Note that any relevant exclusion will need to be explicitly considered during interpretation and can lead to limitations for the further use of the data (in case of an LCI study or data set) and in limitations for the conclusions and recommendations (in case of an LCA study).

3. SHALL—**Requirements for LCIA methods**: All included LCIA methods shall meet the following requirements[21]:

3.1. They should be internationally accepted and preferably additionally be endorsed by a governmental body of the relevant region where the decision is to be supported (Situation A, B) or where the reference of the accounted system is located (Situation C).
3.2. They shall be scientifically and technically valid, as far as possible; the extent of this fact shall be documented.
3.3. They shall have no relevant gaps in coverage of the impact category they relate to, as far as possible; otherwise the gap shall be approximated, reported and explicitly be considered in the results interpretation,
3.4. They shall be based upon a distinct identifiable environmental mechanism or reproducible empirical observation,
3.5. They shall be related exclusively to elementary flows (i.e. interventions between the technosphere and the ecosphere) during normal and abnormal operating conditions, but excluding accidents, spills and the like. [ISO!]
3.6. They shall be free of double-counting across included characterisation factors, as far as possible and unless otherwise required by the goal of the study, and
3.7. They shall be free of value choices and assumptions, as far as possible; these shall be appropriately documented and if relevant, they shall explicitly be considered in the results interpretation.

The development or identification of LCIA methods that are prepared to meet these requirements is supported with the separate guidance document "Framework and requirements for Life Cycle Impact Assessment (LCIA) models and indicators".

Note that for use in comparative assertion studies any used LCIA method and factor may need to undergo a review under ISO in order to be eligible.

4. SHOULD—**Default impact categories and category endpoints**: The selected LCIA methods in their entirety should by default cover all of the following impact categories and provide characterisation factors on midpoint level.

[21]Under the ILCD, recommendations are under preparation on a complete set of such LCIA methods that provide characterisation factors for the ILCD reference elementary flows. These will relate to European and/or global scope, depending on their applicability.

It is recommended that they also provide modelled category endpoint factors that are coherent with the midpoint level and that cover all relevant damages to the three following areas of protection:

4.1. **Impact categories ("midpoint level")**: Climate change, (Stratospheric) Ozone depletion, Human toxicity, Respiratory inorganics, Ionising radiation, (Ground-level) Photochemical ozone formation, Acidification (land and water), Eutrophication (land and water), Ecotoxicity (freshwater, marine, terrestrial), Land use, Resource depletion (of minerals, fossil and renewable energy resources, water, …). [ISO!]

4.2. **Category endpoints ("endpoint level")**: Damage to human health, Damage to ecosystem, Depletion of natural resources. These relate to the three areas of protection "Human health", "Natural environment" and "Natural resources", respectively. [ISO+]

5. SHOULD—**Location and time-generic LCIA**: The LCIA methods should by default be location-generic and time-generic (but see later provision on derived LCIA methods). [ISO!]
6. MAY—**LCIA methodologies**: It is recommended to select available LCIA methodologies that provide a complete set of single LCIA methods, rather than selecting and combining individual LCIA methods. [ISO!]
7. SHOULD—**Excluding impact categories?** Exclusions of any of the above impact categories should be justified as being not relevant for the analysed system(s). This can be done based on experience gained from detailed, complete studies for sufficiently similar systems and/or system group specific/Product Category Rule (PCR) type guidance documents. [ISO+]
8. SHALL—**Adding impact categories?** Check for the specific LCI/LCA study whether next to the default impact categories given above, additional, relevant environmental impacts[22] need to be included in accordance with the goal and scope. If so, identify or develop[23] the relevant LCIA methods to be applied. Note that these shall meet the same requirements as the other included LCIA methods (see above).
9. SHOULD—**Impacts outside the scope of LCA**: Impacts that are outside the LCA frame[24] but for which scientific evidence exists that they are relevant for

[22]Examples are Noise, Desiccation/Salination, Littering of land and sea, etc.

[23]ISO 14044 requires that all relevant impacts are to be covered. In practice of performing LCA studies, the development of new LCIA methods is a rare case. The separate guidance document "Development of Life Cycle Impact Assessment (LCIA) models, methods and factors" supports LCIA method developers in this step.

[24]The inventory related to impacts that are outside the frame of LCA shall not be mixed with the inventory for LCA impacts, i.e. need separate inventorying as separate items outside the general Inputs/Outputs inventory. The LCA frame covers potential impacts on the named three areas of protection that are caused by interventions between Technosphere and Ecosphere during normal and abnormal operation, i.e. Accidents, indoor and workplace exposure, as well as impacts related to direct application or ingestion of products to humans shall not be mixed but be modelled and inventoried separately.

the analysed or compared system(s) should be clearly and individually be identified, including in the Summary and Executive summary of the report/data set. Their brief description should be foreseen in the further documentation. If it is foreseen to include them quantitatively, this requires potentially different modelling and analysis approaches and guidance. This should be done jointly with the LCA study, as far as possible, to ensure coherence, but inventory, impact assessment, etc. shall be kept separately for clear interpretation. [ISO!]

Note that this step is often possible only after the first or second iteration of LCI data collection and modelling, impact assessment and interpretation.

10. SHOULD—**Missing characterisation factors**: If a characterisation factor is missing for an elementary flow of the analysed inventory, and that flow is known to contribute significantly to one or more of the included impact categories, considering the goal and scope of the LCI/LCA study: [ISO+]

10.1. Check the potential importance of the missing characterisation factor by assuming a conservative value or reasonably worst-case value based on chemical, physical, biological and/or other similarity to other elementary flows, which contribute to the same impact category/ies in question.

Note that this procedure requires expert knowledge of an LCIA method developer, especially on fate and exposure modelling to be able to judge which similarities to consider and how; a good chemical and environmental sciences understanding is equally required.

10.2. Apply the assumed characterisation factor(s) to that elementary flow and investigate whether the total result for the affected impact category/ies is changed to a relevant degree (i.e. depending on the required completeness, accuracy and precision).

10.3. If with this approach the contribution from this elementary flow cannot be classified as being not relevant, it should be attempted to get a more accurate and precise value for the missing characterisation factor and use that one for the further work.

Note that this factor will have to fulfil the same conditions as other factors of the respective impact category/method.

10.4. If the latter is not possible or the whole provision is not feasible (e.g. for cost or timing reasons), the fact of a missing relevant characterisation factor shall be reported and the potential influence of the missing factor shall be considered when reporting the achieved data quality and (for LCA studies) in the interpretation of the results.

10.5. If the conservative or reasonably worst-case value does not show a relevant contribution from that elementary flow, the missing characterisation factor can be disregarded. It is recommended to report the fact of a "missing factor" nevertheless and marked as "missing unimportant", at least for those flows that lack relevance but are not fully negligible.

Note that this step is often only possible after the first or second iteration of LCI data collection and modelling, impact assessment and interpretation.

11. SHALL—**Location and time non-generic LCIA methods**: The potential use of LCIA methods that have been derived from the original, location-generic and time-generic ones (i.e. being not generic but, e.g. spatially or otherwise further differentiated or modified) shall be justified along the goal and scope of the study. It shall be demonstrated that significantly different LCIA results are obtained than with the generic methods. The non-generic methods have to meet the other applicable requirements for selected LCIA methods. [ISO!]

Note that this step is often only possible after the first or second iteration of LCI data collection and modelling, impact assessment and interpretation.

Note that for comparative LCA studies also the appropriateness of generic LCIA methods shall be discussed in the interpretation phase of the study. If a further differentiation can be argued or approximated to lead to significantly different results, this finding may limit the conclusions and recommendations that can be drawn from the study.

Note that LCIA results calculated from non-generic LCIA methods are later to be presented separately from the generic ones and discussed jointly.

Normalisation and weighting:

12. SHALL—**Cut-off criteria**: Normalisation and weighting may have been used for defining the cut-off rules. [ISO!]
13. MAY—**Results interpretation**: Normalisation and weighting are in addition optional steps under ISO 14044:2006 that are recommended to support the results interpretation.

Note that the normalisation and weighting shall be made in accordance with the intended application of the LCI/LCA study.

Note that if the study includes a comparative assertion to be disclosed to the public, quantitative weighting of the published indicator results is not permitted.

14. SHALL—**Consistency between cut-off and interpretation**: If used in support of results interpretation, the same normalisation and weighting set shall be used as for the cut-off rules. [ISO!]
15. SHALL—**Requirements for selecting normalisation basis and weighting set**: If used for defining the cut-off and/or in support of the interpretation of the results of the study, select a suitable normalisation basis and weighting set,[25] along the following rules: [ISO!]

15.1. **Normalisation basis**:

15.1.1. As normalisation basis the annual total environmental inventory globally should be preferred. Alternatively the territory-based or consumption-based annual total environmental inventory of the country or region should be used where the supported decisions are made (Situations A, B) or in which the accounting reference is located (Situation C). It is recommended to prefer the average citizen as normalisation basis instead of the global, regional or country total (i.e. the global, regional or country total divided by the number of citizen[26]).

15.1.2. Ensure the relevance of the selected normalisation basis for the intended applications and target audience.

15.1.3. Ensure a high degree of completeness and precision of the overall environmental impact covered and a similar degree of completeness and precision for all covered impact categories.

15.1.4. Ensure a proper link with the used LCIA methods, i.e. relate to the same impact categories/areas of protection and use to a sufficient degree the same elementary flows.

15.1.5. Ensure technical compatibility with the to-be-used weighting set, i.e. relate to the same impact categories/areas of protection.

15.1.6. As year for the normalisation basis the year should be used for which the latest data are available that meet the above requirements.

15.2. **Weighting set**:

15.2.1. The weighting set should represent the normative and other values globally or of the country or region where the supported decisions are made

[25]The development of governmentally supported corresponding normalisation and weighting data in the different regions and countries or globally would be beneficial.

[26]This brings the values of the normalised impacts for goods and services down to a better communicable and interpretable level (typical value range 10–0.00001 instead of 1E−7 to 1E−14).

(Situations A, B), or the reference of the accounting (Situation C). The weighting set should preferably be endorsed by a governmental body of the country or region where the decision is to be supported (Situation A, B) or where the reference of the accounted system is located (Situation C).

15.2.2. Ensure the relevance of the selected weighting set to the intended applications and target audience.

15.2.3. The weighting set shall correctly refer to the used normalisation basis and to the midpoint level or endpoint level indicators of the used LCIA methods, as applied.

15.3. **Extension for added impact categories**: If in the course of the study a non- default impact category has been additionally included, corresponding data for the normalisation basis and a weighting factor shall be additionally provided and used.[27]

Documentation of selected LCIA methods, and of decision/selection of normalisation and weighting:

16. SHALL—**Verifiable documentation of decision on LCIA methods, impact level, normalisation and weighting**: Decide and document now, during the initial scope definition, bindingly on: [ISO!]

16.1. the LCIA methods to be applied by default,

16.2. the selected impact level to be used for reporting and interpretation (i.e. midpoint and/or endpoint level), and if foreseen to be used,

16.3. the specific normalisation and weighting sets to be used for cut-off and for interpretation.

16.4. These decisions shall be documented or published in an appropriate form and way that allows the critical reviewer to later verify the date when these decisions have been made.

16.5. **Permissible adjustments**: Adjustments of these decisions shall only be possible:

16.5.1. If impact categories are added in line with the goal of the study and meeting the related provisions for their addition given more above. This shall result exclusively in an addition to the already selected LCIA methods, normalisation basis and weighting set for the added impact categories.

16.5.2. If using non-generic LCIA methods upon justification as indicated more above. This shall result exclusively in a differentiation of the already selected, generic LCIA methods, unless a best attainable consensus can be found among involved stakeholders on selection of another set of already available non-generic LCIA methods. The normalisation basis and weighting set shall remain unchanged.

[27]This is not required for use of non-generic LCIA methods and for additionally included single elementary flows/characterisation factors, unless this would relevantly change the results, what by default can be assumed not to be the case.

37.3.7 Special Requirements for System Comparisons

When two or more products are compared in an LCA, it is essential that the product systems are modelled consistently in terms of both methodological choices and choices on data to represent the two systems. A qualified consideration of uncertainties is also important when the systems are compared to decide whether one is preferable to the other from an environmental point of view. Particularly strict requirements must be met for comparative assertions disclosed to the public. The provisions from the ILCD guideline for comparisons between systems are the following:

Comparisons between systems (Provisions 6.10)

These provisions are mandatory (shall) only for comparative LCA studies that analyse more than one system or system variants. It is recommended to also apply them analogously to non-comparative LCA studies that include a system internal contribution/weak point analysis.

These provisions also apply to LCI studies and data sets that are intended to be used in context of comparative studies (e.g. as background data).

For all comparative studies

1. SHALL—**Non-assertive, comparative studies**: The ISO 14044:2006 provisions for comparative assertions shall also be applied to non-assertive, comparative studies. Both types together are grouped under the term "comparisons" here. [ISO!]
2. SHALL—**Consistency**: All elements of the scope definition shall be addressed consistently for all systems to be compared, as far as possible. Otherwise, the lack of consistency shall be reported and be considered explicitly when interpreting the results, giving conclusions or recommendations. Especially:

2.1. **LCI model**: The compared system models shall be constructed in an analogous way applying the same rules for system boundaries, LCI modelling principles and method approaches.

2.2. **Assumptions**: Methodological and data assumptions shall be made in an analogous way.

2.3. **Data quality**: The achieved completeness, accuracy and precision of the data shall be sufficiently similar for the compared systems.

3. SHALL—**Uncertainty and accuracy calculations**: Calculations on the stochastic uncertainty and accuracy shall support this analysis. This is not required if uncertainty calculations have already been used to derive the reasonably best and worst case scenarios.

4. SHALL—**Completeness/cut-off**: The cut-off % that has been defined in the study shall also be met for mass and energy, next to for the overall environmental impact.
5. SHALL—**Excluding identical parts**: If included processes/systems of the compared systems are identical for all alternatives, they may be left out of all models. Included processes/systems that are similar but not identical shall remain in the model, but their partial correlation shall be considered when interpreting differences. [ISO+]

Note that the intended applications may not permit to leave out even identical parts.

Note that even apparently identical parts may only be left out of the comparison if they are truly identical.

For example, the same amount of the same aluminium alloy used in the same component of two alternative models may be left out. This shall not be done if the alloy is used in different components of these models, as the inventories of the alloys are only partly correlated in the second case.

6. SHALL—**LCIA to be performed**: A Life Cycle Impact Assessment shall be performed for LCI or LCA studies intended to support comparative studies that are intended to be published.
7. SHALL—**Impact coverage limitations (e.g. Carbon footprint)**: Comparison studies based on selected indicators or impact categories (e.g. Carbon footprint-based comparisons) shall highlight that the comparison is not suitable to identify environmental preferable alternatives, as it only covers the considered impact(s) (e.g. Climate change). This applies unless it can be sufficiently demonstrated that the compared alternatives do not differ in other relevant environmental impacts to a degree that would change the conclusions and/or recommendations of the comparison if those other impacts would be included in the analysis. Such demonstration should draw on robust approximations for the analysed system and/or robust information derived from detailed and complete LCA studies available for sufficiently similar systems. System/product-group specific guidance document and Product Category Rules (PCR) may provide such robust information. The above shall be investigated in any case and if other environmental impacts were identified as being relevant in the above sense, they shall be named in the report. [ISO!]

For studies on systems with similar functional units:
Comparisons shall be made based on the system's reference flows.

8. SHALL—**Functional equivalence**: The compared systems shall have the same (or only insignificantly different) functional unit in terms of both the primary function and possible secondary functions, as far as possible. In the case that

some of the aspects of the functional unit(s) differ significantly between the systems, it shall be ensured that:

8.1. either the functions that the compared systems provide are still seen as sufficiently comparable by the main stakeholders affected by the LCA study,

8.2. or the sufficient comparability is to be achieved by the respective method approaches for consequential modelling or attributional modelling,[28] as to be applied for the respective Situation. For consequential modelling this approach is system expansion.

9. SHOULD—**Selection of compared alternatives**: The study should include—next to the foreseen alternatives—potentially environmentally better market relevant and available alternatives, as otherwise the study would be considered misleading. If such alternatives are not included, this shall later be highlighted in a prominent place of the conclusions and recommendations, as well as in the executive and technical summary chapters of the report, pointing to this fact. [ISO+]
10. SHOULD—**Selection of production, operation and use scenarios**: To ensure a fair comparison, the chosen functional unit should reflect well-justified typical or average production/operation/use scenarios; it shall be agreed with the affected stakeholders in the best attainable consensus. If a typical or otherwise specific scenarios need to be compared in line with to the goal definition, compared, this fact shall later be highlighted in a prominent place of the conclusions and recommendations and executive summary chapter of the report, pointing to this fact. [ISO!]
11. SHOULD—**Modelling replacements over time**: For cases where a system (e.g. a product) needs to be replaced to meet the required duration of performance of the compared functional unit, the replacement should consider that potentially a newer model or system in general will replace the initially used model. This is unless a different agreement can be achieved among the affected stakeholders. This provision analogously relates to the need of repeating a service.
12. SHALL—Indicative only. **Situation A—Assumption scenarios and uncertainty calculation**: For comparative micro-level studies (Situation A): each compared scenario shall be complemented with assumption scenarios of reasonably best and reasonably worst cases. This can be optionally extended to further assumption scenarios within the reasonably best and worst cases. Uncertainty calculation shall be performed, unless such has already been used to derive the reasonably best and worst-case scenarios. The interested parties shall be involved in achieving a best attainable consensus on the definition of the reasonably best and reasonably worst assumption scenarios. The

[28]Comparisons also can occur in accounting type studies (e.g. across product groups in basket-of-product type of studies), while these shall not be used for decision support that would lead to e.g. purchases or policy measures based on superiority or inferiority of the compared alternatives.

assumption scenarios can in principle vary all methods, data and assumptions except for the "shall" provisions.

13. SHALL—Indicative only. **Situation B—Assumption scenarios and uncertainty calculation**: For comparative meso/macro-level studies of Situation B: the scenarios for each of the analysed alternatives shall apply the modelling guidance of Situation A, except for process that are affected by large-scale consequences of the analysed decision. The assumption scenarios can in principle vary all methods, data and assumptions including the "shall" provisions, but excluding the shall provisions of ISO 14040 and 14044.
14. SHALL—**Involvement of interested parties in review** [ISO!].

37.3.8 Needs for Critical Review

The only strict requirement is to decide, based on the goal definition and intended application of the LCA, whether a critical review shall be performed, and if so which type of critical review:

Identifying critical review needs (Provisions 6.11)

1. SHALL—**Review?** Decide whether a critical review shall be performed and if so: [ISO!]

1.1. **Review type**: Decide along the provisions of the separate document "Review schemes for Life Cycle Assessment (LCA)" which type of review is to be performed as minimum.

> Note that an accompanying review can be beneficial. For Situation B, it can moreover help to organise the best attainable consensus among interested parties, which is required for certain scope decisions.

1.2. **Reviewer(s)**: It is recommended to decide at this point, who is/are the reviewer(s). The minimum requirements on reviewer qualification are discussed in Chap. 13 of this book, which also gives an overview of the review requirements.

37.3.9 Planning Reporting of Results

The level of reporting that is required by the intended application of the LCA must be determined already at the onset of the study to ensure that the data needed for the reporting is produced during the study.

Planning reporting (Provisions 6.12)

1. SHALL—Reflecting on the main type of deliverable (i.e. study or data set) and in line with the decision on the target audience(s) and intended application(s), decide on form and level of reporting:

1.1. **Form of reporting**: Decide which form(s) of reporting shall be used to meet the need of the intended application(s) and target audience(s): [ISO!]

1.1.1. detailed report (including non-technical executive summary),

1.1.2. data set,

1.1.3. data set plus detailed report, or

1.1.4. non-technical executive summary (with references to the full report and review reports, if review has been performed).

1.1.5. The electronic ILCD LCA report template and LCI data set format should be foreseen to be used for reporting.

Confidential information can be documented in a separate, complementary report that is not published but only made available to the reviewers under confidentiality.

Note that any form of reporting, also more condensed ones, shall ensure that the contained information cannot easily and unintentionally be misunderstood or misinterpreted beyond what is supported by the study.

1.2. **Level of reporting**: Decide which level of reporting shall be used in accordance with the defined goal. The main levels are:

1.2.1. internal

1.2.2. external (but limited, well defined recipients)

1.2.3. third-party report, publicly accessible

1.2.4. report on comparisons, publicly accessible

37.4 Inventory Analysis

The inventory analysis is discussed in Chap. 9 of this book. It is the third phase of the LCA, where the product system is modelled and elementary flow data is collected for all the processes in the system and scaled according to the reference flow of the study. The resulting life cycle inventory is the basis of the subsequent impact assessment.

The inventory analysis comprises the following six steps:

1. Identifying processes for the LCI model
2. Planning and collecting data
3. Constructing and quality checking unit processes
4. Constructing LCI model and calculating LCI results
5. Preparing the basis for uncertainty management and sensitivity analysis
6. Reporting

37.4.1 Identifying Processes for the LCI Model

Different approaches are taken for identifying processes for the product system depending on whether an attributional or a consequential modelling approach is taken, as described in Sect. 9.2. The provisions and actions of the ILCD guideline are the following:

Identifying processes in attributional modelling (Provisions 7.2.3)

Applicable to Situation A and C, as well as the life cycle model(s) of Situation B, except for those process steps that are affected by large-scale consequences. Also applicable to the assumption scenarios under Situation B for which it has been decided to apply attributional modelling.

Fully applicable for LCI results, partly terminated systems, LCIA results and LCA studies (and for unit processes only to complete the system model for completeness check and precision approximation).

For black box unit processes as deliverable, only those processes that are foreseen to be included are to be identified, as are the product and waste flows that enter or leave the unit process.

For single operation unit processes only the product and waste flows that enter or leave the unit process are to be identified and specified; the named technical flow diagram in that case only consists of one process plus product and waste flows.

1. SHALL—**Identifying processes within the system boundary**: All quantitatively relevant processes shall be identified that are to be attributed to the analysed system(s) and that lay within the system boundary: [ISO+]

1.1. **Start from central process**: This identification should start from the system's functional unit or the reference flow (i.e. from the central process of the foreground system or the analysed system itself).

1.2. **Foreground system**: Stepwise it should be expanded to the entire foreground system. Following a descriptive "supply chain—use—end-of-life" logic it shall as far as possible identify all relevant product and waste flows (or their functional units) that cross the border to or from the background system.

1.3. **Background system**: The processes in the background system shall be identified in the same "supply chain—use—end-of-life" logic as applied in the foreground system.

> Note that it is established practice to embed the foreground system into a third-party or in-house developed general background system of LCI results and/or unit processes. That means that in practice the identification described above ends with the identification of the product and waste flows that connect the foreground system with the background system. Systems or processes that would be missing in such a general background system are for a given case collected or obtained from third parties as required for the analysed system.

1.4. **Justify and document exclusions**: Any exclusion of relevant individual processes or activity types shall be justified using the cut-off criteria. This can build on previous experience including as detailed in related system/product-group specific guidance documents or Product Category Rules (PCRs). In principle all processes are to be inventoried that are to be attributed to the system, as far as they relevantly contribute to the overall environmental impact of the analysed system. This includes in principle—depending on the included life cycle stages and the system boundary in general—activities such as, e.g. mining, processing, manufacturing, use, repair and maintenance, transport, waste treatment and other purchased services linked to the analysed system, such as, e.g. cleaning and legal services, marketing, production and decommissioning of capital goods, operation of premises such as retail, storage, administration offices, staff commuting and business travel, etc.

> Note that individual processes within the background system may need to be identified as well—in context of identifying sensitive issues or if required to meet the specific goal of the study.
>
> The requirements regarding technological, geographical and time-related representativeness of the scope definition shall be met.
>
> Note that the resulting initial list of processes, product and waste flows typically will need a refinement in view of the results of the completed initial life cycle model, impact assessment and interpretation.

Identifying processes in consequential modelling (Provisions 7.2.4)

Applicable for those processes in Situation B that have large-scale consequences, and for use in assumption scenarios in Situation B (if consequential elements are included in those).

Fully applicable to all types of deliverables, except for unit processes.

Expertise [ISO+]

1. SHOULD—Required expertise: Experts in the following domains should be involved in the study, especially for identifying and modelling large-scale consequences:

1.1. technology development forecasting (e.g. learning curves, experience curves),
1.2. scenario development,
1.3. market cost and market forecasting
1.4. technology cost modelling, and
1.5. general-equilibrium and partial-equilibrium modelling

2. SHOULD—Policy scenario experts required?: The involvement of domain experts for policy scenarios is recommended regarding their function as setting constraints. In the case policy scenarios are explicitly analysed in the study, such experts should be involved.

Identifying consequences and constraints to be considered [ISO+]

3. SHALL—**Modelled consequences**: Identify among the following ones those consequences that will be modelled; this step may be taken separately case for each process. Their potential exclusion shall justified by demonstrating at least argumentative/semi-quantitative that they are not relevant for the results; otherwise the exclusion shall be considered when reporting achieved accuracy (in case of data sets) and when interpreting the results (in case of LCA studies):

3.1. **Primary market consequences**:

3.1.1. SHALL—(a) Processes that are operated as direct market consequence of the decision to meet the additional demand of a product (i.e. "consequential modelling of direct consequences; applied for the full system"). This includes among many others also indirect land use effects.

3.1.2. SHALL—(b) Processes that supersede/complement not required co-functions of multifunctional processes that are within the system boundary (i.e. "solving multifunctionality by substitution", reducing the system boundary to exclude the not required function(s)).

3.2. **Secondary market consequences**:

3.2.1. SHOULD—Increased demand for a co-product if its market price is reduced.

3.2.2. SHOULD—Incentive-effects on a process to increase its efficiency due to a higher price for its product(s).

3.2.3. SHOULD—Decreased demand for competing products of a co-product due to the decreased price of the co-product.
3.2.4. SHOULD—Consumer behaviour changes
3.2.5. SHOULD—Further consequences should only be included if explicitly addressed in the goal of the study.
4. SHALL—**Constraints**: Identify the constraints that will be included in the model and that may partly or fully prevent that the marginal process mix as identified along the primary and secondary consequences can directly be used in the system model. The likely specific effect of any included constraint shall be considered when identifying the effective marginal process (es). Their potential exclusion shall be justified by demonstrating at least argumentative/semi-quantitative that they are not relevant for the results; otherwise the exclusion shall be considered when reporting achieved accuracy (in case of data sets) and when interpreting the results (in case of LCA studies). The following constraints should be considered:
4.1. Existing long-term supply-contracts or co-operations that cannot easily be changed.
4.2. High costs that act as a barrier (e.g. limited mobility of some products due to high transport costs).
4.3. Existing or expected political measures/legal constraints that stimulate perceived positive developments or counteract perceived negative developments. (e.g. a political binding target of X % of energy carrier Y in the fuel mix means that energy carrier X is already pre-set and cannot be assumed to be a long-term marginal product in consequence of the analysed decision.)
4.4. Non-scalability of supply of products or natural resources; including of fully used, dependent co-products of joint production.
4.5. Monopolies, i.e. lack of choice of the supplier or technology.
4.6. It is recommended to also consider other constraints in place or expected to be in place that increase, decrease or block a primary or secondary consequence.

Identifying the mix of superseded processes/systems [ISO+]

5. SHOULD—**Stepwise identification of the mix of superseded processes/systems**: Identify the processes/systems within the system boundary that are superseded as consequence of the analysed decision on the investigated system(s). For each process the following steps should be applied, starting from the system's functional unit or reference flow to the entire foreground system and following the identified consequences and constraints of a theoretical "supply chain—use—end-of-life" logic to

include identifying as minimum all product and waste flows (or their functional units) that cross the border to the background system[29]:

5.1. **Primary market consequence and the size of the effect**: First step—consider the primary market consequence and the size of the effect:

5.1.1. Identify the processes that are assumed to be additionally operated or taken out of operation as primary market consequence of the analysed decision and the directly related additional or reduced demand for a function/product, considering the following:

5.1.2. Size of effect:, EITHER

5.1.2.1. "small"—affecting only the extent of operation of one or more existing processes—the short-term marginal process(es) are the ones that should be assumed to be superseded, OR

5.1.2.2. "big"—resulting in additionally installed or de-installed capacity -> the long-term marginal processes are the ones that should be assumed to be superseded.

5.1.2.3. The effect should generally be considered "small", if the annual amount of additional demand or supply is smaller than the average percentage of annual replacement of capacity of the annual supply of that function or system in the given market; if that average percentage is over 5%, 5% should be used instead. Otherwise, it is "big". The percentage is for orientation only and can be for a given case changed to be smaller or bigger upon the argumentation that the change in demand or supply is directly triggering changes in demand and not only via a marginal accumulative effect in contribution to the general market demand/signal.

5.2. **Secondary consequences and constraints**: Second step—consider secondary consequences and constraints:

5.2.1. If the size of the effect of the primary market consequence is "small", check whether the secondary consequences and constraints in the market counteract the primary consequence (rebound), so that the net effect of the consequences is so small that it is not significantly different from being zero. In that case, the "short-term marginal" is best represented by the "average market consumption mix" of the processes/systems (but see next sub-provision).

5.2.2. For the specific case of multifunctionality, a key constraint occurs if the required co-function is an already fully used, dependent co-function of a joint production process (e.g. copper ore mining with silver as dependent but fully used co-product, egg-laying chicken with the dependent co-"product" chicken being fully used for human food or animal fodder), as additional demand cannot be met by additional supply on a net basis. In that case, the required function/product will have to be produced in another

[29] It depends on the chosen background system model solution whether the processes of the background system also need to be individually identified or whether—if embedding the foreground system into an existing background system—this work has been already done.

way (e.g. for the above examples: silver from silver mine, or meat-chicken directly raised for food or fodder).

5.2.3. If the size of the effect of the primary market consequence is "big", check next whether secondary consequences and market constraints counteract the primary consequence, so that the net overall effect is not "big" but "small".

5.2.4. For those processes that are still facing "big" effects, explicitly consider that the affected processes might have been changed by the secondary consequences and constraints. This has to be analysed specifically to correctly identify the final effect/superseded processes.

5.3. **Market situation and the cost-competitiveness**: Third step—market situation and the cost-competitiveness of alternatives:

5.3.1. Market direction, EITHER

5.3.1.1. a "growing, stable, slightly declining market" (i.e. declining less than the average equipment replacement rate, OR

5.3.1.2. a "strongly declining market" (i.e. declining faster than the average equipment replacement rate).

The above named average displacement rate in % is obtained by dividing 100 years by the average or typical life time of the capital equipment, expressed in years.

5.3.2. Based on this: analyse whether the extent of additional demand or supply for the effect "big" is changing the direction of the market, i.e. from a "strongly declining" market to a "slightly declining, stable, or growing" market OR vice versa.

5.3.3. If this is NOT the case, the affected processes/systems are always the "long-term marginal" processes/systems.

5.3.4. For all "small" and "big" cases in addition the cost-competitiveness of alternative processes/systems is relevant:

5.3.4.1. If the market is "growing, stable or slightly declining", the "short-term marginal" (for "small" effects) and the "long-term marginal" (for "big" effects) are the most cost-competitive processes/systems.

5.3.4.2. If the market is "strongly declining", the "short-term marginal" (for "small" effects) and the "long-term marginal" (for "big" effects) are the "least cost-competitive" processes/systems.

5.3.5. If in contrast the market direction IS changing, both the least and the most cost-competitive processes/systems are superseded and their specific type and share needs to be identified individually, drawing on the other provisions of this chapter.

5.4. **Identifying the mix of processes/systems**: Final step—identifying the mix of "short-term" or "long-term" marginal processes/systems:

5.4.1. In the consequential model, not only one single, short-term or long-term marginal process should be modelled but a mix of the most likely marginal processes, given the high uncertainty of market price forecasts and the often large differences of the environmental profiles among alternative marginal processes. To restrict the model to a single marginal process or system is only justifiable if there are no other, similarly cost- competitive processes or systems and hence the use of a single one is more appropriate.
5.4.2. The final amount of function (process or system) that is superseded shall be approximated considering the combined effect of primary and secondary consequences and constraints.

Note that in case the market direction has changed as consequence of the analysed decision, the superseded processes are a specific combination of the least cost-competitive ones and partly the most cost.

Further provisions, comments and recommendation on documentation (7.2.4.5) [ISO+]

6. SHALL—Observe that:
6.1. **Part-system and system–system relationships**: These need special attention (e.g. for energy related products) and correct inventorying. Note that these cases are modelled identically in attributional modelling.
6.2. **Individual processes within the background system**: These may need to be identified as well when identifying significant issues or if required to meet the specific goal of the study.
6.3. **Meet representativeness requirements**: The requirements regarding technological, geographical and time-related representativeness shall be met.
7. SHOULD—**Indirect land use changes**: The appropriate way how to consider indirect land use changes should be developed. If done this shall be in line with the general provisions on consequential modelling. This is unless specific provisions would be published under the ILCD. Such provisions might be part of a future supplement.
8. MAY—Schematic consequential model diagram: It is recommended using the system boundary scheme for overview. Schematic decision-consequence and flow diagrams of the most relevant consequences and marginal processes of the system(s) may be used to document the main identified consequences and constraints and the resulting resource bases, technologies, affected markets, etc. This can serve as basis for a data collection planning and later documentation.

Note again, that any exclusion of individual processes or activity types shall be justified using the cut-off criteria. In principle all processes are to be inventoried that are operated in consequence of the analysed decision. This includes in principle—depending on the system boundary—activities such as, e.g. mining, processing, manufacturing, use, repair and maintenance, transport, waste treatment and other purchased services such as, e.g. cleaning and legal services, marketing, production and decommissioning of capital goods, operation of premises such as retail, storage, administration offices, staff commuting and business travel, etc.

9. MAY—Initial processes' description: It is recommended to also provide an initial description of the identified unit processes of the foreground system and the detailed functional units of those product and waste flows that link it to the background system. This should complement the documentation of the consequences and constraints and be completed with details during the iterations of the LCI work.

Solving multifunctionality of processes and systems [ISO!]

10. SHALL—**Subdivision and virtual subdivision**: Subdivision and virtual subdivision shall be applied in preference to substitution.[30]
11. SHALL—**Combined production**: For cases of truly combined production, the determining physical causality (i.e. the first of the two steps of allocation under attributional modelling) equally applies analogously.
12. SHALL—**Joint production**: For joint production, substitution as a special case of system expansion is the preferred solution to multifunctionality. This shall be done as follows:

12.1. The same provisions shall apply as for general consequential modelling of the system.

12.2. Note the specific constraint for already fully used, dependent co-products of joint production: since their production cannot be increased with that same multifunctional process/technology, their additional provision cannot be modelled. Instead, alternative routes need to be modelled for their supply.

12.3. If for the not required co-function functionally equivalent alternative processes/systems are operated/provided in a commercially relevant extent, the not required co-function shall be substituted with the mix of the superseded marginal processes (excluding the substituted process-route, if quantitatively relevant). Differences in functionality between superseding and superseded function shall be considered by correction of the actually superseded amount of the superseded process(es) or by market price

[30]Observe that virtual subdivision shall not be done if it "cuts" through physically not separable joint processes, as this would distort the substitution.

correction of the superseded process(es)' inventory (if the superseded amount is not known in sufficient detail).

12.4. If such alternative processes/systems do not exist[31] or are not operated in a commercially relevant extent, the provided function in a wider sense should be used for substitution.[32]

Note that the substituted processes or products may also have secondary functions. This can theoretically lead to the problem of an eternally self-referring and/or very extensive, multiple extended system. As the amount of these secondary functions and their relevance within the overalls system goes down with each process step, this problem can be avoided/reduced by applying the cut-off rules.

Substitution for multifunctional processes and systems in reuse/recycling/ recovery [ISO!]

13. SHALL—**Recycling, recovery, reuse, further use**: Substitution shall be applied for cases of recycling, recovery, reuse, further use:

13.1. **Applying general rules to these cases**: Substitution of products recycled or recovered from end-of-life product and waste treatment follows the same rules as for the general cases of multifunctionality. They shall be applied for all cases of waste and end-of-life treatment (i.e. "closed loop" and of "open loop—same primary route" and "open loop—different primary route"). Subdivision and virtual subdivision shall be applied in preference to substitution.

13.2. **Specific aspects and steps (true joint process, interim processes to secondary good, recyclability, …)**: Specific for reuse/recycling/recovery is that interim treatment steps occur more regularly and that often no truly equivalent alternative process/system exist.[33] In this context, also the true joint process of the secondary good is to be identified. Finally, the steps of reuse/recycling/recovery need to be modelled explicitly until the secondary good is obtained that is actually superseding an alternative process/system.

---

[31]E.g. for wheat grain production, many refinery products, etc.

[32]E.g. as for NaOH apart from NaCl electrolysis, or if for a mobile phone the individual function SMS would not be available as commercially relevant, separate consumer product. NaOH provides the general function of neutralising agent and hence other, technically equivalent and competing neutralising agents, KOH, $Ca(OH)_2$, Na_2CO_3, etc. can be assumed to be superseded. For the case of wheat grain and straw production: instead of straw, other dry biomass (e.g. Miscanthus grass, wood for heating, etc.) provides equivalent functions and can be assumed to be superseded.

[33]This is as secondary goods often have distinctly different properties from primary produced goods (e.g. recycled aged plastics vs. primary plastics), what makes a clear assignment to the equivalent or most similar process/system more difficult.

The actual mix of superseded processes shall be identified for the given case and along the following steps:

13.2.1. The true joint process of the secondary good is that process step in the product's life cycle that provides the good with the closest technical similarity to the secondary good; the thereby identified primary good shall not have a lower market value than the secondary good.[34]

13.2.2. The recyclability substitution approach shall be used for substitution. That implies that all interim waste management, treatment, transport, etc. steps are to be modelled and assigned to the analysed system including the step that is producing the valuable co-function (e.g. secondary metal bar).

13.2.3. The amount/degree of recyclability shall refer to the actually achieved recyclability, i.e. accounting for all kinds of losses, e.g. loss due to incomplete collection, sorting, recovery, during recycling processing, rejection, etc. In short, the recyclability is the %[35] of the amount of end-of-life product or waste that is found in the secondary good(s). For practical reasons and for long-living products this should per convention be the currently achieved recyclability for this product (or for new/projected products the achieved recyclability of comparable products in the same market). This can be another reference if the goal of the study explicitly relates to recyclability scenarios.

13.2.4. The superseded process(es)/system(s) shall be identified applying the general consequential modelling guidance as detailed in the above provisions.[36]

---

[34]This serves to avoid a potentially misleading upscaling of the superseded function's inventory in case of applying market value correction when correcting for the functional differences.

[35]Note that this % needs to relate to the appropriate property and unit of the secondary good, e.g. Mass in kg for recycled materials, Lower calorific value in MJ for recovered energy, Pieces in number for reused parts, etc.

[36]That means that the earlier named constraint for already fully used, dependent co-products of joint production also applies here: since the production of e.g. a recycled metal as dependent co-product cannot be increased with that same multifunctional process/technology (i.e. by producing more e.g. metal goods, what is of course not happening), its additional provision via primary production cannot be assumed. Instead, alternative routes need to be modelled for the supply of the recycled metal. As stated for the general case, the determining co-product shall not be substituted. The following example explains what that means and why for "closed loop" and "open loop - same primary route" cases nevertheless the primary production is to be substituted: Example: the determining co- product of primary and secondary metal is the primary metal. The secondary metal, after recycling, is the dependent co-product. If this one is fully used in the same or other products and from the perspective of the metal product made of primary metal, recyclability substitution is applied, substituting the secondary good by primary metal. From the perspective of the user of the secondary good "recycled metal", the metal primary production shall not be substituted, but alternative ways of supplying the recycled metal shall be modelled. This alternative way is, however—what makes this case apparently specific—the primary production of that metal as this is the only way to increase the availability of the required metal on a net basis. Hence in both cases, primary production is to be substituted, but for different reasons.

13.2.5. Also here not one marginal process should be used but the average inventories of several of the potential marginal processes.
13.2.6. For application-unspecific secondary goods, any reduced technical properties of the secondary good should be corrected in the accredited inventory by using the market price ratio (value correction) of the secondary good to the primary produced replaced function.
13.2.7. For application-specific uses of the secondary goods, sufficient functional equivalence with the superseded good shall be ensured and the credited inventory be reduced to the amount that is effectively superseded. In the case, this cannot be determined, the market price ratio (value correction) shall be applied as in the application-unspecific case.
13.2.8. Especially for the case of "open loop—different primary route" in addition it is to be checked whether commercially relevant alternative processes are operated. Otherwise, the provisions for the general case of solving multifunctionality under consequential modelling shall be applied.
13.2.9. The other guidance aspects of this chapter on identifying the superseded processes (e.g. constraints, secondary consequences, etc.) apply analogously.

Note that for scenario formation in comparisons, the various primary and secondary consequences and constraints should be varied jointly when defining "reasonably best case" and "reasonably worst-case" scenarios.

In the inventory analysis, the treatment of multifunctional processes deserves special attention and should be decided before collecting the process data to ensure that the relevant processes are considered. A detailed guidance is given on this aspect in the ILCD Guidelines. Regardless the goal situation, the first choice is to subdivide a multifunctional process into monofunctional processes if that is possible.

Avoiding allocation by subdivision or virtual subdivision (Provisions 7.9.2)

Applicable to Situation C2. Applicable to cases of Situation A, B, C1 only if subdivision, virtual subdivision and substitution/system expansion were not possible or feasible, as identified along the specific provisions for these Situations.

Applicable only to attributional modelling, unless in consequential modelling substitution is not possible or feasible.

1. SHALL—**Analyse whether allocation can theoretically be avoided by subdivision**: Investigate whether the analysed unit process is a black box unit process: does it contain other physically distinguishable sub-process steps and

is it theoretically possible to collect data exclusively for those sub-processes? Next, check whether subdivision can solve the multifunctionality of this black box unit process: can a process or process-chain within the initial black box unit process be identified and modelled separately that provides only the one required functional output?

2. SHALL—**Aim at avoiding allocation by subdivision or virtual subdivision**: Based on the outcome, the following steps shall be followed:

2.1. **Subdivision**: If it is possible to collect data exclusively for those included processes that have only the one, required functional output: inventory data should be collected only for those included unit processes.

2.2. **Partial subdivision**: If this is not possible (i.e. the analysed unit process contains multifunctional single operation unit processes that are attributed to the required functional output) or not feasible (e.g. for lack of access or cost reasons): inventory data should be collected separately for at least some of the included unit processes, especially for those that are main contributors to the inventory and that cannot otherwise (e.g. by virtual subdivision—see later provision) clearly be assigned to only one of the co-functions. [ISO+]

2.3. **Virtual subdivision**: It should be checked whether it is possible by reasoning to virtually partly or fully subdivide the multifunctional process based on process/technology understanding. This is the case wherever a quantitative relationship can be identified and specified that exactly relates the types and amounts of a flow with at least one of the co-functions/reference flow(s) (e.g. the specific mechanical parts or auxiliary materials in a manufacturing plant that are only used for the analysed product can be clearly assigned to that product by subdividing the collected data). For those processes where this can be done, a virtual subdivision should be done, separating included processes as own unit processes. [ISO+]

2.4. **Justify need for allocation and document potential distortion**: If the preceding sub-steps are not possible and a real or virtual separation is not feasible, allocation is the approach that shall be applied (see next provision). In addition and only if subdivision is theoretically possible but was not performed, it should be demonstrated/argued at least via quantitative approximation or reasoning that the decision for allocation does not lead to relevant differences in the resulting inventory, compared to a subdivision. If it leads to relevant differences, the respective cases shall be documented and shall later be explicitly considered when assessing the achieved accuracy of data sets and when interpreting the final results of LCA studies, respectively. [ISO!]

If subdivision is not possible, allocation is the choice for Situation C2 and for those cases in Situation A, B and C, where subdivision, virtual subdivision and substitution/system expansion was not possible or feasible.

Solving multifunctionality by allocation (Provisions 7.9.3)

1. SHALL—**Share inventory between co-functions by allocation**: If allocation is to be done, the environmental burden of the concerned processes shall be shared between the co-function(s) of the process or system by allocation.
2. SHALL—**Differentiate multifunctional processes and multifunctional products**:
   These two cases shall be differentiated [ISO!].
3. SHALL—**Two-step procedure for multifunctional processes**: The following two-step procedure[37] shall be applied [ISO!]:

3.1. **First step and criterion "determining physical causality"**: As first criterion, the "determining physical causal relationships" between each non-functional flow and the co-functions of the process shall be identified and used as allocation criterion. This relationship is the one that determines the way in which quantitative changes of the products or functions delivered by the system change the other inputs and outputs. Within this step, process-related inventory flows (e.g. spontaneous NOx in incineration, consumption of auxiliary materials) should be differentiated from function (product) related inventory flows (e.g. the NOx from the nitrogen in the incinerated fuel, materials or parts ending up at least partly in the co-products).

Note that often a combined, multiple allocation of the different non-functional flows to the co-functions is necessary, applying different criteria for the different flows.

Note also that the preceding step of virtual subdivision is applying the same logic as physical causality allocation.

3.2. **Checklist for "determining physical causality" criteria**: If this is not possible or for any remaining inventory items, the following list gives guidance which criteria should be analysed by default whether they are the "determining physical causal relationship" to be used for allocation in different cases of co-servicing and co- production processes:

3.2.1. **Services**:

- Goods transport: time or distance AND mass or volume (or in specific cases: pieces) of the transported good
- Personal transport: time or distance AND weight of passengers

[37]The need is seen to develop supplementing practice-manuals in line with the ILCD and with explicit allocation- criteria/rules for main process and product groups, to further enhance practicability and reproducibility. This could follow the same general logic as applied when developing Product Category Rules (PCR) in support of Environmental Product Declarations (EPD).

- Staff business travel: added value of system
- Staff commuting: added value of system
- Retailing: time (duration) of shelf-life AND mass or volume of good
- Storage and shelter, i.e. buildings and other three-dimensional infrastructure: time (duration) of use AND volume of good OR area occupied by the good
- Storage and other functions provided by places and other two-dimensional infrastructure: time (duration) of use AND area occupied by the good
- Transport and communication on roads, railways, pipes, cables and other one-dimensional infrastructure: time (duration) AND intensity (e.g. road wearing impact by vehicles of different weight) OR bandwidth of use.
- Heating/cooling of space (keeping a temperature): time (duration of heating/cooling) AND area or volume heated/cooled (depending whether the space is used by area such as in offices, or by volume such as in staple storage halls or retail freezers)
- Heating/cooling of goods (reaching a target temperature): heat capacity of good
- Private administration services: person time or cost charged for admin services OR market value of sales
- Public administration services: person time or cost charged for admin services OR number of cases serviced
- Cleaning services (of objects of similar cleaning technologies): surface area cleaned (or as fallback option: time (duration) of cleaning)
- Guarding services: share of product's value among guarded products AND/OR the production/provision facilities' value of the product among guarded site/object, depending what is the purpose of the guarding
- Marketing services: share of product implicitly or explicitly addressed by marketing (e.g. corporate marketing: share of product's value in corporate turnover)
- Teaching/training services: person time (duration) of training AND number of individuals taught/trained
- R&D services (of objects of similar R&D): person time OR cost charged for R&D services

3.2.2. **Production processes**:

- Extraction processes: for process-related flows the market value, for product-related flows the specific physical properties of the co-products
- Chemical conversion and waste processing (including incineration): quantitative change of the to-be-allocated flows in dependency of quantitative changes in the products or functions delivered by the system. If unknown: the chemical or physical properties that determine the amount of the other flows

- Manufacturing (including physical transformation processes) and mechanical waste processing: length, surface, volume, or mass OR number of items OR time of processing
- General processes by other capital goods' input directly to multifunctional processes (e.g. the processing machines themselves, but not buildings, etc.): time (duration) of use OR mass, volume, length of produced good

3.3. **Justify selection from checklist**: In the case alternatives are given in the above provisions, the chosen alternative shall be concisely justified.

3.4. **Justify other criteria**: If another specific relationship is applied that is not listed above, that choice shall be concisely justified including explaining why none of the default provisions is applicable or the most suitable ones, along the guidance given in the text.

3.5. **Justify non-existence of determining physical causality**: If a "determining physical causal relationships" does not exist (i.e. it is not in the above list and no other can be identified), this shall be concisely justified. Only in that case the second allocation step should be applied (see below); otherwise, the resulting lack of accuracy and potential distortion is to be documented and explicitly be considered in the results interpretation.

4. SHOULD—**Second step and criterion "market price"**: As second, general allocation criterion for multifunctional processes, the market price of the co-functions should be applied. If this is done, the price shall refer to the specific condition and at the point the co-functions leave or enter[38] the multifunctional unit process or are provided. This means for processes that the known, calculated or approximated market price shall relate to, e.g. the specific technical characteristics in quantity and quality such as purity, compressed or not, packaged or not, etc. as well as bulk or small amounts, etc. at the point it leaves the process. If this cannot be done, the resulting lack in accuracy and potential distortion of the results shall be documented and be considered in the results interpretation.

5. SHOULD—**Two-step procedure for multifunctional products** (e.g. consumer products): The following two-step procedure shall be applied: [ISO!]

5.1. **First step and criterion "determining physical causality"**: As first criterion, the "determining physical causal relationships" between each non-functional flow and the co-functions of the product should be identified and applied. The above guidance for multifunctional flows can be applied analogously.

5.2. **Use virtual subdivision principle to perform explicit allocation**: As an initial step, analogously as above for multifunctional processes, the logic of virtual subdivision should be applied to virtually subdivide the multifunctional product.

[38]"Enter" in case of waste and end-of-life treatment services.

5.3. **Second step and criterion "QFD" or "market price"**:

5.3.1. **Preferred second criterion—Quality Function Deployment**: If the above cannot be done, the Quality Function Deployment (QFD) should be used to identify the relevance of the co-function from the user's perspective. If a QFD does not exist and cannot be developed (e.g. due to cost or timing reasons), the second, general allocation criterion of "market price" of equivalent products for the single co-functions can and shall be applied.

5.3.2. **Alternative second criterion—market price**: If the QFD is not feasible, allocation by market price should be done in analogy to the preceding case for multifunctional processes. For products, the representative price of products that provide an equivalent to each single function should be used to allocate among the co-functions of the multifunctional product. [ISO+]

6. SHALL—**Attributional modelling of reuse, recycling and recovery**: The following provisions shall be applied in attributional modelling of recycling and related: [ISO!]

6.1. **Follow general rules for multifunctionality, observing specific aspects**: Allocation of products from end-of-life product and waste treatment shall apply the same general rules as other cases of multifunctionality, with two specific aspects:

6.1.1. **Dealing with waste and end-of-life products of negative market value that generate secondary goods**: Specific is firstly that in case the market value of the end-of-life product or waste is below zero (e.g. soiled post-consumer packaging waste), the appropriate process step at the system boundary to the next life cycle is to be identified, i.e. where the allocation is to be applied. This process step is that one where the valuable co-function is created after one or more initial treatment processes have taken place (e.g. sorted plastic fraction of the above waste).

6.1.2. **True joint process to be identified**: Specific is secondly that for end-of-life products and waste the true joint process is to be identified, which is separated by various, e.g. manufacturing steps from the step where the end-of-life product occurs:

6.1.2.1. For waste or end-of-life products with a market price equal or above zero, the true joint process is that process earlier in the life cycle of the system, where the good (e.g. an aluminium bar) is technically approximately equivalent to the secondary good of the waste or end-of-life product (e.g. aluminium scrap from construction demolishing). Note that for "open loop —different primary route" recycling this step might necessarily involve abstraction to the basic properties of the two products. These two products that have been identified as described above are then considered co-products of the true joint process.

6.1.2.2. For waste and end-of-life products with a market value below zero, the true joint process is that one, which produces that product that is about equivalent to the first valuable product that is produced from the initial waste treatment processes, as described in the preceding provision. These two products that have been identified as described above are then considered co-products of the true joint process.

6.1.2.3. In the case of multiple functions from the waste or end-of-life product (e.g. a complex consumer product is discarded for recycling of its many materials and for energy recovery), there is each one true joint process for each of them that shall be identified.

6.2. **Provisions**: The following provisions can be derived that shall be applied, differentiating between waste/end-of-life products with negative and positive market value:

6.2.1. **Negative market value**: If the market price of the waste/end-of-life product is below zero:

6.2.1.1. The waste/end-of-life management/treatment processes until excluding the process where the pre-treated waste crosses the "zero market value" border (i.e. when a process is generating a function with positive market value) shall be allocated exclusively to the first system. In the case the exact process step or the waste and/or secondary good properties cannot be clearly identified, the resulting lack of accuracy shall be reported and later be considered in the results interpretation

6.2.1.2. Subsequently, the two-step allocation procedure shall be applied between the valuable secondary good and its co-product from the true joint process (i.e. see the next provision). This involves a second, additional allocation exclusively of the inventory of that process step that has produced the first valuable product after the initial waste treatment steps, as follows:

6.2.1.3. The inventory exclusively of the process step that produces a valuable product (secondary good) should be allocated with the market value criterion between the secondary good(s) and the (potentially pre-treated) waste/end-of-life product that enters this process step. The burdens that are allocated to the pre-treated waste/end-of-life product belong to the first system, the ones assigned to the secondary good(s) to the second system (s). Note that the market value of the pre-treated waste/end-of-life product is below zero and that hence the absolute value of its (negative) market price[39] should be used when calculating the allocation key; the rest of the allocation calculation is the same.

6.2.1.4. After that, the two-step allocation is applied between the valuable secondary good and the true joint process, as follows in the next provision, i.e. analogous to the case when the waste or end-of-life product have a positive market price.

6.2.2. **Market value equal or above zero**: If the market price of the waste/end-of-life product is equal or above zero, the two-step allocation procedure shall directly be applied between the process step that generates the waste or end-of-life product and the true joint process. The following procedure shall be applied:

---

[39]E.g. if the market value/gate fee is "−1 US$" this would be "1 US$".

6.2.2.1. As first criterion, the "determining physical causal relationships" between each non-functional flow and the co-functions of the process shall be identified and applied. This is worked out as follows:

6.2.2.2. Two sub-cases are to be differentiated: the first one is where the secondary good is undergoing none or limited changes in the inherent properties (e.g. metal recycling, fibre recycling) and the second one is where it undergoes relevant changes in the inherent properties (e.g. energy recovery from mixed polymer waste). The first sub-case applies to all "closed loop" and "open loop—same primary route" situations. The second sub-case applies to all "Open loop—different primary route" situations.

6.2.2.3. For the first sub-case, the total number of cycles and the therefrom derived total amount of uses (considering the loss at each cycle; concept see text) is determined and used for allocation across the many uses including the initial production up to the true joint process. In result the following formula can be developed for an infinite number of loops (considering the losses at each loop):

6.2.2.4. $e = (P + W) * (1 - r) + R * r$

with

- e average LCI per unit of material, part, or energy carrier
- r average recycling rate [0…1), incorporating both collection efficiencies and processing efficiencies
- P LCI of primary production per unit of material, part, or energy carrier
- W LCI of final waste management per unit of discarded material, part, or energy carrier
- R LCI of effort for reuse/recycling/recovery per unit of material, part, or energy carrier

6.2.2.5. The allocation formula is to consider in addition the change in the inherent properties of the secondary good.

6.2.2.6. If the above cannot be done because information that is required for applying the formula cannot be obtained or at least approximated, the second step of "market value" allocation needs to be applied. In that case, it must be detailed and justified why the above cannot be applied. It shall be also demonstrated that the market value allocation is not disfavouring any competitor product, if the results are intended to be used for comparisons.

6.2.2.7. For the second sub-case, i.e. where the recycled/recovered/reused good undergoes relevant changes in the inherent properties, the true joint process is the one along the production chain that produces the minimum required quality[40] of the good to generate the secondary good. (E.g. in case of soiled low value LDPE post-consumer plastic waste that is incinerated to recover the energy: As the LDPE is incinerated and basically only the lower calorific value is of interest, the minimum required good is even before the production of the LDPE—the crude oil (incl. transport to the country of LDPE production) is meeting the minimum requirements in this case.) Based on this, the general two-step allocation procedure shall be applied between the secondary good and the function(s) or the true joint process.

6.2.2.8. If several functions are generated from the waste/end-of-life product (e.g. different metals recovered), this shall be done individually with each of the true joint processes.

7. SHALL—**System-wide consistent application of allocation**: Consistency shall be ensured as far as possible, using the same allocation criteria for the different co-functions of any specific process and across all similar processes within the system boundary. Otherwise, the lack of consistency and its effect on accuracy, precision and completeness shall be considered when stating the quality of a data set or when interpreting the results of an LCA study, respectively.
8. SHALL—**100% rule**: The sum of the inventories allocated to all co-products shall be equal to the inventory of the system before allocation was done.

37.4.2 Planning Data Collection

The planning of the data collection has the purpose of balancing the invested effort against the relevance of the respective data and information in order to avoid wasting time on collecting high quality data that have a low relevance for the LCA results and/or spend too little time on collecting high quality data where it is highly relevant for the results.

Planning data collection (Provisions 7.3)

1. SHALL—**Identify newly required, study-specific unit processes**: Identify for which processes of the analysed system new, study-specific unit processes have to be developed with producer or operator specific primary and secondary data. This is typically the case for the entire foreground system (including for those parts of existing or planned contractual relationships). The use of technical process or flow diagrams is recommended.
2. SHALL—**Average and generic data**: Identify for which parts of the analysed system the use of average or generic LCI data sets is more appropriate. Note that

[40]Note that this provision ensures fulfilling the ISO 14044 provision on considering the change in inherent properties of the secondary good.

for a given case, average or generic data may be more accurate, complete and precise also for some processes of the foreground system. If such will be used, this shall be justified.

3. MAY—**Identify data and information sources**: It is recommended to systematically identify sources for the required data and information. This includes considering working for the background system primarily with LCI results or with unit process data sets, which both have advantages and disadvantages that are for the given case to be evaluated. Combinations are possible if the data is consistent. Among the LCI data sources, primary and secondary sources can be differentiated. Guiding principle should be the availability and quality of the most appropriate data. Working with well-documented and already reviewed data sets is recommended. This supports a correct use of the data sets, a sound documentation of the analysed system and its review. [ISO+]
4. MAY—**SI units**: It is recommended to aim at collecting data in the Système international d'unités (SI) units, to minimise conversion efforts and potential errors. [ISO+]

> Note that SI units shall be used for reporting.

5. SHOULD—**Multi-annual or generic data to be preferred?** Evaluate along the goal of the study whether multi-annual average data or generic data should be preferred over annual average data as better representing the process/system. This applies for processes with strong inter-annual variations (e.g. agriculture; producer-specific data in general), to ensure sufficient time-related representativeness. [ISO+]
6. MAY—**Relevance-steered data collection**: It is recommended to steer the effort for data collection by the relevance of the respective data and information. Building on existing experience that sufficiently reflects the analysed process or system and that is of high quality is an essential guide. Product Category Rules (PCR) and product-group specific guidance documents can represent this experience. The following is meant to help focusing data collection efforts. The initial data quality and data set quality requirements as identified in the scope definition may need to be fine-tuned/adjusted in subsequent loops as follows: [ISO+]

6.1. For the identification of quantitative LCI data quality needs, determine/estimate the accuracy, completeness and precision of the LCIA results that is required by the intended application (e.g. to allow identifying significant differences among compared alternative products).

6.2. Translate these requirements to related requirements at the level of elementary flows by taking into account the impact potentials of the individual elementary flows and by disregarding the uncertainties/inaccuracies associated with the characterisation factors.

6.3. Use these requirements on the elementary flows to determine the maximum permissible uncertainty, inaccuracy and incompleteness of the overall inventory of the to-be-collected or purchased processes' or systems' inventories.

Note that this includes systematic uncertainties from LCI methods and models applied and from assumptions made when setting up the system model.

6.4. Use this information as indicative guidance on quality requirements in the collection or purchase of inventory data (i.e. unit process or LCI results and similar data sets). For secondary LCI data sets, it is recommended to consider the following additional quality aspects: appropriate documentation, the use of compatible elementary flows and nomenclature, methodological consistency and a completed qualified external review.

37.4.3 Constructing and Quality Checking Unit Processes

The following provisions have the purpose to guide the construction of unit processes and ensure that the process data is appropriate to represent the performance of the model that it represents.

Types of input and output flows to collect (Provisions 7.4.2.4)

1. SHALL—**Types of input and output flows**: Quantitative data of all relevant inputs and outputs that are associated with the unit process shall be collected/modelled, as far as possible. Where not possible, the gaps shall be documented and if they cannot be overcome be considered when reporting the achieved data quality and when interpreting results of a study. These flows typically include, if relevant for the modelled process/system:

1.1. Input of "consumed" products (i.e. materials, services, parts, complex goods, consumables, etc.), as product flows.

1.2. Input of wastes (only in case of waste servicing processes), as waste flows.

1.3. Input of resources from nature (i.e. from ground, water, air, biosphere, land, etc. and with possible further sub-compartment specifications as required by the impact assessment methodology to be applied), as elementary flows.

1.4. Emissions to air, water and soil (with possible further sub-compartment specifications as required by the impact assessment methodology to be applied), as elementary flows

1.5. Other input and output side interventions with the ecosphere (if required by the applied LCIA methods), as elementary flows.

1.6. Output of wastes (e.g. solid, liquid, gaseous waste for waste management within the technosphere[41]), as waste flows.
1.7. Output of valuable goods and services provided by the process, as product flows.

Data and information types for specific, future and generic data sets (Provisions 7.4.2.5)

1. SHOULD—**Raw data types**: Raw data types that should be used for the process, as required: [ISO+]

1.1. **Measured data** collected by/at process operators should be preferred if possible and appropriate. Measurements are not only physical measurements of, e.g. emissions but also other specific information for the operated process such as, e.g. bills and consumption lists, stock/inventory changes and similar.
1.2. **Element composition and energy content** of product and waste flows. This data should later be inventoried as flow property information for these flows to support interim quality control, review and improving data quality.
1.3. **Various other data** can be helpful (also for crosschecks) or even necessary (to fill gaps). These are, e.g. recipes and formulations, part lists, patents, process engineering models, stoichiometric models, process and product specifications and testing reports, legal limits, market shares and sizes, data of similar processes, BAT reference documents, etc.
1.4. **Use stage information**: For modelling the use stage of consumer products and initial waste management, it is recommended to use surveys and studies that analyse the average or typical user behaviour to complement product specifications and user manuals. Information provided in product category rules (PCR) can be supporting.

Representativeness regarding operation conditions (Provisions 7.4.2.7)

1. SHALL—**Full operational cycle of the process, if required**: The collected inventory data for a specific process shall as far as possible and required to meet the goal represent the full operational cycle of the process. This includes all quantitatively relevant steps such as, e.g. preparation, start, operation, closure, standby and cleaning as well as maintenance and repair of the process/system and under normal and abnormal operating conditions. This is unless the data set is meant to represent only a partial cycle. The above applies analogously also to services. The achieved representativeness of the data shall be documented.
2. SHOULD—**One full year as data basis**: For measured data of operated processes, data for at least one full year should be used as basis for deriving representative average data. A sufficient number of samples should be taken and the uncertainty be considered when reporting the precision.

---

[41]The emissions resulting from waste that is directly discarded into the environment shall be modelled as part of the LCI model, with the processes considered to be part of the technosphere.

3. SHOULD—**For parameterised processes**: The mathematical relations should represent the relevant changes of the inventory in dependency of the influential parameters, which can be, e.g. technical, management, or others. This can include quantitative and qualitative relationships between inventory flows. [ISO+]

Note that the mathematical model and its relevant assumptions and limitations later will need to be documented as well.

Guidance is also given with hints on how to perform quality control of the collected data for a unit process.

Interim quality control (Provisions 7.4.2.11)

Many of the following provisions on interim quality control are only recommendations, but the same controls may be part of a subsequent mandatory external review.

General approach

1. SHALL—**Validity check**: A validity check of the collected data shall be performed during the process of data collection and unit process development, to confirm that the data is in line with the goal and scope requirements. The following provisions provides related operational recommendations on this requirement:
2. MAY—**Interim quality control as review along "interpretation" provisions**: For the interim quality control on the unit process level, it is recommended to apply the data quality related technical aspects of the critical review regarding the scope and methods of review together with the guidance on interpretation (especially significant issues, sensitivity check, completeness check and consistency check). These steps can, however, be done in a less formal way. Among others, the following may be done at this point: [ISO+]

2.1. **All relevant flows?** Does the unit process inventory include all relevant product, waste and elementary flows that would be expected based on, e.g. the input of processed materials, of the nature of transformations occurring in the process, and/or based on experience gained with similar processes? Reflect the required technological, geographical and time-related representativeness.

2.2. **Flow amounts are proportionate?** Are the amounts of the individual flows and of the chemical elements, energy and parts in the input and output in expected proportion to each other?

2.3. **Support control by impact assessment**: Controls may also be based on impact assessment results for the process as well as for the whole system. They may reveal errors in the inventory results through showing unexpected high or low values of contributing elementary flows. Compare the LCIA results with data of the same or similar processes/systems from other sources to identify possible problems. Make sure the other sources are of high quality and especially high completeness.

2.4. **Method consistency?** On the system level, carefully check that methods have been applied consistently. This especially applies if combining data from different sources.

2.5. **Follow up on discrepancies**: Check and explain or correct any observed discrepancies in the inventory data by consulting additional data sources or technical experts for the analysed process.

2.6. **Report on findings**: It is recommended providing for the unit process data set at least a brief internal quality control report on the above findings.

2.7. **Reflect findings in data set quality indicators**: Make sure that the data set documentation appropriately describes the process and the identified accuracy, precision and completeness as well as any limitations.

Obtaining better unit process data

3. SHALL—**Dealing with initially missing data**: The potential importance of initially missing data shall be checked in the following way and relevant gaps shall be filled if possible and as detailed below: [ISO!]

3.1. SHOULD—**Identify relevance of initially missing data**: A reasonable worst case or at least conservative value for the missing data should be used in a first screening to see if they may influence the overall results of the LCI/LCA study. This reasonable worst case or conservative value may be derived by inference from knowledge of similar or related processes or from correlation or calculation from other flows of the process. This includes identifying and inventorying flows that were initially not known to occur in the analysed process but that could not be excluded entirely.

3.2. SHOULD—**Dealing with relevant, initially missing data**: If this screening shows that the missing data may be of importance, in further iterations of the LCA work it should be attempted to first identify whether the flow is actually occurring in the analysed process and if so to get the yet missing data. As second option sufficiently good estimates should be obtained. As third option, if also that is not possible, the gap should be kept and reported. (Details see separate provisions more below):

3.3. SHALL—**Filling data gaps with estimates of defined and minimum quality**:

3.3.1 SHALL—For each newly modelled unit process any initially missing data should be documented in a transparent and consistent way. At the end of the iterative steps of improving the data set, the finally missing data and the potential use of data estimates to fill data gaps shall be documented in a transparent and consistent way.

3.3.2. MAY—For judging the relevance of an initial data gap, it is necessary to approximate the achieved accuracy, completeness and precision of the overall environmental impact on system level. This necessarily needs that the subsequent steps of modelling the life cycle and calculating LCI results and LCIA results need to be done first. It is recommended to do this in parallel to developing the unit process data set. For unit processes this means completing the life cycle model around the unit process with background data. Any limited completeness in the used background data shall be not considered when calculating the achieved degree of completeness for the unit process for the final reporting.

3.3.3. MAY—For filling data gaps for single flows estimate data (sets) may be considered to be used. Such may be, e.g.:

3.3.3.1 generic or average data for missing specific data,

3.3.3.2. average data of a group of similar products for missing inventory data for other, not yet analysed products of that group,

3.3.3.3. correlation with other, more complete and high quality data for the same or similar process but from other data sources (e.g. industry average data for improving a producer-specific process),

3.3.3.4. justified judgements of technical experts/process operators.

3.3.4. SHALL—Data gaps shall generally be filled with methodologically consistent data. Gaps of low relevance may also be filed with methodologically not fully but sufficiently consistent data sets while being developed along the guidance of this document and meeting the overall quality requirements as detailed below.

3.3.5. SHALL—Only data that increase the overall quality of the final inventory of the analysed system shall be used to fill data gaps. That means that the individual data/data set's overall quality (i.e. combined accuracy, precision, completeness and methodological appropriateness and consistency) shall be equivalent to at least the "Data estimate" quality level.

Note that this shall include both the quality of the used data estimate and of the amount of the flow. That semi-quantitative approximation of the integrated data estimate plus flow amount quality shall be based at least on an individually, briefly justified expert judgement, explicitly considering the named shortcomings; this may be supported by uncertainty calculation and quantitative calculation of data accuracy.

Dealing with remaining unit process data gaps/missing data

4. SHALL—**Document remaining data gaps**: If data estimates cannot be made available that would meet the above requirements, the data gap shall be kept and be documented instead. The following provisions are made: [ISO!]

4.1. **Missing qualitative information for a unit process inventory item**: The respective flow should be created and used in the regular inventory only if it is a product or waste flow. Little specified elementary flows (e.g. "Metals to air") shall not be kept in the regular inventory but this information shall be documented in another way. This can be either as clearly marked flows that shall not be combined with the elementary flows of the regular inventory when aggregating the data sets of the analysed system, The flows can be marked, e.g. as "missing important" or "missing unimportant", as applicable (see more below), and be excluded from the aggregation. Or they can be documented exclusively in the descriptive information of the data set (e.g. as attached lists).

4.2. **Missing quantitative information for a unit process inventory item**: The flow should be inventoried. If no quantitative information can be given, this has to be documented by marking the flow as "missing important" to avoid misleading readers, as the true value is not zero. The omission must be explicitly addressed and considered in the interpretation of the results. If a conservative estimate for a missing data fails to show any quantitative importance, a zero value may be entered for this data, but marking it as "missing unimportant". If a mean value or a wide range of values (Min and Max) can be given, this should be entered in the inventory. Uncertainty information such as standard deviation and distribution type should be given if possible and if this information has sufficient precision. For both the above cases, the values shall not be aggregated when calculating LCI results. This can be achieved, e.g. by marking theses inventory items as "missing important" or "missing unimportant", as applicable (see more below), and excluding such flows from the aggregation. Or they can be documented exclusively in the descriptive information of the data set (e.g. as attached lists).

4.3. **Missing qualitative and quantitative information**: See preceding two points that are to be combined.

4.4. **Missing LCI data for processes/systems in the background system**: When aggregating the unit processes of the analysed system to LCI results, product and waste flows for which background data of sufficient quality is not available, these flows shall remain in the aggregated inventory, i.e. making the data set a "partly terminated system". The user of such data shall be explicitly informed in a prominent place that these parts of the system need to be still completed or the gap be considered in the further use and interpretation.

Note that any kind of worst case or conservative data and assumptions shall not be kept in the inventory of LCI data that are foreseen to be applicable for comparisons, unless the representing process operators or system producers themselves wish so (e.g. to align LCI data reporting with other values reported on, e.g. site or company level). Note that reasonably worst-case data may, however, be used for scenarios and for checking the robustness of comparisons when doing the sensitivity analysis.

Note the specific requirements for product comparisons such as on, e.g. the consistency of methods, data quality, and assumptions across the compared alternatives.

Guidance is offered on the handling of potentially problematic types of substance flows in the inventory analysis of processes—measurement indicators and groups of elementary flows, ionic compounds, airborne particle emissions, resource uses and energy use indicators.

Emission of measurement indicators and elementary flow groups (Provisions 7.4.3.3)

1. SHALL—**Measurement indicator and substance group elementary flows**: These shall be inventoried as follows: [ISO!]

1.1. **Avoid indicators and flow groups; with permissible exceptions**: Measurement indicator and substance group elementary flows shall be avoided in the inventory by splitting them up to single substances. Exclusively the following exceptions are permissible, while they should be split as well: COD59, BOD, AOX, VOC, NMVOC, PAHs, PCBs, TOC, DOC, Nitrogen in Nitrogen compounds (excluding N2, N20), Phosphorus in Phosphorus compounds, Dioxins (measured as 2,3,7,8-TCDD human toxicity equivalents).

1.2. **Restrictions on partial splitting**: A partial splitting up of measurement indicators and substance group flows should be avoided. This is except for singling out exclusively elementary flows that have higher impacts than the average of the indicator/group and that should be singled out. Partial splits with singling out elementary flows with less than average impacts shall not be done. If singling out single substance elementary flows from the above indicators/flow groups, only the remainder amount of the indicator or flow group shall be inventoried.

1.3. **No double-counting**: Double-counting across the above indicators/flow groups and with the contained individual substances shall be avoided (i.e. correct is to inventory either "BOD" or "COD"[42]; either "VOC" or "NMVOC" plus "Methane"; either "Nitrate" plus "Ammonia" plus … or "Nitrogen in Nitrogen compounds"; etc.).

1.4. **Document composition**: If measured composition information of a split measurement indicator or substance flow group is not available, an assumed composition can be used. Approach and assumptions shall be documented.

> Note that the composition of a measurement indicator or substance flow group can often be derived without direct measurement from process know-how (e.g. processed materials, educts, etc.) or those of sufficiently similar process can be considered.[43]

[42]COD = Chemical oxygen demand, BOD = Biological oxygen demand, AOX = Adsorbable organic halogenated compounds, VOC = Volatile organic compounds, NMVOC = Non-methane volatileorganic compounds, PAH = Polycyclic aromatic hydrocarbons, PCB = Polychlorinated biphenyls, TOC = Total organic carbon, DOC = Dissolved organic carbon.

[43]Default-composition tables for different process-types and industries mightbe developed in PCR-type or sector- specific guidance documents.

1.5. **Do not combine measured flows**: Individually measured substances shall not be integrated/combined into measurement indicators and elementary flow groups but be inventoried individually.

Emission of ionic compounds (Provisions 7.4.3.4)

1. SHALL—**Inventory easily water-soluble salts as ions**: For data sets as deliverables, emissions to air, water, or soil of easily water-soluble ionic compounds (salts) shall be inventoried as separate ions, unless the selected LCIA methods would require otherwise. As convention, the limit is set at a solubility in water at 20 °C of 10 μg/l, above which the ions shall be inventoried separately, below which the compound shall be inventoried. This applies unless the selected LCIA method requires otherwise. [ISO!]

Emission of particles to air (Provisions 7.4.3.5)

1. SHALL—**Inventory only poorly water-soluble compounds as particles**: Particulate matter (PM) emissions to air shall include only poorly water-soluble compounds below a solubility in water at 20 °C of 10 μg/l, as far as feasible. Expert judgement may be needed to identify the composition of the particles. [ISO!]
2. SHOULD—**Differentiate particle size classes**: Particles should be reported split up by particle size class <0.2, 0.2–2.5, 2.5–10, >10 μm if the information is available. <10 μm may be used alternatively is a more differentiated information below 10 μm is not available. This applies unless the selected LCIA method requires otherwise. [ISO!]
3. SHALL—**Inventory particles additionally as the substances they are composed of**: Particles shall be inventoried as both PM and additionally as elementary flows of their environmentally relevant components (e.g. metals contributing to cancer effects), i.e. double counting their mass in the inventory, as far as possible. This applies analogously to other emissions with additive action schemes. [ISO!]

Resource elementary flows (Provisions 7.4.3.6)

1. SHALL—**Provisions for inventorying resource elementary flows**: Resource elementary flows shall be inventoried as follows, with exceptions only if necessary to meet the need of the applied LCIA method: [ISO!]

1.1. **Energy resources**:

1.1.1. **Non-renewable**: These shall be inventoried as type of energy resource and in few cases (only primary, secondary, tertiary crude oil and open pit or underground mining of hard coal) these should be differentiated exclusively by resource extraction type, if this information is available (e.g. "Crude oil, secondary extraction" but not "Crude, Tia Juana Light"; "Hard coal, underground" but not "Hard coal, Western Germany; 39.4 MJ/kg"). The energy/mass relationship shall be provided for all energy resource flows except for nuclear ores. The energy content shall be expressed in the Lower calorific value of the water-free resource, measured in the reference unit MJ. See also separate document "Nomenclature and other conventions".

Note that peat, biomass of primary forests, and some other biogenic energy resources are "non-renewable".

1.1.2. **Renewable**: Renewable energy resources shall be inventoried as the amount of usable energy extracted from nature, e.g. for solar electricity and heat this relates to the amount of electricity and/or heat captured by the solar cells (i.e. not the total solar energy, but what is delivered directly by the cells as electricity and/or usable heat). For biomass from nature this is the amount physically embodied, measured as Lower calorific value, however, of the water-free substance (i.e. measured as if the, e.g. wood would be oven-dry). Note that biomass from fields and managed forests is no elementary flow. In that case, the named energy resources shall be inventoried directly as the respective elementary flows, e.g. "Solar energy" as "Renewable energy resources from air", expressed as Lower calorific value and measured in the reference unit MJ.

1.2. **Avoid geographical differentiation**: Resources shall not be inventoried geographically differentiated (i.e. "Lignite" but not "Lignite, Eastern Germany"). This applies unless the selected LCIA method requires otherwise.

1.3. **Chemical element resources**: Resources for production of metals or other chemical elements should be inventoried as chemical element (e.g. "Iron—Resources from ground" elementary flow).

1.4. **Functional/material resources**: These shall be inventoried as target material resource (e.g. "Schist", "Lime stone", "Anhydrite"). Few exceptions exist where the mineral itself is in industry understood to be the target good; these are reflected in the ILCD reference elementary flows (e.g. "Rock salt", etc.). Other exceptions and exclusively for resources not included in the ILCD reference elementary flows shall be justified by following analogous logic.

1.5. **Flows for completing mass balance**: For completion of the mass balance, a complementary amount of "Inert rock", "Water", or "Air" (or other, as applicable) shall be inventoried for extracted resources (e.g. 0.96 kg "Inert rock" in case of mining 1 kg copper ore with 4% copper content).

1.6. **No minerals or ore bodies**: Inventorying of other minerals (unless these are functional/material resources such as "Granite") or of specific ore, bodies shall not be done (i.e. "Copper", but not "Malachite" and not "Sulphidic copper–silver ore (3.5% Cu; 0.20% Ag)").

Note that when applying the above rules double-counting shall be avoided. Newly created elementary flows shall be checked whether they require carrying a characterisation factor for the applied LCIA method.

2. SHALL—**Land use and transformation**: Direct land use and land transformation shall be inventoried along the needs of the applied LCIA method (if included in the impact assessment).[44]
3. SHALL—**Emissions from land use and transformation**: If land use and/or land transformation are modelled, carbon dioxide and other emissions and related effects should be modelled as follows: [ISO!]

3.1. **Soil organic carbon changes from land use and transformation**: For CO_2 release from or binding in soil organic carbon (SOC) caused by land use and land transformation, the use of the most recent IPCC CO_2 emission factors shall be used, unless more accurate, specific data is available.

3.2. **Land use and transformation related CO_2 emissions from biomass and litter**: For virgin forests and for soil, peat, etc. of all land uses shall be inventoried as "Carbon dioxide (fossil)". Emissions from biomass and litter of secondary forests shall be inventoried as "Carbon dioxide (biogenic)". This applies unless the selected LCIA method requires otherwise.

3.3. **Nutrient losses**: Emissions of nutrients shall be modelled explicitly as part of the land management process.

3.4. **Other emissions**: Other emissions in result of land transformation (e.g. emissions from biomass burning, soil erosion, etc.) should be measured or modelled for the given case or using authoritative sources.

4. MAY—**Water use**: It is recommended to differentiate at least: [ISO+]

4.1. on the input side: surface freshwater, renewable groundwater, fossil/deep ground water, sea water

4.2. on the output side: Emission/discharge of water in liquid form emission in form of steam

4.3. other water quality changes, especially by chemical substances shall be inventoried as separate elementary flows.

Modelling waste treatment (Provisions 7.4.4.2)

1. SHALL—**Waste and end-of-life product deposition**: This shall be modelled as follows: [ISO!]

[44]While this document has been finalised no established and globally applicable practice was available, but several approaches with either only regional applicability or lack of practice experience. These work with fundamentally different inventorying approaches. Any specific recommendation or requirement on inventorying land use and conversion would be implemented and published via revised ILCD reference elementary flows and recommended LCIA methods, and/or a revision of this document.

2. **Model waste management completely**: Waste and wastewater treatment shall be modelled consistently to the boundary between technosphere and ecosphere; otherwise, this shall be clearly documented and be explicitly considered in later interpretation. This modelling includes all treatment steps up to and including disposal of any remaining waste-to-waste deposits or landfills and inventorying the emissions from these sites to/from the ecosphere. Two exceptions are radioactive wastes and wastes in underground deposits (e.g. mine filling), which should be kept as specific waste flows in the inventory, unless detailed, long-term management and related interventions have been entirely modelled also for these.
3. **Modelling discarding of goods into nature**: For unmanaged landfilling, discharge and littering (i.e. discarding goods individually into nature) the related individual interventions that enter the ecosphere shall be modelled as part of the LCI model. This also applies analogously to other interventions than emissions, if the used LCIA method covers such. The littered/landfilled good should be additionally inventoried as reminder flow.
4. **Modelling waste as output**: Waste flows should be modelled following the material flow logic. That means inventorying the waste on the output side of those processes where it is generated (e.g. production waste or end-of-life product as output of the use stage). For waste management processes that means that the waste flows should accordingly be modelled on the input side if the process, with any potentially produced secondary goods and remaining wastes being on the output side. This eases mass and element balancing. For cost calculation purposes, the cost of the waste treatment service may be assigned to the waste flow as additional flow property.

Secondary LCI data sets originating in other LCA studies or LCI unit process databases are often extensively used in the inventory analysis, in particular for the modelling of the background system. Guidance is offered on how to select such secondary LCI data sets in order to meet the requirements from the goal and scope definition and ensure consistency in the study.

Selecting secondary LCI data sets (Provisions 7.6)

1. SHALL—**Use consistent secondary data sets**: The secondary data (generic, average or specific data sets) to be used in the system model shall be methodologically sufficiently consistent among each other and with the primary data sets that were specifically collected.
2. SHOULD—**Quality-oriented selection of secondary data sets**: Secondary data sets should be selected according to their data quality in a stricter sense, i.e. their technological, geographical and time-related representativeness, completeness and precision. Their reference flow(s) and/or functional unit(s) should moreover be sufficiently representative for the specific processes, good or service that they are meant to represent in the analysed system.
3. MAY—**Prefer pre-verified data sets**: It is recommended to give preference to already critically reviewed data sets ("pre-verified data") as this limits the effort for an review of the analysed system: only the appropriate use of these data sets in the analysed system needs to be reviewed. [ISO+]

4. MAY—**Prefer well-documented data sets**: It is recommended to give preference to data sets that are supported by a comprehensive and efficiently organised documentation. This allows the modeller (and later a reviewer) to judge the data set's quality and its appropriateness for the analysed system. [ISO+]

The combined use of data from different sources is facilitated by using either single operation unit process data set background systems that can be adjusted/remodelled by the user to be consistent with the analysed system, or by using LCI results data sets that are consistent with the methodology applied in the analysed system.

37.4.4 Constructing LCI Model and Calculating LCI Results

When all unit processes have been constructed or collected from LCI databases the LCI model can be constructed, using the unit processes as building blocks and scaling them according to the reference flow of the study.

Modelling the system (Provisions 7.8)

1. SHALL—**Scale inventories correctly**: The inventories of all processes within the system boundary shall be correctly scaled to each other and to the functional unit(s) and/or reference flow(s) of the analysed system.[45]
2. SHALL—**Complete system model**: No quantitatively relevant product or waste flows shall be left unmodelled/unconnected, with exception of the reference flow(s) that quantitatively represent(s) the system's functional unit. Otherwise these flows shall be clearly documented and the resulting lack of accuracy and completeness be considered in the interpretation of results. [ISO!]

Note that for unit processes all and for partly terminated systems, selected inventories of the corresponding products and/or wastes modelling processes are intentionally left out of the system boundary. Their systems are nevertheless completed, while only for applying the cut-off rules.

[45]This can be visualised by having all processes connected with each other via their reference flows of interim products and wastes, in the correct amounts. Starting from central process and the amount(s) of the system's functional unit(s) or reference flow(s), all other processes are stepwise, relatively scaled. LCA software with graphical modelling interface shows the system in this way and/or the user is modelling the system explicitly by connecting the processes on that interface. Depending on the modelling approach implemented in the software, other mechanisms can be found that serve the same scaling purpose.

3. SHALL—**Set parameter values**: Set the parameter values to the required values in all used parameterised process data sets, if any. [ISO+]
4. MAY—**Perform another round of interim quality control**: It is recommended to pre-check during modelling whether the data set or system is properly modelled and meets the quality requirements as identified/fine-tuned in the scope phase; the provisions for interim quality control of unit processes apply analogously. For filling initial data gaps of included processes and systems, estimate data sets may be considered to be used. Such may be, e.g. [ISO+]

4.1. generic or average data sets for missing specific processes/systems,

4.2. average data sets of a group of similar processes or systems (e.g. products) for missing processes/systems for other, not yet analysed processes or systems of that group

4.3. correlation with other, more complete and high quality process data sets for the same or similar process but from other data sources (e.g. industry average data for improving a producer-specific process).

5. SHALL—**Use consistent data to fill data gaps**: Data gaps shall be filled with methodologically consistent data sets, while gaps of low relevance may also be filed with methodologically not fully but sufficiently consistent data sets while being developed along the guidance of this document and meeting the overall quality requirements as detailed below. [ISO!]
6. SHALL—**Use sufficiently quality LCI data sets to fill gaps**: Only data and data sets that increase the overall quality of the final inventory of the analysed system shall be used to fill data gaps. That means that the individual data or data set's quality shall be equivalent to at least the "Data estimate" quality level. Remaining data gaps shall be reported. [ISO!]

Note that both the approach(es) used to fill initial data gaps and the resulting lack of representativeness, precision and methodological consistency of the whole data set is later to be clearly documented and explicitly considered when declaring the achieved data set quality or when drawing conclusions or recommendations from an LCA study.

Note that the final check on the achieved overall environmental completeness/cut-off is detailed in Sect. 37.6.2.

Note that decisions on any omissions of life cycle stages, types of activities, individual processes or elementary flows must be clearly reported and should be justified by the fact that they do not contribute significantly to the LCI results in view of the intended application(s) of the outcome of the LCI/LCA study. Otherwise they need to be reported and considered when declaring the achieved data set quality and/drawing conclusions and recommendations from the study.

Calculating LCI results (Provisions 7.10)

1. SHALL—**Apply calculation procedures consistently**: The same calculation procedures shall be applied consistently throughout the analysed system(s) when aggregating the processes within the system boundary for obtaining the LCI results.
2. SHALL—**Calculate and aggregate the inventory data of the system(s)**: (If the model is correctly prepared, the first two following sub-bullets can be skipped):

2.1. Determine for each process within the system boundary how much of its reference flow is required for the system to deliver its functional unit(s) and/or reference flows(s) (i.e. the extent to which the process is involved in the system).

2.2. Scale the inventory of each process accordingly. This way it relates to the functional unit(s) and/or reference flow(s) of the system.

Note that if parameterised process data sets are used in the system model, the parameter values are to be set before scaling and aggregation.

2.3. The correctly scaled inventories of all processes within the system boundary shall be aggregated (summed up) for that system.

2.4. If the intended application of the results requires a location non-generic impact assessment, aggregation of the elementary flows above the required location type or level (e.g. the level of a single site/plant, a region, a country, an environmental sub-compartment, etc.) should be avoided in the LCI results calculation. The same applies for other differentiations (e.g. of environmental sub-compartments or archetypes of emission situations) if those are required for the intended application and impact assessment methods to be used. [ISO +]

2.5. If the disaggregated data cannot be publicly disclosed (e.g. for confidentiality reasons), it is recommended to foresee performing the impact assessment on the disaggregated level and providing the LCIA results together with the aggregated LCI results. [ISO+]

Note that also in this case (as in all cases) the reviewers shall have (at least confidential) access to all underlying data.

2.6. If the disaggregated data cannot be publicly disclosed (e.g. for confidentiality reasons), it is recommended to foresee performing the impact assessment on the disaggregated level and providing the LCIA results together with the aggregated LCI results. [ISO+]

Note that also in this case (as in all cases) the reviewers shall have (at least confidential) access to all underlying data.

3. SHOULD—**Ensure that reference flow(s) is/are only product and waste flow(s)**: Note that after aggregation, the reference flow(s) is/are the only product and/or waste flow(s) that should remain in the LCI results inventory, with two exceptions:

3.1. **For partly terminated systems**: The inventories of selected products and/or waste flows were left out of the system boundary—typically intentionally—and the flows are kept in the inventory. Note, however, that for the purpose of quantifying the achieved completeness via the cut-off rules of environmental impact, also these selected product and waste flows are to be considered via integrating the inventories of the respective production and waste treatment processes.

3.2. **For radioactive waste and waste in underground waste deposits (e.g. mine filling)**: These waste flows can be kept in the inventory for direct use in interpretation.

4. SHALL—**Highlight and explicitly consider remaining non-functional product or waste flows**: Any product and waste flows that remain in the inventory and that are non-functional flows shall be highlighted in the report and/or data set: Either they require to be modelled when later using the data set (e.g. by complementing the data set with a yet missing background LCI data set for, e.g. a specific chemical consumed, or modelling the management/treatment of a specific waste). Alternatively, this gap/missing data needs to be explicitly considered in subsequent interpretation and conclusions drawn.

37.5 Impact Assessment

The impact assessment is discussed in Chap. 10 of this book. It is the fourth phase of the LCA, where the elementary flows of the life cycle inventory are translated into potential contributions to the different impact categories that are modelled in the study. The impact assessment consists of five steps:

1. Selection of impact categories, category indicators and characterisation models
2. Classification—assigning LCI results to impact categories
3. Characterisation—calculating category indicator results for the inventory flows
4. Normalisation—expressing LCIA results for the product system relative to those of a reference system
5. Weighting—prioritising or assigning weights to the each impact category

The first three steps are mandatory for any LCA according to the ISO 14040 standard, while the last two steps are optional.

1.2. MAY—**Normalisation use in cut-off quantification**: For quantification of the achieved completeness/cut-off, in a first step the indicator results for the different impact categories may be normalised by expressing them relative to a common reference, the normalisation basis ("normalisation"). [ISO+]

The decision whether to include normalisation in the cut-off has been made in the scope definition where the specific normalisation basis has also been identified.

2. SHALL—**Calculate normalised LCIA results per impact category**: If normalisation is applied, the "normalised LCIA results" shall be calculated by dividing the LCIA results by the normalisation basis. This shall be done separately for each impact category (for midpoint level approaches) or area of protection (for endpoint level approaches).

Note that normalised results shall not directly be summed up across different impact categories, as this would imply an even weighting of all impact categories. This is unless this even weighting is intended and identified explicitly as weighting when communicating the results.

37.5.3 Weighting

Weighting is the last and optional step of the LCIA and it may be used in order to allow aggregation of multiple impact category results into one score or help identify the impact scores that are most important based on a preselected set of values.

Weighting (Provisions 8.4)

1. Weighting is mainly applied for two purposes:

1.1. MAY—**Weighting to support interpretation**: In support of the interpretation of the results of the study, as an additional, optional element one may perform a "weighting" or other valuation of the—method-wise normalised or not normalised—indicator results.

The decision whether to include weighting in the interpretation has been made in the scope definition.

1.2. MAY—**Weighting use in cut-off quantification**: For quantification of the achieved completeness/cut-off, as second[47] step the normalised indicator results for the different impact categories may be weighted across the indicators ("weighting"). [ISO+]

> The decision whether to include weighting in the cut-off has been made in the scope definition where the specific weighting set has also been identified

2. SHALL—**Calculate weighted LCIA results per impact category**: If weighting is applied, to obtain "weighted LCIA results", the (typically normalised) LCIA results shall be multiplied by the weighting set, separately for each impact category (for midpoint level approaches and in case of having calculated category-wise endpoint results) or Area of protection (for endpoint results that cover each a whole area of protection). The resulting weighted LCIA results can be summed up across the impact categories or areas of protection, respectively.
3. SHALL—**No weighting in published comparative assertions**: Weighting shall not be used in studies leading to comparative assertions intended to be disclosed to the public.

> Note that the setting or selection of weighting factors necessarily involves value choices.

37.6 Interpretation

The interpretation is discussed in Chap. 12 of this book. It is the fifth phase of the LCA, where the results of the other phases are considered together and analysed in the light of the uncertainties of the applied data and the assumptions that have been made and documented throughout the study. The interpretation involves the following five activities:

1. Identification of significant issues
2. Completeness check
3. Sensitivity analysis
4. Consistency check
5. Conclusions, limitations and recommendations

---

[47]Note that some weighting methods work without a separate, preceding normalisation, as the normalisation is part of the weighting step.

37.6.1 *Identification of Significant Issues*

The process of interpretation starts with identification of potentially significant issues in the goal and scope definition, inventory analysis and impact assessment phases, understood as methodological choices, assumptions and specific data used in the model that have a potential to influence the outcome of the study in a significant way.

Identification of significant issues (Provisions 9.2)

1. SHALL—**Identify significant issues**: These can be among the following:
1.1. **Inventory items**: Main contributing "key" life cycle stages, processes, product, waste and elementary flows, parameters. This part is also known as weak point analysis or gravity analysis. Use contribution analysis techniques.
1.2. **Impact categories**: Main contributing "key" impact categories (only identifiable if weighting was applied). Use contribution analysis techniques.
1.3. **Modelling choices and method assumptions**: Relevant modelling choices, such as applied allocation criteria/substitution approaches in the inventory analysis, assumptions made when collecting and modelling inventory data for key processes and flows, selecting secondary data, systematic choices on technological, geographical and time-related representativeness, methodological consistency, extrapolations, etc. Use scenario analysis techniques.
1.4. **Commissioner and interested parties**: The influence of the commissioner and interested parties on decisions in goal and scope definition, modelling choices, weighting sets and the like. Discuss influences on final results and recommendations [ISO!]

Note: For analysing the significant issues of unit processes and partly terminated systems, complete the system model as appropriate (e.g. cradle-to-gate) with a background system before the contribution analysis is done. Focus the contribution analysis to the unit process/partly terminated system itself (i.e. the significant flows, assumptions, parameters, processes, etc. within the original system boundary).

The identification of significant issues is followed by an evaluation of these issues through a check of completeness and consistency in the handling of the issues and an analysis of the sensitivity of the outcome of the study to the significant issues. The outcome of the evaluation is used to inform previous methodological phases on the needs for strengthening the data basis of the study, driving the iterative approach that is inherent to LCA.

37.6.2 Completeness Check

Completeness checks are performed for the inventory and the impact assessment in order to determine the degree to which the available data is complete for the processes and impacts, which were identified as significant issues.

Completeness check (Provisions 9.3.2)

1. SHALL—**Evaluate LCI model completeness (cut-off)**: The cut-off rules as defined in the scope phase shall be systematically applied to ensure that the final data set inventory/ies meets the predefined or goal-derived data quality requirements. Evaluate the completeness of the inventory data in relation to the initially defined cut-off criteria in terms of:

1.1. **Process coverage**: Coverage of all relevant processes in the system

1.2. **Elementary flow coverage**: Coverage of all relevant elementary flows in the inventories for the processes of the system (and in particular the key processes identified under Significant issues), that have characterisation factors for the relevant impact categories (according to the goal of the LCI/LCA study)

1.3. **Operationalise cut-off approximation**: The cut-off criteria/approach and percentage as defined in the scope phase shall be used. This may be operationalised using stepwise the following cut-off rules for flow properties, pre-checking property by property the achieved completeness across all flow types and balancing the aggregated numbers in the inputs against those of the outputs: [ISO+]

1.3.1. **For product flows**: "mass" (of individual key chemical elements), "energy content", "market value" (or "production/provision cost", especially for purchased services).

1.3.2. **For waste flows**: "mass" (of individual key chemical elements), "energy content", "treatment cost".

1.3.3. **For elementary flows**: "mass" (of individual key chemical elements and only for the environmentally relevant flows, i.e. excluding not or less relevant flows such as, e.g. incineration air consumed and waste steam leaving the process as emission to air), "energy content".

1.4. **Cut-off for comparative assertions**: The cut-off shall always be met also by mass and energy, in addition to environmental impact.

1.5. **Additional relevance criteria for elementary and waste flows**: Also those emissions and wastes should be include in the data collection that have a low mass and energy content but a known relevance for the respective type of processes or industry (using, e.g. legal limits and expert judgement). [ISO +]

1.6. **Approximating the 100% value**: The 100% reference of completeness may be approximated by using "best approximation" values for all initially missing information and data, using among others information from similar processes and expert judgement. This missing information and data can be especially: [ISO+]

1.6.1. kind and quantity of initially missing flows,
1.6.2. element composition and energy content of all flows that relevantly contribute to the total mass of the flows,
1.6.3. cost of all goods and services that relevantly contribute to the total production cost and production value
1.6.4. environmental impact of yet missing background data sets for consumed goods and services.

1.7. **Estimating precision of 100% value approximation**: The precision of the 100% approximation may be judged from analysing the share of the different quality levels of the data that make up the inventory: a higher share of low quality data also makes the 100% approximation less precise. [ISO+]
1.8. **Completeness of impact**: As last step, and using the quantitative cut-off value decided upon in the scope definition, approximate the achieved degree of completeness/cut-off. [ISO+]
1.9. **Leaving out negligible flows**: It is an option to leave out negligible flows that jointly make up less than 10% of the share of impact that is cut-off (e.g. if the completeness is 95%, 5% are cut-off. 10% of these 5% are 0.5% that are considered negligible.) It is recommended, however, to not leave them out. [ISO+]

Note that the LCIA methods and (potentially) normalisation and weighting for use in defining the cut-off was decided in the scope phase.

Note that for unit processes and partly terminated systems the completeness is to be judged in relation to the unit process and partly terminated system itself, i.e. any lack of completeness of other processes that were added exclusively to complete the system model for the completeness check shall be disregarded when quantifying the achieved completeness.

2. SHOULD—**Improve completeness, if needed**: In the case of insufficient completeness, the inventory analysis (and sometimes the impact assessment) phases should be revisited to increase the degree of completeness. It is recommended to focus on the key life cycle stages, processes and flows identified as significant issues. This improvement of the LCI data is, however, to be started by potentially fine-tuning or revising goal and scope, i.e. with a complete iteration.
3. SHALL—**Report final completeness; potentially revise scope or goal**: If the aimed at completeness has been achieved, or if it cannot be increased further, the finally achieved degree of completeness shall be reported (as % degree of completeness/cut-off). For LCA studies, it shall be considered when later formulating the limitations in the conclusions and recommendations. If the aimed

at or necessary completeness cannot be achieved, it shall be decided whether the scope or even the goal needs to be revised or redefined.

37.6.3 Sensitivity Analysis

Sensitivity analysis has the purpose of identifying the key processes and most important elementary flows as those elements which contribute most to the overall impacts from the product system.

Sensitivity check (of accuracy and precision) (Provisions 9.3.3)

1. SHALL—**Check sensitivity of results**: Check to what extent the accuracy and precision of the overall results meets the requirements posed by the intended applications. Aim at improving it to the required level, as follows:

1.1. **Sensitivity of significant issues**: Identify the most sensitive among the significant issues identified earlier and analyse the sensitivity of these for the overall results, along with their stochastic and systematic uncertainty estimates. The outcome is determining for the accuracy and precision of the overall results and the strength of the conclusions, which can be drawn from the LCI/LCA study and must be reported together with these. Be aware that calculated uncertainty figures may not include the often-determining systematic uncertainties caused by model assumptions, data gaps and lack of accuracy.

1.1.1. **Sensitivity of LCI items**: Evaluate the sensitivity of the LCIA results (or weighted LCIA results, if applied) to key flows, process parameter settings, flow properties and other data items such as recyclability, life-time of goods, duration of services steps and the like. Assess how sensitive inventory items influence the data representativeness, and precision. [ISO!]

1.1.2. **Sensitivity of LCIA factors**: Evaluate the sensitivity of the LCIA results (or weighted LCIA results, if applied) considering the often widely differing uncertainty of the results due to uncertainties in the impact assessment (e.g. Human toxicity, Ecotoxicity, etc. with high uncertainties and Global warming, Acidification, etc. with lower uncertainty). [ISO!]

1.1.3. **Sensitivity of modelling choices and assumptions**: Evaluate the sensitivity of the LCIA results (or weighted LCIA results, if applied) to different modelling choices and method assumptions ("method issues"), e.g. quantitative and qualitative aspects of the functional unit, superseded processes, allocation criteria, etc. [ISO!]

1.2. **Improve robustness of sensitive issues data, parameters, impact factors, assumptions, etc. as possible**: In the case of lack of quality for some of the significant issues, revisit the inventory analysis and/or the impact assessment phases to improve the concerned data (for data issues), impact factors (for LCIA issues), or try to qualify and discuss the sensitive assumption or choice (for method issues). As for data completeness, also the improvement of the LCI data precision is, however, to be started by potentially fine-tuning or revising goal and scope, i.e. with a complete iteration.

1.3. **Report final achievements; potentially revise scope or goal**: If the certainty of key issues does not meet the needs, or if it cannot be increased to obtain the accuracy and precision that is required by the application of the LCI/LCA study, it shall be decided whether the scope or even the goal needs to be revised or redefined. This shall be reported and for LCA studies later be considered when formulating the limitations in the conclusions and recommendations from the LCA.

37.6.4 Consistency Check

A consistency check is performed to investigate whether the assumptions, methods and data, which have been applied in the study, are consistent with the goal and scope and whether they have been performed consistent for the compared product systems in comparative studies.

Consistency check (Provisions 9.3.4)

For partly terminated systems, LCI results and LCIA results data sets these provisions serve in addition to ensure method consistency across the processes of the model.

For LCA studies, they serve in addition to ensure method consistency across the models of the compared systems.

1. SHALL—**Data quality sufficiently consistent?** Check whether any differences in data quality per se (i.e. accuracy, completeness and precision) and in the selected data sources for the different processes in the system(s) are consistent with the goal and scope of the study. This is especially relevant for comparative studies.
2. SHALL—**Method choices consistent?** Check whether all methodological choices (e.g. LCI modelling principles, allocation criteria or system expansion/substitution approach, system boundary, etc.) are consistent with the goal and scope of the study including the intended applications and target audience. This shall be judged by checking whether the method provisions have been met that are given in relation to the applicable Situation A, B, or C1/C2. [ISO!]

Note that method consistency applies on both unit process level (i.e. consistent approach to develop unit process from raw data) and system level (i.e. consistently modelling the system). This aspect is especially relevant when combining data from different sources.

3. SHALL—**Consistent impact assessment?** Check whether the steps of impact assessment (including normalisation and weighting, if included) have been consistently applied and in line with goal and scope.
4. SHALL—**Evaluate relevance of inconsistencies**: Evaluate the relevance/significance of any identified inconsistencies (as above) for the results and document them, including when reporting the achieved method consistency and appropriateness. For LCA studies additionally consider these findings when drawing conclusions or recommendations from the results.

37.6.5 Conclusions, Limitations and Recommendations

Building on the outcome of the other elements of the interpretation, and drawing on the main findings from the earlier phases of the LCA, the final element of the interpretation has to draw conclusions and identify limitations of the study, and develop recommendations to the intended audience in accordance with the goal definition and the intended applications of the results.

Conclusions, limitations and recommendations (Provisions 9.4)

Note the limitations for Situation C1 and C2 studies in their use for direct decision support. These provisions apply only to comparative and non-comparative LCA studies.

1. SHALL—**Analyse the results in a systems perspective**: Separately analyse and jointly discuss the results obtained in the main system(s) model(s) and—if performed—with the corresponding reasonably worst and best case assumption scenarios and possibly further assumption scenarios. Integrate the results of any potentially performed uncertainty calculations into the analysis. [ISO!]

1.1. **Items that require special or separate analysis**:

1.1.1. **Non-generic LCIA**: Separately analyse and jointly discuss the results obtained with the default LCIA methods and those obtained including any potential additional or modified/non-generic (e.g. spatially or otherwise differentiated) LCIA methods.

1.1.2. **Long-term emissions**: Separately analyse and jointly discuss the results for interventions within the first 100 years from the time of the study and those beyond that time limit.

1.1.3. **Carbon storage and delayed emissions**: Only if such is included in line with an explicit goal requirement: Separately analyse and jointly discuss the results including and excluding carbon storage and delayed emissions/reuse/recycling/reuse credits.

1.2. **Draw conclusions, if foreseen**: Take into account the findings of the earlier elements of the interpretation phase. Draw conclusions in accordance with the goal defined for the LCA study and with the definitions of the scope, in particular those related to data quality requirements, and with the predefined assumptions and known limitations in the methodology and its application in the LCA. Consider all assumptions and related limitations that were noted down in the course of the study.

1.3. **Address impacts outside the LCA scope, if any**: Name any potential or actual effects on the three areas of protection that are based on other mechanisms than those covered by LCA (e.g. accidents, direct application of products to humans, etc.) and that are considered relevant by the interested parties. Clarify that these are outside the scope of LCA.

Note that within the ILCD Handbook, not quantified effects outside the scope of LCA cannot be explicitly or implicitly assessed regarding their relevance in comparison to the LCA results.[48]

1.4. **Conclusions for comparisons**: Differences in data quality and methodological choices between compared systems shall be consistent with the goal and scope of the study, especially:

1.4.1. The functional unit of the compared alternatives shall be sufficiently similar to allow for comparisons, especially in view of stakeholders and potential users.

1.4.2. The setting of system boundaries shall be consistently applied to all systems.

1.4.3. The inventory data should be of comparable quality (i.e. accuracy, completeness, precision, methodological consistency) for all compared alternatives.

---

[48]Effects outside the scope of LCA may be—if available and quantified in a comparable manner (e.g. quantitatively related to the functional unit, considering the whole life cycle etc.)—integrated with LCA results in an additional evaluation and report beyond the scope of LCA and outside the scope of the ILCD. This should consider the relative accuracy and precision of the different approaches and effects.

1.4.4. The steps of impact assessment shall be consistently applied for all systems.

1.4.5. The significance of any above identified inconsistencies to the results of the comparison shall be evaluated and considered when drawing conclusions and giving recommendations from the results.

2. SHALL—**Recommend strictly based on conclusions and limitations**:

2.1. Base any recommendations made in the LCA study exclusively on these conclusions and respecting the limitations. Derive recommendations unambiguously and in a stepwise logical and reasonable consequence of the conclusions. Do so in accordance with the defined goal of the LCA study and specially the intended applications and target audience.

2.2. Recommendations shall be made in a conservative way, only based on significant findings. Any relevant limitations found during the study are to be stated explicitly and clearly in the key message of the LCA study including in the executive summary. [ISO!]

2.3. Special care must be taken to avoid misinterpretations also by a non-technical audience, to avoid interpretation beyond the scope of the LCA study and beyond what is supported by its outcome.

2.4. Equality of compared alternatives shall not be stated, unless it has been shown to be significant: the lack of significant differences alone shall not be misinterpreted as equality of the analysed options. It shall only be stated that with the given data restrictions and/or uncertainties or other causes no significant differences could be identified. [ISO!]

3. SHALL—**Comparisons of systems with dominant subjective preference**: The results and recommendations of comparative studies on not objectively comparable alternatives (e.g. personal services, fashion items, jewellery) shall be presented with the explicit statement that comparability is not assumed per se, but lies with the individual preference and judgement. [ISO!]

4. SHALL—**Conclusions on basket-of-product type of studies**: For studies that analyse several processes or systems in a non-competitive manner, i.e. processes/systems that perform clearly different functions (e.g. basket-of-products, identifying priority products) it shall be clearly reported that no comparability exists in terms of preferability among the processes/systems.

References

EC-JRC European Commission-Joint Research Centre—Institute for Environment and Sustainability.: International Reference Life Cycle Data System (ILCD) Handbook—General guide for Life Cycle Assessment—Detailed guidance. First edition March 2010. EUR 24708 EN. Luxembourg. Publications Office of the European Union (2010)

ISO 14001:2015.: Environmental Management Systems—Requirements with Guidance for Use. The International Organization for Standardization, Geneva

ISO 14015:2001.: Environmental Management—Environmental Assessment of Sites and Organizations (EASO). The International Organization for Standardization, Geneva

ISO 14024:1999.: Environmental labels and declarations—type I environmental labelling—principles and procedures. The International Organization for Standardization, Geneva
ISO 14025:2006.: Environmental Labels and Declarations—Type III Environmental Declarations—Principles and Procedures. The International Organization for Standardization, Geneva
ISO 14031:2013.: Environmental Management—Environmental Performance Evaluation—Guidelines. The International Organization for Standardization, Geneva
ISO 14040:2006.: Environmental Management—Life Cycle Assessment—Principles and Framework. The International Organization for Standardization, Geneva
ISO 14044:2006.: Environmental Management—Life Cycle Assessment—Requirements and Guidelines. The International Organization for Standardization, Geneva
ISO/TR 14062:2002.: Environmental management—Integrating Environmental Aspects into Product Design and Development. The International Organization for Standardization, Geneva

Author Biographies

Michael Z. Hauschild involved in development of LCIA methodology since the early 1990s. Has led several SETAC and UNEP/SETAC working groups and participated in the development of the ISO standards and the ILCD methodological guidelines. Main LCA interests are chemical impacts, spatial differentiation and science-based boundaries in LCIA.

Anders Bjørn part of the LCA community since the early 2010s. Main focus is interpretations of sustainability and integration of sustainability targets in LCA to enable absolute sustainability assessments.

Part V
Annexes

Chapter 38
Report Template

Anders Bjørn, Alexis Laurent and Mikołaj Owsianiak

Abstract To ensure consistent reporting of life cycle assessment (LCA), we provide a report template. The report includes elements of an LCA study as recommended but the ILCD Handbook. Illustrative case study reported according to this template is presented in Chap. 39.

The following LCA report template presents the structure and summarised content of an LCA report that is recommended in the ILCD Handbook[1]. The content should be covered by any LCA report, but the proposed sub-subsections (level 3) may be merged or divided and their sequence may be changed if appropriate. The ILCD Handbook operates with three reporting levels:

1. Internal use by commissioner of study.
2. Third party, i.e. a limited, well-defined list of recipients with at least one organisation that has not participated in the study.
3. Comparative studies to be disclosed to the public.

There are no formal ILCD reporting requirements for level 1, but ILCD recommends following the requirements of level 2. More requirements apply to level 3 than to level 2. The following template applies to both level 2 and level 3 studies (special requirements for level 3 are highlighted here for the reader). The reporting template has been used in the reporting of the illustrative case on window frames in Chap. 39.

[1]EC-JRC (2010) European Commission—Joint Research Centre—Institute for Environment and Sustainability: International Reference Life Cycle Data System (ILCD) Handbook—General guide for Life Cycle Assessment—Detailed guidance. First edition March 2010. EUR 24708 EN. Luxembourg. Publications Office of the European Union.

A. Bjørn · A. Laurent · M. Owsianiak
Division for Quantitative Sustainability Assessment, Department of Management Engineering, Technical University of Denmark, Kongens Lyngby, Denmark

A. Bjørn (✉)
CIRAIG, Polytechnique Montréal, 3333 Chemin Queen-Mary, Montreal, QC, Canada
e-mail: anders.bjoern@polymtl.ca

M.Z. Hauschild et al. (eds.), *Life Cycle Assessment*,
DOI 10.1007/978-3-319-56475-3_38

38.1 Executive Summary

Summarises in a non-technical language the key elements of the goal and scope of the studied system (including the definition of functional unit), the main results from the inventory analysis (including data sources for the foreground system) and the impact assessment components and major conclusions and/or recommendations made.

38.2 Technical Summary

Summarises in a technical language the goal and scope, with relevant limitations and assumptions, a flow diagram of the studied system, the main results from the inventory and impact assessment components (including data sources, major assumptions, and key figures or tables) with consideration of the sensitivity and uncertainty analyses and conclusions and/or recommendations made.

38.3 Main Report

38.3.1 Goal Definition

38.3.1.1 Intended Application(s)

States in a precise and unambiguous way the intended application (e.g. comparison of specific goods or services, weak point analysis of a specific product or greening the supply chain).

38.3.1.2 Method Assumptions and Impact Limitations

States limitations of the study's usability due to (1) choice of methods implying, for example, a limited representation of temporal/spatial variations, and (2) limited impact coverage, e.g. only performing carbon footprinting.

38.3.1.3 Reasons for Carrying Out the LCA Study and Decision Context

Explains why the study was commissioned and classifies its decision context to Situation A, B, C1 or C2.

38.3.1.4 Target Audience

States to whom the results of the study are intended to be communicated in order to help identifying the requirements to a critical review (if any) and the appropriate form and technical level of reporting.

38.3.1.5 Comparative Assertions to be Disclosed to the Public

States if the LCA study includes a comparative assertion intended to be disclosed to the public, since stricter reporting requirements exist for such studies compared to studies only intended for internal use by the commissioner or for specified third parties.

38.3.1.6 Commissioner of the LCA Study and Other Influential Actors

Identifies commissioner(s), including co-financers, and other actors having an influence on the study.

38.3.2 *Scope Definition*

38.3.2.1 Deliverables

Lists the required deliverables of the study (typically LCI and LCIA results), which should be derived from the intended application (Sect. 3.1.1).

38.3.2.2 Function, Functional Unit and Reference Flows

States the function(s) of the assessed product system(s), associated functional unit ("what", "how much", "how well", "where" and "how long/many times"?) and derived reference flow(s), to which all other flows quantitatively relate.

38.3.2.3 LCI Modelling Framework

Based on the identified decision context (Sect. 3.1.1), it is specified whether an attributional or consequential modelling framework (or a combination of the two) is to be followed in the inventory analysis.

38.3.2.4 System Boundaries and Completeness Requirements

Presents system boundaries, including a schematic representation and associated choices of completeness requirements, that are used to determine which processes should be within the system boundaries and which may be excluded.

38.3.2.5 Representativeness of LCI Data

Describes the requirements of inventory data with respect to spatial, temporal and technological representativeness in accordance with the goal of the study.

38.3.2.6 Basis for Impact Assessment

Describes the selection of impact categories and corresponding LCIA models in accordance with the goal of the study for which compatible inventory data is collected.

38.3.2.7 Requirements for Comparative Studies

Only applicable to level 3 studies according to ISO standards. However, in line with ILCD requirements, we recommend to include this section for all comparative studies. It presents special considerations for comparative studies related to, e.g. data quality requirements, exclusion of identical processes and interpretation in light of affected stakeholders.

38.3.2.8 Critical Review Needs

Describes the requirements for critical review (for example in case of public disclosure), the form of the review and who is eligible to perform it.

38.3.3 Life Cycle Inventory Analysis

38.3.3.1 LCI Model at System Level

Presents one or more flow diagrams that clearly describe the details of the foreground system and links to the background system, and all major inputs and outputs. For simple systems, such a diagram may already have been presented in sufficient detail under "System boundaries and completeness requirements" (Sect. 3.2.4), in which case, it should be skipped here. For complex systems, the use of an Annex can be useful to detail graphically the different parts of the system (so as not to overload the main report); cross-references to the Annex should be made when describing the system modelling and inventory building.

38.3.3.2 Data Collection

Presents in a table format a synthetic overview (1 page max.) of the data sources in terms of specificity, type, source and access. This part can provide the reader with a brief overview of the data collected and processed in the modelling of the system. Dividing the table between the different life cycle stages can help structuring this overview and outline the different elements to address in Sect. 3.3.3.

38.3.3.3 System Modelling Per Life Cycle Stage

Details in plain text for each life cycle stage (raw materials stage, production stage, use stage, disposal stage, etc.) the data collected and their further treatment, including main assumptions and evaluations of their reliability (based on data sources used) and uncertainty (to prepare the basis of sensitivity and uncertainty analyses). For clarity, a division of this section into subsections is advocated: a subsection addressing each life cycle stage, and possibly others focusing on key modelling aspects (e.g. electricity supply modelling, when complex scenarios are analysed). From this section, the reader should be able to get a clear and comprehensive overview of the system modelling over its entire life cycle, including all the major assumptions and underlying uncertainties. To make the study fully reproducible, it is recommended to complement this part with an Annex detailing all the assumptions (including minor assumptions), calculations made, detailed unit process data collected for the foreground system, detailed data for modelling of the background system (e.g. detailed scenario description for the disposal stages or for energy systems), and any other relevant information for the system modelling. Cross-references to those different elements in Annex should be made in the main report.

38.3.3.4 Calculated LCI Results

Contains the final LCI results at a level of aggregation required for the intended applications. This section is particularly relevant for studies only aiming at deriving LCI results, where the LCI results should provide a comprehensive listing of elementary flows for the entire system. For studies performing an impact assessment, this section may still be relevant to capture the resulting LCI results for the foreground system that may not be available in existing LCI databases and may thus serve as input data for other LCA studies. Tables detailing these LCI results for each process should be documented in Annex and cross-references should be done in this brief section.

38.3.3.5 Basis for Sensitivity and Uncertainty Analyses

Describes how sensitivity and uncertainty analyses are conducted. Sensitivity and uncertainty analyses should be documented by listing the different parameters tested and describing their uncertainty characteristics such as statistical distribution and variance (for example in a table format), and the methods adopted to conduct these analyses. This section can then be referenced in other sections of the report (e.g. sensitivity check) as the place to find all the methodological details for the sensitivity and uncertainty analyses.

38.3.4 *Life Cycle Impact Assessment*

Presents characterised, normalised (optional) and weighted (optional) results graphically or in table format at a level and type of aggregation that reflects the goal and scope of the study (e.g. results split per life cycle stage, product flows or elementary flows of interest, as relevant). It is often beneficial to merge this section with the section "Interpretation" (see Sect. 3.5) to improve the readability of the report. The results can then be displayed alongside their interpretation and associated intermediate conclusions.

38.3.5 *Interpretation*

38.3.5.1 Significant Issues

Identifies the issues (e.g. methodological choices and assumptions, inventory data, and characterisation or normalisation factors), which have the potential to change the final results of the LCA. The identification of issues often relies on contribution analyses, at a process level (which process contributes the most to the impact indicators?) and/or at a substance level (which substance is the largest contributor to a given impact indicator result?). Such analysis should be documented here.

38.3.5.2 Sensitivity and Uncertainty Analyses Checks

Quantify the sensitivity of results to different elements of the LCA based on information about the uncertainties of significant issues among inventory data, impact assessment data and methodological assumptions and choices. Calculate the uncertainty of the study's central conclusions (e.g. how certain is it that Product A has a lower environmental impact than Product B?). Cross-references to the Sect. 3.3.5 can be made to ease the readability and understanding of the uncertainty and sensitivity analyses.

38.3.5.3 Completeness and Consistency Checks

Consists in two separate sub-sections that briefly document that both completeness and consistency checks have been done. The completeness check aims to ensure, in light of the goals and scope of the study, a sufficient completeness of the inventory and impact assessment data that have been identified as necessary for performing the LCA study. The consistency check investigates whether the assumptions, methods and data have been applied in a consistent way, e.g. consistent choice of the LCI modelling framework, well-argued setting of system boundaries, consistence across all LCIA steps (characterisation, normalisation, weighting), etc.

38.3.6 Conclusions, Limitations and Recommendations

Presents conclusions encompassing the entire study, limitations for the usability of these conclusions, based on the above interpretation sections, and resulting recommendations, based on the intended application of the study.

Annex (Public)

Serves to document elements that would inappropriately interrupt the reading flow of the main part of the report and should include questionnaire/data collection template and raw data, full list of all assumptions and full LCI details and results. Cross-references to the different elements of the Annex should be made in the main report. Note that only non-confidential data should be reported in this appendix.

Annex (Confidential)

This Annex is complementary to the main Annex (public), and only applicable in case confidential data were used in the study. This confidential Annex contains all those data and information that are sensitive or proprietary and cannot be made externally available (only available to the critical reviewers under confidentiality agreements).

References

List of all sources referred to in the report

Author Biographies

Anders Bjørn part of the LCA community since the early 2010s. Main focus is interpretations of sustainability and integration of sustainability targets in LCA to enable absolute sustainability assessments.

Alexis Laurent working with LCA since 2010 with a strong LCIA focus, particularly on normalisation aspects. Main LCA interests include development of LCIA methods, LCA applications and footprinting of large-scale systems for policy-making (e.g. nations, sectors), and LCA applied to various technology domains, including energy systems.

Mikołaj Owsianiak involved in development and application of life cycle impact assessment methods in sustainability assessment of technologies. Has worked on issues associated with: soils (remediation), metals (toxic impact assessment), biodiesel (fate in the environment), and carbonaceous materials (biochar and hydrochar).

Chapter 39
Illustrative Case Study: Life Cycle Assessment of Four Window Alternatives

Mikołaj Owsianiak, Anders Bjørn, Heidi B. Bugge, Sónia M. Carvalho, Leise Jebahar, Jon Rasmussen, Caroline M. White and Stig Irving Olsen

Abstract This report serves as an example report on how to perform an LCA according to the guidance given in Chap. 37 and how to structure the report according to the reporting template in Chap. 38. The goals of the LCA were (i) to perform a benchmarking of a prototype wood/composite (W/C) window made out of glass fibre against three alternative window types currently offered in the market (made of wood (W), wood/aluminium (W/ALU), and PVC) and (ii) to identify environmental hotspots for each window system.

39.1 Executive Summary

Nor-win, a Danish-based windows manufacturer, commissioned an LCA study with the goals (i) to perform a benchmarking of a prototype wood/composite (W/C) window made out of glass fibre against three alternative window types currently offered in the market (made of wood (W), wood/aluminium (W/ALU), and PVC) and (ii) to identify environmental hotspots for each window system. The compared windows differ regarding their ability to prevent heat from escaping the building (insulation performance).

The four window types are compared on the basis of their main function, which is allowing daylight inside the building. The functional unit is thus "Allow daylight

M. Owsianiak (✉) · A. Bjørn · S.M. Carvalho · L. Jebahar · J. Rasmussen · C.M. White · S.I. Olsen
Division for Quantitative Sustainability Assessment, Department of Management Engineering, Technical University of Denmark, Kongens Lyngby, Denmark
e-mail: miow@dtu.dk

A. Bjørn
CIRAIG, Polytechnique Montréal, 3333 Chemin Queen-Mary, Montréal, QC, Canada

H.B. Bugge
Ecolabelling Denmark, Danish Standards Foundation, Göteborg Plads 1, Nordhavn, Denmark

M.Z. Hauschild et al. (eds.), *Life Cycle Assessment*,
DOI 10.1007/978-3-319-56475-3_39

into a residential building through a physical barrier, equivalent to light being transmitted through an area of 1.82 m^2 with visible light transmittance of at least 0.6, for 20 years".

System boundaries comprise all life cycle stages from cradle to grave, including transportation and provision of utilities (electricity or heat). The manufacturing technology for all four windows, including the elements composing the pane and the frame, represent the technology, which is currently used by Nor-win and its suppliers. Nor-win is the provider of primary data used in the product system model. Generic databases were used for background processes and for some foreground processes for which no primary data could be retrieved from the suppliers.

The results show that for most impact categories the impact scores follow the order W/C < W = W/ALU < PVC. The W/C window has the lowest environmental impact in all 14 impact categories, while the PVC window system has the highest impact for 10 impact categories. For nearly all the non-toxicity-related impact categories, the life cycle impacts of the four windows correspond to approximately 10% of the total annual average impacts of an average EU27 citizen in the year 2010.

The main contributor of environmental impacts is the generation of indoor heating to compensate for heat losses through the window. The contribution of this process to the total life cycle impact is for all four windows around 90% for climate change, freshwater eutrophication, or resource depletion, and above 50% for nearly all other impact categories. The manufacturing stage is relevant for impacts on stratospheric ozone depletion, and ionising radiation (human health) across all windows, for impacts on freshwater ecotoxicity and human toxicity (PVC window), and land use (W window).

Several assumptions had to be made for the modelling of the product systems. While most of them were not found to be important for our conclusions, the modelling of chromium steel and galvanised steel using the same processes may influence impact scores in the categories human health (cancer effects) and freshwater ecotoxicity due to associated differences in emissions of chromium (VI) and zinc (II). The impact scores were the most sensitive to the insulation capacity (*U*-value) of the window, and the ranking of the window alternatives does not change when the EU27 heat mix is used instead of the Danish mix.

Overall, results show that there is a trade-off between types of material used and improved insulation properties of windows. The use of glass fibre-based composite in the W/C has some contribution (up to 12%) to total impacts, depending on the impact category, but the use of the composite substantially improves insulation properties causing an overall reduction in environmental impacts, and leading to the superiority of this window type. The design of windows to ensure better environmental performance should focus on optimising insulation properties of windows, which can be done either by improving the design of the frame or by introducing an additional pane that helps improving insulation properties of the whole window.

39.2 Technical Summary

Nor-win, a Danish-based windows manufacturer, wishes to position itself as a proactive company on the market in terms of environmental sustainability, with the ambition to attract customers demanding more environmentally friendly products. For this purpose, an LCA study was commissioned with the goals (i) to perform a benchmarking of a prototype of a wood/composite (W/C) window made out of glass fibre against three window types currently offered in the market (made of wood (W), wood/aluminium (W/ALU), and PVC) and (ii) to identify environmental hotspots for each window system. The deliverables include (i) detailed life cycle inventory model of the compared systems, including unit process data; (ii) life cycle impact assessment results (in both characterised and normalised forms). We follow EU-recommended practice for characterisation modelling at midpoint to quantify life cycle impacts of four window alternatives, referred to as ILCD. Normalisation was carried out using a set of normalisation references for the year 2010.

The functional unit is "Allow daylight into a residential building through a physical barrier, equivalent to light being transmitted through an area of 1.82 m^2 with visible light transmittance of at least 0.6, for 20 years". Decision context is micro-level, product or process-related decision support studies, i.e. situation A of the ILCD guideline (EC-JRC 2010). Consequently, the attributional principle was chosen as LCI modelling principle.

Major properties of the four window alternatives are presented in Table 39.1. The window's ability to prevent heat from escaping the building is described by its heat transfer coefficient—the *U*-value (W m^{-2} K^{-1}). The fraction of light that enters through the window into the building is characterised by its visible light transmittance (T_{vis}).

Table 39.1 Major properties of the four window alternatives

Properties	Window type			
	W	W/ALU	PVC	W/C
Frame material	Mainly wood	Mainly wood and aluminium	Mainly polyvinyl chloride and galvanised steel	Mainly wood and polyamide/glass fibre composite
Glass material	2-layered, coated, sealed with silicone, filled with argon	2-layered, coated, sealed with silicone, filled with argon	2-layered, coated, sealed with silicone, filled with argon	2-layered, coated, sealed with silicone, filled with argon
U-value (W m^{-2} K^{-1})	1.29	1.31	1.36	1.08
T_{vis} (fraction)	0.8	0.8	0.8	0.8
Glass dimensions (m)	1.23 × 1.48	1.23 × 1.48	1.23 × 1.48	1.23 × 1.48

Wood (*W*), wood/aluminium (*W/ALU*), polyvinyl chloride (*PVC*) and wood/composite (*W/C*)

The analysis comprises all life cycle stages from cradle to grave, including transportation and provision of utilities (electricity or heat). The manufacturing technology for all four windows, including the elements composing the pane and the frame, represent the technology, which is currently used by Nor-win and its suppliers. Nor-win is the supplier of primary data used to model the LCI. Ecoinvent and Plastics Europe databases were used for background processes and as source of data for some foreground processes for which no primary data could be retrieved from the suppliers (PlasticsEurope Database 2016; Ecoinvent 2010).

The major assumptions made in inventory modelling include (i) the wood-based windows are sold mainly in Scandinavian countries and Germany, but the use and disposal stages for all windows are modelled using data from processes representative for Denmark; (ii) windows are used only in buildings that are heated by district heating (in Denmark district heating delivers ca. 55% of the total heat demand for buildings); (iii) processes used to model the district heat mix technologies were representative for Norway (except for incineration of bio-waste, which was representative for Switzerland); (iv) energy used for operation of the manufacturing and disassembly facilities for the W/C window is assumed equal to those for other windows, while energy requirements for window disassembly are assumed equal to 1 MJ per kg of dismounted window; (v) losses of materials during production are not considered and all recycled materials replace virgin materials in

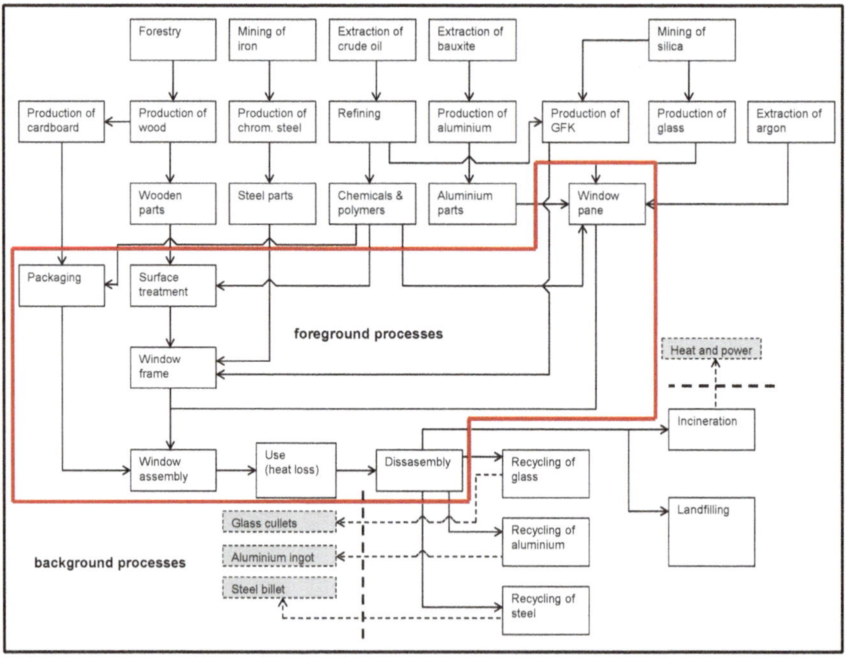

Fig. 39.1 Flow diagram for the wood/composite window (W/C) product system. *Red line* indicates foreground processes. *Grey boxes* indicate avoided processes

the market at a 1:1 ratio without considering any loss of material functionality in the recycling; and (vi) production of chromium steel and galvanised steel was modelled using the same process, while landfilling of EDPM rubber was modelled as that of polypropylene. The flow diagram of the prototype W/C window is presented in Fig. 39.1.

All four window alternatives have impacts within the same order of magnitude, and for most impact categories the impact scores follow the order W/C < W = W/ALU < PVC (Table 39.2). The W/C window has the lowest environmental impact in all 14 impact categories, while the PVC window system has the highest impact scores for 10 impact categories. For nine out of these 10 impact categories the differences in impact scores between the W/C and PVC windows are deemed statistically significant (i.e. the calculated 95% probability ranges of the impact scores do not overlap). For nearly all the non-toxicity-related impact categories, the life cycle impacts of the four windows correspond to approximately 0.1 person equivalents.

Process contribution analysis showed that the main driver of environmental impacts was the production of house heating to compensate for heat losses through the window. The contribution of this process to the total impact is around 90% for climate change, freshwater eutrophication, or resource depletion across all four windows, and above 50% for nearly all other impact categories. The manufacturing stage was relevant for impacts on stratospheric ozone depletion, and ionising radiation (human health) across all windows, and for the impacts on freshwater ecotoxicity, human toxicity (PVC window), and land use (W window). Most of the assumptions made when modelling the LCI were not found to be important for the conclusions, except that modelling of chromium steel and galvanised steel using the same processes may influence impact scores in human health (cancer effects) and freshwater ecotoxicity due to associated differences in emissions of chromium (VI) and zinc (II). The impact scores were the most sensitive to the *U*-value of the window, and the ranking of window alternatives does not change when the EU27 heat mix is used instead of the Danish mix.

Overall, the major conclusions of this LCA are:

I. The W/C window performs significantly better compared to its alternatives in all 14 impact categories. The W/C window is thus the preferable option from an environmental perspective.

II. The PVC window is the least preferred option, as it performs the worst in 11 out of 14 impact categories. This conclusion, however, might change if land use, freshwater ecotoxicity and human health (non-cancer) (where the W window performs significantly worse) are given a higher weight than the rest of the impact categories.

III. The overall environmental performance of the windows is mainly determined by the demand for heat to compensate for heat losses through the window during its use stage. This is true for nearly all impact categories. The *U*-value determines demand for heat, and can thus be considered a key environmental performance indicator of windows.

Table 39.2 Characterised impacts and accompanying 95% probability ranges from Monte Carlo simulations, expressed in category-specific units for each window alternative

Impact category	Unit	Impact score (95% probability range)			
		W	W/ALU	PVC	W/C
Climate change	kg CO_2 eq.	1162 (1134–1189)	1158 (1129–1188)	1232 (1203–1260)	978 (933–1023)
Stratospheric ozone depletion	kg CFC-11 eq.	1.9e−5 (1.8e−5–1.9e−5)	1.6e−5 (1.5e−5–1.6e−5)	1.6e−5 (1.5e−5–1.6e−5)	1.4e−5 (1.4e−5–1.4e−5)
Photochemical ozone formation	kg NMVOC eq.	1.59 (1.55–1.63)	1.57 (1.53–1.61)	1.72 (1.67–1.76)	1.33 (1.26–1.4)
Terrestrial acidification	AE	2.00 (1.95–2.04)	2.00 (1.95–2.05)	2.31 (2.26–2.36)	1.67 (1.6–1.75)
Terrestrial eutrophication	AE	7.14 (6.94–7.35)	7.04 (6.83–7.24)	7.79 (7.56–8.02)	5.96 (5.61–6.32)
Freshwater eutrophication	kg P eq.	0.042 (0.041–0.043)	0.043 (0.041–0.044)	0.046 (0.044–0.047)	0.035 (0.033–0.037)
Marine eutrophication	kg N eq.	0.65 (0.63–0.67)	0.62 (0.60–0.64)	0.68 (0.66–0.7)	0.54 (0.51–0.57)
Freshwater ecotoxicity	CTU_e	2675 (2605–2745)	1755 (1706–1805)	1852 (1809–1895)	1545 (1461–1630)
Human toxicity (cancer)	CTU_h	2e−5 (1.9e−5–2e−5)	1.8e−5 (1.8e−5–1.9e−5)	3.4e−5 (3.3e−5–3.5e−5)	1.6e−5 (1.5e−5–1.7e−5)
Human toxicity (non-cancer)	CTU_h	1.5e−4 (1.5e−4–1.6e−4)	1.3e−4 (1.2e−4–1.3e−4)	1.3e−4 (1.3e−4–1.3e−4)	1.0e−4 (9.9e−5–1.1e−4)
Particulate matter formation	kg $PM_{2.5}$ eq. to air	0.085 (0.083–0.087)	0.082 (0.080–0.084)	0.116 (0.114–0.119)	0.070 (0.067–0.073)
Ionising radiation (human health)	kBq U235 eq.	7.69 (7.56–7.81)	7.99 (7.86–8.12)	8.63 (8.49–8.77)	6.26 (6.07–6.45)
Land use	kg C year	657 (646–668)	405 (399–410)	386 (384–387)	364 (351–377)
Resource depletion (minerals, fossils)	kg Sb eq.	0.0072 (0.0070–0.0073)	0.0074 (0.0072–0.0076)	0.0081 (0.0080–0.0083)	0.0063 (0.0060–0.0066)

W Wood, *W/ALU* wood/aluminium, *PVC* polyvinyl chloride, *W/C* wood/composite
The probability ranges represent the modelled inventory uncertainty, as the uncertainties in the characterisation factors were not known

IV. In addition to processes for generation of heat, other environmental hotspots in the product systems are: production of timber and paint for the W window; the injection moulding process of PVC and production of steel in the PVC window.
V. The use of glass fibre-based composite has some contribution (up to 12%) to total impacts, depending on the impact category, but cannot be considered a hot spot given that the composite substantially improves insulation properties causing an overall reduction in environmental impacts.
VI. Similarly, the use of 3-layered glass instead of 2-layered improves insulation properties resulting in an overall reduction in environmental impacts with the respective heating mix.
VII. The trade-off between impacts from the material used and the improved insulation properties that the material may give the window has to be considered when assessing environmental performance of windows.

Recommendations are given to the commissioner to support eco-design of the new window and greening of the whole value chain:

A. The design of windows to ensure better environmental performance should focus on optimising insulation properties of windows. This can be done by introducing a 3-layered pane, or improving the design of the frame. If the latter is considered, the choice of frame material is important and in each case where new frame material is used in the design of a frame we recommend evaluating (using tools like LCA) whether environmental benefits achieved by improved insulation properties are really sufficient to outweigh potential environmental burden from the use of novel materials. Indeed, if the heat mix changes substantially within the lifetime of the window this could potentially move the hot spots from the use stage to manufacturing and end-of-life stages in which case our recommendations for design of the windows might not hold.
B. Selection of new materials for frame design should consider functional properties of materials in a window design context, i.e. the focus should not be on selection of materials that perform environmentally best per unit mass of the materials, but on selection of materials that perform best considering insulation properties and the amount applied when used in the frame.
C. For the existing W-based windows, improvement potentials lie in selection of paints with lower environmental impact. For the paint applied for maintenance in the use stage, this may be outside the influence of the producer, because it is the window users who will select the type of paint. Our recommendation is to provide information to the users about recommended types of paint.
D. Finally, we recommend to phase-out the PVC window as the option with likely the highest environmental burden overall. If this is not possible, we recommend its redesign through the introduction of a 3-layered pane to improve its insulation properties. Further improvement potentials for the PVC window system lie mainly in selection of cleaner technology for production of PVC frame elements.

The major limitations of the LCA are:

1. Our findings about major drivers of environmental impacts apply to windows where crystal glass is used in the panes with a relatively large (>0.6) visible light transmittance coefficient. They are not thought to be applicable for windows, which change their transparency in response to light intensity (e.g. photochromic windows) where the need for electricity to provide lighting indoor may become an important factor contributing to impacts in the use stage.
2. The disregard of changes in heat mix and heat demand in the future and potential development of more efficient heat supply technologies is another potential limitation. It is uncertain to what extent these will become effective within the time frame of the study (25–30 years). If such is the case, impacts from the manufacturing stage or disposal will become more important in the future (if there is no development of cleaner manufacturing and waste management technologies, which also is uncertain). They may change both the ranking of window alternatives and recommendations given to the commissioner. We expect, however, that in a 20-year time horizon, the use stage will likely remain the most important contributor to total impacts from the window product system, and efforts to design windows with low *U*-values should continue.

39.3 Main Report

Nor-win, a Danish-based windows manufacturer, produces windows for use in residential buildings in Scandinavia (mainly Denmark and Sweden), and some Western European countries (mainly Germany). Existing Nor-win windows on the market are dominated by windows made of wood, a combination of wood and aluminium, or polyvinyl chloride (PVC). Nor-win is currently designing a new type of window made of wood and a composite (glass fibre-reinforced polyamide) to be introduced to the market in 1–2 years, starting with the home market in Denmark. The new window is expected to gain a share of 20–30% of the total current market share of Nor-win. It differs from the existing windows with respect to heat insulation properties, which are improved by combining wood with glass fibre-reinforced plastic in the construction of window frame. The new window is thus expected to have a lower overall environmental impact compared to earlier products from the company. However, a quantitative, life cycle based assessment has not been done yet.

39.3.1 Goal Definition

39.3.1.1 Intended Applications

The study aims to perform a benchmarking for internal use of a wood/composite (W/C) window against three window types made of wood (W), wood/aluminium

(W/ALU), and PVC (PVC) used in Danish residences. With this regard, this study is a comparative case study. However, given that Nor-win will use the results as guidance for the ongoing design of the new window, environmental hotspots for each window system will also be identified. One of the aims of this study is to quantify the trade-off that may occur when potentially reduced environmental impacts from better insulation properties are achieved at the expense of increased impacts from higher demand for materials needed for manufacturing. Overall, the results of this LCA are intended to be used to initiate a greening of the value chains of the four window alternatives.

39.3.1.2 Method Assumptions and Impact Limitations

We follow best, EU-recommended practice for characterisation modelling to quantify life cycle impacts of four window alternatives, referred to as ILCD (EC-JRC 2010; Hauschild et al. 2013). However, even best practice has limitations. The recommended methods currently do not allow for consistent spatially explicit impact assessment. Given that Nor-win operates in Northern Europe, those impacts which occur within this region may be subject to bias, if global-generic characterisation factors are used (Scandinavian soils are for instance quite sensitive to acidification compared to an average European soil). Further, in ILCD, no method has been recommended for dealing with terrestrial and marine ecotoxicity. Thus, impacts on terrestrial and marine ecosystems stemming from emissions of some important stressors, like metals, are not considered in this assessment. In addition, due to insufficient quality of the inventory data, we had to exclude the impact category water use from the set of ILCD methods.

Normalisation was done to relate impact scores to background activities of the society in Europe. However, normalisation references are thought to be underestimated for the toxicity-related impact categories due to the insufficient knowledge of total emissions of the thousands of different chemicals with toxicity potentials in Europe, resulting in overestimation of normalised impact scores for the freshwater ecotoxicity and human toxicity (both cancer and non-cancer effects) impact categories (Laurent et al. 2011).

39.3.1.3 Reasons for Carrying Out the LCA Study and Decision Context

Nor-win wishes to position itself as a proactive company on the market in terms of environmental sustainability, with the ambition to attract customers demanding more environmentally friendly products. For that reason, Nor-win is considering to apply for a Nordic Ecolabel for selected windows in its portfolio, to be used for marketing purposes. Furthermore, the company expects that the LCA study will provide valuable information to be incorporated at the early stages in the development of the new composite window.

The decision context is situation A as the decisions taken by Nor-win stakeholders will primarily have an internal influence (i.e. in supporting the eco-design of the new type of window) and will not result in structural consequences on the market because the share of Nor-win of total window market in Europe is small (ca. 3%). This is mainly because Nor-win is a relatively minor customer for its suppliers and the market production capacity will not be influenced if changes in the choice of suppliers are introduced based on the LCA results.

39.3.1.4 Target Audience

The target audience is environmental and design departments at Nor-win. The company has limited knowledge about life cycle concepts and has neither conducted nor commissioned an LCA before. However, the company has recently employed a designer who is familiar with eco-design principles.

39.3.1.5 Comparisons Intended to Be Disclosed to the Public

This comparative LCA study is not intended to be a comparative assertion disclosed to the public.

39.3.1.6 Commissioner of the LCA Study and Other Influential Actors

This study is commissioned and fully financed by Nor-win. The team carrying out the LCA includes an employee from Ecolabelling Denmark at Danish Standard (Dansk Standard, DS), an organisation that is monitoring and developing ecolabel standards for various products, including windows. Ecolabelling Denmark is thus an influential actor.

39.3.2 Scope Definition

39.3.2.1 Deliverables

This is both a comparative LCA study (for internal use) and an environmental hotspot analysis carried out for each window alternative (see Sect. 2.1). The deliverables include (i) detailed life cycle inventory model of the compared systems, including unit process data; and (ii) life cycle impact assessment results (in both characterised and normalised forms).

39.3.2.2 Function, Functional Unit, and Reference Flows

Function. The four window types are made of different materials and are compared on the basis of their main function, which is allowing light inside the building. The fraction of light that enters through the window into the building is characterised by a parameter called visible light transmittance (T_{vis}). T_{vis} can vary between windows depending on the type and properties of the windowpane, but for windows with panes made of crystal glass it is generally not lower than 0.6. The visible light transmittance is a relevant parameter to consider because light transmittance properties of some windows (like the photochromic or electrochromic ones) may vary, depending on other factors (e.g. light intensity), in which case increased need for indoor lighting should be considered when modelling life cycle inventories. This was not considered relevant in our case study where all four windows have pane made out of crystal glass of constant (and relatively high, $T_{vis} > 0.6$) light transmittance properties. The secondary function of the window (i.e. its ability to transfer heat) was considered by crediting the system for the heat loss in the use stage (as will be explained in detail in Sect. 39.2.3). While all windows must allow daylight into a building (obligatory property), they may differ in some of the positioning properties (Table 39.3).

The window's ability to conduct heat is described by its heat transfer coefficient, the *U*-value (W m^{-2} K^{-1}). The *U*-value is a measure of how well a *window* prevents heat from escaping the building. The lower the *U*-value the lower is the heat loss. The *U*-value depends on the type of material and design of the window frame, and properties (thickness, coating, and number of layers) of the windowpane and on the properties of the material between the panes. The *U*-value is usually measured and provided by the manufacturers. The *U*-value is thus an important parameter because any heat loss through the window must be compensated by providing extra heating to the indoor environment that the window shields. It is estimated that losses through windows can account for 25% of the total heat loss in a residential building (Natural Resources Canada 2015). Note that both window frame and windowpane can be characterised by a *U*-value. The *U*-value used here refers the window as a whole. Major properties of the four window alternatives are presented in Table 39.4.

Table 39.3 Obligatory and positioning properties of windows in this case study

Obligatory property	Positioning properties
• Allow daylight into a building through a physical barrier	• Thermal and noise insulation • Allow ventilation between indoor and outdoor • Provide aesthetic functionality to the building • Protection against breaking into the building

Table 39.4 Major properties of the four window alternatives

Properties	Window type			
	W	W/ALU	PVC	W/C
Frame material	Mainly wood	Mainly wood and aluminium	Mainly polyvinyl chloride and galvanised steel	Mainly wood and polyamide/glass fibre composite
Glass material	2-layered, coated, sealed with silicone, filled with argon	2-layered, coated, sealed with silicone, filled with argon	2-layered, coated, sealed with silicone, filled with argon	2-layered, coated, sealed with silicone, filled with argon
U-value ($W\ m^{-2}\ K^{-1}$)	1.29	1.31	1.36	1.08
T_{vis} (fraction)	0.8	0.8	0.8	0.8
Glass dimensions (m)	1.23 × 1.48	1.23 × 1.48	1.23 × 1.48	1.23 × 1.48

Wood (*W*), wood/aluminium (*W/ALU*), PVC or wood/composite (*W/C*)

Table 39.5 Life times and reference flows needed to fulfil the functional unit for the four window alternatives

Property/reference flow	Window type			
	W	W/ALU	PVC	W/C
Life time of the window frame (years)	40	40	30	40
Life time of the window pane (years)	20	20	20	20
Reference flows (numbers)				
Window frame	0.5	0.5	0.67	0.5
Window pane	1	1	1	1
Packaging	1	1	1	1
Paint for window frame	8	0	0	0

Functional unit. All four windows are compared on the basis for the following functional unit: "Allow daylight into a residential building through a physical barrier, equivalent to light being transmitted through an area of 1.82 m^2 with visible light transmittance of at least 0.6, for 20 years". This definition allows for a fair comparison between windows with different *U*-values. Note, however, that it does not allow for a comparison with an empty hole in the house (which although allows daylight into a building, is not a physical barrier).

Reference flows. Considering the different life times for window frames and panes, the reference flows in Table 39.5 are derived as needed to provide the service defined in the functional unit. It is assumed that the pane will be changed once in windows' lifetime (specifically, after 20 years) and that the window is used for the time equal to the lifetime of the frame. Please note that reference flows for pane and frame would be equal if the whole window is to be replaced after 20 years equal to the expected life time of the pane (and irrespective of the life time of the frame). Whether this is the case will depend on factors like trends in design, the overall state of the building, or the frequency of change in the ownership of the apartment where the window is installed. These factors were not considered here.

39.3.2.3 LCI Modelling Framework

Nor-win's introduction of a new window and improvements in heat insulation properties of existing windows are not expected to have large structural changes on the market (like installation of new composite factories or decommissioning of existing heat pumps). Thus, the decision context is micro-level, product or process-related decision support studies, i.e. situation A in the ILCD Guideline (EC-JRC 2010), suggesting that the attributional principle be chosen as LCI modelling framework. This implies that the systems are modelled depicting existing value chains, i.e. using current Danish electricity and heating mix, and Danish recycling rates for end-of-life scenarios. Consistently with the micro-level decision context, system expansion was done to credit for the heat loss in the use stage using average Danish data. Note, that we applied system expansion (through crediting) using average processes in this attributional approach, consistently with both ILCD and the ISO hierarchy to solving multifunctionality, although system expansion using marginal processes has traditionally been considered for the consequential approach to inventory modelling (allocation has traditionally been used for the attributional approach).

Apart from the secondary function of the window (i.e. its ability to transfer heat), other processes downstream can also have secondary functions or co-products, namely recycling operations (producing recycled steel, aluminium, glass, or recycled PVC). As system expansion is the preferred approach to solving multifunctionality, materials produced from recycled content are credited using virgin materials, i.e. aluminium ingots, steel billet, glass cullets (for all windows), and additionally PVC granulate mix (for PVC window), where all virgin materials and PVC granulate are produced using average technologies. Similarly, incineration of some materials in the end-of-life stage produces heat and power, which is credited using average Danish heat and power mixes. Secondary functions in upstream processes also exist (e.g. naphtha cracking, waste incineration), but these secondary functions had to be handled according to how the database (i.e. ecoinvent v.2.1) did it, so using allocation.

39.3.2.4 System Boundaries and Completeness Requirements

System boundaries. The analysis includes all life cycle stages from cradle to grave (Fig. 39.2). Processes include for all product systems raw material extraction, primary and secondary material production and upstream processes such as mining of metal ores and extraction of crude oil within the system boundaries. Raw materials are transported to metal smelters or refineries to produce virgin metals, fuels and plastics. Similarly, forestry is included to produce wood. These materials are used to produce specific parts from which window components (window frame and windowpane) are made. Assembled windows are packaged and transported to retailers, and from there to residential buildings, in which they are mounted and provide their function. The use stage includes maintenance painting of the wood window (W), and cleaning of all windows. At the end of life, windows are dismantled and transported to a waste handling facility for disassembly. Steel, aluminium and glass are mainly recycled. Landfilling and incineration with heat recovery apply to cardboard, wood, and plastics, and to the remaining, non-recycled fractions of steel, aluminium and glass.

Completeness requirements. As the LCA includes a hotspot analysis, no processes should ideally be excluded from the system boundaries based on their similarity between the four window systems. Yet, we excluded:

(i) Cleaning of windowpane. Although potentially interesting for the hot spot analysis, Nor-win has little control of the cleaning process (e.g. frequency of the cleaning, detergent type). Thus, the inclusion of cleaning is not that important for the goal of the study (that is, to support guidance for the ongoing design of the new window and green supply chain for the existing windows).

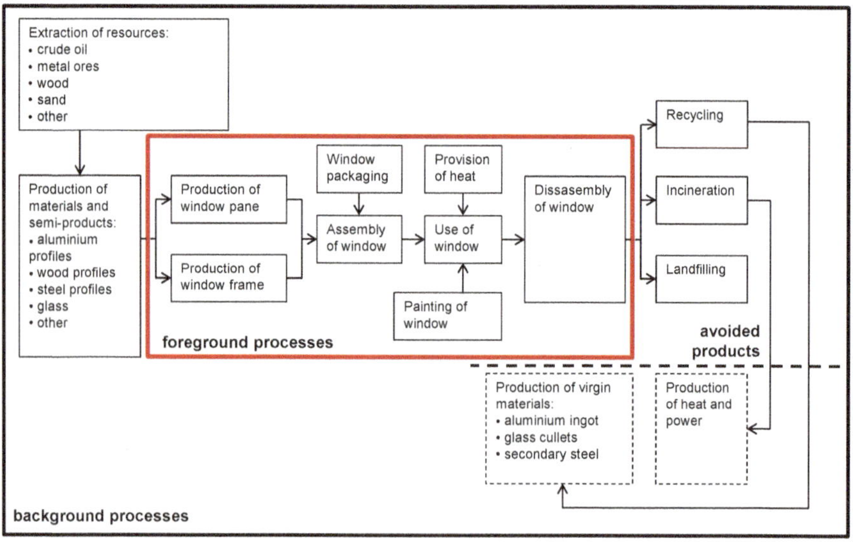

Fig. 39.2 System boundaries for the product systems of the four window alternatives. Transportation and provision of utilities other than heat (e.g. electricity) are included inside the system boundaries but not shown in the figure in order to make it more legible

(ii) Capital equipment such as buildings or machines, unless already integrated in aggregated unit processes of the background system, is excluded. This is common practice in process-based LCA.
(iii) Materials contributing to less than 5% of the total mass of the window are cut-off, excepting substances for surface treatment of the window frame, which are expected to be toxic and hence potentially contribute significantly to some impact scores even with much smaller quantities applied.

39.3.2.5 Representativeness of LCI Data

Technological representativeness. The manufacturing technology for all four windows, including the elements composing the pane and the frame, should (ideally) represent the technology that is currently used by Nor-win and its primary suppliers. This technology is characterised by relatively high efficiency (in terms of material output per day), mainly due to the use of modern (<5 years old) machines and production lines and the employment of relatively new (<7 years old) technological solutions (like those used for impregnating and painting wooden frame, or painting aluminium). Thus, the data for window manufacturing should primarily come from Nor-win and its suppliers. Alternatively, other Scandinavian or European window manufacturers (and European suppliers) that use relatively modern technology can be used as source of data for manufacturing to compensate for missing data.

Data for background processes, like extraction of metal ores and production of raw metals, or extraction of fossils, should ideally represent the average technology currently used globally. It is sufficient that this data comes from generic databases.

Geographical representativeness. Geographical coverage is similar for all four windows (Table 39.6). Wood for the frame originates from forest in Finland, while the pane together with the glass is produced in Sweden. Polyvinyl chloride for the PVC window is produced in Germany. Thus, in the absence of Nor-win and supplier-specific data, they should originate from associations and companies located in Europe. End-of-life data should be the average for the main market, which is either Denmark (W, W/ALU and W/C) or Germany (PVC).

Temporal representativeness. The data for manufacturing processes should be representative for windows produced from 2015 to 2020, i.e. a 5-year time horizon for window manufacturing (the product development takes about 1 year and is not considered important). The average window lifetime is assumed to be 30 (PVC) or 40 years (W, W/ALU and W/C). However, to comply with the definition of the functional unit, the use stage and end-of-life processes should (ideally) be representative for the 25–30 year time horizon, over which the products will be in use or disposed of. Figure 39.3 shows temporal frames of the windows.

Table 39.6 Geographical scope for life cycle stages and central unit processes in the window frames case study

Stage	Window type			
	Wood	Wood/aluminium	PVC	Wood/composite
Materials	Metal ores: not known			
	Crude oil: Norway, Russia, Middle East			
	Forestry: Finland		–	–
Manufacturing	Glass pane: Sweden			
	Wood frame: Scandinavia	Wood/aluminium frame: Scandinavia	PVC frame: Scandinavia	Composite frame: Scandinavia
	Other elements: mainly Europe	Other elements: mainly Europe	Other elements: mainly Europe	Other elements: mainly Europe
	Assembly: Denmark			
Use (heat supply)	Mainly Scandinavia, Germany	Mainly Scandinavia, Germany	Mainly Germany	Mainly Scandinavia
Disposal	The same as the use stage			

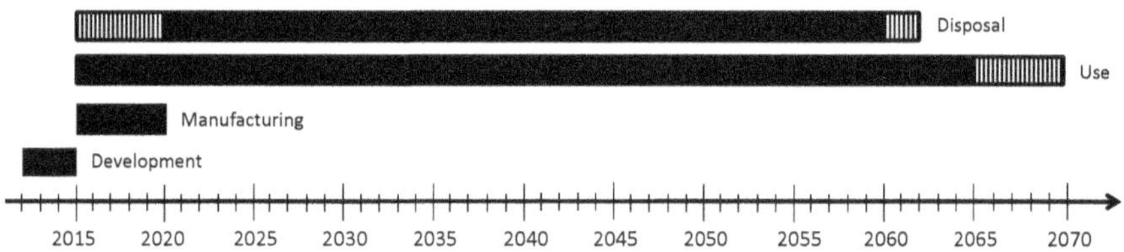

Fig. 39.3 Temporal scope of the W, W/ALU and W/C windows expressed for different life cycle stages. Manufacturing starts in 2015 and continues for 5 years, thus the overall time horizon for the use stage is 5 years longer than the 20-year duration of the use stage for individual window (indicated with a *black-white pattern*). Similarly, the temporal horizon for the disposal stage may start right after the first window had been produced, and end 2 years after end of the use stage (as indicated with the *black-white pattern*). The temporal scope for the use and disposal of the PVC window is 10 years shorter compared to the three other windows (not shown). Note that the temporal scope looks at the time horizon of the window life cycle (i.e. 40 years), regardless of the duration considered in the functional unit (in our study equal to 20 years)

39.3.2.6 Basis for Impact Assessment

ILCD's recommended practice for characterisation modelling is employed as life cycle impact assessment method (EC-JRC 2012; Hauschild et al. 2013). The ILCD is a combination of state of the art methods for LCIA (as of 2009). Analysis of the sensitivity of the results to alternative LCIA methods was not deemed necessary since the results are for internal use only. Modelling impacts at midpoint is considered sufficient given the goal of the study (comparative assessment and identification of hotspots). Normalisation was performed using the set of normalisation references presented for year 2010. The following impact categories are included in the ILCD: climate change (unit: kg CO_2 eq.), ozone depletion (kg CFC-11 eq.) photochemical ozone formation (kg NMVOC eq.), terrestrial acidification (AE, accumulated exceedance), terrestrial eutrophication (AE, accumulated exceedance), freshwater eutrophication (kg P eq.), marine eutrophication (kg N eq.), freshwater ecotoxicity (CTU_e, comparative toxic unit for ecosystems), ionising radiation (human health, in kBq U235 eq.), particulate matter/(kg $PM_{2.5}$ eq. to air), human toxicity (cancer effects, in CTU_h for human health), human toxicity (non-cancer effects, in CTU_h for human health), land use (kg C year), and resource depletion (mineral and fossils, in kg Sb eq.).

Product systems were modelled in GaBi, version 4.3 (PE International, Germany; renamed to thinkstep). Because the ILCD LCIA method was not implemented in GaBi at the time of the study, characterisation factors for the ILCD methods (version 1.0.3, 01 March 2012) were downloaded from the Life Cycle website of the European Commission (http://lct.jrc.ec.europa.eu/assessment) and were imported into the software. For those impact categories where ReCiPe 2008 (Goedkoop et al. 2009) is the recommended method (6 categories in total), impact scores were calculated using the original set of ReCiPe (version 1.05) characterisation factors as implemented in GaBi. Normalisation references are for the EU27 in the reference year 2010 as presented in Benini et al. (2014). The LCIA methods and normalisation factors are presented in Annex, Sect. 39.4.1.

39.3.2.7 Requirements for Comparative Studies

Although requirements for a comparative study (like quality requirements, exclusion of identical processes and interpretation in light of affected stakeholders) are only applicable to studies reported at level 3 (that is, comparative studies to be disclosed to the public), we note that in our case study comparison has been made using the same functional unit, system boundaries omit common processes only (i.e. window washing), and data quality is the same between the compared windows (e.g. primary data come from manufacturer). Thus, the comparison between the four window systems is fair and requirements for comparative studies are met. The readers should note, however, that although we quantified inventory uncertainties, comparison could not be made taking into account correlation between inventory uncertainties of those inventory processes which are the same for the compared window systems (as will be explained in detail in Sect. 39.3.5). Thus, in some cases there may be statistically significant difference in impact scores between window systems, even if that is not apparent in our analysis. On the other hand, uncertainties in background processes were not considered, which may, at least partially, outweigh a potential decrease in uncertainty due to correlations.

39.3.2.8 Critical Review Needs

This is a comparative study but since it is not intended for disclosure to the public, there is no obligation for a critical review by a third-party panel.

39.3.3 *Life Cycle Inventory Analysis*

39.3.3.1 LCI Model at System Level

Flow diagrams show the product systems of the four windows in Figs. 39.4, 39.5, 39.6 and 39.7. Comparison between the flow diagrams shows that many processes are the same for the four windows (e.g. mining of silica and production of glass, or mining of iron and production of chromium steel). Yet, magnitude of flows often varies between the systems (not shown).

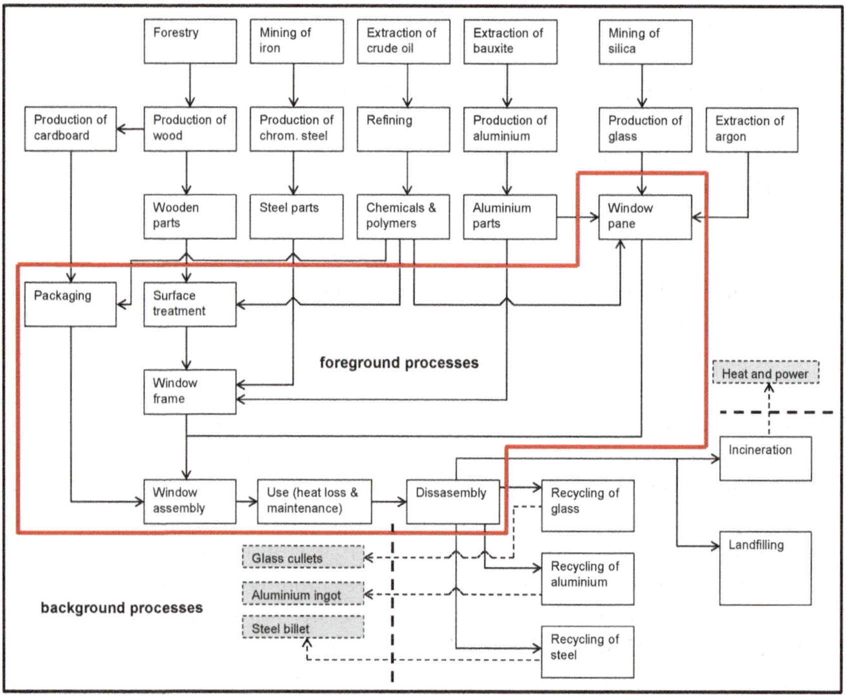

Fig. 39.4 Product system of the wood window (W). *Red line* indicates foreground processes. *Grey boxes* indicate avoided processes

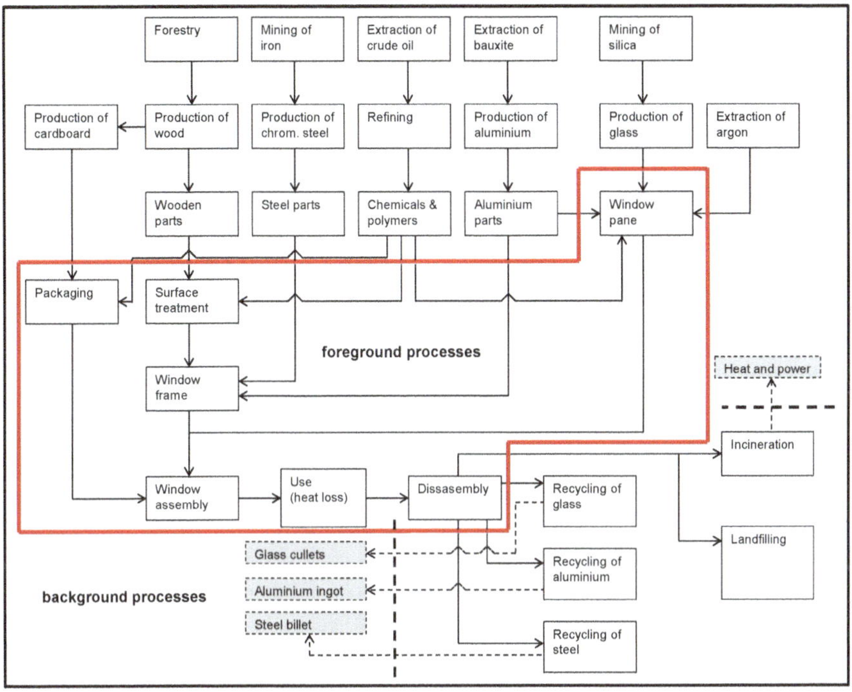

Fig. 39.5 Product system of the wood/aluminium window (W/ALU). *Red line* indicates foreground processes. *Grey boxes* indicate avoided processes

39.3.3.2 Data Collection

Data used to model life cycle inventories for the foreground systems were collected from two sources: (i) Nor-win, who provided primary data related mainly to energy use in the manufacturing and bills of materials and (ii) ecoinvent and Plastics Europe databases for foreground processes where primary data could not be achieved (PlasticsEurope Database 2016; Ecoinvent 2010). The primary data from Nor-win meet the quality requirements given in Sect. 39.3.5. The data are synthesised in Table 39.7.

39.3.3.3 System Modelling Per Life Cycle Stage

Below, we present details of the system modelling, the data collected and treatment, and major assumptions. The full list of major and minor assumptions is given in the report Annex, Sect. 39.4.3.

Materials stage. Bills of activities required to produce one window are given in Table 39.7, with details on the bill of materials presented in Annex, Sect. 39.4.2 (Table 39.12). Amounts of materials in each window are provided by Nor-win.

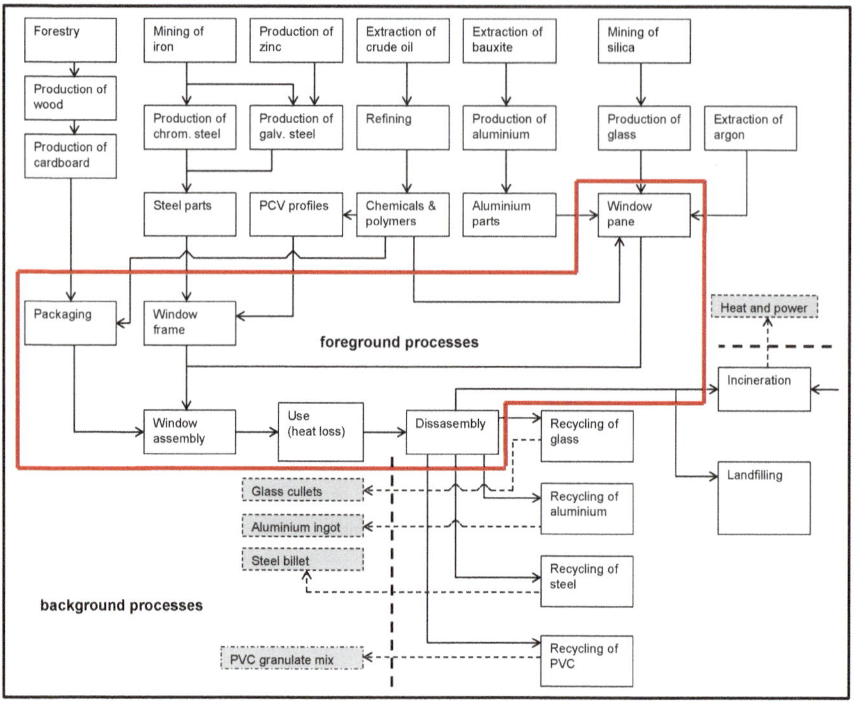

Fig. 39.6 Product system of the polyvinyl chloride window (PVC). *Red line* indicates foreground processes. *Grey boxes* indicate avoided processes

Data for W, W/ALU and PVC windows are precise, because these windows are already on the market and detailed information is available. Data for the W/C window are considered sufficiently accurate to be used in modelling, because the prototype of the window has been produced. Note that based on the outcome of this study, the W/C window may be redesigned, bringing about a change in amounts of some materials in which case the LCA may have to be updated with the new numbers. It is not expected that this change will be higher than 5% for any window frame material. No major assumptions were made for the materials stage.

Manufacturing stage. Data on electricity use for production come from measurements of the actual processes and are provided by Nor-win. These data are of high quality and are considered certain. Data on electricity requirements for assembly of the W/C window frame are less certain, and are initial estimates provided by Nor-win. The three major assumptions made in the production stage are (i) losses of materials during production are not considered, (ii) energy used for operation of the manufacturing and window disassembly facilities for the W/C window is assumed equal to numbers for other windows, and (iii) energy requirements for window disassembly are assumed equal to 1 MJ per 1 kg of dismounted window.

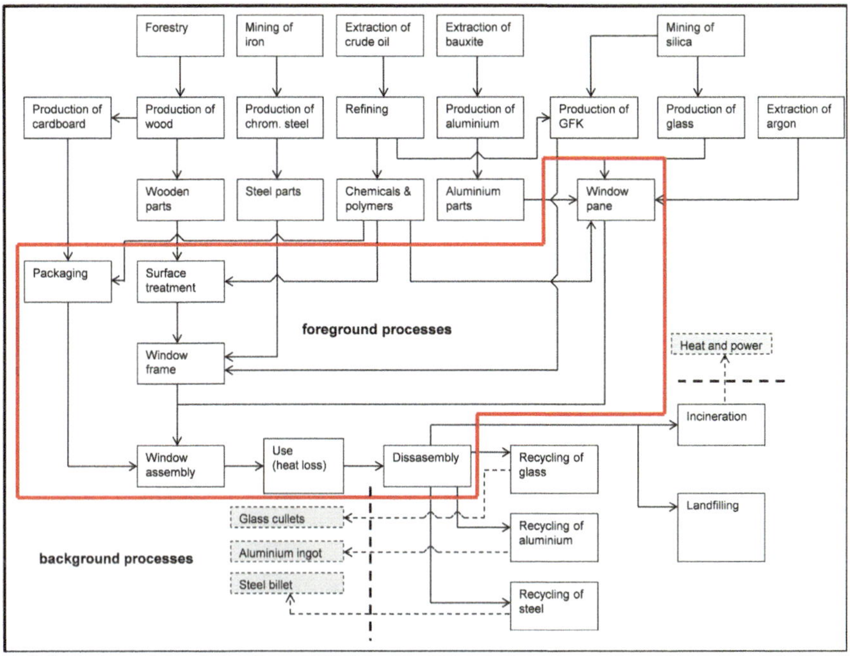

Fig. 39.7 Product system of the wood/composite window (W/C). *Red line* indicates foreground processes. *Grey boxes* indicate avoided processes

Use stage. Data on heat use during the use stage are calculated using the *U*-values and average temperature difference between outdoor and indoor environment (Table 39.7). Several assumptions were made for the use stage. First, we modelled the heat loss based on the annual average temperatures indoor and outdoor, without considering the temperature dynamics during the year. Second, we assumed that there is no shift in the source of heat (e.g. towards wind-driven electricity) over the lifetime of the window. Third, we assumed that the windows are used only in buildings to which heat is provided by district heating. In Denmark, district heating was estimated to deliver 55% of the total heat demand for buildings in 2010 (Dyrelund and Lund 2009). Fourth, processes used to model the district heat mix technologies were representative for Switzerland, or were based on European processes for the generation of heat from the sources included in the study, as no Danish processes were available in ecoinvent 2.2. The fifth assumption is that processes for generation of heat from incineration of straw and non-renewable waste in the Danish heat mix were modelled as incineration of bio-waste, while heat generation from biomass in the EU27 heat mix was modelled as incineration of bio-waste combined with combustion of wood pellets (50:50).

Disposal stage. Waste treatment options are based on the data retrieved from Eurostat (2016). Glass, aluminium and steel are mainly recycled, and wood is

Table 39.7 Metadata including model parameters and data sources for foreground processes for the four window alternatives

Parameter	Value				Unit	Note	Source
	W	W/ALU	PVC	W/C			
Materials							
Frame materials	See Annex, Sect. 39.4.2, Table 39.12				kg	In addition to materials presented in Table 39.12, other materials used to produce the frames include acrylic binder, triethylene glycol and wood preservative which are used in small amounts (<1% of total frame mass). Bills of materials are retrieved from the producer	Measured
Pane materials	See Annex, Sect. 39.4.2, Table 39.12				kg	In addition to materials presented in Table 39.12, synthetic rubber (ethylene propylene diene monomer, EDPM is used in smaller amounts (<2–6% of total pane mass). Bills of materials are retrieved from the producer	Measured
Thickness of glass underlying window pane	5				mm	Nominal thickness as provided by the producer	Measured
Packaging	See Annex, Sect. 39.4.2, Table 39.12				kg	Packaging is made of polyethylene and recycled cardboard. Bills of materials are retrieved from the producer and are presented in Table 39.12	Measured
Manufacturing							
Electricity for production of window frame	15	21	24	30	MJ	The data come from measurements of the actual processes and are provided by the producer, apart from the W/C window for which the producer assumed that electricity required for assembly of the frame for the W/C window is 2 times higher than for the W window	Measured or assumed
Electricity for production of window pane	5	5	5	5	MJ	The data come from measurements of the actual processes and are provided by the producer	Measured
Electricity for mounting window frame and window pane into a window (MJ)	38	40	46	50	MJ	The data come from measurements of the actual processes and are provided by the producer, apart from the W/C window for which the producer assumed that electricity required for mounting of thee W/C window is 1.3 times higher than for the W window	Measured or assumed

(continued)

Table 39.7 (continued)

Parameter	Value				Unit	Note	Source
	W	W/ALU	PVC	W/C			
Electricity for operation of the manufacturing facility	80	80	80	80	MJ	The data is provided by the producer for the plant and is scaled to one window basing on the electricity bills and production capacity while taking into account electricity used for production and mounting	Measured
Heat demand							
Heat loss per year	741	752	781	620	MJ/year	Heat loss is directly proportional to the *U*-value and the window area and to the temperature difference between indoor and outdoor environment, and is calculated using the formula $\Phi_t = U \cdot A \cdot (\theta_i - \theta_e)$ where Φ_t [W] is the heat loss, *U* [W m^{-2} K^{-1}] is the *U*-value, *A* [m^2] is the heat exchange area; θ_{i} is the indoor temperature (K), and θ_{e} is the outdoor temperature (K)	Calculated
Indoor temperature	17	17	17	17	°C	Annual average indoor temperature in residential buildings	Assumed
Outdoor temperature	7.7	7.7	7.7	7.7	°C	Annual average outdoor temperature in Denmark, calculated based on the temperature data retrieved for year 2014 from DMI (2016)	Calculated
Maintenance of window							
Painting needed for painting of 1 window	0.6	0	0	0	kg	Approximately 1 L of paint is needed for a window frame area of 1 m^2 (based on data from a single paint producer)	Assumed
Transportation distances and means							
Elements composing window frame and window pane	1240	1110	830	310	km	Distance between suppliers of elements underlying window frame and windowpane and location of the producer. Calculated using Google maps. Transport by truck 34–40 t, EURO4	Calculated

(continued)

Table 39.7 (continued)

Parameter	Value				Unit	Note	Source
	W	W/ALU	PVC	W/C			
Window from producer to warehouse	300	300	300	300	km	Distance between producer and warehouse. Calculated using Google maps. Transport by truck 34–40 t, EURO4	Calculated
Window from warehouse to residential building	100	100	100	100	km	Distance between the warehouse and residential building including retail. The location of final user is unknown and distance had to be assumed. Transport by truck 12–14 t, EURO3	Assumed
Window from residential building to disposal/recycling site	100	100	100	100	km	Distance between the residential building and disposal/recycling site. The locations are unknown and thus the distance had to be assumed. Transport by truck 12–14 t, EURO3	Assumed
Packaging from residential building to the disposal site	85	85	85	85	km	Distance between the warehouse and residential building including retail. The location of final user is unknown and distance had to be assumed. Transport by truck 12–14 t, EURO3	Assumed
Disassembly and disposal							
Electricity for disassembly of window	90	77	91	76	MJ	It is assumed that electricity consumption is equal to 1 MJ per 1 kg of dismounted window	Assumed
Waste treatment options	See Sect. 39.4.2, Table 39.13				%	Disposal according to the Danish waste policy (Eurostat 2016). Treatment rates are presented in Annex, Sect. 39.4.2, Table 39.12	Measured

Wood (*W*), wood/aluminium (*W/ALU*), PVC or wood/composite (*W/C*)

Note that the values are scaled to one window used for one year, not to the functional unit of the window systems incinerated. Other materials are mainly incinerated, or landfilled. PVC is technically recyclable but not to the extent as for other plastics (30%). The remaining part of PVC is landfilled. The composite (glass fibre/polyamide) is technically difficult to recycle, and is assumed 100% incinerated. Details of end-of-life options are presented in Annex, Sect. 39.4.2 (Table 39.13). The two major assumptions are (i) all recycled materials replace virgin materials in the market, i.e. glass cullets, aluminium ingot, steel billet, and PVC granulate mix, at a 1:1 ratio, i.e. without considering any loss of material functionality in the recycling; and (ii) although the wood-based windows are sold mainly in Scandinavian countries and Germany, the use and disposal stages for all windows are modelled using data from processes representative for Denmark, e.g. Danish heating and electricity mixes and waste management systems.

Transportation. Transportation distances and means are either provided by Nor-win or assumed. The data provided by Nor-win are considered sufficiently accurate, whereas the assumed data are considered uncertain.

39.3.3.4 Basis for Sensitivity and Uncertainty Analyses

To test the influence of the assumptions made on the results of the LCA, sensitivity analyses were performed, followed by uncertainty and variability analyses.

Sensitivity analyses. First, to identify which of the parameters influence impact scores the most, and to provide a basis for uncertainty and variability analysis, we calculated normalised sensitivity coefficients ($X_{\mathrm{IS},k}$), according to Eq. 39.1 (e.g. Prommer et al. 2006):

$$X_{\mathrm{IS},k} = \frac{\Delta\mathrm{IS}/\mathrm{IS}}{\Delta a_k/a_k} \tag{39.1}$$

where $X_{\mathrm{IS},k}$ is the normalised sensitivity coefficient of impact score (IS) for perturbance of a parameter k, a_k is the default value of parameter k, Δa_k is the perturbation of parameter a_k, IS is the calculated impact score for parameter value a_k, and ΔIS is the change of the impact score that results from the perturbation of parameter a_k. The following parameters were tested: amount of wood, aluminium, steel composite (W/C window only), and PVC (PVC window only) in the window frame, the amount of glass in the pane, amount of paint for manufacturing, electricity needed for assembly, U-value, and transportation distance from Nor-win to retailers. All input parameters were perturbed by 10%, which is a realistic range around the expected values. $X_{\mathrm{IS},k}$ equal to 1 means that a 10% increase in parameter value brings about a 10% increase in the impact score. Generally, a parameter is considered to have medium sensitivity if $X_{\mathrm{IS},k} > 0.3$, and large sensitivity if $X_{\mathrm{IS},k} > 0.5$. In this study, a parameter is considered important when $X_{\mathrm{IS},k} > 0.3$.

Second, in addition to testing sensitivity to individual parameters through computation of normalised sensitivity coefficients, perturbing each parameter at once, a separate sensitivity check was done, where several parameters expected to be important were perturbed at once. The overview of the two sensitivity scenarios considered is given in Table 39.8. Scenario 1 reflects a situation where the window is used by an average European residence rather than a Danish residence. Scenario 2 reflects the situation where a 3-layered windowpane is used instead of a 2-layered one, which improves insulation properties of the whole window (without any considerable influence on visible light transmission properties). This scenario was included to identify potential for improvements of existing and new windows. Note, that over the coming 20 years we may see a shift in the heat source (e.g. towards wind-driven electricity) but it is uncertain to what extent these will become effective within the time frame of the study (25–30 years). On the other hand, we may also witness the development of cleaner manufacturing and waste management technologies in 20 years (which is also uncertain). Thus, the potential change in heat mix and change in manufacturing and waste management systems were not considered in the sensitivity analysis.

Uncertainty and variability analysis. Parameter uncertainties stem from the imprecision in knowledge about the actual value of a parameter, e.g. electricity use during window assembly. By contrast, variability is the inherent variance that will exist between similar processes depending on technological level and spatial location, e.g. transportation distance from factory to retail Steinmann et al. (2014).

Table 39.8 Sensitivity scenarios and corresponding model parameters

Sensitivity parameters	Baseline scenario	Sensitivity scenario	
		Scenario 1	Scenario 2
Use location[a]	DK	EU27	DK
Disposal route[b,e]			
Heat mix[c,f]			
Electricity mix[d,g]			
Pane design	2-layered	2-layered	3-layered[h]

[a]*DK* Denmark; *EU27* European Union's 27 member states

[b]Please see Annex, Sect. 39.4.2 (Table 39.13) for details of end-of-life options in DK and EU27

[c]Danish heating mix in 2010 was based on: natural gas (24%), coal (23%), straw (8%), wood chips (12%), wood pellets (10%), non-renewable waste (17%), oil (2%), and other sources (4%) (Energynet 2012)

[d]Danish electricity mix in 2010 as based on: hard coal (36%), natural gas (14%), wind power (15%), oil (2%), import from Sweden (14%), Norway (10%), Germany (3%), and other sources (6%) (Ecoinvent 2010)

[e]Compared to Danish disposal routes the EU27 disposal routes in 2010 is characterised by lower frequency of recycling and/or incineration, and increased frequency of landfilling (Eurostat 2016). The disposal options are summarised in Annex, Sect. 39.4.2 (Table 39.13)

[f]EU27 heat mix in 2010 was based on: natural gas (57%), oil (21%), biomass (13%), and coal (9%) (Connolly et al. 2012)

[g]EU27 electricity mix was 2010 is based on: nuclear power (28%), coal and peat (27%), natural gas (27%), hydropower (11%), wind power (4%), oil (3%), biofuels (3%), and non-renewable waste (7%) (Ecoinvent 2010)

[h]3-layered windows have improved insulation properties thanks to smaller *U*-values, which were reduced by 25% for the W, W/ALU, and PVC windows, and by 30% for the W/C window

Table 39.9 Uncertain or variable parameters included in the Monte Carlo simulation and the associated relative standard deviation, expressed in percentage

Uncertain or variable parameter	Mean (relative standard deviation)[a]			
	W	W/ALU	PVC	W/C
Amount of wood in the frame[b]	30 (1%)	9.2 (1%)	0 (0%)	9.2 (2.5%)
Amount of steel in the frame[b]	0.5 (1%)	1.2 (1%)	15.1 (1%)	1.2 (2.5%)
Amount of glass in the pane[c]	56 (0.5%)	56 (0.5%)	56 (0.5%)	56 (0.5%)
U-value of the window[d]	1.29 (1.5%)	1.31 (1.5%)	1.36 (1.5%)	1.08 (3%)

[a]Relative standard deviation (also known as coefficient of variation, CV) is equal to sample standard deviation divided by sample mean, expressed in percentage. Sample standard deviation was estimated using an empirical rule that the sample standard deviation is equal to one fourth of the whole parameter range (equal to the difference between maximum and minimum value)
[b]Change in amounts of wood and steel in the frame depend mainly on losses in the production, and are expected to be maximum 2% for W, W/ALU and PVC windows, and 5% for the W/C window, because of the ongoing development of the latter. These values are realistic values provided by Nor-win based on the information retrieved from suppliers
[c]Change in the amount of glass is expected to be by maximum 1%. Again, this value was provided by Nor-win
[d]Although the *U*-value is considered as an inherent property of a window, the actual amount of heat exchanged depends on other factors, like the quality of the work during window installation, type and quality of insulation used to install the window in the wall, or type and properties of walls. To account for this variability, a maximum change in the *U*-value of 3% was used for W, W/ALU and PVC windows, based on the information from Nor-win. For the W/C window, 6% was used to calculate minimum and maximum *U*-values (again, because it is ongoing product development)

Here, parameter uncertainty was assessed together with variability by means of a Monte Carlo simulation. Only parameters that were found important ($X_{IS,k} > 0.3$) in the sensitivity analysis, for any of the considered impact categories for either window design option, were considered. In total, four parameters where considered (Table 39.9). They were assigned relative standard deviations derived from the expected range of parameter values. The uncertainty ranges and number of uncertain parameters is higher for the W/C window, because this window is still in under development and very accurate bills of materials and performance parameters (*U*-value) are not known. We assumed normal distributions of all parameters mainly because this is one of two types of distribution implemented in our version of the software, GaBi v. 4.3 (the other being equal distribution). Other distribution types (e.g. lognormal) can be used if found more appropriate, provided that such is possible in the modelling software employed. Uncertainties in the background processes were not considered as they were not known and the unit process database did not include them at the time of the study. Differences in impact scores between the compared systems were considered significant if the calculated 95% probability ranges of the impact scores from 1000 iterations did not overlap.

Although not deemed necessary in this case study, all other flows and parameters could be ascribed to standard deviations, supporting a more comprehensive uncertainty analysis. In such cases, standard deviations for each flow in foreground processes could be computed using the Pedigree matrix approach (Ciroth 2013). Uncertainties in the background processes should be considered based on standard deviations already assigned to flows in processes of the considered unit process database. Newer versions of the database offer such features.

The calculated probability ranges represent the modelled inventory uncertainty, but we did not account for covariation between processes that occur in some or all of the compared window systems (e.g. production of heat for the use stage), leading to correlations between uncertainties of those inventory processes. The employed modelling software (GaBi v. 4.3) did not allow taking this into account, but it would have reduced the uncertainty in comparison between the systems (see Sect. 11.4.2). Thus, in some cases there may be statistically significant difference in impact scores, even though that is not revealed by our analysis. On the other hand, uncertainties in background processes were also not considered in our case study, which would increase the uncertainty in the results and may, to some extent, counterbalance this effect. In addition, the characterisation and normalisation factors applied in the impact assessment are accompanied by uncertainties but these were not known to us and we were therefore unable to take them into account in our uncertainty analysis. They are expected to be equal to or higher than the inventory uncertainties.

39.3.3.5 Calculated LCI Results

Unit processes and life cycle inventories showing elementary flows for each window product system are documented in Annex, Sect. 39.4.4 (Tables 39.15, 39.16, 39.17, 39.18, 39.19, 39.20, 39.21, 39.22, 39.23, 39.24, 39.25, 39.26, 39.27, 39.28, 39.29, 39.30, 39.31, 39.32, 39.33 and 39.34).

39.3.4 Life Cycle Impact Assessment

Characterised results. The life cycle impacts are listed in characterised form in Table 39.10. All four window alternatives have impacts within the same order of magnitude. For most impact categories the impact scores follow the order W/C < W = W/ALU < PVC. Ranking of window systems normalised internally to the W window (equal to 100% of impact) is presented in Fig. 39.8. The W/C window has the lowest environmental impact in all 14 impact categories, while the PVC window system has the highest impact scores for 11 impact categories. For these 11 impact categories, the differences in impact scores between the W/C and PVC windows are statistically significant (the calculated 95% probability ranges of the impact scores do not overlap). The PVC window performs better in land use impacts with a significantly lower impact compared to the W and W/ALU window systems, but still slightly higher compared to the W/C window system. By contrast, the W window system performs significantly worse than the other window systems

Table 39.10 Characterised impacts and accompanying 95% probability ranges from Monte Carlo simulations for each window alternative

Impact category	Unit	Impact score (95% probability range)			
		W	W/ALU	PVC	W/C
Climate change	kg CO_2 eq.	1162 (1134–1189)	1158 (1129–1188)	1232 (1203–1260)	978 (933–1023)
Stratospheric ozone depletion	kg CFC-11 eq.	1.9e−5 (1.8e−5–1.9e.5)	1.6e−5 (1.5e−5–1.6e−5)	1.6e−5 (1.5e−5–1.6e−5)	1.4e−5 (1.4e−5–1.4e−5)
Photochemical ozone formation	kg NMVOC eq.	1.59 (1.55–1.63)	1.57 (1.53–1.61)	1.72 (1.67–1.76)	1.33 (1.26–1.4)
Terrestrial acidification	AE	2.00 (1.95–2.04)	2.00 (1.95–2.05)	2.31 (2.26–2.36)	1.67 (1.6–1.75)
Terrestrial eutrophication	AE	7.14 (6.94–7.35)	7.04 (6.83–7.24)	7.79 (7.56–8.02)	5.96 (5.61–6.32)
Freshwater eutrophication	kg P eq.	0.042 (0.041–0.043)	0.043 (0.041–0.044)	0.046 (0.044–0.047)	0.035 (0.033–0.037)
Marine eutrophication	kg N eq.	0.65 (0.63–0.67)	0.62 (0.60–0.64)	0.68 (0.66–0.7)	0.54 (0.51–0.57)
Freshwater ecotoxicity	CTU_e	2675 (2605–2745)	1755 (1706–1805)	1852 (1809–1895)	1545 (1461–1630)
Human toxicity (cancer)	CTU_h	2e−5 (1.9e−5–2e−5)	1.8e−5 (1.8e−5–1.9e−5)	3.4e−5 (3.3e−5–3.5e−5)	1.6e−5 (1.5e−5–1.7e−5)
Human toxicity (non-cancer)	CTU_h	1.5e−4 (1.5e−4–1.6e−4)	1.3e−4 (1.2e−4–1.3e−4)	1.3e−4 (1.3e−4–1.3e−4)	1.0e−4 (9.9e−5–1.1e−4)
Particulate matter formation	kg $PM_{2.5}$ eq. to air	0.085 (0.083–0.087)	0.082 (0.080–0.084)	0.116 (0.114–0.119)	0.070 (0.067–0.073)
Ionising radiation (human health)	kBq U235 eq.	7.69 (7.56–7.81)	7.99 (7.86–8.12)	8.63 (8.49–8.77)	6.26 (6.07–6.45)
Land use	kg C year	657 (646–668)	405 (399–410)	386 (384–387)	364 (351–377)
Resource depletion (minerals, fossils)	kg Sb eq.	0.0072 (0.0070–0.0073)	0.0074 (0.0072–0.0076)	0.0081 (0.0080–0.0083)	0.0063 (0.0060–0.0066)

W/ALU Wood/aluminium, *PVC* polyvinyl chloride, *W/C* wood/composite

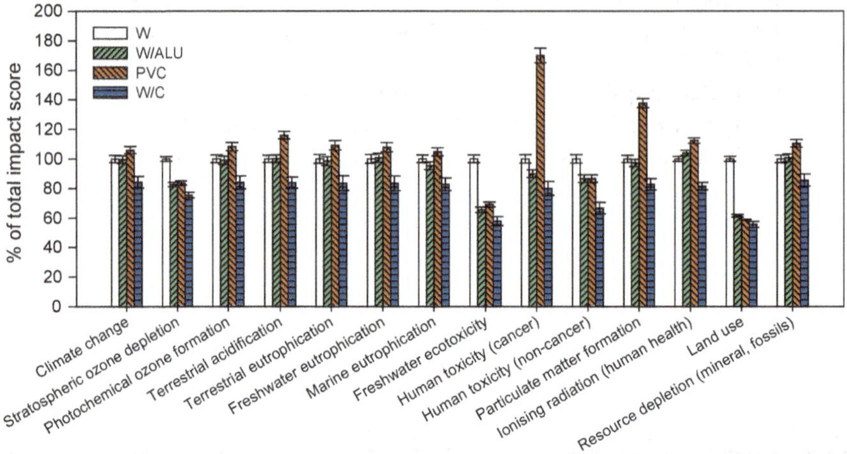

Fig. 39.8 Ranking of the four window options with impact scores scaled to those of the W window (equal to 100% of total impact). Whiskers represent inventory uncertainty stemming from uncertainty and variability in model parameters presented in Table 39.8

for the impact categories freshwater ecotoxicity, human toxicity (non-cancer), stratospheric ozone depletion, and land use. The W and W/ALU window systems rank as second or third for 10 out of 14 impact categories, but for these alternatives, differences between impact scores are only statistically significant in the ionising radiation impact category.

Normalised results. Figure 39.9 shows the normalised results. The common unit for indicator scores is person equivalents (pe) representing the annual impact of an average person in the European Union (EU27) in 2010. For nearly all the non-toxicity impact categories, like climate change, the life cycle impacts of the four windows correspond to approximately 10% of the total annual average impacts of an average EU27 citizen in the year 2010. Much smaller normalised impact scores are seen for stratospheric ozone depletion. Normalised results are somewhat higher for freshwater ecotoxicity and human toxicity impact categories (scoring up to 1 PE for cancer effects), but are smaller for respiratory effects and ionising radiation impacts on human health (around or below 0.1 PE). Normalised impact scores are the highest for human toxicity (cancer), equal to ca. 0.5 PE, but are small for land use (below 0.001 PE).

39.3.5 Interpretation

Before providing final recommendations to the commissioner of the study, it is necessary to interpret the results of the LCA.

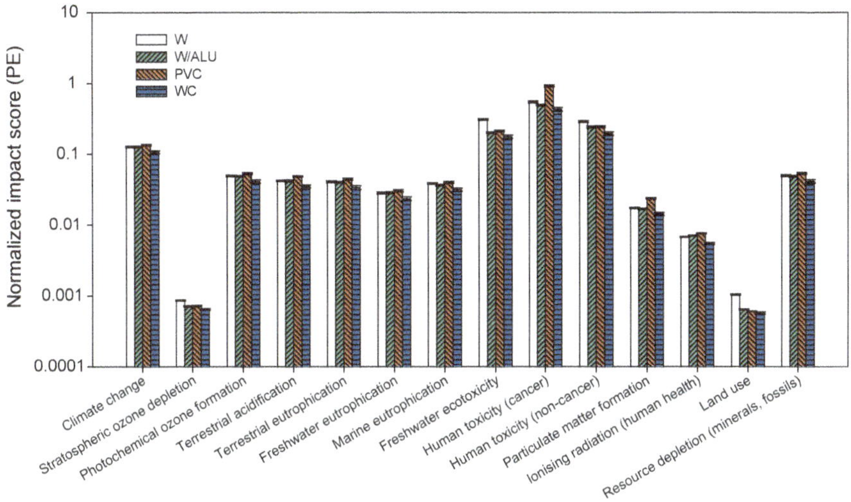

Fig. 39.9 Normalised impacts and accompanying 95% confidence intervals ($\log_{10}$-scale) in person equivalents (pe) for each window system (*W* wood, *W/ALU* wood/aluminium, *PVC* polyvinyl chloride, *W/C* wood/composite). Whiskers represent inventory uncertainty stemming from uncertainty and variability in model parameters presented in Table 39.8

39.3.5.1 Significant Issues

Process contribution analysis. To explain differences in window ranking and identify hot spots, a process contribution analysis was conducted, i.e. identifying the processes with the largest environmental burden.

Figure 39.10 shows that the main driver of environmental impacts is the production of residential heating to compensate for heat losses through the window. The contribution of this process to total impact is around 90% for climate change, freshwater eutrophication, or resource depletion, and is above 50% for most other impact categories (apart from ozone depletion and ionising radiation, where the contribution is smaller). This trend is consistent across all four window systems. Across all window systems, climate change impacts from the use stage due to combustion of fossil coal and natural gas, which constitute 25 and 31% of total Danish heating mix, respectively. The use of fossil fuels in the use stage is also the major driver of impacts related to depletion of resources. For other impact categories where the use stage is important (>50% of total impact score), however, the major driver of environmental impact is the use of other fuels like wood, straw and bio-waste. These processes are important for the impact categories terrestrial and freshwater eutrophication, and all the toxicity related impact categories.

Although the use stage is the main driver for the above-mentioned impact categories, for some impact categories the differences between window systems can sometimes be attributed to differences in material composition of the window. The manufacturing stage is important (>50% of total impacts) for impacts on

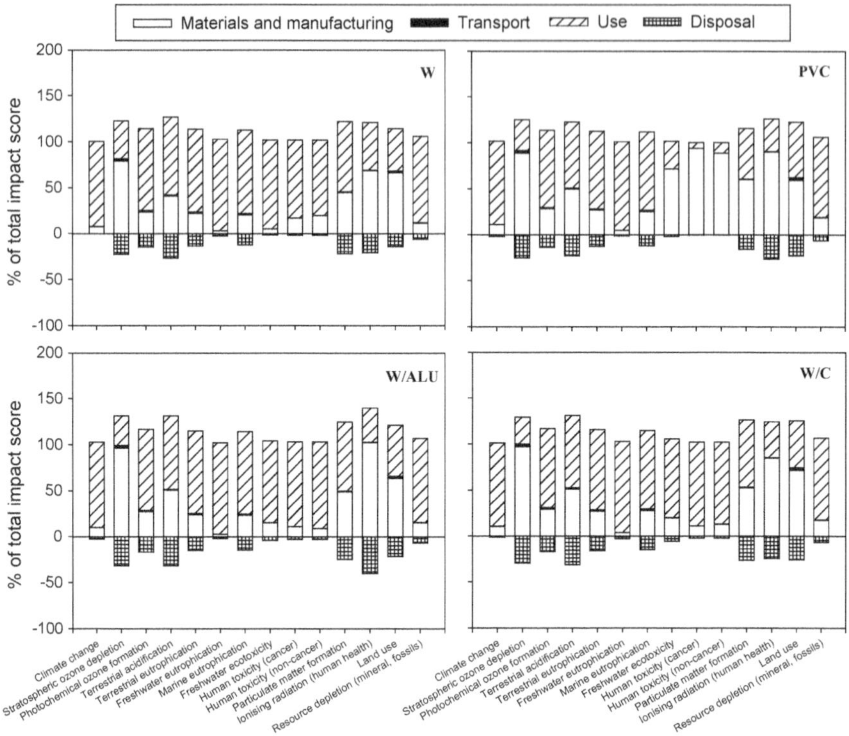

Fig. 39.10 Contribution of individual life cycle stages to total impact for each impact category for the four window systems (*W* wood, *W/ALU* wood/aluminium, *PVC* polyvinyl chloride, *W/C* wood/composite)

stratospheric ozone depletion, and ionising radiation (human health) across all windows, and for impacts on freshwater ecotoxicity and human toxicity. In addition, the manufacturing processes overall contribute to impacts on land use (around 40% of total impact) and to some extent also to the remaining impact categories (with contributions from 10 to 30%), reflecting that the materials used in the windows are considered part of the manufacturing stage. Substantial contribution to land use impacts in the W window is thus from the production of glue laminated timber. Impacts in these categories are also caused by production of alkyd paint (18 and 13% of total impact, respectively). In addition, the alkyd paint shows contribution of the same order of magnitude for four other impact categories, i.e. aquatic acidification, ionising radiation, ozone layer depletion and photochemical ozone formation. For the W/ALU window system, considerable impacts are caused by production of aluminium. This process contributes substantially to terrestrial acidification and stratospheric ozone depletion (20–23% of total impacts). Note that the introduction of aluminium has negative influence on the window performance in those impact categories that are determined by the use stage, because insulating properties are slightly worse than for the W window. Yet, the overall differences in impact scores are not statistically significant. Environmental impacts in the PVC window system in the manufacturing stage originate mainly from the PVC injection moulding process and production of steel. Injection moulding contributes substantially to impacts on human health (42 and 34% for carcinogens and non-carcinogens, respectively) while 94% of total impacts on mineral depletion is caused by the need for chromium; this, however, is not apparent in Fig. 39.10 because the resource depletion impact category is driven by the use of fossils. Given that insulation properties of the PVC window are not improved when PVC and steel are used in the window frame (they even decrease), the PVC window performs the worst among considered alternatives. An exception is an impact on land use, in which the PVC windows performs nearly as good as the best W/C window, which is mainly due to no use of wood in the PVC window. For the W/C window, the composite contributes to some extent (up to 12%) to some impact categories, but environmental benefits are obtained due to improved insulation properties. Across all windows, production of flat glass is a considerable contributor (>25% of total impact) to impacts on ionising radiation, stratospheric ozone depletion, and respiratory effects. In addition, silicone used as insulating material in the pane contributes substantially to ionising radiation and ozone layer depletion (15 and 32%, respectively).

The disposal stage is less important across all windows and impact categories, with contribution from 1 to 20% of the total impacts, depending on the impact category. Benefits are mainly due to recycling of materials, like aluminium in the W/ALU window system. Transportation is not seen as substantial for any impact category, irrespective of the window system.

Substance contribution analysis. To provide further insights into the causes of environmental impacts from the window product systems, the contribution analysis was also conducted at the level of elementary flows, identifying the individual substances that cause the largest environmental burden. The analysis was carried out for the W window system only, because for most impact categories the drivers of environmental impacts are expected to the same across windows. However, differences in contributing substances between the W window and the alternative design options are also discussed, when found important for the interpretation of results.

Climate change impacts are mainly driven by emissions of CO_2, which contributes to 99% of the total impacts. This contribution is mainly due to emissions from processes associated with generation of heat. Emissions of other substances from the generation of heat drive impact scores for several other impact categories. Potential impacts of photochemical ozone formation on human health are mainly due to emissions of nitrogen oxides (NO_x), which account for 95% of the total impact. Note, that the current implementation of characterisation factors into the modelling software employed omits potential contribution from unspecified emissions of non-methane volatile organic compounds (NMVOC), which are also reported in life cycle inventories (see Annex, Sect. 39.4.4, Table 39.34) and would be expected to contribute to photochemical ozone formation. Ammonia (NH_3), nitrogen oxides (NO_x) and sulphur dioxide (SO_2) are the substances that dominate the acidification and eutrophication impacts in terrestrial ecosystems, whereas eutrophication impacts in freshwater and marine ecosystems are mainly due to emissions of NO_x and phosphorus (P). By contrast, toxic impacts in freshwater ecosystems are dominated by emissions of metals (again, stemming mainly from processes associated with generation of heat), namely zinc (II) and copper (II).

For all window systems except the PVC system, the use stage is also the main contributor to the human health impact categories (carcinogens and non-carcinogens). Again, production of heat from incineration of fossil fuels and biomass, and the associated emissions of metals, are the major contributors to human health impacts; arsenic (V) and zinc (II) emitted to freshwater drive toxic impact scores for non-cancer effects, while chromium (VI) emitted to freshwater is the major driver of cancer effects. By contrast, for the PVC window, human health impacts (cancer and non-cancer effects) are mainly driven by substances associated with production of steel in the manufacturing stage. Potential impacts on depletion of resources also vary between windows when only mineral resources are considered (e.g. impacts of the PVC window are dominated by the need for chromium in production of the PVC window frame), but altogether (combining impact scores from depletion of fossils and minerals) this impact category is dominated by the depletion of fossils.

39.3.5.2 Sensitivity and Uncertainty Checks

The assumptions and choices that had to be made when modelling window systems can potentially influence conclusions from the study and they were systematically compiled in Table 39.14 of the Annex, Sect. 39.4.3. To determine the extent of this potential influence, we first identified individual parameters that are important for the results. Annex, Sect. 39.4.5, Tables 39.35, 39.36, 39.37, and 39.38 gives details of normalised sensitivity coefficients. Next, we compared the baseline and the two sensitivity scenarios with all uncertain parameters perturbed at once. Thereby, we found that many of the assumptions presented in Table 39.34 did not influence the results in terms of ranking or identification of hot spots to the extent that would change our conclusions.

The influence of heat loss. The parameters involved in the modelling of the heat loss compensation are important because their uncertainty can potentially change the results of the comparative part of the LCA (which window performs best?) and the results of the weak point analysis (what are the most environmentally harmful parts of the product life cycle?). Such parameters are the modelled heat loss, the assumed heat mix, the LCI processes used to model the heat mix technologies and the relevant characterisation factors and normalisation references involved in the impact assessment. This was confirmed in sensitivity and uncertainty analyses; impact scores are the most sensitive to the *U*-value of the window, and furthermore this parameter is the dominant driver of difference in impact scores between the compared window systems. Indeed the differences in impact scores between W and

W/ALU windows are in most cases are not statistically significant when uncertainties in *U*-values are considered. The assumption about using average indoor and outdoor temperatures when calculating the heat loss was not tested in the sensitivity analysis but is not expected to change our conclusions about which window performs best as heat loss is a linear function of the temperature difference. Similarly, it would not change our conclusion about hot spots; if higher temperature difference was considered (e.g. corresponding to winter temperatures), the contribution of heat to total impact scores would increase due to higher demand for heat.

The influence of materials and production. Out of all assumptions in the materials and production stages, the most important one is about modelling of chromium steel and galvanised steel using the same processes (for chromium steel). This assumption may influence impact scores in human health (cancer effects) and freshwater ecotoxicity, where impact scores might be overestimated (because production of chromium steel is associated with toxic emissions of chromium (VI). In contrast, impacts in human health (non-cancer effects) are expected to increase if process for galvanised steel had been used, due to expected increase in emissions of toxic zinc (II). The contribution of electricity requirements in window manufacturing and disassembly is for most impact categories too small to influence our comparison, and the same is the case for assumptions on transportation distances in these life cycle stages. The exclusion of painting activity (but not production of paint) is also not expected to be important for the result, because impacts are mainly expected to stem from transportation of paint from retailer to the housing (which is small relative to other impacts from the window product systems).

The influence of disposal. Assumptions about incineration and landfilling processes for some materials are not expected to influence our conclusions, given that the contribution of disposal to total impact is relatively small (10–15%, depending on the impact category). The inclusion of landfilling of copper and zinc used in window frames could potentially influence impact scores for the toxicity-related impact categories (where both copper and zinc are characterised as very toxic), but the amounts of these metals is very small compared to emissions from production of heat in the use stage. For the same reason, omitting of disposal of wood preservative and acrylic binder in the window frame is not expected to change impact scores.

Comparison between the baseline and the two sensitivity scenarios. Figure 39.11 shows the comparison between the baseline scenario and the two sensitivity scenarios. When EU27 average heat mix is used (along with and EU27 electricity mix and EU27 average disposal scenarios), impact scores generally increase compared to the base scenario, apart from the three eutrophication impact categories, and freshwater ecotoxicity. This is because the European heating mix mainly relies on natural gas (57%), with smaller contribution from coal and biomass compared to the Danish mix. On the other hand, a larger proportion of natural gas and oil (57 and 21%, respectively), results in considerably higher impacts in other impact categories. We also tested a scenario where windowpanes are changed into 3-layered ones, causing a decrease in *U*-value thereby improving insulation properties of the window. The results show that additional environmental impacts from the extra layer of glass are generally compensated for by the reduced heat loss in the use stage, and the

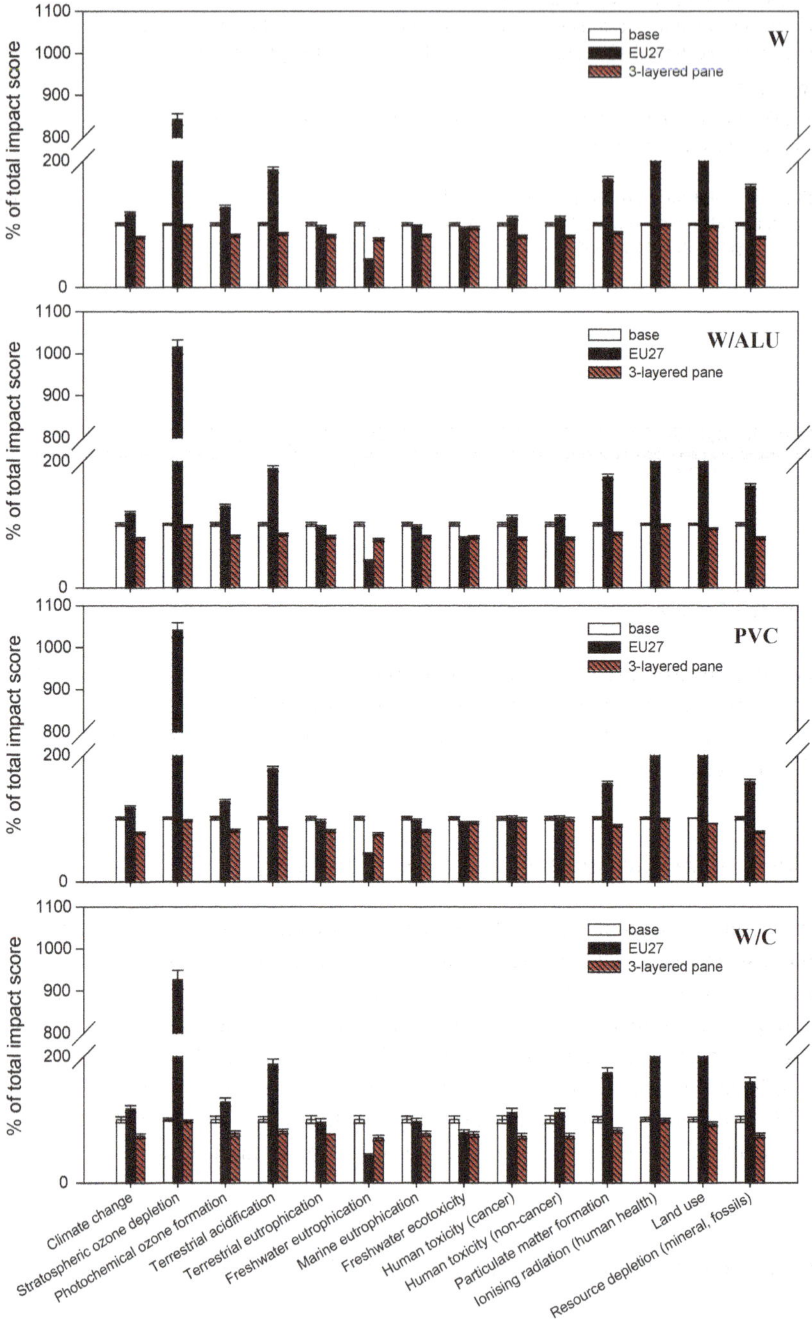
% of total impact score
base
EU27
3-layered pane
W
W/ALU
PVC
W/C
Climate change
Stratospheric ozone depletion
Photochemical ozone formation
Terrestrial acidification
Terrestrial eutrophication
Freshwater eutrophication
Marine eutrophication
Freshwater ecotoxicity
Human toxicity (cancer)
Human toxicity (non-cancer)
Particulate matter formation
Ionising radiation (human health)
Land use
Resource depletion (mineral, fossils)

◂**Fig. 39.11** Comparison between the baseline and two sensitivity scenarios: (i) where EU27 electricity and heating mix and EU27 disposal options are used for each window, and (ii) where 3-layered windowpane used instead of 2-layered one. *W* wood, *W/ALU* wood/aluminium, *PVC* polyvinyl chloride, *W/C* wood/composite. Whiskers represent inventory uncertainty stemming from uncertainty and variability in model parameters presented in Table 39.8

overall life cycle impact are smaller compared to the base scenario by up to 20%. High increase for stratospheric ozone depletion is most likely an artefact related to the use of relatively old processes for generation of heat from natural gas and oil in the EU27 system, since ozone-depleting substances have been largely banned for at least a decade. Despite these differences in impact scores, the ranking of window options generally does not change irrespective of the analysed scenarios (Fig. 39.12). As whiskers do not overlap, the results can be considered statistically significant, although we repeat that neither were uncertainties in background processes considered (which would increase the overall inventory uncertainty), nor could correlation between uncertainties in processes that are the same be addressed (which would have an opposite effect). It is, however, clear that if the heat mix changes substantially within the lifetime of the window as a consequence of the decarbonisation of our energy systems, the hot spots may move from the use stage to manufacturing and end-of-life stages and this would change the ranking of the alternatives and also the recommendations for design of the windows.

Uncertainties in characterisation factors and sensitivity to LCIA method chosen. All characterisation factors in ILCD (just as in any other LCIA method) are associated with uncertainties, meaning that the contribution to impacts of different modelled elementary flows and processes (such as heating) display varying uncertainties across impact categories. Although the uncertainties in characterisation factors were not considered in this study (they are rarely even known today), we expect that the uncertainty in characterisation factors will result in lack of statistical significance of difference in impact scores for freshwater ecotoxicity and human toxicity across all four windows. These are the impact categories where the uncertainties in individual characterisation factors are the highest (up to a few orders of magnitude) (Rosenbaum et al. 2008).

Sensitivity of the results to the chosen LCIA methods is also not considered in this LCA report (because the results are for internal use only). Such a sensitivity analysis could reveal that window ranking generally does not change for most impact categories because it is a few processes, associated with the production of heat, that are driving the main environmental impacts and there is large difference in demand for heat between the compared windows. This is expected to be the case for climate change and acidifying and eutrophying emissions where the driving elementary flows are very similar between different impact assessment methods. However, this may not be the case for freshwater and human toxicity, where impact scores can be sensitive to the inclusion of one or few substances with high characterisation factors, depending on the method, as for these impact categories up to 12 orders of magnitude between characterisation factors are observed (Rosenbaum et al. 2008).

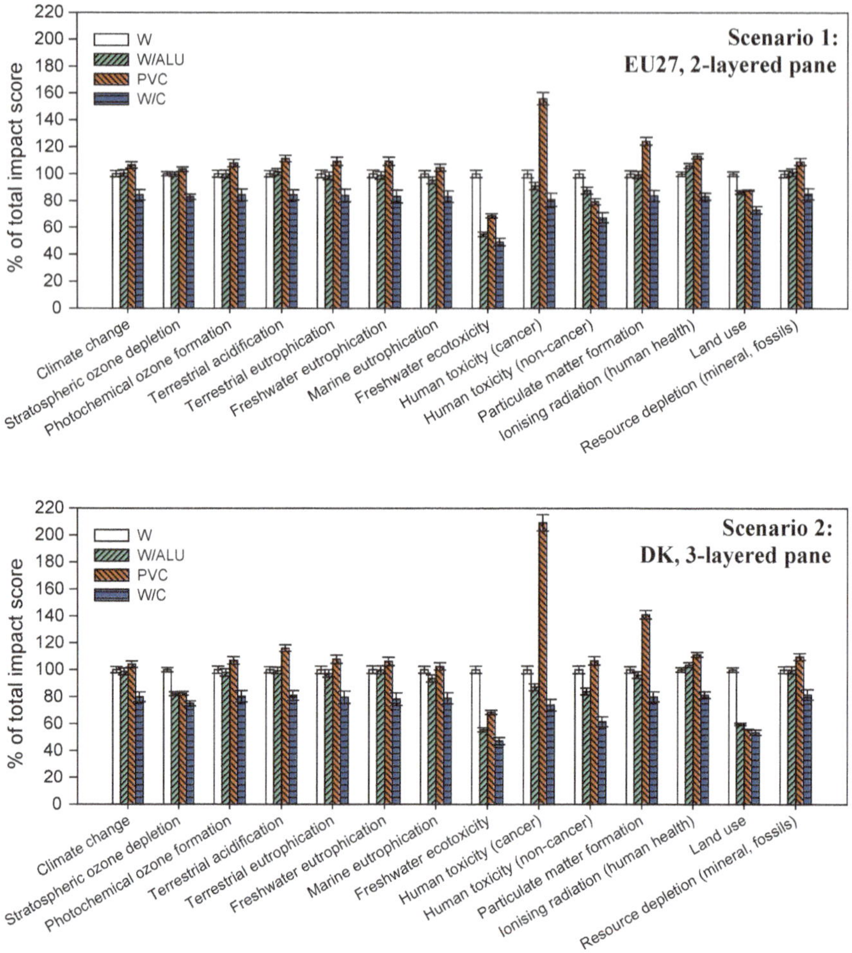

Fig. 39.12 Ranking of four window options where impact scores are scaled to those of W window (equal to 100% of total impact) for the two sensitivity scenarios presented in Fig. 39.11. *W* wood, *W/ALU* wood/aluminium, *PVC* polyvinyl chloride, *W/C* wood/composite. Whiskers represent inventory uncertainty stemming from uncertainty and variability in model parameters presented in Table 39.8

39.3.5.3 Completeness and Consistency Checks

Completeness check. The cut-off rules have been consistently applied across the whole life cycle for all four window alternatives in order to ensure the completeness of the study. However, two processes had to be left out when modelling life cycle inventories due either to difficulties in finding and approximating data, or they were not thought to be important initially. First, we did not include the coating of glass in the windowpane, where the current Nor-win technology uses nanomaterials because of limited information about input and output flows from nanomaterial production. This is expected to result in underestimation of human health and ecotoxicity impacts (some of the nanomaterials used by Nor-win are recognised to be toxic), and furthermore production of nanomaterials will to some extent contribute to total impact scores for other impact categories (Jolliet et al. 2014). We estimate that this contribution will not be larger than 1–2% of total impact scores for all impact categories, apart from the three toxicity-related impact categories where our rough estimate is 2.5–5% contribution. Second, we assumed no loss in material functionality in recycling of PVC (for metals and glass this assumption is expected to hold), nor did we assume material loss during recycling or production of the materials. Assuming that 10% increase in material is sufficient to cover this, total impact scores are expected to be higher by roughly 1–5%, depending on the impact category and contribution of manufacturing and disposal to total impacts. Finally, we did not include capital equipment for foreground processes. The contribution of capital equipment can be 10–30%, depending on the type of sector (Frischknecht et al. 2007). Given that contribution to overall impact from the materials and production stages is around 30% (although this number varies between windows and impact categories, see Fig. 39.10), the contribution of capital equipment is expected to be equal to ca. 10% to total impact score. Overall, we estimate that the calculated impact scores represent 75–85% of the actual total impacts.

Consistency check. The major source of inconsistency in data quality is the limited knowledge of performance parameters of the prototype W/C window (like the *U*-values), and we took this into account in the uncertainty and variability analysis. The major source of inconsistency in the applied life cycle impact assessment method is missing characterisation factors for some of the flows, due to incorrect implementation of life cycle impact assessment methods into the modelling software employed. This inconsistency is not expected to change impact scores to an extent that would change our conclusion about window ranking or major drivers of environmental impacts, since the majority of input and output flows are the same for all four windows (see Annex, Sect. 39.4.4, Table 39.34). Cut-off criteria were applied consistently across the four window product systems and the same processes were omitted. Other assumptions, methods and data (like the attributional principle with credits given to recycling, or the sources and quality of primary and secondary data) have also been applied consistently to all four window options.

39.3.6 Conclusions, Limitations and Recommendations

Conclusions:

I. The W/C window performs significantly better compared to its alternatives in all 14 impact categories. The W/C window is thus the preferable option from an environmental perspective.

II. The PVC window is the least preferred option, as it performs the worst in 11 out of 14 impact categories. This conclusion, however, might change if land use, freshwater ecotoxicity and human health (non-cancer) (where the W window performs significantly worse) are given a higher weight than the rest of the impact categories.
III. The overall environmental performance of the windows is mainly determined by the demand for heat to compensate for heat losses through the window during its use stage. This is true for nearly all impact categories. The U-value determines demand for heat, and can thus be considered a key environmental performance indicator of windows.
IV. In addition to processes for generation of heat, other environmental hotspots in the product systems are: production of timber and paint for the W window; the injection moulding process of PVC and production of steel in the PVC window.
V. The use of glass fiber based composite has some contribution (up to 12%) to total impacts, depending on the impact category, but cannot be considered a hotspot given that the composite substantially improves insulation properties causing an overall reduction in environmental impacts.
VI. Similarly, the use of 3-layered glass instead of 2-layered improves insulation properties resulting in an overall reduction in environmental impacts with the respective heating mix.
VII. The trade-off between impacts from the material used and the improved insulation properties that the material may give the window has to be considered when assessing environmental performance of windows.

Limitations:

The major limitations of the LCA are:

1. Our findings about major drivers of environmental impacts apply to windows where crystal glass is used in the panes with a relatively large (>0.6) visible light transmittance coefficient. They are not thought to be applicable for windows, which change their transparency in response to light intensity (e.g. photochromic windows) where the need for electricity to provide lighting indoor may become an important factor contributing to impacts in the use stage.
2. The disregard of changes in heat mix and heat demand in the future and potential development of more efficient heat supply technologies is another potential limitation. It is uncertain to what extent these will become effective within the time frame of the study (25–30 years). If such is the case, impacts from the manufacturing stage or disposal will become more important in the future (if there is no development of cleaner manufacturing and waste management technologies, which also is uncertain). They may change both the ranking of window alternatives and recommendations given to the commissioner. We expect, however, that in a 25–30 year time horizon the use stage will likely remain the most important contributor to total impacts from the window

product system, and efforts to design windows with low *U*-values should continue.

Recommendations:

Recommendations are given to the commissioner to support eco-design of the new window and greening of the whole value chain:

A. The design of windows to ensure better environmental performance should focus on optimising insulation properties of windows. This can be done by introducing a 3-layered pane, or improving the design of the frame. If the latter is considered, the choice of frame material is important and in each case where new frame material is used in the design of a frame we recommend evaluating (using tools like LCA) whether environmental benefits achieved by improved insulation properties are really sufficient to outweigh potential environmental burden from the use of novel materials. Indeed, if the heat mix changes substantially within the lifetime of the window this could potentially move the hotspots from the use stage to manufacturing and end-of-life stages in which case our recommendations for design of the windows might not hold.
B. Selection of new materials for frame design should consider functional properties of materials in a window design context, i.e. the focus should not be on selection of materials that perform environmentally best per unit mass of the materials, but on selection of materials that perform best considering insulation properties and the amount applied when used in the frame.
C. For the existing W-based windows, improvement potentials lie in selection of paints with lower environmental impact. For the paint applied for maintenance in the use stage, this may be outside the influence of the producer, because it is the window users who will select the type of paint. Our recommendation is to provide information to the users about recommended types of paint.
D. Finally, we recommend to phase-out the PVC window as the option with likely the highest environmental burden overall. If this is not possible, we recommend its redesign through the introduction of a 3-layered pane to improve its insulation properties. Further improvement potentials for the PVC window system lie mainly in selection of cleaner technology for production of PVC frame elements.

39.4 Annex (Public)

39.4.1 Life Cycle Impact Assessment Methods and Normalisation Factors

See Table 39.11.

Table 39.11 ILCD methods and normalisation factors for the impact categories considered in this study (EC-JRC 2011)

Impact category	Indicator	Unit	Model reference	Normalisation factor [unit/person/year]
Climate change	Radiative forcing as Global Warming Potential, 100 years horizon (GWP100)	kg CO_2 eq.	Baseline model of 100 years of the IPCC	9.10E+03
Ozone depletion	Ozone Depletion Potential (ODP)	kg CFC-11 eq.	Steady-state ODPs 1999 as in WMOassessment	2.16E−02
Human toxicity, cancer effects	Comparative Toxic Unit for humans	CTU_h	USEtox model (Rosenbaum et al. 2008)	3.68E−05
Human toxicity, non-cancer effects	Comparative Toxic Unit for humans	CTU_h	USEtox model (Rosenbaum et al. 2008)	5.32E−04
Particulate matter	Intake fraction for fine particles	kg $PM_{2.5}$ eq.	RiskPoll model (Rabl and Spadaro 2004) and Greco et al. (2007)	4.82E+00
Ionising radiation (human health)	Human exposure efficiency relative to U235	kg U235 eq.	Human health effect model as developed by Dreicer et al. (1995), Frischknecht et al. (2000)	1.13E+03
Photochemical ozone formation	Tropospheric ozone concentration increase	kg NMVOC eq.	LOTOS-EUROS (Van Zelm et al. 2008) as applied in ReCiPe	3.18E+01
Acidification	Accumulated Exceedance	mol H^+ eq.	Accumulated exceedance (Seppälä et al. 2006; Posch et al. 2008)	4.72E+01
Terrestrial eutrophication	Accumulated Exceedance	mol N eq.	Accumulated exceedance (Seppälä et al. 2006; Posch et al, 2008)	1.74E+02
Freshwater eutrophication	Residence time of nutrients in freshwater compartment (P)	kg P eq.	EUTREND model (Struijs et al. 2009) as implemented in ReCiPe	1.48E+00
Marine eutrophication	Residence time of nutrients in marine compartment (N)	kg N eq.	EUTREND model (Struijs et al. 2009) as implemented in ReCiPe	1.68E+01

(continued)

Table 39.11 (continued)

Impact category	Indicator	Unit	Model reference	Normalisation factor [unit/person/year]
Freshwater ecotoxicity	Comparative Toxic Unit for ecosystems	CTU_e	USEtox model (Rosenbaum et al. 2008)	8.71E+03
Land use	Soil Organic Matter	kg C deficit	Model based on Soil organic matter (SOM) (Milà i Canals et al. 2007)	6.30E+05
Water resource depletion	Water use related to local scarcity of water	kg water eq.	Model for water consumption as in Swiss Ecoscarcity (Frischknecht et al. 2006)	7.89E+01
Mineral fossil and renewable resource depletion	Scarcity	kg Sb eq.	CML 2002 (Guinée et al. 2002)	1.00E−01

Normalisation factors are for EU27 for the reference year 2010 as presented in Benini et al. (2014)

39.4.2 Bills of Materials and End-of-Life Options

See Tables 39.12 and 39.13.

Table 39.12 Amounts of materials (in kg) required to produce one window

Material	Window type			
	W	W/ALU	PVC	W/C
Window frame				
Heartwood	30	9.2	–	9.2
Polyvinyl chloride (PVC)	–	–	14	–
Composite	–	–	–	3.9
Aluminium	0.2	4.6	–	–
Galvanised steel	–	–	10	–
Chromium steel	0.5	1.2	5.1	1.2
Acrylic binder	0.168	0.056	–	0.056
Triethylene glycol	0.00427	0.00142	–	0.00142
Wood preservative	0.000525	0.000175	–	0.000175
Window pane				
Glass	56	56	56	56
Aluminium	0.4	0.4	0.4	0.4
Argon	0.06	0.06	0.06	0.06
Synthetic rubber (EDPM)	1	3.6	3.6	3.6
Silicone	1.4	1.4	1.4	1.4
Window packaging				
Polyethylene	0.2	0.2	0.2	0.2
Cardboard	1	1	1	1

Note, that the amounts are not scaled to the functional unit

Table 39.13 End-of-life options for window materials (percentage recycled/incinerated/landfilled) in Denmark and EU27 (given in brackets)

ID	Material	Window type			
		W	W/ALU	PVC	W/C
Window frame	Heartwood	DK: 90.5/9/0.5 EU27: 47/52/1		Not relevant	DK: 90.5/9/0.5 EU27: 47/52/1
	Polyvinyl chloride (PVC)	Not relevant	Not relevant	DK: 93/5/2 EU27: 76/17/7	Not relevant
	Composite	Not relevant	Not relevant	Not relevant	DK: 93/5/2 EU27: 76/17/7
	Aluminium	DK: 99.987/0.01/0.003 EU27: 99.885/0/0.115		Not relevant	Not relevant
	Steel (galvanised and chromium)	DK: 99.987/0.01/0.003 EU27: 99.885/0/0.115			
Window pane	Glass	99.853/0.018/0.129			
	Aluminium	DK: 99.987/0.01/0.003 EU27: 99.885/0/0.115			
	Synthetic rubber (EDPM)	99.39/0.604/0.06			
	Silicone	99.39/0.604/0.06			
Window packaging	Polyethylene	DK: 93/5/2 EU27: 76/17/7			
	Cardboard	DK: 99.32/0.59/0.09 EU27: 98.73/0.93/0.34			

Data from Eurostat (2016) for the reference year 2014. The options are based on the data retrieved from Eurostat for the categories: metal wastes (mixed ferrous and non-ferrous), glass wastes, paper and cardboard wastes, rubber wastes, plastic wastes, and wood wastes

39.4.3 List of Assumptions

See Table 39.14.

Table 39.14 List of assumptions

Assumptions	Window type			
	W	WA	PVC	W/C
Heat loss is based on the annual average temperatures indoor and outdoor in Denmark (7 and 17 °C, respectively), without considering the dynamics of temperature change during the year	x	x	x	x
Windows are only used in places covered by district heating	x	x	x	x
Processes for generation of heat from incinerating straw and incinerating of non-renewable waste in the Danish heat mix were modelled as incineration of bio-waste	x	x	x	x
Heat generation from biomass in the EU27 heat mix was modelled as incineration of bio-waste combined with combustion of wood pellets (50:50)	x	x	x	x
Energy consumption for window assembly covers all processes in the factory	x	x	x	x
Chromium steel and galvanised steel are modelled using the same process			x	
Painting activity of window frame is not modelled (but production of the paint is)	x			
Energy used in disassembly in end of life assumed equal to 1 MJ per 1 kg of window	x	x	x	x
PVC is 30% recycled, and 70% landfilled			x	
Disposal of wood preservative and acrylic binder is not modelled	x	x		
Incineration of Aluminium is modelled as municipal solid waste (MSW)	x	x	x	x
Landfilling of EDPM rubber is modelled as polypropylene (PP)	x	x	x	x
Incineration of silicone is modelled as incineration of plastic mixture	x	x	x	x
Argon from window pane is released to the atmosphere during window disassembly	x	x	x	x
Landfilling of copper and zinc in window frame is not modelled	x	x	x	x
Transportation distances are the same for all windows in the distribution stage	x	x	x	x
Transportation distances are the same for all windows in the end-of-life stage	x	x	x	x
Packaging is the same for all windows	x	x	x	x

Table 39.15 Inventory of the unit process “Use of window, U, MIOW”

Activity	W	W/ALU	PVC	W/C	Unit	Source/note
Output (main product or function)						
Window use, U, MIOW	0.5	0.5	0.67	0.5	p	Process output
Other outputs (waste to treatment)						
Disassembly of window, U, MIOW	0.5	0.5	0.67	0.5	p	See Table 39.19
DE: polyethylene, incineration (PE, Adapted)	1	1	1	1	kg	PlasticsEurope
DE: paper/cardboard, incineration (Adapted)	0.2	0.2	0.2	0.2	kg	PlasticsEurope
Inputs (materials, energy, resources)						
Assembly and packaging of window, U, MIOW	0.5	0.5	0.67	0.5	p	See Table 39.16
Production of window pane, 2-layered, U, MIOW	1	1	1	1	p	See Table 39.18
DK: heat mix	14,820	15,040	15,620	12,400	MJ	See Table 39.32
RER: alkyd paint, white, 60% in solvent, at plant	4.8	0	0	0	kg	ecoinvent, v. 2.2
GLO: truck PE <u-so> technology mix, diesel driven, Euro4, cargo |>34–40 t total cap. /27 t payload capacity	27.3	22.5	26.8	22.4	tkm	ecoinvent, v. 2.2
GLO: truck PE <u-so> technology mix, diesel driven, Euro3, cargo |>12–14 t total cap. /9.3 t payload capacity	9.1	7.5	8.9	7.5	tkm	ecoinvent, v. 2.2
RER: diesel, low-sulphur, at regional storage	0.703	0.581	0.69	0.803	kg	ecoinvent, v. 2.2

All outputs and inputs are scaled to the functional unit of the window systems, with windows used for 20 years

Table 39.16 Inventory of the unit process "Assembly and packaging of window, U, MIOW"

Activity	W	W/ALU	PVC	W/C	Unit	Source/note
Output (main product or function)						
Window assembled and packed, U, MIOW	1	1	1	1	p	Process output
Inputs (materials, energy, resources)						
Production of window frame, U, MIOW	1	1	1	1	p	See Table 39.17
Production of window pane, 2-layered, U, MIOW	1	1	1	1	p	See Table 39.18
DK: electricity, production mix DK	118	120	126	130	MJ	ecoinvent, v. 2.2
DK: heat mix	50	50	50	50	MJ	See Table 39.32
RER: corrugated board base paper, kraftliner, at plant	1	1	1	1	kg	ecoinvent, v. 2.2
RER: polyethylene film (PE-LD)	0.2	0.2	0.2	0.2	kg	PlasticsEurope

Note that inputs and outputs are not scaled to the functional unit of the window systems

39.4.4 Unit Processes and LCI Results

See Tables 39.15, 39.16, 39.17, 39.18, 39.19, 39.20, 39.21, 39.22, 39.23, 39.24, 39.25, 39.26, 39.27, 39.28, 39.29, 39.30, 39.31, 39.32, 39.33 and 39.34.

39.4.5 Normalised Sensitivity Coefficients

Normalised sensitivity coefficients were computed for the perturbance of the following parameters: amount of wood, aluminium, steel (W/C window only), and PVC (PVC window only) in the window frame, the amount of glass in the pane, amount of paint, electricity needed for assembly, *U*-value, and transportation distance from Nor-win to retailers. Thereby, we found that impact scores are most sensitive to *U*-value, and three other parameters (amount of wood and steel in the frame, and amount of glass in the pane). The normalised sensitivity coefficients for these four parameters are presented in Tables 39.35,39.36, 39.37 and 39.38.

39.5 Annex (Confidential)

No confidential data were used in the study.

Table 39.17 Inventory of the unit process "Production of window frame, U, MIOW"

Activity	W	W/ALU	PVC	W/C	Unit	Source/note
Output (main product or function)						
Window frame, U, MIOW	1	1	1	1	p	Process output
Inputs (materials, energy, resources)						
RER: glued laminated timber, outdoor use, at plant [benefication]	0.06	0.0184	0	0.0184	m^3	ecoinvent, v. 2.2
GLO: Truck PE <u-so> technology mix, diesel driven, Euro4, cargo |>34–40 t total cap. /27 t payload capacity	30	9.2	11.2	10.37	tkm	ecoinvent, v. 2.2
RER: diesel, low-sulphur, at regional storage	0.419	0.129	0.157	0.129	kg	ecoinvent, v. 2.2
RER: Aluminium extrusion profile <agg>	0.2	4.6	0	0	kg	ecoinvent, v. 2.2
RER: chromium steel product manufacturing, average metal working	0.5	1.2	15.1	1.2	kg	ecoinvent, v. 2.2
RER: triethylene glycol, at plant [organics]	0.00427	0.00131	0	0.00131	kg	ecoinvent, v. 2.2
RER: acrylic binder, 34% in H_2O, at plant [manufacturing] <agg>	0.168	0.0515	0	0.0515	kg	ecoinvent, v. 2.2
RER: wood preservative, organic salt, Cr-free, at plant [manufacturing] <agg>	0.000525	0.000161	0	0.000161	kg	ecoinvent, v. 2.2
DK: electricity, production mix DK	15	21	24	30	kg	ecoinvent, v. 2.2
RER: zinc coating, coils	0	0	0.151	0	m^2	ecoinvent, v. 2.2
RER: polyvinylchloride injection moulding part (PVC) PlasticsEurope	0	0	14	0	kg	PlasticsEurope
RER: glass fibre-reinforced plastic, polyamide, injection moulding, at plant	0	0	0	3.9	kg	ecoinvent, v. 2.2

Note that inputs and outputs are not scaled to the functional unit of the window systems

Table 39.18 Inventory of the unit process "Production of window pane, 2-layered, U, MIOW"

Activity	W	W/ALU	PVC	W/C	Unit	Source/note
Output (main product or function)						
Window pane, 2-layered, U, MIOW	1	1	1	1	p	Process output
Inputs (materials, energy, resources)						
REE: flat glass, coated, at plant	56	56	56	56	kg	ecoinvent, v. 2.2
GLO: truck PE <u-so> technology mix, diesel driven, Euro4, cargo |>34–40 t total cap. /27 t payload capacity	14	14	14	14	tkm	ecoinvent, v. 2.2
RER: diesel, low-sulphur, at regional storage	0.196	0.196	0.196	0.196	kg	ecoinvent, v. 2.2
RER: aluminium extrusion profile <agg>	0.4	0.4	0.4	0.4	kg	ecoinvent, v. 2.2
DE: polypropylene-EPDM granulate mix PE	1	1	1	1	kg	ecoinvent, v. 2.2
RER: silicone product, at plant	1.4	1.4	1.4	1.4	kg	ecoinvent, v. 2.2
DE: argon (gaseous)	0.06	0.06	0.06	0.06	kg	ecoinvent, v. 2.2
DK: electricity, production mix DK	5	5	5	5	MJ	ecoinvent, v. 2.2

Note that inputs and outputs are not scaled to the functional unit of the window systems

Table 39.19 Inventory of the unit process "Disassembly of window, U, MIOW"

Activity	W	W/ALU	PVC	W/C	Unit	Source/note
Output (main product or function)						
Window disassembled, U, MIOW	1	1	1	1	p	Process output
Other outputs (waste to treatment)						
Disposal of aluminium, U, MIOW	0.6	5	0.4	0.4	kg	See Table 39.20
Disposal of wood, U, MIOW	30.2	9.25	0	9.25	kg	See Table 39.26
Disposal of EPDM, U, MIOW						
Disposal of silicone, U, MIOW	1.4	1.4	1.4	1.4	kg	See Table 39.22
Disposal of steel, U, MIOW	0.5	1.2	15.1	1.2	kg	See Table 39.28
Disposal of polyvinyl chloride, U, MIOW	0	0	14	0	kg	See Table 39.30
Disposal of composite, U, MIOW	0	0	0	0	kg	See Table 39.31
Disposal of glass, U, MIOW	56	56	56	56	kg	See Table 39.23
Inputs (materials, energy, resources)						
DK: electricity, production mix DK	50	50	50	50	MJ	ecoinvent, v. 2.2
GLO: truck PE <u-so> technology mix, diesel driven, Euro4, cargo |>34–40 t total cap. /27 t payload capacity	4.485	3.695	4.4	3.67	tkm	ecoinvent, v. 2.2

Note that inputs and outputs are not scaled to the functional unit of the window systems

Table 39.20 Inventory of the unit process "Disposal of aluminium, U, MIOW"

Activity	W	W/ALU	PVC	W/C	Unit	Source/note
Output (main product or function)						
Aluminium recycling, U, MIOW	0.88	0.88	0.88	0.88	kg	See Table 39.21
Inputs (materials, energy, resources)						
Disposal of aluminium, U, MIOW	1	1	1	1	kg	Process input
CH: disposal, aluminium, 0% water, to municipal incineration	0	0	0	0	kg	ecoinvent, v. 2.2
CH: disposal, wood untreated, 20% water, to sanitary landfill	0.12	0.12	0.12	0.12	kg	ecoinvent, v. 2.2

Note that inputs and outputs are not scaled to the functional unit of the window systems

Table 39.21 Inventory of the unit process "Aluminium recycling, U, MIOW"

Activity	W	W/ALU	PVC	W/C	Unit	Source/note
Output (avoided product or function)						
DE: Aluminium ingot mix (Inverted)	0.97	0.97	0.97	0.97	kg	ecoinvent, v. 2.2; inverted process
Inputs (materials, energy, resources)						
Aluminium recycling, U, MIOW	1	1	1	1	kg	Process input

Note that inputs and outputs are not scaled to the functional unit of the window systems

Table 39.22 Inventory of the unit process "Disposal of EPDM, U, MIOW"

Activity	W	W/ALU	PVC	W/C	Unit	Source/note
Output (main product or function)						
RER: EPDM seal PE, p-agg	1	1	1	1	kg	ecoinvent, v. 2.2
Inputs (materials, energy, resources)						
Disposal of EPDM, U, MIOW	1	1	1	1	kg	Process input
CH: disposal, polypropylene, 15.9% water, to sanitary landfill	0	0	0	0	kg	ecoinvent, v. 2.2

Note that inputs and outputs are not scaled to the functional unit of the window systems

Table 39.23 Inventory of the unit process "Disposal of glass, U, MIOW"

Activity	W	W/ALU	PVC	W/C	Unit	Source/note
Inputs (materials, energy, resources)						
Disposal of glass, U, MIOW	1	1	1	1	kg	Process input
CH: disposal, building, glass pane (in burnable frame), to sorting plant, U, MIOW	1	1	1	1	kg	See Table 39.24

Note that inputs and outputs are not scaled to the functional unit of the window systems

Table 39.24 Inventory of the unit process "CH: disposal, building, glass pane (in burnable frame), to sorting plant, U, MIOW"

Activity	W	W/ALU	PVC	W/C	Unit	Source/note
Output (main product or function)						
CH: disposal, building, glass pane (in burnable frame), to sorting plant, U, MIOW	1	1	1	1	kg	Process output
Other outputs (avoided product or function)						
RER: flat glass, uncoated, at plant (inverted)	0.9	0.9	0.9	0.9	kg	ecoinvent, v. 2.2; inverted process
Inputs (materials, energy, resources)						
CH: disposal, building, glass pane (in burnable frame), to sorting plant	0.1	0.1	0.1	0.1	kg	ecoinvent, v. 2.2
CH: disposal, glass, 0% water, to inert material landfill	0	0	0	0	kg	ecoinvent, v. 2.2
RER: glass, cullets, sorted, at sorting plant	0.0071	0.0071	0.0071	0.0071	kg	ecoinvent, v. 2.2
RER: excavation, hydraulic digger	7.9E −06	7.9E −06	7.9E −06	7.9E −06	m^3	ecoinvent, v. 2.2
CH: electricity, low voltage, at grid	0.00014	0.00014	0.00014	0.00014	MJ	ecoinvent, v. 2.2
CH: sorting plant for construction waste	1.8E −12	1.8E −12	1.8E −12	1.8E −12	p	ecoinvent, v. 2.2

Note that inputs and outputs are not scaled to the functional unit of the window systems

Table 39.25 Inventory of the unit process "Disposal of silicone, U, MIOW"

Activity	W	W/ALU	PVC	W/C	Unit	Source/note
Inputs (materials, energy, resources)						
Disposal of silicone, U, MIOW	1	1	1	1	kg	Process input
CH: disposal, plastics, mixture, 15.3% water, to municipal incineration	1	1	1	1	kg	ecoinvent, v. 2.2

Note that inputs and outputs are not scaled to the functional unit of the window systems

Table 39.26 Inventory of the unit process "Disposal of wood, U, MIOW"

Activity	W	W/ALU	PVC	W/C	Unit	Source/note
Output (main product or function)						
Wood incineration, U, MIOW	1	1	1	1	kg	See Table 39.27
Inputs (materials, energy, resources)						
Disposal of wood, U, MIOW	1	1	1	1	kg	Process input

Note that inputs and outputs are not scaled to the functional unit of the window systems

Table 39.27 Inventory of the unit process "Wood incineration, U, MIOW"

Activity	W	W/ALU	PVC	W/C	Unit	Source/note
Output (main product or function)						
DE: wood (natural) in municipal waste incineration PE, p-agg	0.92	0.92	0.92	0.92	kg	PlasticsEurope
Inputs (materials, energy, resources)						
Wood incineration, U, MIOW	1	1	1	1	kg	Process input
CH: disposal, aluminium, 0% water, to municipal incineration	0.00022	0.00022	0.00022	0.00022	kg	ecoinvent, v. 2.2
CH: disposal, copper, 0% water, to municipal incineration	8.8E−05	8.8E−05	8.8E−05	8.8E−05	kg	ecoinvent, v. 2.2
CH: disposal, zinc in car shredder residue, 0% water, to municipal incineration	0.0041	0.0041	0.0041	0.0041	kg	ecoinvent, v. 2.2
CH: disposal, paint, 0% water, to municipal incineration	0.074	0.074	0.074	0.074	kg	ecoinvent, v. 2.2

Note that inputs and outputs are not scaled to the functional unit of the window systems

Table 39.28 Inventory of the unit process "Disposal of steel, U, MIOW"

Activity	W	W/ALU	PVC	W/C	Unit	Source/note
Output (main product or function)						
Steel recycling, U, MIOW	0.88	0.88	0.88	0.88	kg	See Table 39.29
Inputs (materials, energy, resources)						
Disposal of steel, U, MIOW	1	1	1	1	kg	Process input
CH: disposal, steel, 0% water, to municipal incineration	0	0	0	0	kg	ecoinvent, v. 2.2
CH: disposal, steel, 0% water, to inert material landfill	0.12	0.12	0.12	0.12	kg	ecoinvent, v. 2.2

Note that inputs and outputs are not scaled to the functional unit of the window systems

Table 39.29 Inventory of the unit process "Steel recycling, U, MIOW"

Activity	W	W/ALU	PVC	W/C	Unit	Source/note
Output (avoided product or function)						
DE: steel billet PE (inverted)	1	1	1	1	kg	PlasticsEurope, inverted process
Inputs (materials, energy, resources)						
Steel recycling, U, MIOW	1	1	1	1	kg	Process input

Note that inputs and outputs are not scaled to the functional unit of the window systems

Table 39.30 Inventory of the unit process "Disposal of polyvinyl chloride, U, MIOW"

Activity	W	W/ALU	PVC	W/C	Unit	Source/note
Output (main product or function)						
DE: polyvinyl chloride (PVC) PE, p-agg	0	0	0.7	0	kg	PE
Output (avoided product or function)						
DE: polyvinylchloride granulate mix (S-PVC) PE (inverted)	0	0	0.3	0	kg	PE, inverted process
Inputs (materials, energy, resources)						
Disposal of PVC, U, MIOW	0	0	1	0	kg	Process input
CH: disposal, polyvinyl chloride, 0.2% water, to sanitary landfill	0	0	0	0	kg	ecoinvent, v. 2.2

Note that inputs and outputs are not scaled to the functional unit of the window systems

Table 39.31 Inventory of the unit process "Disposal of composite, U, MIOW"

Activity	W	W/ALU	PVC	W/C	Unit	Source/note
Output (main product or function)						
RER: polyamide (PA) 6.6 GF ELCD/PE-Gabi p-agg	0	0	0	1	kg	PE
Inputs (materials, energy, resources)						
Disposal of composite, U, MIOW	0	0	0	1	kg	Process input

Note that inputs and outputs are not scaled to the functional unit of the window systems

Table 39.32 Inventory of the unit process "Heat, DK, SERF"

Activity	Value	Unit	Source/note
Output (main product or function)			
Heat, DK, SERF	1	MJ	Process output
Inputs (materials, energy, resources)			
CH: heat, wood pellets, at furnace 50 kW	0.0980	MJ	ecoinvent, v. 2.2
CH: heat, softwood chips from industry, at furnace 50 kW	0.1257	MJ	ecoinvent, v. 2.2
RER: heat, heavy fuel oil, at industrial furnace 1 MW	0.0235	MJ	ecoinvent, v. 2.2
RER: heat, natural gas, at industrial furnace >100 kW	0.2451	MJ	ecoinvent, v. 2.2
RER: heat, hard coal briquette, at stove 5–15 kW	0.2344	MJ	ecoinvent, v. 2.2
CH: heat, bio-waste, at waste incineration plant, allocation price	0.25	MJ	ecoinvent, v. 2.2
CH: heat, at cogen, biogas agricultural mix, allocation exergy	0.0195	MJ	ecoinvent, v. 2.2
CH: heat, at heat pump 30 kW, allocation exergy	0.0005	MJ	ecoinvent, v. 2.2
CH: heat, at solar + gas heating, tube collector, one-family house, combined system	0.0033	MJ	ecoinvent, v. 2.2

Note that inputs and outputs are not scaled to the functional unit of the window systems. The Danish heat mix is based on data from Energynet (2012). Processes for generation of heat from incinerating straw (0.077 MJ/MJ heat output) and incinerating of non-renewable waste (0.173 MJ/MJ heat output) are modelled as incineration of bio-waste

Table 39.33 Inventory of the unit process "Heat, EU27, MIOW"

Activity	Value	Unit	Source/note
Output (main product or function)			
Heat, EU27, MIOW	1	MJ	Process output
Inputs (materials, energy, resources)			
CH: heat, wood pellets, at furnace 50 kW	0.065	MJ	ecoinvent, v. 2.2
RER: heat, heavy fuel oil, at industrial furnace 1 MW	0.21	MJ	ecoinvent, v. 2.2
RER: heat, natural gas, at industrial furnace >100 kW	0.57	MJ	ecoinvent, v. 2.2
RER: heat, hard coal briquette, at stove 5–15 kW	0.09	MJ	ecoinvent, v. 2.2
CH: heat, bio-waste, at waste incineration plant, allocation price	0.065	MJ	ecoinvent, v. 2.2

Note that inputs and outputs are not scaled to the functional unit of the window systems. The EU27 heat mix is based on data from Conolly et al. (2012)

Table 39.34 LCI results (elementary flows for each window product system)

Substance	W	W/ALU	PVC	W/C
Emission to air				
1,1,1-trichloroethane	8.33E−11	1.22E−10	1.25E−10	6.49E−11
1-butanol	1.10E−12	5.10E−13	5.17E−13	4.57E−13
Acenaphthene	9.08E−11	8.27E−11	1.21E−10	8.30E−11
Acetaldehyde (ethanal)	3.62E−04	3.61E−04	3.70E−04	3.03E−04
Acetic acid	7.42E−04	6.45E−04	6.28E−04	5.81E−04
Acetone (dimethylcetone)	9.99E−05	1.02E−04	1.02E−04	8.42E−05
Acetonitrile	7.05E−08	6.35E−08	6.58E−08	6.21E−08
Acrolein	4.24E−08	3.26E−08	5.05E−08	2.75E−08
Acrylic acid	1.70E−08	7.91E−09	7.98E−09	7.09E−09
Aldehyde (unspecified)	1.61E−06	1.51E−06	3.28E−07	7.39E−06
Alkane (unspecified)	2.51E−02	4.60E−03	1.02E−02	3.97E−03
Alkene (unspecified)	8.42E−03	8.44E−03	8.73E−03	6.94E−03
Aluminium	1.56E−02	1.60E−02	1.80E−02	1.31E−02
Ammonia	6.25E−02	5.94E−02	7.32E−02	4.94E−02
Ammonium	7.86E−11	1.22E−10	2.85E−11	9.58E−11
Ammonium carbonate	9.60E−08	3.32E−08	4.78E−08	3.41E−08
Ammonium nitrate	4.55E−11	4.89E−11	3.64E−11	4.02E−11
Anthracene	1.98E−09	2.18E−09	2.11E−09	1.79E−09
Antimony	5.81E−06	7.24E−06	7.61E−06	6.32E−06
Aromatic hydrocarbons (unspecified)	9.28E−05	7.22E−05	3.22E−04	1.12E−04
Arsenic (+V)	3.53E−05	3.66E−05	4.83E−05	2.84E−05
Arsenic trioxide	2.27E−12	3.45E−12	2.41E−12	2.80E−12
Barium	7.00E−05	8.79E−05	7.78E−05	6.89E−05

(continued)

Table 39.34 (continued)

Substance	W	W/ALU	PVC	W/C
Benzal chloride	7.62E−15	3.04E−14	1.99E−14	5.85E−15
Benzaldehyde	1.29E−08	6.91E−09	1.59E−08	6.26E−09
Benzene	3.97E−03	3.91E−03	4.15E−03	3.32E−03
Benzo(a)anthracene	9.99E−10	1.10E−09	1.06E−09	9.00E−10
Benzo(a)pyrene	1.98E−06	2.21E−06	2.15E−06	1.59E−06
Benzo(ghi)perylene	8.91E−10	9.80E−10	9.46E−10	8.03E−10
Benzofluoranthene	1.78E−09	1.96E−09	1.89E−09	1.61E−09
Beryllium	3.57E−07	3.89E−07	4.30E−07	2.97E−07
Boron	1.26E−03	1.23E−03	1.40E−03	1.04E−03
Boron compounds (unspecified)	6.89E−04	7.32E−04	7.25E−04	5.80E−04
Boron trifluoride	1.42E−15	6.22E−16	6.24E−16	5.62E−16
Bromine	5.25E−04	5.37E−04	5.58E−04	4.38E−04
Butadiene	2.88E−10	2.12E−10	1.38E−10	1.78E−10
Butane	2.20E−03	2.28E−03	2.21E−03	1.95E−03
Butane (*n*-butane)	1.18E−03	1.22E−03	1.24E−03	9.94E−04
Butanone (methyl ethyl ketone)	3.06E−05	1.42E−05	1.44E−05	1.28E−05
Butene	1.94E−05	1.37E−05	1.75E−05	2.17E−05
Butylene glycol (butane diol)	3.53E−10	1.65E−10	1.67E−10	1.47E−10
butyrolactone	1.02E−10	4.76E−11	4.82E−11	4.27E−11
Cadmium (+II)	8.00E−06	7.30E−06	8.45E−06	6.08E−06
Carbon dioxide	8.39E+02	8.36E+02	8.90E+02	7.10E+02
Carbon dioxide (biotic)	2.47E+02	2.46E+02	2.56E+02	2.03E+02
Carbon dioxide (biotic)	4.94E−01	4.76E−01	4.69E−01	4.15E−01
Carbon dioxide, land transformation	2.68E−03	2.18E−03	3.00E−03	2.16E−03

(continued)

Table 39.34 (continued)

Substance	W	W/ALU	PVC	W/C
Carbon disulphide	3.22E−04	3.66E−04	3.59E−04	2.27E−04
Carbon monoxide	3.35E−01	3.22E−01	2.82E−01	2.57E−01
Carbon monoxide (biotic)	2.18E−01	2.18E−01	2.25E−01	1.79E−01
Carbon tetrachloride (tetrachloromethane)	5.56E−07	1.21E−07	1.66E−07	1.05E−07
Chloride (unspecified)	2.75E−05	5.97E−05	2.88E−05	2.75E−05
Chlorine	5.43E−04	5.27E−04	1.74E−03	4.37E−04
Chloromethane (methyl chloride)	5.16E−05	3.29E−05	3.29E−05	3.29E−05
Chlorosilane, trimethyl-	3.06E−10	1.42E−10	1.43E−10	1.27E−10
Chromium (+III)	1.20E−08	1.28E−08	1.26E−08	1.03E−08
Chromium (+VI)	2.93E−06	4.09E−06	4.94E−05	3.98E−06
Chromium (unspecified)	1.33E−04	1.77E−04	1.99E−03	1.72E−04
Chrysene	2.45E−09	2.70E−09	2.60E−09	2.21E−09
Cobalt	1.05E−05	1.05E−05	3.20E−05	9.73E−06
Copper (+II)	1.17E−04	1.08E−04	2.09E−04	9.33E−05
Cumene (isopropylbenzene)	2.52E−05	1.38E−05	1.64E−05	3.48E−05
Cyanide (unspecified)	3.03E−03	3.09E−03	3.25E−03	2.54E−03
Cycloalkanes (unspec.)	5.15E−06	2.46E−07	3.28E−07	3.89E−06
Cyclohexane (hexahydro benzene)	1.68E−07	1.95E−07	1.27E−07	1.39E−07
Dibenz(a)anthracene	5.55E−10	6.11E−10	5.90E−10	5.00E−10
Dichlorobenzene (o-DCB; 1,2-dichlorobenzene)	4.75E−08	2.21E−08	2.24E−08	1.98E−08
Dichloroethane (1,2-dichloroethane)	4.74E−06	3.03E−06	8.26E−06	3.55E−05
Dichloroethane (ethylene dichloride)	1.01E−09	1.01E−09	4.53E−04	1.01E−09
Dichloromethane (methylene chloride)	2.79E−09	2.73E−09	2.93E−06	9.60E−09
Diethylamine	1.97E−15	3.04E−15	7.13E−16	2.40E−15

(continued)

Table 39.34 (continued)

Substance	W	W/ALU	PVC	W/C
Dioxins (unspec.)	1.95E−14	2.89E−14	2.74E−09	2.46E−14
Dust (>PM_{10})	2.68E−01	2.57E−01	2.89E−01	2.19E−01
Dust (PM_{10})	2.76E−03	3.21E−03	1.98E−02	2.40E−03
Dust ($PM_{2,5}$–PM_{10})	2.45E−02	1.66E−02	3.42E−02	1.83E−02
Dust ($PM_{2.5}$)	6.70E−02	6.42E−02	8.51E−02	5.52E−02
Dust (unspecified)	3.19E−02	4.03E−02	3.41E−02	2.71E−02
Emissions to air	4.94E+03	5.01E+03	5.08E+03	4.16E+03
Ethane	6.12E−03	6.76E−03	6.37E−03	5.71E−03
Ethanol	1.79E−04	1.92E−04	1.90E−04	1.57E−04
Ethene (ethylene)	2.36E−04	1.75E−04	4.25E−04	1.71E−04
Ethine (acetylene)	1.36E−04	1.25E−04	1.37E−04	1.24E−04
Ethyl benzene	7.63E−04	8.04E−04	7.95E−04	6.36E−04
Ethyl cellulose	6.19E−08	2.87E−08	2.90E−08	2.57E−08
Ethylene acetate (ethyl acetate)	3.06E−05	1.42E−05	1.44E−05	1.28E−05
Ethylene oxide	4.42E−07	2.31E−07	2.30E−07	4.25E−07
Ethylenediamine	6.52E−10	6.50E−10	6.58E−10	6.50E−10
Exhaust	3.54E+03	3.58E+03	3.61E+03	2.97E+03
Fluoranthene	6.46E−09	7.11E−09	6.86E−09	5.82E−09
Fluorene	2.05E−08	2.26E−08	2.18E−08	1.85E−08
Fluoride	6.62E−05	3.09E−04	5.60E−05	5.62E−05
Fluorine	1.26E−04	1.25E−04	1.36E−04	1.08E−04
Formaldehyde (methanal)	2.81E−03	2.34E−03	2.20E−03	1.97E−03
Formic acid (methane acid)	5.10E−07	4.42E−07	4.58E−07	4.31E−07
Furan	1.34E−07	1.21E−07	1.25E−07	1.18E−07

(continued)

Table 39.34 (continued)

Substance	W	W/ALU	PVC	W/C
Group NMVOC to air	2.74E−01	2.30E−01	2.40E−01	2.05E−01
Group PAH to air	5.33E−05	7.66E−05	5.71E−05	4.82E−05
Halogenated hydrocarbons (unspecified)	1.93E−07	1.93E−07	1.39E−04	1.93E−07
Halogenated organic emissions to air	4.40E−04	4.85E−04	2.23E−03	4.06E−04
Halon (1211)	5.52E−07	4.23E−07	5.58E−07	3.93E−07
Halon (1301)	5.67E−07	4.17E−07	5.54E−07	3.69E−07
Heavy metals to air	2.45E−03	2.73E−03	5.25E−03	2.07E−03
Heavy metals to air (unspecified)	7.49E−10	6.98E−10	−2.10E−08	2.78E−10
Helium	5.92E−05	4.77E−05	5.33E−05	4.14E−05
Heptane (isomers)	2.05E−04	1.56E−04	1.73E−04	1.38E−04
Hexachlorobenzene (Perchlorobenzene)	8.38E−07	8.51E−07	9.04E−07	7.00E−07
Hexafluorosilicates	1.22E−06	9.74E−07	1.19E−06	9.21E−07
Hexamethylene diamine (HMDA)	4.53E−12	6.95E−12	2.48E−12	5.52E−12
Hexane (isomers)	1.21E−03	4.15E−04	4.93E−04	3.79E−04
Hydrocarbons (unspecified)	9.01E−04	9.01E−04	3.36E−02	9.01E−04
Hydrocarbons, aromatic	1.33E−04	1.10E−04	1.88E−04	9.60E−05
Hydrocarbons, chlorinated	3.25E−04	3.09E−04	2.17E−04	2.45E−04
Hydrogen	2.51E−03	2.17E−03	4.33E−02	2.33E−03
Hydrogen arsenic (arsine)	1.89E−10	2.86E−10	2.00E−10	2.32E−10
Hydrogen bromine (hydrobromic acid)	1.47E−08	1.54E−08	8.83E−09	1.91E−08
Hydrogen chloride	7.40E−03	8.06E−03	1.83E−02	6.71E−03
Hydrogen cyanide (prussic acid)	4.41E−08	7.64E−08	−1.41E−07	2.97E−08
Hydrogen fluoride	1.72E−03	1.98E−03	2.08E−03	1.51E−03
Hydrogen iodide	1.52E−11	1.58E−11	8.71E−12	2.01E−11

(continued)

Table 39.34 (continued)

Substance	W	W/ALU	PVC	W/C
Hydrogen phosphorous	5.23E−09	2.82E−08	4.18E−09	4.18E−09
Hydrogen sulphide	−2.82E−05	−2.63E−03	1.71E−04	−2.80E−05
Indeno(1,2,3-cd)pyrene	6.63E−10	7.29E−10	7.04E−10	5.97E−10
Inorganic emissions to air	1.39E+03	1.42E+03	1.47E+03	1.18E+03
Iodine	1.27E−05	9.96E−06	1.65E−05	9.97E−06
Iron	1.05E−04	9.55E−05	2.77E−04	8.44E−05
Isocyanide acid	1.20E−03	3.67E−04	5.38E−07	3.67E−04
Isoprene	6.21E−09	5.59E−09	5.80E−09	5.47E−09
Lanthanides	1.39E−10	1.63E−10	1.09E−10	1.18E−10
Lead (+II)	1.62E−04	1.74E−04	2.80E−04	1.31E−04
Lead dioxide	5.85E−12	9.28E−12	4.08E−12	5.13E−12
Magnesium	2.28E−10	8.97E−10	5.89E−10	1.75E−10
Manganese (+II)	4.58E−04	4.57E−04	4.96E−04	3.77E−04
Mercaptan (unspecified)	5.80E−07	1.09E−06	2.97E−05	9.46E−07
Mercury (+II)	1.66E−05	1.62E−05	2.24E−05	1.32E−05
Metals (unspecified)	7.82E−06	1.81E−05	3.41E−04	1.81E−05
Methacrylate	1.93E−08	8.98E−09	9.05E−09	8.04E−09
Methane	1.86E+00	1.89E+00	2.25E+00	1.63E+00
Methane (biotic)	9.46E−03	9.04E−03	1.08E−02	8.00E−03
Methanol	7.97E−04	4.35E−04	4.05E−04	4.19E−04
Methyl amine	3.69E−11	1.72E−11	1.74E−11	1.54E−11
Methyl borate	6.53E−15	3.03E−15	3.06E−15	2.71E−15
Methyl bromide	1.65E−15	6.86E−15	4.46E−15	1.26E−15
Methyl formate	7.50E−11	3.48E−11	3.51E−11	3.12E−11

(continued)

Table 39.34 (continued)

Substance	W	W/ALU	PVC	W/C
Methyl tert-butylether	1.94E−06	1.00E−06	6.17E−07	9.31E−07
Molybdenum	1.86E−06	1.63E−06	2.03E−06	1.43E−06
Monoethanolamine	1.13E−05	1.07E−05	1.09E−05	1.07E−05
Naphthalene	2.08E−07	2.29E−07	2.21E−07	1.88E−07
Nickel (+II)	8.04E−05	6.70E−05	1.04E−04	5.96E−05
Nitrate	1.05E−07	1.13E−07	1.59E−07	8.70E−08
Nitrogen (atmospheric nitrogen)	3.42E−02	3.74E−02	3.24E−02	2.93E−02
Nitrogen dioxide	9.95E−04	9.95E−04	7.04E−02	9.95E−04
Nitrogen monoxide	1.99E−09	2.17E−09	2.07E−09	1.72E−09
Nitrogen oxides	1.48E+00	1.46E+00	1.53E+00	1.24E+00
Nitrous oxide (laughing gas)	9.64E−02	9.35E−02	9.73E−02	7.88E−02
NMVOC (unspecified)	2.01E−01	1.77E−01	1.79E−01	1.60E−01
Octane	6.50E−06	1.14E−05	8.01E−06	9.83E−06
Organic chlorine compounds	6.61E−10	6.63E−10	9.66E−05	6.59E−10
Organic emissions to air (group VOC)	2.15E+00	2.13E+00	2.54E+00	1.85E+00
Other emissions to air	3.54E+03	3.59E+03	3.61E+03	2.98E+03
Oxygen	2.06E−02	2.45E−02	1.96E−02	1.87E−02
Ozone	2.05E−04	1.54E−04	2.46E−04	1.49E−04
Palladium	−1.13E−14	−6.16E−14	−9.03E−15	−9.01E−15
Particles to air	4.09E−01	3.97E−01	4.80E−01	3.35E−01
Pentachlorobenzene	2.05E−06	2.09E−06	2.18E−06	1.71E−06
Pentachlorophenol (PCP)	4.09E−07	3.76E−07	4.67E−07	3.32E−07
Pentane (*n*-pentane)	4.00E−03	3.84E−03	4.10E−03	3.23E−03
Phenanthrene	6.55E−08	7.20E−08	6.95E−08	5.90E−08

(continued)

Table 39.34 (continued)

Substance	W	W/ALU	PVC	W/C
Phenol (hydroxy benzene)	4.46E−06	3.80E−06	6.54E−06	1.25E−05
Phosphorus	1.69E−03	1.71E−03	1.78E−03	1.41E−03
Platinum	1.70E−11	1.58E−11	2.28E−11	1.64E−11
Polychlorinated biphenyls (PCB unspecified)	3.81E−08	3.31E−08	6.03E−08	2.97E−08
Polychlorinated dibenzo-*p*-dioxins (2,3,7,8—TCDD)	6.50E−09	6.63E−09	6.91E−09	5.44E−09
Polycyclic aromatic hydrocarbons (PAH)	5.11E−05	7.41E−05	5.46E−05	4.63E−05
Propane	8.26E−03	9.58E−03	8.81E−03	8.00E−03
Propanol (iso-propanol; isopropanol)	6.59E−06	3.06E−06	3.08E−06	2.74E−06
Propene (propylene)	1.27E−04	1.12E−04	1.46E−04	1.19E−04
Propionaldehyde	1.32E−08	7.18E−09	1.64E−08	6.53E−09
Propionic acid (propane acid)	5.10E−06	4.23E−06	5.75E−06	3.82E−06
Propylene oxide	1.40E−06	1.12E−06	1.15E−06	1.00E−06
R 11 (trichlorofluoromethane)	4.22E−07	1.16E−06	2.11E−07	5.00E−07
R 113 (trichlorofluoroethane)	8.09E−10	3.76E−10	3.79E−10	3.36E−10
R 114 (dichlorotetrafluoroethane)	7.19E−07	1.42E−06	5.70E−07	7.41E−07
R 116 (hexafluoroethane)	2.44E−06	8.08E−06	2.11E−06	1.88E−06
R 12 (dichlorodifluoromethane)	3.98E−06	2.73E−06	2.72E−06	3.85E−06
R 13 (chlorotrifluoromethane)	5.70E−08	1.57E−07	2.85E−08	6.76E−08
R 134a (tetrafluoroethane)	2.07E−08	1.57E−08	2.29E−08	1.55E−08
R 152a (difluoroethane)	1.91E−08	1.51E−08	2.42E−08	1.51E−08
R 21 (dichlorofluoromethane)	1.21E−11	6.81E−12	7.00E−12	6.56E−12
R 22 (chlorodifluoromethane)	2.19E−06	1.85E−06	2.17E−06	1.59E−06
R 23 (trifluoromethane)	3.84E−09	2.17E−09	2.23E−09	2.09E−09
Radioactive emissions to air	5.07E−06	4.93E−06	6.27E−06	4.09E−06

(continued)

Table 39.34 (continued)

Substance	W	W/ALU	PVC	W/C
Rhodium	−1.09E−14	−5.95E−14	−8.71E−15	−8.69E−15
Scandium	1.90E−08	1.69E−08	6.06E−08	1.52E−08
Selenium	7.02E−05	7.31E−05	7.47E−05	5.87E−05
Silicium tetrafluoride	2.69E−08	1.76E−08	1.80E−08	1.72E−08
Silver	3.57E−09	2.90E−09	9.33E−09	2.93E−09
Sodium chlorate	1.21E−07	1.10E−07	1.16E−07	1.05E−07
Sodium dichromate	3.44E−07	3.37E−07	4.66E−07	2.83E−07
Sodium formate	4.91E−07	4.87E−07	4.91E−07	4.86E−07
Sodium hydro	1.71E−07	7.94E−08	8.02E−08	7.11E−08
Steam	3.05E+02	3.34E+02	3.22E+02	2.62E+02
Strontium	5.10E−06	4.19E−06	1.19E−05	3.95E−06
Styrene	5.06E−07	4.23E−07	7.34E−07	5.74E−07
Sulphate	1.99E−10	2.91E−10	2.99E−10	1.55E−10
Sulphur dioxide	5.44E−01	5.62E−01	6.95E−01	4.62E−01
Sulphur hexafluoride	2.85E−06	2.06E−06	3.11E−06	2.03E−06
Sulphuric acid	1.15E−07	1.56E−07	6.24E−08	8.50E−08
Tellurium	1.60E−09	1.70E−09	1.68E−09	1.37E−09
Terpenes	5.87E−08	5.29E−08	5.49E−08	5.17E−08
Tetrachloroethene (perchloroethylene)	1.88E−05	4.51E−05	7.56E−04	4.51E−05
Tetrafluoromethane	2.14E−05	7.25E−05	1.88E−05	1.67E−05
Thallium	1.21E−07	1.15E−07	7.95E−08	9.26E−08
Thorium (Th230)	4.10E−09	3.44E−09	5.17E−09	3.05E−09
Thorium (Th232)	6.85E−07	5.43E−07	1.18E−06	5.28E−07
Tin (+IV)	9.88E−05	1.01E−04	1.12E−04	9.27E−05

(continued)

Table 39.34 (continued)

Substance	W	W/ALU	PVC	W/C
Tin oxide	5.09E−13	8.08E−13	3.55E−13	4.46E−13
Titanium	5.62E−06	5.06E−06	1.80E−05	4.58E−06
Toluene (methyl benzene)	2.23E−03	2.22E−03	2.32E−03	1.83E−03
Trichloromethane (chloroform)	3.33E−08	1.98E−08	2.40E−08	1.88E−08
Trimethylbenzene	4.96E−12	7.87E−12	3.46E−12	4.34E−12
Uranium (total)	4.38E−06	4.38E−06	5.09E−06	3.56E−06
Used air	7.05E+00	1.56E+01	−1.50E+00	1.55E+01
Vanadium (+III)	1.14E−04	1.14E−04	1.15E−04	8.98E−05
Vinyl chloride (VCM; chloroethene)	2.91E−06	2.07E−06	5.86E−04	1.84E−05
VOC (unspecified)	3.92E−05	4.23E−05	6.96E−04	4.07E−05
Wood (dust)	1.88E−10	2.98E−10	1.31E−10	1.65E−10
Xylene (dimethyl benzene)	3.37E−03	3.51E−03	3.57E−03	2.82E−03
Xylene (meta-xylene; 1,3-dimethylbenzene)	2.97E−04	2.96E−04	3.07E−04	2.44E−04
Zinc (+II)	1.02E−03	1.27E−03	1.40E−03	8.37E−04
Zinc oxide	1.02E−12	1.62E−12	7.10E−13	8.92E−13
Zinc sulphate	4.54E−09	6.99E−09	4.81E−09	5.68E−09
Emissions to freshwater				
1,2-dibromoethane	3.95E−11	4.57E−11	2.97E−11	3.27E−11
1-butanol	1.11E−07	5.16E−08	5.22E−08	4.63E−08
Acenaphthene	4.75E−09	4.54E−09	4.29E−09	3.98E−09
Acenaphthylene	6.41E−10	9.52E−10	6.95E−10	8.26E−10
Acetaldehyde (Ethanal)	2.03E−07	9.43E−08	9.53E−08	8.45E−08
Acetic acid	2.31E−05	1.56E−05	9.00E−06	1.72E−05
Acetone (dimethylcetone)	1.10E−10	4.32E−10	2.84E−10	8.43E−11

(continued)

Table 39.34 (continued)

Substance	W	W/ALU	PVC	W/C
Acid (calculated as H +)	9.13E−05	4.21E−04	2.07E−04	1.18E−04
Acrylic acid	4.03E−08	1.87E−08	1.89E−08	1.68E−08
Acrylonitrile	2.66E−10	4.08E−10	1.45E−10	3.24E−10
Adsorbable organic halogen compounds (AOX)	3.90E−05	3.41E−05	2.47E−05	3.23E−05
Alkane (unspecified)	7.90E−05	5.50E−05	6.46E−05	4.90E−05
Alkene (unspecified)	7.29E−06	5.08E−06	5.96E−06	4.52E−06
Aluminium (+III)	6.72E+00	7.10E+00	7.12E+00	5.66E+00
Aluminium (+III)	1.82E−03	2.45E−03	7.38E−03	1.80E−03
Ammonia	4.43E−07	4.70E−07	1.16E−07	3.68E−07
Ammonium/ammonia	1.64E−05	1.25E−05	2.79E−03	1.34E−05
Ammonium/ammonia	1.79E−03	1.67E−03	3.10E−03	5.15E−03
Analytical measures to freshwater	4.23E+00	3.85E+00	5.21E+00	3.25E+00
Anthracene	1.56E−09	2.98E−09	1.87E−09	2.49E−09
Antimony	3.82E−05	1.99E−05	6.67E−05	7.28E−05
Antimony	2.16E−05	1.15E−05	2.03E−05	4.03E−05
Aromatic hydrocarbons (unspecified)	3.21E−04	2.25E−04	2.78E−04	2.00E−04
Arsenic (+V)	9.48E−04	9.66E−04	1.02E−03	7.93E−04
Arsenic (+V)	8.62E−04	8.71E−04	9.34E−04	7.20E−04
Barium	4.92E−04	4.00E−04	2.07E−03	4.01E−04
Barium	5.56E−04	4.25E−04	9.33E−04	3.73E−04
Benzene	9.88E−05	6.16E−05	7.41E−05	1.31E−04
Benzo(a)anthracene	1.27E−10	2.53E−10	1.62E−10	2.24E−10
Benzofluoranthene	5.13E−11	1.14E−10	7.24E−11	1.07E−10
Beryllium	3.44E−06	2.66E−06	7.92E−06	2.64E−06

(continued)

Table 39.34 (continued)

Substance	W	W/ALU	PVC	W/C
Beryllium	1.12E−08	1.19E−08	1.43E−08	9.18E−09
Biological oxygen demand (BOD)	1.43E−01	9.34E−02	1.17E−01	8.63E−02
Boron	6.11E−03	6.12E−03	6.58E−03	5.10E−03
Boron	2.23E−03	2.28E−03	2.38E−03	1.88E−03
Bromate	6.12E−04	5.08E−04	5.42E−04	4.18E−04
Bromine	9.04E−04	9.18E−04	1.44E−03	7.70E−04
Bromine	4.69E−03	4.63E−03	4.93E−03	3.89E−03
Butene	1.86E−07	2.98E−07	3.77E−06	2.31E−05
Butylene glycol (butane diol)	1.41E−10	6.58E−11	6.67E−11	5.90E−11
butyrolactone	2.45E−10	1.14E−10	1.16E−10	1.02E−10
Cadmium (+II)	2.15E−05	2.16E−05	4.16E−05	1.81E−05
Cadmium (+II)	1.19E−05	1.22E−05	1.17E−05	9.81E−06
Calcium (+II)	1.74E+01	1.78E+01	1.86E+01	1.46E+01
Calcium (+II)	2.90E−01	2.90E−01	2.71E−01	2.46E−01
Carbon, organically bound	3.78E−05	6.39E−05	4.17E−05	5.21E−05
Carbonate	4.50E−04	6.78E−04	7.71E−03	7.18E−04
Cesium	6.08E−07	4.23E−07	4.97E−07	3.77E−07
Chemical oxygen demand (COD)	3.71E+00	3.61E+00	4.79E+00	3.00E+00
Chlorate	4.68E−03	3.88E−03	6.37E−03	3.21E−03
Chloride	9.62E−02	8.47E−02	3.11E+00	7.03E−02
Chloride	7.65E+00	7.62E+00	8.42E+00	6.34E+00
Chlorinated hydrocarbons (unspecified)	8.88E−08	2.13E−07	3.58E−06	2.13E−07
Chlorine (dissolved)	5.85E−05	1.51E−04	7.49E−05	6.81E−05
Chlorobenzene	9.81E−07	4.57E−07	4.63E−07	4.10E−07

(continued)

Table 39.34 (continued)

Substance	W	W/ALU	PVC	W/C
Chloromethane (methyl chloride)	1.74E−08	1.90E−08	−5.10E−09	1.49E−08
Chlorous dissolvent	1.11E−05	7.31E−06	7.49E−06	8.47E−06
Chromium (+III)	1.40E−07	3.34E−07	7.86E−08	1.58E−07
Chromium (+VI)	1.26E−03	1.13E−03	1.46E−03	9.95E−04
Chromium (+VI)	4.39E−04	3.85E−04	4.94E−04	3.39E−04
Chromium (unspecified)	1.15E−05	1.65E−05	1.00E−04	1.44E−05
Chrysene	5.36E−10	1.09E−09	7.00E−10	9.78E−10
Cobalt	4.27E−03	4.11E−03	6.06E−03	3.42E−03
Cobalt	2.95E−05	6.63E−06	2.68E−05	6.44E−06
Copper (+II)	1.42E−02	1.35E−02	1.56E−02	1.13E−02
Copper (+II)	2.49E−05	2.39E−05	7.19E−05	4.95E−05
Cresol (methyl phenol)	4.50E−10	4.65E−10	−3.99E−10	3.44E−10
Cumene (isopropylbenzene)	6.06E−05	3.31E−05	3.95E−05	8.35E−05
Cyanide	2.20E−05	2.53E−05	2.65E−05	1.94E−05
Dichloroethane (ethylene dichloride)	1.47E−06	5.60E−07	1.92E−05	6.48E−07
Dichloromethane (methylene chloride)	1.04E−05	6.90E−06	8.37E−06	6.24E−06
Dichloropropane	3.63E−15	5.58E−15	1.99E−15	4.43E−15
Dichromate	1.25E−06	1.22E−06	1.70E−06	1.03E−06
Dissolved organic carbon, DOC (ecoinvent)	1.40E+00	1.38E+00	2.20E+00	1.14E+00
Ecoinvent long-term to freshwater	2.02E+01	2.06E+01	2.15E+01	1.69E+01
Emissions to freshwater	3.55E+01	3.53E+01	3.90E+01	2.93E+01
Ethanol	2.56E−07	1.19E−07	1.20E−07	1.06E−07
Ethene (ethylene)	6.84E−05	2.64E−05	1.52E−05	4.03E−05
Ethyl benzene	1.47E−05	1.04E−05	1.21E−05	9.23E−06

(continued)

Table 39.34 (continued)

Substance	W	W/ALU	PVC	W/C
Ethylene acetate (ethyl acetate)	1.74E−11	8.10E−12	8.19E−12	7.25E−12
Ethylene oxide	5.11E−08	2.92E−08	2.44E−08	2.83E−08
Ethylenediamine	1.58E−09	1.58E−09	1.59E−09	1.58E−09
Fatty acids (calculated as total carbon)	2.24E−03	1.56E−03	1.83E−03	1.39E−03
Fluoranthene	1.59E−10	3.41E−10	1.95E−10	2.80E−10
Fluoride	4.02E−01	4.13E−01	4.26E−01	3.38E−01
Fluorine	1.36E−07	1.48E−07	1.20E−07	1.25E−07
Formaldehyde (methanal)	7.18E−05	1.95E−05	1.57E−05	2.17E−05
Freshwater	1.94E+01	1.97E+01	2.57E+01	1.61E+01
Halogenated organic emissions to freshwater	2.43E−05	1.56E−05	1.18E−04	1.61E−05
Heavy metals to freshwater	2.37E−02	2.17E−02	3.04E−02	1.83E−02
Heavy metals to water (unspecified)	2.87E−07	3.04E−07	2.51E−07	2.40E−07
Hexafluorosilicates	2.19E−06	1.75E−06	2.13E−06	1.66E−06
Hexane (isomers)	4.94E−11	5.14E−11	−4.33E−11	3.79E−11
Hydrocarbons (unspecified)	8.69E−05	7.86E−05	3.77E−04	1.40E−04
Hydrocarbons to freshwater	4.59E−02	2.92E−02	3.60E−02	2.68E−02
Hydrogen chloride	1.21E−08	1.29E−08	1.24E−08	1.04E−08
Hydrogen fluoride (hydrofluoric acid)	1.23E−09	2.42E−09	−9.55E−09	1.96E−09
Hydrogen peroxide	2.22E−06	1.99E−06	2.08E−06	1.97E−06
Hydrogen sulphide	2.12E−03	2.15E−03	2.76E−03	1.77E−03
Hydrogen sulphide	2.66E−06	2.23E−06	3.36E−06	1.99E−06
Hydroxide	2.36E−04	1.27E−03	1.89E−04	1.88E−04
Hypochlorite	8.90E−06	7.61E−06	1.17E−05	7.73E−06
Inorganic emissions to freshwater	1.09E+01	1.07E+01	1.20E+01	8.93E+00

(continued)

Table 39.34 (continued)

Substance	W	W/ALU	PVC	W/C
Inorganic salts and acids (unspecified)	7.77E−04	1.86E−03	3.13E−02	1.86E−03
Iodide	7.65E−11	5.87E−11	2.88E−10	6.55E−11
Iodide	1.08E−04	9.05E−05	1.01E−04	7.74E−05
Iron	2.64E−01	2.66E−01	3.17E−01	2.21E−01
Iron	1.92E−02	1.77E−02	2.57E−02	1.49E−02
Lead (+II)	2.63E−03	1.81E−03	1.91E−03	1.57E−03
Lead (+II)	2.57E−05	3.13E−05	5.79E−05	2.32E−05
Lithium	1.18E−05	4.65E−05	3.05E−05	9.06E−06
Magnesium (+III)	2.24E+00	2.27E+00	2.38E+00	1.87E+00
Magnesium (+III)	3.90E−02	3.88E−02	4.12E−02	3.21E−02
Magnesium chloride	−3.07E−09	−1.68E−08	−2.46E−09	−2.45E−09
Manganese (+II)	2.33E−02	2.33E−02	2.52E−02	1.93E−02
Manganese (+II)	6.06E−04	6.05E−04	6.33E−04	4.98E−04
Mercury (+II)	4.11E−06	4.01E−06	1.09E−05	3.70E−06
Mercury (+II)	2.19E−06	2.09E−06	2.62E−06	1.74E−06
Metal ions (unspecific)	3.41E−02	2.68E−03	1.02E−02	2.46E−03
Metal ions (unspecific)	0.00E+00	0.00E+00	−3.56E−05	0.00E+00
Metals (unspecified)	1.88E−05	4.18E−05	1.15E−03	4.18E−05
Methanol	6.61E−04	6.35E−04	6.29E−04	5.82E−04
Methyl acrylate	3.78E−07	1.75E−07	1.77E−07	1.57E−07
Methyl amine	8.84E−11	4.12E−11	4.17E−11	3.69E−11
Methyl isobutyl ketone	4.60E−11	1.81E−10	1.19E−10	3.54E−11
Methyl tert-butylether	3.04E−08	1.60E−08	1.01E−08	1.49E−08
Methylformate	2.99E−11	1.39E−11	1.40E−11	1.24E−11

(continued)

Table 39.34 (continued)

Substance	W	W/ALU	PVC	W/C
Molybdenum	3.17E−04	2.98E−04	3.01E−04	2.47E−04
Molybdenum	1.01E−04	8.97E−05	1.00E−04	7.74E−05
Naphthalene	5.99E−08	1.12E−07	7.28E−08	9.66E−08
n-Butyl acetate	1.44E−07	6.71E−08	6.78E−08	6.01E−08
Neutral salts	1.60E−09	1.07E−08	−4.77E−08	6.87E−10
Nickel (+II)	5.30E−03	5.47E−03	1.22E−02	4.62E−03
Nickel (+II)	4.41E−05	4.63E−05	1.04E−04	6.68E−05
Nitrate	1.21E−01	1.24E−01	1.29E−01	1.02E−01
Nitrate	1.66E−01	6.13E−02	6.94E−02	9.12E−02
Nitrite	8.93E−07	6.79E−07	1.52E−04	7.30E−07
Nitrite	1.97E−05	2.02E−05	9.74E−05	1.84E−05
Nitrogen	1.37E−03	1.06E−03	1.59E−03	1.09E−03
Nitrogen organic bounded	2.68E−05	2.03E−05	4.57E−03	2.19E−05
Nitrogen organic bounded	2.05E−04	2.63E−04	3.59E−04	1.81E−04
Oil (unspecified)	4.17E−02	2.62E−02	3.23E−02	2.38E−02
Organic chlorine compounds (unspecified)	2.81E−09	2.81E−09	5.86E−06	2.81E−09
Organic compounds (dissolved)	2.29E−06	2.29E−06	1.26E−04	2.29E−06
Organic compounds (unspecified)	3.31E−09	3.31E−09	1.47E−03	3.31E−09
Organic emissions to freshwater	4.60E−02	2.93E−02	3.78E−02	2.69E−02
Particles to freshwater	1.48E−01	1.68E−01	2.16E−01	1.72E−01
Phenol (hydroxy benzene)	6.40E−05	4.77E−05	5.88E−05	5.23E−05
Phosphate	1.17E−01	1.18E−01	1.25E−01	9.79E−02
Phosphate	7.05E−04	3.82E−04	1.26E−03	3.31E−04
Phosphorus	1.73E−04	8.25E−05	4.12E−04	1.37E−04

(continued)

Table 39.34 (continued)

Substance	W	W/ALU	PVC	W/C
Polychlorinated dibenzo-p-dioxins (2,3,7,8—TCDD)	1.80E−13	1.80E−13	6.17E−08	1.80E−13
Polycyclic aromatic hydrocarbons (PAH, unspec.)	4.84E−06	4.86E−06	3.84E−06	3.46E−06
Potassium	2.48E+00	2.52E+00	2.63E+00	2.08E+00
Potassium	5.19E−01	5.27E−01	5.51E−01	4.35E−01
Propene	2.88E−05	1.70E−05	2.41E−05	6.62E−05
Propylene oxide	3.37E−06	2.70E−06	2.76E−06	2.41E−06
Rubidium	8.55E−06	5.79E−06	6.88E−06	5.19E−06
Scandium	5.89E−06	4.50E−06	8.90E−06	4.41E−06
Scandium	1.40E−06	1.07E−06	2.01E−06	1.05E−06
Selenium	2.87E−04	2.92E−04	3.10E−04	2.41E−04
Selenium	1.42E−04	1.44E−04	1.51E−04	1.19E−04
Silicate particles	−2.87E−09	−6.82E−09	−1.21E−07	−7.04E−09
Silicon dioxide (silica)	9.30E−21	9.30E−21	2.41E−19	9.30E−21
Silver	1.46E−07	1.20E−07	7.49E−07	1.88E−07
Silver	6.25E−07	5.22E−07	5.59E−07	3.99E−07
Sodium (+I)	1.05E+00	1.06E+00	1.13E+00	8.76E−01
Sodium (+I)	6.63E−01	5.79E−01	8.14E−01	4.96E−01
Sodium chloride (rock salt)	3.29E−10	1.75E−09	−5.74E−10	3.39E−10
Sodium formate	1.18E−06	1.17E−06	1.18E−06	1.17E−06
Sodium hypochlorite	7.30E−09	2.43E−08	−4.24E−07	5.77E−09
Soil loss by erosion into water	1.56E−08	4.54E−08	2.09E−08	3.93E−08
Solids (dissolved)	9.07E−02	8.25E−02	2.22E−01	9.13E−02
Solids (suspended)	3.96E+00	3.92E+00	5.63E+00	3.28E+00
Solids (suspended)	1.48E−01	1.68E−01	2.15E−01	1.72E−01

(continued)

Table 39.34 (continued)

Substance	W	W/ALU	PVC	W/C
Strontium	3.98E−04	3.10E−04	1.13E−03	3.09E−04
Strontium	1.91E−03	1.62E−03	1.76E−03	1.38E−03
Sulphate	3.58E+00	3.59E+00	3.80E+00	2.97E+00
Sulphate	1.10E+00	1.12E+00	1.29E+00	9.30E−01
Sulphide	7.68E−05	1.31E−04	9.06E−05	1.14E−04
Sulphite	3.12E−04	3.13E−04	3.43E−04	2.65E−04
Sulphur	2.32E−04	8.72E−05	1.18E−04	7.93E−05
Sulphuric acid	1.56E−06	1.66E−06	1.60E−06	1.34E−06
Suspended solids, unspecified	−1.91E−20	−5.54E−20	−3.19E−20	−1.06E−20
Thallium	2.74E−06	2.65E−06	5.83E−06	2.28E−06
Thallium	9.72E−08	8.37E−08	1.28E−07	8.46E−08
Tin (+IV)	3.56E−03	3.52E−03	6.24E−03	2.91E−03
Tin (+IV)	1.14E−06	1.13E−06	1.68E−06	1.09E−06
Titanium	5.92E−05	6.31E−06	1.01E−05	5.80E−06
Toluene (methyl benzene)	7.16E−05	5.16E−05	6.11E−05	4.57E−05
Total dissolved organic bounded carbon	1.41E−01	3.07E−02	3.73E−02	2.80E−02
Total organic bounded carbon	1.44E−01	3.37E−02	4.14E−02	3.83E−02
Trichloromethane (chloroform)	2.26E−09	1.05E−09	1.06E−09	9.41E−10
Tungsten	3.26E−06	2.50E−06	4.43E−06	2.47E−06
Tungsten	1.98E−06	1.52E−06	2.67E−06	1.50E−06
Vanadium (+III)	1.07E−03	9.44E−04	1.66E−03	7.94E−04
Vanadium (+III)	8.97E−06	8.31E−06	1.11E−05	7.14E−06
Vinyl chloride (VCM; chloroethene)	1.71E−07	1.10E−07	7.84E−05	1.13E−07
VOC (unspecified)	2.17E−04	1.51E−04	1.79E−04	1.35E−04

(continued)

Table 39.34 (continued)

Substance	W	W/ALU	PVC	W/C
Xylene (isomers; dimethyl benzene)	5.89E−05	4.21E−05	4.80E−05	3.73E−05
Xylene (meta-Xylene; 1,3-Dimethylbenzene)	3.32E−10	1.31E−09	8.61E−10	2.55E−10
Xylene (ortho-xylene; 1,2-Dimethylbenzene)	2.42E−10	9.54E−10	6.27E−10	1.86E−10
Zinc (+II)	3.46E−02	1.27E−02	4.05E−03	1.23E−02
Zinc (+II)	1.78E−04	1.43E−04	2.18E−04	1.21E−04
Emissions to agricultural soil				
2,4-dichlorophenoxyacetic acid (2,4-D)	2.37E−08	2.13E−08	2.21E−08	2.08E−08
Aclonifen	1.38E−05	1.49E−07	8.68E−08	1.37E−07
Aldrin	4.38E−10	2.04E−10	2.05E−10	1.82E−10
Aluminium	6.80E−03	6.79E−03	7.08E−03	5.60E−03
Antimony	1.65E−07	1.65E−07	1.65E−07	1.65E−07
Arsenic (+V)	2.26E−06	2.26E−06	2.34E−06	1.87E−06
Atrazine	1.15E−10	5.34E−11	5.39E−11	4.78E−11
Barium	3.16E−06	3.16E−06	3.16E−06	3.16E−06
Benomyl	1.51E−10	1.36E−10	1.41E−10	1.33E−10
Bentazone	7.05E−06	7.60E−08	4.43E−08	6.98E−08
Cadmium (+II)	5.35E−06	4.66E−06	4.83E−06	3.85E−06
Carbetamide	2.50E−06	3.17E−08	2.06E−08	2.95E−08
Carbofuran	8.26E−08	7.44E−08	7.72E−08	7.28E−08
Carbon (unspecified)	4.22E−03	4.20E−03	4.91E−03	3.48E−03
Chlorine	1.04E−03	1.04E−03	1.08E−03	8.58E−04
Chlorothalonil	4.78E−06	4.78E−06	4.81E−06	4.76E−06
Chromium (unspecified)	6.94E−05	6.39E−05	6.63E−05	5.27E−05
Cobalt	5.87E−06	5.87E−06	6.10E−06	4.84E−06

(continued)

Table 39.34 (continued)

Substance	W	W/ALU	PVC	W/C
Copper (+II)	4.97E−05	5.40E−05	5.68E−05	4.46E−05
Cypermethrin	6.62E−08	1.13E−08	1.15E−08	1.11E−08
Different pollutants	1.48E−01	1.48E−01	1.53E−01	1.22E−01
Emissions to agricultural soil	2.36E−01	2.14E−01	2.29E−01	1.80E−01
Fenpiclonil	6.65E−07	1.93E−07	1.92E−07	1.92E−07
Glyphosate	7.05E−07	6.56E−07	6.75E−07	5.88E−07
Heavy metals to agricultural soil	1.54E−02	1.53E−02	1.63E−02	1.26E−02
Inorganic emissions to agricultural soil	1.41E−02	1.40E−02	1.46E−02	1.16E−02
Iron	7.65E−03	7.63E−03	8.33E−03	6.30E−03
Lead (+II)	2.29E−05	2.18E−05	2.28E−05	1.81E−05
Linuron	1.06E−04	1.15E−06	6.69E−07	1.05E−06
Mancozeb	6.20E−06	6.20E−06	6.24E−06	6.18E−06
Manganese (+II)	6.51E−03	6.51E−03	6.74E−03	5.36E−03
Mercury (+II)	1.27E−07	3.69E−08	4.22E−08	3.08E−08
Metaldehyde	4.72E−07	7.20E−09	5.10E−09	6.78E−09
Metolachlor	7.71E−04	8.30E−06	4.84E−06	7.63E−06
Metribuzin	2.18E−07	2.18E−07	2.20E−07	2.18E−07
Molybdenum	1.21E−06	1.21E−06	1.27E−06	9.99E−07
Napropamide	8.36E−07	1.27E−08	9.03E−09	1.20E−08
Nickel (+II)	1.74E−05	1.83E−05	1.90E−05	1.51E−05
Oil (unspecified)	5.35E−02	3.31E−02	3.96E−02	3.01E−02
Orbencarb	1.18E−06	1.18E−06	1.19E−06	1.18E−06
Organic emissions to agricultural soil	5.77E−02	3.73E−02	4.45E−02	3.36E−02
Other emissions to agricultural soil	1.49E−01	1.48E−01	1.53E−01	1.22E−01

(continued)

Table 39.34 (continued)

Substance	W	W/ALU	PVC	W/C
Pesticides to agricultural soil	9.18E−04	2.31E−05	1.91E−05	2.22E−05
Phosphorus	3.19E−03	3.19E−03	3.30E−03	2.63E−03
Pirimicarb	6.67E−07	7.19E−09	4.19E−09	6.61E−09
Strontium	2.72E−08	2.38E−08	2.73E−08	2.12E−08
Sulphur	3.02E−03	3.02E−03	3.18E−03	2.49E−03
Sulphuric acid	2.21E−11	1.03E−11	1.03E−11	9.19E−12
Tebutam	1.98E−06	3.02E−08	2.14E−08	2.84E−08
Teflubenzuron	1.46E−08	1.46E−08	1.47E−08	1.45E−08
Thiram	2.67E−10	2.41E−10	2.50E−10	2.36E−10
Tin (+IV)	3.57E−07	3.55E−07	4.20E−07	3.51E−07
Titanium	4.49E−04	4.49E−04	4.65E−04	3.70E−04
Vanadium (+III)	1.29E−05	1.29E−05	1.33E−05	1.06E−05
Zinc (+II)	5.59E−04	5.43E−04	5.65E−04	4.48E−04
Emissions to industrial soil				
Aluminium	4.12E−04	2.72E−04	3.31E−04	2.46E−04
Aluminium (+III)	7.37E−07	1.52E−06	9.21E−07	1.28E−06
Ammonia	3.45E−04	6.98E−04	4.33E−04	5.95E−04
Arsenic (+V)	1.65E−07	1.10E−07	1.33E−07	9.90E−08
Barium	2.06E−04	1.36E−04	1.65E−04	1.23E−04
Bromide	9.86E−08	2.03E−07	1.25E−07	1.73E−07
Cadmium (+II)	1.41E−08	1.24E−08	−6.97E−08	1.02E−08
Calcium (+II)	1.66E−03	1.10E−03	1.33E−03	9.91E−04
Carbon (unspecified)	1.24E−03	8.17E−04	9.92E−04	7.39E−04
Chloride	1.15E−04	2.37E−04	1.46E−04	2.02E−04

(continued)

Table 39.34 (continued)

Substance	W	W/ALU	PVC	W/C
Chlorine	2.45E−02	2.31E−02	3.62E−02	2.03E−02
Chromium (+III)	1.12E−10	1.21E−10	1.17E−10	9.63E−11
Chromium (+VI)	1.01E−05	8.42E−06	2.51E−05	7.67E−06
Chromium (unspecified)	2.78E−06	2.76E−06	2.53E−06	2.42E−06
Cobalt	1.15E−08	2.37E−08	1.45E−08	2.02E−08
Copper (+II)	7.43E−06	6.20E−06	1.66E−05	5.72E−06
Different pollutants	4.72E−05	3.15E−05	4.08E−05	2.85E−05
Emissions to industrial soil	1.90E−01	1.90E−01	2.18E−01	1.57E−01
Fluoride	1.53E−01	1.56E−01	1.61E−01	1.27E−01
Glyphosate	2.36E−06	2.00E−06	2.08E−06	1.92E−06
Heavy metals to industrial soil	5.69E−03	4.93E−03	6.12E−03	4.65E−03
Inorganic emissions to industrial soil	1.83E−01	1.84E−01	2.10E−01	1.52E−01
Iron	5.38E−03	4.42E−03	5.74E−03	4.21E−03
Lead (+II)	5.66E−07	5.11E−07	5.24E−07	4.55E−07
Magnesium (+III)	3.31E−04	2.19E−04	2.65E−04	1.98E−04
Manganese (+II)	1.66E−05	1.12E−05	1.34E−05	1.01E−05
Mercury (+II)	1.39E−11	2.78E−11	1.74E−11	2.36E−11
Nickel (+II)	4.07E−07	6.12E−07	4.28E−07	5.06E−07
Oil (unspecified)	3.21E−04	2.24E−04	2.26E−04	1.98E−04
Organic emissions to industrial soil	1.56E−03	1.04E−03	1.22E−03	9.37E−04
Other emissions to industrial soil	4.95E−05	3.35E−05	4.29E−05	3.04E−05
Pesticides to industrial soil	2.36E−06	2.00E−06	2.08E−06	1.92E−06
Phosphorus	5.64E−05	8.56E−05	6.13E−05	7.37E−05
Potassium (+I)	2.28E−04	2.67E−04	2.20E−04	2.32E−04

(continued)

Table 39.34 (continued)

Substance	W	W/ALU	PVC	W/C
Sodium (+I)	1.11E−03	1.13E−03	9.47E−03	1.07E−03
Strontium	2.24E−04	4.45E−04	2.78E−04	3.80E−04
Sulphate	1.09E−05	2.22E−05	1.36E−05	1.88E−05
Sulphide	6.56E−05	1.33E−04	8.17E−05	1.13E−04
Sulphur	2.47E−04	1.63E−04	1.98E−04	1.48E−04
Zinc (+II)	4.51E−05	3.93E−05	4.10E−05	3.50E−05
Emissions to seawater				
Acenaphthene	5.53E−07	5.96E−07	5.97E−07	4.97E−07
Acenaphthylene	2.10E−07	2.26E−07	2.26E−07	1.88E−07
Acetic acid	1.04E−07	1.96E−07	1.41E−07	1.82E−07
Adsorbable organic halogen compounds (AOX)	9.38E−08	5.75E−08	6.98E−08	5.22E−08
Alkane (unspecified)	3.14E−05	1.96E−05	2.38E−05	1.78E−05
Alkene (unspecified)	2.90E−06	1.81E−06	2.20E−06	1.65E−06
Aluminium (+III)	8.43E−05	6.02E−05	7.50E−05	5.52E−05
Ammonia	4.59E−08	4.73E−08	−4.07E−08	3.51E−08
Ammonium/ammonia	2.93E−05	1.77E−05	2.15E−05	1.60E−05
Analytical measures to sea water	2.64E+00	2.58E+00	3.67E+00	2.13E+00
Anthracene	1.22E−07	1.33E−07	1.32E−07	1.10E−07
Aromatic hydrocarbons (unspecified)	1.36E−04	8.51E−05	1.03E−04	7.73E−05
Arsenic (+V)	1.42E−06	1.45E−06	1.10E−06	1.17E−06
Barium	4.15E−04	3.63E−04	3.83E−04	3.13E−04
Barytes	4.18E−03	2.83E−03	3.55E−03	2.59E−03
Benzene	1.18E−04	1.17E−04	1.20E−04	9.79E−05
Benzo(a)anthracene	1.25E−07	1.35E−07	1.35E−07	1.13E−07

(continued)

Table 39.34 (continued)

Substance	W	W/ALU	PVC	W/C
Benzofluoranthene	1.41E−07	1.52E−07	1.52E−07	1.26E−07
Beryllium	3.39E−07	3.79E−07	3.70E−07	3.18E−07
Biological oxygen demand (BOD)	1.19E+00	1.16E+00	1.43E+00	9.59E−01
Boron	2.06E−06	1.27E−06	1.48E−06	1.15E−06
Bromine	1.69E−04	1.06E−04	1.28E−04	9.60E−05
Cadmium (+II)	1.51E−06	1.60E−06	1.39E−06	1.28E−06
Calcium (+II)	1.55E−02	7.29E−03	8.32E−03	6.79E−03
Carbonate	1.25E−02	1.42E−02	1.36E−02	1.19E−02
Cesium	2.42E−07	1.51E−07	1.83E−07	1.37E−07
Chemical oxygen demand (COD)	2.89E−02	2.01E−02	2.43E−02	1.80E−02
Chloride	1.10E+00	1.19E+00	1.16E+00	9.99E−01
Chlorous dissolvent	3.82E−13	1.86E−13	1.10E−13	1.71E−13
Chromium (unspecified)	1.28E−06	2.37E−06	1.31E−06	1.59E−06
Chrysene	7.11E−07	7.66E−07	7.67E−07	6.39E−07
Cobalt	5.94E−06	6.63E−06	6.48E−06	5.56E−06
Copper (+II)	4.29E−06	4.29E−06	3.74E−06	3.50E−06
Cresol (methyl phenol)	3.46E−10	3.56E−10	−3.07E−10	2.65E−10
Cyanide	2.77E−06	1.41E−06	1.53E−06	1.27E−06
Emissions to sea water	4.03E+00	3.90E+00	4.98E+00	3.25E+00
Ethyl benzene	3.36E−05	3.26E−05	3.41E−05	2.72E−05
Fatty acids (calculated as total carbon)	1.38E−03	8.65E−04	1.05E−03	7.87E−04
Fluoranthene	1.46E−07	1.58E−07	1.58E−07	1.31E−07
Fluoride	1.08E−04	4.22E−05	4.58E−05	4.00E−05
Glutaraldehyde	5.16E−07	3.50E−07	4.38E−07	3.20E−07
Halogenated organic emissions to sea water	3.82E−13	1.86E−13	1.10E−13	1.71E−13

(continued)

Table 39.34 (continued)

Substance	W	W/ALU	PVC	W/C
Heavy metals to sea water	9.17E−04	6.89E−04	7.79E−04	6.09E−04
Hexane (isomers)	3.78E−11	3.89E−11	−3.35E−11	2.89E−11
Hydrocarbons (unspecified)	7.85E−05	5.30E−05	6.65E−05	4.86E−05
Hydrocarbons to sea water	1.15E−02	8.14E−03	9.72E−03	7.26E−03
Hypochlorite	9.27E−06	7.93E−06	1.22E−05	8.06E−06
Inorganic emissions to sea water	1.35E+00	1.28E+00	1.26E+00	1.08E+00
Iodide	2.42E−05	1.51E−05	1.83E−05	1.37E−05
Iron	9.41E−05	8.97E−05	8.95E−05	7.58E−05
Lead (+II)	2.58E−06	1.96E−06	2.11E−06	1.71E−06
Magnesium	1.34E−03	8.38E−04	1.01E−03	7.61E−04
Manganese (+II)	1.86E−05	1.53E−05	1.65E−05	1.33E−05
Mercury (+II)	5.02E−08	5.30E−08	5.24E−08	4.45E−08
Methanol	6.30E−06	4.46E−06	5.89E−06	3.99E−06
Methyl tert-butylether	1.62E−06	9.82E−07	1.19E−06	8.92E−07
Molybdenum	5.11E−08	3.11E−08	3.75E−08	2.83E−08
Naphthalene	1.61E−05	1.75E−05	1.74E−05	1.46E−05
Nickel (+II)	3.68E−06	3.92E−06	3.65E−06	3.26E−06
Nitrate	1.33E−04	1.02E−04	1.41E−04	9.56E−05
Nitrite	1.56E−06	1.19E−06	1.91E−06	1.18E−06
Nitrogen	1.17E−06	7.45E−07	9.09E−07	6.75E−07
Nitrogen organic bounded	5.62E−05	2.75E−05	3.35E−05	2.55E−05
Oil (unspecified)	9.32E−03	6.58E−03	7.91E−03	5.87E−03
Organic emissions to sea water	1.15E−02	8.16E−03	9.74E−03	7.27E−03
Other emissions to sea water	1.78E−06	9.29E−07	1.29E−06	8.62E−07
Particles to sea water	3.82E−02	3.54E−02	3.72E−02	2.99E−02

(continued)

Table 39.34 (continued)

Substance	W	W/ALU	PVC	W/C
Pesticides to sea water	1.78E−06	9.29E−07	1.29E−06	8.62E−07
Phenol (hydroxy benzene)	1.74E−04	1.77E−04	1.79E−04	1.49E−04
Phosphate	3.52E−04	1.14E−04	1.16E−04	1.11E−04
Phosphorus	2.00E−06	1.22E−06	1.49E−06	1.11E−06
Polycyclic aromatic hydrocarbons (PAH, unspec.)	1.92E−06	1.20E−06	1.46E−06	1.09E−06
Potassium	1.02E−03	6.38E−04	7.74E−04	5.80E−04
Selenium	7.64E−08	4.64E−08	5.63E−08	4.22E−08
Silver	1.45E−07	9.10E−08	1.10E−07	8.26E−08
Sodium (+I)	7.46E−02	4.69E−02	5.67E−02	4.25E−02
Solids (suspended)	3.82E−02	3.54E−02	3.72E−02	2.99E−02
Strontium	4.40E−04	2.75E−04	3.33E−04	2.50E−04
Sulphate	1.35E−01	1.47E−02	1.48E−02	1.33E−02
Sulphide	2.01E−03	2.31E−03	2.20E−03	1.94E−03
Sulphur	3.78E−06	2.40E−06	2.89E−06	2.17E−06
Tin (+IV)	4.71E−10	4.85E−10	−4.18E−10	3.60E−10
Titanium	2.08E−08	1.50E−08	1.85E−08	1.37E−08
Toluene (methyl benzene)	1.14E−04	1.05E−04	1.10E−04	8.81E−05
Total dissolved organic bounded carbon	8.89E−03	5.95E−03	7.29E−03	5.36E−03
Total organic bounded carbon	1.41E+00	1.39E+00	2.21E+00	1.15E+00
Tributyltinoxide	1.78E−06	9.29E−07	1.29E−06	8.62E−07
Triethylene glycol	5.35E−06	3.82E−06	5.04E−06	3.43E−06
Vanadium (+III)	4.22E−06	4.64E−06	4.55E−06	3.90E−06
VOC (unspecified)	8.45E−05	5.28E−05	6.41E−05	4.80E−05
Xylene (isomers; dimethyl benzene)	4.32E−05	3.56E−05	3.80E−05	3.12E−05
Zinc (+II)	3.40E−04	2.81E−04	3.15E−04	2.47E−04

Table 39.35 Normalised sensitivity coefficients computed for 10% perturbance of amount of wood in the frame

Impact category	Normalised sensitivity coefficient			
	W	W/ALU	PVC	W/C
Land use	*8.8E−01*	*7.5E−01*	0.0E+00	*7.3E−01*
Climate change	5.4E−02	1.7E−02	0.0E+00	1.9E−02
Freshwater ecotoxicity	6.3E−02	2.0E−02	0.0E+00	2.3E−02
Freshwater eutrophication	−7.3E−03	−2.2E−03	0.0E+00	−2.7E−03
Human toxicity (cancer)	2.0E−03	6.2E−04	0.0E+00	7.5E−04
Ionising radiation (human health)	7.8E−02	2.3E−02	0.0E+00	2.6E−02
Human toxicity (non-cancer)	6.1E−02	1.9E−02	0.0E+00	2.3E−02
Marine eutrophication	9.1E−03	2.9E−03	0.0E+00	3.2E−03
Resource depletion (minerals, fossils)	3.2E−02	9.8E−03	0.0E+00	8.5E−03
Stratospheric ozone depletion	2.4E−02	7.6E−03	0.0E+00	6.5E−03
Particulate matter formation	2.1E−02	6.3E−03	0.0E+00	7.0E−03
Photochemical ozone formation	1.2E−02	3.7E−03	0.0E+00	4.3E−03
Terrestrial acidification	1.7E−02	5.0E−03	0.0E+00	5.6E−03

Values >0.3 are in italics

Table 39.36 Normalised sensitivity coefficients computed for 10% perturbance of amount of steel in the frame

Impact category	Normalised sensitivity coefficient			
	W	W/ALU	PVC	W/C
Land use	1.3E−04	9.0E−04	6.3E−02	8.8E−04
Climate change	1.0E−03	2.6E−03	4.3E−02	2.8E−03
Freshwater ecotoxicity	5.1E−03	1.3E−02	1.9E−01	1.5E−02
Freshwater eutrophication	1.8E−04	4.4E−04	7.9E−03	5.3E−04
Human toxicity (cancer)	2.4E−04	5.6E−04	1.0E−02	6.8E−04
Ionising radiation (human health)	6.7E−03	1.6E−02	2.8E−01	1.8E−02
Human toxicity (non-cancer)	6.0E−03	1.5E−02	2.2E−01	1.7E−02
Marine eutrophication	5.1E−04	1.2E−03	2.3E−02	1.4E−03
Resource depletion (minerals, fossils)	3.5E−02	8.4E−02	*6.4E−01*	7.3E−02
Stratospheric ozone depletion	2.7E−03	6.6E−03	1.2E−01	5.7E−03
Particulate matter formation	2.3E−03	5.4E−03	9.6E−02	6.0E−03
Photochemical ozone formation	5.6E−04	1.3E−03	2.4E−02	1.5E−03
Terrestrial acidification	1.4E−03	3.5E−03	7.0E−02	3.8E−03

Values >0.3 are in italics

Table 39.37 Normalised sensitivity coefficients computed for 10% perturbance of amount of glass in the pane

Impact category	Normalised sensitivity coefficient			
	W	W/ALU	PVC	W/C
Land use	6.2E−03	1.7E−02	6.3E−02	2.5E−02
Climate change	2.3E−02	2.3E−02	2.1E−02	3.8E−02
Freshwater ecotoxicity	8.8E−03	9.1E−03	7.3E−03	1.6E−02
Freshwater eutrophication	1.6E−03	1.6E−03	1.5E−03	2.9E−03
Human toxicity (cancer)	2.6E−03	2.5E−03	2.4E−03	4.6E−03
Ionising radiation (human health)	1.1E−01	1.1E−01	9.8E−02	1.8E−01
Human toxicity (non-cancer)	9.1E−03	9.4E−03	7.3E−03	1.7E−02
Marine eutrophication	2.8E−02	2.9E−02	2.7E−02	4.8E−02
Resource depletion (minerals, fossils)	*4.5E−01*	*4.5E−01*	1.8E−01	*5.9E−01*
Stratospheric ozone depletion	6.2E−02	6.4E−02	6.2E−02	8.3E−02
Particulate matter formation	5.4E−02	5.4E−02	4.6E−02	9.0E−02
Photochemical ozone formation	3.4E−02	3.4E−02	3.1E−02	5.8E−02
Terrestrial acidification	5.7E−02	5.7E−02	4.8E−02	9.5E−02

Values >0.3 are in italics

Table 39.38 Normalised sensitivity coefficients computed for 10% perturbance of *U*-values

Impact category	Normalised sensitivity coefficient			
	W	W/ALU	PVC	W/C
Land use	*8.2E−01*	*8.6E−01*	*7.9E−01*	*7.7E−01*
Climate change	*9.0E−01*	*9.5E−01*	*7.9E−01*	*9.2E−01*
Freshwater ecotoxicity	*1.0E+00*	*1.0E+00*	*9.9E−01*	*1.0E+00*
Freshwater eutrophication	*9.9E−01*	*9.9E−01*	*9.9E−01*	*9.9E−01*
Human toxicity (cancer)	*5.7E−01*	*5.7E−01*	*5.4E−01*	*5.2E−01*
Ionising radiation (human health)	*9.0E−01*	*9.5E−01*	*7.7E−01*	*9.2E−01*
Human toxicity (non-cancer)	*9.0E−01*	*9.4E−01*	*9.0E−01*	*8.7E−01*
Marine eutrophication	*3.8E−01*	*3.9E−01*	1.6E−01	2.8E−01
Resource depletion (minerals, fossils)	*5.5E−01*	*5.7E−01*	*5.8E−01*	*4.1E−01*
Stratospheric ozone depletion	*8.3E−01*	*8.5E−01*	*7.5E−01*	*7.7E−01*
Particulate matter formation	*9.1E−01*	*9.2E−01*	*8.8E−01*	*8.7E−01*
Photochemical ozone formation	*8.2E−01*	*8.3E−01*	*7.3E−01*	*7.6E−01*
Terrestrial acidification	*8.2E−01*	*8.6E−01*	*7.9E−01*	*7.7E−01*

Values >0.3 are in italics

References

Benini, L., Mancini, L., Sala, S., Schau, E., Manfredi, S., Pant, R.: Normalisation method and data for environmental footprints. Report EUR 26842 EN. Ispra (2014) doi:10.2788/16415

Ciroth, A.: Refining the pedigree matrix approach in ecoinvent: towards empirical uncertainty factors. LCA Discussion Forum Zürich, 13 Sept (2013)

Connolly, D., Mathiesen, B.V., Østergaard, P.A., Möller, B., Nielsen, S., Lund, H., et al.: Heat Roadmap Europe 1: First Pre-Study for the EU27 (2012)

DMI: Danish Meteorological Institute. Available at http://www.dmi.dk/dmi/index/ (Accessed May 2016) (2016)

Dreicer, M., Tort, V., Manen, P.: ExternE, externalities of energy, vol. 5 9 Nuclear, Centre d'étude sur l'Evaluation de la Protection dans le domaine 10 nucléaire (CEPN). In: European Commission DGXII (ed) Science, 11 Research and development JOULE, Luxembourg (1995)

Dyrelund, A., Lund, H.: Danish heating sector can be CO_2 neutral before 2030. EuroHeat&Power **6**, 34–37 (2009)

EC-JRC: ILCD Handbook: General Guide for Life Cycle Assessment—Provisions and Action Steps, 1st edn. European Commission, Joint Research Centre, Institute for Environment and Sustainability, Ispra, Italy (2010)

EC-JRC: Recommendations based on existing environmental impact assessment models and factors for life cycle assessment in European context. ILCD Handbook—International Reference Life Cycle Data System, European Union EUR24571EN. ISBN: 978-92-79-17451-3. http://lct.jrc.ec.europa.eu/assessment/assessment/projects#consultation_impact (2011). Accessed Jan 2012

EC-JRC: Characterization factors of the ILCD recommended life cycle impact assessment methods. Database and Supporting Information. EUR25167. http://lct.jrc.ec.europa.eu/ (2012)

Ecoinvent: Ecoinvent Data v2.2.Final reports Ecoinvent v2.2 No. 1–25. Swiss Centre for Life Cycle Inventories, Dübendorf (2010)

Energynet: Technology Data for Energy Plants—Generation of Electricity and District Heating, Energy Storage and Electricity Carrier Generation and Conversion. doi:ISBN: 978-87-7844-940-5 (2012)

Eurostat: Treatment of waste by waste category, hazardousness and waste operations [WWW Document]. URL http://ec.europa.eu/eurostat/web/products-datasets/-/env_wastrt (Accessed 19 Sept 2016) (2016)

Frischknecht, R., Braunschweig, A., Hofstetter, P., Suter, P.: Modelling human health effects of radioactive releases in life cycle impact assessment. Environ. Impact Assess. Rev. **20**, 159–189 (2000)

Frischknecht, R., Steiner, R., Jungbluth, N.: The Ecological Scarcity Method—Eco-Factors: A method for impact assessment in LCA. 2009, Federal Office for the Environment FOEN: Zürich und Bern [WWW Document] (2006)

Frischknecht, R., Althaus, H., Bauer, C., Doka, G., Heck, T., Jungbluth, N., Kellenberger, D., Nemecek, T.: The environmental relevance of capital goods in life cycle assessments of products and services. Int. J. Life Cycle Assess. **12**, 7–17 (2007)

Goedkoop, M.J., Heijungs, R., Huijbregts, M., De Schryver, A., Struijs, J., Van Zelm, R.:ReCiPe 2008. A life cycle impact assessment method which comprises harmonised category indicators at the midpoint and the endpoint level. First edition report I: characterization, 6 Jan 2009. http://www.lcia-recipe.net/. Accessed July 2013 (2009)

Greco, S.L., Wilson, A.M., Spengler, J.D., Levy, J.I.: Spatial patterns of mobile source particulate matter emissions-to-exposure relationships across the United States. Atmos. Environ. **41**, 1011–1025 (2007)

Guinée, J.B., Gorrée, M., Heijungs, R., Huppes, G., Kleijn, R., de Koning, A., Van Oers, L., Wegener Sleeswijk, A., Suh, S., Udo de Haes, H., De Bruijn, J.A., Van Duin, R., Huijbregts, M.A.J.: Handbook on Life Cycle Assessment: Operational Guide to the ISO Standards. Series: Eco-efficiency in industry and science. Kluwer Academic Publishers, Dordrecht (Hardbound, ISBN 1-4020-0228-9; Paperback, ISBN 1-4020-0557-1) (2002)

Hauschild, M.Z., Goedkoop, M., Guinee, J., Heijungs, R., Huijbregts, M., Jolliet, O., Margni, M., De Schryver, A., Humbert, S., Laurent, A., Sala, S., Pant, R.: Identifying best existing practice for characterization modeling in life cycle impact assessment. Int. J. Life Cycle Assess. **18**, 683–697 (2013)

ISO: ISO 14044:2006 Environmental Management—Life Cycle Assessment—Requirements and Guidelines. International Standards Organization (2006)

Jolliet, O., Rosenbaum, R.K., Laurent, A.: Life cycle risks and impacts of nanotechnologies. Chapter 11 In: Malsch I., Emond C. (eds.) Nanotechnology and Human Health, CRC Press pp. 213–278 (2014). Print ISBN: 978-0-8493-8144-7, eBook ISBN: 978-0-8493-8145-4. https://doi.org/10.1201/b15341-18

Laurent, A., Lautier, A., Rosenbaum, R.K., Olsen, S.I., Hauschild, M.Z.: Normalisation references for Europe and North America for application with USEtox characterisation factors. Int. J. Life Cycle Assess. **16**(8), 728–738 (2011)

Milà i Canals, L., Romanyà, J., Cowell, S.: Method for assessing impacts on life support functions (LSF) related to the use of "fertile land" in life cycle assessment (LCA). J. Clean. Prod. **15**, 1426–1440 (2007)

Natural Resources Canada: Improving window energy efficiency. URL http://oee.nrcan.gc.ca/sites/oee.nrcan.gc.ca/files/pdf/residential/personal/documents/windows-eng.pdf (Accessed June 2016) (2015)

PlasticsEurope Database. URL: http://www.plasticseurope.org/. Accessed 21 Nov 2016

Posch, M., Seppälä, J., Hettelingh, J.P., Johansson, M., Margni, M., Jolliet, O.: The role of atmospheric dispersion models and ecosystem sensitivity in the determination of characterisation factors for acidifying and eutrophying emissions in LCIA. Int. J. Life Cycle Assess. **13**, 477–486 (2008)

Prommer, H., Tuxen, N., Bjerg, P.L.: Fringe-controlled natural attenuation of phenoxy acids in a landfill plume: integration of field-scale processes by reactive transport modeling. Environ. Sci. Technol. **40**, 4732–4738 (2006)

Rabl, A., Spadaro, J.V.: The RiskPoll software, version is 1.051 (dated August 2004). www.arirabl.com (2004). Accessed Jan 2012

Rosenbaum, R.K., Bachmann, T.M., Gold, L.S., Huijbregts, M.A.J., Jolliet, O., Juraske, R., Koehler, A., Larsen, H.F., MacLeod, M., Margni, M., McKone, T.E., Payet, J., Schuhmacher, M., van de Meent, D., Hauschild, M.Z.: USEtox-the UNEP-SETAC toxicity model: recommended characterisation factors for human toxicity and freshwater ecotoxicity in life cycle impact assessment. Int. J. Life Cycle Assess. **13**, 532–546 (2008) doi:10.1007/s11367-008-0038-4

Seppälä, J., Posch, M., Johansson, M., Hettelingh, J.P.: Country-dependent characterisation factors for acidification and terrestrial eutrophication based on accumulated exceedance as an impact category indicator. Int. J. Life Cycle Assess. **11**, 403–416 (2006)

Steinmann, Z.J.N., Hauck, M., Karuppiah, R., Laurenzi, I.J., Huijbregts, M.A.J.: Int. J. Life Cycle Assess. **19**, 1146–1155 (2014)

Struijs, J., Beusen, A., van Jaarsveld, H., Huijbregts, M.A.J.: ReCiPe 2008 A life cycle impact assessment method which comprises harmonised category indicators at the midpoint and the endpoint level. Report I: Characterisation factors, 1st edn. (2009)

The ecoinvent database, v2.2; ecoinvent: Zürich (2010)

Van Zelm, R., Huijbregts, M.A.J., Den Hollander, H.A., Van Jaarsveld, H.A., Sauter, F.J., Struijs, J., Van Wijnen, H.J., Van de Meent, D.: European characterization factors for human health damage of PM10 and ozone in life cycle impact assessment. Atmos. Environ. **42**, 441–453 (2008)

Disclaimer

This report is based on an LCA that was delivered as part of the requirements to pass the MSc course "Life Cycle Assessment of Products and Systems", given at the Department of Management Engineering of the Technical University of Denmark. The report has been reworked somewhat to serve as an example report to illustrate to students how to perform and how to structure the report on an LCA according to the requirements of ISO 14044:2006 (ISO 2006) and the reporting template in Chap. 38 from the ILCD Handbook (EC-JRC 2010). The reader should note that it is not the intention to provide an example of "the perfect LCA study" or "the perfect LCA report". The results of this LCA should not be directly used to inform a choice between windows, not even in Denmark. As a result of students collaborating in project teams of 5–6 members during one semester (~ 13 weeks, 10 ECTS MSc course), this is primarily the result of a well-achieved learning experience from LCA beginners. Its main purpose is to illustrate reporting, not good or best LCA practice, which is why many details are not necessarily handled the way they should be according to part II of the book, because there are many constraints on what can be achieved in one semester of learning LCA. LCA studies and reports produced by experienced LCA professionals can have a wide range of different structures and follow different emphases depending on the goal of the study.

Author Biographies

Mikołaj Owsianiak Involved in development and application of life cycle impact assessment methods in sustainability assessment of technologies. Has worked on issues associated with: soils (remediation), metals (toxic impact assessment), biodiesel (fate in the environment), and carbonaceous materials (biochar and hydrochar).

Anders Bjørn Part of the LCA community since the early 2010s. Main focus is interpretations of sustainability and integration of sustainability targets in LCA to enable absolute sustainability assessments.

Heidi B. Bugge Working with life cycle thinking in relation to ecolabeling since mid-2000s. LCA experience mostly by applying LCAs in developing ecolabel criteria for the Nordic Swan. Here focusing on how to use environmental hot spots in the life cycle to set relevant and steerable ecolabel requirements.

Sónia M. Carvalho Former student at the Technical University of Denmark.

Leise Jebahar Engineer specialised in LCA applied on circular economy, focusing mainly on the effect of different end-of-life scenarios and cascading and is working with waste management concerning waste from households, industries and the built environment.

Jon Rasmussen Former student at the Technical University of Denmark.

Caroline M. White Former student at the Technical University of Denmark.

Stig Irving Olsen LCA expert both as researcher and as consultant. Involved in the development of LCA methodologies since mid 1990s. Contributed to UNEP/SETAC working groups on LCIA methodology. Main LCA interest is human toxicity impacts, emerging technologies, and decision making.

Chapter 40
Overview of Existing LCIA Methods—Annex to Chapter 10

Ralph K. Rosenbaum

Abstract The chapter gives an overview and a systematic comparison of a selection of the most used Life Cycle Impact Assessment (LCIA) methods, focusing on methods that have been implemented and made available in LCA software. Currently available midpoint and endpoint characterisation methodologies are presented and their specific properties are qualitatively compared in detailed tables.

Learning objectives
After studying this chapter the reader should be able to:

- Name and summarise the LCIA methods most relevant in current LCA practice
- Identify and distinguish their main features, properties, advantages and limitations
- Select one or several adequate LCIA method(s) for a given goal and scope definition
- Discuss the (apparent) developments from earlier LCIA methods to the current state-of-the-art

---

The contents of this chapter have been modified from Rosenbaum, R.K.: Selection of impact categories, category indicators and characterisation models in goal and scope definition, appearing as Chapter 2 of Curran MA (ed.) LCA Compendium—The Complete World of Life Cycle Assessment—Goal and scope definition in Life Cycle Assessment pp. 63–122. Springer, Dordrecht (2017). Most notably, the LCIA method comparison tables have been updated for IMPACT World+ and LC-Impact.

R.K. Rosenbaum (✉)
IRSTEA, UMR ITAP, ELSA Research Group and ELSA-PACT—Industrial Chair
for Environmental and Social Sustainability Assessment, 361 rue Jean-François Breton,
BP 5095, 34196, Montpellier, France
e-mail: ralph.rosenbaum@irstea.fr

M.Z. Hauschild et al. (eds.), *Life Cycle Assessment*,
DOI 10.1007/978-3-319-56475-3_40

40.1 Introduction

An essential element in the choice of category indicators or entire LCIA methods is sufficient knowledge and overview of their (most important) characteristics. However, this knowledge is not easy to come by or readily available in the literature, without having to study the documentation of each method respectively. A profound comparison of existing LCIA methods was performed by Hauschild et al. (2013) for the establishment of recommended LCIA models for the European context. The results can be found in the ILCD handbook on LCIA recommended practice in Europe (EC-JRC 2011) and will provide some helpful guidance, including for the non-European context, as it contains both facts and evaluative expert judgements on the models, with only the latter being partially specific to the European context. Taking Hauschild et al.'s work as a starting point, the following tables provide a complete and updated qualitative comparison of widely used LCIA methods available in current LCA software.

Only models integrated into LCIA methods (and thus readily available for practitioners in LCA software and databases) are represented here with the exception of the latest methods IMPACT World+ and LC-Impact, which by the time of writing (mid 2017) were not yet fully implemented into LCA software but readily available to be imported manually (see respective websites for further information). It is worth mentioning that the authors of the LC-Impact method intend to provide both midpoint and endpoint characterisation factors (CFs). So far, endpoint CFs have been published, while midpoint CFs are not yet available but foreseen for later publication and thus not included in Table 40.1. During the finalisation of this chapter, a major update of the ReCiPe 2008 method, called ReCiPe 2016, has been published but could unfortunately not be included in the update of the comparison tables, so that only the latest version of ReCiPe 2008 (from 2013) is described in Tables 40.1 and 40.2. The Japanese LCIA method LIME has been updated to version 3.0, but to the author's knowledge no documentation in another language than Japanese is available, which is why only version 2.0 is covered here.

Further models (published but not yet integrated into LCIA methods) are discussed in the ILCD handbooks on LCIA (EC-JRC 2010, 2011) and of course in current scientific literature. Models not based on mechanistic cause-effect chain modeling, such as regulatory-based distance-to-target approaches like the Swiss Eco-scarcity method (Frischknecht et al. 2006) or the MEEuP approach based on emission limit values (Kemna et al. 2005) were also excluded from this overview. Such approaches require specific interpretation, different from cause-effect-based methods, due to their non-mechanistic and often policy-priority-based nature. If a potential environmental impact is expressed based on its difference to a political target, the resulting impact score will essentially represent the importance of an emission or resource extraction relative to established political target limits but not necessarily relative to its environmental relevance (depending on how closely political targets are related to environmental issues). This is because political targets are established based on a number of influences and lobbies, and will vary substantially from one country to another, especially on a global scale. Compared to a mechanistic modelling approach, political targets are not comprehensive enough and will only cover a selected number of known issues. In other words, a number of potentially important environmental issues may not (yet) be represented in politically set targets and will therefore not be evaluated when applying a distance-to-target approach. Political thresholds for some emissions, e.g. toxic chemicals, may furthermore be established based on risk measures instead of best (average) estimates or potential impacts and thus contain (inconsistent) safety values and other potential biases. However, in specific cases, where the goal of a study is to evaluate the environmental profile of a product, service or organisation, or the consequences of an environmental policy towards their relevance regarding political targets, such a method is a meaningful choice among available LCIA methods.

An important point to keep in mind is that the implementation of a given LCIA method may vary from one software to another (due to the need to adapt the LCIA method to the architecture and structure of the software, which may in some cases involve a re-interpretation) and not always all options proposed by LCIA developers may be implemented in each software. Hence, depending on which software you are using, you may find smaller (only in rare cases larger) deviations when it comes to implemented archetypes and other details and options. The descriptions in Tables 40.1 and 40.2 are mostly based on the original proposals by the LCIA method developers. The content of these tables is restricted to facts, while judgements on quality etc. were excluded as far as possible. For a further evaluation including expert judgements, e.g. on scientific validity, environmental relevance, or stakeholder acceptance, the reader is referred to the ILCD handbook on LCIA (EC-JRC 2010, 2011). Given the large amount of information contained in these tables, mistakes cannot be excluded, but as much information as possible has been verified in the original documentation of the methods (and if required corrected when taken from the ILCD handbook, which contains a number of small errors in the method descriptions). LCIA methods are under constant improvement and may be updated and corrected over time. Consequently, the information contained in Tables 40.1 and 40.2 is a snapshot of the situation and available information by the time of writing of this chapter (mid 2017) and is likely to change over time.

Tables 40.1 and 40.2 contain a qualitative comparison of a number of specific properties of available LCIA methods. Each column represents an LCIA method, while the rows are structured by impact category. This allows easy identification of the differences (and similarities) in these properties per impact category among methods and choosing the most suitable one for a given goal and scope. Table 40.1 lists the most important midpoint characterisation methods, while Table 40.2 contains methods providing endpoint or damage assessment characterisation factors. A number of methods were published before 2000, but are not included in this overview as they have been replaced by their authors with newer versions or must be considered outdated and obsolete for today's LCA practice. As a support to using and interpreting the tables, a brief description of each property reported in the tables and of its meaning are given hereafter (note that all properties may not necessarily apply to each impact category):

- Aspects/diseases/ecosystems considered: lists which kinds of impacts are considered, e.g. which kinds of resources (for resource use), or which kinds of diseases (for human health), or which ecosystems out of freshwater, marine water, and terrestrial ecosystems are covered by a method.
- Characterisation model: gives the name (if applicable) and points to the main reference(s) for the corresponding characterisation model used to calculate the characterisation factors for a given impact category and LCIA method.
- Human health effects: details about which kind of health effects were included.
- Ecosystem effects: details about which kind of effects on ecosystems were included.
- Biotic resources effects: consideration of potential impacts on biotic resources is still a rare property, but is included in some methods and may be an important point for some studies.
- Fate modeling: details about how the modeling of the distribution of an emission in the environment is considered (the concept of fate may also be applied for modeling a part of the cause-effect chain of a resource extraction instead of an emission).
- Exposure modeling: details about how the transfer of a substance from the environment into a given target (e.g. human population or an ecosystem) is considered (the concept of exposure may also be applied for modeling a part of the cause-effect chain of a resource extraction instead of an emission).
- Effect modeling: details about how the effect(s) of a substance transferred from the environment into a given target (e.g. human population or an ecosystem) is considered (the concept of exposure may also be applied for modeling a part of the cause-effect chain of a resource extraction instead of an emission).
- Marginal/average: these terms are used in different ways and meanings in the LCA context; here they describe two different impact modeling principles or choices: a marginal impact modeling approach represents the additional impact per additional unit emission/resource extraction within a product system on top of an existing background impact which is not coming from the modelled product system. This allows e.g. considering non-linearities of impacts depending on local conditions like high or low background concentrations to which the product systems adds an additional emission/resource extraction). An average impact modeling approach is strictly linear and represents an average impact independent from existing background impacts, which is similar to dividing the overall effect by the overall emissions.
- Emission compartment(s): for which emission compartment(s) the method provides characterisation factors.

- Time horizon: details on the time horizon(s) used to calculate potential impacts. A prominent example are the GWP-time horizons of 20, 100, and until IPCC (2007a) also 500 years. The essential difficulty with time horizons is that a short time horizon may exclude an important amount of future potential impacts from the assessment (risking violating the sustainability principle of inter-generational equality). Whereas, a long time horizon may "dilute" large short-term impacts over a longer time (i.e. making them look smaller), which would give a small but permanently continuing impact a similar impact potential than that of a large impact occurring within a short time. In other words, it would give the same importance to a large impact within one generation as to a small impact affecting several generations of humans for example. An important and widely ignored issue in current LCA practice is the inconsistency among time horizons between different impact categories, with some representing 100 years and others several hundreds to even thousands of years. An inconsistency that, in principle, would disallow adding up endpoint scores into areas of protection or normalising and weighting midpoint scores. Its importance, however, needs further study and most likely it is far from being a large source of uncertainty relative to other issues in LCA.
- Region modelled/valid: details on which region(s) has been modelled (i.e. which region is represented by the parameters used in the characterisation model). A model may either represent one or several specific region(s) (the larger the region, the more averaging is applied and the less specific the model is representing a region) or a global (or sometimes continental) average, also referred to as generic.
- Level of spatial differentiation: if the characterisation model represents more than one region, it is spatially (or geographically) differentiated. The level of differentiation may range from coarse (e.g. continental, sub-continental, countries, etc.) to fine (e.g. small grid-cells of a few km or sub-watersheds). The finer the spatial differentiation, the better a model captures variability of local conditions which may influence potential impacts by up to several orders of magnitude for some impact categories, such as toxicity or water consumption.
- Number of substances/land use types/resources: the more substances or land-use types/resources are covered by a method, the more likely it will consider all important (=highly contributing to impact) emissions or resource extractions of a product system. A missing characterisation factor for any given elementary flow automatically leads to its omission in the impact profile.
- Unit: the dimension of the indicator.
- "n/a" means that information was not available or that a property is not applicable.

Not all these properties may be of equal relevance for choosing an LCIA method for each practitioner or study, but are intended to represent the most relevant and fact-based properties.

Table 40.1 Detailed characteristics of available midpoint characterisation methods [extended and updated from ILCD handbook on LCIA (EC-JRC 2010, 2011; Sala et al. 2016)]

		CML-IA	TRACI 1.0	IMPACT 2002+	EDIP 2003	ReCiPe 2008	TRACI 2	ILCD/PEF/OEF	IMPACT World+
	Reference	Guinée et al. (2002)	Bare et al. (2003)	Jolliet et al. (2003)	Hauschild and Potting (2005); Potting and Hauschild (2005)Hauschild and Wenzel (1998)	Goedkoop et al. (2012)	Bare (2011)	EC-JRC (2011)	Bulle et al. (in review)
	Website	https://cml.leiden.edu/software/data-cmlia.html	https://epa.gov/chemical-research/tool-reduction-and-assessment-chemicals-and-other-environmental-impacts-traci	https://www.quantis-intl.com/pdf/IMPACT2002_UserGuide_for_vQ2.21.pdf		https://lcia-recipe.net	https://epa.gov/chemical-research/tool-reduction-and-assessment-chemicals-and-other-environmental-impacts-traci	https://eplca.jrc.ec.europa.eu/?page_id=86	https://impactworldplus.org
	Update of	CML 1992			EDIP 97	CML-IA	TRACI 1.0		IMPACT 2002+, EDIP 2003, and LUCAS
	Uncertainties quantified								Partially, spatial variability for each CF[1]
	Normalisation factors	Netherlands (1997), Europe (1995), World (1995), production-based) (Huijbregts et al. 2003)	USA and USA+Canada (production-based) reference year 2005 (Lautier et al. 2010)	Western Europe (production-based)	Europe (production-based), reference year 2004 (Laurent et al. 2011)	Europe, global (production-based), reference year 2000 (Sleeswijk et al. 2008)	USA and USA+Canada (production-based) reference year 2008 (Ryberg et al. 2014)	European, global (production-based)	No Normalization recommended at the midpoint level
	Weighting scheme	Panel-based (~80 replies), deriving cultural perspectives: Individualist, Hierarchist, and Egalitarian based on Hofstetter (1998)			For emissions based on political targets for 2000, expressed as Political Target Person-Equivalents (PET), for resource consumption based on person reserves (PR) for known reserves in 1990	Panel-based (~80 replies), deriving cultural perspectives: Individualist, Hierarchist, and Egalitarian based on Hofstetter (1998)			No weighting recommended at the midpoint level
Climate change	**Characterisation model**	GWP[2] from IPCC[3] (2007a)	GWP[2] from IPCC[3] (2001)	GWP[2] from IPCC[3] (2001)	GWP[2] from IPCC[3] (2001)	GWP[2] from IPCC[3] (2007a)	GWP[2] from IPCC[3] (2007a)	GWP[2] from IPCC[3] (2007a)	GWP[2] and GTP[4] from IPCC[3] 2013 (Myhre et al. 2013)
	Indicator	Radiative forcing	Radiative forcing	Radiative forcing	Radiative forcing	Radiative forcing	Radiative forcing	Radiative forcing	Radiative forcing
	Marginal/average	Average	Average	Average	Average	Average	Average	Average	Average
	Time horizon(s) [y]	100	100	500	100	100	100	100	GWP100 for short-term, GTP100 for long-term impacts
	Region modelled	Global	Global	Global	Global	Global	Global	Global	Global
	No. of substances	70	78	78	78	70	70	70	212
	Unit	CO_2 equivalents	CO_2 equivalents	CO_2 equivalents	CO_2 equivalents	CO_2 equivalents	CO_2 equivalents	CO_2 equivalents	CO_2 equivalents
Ozone depletion	**Characterisation model(s)**	ODP[5] from WMO[6] (2003)	ODP[5] from WMO[6] (1999)	ODP[5] from WMO[6] (2003)	ODP[5] from WMO[6] (2003)	ODP[5] from WMO[6] (2003)	ODP[5] from WMO[6] (2003)	ODP[5] from WMO[6] (2003)	ODP[5] from WMO[6] (2014)
	Indicator	Global degradation in stratospheric O_3 concentration	Global degradation in stratospheric O_3 concentration	Global degradation in stratospheric O_3 concentration	Global degradation in stratospheric O_3 concentration	Global degradation in stratospheric O_3 concentration	Global degradation in stratospheric O_3 concentration	Global degradation in stratospheric O_3 concentration	Global degradation in stratospheric O_3 concentration
	Marginal/average	Average	Average	Average	Average	Average	Average	Average	Average
	Time horizon(s) [y]	Infinite	Infinite	Infinite	Infinite	Infinite	Infinite	Infinite	Infinite
	Region modelled	Global	Global	Global	Global	Global	Global	Global	Global
	No. of substances	20	20	20	20	20	20	20	30
	Unit	[7]CFC-11 equivalents	[7]CFC-11 equivalents	[7]CFC-11 equivalents	[7]CFC-11 equivalents	[7]CFC-11 equivalents	[7]CFC-11 equivalents	[7]CFC-11 equivalents	[7]CFC-11 equivalents

(continued)

Table 40.1 (continued)

		CML-IA	TRACI 1.0	IMPACT 2002+	EDIP 2003	ReCiPe 2008	TRACI 2	ILCD/PEF/OEF	IMPACT World+
Particulate matter formation / Respiratory inorganics	Characterisation model	PM[10] emissions to air considered in human toxicity impact category	De Hollander et al. (1999)	Hofstetter (1998)		van Zelm et al. (2008)	Same as TRACI 1.0	[8]Humbert (2009)	Humbert et al. (2011) for fate and exposure and Gronlund et al. (2015) for effects
	Fate/exposure modeling		CALPUFF model, mechanistic, closed, multimedia LCA model	Empirical data		Mechanistic, option for low, undefined and high stack emission		[8]Humbert (2009) (USEtox (Rosenbaum et al. 2008) is used for primary PM and CO, Greco et al. (2007) for secondary PM from SO_2 and NOx, Van Zelm et al. (2008) for secondary PM from NH_3 and RiskPoll (Rabl and Spadaro 2004) to differentiate among high-stack, low-stack, and ground-level emissions of primary PM for urban and rural conditions, respectively)	Mechanistic (same as in ILCD/PEF/OEF but with updated parameters), Humbert et al. (2011)
	Effect modeling		endpoint-based indicator, dose-response (chronic mortality, acute mortality, acute respiratory morbidity, acute cardiovascular morbidity). 10.9 DALY per case mortality is used	Dose-response (chronic mortality, acute mortality, acute respiratory morbidity, acute cardiovascular morbidity)		Endpoint-based indicator, dose-response (chronic mortality, acute mortality, acute respiratory morbidity, acute cardiovascular morbidity); chronic bronchitis not considered		Epidemiological studies (dose-response (chronic mortality, acute mortality, acute respiratory morbidity, acute cardiovascular morbidity)	Epidemiological studies (dose-response (chronic mortality, acute mortality, acute respiratory morbidity, acute cardiovascular morbidity) Gronlund et al. (2015)
	Marginal/average		Marginal and Average give same results	Marginal and Average give same results		Marginal and Average give same results		Marginal and Average give same results	Marginal and Average give same results
	Emission compartment(s)		Air (point source (proxy for high stack) and mobile source (proxy for low stack)	Air		Air (rural, urban, and undefined)		Air (high stack, low stack, ground-level, undefined; remote, rura, urban and undefined environnement)	Air (high stack, low stack, ground-level, undefined; remote, rura, urban and undefined environnement)
	Time horizon, discounting		Infinite	Infinite		No timeframe, no discounting		No timeframe, 3% discounting for future cost	No timeframe, no discounting
	Region modelled		USA	Europe		Europe		Generic	Generic and continental
	Level of spatial differentiation					Urban and rural archetypes		High stack, low stack, ground-level, undefined; remote, rura, urban and undefined environnement	High stack, low stack, ground-level, undefined; remote, rural, urban and undefined environnement
	No. of substances		7 (TSP[9], PM10, PM2.5, NO, NO_2 NO_x/nitrate and SO_2/sulfate)[10]	6 (primary PM10 and PM 2.5, SO_2, NO_2, secondary PM10 from NO_2, SO_2)[10]		4 (primary PM10; secondary PM10 from NH_3, NO_x, SO_2)[10]		5 (primary PM2.5, primary PM10, secondary PM from NO_x and SO_2, and CO)[10]	5 (primary PM2.5, primary PM10, secondary PM from NO_x and SO_2, and CO)[10]
	Unit		PM10 equivalents[10]	PM2.5 equivalents[10]		PM10 equivalents[10]	PM2.5 equivalents[10]	PM2.5 equivalents[10]	PM2.5 equivalents[10]

(continued)

Table 40.1 (continued)

		CML-IA	TRACI 1.0	IMPACT 2002+	EDIP 2003	ReCiPe 2008	TRACI 2	ILCD/PEF/OEF	IMPACT World+
Photochemical ozone formation / Respiratory organics	Characterisation model	Derwent et al. (1998), Scenario reflecting realistic worst case for British Isles	Norris (2003)	POCP[11] from Jenkin et al. (1999)	Hauschild et al. (2006)	van Loon et al. (2007), Vautard et al. (2007), vanZelm et al. (2008)	Based on TRACI 1.0 but updated	Recommended: LOTOS-EUROS fate model as used in ReCiPe, resulting indicator: tropospheric ozone concentration increase (no effects)	van Loon et al. (2007), Vautard et al. (2007), vanZelm et al. (2008)
	Fate modeling	POCP[11], detailed fate modeling of individual VOCs[12]	Change in Maximum Incremental Reactivity (MIR) (Carter 2000), ASTRAP model for source-receptor matrix, factor for NO_x based on US average estimate	POCP[11] from Jenkin et al. (1999)	Regression model derived from RAINS	LOTOS-EUROS model 2007	Updated MIR (Maximum Incremental Reactivity) based on Carter (Bare 2011)		LOTOS-EUROS model 2007
	Exposure modeling			Population densities and average daily inhalation	Increased exposure of humans or vegetation above critical threshold, population density	Population densities within grid cells, atmospheric concentration of O_3 and average daily inhalation	Similar to TRACI 1.0		Population densities within grid cells, atmospheric concentration of O_3 and average daily inhalation
	Effect modeling		Human health only, linear dose-response model		Exceedance for exposure above critical level for humans and vegetation based on WHO[13] guidelines (AOT60 for human health, AOT40 for vegetation)	Human health only, linearity assumed with no threshold (based on WHO[13] recommendation)			Human health only, linearity assumed with no threshold (based on WHO[13] recommendation)
	Marginal/average	Marginal (ΔO_3 per marginal ΔVOC[12])	Marginal for fate, average for effect	Marginal and Average give same results	Marginal increase in long term ozone levels	Marginal (linearity checked up to 10% increase)			Marginal (linearity checked up to 10% increase)
	Emission compartment(s)	Air	Air	Air	Air	Air			Air
	Region modelled/valid	North Western Europe	USA	Europe	Europe	Europe			Europe
	Level of spatial differentiation		US states		European countries	Global generic			Global generic
	No. of substances	127 (VOCs[12], CH_4, SO_2, NO, NO_2 and CO)	~580	~130	4 (nmVOC[14], CH_4, No_x and CO)	2 (nmVOC[14] and NO_x)	~1200		134
	Unit	C_2H_4 equivalents	NO_x equivalents	ethylene equivalents	person*ppm*h and m^2*ppm*h	kg nmVOC[14] equivalents	NO_x equivalents		kg nmVOC[14] equivalents

(continued)

Table 40.1 (continued)

		CML-IA	TRACI 1.0	IMPACT 2002+	EDIP 2003	ReCiPe 2008	TRACI 2	ILCD/PEF/OEF	IMPACT World+
Ionizing radiation	Characterisation model	Frischknecht et al. (2000)		Same as CML-IA, but without effect model, indicator: human exposure level		Same as CML-IA, but without effect/damage model, indicator: human exposure level		Recommended: Frischknecht et al. (2000) for human health but excluding damage assessment (only effect) Interim use of method by Garnier-Laplace et al. (2008; 2009) for ecosystem impacts possible	Same as CML-IA, but without effect model, indicator: human exposure level Garnier-Laplace et al. (2008; 2009) for ecosystem impacts
	Fate modeling	Dreicer et al. (1995), using routine atmospheric and liquid discharges into rivers in French nuclear fuel cycle including surrounding conditions. For globally dispersed radionuclides simplified models are used.							
	Exposure modeling	Effective dose [Sv][15] via inhalation, ingestion, external irradiation, based on human body equivalence factors for α-, β-, γ-radiation, and neutrons							
	Effect modeling	Dose-response functions directly based on human subjects exposed in Nagasaki and Hiroshima (extrapolated to low-dose exposure)							
	Time horizon(s) [y]	100, 100000							
	Emission compartment(s)	Air, water							
	Region modelled/valid	Global, Europe (fate based on French conditions)							
	Level of spatial differentiation								
	No. of substances	31 (21 radionuclides to outdoor air, 13 to water, 15 to ocean)							26 for human health and 13 for ecosystem quality impacts
	Unit	DALY		C-14 Bq equivalents[16]		man.Sv/kBq [15, 16]			Bq C-14 equivalents[16]

(continued)

Table 40.1 (continued)

		CML-IA	TRACI 1.0	IMPACT 2002+	EDIP 2003	ReCiPe 2008	TRACI 2	ILCD/PEF/OEF	IMPACT World+
Human toxicity	Diseases considered	Cancer/non-cancer	Cancer/non-cancer	Cancer/non-cancer	Cancer/non-cancer	Cancer/non-cancer	Cancer/non-cancer	Cancer/non-cancer	Cancer/non-cancer
	Characterisation model	USES-LCA 1.0 (Huijbregts et al. 2000)	CalTOX 4.0 (Hertwich et al. 2001; McKone et al. 2001)	IMPACT 2002 (Pennington et al. 2005)	EDIP (Hauschild and Potting 2005)	USES-LCA 2.0 (van Zelm et al. 2009)	USEtox 1.0 (Rosenbaum et al. (2008; 2011))	USEtox 1.0 (Rosenbaum et al. (2008; 2011))	USEtox 2.0 (https://usetox.org) Rosenbaum et al. (2008;2011)
	Fate modeling	Mechanistic, nested, multimedia, mass-balance model (developed for LCA)	Mechanistic, closed, multimedia, mass-balance model (developed for ERA)	Mechanistic, nested, multimedia, mass-balance model (developed for LCA)	Key property, partial fate	Mechanistic, nested, multimedia, mass-balance model (developed for LCA)	Mechanistic, nested, multimedia, mass-balance model (developed for LCA)	Mechanistic, nested, multimedia, mass-balance model (developed for LCA)	Mechanistic, nested, multimedia, mass-balance model (developed for LCA)
	Exposure modeling	Inhalation, various ingestion pathways	Inhalation, various ingestion pathways, dermal uptake	Inhalation, various ingestion pathways	Inhalation, various ingestion pathways	Inhalation, various ingestion pathways	Inhalation, various ingestion pathways	Inhalation, various ingestion pathways	Inhalation including indoor exposure (Hellweg et al. 2009; Wenger et al. 2012; Rosenbaum et al. 2015), various ingestion pathways (Rosenbaum et al. 2011) including pesticide residues in food from Fantke et al. (2011a; 2011b; 2013)
	Effect modeling	[17]ED50	linear slope factor based on RfD[18]	[19]ED10	[19]ED10	[17]ED50	[17]ED50	[17]ED50	[17]ED50 (Rosenbaum et al. 2011)
	Marginal/average	Marginal (non-linear effect factor)	Marginal and Average give same results	Marginal and Average give same results	Marginal and Average give same results	Marginal (non-linear effect factor)	Marginal and Average give same results	Marginal and Average give same results	Marginal and Average give same results
	Time horizon	Infinite, 100 years for metals	Infinite	Infinite	Infinite	Infinite, 100 years for metals or all substances in Individualist perspective	Infinite	Infinite	Infinite
	Emission compartment(s)	Rural air, urban air, freshwater, agricultural soil, natural soil, industrial soil	Air, freshwater, soil	Air, freshwater, marine water, soil	Air, freshwater, soil	Rural air, urban air, freshwater, agricultural soil, natural soil, industrial soil	Rural air, urban air, water, agricultural soil, natural soil	Rural air, urban air, water, agricultural soil, natural soil	Rural air, urban air, water, agricultural soil, natural soil
	Region modelled	Europe, version available for various continents	USA	Europe	Europe	Europe	Generic	Generic	Generic global average + 9 parameterised sub-continents
	Level of spatial differentiation								sub-continental level (Kounina et al. 2014)
	No. of substances	~860	~380	~800	~180	~1000	~1250	~1250	~1250
	Unit	1,4-DCB equivalents[20]	2,4-D equivalents[21]	chloroethylene equivalents	m^3 (volume of poisoned compartment)	1,4-DCB equivalents[20]	cases or CTU_h[22]	cases or CTU_h[22]	cases or CTU_h[22]

(continued)

Table 40.1 (continued)

		CML-IA	TRACI 1.0	IMPACT 2002+	EDIP 2003	ReCiPe 2008	TRACI 2	ILCD/PEF/OEF	IMPACT World+
Ecotoxicity	Ecosystems considered	Freshwater, freshwater sediment, marine, marine sediment, terrestrial	Freshwater, terrestrial	Freshwater, marine, terrestrial	Freshwater, terrestrial	Freshwater, marine, terrestrial	Freshwater	Freshwater	Freshwater, interim factors for marine and terrestrial
	Characterisation model	USES-LCA 1.0 (Huijbregts et al. 2000)	CalTOX 4.0 (Hertwich et al. 2001; McKone et al. 2001; Bare et al. 2003)	IMPACT 2002 (Pennington et al. 2005)	EDIP1997, combined with site dependent factors (Tørsløv et al. 2005)	USES-LCA 2.0 (van Zelm et al. 2009)	USEtox 1.0 (Rosenbaum et al. 2008; Henderson et al. 2011)	USEtox 1.0 (Rosenbaum et al. 2008; Henderson et al. 2011)	USEtox 2.0 (https://usetox.org) (Rosenbaum et al. 2008; Henderson et al. 2011)
	Fate/exposure modeling	Mechanistic, nested, multimedia, mass-balance model (developed for LCA)	Mechanistic, closed, multimedia, mass-balance model (developed for ERA)	Mechanistic, nested, multimedia, mass-balance model (developed for LCA)	Key property, partial fate	Mechanistic, nested, multimedia, mass-balance model (developed for LCA)	Mechanistic, nested, multimedia, mass-balance model (developed for LCA)	Mechanistic, nested, multimedia, mass-balance model (developed for LCA)	Mechanistic, nested, multimedia, mass-balance model (developed for LCA)
	Effect modeling	Most sensitive species	Most sensitive species	Average toxicity	Most sensitive species	Average toxicity	Average toxicity	Average toxicity	Average toxicity
	Marginal/average	Marginal and Average give same results	Marginal and Average give same results	Marginal and Average give same results	Marginal and Average give same results	Marginal (non-linear effect factor)	Marginal and Average give same results	Marginal and Average give same results	Marginal and Average give same results
	Time horizon	Infinite	Infinite	Infinite	Infinite	Infinite, 100 years for metals	Infinite	Infinite	Infinite
	Emission compartment(s)	Air, freshwater, marine water, agricultural soil, industrial soil	Air, freshwater	Air, freshwater, soil	Air, freshwater, soil	Air, freshwater, marine water, agricultural soil, natural soil	Air, freshwater, marine water, agricultural soil, natural soil	Air, freshwater, marine water, agricultural soil, natural soil	Air, freshwater, marine water, agricultural soil, natural soil
	Region modelled	Europe	USA	Europe	Generic	Europe	Generic	Generic	Generic global average + 9 parameterised sub-continents
	Level of spatial differentiation								sub-continental level (Kounina et al. 2014)
	No. of substances	~170	~160	~430	~190	~2650	~2550	~2550	~2550
	Unit	1,4-DCB equivalents[20]	2,4-D equivalents[21]	triethylene-glycol equivalents	m^3 (volume of poisoned compartment)	1,4-DCB equivalents[20]	PAF[24] in [m^3*day] or CTU_e[24]	PAF[23] in [m^3*day] or CTU_e[24]	PAF[23] in [m^3*day] or CTU_e[24]

(continued)

Table 40.1 (continued)

		CML-IA	TRACI 1.0	IMPACT 2002+	EDIP 2003	ReCiPe 2008	TRACI 2	ILCD/PEF/OEF	IMPACT World+
Eutrophication	**Ecosystems considered**	Freshwater, terrestrial	Freshwater	Freshwater	Freshwater, terrestrial	Freshwater, marine		Freshwater, marine, terrestrial	Freshwater, marine
	Fate modeling	No mechanistic fate model, mineralisation with full release of bioavailable nutrients, Redfield ratio assumed for N/P ratio, no advection, no distinction between sensitive and insensitive recipients	Mechanistic (ASTRAP for atmospheric transport, WSAG topological network for water transport) except for NH_3 (NO_x matrices have been used as placeholder), no advection, no N or P removal, distinction between sensitive and insensitive recipients		Atmospheric fate and transport (RAINS model pre 2000 with detailed transport and fate model (applying EMEP)), no advection, removal of nutrients in hydrological cycle (using CARMEN model as fixed removal ratio for N- and P-compounds in different emission scenarios)	Mechanistic fate model EUTREND for atmospheric and CARMEN for waterborne emissions (Struijs et al. 2009), no advection leaving Europe, fixed removal ratio for N and P in different emission scenarios		Recommendation for freshwater and marine: EUTREND model (Struijs et al. 2009) as implemented in ReCiPe [kg P to freshwater and kg N to freshwater for freshwater and marine respectively] Recommendation for terrestrial: Accumulated Exceedance based on Seppälä et al. (2006), Posch et al. (2008), not included in any other LCIA method	Freshwater: fate transport based on Helmes et al. (2012) at the 0.5°x0.5° resolution scale Marine : GEOSchem model (from NASA http://map.nasa.gov/GEOSCHEM.html) for atmospheric fate and transport until coastal zones at the 2°x2.5° resolution scale
	Exposure modeling		Distinction of N- and P-limited recipients		No distinction between freshwaters and marine waters, no critical levels in water, exceedance of critical load in soil considered	Distinction between exposure of N- and P-limited systems		Distinction between exposure of N- and P-limited systems	Distinction between exposure of N- (marine) and P-limited (freshwater) systems
	Effect modeling			Same as CML-IA, but distinguishing P-limited watershed, N-limited watershed, and undefined watershed (modeled as a 50% P-limited and a 50% N-limited watershed)		Linear dose-response relationship according to limiting nutrient	Same as TRACI 1.0 but with additional substances covered		
	Marginal/average	Marginal	Average		Marginal	Marginal		Marginal	Marginal
	Emission compartment(s)	Air, freshwater, marine water, soil	Air, freshwater		Air, freshwater	Air, freshwater, soil		Air, freshwater, marine water	Air, freshwater, marine water
	Time horizon	Infinite	Infinite		Infinite	Infinite		Infinite	Infinite
	Region modelled/valid	Global generic	USA		Europe	Europe		Europe	Global
	Level of spatial differentiation		US states		European countries	Spatial differentiation not found important for aquatic eutrophication (factor 3-7 between European countries)		Generic, countries for NH_3 and NO_2	CFs[a] available at global, continental, country, and fine scale (0.5°x0.5° resolution scale for freshwater eutrophication 2°x2.5° for marine eutrophication)
	No. of substances	13 (8 N-compounds, 4 P compounds, COD[25])	11 (to air: NO, NO_2, NO_x, NH_3 to water: NH_4^+, N, NO_3^-, PO_4^{3-}, P, COD[25])		Freshw.: 12 (5 to air, 7 to water; 9 N-comp, 3 P-comp); Terrestrial: SO_2, SO_3, H_2SO_4, H_2S, NO, NO_2, NO_x, HNO, NH_3, HCL, HF	4 (N-total, P-total, NO_x and NH_3, but differentiation of N-total and P-total according to source)		Freshwater: P, H_3PO_4, P-total; marine: NH_3, NH_4^+, NO_3^-, NO_2^-, NO_2, NO, N-total; terrestrial: NH_3, NH_4^+, NO_2, NO, NO_3^- (to air)	8 for freshwater eutrophication, 16 for marine eutrophication
	Unit	PO_4^{3-} equivalents	PO_4^{3-} equivalents		Freshwater: NO_3^- equivalents; terrestrial: m^2 unprotected ecosystem	kg P to freshwater, kg N to freshwater (for marine eutrophication)		kg P to freshwater, kg N to freshwater (for marine eutrophication), mol N-equivalents for terrestrial	kg PO_4 P-lim equivalents for freshwater eutrophication, kg N N-lim equivalents for marine eutrophication

(continued)

Table 40.1 (continued)

		CML-IA	TRACI 1.0	IMPACT 2002+	EDIP 2003	ReCiPe 2008	TRACI 2	ILCD/PEF/OEF	IMPACT World+
Acidification	Ecosystems considered	Terrestrial	Terrestrial		Terrestrial	Terrestrial (forest soil)		Terrestrial	Terrestrial, freshwater, marine
	Characterisation model	Huijbregts et al.(2001)	Norris (2003)		Potting et al. (1998)	van Zelm et al. (2007), EUTREND, SMART 2		Seppälä et al. (2006) and Posch et al. (2008)	Roy et al. (2012a; 2012b; 2014)
	Fate modeling	Mechanistic atmospheric fate model (RAINS model dated before 2000), linear increase in sensitive area change according to emission scenario	Atmospheric fate and deposition on land (ASTRAP model dated before 2000), no soil sensitivity to acidifying deposition		Mechanistic atmospheric fate model, linear increase in sensitive area change, sensitive areas considered, incl. with limited buffer capacity, discounting for deposition in area above critical load	Mechanistic atmospheric fate model (EUTREND model) including deposition on land (SMART 2 model)		Mechanistic atmospheric fate model including deposition on land (EMEP model)	GEOSchem model (from NASA) for atmospheric fate and transport at the 2°x2.5° resolution scale for both terrestrial and freshwater acidification + receiving environment fate model for freshwater acidification
	Exposure modeling	Sensitive areas considered, incl. with limited buffer capacity, acidification potential modelled with slope inversly proportional to critical load and applied above and below it				Sensitive areas consider magnitude of deposition above critical load and areas with limited buffer capacity for forests (extrapolated to other ecosystems)		Linear increase (sensitive area change), sensitive areas, incl. with limited buffer capacity, acidification potential modelled as the exceedence above the critical load, potency	Terrestrial acidification: Soil sensitivity factor giving the change in soil solution H^+ concentration due to a change in the atmospheric deposits of pollutant using the PROFILE geochemical steady-state model
	Effect modeling	Dose-response slope based on a midpoint hazard risk ratio, similar to PEC / PNEC, and applied above and below critical load (validity still need to be verified as some doubts exist on the relevance of the dose-response curve, as the slope over buffer capacity depends on the buffer capacity itself		Same as CML-IA			Same as TRACI 1.0 but with additional substances covered		
	Marginal/average	Marginal	Average		Marginal	Marginal		Marginal	Marginal
	Emission compartments	Air, freshwater, soil	Air		Air	Air		Air	Air
	Time horizon(s) [y]	1990 and 2010 emission scenario	present		1990 and 2010 emission scenario	20, 50, 100 and 500		2002 and 2010 emission scenario	2010 emission scenario
	Region modelled	Europe	North America		Europe	Europe + country specific validity		Europe	Global
	Level of spatial differentiation	European countries	US counties		European countries			European countries	CFs[1] available at global, continental, country, and fine scale (2° x 2.5° resolution scale)
	No. of substances	4 (NH_3, NO_x (NO_2 and NO), SO_2)	8 (H_2S, SO_x, NO, NO_2, NO_x, HCL, HF, NH_3)		11 (SO_2, SO_3, H_2SO_4, H_2S, NO, NO_2, NO_x, HNO, NH_3, HCL, HF)	4 (NO_x (NO, NO_x), NH_3, SO_2)		4 (SO_2, NO_x (NO, NO_2), NH_3)	15
	Unit	kg SO_2 equivalents	H^+ equivalents		kg SO_2 equivalents	BS[26] [m^2 * y]		mol H^+ equivalents	kg SO_2 equivalents

(continued)

Table 40.1 (continued)

		CML-IA	TRACI 1.0	IMPACT 2002+	EDIP 2003	ReCiPe 2008	TRACI 2	ILCD/PEF/OEF	IMPACT World+
Land use	Aspects considered			Based on Eco-Indicator 99 endpoints: obtained by dividing damage factors from Eco-Indicator 99 by damage factor of organic arable land from Eco-Indicator 99		Land surface used		Erosion resistance, mechanical filtration, groundwater replenishment, biotic production; for PEF/OEF recommended to provide net blue water consumption as additional environmental information	Biodiversity, Ecosystem services: erosion resistance capacity/potential (ERP), freshwater recharge potential (FWRP), mechanical filtration potential (MWFP), chemical filtration potential (CWFP), biotic production potential (BPP)
	Characterisation model					None, hence no distinction8 of different species composition between land use types		LANCA 2.0 (Bos et al. 2016)	Biodiversity: de Baan et al. (2013a; 2013b) Ecosystem services: updated models from Saad et al. (2011; 2013). and Brandão & Milà i Canals (2013)
	Marginal/average					Average		Not described	Not described
	Region modelled					Global generic		Global	Global
	Level of spatial differentiation							Country, world default and local (site-specific)	16 WWF biomes for biodiversity; 36 Holdridge lifezones for ERP, MWPP, PCWPP, FWRP; 12 IPCC climate zones for BPP
	No. of land use types					3 (2 occupation (competition in agricultural area, competition in urban area) and 1 transformation)		Up tp 58, although many highly correlated	8 for biodiversity; 36 for ERP, MWPP, PCWPP, FWRP; 8/26 for BPP
	Unit			m^2 equivalents of organic arable land * y		m^2 occupation or transformation		Erosion resistance in kg/m² year, mechanical filtration in m³/m² year, groundwater replenishment in m³/m² year, biotic production in kg/m² year	Biodiversity in ha equivalents of arable land * y, ERP in ton/ha/y, FWRP in mm/y, MWFP in cm/d, CWFP in cmol/kg_{soil}, BPP in tC/ha/y

(continued)

Table 40.1 (continued)

		CML-IA	TRACI 1.0	IMPACT 2002+	EDIP 2003	ReCiPe 2008	TRACI 2	ILCD/PEF/OEF	IMPACT World+
Resources	**Aspects considered**	Abiotic: amount of non-renewable resources (fossil fuels and minerals) extracted			Abiotic: amount of non-renewable resources (fossil fuels and minerals) extracted Biotic: wood extraction	Abiotic: non-renewable resources (fossil fuels and minerals), based on change in availability of high grade minerals and fossil resources, mining of bulk res. covered by land use, water only as inventory parameter Biotic: use of agricultural, silvicultural biotic resources covered by land use		1) Abiotic resource depletion; 2) Cumulative Energy Demand (CED); as additional information in PEF: 3) Abiotic and biotic resource intensity (i.e. mass accounting of biotic and abiotic resources; 4) Criticality focusing on economic and social aspects	Fossil fuels and mineral resources, approaches based on dissipation of resource functionality (instead of depletion of stocks) assuming that extraction does not contribute to functionality loss and therefore only dissipation of a resource has an impact
	Characterisation model	Guinée and Heijungs (1995) based on extraction rates and using ultimate reserves, average rock is used	Same as Eco-Indicator 99 endpoint for fossil use only	Based on Eco-Indicator 99 endpoints: obtained by dividing damage factors from Eco-Indicator 99 by damage factor of iron in ore from Eco-Indicator 99	Hauschild and Wenzel (1998): amount of resource extracted is compared to 1990 extraction levels in person equivalents (fraction of resource which can be exploited economically is used (can be much smaller than the ultimate reserves))	Kirkham and Rafer (2003)	Same as TRACI 1.0	Abiotic resource depletion: Anthropogenic stock extended Abiotic Depletion Potential (AADP) model (Schneider et al. 2015); CED as in Ecoinvent (Hischier et al. 2010); Criticality based on Supply risk_JRC (Mancini et al. 2016)	Fossil fuels: primary energy content Minerals: Material competition scarcity index from de Bruille (2014)
	Time horizion	Infinite			For renewables annual regeneration used to determine supply horizon	Minerals: infinite Fossil fuels: before and after 2030		Infinite	Infinite
	Region modelled	Global			Global	Global		Global	Global
	No. of resource types	82 elements			36	20 (minerals) + 34 (fossil fuels)		n/a	n/a
	Unit	Abiotic Depletion Potential (ADP) [dimensionless]		kg equivalents of iron in ore	Person reserve - quantity of resource available per person (according to economically exploitable reserve)	Marginal increase of extraction costs		AADP in t Sb-eq/t; CED in MJ; resource intensity in kg; criticality dimensionless	Fossil fuels: MJ deprived, Minerals: kg deprived
Water use	**Aspects considered**							Water deprivation	Water deprivation
	Characterisation model(s)							AWARE model representing relative Available WAter REmaining per area in a watershed after demand of humans and aquatic ecosystems is met; it assesses potential of water deprivation to humans or ecosystems under assumption that the less water remaining available per area, the more likely another user will be deprived (Boulay et al. 2017).	AWARE model representing relative Available WAter REmaining per area in a watershed after demand of humans and aquatic ecosystems is met; it assesses potential of water deprivation to humans or ecosystems under assumption that the less water remaining available per area, the more likely another user will be deprived (Boulay et al. 2017).
	Marginal/average							Not described	Not described
	Region modelled							Global	Global
	Level of spatial differentiation							Country, watershed, global average consumption-weighted; for agricultural, non-agricultural and unknown water use	Country, watershed, global average consumption-weighted; for agricultural, non-agricultural and unknown water use
	Unit							m^3 world eq. deprived	m^3 world eq. m^3 deprived

(continued)

Table 40.1 (continued)

[1]Characterisation Factor
[2]Global Warming Potential
[3]Intergovernmental Panel on Climate Change
[4]Global Temperature Potential
[5]Ozone Depletion Potential
[6]World Meteorological Organisation
[7]Chlorofluorocarbon
[8]The information given in ILCD handbook and related documents is incorrect. The correct models are given in Humbert (2009)—the ILCD recommended approach—and are reflected here as well
[9]Total Suspended Particulate matter
[10]PM—Particulate Matter (with diameters up to 2.5 and 10 μm respectively)
[11]Photochemical Ozone Creation Potential
[12]Volatile Organic Compounds
[13]World Health Organisation
[14]Non-Methane Volatile Organic Compounds
[15]Sievert, unit of ionizing radiation dose
[16]Becquerel, unit of radioactivity (1 Bq = 1 disintegration per second)
[17]Effective Dose affecting 50% of tested individuals
[18]Reference Dose (US-EPA's acceptable daily oral exposure to the human population likely to be without risk of deleterious effects during a lifetime)
[19]Effective Dose affecting 10% of tested individuals
[20]Dichlorobenzene
[21]Dichlorophenoxyacetic acid
[22]Comparative Toxic Unit for humans
[23]Potentially Affected Fraction of species (not an actual unit but a fraction)
[24]Comparative Toxic Unit for ecosystems
[25]Chemical Oxygen Demand
[26]Base Saturation

Table 40.2 Detailed characteristics of available endpoint characterisation methods [extended and updated from ILCD handbook on LCIA (EC-JRC 2010, 2011; Sala et al. 2016)]

	EPS 2000	Eco-Indicator99	IMPACT 2002+	LIME 1.0 (2003)	LIME 2.0 (2008)[1]	ReCiPe 2008	ILCD/PEF/OEF	IMPACT World+	LC-Impact
Reference	Steen (1999)	Goedkoop and Spriensma (2000)	Jolliet et al. (2003)	Itsubo and Inaba (2003)	Itsubo and Inaba (2012)	Goedkoop et al. (2012)	EC-JRC (2011)	Bulle et al. (in review)	Verones et al. (in preparation), Verones et al. (2016a)
Website	https://cpmdatabase.cpm.chalmers.se/AboutDatabase_2.htm	https://pre-sustainability.com/eco-indicator-99-manuals	https://www.quantis-intl.com/pdf/IMPACT2002_UserGuide_for_vQ2.21.pdf		https://lca-forum.org/english	https://lcia-recipe.net	https://eplca.jrc.ec.europa.eu/?page_id=86	https://impactworldplus.org	https://lc-impact.eu
Update of					LIME 1.0	Eco-Indicator99		IMPACT 2002+	in part: ReCiPe 2008
Uncertainties quantified					full uncertainty information based on Monte Carlo Analysis			Partially, spatial variability for each CF[2]	Qualitative, if possible quantitative
Normalisation factors	Integrated in CFs[2]	Western Europe (production-based)	Western Europe (production-based)	Integrated in CFs[2]	Integrated in CFs[2]	Europe, global (production-based)	European, global (production-based)	Global	Global
Weighting scheme	Integrated in CFs[2]	Panel-based cultural perspectives: Individualist, Hierarchist, and Egalitarian		Willingness to pay [Yen] (Conjoint analysis of interview survey of ~400 Japanese people (Kanto region) on opinions about environmental policy)	Same as LIME 1.0, but for ~1000 people from all over Japan and with improved statistical analysis	Cultural perspectives: Individualist, Hierarchist, and Egalitarian		No recommended weighing factors, but STEPWISE factors (Weidema et al. 2006) are compatible and thus optional	Additive impacts, including amongst others different time horizons (CF^2 core and CF^2 extended)
Areas of protection	Human health [YOLL][3], biodiversity [NEX][4], abiotic resources [kg], ecological productivity [kg]	Human health [DALY][5], ecosystem quality [PDF in km^2*y][6], resources [MJ surplus energy]	Human health [DALY][5], ecosystem quality [PDF in m^2*y][6], resources [MJ surplus energy], climate change [CO_2 eq]	Human health [DALY][5], social welfare [Yen], biodiversity [EINES][7], primary production [kg-DW][8]	Human health [DALY][5], social assets [Yen], biodiversity [EINES][7], primary production [kg-DW][8]	Human health [DALY][5], Ecosystems [species*y], resources [surplus cost]		Human health [DALY][5], Ecosystems [PDF in m^2*y][6], Resources and ecosystem services [$]	Human Health [DALY][5], Ecosystems incl. vulnerability of species thus represent global extinctions [PDF][6], Resources [Surplus Ore Potential].

(continued)

Table 40.2 (continued)

		EPS 2000	Eco-Indicator99	IMPACT 2002+	LIME 1.0 (2003)	LIME 2.0 (2008)[1]	ReCiPe 2008	ILCD/PEF/OEF	IMPACT World+	LC-Impact
Climate change	**Characterisation model(s)**	(IPCC 1990)	GWP[9] from IPCC[10] (1995)	only midpoint CF	GWP[9] from IPPC[10] (2001)	GWP[9] from IPPC[10] (2007b)	GWP[9] from IPCC[10] (2007b), de Schryver et al. (2009)	No recommendation, but interim use of ReCiPe 2008 model	aGTP[11] from IPCC[10] 2013 (Myhre et al. 2013), de Schryver et al. (2009) with corrected values for "semi natural areas"	GWP[9] from IPCC[10] 2013 (Myhre et al. 2013), de Schryver et al. (2009; 2011), Hanafiah et al. (2011)
	Human health effects	Thermal stress, flooding, malaria, starvation (no diarhea)	Thermal stress, vector borne diseases, flooding (no starvation)		Heat and cold stress, vector borne diseases, malnutrition, natural disasters	Heat and cold stress, malaria, dengue, disaster damage, malnutrition, hunger	Cardiovascular diseases, malaria, diarrhoea and malnutrition, flooding		Cardiovascular diseases, malaria, diarrhoea and malnutrition, flooding	Cardiovascular diseases, malaria, diarrhoea and malnutrition, flooding
	Ecosystem effects						Species loss based on global assessment of ecosystem studies in different regions and different species groups		Species loss based on global assessment of ecosystem studies in different regions and different species groups	Species loss incl. all species for terr. (De Schryver et al. 2009) and fish as proxy for aqu. (Hanafiah et al. 2011)
	Biotic resources effects	Decrease in crop production, increase wood production and species loss			Links to crop production and fish production	Agricultural production, energy consumption, land disappearance			Ecosystem service loss (interim)	
	Marginal/average	Average	Marginal		Average	Average	Marginal		Marginal	Marginal
	Time horizon(s) [y]	100	200		100	100	100		100, [100-500]	100, 1000
	Region modelled	Global	Global		Global	Global	Global		Global	Global
	Level of spatial differentiation									
	No. of substances	106	38		n/a	n/a	70		211	207
	Unit	Willingness to pay (WTP)	DALY[5]		DALY[5] and Yen (crop loss)	DALY[5] and Yen	DALY[5] and PDF[6] in $[m^2*y]$		DALY[5], PDF[6] in $[m^2*y]$ and $	DALY[5], PDF[6] in [yr/kg]
Ozone depletion	**Characterisation model(s)**	ODP[12] from WMO[13] (1999)	ODP[12] from WMO[13] (1999) and data from TOMS satelite	ODP[12] from WMO[13] (2003) and US-EPA	ODP[12] from WMO[13] (1999)	Same as LIME 1.0	ODP[12] from WMO[13] (2003)	No recommendation, but interim use of ReCiPe 2008 model	ODP[12] from WMO[13] (2014)	ODP[12] from WMO[13] (2011), De Schryver et al. (2011)
	Human health effects	Skin cancer	Skin cancer and cataract	Same as Eco-Indicator 99 (Egalitarian scenario), but with updated ODP	Cataract and skin cancer (malignant melanoma, basal cell carcinoma, squamous cell carcinoma)		Skin cancer and cataract (latter not for individualist perspect.) based on the AMOUR 2.0 model (van Dijk et al. 2008; Den Outer et al. 2008)		Skin cancer and cataract based on the AMOUR 2.0 model (van Dijk et al. 2008; Den Outer et al. 2008)	Skin cancer, cataract
	Biotic resources effects				Net primary productivity for coniferous forests, agriculture (soybean,rice,greenpea, mustard) and phytoplancton at high latitudes					
	Marginal/average	Average	Marginal		Marginal		Marginal		Average	Average
	Time horizon(s) [y]	Infinite	500		Infinite		Infinite		Integrated from 2007 to 2100	100, infinite
	Region modelled	Global	Global		Global		Global		Global	Global
	Level of spatial differentiation									
	No. of substances	20	25	95	19		22		30	22
	Unit	DALY[5] as monetary value via willingness to pay	DALY[5]	DALY[5]	DALY[5], NPP[14] and kg Wood		DALY[5]		DALY[5]	DALY[5]

(continued)

Table 40.2 (continued)

		EPS 2000	Eco-Indicator99	IMPACT 2002+	LIME 1.0 (2003)	LIME 2.0 (2008)[1]	ReCiPe 2008	ILCD/PEF/OEF	IMPACT World+	LC-Impact
Particulate matter formation / Respiratory inorganics	Characterisation model		Hofstetter (1998)	Same as Eco-Indicator 99	Hofstetter (1998), Ikeda (2001)	Hofstetter (1998), Ikeda (2001), Pope et al. (2002)	van Zelm et al. (2008)	Recommendation: DALY[5] based on Humbert (2009)	Humbert et al. (2011) for fate and exposure and Gronlund et al. (2015) for effects	van Zelm et al. (2016)
	Fate/exposure modeling		Mechanistic		Mechanistic	Mechanistic atmospheric fate (plume model and puff model) distinguishing chimneys and vehicles source types	Mechanistic, distinguishing low, undefined and high stack emission		Mechanistic (same as in ILCD/PEF/OEF but with updated parameters)	TM5-FASST, a global chemistry transport model
	Effect modeling		Dose-response (chronic mortality, acute mortality, acute respiratory morbidity, acute cardiovascular morbidity, etc.)		Dose-response (chronic mortality, acute mortality, acute respiratory morbidity, acute cardiovascular morbidity, etc.)	Similar to LIME 1.0 but disease and death rates were increased for some substances/endpoints according to newer data from Pope et al. (2002)	Dose-response (chronic mortality, acute mortality, acute respiratory morbidity, acute cardiovascular morbidity) ; chronic bronchitis not considered		Dose-response, mass-based, surface-based, and number-based	Updated region-specific effect and damage data
	Marginal/average		Marginal and Average give same results		Marginal and Average give same results	Marginal and Average give same results	Marginal and Average give same results		Marginal and Average give same results	Linear
	Emission compartment		Air		Air	Air (from chimney, from automobile)	Air (for low, undefined and high stack emission)		Air (high stack, low stack, ground-level, undefined; remote, rura, urban and undefined environnement)	Air
	Time horizon, discounting		No timeframe, no discounting		Infinite	Same as LIME 1.0	No timeframe, no discounting		No timeframe, no discounting	No timeframe, no discounting
	Region modelled		Generic continent		Average atmospheric conditions for 7 regions of Japan		Europe		Generic and continental	World
	Level of spatial differentiation				7 Japanese regions				high stack, low stack, ground-level, undefined; remote, rura, urban and undefined environnement	56 regions
	No. of substances		4 (primary PM10; secondary PM10 from NH_3, NO_x, SO_2; no values for PM2.5)[15]		6 (primary PM10 and PM 2.5, SO_2, NO_2, secondary PM10 from NO_2, SO_2)[15]		4 (primary PM10; secondary PM10 from NH_3, NO_x, SO_2)[15]		5 (primary PM2.5, primary PM10, secondary PM from NO_x and SO_2, and CO)[15]	PM2.5 (from PM2.5, NH_3, NO_x and SO_2)
	Unit		DALY[5]		DALY[5]		DALY[5]		DALY[5]	DALY[5]

(continued)

Table 40.2 (continued)

		EPS 2000	Eco-Indicator99	IMPACT 2002+	LIME 1.0 (2003)	LIME 2.0 (2008)[1]	ReCiPe 2008	ILCD/PEF/OEF	IMPACT World+	LC-Impact
Photochemical ozone formation / respiratory organics	**Human health effects**	Morbidity and severe morbidity	Acute mortality	Same as Eco-Indicator 99 (Egalitarian scenario)	Acute mortality, morbidity, no consideration of NO_x	Same as LIME 1.0	Only acute mortality, chronic effects disregarded due to lack of empirical evidence	Recommendation: ReCiPe 2008 model	Constant damage factor for acute mortality, chronic effects disregarded due to lack of empirical evidence	van Zelm et al. (2016)
	Ecosystem effects				All relevant mechanisms, no consideration of NOx					Following Van Goethem et al. (2013b) but for the World
	Characterisation model	POCP[16] (Lindfors et al. 1994)	Hofstetter (1998)		Ozone conversion equivalency factors (OCEF), Schere & Demerjian, (1984), corrected by Uno and Wakamatsu (1992)		van Loon et al. (2007), Vautard et al. (2007), van Zelm et al. (2008)		van Loon et al. (2007), Vautard et al. (2007), van Zelm et al. (2008)	Van Zelm et al. (2008) Van Goethem et al. (2013a; 2013b), van Zelm et al. (2016)
	Fate modeling	Based on old Swedish version of POCP[16] (Lindfors et al. 1994)	POCP[16] from Jenkin et al. (1999)		Average atmospheric conditions for 7 regions of Japan		LOTOS-EUROS model 2007		LOTOS-EUROS model 2007	TM5-FASST, a global chemistry transport model
	Exposure modeling		Population densities and average daily inhalation		Population distribution for human health, integrated in dose-response model for other AOPs[17]		Population densities within grid cells, atmospheric concentration of O_3 and average daily inhalation		Population densities within grid cells, atmospheric concentration of O_3 and average daily inhalation	Updated region-specific effect and damage data Van Goethem et al. (2013a; 2013b)
	Effect modeling	Based on global average situation	Linear exposure-response functions		Linear exposure-response functions, also considering crops and natural vegetation, wood		Linearity assumed, no threshold and only including acute effects		Human health only, linearity assumed with no threshold (based on WHO[18] recommendation)	
	Marginal/average	Average	Marginal and Average give same results		Marginal (ΔO_3 per marginal ΔVOC[19] for 8 VOC[19] archetypes)		Marginal (linearity checked up to 10% increase)		Marginal (linearity checked up to 10% increase)	Linear
	Emission compartment	Air	Air		Air		Air		Air	Air
	Region modelled/valid	Global damage data, European fate data (POCP[16])	Europe		Japan		Europe		Europe	World
	Level of spatial differentiation									56 regions
	No. of substances	67 (nmVOCs[20] and No_x)	130		8 archetypes from which individual VOCs[19] can be estimated		2 (nmVOC[20] and NO_x) but factors for individual VOCs[19] can be determined applying POCPs[16], e.g. from CML-IA		134	Ozone (from NO_x and NMVOC)
	Unit	DALY[5] converted to monetary value using willingness to pay	DALY[5]		DALY[5]		DALY[5]		DALY[5]	Human health [DALY][5] and Ecosystems [PDF][6]

(continued)

Table 40.2 (continued)

		EPS 2000	Eco-Indicator99	IMPACT 2002+	LIME 1.0 (2003)	LIME 2.0 (2008)[1]	ReCiPe 2008	ILCD/PEF/OEF	IMPACT World+	LC-Impact
Ionizing radiation	Characterisation model		Frischknecht et al. (2000)							
	Fate modeling		Dreicer et al. (1995), using routine atmospheric and liquid discharges into rivers in French nuclear fuel cycle including surrounding conditions. For globally dispersed radionuclides simplified models are used.							
	Exposure modeling		Effective dose [Sv][21] via inhalation, ingestion, external irradiation, based on human body equivalence factors for α-, β-, γ-radiation, and neutrons	Same as Eco-Indicator 99 (Egalitarian scenario)			Same as Eco-Indicator 99	No recommendation, but interim use of Frischknecht et al. (2000) as used in Eco-Indicator 99, IMPACT 2002+, ReCiPe	Same as CML-IA, but without effect model, indicator: human exposure level Garnier-Laplace et al. (2008; 2009) for ecosystem impacts	Same as CML-IA, complemented with De Schryver et al. (2011)
	Effect modeling		Dose-response functions directly based on human subjects exposed in Nagasaki and Hiroshima (extrapolated to low-dose exposure)							
	Time horizon(s) [y]		100, 100000							
	Emission compartment		Air, water							
	Region modelled/valid		Global, Europe (fate based on French conditions)							
	No. of substances		31 (21 radionuclides to outdoor air, 13 to water, 15 to ocean)						26 for human health and 13 for ecosystem quality impacts	63
	Unit		DALY[5]						DALY[5], PDF[6] in [m^2*y]	DALY[5]

(continued)

Table 40.2 (continued)

		EPS 2000	Eco-Indicator99	IMPACT 2002+	LIME 1.0 (2003)	LIME 2.0 (2008)[1]	ReCiPe 2008	ILCD/PEF/OEF	IMPACT World+	LC-Impact
Human toxicity	**Diseases considered**	Cancer/non-cancer	Cancer	Cancer/non-cancer	Cancer, oral chronic non-cancer diseases	Cancer, oral and inhalatory chronic non-cancer diseases, sick house syndrome via indoor exposure to formaldehyde, NO_x, SO_2, PM	Cancer/non-cancer	Recommendation for cancer effects: USEtox model (Rosenbaum et al. 2008) combined with DALYs[5] from Huijbregts et al. (2005) No recommendation for non-cancer effects, but interim use of USEtox model (Rosenbaum et al. 2008) combined with severity factors from Huijbregts et al. (2005)	Cancer/non-cancer	Same as IMPACT World+
	Characterisation model	None	EUSES 1.0 (EC and RIVM 1996)	IMPACT 2002 (Pennington et al. 2005)	Modified version of IMPACT 2002 model	Modified version of IMPACT 2002 model	USES-LCA 2.0 (van Zelm et al. 2009)		USEtox 2.0 (www.usetox.org)	
	Fate modeling	None	Mechanistic, mass-balance model (developed for risk assessment)	Mechanistic, mass-balance model (developed for LCA)			Mechanistic, mass-balance model (developed for LCA)		Mechanistic, mass-balance model (developed for LCA) (Rosenbaum et al. 2007; Rosenbaum et al. 2008; Henderson et al. 2011; Fantke et al. 2013)	
	Exposure modeling	some pathways missing	inhalation, various ingestion pathways	inhalation, various ingestion pathways			inhalation, various ingestion pathways		Inhalation including indoor exposure (Hellweg et al. 2009; Wenger et al. 2012; Rosenbaum et al. 2015), various ingestion pathways (Rosenbaum et al. 2011) including pesticide residues from Fantke et al. (2011a; 2011b; 2013)	
	Effect modeling	RfD[22], safety factors	Unit-Risk concept	ED10[23]	Unit-risk concept for cancer ED10[23] for oral chronic diseases	Improved models and data, including inhalation exposure to heavy metals based on epidemiological data, sick house syndrome effect modeling via indoor inhalation volume-response curve based on clinic survey	ED50[24]		ED50[24] (Rosenbaum et al. 2011) combined with severity factors from Huijbregts et al. (2005)	
	Marginal/average	Average, Sweden, Europe or USA taken as reference	Marginal and Average give same results	Marginal and Average give same results	Similar to IMPACT 2002+	Similar to IMPACT 2002+	Marginal (non-linear effect factor)		Marginal and Average give same results	
	Time horizon	n/a	Infinite	Infinite			Infinite, 100 years for metals or all substances in Individualist perspective		Infinite, ≤100 y, >100 y for metals	
	Emission compartment	n/a	Air, water, soil	Air, water, soil			Rural air, urban air, water, agricultural soil, natural soil		rural air, urban air, indoor air, agricultural soil, natural soil, freshwater, marine water	
	Region modelled	Global average, calculated based on Swedish data	Europe	Europe	Japan	Japan	Europe		global average, 16 sub-continental zones, 8 continental zones	
	Level of spatial differentiation								sub-continental level (Kounina et al. 2014)	
	No. of substances	~57	~10	~800	~800	~1000	~1000		~1250	
	Unit	DALY[5] converted to monetary value	DALY[5]	DALY[5]	DALY[5]	DALY[5]	DALY[5]		DALY[5]	

(continued)

Table 40.2 (continued)

		EPS 2000	Eco-Indicator99	IMPACT 2002+	LIME 1.0 (2003)	LIME 2.0 (2008)[1]	ReCiPe 2008	ILCD/PEF/OEF	IMPACT World+	LC-Impact
Ecotoxicity	Ecosystems considered	n/a	Freshwater, terrestrial	Freshwater, marine, terrestrial	Freshwater, terrestrial	Freshwater, terrestrial	Freshwater, marine, terrestrial		Freshwater, marine, terrestrial	
	Characterisation model	None (red list species pot. threatened by chemicals)	EUSES 1.0 (EC and RIVM 1996)	IMPACT 2002 (Pennington et al. 2005)			USES-LCA 2.0 (van Zelm et al. 2009)		USEtox 2.0 (www.usetox.org)	
	Fate/exposure modeling	None	Mechanistic, mass-balance model (developed for risk assessment)	Mechanistic, mass-balance model (developed for LCA)			Mechanistic, mass-balance model (developed for LCA)		Mechanistic, mass-balance model (developed for LCA) (Rosenbaum et al. 2007; Rosenbaum et al. 2008; Henderson et al. 2011)	
	Effect modeling	None	Most sensitive species	Average toxicity	Modified version of IMPACT 2002 model	Modified version of IMPACT 2002 model	Average toxicity		Average toxicity (Henderson et al. 2011) with generic severity factor (PAF-PDF) of 0.5 (Jolliet et al. 2003)	
	Marginal/average	Marginal and Average give same results	Marginal and Average give same results	Marginal and Average give same results			Marginal (non-linear effect factor)	No recommendation	Marginal and Average give same results	Same as IMPACT World+, but with severity factor (PAF-PDF) of 1 and including vulnerability factors for taking species' vulnerability into account (leading to global extinctions)
	Time horizon	n/a	Infinite	Infinite			Infinite, 100 years for metals		Infinite, ≤100 y and >100 y for metals	
	Emission compartment	n/a	Air, water, soil	Air, water, soil			Air, water, agricultural soil, natural soil		Air, freshwater, marine water, agricultural soil, natural soil	
	Region modelled	Global average	Europe	Europe	Japan	Japan	Europe		global average, 16 sub-continental zones, 8 continental zones	
	Level of spatial differentiation								sub-continental level (Kounina et al. 2014)	
	No. of substances	~45	~45	~430	n/a	n/a	~2650		~3100	
	Unit	NEX[4]	PDF[6] in [m^2*y]	PDF[6] in [m^2*y]	EINES[25]	EINES[25]	PDF[6] in [y/t]		PDF[6] in [m^2*y]	

(continued)

Table 40.2 (continued)

		EPS 2000	Eco-Indicator99	IMPACT 2002+	LIME 1.0 (2003)	LIME 2.0 (2008)[1]	ReCiPe 2008	ILCD/PEF/OEF	IMPACT World+	LC-Impact
Eutrophication	Ecosystems considered	Freshwater, terrestrial	Terrestrial (incl. wetlands and swamps)	Freshwater	Marine (only benthic communities in coastal water)		Freshwater		Freshwater, marine	Freshwater, marine
	Fate modeling	No mechanistic fate model, empirical data used, only distinction between deposition on land and water for airborne emissions	No atmospheric fate model, assuming 60% of emissions deposited on European natural soil (no advection leaving Europe)	No mechanistic fate model, Redfield ratio assumed for N/P	No leaching of N from land, no N or P removal processes, estuarine circulation modelled		Mechanistic fate model EUTREND for atmospheric and CARMEN for waterborne emissions (Struijs et al. 2009), fixed removal ratio for N and P in different emission scenarios		Freshwater: fate transport based on Helmes et al. (2012) at the 0.5°x0.5° resolution scale Marine: GEOSchem model (from NASA http://map.nasa.gov/GEOS_CHEM.html) for atmospheric fate and transport until coastal zones at the 2°x2.5° resolution scale	Freshwater: fate transport based on Helmes et al. (2012) Marine: Cosme and Hauschild (2017)
	Exposure modeling	Effects assumed for 90% of N emissions exposing N-limited ecosystems and 10% of P emissions reaching P-limited ecosystems	Change in soil nutrients for a marginal deposition increase based on Dutch conditions	Mineralisation with full release of nutrients in biologically available form assumed, freshwater ecosystems P-limited	Empirical data for four Japanese bays		Distinction between exposure of N- and P-limited systems		Distinction between exposure of N- and P-limited systems	Only P-limited systems for freshwater, only N-limited systems for marine
	Effect modeling	No dose-response	Linear dose-response relationship, endpoint as impacts on higher plants in the Netherlands	Linear dose-response relationships for different classes of European P-limited freshwater ecosystems	Linear dose-response relationships for N and P based on empirical observations on disappearance of benthic species due to oxygen depletion	Same as LIME 1.0	Linear dose-response relationship according to limiting nutrient	No recommendation, but interim use of ReCiPe 2008 model for freshwater eutrophication	Freshwater = P limited Generic effect factor from Tirado-Seco (2005) Marine = N limited Generic, empirical effect factor	Freshwater: Azevedo et al. (2013a) Marine: Cosme and Hauschild (2016)
	Marginal/average	Average	Marginal	Average	Marginal		Marginal		Marginal	Both linear/average
	Emission compartment	Air, water	Air	Air, water, soil	Air, water		Air, water, soil		Air, freshwater, ocean	FW: soil, water Marine: air, surface water, groundwater, ocean
	Time horizon	Infinite	Infinite	Infinite	Infinite		Infinite		Infinite	Not relevant
	Region modelled/valid	Global average (extrapolated from Swedish data)	Europe + country specific validity/ Dutch validity for midpoint-endpoint modeling	Europe	Japan		Europe		Global	Global
	Level of spatial differentiation								CFs[7] available at global, continental, country, and fine scale (0.5°x0.5° resolution scale for freshwater eutrophication 2°x2.5° for marine eutrophication	Freshwater: 0.5x0.5 arc degrees, countries, continents, global average Marine: 215 country to LME values, countries, regions, world average
	No. of substances	5 (1 to air, 4 to water)	4 (NH_3, NO, NO_2, No_x)	4 P compounds and COD[26] from CML	5 (2 to air, 3 to water)		4 (N-total, P-total, NO_x and NH_3)		8 for freshwater eutrophication, 16 for marine eutrophication	Total P for freshwater NH_3, NO_x to air and N to surface water, ground water and ocean for marine
	Unit	NEX[4]	PDF[6] in [m^2*y]	PDF[6] in [m^2*y]	NPP[14](Loss of Benthic biomass)		PDF[6] in [y/t]		PDF[6] in [m^2*y]	PDF[6] for freshwater and marine

(continued)

Table 40.2 (continued)

		EPS 2000	Eco-Indicator99	IMPACT 2002+	LIME 1.0 (2003)	LIME 2.0 (2008)[1]	ReCiPe 2008	ILCD/PEF/OEF	IMPACT World+	LC-Impact
Acidification	Ecosystems considered		Terrestrial (incl. wetlands and swamps)		Terrestrial	Terrestrial	Terrestrial		Terrestrial, freshwater, marine	Terrestrial
	Characterisation model		SMART, MOVE		Hayashi et al. (2004)	Hayashi et al. (2004) and others	van Zelm et al. (2007), EUTREND, SMART 2		Roy et al. (2012a; 2012b; 2014), Azevedo et al. (2015)	Same as IMPACT World+ complemented by Azevedo et al. (2013b)
	Fate modeling		No atmospheric fate model, assuming 60% of emissions deposited on European natural soil (no advection leaving Europe)		Atmospheric fate model (mixing mechanistic and empirical approaches - Euler-type model for SO_2 and No_x, empirical values for rest of chemicals with emission and deposition data between 1991-1997) including deposition on land, domestic average	Similar to LIME 1.0 but dividing Japan into 6 zones, including geographical fluctuation, calculating source attribution by zone based on the ratio of terrestrial area and marine area	Mechanistic atmospheric fate model (EUTREND model) including deposition on land (SMART 2 model)		GEOSchem model (NASA http://map.nasa.gov/GEOS_CHEM.html) for atmospheric fate and transport at the 2°x2.5° resolution scale for both terrestrial and freshwater acidification + receiving environment fate model for freshwater acidification	Same as IMPACT World+
	Exposure modeling		Change in soil pH for a marginal deposition increase based on Dutch conditions		Sensitive areas consider magnitude of deposition above critical load and areas with limited buffer capacity for representative soils in Japan	Naturally produced SO_2 excluded from emissions used for calculation of source attribution of SO_2	Sensitive areas consider magnitude of deposition above critical load and areas with limited buffer capacity for forests (extrapolated to other ecosystems)		Terrestrial: soil sensitivity factor giving the change in soil solution H^+ conc. due to a change in the atmos. deposits of pollutant using the PROFILE geochemical steady-state model (Roy et al. 2012a)	
	Effect modeling		Linear dose-response relationship, endpoint as impacts on higher plants in the Netherlands	Same as Eco-Indicator 99 (Egalitarian scenario)	Change in NPP of forestry plant species per change H+ deposition: [NPP] / [H^+_dep], linear function between H+ deposition and NPP, latter calculated as a function of soil acidification and Al_3^+ concentration	Similar to LIME 1.0 but updated modeling and added biodiversity of terrestrial (plant) ecosystems	Change in Potentially Not Occurring Fraction of forestry plant species per change in Base Saturation: [PNOF] / [BS], linear function for BS > 0.15 based on 240 vascular plants species	No recommendation, but interim use of ReCiPe 2008 model	Terrestrial: site specific effect factor based on biome regr. models from Azevedo et al. (2013b) Freshwater: location-specific effect factor linking change in H^+ concentration in a lake to change in potentially disappeared fraction (Roy et al. 2014) Marine: dose response curve for pH effect on different calcifying organisms using available data to assess EC50[27] (Azevedo et al. 2015)	Terrestrial: same as IMPACT World+
	Marginal/average		Marginal		Marginal		Marginal		Marginal	
	Emission compartment		Air		Air		Air		Air	
	Time horizon(s) [y]		Infinite		Infinite	Same as LIME 1.0	20, 50, 100 and 500		Infinite	Same as IMPACT World+
	Region modelled		Europe + country specific validity/ Dutch validity for midpoint-endpoint modeling		Japan		Europe + country specific validity/ Dutch validity for midpoint-endpoint modeling		Global	
	Level of spatial differentiation								CFs[2] available at global, continental, country, and fine scale (2° x 2.5° resolution scale)	
	No. of substances		7 (SO_2, SO_3, SO_x, NH_3, NO, NO_2, No_x)		5 (SO_2, NO, NO_2, HCL, NH_3)	Same as LIME 1.0	4 (No_x (NO, NO_x), NH_3, SO_2)		15	3 (SO_x, NO_x, NH_3)
	Unit		PDF[6] in [m^2*y]		NPP[14]	NPP[14] and Yen	PDF[6] in [m^2*y]	species*y	PDF[6] in [m^2*y]	PDF[6]

(continued)

Table 40.2 (continued)

		EPS 2000	Eco-Indicator99	IMPACT 2002+	LIME 1.0 (2003)	LIME 2.0 (2008)[1]	ReCiPe 2008	ILCD/PEF/OEF	IMPACT World+	LC-Impact
Land use	**Aspects considered**	biodiversity (only red list species) and primary wood productivity of forest	Biodiversity		Biodiversity and net primary production of an area		Biodiversity		Biodiversity, Ecosystem service loss: erosion resistance capacity/potential (ERP), freshwater recharge potential (FWRP), mechanical filtration potential (MWFP), chemical filtration potential (CWFP), biotic production potential (BPP)	Biodiversity global extinction of species, vulnerability of species communities considered taxa-specific
	Characterisation model	(Jarvinen and Miettinen 1987)	Koellner (2001), Species richness (based on vascular plants), species-area relationship		NPP based on Chikugo model. Biodiversity based on red species list and life expectancy (study by Matsuda et al. (2003))	Similar to LIME 1.0 but with updated data covering 30 types of vegetation	Koellner (2001), Species richness (based on vascular plants), species-area relationship		Biodiversity: de Baan et al. (2013a; 2013b) Ecosystem services: updated models from Saad et al. (2011; 2013), and Brandão & Milà i Canals (2013)	Chaudhary et al. (2015)
	Marginal/average	Not described	Marginal		Marginal		Marginal		Not described	Marginal and average
	Region modelled	Global, but based only on Swedish data	Mid Europe, based on Swiss data	Same as Eco-Indicator 99 (Egalitarian scenario)	Japan, for NPP[14], overseas areas also considered		North West-Europe, based on British and Swiss data	No recommendation due to models' immaturity, but interim use of Chaudhary et al. (2015) model	Global	Global
	Level of spatial differentiation								16 WWF biomes for biodiversity; 36 Holdridge lifezones for ERP, MWPP, PCWPP, FWRP; 12 IPCC climate zones for BPP	804 terrestrial ecoregions
	No. of land use types	3 (arable land, forest for forest, and forest for roads)	16 land occupations + 11 land conversions		n/a	9	18 land occupations (including 3 intensities for arable areas) + 4 land conversions		8 for biodiversity; 36 for ERP, MWPP, PCWPP, FWRP; 8/26 for BPP	6 (each for occupation and transformation)
	Unit	For biodiversity: NEX[4]; for primary production: 1kg dry wood	PDF[6] in [m^2*y]		EINES[7] and NPP[14]	EINES[7] and NPP[14]	PDF[6] in [m^2*y]		Biodiversity affecting Ecosystem quality endpoint in PDF[6] in [m^2*y], Ecosystem services (FWRP, MWFP, CWFP, BPP) affecting all three endpoints given in DALY[5], PDF[6] in [m^2*y], $, Ecosystem service ERP affecting Resources and ecosystem services in $	species-eq or PDF[6] (aggregated across taxa)

(continued)

Table 40.2 (continued)

		EPS 2000	Eco-Indicator99	IMPACT 2002+	LIME 1.0 (2003)	LIME 2.0 (2008)[1]	ReCiPe 2008	ILCD/PEF/OEF	IMPACT World+	LC-Impact
Resources	**Aspects considered**	Abiotic: non-renewable resources (fossil fuels and minerals) based on resource in average rock, water extraction (irrigation, drinking) Biotic: wood extraction, fish and meat extraction	Abiotic: non-renewable resources (fossil fuels and minerals), based on (long term) change in availability of high grade minerals and fossil resources, mining of bulk resources covered by land use Biotic: use of agricultural, silvicultural biotic resources covered by land use	Abiotic: non-renewable resources: minerals as modelled in Eco-Indicator 99 and fossil fuels modelled in a specific way (no stock size included), mining of bulk resources covered by land use Biotic: use of agricultural, silvicultural biotic resources covered by land use	Abiotic: minerals, fossil resources, fossil fuels Biotic: living resources (terrestrial ecosystem)	Same as LIME 1.0 but with improved models and data	Abiotic: non-renewable resources (fossil fuels and minerals), based on change in availability of high grade minerals and fossil resources, mining of bulk resources covered by land use, water only as inventory parameter Biotic: use of agricultural, silvicultural biotic resources covered by land use	No recommendation due to models' immaturity, but interim use of Surplus Cost Potential SCP model (Vieira et al. 2016) of mining and milling activities including all future metal extractions and the operating mining costs accounting for co-production; assuming that mines with lower operating costs are explored first; costs assumed as measure of depletion; unit of SCP is USD2013/kg	Endpoint: resources and ecosystem services affected due to functional deprivation of fossil fuels and mineral resources, approaches based on dissipation of resource functionality (instead of depletion of stocks) assuming that extraction does not contribute to functionality loss and therefore only dissipation of a resource has an impact	Mineral resources
	Characterisation model	Numerous sources from 1990-2000, not all input data is traceable	(de Vries 1988) (Müller-Wenk 1998)	For minerals see Eco-Indicator 99; for fossils: Ecoinvent database as of 2003	n/a		Kirkham and Rafel (2003)		Functionality specific effect factor for fossil fuels from Fatemi (2012) and for minerals from de Bruille (2014)	Vieira et al. (2012; 2017)
	Time horizon	Long time frame, weighting/Normalization based on future technologies	5x the historical extraction before 1990	Minerals: 5x the historical extraction before 1990. Fossil fuels: infinite	n/a		Minerals: infinite Fossil fuels: before and after 2030		n/a	n/a
	Region modelled	Global	Global	Global	n/a		Global		Global	Global
	No. of resource types	67 (minerals) + 3 (fossils) + 3 (biotic) + 1 (water)	12 (minerals) + 9 (fossil fuels)	22 (minerals) + 9 (fossil fuels)	n/a		20 (minerals) + 34 (fossil fuels) + 5 (water use)		n/a	18 minerals
	Unit	kg of element or resource used	Surplus energy [MJ] needed to extract one kg extra element or one MJ fuel.	Surplus Energy and primary energy [MJ]	Yen, EINES[7], and NPP[14]		Marginal increase of extraction costs		$	Surplus Ore Potential

(continued)

Table 40.2 (continued)

		EPS 2000	Eco-Indicator99	IMPACT 2002+	LIME 1.0 (2003)	LIME 2.0 (2008)[1]	ReCiPe 2008	ILCD/PEF/OEF	IMPACT World+	LC-Impact
Water use	Aspects considered								Human health, ecosystem quality, resources and ecosystem services	Human health , ecosystem quality
	Characterisation model(s)								Boulay et al. (2011), van Zelm et al. (2010), Hanafiah et al. (2011), Verones et al. (2010), Humbert and Maendly (2008)	Pfister et al. (2009; 2014), Verones et al. (2016b)
	Ecosystem effects								Terrestrial: van Zelm et al. (2010), regionalised using likelihood of shallow vs. deep groundwater extr. Freshwater: Hanafiah et al. (2011) as global average value, model from Verones et al. (2010) for thermal pollution, and model from Humbert and Maendly (2008) for water stream use	Terrestrial: Plants and aquatic: Verones et al. (2016b)
	Human health effects								Human health based on Boulay et al. (2011), representing water stress, i.e. level of competition among users (anthropogenic and ecosystems) due to physical stress of the resource, addressing seasonal variations, and distinguishing between surface and groundwater	Pfister et al. (2009; 2014)
	Region modelled								Global	Global
	Level of spatial differentiation								CFs[2] available at global, continental, country, and for human health also at fine scale (watershed resolution scale)	CFs[2] for human health: 11050 watersheds, countries, global average CFs[2] for ecosystems: 0.05x0.05 arc degrees, coutries, global average
	Unit								DALY[5], PDF[6] in [m^2*y], $	DALY[5], PDF[6]
Rare or emerging impact categories	Noise					From transport applying fate (increase in noise levels due to additional cars/km - distinguishing day/night and small/large cars), exposure (distribution of exposed population with background levels), effect (volume-response curve based on social survey), and damage due to sleep disorder and conversation disorder				
	Waste					User cost, biodiversity of terrestrial ecosystems				

(continued)

Table 40.2 (continued)

[1]Version 3.0 of LIME, which is currently not documented in english but already available, includes, among other, a water use characterization model, and focuses on global coverage for many impact categories
[2]Characterization Factor
[3]Years of Life Lost (actual unit is [year])
[4]Normalized Extinction of species
[5]Disability Adjusted Life Years (actual unit is [year])
[6]Potentially Disappeared Fraction of species (not an actual unit but a fraction of 1)
[7]Expected Increase in Number of Extinct species
[8]Dry Weight
[9]Global Warming Potential (a measure of infrared radiative forcing in [W year/m^2] or in CO_2-eq if normalised to CO_2)
[10]Intergovernmental Panel on Climate Change
[11]Absolute Global Temperature Potential
[12]Ozone Depletion Potential
[13]World Meteorological Organisation
[14]Net Primary Productivity
[15]PM—Particulate Matter (with diameters up to 2.5 and 10 μm respectively)
[16]Photochemical Ozone Creation Potential
[17]Area of Protection
[18]World Health Organisation
[19]Volatile Organic Compounds
[20]Non-Methane Volatile Organic Compounds
[21]Sievert, unit of ionizing radiation dose
[22]Reference Dose (US-EPA's acceptable daily oral exposure to the human population likely to be without risk of deleterious effects during a lifetime)
[23]Effective Dose affecting 10% of tested individuals
[24]Effective Dose affecting 50% of tested individuals
[25]Expected increase in Number of Extinct Species
[26]Chemical Oxygen Demand
[27]Effective Concentration affecting 50% of individuals

Acknowledgements The author is very grateful to Cécile Bulle (CIRAIG/UQAM) and Francesca Verones (NTNU) who provided very helpful and valuable inputs to this chapter.

References

Azevedo, L.B., Henderson, A.D., van Zelm, R., Jolliet, O., Huijbregts, M.A.J.: Assessing the importance of spatial variability versus model choices in life cycle impact assessment: the case of freshwater eutrophication in Europe. Environ. Sci. Technol. **47**, 13565–13570 (2013a). doi:10.1021/es403422a

Azevedo, L.B., van Zelm, R., Hendriks, A.J., Bobbink, R., Huijbregts, M.A.J.: Global assessment of the effects of terrestrial acidification on plant species richness. Environ. Pollut. **174**, 10–15 (2013b). doi:10.1016/j.envpol.2012.11.001

Azevedo, L.B., De Schryver, A.M., Hendriks, A.J., Huijbregts, M.A.J.: Calcifying species sensitivity distributions for ocean acidification. Environ. Sci. Technol. **49**, 1495–1500 (2015). doi:10.1021/es505485m

Bare, J.: TRACI 2.0: the tool for the reduction and assessment of chemical and other environmental impacts 2.0. Clean Technol. Environ. Policy **13**, 687–696 (2011). doi:10.1007/s10098-010-0338-9

Bare, J.C., Norris, G.A., Pennington, D.W., McKone, T.: TRACI: the tool for the reduction and assessment of chemical and other environmental impacts. J. Ind. Ecol. **6**, 49 (2003)

Bos, U., Horn, R., Beck, T., Lindner, J., Fischer, M.: LANCA®—Characterization Factors for Life Cycle Impact Assessment, Version 2.0. Fraunhofer Verlag, Stuttgart, Germany (2016)

Boulay, A.-M., Bare, J., Benini, L., Berger, M., Lathuilliere, M., Manzardo, A., Margni, M., Motoshita, M., Núñez, M., Pastor, A.V., Ridoutt, B.G., Oki, T., Worbe, S., Pfister, S.: The WULCA consensus characterization model for water scarcity footprints: assessing impacts of water consumption based on available water remaining (AWARE). Int. J. Life Cycle Assess. In Press (2017). doi:10.1007/s11367-017-1333-8

Boulay, A.-M., Bulle, C., Bayart, J.-B., Deschenes, L., Margni, M.: Regional characterization of freshwater use in LCA: modeling direct impacts on human health. Environ. Sci. Technol. **45**, 8948–8957 (2011)

Brandão, M., i Canals, L.M.: Global characterisation factors to assess land use impacts on biotic production. Int. J. Life Cycle Assess. **18**, 1243–1252 (2013). doi:10.1007/s11367-012-0381-3

Bulle, C., Margni, M., Kashef-Haghighi, S., Boulay, A.-M., Bourgault, G., De Bruille, V., Cao, V., Fantke, P., Hauschild, M.Z., Henderson, A., Humbert, S., Kounina, A., Laurent, A., Levasseur, A., Liard, G., Patouillard, L., Rosenbaum, R.K., Roy, P.-O., Shaked, S., Jolliet, O.: IMPACT World+: A Globally Regionalized Life Cycle Impact Assessment Method. In Review (2017)

Carter, W.: Updated Maximum Incremental Reactivity Scale for Regulatory Applications. Sacramento, CA (2000)

Chaudhary, A., Verones, F., de Baan, L., Hellweg, S.: Quantifying land use impacts on biodiversity: combining species-area models and vulnerability indicators. Environ. Sci. Technol. **49**, 9987–9995 (2015). doi:10.1021/acs.est.5b02507

Cosme, N., Hauschild, M.Z.: Effect factors for marine eutrophication in LCIA based on species sensitivity to hypoxia. Ecol. Indic. **69**, 453–462 (2016). doi:10.1016/j.ecolind.2016.04.006

Cosme, N., Hauschild, M.Z.: Characterization of waterborne nitrogen emissions for marine eutrophication modelling in life cycle impact assessment at the damage level and global scale. Int. J. Life Cycle Assess. (2017). doi:10.1007/s11367-017-1271-5

de Baan, L., Alkemade, R., Koellner, T.: Land use impacts on biodiversity in LCA: a global approach. Int. J. Life Cycle Assess. **18**, 1216–1230 (2013a). doi:10.1007/s11367-012-0412-0 de Baan, L., Mutel, C.L., Curran, M., Hellweg, S., Koellner, T.: Land use in life cycle assessment: global characterization factors based on regional and global potential species extinction. Environ. Sci. Technol. **47**, 9281–9290 (2013b). doi:10.1021/es400592q

De Bruille, V.: Impact de l'utilisation des ressources minérales et métalliques dans un contexte cycle de vie : une approche fonctionnelle. Ecole Polytechnique Montreal (2014)

De Hollander, A.E.M., Melse, J.M., Lebret, E., Kramers, P.G.N.: An aggregate public health indicator to represent the impact of multiple environmental exposures. Epidemiology **10**, 606–617 (1999)

De Schryver, A.M., Brakkee, K.W., Goedkoop, M.J., Huijbregts, M.A.J.: Characterization factors for global warming in life cycle assessment based on damages to humans and ecosystems. Environ. Sci. Technol. **43**, 1689–1695 (2009)

De Schryver, A.M., van Zelm, R., Humbert, S., Pfister, S., McKone, T.E., Huijbregts, M.A.J.: Value choices in life cycle impact assessment of stressors causing human health damage. J. Ind. Ecol. **15**, 796–815 (2011). doi:10.1111/j.1530-9290.2011.00371.x

de Vries, D.: Sustainable Resource Use: Optimal Depletion within a Geostatistical Framework. Groningen, The Netherlands (1988)

Den Outer, P.N., van Dijk, A., Slaper, H.: Validation of Ultraviolet Radiation Budgets Using Satellite Observations from the OMI Instrument. Bilthoven, The Netherlands (2008)

Derwent, R.G., Jenkin, M.E., Saunders, S.M., Pilling, M.J.: Photochemical ozone creation potentials for organic compounds in Northwest Europe calculated with a master chemical mechanism. Atmos. Environ. **32**, 2429–2441 (1998)

Dreicer, M., Tort, V., Manen, P.: ExternE, Externalities of Energy. European Commission DGXII, Science, Research and development JOULE, Luxembourg (1995)

EC, RIVM: EUSES—The European Union System for the Evaluation of Substances. Available from the European Chemicals Bureau, Ispra, Italy (1996)

EC-JRC: International Reference Life Cycle Data System (ILCD) Handbook—Recommendations for Life Cycle Impact Assessment in the European context, 1st edn. European Commission, Joint Research Centre, Institute for Environment and Sustainability, Ispra, Italy (2011)

EC-JRC: International Reference Life Cycle Data System (ILCD) Handbook: Analysis of existing Environmental Impact Assessment methodologies for use in Life Cycle Assessment. European Commission, Joint Research Centre, Institute for Environment and Sustainability, Ispra, Italy (2010)

Fantke, P., Charles, R., De Alencastro, L.F., Friedrich, R., Jolliet, O.: Plant uptake of pesticides and human health: dynamic modeling of residues in wheat and ingestion intake. Chemosphere **85**, 1639–1647 (2011a). doi:10.1016/j.chemosphere.2011.08.030

Fantke, P., Juraske, R., Antón, A., Friedrich, R., Jolliet, O.: Dynamic multicrop model to characterize impacts of pesticides in food. Environ. Sci. Technol. **45**, 8842–8849 (2011b). doi:10.1021/es201989d

Fantke, P., Wieland, P., Wannaz, C., Friedrich, R., Jolliet, O.: Dynamics of pesticide uptake into plants: From system functioning to parsimonious modeling. Environ. Model Softw. **40**, 316–324 (2013). doi:10.1021/es303525x

Fatemi, F.: A novel methodology for the assessment of the direct and indirect impacts associated with the depletion of resources in life cycle assessment. Ecole Polytechnique Montreal (2012)

Frischknecht, R., Steiner, R., Jungbluth, N.: The ecological scarcity method—eco-factors 2006: a method for impact assessment in LCA (2009)

Frischknecht, R., Braunschweig, A., Hofstetter, P., Suter, P.: Modelling human health effects of radioactive releases in life cycle impact assessment. Environ. Impact Assess. Rev. **20**, 159–189 (2000)

Garnier-Laplace, J.C., Beaugelin-Seiller, K., Gilbin, R., Della-Vedova, C., Jolliet, O., Payet, J.: A screening level ecological risk assessment and ranking method for liquid radioactive and chemical mixtures released by nuclear facilities under normal operating conditions. In: Proceedings of International Conference Radioecology Environmental Protection 15–20 June, Bergen (2008)

Garnier-Laplace, J.C., Beaugelin-Seiller, K., Gilbin, R., Della-Vedova, C., Jolliet, O., Payet, J.: A screening level ecological risk assessment and ranking method for liquid radioactive and chemical mixtures released by nuclear facilities under normal operating conditions. Radioprotection **44**, 903–908 (2009). doi:10.1051/radiopro/20095161

Goedkoop, M., Heijungs, R., Huijbregts, M.A.J., De Schryver, A., Struijs, J., van Zelm, R., Ministry of Housing SP and E (VROM): ReCiPe 2008: a life cycle impact assessment method which comprises harmonised category indicators at the midpoint and the endpoint level, 1st edn. Ministry of Housing, Spatial Planning and Environment (VROM) (2012)

Goedkoop, M., Spriensma, R.: The Eco-Indicator 99: A Damage Oriented Method for Life Cycle Assesment, Methodology Report. Pré Consultants, Amersfoort (2000)

Greco, S.L., Wilson, A.M., Spengler, J.D., Levy, J.I.: Spatial patterns of mobile source particulate matter emissions-to-exposure relationships across the United States. Atmos. Environ. **41**, 1011–1025 (2007)

Gronlund, C., Humbert, S., Shaked, S., O'Neill, M., Jolliet, O.: Characterizing the Burden of disease of particulate matter for life cycle impact assessment. Air Qual. Atmos. Health **8**, 29–46 (2015). doi:10.1007/s11869-014-0283-6

Guinée, J.B., Heijungs, R.: A proposal for the definition of resource equivalency factors for use in product life-cycle assessment. Environ. Toxicol. Chem. **14**, 917–925 (1995). doi:10.1002/etc.5620140525

Guinée, J.B., Gorrée, M., Heijungs, R., Huppes, G., Kleijn, R., van Oers, L., Wegener Sleeswijk, A., Suh, S., Udo de Haes, H.A., de Bruijn, H., van Duin, R., Huijbregts, M.A.J.: Handbook on Life Cycle Assessment: Operational Guide to the ISO Standards. Kluwer, Dordrecht (2002)

Hanafiah, M.M., Xenopoulos, M.A., Pfister, S., Leuven, R.S.E.W., Huijbregts, M.A.J.: Characterization factors for water consumption and greenhouse gas emissions based on freshwater fish species extinction. Environ. Sci. Technol. **45**, 5272–5278 (2011)

Hauschild, M.Z., Potting, J.: Spatial Differentiation in Life Cycle Impact Assessment: the EDIP2003 Methodology. The Danish Ministry of the Environment, Environmental Protection Agency, Copenhagen (2005)

Hauschild, M., Wenzel, H.: Environmental Assessment of Products, Volume 2: Scientific Background. Kluwer, Hingham (1998)

Hauschild, M.Z., Potting, J., Hertel, O., Schöpp, W., Bastrup-Birk, A.: Spatial differentiation in the characterisation of photochemical ozone formation: the EDIP2003 methodology. Int. J. Life Cycle Assess. **11**, 72–80 (2006)

Hauschild, M., Goedkoop, M., Guinée, J., Heijungs, R., Huijbregts, M., Jolliet, O., Margni, M., Schryver, A., Humbert, S., Laurent, A., Sala, S., Pant, R.: Identifying best existing practice for characterization modeling in life cycle impact assessment. Int. J. Life Cycle Assess. **18**, 683–697 (2013). doi:10.1007/s11367-012-0489-5

Hayashi, K., Okazaki, M., Itsubo, N., Inaba, A.: Development of damage function of acidification for terrestrial ecosystems based on the effect of aluminum toxicity on net primary production. Int. J. Life Cycle Assess. **9**, 13–22 (2004)

Hellweg, S., Demou, E., Bruzzi, R., Meijer, A., Rosenbaum, R.K., Huijbregts, M.A.J., McKone, T.E.: Integrating indoor air pollutant exposure within life cycle impact assessment. Environ. Sci. Technol. **43**, 1670–1679 (2009). doi:10.1021/es8018176

Helmes, R.K., Huijbregts, M.J., Henderson, A., Jolliet, O.: Spatially explicit fate factors of phosphorous emissions to freshwater at the global scale. Int. J. Life Cycle Assess. **17**, 646–654 (2012). doi:10.1007/s11367-012-0382-2

Henderson, A., Hauschild, M.Z., Van de Meent, D., Huijbregts, M.A.J., Larsen, H.F., Margni, M., McKone, T.E., Payet, J., Rosenbaum, R.K., Jolliet, O.: USEtox fate and ecotoxicity factors for comparative assessment of toxic emissions in life cycle analysis: sensitivity to key chemical properties. Int. J. Life Cycle Assess. **16**, 701–709 (2011). doi:10.1007/s11367-011-0294-6

Hertwich, E., Matales, S.F., Pease, W.S., McKone, T.E.: Human toxicity potentials for life-cycle assessment and toxics release inventory risk screening. Environ. Technol. Chem. **20**, 928–939 (2001)

Hischier, R., Weidema, B., Althaus, H.-J., Bauer, C., Doka, G., Dones, R., Frischknecht, R., Hellweg, S., Humbert, S., Jungbluth, N., Köllner, T., Loerincik, Y., Margni, M., Nemecek, T.: Implementation of Life Cycle Impact Assessment Models. Ecoinvent report No. 3, v2.2. Dübendorf, Switzerland (2010)

Hofstetter, P.: Perspectives in Life Cycle Impact Assessment: A Structure Approach to Combine Models of the Technosphere. Kluwer, Dordrecht (1998)

Huijbregts, M.A.J., Thissen, U., Guinée, J.B., Jager, T., Kalf, D., van de Meent, D., Ragas, A.M.J., Wegener Sleeswijk, A., Reijnders, L.: Priority assessment of toxic substances in life cycle assessment. Part I: calculation of Toxicity potentials for 181 substances with the nested multi-media fate, exposure and effects model USES-LCA. Chemosphere **41**, 541–573 (2000)

Huijbregts, M.A.J., Verkuijlen, S.W.E., Heijungs, R., Reijnders, L.: Spatially explicit characterization of acidifying and eutrophying air pollution in life-cycle assessment. J. Ind. Ecol. **4**, 75–92 (2001)

Huijbregts, M.A.J., Breedveld, L., Huppes, G., de Koning, A., van Oers, L., Suh, S.: Normalisation figures for environmental life-cycle assessment: the Netherlands (1997/1998), Western Europe (1995) and the world (1990 and 1995). J. Clean. Prod. **11**, 737–748 (2003). doi:10.1016/S0959-6526(02)00132-4

Huijbregts, M.A.J., Rombouts, L.J.A., Ragas, A.M.J., Van de Meent, D.: Human-toxicological effect and damage factors of carcinogenic and noncarcinogenic chemicals for life cycle impact assessment. Integr. Environ. Assess Manag. **1**, 181–192 (2005)

Humbert, S.: Geographically Differentiated Life-Cycle Impact Assessment of Human Health. University of California, Berkeley (2009)

Humbert, S., Maendly, R.: Characterization factors for damage to aquatic biodiversity caused by water use especially from dams used for hydropower. In: SETAC North America 29th Annual Meeting Tampa, Florida, USA (2008)

Humbert, S., Marshall, J.D., Shaked, S., Spadaro, J.V., Nishioka, Y., Preiss, P., McKone, T.E., Horvath, A., Jolliet, O.: Intake fractions for particulate matter: recommendations for life cycle assessment. Environ. Sci. Technol. **45**, 4808–4816 (2011)

Ikeda, Y.: Establishment of Comprehensive Measures for Control of the Amount of Air Pollutant Emissions in the Whole East Asia, Report on Research Results for FY1997 to FY 2000 Scientific Research Subsidies (2001)

IPCC: Climate Change: The IPCC Scientific Assessment. Cambridge University Press, Cambridge (1990)

IPCC: Climate Change 1995: The Science of Climate Change. Cambridge University Press, Camridge (1995)

IPCC: Climate Change 2007—The Physical Science Basis. Intergovernmental Panel on Climate Change (2007a)

IPCC: Climate Change 2007—Synthesis Report. Intergovernmental Panel on Climate Change (2007b)

IPCC: Third Assessment Report. The Scientific Basis (2001)

Itsubo, N., Inaba, A.: LIME 2: Life-Cycle Impact Assessment Method Based on Endpoint Modeling. Life Cycle Assessment Society of Japan (2012)

Itsubo, N., Inaba, A.: A new LCA method: LIME has been completed. Int. J. Life Cycle Assess. **8**, 305 (2003)

Jarvinen, O., Miettinen, K.: “Sista paret ut” Naturskyddsforeningen (1987)

Jenkin, M., Hayman, G.: Photochemical ozone creation potentials for oxygenated volatile organic compounds: sensitivity to variations in kinetic and mechanistic parameters. Atmos. Environ. **33**, 1275–1293 (1999)

Jolliet, O., Margni, M., Charles, R., Humbert, S., Payet, J., Rebitzer, G., Rosenbaum, R.K.: IMPACT 2002+: a new life cycle impact assessment methodology. Int. J. Life Cycle Assess. **8**, 324–330 (2003). doi:10.1007/BF02978505

Kemna, R., Van Elburg, M., Li, W., Van Holsteijn, R.: MEEUP: Methodology Report. EC, Brussels (2005)

Kirkham, R.V., Rafer, A.B.: Selected World Mineral Deposits Database (2003)

Köllner, T: Land Use in Product Life Cycles and Its Consequences for Ecosystem Quality. University St. Gallen (2001)
Kounina, A., Margni, M., Shaked, S., Bulle, C., Jolliet, O.: Spatial analysis of toxic emissions in LCA: a sub-continental nested USEtox model with freshwater archetypes. Environ. Int. **69**, 67–89 (2014). doi:10.1016/j.envint.2014.04.004
Laurent, A., Olsen, S., Hauschild, M.: Normalization in EDIP97 and EDIP2003: updated European inventory for 2004 and guidance towards a consistent use in practice. Int. J. Life Cycle Assess. **16**, 401–409 (2011). doi:10.1007/s11367-011-0278-6
Lautier, A., Rosenbaum, R.K., Margni, M., Bare, J., Roy, P.-O., Deschênes, L.: Development of normalization factors for Canada and the United States and comparison with European factors. Sci. Total Environ. **409**, 33–42 (2010). doi:10.1016/j.scitotenv.2010.09.016
Lindfors, L.G., Christiansen, K., Hoffman, L., Virtanen, Y., Juntilla, V., Leskinen, A., Hanssen, O.-J., Rønning, A., Ekvall, T., Finnveden, G.: LCA-Nordic, Technical report No 10, Tema Nord 1995:503. Denmark, Copenhagen (1994)
Mancini, L., Benini, L., Sala, S.: Characterization of raw materials based on supply risk indicators for Europe. Int. J. Life Cycle Assess. In Press (2016)
Matsuda, H., Serizawa, S., Ueda, K., Kato, T., Yahara, T.: Assessing the impact of the Japanese 2005 World exposition project on vascular plants' risk of extinction. Chemosphere **53**, 325–336 (2003). doi:10.1016/S0045-6535(03)00013-4
McKone, T.E., Bennett, D.H., Maddalena, R.L.: CalTOX 4.0 technical support document, vol. 1. Lawrence Berkeley National Laboratory, Berkeley, CA (2001)
Müller-Wenk, R.: Depletion of Abiotic Resources Weighted on the Base of "Virtual" Impacts of Lower Grade Deposits in Future. St. Gallen, Switzerland (1998)
Myhre, G., Shindell, D., Bréon, F.-M., Collins, W., Fuglestvedt, J., Huang, J., Koch, D., Lamarque, J.-F., Lee, D., Mendoza, B., Nakajima, T., Robock, A., Stephens, G., Takemura, T., Zha, H. (2013) Anthropogenic and natural radiative forcing. In: Stocker, T.F., Qin, D., Plattner, G.-K., et al. (eds.) Climate Change 2013 Physics Science Basis. Contribution Working Group I to Fifth Assessment Report. Intergovernmental Panel on Climater Change. Cambridge University Press, Cambridge, UK, pp. 659–740
Norris, G.A.: Impact characterization in the tool for the reduction and assessment of chemical and other environmental impacts: methods for acidification, eutrophication, and ozone formation. J. Ind. Ecol. **6**, 79–101 (2003)
Pennington, D.W., Margni, M., Ammann, C., Jolliet, O.: Multimedia fate and human intake modeling: spatial versus nonspatial insights for chemical emissions in Western Europe. Environ. Sci. Technol. **39**, 1119–1128 (2005)
Pfister, S., Bayer, P.: Monthly water stress: spatially and temporally explicit consumptive water footprint of global crop production. J. Clean. Prod. **73**, 52–62 (2014)
Pfister, S., Koehler, A., Hellweg, S.: Assessing the environmental impacts of freshwater consumption in LCA. Environ. Sci. Technol. **43**, 4098–4104 (2009)
Pope, C.A., Burnett, R.T., Thun, M.J., Calle, E.E., Krewski, D., Ito, K., Thurston, G.D.: Lung cancer, cardiopulmonary mortality, and long-term exposure to fine particulate air pollution. J. Am. Med. Assoc. **287**, 1132–1141 (2002)
Posch, M., Seppälä, J., Hettelingh, J.-P., Johansson, M., Margni, M., Jolliet, O.: The role of atmospheric dispersion models and ecosystem sensitivity in the determination of characterisation factors for acidifying and eutrophying emissions in LCIA. Int. J. Life Cycle Assess. **13**, 477–486 (2008). doi:10.1007/s11367-008-0025-9
Potting, J., Hauschild, M.: Background for spatial differentiation in life cycle impact assessment: the EDIP2003 methodology. Copenhagen (2005)
Potting, J., Schöpp, W., Blok, K., Hauschild, M.Z.: Comparison of the acidifying impact from emissions with different regional origin in life-cycle assessment. J. Hazard. Mater. **61**, 155–162 (1998)
Rabl, A., Spadaro, J.V.: The RiskPoll Software, Version 1.051 (2004)

Rosenbaum, R.K., Margni, M., Jolliet, O.: A flexible matrix algebra framework for the multimedia multipathway modeling of emission to impacts. Environ. Int. **33**, 624–634 (2007). doi:10.1016/j.envint.2007.01.004

Rosenbaum, R.K., Bachmann, T.M.K., Gold, L.S., Huijbregts, M.A.J., Jolliet, O., Juraske, R., Koehler, A., Larsen, H.F., MacLeod, M., Margni, M., McKone, T.E., Payet, J., Schuhmacher, M., Van de Meent, D., Hauschild, M.Z.: USEtox: the UNEP/SETAC-consensus model: recommended characterisation factors for human toxicity and freshwater ecotoxicity in life cycle impact assessment. Int. J. Life Cycle Assess. **13**, 532–546 (2008). doi:10.1007/s11367-008-0038-4

Rosenbaum, R.K., Huijbregts, M.A.J., Henderson, A., Margni, M., McKone, T.E., Van de Meent, D., Hauschild, M.Z., Shaked, S., Li, D.S., Slone, T.H., Gold, L.S., Jolliet, O.: USEtox human exposure and toxicity factors for comparative assessment of toxic emissions in life cycle analysis: sensitivity to key chemical properties. Int. J. Life Cycle Assess. **16**, 710–727 (2011). doi:10.1007/s11367-011-0316-4

Rosenbaum, R.K., Meijer, A., Demou, E., Hellweg, S., Jolliet, O., Lam, N., Margni, M., McKone, T.E.: Indoor air pollutant exposure for life cycle assessment: regional health impact factors for households. Environ. Sci. Technol. **49**, 12823–12831 (2015). doi:10.1021/acs.est.5b00890

Roy, P.-O., Deschênes, L., Margni, M.: Life cycle impact assessment of terrestrial acidification: modeling spatially explicit soil sensitivity at the global scale. Environ. Sci. Technol. **46**, 8270–8278 (2012a). doi:10.1021/es3013563

Roy, P.-O., Huijbregts, M., Deschênes, L., Margni, M.: Spatially-differentiated atmospheric source–receptor relationships for nitrogen oxides, sulfur oxides and ammonia emissions at the global scale for life cycle impact assessment. Atmos. Environ. **62**, 74–81 (2012b). doi:10.1016/j.atmosenv.2012.07.069

Roy, P.-O., Deschênes, L., Margni, M.: Uncertainty and spatial variability in characterization factors for aquatic acidification at the global scale. Int. J. Life Cycle Assess. **19**, 882–890 (2014). doi:10.1007/s11367-013-0683-0

Ryberg, M., Vieira, M.D.M., Zgola, M., Bare, J., Rosenbaum, R.K.: Updated US and Canadian normalization factors for TRACI 2.1. Clean Technol. Environ. Policy **16**, 329–339 (2014). doi:10.1007/s10098-013-0629-z

Saad, R., Margni, M., Koellner, T., Wittstock, B., Deschênes, L.: Assessment of land use impacts on soil ecological functions: development of spatially differentiated characterization factors within a Canadian context. Int. J. Life Cycle Assess. **16**, 198–211 (2011)

Saad, R., Koellner, T., Margni, M.: Land use impacts on freshwater regulation, erosion regulation, and water purification: a spatial approach for a global scale level. Int. J. Life Cycle Assess. **18**, 1253–1264 (2013). doi:10.1007/s11367-013-0577-1

Sala, S., Benini, L., Castellani, V., Vidal-Legaz, B., Pant, R.: Environmental Footprint: Update of Life Cycle Impact Assessment Methods; DRAFT for TAB (Status: May 2, 2016) Resources, Water, Land. Ispra, Italy (2016)

Schere, K.L., Demerjian, K.L.: User's Guide for the Photochemical Box Model (PBM) (1984)

Schneider, L., Berger, M., Finkbeiner, M.: Abiotic resource depletion in LCA: background and update of the anthropogenic stock extended abiotic depletion potential (AADP) model. Int. J. Life Cycle Assess. **20**, 709–721 (2015). doi:10.1007/s11367-015-0864-0

Seppälä, J., Posch, M., Johansson, M., Hettelingh, J.P.: Country-dependent characterisation factors for acidification and terrestrial eutrophication based on accumulated exceedance as an impact category indicator. Int. J. Life Cycle Assess. **11**, 403–416 (2006)

Sleeswijk, A.W., van Oers, L.F.C.M., Guinée, J.B., Struijs, J., Huijbregts, M.A.J.: Normalisation in product life cycle assessment: an LCA of the global and European economic systems in the year 2000. Sci. Total Environ. **390**, 227–240 (2008). doi:10.1016/j.scitotenv.2007.09.040

Steen, B.: A systematic approach to environmental priority strategies in product development (EPS). Version 2000: models and data of the default method. Centre for Environmental assessment of products and material systems. Chalmers University of Technology, Technical Environmental Planning, Gothenburg, Sweden (1999)

Struijs, J., Beusen, A., Van Jaarsveld, H., Huijbregts, M.A.J.: Aquatic eutrophication. ReCiPe 2008 A life cycle impact assessment method which comprises Harmon. Category indicator midpoint endpoint level. Rep. I Characterisation factors (2009)

Tirado-Seco, P.: Development of damage functions for aquatic eutrophication in life cycle assessment. Université de Genève (2005)

Tørsløv, J., Hauschild, M.Z., Rasmussen, D.: Ecotoxicity. The Danish Ministry of the Environment, Environmental Protection Agency, Copenhagen (2005)

Uno, I., Wakamatsu, S.: Analysis of winter high-concentration NO_2 pollution by the photochemical box model. J. Jpn. Soc. Air Pollut. **27**, 246–257 (1992)

van Dijk, A., Den Outer, P.N., Slaper, H.: Climate and Ozone change Effects on ultraviolet radiation and Risks (COEUR) using and validating earth observations. Bilthoven, The Netherlands (2008)

van Goethem, T.M.W.J., Azevedo, L.B., van Zelm, R., Hayes, F., Ashmore, M.R., Huijbregts, M. A.J.: Plant species sensitivity distributions for ozone exposure. Environ. Pollut. **178**, 1–6 (2013a). doi:10.1016/j.envpol.2013.02.023

van Goethem, T.M.W.J., Preiss, P., Azevedo, L.B., Roos, J., Friedrich, R., Huijbregts, M.A.J., van Zelm, R.: European characterization factors for damage to natural vegetation by ozone in life cycle impact assessment. Atmos. Environ. **77**, 318–324 (2013b). doi:10.1016/j.atmosenv.2013.05.009

Van Loon, M., Vautard, R., Schaap, M., Bergstrom, R., Bessagnet, B., Brandt, J., Builtjes, P., Christensen, J.H., Cuvelier, K., Graf, A., Jonson, J., Krol, M., Langner, J., Roberts, P., Rouil, L., Stern, R., Tarrason, L., Thunis, P., Vignati, E., White, L., Wind, P.: Evaluation of long-term ozone simulations from seven regional air quality models and their ensemble average. Atmos. Environ. **41**, 2083–2097 (2007)

van Zelm, R., Huijbregts, M.A.J., van Jaarsveld, H.A., Reinds, G.J., de Zwart, D., Struijs, J., van de Meent, D.: Time horizon dependent characterization factors for acidification in life-cycle assessment based on forest plant species occurrence in Europe. Environ. Sci. Technol. **41**, 922–927 (2007). doi:10.1021/es061433q

van Zelm, R., Huijbregts, M.A.J., den Hollander, H.A., van Jaarsveld, H.A., Sauter, F.J., Struijs, J., van Wijnen, H.J., van de Meent, D.: European characterization factors for human health damage of PM10 and ozone in life cycle impact assessment. Atmos. Environ. **42**, 441–453 (2008). doi:10.1016/j.atmosenv.2007.09.072

van Zelm, R., Huijbregts, M.A.J., Van de Meent, D.: USES-LCA 2.0-a global nested multi-media fate, exposure, and effects model. Int. J. Life Cycle Assess. **14**, 282–284 (2009)

van Zelm, R., Schipper, A.M., Rombouts, M., Snepvangers, J., Huijbregts, M.A.J.: Implementing groundwater extraction in life cycle impact assessment: characterization factors based on plant species richness for the Netherlands. Environ. Sci. Technol. **45**, 629–635 (2010). doi:10.1021/es102383v

van Zelm, R., Preiss, P., van Goethem, T., Van Dingenen, R., Huijbregts, M.: Regionalized life cycle impact assessment of air pollution on the global scale: damage to human health and vegetation. Atmos. Environ. **134**, 129–137 (2016). doi:10.1016/j.atmosenv.2016.03.044

Vautard, R., Builtjes, P.J.H., Thunis, P., Cuvelier, C., Bedogni, M., Bessagnet, B., Honore, C., Moussiopoulos, N., Pirovano, G., Schaap, M., Stern, R., Tarraso, L., Wind, P.: Evaluation and intercomparison of ozone and PM10 simulations by several chemistry transport models over four European cities within the city delta project. Atmos. Environ. **41**, 173–188 (2007)

Verones, F., Hanafiah, M.M., Pfister, S., Huijbregts, M.A.J., Pelletier, G.J., Koehler, A.: Characterization factors for thermal pollution in freshwater aquatic environments. Environ. Sci. Technol. **44**, 9364–9369 (2010). doi:10.1021/es102260c

Verones, F., Hellweg, S., Azevedo, L., Chaudhary, A., Cosme, N., Fantke, P., Goedkoop, M., Hauschild, M., Laurent, A., Mutel, C., Pfister, S., Ponsioen, T., Steinmann, Z., Van Zelm, R., Vieira, M., Huijbregts, M.: LC-IMPACT Version 0.5—A Spatially Differentiated Life Cycle Impact Assessment Approach (2016a). http://www.lc-impact.eu

Verones, F., Pfister, S., van Zelm, R., Hellweg, S.: Biodiversity impacts from water consumption on a global scale for use in life cycle assessment. Int J Life Cycle Assess. 1–10 (2016b). doi:10.1007/s11367-016-1236-0

Vieira, M.D.M., Goedkoop, M.J., Storm, P., Huijbregts, M.A.J.: Ore grade decrease as life cycle impact indicator for metal scarcity: the case of copper. Environ. Sci. Technol. **46**, 12772–12778 (2012). doi:10.1021/es302721t

Vieira, M.D.M., Ponsioen, T.C., Goedkoop, M.J., Huijbregts, M.A.J.: Surplus cost potential as a life cycle impact indicator for metal extraction. Resources **5**, 2 (2016). doi:10.3390/resources5010002

Vieira, M.D.M., Ponsioen, T.C., Goedkoop, M.J., Huijbregts, M.A.J.: Surplus ore potential as a scarcity indicator for resource extraction. J. Ind. Ecol. **21**, 381–390 (2017). doi:10.1111/jiec.12444

Weidema, B., Hauschild, M., Jolliet, O.: Preparing characterisation methods for endpoint impact assessment (previously submitted under the title: “Stepwise 2006: a new environmental impact assessment method”) (2006)

Wenger, Y., Li, D.S., Jolliet, O.: Indoor intake fraction considering surface sorption of air organic compounds for life cycle assessment. Int. J. Life Cycle Assess. **17**, 919–931 (2012). doi:10.1007/s11367-012-0420-0

WMO: Scientific Assessment of Ozone Depletion: 1998, Global Ozone Research and Monitoring Project–Report No. 44. Geneva, Switzerland (1999)

WMO: Scientific Assessment of Ozone Depletion: 2002, Global Ozone Research and Monitoring Project–Report No. 47. Geneva, Switzerland (2003)

WMO: Scientific Assessment of Ozone Depletion: 2010, Global Ozone Research and Monitoring Project–Report No. 52. Geneva, Switzerland (2011)

WMO: Scientific Assessment of Ozone Depletion: 2014, World Meteorological Organization, Global Ozone Research and Monitoring Project Report No. 55. Geneva, Switzerland (2014)

Author Biography

Ralph K. Rosenbaum LCA expert and environmental modeller focusing on LCIA development since early 2000s. Contributed to several UNEP/SETAC working groups towards global harmonisation of LCA methodology. Interested in LCIA modelling of emissions and water/soil resource use, operationalisation of uncertainty management and spatial differentiation.

Glossary

Term	Definition	Reference
Allocation	Partitioning the input or output flows of a process or a product system between the product under study and one or more other product systems	ISO 14044
Ancillary input	Material input that is used by the unit process producing the product, but does not constitute part of the product	ISO 14044
Area of protection	A cluster of category endpoints of recognisable value to society, viz. human health, natural resources, natural environment and sometimes man-made environment	Guinée et al. (2002)
Attributional modelling (or descriptive book keeping)	LCI modelling frame that inventories the inputs and output flows of all processes of a system as they occur. Modelling process along an existing supply chain is of this type	ILCD, LCI
By-product	See also co-product. A product from a process that is not the reason why the process is run and that usually has lower value than the main product of the process	Own definition
Category endpoint	Attribute or aspect of natural environment, human health, or resources, identifying an environmental issue giving cause for concern	ISO 14044
Category indicator	See impact category indicator	
Cause–effect chain	See environmental mechanism	
Cause–effect network	See environmental mechanism	
Characterisation	A step of the impact assessment, in which the environmental interventions assigned qualitatively to a particular impact category (in classification) are quantified in terms of a common unit for that category, allowing aggregation into one figure of the indicator result	Guinée et al. (2002)

(continued)

M.Z. Hauschild et al. (eds.), *Life Cycle Assessment*,
DOI 10.1007/978-3-319-56475-3

(continued)

Term	Definition	Reference
Characterisation factor	Factor derived from a characterization model which is applied to convert an assigned life cycle inventory analysis result to the common unit of the category indicator. *Note*: The common unit allows calculation of the category indicator result	ISO 14044
Characterisation model	Reflect the environmental mechanism by describing the relationship between the LCI results, category indicators and, in some cases, category endpoint(s). The characterization model is used to derive the characterization factors	ISO 14044
Classification	A step of impact assessment, in which environmental interventions are assigned to predefined impact categories on a purely qualitative basis	Guinée et al. (2002)
Co-function	Any of two or more functions provided by the same unit process or system	ILCD, LCI
Comparative assertion	Environmental claim regarding the superiority or equivalence of one product versus a competing product that performs the same function	ISO 14044
Completeness check	Process of verifying whether information from the phases of a life cycle assessment is sufficient for reaching conclusions in accordance with the goal and scope definition	ISO 14044
Consequential modelling	LCI modelling principle that identifies and models all processes in the background system of a system in consequence of decisions made in the foreground system	ILCD, LCI
Consistency check	Process of verifying that the assumptions, methods and data are consistently applied throughout the study and are in accordance with the goal and scope definition performed before conclusions are reached	ISO 14044
Co-product	Any of two or more products coming from the same unit process or product system	ISO 14044
Critical review	Process intended to ensure consistency between a life cycle assessment and the principles and requirements of the International Standards on life cycle assessment	ISO 14044
Cut-off criteria	Specification of the amount of material or energy flow or the level of environmental significance associated with unit processes or product system to be excluded from a study	ISO 14044

(continued)

(continued)

Term	Definition	Reference
Damage approach	See endpoint method	
Data quality	Characteristics of data that relate to their ability to satisfy stated requirements	ISO 14044
Eco-efficiency	Ratio between the value created and the environmental impact caused by an activity	Own definition
Ecosphere	The biosphere of the earth, especially when the interaction between the living and non-living components is emphasised	Oxford Dictionary of English
Ecosystem quality	Area of protection "Ecosystem Quality" that deals with damages on the intrinsic value of natural ecosystems. See also natural environment	Verones et al. (2017)
Elementary flow/elementary exchange	Material or energy entering the system being studied that has been drawn from the environment without previous human transformation, or material or energy leaving the system being studied that is released into the environment without subsequent human transformation	ISO 14044
Emission	See release	
End-of-life product	Product at the end of its useful life that will potentially undergo reuse, recycling, or recovery	ILCD, LCI
Endpoint	See category endpoint	
Endpoint method/model/indicator	The category endpoint is an attribute or aspect of natural environment, human health, or resources, identifying an environmental issue giving cause for concern (ISO 14040). Hence, endpoint method (or damage approach)/model is a characterisation method/model that provides indicators at the level of Areas of Protection (natural environment's ecosystems, human health, resource availability) or at a level close to the areas of protection level	ILCD, LCIA
Energy flow	Input to or output from a unit process or product system, quantified in energy units. *Note*: Energy flow that is an input may be called an energy input; energy flow that is an output may be called an energy output	ISO 14044
Environmental aspect	Element of an organisation's activities, products or services that can interact with the environment	ISO 14044

(continued)

(continued)

Term	Definition	Reference
Environmental impact	Potential impact on the natural environment, human health or the depletion of natural resources, caused by the interventions between the technosphere and the ecosphere as covered by LCA (e.g. emissions, resource extraction, land use)	ILCD, LCI
Environmental indicator	An environmental indicator can be a measurable feature or features that provide managerially and scientifically useful evidence of the environment and ecosystem quality or reliable evidence of trends in quality. Thus, environmental indicators must be measurable with available technology, scientifically valid for assessing or documenting ecosystem quality, and useful for providing information for management decision-making. Indicators can be used to: (1) compare current conditions with desired performance; (2) show trends over time, to allow comparisons between different regions; (3) help judge the sustainability of current practices; and (4) define and publicise new standards and measures for assessing progress toward a sustainable future	JRC
Environmental intervention	A human intervention in the environment, either physical, chemical or biological; in particular resource extraction, emissions (incl. noise and heat) and land use; the term is thus broader than "elementary flow"	Guinée et al. (2002)
Environmental mechanism	System of physical, chemical and biological processes for a given impact category, linking the life cycle inventory analysis results to category indicators and to category endpoints by means of a characterisation model	ISO 14044
Environmental life cycle costing	Assessment of all costs associated with the life cycle of a product that are directly covered by any one or more of the actors in the product life cycle with complementary inclusion of externalities that are anticipated to be internalised in the decision-relevant future (Hunkeler et al. 2008). The analysis is performed consistent with the system boundaries of the environmental LCA	Hunkeler et al. (2008)
Environmental process	A physical, chemical or biological process in the environment system that is identified as part of the causal chain linking a	Guinée et al. (2002)

(continued)

(continued)

Term	Definition	Reference
	particular environmental intervention to a particular impact, e.g. pollution leaching or bioaccumulation; for a given impact category, the environmental processes together form the environmental mechanism	
Environmental (impact) profile	The result of the characterisation step showing the indicator results for all the predefined impact categories, supplemented by any other relevant information	Guinée et al. (2002)
Environmental relevance	Degree of linkage between category indicator result and category endpoints	ISO 14044
Environmentally extended input–output analysis	Linking environmental impacts to economic demand through the use of economic input–output tables originally developed for macroeconomic systems analysis and planning by combining them with tables that describe how much direct environmental impacts each economic sector causes per economic output during a year of production	Own definition
Evaluation	Element within the life cycle interpretation phase intended to establish confidence in the results of the life cycle assessment. *Note*: Evaluation includes completeness check, sensitivity check, consistency check, and any other validation that may be required according to the goal and scope definition of the study	ISO 14044
Extraction	Withdrawal of a biotic or abiotic resource from the environment in a unit process, considered as an environmental intervention	Guinée et al. (2002)
Feedstock energy	Heat of combustion of a raw material input that is not used as an energy source to a product system, expressed in terms of higher heating value or lower heating value. *Note*: Care is necessary to ensure that the energy content of raw materials is not counted twice	ISO 14044
Functional unit	Quantified performance of a product system for use as a reference un	ISO 14044
Grouping	Sorting and possibly ranking of the impact categories	ISO 14044
Impact assessment	See life cycle impact assessment	
Impact category	Class representing environmental issues of concern to which life cycle inventory analysis results may be assigned	ISO 14044
Impact category indicator	Quantifiable representation of an impact category	ISO 14044

(continued)

(continued)

Term	Definition	Reference
Impact pathway	Cause–effect chain of an environmental mechanism linking interventions through midpoint impacts to damages to areas of protection	Own definition
Impact score	See indicator result	
Indicator result	The numerical result of the characterisation step for a particular impact category, e.g. 12 kg CO_2-equivalents for climate change	Guinée et al. (2002)
Input	Product, material or energy flow that enters a unit process. *Note*: Products and materials include raw materials, intermediate products and co-products	ISO 14044
Interested party	Individual or group concerned with or affected by the environmental performance of a product system, or by the results of the life cycle assessment	ISO 14044
Intermediate flow	Product, material or energy flow occurring between unit processes of the product system being studied	ISO 14044
Intermediate product	Output from a unit process that is input to other unit processes that require further transformation within the system	ISO 14044
Interpretation	See life cycle interpretation	
Inventory analysis	See life cycle inventory analysis	
Inventory table	See life cycle inventory analysis results	
Land occupation	The unavailability of a given plot of land for alternative uses for a certain period of time	Guinée et al. (2002)
Land transformation	The change in the quality of a given plot of land due to a particular mode of human use, measured in terms of changes in biodiversity and life support functions	Guinée et al. (2002)
Land use	The impact category land use reflects the damage to ecosystems due to the effects of occupation and transformation of land. Examples of land use are agricultural production, mineral extraction and human settlement. Occupation of land can be defined as the maintenance of an area in a particular state over a particular time period. Transformation is the conversion of land from one state to another state, e.g. from its original state to an altered state or from an altered state to another altered state	ILCD-LCIA
LCIA method	Collection of individual characterisation models (each addressing their separate impact category)	Hauschild et al. (2013)

(continued)

(continued)

Term	Definition	Reference
Life cycle	Consecutive and interlinked stages of a product system, from raw material acquisition or generation from natural resources to final disposal	ISO 14044
Life cycle assessment (LCA)	Compilation and evaluation of the inputs, outputs and the potential environmental impacts of a product system throughout its life cycle	ISO 14044
Life cycle impact assessment (LCIA)	Phase of life cycle assessment aimed at understanding and evaluating the magnitude and significance of the potential environmental impacts for a product system throughout the life cycle of the product	ISO 14044
Life cycle impact category indicator	See category indicator	
Life cycle initiative (UNEP/SETAC LCI)	An international partnership to enable users around the world to put life cycle thinking into effective practice. Launched in 2002 by The United Nations Environment Programme (UNEP) and the Society for Environmental Toxicology and Chemistry (SETAC)	www.lifecycleinitiative.org
Life cycle interpretation	Phase of life cycle assessment in which the findings of either the inventory analysis or the impact assessment, or both, are evaluated in relation to the defined goal and scope in order to reach conclusions and recommendations	ISO 14044
Life cycle inventory analysis (LCI)	Phase of life cycle assessment involving the compilation and quantification of inputs and outputs for a product throughout its life cycle	ISO 14044
Life cycle inventory analysis result	Outcome of a life cycle inventory analysis that catalogues the flows crossing the system boundary and provides the starting point for life cycle impact assessment	ISO 14044
Life cycle phase	Major methodological element of LCA. The four phases of an LCA are: Goal and scope definition, life cycle inventory analysis, life cycle impact assessment, life cycle interpretation	Own definition based on ISO 14044
Life cycle stage	A stage in the life time of the product/service. Defined by the LCA practitioner, but often considered as stages are: raw materials extraction, manufacturing, distribution, use and disposal	Own definition based on ISO 14044

(continued)

(continued)

Term	Definition	Reference
Marginal process	Process that is affected (employed or taken out of use) as a response to an increase or decrease in the demand of a product, respectively	Own definition
Midpoint indicator	Impact category indicator located somewhere along the impact pathway between emission and category endpoint	Hauschild and Huijbregts (2015)
Midpoint method/approach	The midpoint method is a characterisation method that provides indicators for comparison of environmental interventions at a level of cause–effect chain between emissions/resource consumption and the endpoint level	ILCD, LCIA
Multifunctional process	Process or system that performs more than one function. Examples: Processes with more than one product as output (e.g. NaOH, Cl_2 and H_2 from chloralkali electrolysis) or more than one waste treated jointly (e.g. mixed household waste incineration with energy recovery). See also: "Allocation" and "System expansion"	ILCD, LCI
Natural environment	Area of protection that addresses impacts to ecosystems and landscapes. See also ecosystem quality	ILCD-LCIA
Normalisation	Calculation of the magnitude of category indicator results relative to some reference information	ISO 14044
Normalisation result/factor/reference	See normalised environmental profile	
Normalised environmental profile	The result of the normalisation step: a table showing the normalised indicator results for all the selected impact categories, supplemented by any other relevant information	Guinée et al. (2002)
Normalised indicator result	The numerical result of normalisation for a particular impact category, e.g. 0.02 year for climate change	Guinée et al. (2002)
Obligatory property	Feature that the product must possess for any user to perceive it as a product valid for fulfilling the desired function. May also include legally required features. Can usually be expressed in technical terms	Own definition
Output	Product, material or energy flow that leaves a unit process. *Note*: Products and materials include raw materials, intermediate products, co-products and releases	ISO 14044

(continued)

(continued)

Term	Definition	Reference
Positioning property	Optional feature of a product, which can be used to position it as more attractive to the consumer in the competition with other similar products. In contrast to obligatory properties (see this), positioning properties often vary from consumer to consumer	Own definition
Potential impact	Relative performance indicators which can be the basis of comparisons and optimisation of the system or product	Hauschild and Huijbregts (2015)
Problem-oriented approach	See midpoint approach	
Process	Set of interrelated or interacting activities that transforms inputs into outputs	ISO 14044
Process energy	Energy input required for operating the process or equipment within a unit process, excluding energy inputs for production and delivery of the energy itself	ISO 14044
Product	Any goods or service. *Note 1*: The product can be categorised as follows: – services (e.g. transport) *Note 2*: Services have tangible and intangible elements. Provision of a service can involve, for example, the following: – an activity performed on a customer-supplied tangible product (e.g. automobile to be repaired) – an activity performed on a customer-supplied intangible product (e.g. the income statement needed to prepare a tax return) – the delivery of an intangible product (e.g. the delivery of information in the context of knowledge transmission) – the creation of ambience for the customer (e.g. in hotels and restaurants) Software consists of information and is generally intangible and can be in the form of approaches, transactions or procedures. Hardware is generally tangible and its amount is a countable characteristic. Processed materials are generally tangible and their amount is a continuous characteristic: – software (e.g. computer program, dictionary) – hardware (e.g. engine mechanical part) – processed materials (e.g. lubricant)	ISO 14044

(continued)

(continued)

Term	Definition	Reference
Product category rules	Set of specific rules, requirements and guidelines for developing type III environmental declarations for one or more product categories	ISO 14025
Product environmental footprint	Result of a product environmental footprint study based on the product environmental footprint method	2013/179/EU
Product environmental footprint method	General method to measure and communicate the potential life cycle environmental impact of a product. In EU this methods is detailed in Annex II of COMMISSION RECOMMENDATION of 9 April 2013 on the use of common methods to measure and communicate the life cycle environmental performance of products and organisations	2013/179/EU
Product flow	Products entering from or leaving to another product system	ISO 14044
Product system	Collection of unit processes with elementary and product flows, performing one or more defined functions, and which models the life cycle of a product	ISO 14044
Raw material	Primary or secondary material that is used to produce a product. *Note*: Secondary material includes recycled material	ISO 14044
Recycling, reuse recovery	Recovery is any form of recovering value from a waste stream whether in the form of material value (i.e. recycling) or recovery of energy content through incineration. "Recycling" means any recovery operation by which waste materials are reprocessed into products, materials or substances whether for the original or other purposes. It includes the reprocessing of organic material but does not include energy recovery and the reprocessing into materials that are to be used as fuels or for backfilling operations Reuse is a form of waste prevention since the product re-use avoids the need for the manufacture of a new product. A simple example is the direct re-use of containers, bricks or other materials on site	Based on EUR 24916 EN (2011)
Reference flow	Measure of the outputs from processes in a given product system required to fulfil the function expressed by the functional unit'	ISO 14044
Releases	Emissions to air and discharges to water and soil	ISO 14044

(continued)

(continued)

Term	Definition	Reference
Scenario	Generally, scenarios are different, more or less realistic, descriptions of actions or situations in the future based on certain assumptions and factors	Own definition
Secondary good	Secondary material, recovered energy, reused part or similar as the product of a reuse, recycling, recovery, refurbishing or similar process	ILCD, LCI
Secondary function	Unintended functions that usually have low or no relevance to the users of a product, meaning that they are not contributing to the obligatory or positioning properties	Own definition
Sensitivity analysis	Systematic procedures for estimating the effects of the choices made regarding methods and data on the outcome of a study	ISO 14044
Sensitivity check	Process of verifying that the information obtained from a sensitivity analysis is relevant for reaching the conclusions and giving recommendations	ISO 14044
Subcategory	A subdivision of an impact category, e.g. freshwater aquatic ecotoxicity as a subcategory of ecotoxicity	Guinée et al. (2002)
Substitution	Solving multifunctionality of processes and products by expanding the system boundaries and substituting the not required function with an alternative way of providing it, i.e. the process(es) or product (s) that the not required function supersedes. Effectively the life cycle inventory of the superseded process(es) or product(s) is subtracted from that of the analysed system, i.e. it is "credited". Substitution is a special (subtractive) case of applying the system expansion principle	ILCD, LCI
System	Any good, service, event, basket-of-products, average consumption of a citizen, or similar object that is analysed in the context of the LCA study	ILCD, LCI
System boundary	Set of criteria specifying which unit processes are part of a product system	ISO 14044
System expansion	Adding specific processes or products and the related life cycle inventories to the analysed system. Used to make several multifunctional systems with an only partly equivalent set of functions comparable within LCA	ILCD, LCI

(continued)

(continued)

Term	Definition	Reference
Technosphere	The sphere or realm of human technological activity; the technologically modified environment	Oxford Dictionary of English
Transparency	Open, comprehensive and understandable presentation of information	ISO 14044
Uncertainty analysis	Systematic procedure to quantify the uncertainty introduced in the results of a life cycle inventory analysis due to the cumulative effects of model imprecision, input uncertainty and data variability. *Note*: Either ranges or probability distributions are used to determine uncertainty in the results	ISO 14044
Unit process	Smallest element considered in the life cycle inventory analysis for which input and output data are quantified	ISO 14044
Waste	Substances or objects which the holder intends or is required to dispose of (ISO 14044, 2008) Output with zero or negative value. The moment it gets a value, it turns into a co-product or secondary function and system expansion or allocation become relevant	Own definition
Weighting	Converting and possibly aggregating indicator results across impact categories using numerical factors based on value-choices; data prior to weighting should remain available	Guinée et al. (2002)
Weighting factor	A factor obtained with a weighting method and used to express a particular (normalised) indicator result in terms of the common unit of the weighting result	Guinée et al. (2002)
Weighting result	The numerical part of the result of weighting and aggregation of all (normalised) indicator results, e.g. 0.08 year (*Note*: the result may be expressed as more than one numerical value)	Guinée et al. (2002)

References

2013/179/EU: COMMISSION RECOMMENDATION of 9 April 2013 on the use of common methods to measure and communicate the life cycle environmental performance of products and organisations (2013/179/EU)

EUR 24916 EN Supporting Environmentally Sound Decisions for Waste Management—A Technical Guide to Life Cycle Thinking (LCT) and Life Cycle Assessment (LCA) for Waste Experts and LCA Practitioners (2011)

Guinée, J.B. (ed.), Gorrée, M., Heijungs, R., Huppes, G., Kleijn, R., de Koning, A., Van Oers, L., Wegener Sleeswijk, A., Suh, S., Udo de Haes, H.A., De Bruijn, J.A., Van Duin, R., Huijbregts, M.A.J.: Handbook on Life Cycle Assessment: Operational Guide to the ISO Standards. In: Eco-Efficiency in Industry and Science. Kluwer Academic Publishers. Dordrecht (Hardbound, ISBN 1-4020-0228-9; Paperback, ISBN 1-4020-0557-1) (2002)

Hauschild, M.Z., Goedkoop, M., Guinée, J., et al.: Identifying best existing practice for characterization modeling in life cycle impact assessment. Int. J. Life Cycle Assess. **18**, 683 (2013). doi:10.1007/s11367-012-0489-5

Hauschild, M.Z., Huijbregts, M.A.J. (eds.): Life Cycle Impact Assessment. Springer, Netherlands (2015). doi:10.1007/978-94-017-9744-3

Hunkeler, D.D., Lichtenvort, K., Rebitzer, G., Ciroth, A.: Environmental life cycle costing. CRC Press, Pensacola, Florida (2008)

ILCD, LCI: ILCD Handbook: Specific guide for Life Cycle Inventory (LCI) Data Sets, 1st edn. (2010)

ILCD, LCIA: ILCD Handbook: Recommendations for Life Cycle Impact Assessment in the European Context, 1st edn (2011), and ILCD Handbook: Framework and Requirements for Life Cycle Impact Assessment models and Indicators, 1st edn. (2010)

ISO 14025: Environmental labels and declarations—type III environmental declarations—principles and procedures (2010)

ISO 14044: Environmental management—life cycle assessment—requirements and guidelines (2008)

Verones, F., Bare, J., Bulle, C., Frischknecht, R., Hauschild, M., Hellweg, S., Henderson, A., Jolliet, O., Laurent, A., Liao, X., Lindner, J.P., de Souza, D., Michelsen, O., Patouillard, L., Pfister, S., Posthuma, L., Prado, V., Ridoutt, B., Rosenbaum, R.K., Sala, S., Ugaya, C., Vieira, M., Fantke, P.: LCIA framework and cross-cutting issues guidance within the UNEP-SETAC life cycle initiative (2017) (in review)

Index

M.Z. Hauschild et al. (eds.), *Life Cycle Assessment*,
DOI 10.1007/978-3-319-56475-3

D

E

P

GPSR Compliance

The European Union's (EU) General Product Safety Regulation (GPSR) is a set of rules that requires consumer products to be safe and our obligations to ensure this.

If you have any concerns about our products, you can contact us on ProductSafety@springernature.com

In case Publisher is established outside the EU, the EU authorized representative is:

Springer Nature Customer Service Center GmbH
Europaplatz 3
69115 Heidelberg, Germany

Batch number: 10426266

Printed by Printforce, the Netherlands